Diagnosis and Treatment of Traumatic Brain Injury

Diagnosis and Treatment of Traumatic Brain Injury

The Neuroscience of Traumatic Brain Injury

Edited by

Rajkumar Rajendram
Department of Medicine, King Abdulaziz Medical City, King Abdulaziz International Medical Research Center, Ministry of National Guard - Health Affairs, Riyadh, Saudi Arabia

College of Medicine, King Saud bin Abdulaziz University of Health Sciences, Riyadh, Saudi Arabia

Victor R. Preedy
King's College London, London

Colin R. Martin
Institute for Health and Wellbeing, University of Suffolk, United Kingdom

ELSEVIER

ACADEMIC PRESS
An imprint of Elsevier

Academic Press is an imprint of Elsevier
125 London Wall, London EC2Y 5AS, United Kingdom
525 B Street, Suite 1650, San Diego, CA 92101, United States
50 Hampshire Street, 5th Floor, Cambridge, MA 02139, United States
The Boulevard, Langford Lane, Kidlington, Oxford OX5 1GB, United Kingdom

ISBN 978-0-12-823347-4

SET ISBN 978-0-323-99197-1

For information on all Academic Press publications
visit our website at https://www.elsevier.com/books-and-journals

Publisher: Nikki Levy
Acquisitions Editor: Natalie Farra
Editorial Project Manager: Timothy Bennett
Production Project Manager: Paul Prasad Chandramohan
Cover Designer: Mark Rogers

Typeset by STRAIVE, India

Dedication

I dedicate this book to my wonderful daughter, Dr. Caragh Brien, of whom I am so incredibly proud.
Colin R. Martin

Contents

Section B
Clinical features of traumatic brain injury

Section E
Rehabilitation in traumatic brain injury

Contributors

Numbers in parentheses indicate the pages on which the authors' contributions begin.

Andrew Y. Abashkin (301, 411), Department of Neurotrauma, Trauma Center Level II, Urgent Care Hospital, Dzerzhinsk, Russian Federation

Samar Abdelhady (239), Faculty of Medicine, Alexandria University, Alexandria, Egypt

Reem Abedi (39), Department of Biochemistry and Molecular Genetics, Faculty of Medicine, American University of Beirut, Beirut, Lebanon

Darya I. Agarkova (301, 411), Department of Ophthalmology, Trauma Center Level I, Regional Clinical Hospital named after NA Semashko, Nizhny Novgorod, Russian Federation

Hasan Alam (477), Department of Surgery, Feinberg School of Medicine, Northwestern Memorial Hospital, Chicago, IL, United States

George Alexiou (289), Department of Neurosurgery, University Hospital of Ioannina, Ioannina, Greece

Georgios Alexiou (233), Department of Neurosurgery, University Hospital of Ioannina, Ioannina, Greece

Garzon Heredia Alicia (517), Department of Psychology des Pays de la Loire, SFR Confluences, University of Angers, University of Nantes, Maison de la recherche Germaine Tillion, Angers Cedex 01; Department of Neurology, CHU Angers, Angers, France

Marta Aloisi (205), Post-Coma Unit, IRCCS Santa Lucia Foundation, Rome, Italy

Sudharsana Rao Ande (3), Department of Radiology, Rady Faculty of Health Sciences, University of Manitoba, Winnipeg, MB, Canada

Karim Asehnoune (111), Surgical Intensive Care Unit, Hotel-Dieu, University Hospital of Nantes; Anti-Infectious Therapeutics, Health Research Institute 2, Medical University of Nantes, Nantes, France

Ohanes Ashekyan (239), Department of Biochemistry and Molecular Genetics, Faculty of Medicine, American University of Beirut, Beirut, Lebanon

Eva Azicnuda (531), IRCCS Fondazione Santa Lucia, Rome, Italy

Karen M. Barlow (151), Acquired Brain Injury in Children Research Program, Child Health Research Centre, Faculty of Medicine, The University of Queensland; Dr. Paul Hopkins Chair of Paediatric Rehabilitation in Acquired Brain Injury, Queensland Children's Hospital, South Brisbane, QLD, Australia

Carleen Batson (169, 181), Department of Human Anatomy and Cell Science, Rady Faculty of Health Sciences, University of Manitoba, Winnipeg, MB, Canada

Eslam Belal (239), Faculty of Medicine, Alexandria University, Alexandria, Egypt

Carmela Belardo (463), Department of Experimental Medicine, Pharmacology Division, University of Campania "L. Vanvitelli", Naples, Italy

Francesco Biroli (15), FROM Research Foundation, Papa Giovanni XXIII Hospital, Neuroscience Area, Bergamo, Italy

Umberto Bivona (71, 531), Post-Coma Unit; IRCCS Fondazione Santa Lucia, Rome, Italy

Serena Boccella (463), Department of Experimental Medicine, Pharmacology Division, University of Campania "L. Vanvitelli", Naples, Italy

Marwan Bouras (111), Surgical Intensive Care Unit, Hotel-Dieu, University Hospital of Nantes; Anti-Infectious Therapeutics, Health Research Institute 2, Medical University of Nantes, Nantes, France

Denis E. Bragin (301, 411), Lovelace Biomedical Research Institute; University of New Mexico, School of Medicine, MSC 105615 Neurosurgery, Albuquerque, NM, United States

Rocco Salvatore Calabrò (497), IRCCS Centro Neurolesi "Bonino-Pulejo", Messina, Italy

C.S. Carabias (251), Neurotraumatology and Subarachnoid Hemorrhage Research Unit, Area 8: Neuroscience and Mental Health, Research Institute Hospital 12 de Octubre (i+12); Department of Neurosurgery, Hospital Universitario 12 de Octubre, Madrid, Spain

Lucia Cattin (191), Department of Anesthesiology and Intensive Care, San Bortolo Hospital, Vicenza, Italy

Siddharth Chavali (367), Institute of Neurosciences, AIG Hospitals, Hyderabad, India

Helena C. Chui (87), Department of Neurology, Keck School of Medicine, University of Southern California, Los Angeles, CA, United States

Maria Paola Ciurli (71), Neuropsychology Unit, IRCCS Fondazione Santa Lucia, Rome, Italy

Vallat-Azouvi Claire (517), Department of Fonctionnement et Dysfonctionnement Cognitifs : les âges de la vie (DYSCO), University of Paris 8, Saint-Denis, France

Marianna Contrada (205), Post-Coma Unit, IRCCS Santa Lucia Foundation, Rome, Italy

Alvaro Cordoba (357, 379, 437), Neurosurgery and Intensive Care Unit ASESP; Medical University of Vienna, Vienna, Austria; ASESP Neurosurgical Department ICU Unit, Montevideo, Uruguay

Matheus Fernandes de Oliveira (315), Neurosurgery Department, Public Servant State Hospital; Neurosurgery Department, DFV Neuro, São Paulo, Brazil

Silvia De Rosa (191), Department of Anesthesiology and Intensive Care, San Bortolo Hospital, Vicenza, Italy

Agilandeswari Devarajan (485), Research and Development Department, Synapsesai India Private Limited, Indore, Madhya Pradesh, India

Vipin V. Dhote (485), VNS Group of Institutions, Neelbud, Bhopal, Madhya Pradesh, India

Michael Dobrzeniecki (301, 411), Department of Neurologic Diseases, Privolzhsky Research Medical University, Nizhny Novgorod, Russian

João Gustavo Rocha Peixoto dos Santos (327), Neurosurgery Department, Santa Casa de Maceio, Maceio, Brazil

Ali Eid (39), Faculty of Medicine, Department of Pharmacology and Toxicology, American University of Beirut, Beirut, Lebanon

Maya El Dor (39), PRASE, Lebanese University, Beirut, Lebanon

Evert A. Eriksson (449), Medical University of South Carolina, Charleston, SC, United States

Michele Rechia Fighera (397), Department of Neuropsychiatry, Laboratory of Experimental and Clinical Neuropsychiatry, Health Sciences Center, Medical School, Federal University of Santa Maria, Santa Maria, RS, Brazil

Rita Formisano (205), Post-Coma Unit, IRCCS Santa Lucia Foundation, Rome, Italy

Bernardina Frache (357, 379), Intensive Care Unit, Sanatorio Americano, Montevideo, Uruguay

Logan Froese (169, 181), Biomedical Engineering, Faculty of Engineering, University of Manitoba, Winnipeg, MB, Canada

Ana Flavia Furian (397), Neurotoxicity and Psychopharmacology Laboratory, Health Sciences Center, Federal University of Santa Maria, Santa Maria, RS, Brazil

Elena Romana Gasenzer (455), Department Physician Assistance, Carl Remigius Medical School, Frankfurt a.M., Germany

Alwyn Gomez (169, 181), Section of Neurosurgery, Department of Surgery; Department of Human Anatomy and Cell Science, Rady Faculty of Health Sciences, University of Manitoba, Winnipeg, MB, Canada

Paolo Gritti (15), Department of Anesthesia and Critical Care Medicine, Papa Giovanni XXIII Hospital, Bergamo, Italy

Francesca Guida (463), Department of Experimental Medicine, Pharmacology Division, University of Campania "L. Vanvitelli", Naples, Italy

Aysa Hacioglu (339), Department of Endocrinology, Erciyes University Medical School, Kayseri, Turkey

Muhammad Ali Haidar (39, 239), Department of Biochemistry and Molecular Genetics, Faculty of Medicine, American University of Beirut, Beirut, Lebanon

Steven D. Hicks (277), Department of Pediatrics, Penn State College of Medicine, Hershey, PA, United States

Makoto Higuchi (123), Department of Functional Brain Imaging Research, National Institute of Radiological Sciences, National Institutes for Quantum and Radiological Science and Technology, Chiba, Japan

Shih-Wei Huang (223), Department of Physical Medicine and Rehabilitation, Shuang Ho Hospital, Taipei Medical University, New Taipei City, Taiwan ROC

Monica Iannotta (463), Department of Experimental Medicine, Pharmacology Division, University of Campania "L. Vanvitelli", Naples, Italy

Rosmara Infantino (463), Department of Experimental Medicine, Pharmacology Division, University of Campania "L. Vanvitelli", Naples, Italy

Andrei Irimia (87), Ethel Percy Andrus Gerontology Center, Leonard Davis School of Gerontology; Corwin D. Denney Research Center, Department of Biomedical Engineering, University of Southern California, Los Angeles, CA, United States

Hawraa Issa (39, 239), College of Public Health, Phoenicia University, Zahrani, Lebanon

Kartik K. Iyer (151), Acquired Brain Injury in Children Research Program, Child Health Research Centre, Faculty of Medicine, The University of Queensland, South Brisbane, QLD, Australia

Ayhan Kanat (455), Department of Neurosurgery, Medical Faculty, Recep Tayyip Erdogan University, Rize, Turkey

Min-Gu Kang (137), Department of Physical Medicine and Rehabilitation, Dong-A University College of Medicine, Busan, Korea

Zuleyha Karaca (339), Department of Endocrinology, Erciyes University Medical School, Kayseri, Turkey

Fahrettin Kelestimur (339), Department of Endocrinology, Faculty of Medicine, Yeditepe University, Kosuyolu Hospital, Istanbul, Turkey

Michael Kemp (477), Department of Surgery, University of Michigan, Ann Arbor, MI, United States

Doo Young Kim (507), Department of Physical Medicine and Rehabilitation, Catholic Kwandong University International St Mary's Hospital, Incheon, Republic of Korea

Eunkyung Kim (137), Department of Rehabilitation Medicine; Biomedical Research Institute, Seoul National University Hospital, Seoul, Korea

Andrew Kiragu (421), Division of Critical Care, Department of Pediatrics, University of Minnesota; Division of Critical Care, Department of Pediatrics, Hennepin Healthcare; Children's Minnesota, Minneapolis, MN, United States

Firas H. Kobeissy (39, 239), Department of Biochemistry and Molecular Genetics, Faculty of Medicine, American University of Beirut, Beirut, Lebanon; Program for Neurotrauma, Neuroproteomics & Biomarkers Research, Departments of Emergency Medicine, Psychiatry, Neuroscience and Chemistry, University of Florida, Gainesville, FL, United States

Artem A. Kopylov (411), Department of Neurosurgery, Trauma Center Level I, Regional Clinical Hospital named after NA Semashko, Nizhny Novgorod, Russian Federation

Svyatoslav B. Korolev (411), Department of Traumatology and Neurosurgery, Privolzhsky Research Medical University, Nizhny Novgorod, Russia

Hanayama Kozo (27), Department of Rehabilitation Medicine, Kawasaki Medical School, Okayama, Japan

A. Lagares (251), Neurotraumatology and Subarachnoid Hemorrhage Research Unit, Area 8: Neuroscience and Mental Health, Research Institute Hospital 12 de Octubre (i+12); Department of Neurosurgery, Hospital Universitario 12 de Octubre; Department of Surgery, Faculty of Medicine, Complutense University of Madrid (UCM), Madrid, Spain

E. Meng Law (87), iBRAIN Research Laboratory, Departments of Neuroscience, Computer Systems and Electrical Engineering, Monash University; Department of Radiology, The Alfred Health Hospital, Melbourne, VIC, Australia

Yu-Hao Lee (223), Department of Physical Medicine and Rehabilitation, Shuang Ho Hospital, Taipei Medical University, New Taipei City, Taiwan ROC

Gustavo Fernandes Leobas (315), Neurosurgery Department, Public Servant State Hospital, São Paulo, Brazil

Gabriel Corrêa Lima (397), Exercise Biochemistry Laboratory, Center of Physical Education and Sports, Federal University of Santa Maria—UFSM, Santa Maria, RS, Brazil

Tsan-Hon Liou (223), Department of Physical Medicine and Rehabilitation, Shuang Ho Hospital, Taipei Medical University, New Taipei City, Taiwan ROC

Livio Luongo (463), Department of Experimental Medicine, Pharmacology Division, University of Campania "L. Vanvitelli", Naples, Italy

Maria Grazia Maggio (497), Cognitive Rehabilitation and Relational Psychotherapy, Messina, Sicily, Italy

Sabatino Maione (463), Department of Experimental Medicine, Pharmacology Division, University of Campania "L. Vanvitelli", Naples, Italy

Dmitry S. Martynov (411), Nizhny Novgorod State Technical University n.a. R.E. Alekseev, Nizhny Novgorod, Russian Federation

Dimitrios Metaxas (233), Department of Neurosurgery, University Hospital of Ioannina, Ioannina, Greece

Masaru Mimura (123), Department of Neuropsychiatry, Keio University School of Medicine, Tokyo, Japan

Muthu Kumaradoss MohanMarugaRaja (485), Department of Pharmacognosy and Phytochemistry, Parul Institute of Pharmacy & Research, Parul University, Vadodara, Gujarat, India

Stefania Mondello (239), Department of Biomedical and Dental Sciences and Morphofunctional Imaging, University of Messina, Messina, Italy

Ryuta Nakae (101), Department of Emergency and Critical Care Medicine, Nippon Medical School, Tokyo, Japan

Anastasios Nasios (289), Department of Neurosurgery, University Hospital of Ioannina, Ioannina, Greece

Leila Nasrallah (39), Department of Biochemistry and Molecular Genetics, Faculty of Medicine, American University of Beirut, Beirut, Lebanon

Byung-Mo Oh (137), Department of Rehabilitation Medicine, Seoul National University Hospital; Department of Rehabilitation Medicine, Seoul National University College of Medicine; Institute on Aging, Seoul National University, Seoul; National Traffic Injury Rehabilitation Hospital, Yangpyeong; Neuroscience Research Institute, Seoul National University College of Medicine, Seoul, Korea

M.S. Oliveira (397), Neurotoxicity and Psychopharmacology Laboratory, Health Sciences Center, Federal University of Santa Maria, Santa Maria, RS, Brazil

Oumaima Outani (239), Faculty of Medicine and Pharmacy of Rabat, Muhammad-V University, Rabat, Morocco

Wellingson Silva Paiva (327), Neurology Department, Hospital das Clínicas, University of Sao Paulo, Sao Paulo, Brazil

Enza Palazzo (463), Department of Experimental Medicine, Pharmacology Division, University of Campania "L. Vanvitelli", Naples, Italy

Vinood B. Patel (543), School of Life Sciences, University of Westminster, London, United Kingdom

Allain Philippe (517), Department of Psychology des Pays de la Loire, SFR Confluences, University of Angers, University of Nantes, Maison de la recherche Germaine Tillion, Angers Cedex 01; Department of Neurology, CHU Angers, Angers, France

Azouvi Philippe (517), AP-HP, Raymond Poincare Hospital, Service de MPR, Garches; INSERM DevPsy Team, Centre de Recherche en Épidémiologie et Santé des Populations, UMR 1018, Paris-Saclay University, UVSQ, Versailles, France

Victor R. Preedy (543), Diabetes and Nutritional Sciences Research Division, Faculty of Life Science and Medicine, King's College London, London, United Kingdom

Sung-Bom Pyun (507), Department of Physical Medicine and Rehabilitation; Department of Biomedical Sciences, College of Medicine; Brain Engineering Convergence Research Center, Korea University, Seoul, Republic of Korea

Rajkumar Rajendram (543), College of Medicine, King Saud bin Abdulaziz University for Health Sciences; Department of Medicine, King Abdulaziz Medical City, King Abdullah International Medical Research Center, Ministry of National Guard Health Affairs, Riyadh, Saudi Arabia; Stoke Mandeville Hospital, Buckinghamshire Healthcare NHS Trust, Aylesbury, United Kingdom

Leonardo Magno Rambo (397), Exercise Biochemistry and Physiology Laboratory, Biochemistry Graduate Program, Federal University of Pampa, Uruguaiana, RS, Brazil

Marta Rapiti (531), IRCCS Fondazione Santa Lucia, Rome, Italy

Girija Prasad Rath (367), Department of Neuroanaesthesiology and Critical Care, Neurosciences Centre, All India Institute of Medical Sciences (AIIMS), New Delhi, India

Mohammad Amine Reslan (39), Department of Biochemistry and Molecular Genetics, Faculty of Medicine, American University of Beirut, Beirut, Lebanon

Flavia Ricciardi (463), Department of Experimental Medicine, Pharmacology Division, University of Campania "L. Vanvitelli", Naples, Italy

Luca Ricciardi (191), Dipartimento NESMOS, UOC di Neurochirurgia, AOU Sant'Andrea, Sapienza, Roma, Italy

Dimitrios Rizos (233), Department of Intensive Care Unit, "Hatzikosta" General Hospital, Ioannina, Greece

Jose Maria Dominguez Roldan (357), Intensive Care Unit Hospital Virgen del Rocio, Sevilla, Spain

Antoine Roquilly (111), Surgical Intensive Care Unit, Hotel-Dieu, University Hospital of Nantes; Anti-Infectious Therapeutics, Health Research Institute 2, Medical University of Nantes, Nantes, France

Kenneth A. Rostowsky (87), Ethel Percy Andrus Gerontology Center, Leonard Davis School of Gerontology, University of Southern California, Los Angeles, CA, United States

Luiz Fernando Freire Royes (397), Exercise Biochemistry Laboratory, Center of Physical Education and Sports, Federal University of Santa Maria—UFSM, Santa Maria, RS, Brazil

Yasmine Samir (239), Departments of Emergency Medicine, Psychiatry, Neuroscience and Chemistry, Faculty of Science, Alexandria University, Alexandria, Egypt

Alba Scerrati (191), Department of Morphology, Surgery and Experimental Medicine, University Hospital S. Anna, Ferrara, Italy

Zhanna B. Semenova (301), Department of Neurosurgery, Children's Clinical and Research Institute of Emergency Surgery and Trauma, Moscow, Russian Federation

Zaynab Shakkour (239), Department of Biochemistry and Molecular Genetics, Faculty of Medicine, American University of Beirut, Beirut, Lebanon

Jai Jai Shiva Shankar (3), Department of Radiology, Rady Faculty of Health Sciences, University of Manitoba, Winnipeg, MB, Canada

Anatoly Y. Sheludyakov (301, 411), Department of Neurosurgery, Trauma Center Level I, City Clinical Hospital, Nizhny Novgorod, Russia

Daniela Silvestro (531), IRCCS Fondazione Santa Lucia, Rome, Italy

Tina Slusher (421), Division of Critical Care, Department of Pediatrics, University of Minnesota; Division of Critical Care, Department of Pediatrics, Hennepin Healthcare, Minneapolis, MN, United States

Bedriye Müge Sönmez (57), Department of Emergency Medicine, Univercity of Health Sciences, Dışkapı Yıldırım Beyazıt Education and Research Hospital, Ankara, Turkey

Anselmo Alves Boa Sorte, Jr (315), Neurosurgery Department, Public Servant State Hospital, São Paulo, Brazil

Athena Stein (151), Acquired Brain Injury in Children Research Program, Child Health Research Centre, Faculty of Medicine, The University of Queensland, South Brisbane, QLD, Australia

Maha Tabet (39), Molecular, Cellular and Developmental biology department (MCD), Centre de Biologie Integrative (CBI), University of Toulouse, CNRS, UPS, Toulouse, France

Kenji Tagai (123), Department of Functional Brain Imaging Research, National Institute of Radiological Sciences, National Institutes for Quantum and Radiological Science and Technology, Chiba, Japan

Keisuke Takahata (123), Department of Functional Brain Imaging Research, National Institute of Radiological Sciences, National Institutes for Quantum and Radiological Science and Technology, Chiba; Department of Neuropsychiatry, Keio University School of Medicine, Tokyo, Japan

Hiraoka Takashi (27), Department of Rehabilitation Medicine, Kawasaki Medical School, Okayama, Japan

Christopher S. Thomas (449), Department of Surgery Education Office, Medical University of South Carolina, Charleston, SC, United States

Alexey O. Trofimov (301, 411), Department of Neurologic Diseases, Privolzhsky Research Medical University; Department of Neurosurgery, Trauma Center Level I, City Clinical Hospital, Nizhny Novgorod, Russia

Xenia A. Trofimova (411), Department of Neurologic Diseases, Privolzhsky Research Medical University, Nizhny Novgorod, Russia

Dolores Villalobos (71), Laboratory of Cognitive and Computational Neuroscience, Center of Biomedical Technology, Technical University of Madrid, Complutense University of Madrid; Department of Experimental Psychology, School of Psychology. Complutense University; The European Centre of Neuroscience, Madrid, Spain

Spyridon Voulgaris (233, 289), Department of Neurosurgery, University Hospital of Ioannina, Ioannina, Greece

Glenn Wakam (477), Department of Surgery, University of Michigan, Ann Arbor, MI, United States

Andrew Wu (421), Division of Critical Care Medicine, Department of Anesthesiology, Critical Care, and Pain Medicine, Boston Children's Hospital, Boston, MA, United States

Olga Ygropoulou (233), Department of Intensive Care Unit, "Hatzikosta" General Hospital, Ioannina, Greece

Shoji Yokobori (101), Department of Emergency and Critical Care Medicine, Nippon Medical School, Tokyo, Japan

Rosalia Zangari (15), FROM Research Foundation, Papa Giovanni XXIII Hospital, Neuroscience Area, Bergamo, Italy

Frederick A. Zeiler (169, 181), Section of Neurosurgery, Department of Surgery; Department of Human Anatomy and Cell Science, Rady Faculty of Health Sciences; Biomedical Engineering, Faculty of Engineering; Centre on Aging, University of Manitoba, Winnipeg, MB, Canada; Division of Anaesthesia, Department of Medicine, Addenbrooke's Hospital, University of Cambridge, Cambridge, United Kingdom

Kazem Zibara (39), PRASE, Lebanese University, Beirut, Lebanon

Preface

Traumatic brain injury has complex etiology and may arise as a consequence of physical abuse, violence, war, vehicle collisions, working in the construction industry, and sports. The variable nature of traumatic brain injury is evidenced by the fact that there are a number of classifications. For example, traumatic brain injury can be acute or chronic, penetrating or nonpenetrating. There are also several other, more extensive classifications based on the severity of the brain injury, its anatomical location, the level of consciousness, or predicted outcomes.

Classifications and diagnosis go hand in hand. There are many scoring systems and scales. Biological markers have also been used to diagnose brain injury and predict outcomes. Features of traumatic brain injury also vary; locations, the presence of edema, and white matter damage all contribute to the diversity of traumatic brain injury. While physical damage often leaves visible scars, the psychosocial effects, though "invisible," can be long-lasting and may even have greater impact on patients, their family, and their work colleagues. These changes all require adequate assessment, diagnosis, classification, treatment, and rehabilitation. Clinical evaluations of patients may include the use of biochemical markers and electroencephalography alongside imaging studies (e.g., plain X-ray, computed tomography, and magnetic resonance imaging). Treatments vary depending on the anatomical site of the injury and the extent of the damage. Clinical management includes the use of pharmacological agents, control of raised intracranial pressure, and neurosurgery to control the myriad of symptoms that occur in traumatic brain injury. The morbidity and mortality of traumatic brain injury are high, and in the initial phase after the injury, the provision of organ support within intensive therapy units is often required. However, subsequent rehabilitation of survivors will include physiotherapy, speech therapy, occupational therapy, treatments for depression or posttraumatic stress, family counseling, and follow-up investigations for secondary conditions.

It can be argued that traumatic brain injury is one of the most complex conditions to treat. There is an ongoing dialog between the established information of the day and new scientific discoveries. The characterization, diagnosis, treatment, and rehabilitation of individuals with traumatic brain injury revolve around an endless pool of scientific information. However, to adequately treat such patients, diverse skill sets are needed. To achieve the best outcomes, timely, effective, coordinated, evidence-based multidisciplinary input is required from the moment the injury occurs through to the point of discharge. Herein lies a difficulty: Experts in one field are not necessarily well versed in other fields. For example, experts in speech therapy may not necessarily be knowledgeable about the diagnosis and treatment of traumatic brain injury and *vice versa*. This is compounded by the fact that much of the information relating to traumatic brain injury is spread across different scientific domains and is often presented by experts for experts rather than those new to the field.

The effects of traumatic injury on the brain include not only the consequences of physical forces or agents (e.g., penetrating objects, projectiles and head-surface contact) but also the ensuing physiological and biochemical events. These include changes to the following: cranial structures, the blood-brain barrier, the physiological architecture of blood vessels, blood flow (resulting in regional hypoxia or ischemia), intracranial pressure dynamics, oxidative balance, neurotransmitter function, and the ionic milieu. Other effects may include edema, extradural hematoma, and cellular biochemistry. Cellular biochemistry itself will include alterations in gene expression, apoptotic cascades, protein turnover, and subcellular organelles. Some changes in cellular biochemistry are very similar to those occurring in modeling systems and also in nontraumatic brain injuries (e.g., due to toxic agents, aging, or neurodegenerative diseases).

Traumatic brain injury thus represents a transition between the initiating insults, physiological processes, and altered cellular and tissue biochemistry. This complex milieu ultimately results in psychosocial and behavioral changes. However, understanding or connecting these different processes is problematic as hitherto there has never been a single source material that covers all of these aspects yet is suitable for the expert and the novice alike. Furthermore, hardly any of the existing scientific text dealing with changes in traumatic brain injury has been described in relation to their applicability to other areas of neuroscience, i.e., translational information in its broadest sense. A spectrum of knowledge is necessary to fully understand diseases and processes. This is addressed in the two-volume set *The Neuroscience of Traumatic Brain Injury* comprising the 2 books:

Diagnosis and Treatment of Traumatic Brain Injury
Cellular, Molecular, Physiological, and Behavioral Aspects of Traumatic Brain Injury

This book, *Diagnosis and Treatment of Traumatic Brain Injury*, comprises six sections:

- *Setting the scene and introductory chapters*
- *Clinical features of traumatic brain injury*
- *Diagnosis and evaluation*
- *Treatments: Experimental and clinical*
- *Rehabilitation in traumatic brain injury*
- *Resources*

Each chapter has the following sections:

- *An abstract (published online)*
- *Key facts*
- *Mini-dictionary of terms*
- *Applications to other areas of neuroscience*
- *Summary points*

The sections *Key facts*, *Mini-dictionary of terms*, and *Summary points* enable the reader to cross the trans-intellectual and trans-disciplinary divides. The section *Applications to other areas of neuroscience* relates to the translational aspects of the chapter and the applicability of the information.

The Neuroscience of Traumatic Brain Injury is designed for research and teaching purposes. It is suitable for neurologists, surgeons, trauma specialists, psychologists, health scientists, public health workers, doctors, pharmacologists, and research scientists. It is valuable as a personal reference book and also for academic libraries that cover the domains of trauma and neurology. Contributions are from leading national and international experts including those from world-renowned institutions. The book is suitable for undergraduate and postgraduate students as well as lecturers and academic professors.

Rajkumar Rajendram, Victor R. Preedy, and Colin R. Martin
(Editors)

Section A

Setting the scene and introductory chapters

Chapter 1

Traumatic brain injury and in-hospital mortality- CT perfusion and beyond

Sudharsana Rao Ande and Jai Jai Shiva Shankar
Department of Radiology, Rady Faculty of Health Sciences, University of Manitoba, Winnipeg, MB, Canada

List of abbreviations

CBV cerebral blood volume
CBF cerebral blood flow
CTA computed tomography angiography
CTP computed tomography perfusion
DTI diffusion tensor imaging
FA fractional anisotropy
GCS Glasgow Coma Scale
GOSe glasgow outcome scale
ICU intensive care unit
TBI traumatic brain injury

Introduction

Traumatic brain injury (TBI) is a major health concern all over the world. It is responsible for social, economic, and health problems related to premature death and long-term disabilities. TBI can result in cognitive deficits, endocrine, psychiatric disorders affecting the capacity to work and impair the quality of life of the patients (Areas et al., 2019). Please see Table 1 for causative factors and consequences of TBI. The severity of the TBI is most commonly classified as mild, moderate, and severe according to the Glasgow Coma Scale (GCS) (Teasdale & Jennett, 1974). Please see Table 2 for the details on how to classify TBI based on the GCS scale. Severe TBI is a clinical emergency during which trauma teams work effectively to provide appropriate level of care to the patients. Severe TBI is the principal cause of post-injury hospitalization, disability, and death throughout the world and has been called a "deadly killer" or "silent epidemic" (Narayan et al., 2002; Shankar et al., 2020). TBI represents the hefty financial expenditure incurred by the various medical insurance companies for rehabilitation in Canada and around the world (National Scientific Advisory Committee, Identifying Priorities for Research and Capacity Development in Injury as a Multi-Institute Strategic Initiative, 2004). TBI is predicted to be the most prevalent and costliest (accounting total indirect cost of over $8 billion) neurological condition in Canada through the year 2031 (National Scientific Advisory Committee, Identifying Priorities for Research and Capacity Development in Injury as a Multi-Institute Strategic Initiative, 2004). Each year, over 50 million people suffer from TBI throughout the world, with an annual estimated cost for TBI of US$ 400 billion to the global economy (Maas et al., 2017). Despite very resource-intense care including multiple complicated surgeries, up to 50% of these patients die within the first 48 h of hospital admission (National Collaborating Centre for Acute Care (UK), 2007; Trauma Registry Report on Injury in Nova Scotia: Provincial Report 2012–2013 Data Year.; 2014., 2014). Early *in-hospital* mortality (<48 h) is likely dependent on the "pre-injury" and more likely on the "injury" factors (Maas et al., 2017; McMillan, Bigler, Teasdale, Ponsford, & Murray, 2018). It is possible that a significant percentage of these patients have permanent brain damage at hospital admission, including brain death. The gold standard to diagnose brain death is a clinical examination (Shemie et al., 2014), which is complicated to elicit because of the nature of the injury as well as patients' life support and sedative medications. Thus, patients receive elaborate resource-intense treatment despite a lack of clear prognoses. Needless to mention, there are several outcome scales proposed for the final neurological outcome (Maas et al., 2017). Existence of numerous outcome scales reflects the complexity of long-term

Diagnosis and Treatment of Traumatic Brain Injury. https://doi.org/10.1016/B978-0-12-823347-4.00035-X

TABLE 1 Causative factors and consequences of TBI.

Causative factors TBI	Consequences of TBI
Motor vehicle accidents	Loss or decreased consciousness
Assaults	Loss of memory of events happened before or after TBI. Loss of reasoning
Unintentional Trauma	Neurological deficits such as muscle weakness, loss of vision, and slurred speech
Struck by/against an object	Mental state such as disorientation, slow thinking, or difficulty on concentrating
	Difficulty on communication, expression, and understanding
	Depression, anxiety, changes in the personality, aggression, acting out, and inappropriate social behavior
	TBI can increase the risk for other neurological disorders such as Alzheimer's disease, Parkinson's disease, and Epilepsy

TABLE 2 Glasgow Coma Scale description.

Category	Response	Points
Eye opening	Spontaneously To speech To pain No response	4 3 2 1
Best Verbal Response	Orientated Confused (Disoriented) Inappropriate words Incomprehensive sounds No response	5 4 3 2 1
Best Motor Response	Obeys Localize Withdrawal Abnormal flexion Abnormal extension No response	6 5 4 3 2 1
Severe brain injury—score of 3–8 Moderate brain injury—Score of 9–12 Mild brain injury—score of 13–15		

outcome assessment in patients with severe TBI. The most commonly used is an eight-point extended Glasgow outcome scale (GOSe) (Table 3). GOSe score of 1 (death), 2 (vegetative state), or 3 (severe neurological deficit) are considered to be a poor neurological outcome (Wilson, Pettigrew, & Teasdale, 1998). This outcome assessment is further complicated by pre-injury and injury factors as well as patients' response at various stages. Unavailability of definitive, reliable, validated, and prompt triage tool to diagnose brain death in patients with severe TBI, warrants the urgent need of research in this field. This will help reduce resource intensive but futile care in those who are already dead at the time of hospital admission and may facilitate the precious gift of organ donation.

Neuroimaging for TBI

Neuroimaging plays a crucial role in the evaluation, management, diagnosis, and prognosis of TBI. Here are some of the neuroimaging techniques that are used in the diagnosis and prognosis of TBI (Table 4).

TABLE 3 Extended Glasgow Outcome Scale (GOSE).

1	Death	D	
2	Vegetative state	VX	Condition of unawareness with only reflex responses but with periods of spontaneous eye opening
3	Lower severe disability	SD-	Patient who is dependent for daily support for mental or physical disability, usually a combination of both. If the patient can be left alone for more than 8h at home it is upper level of SD, if not then it is low level of SD
4	Upper severe disability	SD+	
5	Lower moderate disability	MD-	Patients have some disability such as aphasia, hemiparesis or epilepsy and/or deficits of memory or personality but are able to look after themselves. They are independent at home but dependent outside. If they are able to return to work even with special arrangement it is upper level of MD, if not then it is low level of MD
6	Upper moderate disability	MD+	
7	Lower good recovery	GR-	Resumption of normal life with the capacity to work even if pre-injury status has not been achieved. Some patients have minor neurological or psychological deficits. If these deficits are not disabling then it is upper level of GR, if disabling then it is lower level of GR
8	Upper good recovery	GR+	

TABLE 4 List of some clinical tools used in Traumatic Brain Injury.

Clinical tools	Diagnoses
Plain head CT	Plain head CT accurately and promptly diagnoses structural abnormalities such as skull fractures, intracranial bleeds, brain contusions and brain herniation. But does not provide any functional information about the brain
CT angiography	CTA provides anatomic information about the brain blood vessels such as injury or traumatic occlusion of the blood vessels
CT perfusion	It is an advanced CT scan that provides both functional and anatomic information about the brain, and it is mainly used in triage patients with acute stroke. This imaging technique quantifies perfusion parameters such as cerebral blood flow and cerebral blood volume
Magnetic resonance perfusion	Magnetic resonance perfusion is non-invasive and can be used to detect intracranial arterial blood flow. It can also detect perfusion parameters of affected brain tissues such as cerebral blood flow and cerebral blood volume
Diffusion tensor imaging	Diffusion tensor imaging (DTI) is a promising technique that provides vital information on the connections within the brain. It provides information on microstructural changes and structural connectivity of the brain

Plain head computed tomography (CT)

Computed tomography was introduced in 1970 and from then on, it has revolutionized the assessment of head injuries including traumatic brain injuries (Rosenblum, Hoff, Norman, Weinstein, & Pitts, 1978). Plain CT of the head is the standard imaging test performed as the part of standard care at admission and during the hospital stay for patients with severe TBI (Mayer & Kossoff, 1999). It is a fast and readily available test and requires no contrast medium. Plain CT accurately and promptly diagnoses structural abnormalities such as skull fractures, intracranial bleeds, brain contusions, and brain herniation (Fig. 1). But plain head CT does not provide any functional information about the brain, does not assess intracranial blood flow and no information on brain death is presumed by CT (Astrup, 1982; Corral et al., 2009; Huang et al., 2011; Mayer & Kossoff, 1999; Mushkudiani et al., 2008; Soustiel, Mahamid, Goldsher, & Zaaroor, 2008). So plain head CT imaging test even though it provides valuable information is not reliable in determining TBI.

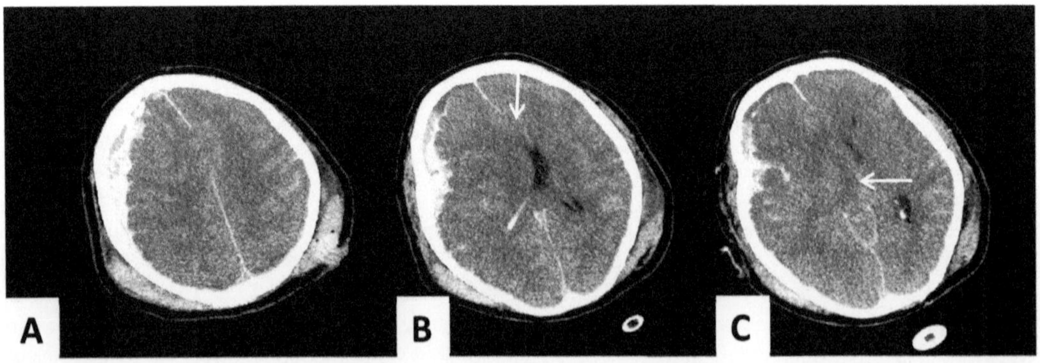

FIG. 1 Plain head CT images of a patient with severe traumatic brain injury showing large right subdural hematoma causing significant mass effect, midline shift (arrow in B) and right uncal herniation (arrow in C).

CT angiography (CTA)

CTA is a readily available, practical, and omnipresent imaging test. However, the CT angiography protocol for confirmation of brain death varies considerably among different centers, which makes it difficult for it to be accepted as the gold standard test (Greer, Varelas, Haque, & Wijdicks, 2008). Systematic reviews of the literature have concluded that the sensitivity of CTA is inadequate to confirm brain death and that clinicians should be wary when using it as an ancillary test (Kramer & Roberts, 2014; Taylor et al., 2014). Moreover, CTA provides anatomic, but not functional, information about the brain (Evans et al., 2010; Shankar & Vandorpe, 2013). The presence of contrast medium in the different segments of the intracranial arteries can be analyzed on CTA images by use of 4-, 7-, and 10-point scales. A 4-point scale is used to assess lack of opacification of the M4 segment of the bilateral middle cerebral arteries and bilateral internal cerebral veins (Dupas et al., 1998). A 7-point scale is used to assess lack of opacification of the M4 segments of the bilateral middle cerebral arteries, the A2 segments of the bilateral anterior cerebral arteries, bilateral internal cerebral veins, and the vein of Galen (Leclerc et al., 2006). A 10-point scale is used to assess the basilar artery and the P2 segments of the bilateral posterior cerebral arteries in addition to the assessments in the 7-point scale (Sawicki et al., 2013) (Fig. 2G). Fig. 2 shows images of a patient with severe traumatic brain injury with plain CT head done at the time of admission showing no obvious abnormality. CT angiogram shows filling of intracranial vessels. MRI done 2 days later showed foci of restricted diffusion and multiple foci of microhemorrhages on susceptibility weighted images suggestive likely of diffuse axonal injury.

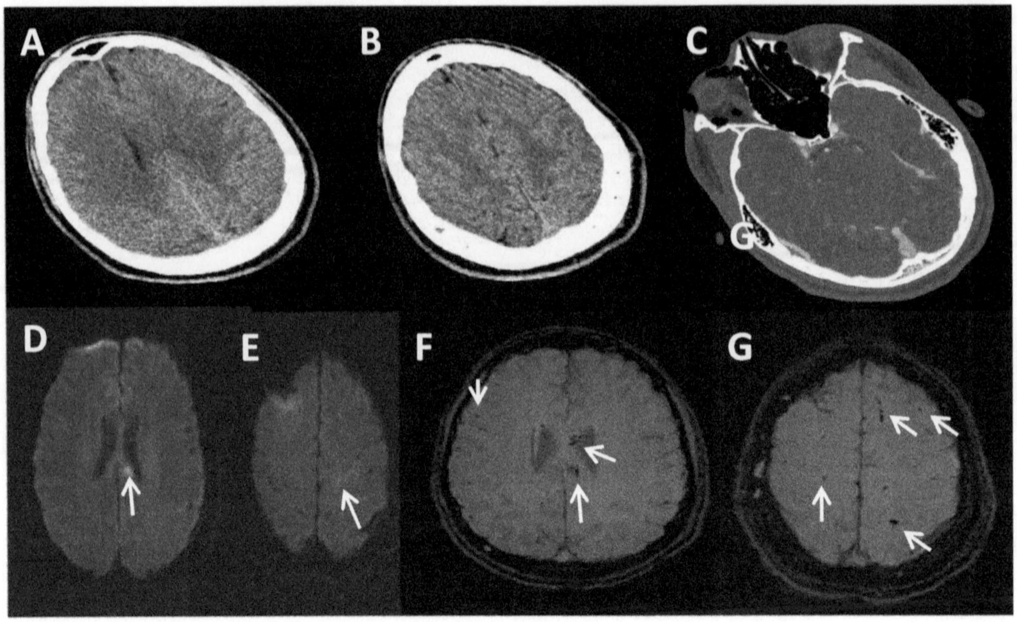

FIG. 2 A patient with severe traumatic brain injury with CT perfusion done at the time of admission showed absent cerebral Blood Flow (A, B, C) and cerebral Blood Volume (D, E, F) at the level of medulla (A, D), pons (B, E) and brainstem (C, F). The extracranial flow continued to be preserved.

CT perfusion

It is an advanced CT scan that provides both functional and anatomic information about the brain, and it is mainly used in triage patients with acute stroke and in TBI (Albers et al., 2017; Du & Shankar, 2017; Nogueira et al., 2018; Shankar, Langlands, Doucette, & Phillips, 2016; Shankar & Lum, 2011; Wilson & Shankar, 2014). In this technique, the temporal change in contrast density is used to quantify perfusion parameters such as cerebral blood flow (CBF) and cerebral blood volume (CBV) which indicate areas of ischemic or infarcted brain (area of "dead brain"). In a healthy person CBF is autoregulated and CTP is very sensitive in detecting CBF and can detect decreased perfusion as low as 2% to 3% in CBF and 2% in CBV (Uwano et al., 2012). If CTP is unable to detect CBF, or if CTP shows markedly decreased CBF (<5 mL/100 g/min), brain death can be confirmed because cellular viability is not possible at such a low CBF. Moreover, CTP with approximately whole-brain coverage can also provide data for CTA. We were the *first* to describe the CTP features of brain death as the most sensitive and specific imaging test for confirmation of brain death in intensive care unit (ICU) patients (MacDonald, Stewart-Perrin, & Shankar, 2018; Shankar & Vandorpe, 2013). Furthermore, we provided the astonishing evidence of using CTP to predict *in-hospital* mortality in comatose cardiac arrest patients (Shankar, Stewart-Perrin, Quraishi, Bata, & Vandorpe, 2018). Fig. 3 depicts images of a patient with severe traumatic brain injury with CT perfusion done at the time of admission showed preserved cerebral blood flow and cerebral blood volume at the level of medulla, pons and brainstem. Fig. 4 shows another patient with severe traumatic brain injury with CT perfusion done at the time of admission showed absent cerebral blood flow and cerebral blood volume at the level of medulla, pons, and brainstem. The extracranial flow continued to be preserved. There are some limitations with the use of CT perfusion in TBI. The radiation dose used in the perfusion must be kept low as it may affect the patients. In some patients, the contrast material used in CT perfusion can result in renal impairment or allergic reaction. Therefore, institutions using CT perfusion have to keep these limitations in mind.

Magnetic resonance perfusion (MR perfusion)

MRP is noninvasive and can be used to detect intracranial arterial blood flow. It can also detect perfusion parameters of affected brain tissues such as cerebral blood flow and cerebral blood volume. There are not many reports in the

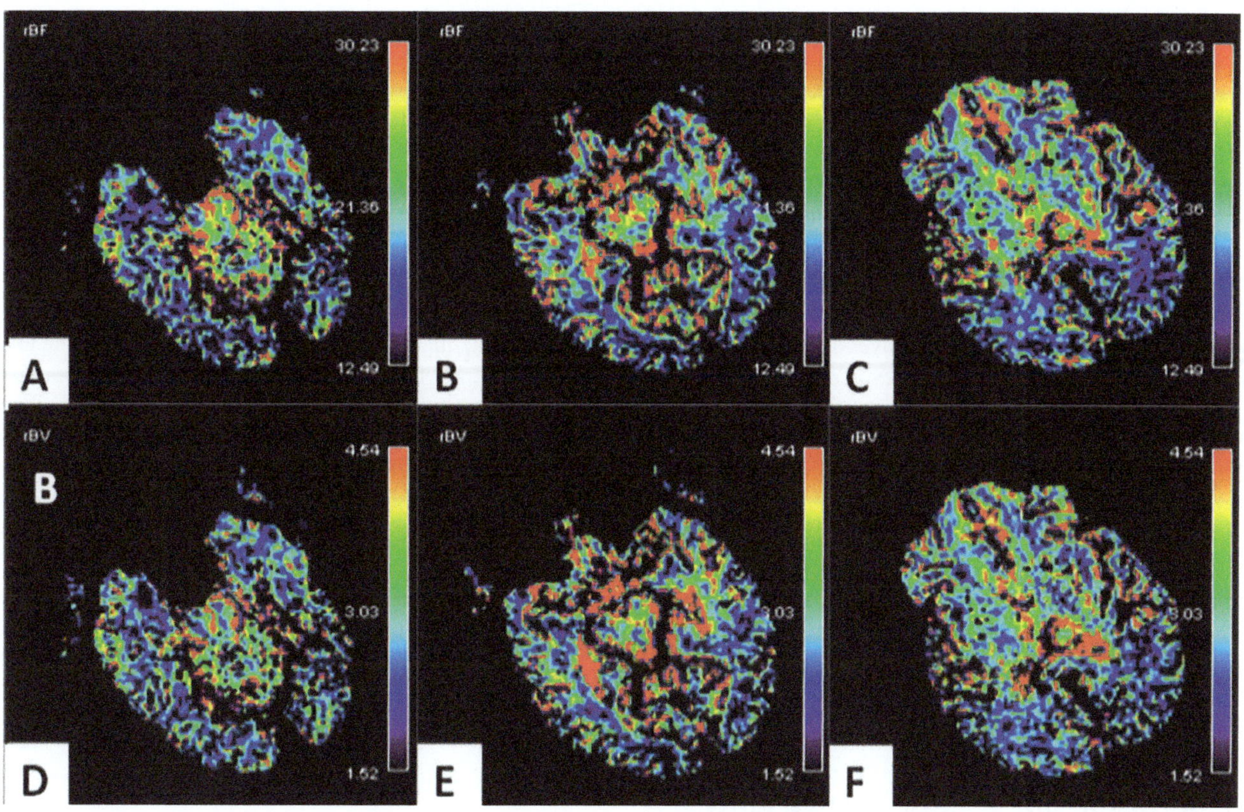

FIG. 3 A patient with severe traumatic brain injury with CT perfusion done at the time of admission showed preserved cerebral Blood Flow (A, B, C) and cerebral Blood Volume (D, E, F) at the level of medulla (A, D), pons (B, E) and brainstem (C, F).

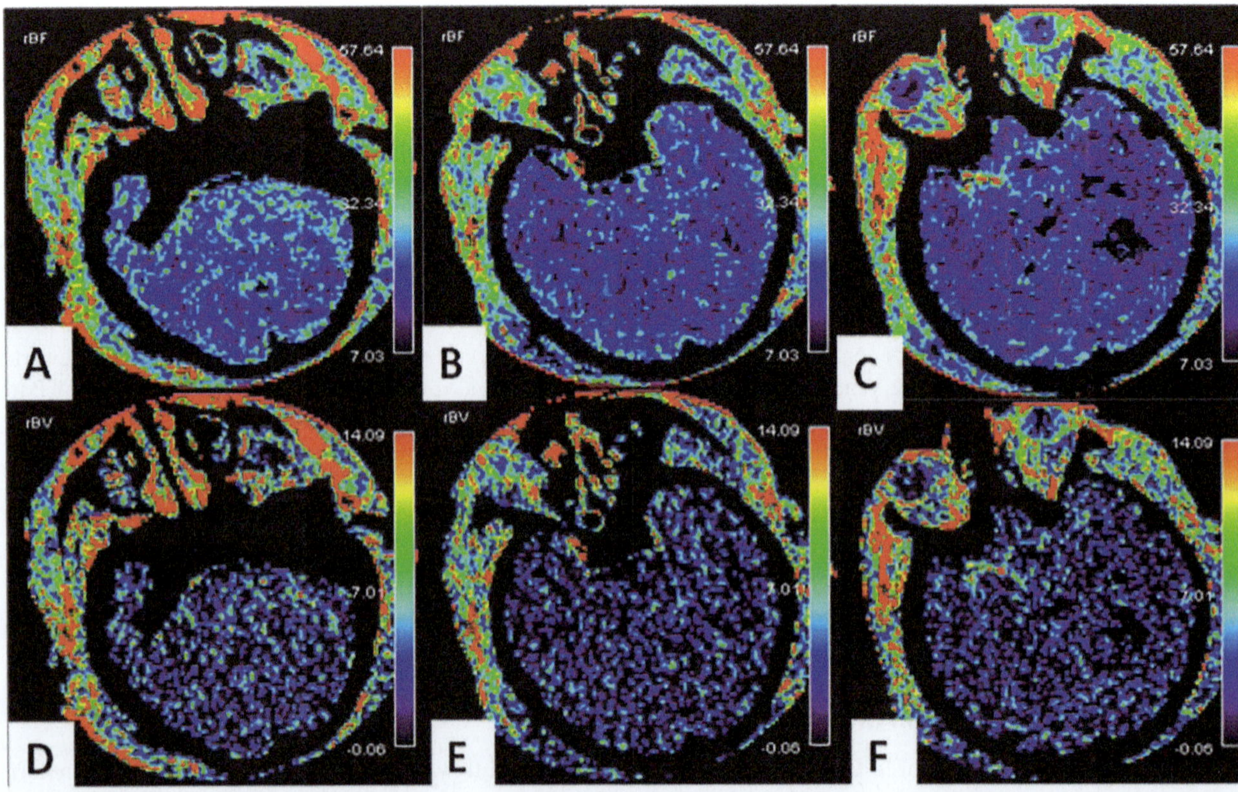

FIG. 4 A patient with severe traumatic brain injury with plain CT head (A, B) done at the time of admission showing no obvious abnormality. CT angiogram (C) shows filling of intracranial vessels. MRI done 2 days later showed foci of restricted diffusion (arrows, D, E) and multiple foci of micro-hemorrhages on susceptibility weighted images (arrows, F, G) suggestive likely of diffuse axonal injury.

literature that used MR perfusion as an ancillary imaging tool in TBI and more research studies are needed to establish the reliability of this technique in the clinical setting.

Diffusion tensor imaging

Diffusion tensor imaging (DTI) is a promising technique that provides vital information on the connections within the brain. It provides information on microstructural changes and structural connectivity of the brain. Two main DTI metrics include fractional anisotropy (FA) and mean diffusivity (MD). FA calculates the anisotropy within a voxel and MD quantifies the average magnitude of diffusion over all sampled directions. Normal white matter of the brain is highly anisotropic and thus has high FA and low MD. However, during traumatic axonal injury FA and MD can be altered. Wallace and colleagues conducted a systematic review and meta-analysis recently. They included studies of DTI changes in mild, moderate, and severe TBI, which showed a variety of brain regions involved (Wallace, Mathias, & Ward, 2018). In the analysis of the mild TBI subgroup, 88% of the all of the regions examined had significantly lower FA values when compared to the controls. In the analysis of the moderate–severe TBI subgroup, 92% of all of the regions were found to have lower FA values compared with the controls. When Wallace and colleagues looked at MD, 95% of all of the regions examined were higher in MD in the mild TBI group compared to the controls. In moderate–severe TBI, 100% of all of the regions examined were higher in MD in the TBI group compared to the controls (Wallace et al., 2018). There are some limitations in use of DTI in TBI patients. There is not enough evidence in the literature to suggest that DTI can be used to diagnose mild TBI in individual patients (Niogi & Mukherjee, 2010; Shenton et al., 2012; Ware et al., 2017). In addition, findings of decreased fractional anisotropy lack specificity. This is because fractional anisotropy alterations can be seen in variety of other neurological conditions especially that affect the white matter (Sbardella, Tona, Petsas, & Pantano, 2013).

CT perfusion and in-hospital mortality

Patients with TBI when presented to the hospital, several of them might have had fatal injuries and may not be recognized at the hospital admission because of inadequate clinical examination. Hence having a reliable imaging test that predicts in-hospital mortality in patients with severe TBI is extremely important. CT perfusion can be a reliable diagnostic imaging tool as it predicts functional aspects of the brain including cerebral blood flow and cerebral blood volume. There are only few clinical studies that are present in the literature, which shows that CT perfusion can be a valid and reliable diagnostic tool that rightly predicts the parameters of brain death in patients with severe TBI. Here in, we present some of those clinical studies.

CTP has rarely been used for patients with severe TBI due to lack of adequate reliable evidence on the efficacy of CTP in this patient population. Wintermark and colleagues described one of the first uses of CTP in patients with severe TBI upon hospital admission (Wintermark et al., 2004). They assessed the prognostic value of admission perfusion computed tomography in patients with severe head trauma (Wintermark et al., 2004). In this prospective study, they included 130 patients with severe trauma, aged 19 to 86 years, admitted with a Glasgow Coma Scale score of 8 or less. Patients underwent CT perfusion as part of their admission CT survey. Clinical data, unenhanced cerebral CT findings, and perfusion CT scans were evaluated with respect to the Glasgow Outcome Scale (GOS) score at 3 months. CT perfusion features were evaluated in patients with intracranial hypertension, cerebral contusions, and juxtadural hematomas. Ordered logistic regression was used to determine risk factors for an unfavorable GOS score at 3 months. Their results pointed out that CT perfusion is more sensitive than conventional unenhanced CT in the detection of cerebral contusions. CT perfusion featured specific patterns with respect to patient outcome, with normal brain perfusion or hyperemia in patients with favorable outcome, and oligemia in patients with unfavorable outcome. The number of arterial territories with low regional cerebral blood volume at CT perfusion was an independent prognostic factor ($P=0.008$), as were mean arterial pressure at the scene of accident ($P=0.083$), base excess at admission ($P=0.002$), presence of skull fractures ($P=0.041$), and signs of herniation ($P=0.013$) at admission unenhanced cerebral CT. CT perfusion also showed a range of brain perfusion alterations in patients with juxtadural collections, cerebral edema, or intracranial hypertension. Authors have concluded that CT perfusion in patients with severe head trauma provides independent prognostic information regarding functional outcome (Wintermark et al., 2004).

In 2012, Kaloostian and colleagues measured quantitative cortical mantle cerebral blood flow using xenon computed tomography within the first 12 h after severe TBI (Kaloostian et al., 2012). Outcome measures were assessed by Glasgow Outcome Scale score at discharge and 1, 3, and 6 months after discharge. The results of this study suggest that quantitative cerebral cortical mantle cerebral blood flow measured within the first 6 and 12 h after TBI predicts 6-month outcome (Kaloostian et al., 2012). Quantitative cortical mantle CBF differentiated GOS 1 and GOS 2 (dead or vegetative state) from GOS 3 to 5 (severely disabled to good recovery; $P<0.001$). Receiver operating characteristic (ROC) curve analysis for percent total normal plus hyperemic flow volume (TNHV) predicting GOS 3 to 5 outcome at 6 months for CBF measured <6 and <12 h after injury showed ROC area under the curve (AUC) cut-scores of 0.92 and 0.77, respectively. These findings give evidence that measurement of CBF at hospital admission can be useful in guiding TBI patient care.

We published a recent report on assessing the feasibility and safety of CT perfusion in patients with severe TBI at hospital admission and to examine whether early in-hospital mortality could be characterized with CT perfusion (Shankar et al., 2020). To our knowledge, ours is the first study in the literature to characterize early in-hospital mortality of patients with severe TBI on the basis of admission CT perfusion findings. In our study, a total of 19 patients with severe TBI were enrolled. No complications associated with CT perfusion were observed. In the first 48 h after admission, 4 patients died. The results of our study point out that CT perfusion showed 75% sensitivity, 100% specificity, 100% positive predictive value (PPV), and 93.75% negative predictive value (NPV) (Shankar et al., 2020). In conclusion, the findings of our study point out that admission CT perfusion with severe TBI is feasible and safe. Admission CT perfusion findings can correctly classify early in-hospital mortality in the first 48 h of hospital admission. These results are in keeping with what we have previously reported for PCT characteristics of brain death (Shankar et al., 2018; Shankar & Vandorpe, 2013).

In a case series by Bendinelli et al., authors hypothesized that in severe TBI patients early CT perfusion may provide additional information beyond the non-contrast CT (Bendinelli, Bivard, Nebauer, Parsons, & Balogh, 2013). They did subset analysis of an ongoing prospective observational study on trauma patients with severe TBI who did not require craniectomy and deteriorated or failed to improve neurologically during the first 48 h from trauma. Subsequently to follow-up non-contrast CT, a CT perfusion was obtained. CT perfusion altered clinical management in three patients (10%), who were diagnosed with massive and unsurvivable strokes despite minimal changes on unenhanced head CT. Authors concluded that when compared to unenhanced head CT, CT perfusion provided additional diagnostic information in 60% of patients with severe TBI. CT perfusion altered clinical management in 10% of patients (Bendinelli et al., 2013).

Another recent report by Bendinelli et al., pointed out that following severe TBI, CT perfusion performed at the time of the first follow-up plain CT is extremely valuable tool for the early outcome prediction in severe TBI patients (Bendinelli et al., 2017). From the findings of the study, authors concluded that CT perfusion can have a potential impact on management and it is cost effective in patients with severe TBI. A total of 50 patients were investigated with CT perfusion and out of them 30 patients (50.8%) had an unfavorable outcome. GCS was the moderate predictor of unfavorable outcome with AUC = 0.74, CT perfusion showed greater predictive ability with AUC = 0.92 and AUC for the presence of ischemia is 0.81. Authors concluded that CT perfusion is extremely valuable tool for early outcome prediction in patients with severe TBI.

In a more recent study, authors prospectively evaluated whole-brain CT perfusion on admission to predict delayed cerebral ischemia (DCI) in patients with aneurysmal subarachnoid hemorrhage (aSAH) (Dong et al., 2019). Out of 252 consecutive patients enrolled in this study, 191 patients were evaluated and 57 of whom (29.8%) developed DCI during hospitalization. Patients with diffused hypoperfusion had the highest incidence rate of DCI (43%, 46/107). Mean TMax produced the largest area under the curve of 0.726 (95% confidence interval [CI] 0.638–0.814), and a cut off value of 2.240s provided sensitivity of 73.7% and specificity of 71.6% for early prediction of developing DCI. Glasgow Coma Scale score (odds ratio [OR]=0.716, 95% CI 0.565–0.908, P=0.006), cerebral vasospasm (OR=6.117, 95% CI 1.427–26.223, P=0.015), hydrocephalus (OR=3.795, 95% CI 1.327–10.858, P=0.013), and qualitative CTP analysis (OR=3.383, 95% CI 1.686–6.789, P=0.001) were all significant independent predictors of DCI. From these results, authors concluded that whole-brain CTP within 24h of admission can qualitatively and quantitatively detect abnormal cerebral perfusion. It is possible to predict the risk of developing DCI after aSAH when the TMax value is larger than 2.240s (Dong et al., 2019).

In another interesting study, authors investigated if data obtained from a CT perfusion on admission could correlate to outcomes for the patients, including the patient's length of stay in the hospital and their initial and final GCS, as well as the modified Rankin Scale (mRS) on discharge. They included patients fulfilling the following inclusion criteria: over the age of 18 with mild, moderate, or severe TBI. Patients admitted with a diagnosis of TBI had CT perfusion performed within 48h of admission (Patchana et al., 2019). Preliminary data were obtained on six patients exhibiting TBI, ranging from mild to severe. The mean GCS of the patient cohort on admission was eight, with the most common mechanism of injury found to be falls (50%) and motor vehicle accidents (50%). Cerebral blood volume seemed to increase with Rankin value (Pearson's correlations coefficient=0.43 but was statistically insignificant (P=0.21)). Cerebral blood flow was found to be correlated with CBV, and both increased with Rankin score (Pearson's correlation coefficient=0.56) but were statistically insignificant (P=0.27). Authors concluded that CT perfusion might provide information that can be correlated clinically to patient outcomes (Patchana et al., 2019). This study shows that CBF and CBV may serve as useful indicators for prognostication of TBI patients.

The above studies assessed functional outcomes in patients with TBI by using CT perfusion at hospital admission. However, some of those studies have their own limitations. For example, Bendinelli et al performed CTP in patients with severe TBI when they did not improve neurologically during the first 48h after trauma (Bendinelli et al., 2013). They found low perfusion in one third of their patients and CTP altered clinical management in 10% of their patients who were diagnosed with massive and fatal strokes despite minimal changes on plain CT of head. This study used CTP with limited brain coverage due to limitation of CTP technology at the time of these studies. This limited coverage of the brain might have missed important findings in other parts of the brain. These studies addressed the neurological outcome at the end of hospital stay but failed to specify any association of CTP results with in-hospital mortality. Our study did not have those limitations and we have shown that admission CT perfusion of patients with severe TBI is feasible and safe (Shankar et al., 2020). Admission CT perfusion findings helped in correctly classifying early in-hospital mortality in the first 48h of hospital admission.

Applications to other areas of neuroscience

Neuroimaging techniques such as plain head CT, CT angiogram, CT perfusion, and MRI play an important role in diagnosis and prognosis of TBI. CT perfusion is used for the detection, and assessing the severity of TBI. The same imaging techniques can be used in diagnosis and prognosis of other neurological diseases such as brain tumors, Parkinson's disease, epilepsy, stroke, brain death, and variety of headaches. Perfusion imaging can also be used for detection of brain death as both TBI and brain death are inter-linked. Imaging studies that are being conducted for TBI patients can also be used for other relevant neurological disorders such as acute ischemic stroke and other brain hematomas. In acute ischemic stroke, CT perfusion can help in the diagnosis and selection of patients for thrombolytic therapy (Harrigan, Leonardo, Gibbons, Guterman, & Hopkins, 2005). In aneurysmal subarachnoid hemorrhage patients, CT perfusion can be used to identify the delayed cerebral ischemia and to guide medical and endovascular therapy for vasospasm (Harrigan et al., 2005). CT perfusion is also used in the diagnosis of brain tumors. Perfusion imaging in brain tumors helps in treatment planning process, to optimize decision-making in treatment and to enable monitoring and prognosis. Perfusion imaging can also help in differentiating malignant brain lesions, distinguishing true progression from post-treatment effects, and predicting survival after treatment therapies (Eilaghi et al., 2016). Technological improvements should improve accuracy and expand the anatomic area imaged by CT perfusion. Easy availability of CT perfusion should increase the usage of this technology and broaden the range of clinical applications (Harrigan et al., 2005).

Mini-dictionary of terms

CT: It is a fast and readily available imaging test and requires no contrast medium. CT can accurately and promptly diagnoses structural abnormalities such as skull fractures, intracranial bleeds, brain contusions, and brain herniation.

CTA: CT Angiography is a minimally invasive imaging technique wherein we can assess the health of veins and arteries.

CTP: It is a non-invasive imaging test that provides detailed information on the flow of blood to the brain.

Cerebral Blood Flow: It is defined as the volume of blood that flows per unit mass and per unit time in the brain tissue. It is expressed in mL/100 g/min.

Cerebral Blood Volume: It is defined as the volume of blood that is present at a given point of time in the brain tissue. It is expressed in mL/100 g.

Fractional Anisotropy: It is a common measurement used in DTI studies. It has a scale value of 0 to 1 that describes the degree of anisotropy of diffusion process. The value 0 represents isotropic diffusion and 1 represents anisotropic diffusion.

Glasgow Coma Scale: It is a neurological scale used to describe the extent of consciousness in patients with traumatic brain injury.

Glasgow Outcome Scale: It is a global scale to describe functional outcomes of the patients. It is divided into five categories: Dead, Vegetative State, Severe Disability, Moderate Disability, or Good Recovery.

Extended Glasgow Outcome Scale: It provides more detailed categorization of Severe Disability, Moderate Disability, and Good Recovery into a lower and upper category.

Modified Rankin Scale: It is a commonly used scale to determine the degree of dependence or disability in the daily activities of patients with TBI, stroke or other neurological disorders.

Traumatic Brain Injury: It is an injury caused to the brain because of bump, blow, or jolt or penetrating head injury.

Key facts of TBI

The severity of the TBI is most commonly classified as mild, moderate and severe according to the Glasgow Coma Scale.

Neuroimaging plays a crucial role in the evaluation, management, diagnosis, and prognosis of TBI.

Plain head CT accurately and promptly diagnoses structural abnormalities such as skull fractures, intracranial bleeds, brain contusions, and brain herniation.

CT perfusion imaging technique can quantify cerebral blood volume and cerebral blood flow.

Diffusion tensor imaging is a promising technique that provides vital information on the connections within the brain.

There are reports in the literature which show that CT perfusion is a very useful method in assessing the diagnosis of TBI patients.

CT perfusion possess eminent potential to become standard imaging technique in determination of in-hospital mortality in severe TBI patients.

Summary points

- Traumatic brain injury is a major catastrophic event occurring primarily in young and healthy population who previously had excellent quality of life.
- It is estimated that more than 1.7 million cases of TBI occurs in the United States every year and approximately 5.4 billion people live with disabilities because of TBI in the United States alone.
- TBI can be detected using various imaging techniques such as plain head CT, CT angiogram, CT perfusion, and MRI. Plain head CT and CT angiogram have their own disadvantages, as they do not provide any functional aspects of the brain.
- CTP provides both anatomical and functional aspects of the brain, and it is very useful in detection of TBI and brain death.
- In this chapter, we provide latest evidences on using CT perfusion at hospital admission and its relationship with in-hospital mortality rate.
- CTP can be a reliable and prompt triage tool to diagnose brain death in patients with severe TBI. This will help reduce resource intensive but futile care in those who are already dead at the time of hospital admission and may facilitate the precious gift of organ donation.

References

Albers, G. W., Lansberg, M. G., Kemp, S., Tsai, J. P., Lavori, P., Christensen, S., ... Marks, M. P. (2017). A multicenter randomized controlled trial of endovascular therapy following imaging evaluation for ischemic stroke (DEFUSE 3). *International Journal of Stroke: Official Journal of the International Stroke Society*, *12*(8), 896–905. https://doi.org/10.1177/1747493017701147.

Areas, F. Z., Schwarzbold, M. L., Diaz, A. P., Rodrigues, I. K., Sousa, D. S., Ferreira, C. L., ... Walz, R. (2019). Predictors of hospital mortality and the related burden of disease in severe traumatic brain injury: A prospective multicentric study in Brazil. *Frontiers in Neurology*, *10*, 432. https://doi.org/10.3389/fneur.2019.00432.

Astrup, J. (1982). Energy-requiring cell functions in the ischemic brain: Their critical supply and possible inhibition in protective therapy. *Journal of Neurosurgery*, *56*, 482–497.

Bendinelli, C., Bivard, A., Nebauer, S., Parsons, M., & Balogh, Z. (2013). Brain CT perfusion provides additional useful information in severe traumatic brain injury. *Injury*, *44*, 1208–1212.

Bendinelli, C., Cooper, S., Evans, T., Bivard, A., Pacey, D., Parson, M., & Balogh, Z. J. (2017). Perfusion abnormalities are frequently detected by early CT perfusion and predict unfavourable outcome following severe traumatic brain injury. *World Journal of Surgery*, *41*(10), 2512–2520. https://doi.org/10.1007/s00268-017-4030-7.

Corral, L., Herrero, J., Monfort, J., Ventura, J., Javierre, C., Juncadella, M., ... Gabarrós, A. (2009). First CT findings and improvement in GOS and GOSE scores 6 and 12 months after severe traumatic brain injury. *Brain Injury*, *23*(5), 403–410.

Dong, L., Zhou, Y., Wang, M., Yang, C., Yuan, Q., & Fang, X. (2019). Whole-brain CT perfusion on admission predicts delayed cerebral ischemia following aneurysmal subarachnoid hemorrhage. *European Journal of Radiology*, *116*, 165–173. https://doi.org/10.1016/j.ejrad.2019.05.008.

Du, E. H. Y., & Shankar, J. J. S. (2017). Rapid endovascular treatment of acute ischemic stroke: What a general radiologist should know. *Canadian Association of Radiologists Journal = Journal l'Association Canadienne Des Radiologistes*, *68*(2), 154–160. https://doi.org/10.1016/j.carj.2016.10.002.

Dupas, B., Gayet-Delacroix, M., Villers, D., Antonioli, D., Veccherini, M. F., & Soulillou, J. P. (1998). Diagnosis of brain death using two-phase spiral CT. *AJNR. American Journal of Neuroradiology*, *19*(4), 641–647.

Eilaghi, A., Yeung, T., d'Esterre, C., Bauman, G., Yartsev, S., Easaw, J., ... Frayne, R. (2016). Quantitative perfusion and permeability biomarkers in brain cancer from tomographic CT and MR images. *Biomarkers in Cancer*, *8*(Suppl 2), 47–59. https://doi.org/10.4137/BIC.S31801.

Evans, J. A., van Wessem, K. J. P., McDougall, D., Lee, K. A., Lyons, T., & Balogh, Z. J. (2010). Epidemiology of traumatic deaths: Comprehensive population-based assessment. *World Journal of Surgery*, *34*(1), 158–163. https://doi.org/10.1007/s00268-009-0266-1.

Greer, D., Varelas, P., Haque, S., & Wijdicks, E. (2008). Variability of brain death determination guidelines in leading US neurologic institutions. *Neurology*, *70*(4), 284–289.

Harrigan, M. R., Leonardo, J., Gibbons, K. J., Guterman, L. R., & Hopkins, L. N. (2005). CT perfusion cerebral blood flow imaging in neurological critical care. *Neurocritical Care*, *2*(3), 352–366. https://doi.org/10.1385/NCC:2:3:352.

Huang, A., Lee, C.-W., Hsieh, H.-J., Yang, C.-C., Tsai, Y.-H., Tsuang, F.-Y., ... Tsai, J.-C. (2011). Early parenchymal contrast extravasation predicts subsequent hemorrhage progression, clinical deterioration, and need for surgery in patients with traumatic cerebral contusion. *Journal of Trauma-Injury Infection & Critical Care*, *71*(6), 1593–1599.

Kaloostian, P., Robertson, C., Gopinath, S. P., Stippler, M., King, C. C., Qualls, C., ... Nemoto, E. M. (2012). Outcome prediction within twelve hours after severe traumatic brain injury by quantitative cerebral blood flow. *Journal of Neurotrauma*, *29*(5), 727–734. https://doi.org/10.1089/neu.2011.2147.

Kramer, A. H., & Roberts, D. J. (2014). Computed tomography angiography in the diagnosis of brain death: A systematic review and meta-analysis. *Neurocritical Care*, *21*(3), 539–550. https://doi.org/10.1007/s12028-014-9997-4.

Leclerc, X., Taschner, C. A., Vidal, A., Strecker, G., Savage, J., Gauvrit, J. Y., & Pruvo, J. P. (2006). The role of spiral CT for the assessment of the intracranial circulation in suspected brain-death. *Journal of Neuroradiology. Journal de Neuroradiologie*, *33*(2), 90–95.

Maas, A. I. R., Menon, D. K., Adelson, P. D., Andelic, N., Bell, M. J., Belli, A., ... InTBIR Participants and Investigators. (2017). Traumatic brain injury: Integrated approaches to improve prevention, clinical care, and research. *The Lancet. Neurology*, *16*(12), 987–1048. https://doi.org/10.1016/S1474-4422(17)30371-X.

MacDonald, D., Stewart-Perrin, B., & Shankar, J. J. S. (2018). The role of neuroimaging in the determination of brain death. *Journal of Neuroimaging: Official Journal of the American Society of Neuroimaging*, *28*(4), 374–379. https://doi.org/10.1111/jon.12516.

Mayer, S., & Kossoff, S. (1999). Withdrawal of life support in the neurological intensive care unit. *Neurology*, *52*(8), 1602.

McMillan, T., Bigler, E. D., Teasdale, G., Ponsford, J., & Murray, G. D. (2018). Outcome assessment after traumatic brain injury. *The Lancet. Neurology*, *17*(4), 299. https://doi.org/10.1016/S1474-4422 (18)30041-3.

Mushkudiani, N., Hukkelhoven, C., Hernández, A., Murray, G., Choi, S., Maas, A., & Steyerberg, E. (2008). A systematic review finds methodological improvements necessary for prognostic models in determining traumatic brain injury outcomes. *Journal of Clinical Epidemiology*, *61*(4), 331–343.

Narayan, R. K., Michel, M. E., Ansell, B., Baethmann, A., Biegon, A., Bracken, M. B., ... Yurkewicz, L. (2002). Clinical trials in head injury. *Journal of Neurotrauma*, *19*(5), 503–557. https://doi.org/10.1089/089771502753754037.

National Collaborating Centre for Acute Care (UK). (2007). *Head injury: Triage, assessment, investigation and early management of head injury in infants, children and adults*. National Collaborating Centre for Acute Care (UK). http://www.ncbi.nlm.nih.gov/books/NBK53036/.

National Scientific Advisory Committee. (2004). *Identifying priorities for research and capacity development in injury as a multi-institute strategic initiative*. Canadian Institute of Health Research.

Niogi, S. N., & Mukherjee, P. (2010). Diffusion tensor imaging of mild traumatic brain injury. *The Journal of Head Trauma Rehabilitation*, *25*(4), 241–255. https://doi.org/10.1097/HTR.0b013e3181e52c2a.

Nogueira, R. G., Jadhav, A. P., Haussen, D. C., Bonafe, A., Budzik, R. F., Bhuva, P., … DAWN Trial Investigators. (2018). Thrombectomy 6 to 24 hours after stroke with a mismatch between deficit and infarct. *The New England Journal of Medicine*, *378*(1), 11–21. https://doi.org/10.1056/NEJMoa1706442.

Patchana, T., Dorkoski, R., Zampella, B., Wiginton, J. G., Sweiss, R. B., Menoni, R., & Miulli, D. E. (2019). The use of computed tomography perfusion on admission to predict outcomes in surgical and nonsurgical traumatic brain injury patients. *Cureus*, *11*(7). https://doi.org/10.7759/cureus.5077, e5077.

Rosenblum, M. L., Hoff, J. T., Norman, D., Weinstein, P. R., & Pitts, L. (1978). Decreased mortality from brain abscesses since advent of computerized tomography. *Journal of Neurosurgery*, *49*(5), 658–668. https://doi.org/10.3171/jns.1978.49.5.0658.

Sawicki, M., Bohatyrewicz, R., Safranow, K., Walecka, A., Walecki, J., Rowinski, O., … Czarnecka, A. (2013). Dynamic evaluation of stasis filling phenomenon with computed tomography in diagnosis of brain death. *Neuroradiology*, *55*(9), 1061–1069. https://doi.org/10.1007/s00234-013-1210-5.

Sbardella, E., Tona, F., Petsas, N., & Pantano, P. (2013). DTI measurements in multiple sclerosis: Evaluation of brain damage and clinical implications. *Multiple Sclerosis International*, *2013*, 671730. https://doi.org/10.1155/2013/671730.

Shankar, J. J. S., Green, R., Virani, K., Wong, H., Eddy, K., & Vandorpe, R. (2020). Admission perfusion CT for classifying early in-hospital mortality of patients with severe traumatic brain injury: A pilot study. *AJR. American Journal of Roentgenology*, *214*(4), 872–876. https://doi.org/10.2214/AJR.19.21599.

Shankar, J., Langlands, G., Doucette, S., & Phillips, S. (2016). CT perfusion in acute stroke predicts final infarct volume-inter-observer study. *Canadian Journal of Neurological Sciences*, *43*, 93–97.

Shankar, J., & Lum, C. (2011). Whole brain CT perfusion on a 320-slice CT scanner. *Indian Journal of Radiology and Imaging*, *21*(3), 209–214.

Shankar, J. J. S., Stewart-Perrin, B., Quraishi, A.-U.-R., Bata, I., & Vandorpe, R. (2018). Computed tomography perfusion aids in the prognostication of comatose Postcardiac arrest patients. *The American Journal of Cardiology*, *121*(7), 874–878. https://doi.org/10.1016/j.amjcard.2017.12.027.

Shankar, J. J. S., & Vandorpe, R. (2013). CT perfusion for confirmation of brain death. *AJNR. American Journal of Neuroradiology*, *34*(6), 1175–1179. https://doi.org/10.3174/ajnr.A3376.

Shemie, S., Hornby, L., Baker, A., Teitelbaum, J., Torrance, S., Young, K., … Noel, L. (2014). International guideline development for the determination of death. *Intensive Care Medicine*, *40*(6), 788–797.

Shenton, M. E., Hamoda, H. M., Schneiderman, J. S., Bouix, S., Pasternak, O., Rathi, Y., … Zafonte, R. (2012). A review of magnetic resonance imaging and diffusion tensor imaging findings in mild traumatic brain injury. *Brain Imaging and Behavior*, *6*(2), 137–192. https://doi.org/10.1007/s11682-012-9156-5.

Soustiel, J., Mahamid, E., Goldsher, D., & Zaaroor, M. (2008). Perfusion-CT for early assessment of traumatic cerebral contusions. *Neuroradiology*, *50*(2), 189–196.

Taylor, T., Dineen, R. A., Gardiner, D. C., Buss, C. H., Howatson, A., & Pace, N. L. (2014). Computed tomography (CT) angiography for confirmation of the clinical diagnosis of brain death. *The Cochrane Database of Systematic Reviews*, *3*. https://doi.org/10.1002/14651858.CD009694.pub2, CD009694.

Teasdale, G., & Jennett, B. (1974). Assessment of coma and impaired consciousness. A practical scale. *Lancet (London, England)*, *2*(7872), 81–84. https://doi.org/10.1016/s0140-6736(74)91639-0.

Trauma Registry Report on Injury in Nova Scotia: Provincial Report 2012–2013 Data Year, 2014.; 2014.

Uwano, I., Kudo, K., Sasaki, M., Christensen, S., Østergaard, L., Ogasawara, K., & Ogawa, A. (2012). CT and MR perfusion can discriminate severe cerebral hypoperfusion from perfusion absence: Evaluation of different commercial software packages by using digital phantoms. *Neuroradiology*, *54*, 467–474.

Wallace, E. J., Mathias, J. L., & Ward, L. (2018). Diffusion tensor imaging changes following mild, moderate and severe adult traumatic brain injury: A meta-analysis. *Brain Imaging and Behavior*, *12*(6), 1607–1621. https://doi.org/10.1007/s11682-018-9823-2.

Ware, J. B., Hart, T., Whyte, J., Rabinowitz, A., Detre, J. A., & Kim, J. (2017). Inter-subject variability of axonal injury in diffuse traumatic brain injury. *Journal of Neurotrauma*, *34*(14), 2243–2253. https://doi.org/10.1089/neu.2016.4817.

Wilson, J., Pettigrew, L., & Teasdale, G. (1998). Structured interviews for the Glasgow outcome scale and the extended Glasgow outcome scale: Guidelines for their use. *Journal of Neurotrauma*, *15*(8), 573–585.

Wilson, C., & Shankar, J. (2014). Diagnosing vasospasm after subarachnoid hemorrhage: CTA and CTP. *The Canadian Journal of Neurological Sciences*, *41*(3), 314–319.

Wintermark, M., van Melle, G., Schnyder, P., Revelly, J.-P., Porchet, F., Regli, L., … Chioléro, R. (2004). Admission perfusion CT: Prognostic value in patients with severe head trauma. *Radiology*, *232*, 211–220.

Chapter 2

Predictors of outcome in moderate and severe traumatic brain injury

Rosalia Zangari[a], Paolo Gritti[b], and Francesco Biroli[a]

[a]*FROM Research Foundation, Papa Giovanni XXIII Hospital, Neuroscience Area, Bergamo, Italy,* [b]*Department of Anesthesia and Critical Care Medicine, Papa Giovanni XXIII Hospital, Bergamo, Italy*

List of abbreviations

ADAPT	approaches and decisions in acute pediatric TBI trial
AR	autoregulation
CPP	cerebral perfusion pressure
CSF	cerebral spinal fluid
CENTER-TBI	collaborative European neurotrauma effectiveness research in TBI
CREACTIVE	collaborative research on acute traumatic brain injury in intensive care medicine in Europe
CER	comparative effectiveness research
CT	computed tomography
CRASH	corticosteroid randomization after significant head injury
GCS	Glasgow Coma Scale
GOS	Glasgow Outcome Scale
GOSE	Glasgow Outcome Scale-extended
GFAP	glial fibrillary acid protein
HICs	high-income countries
MPACT	international mission for prognosis and analysis of clinical trials in TBI
ICP	intracranial cerebral pressure
ICU	intensive care unit
LMICs	low- and middle-income countries
MAP	mean arterial pressure
NF-L	neurofilament-light
NSE	neuron-specific enolase
pTBI	pediatric TBI
PM	predictive model
PRx	pressure reactivity index
RCTs	randomized controlled trials
tau	tau protein
TRACK-TBI	transforming research and clinical knowledge in TBI
UCH-L1	ubiquitin C-terminal hydrolase-l1

Introduction

Traumatic brain injury (TBI) is one of the leading causes of disability and mortality worldwide. Much research is undertaken to improve outcome after TBI; however, no significant improvements in prognosis have been noted in recent years. Because of disease heterogeneity, including variability of clinical course and differences in management standards among different countries, prediction of outcome after TBI remains challenging. Accurate prediction of outcome after moderate-to-severe TBI is essential to guide clinical management, patient course organization, and family expectations. In clinical

Diagnosis and Treatment of Traumatic Brain Injury. https://doi.org/10.1016/B978-0-12-823347-4.00001-4

practice, reliable assessment of prognosis could become important when making decisions, as clinicians treating patients often make therapeutic decisions based on personal judgment.

Outcomes after TBI are influenced by many factors, including patient demographics, injury characteristics, and post-injury factors (named predictors). For accurate outcome forecasting, predictors need to be considered jointly in a statistical model (named prognostic/predictive model, PM).

The International Mission for Prognosis and Analysis of Clinical Trials in TBI (IMPACT) and Corticosteroid randomization after significant head injury (CRASH) are frequently used for dealing with the heterogeneity of TBI and define more precisely the prognostic value of many known predictors [i.e., age, Glasgow Coma Scale (GCS) or GCS-motor score and pupillary reactivity].

Along with the heterogeneity of patient- or injury-related factors, the differences in outcome measures and research methodology (e.g., mortality or outcome measured at different time points), as well as differences between the standards of care across centers [e.g., models not being generalizable to different settings, such as for high-income countries (HICSs) compared with low and middle-income countries (LMICs)] complicate prognostic research in TBI. This highlights the need to confirm or to update the current state-of-the-art in prognostic research in TBI and to research new potential predictors.

To this end, an international collaborative effort (Comparative Effectiveness Research, CER) is underway to update the available knowledge. The proposed approach embraces heterogeneity, not treating it as an obstacle to overcome, but as an advantage that can be leveraged to facilitate outcome prognostication.

The chapter aims to present a brief overview of the main predictors and available PMs developed for predicting subsequent outcomes in moderate–severe TBI, offering a review of the literature, general considerations, and the authors' personal experience. Most of these data come from HICSs, but evolving evidence from LMICs are also included.

Specific sections are dedicated to exploring the use of predictors and providing methodological tips for describing the relationship between predictors and outcomes.

Predictors and outcome measures

Predictors are conventionally defined as variables recorded at admission to the emergency department or during the acute phase of ICU treatment (acute predictors). IMPACT and CRASH have been important to confirm the core set of acute predictors of unfavorable outcome after moderate–severe TBI: older age, less responsive pupils, and lower GCS (motor) score (Table 1). It should be noted that these acute predictors only explain ~35% of the variance in outcomes (Maas, Lingsma, & Roozenbeek, 2015).

In the days after injury, TBI evolves in a varied spectrum of secondary injuries. The role of these sub-acute predictors of outcome in moderate and severe TBI remains less studied, with only a few studies focusing on sub-acute clinical variables as possible outcome predictors. Knowing sub-acute clinical variables could be significant in guiding rehabilitation and the post-acute course (Gritti et al., 2019).

For an overview of acute and sub-acute predictors of outcome, see Fig. 1. This overview is not at all meant to be complete; the interested reader is encouraged to refer to the reported literature for more in-depth information.

Regarding outcome measures, mortality and functional outcome, as assessed by the Glasgow Outcome Scale and its Extended version (GOS/GOSE), are usually considered (Wilson, Pettigrew, & Teasdale, 1998).

Mortality is evaluated as in-hospital and late mortality. The GOS or GOSE are dichotomized into favorable [good recovery (score > 4)] versus unfavorable [severe disability, vegetative state, and mortality (score < or ≤ 4, respectively)].

The time of outcome assessment varies considerably, from discharge to 6–12 months or later. GOSE has been criticized for its failure to capture the multifaceted effects of TBI, especially in the cognitive dimension. Composite outcomes, including batteries of both functional and cognitive measures (e.g., domains: episodic memory, information processing speed, and executive functions), have been proposed as secondary outcomes to overcome this limitation, because of their better representation of the spectrum of long-term TBI outcomes (Alali et al., 2015).

Acute predictors

Age is one of the strongest and best-documented predictors of TBI, with a large body of evidence demonstrating that the outcome is poorer for older age groups. Some debate exists, however, whether age is a stronger outcome predictor than GCS. CRASH results identify older age in HICs and lower GCS in LMICs as stronger outcome predictors (MRC CRASH Trial Collaborators et al., 2008). The average age in TBI is increasing in HICs, which could be explained by higher life expectancy and advances in health care. However, older people are also more susceptible to an unfavorable outcome, calling for further research; it is not only a clinical problem but is also a social and ethical problem.

TABLE 1 An overview of IMPACT and CRASH.

Characteristics	IMPACT	CRASH
Endpoints	6-Month mortality 6-Month Glasgow Outcome Scale	14-Day mortality 6-Month Glasgow Outcome Scale
Population, *(n)*	9205	10,008
Enrollment period, *(y)*	1984–1995	1999–2004
Study type	8 Clinical trials + 3 observational studies (retrospective)	1 Clinical trial (prospective)
Setting/applicability	High-Income Countries	High-Income Countries Low/Middle-Income Countries
Online calculator	http://tbi-impact.org/	http://www.crash2.lshtm.ac.uk/Risk%20 calculator/index.html
Core model		
Age, (y)	14–99	≤40, 40–99
Glasgow Coma Scale[a]		3–14
Motor score	✓ None/extension ✓ Abnormal flexion ✓ Normal flexion ✓ Localizes/obeys ✓ Untestable/missing	
Pupil reactivity to light	Pupils: both, one, or none	Pupils: both, one, or none
Extracranial injury		Y/N
CT model		
Hypoxia	Y or suspected/N	
Hypotension	Y or suspected/N	
CT abnormalities	Marshall CT classification (I, II, III–IV, and V–VI)	Petechial hemorrhages; obliteration of 3rd or basal cisterns; midline shift; non-evacuated hematoma
Subarachnoid hemorrhage	Y/N	Y/N
Epidural hematoma	Y/N	
Laboratory model (threshold values, *mmol/L*)		
Glucose	< 6, 6–8.9, 9–11.9, 12–14.9 or ≥ 15	
Hemoglobin	< 9, 9–11.9, 12–14.9 or ≥ 15	

[a]*The IMPACT database includes patients with moderate and severe TBI Glasgow Coma Scale ≤ 12 (variable not included in the model.*
The table summarizes the main characteristics and predictors of IMPACT and CRASH studies (MRC CRASH Trial Collaborators et al., 2008; Steyerberg et al., 2008). *Gray* color: variables not applicable in the models. Computed Tomography (CT).

Age

With an aging population, TBI is being seen more frequently in elderly patients (Maas, Marmarou, Murray, Teasdale, & Steyerberg, 2007). Due to associated co-morbidities (such as diabetes, hypertension, and anticoagulant/antiplatelet treatment), elderly patients sustaining severe TBI have been reported to expect increased mortality rates or unfavorable recovery compared to younger patients with similar anatomical lesions at presentation. In-hospital mortality after severe TBI has been shown to reach up to 30%–40% in patients over 65–70 years old (Gritti et al., 2019; MRC CRASH Trial Collaborators et al., 2008; Walder et al., 2013). The use of anticoagulant and antiplatelet treatments in elderly patients explain why acute subdural hematoma is frequently seen in elderly patients showing more severe injury with potentially worse outcome (Gritti et al., 2019).

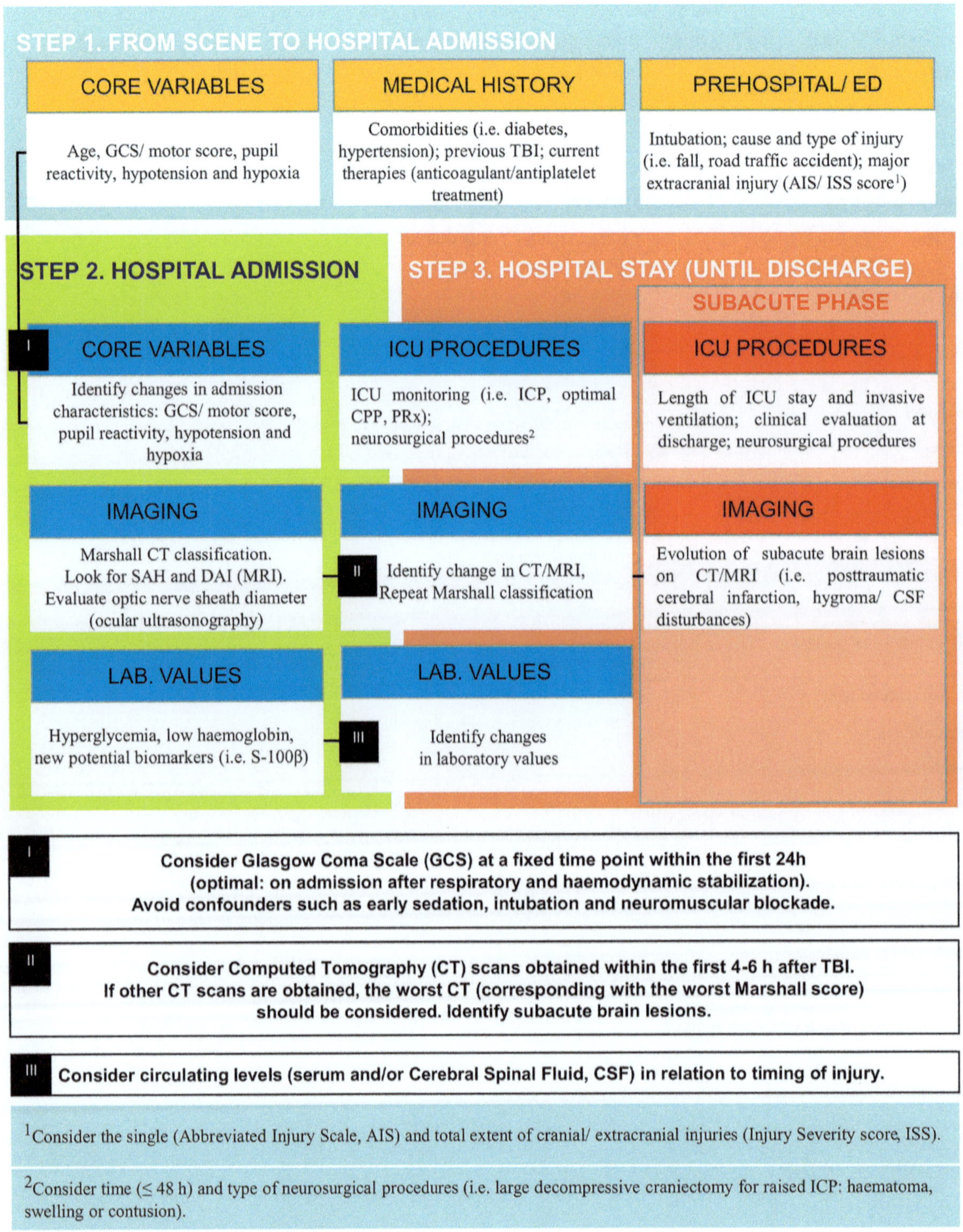

FIG. 1 Predictors to be considered in developing a predictive model in TBI. Predictors can be divided into 3 building-blocks of patient course. The figure serves an overview of the most common predictors of outcome and is not meant to be complete. Some variables are not discussed in the chapter but associated with outcome on literature. ED (emergency department). ICU (intensive care unit). Magnetic resonance imaging (MRI). Diffuse axonal injury (DAI). Subarachnoid hemorrhage (SAH). Cerebral perfusion pressure (CPP). Intracranial cerebral pressure (ICP). Pressure reactivity index (PRx).

The association between age and outcome is usually analyzed using threshold values, varying from 30 to 60 years of age. Only a few studies examine age as a continuous variable, and report a change around age 30–40 years, above which outcome becomes increasingly poorer (Lingsma, Roozenbeek, Steyerberg, Murray, & Maas, 2010; Maas, Lingsma, & Roozenbeek, 2015).

The age–gender relationship may also associate with outcome in TBI patients; however, its specific dynamics remain mostly unknown. Gender differences, such as estrogen and progesterone influences and hormonal treatment, have raised interest in explaining the possible positive effect in young women after TBI (Mushkudiani et al., 2008).

GCS

GCS is the common scale to assess the level of consciousness after injury. According to the GCS, TBI is classified as mild (13–15 GCS), moderate (9–12 GCS), and severe (3–8 GCS). Debate exists over whether GCS 13 should be classified as moderate or mild TBI (Steyerberg et al., 2019).

The score is often obtained on the scene, at the emergency department or after stabilization, and counts the lowest GCS score in patients with subsequent neurologic deterioration (within the first 24 h) (Gritti et al., 2019; Lingsma et al., 2010; Marmarou et al., 2007). With the current frequent use of early sedation, intubation and neuromuscular blockade to prevent secondary brain damages, accurate assessment of the motor scale may be confounded, leading to erroneously recording the motor score as being absent (Hukkelhoven et al., 2005; Stocchetti et al., 2004).

The greater use of sedation in LMICs could explain the different predictive power acquired by GCS in LMICs, as previously discussed (MRC CRASH Trial Collaborators et al., 2008).

GCS is most accurate at predicting outcome when combined with age and pupillary response. Age affects the relationship between the GCS score and anatomical severity. Elderly TBI patients have better GCS scores than younger patients with similar anatomical lesions. Although several potential explanations exist, the larger intracranial space due to cerebral atrophy could be the most likely reason (Salottolo, Levy, Slone, Mains, & Bar-Or, 2014).

PMs include either GCS or motor score, or both. When the evaluation is difficult (i.e., due to sedation or intubation), the GCS should be recorded as untestable (Hukkelhoven et al., 2005).

In pediatric TBI patients (≤2 years), a modified GCS score is used, adapted to the child's reaction to stimulation (online calculator: http://www.mdcalc.com/pediatric-glasgow-coma-scale-pgcs).

Abnormal pupillary response

Acute dilatation of the pupil and unresponsiveness to light is considered a concerning sign strongly associated with unfavorable prognosis. Acute abnormal pupillary reactions after TBI may be the result of uncal herniation with third cranial nerve compression and/or subsequent brain stem involvement (Maas, Lingsma, & Roozenbeek, 2015; Marmarou et al., 2007; MRC CRASH Trial Collaborators et al., 2008). Likewise, small unreactive pupils may be caused by the effects of deep sedation (Hukkelhoven et al., 2005) or direct injury to the orbit. The abnormal pupillary reaction is defined as anisocoria or mydriasis on scene (Gritti et al., 2019; Maas, Lingsma, & Roozenbeek, 2015). Pupillary reactivity seems to be less sensitive to changes over time and more informative when related to in-hospital mortality (Gritti et al., 2019).

Hypoxia and hypotension

TBI causes fluctuations in vital signs, such as blood pressure, pulse rate, respiratory rate, and temperature. Most of these parameters are evaluated on scene or after the patient arrives in the emergency department. Hypotension and hypoxia following TBI are recognized as significant causes of secondary insult associated with unfavorable outcome. Hypotension is indicated by a systolic blood pressure below 90 mmHg, and hypoxia is considered at a pO_2 below 60 mmHg or oxygen saturation 90% (Hukkelhoven et al., 2005; Maas, Lingsma, & Roozenbeek, 2015). The occurrence of cardiac/respiratory arrest, the pre-hospital airway and hemodynamic management should be taken into account when considering the prognostic effects of these predictors (Gritti et al., 2019).

CT findings

Computed tomography (CT) provides an objective assessment of structural damage to the brain and associated outcome. The most common classification used after TBI, the Marshall classification, systematically groups a spectrum of abnormalities—including diffuse injury, midline shift, obliterated basal cisterns, and mass lesions—into six categories. The Rotterdam classification integrates traumatic subarachnoid hemorrhage (SAH) and seems to have a stronger predictive value than the Marshall classification (Maas, Hukkelhoven, Marshall, & Steyerberg, 2005). Overall, the prognosis in patients with mass lesions is better for an epidural hematoma, and poorer for an acute subdural hematoma. Presence of midline shift, SAH, obliteration of the basal cisterns or third ventricle, and non-evacuated hematoma are associated with unfavorable outcome (Gritti et al., 2019; Maas et al., 2007; Mattioli et al., 2003; Murray et al., 2007; Servadei et al., 2000).

For an accurate predictor definition, admission CT scans obtained within the first 4–6 h after the injury should be considered. If other CT scans are performed, the worst CT (corresponding with the worst CT score) should be considered and used in the prognostic analyses (Maas, Steyerberg, et al., 2007). PMs usually include the presence of mass lesions and/or CT classification. The variables are categorized as follows: midline shift as $\leq$ or >5 mm; the basal cisterns as normal or abnormal (compressed or absent); the presence and type of mass lesions (extradural, subdural or intracerebral) as $\leq$ vs >25 mL in volume; epidural hematomas as present or absent; intra-ventricular blood and SAH as present or absent. Based on the amount of bleeding, SAH should be categorized as mild (defined as a small amount of blood at 1 or 2 sites) moderate (>2 sites) and severe (>2 sites filled bilaterally with blood or clots that have expanded to the original size of the cistern or fissure; Maas, Lingsma, & Roozenbeek, 2015).

Sub-acute predictors

Late neuroimaging features and clinical variables are possible predictors of outcome after TBI. Although early assessment of injury is crucial in decision-making, following clinical and neuroimaging features are more informative. Sub-acute factors, integrating the information of acute predictors, could be used as refinements to optimize the prognosis and guide the rehabilitation phase. In particular, a systematic neurologic evaluation at ICU discharge, (i.e., Oxford Handicap Scale), the presence of late brain lesions and cerebral spinal fluid (CSF; e.g., posttraumatic cerebral infarction, hydrocephalus and hygroma), the length of stay in the ICU and length of invasive ventilation (days) could be considered as late predictors (Gritti et al., 2019; Perel, Edwards, Shakur, & Roberts, 2008; Perel, Edwards, Wentz, & Roberts, 2006).

ICU neuromonitoring

ICU neuromonitoring for severe TBI aims to reduce secondary brain damage due to mass lesions and swelling. The rationale for monitoring is to continuously check the status of the intracranial compartment of the brain and its perfusion. The main clinical targets are to avoid intracranial hypertension and to keep an adequate cerebral blood flow sufficient for cell metabolism (autoregulation, AR), by control of intracranial pressure (ICP) and cerebral perfusion pressure (CPP, usually ranges approximately <60–70 mmHg).

Even though the effect of ICP monitoring on outcomes remains a controversial issue (Aiolfi et al., 2017), the documented relationship between intracranial hypertension and unfavorable outcomes has led to the widespread use of ICP >20 mmHg (for 15 min within a 1-h period) as the threshold for therapeutic interventions to lower the intracranial pressure (Yuan et al., 2015).

The pressure reactivity index (PRx), along with an optimal CPP, have been proposed as principal characters of AR. PRx is derived from ICP and mean arterial pressure (MAP), where a negative to zero value of the PRx implies intact pressure reactivity (detailed in the next chapters). Disturbances of AR are associated with unfavorable outcomes (Liu et al., 2015; Riemann et al., 2020; Svedung Wettervik, Howells, Enblad, & Lewén, 2019). However, despite the documented relationship between autoregulation and outcome, available ICU therapies and techniques could be still inadequate in modulating cerebrovascular reactivity, resulting in a non-improvement of outcome prediction.

Biomarkers

Several laboratory variables, routinely recorded on admission following TBI, have been associated with outcome. Among them, glycemia and prothrombin time show a positive linear relationship with outcome (increasing values associated with an unfavorable outcome), while hemoglobin and platelets show a negative relationship (Gritti et al., 2019; Van Beek et al., 2007). Glycemia and hemoglobin are included in IMPACT (Table 1).

More recently, biomarkers of neuronal tissue damage have been proposed as potential adjuvants to outcome prediction, including S-100β protein, neuron-specific enolase (NSE), Neurofilament-light (NF-L), glial fibrillary acid protein (GFAP), ubiquitin C-terminal hydrolase-L1 (UCH-L1), tau protein, tumor necrosis factor-alpha (TNF-α), interleukin-6 (IL-6), and myelin basic protein (MBP) (Maas, Lingsma, & Roozenbeek, 2015; Thelin et al., 2017).

To accurately demonstrate the association of a biomarker concentration with outcome, it is of importance to assess the circulating temporal profile of the biomarkers (serum and/or CSF levels) concerning the timing of the injury. For instance, UCH-L1, S-100β, and tau protein have a short half-life (hours), GFAP and NSE remain elevated for days and NF-L for several weeks. Another critical factor is the availability of biomarkers in the brain compartment. For instance, S-100β can be released from non-cerebral tissues into the brain in the early phase after injury and detected into the bloodstream after the breakdown of the blood brain barrier. In contrast, GFAP and NF-L are brain-specific biomarkers (E. Thelin et al., 2019).

The combination of different biomarkers, reflecting various aspects of the pathophysiological processes of TBI, could be a crucial strategy for improving the prediction of PMs (E. Thelin et al., 2019).

Pediatric TBI

Even though plasticity offers greater potential for recovery and reorganization, prediction of outcome following pediatric TBI (pTBI) can be difficult in a developing brain. As for adult TBI, mortality and GOSE-Peds (the pediatric version of the original GOSE interview) fail to describe the range of pTBI outcomes. Neurological disabilities, cognitive, and neuropsychological tests must be prioritized to define the burden of illness in the pediatric population. However, no clear gold standard exists, limiting the applicability of available measures, that often should be changed according to patient age (McCauley et al., 2012). IMPACT and CRASH are derived from predominately adult populations; in the development of these models, only patients as young as 14 years of age are included. Although their validity in the pediatric setting is questionable, potential benefits of adapting adult models by reweighing predictors using pediatric patients have been demonstrated (Gritti et al., 2019; Huth, Slater, Waak, Barlow, & Raman, 2020). However, deriving models directly from pediatric patients can further improve outcome prediction. The Approaches and Decisions in Acute Pediatric TBI Trial (ADAPT) and the CENTER-TBI seek to integrate a wide array of variables to predict physical and neurological recovery (Maas, Menon, et al., 2015; Sarnaik et al., 2018). Ideally, the novel insight produced by these studies will lead to the formulation of new clinical tools for the routine assessment of the risk for pediatric patient complications.

Prognostic models

The use of PMs is becoming increasingly important in TBI research as a tool to compare outcomes across institutions, healthcare systems, and countries, as provided by CER studies (detailed below), and for a better design and baseline risk stratification in clinical trials (Roozenbeek, Maas, Lingsma, et al., 2009). The failure of numerous clinical trials in TBI (RCTs) has been attributed to the broad heterogeneity and patient case-mix. To limit heterogeneity, RCTs often apply strict enrollment criteria (i.e., excluding patients with unfavorable prognosis, such as patients >65 years old) and require large study populations from multiple centers. These criteria are perhaps unrealistic from a clinical perspective and decreases generalizability (Roozenbeek, Maas, Marmarou et al., 2009).

Based on a recent systematic review, 67 PMs for functional outcome after moderate-to-severe TBI have been developed (Dijkland et al., 2020).

IMPACT and CRASH currently dominate the field of predictive modeling in TBI (Table 1).

IMPACT brought together data from 8 RCTs and 3 observational studies, mainly from the US and Europe (Steyerberg et al., 2008). IMPACT estimates 14-day mortality, as well as mortality and outcome at 6-months and includes 3 models of complexity (core, extended, and laboratory model).

CRASH is the largest trial ever conducted in TBI (MRC CRASH Trial Collaborators et al., 2008). CRASH estimates the 14-day mortality and 6-month outcome and includes 2 models of increasing complexity (core and extended model).

Both models are set up for moderate-severe TBI; however, CRASH also captures mild cases (GCS 13–14; Lingsma et al., 2015).

Epidemiological considerations: the importance of outcome prediction considering between-country differences

The majority of the literature on TBI comes from HICs in North America and Europe (~90%), while only a few studies are developed using populations from LMICs (Perel et al., 2006). It is noteworthy that only 2% of the PMs included patients from LMICs, considering that most of the burden (90%) occurs in these. Nevertheless, it may be inappropriate to generalize from models for HICs to lower settings, where adverse outcomes occur more frequently, and there is less access to treatment facilities and rehabilitation programs. It is reasonable to consider that the differences between HICs and LMICs may affect the accuracy of a PM (external validation).

CRASH, as compared with IMPACT, performs well when combining data from HICs and LMICs. This effect is probably the result of the CRASH model being developed based on patients (75%) from LMICs (Steyerberg et al., 2008). Along with common predictors, the CRASH considers the level of income in a country. It shows that the strength of outcome predictors varies according to whether patients are from high or low-middle–income countries (MRC CRASH Trial Collaborators et al., 2008). Given the lack of appropriate guidelines, the potential impact of risk identification strategies specifically designed for these countries to support high-quality care in lower settings may be substantial. To achieve this goal, specialty training in trauma among the healthcare workforce (Calero-Martinez et al., 2020) and trauma registries to derive appropriate guidelines are required, which can be a significant challenge in countries with scarce resources.

Approaches to predictive models in TBI: Discordance between clinical research and clinical management of TBI

CRASH and IMPACT become extremely useful for dealing with the heterogeneity of the TBI population and discriminate well between low- and high-risk patients. These extensive analyses have defined and quantified more precisely the predictive value of known variables and yielded new predictors in moderate-severe TBI.

However, external validations of these models show substantial variation in performance (Dijkland et al., 2020; Roozenbeek et al., 2012).

The latest findings from the IMPACT studies have shown that the risk of unfavorable outcomes could vary by up to 3 times across centers (Lingsma et al., 2011). The inter-center variations in clinical care could produce substantial noise in results, explaining why it is difficult for a PM to make useful predictions.

CER studies (large-scale observational studies) are designed to overcome these limitations (Maas et al., 2012); taking advantage of heterogeneity in care to generate evidence, and by analyzing current guideline adherence, treatment variation, and therapies associated with the best outcomes in real-world situations (Steyerberg et al., 2019). The concept of an international cooperative effort to improve health care and lessen the global burden of TBI, as recommended in the InTBIR (International Initiative for Traumatic Brain Injury Research), has resulted in a rising number of linked CER projects. Possibilities for updating the IMPACT and CRASH models are currently being evaluated in various studies [including the Collaborative European NeuroTrauma Effectiveness Research in TBI (CENTER-TBI), Transforming Research and Clinical Knowledge in TBI (TRACK-TBI) and Collaborative REsearch on ACute Traumatic Brain Injury in intensiVe Care Medicine in Europe (CREACTIVE)] (Maas, Menon, et al., 2015; Roozenbeek, Maas, Marmarou, et al., 2009).

In this re-orientation of clinical research in TBI, some aspects should be considered (Maas et al., 2012; Timmons & Toms, 2012). With a focus on clinical management of TBI, observational studies tend to have much broader inclusion criteria than RCTs, where patients with the most extreme prognosis are often excluded. On the other hand, the epidemiological shift of TBI patients towards older and comorbid populations might increase rates of unfavorable outcomes in observational studies. Thus, a comparison between RCTs and observational studies should be interpreted with great caution.

Recommendation for developing a predictive model in TBI

A recent review of available methodological issues provides tentative guidelines for prognostic research on TBI (Retel Helmrich, Lingsma, Turgeon, Yamal, & Steyerberg, 2020). Predictive modeling should consider 4 recommendations: the study population, outcome measures and predictors, model development and validation (Mushkudiani et al., 2008). The study population includes sample size (number of patients), inclusion criteria, and period of data collection. Mortality and GOS/GOSE are commonly used as primary endpoints (dichotomized as previously described), but the time of assessment varied considerably among studies (from discharge to > 1-year post-injury). Predictors should be statistically associated with the outcome in univariate analysis; tested as continuous (i.e., age, the continuous variable should not be collapsed into categories), binary (i.e., gender: male versus female) or categorical variables (i.e., threshold values of hyperglycemia). Missing values on considered predictors should be estimated; imputation is recommended to be more efficient than dropping cases (Hukkelhoven et al., 2005; Mushkudiani et al., 2008).

It is recommended to start with selected predictors, known from the literature, after which variables from the own population (fig. 1) are included (Mushkudiani et al., 2008). A good PM should consist of no more than 5–7 predictors, to avoid over-fitting and decreases clinical applicability (Hukkelhoven et al., 2005; Mushkudiani et al., 2008). The most common and easy-to-use method to create a PM is logistic regression, but other promising modeling techniques are gaining increased attention (i.e., machine learning). Internal validation and external validation are essential to provide evidence that the model accurately predicts outcomes and to support its generalizability to other settings (Maas, Lingsma, & Roozenbeek, 2015).

Applications to other areas of neuroscience

In this chapter, the current state-of-the-art on predictors and PMs for outcome after moderate and severe TBI has been reviewed. The importance of outcome prognosis is seen in other acute brain injuries, such as ischemic stroke. Even though there are significant differences in these two pathologies, both conditions are among the leading causes of disability in elderly patients. Both diseases benefit from an accurate outcome prognosis for the same reasons: (1) the information could be used to select appropriate treatments, to improve rehabilitation and discharge planning; (2) the information could be used to establish priorities and guide the allocation of resources.

While many prognostic models for TBI have been published, there is a need for external validation of the existing models for ischemic stroke. Most of the models are developed from elderly patients with moderately severe ischemic stroke, mainly in HICs (Jampathong, Laopaiboon, Rattanakanokchai, & Pattanittum, 2018). Conversely, knowledge of the biological mechanisms involved after TBI is lacking. The integration of protein biomarkers with genetic profiling could be essential to determine potential mechanisms driving impaired cerebrovascular reactivity after TBI (Maas, Lingsma, & Roozenbeek, 2015).

Continued development, refinement, and validation of PMs and the application of new modeling techniques, such as machine learning and big data analysis, should be a common frontier in outcome prediction. This will allow for a wide array of clinical, laboratory and imaging variables to be interpreted rapidly.

Mini-dictionary of terms

Acute predictors. Risk factors/clinical variables recorded at admission to the emergency department or during the acute phase of intensive treatment and statistically associated with the outcome.

Biomarkers. Proteins exhibiting different circulating temporal dynamics in response to a pathophysiological process.

Cerebral autoregulation. The mechanism that maintains a constant cerebral blood flow.

Comparative Effectiveness Research. Observational study design. Main characteristics: adjustment for confounding data, reflects real-world clinical practice.

CRASH. A double-blind, randomized placebo-controlled trial of the effect of early steroid administration on death and disability after TBI.

External validation. Method used to determine the ability of a model to predict outcome in a study population that is different from that in which the model has been developed. The performance is assessed in terms of discrimination and calibration.

Logistic regression analysis. The technique used to examine a series of predictors (variables used as categorical, ordinal, or continuous) to determine those that best predict a specific outcome (dependent variable), presented as a dichotomous variable (e.g., favorable versus unfavorable outcome).

Predictive model. A model that could be used to distinguish patients at high risk for an unfavorable outcome, based on ≥ 2 predictors.

Randomized Controlled Trials. Main characteristics: minimization of cofounding data (i.e., inclusion criteria), does not reflect real-world clinical practice.

Sub-acute predictors. Clinical variables related to secondary injuries/events and statistically associated with outcome.

Key facts of potential strategies in TBI outcome research

Integrating information about acute factors with sub-acute predictors (neuroimaging features and late clinical variables) could help to optimize outcome prediction.

Continuous cerebrovascular reactivity monitoring in moderate-severe TBI (autoregulation) is a potential pathway to improving outcome.

The inclusion of circulating biomarkers involved in the pathophysiological processes of TBI is an important strategy for improving outcome prediction.

Deriving models directly from pediatric TBI patients can further improve outcome prediction.

The integration of protein biomarkers with genetic profiling could be important in determining potential mechanisms driving impaired cerebrovascular reactivity after TBI.

Along with the conventional methods, the application of new modeling techniques, such as machine learning and big data analysis, could improve outcome prediction.

Summary points

- Accurate prediction of outcome after moderate-to-severe TBI is essential to guide clinical management, patient course organization and family expectations.
- In clinical practice, reliable assessment of prognosis is important when making decisions, as clinicians treating patients often make therapeutic decisions based on their assessment.
- IMPACT and CRASH currently dominate the field of predictive modeling in TBI [core set: age, Glasgow Coma Scale (GCS) or motor score and pupillary reactivity at admission].
- Aging is a strong predictor of unfavorable outcome.
- GCS is most accurate at predicting outcome when combined with age and pupillary response.
- Hypoxia, hypotension, computed tomography characteristics and laboratory parameters (such as glucose, hemoglobin) are considered as an extension of the core of predictors.
- Most of the predictive models are being developed in developed countries, making inappropriate the generalizability of their results to lower settings.
- The inter-center variability produced by external validation of IMPACT and CRASH set a context for future updating of these models through a collaborative effort using large-observational studies.
- Along with the core of predictors, computed tomography characteristics and second insults, sub-acute predictors, intensive monitoring measures, and biomarkers are relevant factors to include in models to predict outcome.

References

Aiolfi, A., Benjamin, E., Khor, D., Inaba, K., Lam, L., & Demetriades, D. (2017). Brain trauma foundation guidelines for intracranial pressure monitoring: Compliance and effect on outcome. *World Journal of Surgery*, *41*(6), 1543–1549. https://doi.org/10.1007/s00268-017-3898-6.

Alali, A. S., Vavrek, D., Barber, J., Dikmen, S., Nathens, A. B., & Temkin, N. R. (2015). Comparative study of outcome measures and analysis methods for traumatic brain injury trials. *Journal of Neurotrauma*, *32*(8), 581–589. https://doi.org/10.1089/neu.2014.3495.

Calero-Martinez, S. A., Matula, C., Peraud, A., Biroli, F., Fernández-Alén, J., Bierschneider, M., … Rubiano, A.M. (2020). Development and assessment of competency-based neurotrauma course curriculum for international neurosurgery residents and neurosurgeons. *Neurosurgical Focus*, *48*(3). https://doi.org/10.3171/2019.12.FOCUS19850, E13.

Dijkland, S. A., Foks, K. A., Polinder, S., Dippel, D. W. J., Maas, A. I. R., Lingsma, H. F., & Steyerberg, E. W. (2020). Prognosis in moderate and severe traumatic brain injury: A systematic review of contemporary models and validation studies. *Journal of Neurotrauma*, *37*(1), 1–13. https://doi.org/10.1089/neu.2019.6401.

Gritti, P., Zangari, R., Carobbio, A., Zucchi, A., Lorini, F. L., Ferri, F., … Biroli, F. (2019). Acute and subacute outcome predictors in moderate and severe traumatic brain injury: A retrospective monocentric study. *World Neurosurgery*, *128*, e531–e540. https://doi.org/10.1016/j.wneu.2019.04.190.

Hukkelhoven, C. W. P. M., Steyerberg, E. W., Habbema, J. D. F., Farace, E., Marmarou, A., Murray, G. D., … Maas, A.I.R. (2005). Predicting outcome after traumatic brain injury: Development and validation of a prognostic score based on admission characteristics. *Journal of Neurotrauma*, *22*(10), 1025–1039. https://doi.org/10.1089/neu.2005.22.1025.

Huth, S. F., Slater, A., Waak, M., Barlow, K., & Raman, S. (2020). Predicting neurological recovery after traumatic brain injury in children: A systematic review of prognostic models. *Journal of Neurotrauma*, *37*(20), 2141–2149. https://doi.org/10.1089/neu.2020.7158.

Jampathong, N., Laopaiboon, M., Rattanakanokchai, S., & Pattanittum, P. (2018). Prognostic models for complete recovery in ischemic stroke: A systematic review and meta-analysis. *BMC Neurology*, *18*. https://doi.org/10.1186/s12883-018-1032-5.

Lingsma, H. F., Roozenbeek, B., Li, B., Lu, J., Weir, J., Butcher, I., … Steyerberg, E.W. (2011). Large between-center differences in outcome after moderate and severe traumatic brain injury in the international mission on prognosis and clinical trial design in traumatic brain injury (IMPACT) study. *Neurosurgery*, *68*(3), 601–608. https://doi.org/10.1227/NEU.0b013e318209333b.

Lingsma, H. F., Roozenbeek, B., Steyerberg, E. W., Murray, G. D., & Maas, A. I. R. (2010). Early prognosis in traumatic brain injury: From prophecies to predictions. *The Lancet. Neurology*, *9*(5), 543–554. https://doi.org/10.1016/S1474-4422(10)70065-X.

Lingsma, H. F., Yue, J. K., Maas, A. I. R., Steyerberg, E. W., Manley, G. T., & TRACK-TBI Investigators. (2015). Outcome prediction after mild and complicated mild traumatic brain injury: External validation of existing models and identification of new predictors using the TRACK-TBI pilot study. *Journal of Neurotrauma*, *32*(2), 83–94. https://doi.org/10.1089/neu.2014.3384.

Liu, X., Czosnyka, M., Donnelly, J., Budohoski, K. P., Varsos, G. V., Nasr, N., … Smielewski, P. (2015). Comparison of frequency and time domain methods of assessment of cerebral autoregulation in traumatic brain injury. *Journal of Cerebral Blood Flow and Metabolism: Official Journal of the International Society of Cerebral Blood Flow and Metabolism*, *35*(2), 248–256. https://doi.org/10.1038/jcbfm.2014.192.

Maas, A. I. R., Hukkelhoven, C. W. P. M., Marshall, L. F., & Steyerberg, E. W. (2005). Prediction of outcome in traumatic brain injury with computed tomographic characteristics: A comparison between the computed tomographic classification and combinations of computed tomographic predictors. *Neurosurgery*, *57*(6), 1173–1182. discussion 1173-1182 https://doi.org/10.1227/01.neu.0000186013.63046.6b.

Maas, A. I. R., Lingsma, H. F., & Roozenbeek, B. (2015). Predicting outcome after traumatic brain injury. *Handbook of Clinical Neurology*, *128*, 455–474. https://doi.org/10.1016/B978-0-444-63521-1.00029-7.

Maas, A. I. R., Marmarou, A., Murray, G. D., Teasdale, S. G. M., & Steyerberg, E. W. (2007). Prognosis and clinical trial design in traumatic brain injury: The IMPACT study. *Journal of Neurotrauma*, *24*(2), 232–238. https://doi.org/10.1089/neu.2006.0024.

Maas, A. I. R., Menon, D. K., Lingsma, H. F., Pineda, J. A., Sandel, M. E., & Manley, G. T. (2012). Re-orientation of clinical research in traumatic brain injury: Report of an international workshop on comparative effectiveness research. *Journal of Neurotrauma*, *29*(1), 32–46. https://doi.org/10.1089/neu.2010.1599.

Maas, A. I. R., Menon, D. K., Steyerberg, E. W., Citerio, G., Lecky, F., Manley, G. T., … CENTER-TBI Participants and Investigators. (2015). Collaborative European NeuroTrauma effectiveness research in traumatic brain injury (CENTER-TBI): A prospective longitudinal observational study. *Neurosurgery*, *76*(1), 67–80. https://doi.org/10.1227/NEU.0000000000000575.

Maas, A. I. R., Steyerberg, E. W., Butcher, I., Dammers, R., Lu, J., Marmarou, A., … Murray, G.D. (2007). Prognostic value of computerized tomography scan characteristics in traumatic brain injury: Results from the IMPACT study. *Journal of Neurotrauma*, *24*(2), 303–314. https://doi.org/10.1089/neu.2006.0033.

Marmarou, A., Lu, J., Butcher, I., McHugh, G. S., Murray, G. D., Steyerberg, E. W., … Maas, A.I.R. (2007). Prognostic value of the Glasgow coma scale and pupil reactivity in traumatic brain injury assessed pre-hospital and on enrollment: An IMPACT analysis. *Journal of Neurotrauma*, *24*(2), 270–280. https://doi.org/10.1089/neu.2006.0029.

Mattioli, C., Beretta, L., Gerevini, S., Veglia, F., Citerio, G., Cormio, M., & Stocchetti, N. (2003). Traumatic subarachnoid hemorrhage on the computerized tomography scan obtained at admission: A multicenter assessment of the accuracy of diagnosis and the potential impact on patient outcome. *Journal of Neurosurgery*, *98*(1), 37–42. https://doi.org/10.3171/jns.2003.98.1.0037.

McCauley, S. R., Wilde, E. A., Anderson, V. A., Bedell, G., Beers, S. R., Campbell, T. F., … Pediatric TBI Outcomes Workgroup. (2012). Recommendations for the use of common outcome measures in pediatric traumatic brain injury research. *Journal of Neurotrauma*, *29*(4), 678–705. https://doi.org/10.1089/neu.2011.1838.

MRC CRASH Trial Collaborators, Perel, P., Arango, M., Clayton, T., Edwards, P., Komolafe, E., … Yutthakasemsunt, S. (2008). Predicting outcome after traumatic brain injury: Practical prognostic models based on large cohort of international patients. *BMJ (Clinical Research Ed.)*, *336*(7641), 425–429. https://doi.org/10.1136/bmj.39461.643438.25.

Murray, G. D., Butcher, I., McHugh, G. S., Lu, J., Mushkudiani, N. A., Maas, A. I. R., … Steyerberg, E.W. (2007). Multivariable prognostic analysis in traumatic brain injury: Results from the IMPACT study. *Journal of Neurotrauma*, *24*(2), 329–337. https://doi.org/10.1089/neu.2006.0035.

Mushkudiani, N. A., Hukkelhoven, C. W. P. M., Hernández, A. V., Murray, G. D., Choi, S. C., Maas, A. I. R., & Steyerberg, E. W. (2008). A systematic review finds methodological improvements necessary for prognostic models in determining traumatic brain injury outcomes. *Journal of Clinical Epidemiology*, *61*(4), 331–343. https://doi.org/10.1016/j.jclinepi.2007.06.011.

Perel, P., Edwards, P., Shakur, H., & Roberts, I. (2008). Use of the Oxford handicap scale at hospital discharge to predict Glasgow Outcome Scale at 6 months in patients with traumatic brain injury. *BMC Medical Research Methodology*, *8*, 72. https://doi.org/10.1186/1471-2288-8-72.

Perel, P., Edwards, P., Wentz, R., & Roberts, I. (2006). Systematic review of prognostic models in traumatic brain injury. *BMC Medical Informatics and Decision Making*, *6*, 38. https://doi.org/10.1186/1472-6947-6-38.

Retel Helmrich, I. R. A., Lingsma, H. F., Turgeon, A. F., Yamal, J.-M., & Steyerberg, E. W. (2020). Prognostic research in traumatic brain Injury: Markers, modeling, and methodological principles. *Journal of Neurotrauma*. https://doi.org/10.1089/neu.2019.6708.

Riemann, L., Beqiri, E., Smielewski, P., Czosnyka, M., Stocchetti, N., Sakowitz, O., … the CENTER-TBI High Resolution ICU (HR ICU) Sub-Study Participants and Investigators. (2020). Low-resolution pressure reactivity index and its derived optimal cerebral perfusion pressure in adult traumatic brain injury: A CENTER-TBI study. *Critical Care*, *24*(1), 266. https://doi.org/10.1186/s13054-020-02974-8.

Roozenbeek, B., Lingsma, H. F., Lecky, F. E., Lu, J., Weir, J., Butcher, I., … International Mission on Prognosis Analysis of Clinical Trials in Traumatic Brain Injury (IMPACT) Study Group, Corticosteroid Randomisation After Significant Head Injury (CRASH) Trial Collaborators, & Trauma Audit and Research Network (TARN). (2012). Prediction of outcome after moderate and severe traumatic brain injury: External validation of the international mission on prognosis and analysis of clinical trials (IMPACT) and corticoid randomisation after significant head injury (CRASH) prognostic models. *Critical Care Medicine*, *40*(5), 1609–1617. https://doi.org/10.1097/CCM.0b013e31824519ce.

Roozenbeek, B., Maas, A. I. R., Lingsma, H. F., Butcher, I., Lu, J., Marmarou, A., … IMPACT Study Group. (2009). Baseline characteristics and statistical power in randomized controlled trials: Selection, prognostic targeting, or covariate adjustment? *Critical Care Medicine*, *37*(10), 2683–2690. https://doi.org/10.1097/ccm.0b013e3181ab85ec.

Roozenbeek, B., Maas, A. I. R., Marmarou, A., Butcher, I., Lingsma, H. F., Lu, J., … IMPACT Study Group. (2009). The influence of enrollment criteria on recruitment and outcome distribution in traumatic brain injury studies: Results from the impact study. *Journal of Neurotrauma*, *26*(7), 1069–1075. https://doi.org/10.1089/neu.2008.0569.

Salottolo, K., Levy, A. S., Slone, D. S., Mains, C. W., & Bar-Or, D. (2014). The effect of age on Glasgow coma scale score in patients with traumatic brain injury. *JAMA Surgery*, *149*(7), 727–734. https://doi.org/10.1001/jamasurg.2014.13.

Sarnaik, A., Ferguson, N. M., O'Meara, A. M. I., Agrawal, S., Deep, A., Buttram, S., … Investigators of the ADAPT Trial. (2018). Age and mortality in pediatric severe traumatic brain injury: Results from an international Study. *Neurocritical Care*, *28*(3), 302–313. https://doi.org/10.1007/s12028-017-0480-x.

Servadei, F., Nasi, M. T., Giuliani, G., Cremonini, A. M., Cenni, P., Zappi, D., & Taylor, G. S. (2000). CT prognostic factors in acute subdural haematomas: The value of the «worst» CT scan. *British Journal of Neurosurgery*, *14*(2), 110–116. https://doi.org/10.1080/02688690050004525.

Steyerberg, E. W., Mushkudiani, N., Perel, P., Butcher, I., Lu, J., McHugh, G. S., … Maas, A.I.R. (2008). Predicting outcome after traumatic brain injury: Development and international validation of prognostic scores based on admission characteristics. *PLoS Med.*, *5*(8), e165. discussion e165 https://doi.org/10.1371/journal.pmed.0050165.

Steyerberg, E. W., Wiegers, E., Sewalt, C., Buki, A., Citerio, G., Keyser, V. D., … Zoerle, T. (2019). Case-mix, care pathways, and outcomes in patients with traumatic brain injury in CENTER-TBI: A European prospective, multicentre, longitudinal, cohort study. *The Lancet Neurology*, *18*(10), 923–934. https://doi.org/10.1016/S1474-4422(19)30232-7.

Stocchetti, N., Pagan, F., Calappi, E., Canavesi, K., Beretta, L., Citerio, G., … Colombo, A. (2004). Inaccurate early assessment of neurological severity in head injury. *Journal of Neurotrauma*, *21*(9), 1131–1140. https://doi.org/10.1089/neu.2004.21.1131.

Svedung Wettervik, T., Howells, T., Enblad, P., & Lewén, A. (2019). Temporal neurophysiological dynamics in traumatic brain injury: Role of pressure reactivity and optimal cerebral perfusion pressure for predicting outcome. *Journal of Neurotrauma*, *36*(11), 1818–1827. https://doi.org/10.1089/neu.2018.6157.

Thelin, E., Al Nimer, F., Frostell, A., Zetterberg, H., Blennow, K., Nyström, H., … Nelson, D.W. (2019). A serum protein biomarker panel improves outcome prediction in human traumatic brain injury. *Journal of Neurotrauma*, *36*(20), 2850–2862. https://doi.org/10.1089/neu.2019.6375.

Thelin, E. P., Zeiler, F. A., Ercole, A., Mondello, S., Büki, A., Bellander, B.-M., … Nelson, D.W. (2017). Serial sampling of serum protein biomarkers for monitoring human traumatic brain injury dynamics: A systematic review. *Frontiers in Neurology*, *8*. https://doi.org/10.3389/fneur.2017.00300.

Timmons, S. D., & Toms, S. A. (2012). Comparative effectiveness research in neurotrauma. *Neurosurgical Focus*, *33*(1). https://doi.org/10.3171/2012.6.FOCUS1296, E3.

Van Beek, J. G. M., Mushkudiani, N. A., Steyerberg, E. W., Butcher, I., McHugh, G. S., Lu, J., … Maas, A.I.R. (2007). Prognostic value of admission laboratory parameters in traumatic brain injury: Results from the IMPACT study. *Journal of Neurotrauma*, *24*(2), 315–328. https://doi.org/10.1089/neu.2006.0034.

Walder, B., Haller, G., Rebetez, M. M. L., Delhumeau, C., Bottequin, E., Schoettker, P., … Osterwalder, J. (2013). Severe traumatic brain injury in a high-income country: An epidemiological study. *Journal of Neurotrauma*, *30*(23), 1934–1942. https://doi.org/10.1089/neu.2013.2955.

Wilson, J. T., Pettigrew, L. E., & Teasdale, G. M. (1998). Structured interviews for the Glasgow Outcome Scale and the extended Glasgow Outcome Scale: Guidelines for their use. *Journal of Neurotrauma*, *15*(8), 573–585. https://doi.org/10.1089/neu.1998.15.573.

Yuan, Q., Wu, X., Sun, Y., Yu, J., Li, Z., Du, Z., … Hu, J. (2015). Impact of intracranial pressure monitoring on mortality in patients with traumatic brain injury: A systematic review and meta-analysis. *Journal of Neurosurgery*, *122*(3), 574–587. https://doi.org/10.3171/2014.10.JNS1460.

Chapter 3

Thirty years post-injury: Impact of traumatic brain injury on later Alzheimer's disease

Hiraoka Takashi and Hanayama Kozo
Department of Rehabilitation Medicine, Kawasaki Medical School, Okayama, Japan

List of abbreviations

Aβ	amyloid-β
AD	Alzheimer's disease
ApoE	apolipoprotein E
CTE	chronic traumatic encephalopathy
DAI	diffuse axonal injury
HBD	higher brain dysfunction
PET	positron emission tomography
tau	tubulin-associated unit
TBI	traumatic brain injury

Introduction

In recent years, the incidence of traumatic brain injury (TBI) has increased worldwide. According to a 2010 report, approximately 1.7 million cases of TBI occur yearly in the United States (Faul, Wald, Xu, & Coronado, 2010). In Japan, an estimated 300,000 cases of TBI occur yearly, of which more than 10,000 are believed to constitute severe TBI (Hiraoka, 2019). TBI that involves a loss of consciousness that lasts for more than 24 h is categorized as severe TBI; this type is primarily caused by traffic accidents and falls. On the other hand, TBI that involves short-term loss of consciousness is categorized as mild TBI (also known as a concussion); this type of TBI is often caused by contact sports, including boxing and American football. In general, immediately after a severe TBI, patients tend to exhibit signs of higher brain dysfunction (HBD) as an after-effect, including attention, memory, and social behavior disorders, as well as executive dysfunction (Hiraoka, 2019). In most cases, these HBDs do not progress beyond the sub-acute period of TBI. On the other hand, under the names "punch-drunk syndrome" (Martland, 1928) and "dementia pugilistica" (Castellani & Perry, 2017; Millspaugh, 1937), mild repetitive TBI has long been known to involve late-onset cognitive/mental impairment. Further, irrespective of whether or not the injury in question is isolated, repetitive, severe, or mild, several publications have discussed the possibility that once an individual suffers a blow to the head, delayed-onset neuropsychiatric problems can occur anywhere from several years to several decades later (Hiraoka, 2019; LoBue et al., 2018; Smith, Johnson, & Stewart, 2013). While the current mainstream theory regarding these "delayed-onset encephalopathies" is that they are a type of tauopathy caused by the over-accumulation of tubulin-associated unit (tau) protein in the brain, the specific details of the mechanisms through which TBI causes delayed-onset cognitive decline or psychological symptoms remain largely unclear. Elucidation of the pathology of delayed-onset cognitive decline/psychiatric symptom onset will be beneficial not only for the diagnosis and treatment of patients with TBI but also for the improvement of the quality of life of patients with TBI. Here, we will discuss two emblematic delayed-onset pathologies often seen after TBI: Alzheimer's disease (AD) and chronic traumatic encephalopathy (CTE), the latter of which has recently garnered much attention.

Diagnosis and Treatment of Traumatic Brain Injury. https://doi.org/10.1016/B978-0-12-823347-4.00022-1

Understanding traumatic brain injury as a risk factor for Alzheimer's disease

Risk factors for AD are easy to understand when divided into three levels. The first are risk factors for which preventive intervention is not possible, namely age and gene mutations. The second are medical level risk factors, which include the topic of this chapter, TBI. The third are lifestyle-related risk factors which can be improved to some extent through an individual's efforts (Table 1).

The major after-effects of TBI are shown in Table 2. Among these, it is essential to focus on "mild repetitive TBI" and "severe isolated TBI" as levels of "severity" and "repetition" that lead TBI to be a risk factor for cognitive decline, including AD, in the chronic phase. These two types of TBI are summarized below.

Similarities in the long-term prognosis of mild repetitive and severe isolated traumatic brain injury

While mild repetitive TBI and severe isolated TBI are characterized by vastly different mechanisms of injury and acute pathology/progress, they share a commonality in that both can lead to cognitive decline in the long term (Mez et al., 2017;

TABLE 1 Reported risk and protective factors for Alzheimer's disease.

Risk factor level	Risk factor	Protective factor
Protective intervention not possible	• Age • Apolipoprotein E gene • Down syndrome • Mutations such as presenilin gene • Family history	• None
Medical level	• **Traumatic brain injury** • Depression • Diabetes • Hypertension & hypotension • Surgery under general anesthesia • Arteriosclerosis • Hyperhomocysteinemia • Hyperlipidemia • Tooth loss • Hypothyroidism • Congestive heart failure, atrial fibrillation, thrombotic episodes • Hypercholesterolemia • Abnormal cerebral hemodynamics	• Antihypertensive drugs • Statins • Non-steroidal anti-inflammatory drugs
Lifestyle-related	• Smoking • Alcoholism • High saturated fat intake • Lower education level	• Exercise • Mentally stimulating activities • Appropriate alcohol consumption • Social connections • Higher education level

Traumatic brain injury is listed as one risk factor. Of the three risk factor levels above, "risk factors for which preventive intervention is not possible" are considered to have the most control over Alzheimer's disease onset.

TABLE 2 Major traumatic brain injury after-effects.

After-effects of cerebral contusion/intracranial hemorrhage	• Persistent coma • Minimal brain dysfunction • Traumatic hydrocephalus, etc.
After-effects of bone fracture	• Cranial nerve palsy • Cerebrospinal fluid leakage • Meningitis • Pneumocranium/pneumocephalus • Growing skull fracture
Infection	• Brain abscess • Meningitis • Subdural or epidural abscess
Cerebrovascular disorders	• Traumatic carotid-cavernous sinus fistula • Traumatic cerebral aneurysm
Chronic subdural hematoma/hygroma	
Post-traumatic epilepsy	
Post-concussion syndrome	• General malaise including headaches, dizziness, fatigue
Alzheimer's disease/chronic traumatic encephalopathy	

It is essential to focus on "mild repetitive TBI" and "severe isolated TBI" as levels of "severity" and "repetition" that lead TBI to be a risk factor for cognitive decline, including AD/CTE, in the chronic phase.
AD: Alzheimer's disease; *CTE*: chronic traumatic encephalopathy; *TBI*: traumatic brain injury.

Smith et al., 2013). In particular, late-onset neurodegenerative disorder associated with mild repetitive TBI, characterized by cognitive decline, has recently been defined as CTE, and has been the subject of much attention (Mez et al., 2017; Mez, Solomon, Daneshvar, Stein, & McKee, 2016; Omalu et al., 2005, 2006; Omalu, Fitzsimmons, Hammers, & Bailes, 2010). Similarly, progressive cognitive decline has also been reported in the chronic phase of severe isolated TBI, and has been linked to an increased risk of developing AD and other neurodegenerative diseases (Hiraoka, 2019; Johnson, Stewart, & Smith, 2013; Smith et al., 2013). Furthermore, the occurrence of psychological symptoms in the chronic phase is a characteristic feature of both mild repetitive TBI (McKee et al., 2013) and severe isolated TBI (Max, 2014; Max et al., 2013, 2015). Late-onset schizophreniform disorders have also been reported in a number of clinical studies of severe TBI (Deighton et al., 2016; Nielsen, Mortensen, O'Callaghan, Mors, & Ewald, 2002). The mean delay in the onset of psychosis following severe TBI is more than 5 years (Achté, Jarho, Kyykkä, & Vesterinen, 1991; Fujii & Ahmed, 2002; Nielsen et al., 2002). Such a long interval suggests the involvement of chronic neurodegenerative changes in the onset of post-TBI psychosis. Thus, cognitive and neuropsychiatric issues are commonly encountered long-term after-effects of both mild repetitive and severe isolated TBI. AD and CTE—emblematic examples of the various post-TBI, chronic-phase pathologies that present as a variety of cognitive and psychological symptoms—are typified by similar characteristics.

Post-traumatic brain injury Alzheimer's disease

Many previous studies reference TBI as a possible trigger or accelerator of AD onset (De Silva & Faraci, 2016; Fleminger, Oliver, Lovestone, Rabe-Hesketh, & Giora, 2003; Fujii & Ahmed, 2002; Ikonomovic et al., 2004; Lye & Shores, 2000; Mortimer et al., 1991; Perry et al., 2016; Ramos-Cejudo et al., 2018; Roberts, Gentleman, Lynch, & Graham, 1991; Roberts et al., 1994; Salib & Hillier, 1997; Schofield et al., 1997; Smith et al., 2013; Washington, Morffy, Parsadanian, Zapple, & Burns, 2014) and the National Academy of Medicine has concluded that TBI is a risk factor for AD. AD is an irreversible, progressive disorder of the brain that gradually impairs patients' memory and other cognitive functions, ultimately preventing them from completing even the most basic activities of daily life. Pathologically, the presence of cerebral amyloid plaques and neurofibrillary tangles are the characteristic signs of AD. A variety of different hypotheses concerning the pathological

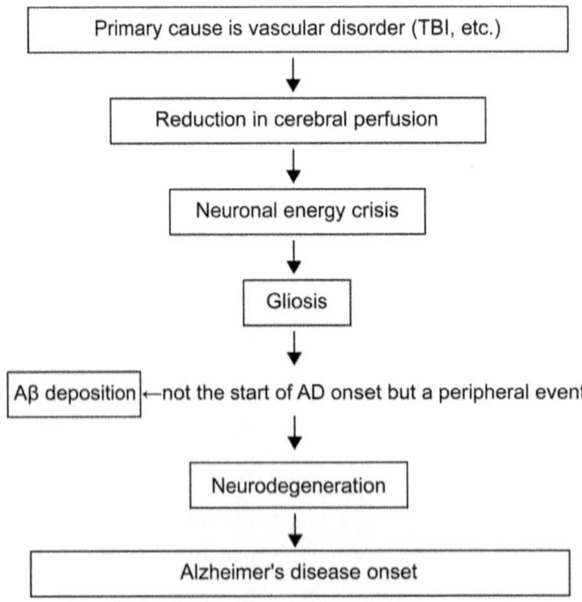

FIG. 1 Illustration of the vascular hypothesis of traumatic brain injury-induced Alzheimer's disease. The vascular hypothesis proposes that structural and functional disorders of the cerebral vasculature brought about primarily by various vascular disorders are the basic cause of AD. Aβ aggregation is considered a peripheral event in this process. *Aβ*: amyloid-β; *AD*: Alzheimer's disease; *TBI*: traumatic brain injury.

mechanisms underlying AD have been published. Chief among these are the amyloid cascade (Hardy & Selkoe, 2002; Selkoe, 1991), vascular (de la Torre, 2004; de la Torre & Mussivand, 1993), cognitive reserve (Katzman, 1993), and glucocorticoid cascade hypotheses (Sapolsky, Krey, & McEwen, 1986). The most promising of these is the amyloid cascade hypothesis. This hypothesis holds that amyloid-β (Aβ) accumulates outside of neurons in the brain, forming senile plaques, which cause the formation of tau protein aggregates. These aggregates subsequently cause neurofibrillary tangles, which ultimately cause the death of the neurons in which they form. In other words, the hypothesis supposes that AD is a tauopathy and that Aβ sits at the top of the cascade that ultimately leads to disease onset. In contrast, the vascular hypothesis (Fig. 1), formulated as a critical response to the amyloid cascade hypothesis, holds that vascular damage, which can be caused by a number of factors, is the source of AD onset; here, Aβ aggregation is a peripheral downstream consequence of this damage. Considering the relationship between TBI and AD onset, if one were to consider this hypothesis, cerebrovascular damage caused by TBI should be considered a significant risk factor for AD. The fact that pure AD is quite rare in the clinical setting and that the idea of AD with cerebrovascular disease (CVD)/mixed AD has come to be far more accepted, does not contradict this possibility. This has led to further suppositions of mechanisms in which cerebrovascular damage due to any cause has a complementary or interdependent effect on AD onset (Perrotta, Lembo, & Carnevale, 2016). Conversely, there are also negative reports that contest that TBI is a risk factor for the development of AD (Mehta et al., 1999). Thus, whether TBI is a risk factor for AD is still a matter of debate, and many aspects of its onset/mechanism of progress remain unclear.

Amyloid-β pathology in traumatic brain injury

In recent years, amyloid positron emission tomography (PET) has been used to visualize the accumulation of cerebral Aβ while patients are still alive (Herholz & Ebmeier, 2011; Mintun et al., 2006; Okamura, Harada, Kudo, & Yanai, 2015), yielding a host of new findings.

Evidence regarding Aβ has to be interpreted with regard to the time scale of Aβ plaque formation. Autopsies of patients with isolated severe-to-moderate TBI that died in the acute post-injury period revealed Aβ plaques in up to 30% of cases (Ikonomovic et al., 2004; Lye & Shores, 2000; Roberts et al., 1991, 1994). These reports describe plaques appearing within a few hours of injury. While the presence of Aβ pathology in the chronic phase of moderate-to-severe TBI several years after injury has been reported (Johnson, Stewart, & Smith, 2012), there are also reports detailing that, as time passes after TBI occurrence, Aβ deposits are less frequently observed (Chen, Johnson, Uryu, Trojanowski, & Smith, 2009; Kawai, Shishido, Hatakeyama, & Tamiya, 2017). Chen and colleagues questioned the post-TBI amyloid cascade hypothesis of AD onset by pointing out that while Aβ deposits are less frequently observed as time passes from a TBI occurrence, the Aβ-degrading enzyme neprilysin is more frequently observed (Chen et al., 2009). Kawai presented a similar opinion, describing a possible mechanism for TBI-induced Aβ deposition and the progressive cognitive decline seen in older individuals that hinges on the following: despite the fact that endogenous Aβ removal mechanisms operate normally in cases of TBI in young individuals, in older individuals, neprilysin expression drops significantly (Hellström-Lindahl, Ravid, & Nordberg, 2008; Kawai et al., 2017). A report describing a significant increase in Aβ deposition rate as age at injury increases serves as further evidence of the effect of age at TBI onset on Aβ accumulation (Roberts et al., 1994).

It has been reported that Aβ deposition is accelerated if the ε4-type allele of apolipoprotein E (ApoE) is present (Nicoll, Roberts, & Graham, 1995). My own work (Hiraoka, 2019), as well as that of Chen et al. (2009), Kawai et al. (2017), Roberts et al. (1994), and Nicoll et al. (1995), suggest that many cases involve amyloid PET scans that are Aβ(−), constitute isolated severe TBIs after which more than 30 years had passed, and are ApoE phenotypes of the ε3/3 type (Table 3, Figs. 2 and 3). This aligns well with these earlier reports.

A report in which patients with TBI were compared against age-matched controls with no history of TBI indicates that patients with TBI have more widespread Aβ plaques (Huber, Gabbert, Kelemen, & Cervos-Navarro, 1993). Further, reports have also indicated the possible involvement of diffuse axonal injury (DAI) in post-TBI Aβ formation (Adams et al., 1989; Adams, Graham, Murray, & Scott, 1982; Johnson et al., 2013) in pigs (Smith et al., 1999) and humans (Smith, Graham, Murray, & Nicoll, 2003; Smith, Chen, Iwata, & Graham, 2003; Chen et al., 2009). These Aβ pathologies lend further support to the notion that TBI is a risk factor for AD.

Thus, while there are a number of reports on post-TBI Aβ deposition and its effect on AD onset, much is still unclear about the pathological mechanisms at play in these conditions.

Tubulin-associated unit pathology in traumatic brain injury

In recent years, PET scans have made the non-invasive in vivo detection of tau proteins possible, and the number of related reports has increased accordingly (Takahata et al., 2019; Xia et al., 2013). The spatial distribution of tau accumulation following severe TBI is reported to resemble that following mild repetitive TBI (Johnson, Stewart, Arena, & Smith, 2017). In terms of the time scale, unlike Aβ-related pathologies, tau lesions in severe isolated TBI have been reported to appear at some point after the injury occurred (Kawai et al., 2017). A study in which tau immunostaining was performed in the autopsied brains of patients who had died soon after injury found no difference between the injured and control group (Smith, Chen, et al., 2003; Smith, Graham, et al., 2003). However, another study found neurofibrillary tangles in 30% of patients who had survived for several years after their injuries (Johnson et al., 2012), and reports discussing the accumulation of fibrous tau in the chronic phase of severe TBI are not uncommon (Shively, Scher, Perl, & Diaz-Arrastia, 2012). It seems quite likely that tau accumulation is implicated in TBI-induced delayed-onset neuropsychiatric symptoms, regardless of the type of injury. However, it is not yet clear whether tau accumulation due to TBI affects later AD onset. Recently, it has been reported that the tau microstructure observed in CTE differs morphologically from that seen in AD (Falcon et al., 2019); this finding may prove vital in the differential diagnosis of post-TBI AD and CTE.

Applications to other areas of neuroscience

In recent years, chronic traumatic encephalopathy has received much attention as an emblematic post-traumatic brain injury delayed-onset pathology (Mez et al., 2016, 2017; Omalu et al., 2006, 2005, 2010). Here, we have depicted chronic traumatic encephalopathy as a delayed-onset pathology of traumatic brain injury, and outlined how it is believed to involve pathological mechanisms different from those underlying Alzheimer's disease.

TABLE 3 Case characteristics and amyloid-β deposition amyloid-β was negative in both cases and no association with Alzheimer's disease was found (Hiraoka, 2019).

Case	Age (y)/gender	Time from injury	WAIS-III FIQ	WAIS-III VIQ	WAIS-III PIQ	FIM cognitive score	Amyloid-β deposition	Apolipoprotein E phenotype
1	56/M	38	72	76	72	16/35	negative	ε3/3
2	48/F	31	72	66	84	24/35	negative	ε3/3

FIQ: full intelligence quotient; *FIM*: functional independence measure; *PIQ*: performance intelligence quotient; *VIQ*: verbal intelligence quotient; *WAIS-III*: Wechsler Adult Intelligence Scale Third Edition.

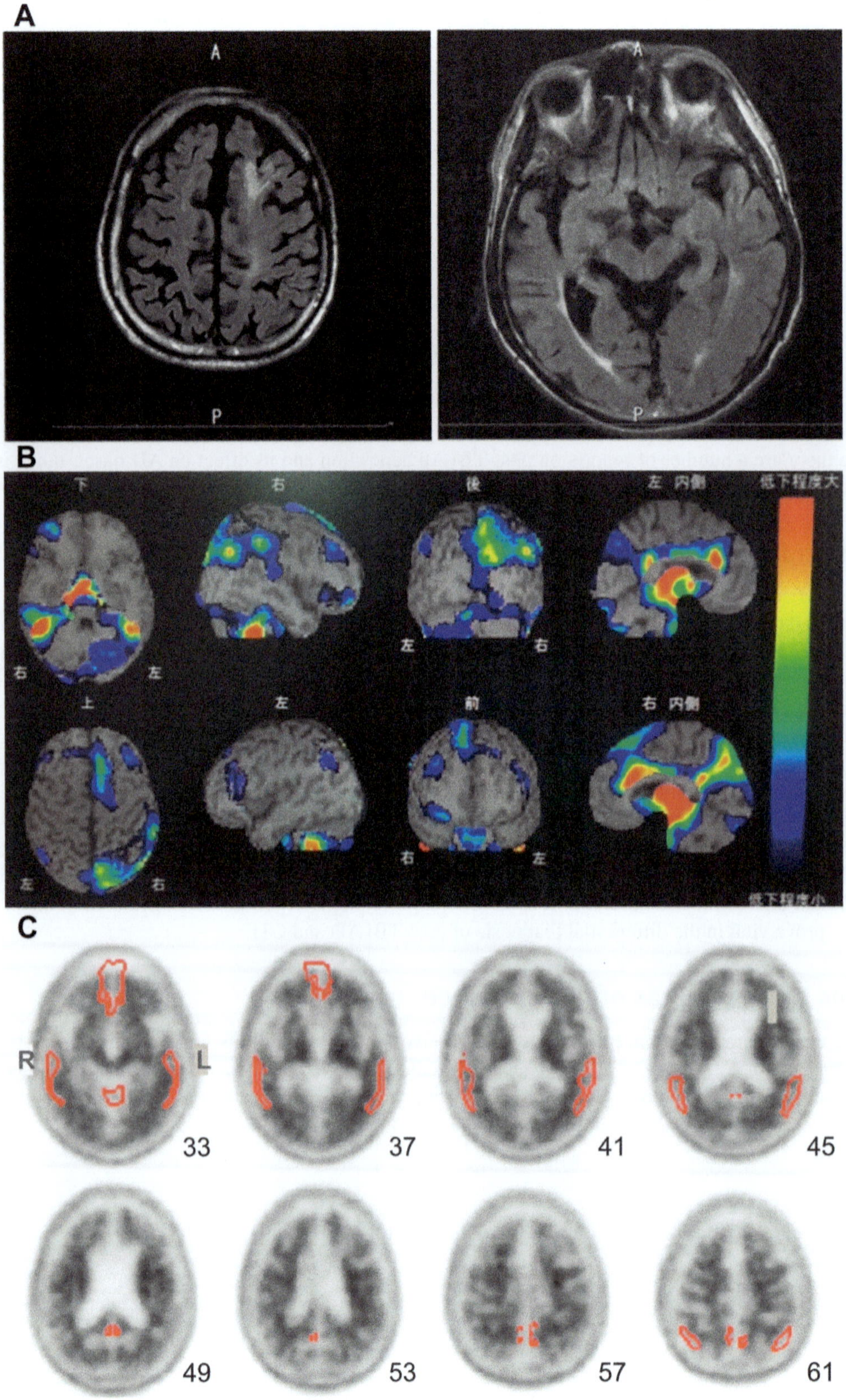

FIG. 2 Case 1 MRI, SPECT, and Amyloid PET findings (Hiraoka, 2019). (A) Case 1 MRI: The brain MRI revealed contusions and atrophy in the left frontal lobe. (B) Case 1 SPECT: SPECT showed decreased blood flow to both frontal lobes, the right parietal lobe, the occipital lobe, and the left cerebellum. (C) Case 1 Amyloid PET: The amyloid PET was Aβ-negative. The portion outlined in *red* denotes the mean ROI. *Aβ*: amyloid-β, *MRI*: magnetic resonance imaging, *PET*: positron emission tomography, *SPECT*: single-photon emission computed tomography.

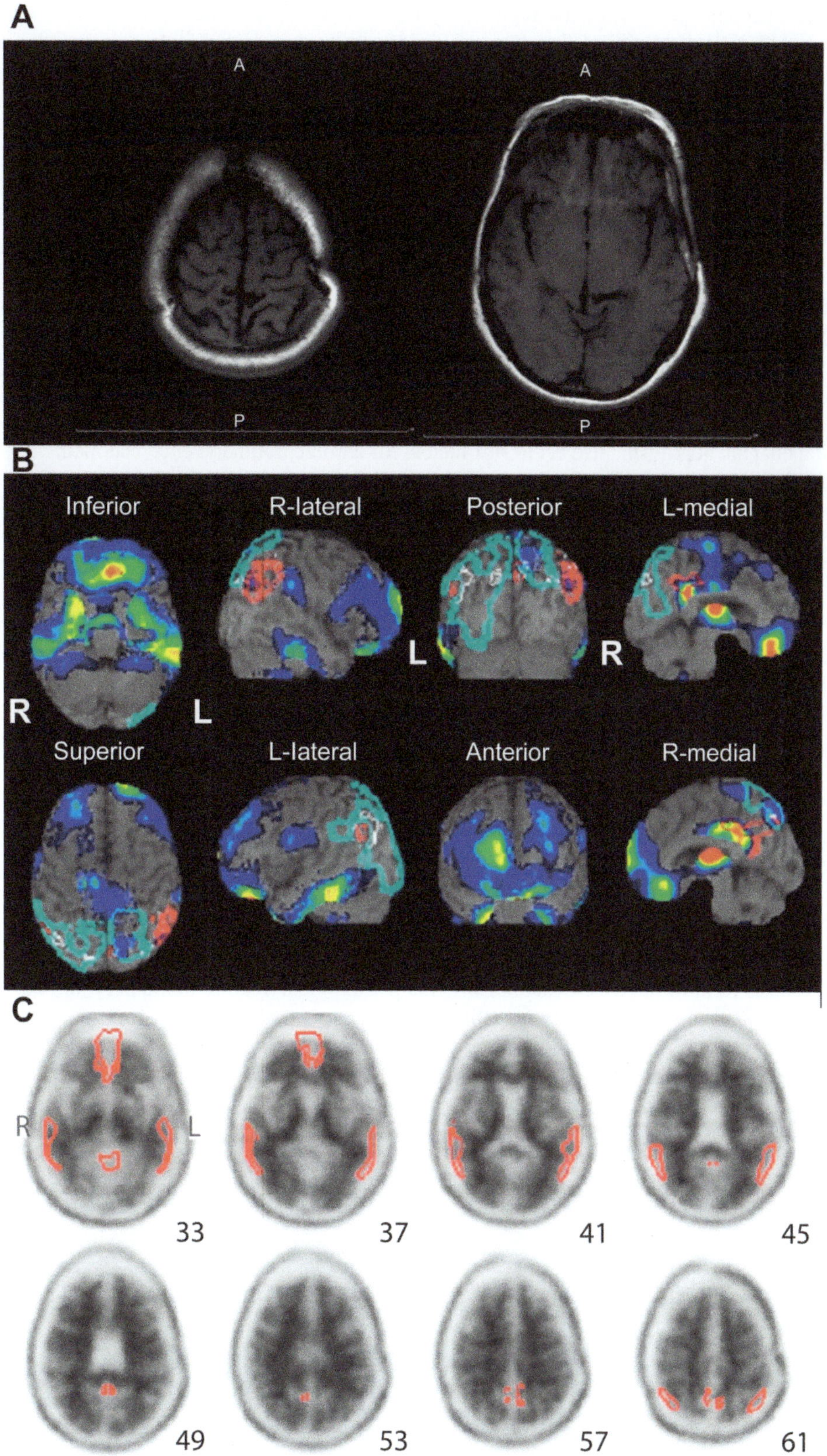

FIG. 3 Case 2 MRI, SPECT, and Amyloid PET findings (Hiraoka, 2019). (A) Case 2 MRI: The brain MRI revealed atrophy and old traumatic changes in the right frontal lobe, parietal lobe, and splenium of the corpus callosum. (B) Case 2 SPECT: SPECT did not show clear Alzheimer's disease patterns, and a reduction in blood flow to both frontal lobes, particularly to the right frontal lobe, was observed. (C) Case 2 Amyloid PET: The amyloid PET was Aβ-negative. *Aβ*: amyloid-β, *MRI*: magnetic resonance imaging, *PET*: positron emission tomography, *SPECT*: single-photon emission computed tomography.

Chronic traumatic encephalopathy was first reported by the New Jersey pathologist Harrison Martland in 1928. His research touched on the clinical aspects of punch-drunk syndrome; a progressive form of neurological decline caused by repeated brain injury suffered by boxers (Martland, 1928). This pathology was also known as dementia pugilistica (Millspaugh, 1937), which refers to the psychological symptoms of cognitive decline and personality shifts seen in boxers that had taken repeated fierce blows to the head over many years. However, it was soon realized that this condition occurred in arenas outside of boxing; thus, in recent years, it came to be known as chronic traumatic encephalopathy (Critchley, 1957; Mez et al., 2016; Omalu et al., 2010). Mild traumatic brain injury does not leave one with apparent after-effects immediately after injury. However, as brain trauma is repeatedly suffered, neurodegenerative mechanisms are believed to be induced, ultimately leading to the onset of chronic traumatic encephalopathy. Aggression, impatience, confusion, depression, suicidal tendencies, and other psychological symptoms as well as memory impairment and other forms of cognitive decline are all seen in chronic traumatic encephalopathy. As the disease progresses, general cognitive decline (dementia), gait disturbance, Parkinsonism, and a lisp can occur. Eventually, it becomes quite difficult to distinguish Alzheimer's disease from chronic traumatic encephalopathy (Stern et al., 2013; Takahata, Mimura, Shimada, Higuchi, & Suhara, 2015). The more brain trauma is suffered, the faster onset occurs, and the condition itself tends to become more severe. It has been reported that Alzheimer's disease risk factors, including the apolipoprotein ε4 gene and changes in the amyloid-β-degrading enzyme neprilysin, also act as risk factors for chronic traumatic encephalopathy (DeKosky, Blennow, Ikonomovic, & Gandy, 2013), highlighting the commonalities between the two conditions. While phosphorylated tubulin-associated unit-positive neurofibrillary tangles across a wide area of the frontal and temporal lobes are characteristic of both conditions, their distribution is different in chronic traumatic encephalopathy (Katsumoto, Takeuchi, & Tanaka, 2019; McKee et al., 2013, 2016; Omalu et al., 2005), where they tend to be distributed in gyri valleys and in the vicinity of blood vessels. Furthermore, in many cases, transactive response DNA-binding protein 43 inclusion bodies can be seen inside neurons. With respect to amyloid-β deposition, cases with senile plaques are comparatively less common, and are seen mostly in older adults, if at all (McKee et al., 2013). Currently, there is no treatment for chronic traumatic encephalopathy, with prevention of traumatic brain injury being the only option.

Mini-dictionary of terms

Traumatic brain injury: Traumatic brain injury is a collective name for the variety of symptoms that result from suffering a physical blow to the head. Traffic accidents are the most common cause of traumatic brain injury, followed by falls, sports, and physical labor. Traumatic brain injury is categorized into a severe type, which involves loss of consciousness that lasts for more than 24 h, and mild traumatic brain injury, which either does not involve a loss of consciousness or only involves short-term losses.

Alzheimer's disease: Alzheimer's disease is a form of dementia that involves impaired memory, thinking, and behavior. Dementia is a general term for an illness state in which a pathological decline in memory and other cognitive functions, significant enough to interfere with daily life, is observed. Alzheimer's disease comprises 60%–80% of all dementia cases. It is believed to be caused by a variety of different factors, one of which is referred to as traumatic brain injury. Pathologically, the two most characteristic signs of Alzheimer's disease are amyloid plaques and neurofibrillary tangles in the brain.

Chronic traumatic encephalopathy: Chronic traumatic encephalopathy is a neurodegenerative disorder caused by traumatic brain injury; neuropathologically, it is a tauopathy, which involves the over-accumulation of tubulin-associated unit protein in the brain. At present, there is no definitive way to diagnose the disease while a patient is still alive; diagnoses are therefore made via autopsy. Typically, it occurs in boxers, American football players, retired service men, etc.—individuals that have experienced multiple concussions throughout their life. However, isolated severe traumatic brain injury can also lead to chronic traumatic encephalopathy findings. In addition to cognitive impairment, the observed problems include high rates of psychological symptoms and a high risk of suicide.

Amyloid-β: The primary component of amyloid plaques present in the brain of patients with Alzheimer's disease. It is an amino acid peptide implicated in Alzheimer's disease. This peptide is derived from amyloid precursor protein, which is then cleaved by β-secretase and γ-secretase.

Neprilysin: An enzyme known to degrade amyloid-β. Knockout mice lacking neprilysin exhibit Alzheimer's-like behavioral impairment as well as amyloid-β aggregates in the brain. This evidence points to the involvement of neprilysin in the Alzheimer's disease onset process.

Tubulin-associated unit protein: A protein that binds to the microtubules that make up the cytoskeleton of neurons. It is involved in intracellular cytoskeleton formation and material transport. Abnormal tubulin-associated unit phosphorylation and subsequent intracellular accumulation are known to occur in a variety of psychoneurological diseases, including Alzheimer's-type dementia.

Tauopathies are diseases in which tubulin-associated unit protein becomes hyperphosphorylated, thereby abnormally depositing as insoluble aggregates in neurons or glial cells. The most representative of the tauopathies is Alzheimer's disease, in which phosphorylated tubulin-associated unit is abnormally deposited inside cells, where it forms insoluble aggregates called neurofibrillary tangles. Other tauopathies include progressive supranuclear palsy, corticobasal degeneration, Pick's disease, argyrophilic grain disease, senile dementia of the neurofibrillary tangle type, and chronic traumatic encephalopathy.

Positron emission tomography: A technique that enables external visualization of the movement of biological molecules inside a living organism. Positron emission tomography agents labeled with specific radioisotopes are given to patients, and the gamma rays caused by the emission of positrons from these positron emission tomography agents are detected, enabling the 3D localization and quantification of biological molecules in deep parts of the body.

Higher brain dysfunction: A Japan-specific diagnosis/administrative term referring to a particular type of daily life impairment caused by cognitive impairments, including memory disorder, attention disorder, executive dysfunction, and social disorder caused by brain injury, stroke, or other forms of damage to the brain. This state overlaps with the International Classifications of Diseases (ICD) categories F04/06/07.

Key facts of post-traumatic brain injury delayed-onset cognitive and psychiatric disorders

- Described as "punch-drunk syndrome" and "dementia pugilistica," we have long known that cognitive and psychological dysfunctions can occur in the late phase of traumatic brain injury.
- Late-onset schizophreniform disorders have also been reported, with a mean onset delay of psychosis following severe traumatic brain injury of more than 5 years.
- Alzheimer's disease and chronic traumatic encephalopathy are both believed to be post-traumatic brain injury delayed-onset cognitive and psychiatric complications and involve deceptively similar pathologies.
- Alzheimer's disease, first identified in 1901, is an irreversible, progressive disorder of the brain that gradually impairs memory and other cognitive functions.
- The first seminal report on chronic traumatic encephalopathy, in boxers, was published in 1949 by Critchley et al.
- Full-scale research into chronic traumatic encephalopathy began around 2005; it is therefore a pathology only just beginning to be explored, and much of the details surrounding it are unknown.
- Chronic traumatic encephalopathy is known to occur in a variety of different sports, but among these, it is particularly frequent in boxing due to the prevalence of head trauma experienced by practitioners of the sport.

Summary points

- Much remains unknown about the pathologies of delayed-onset after-effects that manifest years after a traumatic brain injury.
- Further advances in positron emission tomography are expected to help detect traumatic brain injury-caused delayed-onset pathologies in alive patients and evaluate underlying pathological mechanisms, thereby improving diagnosis and treatment of traumatic brain injury.
- Here, we focus on the relationship between two emblematic delayed-onset pathologies often seen after traumatic brain injury; Alzheimer's disease and chronic traumatic encephalopathy.
- Both are believed to be post-traumatic brain injury delayed-onset cognitive and psychiatric pathologies.
- Traumatic brain injury was found to be a risk factor for Alzheimer's disease onset.
- Alzheimer's disease onset after a long progression of traumatic brain injury may involve amyloid-β pathologies.
- While it is generally accepted that the so-called "delayed-onset encephalopathies" caused by traumatic brain injury are a type of tauopathy, whether tau accumulation due to traumatic brain injury has an effect on Alzheimer's disease onset is unclear.
- While traumatic brain injury-caused Alzheimer's disease and chronic traumatic encephalopathy are deceptively similar pathologies, they are considered to be different entities.

Acknowledgments

We would like to thank Editage (www.editage.com) for English language editing.

References

Achté, K., Jarho, L., Kyykkä, T., & Vesterinen, E. (1991). Paranoid disorders following war brain damage. *Psychopathology, 24*(5), 309–315.

Adams, J. H., Doyle, D., Ford, I., Gennarelli, T. A., Graham, D. I., & McLellan, D. R. (1989). Diffuse axonal injury in head injury: Definition, diagnosis and grading. *Histopathology, 15*(1), 49–59.

Adams, J. H., Graham, D. I., Murray, L. S., & Scott, G. (1982). Diffuse axonal injury due to nonmissile head injury in humans: An analysis of 45 cases. *Annals of Neurology: Official Journal of the American Neurological Association and the Child Neurology Society, 12*(6), 557–563.

Castellani, R. J., & Perry, G. (2017). Dementia pugilistica revisited. *Journal of Alzheimer's Disease, 60*(4), 1209–1221.

Chen, X. H., Johnson, V. E., Uryu, K., Trojanowski, J. Q., & Smith, D. H. (2009). A lack of amyloid β plaques despite persistent accumulation of amyloid β in axons of long-term survivors of traumatic brain injury. *Brain Pathology, 19*(2), 214–223.

Critchley, M. (1957). Medical aspects of boxing, particularly from a neurological standpoint. *British Medical Journal, 1*, 357–362.

de la Torre, J. C. (2004). Is Alzheimer's disease a neurodegenerative or a vascular disorder? Data, dogma, and dialectics. *The Lancet Neurology, 3*(3), 184–190.

de la Torre, J. C., & Mussivand, T. (1993). Can disturbed brain microcirculation cause Alzheimer's disease? *Neurological Research, 15*(3), 146–153.

De Silva, T. M., & Faraci, F. M. (2016). Microvascular dysfunction and cognitive impairment. *Cellular and Molecular Neurobiology, 36*(2), 241–258.

Deighton, S., Buchy, L., Cadenhead, K. S., Cannon, T. D., Cornblatt, B. A., McGlashan, T. H., … Woods, S. W. (2016). Traumatic brain injury in individuals at clinical high risk for psychosis. *Schizophrenia Research, 174*(1–3), 77–81.

DeKosky, S. T., Blennow, K., Ikonomovic, M. D., & Gandy, S. (2013). Acute and chronic traumatic encephalopathies: Pathogenesis and biomarkers. *Nature Reviews Neurology, 9*(4), 192.

Falcon, B., Zivanov, J., Zhang, W., Murzin, A. G., Garringer, H. J., Vidal, R., … Scheres, S. H. (2019). Novel tau filament fold in chronic traumatic encephalopathy encloses hydrophobic molecules. *Nature, 568*(7752), 420–423.

Faul, M., Wald, M. M., Xu, L., & Coronado, V. G. (2010). *Traumatic brain injury in the United States; emergency department visits, hospitalizations, and deaths 2002–2006*.

Fleminger, S., Oliver, D. L., Lovestone, S., Rabe-Hesketh, S., & Giora, A. (2003). Head injury as a risk factor for Alzheimer's disease: The evidence 10 years on; a partial replication. *Journal of Neurology, Neurosurgery, and Psychiatry, 74*, 857–862.

Fujii, D., & Ahmed, I. (2002). Characteristics of psychotic disorder due to traumatic brain injury: An analysis of case studies in the literature. *The Journal of Neuropsychiatry and Clinical Neuroscience, 14*(2), 130–140.

Hardy, J., & Selkoe, D. J. (2002). The amyloid hypothesis of Alzheimer's disease: Progress and problems on the road to therapeutics. *Science, 297*(5580), 353–356.

Hellström-Lindahl, E., Ravid, R., & Nordberg, A. (2008). Age-dependent decline of neprilysin in Alzheimer's disease and normal brain: Inverse correlation with Aβ levels. *Neurobiology of Aging, 29*(2), 210–221.

Herholz, K., & Ebmeier, K. (2011). Clinical amyloid imaging in Alzheimer's disease. *The Lancet Neurology, 10*(7), 667–670.

Hiraoka, T. (2019). Association of late effects of single, severe traumatic brain injury with Alzheimer's disease using amyloid PET. *Neurocase, 25*(1–2), 10–16.

Huber, A., Gabbert, K., Kelemen, J., & Cervos-Navarro, J. (1993). Density of amyloid plaques in brains after head trauma. *Journal of Neurotrauma, 10*, S180.

Ikonomovic, M. D., Uryu, K., Abrahamson, E. E., Ciallella, J. R., Trojanowski, J. Q., Lee, V. M. Y., … De Kosky, S. T. (2004). Alzheimer's pathology in human temporal cortex surgically excised after severe brain injury. *Experimental Neurology, 190*(1), 192–203.

Johnson, V. E., Stewart, W., Arena, J. D., & Smith, D. H. (2017). Traumatic brain injury as a trigger of neurodegeneration. *Advances in Neurobiology, 15*, 383–400.

Johnson, V. E., Stewart, W., & Smith, D. H. (2012). Widespread tau and amyloid-beta pathology many years after a single traumatic brain injury in humans. *Brain Pathology, 22*(2), 142–149.

Johnson, V. E., Stewart, W., & Smith, D. H. (2013). Axonal pathology in traumatic brain injury. *Experimental Neurology, 246*, 35–43.

Katsumoto, A., Takeuchi, H., & Tanaka, F. (2019). Tau pathology in chronic traumatic encephalopathy and Alzheimer's disease: Similarities and differences. *Frontiers in Neurology, 10*, 980.

Katzman, R. (1993). Education and the prevalence of dementia and Alzheimer's disease. *Neurology, 43*(1), 13–20.

Kawai, N., Shishido, H., Hatakeyama, T., & Tamiya, T. (2017). Cerebral amyloid β deposition and cognitive impairment after traumatic brain injury: Alzheimer's disease secondary to TBI. *Neurotraumatology, 40*, 96–104.

LoBue, C., Cullum, C. M., Didehbani, N., Yeatman, K., Jones, B., Kraut, M. A., & Hart, J., Jr. (2018). Neurodegenerative dementias after traumatic brain injury. *The Journal of Neuropsychiatry and Clinical Neurosciences, 30*(1), 7–13.

Lye, T. C., & Shores, E. A. (2000). Traumatic brain injury as a risk factor for Alzheimer's disease: A review. *Neuropsychology Review, 10*(2), 115–129.

Martland, H. S. (1928). Punch drunk. *Journal of the American Medical Association, 91*(15), 1103–1107.

Max, J. E. (2014). Neuropsychiatry of pediatric traumatic brain injury. *Psychiatric Clinics of North America, 37*(1), 125–140.

Max, J. E., Friedman, K., Wilde, E. A., Bigler, E. D., Hanten, G., Schachar, R. J., … Yang, T. T. (2015). Psychiatric disorders in children and adolescents 24 months after mild traumatic brain injury. *The Journal of Neuropsychiatry and Clinical Neurosciences, 27*(2), 112–120.

Max, J. E., Schachar, R. J., Landis, J., Bigler, E. D., Wilde, E. A., Saunders, A. E., … Levin, H. S. (2013). Psychiatric disorders in children and adolescents in the first six months after mild traumatic brain injury. *The Journal of Neuropsychiatry and Clinical Neurosciences*, *25*(3), 187–197.

McKee, A. C., Cairns, N. J., Dickson, D. W., Folkerth, R. D., Keene, C. D., Litvan, I., … Tripodis, Y. (2016). The first NINDS/NIBIB consensus meeting to define neuropathological criteria for the diagnosis of chronic traumatic encephalopathy. *Acta Neuropathologica*, *131*(1), 75–86.

McKee, A. C., Stein, T. D., Nowinski, C. J., Stern, R. A., Daneshvar, D. H., Alvarez, V. E., … Riley, D. O. (2013). The spectrum of disease in chronic traumatic encephalopathy. *Brain*, *136*(1), 43–64.

Mehta, K. M., Ott, A., Kalmijn, S., Slooter, A. J. C., Van Duijn, C. M., Hofman, A., & Breteler, M. M. B. (1999). Head trauma and risk of dementia and Alzheimer's disease: The Rotterdam study. *Neurology*, *53*(9), 1959.

Mez, J., Daneshvar, D. H., Kiernan, P. T., Abdolmohammadi, B., Alvarez, V. E., Huber, B. R., … Cormier, K. A. (2017). Clinicopathological evaluation of chronic traumatic encephalopathy in players of American football. *Journal of the American Medical Association*, *318*(4), 360–370.

Mez, J., Solomon, T. M., Daneshvar, D. H., Stein, T. D., & McKee, A. C. (2016). Pathologically confirmed chronic traumatic encephalopathy in a 25-year-old former college football player. *Journal of the American Medical Association Neurology*, *73*(3), 353–355.

Millspaugh, J. (1937). Dementia pugilistica. *United States Naval Medical Bulletin*, *35*, 297–303.

Mintun, M. A., Larossa, G. N., Sheline, Y. I., Dence, C. S., Lee, S. Y., Mach, R. H., … Morris, J. C. (2006). [^{11}C] PIB in a nondemented population: Potential antecedent marker of Alzheimer disease. *Neurology*, *67*(3), 446–452.

Mortimer, J. A., Van Duijn, C. M., Chandra, V., Fratiglioni, L., Graves, A. B., Heyman, A., … Shalat, S. L. (1991). Head trauma as a risk factor for Alzheimer's disease: A collaborative re-analysis of case-control studies. *International Journal of Epidemiology*, *20*, S28–S35.

Nicoll, J. A., Roberts, G. W., & Graham, D. I. (1995). Apolipoprotein E ε4 allele is associated with deposition of amyloid β-protein following head injury. *Nature Medicine*, *1*(2), 135–137.

Nielsen, A. S., Mortensen, P. B., O'Callaghan, E., Mors, O., & Ewald, H. (2002). Is head injury a risk factor for schizophrenia? *Schizophrenia Research*, *55*(1–2), 93–98.

Okamura, N., Harada, R., Kudo, Y., & Yanai, K. (2015). Amyloid and tau imaging, its present and future. *Folia Pharmacologica Japaonica*, *146*, 144–149.

Omalu, B. I., DeKosky, S. T., Hamilton, R. L., Minster, R. L., Kamboh, M. I., Shakir, A. M., & Wecht, C. H. (2006). Chronic traumatic encephalopathy in a national football league player: Part II. *Neurosurgery*, *59*(5), 1086–1093.

Omalu, B. I., DeKosky, S. T., Minster, R. L., Kamboh, M. I., Hamilton, R. L., & Wecht, C. H. (2005). Chronic traumatic encephalopathy in a National Football League player. *Neurosurgery*, *57*(1), 128–134.

Omalu, B. I., Fitzsimmons, R. P., Hammers, J., & Bailes, J. (2010). Chronic traumatic encephalopathy in a professional American wrestler. *Journal of Forensic Nursing*, *6*(3), 130–136.

Perrotta, M., Lembo, G., & Carnevale, D. (2016). Hypertension and dementia: Epidemiological and experimental evidence revealing a detrimental relationship. *International Journal of Molecular Sciences*, *17*(3), 347.

Perry, D. C., Sturm, V. E., Peterson, M. J., Pieper, C. F., Bullock, T., Boeve, B. F., … Welsh-Bohmer, K. A. (2016). Association of traumatic brain injury with subsequent neurological and psychiatric disease: A meta-analysis. *Journal of Neurosurgery*, *124*(2), 511–526.

Ramos-Cejudo, J., Wisniewski, T., Marmar, C., Zetterberg, H., Blennow, K., de Leon, M. J., & Fossati, S. (2018). Traumatic brain injury and Alzheimer's disease: The cerebrovascular link. *eBioMedicine*, *28*, 21–30.

Roberts, G. W., Gentleman, S. M., Lynch, A., & Graham, D. I. (1991). βA4 amyloid protein deposition in brain after head trauma. *The Lancet*, *338*(8780), 1422–1423.

Roberts, G. W., Gentleman, S. M., Lynch, A., Murray, L., Landon, M., & Graham, D. I. (1994). Beta amyloid protein deposition in the brain after severe head injury: Implications for the pathogenesis of Alzheimer's disease. *Journal of Neurology, Neurosurgery & Psychiatry*, *57*(4), 419–425.

Salib, E., & Hillier, V. (1997). Head injury and the risk of Alzheimer's disease: A case control study. *International Journal of Geriatric Psychiatry*, *12*(3), 363–368.

Sapolsky, R. M., Krey, L. C., & McEwen, B. S. (1986). The neuroendocrinology of stress and aging: The glucocorticoid cascade hypothesis. *Endocrine Reviews*, *7*(3), 284–301.

Schofield, P. W., Tang, M., Marder, K., Bell, K., Dooneief, G., Chun, M., … Mayeux, R. (1997). Alzheimer's disease after remote head injury: An incidence study. *Journal of Neurology, Neurosurgery, and Psychiatry*, *62*, 119–124.

Selkoe, D. J. (1991). The molecular pathology of Alzheimer's disease. *Neuron*, *6*(4), 487–498.

Shively, S., Scher, A. I., Perl, D. P., & Diaz-Arrastia, R. (2012). Dementia resulting from traumatic brain injury: What is the pathology? *Archives of Neurology*, *69*(10), 1245–1251.

Smith, D. H., Chen, X. H., Iwata, A., & Graham, D. I. (2003). Amyloid β accumulation in axons after traumatic brain injury in humans. *Journal of Neurosurgery*, *98*(5), 1072–1077.

Smith, D. H., Chen, X. H., Nonaka, M., Trojanowski, J. Q., Lee, V. Y., Saatman, K. E., … Meaney, D.F. (1999). Accumulation of amyloid β and tau and the formation of neurofilament inclusions following diffuse brain injury in the pig. *Journal of Neuropathology and Experimental Neurology*, *58*(9), 982–992.

Smith, C., Graham, D. I., Murray, L. S., & Nicoll, J. A. R. (2003). Tau immunohistochemistry in acute brain injury. *Neuropathology and Applied Neurobiology*, *29*(5), 496–502.

Smith, D. H., Johnson, V. E., & Stewart, W. (2013). Chronic neuropathologies of single and repetitive TBI: Substrates of dementia? *Nature Reviews Neurology*, *9*(4), 211–221.

Stern, R. A., Daneshvar, D. H., Baugh, C. M., Seichepine, D. R., Montenigro, P. H., Riley, D. O., … Simkin, I. (2013). Clinical presentation of chronic traumatic encephalopathy. *Neurology*, *81*(13), 1122–1129.

Takahata, K., Kimura, Y., Sahara, N., Koga, S., Shimada, H., Ichise, M., … Umeda, S. (2019). PET-detectable tau pathology correlates with long-term neuropsychiatric outcomes in patients with traumatic brain injury. *Brain*, *142*(10), 3265–3279.

Takahata, K., Mimura, M., Shimada, H., Higuchi, M., & Suhara, T. (2015). Late-onset neurocognitive deficits following traumatic brain injury: chronic traumatic encephalopathy (CTE) and psychotic disorder following TBI (PDF TBI). *Higher Brain Function Research*, *35*, 276–282.

Washington, P. M., Morffy, N., Parsadanian, M., Zapple, D. N., & Burns, M. P. (2014). Experimental traumatic brain injury induces rapid aggregation and oligomerization of amyloid-beta in an Alzheimer's disease mouse model. *Journal of Neurotrauma*, *31*(1), 125–134.

Xia, C. F., Arteaga, J., Chen, G., Gangadharmath, U., Gomez, L. F., Kasi, D., … Mu, F. (2013). [^{18}F] T807, a novel tau positron emission tomography imaging agent for Alzheimer's disease. *Alzheimer's & Dementia*, *9*(6), 666–676.

Chapter 4

Drug interventions and stem cells in traumatic brain injury: Translation from experimental model to bedside

Maha Tabet[a], Mohammad Amine Reslan[b], Muhammad Ali Haidar[b], Hawraa Issa[c], Maya El Dor[d], Reem Abedi[b], Leila Nasrallah[b], Ali Eid[e], Kazem Zibara[d], and Firas H. Kobeissy[b]

[a]*Molecular, Cellular and Developmental biology department (MCD), Centre de Biologie Integrative (CBI), University of Toulouse, CNRS, UPS, Toulouse, France,* [b]*Department of Biochemistry and Molecular Genetics, Faculty of Medicine, American University of Beirut, Beirut, Lebanon,* [c]*College of Public Health, Phoenicia University, Zahrani, Lebanon,* [d]*PRASE, Lebanese University, Beirut, Lebanon,* [e]*Faculty of Medicine, Department of Pharmacology and Toxicology, American University of Beirut, Beirut, Lebanon*

List of abbreviations

ASC	adult stem cells
BBB	blood–brain barrier
BMMNC	bone marrow mononuclear cells
BDNF	brain-derived neurotrophic factor
CCI	controlled cortical impact
COX-2	cyclooxygenase-2
DAI	diffuse axonal injury
ESC	embryonic stem cells
ECFC	endothelial colony-forming cell
EDA	Edaravone
EPC	endothelial progenitor cell
FDA	Food and Drug Administration
GCS	Glasgow Coma Scale
GDNF	glial cell line-derived neurotrophic factor
GH	growth hormone
GPx	glutathione peroxidase
HBOT	hyperbaric oxygen therapy
iPSCs	induced pluripotent stem cells
LPA	lipoic acid
MAPC	multi-potent adult progenitor cells
MitoQ	mitoquinone
MSC	mesenchymal stem cell
mTBI	mild TBI
NAC	*N*-acetylcysteine
NSC	neural stem cells
ROS	reactive oxygen species
SOD	superoxide dismutase
TBI	traumatic brain injury
VAS203	4-amino-tetrahydrobiopterin

Diagnosis and Treatment of Traumatic Brain Injury. https://doi.org/10.1016/B978-0-12-823347-4.00012-9

Introduction

Traumatic brain injury (TBI) is a major health concern defined as an impairment of brain functions caused by an external force. An estimated 69 million people suffer from TBI each year with an incidence of 939 cases per 100,000 with the highest incidence being assessed in the Americas (Dewan et al., 2018). Several factors play a role in defining the severity of the injury and its patterns, including the nature, speed, and duration of the impact. The severity of TBI is classified most commonly according to the Glasgow Coma scale (GCS). GSC measures clinical signs in patients based on the level of consciousness, eye movement, and motor abilities. The most severe cases have low scores on the GCS (3–8), which increases as the severity decreases such that the mild injury has a score of 13–15 (Teasdale & Jennett, 1974). Notably, 81% of the total TBI cases are mild cases (Dewan et al., 2018).

The events that lead to TBI are very heterogeneous. TBI can be initiated by a direct impact to the head. The chief causes of TBI worldwide are considered to be falls, vehicle accidents, war assaults, and sports-related injuries (Bushnik, Hanks, Kreutzer, & Rosenthal, 2003). The pathophysiological changes of TBI are reflected in patients as pathological conditions including seizures, sleep disorders, increased susceptibility to neurodegenerative diseases including Alzheimer's disease, and psychiatric problems including depression (Bramlett & Dietrich, 2015).

To date, there are no Food and Drug Administration (FDA)-approved drugs for the treatment of TBI. However, early after an injury, blocking one of the mechanisms leading to the symptoms can be effective in ameliorating its consequences (Mohamadpour, Whitney, & Bergold, 2019). Most commonly, drug interventions were used to alleviate the consequences of TBI; however, these remain of limited efficacy. Thus, other treatment options such as stem cells are under study to test their potential for advancing the treatment of TBI (Tabet et al., 2020). In this chapter, we will shed light on the pathology mechanisms underlying the consequences of TBI and the major drug interventions used. Then, we will discuss stem cells as a treatment option for TBI.

The pathophysiology of TBI

Understanding the pathophysiology of TBI is essential for better treatment. There are two main types of primary injuries resulting from the immediate impact to the head: focal and diffuse brain injuries. These two types can coexist or occur independently (Skandsen et al., 2010). Typically, the impact leads to necrosis of neuronal cells with compromised blood supply and hemorrhage at some layers of the brain (Schmidt, Infanger, Heyde, Ertel, & Stahel, 2004). The hallmark of diffuse brain damage is diffuse axonal injury (DAI) caused by the acceleration/deceleration forces imposed by the impact. This axonal shearing takes place primarily in the white matter tracts of the brain and can persist for months following the injury (Saatman et al., 2008).

The primary injury leads to primary neuronal death and the breakdown of the blood–brain barrier (BBB) which leads to the excessive release of excitatory amino acids such as glutamate from presynaptic terminals (Chamoun, Suki, Gopinath, Goodman, & Robertson, 2010). The subsequent activation of glutamate receptors allows the influx of ions such as Ca^{2+} and Na^{+} into the post-synaptic terminals (Sun et al., 2008), and induces the production of reactive oxygen species (ROS) that aggravate the secondary injury (Girouard et al., 2009). The excessive amount of Ca^{2+} and ROS leads to the impairment of mitochondrial function through the depolarization and permeabilization of the mitochondrial membrane, the inhibition of ATP synthesis, and the activation of cell death pathways via caspases or calpains (Chamoun et al., 2010). Moreover, the disruption of BBB permits the infiltration of immune cells including monocytes and lymphocytes to the site of injury. These secrete pro-inflammatory cytokines such as IL-6 and TNF-α that contribute to the activation of caspases implicated in apoptosis (Morganti-Kossmann, Rancan, Stahel, & Kossmann, 2002). Prolonged inflammation provides grounds for the recruitment of macrophages and the activation of glial cells, mainly microglia and astrocytes (Bye et al., 2007). DAI resulting from the primary impact together with the activation of Ca^{2+}-dependent proteolytic enzymes culminate in prolonged axonal damage marked by myelin degradation and impairment of axonal transport and the subsequent formation of retraction bulbs, axonal swelling, and cell death as illustrated in Fig. 1 (Buki & Povlishock, 2006).

Possible treatments for TBI: Drug interventions

The Food and Drug Administration (FDA) has yet to approve a neuroprotective treatment against the insidious secondary injuries sustained due to TBI (Lamade et al., 2020). That being said, several contenders have been circulating in research circles; and while some have demonstrated diminished efficacy, others have shown promising results (Table 1). Potential neuroprotective interventions often seek to modulate injury cascades—such as cellular stress, oxidative stress, immunotoxicity, and inflammation—which are triggered by secondary brain injury (Lucke-Wold et al., 2018).

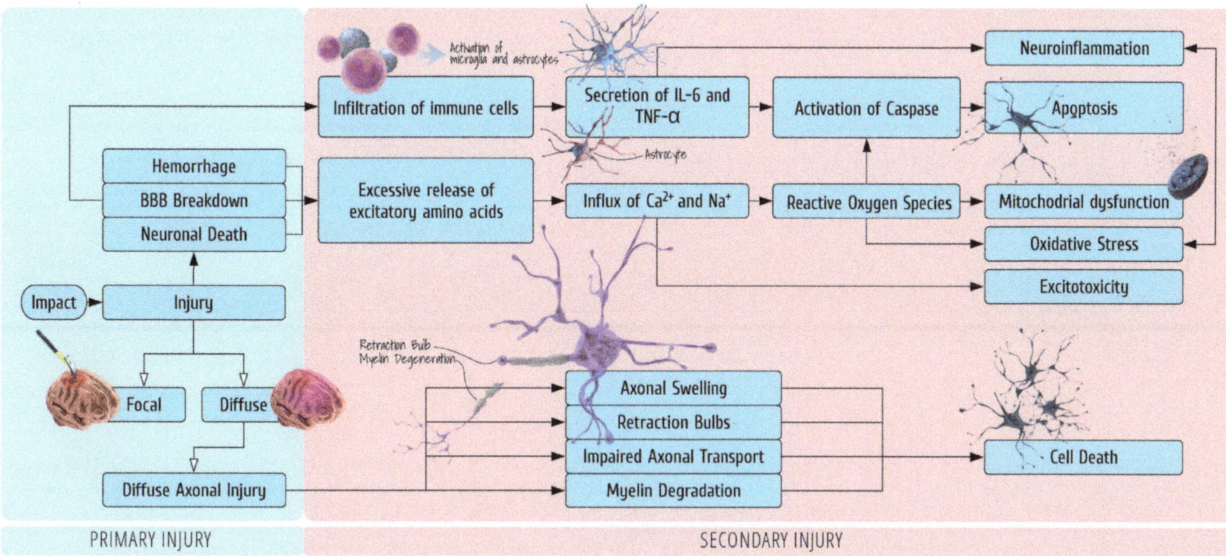

FIG. 1 The Pathophysiology of TBI. The impact on the brain can lead to a focal damage or a diffuse one. The latter leads to diffuse axonal injury. The impact is the primary injury that triggers a cascade of biochemical reactions including excitotoxicity following an ion imbalance, oxidative stress and an excessive production of reactive oxygen species, the infiltration of immune cells and the activation of resident glial cells that lead to neuroinflammation, and mitochondrial dysfunction. The pathological changes that occur following TBI may lead to cell death which is a hallmark of neurodegenerative diseases.

Oxidative stress resulting from the intracellular build-up of reactive ROS can be mitigated by anti-oxidants. For example, *N*-acetylcysteine (NAC) is a thiol possessing anti-oxidant and anti-inflammatory properties. In 20 animal studies, administration of NAC post-TBI significantly promoted cortical sparing, reduced apoptosis, and lessened the levels of biomarkers of both inflammation and oxidative stress. Additionally, considerable improvements in cognition and psychomotor performance were reported. In two randomized clinical trials, however, the reported improvements were less substantial and limited to functional outcome (Bhatti et al., 2017).

Mitoquinone (MitoQ), a potent antioxidant, has manifested neuroprotective activity against a number of neurological disorders in several animal studies. In TBI, its function remains indefinite. However, in a weight-drop mouse model of TBI, MitoQ allowed for improved neurological deficits, reduced brain edema, and lesser neuronal apoptosis in the cortex. MitoQ also stimulated the activity of the antioxidant enzymes superoxide dismutase (SOD) and glutathione peroxidase (GPx) (Zhou et al., 2018). Another antioxidant, Edaravone (EDA), has been investigated with respect to its neuroprotective capabilities post-TBI. In a controlled cortical impact (CCI) model of TBI, mice injected with EDA post-impact demonstrated enhanced neurological function, lesser neuronal apoptosis, and reduced inflammation and oxidative stress (Zhang et al., 2019). In a similar vein, lipoic acid (LPA), a potent antioxidant that suppresses lipid peroxidation, proved effective in abating neuronal cell death and reinforcing the activity of the anti-oxidant enzymes SOD and GPx (Lucke-Wold et al., 2018).

Glutamate antagonists may limit the detrimental secondary injury effects of elevated extracellular glutamate (Lucke-Wold et al., 2018). For example, memantine—an uncompetitive NMDA receptor antagonist—shows promise in providing neuroprotection in TBI. In a fluid percussion TBI model, rats treated with memantine injections demonstrated improved motor and sensory functional outcomes, decreased neuronal apoptosis, astrogliosis, and microgliosis, and a significantly decreased volume of cerebral infarction 72 h after TBI. The post-TBI treatment with memantine also enhanced neuronal survival and alleviated neuronal nitrosative stress in the perilesioned cortex (Wang et al., 2018).

The anomalous intracellular build-up of calcium post-TBI can lead to endoplasmic reticulum stress and eventually cell death. Consequently, calcium-channel blockers have been tested for the treatment of neuropathological consequences partially arising out of excessive Ca^{2+} influx. One such example is SNX-185—a N-type voltage-gated calcium channel blocker. SNX-185 was tested in an in vitro mechanical strain injury model of TBI, and the results showed that cultures treated with SNX-185 had a reduced calcium accumulation 3 h post-injury and a higher number of viable neurons 24 h after injury (Shahlaie et al., 2013).

Cyclooxygenase-2 (COX-2) inhibitors have also been researched for the treatment of secondary neuroinflammation due to TBI. In a focal penetrating TBI model, the inhibition of COX-2 by diclofenac was correlated with reduced apoptosis and

TABLE 1 Drug interventions as possible treatments for TBI.

Potential TBI treatment	Mode of action	Outcome	Limitations	Reference
N-acetylcysteine	Replenishes glutathione synthesis	In animal studies: • Promoted cortical sparing • Reduced apoptosis • Lessened inflammation and oxidative stress • Improved cognition and psychomotor performance	Less substantial improvements in clinical trials	Bhatti et al. (2017)
Mitoquinone	Mitochondria-targeted anti-oxidant	In a mouse model: • Improved neurological deficits • Reduced brain edema • Reduced neuronal death in the cortex • Improved the activity of anti-oxidant enzymes such as SOD and GPx	• Pre-clinical trials are scarce • Has not been tested for the treatment of TBI in a clinical trial	Zhou et al. (2018)
Edaravone	Anti-oxidant	In a mouse model: • Enhanced neurological function • Reduced neuronal apoptosis • Reduced inflammation and oxidative stress	• Short half-life • Pre-clinical trials are scarce • Has not been tested for the treatment of TBI in a clinical trial	Zhang et al. (2019)
Lipoic Acid	Endogenous anti-oxidant; suppresses lipid peroxidation	In a rat model: • Reduced neuronal cell death • Reinforced the activity of the anti-oxidant enzymes SOD and GPx	• Short half-life • Pre-clinical and clinical trials are scarce	Lucke-Wold et al. (2018)
Memantine	Uncompetitive NMDA receptor antagonist	In a rat model: • Improved motor and sensory functional outcomes • Decreased neuronal apoptosis, astrogliosis, and microgliosis • Decreased the volume of cerebral infarction 72 h after TBI • Enhanced neuronal survival • Alleviated neuronal nitrosative stress in the peri-lesioned cortex	• Pre-clinical and clinical trials are scarce	Wang, Wee, Hu, Chio, and Kuo (2018)

SNX-185	N-type voltage-gated calcium channel blocker	In an in vitro model: • Reduced calcium accumulation 3 h post-injury • Increased the number of viable neurons 24 h after injury	• Pre-clinical trials are scarce • Has not been tested for the treatment of TBI in a clinical trial	Shahlaie, Gurkoff, Lyeth, Muizelaar, and Berman (2013)
Diclofenac	Cyclooxygenase-2 inhibitor	In a rat model: • Reduced apoptosis • Reduced the lesion area	• Pre-clinical trials are scarce • Has not been tested for the treatment of TBI in a clinical trial	Dehlaghi Jadid et al. (2019)
Hyperbaric oxygen therapy	Increases oxygen concentration in the body	Exhibits potential in the clinical setting	Surrounded by controversy with respect to its safety, feasibility, and mixed results	Daly et al. (2018)
Growth hormone	Replacement therapy for patients with GH-deficiency	Improve cognitive functions, behavioral functions, activities of daily living (ADL), and quality of life (QoL) functions in patients with GH-deficiency	Similar effects are not as well established in patients with no GH-deficiency	Devesa, Reimunde, Devesa, Barbera, and Arce (2013), Gardner et al. (2015), Moreau et al. (2013)
Rosuvastatin	Reduces inflammatory cytokines	May induce an anti-inflammatory effect and may promote recovery after TBI	Study had a small sample size	Sanchez-Aguilar et al. (2013)
Cannabinoids	Neuroprotective actions	In animal models: • Provide protection against excitotoxicity, calcium influx, free radical formation and neuroinflammation	• Similar effects are not as well established in humans • Clinical trials are insufficient	Biegon (2004)
β-Blockers	Down-regulate the levels of catecholamines	The use of β-blockers before injury may promote the chances of survival	• Clinical trials are insufficient • The mechanism by which β-blockers provide neuroprotection remains unclear	Mohseni, Talving, Wallin, Ljungqvist, and Riddez (2014)
Enzogenol	Anti-oxidant; anti-inflammatory flavonoid-rich extract	May significantly reduce cognitive failures after 6 weeks	Study had a small sample size and was conducted among patients with mTBI	Theadom et al. (2013)
Cerebrolysin	Nootropic	May improve cognitive function, notably long-term memory and drawing function	Study had a small sample size and was conducted among patients with mTBI	Chen, Wei, Tsaia, Chen, and Cho (2013)
VAS203	Nitric oxide synthase inhibitor	Can provide neuroprotection after TBI	• Study had a small sample size • Drug increases the risk of acute kidney injury at high doses	Stover et al. (2014)

lesion area. However, the mechanisms by which COX-2 and COX-2 inhibitors affect post-TBI neuroinflammation remain unclear. For example, post-traumatic COX-2 expression has been found to be higher in male rats than female rats. This gender dichotomy remains to be explained (Dehlaghi Jadid et al., 2019).

The treatment of TBI-induced brain ischemia via hyperbaric oxygen therapy (HBOT)—a treatment that increases oxygen concentration in the body—has been repeatedly investigated. A systematic review of 22 pre-clinical and eight clinical studies that administered HBOT within 30 days of TBI concluded that the therapy exhibits a promising potential in the clinical setting. Results from the pre-clinical studies confirm the positive effects of the treatment with regard to several outcome measures. The clinical studies also reported positive results with a significant improvement in physiological outcome measures and the absence of cerebral or pulmonary toxicity (Daly et al., 2018).

The supplementation of certain vitamins and minerals contributed to positive results in several pre-clinical models of mild TBI (mTBI). For instance, in a fluid percussion model of TBI, vitamin E supplementation promoted a reduction in oxidative stress and an improvement in learning and memory. In a different model, vitamin D—in combination with progesterone—allowed for the amelioration of both inflammation and neuronal cell death. Similarly, when administered alongside progesterone, nicotinamide lead to the betterment of recovery in rats with mTBI. The sustained infusion of nicotinamide afforded additional neuroprotection. It should be noted however that vitamins are more likely to be utilized as complementary supplements that boost the performance of other viable therapeutic agents (Lucke-Wold et al., 2018).

Various other pharmacological agents have also demonstrated significant neuroprotective effects after TBI and warrant further research. These include, but are not limited to, growth hormone (GH), statins, marijuana, β-blockers, Enzogenol, Cerebrolysin, and nitric oxide synthase inhibitors. In GH-deficient TBI patients, GH replacement therapy was shown to improve cognitive functions, behavioral functions, activities of daily living, and quality of life functions (Gardner et al., 2015; Moreau et al., 2013). However, these results are not as well-established in TBI patients who are not GH-deficient (Devesa et al., 2013). In a randomized controlled trial, Rosuvastatin (a statin medication) exhibited promising results in lowering the levels of pro-inflammatory mediators and improving functional recovery after TBI (Sanchez-Aguilar et al., 2013). While the neuroprotective effects of marijuana and other cannabinoid analogs have been established in animal studies, the same effects have not yet been verified in human patients with TBI. In animal TBI models, cannabinoid analogs were able to reduce neurotoxicity caused by glutamate and lessen the damage due to free radicals, as well as reduce inflammation (Biegon, 2004). In the case of β-blockers, a retrospective chart review study that included 662 cases of severe TBI revealed that the use of β-blockers before injury promoted the chances of survival (Mohseni et al., 2014). Nevertheless, this study was not randomized and does not provide information on the mechanism by which β-blockers provide neuroprotection. It is speculated that β-blockers exert their neuroprotective effect by down-regulating the levels of catecholamines which increase tremendously after TBI.

Enzogenol is extracted from the bark of *Pinus radiata*. It is rich in flavonoids and possesses anti-oxidant and anti-inflammatory properties. In a placebo-controlled RCT that recruited 60 adults who had sustained mTBI 3–12 months before the trial, the supplementation of Enzogenol was shown to be safe and well-tolerated. It also resulted in significant reduction in cognitive failures after 6 weeks. Despite these favorable results, this study had a small sample size and was conducted among patients with mTBI only (Theadom et al., 2013). Cerebrolysin—a nootropic drug—was also investigated in a small sample of patients with mTBI. The double-blind placebo-controlled randomized study concluded that Cerebrolysin improves cognitive function, notably long-term memory and drawing function, in mTBI patients (Chen, Wei, et al., 2013). Last but not least, the efficacy of the nitric oxide synthase inhibitor 4-amino-tetrahydrobiopterin (VAS203) in TBI was investigated in a placebo-controlled randomized Phase IIa trial. This study concluded that VAS203 can provide neuroprotection after TBI. It should be noted, however, that the results elucidated in this study were limited by the small sample size. Additionally, VAS203 was associated with a risk of acute kidney injury at high doses (Stover et al., 2014). It is also worth mentioning that several other agents have been tested in the matter of TBI, but with mixed results. These include progesterone, ethanol, barbiturates, and erythropoietin.

It is safe to say that TBI is characterized by extensive etiological heterogeneity, complex mechanisms, and a plethora of unpredictable outcomes. Therefore, focusing on isolated mechanisms and testing neuroprotective agents that target a single dimension of this multi-dimensional disease might not be enough. In this regard, the tendency to neglect the multi-faceted nature of TBI has likely contributed to the failure of numerous TBI trials. Stem-cell therapy is a prospective therapeutic approach that conceivably has the potential to simultaneously target several of the injurious aftereffects of TBI. In theory, stem-cell therapy can induce the up-regulation of endogenous neurogenesis and angiogenesis. It also allows for the restoration of damaged tissues and the amelioration of neuroinflammation, among others. Several pre-clinical and clinical trials provide evidence of the untapped potential of stem cell therapy; however, clinical trials have so far failed to progress to Phase II (Tabet et al., 2020).

Introduction to stem cells and their therapeutic potential

Stem cells are non-specialized cells that can differentiate into specialized cells and have a self-renewal capacity (Takahashi et al., 2007). Stem cells are classified based on potency into totipotent, pluripotent, multi-potent, or unipotent (Singh, Saini, Kalsan, Kumar, & Chandra, 2016). Moreover, stem cells can be classified based on their origin into embryonic stem cells (ESC) or adult stem cells (ASC). Among ASCs, stem cells that persist in differentiated somatic tissues are known as mesenchymal stem cells (MSC) and can be found in several organs, including the bone marrow, liver, heart, kidneys, adipose tissue, and the brain (Fathi & Farahzadi, 2016).

ESCs are pluripotent and are mainly used to study developmental processes and cell therapy where they can be maintained in an undifferentiated form or can be differentiated into other cell lineages (Nishikawa, Jakt, & Era, 2007). On the other hand, ASCs are more specialized and are mainly used in regenerative medicine (Fann, Uomoto, & Katon, 2000). MSCs are the most studied type among ASCs. They are spindle-like in shape with multi-potent properties (Bianco, 2014). When cells are injured, MSCs have the ability to substitute these cells and to secrete various growth factors and cytokines giving them a therapeutic property (Caplan & Dennis, 2006). In addition to their traditional therapeutic properties, MSCs have the advantage to be used as a therapeutic option due to their low immunogenicity, abundance, and ease of supply (Aggarwal & Pittenger, 2005; Yarak & Okamoto, 2010). Interestingly, MSCs can be converted into neural stem cells (NSCs) which makes them feasible for the treatment of neuronal disorders (Khalil, Tiraihi, Soleimani, Baheiraei, & Zibara, 2020)

In order to overcome the limited ability of ASCs to differentiate, a novel approach was developed by Takahashi and Yamanaka (2006). This resulted in what has come to be known as induced pluripotent stem cells (iPSCs), which were formed by reprogramming mouse-derived multi-potent ASCs (Takahashi & Yamanaka, 2006). The researchers introduced a mix of transcription factors (Oct3/4, Sox2, Klf4, and c-Myc) that allowed somatic cells to regain pluripotency. One year later, iPSCs were generated from human ASCs showing promising results (Takahashi et al., 2007), thus proving the importance of stem cell therapy for present and future treatments as shown in Fig. 2.

Stem cells have been used for the treatment of a wide array of diseases with successful clinical trials in the fields of macular degeneration, diabetes, spinal cord injury, and myocardial infarction (Deng-Bryant, Singh, Carrico, & Hall, 2008). Notably, stem cells have been used for the treatment of neurodegenerative diseases such as Alzheimer's disease and amyotrophic lateral sclerosis which are characterized by progressive neuronal death that leads to debilitating neurologic impairments. Pre-clinical and clinical studies investigating the therapeutic potential of stem cells in neurological disorders showed that it is possible to not only delay the progression of such diseases, but also to solve the source of the problem (Shiri et al., 2020; Staff, Jones, & Singer, 2019). Of interest, stem cells have also been used for the treatment of TBI. Numerous preclinical studies showed a promising possibility for translation of stem cell therapy to the bedside with a few completed clinical studies while others being underway to completion.

Pre-clinical studies utilizing stem cells for the treatment of TBI

Stem cells may be used for the treatment of TBI by allowing the replacement of lost cells post-injury (Zhou, Shao, Xu, Wu, & Deng, 2019). Several studies utilized different types of stem cells to investigate their effect on disease pathology and prognosis. To start with, earlier research has confirmed that MSCs can improve neurological function and lead to a better prognosis post-TBI (Djouad et al., 2003). A study on a TBI rat model revealed that MSCs can migrate selectively to the affected tissue and differentiate into neurons and astrocytes to repair the injury site (Wang et al., 2015). In another study in a similar model of rat TBI exploring the immunomodulatory and anti-inflammatory characteristics of MSCs, Zhang et al. demonstrated that the neurological function of TBI rats improved and that brain water content decreased significantly 3 days and 28 days post-administration. The observed effects of MSCs also included decreasing the number of inflammatory cells like microglia, macrophages, neutrophils, and lymphocytes along with reducing pro-inflammatory cytokines and apoptotic cells in the affected cortex (Zhang et al., 2013). Based on the notion that SOD2 is involved in reducing oxidative stress, a study conducted by Shi et al. used MSCs with an overexpression of the SOD2 gene by intravenously transfecting them in TBI mice. This resulted in subsidizing neuroinflammation and sustaining BBB integrity, both of which contributed to improved neurological functions (Shi et al., 2018).

NSCs treatment in TBI has been investigated earlier after transfecting the cells with growth factor genes that promote their proliferation and differentiation. Upon NSC therapy, cognitive and motor functions were improved in the animals, along with a significant elevation in the survival of pyramidal cells in the hippocampus (Philips et al., 2001). Moreover, treatment with NSCs was revealed to be dependent on the timing of injection and transplantation method. Injection through the lateral ventricle showed higher survival rates in vivo as opposed to stereotactic injection, and significantly better effects were observed when the treatment was performed 2 days or one-week post-TBI, instead of 2 weeks (Zhou et al., 2019).

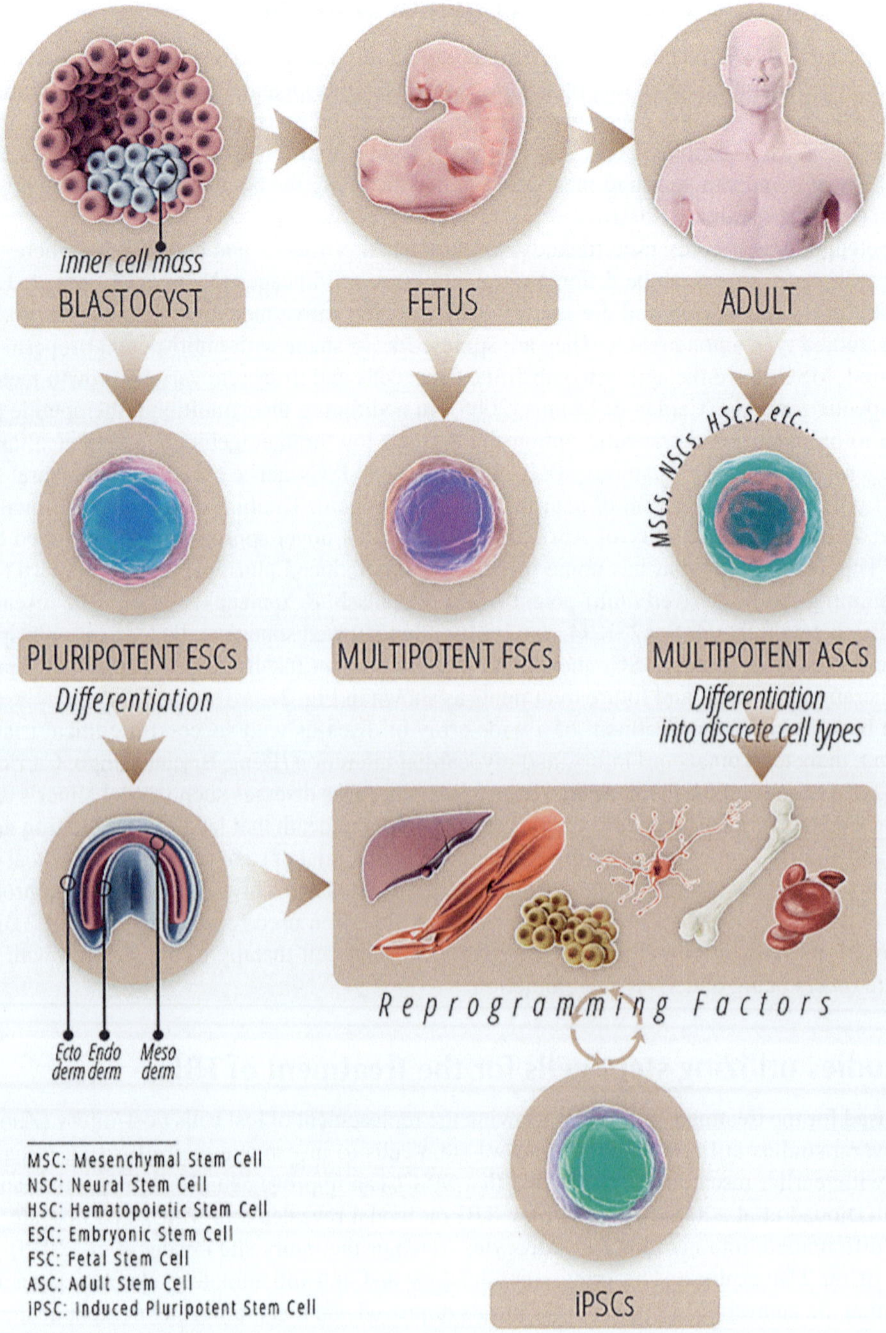

FIG. 2 Potency of stem cells at the different stages of life. The potency of stem cells changes as the organism develops. ESCs are pluripotent and can give rise to the three germ layers where each differentiates into specific cell types. FSCs are more specialized multipotent cells. ASCs, which exist in several tissues, can differentiate into discrete cell types. However, reprogramming factors can be used to help ASCs regain pluripotency as iPSCs.

On the other hand, a drawback with human cell transplantation in TBI models is the long-term survival due to immune host rejection. In an attempt to overcome this, Haus et al. produced an immune-deficient TBI rat model and transplanted it with human NSCs where the experimental results showed long-term cognitive function recovery. Notably, the rat model was proven successful considering that 9% to 25% of NSCs survived for at least 5 months, successfully differentiating into mature neurons, astrocytes, and oligodendrocytes (Haus et al., 2016). Lastly, NSCs exhibited the ability to create a favorable environment for increased neurogenesis in one study where they increased neural progenitor cell proliferation and decreased astrogliosis 7 days post-injury (Nasser et al., 2018).

In other studies, iPSCs were transplanted into the brain injury zone. For instance, the animal model showed improved neurological dysfunction by modulating lncRNA and mRNA expression (Lyu et al., 2017). Other studies confirmed the potential of iPSCs treatment in improving motor and cognitive functions post-TBI by regulating certain regenerative factors such as brain-derived neurotrophic factor (BDNF) and glial cell line-derived neurotrophic factor (GDNF), and increasing the expression of genes like oxytocin and oxytocin receptor (Dunkerson et al., 2014; Wei et al., 2016).

Another category of stem cells that has been shown to participate in endothelial repair after brain injury is endothelial progenitor cells (EPCs) (Zhou et al., 2019). Interestingly, EPCs count in peripheral blood can be used as a marker of BBB integrity, and it is significantly elevated within 24-h post-TBI (Lin, Weisdorf, Solovey, & Hebbel, 2000). Under certain conditions, these cells can develop into endothelial colony-forming cells (ECFC) which were found to take part in repairing BBB after injury and enhancing neovascularization when injected through the ventricles of the TBI animal model (Huang et al., 2013). Guo et al. tried to track the migration of EPCs using BrdU and GFP double labeling and showed that they home to affected tissue post-intravenous injection leading to neurogenesis and angiogenesis in the hippocampus (Guo, Deng, & Wei, 2017). In a study that utilized EPCs derived from adipose tissue, the cells accumulated in the injury site and helped in promoting blood vessel formation, reducing astrocyte numbers, and decreasing inflammation (Xue et al., 2010). Additionally, in a TBI rat model, SPIO-labeled EPCs were injected 6 and 12h after injury. One week later, microvessels increased considerably in the affected area and cerebral blood perfusion was recovered (Chen, Yin, et al., 2013).

Lastly, multi-potent adult progenitor cells (MAPCs) have only been barely investigated, unlike the previously mentioned stem cells that have been more widely studied. MAPCs were introduced in 2002 and are different than MSCs in that they have a low expression of MHC class I proteins and can differentiate into mesenchymal as well as endothelial cells (Jiang et al., 2002; Zhou et al., 2019). In one of the few studies available, 106 MAPCs were injected post-TBI at 2h and 24h. The animals exhibited improved spatial learning, information retention, memory retrieval, and dyskinesia 120days post-injury. This was observed along with maintaining BBB integrity during the acute phase of the injury by up-regulating anti-inflammatory factors illustrated in Table 2 (Bedi et al., 2013).

Clinical studies utilizing stem cells

The clinical studies that investigate the therapeutic potential of stem cells are scarce, and only a few have been completed while others remain underway (Schepici, Silvestro, Bramanti, & Mazzon, 2020). In a double-blinded controlled phase II study (NCT02416492), 61 TBI patients with chronic motor deficits were treated intercranially with modified bone marrow-derived MSCs. This study showed that the treatment was safe and effective as it led to a significant motor improvement 24weeks after infusion. Another phase I/II study (NCT01575470) enrolled 25 patients aged 18–55 with GCS of 5–8 within 24h post-injury. Bone marrow mononuclear cells (BMMNCs) were infused intravenously with increasing concentrations among the groups of the cohort. This intervention allowed for the preservation of critical brain regions and the reduction of pro-inflammatory biomarkers with no significant adverse side events. A completed phase I (NCT00254722) clinical trial recruited 10 patients aged 5–14 with a GSC of 5–8. Autologous bone marrow progenitor cells (6×10^6 cells/kg) were infused intravenously within 36h post-injury. The results of this study are yet to be published. A study by Wang et al. investigated the effect of umbilical cord MSCs on 20 patients. Motor, sensory, and balance functions were assessed 6 months post-treatment showing significant improvements (Wang et al., 2013). Another study that utilized BMMNCs in 29 pediatric patients of severe TBI showed reduces intracranial pressure, organ damage, and neuroinflammation (Liao et al., 2015). Finally, a study by Tian et al. evaluated the efficacy of bone marrow MSCs on 97 subjects. Patients showed enhancement in motor functions and consciousness when cells were transplanted 14days post-transplantation (Tian et al., 2013) (Table 3).

The9 prospective for clinical translation of stem-cell–based therapy is promising, but there are a lot of confounding factors that need to be considered such as the route of administration, conditioning procedure, and immune competence of the recipient; all of which may limit the potential of stem cells as a treatment option for TBI (Tolar et al., 2006).

Limitations of stem cells use

Each type of stem cells is limited by a certain property. For example, ESCs are a matter of ethical debate since they require the destruction of human embryos. ESCs pose a huge concern regarding the possible risk of immunological rejection and tumorigenesis following transplantation. On the other hand, iPSCs do not raise ethical considerations nor the risk of transplant rejection. However, they are poorly efficient and are coupled with high genomic instability. This instability during cell division together with the use of viral infections to direct cell reprogramming can increase the likelihood of tumor formation (Dekmak et al., 2018). On the other hand, ASCs show a restricted differentiation and self-renewal capacity and thus a lower therapeutic potential. Adult NSCs present technical limitations since they are difficult to obtain, grow, and maintain in culture. Importantly, NSCs transplantation is generally linked with immunological rejection. In such case, the administration of immunosuppressive therapy can aggravate the situation since it increases the risk of opportunistic infections. On the other side, MSCs are associated with a modest engraftment rate and healing potential (Dekmak et al., 2018).

TABLE 2 The pre-clinical studies utilizing stem cells as a treatment option for TBI.

Type of Stem Cells	TBI Model and Animals	Injury Location	Results	Mechanism	Reference
Mesenchymal stem cells (MSCs)	Sprague-Dawley rats hit from a height of 20 cm with a weight-drop hitting device of 4.5-mm-diameter cylinder bar that weighs 40 g	Right cortex midway between lamda and bregma	Improved neurological function after 3–28 days	Cells had anti-inflammatory and immunomodulatory by suppressing NF-kB pathway through the secretion of TSG-6	Zhang et al. (2013)
	Balb/c mice (6–8 weeks) hit with a pneurmatic piston impactor tip of 3 mm diameter with a penetration depth of 1.0 mm and a velocity of 4.5 m/s	Right skull	Improved neurological function	SOD took part in decreasing neuroinflammatory response, specifically that of ipsilateral contexin, and maintaining BBB integrity.	Shi et al. (2018)
Neural stem cells (NSCs)	Adult male Wistar rats underwent dura exposure through sealed fitting with a lateral FPI of moderate severity (2.3–2.4 atm)	Left parietal bone centered between the coronal, sagittal, and lambdoid sutures	Increased hippocampus pyramidal cell survival and improved cognition, learning, and motor functions	Secretion of transgenic nerve growth factor	Philips et al. (2001)
	Male Athymic nude rats (NCI RNU −/− homozygous, 10–11 weeks old) hit with a pneumatic controlled cortical impact device with a 5 mm flat metal impactor tip. The velocity was 4.5 m/s. compression was 2.5 mm and the dwell time was 500 ms.	Left cortex, centered over hippocampus	Retrieved cognitive functions on the long term.	Enhanced neuron survival in the hippocampus	Haus et al. (2016)
	C57BL/6 adult mice subject to CCI in the somatosensory area. The impactor had a 1 mm diameter, impact velocity of 4 m/*sec*, depth of 1 mm and a dwell time of 150 ms.	Midway between bregma and lambda	Increased neurogenesis and significantly reduced astrogliosis.	Induced neural progenitor proliferation	Nasser et al. (2018)

Induced pluripotent stem cells (iPSCs)	Adult female common marmosets subjected to spinal cord injury	N/A	Promoted functional recovery	Aided in the regeneration of axons and protected from tissue damage	Kobayashi et al. (2012)
	Male P14 Wistar rats underwent damage to sensorimotor cortex using an electric impact device with a velocity of 3 m/s and depth of 2.0 mm	Midway between lambda and bregma	Improved in social interactions, social novelty, and social transmission of food preference tests	Increased HIF-α and up-regulated regenerative factors like BDNF, GDNF, and vascular endothelial growth factor.	Wei et al. (2016)
	Sprague-Dawley rats subjected to right cerebral cortex injury by a 50 g hammer	2.5 mm away from the sagittal suture and 1.5 mm away from the arcuate suture	Improved neurological function	Modulated the expression of lncRNA and mRNA	Lyu et al. (2017)
	Adult male rats hit with controlled cortical impact device	Medical frontal cortex	Full cognitive function recovery and motor function was improved	N/A	Dunkerson et al. (2014)
Endothelial stem cells (ESCs)	Adult demale BALB/C nude mice injured using a fluid percussion instrument with an impact pressure of 1.5e1.8 atm	2.0 mm posterior from bregma and 1.5 mm lateral to sagittal suture	New vessel formation and improvement in neurological function and BBB integrity	Enhanced angiogenesis	Huang et al. (2013)
	Adult male Wistar rats injured suing fluid percussion device	4.0 mm posterior from bregma and 3.0 mm lateral to the sagittal suture	Improved neurological function post-TBI	Enhanced neurogenesis and angiogenesis in the hippocampus	Guo et al. (2017)
	Wistar rats subjected to injury with a footplate of a 4.5 mm diameter tip at a force of 20 g from a 30 cm height	Left hemisphere, 2 mm anterior and 2.5 mm lateral to the bregma	Improved function and reduced volume deficiency	Stem cells took part in the formation of capillaries and reduced astrocyte proliferation and inflammation	Xue et al. (2010)
Multi-potent adult progenitor cells (MAPCs)	Adult rats injured using an impact with a flat tip of 6 mm diameter at a single impact depth of 2.7 mm and velocity of 5.6 m/s	The midpoint between bregma and lambda, 3 mm lateral to the midline, overlying the temporoparietal cortex	Enhanced spatial learning, information retention and memory retrieval	Treatment showed anti-inflammatory properties	Bedi et al. (2013)

TABLE 3 Clinical studies utilizing stem cells for the treatment of TBI.

Study title	Phase	N	Inclusions	Condition	Administration	Results	References
A Double-Blind, Controlled Phase 2 Study of the Safety and Efficacy of Modified Stem Cells (SB623) in Patients with Chronic Motor Deficit From Traumatic Brain Injury (TBI)	2	61	18–75 years Motor deficits	12 months post-TBI	Intracranial administration of SB623 cells	Significant improvement in motor function after 24 weeks furthermore 5 and 3 severe side events were reported in patients treated with SB623 and control groups respectively	NCT02416492
Treatment of Severe Adult Traumatic Brain Injury Using Bone Marrow Mononuclear Cells	½	25	18–55 years GCS 5–8	36 h after injury	BMMNCs Infusion	No one of participants have shown severe side events in addition to a decrease in the inflammatory response	NCT01575470
Umbilical cord mesenchymal stem cell transplantation significantly improves neurological function in patients with sequelae of traumatic brain injury	3	40	5–48 years GCS 6–10	1–10 years from injury	2 mL of stem cell suspension that containing 1×10^7 SCs, via lumbar puncture	Improvement in neurological function and self-care in TBI patients	Wang et al. (2013)
Autologous bone marrow mononuclear cells reduce therapeutic intensity for severe traumatic brain injury in children	1	29	5–15 years GCS 5–8	$<$24 h post-trauma	6 Million autologous BMMNCs/kg body weight IV within 48 h of injury	Reduction in therapy intensity needed to maintain intracranial pressure and severity of organ damage	Liao et al. (2015)
Autologous bone marrow mesenchymal stem cell therapy in the subacute stage of traumatic brain injury by lumbar puncture	3	97	Complex cerebral lesion	$>$1 month post-TBI	1×10^6 cells/mL were administered by lumbar puncture between the lumbar vertebrae L3/L4 or L4/L5	Bone marrow stem cell therapy is safe and effective on patients with traumatic brain injury with no serious complications or adverse events	Tian et al. (2013)

The optimal protocol design ensuring maximum stem cell activity is debatable and more studies are needed to determine the ideal source of stem cells, the convenient number of transplanted cells, the best time of intervention according to the disease state and the type of injury, and the best route of administration. In general, the intravenous route failed to deliver sufficient amounts of cells to the site of injury as a result of cell distribution to multiple organs and the pulmonary first-pass effect. Also, the administration of stem cells into the carotid artery is correlated with various side effects such as brain infarction, decreased cerebral blood flow, inflammation, and no engraftment to the injury site. Therefore, direct intralesional injection seems to be the optimal therapeutic choice since it is characterized by the functional integration of stem cells to the damaged area. However, this route is invasive and applicable only to a small number of cases with penetrating TBI (Tabet et al., 2020; Zibara et al., 2019).

Application to other fields of neuroscience

In this chapter, we discussed the advancements made using stem cells as a potential treatment for TBI. Stem cells have also been used for the treatment of neurological disorders and neurodegenerative diseases. In ischemic strokes, blocking the blood flow leads to neuronal loss and disruption of synaptic structure. The injection of stem cells in animal models has been shown to be effective in enhancing motor abilities, reducing inflammatory markers, and improving angiogenesis post-stroke (Huang, Wong, Snyder, Hamblin, & Lee, 2014; Nam et al., 2015). Stem cell treatment was successful in improving cognitive functions and reducing amyloid plaque formation in animal models of Alzheimer's disease (Lee, Jin, & Bae, 2009; Moghadam et al., 2009). In addition, stem cell-based therapy was efficient to delay motor impairments and reduce pathological consequences in a rat model of amyotrophic lateral sclerosis (Zalfa et al., 2019). The above mentioned studies are just a glimpse aimed to shed light on the possibility of using stem cells as therapeutic agents for a myriad of neurodegenerative and neurological disorders.

Mini-dictionary of terms

- **Excitotoxicity**: occurs in neurons when the concentration of an excitatory neurotransmitter (e.g., glutamate) increases to harmful levels prompting hazardous overstimulation of receptors.
- **Cell potency**: describes a cell's potential to differentiate into other cell types. Cells with greater potency can differentiate into a greater number of cell types.
- **Retraction bulb/ball**: forms by cause of an abrupt stretching of an axon, which leads to the breaking of its cytoskeleton. The break prevents cellular products from being delivered via axonal transport from proceeding along the axon. Thereafter, they accumulate and produce a swelling that tears the axon, which in turn provokes the axon to drawback towards the cell body, forming a bulb.
- **Biomarker**: is an indicator to assess a biological state or condition.
- **Stem cell niche**: is a specialized microenvironment that not only harbors stem cells, but also governs their fate. The components of a niche may differ between the different types of stem cells.

Key facts of stem cells

- Stem cells have two defining characteristics: unlimited self-renewal and the ability to differentiate.
- With respect to their differentiation potential, stem cells are classified as totipotent (highest potential), pluripotent, multi-potent, oligopotent, or unipotent (lowest potential).
- In some cases, ASCs can remain quiescent for a very long time.
- In response to stress, the quiescent state may be terminated and the cells can resume the processes of proliferation and differentiation.
- The deliberate stimulation of endogenous ASCs is limited.

Summary points

- TBI is a major health concern with heterogeneous underlying pathologic events.
- There is currently no FDA approved drug for TBI with drug interventions showing limited efficacy.
- Stem cell-based therapies have been studied in pre-clinical settings and showed promising results in the treatment of TBI.
- Clinical studies utilizing stem cells as a treatment option are still scarce.
- Further studies are needed to determine the potential of translating stem cell therapy to the bedside.

References

Aggarwal, S., & Pittenger, M. F. (2005). Human mesenchymal stem cells modulate allogeneic immune cell responses. *Blood*, *105*(4), 1815–1822. https://doi.org/10.1182/blood-2004-04-1559.

Bedi, S. S., Hetz, R., Thomas, C., Smith, P., Olsen, A. B., Williams, S., … Cox, C. S., Jr. (2013). Intravenous multipotent adult progenitor cell therapy attenuates activated microglial/macrophage response and improves spatial learning after traumatic brain injury. *Stem Cells Translational Medicine*, *2*(12), 953–960. https://doi.org/10.5966/sctm.2013-0100.

Bedi, S. S., Walker, P. A., Shah, S. K., Jimenez, F., Thomas, C. P., Smith, P., … Cox, C. S., Jr. (2013). Autologous bone marrow mononuclear cells therapy attenuates activated microglial/macrophage response and improves spatial learning after traumatic brain injury. *Journal of Trauma and Acute Care Surgery*, *75*(3), 410–416. https://doi.org/10.1097/TA.0b013e31829617c6.

Bhatti, J., Nascimento, B., Akhtar, U., Rhind, S. G., Tien, H., Nathens, A., & da Luz, L. T. (2017). Systematic review of human and animal studies examining the efficacy and safety of N-acetylcysteine (NAC) and N-acetylcysteine amide (NACA) in traumatic brain injury: Impact on neurofunctional outcome and biomarkers of oxidative stress and inflammation. *Frontiers in Neurology*, *8*, 744. https://doi.org/10.3389/fneur.2017.00744.

Bianco, P. (2014). "Mesenchymal" stem cells. *Annual Review of Cell and Developmental Biology*, *30*, 677–704. https://doi.org/10.1146/annurev-cellbio-100913-013132.

Biegon, A. (2004). Cannabinoids as neuroprotective agents in traumatic brain injury. *Current Pharmaceutical Design*, *10*(18), 2177–2183. https://doi.org/10.2174/1381612043384196.

Bramlett, H. M., & Dietrich, W. D. (2015). Long-term consequences of traumatic brain injury: Current status of potential mechanisms of injury and neurological outcomes. *Journal of Neurotrauma*, *32*(23), 1834–1848. https://doi.org/10.1089/neu.2014.3352.

Buki, A., & Povlishock, J. T. (2006). All roads lead to disconnection?—Traumatic axonal injury revisited. *Acta Neurochirurgica*, *148*(2), 181–193. discussion 193-184 https://doi.org/10.1007/s00701-005-0674-4.

Bushnik, T., Hanks, R. A., Kreutzer, J., & Rosenthal, M. (2003). Etiology of traumatic brain injury: Characterization of differential outcomes up to 1 year postinjury. *Archives of Physical Medicine and Rehabilitation*, *84*(2), 255–262. https://doi.org/10.1053/apmr.2003.50092.

Bye, N., Habgood, M. D., Callaway, J. K., Malakooti, N., Potter, A., Kossmann, T., & Morganti-Kossmann, M. C. (2007). Transient neuroprotection by minocycline following traumatic brain injury is associated with attenuated microglial activation but no changes in cell apoptosis or neutrophil infiltration. *Experimental Neurology*, *204*(1), 220–233. https://doi.org/10.1016/j.expneurol.2006.10.013.

Caplan, A. I., & Dennis, J. E. (2006). Mesenchymal stem cells as trophic mediators. *Journal of Cellular Biochemistry*, *98*(5), 1076–1084. https://doi.org/10.1002/jcb.20886.

Chamoun, R., Suki, D., Gopinath, S. P., Goodman, J. C., & Robertson, C. (2010). Role of extracellular glutamate measured by cerebral microdialysis in severe traumatic brain injury. *Journal of Neurosurgery*, *113*(3), 564–570. https://doi.org/10.3171/2009.12.Jns09689.

Chen, C. C., Wei, S. T., Tsaia, S. C., Chen, X. X., & Cho, D. Y. (2013). Cerebrolysin enhances cognitive recovery of mild traumatic brain injury patients: Double-blind, placebo-controlled, randomized study. *British Journal of Neurosurgery*, *27*(6), 803–807. https://doi.org/10.3109/02688697.2013.793287.

Chen, X., Yin, J., Wu, X., Li, R., Fang, J., Chen, R., … Zhang, W. (2013). Effects of magnetically labeled exogenous endothelial progenitor cells on cerebral blood perfusion and microvasculature alterations after traumatic brain injury in rat model. *Acta Radiologica*, *54*(3), 313–323. https://doi.org/10.1258/ar.2012.120605.

Daly, S., Thorpe, M., Rockswold, S., Hubbard, M., Bergman, T., Samadani, U., & Rockswold, G. (2018). Hyperbaric oxygen therapy in the treatment of acute severe traumatic brain injury: A systematic review. *Journal of Neurotrauma*, *35*(4), 623–629. https://doi.org/10.1089/neu.2017.5225.

Dehlaghi Jadid, K., Davidsson, J., Lidin, E., Hanell, A., Angeria, M., Mathiesen, T., … Gunther, M. (2019). COX-2 inhibition by diclofenac is associated with decreased apoptosis and lesion area after experimental focal penetrating traumatic brain injury in rats. *Frontiers in Neurology*, *10*, 811. https://doi.org/10.3389/fneur.2019.00811.

Dekmak, A., Mantash, S., Shaito, A., Toutonji, A., Ramadan, N., Ghazale, H., … Zibara, K. (2018). Stem cells and combination therapy for the treatment of traumatic brain injury. *Behavioural Brain Research*, *340*, 49–62. https://doi.org/10.1016/j.bbr.2016.12.039.

Deng-Bryant, Y., Singh, I. N., Carrico, K. M., & Hall, E. D. (2008). Neuroprotective effects of tempol, a catalytic scavenger of peroxynitrite-derived free radicals, in a mouse traumatic brain injury model. *Journal of Cerebral Blood Flow and Metabolism*, *28*(6), 1114–1126. https://doi.org/10.1038/jcbfm.2008.10.

Devesa, J., Reimunde, P., Devesa, P., Barbera, M., & Arce, V. (2013). Growth hormone (GH) and brain trauma. *Hormones and Behavior*, *63*(2), 331–344. https://doi.org/10.1016/j.yhbeh.2012.02.022.

Dewan, M. C., Rattani, A., Gupta, S., Baticulon, R. E., Hung, Y.-C., Punchak, M., … Rubiano, A. M. (2018). Estimating the global incidence of traumatic brain injury. *Journal of Neurosurgery*, *130*(4), 1080–1097.

Djouad, F., Plence, P., Bony, C., Tropel, P., Apparailly, F., Sany, J., … Jorgensen, C. (2003). Immunosuppressive effect of mesenchymal stem cells favors tumor growth in allogeneic animals. *Blood*, *102*(10), 3837–3844. https://doi.org/10.1182/blood-2003-04-1193.

Dunkerson, J., Moritz, K. E., Young, J., Pionk, T., Fink, K., Rossignol, J., … Smith, J. S. (2014). Combining enriched environment and induced pluripotent stem cell therapy results in improved cognitive and motor function following traumatic brain injury. *Restorative Neurology and Neuroscience*, *32*(5), 675–687. https://doi.org/10.3233/RNN-140408.

Fann, J. R., Uomoto, J. M., & Katon, W. J. (2000). Sertraline in the treatment of major depression following mild traumatic brain injury. *The Journal of Neuropsychiatry and Clinical Neurosciences*, *12*(2), 226–232. https://doi.org/10.1176/jnp.12.2.226.

Fathi, E., & Farahzadi, R. (2016). Isolation, culturing, characterization and aging of adipose tissue-derived mesenchymal stem cells: A brief overview. *Brazilian Archives of Biology and Technology*, *59*, e16150383. https://doi.org/10.1590/1678-4324-2016150383.

Gardner, C. J., Mattsson, A. F., Daousi, C., Korbonits, M., Koltowska-Haggstrom, M., & Cuthbertson, D. J. (2015). GH deficiency after traumatic brain injury: Improvement in quality of life with GH therapy: Analysis of the KIMS database. *European Journal of Endocrinology*, *172*(4), 371–381. https://doi.org/10.1530/EJE-14-0654.

Girouard, H., Wang, G., Gallo, E. F., Anrather, J., Zhou, P., Pickel, V. M., & Iadecola, C. (2009). NMDA receptor activation increases free radical production through nitric oxide and NOX2. *The Journal of Neuroscience*, *29*(8), 2545–2552. https://doi.org/10.1523/jneurosci.0133-09.2009.

Guo, X. B., Deng, X., & Wei, Y. (2017). Homing of cultured endothelial progenitor cells and their effect on traumatic brain injury in rat model. *Scientific Reports*, *7*(1), 4164. https://doi.org/10.1038/s41598-017-04153-2.

Haus, D. L., López-Velázquez, L., Gold, E. M., Cunningham, K. M., Perez, H., Anderson, A. J., & Cummings, B. J. (2016). Transplantation of human neural stem cells restores cognition in an immunodeficient rodent model of traumatic brain injury. *Experimental Neurology*, *281*, 1–16. https://doi.org/10.1016/j.expneurol.2016.04.008.

Huang, L., Wong, S., Snyder, E. Y., Hamblin, M. H., & Lee, J. P. (2014). Human neural stem cells rapidly ameliorate symptomatic inflammation in early-stage ischemic-reperfusion cerebral injury. *Stem Cell Research & Therapy*, *5*(6), 129. https://doi.org/10.1186/scrt519.

Huang, X. T., Zhang, Y. Q., Li, S. J., Li, S. H., Tang, Q., Wang, Z. T., … Zhang, J. N. (2013). Intracerebroventricular transplantation of ex vivo expanded endothelial colony-forming cells restores blood-brain barrier integrity and promotes angiogenesis of mice with traumatic brain injury. *Journal of Neurotrauma*, *30*(24), 2080–2088. https://doi.org/10.1089/neu.2013.2996.

Jiang, Y., Jahagirdar, B. N., Reinhardt, R. L., Schwartz, R. E., Keene, C. D., Ortiz-Gonzalez, X. R., … Verfaillie, C. M. (2002). Pluripotency of mesenchymal stem cells derived from adult marrow. *Nature*, *418*(6893), 41–49. https://doi.org/10.1038/nature00870.

Khalil, W., Tiraihi, T., Soleimani, M., Baheiraei, N., & Zibara, K. (2020). Conversion of neural stem cells into functional neuron-like cells by MicroRNA-218: Differential expression of functionality genes. *Neurotoxicity Research*. https://doi.org/10.1007/s12640-020-00244-7.

Kobayashi, Y., Okada, Y., Itakura, G., Iwai, H., Nishimura, S., Yasuda, A., … Okano, H. (2012). Pre-evaluated safe human iPSC-derived neural stem cells promote functional recovery after spinal cord injury in common marmoset without tumorigenicity. *PLoS One*, *7*(12). https://doi.org/10.1371/journal.pone.0052787, e52787.

Lamade, A. M., Anthonymuthu, T. S., Hier, Z. E., Gao, Y., Kagan, V. E., & Bayir, H. (2020). Mitochondrial damage & lipid signaling in traumatic brain injury. *Experimental Neurology*, *329*, 113307. https://doi.org/10.1016/j.expneurol.2020.113307.

Lee, J. K., Jin, H. K., & Bae, J. S. (2009). Bone marrow-derived mesenchymal stem cells reduce brain amyloid-beta deposition and accelerate the activation of microglia in an acutely induced Alzheimer's disease mouse model. *Neuroscience Letters*, *450*(2), 136–141. https://doi.org/10.1016/j.neulet.2008.11.059.

Liao, G. P., Harting, M. T., Hetz, R. A., Walker, P. A., Shah, S. K., Corkins, C. J., … Cox, C. S., Jr. (2015). Autologous bone marrow mononuclear cells reduce therapeutic intensity for severe traumatic brain injury in children. *Pediatric Critical Care Medicine*, *16*(3), 245–255. https://doi.org/10.1097/pcc.0000000000000324.

Lin, Y., Weisdorf, D. J., Solovey, A., & Hebbel, R. P. (2000). Origins of circulating endothelial cells and endothelial outgrowth from blood. *The Journal of Clinical Investigation*, *105*(1), 71–77. https://doi.org/10.1172/JCI8071.

Lucke-Wold, B. P., Logsdon, A. F., Nguyen, L., Eltanahay, A., Turner, R. C., Bonasso, P., … Rosen, C. L. (2018). Supplements, nutrition, and alternative therapies for the treatment of traumatic brain injury. *Nutritional Neuroscience*, *21*(2), 79–91. https://doi.org/10.1080/1028415X.2016.1236174.

Lyu, Q., Zhang, Z. B., Fu, S. J., Xiong, L. L., Liu, J., & Wang, T. H. (2017). Microarray expression profile of lncRNAs and mRNAs in rats with traumatic brain injury after A2B5+ cell transplantation. *Cell Transplantation*, *26*(10), 1622–1635. https://doi.org/10.1177/0963689717723014.

Moghadam, F. H., Alaie, H., Karbalaie, K., Tanhaei, S., Nasr Esfahani, M. H., & Baharvand, H. (2009). Transplantation of primed or unprimed mouse embryonic stem cell-derived neural precursor cells improves cognitive function in Alzheimerian rats. *Differentiation*, *78*(2–3), 59–68. https://doi.org/10.1016/j.diff.2009.06.005.

Mohamadpour, M., Whitney, K., & Bergold, P. J. (2019). The importance of therapeutic time window in the treatment of traumatic brain injury. *Frontiers in Neuroscience*, *13*, 07. https://doi.org/10.3389/fnins.2019.00007.

Mohseni, S., Talving, P., Wallin, G., Ljungqvist, O., & Riddez, L. (2014). Preinjury beta-blockade is protective in isolated severe traumatic brain injury. *Journal of Trauma and Acute Care Surgery*, *76*(3), 804–808. https://doi.org/10.1097/TA.0000000000000139.

Moreau, O. K., Cortet-Rudelli, C., Yollin, E., Merlen, E., Daveluy, W., & Rousseaux, M. (2013). Growth hormone replacement therapy in patients with traumatic brain injury. *Journal of Neurotrauma*, *30*(11), 998–1006. https://doi.org/10.1089/neu.2012.2705.

Morganti-Kossmann, M. C., Rancan, M., Stahel, P. F., & Kossmann, T. (2002). Inflammatory response in acute traumatic brain injury: A double-edged sword. *Current Opinion in Critical Care*, *8*(2), 101–105. https://doi.org/10.1097/00075198-200204000-00002.

Nam, H. S., Kwon, I., Lee, B. H., Kim, H., Kim, J., An, S., … Heo, J. H. (2015). Effects of mesenchymal stem cell treatment on the expression of matrix metalloproteinases and angiogenesis during ischemic stroke recovery. *PLoS One*, *10*(12). https://doi.org/10.1371/journal.pone.0144218, e0144218.

Nasser, M., Ballout, N., Mantash, S., Bejjani, F., Najdi, F., Ramadan, N., … Kobeissy, F. (2018). Transplantation of embryonic neural stem cells and differentiated cells in a controlled cortical impact (CCI) model of adult mouse somatosensory cortex. *Frontiers in Neurology*, *9*, 895. https://doi.org/10.3389/fneur.2018.00895.

Nishikawa, S., Jakt, L. M., & Era, T. (2007). Embryonic stem-cell culture as a tool for developmental cell biology. *Nature Reviews. Molecular Cell Biology*, *8*(6), 502–507. https://doi.org/10.1038/nrm2189.

Philips, M. F., Mattiasson, G., Wieloch, T., Björklund, A., Johansson, B. B., Tomasevic, G., … McIntosh, T. K. (2001). Neuroprotective and behavioral efficacy of nerve growth factor—Transfected hippocampal progenitor cell transplants after experimental traumatic brain injury. *Journal of Neurosurgery*, *94*(5), 765. https://doi.org/10.3171/jns.2001.94.5.0765.

Saatman, K. E., Duhaime, A. C., Bullock, R., Maas, A. I., Valadka, A., & Manley, G. T. (2008). Classification of traumatic brain injury for targeted therapies. *Journal of Neurotrauma*, *25*(7), 719–738. https://doi.org/10.1089/neu.2008.0586.

Sanchez-Aguilar, M., Tapia-Perez, J. H., Sanchez-Rodriguez, J. J., Vinas-Rios, J. M., Martinez-Perez, P., de la Cruz-Mendoza, E., … Gordillo-Moscoso, A. (2013). Effect of rosuvastatin on cytokines after traumatic head injury. *Journal of Neurosurgery*, *118*(3), 669–675. https://doi.org/10.3171/2012.12.JNS121084.

Schepici, G., Silvestro, S., Bramanti, P., & Mazzon, E. (2020). Traumatic brain injury and stem cells: An overview of clinical trials, the current treatments and future therapeutic approaches. *Medicina (Kaunas, Lithuania)*, *56*(3), 137. https://doi.org/10.3390/medicina56030137.

Schmidt, O. I., Infanger, M., Heyde, C. E., Ertel, W., & Stahel, P. F. (2004). The role of neuroinflammation in traumatic brain injury. *European Journal of Trauma*, *30*(3), 135–149. https://doi.org/10.1007/s00068-004-1394-9.

Shahlaie, K., Gurkoff, G. G., Lyeth, B. G., Muizelaar, J. P., & Berman, R. F. (2013). Neuroprotective effects of SNX-185 in an in vitro model of TBI with a second insult. *Restorative Neurology and Neuroscience*, *31*(2), 141–153. https://doi.org/10.3233/RNN-120243.

Shi, X., Bai, Y., Zhang, G., Liu, Y., Xiao, H., Liu, X., & Zhang, W. (2018). Effects of over-expression of SOD2 in bone marrow-derived mesenchymal stem cells on traumatic brain injury. *Cell and Tissue Research*, *372*(1), 67–75. https://doi.org/10.1007/s00441-017-2716-7.

Shiri, E., Pasbakhsh, P., Borhani-Haghighi, M., Alizadeh, Z., Nekoonam, S., Mojaverrostami, S., … Kashani, I. R. (2020). Mesenchymal stem cells ameliorate cuprizone-induced demyelination by targeting oxidative stress and mitochondrial dysfunction. *Cellular and Molecular Neurobiology*. https://doi.org/10.1007/s10571-020-00910-6.

Singh, V. K., Saini, A., Kalsan, M., Kumar, N., & Chandra, R. (2016). Describing the stem cell potency: The various methods of functional assessment and in silico diagnostics. *Frontiers in Cell and Developmental Biology*, *4*, 134. https://doi.org/10.3389/fcell.2016.00134.

Skandsen, T., Kvistad, K. A., Solheim, O., Strand, I. H., Folvik, M., & Vik, A. (2010). Prevalence and impact of diffuse axonal injury in patients with moderate and severe head injury: A cohort study of early magnetic resonance imaging findings and 1-year outcome. *Journal of Neurosurgery*, *113*(3), 556–563. https://doi.org/10.3171/2009.9.Jns09626.

Staff, N. P., Jones, D. T., & Singer, W. (2019). Mesenchymal stromal cell therapies for neurodegenerative diseases. *Mayo Clinic Proceedings*, *94*(5), 892–905. https://doi.org/10.1016/j.mayocp.2019.01.001.

Stover, J. F., Belli, A., Boret, H., Bulters, D., Sahuquillo, J., Schmutzhard, E., … Investigators, N. (2014). Nitric oxide synthase inhibition with the antipterin VAS203 improves outcome in moderate and severe traumatic brain injury: A placebo-controlled randomized phase IIa trial (NOSTRA). *Journal of Neurotrauma*, *31*(19), 1599–1606. https://doi.org/10.1089/neu.2014.3344.

Sun, D. A., Deshpande, L. S., Sombati, S., Baranova, A., Wilson, M. S., Hamm, R. J., & DeLorenzo, R. J. (2008). Traumatic brain injury causes a long-lasting calcium (Ca2+)-plateau of elevated intracellular Ca levels and altered Ca2+ homeostatic mechanisms in hippocampal neurons surviving brain injury. *The European Journal of Neuroscience*, *27*(7), 1659–1672. https://doi.org/10.1111/j.1460-9568.2008.06156.x.

Tabet, M., Hasan, H., Abdelhady, S., Mahavadi, A. K., Clervius, H., Nasrallah, L., … Shaito, A. (2020). *Evaluation of evidence: Stem cells as a treatment option for traumatic brain injury*. Chichester: eLS, John Wiley & Sons, Ltd. https://doi.org/10.1002/9780470015902.a0025801.

Takahashi, K., Tanabe, K., Ohnuki, M., Narita, M., Ichisaka, T., Tomoda, K., & Yamanaka, S. (2007). Induction of pluripotent stem cells from adult human fibroblasts by defined factors. *Cell*, *131*(5), 861–872. https://doi.org/10.1016/j.cell.2007.11.019.

Takahashi, K., & Yamanaka, S. (2006). Induction of pluripotent stem cells from mouse embryonic and adult fibroblast cultures by defined factors. *Cell*, *126*(4), 663–676. https://doi.org/10.1016/j.cell.2006.07.024.

Teasdale, G., & Jennett, B. (1974). Assessment of coma and impaired consciousness. A practical scale. *Lancet*, *2*(7872), 81–84.

Theadom, A., Mahon, S., Barker-Collo, S., McPherson, K., Rush, E., Vandal, A. C., & Feigin, V. L. (2013). Enzogenol for cognitive functioning in traumatic brain injury: A pilot placebo-controlled RCT. *European Journal of Neurology*, *20*(8), 1135–1144. https://doi.org/10.1111/ene.12099.

Tian, C., Wang, X., Wang, X., Wang, L., Wang, X., Wu, S., & Wan, Z. (2013). Autologous bone marrow mesenchymal stem cell therapy in the subacute stage of traumatic brain injury by lumbar puncture. *Experimental and Clinical Transplantation*, *11*(2), 176–181. https://doi.org/10.6002/ect.2012.0053.

Tolar, J., O'Shaughnessy, M. J., Panoskaltsis-Mortari, A., McElmurry, R. T., Bell, S., Riddle, M., … Blazar, B. R. (2006). Host factors that impact the biodistribution and persistence of multipotent adult progenitor cells. *Blood*, *107*(10), 4182–4188. https://doi.org/10.1182/blood-2005-08-3289.

Wang, S., Cheng, H., Dai, G., Wang, X., Hua, R., Liu, X., … An, Y. (2013). Umbilical cord mesenchymal stem cell transplantation significantly improves neurological function in patients with sequelae of traumatic brain injury. *Brain Research*, *1532*, 76–84. https://doi.org/10.1016/j.brainres.2013.08.001.

Wang, L., Wang, X., Su, H., Han, Z., Yu, H., Wang, D., … Zhang, J. (2015). Recombinant human erythropoietin improves the neurofunctional recovery of rats following traumatic brain injury via an increase in circulating endothelial progenitor cells. *Translational Stroke Research*, *6*(1), 50–59. https://doi.org/10.1007/s12975-014-0362-x.

Wang, C. C., Wee, H. Y., Hu, C. Y., Chio, C. C., & Kuo, J. R. (2018). The effects of memantine on glutamic receptor-associated nitrosative stress in a traumatic brain injury rat model. *World Neurosurgery*, *112*, e719–e731. https://doi.org/10.1016/j.wneu.2018.01.140.

Wei, Z. Z., Lee, J. H., Zhang, Y., Zhu, Y. B., Deveau, T. C., Gu, X., … Yu, S. P. (2016). Intracranial transplantation of hypoxia-preconditioned iPSC-derived neural progenitor cells alleviates neuropsychiatric defects after traumatic brain injury in juvenile rats. *Cell Transplantation*, *25*(5), 797–809. https://doi.org/10.3727/096368916X690403.

Xue, S., Zhang, H. T., Zhang, P., Luo, J., Chen, Z. Z., Jang, X. D., & Xu, R. X. (2010). Functional endothelial progenitor cells derived from adipose tissue show beneficial effect on cell therapy of traumatic brain injury. *Neuroscience Letters*, *473*(3), 186–191. https://doi.org/10.1016/j.neulet.2010.02.035.

Yarak, S., & Okamoto, O. K. (2010). Human adipose-derived stem cells: Current challenges and clinical perspectives. *Anais Brasileiros de Dermatologia*, *85*(5), 647–656. https://doi.org/10.1590/s0365-05962010000500008.

Zalfa, C., Rota Nodari, L., Vacchi, E., Gelati, M., Profico, D., Boido, M., … Ferrari, D. (2019). Transplantation of clinical-grade human neural stem cells reduces neuroinflammation, prolongs survival and delays disease progression in the SOD1 rats. *Cell Death & Disease*, *10*(5), 345. https://doi.org/10.1038/s41419-019-1582-5.

Zhang, R., Liu, Y., Yan, K., Chen, L., Chen, X.-R., Li, P., … Jiang, X.-D. (2013). Anti-inflammatory and immunomodulatory mechanisms of mesenchymal stem cell transplantation in experimental traumatic brain injury. *Journal of Neuroinflammation*, *10*(1), 871. https://doi.org/10.1186/1742-2094-10-106.

Zhang, M., Teng, C. H., Wu, F. F., Ge, L. Y., Xiao, J., Zhang, H. Y., & Chen, D. Q. (2019). Edaravone attenuates traumatic brain injury through anti-inflammatory and anti-oxidative modulation. *Experimental and Therapeutic Medicine*, *18*(1), 467–474. https://doi.org/10.3892/etm.2019.7632.

Zhou, Y., Shao, A., Xu, W., Wu, H., & Deng, Y. (2019). Advance of stem cell treatment for traumatic brain injury. *Frontiers in Cellular Neuroscience*, *13*(301). https://doi.org/10.3389/fncel.2019.00301.

Zhou, J., Wang, H., Shen, R., Fang, J., Yang, Y., Dai, W., … Zhou, M. (2018). Mitochondrial-targeted antioxidant MitoQ provides neuroprotection and reduces neuronal apoptosis in experimental traumatic brain injury possibly via the Nrf2-ARE pathway. *American Journal of Translational Research*, *10*(6), 1887–1899.

Zibara, K., Ballout, N., Mondello, S., Karnib, N., Ramadan, N., Omais, S., … Kobeissy, F. (2019). Combination of drug and stem cells neurotherapy: Potential interventions in neurotrauma and traumatic brain injury. *Neuropharmacology*, *145*(Pt. B), 177–198. https://doi.org/10.1016/j.neuropharm.2018.09.032.

Chapter 5

Management of traumatic brain injury from the aspect of emergency department and case studies

Bedriye Müge Sönmez

Department of Emergency Medicine, Univercity of Health Sciences, Dışkapı Yıldırım Beyazıt Education and Research Hospital, Ankara, Turkey

List of abbreviations

ABCDEs airway-breathing-circulation-disability-expose
ATLS Advanced Trauma Life Support
BG blood glucose
BP blood pressure
BTF Brain Trauma Foundation
CBF cerebral blood flow
CPP cerebral perfusion pressure
CT computed tomography
ED emergency department
eFAST extended focused assessment with sonography for trauma
ES emergency specialist
GCS Glasgow Coma Scale
Hb hemoglobin
HR heart rate
ICP intracranial pressure
INR international normalized ratio
MAP mean arterial pressure
MRI magnetic resonance imaging
mrTBI moderate traumatic brain injury
mTBI mild traumatic brain injury
NMBA neuromuscular blocking agents
$PaCO_2$ partial carbon dioxide
PaO_2 arterial oxygen pressure
PCC prothrombin complex concentrate
pRBC packet red blood cell
RR respiratory rate
RSI rapid sequence intubation
SBP systolic blood pressure
SCh succinylcholine
SpO_2 oxygen saturation
sTBI severe traumatic brain injury
TBI traumatic brain injury
TT thrombin time
TTM targeted temperature management

Diagnosis and Treatment of Traumatic Brain Injury. https://doi.org/10.1016/B978-0-12-823347-4.00034-8

Introduction

Traumatic brain injury (TBI) constitutes a high proportion of trauma cases admitted to emergency departments (EDs); its clinical manifestations range widely, from mildly altered consciousness to deep coma and death (Galgano et al., 2017; Galvagno, Nahmias, & Young, 2019). Since emergency specialists have a major responsibility for the prevention of secondary injury in TBI, there is a need for a summary of urgent management strategies. In this review, by combining data from both clinical practice and basic research, we provide a summary of the most current emergency treatment of TBI, including novel combination therapies and advanced therapeutics. The first algorithm of priorities in the management of TBI is shown in Table 1.

Emergency department considerations

Both blunt and penetrating injury types can cause traumatic brain injury (Vella, Crandall, & Patel, 2017). Glasgow Coma Scale (GCS) is typically used to rate the severity of TBI, i.e. mild, moderate, or severe. Thus, GCS rating is of utmost importance for the management of TBI in ED. The primary brain injury is irreversible and responsible for brain tissue damage, dysregulated cerebral blood flow (CBF), and perturbations of cerebral metabolism including inflammatory mediator upregulation, increased oxidative stress, and enhanced vasospasm. Altogether, these processes are termed as secondary injury cascades of the brain and, ultimately, cause neuronal cell death, and generalized brain edema to develop (Vella et al., 2017).

An overwhelming majority (75%–90%) of TBI cases admitted to ED are categorized as "mild" TBI (mTBI), which corresponds to a GCS score of 13–15. Moderate brain injury (mrTBI) is present in about 15% of ED cases and corresponds to a GCS score between 9 and 13. Lastly, severe injury (sTBI) corresponding to a GCS score between 3 and 8 affects 10% of patients (Dixon et al., 2020; Foks et al., 2017) (Fig. 1).

Patients with TBI are initially managed in the same manner with other types of trauma, with attention being paid to the Advanced Trauma Life Support (ATLS) principles that dictate securing airway, breathing, and circulation, and then performing a rapid neurologic examination and exposure of the patient while avoiding hypothermia (Henry, 2018). The algorithm of ED care was described in Fig. 1.

TABLE 1 Steps of managing traumatic brain injury.

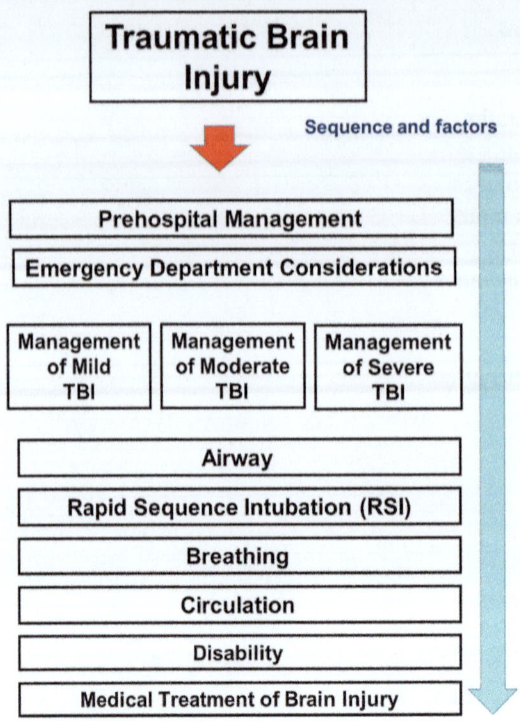

Management of TBI before and in the ED according to GCS categorization.

Management of mild brain injury (GCS score 13–15)

mTBI is caused by non-penetrating head trauma causing a short-lived episode of unconsciousness, which is followed by cognitive impairment accompanied by several other symptoms collectively termed as post-concussion syndrome (Blennow et al., 2016; McInnes, Friesen, MacKenzie, Westwood, & Boe, 2019).

Evaluation of patients with mTBI should begin with a detailed history and neurological examination. One must obtain information about the patient's demographic characteristics, mechanism and time of injury, presence of loss of consciousness (LOC) immediately after post-injury, subsequent level of alertness, amnesia (retrograde/antegrade) and the amnestic interval, headache (severity), vomiting, seizure; patient allergies, medications (particularly about the use of anticoagulants, past medical history, last meal and events leading to injury as called mnemonics AMPLE should also be questioned to assess the degree of concussion-induced amnesia (Henry, 2018).

Since intracranial damage is highly unlikely in these patients, cranial imaging studies, or hospital admission is not always essential (Foks et al., 2017). In cases where making a decision to perform neuroimaging is needed, New Orleans Criteria, National Emergency X-Radiography Utilization Study II clinical criteria, or the Canadian CT Head Rule can be utilized; a repeat non-contrast head CT scan or cranial magnetic resonance imaging (MRI) are the imaging modalities of choice when there are persistent or worsening neurological signs despite a negative initial head CT scan (Smith et al., 2019). Cranial CT should be considered for diagnostic purposes when witnessed LOC occurred for longer than > 5 min; the patient has become amnestic for longer than 30 min; the patient has had witnessed disorientation or is currently disoriented during serial examinations); if she or he has a GCS score below 15 when 2 h have passed since injury; there is suspected open or depressed skull fracture; there is any sign of basilar skull fracture (hemotympanum, raccoon eyes, otorrhea or rhinorrhea, Battle's sign); if the patient vomits more than two times; if there is seizure activity or convulsions; if the victim is older than 65 years of age; if there is a history of using anticoagulants; and the mechanism of injury is deemed dangerous (e.g., motor vehicle accidents involving pedestrians, in-vehicle traffic accidents where the passengers or the driver has been ejected out of the vehicle and falls off heights higher than 3 ft or five stairs) (Webster, Moore, & Stewart, 2017).

Patients should be admitted to the hospital and observed with serial neurological checks when a head CT is needed but unavailable; if there are abnormal signs on CT; if GCS does not normalize by 2 h; and there are signs of intoxication. Patients with abnormal CT and/or neurological examination should be consulted with the neurosurgery department, and CT should be repeated if the neurological status of the patient worsens.

Patients with mTBI can be discharged and instructed to watch for head injury warning rules and return for a follow-up visit if their GCS score is normal and neuroimaging studies (CT or traditional MRI) did not reveal any abnormality, which indicates the absence of any neurological damage (Mayer, Quinn, & Master, 2017). The flowchart of managing mTBI is shown in Fig. 2.

Management of moderate TBI

mrTBI once encompassed patients with TIB and a GCS score of 9–12 but after the demonstration that patients with a GCS of 13 had similar traumatic lesions on cranial CT (Godoy, Rubiano, Rabinstein, Bullock, & Sahuquillo, 2016). Patients should be initially managed in the emergency service according to the ATLS guidelines; then, the GCS score should guide

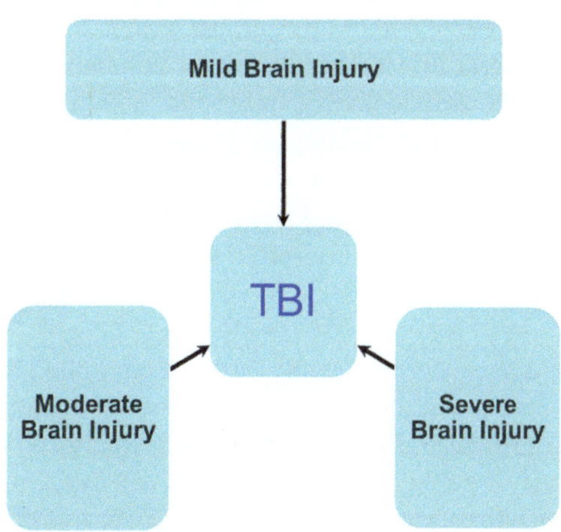

FIG. 1 Traumatic brain injury classification. TBI types according to the GCS.

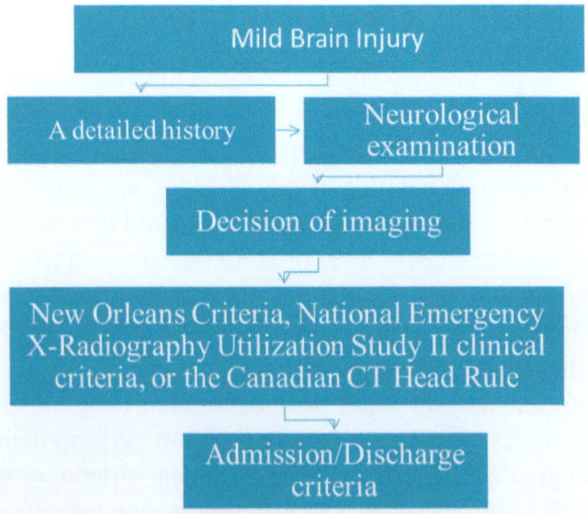

FIG. 2 Steps of managing mild traumatic brain injury. Emergency department considerations of mild TBI.

clinicians whether a head CT is required (Henry, 2018). These patients should be ideally admitted to a healthcare facility with an intensive care unit for the first 12–24 h (Dixon et al., 2020). In these patients, neurological status dictates patient management; with an abnormal admission CT or worsening patient condition necessitates a repeat CT in the first 24 h or worsening of neurological status requires the application of sTBI protocols whereas improving neurological condition allows safe patient discharge (Henry, 2018). Fig. 3 shows the management of mtTBI.

Management of sTBI

Moderate to sTBI is acutely managed on the basis of ATLS criteria combined with serial neurological evaluation, a thorough systematic examination to exclude injury to other systems and, if necessary, to aid in balancing various treatment priorities (airway-breathing-circulation-disability-expose (ABCDEs) and primary survey and resuscitation) (Whitaker-Lea, 2017). According to ATLS, the diagnosis of patients with TBI is a priority to prevent secondary injury, which is mainly caused by hypotension, hypoxia, hypercarbia, and increasing intracranial pressure (ICP). Therefore, the primary treatment targets include ABCs, that is airway protection early in the management, ensuring adequate breathing by supplying extra oxygen, and protect circulation by adequate support so that brain is supplied with adequate oxygen and blood flow (Abdelmalik, Draghic, & Ling, 2019; Dixon et al., 2020). Fig. 4 shows the basis of sTBI management. Treatment of sTBI was described in Fig. 2.

Airway

Airway protection is a pivotal measure in sTBI as affected patients cannot secure their airways due to a depressed level of consciousness (Galvagno et al., 2019). Suspected elevated ICP necessitates rapid sequence intubation (RSI) as the method of choice to ensure securing airways. All patients with a GCS of less than 8–9 should be intubated, as well as those who are unable to protect the airway, who are agitated, who are unable to maintain $SpO_2 > 90\%$ under supplemental oxygen, who need to be transported to another healthcare facility or in-hospital unit, and who are clinically suspected of having cerebral herniation where adequate oxygenation (arterial oxygen pressure (PaO_2) > 60 mmHg) is a concern (Carney et al., 2017; Haydel & Burns, 2021). In-line C-spine stabilization should be maintained during the entire intubation procedure (Ginsburg & Huff, 2021).

Rapid sequence intubation

In patients with elevated ICP, RSI should be performed gently and in a controlled fashion. Prolonged or multiple intubation attempts are not desirable; an appropriate neurologic evaluation is of utmost importance before attempting intubation (Kramer, Lebowitz, Walsh, & Ganti, 2018; Rajajee, Riggs, & Seder, 2017). Patients with elevated ICP need some measures to improve their oxygenation; hence, it is of paramount importance to pre-oxygenate patients to achieve and maintain a sufficiently high level of $SaO_2\%$ before neuromuscular blockade starts to exert its effect (Kovacs & Sowers, 2018; Rajajee et al., 2017).

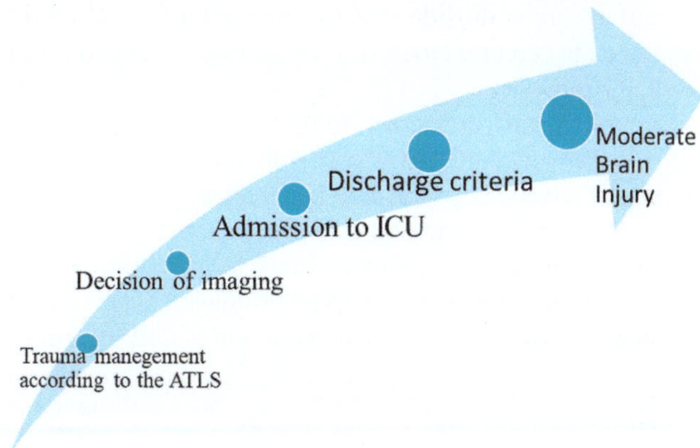

FIG. 3 Steps of managing severe traumatic brain injury. Emergency department considerations of severe TBI.

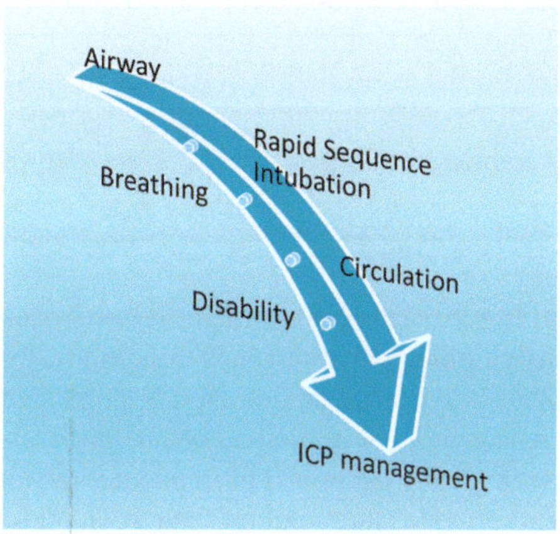

FIG. 4 Steps of managing moderate traumatic brain injury. Emergency department considerations of mild TBI.

Tracheal intubation with laryngoscopy may elicit a reflex sympathetic response that increases ICP while airway manipulation may create direct reflex responses that result in the same effect. The intensity of these responses may be reduced by administering an ultrashort-acting opioid before such procedures; although the target of neuroprotection can be achieved by using fentanyl and its analogs in patients with increased ICP, there is controversy surrounding other pre-medications including atropine, lidocaine, B-blockers, and a de-fasciculation dose of neuromuscular blocking agents (NMBA) (Garvin & Mangat, 2017; Kramer et al., 2018).

Propofol may be a potent blood pressure (BP) reducing agent even after the first bolus dose and therefore is not suitable for patients with TBI; on the other hand, whereas ketamine may cause an increase in ICP and has been traditionally avoided in these patients, recent evidence suggests that it does not affect ICP and even lowers it. Patients with ICP elevation may also benefit from etomidate as an induction agent particularly when a patient is hypotensive or may be hypotensive during the course while methohexital is cardioprotective and can be used safely in patients with elevated ICP (DeWitt & Porter, 2021; Garvin & Mangat, 2017).

Succinylcholine (SCh) is the NMBA of choice for RSI in patients with elevated ICP. For anesthesia induction rocuronium is the most widely used short-acting competitive drug as non-depolarizing NMBAs which are free of the properties causing ICP elevation, although its longer duration of effect than SCh may prevent proper neurologic examinations and management modifications; Sugammadex may be beneficial in these instances (Curley, Ciceri, & Culp, 2020; deBacker, Hart, & Fan, 2017; Renew, Ratzlaff, Hernandez-Torres, Brull, & Prielipp, 2020).

Breathing

Desired levels of respiratory parameters can be well achieved via mechanical ventilation for which intubation is required (Cinotti, Bouras, Roquilly, & Asehnoune, 2018). Currently, sTBI patients are expected to normally ventilate provided that they are free of cerebral herniation. The preferred targets for the parameters of arterial blood gas analysis include a

PaO_2 greater than 60 mmHg and a partial carbon dioxide ($PaCO_2$) pressure of 35–45 mmHg (Carney et al., 2017; Cinotti et al., 2018). $PaCO_2$ is strongly correlated to cerebral blood flow by having direct control of cerebral vascular dilation and contraction, immediately affecting ICP (Carney et al., 2017). The preferred management method should be administering patients 100% oxygen until measuring blood gas parameters, followed by tailoring the fraction of inspired oxygen to the patient's needs (Galvagno et al., 2019). Patients with elevated ICP should not be treated with lung-protective ventilation strategies since cerebral perfusion may be adversely affected by hypercapnia and acidosis; these strategies are permissible when the risk of acute respiratory disease syndrome is high and hypercapnia and acidosis are prevented (Cinotti et al., 2018; Picetti et al., 2020). Hyperventilation can be used as a temporary technique to reduce ICP for patients showing acute worsening of neurological function or signs of herniation; however, prolonged hyperventilation with $PaCO_2 < 25$ mmHg is contraindicated in the first 24 h following injury due to a critical reduction in CBF (Carney et al., 2017).

Circulation

Since maintaining decent BP and CBF as well as preventing hypotension is of utmost importance, the available guidelines has given the emphasis on keeping SBP equal or greater than 100 mmHg among patients aged 50–69 years and equal or greater than 110 mmHg among those aged 15–49 years or older than 70 years (Carney et al., 2017; Jain, Choudhary, & Pandit, 2019). In these patients arterial BP should be monitored invasively on a continuous basis; isotonic fluids, and when indicated, vasopressors and blood products (for the resuscitation of trauma victims) should be administered to maintain BP and hemodynamic goals (Carney et al., 2017; Dixon et al., 2020; Vella et al., 2017). Since hypotonic fluids and fluids with dextrose contents can further deteriorate cerebral edema, they should be avoided (Dixon et al., 2020).

Disability

The disability status of a patient with TBI is evaluated by a rapid and careful neurological examination at the emergency department (Vella et al., 2017). As in other neurological emergency conditions, the initial assessment in TBI typically involves recording the neurological vital signs, i.e. determination of GCS, pupil size, and checking the light reflex for both eyes (Whitaker-Lea, 2017). The clinical management of particularly sTBI involves maintaining ICP within the established goals; therefore, the patient should be neurologically examined at frequent intervals (Le Roux, 2016).

When the initial neurological and other systemic examinations have been performed and the patient hemodynamically stabilized, a secondary investigation should be performed, including a non-contrast-enhanced cranial CT. The first CT should be followed by a subsequent CT when a patient's neurological status worsens without a clear explanation or does not adequately respond to invasive procedures and other resuscitation efforts; in contrast, serial CTs are not indicated for patients with a stable neurological condition for several hours following injury (Whitaker-Lea, 2017). Patients who take oral anticoagulants or are suspected to do so need a follow-up cranial tomographic scan only when there are signs of intracranial hemorrhage in the initial examination. When a decision is made to proceed with a follow-up tomographic head scan, the latter is ideally performed 6–24 h following the trauma (Wiegele et al., 2019).

Medical treatment of brain injury

The established principles for the management of TBI include mitigation of cerebral edema and reduction of ICP; prevention of tissue hypoxia and ischemia; protection of neural tissue by cooling down inflammation and lowering its metabolic demand; treating existing abnormalities of coagulation and preventing the progression of hemorrhage, and adequately addressing systemic complications, the major ones being pulmonary, infectious, nutritional, thromboembolic, musculoskeletal, and neurologic (e.g., seizure) (Davanzo, Sieg, & Timmons, 2017). The steps are summarized in Table 2.

Intravenous fluids

It is known that hypovolemia is unfavorable for patients with TBI; therefore, adequate hydration of these patients is essential, but care should be exercised to avoid hypervolemia and not to administer hypotonic fluids known to worsen cerebral edema, such as dextrose 5%, sodium chloride 0.45%, or Ringer's lactate solution. The initial fluids of choice are isotonic crystalloid fluids and appropriate blood products administered via the intravenous route. They may be supported by vasopressors such as phenylephrine or norepinephrine if BP targets (typically mean arterial pressure (MAP)) have not been achieved (DeWitt & Porter, 2021; Farrokh, Cho, & Suarez, 2019). Since hyponatremia is well known to give rise to cerebral edema, sodium level should be closely monitored (Gankam Kengne & Decaux, 2018; Rajagopal, Swaminathan, Nair, & Joseph, 2017).

TABLE 2 Steps of managing intracranial pressure.

Medical Treatment of Brain Injury
Intravenous Fluids
Correction of Anticoagulation
Rapid Sequence Intubation (RSI)
Tranexamic Acid (TXA)
Transfusion
Body Positioning
Hyperosmolar Agents
Hyperventilation
Temperature management
Sedation and Analgesia
Seizure management
Corticosteroids
Calcium channel blockers

The procedures and medical treatment of regulating intracranial pressure in TBI patients.

Correction of anticoagulation

Chronic anticoagulation is an important issue that should be addressed in the management of TBI. International normalized ratio (INR) should be rapidly determined, and a head CT is taken without a delay when clinically indicated. The Neurocritical Care Society, in its most recent guidelines, recommends that anticoagulation be promptly reversed if the intracranial hemorrhage is present (Frontera et al., 2016). As the bleeding risk is proportional to the INR level, INR should be kept below 1.4 or 1.5 if surgery is contemplated (Macdonald, 2018; Wiegele et al., 2019). When a patient uses novel oral anticoagulant agents, the anticoagulation level should not be checked using standard coagulation tests such as prothrombin time or active partial thromboplastine time; thrombin time (TT) or dilute TT (dTT) may be studied if dabigatran use is known or suspected while xabans' blood-thinning effects can be quantified using anti-Xa calibrated to low molecular weight heparin (Cuker, 2016). Reversal of anticoagulation is generally not indicated in TBI unless there is active bleeding or pathological signs on the initial head CT. When intervention for anticoagulation is considered, it can be accomplished by temporarily withholding anticoagulants; if reversal of anticoagulation with warfarin is needed, fresh frozen plasma, prothrombin complex concentrate, and/or vitamin K can be administered (Macdonald, 2018; Pollack, 2016; Wiegele et al., 2019). The available evidence is not so strong for making any goal-specific recommendations for raising the platelet count in patients with thrombocytopenia (Briggs et al., 2015; Furay et al., 2018).

Tranexamic acid

Tranexamic acid is an anti-fibrinolytic agent but, its benefits must exceed its risk of thrombotic and embolic events including myocardial infarction, stroke, pulmonary embolism, and deep vein thrombosis (Marehbian, Muehlschlegel, Edlow, Hinson, & Hwang, 2017). Coagulopathy complicates TBI through tissue factor activation led by tissue factors released from the brain tissue into the circulation, low platelet count, platelet dysfunction, protein C activation, and hyperfibrinolysis maximally occurring 3h following an injury event (Nakae et al., 2019). Despite some evidence that it may lower mortality after TBI, its efficacy in this indication is currently debated and warrants further research (Marehbian et al., 2017; Yokobori et al., 2020).

Transfusion in TBI

Hemoglobin (Hb) level is a major factor affecting brain oxygenation in sTBI; the widely agreed clinical consensus dictates red blood cell transfusion (RBCT) for a Hb level below 7 g/dL in patients with TBI who are in critical condition (East, Viau-Lapointe, & McCredie, 2018; Lelubre, Bouzat, Crippa, & Taccone, 2016). Although trauma and critical care societies currently urge to keep hemoglobin level between 7 and 9 g/dL in such patients (Rossaint et al., 2016), a global agreement on transfusion decisions is yet to be reached, with making transfusion decisions on a patient by patient basis seems reasonable (East et al., 2018).

Management of elevated ICP

Sufficiently prolonged, the severe elevation of ICP jeopardizes cerebral perfusion (CPP=MAP−ICP) and causes brain ischemia and/or infarction, ultimately resulting in irreversible cerebral injury and death. Therefore, ICP remains the main parameter that should be maintained in certain well-established limits, and every attempt should be made to aggressively recognize and treat an elevated ICP (Khellaf, Khan, & Helmy, 2019; Ravishankar et al., 2020). The guidelines issued by the Brain Trauma Foundation stress the importance of monitoring ICP in sTBI in order to cut in-hospital and 2-week post-injury mortality rates (Carney et al., 2017; Moore et al., 2020).

Body positioning

In TBI, body positioning is an essential and easy-to-apply component of patient management. Head elevation between 30 and 45 degrees may effectively reduce ICP without compromising CBF and CPP while putting the patient's face at midline augments cerebral venous drainage via the internal jugular vein (Galgano et al., 2017; Schizodimos, Soulountsi, Iasonidou, & Kapravelos, 2020). Limiting excess neck flexion and/or rotation, restrictive neck taping, and cough and Valsalva reflexes by avoiding endotracheal suctioning as much as possible are other important body positioning maneuvers to which emphasis should be given (Schizodimos et al., 2020).

Hyperosmolar agents

Hyperosmolar agents such as mannitol and hyperosmolar saline have been routinely administered to patients with elevated ICP and cerebral herniation and effectively lower ICP and brain edema by several mechanisms when intravenously administered, including the induction of temporary hyperosmolarity in the extracellular space, creation of an osmotic pressure difference on the blood-brain barrier, and pulling water out of the brain tissue into the hypertonic intravascular compartment (Carney et al., 2017; Geeraerts et al., 2018; Marehbian et al., 2017). There is no clear recommendation as to which agent should be selected, but it should be remembered that hypertonic saline may be detrimental in chronic hyponatremia and mannitol may aggravate pre-existing hypotension (Marehbian et al., 2017).

Hyperventilation

Hyperventilation is a disputable management strategy in TBI. By creating hypocapnia, hyperventilation raises the pH level of the cerebrospinal fluid, which in turn leads to cerebral vasoconstriction, reduced cerebral blood volume, and ICP lowering. However, when prophylactically applied, particularly for long periods in the first 24–48 h, hyperventilation may paradoxically reduce cerebral perfusion to critically low levels, thereby creating potential brain ischemia and worsened neurological damage. Therefore, hyperventilation should be spared as an acceptable temporary measure for sTBI accompanied by an elevated ICP or acute neurological deterioration. Patients with sTBI but no cerebral herniation or hypercarbia (increased $PaCO_2$) in arterial blood gas analysis should be best managed by allowing them to breathe normally (Carney et al., 2017; Davanzo et al., 2017; Galgano et al., 2017; Godoy, Seifi, Garza, Lubillo-Montenegro, & Murillo-Cabezas, 2017; Wettervik et al., 2020).

Temperature management

In addition to the primary trauma-induced neural cell injury, sTBI triggers a cascade of temperature-sensitive events leading to secondary brain injury, which encompasses cerebral ischemia and hypoxia, inflammatory reaction, metabolic disorders, and blood-brain barrier breakdown (Kaur & Sharma, 2018). Hence, high body temperature is among the main determinants of prognosis since it leads to an increase in brain metabolism, worsens brain edema, and elevates ICP, ultimately resulting in neural injury and even irreversible brain damage (Huang, Zhao, & Pu, 2020). Targeted temperature management (TTM) is a management approach that encompasses the components of targeted hypothermia, maintaining normothermia, and preventing fever (DeWitt & Porter, 2021). Whereas prophylactic or therapeutic hypothermia may mitigate resistant ICP elevation and be associated with more favorable outcomes in patients with TBI, data from meta-analyses are still lacking and no consensus has been reached on this subject. Moreover, mild hypothermia alone is incapable of improving the overall poor survival rate of sTBI; also, it is unclear how patients treated with therapeutic hypothermia fare neurologically in the long term. It is therefore not advisable to universally utilize hypothermia in sTBI (Huang et al., 2020; Kim et al., 2020). If it is to be administered at all, it should be done so in the first 24 h and as a therapeutic intervention, not a prophylactic one (Chen et al., 2019). In line with the view, current guidelines discourage prophylactic use of hypothermia to reduce TBI-induced ICP; however, it remains a top-notch neuroprotective management strategy and it may still be considered as a second option if there is persistent intracranial hypertension despite other maneuvers (Zusman, Kochanek, & Jha, 2020).

Suppression of shivering by medications such as analgesics, sedatives, NMBAs, or non-pharmacologic measures should be added to hypothermia since shivering may offset the favorable effects of TTM due to its adverse effects such as the increases in the rate of metabolism, oxygen consumption, and carbon dioxide production, (DeWitt & Porter, 2021; Jain et al., 2018).

Sedation and analgesia

Patients with sTBI should be optimally sedated and provided with adequate pain relief because both agitation and pain elevate ICP. At times, patients may need to be put in a coma induced by the infusion of sedatives as the last resort when other approaches fail (Galgano et al., 2017; Schizodimos et al., 2020). As guidelines have already pointed, propofol is one of the agents of choice for sedation for controlling ICP, but care must be exercised not to experience its untoward effects, particularly the fatal "propofol-infusion syndrome" (Carney et al., 2017). When propofol alone does not suffice to reach the sedation goal set for the patient, fentanyl and particularly remifentanil can be added to make use of their synergistic effects with propofol (Zusman et al., 2020). Neuromuscular blocking agents (e.g., vecuronium, cisatracurium) have been poorly studied so far, but they can serve for the control of refractory ICP increase when selectively used for very patients with severe agitation, shivering, and respiratory distress (Godoy, Lubillo, & Rabinstein, 2018). BTF guidelines advise using high dose barbiturates, including pentobarbital and thiopental, for deep sedation as part of attempts of cooling down brain metabolism, limiting CBF, and improving cerebral oxygenation. However, hemodynamic stability should be checked before administering these agents and they should be applied when ICP elevations persist despite standard interventions (Carney et al., 2017; DeWitt & Porter, 2021).

Seizure management

Seizures have the potential of worsening intracranial hypertension by augmenting the cerebral metabolic rate of oxygen and CBF (Schizodimos et al., 2020). Despite the fact that the role of TBI severity and anticonvulsant use in the prevention of cerebral injury has not been well studied, TBI guidelines favor the use of prophylactic antiepileptic agents for 1 week after TBI with the goal of preventing early seizure attacks but no compelling evidence exists for the prevention of long-term seizure activity. In accordance with this recommendation, anti-epileptics are typically stopped after 1 week (Carney et al., 2017). The anti-epileptics of choice include phenytoin, fosphenytoin, levetiracetam, and valproate (Carney et al., 2017; DeWitt & Porter, 2021; Stein, Feather, & Napolitano, 2017).

Corticosteroids

BTF strongly discourages the use of corticosteroids in sTBI patients as neuroprotective agents (Carney et al., 2017; Marehbian et al., 2017).

Calcium channel blockers

Calcium channel blocker (nimodipine, ziconotide) use in TBI is a debated topic due to conflicting and mixed study results; hence, believed to reduce delayed injury progression in TBI, particularly nimodipine among others, can be administered to selected high-risk TBI patients provided that hypotension is not a concern (Carlson, Hanggi, Macdonald, & Shuttleworth, 2020; Galgano et al., 2017).

Conclusions and future directions

The initial emergency care of TBI patients is composed of quick and targeted triage, rapid diagnosis of the condition, and rapid patient stabilization if there is a severe brain injury or critical injury or dysfunction of other body parts. These patients should be ideally managed on the basis of existing evidence and in a standardized manner because doing so ensures more favorable outcomes from a global perspective. Unfortunately, the available evidence obtained from animal and human research is of low quality despite recent scientific efforts, rendering recommendations in this field to expert opinions in the absence of a sufficient number of high-quality prospective randomized trials. With the knowledge in mind that a variety of several factors affect both short- and long-term outcomes of patients with TBI, utilization of available resources, diagnostic and therapeutic strategies for TBI should be further studied.

Case 1

A 73-year-old male patient was admitted to the ED with his head hitting the kitchen cabinet door. The history of the trauma, its mechanism, the presence of post-traumatic symptoms, co-morbidities (concussion, migraine/headache, and depression history; bleeding disorders; or anticoagulant medication) were questioned in the patient's history. He had a history of hypertension, cardiac arrhythmia (atrial fibrillation), and mitral valve replacement; Calcium channel blocker and oral anticoagulant use. He did not have any allergies. He stated that his foot caught on the carpet and he hit his head to the open kitchen cabinet door. He had no any other additional trauma. In the neurological examination of the patient; his GCS was 15, pupillary function was normal, and there was approximately 2×2cm superficial incision in the left parietal area and an active bleeding in the form of leakage from the incision site. No acute pathological examination findings were detected in the cervical spine and other system examinations. He had vital signs of BP: 138/87mmHg, heart rate (HR): 83bpm, body temperature: 36.3°C, respiratory rate (RR): 14/min and blood glucose: 115mg/dL. Because scalp lacerations can lead to massive blood loss, direct pressure was applied to control bleeding. According to the criteria of Canada CT head rule, the patient was considered to be at high risk for neurosurgical intervention due to the use of oral anticoagulants and being over 65 years old, and brain CT was planned. Trauma specific blood samples, including the patient's coagulation parameters, were collected. The head CT was revealed no any intracranial pathology and the patient was taken under observation. In the repeated examinations of the patient, his GCS was found to be 13 (eye: 3, verbal: 4, motor: 6) at the second hour under observation, the brain CT was repeated and a 3×2cm parenchymal hematoma was detected in the left parietal lobe on CT. Neurosurgical consultation was requested, the patient was monitored, the bedside was elevated 30 degrees. During the observation in ED, vital signs were as follows: BP 125/82mm/Hg, HR: 80bpm, body temperature: 36.3°C, RR: 14/min, BG: 110mg/dL, SaO_2: 96. Since his GCS: 13, advanced airway intervention was not considered. According to the blood samples, his INR result was 3.4IU so reversal of anticoagulation should be considered and 10mg of IV vitamin K and prothrombin complex concentrate (PCC) were administered. The patient was evaluated by neurosurgery and was admitted to intensive care unit.

Case 2

A 26-year-old female patient was brought to the emergency room by emergency medical service due to an in-vehicle traffic accident. The patient was the driver of the vehicle, hit the stationary truck from behind, was removed from the jammed vehicle, and her seat belt was fastened. At the scene, it was found that the patient was unconscious and her GCS was 7, her vital signs were BP: 80/40mmHg, BPM: 124/min, RR/min, SaO_2: 83, respectively. The patient was admitted to resuscitation area and underwent a primary survey, adhering to the ABCDE priorities. Two large-bore peripheral venous catheters were placed to administer fluid, blood, and plasma. Blood samples for baseline hematologic studies are obtained, including a pregnancy test, blood type and cross matching, blood gases, and/or lactate levels. On primary survey; she had an altered level of consciousness with a GCS score of 7 (E: 1, V: 1, M: 5) suggestive of sTBI. Simultaneously measured vital signs were BP: 80/50mmHg, HR: 138bpm, RR: 12/min, SaO_2: 80 respectively. On inspection, the left hemithorax did not participate in breathing and respiratory sounds were not revealed in the left hemithorax on auscultation. Simultaneous extended focused assessment with sonography for trauma (eFAST) examination revealed pneumothorax in the left hemithorax and free fluid in the sub-phrenic space. Because it is imperative to rapidly achieve cardiopulmonary stabilization in patients with severe brain injury 100% O_2 support was administered and as she was comatose (GCS $<$ 8) she could not maintain her airway integrity so a definitive airway with rapid sequence entubation was considered with cervical spine motion restriction. Because she was hypotensive, ketamine was preferred for induction and succinylcholine was the NMBA of choice. Before entubation, tube thoracostomy for chest decompression was performed because pneumothorax could be liable to convert to a tension pneumothorax as well as it was compromising the patient's ventilation. There was no any contraindication for bladder catheterization so a transurethral catheter was inserted to assess the urine for hematuria and to monitöre urine output. She was in class 3 shock (heart rate > 120 bpm, BP was decreased) with the probability of sTBI, a bolus of 1 L of an isotonic solution and after then because the patient was a female of childbearing age, to prevent sensitization 2 units of type 0 Rh-negative packet red blood cell (pRBC) with 2 units of AB plasma was administered to achieve a BP of ≥ 110 mmHg. After transfusion, BP was 90/60 mmHg, HR was 119 bpm. In neurological examination, right pupil was mydriatic and light reflex could not be obtained and GCS was still 7. The patient was evaluated as to failure to respond to crystalloid and blood administration in the ED, so surgical evaluation was provided immediately for definitive intervention to control probable exsanguinating hemorrhage. The patient was decided to be transferred to operation room for laparoscopic evaluation and obtaining a head CT was decided to be taken after operation as the first priority is to establish the cause of the hypotension and hemodynamic stabilization even the patient has sTBI.

Applications to other areas of neuroscience

In this chapter we have reviewed the management of traumatic brain injury in emergency department. We also interlink the role of emergency physicians in the golden hour of these patients.

The interconnection between neurological injury and factors resulting in hypoxia due to the imbalance between metabolic demand and supply is seen in many other studies. These include stroke, spinal cord injury, and refractory status epilepticus. The pivotal variables preventing hypoxia and hypotension are also critical in the management of spinal cord injury. However, the difference between traumatic brain injury and spinal cord injury is in RSI. In patients suspected spinal cord injury, avoidance of drugs that results bradycardia or hypotension and mitigate this effect by using atropine or/and vasopressors around the RSI is utmost of importance. Also, while steroids should be avoided in traumatic brain injury, in spinal cord injury patients they should not be used routinely but in central nervous system infections they are routinely recommended (DeWitt & Porter, 2021; Henry, 2018). Again, although BP regulation is a cornerstone in the treatment for both traumatic brain injury and stroke; in the management of traumatic brain injury, while significant values for BP are specified, an ideal BP recommendation is not mentioned in stroke, but it is recommended to avoid rapid and sharp reduction in of BP values mostly due to concomitant hypertension (Jain et al., 2019). For emergency specialists, preventing secondary injuries is common and targeted outcome for all neurological acute admissions.

Mini-dictionary of terms

- ABCDE: The priorities of resuscitating an emergent patient in a methodical order.
- Advanced Trauma Life Support: A systematic and evidence-based approach in the acute care of trauma patients.
- Body-positioning: Posture given to the body in traumatic brain injury patients to prevent intracranial pressure increase and to increase oxygenation.
- Intracranial pressure: A combination pressure of intracranial vessels, fluids, and brain tissue.
- Minimum arterial pressure: The minimum blood pressure level required for end-organ perfusion.
- Neuromuscular blocking agent: Drugs which are inhibiting the effects of acetylcholine in the neuromuscular junction and paralyzing the muscles.
- Rapid sequence intubation: An emergency airway management technique to control and secure airway with specific drugs.
- Sedation: An intervention to reduce the patient's anxiety and agitation with specific drugs to perform a medical procedure or in the treatment algorithm.
- Tracheal intubation: A procedure which allows the placement of a tube into the trachea to provide advanced airway in patients.
- Traumatic brain injury: A damage to the brain by external forces which can be blunt or penetrating and disrupts the normal brain function.

Key facts of advanced trauma life support

- Traumatic brain injury is a condition that results in brain damage and disrupts its' normal function after a fall, crash, traffic accident, recreational activities, violence and a leading cause of morbidity and mortality worldwide.
- Emergency departments are typically the first medical contact for patients with traumatic brain injury.
- Adequate resuscitation as preventing hypovolemia and hypoxemia is important in limiting secondary brain injury.

- Resuscitation with Ringer's lactate solution, normal saline, or similar isotonic solutions without hypotonic solutions, especially dextrose is critical.
- Treatment of brain injury with other life-threatening injuries such as hemorrhage is pivotal.
- Not every patient needs imaging, the necessity for this is a situation that needs to be decided by clinical and observation strategies of physicians.

Summary points

- According to the patient's clinical status, traumatic brain injuries can be categorized as mild, moderate, and severe. Neurological vital signs are pivotal in evaluating the patients' severity.
- The major concern for emergency specialists is identifying the patients with a moderate or high risk of neurosurgical intervention requirement.
- Since the time patients spend in the emergency department is the most important part of the intervention, it can be expressed as golden hour.
- Avoidance of secondary injury; preventing hypoxia, hypotension, hypoglycemia and a strict intracranial pressure control are mandatory keys for the acute management of traumatic brain injury in emergency department.
- Practical management of traumatic brain injury should be directed according to the evidence-based recommendation levels in guidelines in a continuum manner.

References

Abdelmalik, P. A., Draghic, N., & Ling, G. S. F. (2019). Management of moderate and severe traumatic brain injury. *Transfusion*, *59*, 1529–1538. https://doi.org/10.1111/trf.15171.

Blennow, K., Brody, D. L., Kochanek, P. M., Levin, H., McKee, A., Ribbers, G. M., … Zetterberg, H. (2016). Traumatic brain injuries. *Nature Reviews Disease Primers*, *2*. https://doi.org/10.1038/nrdp.2016.84. ARTN 16084.

Briggs, A., Gates, J. D., Kaufman, R. M., Calahan, C., Gormley, W. B., & Havens, J. M. (2015). Platelet dysfunction and platelet transfusion in traumatic brain injury. *Journal of Surgical Research*, *193*(2), 802–806. https://doi.org/10.1016/j.jss.2014.08.016.

Carlson, A. P., Hanggi, D., Macdonald, R. L., & Shuttleworth, C. W. (2020). Nimodipine reappraised: An old drug with a future. *Current Neuropharmacology*, *18*(1), 65–82. https://doi.org/10.2174/1570159x17666190927113021.

Carney, N., Totten, A. M., O'Reilly, C., Ullman, J. S., Hawryluk, G. W. J., Bell, M. J., … Ghajar, J. (2017). Guidelines for the management of severe traumatic brain injury, fourth edition. *Neurosurgery*, *80*(1), 6–15. https://doi.org/10.1227/Neu.0000000000001432.

Chen, H., Wu, F., Yang, P., Shao, J., Chen, Q., & Zheng, R. (2019). A meta-analysis of the effects of therapeutic hypothermia in adult patients with traumatic brain injury. *Critical Care*, *23*(1), 396. https://doi.org/10.1186/s13054-019-2667-3.

Cinotti, R., Bouras, M., Roquilly, A., & Asehnoune, K. (2018). Management and weaning from mechanical ventilation in neurologic patients. *Annals of Translational Medicine*, *6*(19), 381. https://doi.org/10.21037/atm.2018.08.16.

Cuker, A. (2016). Laboratory measurement of the non-vitamin K antagonist oral anticoagulants: Selecting the optimal assay based on drug, assay availability, and clinical indication. *Journal of Thrombosis and Thrombolysis*, *41*(2), 241–247. https://doi.org/10.1007/s11239-015-1282-7.

Curley, J. M., Ciceri, D. P., & Culp, W. C. (2020). Sugammadex administration to facilitate timely neurologic examination in the traumatic brain injury patient. *Neurocritical Care*, *32*(3), 880–882. https://doi.org/10.1007/s12028-019-00901-6.

Davanzo, J. R., Sieg, E. P., & Timmons, S. D. (2017). Management of traumatic brain injury. *Surgical Clinics of North America*, *97*(6), 1237–1253. https://doi.org/10.1016/j.suc.2017.08.001.

deBacker, J., Hart, N., & Fan, E. (2017). Neuromuscular blockade in the 21st century management of the critically ill patient. *Chest*, *151*(3), 697–706. https://doi.org/10.1016/j.chest.2016.10.040.

DeWitt, K. M., & Porter, B. A. (2021). Emergency neuropharmacology. *Emergency Medicine Clinics of North America*, *39*(1), 133–154. https://doi.org/10.1016/j.emc.2020.09.008.

Dixon, J., Comstock, G., Whitfield, J., Richards, D., Burkholder, T. W., Leifer, N., … Hynes, E. J. C. (2020). Emergency department management of traumatic brain injuries: A resource tiered review. *African Journal of Emergency Medicine*, *10*(3), 159–166. https://doi.org/10.1016/j.afjem.2020.05.006.

East, J. M., Viau-Lapointe, J., & McCredie, V. A. (2018). Transfusion practices in traumatic brain injury. *Current Opinion in Anaesthesiology*, *31*(2), 219–226. https://doi.org/10.1097/ACO.0000000000000566.

Farrokh, S., Cho, S. M., & Suarez, J. I. (2019). Fluids and hyperosmolar agents in neurocritical care: An update. *Current Opinion in Critical Care*, *25*(2), 105–109. https://doi.org/10.1097/MCC.0000000000000585.

Foks, K., Cnossen, M., Dippel, D., Maas, A., Menon, D. K., van der Naalt, J., … Polinder, S. (2017). Management of mild traumatic brain injury at the emergency department and hospital admission in Europe: A survey of 71 neurotrauma centers participating in the CENTER-TBI study. *European Journal of Neurology*, *24*, 101. Retrieved from <Go to ISI>://WOS:000405530100166.

Frontera, J. A., Lewin, J. J., 3rd, Rabinstein, A. A., Aisiku, I. P., Alexandrov, A. W., Cook, A. M., … Zerfoss, C. L. (2016). Guideline for reversal of antithrombotics in intracranial hemorrhage: A statement for healthcare professionals from the Neurocritical Care Society and Society of Critical Care Medicine. *Neurocritical Care*, *24*(1), 6–46. https://doi.org/10.1007/s12028-015-0222-x.

Furay, E., Daley, M., Teixeira, P. G., Coopwood, T. B., Aydelotte, J. D., Malesa, N., ... Brown, C. V. R. (2018). Goal-directed platelet transfusions correct platelet dysfunction and may improve survival in patients with severe traumatic brain injury. *Journal of Trauma and Acute Care Surgery*, *85*(5), 881–887. https://doi.org/10.1097/Ta.0000000000002047.

Galgano, M., Toshkezi, G., Qiu, X. C., Russell, T., Chin, L., & Zhao, L. R. (2017). Traumatic brain injury: Current treatment strategies and future endeavors. *Cell Transplantation*, *26*(7), 1118–1130. https://doi.org/10.1177/0963689717714102.

Galvagno, S. M., Jr., Nahmias, J. T., & Young, D. A. (2019). Advanced trauma life support((R)) update 2019: Management and applications for adults and special populations. *Anesthesiology Clinics*, *37*(1), 13–32. https://doi.org/10.1016/j.anclin.2018.09.009.

Gankam Kengne, F., & Decaux, G. (2018). Hyponatremia and the brain. *Kidney International Reports*, *3*(1), 24–35. https://doi.org/10.1016/j.ekir.2017.08.015.

Garvin, R., & Mangat, H. S. (2017). Emergency neurological life support: Severe traumatic brain injury. *Neurocritical Care*, *27*, S159–S169. https://doi.org/10.1007/s12028-017-0461-0.

Geeraerts, T., Velly, L., Abdennour, L., Asehnoune, K., Audibert, G., Bouzat, P., ... Association des anesthesistes-reanimateurs pediatriques d'expression Française. (2018). Management of severe traumatic brain injury (first 24 hours). *Anaesthesia, Critical Care & Pain Medicine*, *37*(2), 171–186. https://doi.org/10.1016/j.accpm.2017.12.001.

Ginsburg, J., & Huff, J. S. (2021). Closed head trauma. In *StatPearls*. Treasure Island, FL: StatPearls Publishing.

Godoy, D. A., Lubillo, S., & Rabinstein, A. A. (2018). Pathophysiology and management of intracranial hypertension and tissular brain hypoxia after severe traumatic brain injury an integrative approach. *Neurosurgery Clinics of North America*, *29*(2), 195. https://doi.org/10.1016/j.nec.2017.12.001.

Godoy, D. A., Rubiano, A., Rabinstein, A. A., Bullock, R., & Sahuquillo, J. (2016). Moderate traumatic brain injury: The grey zone of neurotrauma. *Neurocritical Care*, *25*(2), 306–319. https://doi.org/10.1007/s12028-016-0253-y.

Godoy, D. A., Seifi, A., Garza, D., Lubillo-Montenegro, S., & Murillo-Cabezas, F. (2017). Hyperventilation therapy for control of posttraumatic intracranial hypertension. *Frontiers in Neurology*, *8*. https://doi.org/10.3389/fneur.2017.00250. ARTN 250.

Haydel, M. J., & Burns, B. (2021). Blunt head trauma. In *StatPearls*. Treasure Island, FL: StatPearls Publishing.

Henry, S. (2018). ATLS 10th edition offers new insights into managing trauma patients. In *Bulletin of the American College of Surgens*.

Huang, H. P., Zhao, W. J., & Pu, J. (2020). Effect of mild hypothermia on prognosis of patients with severe traumatic brain injury: A meta-analysis with trial sequential analysis. *Australian Critical Care*, *33*(4), 375–381. https://doi.org/10.1016/j.aucc.2019.08.005.

Jain, V., Choudhary, J., & Pandit, R. (2019). Blood pressure target in acute brain injury. *Indian Journal of Critical Care Medicine*, *23*(Suppl. 2), S136–S139. https://doi.org/10.5005/jp-journals-10071-23191.

Jain, A., Gray, M., Slisz, S., Haymore, J., Badjatia, N., & Kulstad, E. (2018). Shivering treatments for targeted temperature management: A review. *Journal of Neuroscience Nursing*, *50*(2), 63–67. https://doi.org/10.1097/Jnn.0000000000000340.

Kaur, P., & Sharma, S. (2018). Recent advances in pathophysiology of traumatic brain injury. *Current Neuropharmacology*, *16*(8), 1224–1238. https://doi.org/10.2174/1570159X15666170613083606.

Khellaf, A., Khan, D. Z., & Helmy, A. (2019). Recent advances in traumatic brain injury. *Journal of Neurology*, *266*(11), 2878–2889. https://doi.org/10.1007/s00415-019-09541-4.

Kim, J. H., Nagy, A., Putzu, A., Belletti, A., Biondi-Zoccai, G., Likhvantsev, V. V., ... Landoni, G. (2020). Therapeutic hypothermia in critically ill patients: A systematic review and meta-analysis of high quality randomized trials. *Critical Care Medicine*, *48*(7), 1047–1054. https://doi.org/10.1097/CCM.0000000000004364.

Kovacs, G., & Sowers, N. (2018). Airway management in trauma. *Emergency Medicine Clinics of North America*, *36*(1), 61–84. https://doi.org/10.1016/j.emc.2017.08.006.

Kramer, N., Lebowitz, D., Walsh, M., & Ganti, L. (2018). Rapid sequence intubation in traumatic brain-injured adults. *Cureus*, *10*(4). https://doi.org/10.7759/cureus.2530, e2530.

Le Roux, P. (2016). Intracranial pressure monitoring and management. In D. Laskowitz, & G. Grant (Eds.), *Translational research in traumatic brain injury*. Boca Raton, FL: CRC Press/Taylor and Francis Group.

Lelubre, C., Bouzat, P., Crippa, I. A., & Taccone, F. S. (2016). Anemia management after acute brain injury. *Critical Care*, *20*(1), 152. https://doi.org/10.1186/s13054-016-1321-6.

Macdonald, R. L. (2018). Management of intracranial hemorrhage in the anticoagulated patient. *Neurosurgery Clinics of North America*, *29*(4), 605–613. https://doi.org/10.1016/j.nec.2018.06.013.

Marehbian, J., Muehlschlegel, S., Edlow, B. L., Hinson, H. E., & Hwang, D. Y. (2017). Medical management of the severe traumatic brain injury patient. *Neurocritical Care*, *27*(3), 430–446. https://doi.org/10.1007/s12028-017-0408-5.

Mayer, A. R., Quinn, D. K., & Master, C. L. (2017). The spectrum of mild traumatic brain injury: A review. *Neurology*, *89*(6), 623–632. https://doi.org/10.1212/Wnl.0000000000004214.

McInnes, K., Friesen, C. L., MacKenzie, D. E., Westwood, D. A., & Boe, S. G. (2019). Mild traumatic brain injury (mTBI) and chronic cognitive impairment: A scoping review (vol 12, e0174847, 2017). *PLoS One*, *14*(6). https://doi.org/10.1371/journal.pone.0218423. ARTN e0218423.

Moore, L., Tardif, P. A., Lauzier, F., Berube, M., Archambault, P., Lamontagne, F., ... Re, C. T. B. I. (2020). Low-value clinical practices in adult traumatic brain injury: An umbrella review. *Journal of Neurotrauma*, *37*(24), 2605–2615. https://doi.org/10.1089/neu.2020.7044.

Nakae, R., Yokobori, S., Takayama, Y., Kanaya, T., Fujiki, Y., Igarashi, Y., ... Yokota, H. (2019). Time course of coagulation and fibrinolytic parameters in patients with traumatic brain injury. *Brain Injury*, *33*, 39. Retrieved from <Go to ISI>://WOS:000466897000091.

Picetti, E., Pelosi, P., Taccone, F. S., Citerio, G., Mancebo, J., Robba, C., & On the Behalf of the ESICM NIC/ARF Sections. (2020). VENTILatOry strategies in patients with severe traumatic brain injury: The VENTILO Survey of the European Society of Intensive Care Medicine (ESICM). *Critical Care*, *24*(1), 158. https://doi.org/10.1186/s13054-020-02875-w.

Pollack, C. V. (2016). Coagulation assessment with the new generation of oral anticoagulants. *Emergency Medicine Journal*, *33*(6), 423–430. https://doi.org/10.1136/emermed-2015-204891.

Rajagopal, R., Swaminathan, G., Nair, S., & Joseph, M. (2017). Hyponatremia in traumatic brain injury: A practical management protocol. *World Neurosurgery*, *108*, 529–533. https://doi.org/10.1016/j.wneu.2017.09.013.

Rajajee, V., Riggs, B., & Seder, D. B. (2017). Emergency neurological life support: Airway, ventilation, and sedation. *Neurocritical Care*, *27*(Suppl. 1), 4–28. https://doi.org/10.1007/s12028-017-0451-2.

Ravishankar, N., Nuoman, R., Amuluru, K., El-Ghanem, M., Thulasi, V., Dangayach, N. S., … Al-Mufti, F. (2020). Management strategies for intracranial pressure crises in subarachnoid hemorrhage. *Journal of Intensive Care Medicine*, *35*(3), 211–218. https://doi.org/10.1177/0885066618813073.

Renew, J. R., Ratzlaff, R., Hernandez-Torres, V., Brull, S. J., & Prielipp, R. C. (2020). Neuromuscular blockade management in the critically ill patient. *Journal of Intensive Care*, *8*, 37. https://doi.org/10.1186/s40560-020-00455-2.

Rossaint, R., Bouillon, B., Cerny, V., Coats, T. J., Duranteau, J., Fernandez-Mondejar, E., … Spahn, D. R. (2016). The European guideline on management of major bleeding and coagulopathy following trauma: Fourth edition. *Critical Care*, *20*, 100. https://doi.org/10.1186/s13054-016-1265-x.

Schizodimos, T., Soulountsi, V., Iasonidou, C., & Kapravelos, N. (2020). An overview of management of intracranial hypertension in the intensive care unit. *Journal of Anesthesia*, *34*(5), 741–757. https://doi.org/10.1007/s00540-020-02795-7.

Smith, L. G. F., Milliron, E., Ho, M. L., Hu, H. H., Rusin, J., Leonard, J., & Sribnick, E. A. (2019). Advanced neuroimaging in traumatic brain injury: An overview. *Neurosurgical Focus*, *47*(6). https://doi.org/10.3171/2019.9.Focus19652. Artn E17.

Stein, D. M., Feather, C. B., & Napolitano, L. M. (2017). Traumatic brain injury advances. *Critical Care Clinics*, *33*(1), 1–13. https://doi.org/10.1016/j.ccc.2016.08.008.

Vella, M. A., Crandall, M. L., & Patel, M. B. (2017). Acute management of traumatic brain injury. *Surgical Clinics of North America*, *97*(5), 1015. https://doi.org/10.1016/j.suc.2017.06.003.

Webster, N. J., Moore, N., & Stewart, F. (2017). *Reducing unnecessary head computed tomography in mild traumatic brain injury (publication no. 10.1097/Tme.0000000000000169)* (pp. 1931–4485). Retrieved Oct-Dec <Go to ISI>://WOS:000415078600010.

Wettervik, T. S., Howells, T., Hillered, L., Nilsson, P., Engquist, H., Lewen, A., … Rostami, E. (2020). Mild hyperventilation in traumatic brain injury-relation to cerebral energy metabolism, pressure autoregulation, and clinical outcome. *World Neurosurgery*, *133*, E567–E575. https://doi.org/10.1016/j.wneu.2019.09.099.

Whitaker-Lea, W. A. (2017). Acute management of moderate-severe traumatic brain injury. *Physical Medicine and Rehabilitation Clinics of North America*, *28*(2), 227. https://doi.org/10.1016/j.pmr.2016.12.002.

Wiegele, M., Schochl, H., Haushofer, A., Ortler, M., Leitgeb, J., Kwasny, O., … Schaden, E. (2019). Diagnostic and therapeutic approach in adult patients with traumatic brain injury receiving oral anticoagulant therapy: An Austrian interdisciplinary consensus statement. *Critical Care*, *23*(1), 62. https://doi.org/10.1186/s13054-019-2352-6.

Yokobori, S., Yatabe, T., Kondo, Y., Kinoshita, K., Japan Resuscitation Council Neuroresuscitation Task Force, & the Guidelines Editorial Committee. (2020). Efficacy and safety of tranexamic acid administration in traumatic brain injury patients: A systematic review and meta-analysis. *Journal of Intensive Care*, *8*, 46. https://doi.org/10.1186/s40560-020-00460-5.

Zusman, B. E., Kochanek, P. M., & Jha, R. M. (2020). Cerebral edema in traumatic brain injury: A historical framework for current therapy. *Current Treatment Options in Neurology*, *22*(3). https://doi.org/10.1007/s11940-020-0614-x. ARTN 9.

Chapter 6

Neuropsychiatric disorders after severe traumatic brain injury: An overview

Dolores Villalobos[a,b,c], Umberto Bivona[d,f], and Maria Paola Ciurli[e]

[a]*Laboratory of Cognitive and Computational Neuroscience, Center of Biomedical Technology, Technical University of Madrid, Complutense University of Madrid, Madrid, Spain,* [b]*Department of Experimental Psychology, School of Psychology. Complutense University, Madrid, Spain,* [c]*The European Centre of Neuroscience, Madrid, Spain,* [d]*Post-Coma Unit, IRCCS Fondazione Santa Lucia, Rome, Italy,* [e]*Neuropsychology Unit, IRCCS Fondazione Santa Lucia, Rome, Italy,* [f]*IRCCS Fondazione Santa Lucia, Rome, Italy*

List of abbreviations

ADL	activities of daily living
AES	The Apathy Evaluation Scale
BIRT-PQs	five Brain Injury Rehabilitation Trust Personality Questionnaires
CBT	cognitive behavioral therapy
DSM-V	Diagnostic and Statistical Manual of Mental Disorders 5
FDA	Food and Drug Administration
GCS	Glasgow Coma Scale
ISA	impaired self-awareness
LOC	loss of consciousness
MDD	major depressive disorder
NPI	neuropsychiatric inventory
PTSD	post-traumatic stress disorder
OAS	The Overt Aggression Scale
OAS-MNR-E	The Overt Aggression Scale-Modified for Neurorehabilitation-Extended
OCD	obsessive-compulsive disorder
PTA	post-traumatic amnesia
PTCS	post-traumatic confusional state
QOL	quality of life
sTBI	severe traumatic brain injury
TMS	transcranial magnetic stimulation

Introduction

Severe traumatic brain injury (sTBI) may cause important neuropsychological and disabling neuropsychiatric symptoms. Indeed, they can include cognitive, mood, motivational, and behavioral symptoms, like changes in personality, which interfere with successful participation in social and vocational roles (Arciniegas & Wortzel, 2014; Ciurli, Formisano, Bivona, Cantagallo, & Angelelli, 2011; Sandhaug, Andelic, Berntsen, Seiler, & Mygland, 2012), as well as with caregivers well-being, for instance in terms of low quality of life (QoL) (D'Ippolito et al., 2018; Formisano et al., 2017) and sexuality changes (Bivona et al., 2016).

Neurobehavioral changes can be differently associated with sTBI, and temporal association does not guarantee causality. Of note, while sTBI may cause higher rates of neuropsychiatric disorders, also neuropsychiatric illness or symptoms are a risk factor for sTBI (Rao, Koliatsos, Ahmed, Lyketsos, & Kortte, 2015; Vassallo, Proctor-Weber, Lebowitz, Curtiss, & Vanderploeg, 2007). The present chapter will focus on the neuropsychiatric sequelae of a sTBI.

Diagnosis and Treatment of Traumatic Brain Injury. https://doi.org/10.1016/B978-0-12-823347-4.00037-3

Behavioral changes could be related to the meaning of the injury, as a reaction to a loss of self-esteem due to physical injury, and loss of mobility, unemployment (McAlister, 2008). Also environmental changes, such as in living situation and caregiving, can have an enormous impact on the behavior and adaptation of the survivor.

The incidence rates of neuropsychiatric disturbances after TBI are really complex and, as such, they have been heterogeneously described in the literature. It depends on the level of TBI severity, the kind of symptoms considered, and how they are assessed. However, the relationship between neuropsychiatric symptoms and sTBI is unequivocal (Stéfan et al., 2016). Behavioral disorders have been classified into three main groups, with different incidence rates: *disruptive primary behaviors by excess* (i.e., agitation 11%–70%, aggression 25%–39%, irritability 29%–71%, substance abuse 2%–26%); *disruptive primary behaviors by default* (apathy 20%–71%); *affective disorders* (depression 12%–76%, anxiety 0.8%–24.5%, post-traumatic stress 11%–18%, obsessive-compulsive disorders 1.2%–30%, psychosis 0.7%); and *suicide attempts* and *suicide* (1%). In particular, in a previous work conducted on a well-selected sample of patients with sTBI (Ciurli et al., 2011) through the Neuropsychiatric Inventory (NPI—Cummings et al., 1994), it has been found a wide range of neuropsychiatric symptoms, such as apathy (42%), irritability (37%), depressed mood (29%), disinhibition (28%), eating disturbances (27%), and agitation (24%).

Many variables have been proposed to account for the differential expression of neuropsychiatric symptoms after sTBI. Probably, the magnitude of risk factors, or complex interplay pattern between them, may better account for this variance. Warriner and Velikonja (2006) proposed a model positing four factors: (a) *anatomo-pathology* of the sTBI; (b) *environmental* variables (support system and interpersonal relationships); (c) *pre-morbid* person variables (personality, pre-existing psychological disturbances, cognitive and educational–occupational status); and (d) *post-injury* person variables (age, physical, cognitive sequelae, emotional response, self-awareness, coping styles) (Fig. 1).

The role of brain injury severity and structural pathophysiologic changes

TBI severity is undoubtedly a main factor. It can range from mild to severe, based on the length of loss of consciousness (LOC) and length of post-traumatic amnesia (PTA), also known as "Post-Traumatic Confusional State" (PTCS; Stuss et al., 1999) or "Post-Traumatic Delirium" (PTD; Ponsford et al., 2014). However, the Glasgow Coma Scale (GCS) (Jennett & Bond, 1975) is probably the most used measure to assess the severity of TBI, associated to LOC and PTA. The table below shows the different criteria according to O'Neil et al. (2013) and Baxendale, Heaney, Rugg-Gunn, and Friedland (2019), to define the severity of TBI (Table 1).

However, regardless of TBI severity, neuropsychiatric disturbances are often persistent (Whelan-Goodinson, Ponsford, Schönberger, & Johnston, 2010): among subjects with moderate to severe TBI, 49% had evidence of a neuropsychiatric illness in the first year following TBI, compared to 34% in mild TBIs and 18% in non-TBIs (Fann et al., 2004). In particular, apathy and disinhibition had been associated with the level of disability. Indeed, persons with TBI and a severe disability

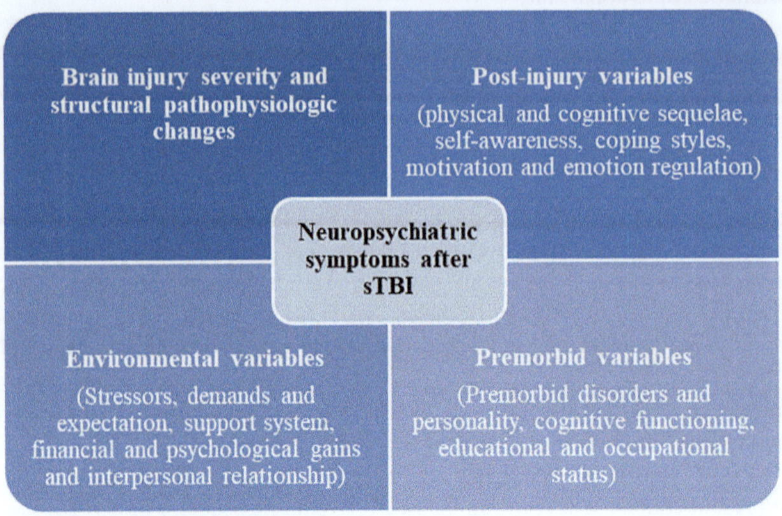

FIG. 1 Different factors of neuropsychiatric symptoms after sTBI. The figure includes the possible variables accounting for the different features of neuropsychiatric symptoms after sTBI, namely the severity of TBI and the structural pathophysiologic changes, as well as some post-injury, environmental and pre-morbid variables.

TABLE 1 Classification systems for determining the severity of head injury according to O'Neil et al. (2013) and Baxendale et al. (2019).

	Glasgow Coma Scale score	Loss of consciousness length	Post-traumatic amnesia length	Structural imaging
Severe	3–8	>24h	7days	Normal or abnormal
Moderate	9–12	30min to 24h	1–7days	Normal or abnormal
Mild	13–15	<30min	<24h	Normal or abnormal

The table shows the different criteria applied to define the severity of TBI, from the GCS score to the neuroimaging data.

had a four times greater risk of developing apathy, and a three times higher risk of developing disinhibition, than patients with lower disability (Ciurli et al., 2011).

Regarding anatomo-pathological variables, a simplified biological explanation for the increased vulnerability of sTBI patients to develop neuropsychiatric disturbances is that brain regions involved in emotional regulation, behavioral control, and higher-order cognitive operations (all aspects related to psychological status) are the brain regions which are most frequently affected by sTBI (Rao et al., 2015). The inferior frontal and temporo-polar regions are usually involved in coup-countercoup injuries from violent impacts to the head. In addition, long tracts are especially vulnerable to impulses associated with rotational acceleration as occurs in vehicle crashes and high-impact falls. All these regions and pathways are essential for higher central nervous system functions (Rao et al., 2015). Moreover, following sTBI, there are acute changes in the neurotransmitters, which lead to neuropsychiatric manifestations by altering the levels of acetylcholine, norepinephrine, dopamine, and serotonin.

The role of post-injury variables

Many variables have been related to the development of neuropsychiatric symptoms after sTBI, such as emotional response to injury, physical and cognitive sequelae, degree of decline in functioning, reaction to changes and ability to adapt and coping styles (Warriner & Velikonja, 2006). Younger patients are at higher risk of psychiatric morbidity than aged over 65, although neurobehavioral symptoms are prevalent in both groups (Deb & Burns, 2007).

Despite neuromotor sequelae of sTBI, showed a pattern of improvement over time (with the greatest improvement occurring between the inpatient and 6-month follow-up assessments), more than one-third of subjects usually shows a neuromotor abnormality 2 years following acute rehabilitation (Walker & Pickett, 2007). Its influence deserves a special consideration since patients with severe neuromotor disabilities lose their autonomy, thus having severe difficulties in readjustment, sadness or anger, loss of self-esteem, which often cause conflicts with caregivers, or substance abuse.

Also cognitive sequelae are common after sTBI, and usually affect attention, memory and executive functions processes (Ahmed et al., 2017). Moreover, due to frontal damage patients with TBI usually present dysexecutive disorders, and deficits related to social cognition, that affect patient's emotions and behaviors (Bird, Castelli, Malik, Frith, & Husain, 2004).

Finally, self-awareness is often impaired after sABI (Prigatano, 2010). According to Prigatano and Klonoff (1998), impaired self-awareness (ISA), unlike denial, can be associated to specific neuropsychiatric patterns. In particular, in a recent work of Bivona et al. (2019) a clear relationship has been found between ISA and both apathy and greater difficulty describing their emotional state, with respect to those with good self-awareness who showed instead higher levels of depression.

Additionally, coping styles, motivation and emotion regulation can affect or even cause the development of neuropsychiatric symptoms and significantly predict outcome at 1 year in participation and social adaptation (Cattran, Oddy, Wood, & Moir, 2011). Accordingly, they could help clinicians to plan targeted rehabilitation programs.

The role of pre-morbid variables

Personality, psychiatric disorder or psychological distress, coping styles, cognitive functioning, educational, and occupational status are the most important pre-morbid variables related with neuropsychiatric disturbance after sTBI (Warriner & Velikonja, 2006).

Pre-morbid disorders are predictive of the development of post-injury mood disorder such as anxiety (Zgaljardic et al., 2015). However, depression and anxiety are often novel, suggesting that significant factors, other than pre-TBI psychiatric status, contribute to post-TBI psychiatric outcome. Female gender, lower education, and pain were also associated with post-injury depression, while unemployment and older age were with anxiety (Whelan-Goodinson et al., 2010).

Personality has been typically studied within the context of resilience and coping with general stressful events, including the "The Big 5" dimensions of personality (neuroticism, extraversion, openness to experience, conscientiousness, and agreeableness). Higher levels of extraversion, openness to experience, conscientiousness, and agreeableness, have been positively associated with mental resilience, whereas high neurotic personality traits show the opposite effect (Sela-Kaufman, Rassovsky, Agranov, Levi, & Vakil, 2013). These personality traits could also influence the possible development of neuropsychiatric symptoms after sTBI and patient's coping strategies. Also pre-morbid personality characteristics provided a robust moderator of injury severity on occupational outcome, being a relevant reserve construct in predicting outcome in this clinical population (Sela-Kaufman et al., 2013).

The role of environmental variables

Stressors, demands and expectation, support system, financial and psychological gains, and quality of interpersonal relationship can affect the development of neuropsychiatric disturbance after sTBI (Warriner & Velikonja, 2006).

Particular attention should be given to the informal caregivers, since they become responsible of care, in addition to the routine of family life. Many caregivers show distress, depression, anxiety, psychosomatic disorders, increased consumption of drugs, financial difficulties, role changes, and poor social adjustment with increasing social isolation, thus requiring medical and mental health care services (D'Ippolito et al., 2018; Lefebvre & Levert, 2012).

A recent review of chronic family stress concluded that behavioral and emotional sequelae post-TBI impose greater stress on family members than injury severity or physical problems. The authors highlighted that depression, anxiety, behavioral and personality problems after TBI are directly associated with greater depression, anxiety, and burden among family members (Zgaljardic et al., 2015).

Methodological issues related to the neuropsychiatric assessment

The diagnosis of neuropsychiatric disorders after sTBI is often challenging. Indeed, some problems concern neuropsychiatric assessment in TBIs.

The diagnostic classification is usually based on the *Diagnostic and Statistical Manual of Mental Disorders* (American Psychiatric Association DSM-5 Task Force, 2013) criteria, and generically separates "primary" syndromes from "secondary" disorders (i.e., associated with general medical conditions); moreover, it does not include phenomenology and specific neuropsychiatric disorders that are very common in populations with TBI (i.e., apathy, irritability or agitation, and disinhibition).

There is limited applicability of traditional diagnostic tools in a neurological population. In fact, different assessment methods have been proposed: self-report questionnaires (e.g., the Beck Depression Inventory; Beck, Steer, & Brown, 1996) and clinical interviews (e.g., Hamilton Depression Rating Scale; Hamilton, 1960), but they can lead to different neuropsychiatric diagnoses, as they can be inappropriate for neurological populations. Indeed, cognitive–behavioral impairments after sTBI (such as impulsivity, reduced insight and impaired concentration, language, and memory deficits) and ISA can hamper the patients' reliability. Moreover, some symptoms (such as irritability, anxiety and sleep difficulties, or impaired concentration and memory) may lead to misdiagnosis since some neurological and cognitive symptoms (e.g., hypoarousal, loss of energy, and neurovegetative disorders) often mimic or alter the clinical presentation of some psychiatric disturbances, which are generally interpreted as somatizations of depression and anxiety (Sumpter & McMillan, 2005).

However, new clinical tools have been developed, and are generally based on information gathered from caregivers, due to the patients' cognitive deficit and ISA. The NPI, an informant-based interview, provides a comprehensive assessment of psychopathology in neurological populations. The NPI was used successfully to describe the neuropsychiatric symptoms after Alzheimer dementia, stroke, and sTBI (see Ciurli et al., 2011, for details). The five Brain Injury Rehabilitation Trust personality questionnaires (BIRT-PQs) is a specific battery of five questionnaires administrable singly or in combination, to assess motivation, impulsivity, disinhibition, social cognition, and emotional regulation (Cattran, Oddy, & Wood, 2011). The Overt Aggression Scale (OAS) (Yudofsky, Silver, Jackson, Endicott, & Williams, 1986) and the Overt Aggression Scale-Modified for Neurorehabilitation-Extended (OAS-MNR-E) (Giles & Mohr, 2007) allow measuring aggressive behavior after sTBI. The Apathy Evaluation Scale (AES) (Marin, Biedrzycki, & Firinciogullari, 1991) quantifies and characterizes apathy in patients with sTBI (Lee, Gleason, & Umucu, 2020).

To better assess neuropsychiatric symptoms we recommend the combined use of clinical observation and some of the above scales. Even if the whole spectrum of personalities characteristics cannot be fully depicted by other persons,

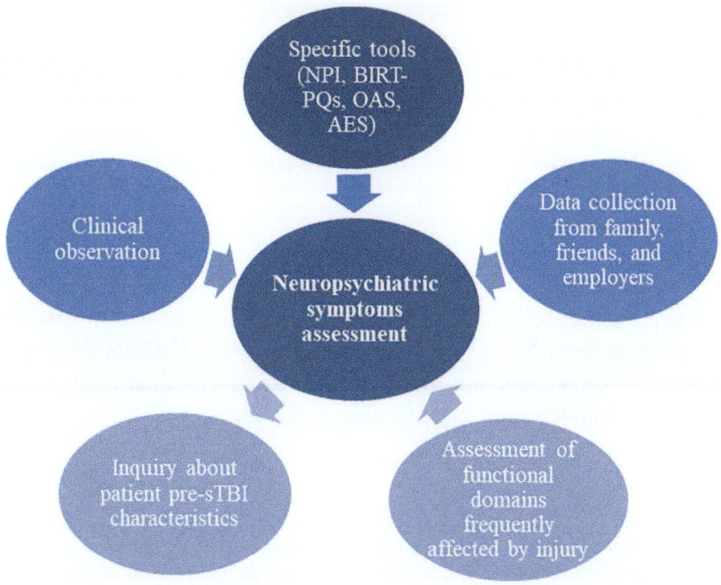

FIG. 2 Comprehensive assessment of neuropsychiatric symptoms. The figure shows the possible determining aspects leading to a comprehensive and effective assessment of neuropsychiatric symptoms after sTBI, from the clinical observation of the patients, to the knowledge of their pre-morbid personality through a careful anamnesis obtained by their relatives.

it's mandatory to collect data also from family members, friends, and employers, that can help to clarify the history and current clinical patients' condition. Moreover, the assessment of the effects of a brain injury should start with a thorough assessment of the patient pre-sTBI characteristics (Sela-Kaufman et al., 2013) as early as possible, since the tendency to over-estimate the role of sTBI in determining neuropsychiatric sequelae can increase over time (McAlister, 2008). It is also important to carefully review the functional domains that are frequently affected by injury, including cognition, personality, mood regulation, speech/language, mobility, and higher order domains such as vocational performance, major role performance within the family or equivalent context (McAlister, 2008) (Fig. 2).

Classification of the main neuropsychiatric disorders

Beyond the cognitive disorders after sTBI, DSM-V allows diagnosing a wide variety of neuropsychiatric disorders due to another medical condition (i.e., the TBI itself), including depressive, bipolar, anxiety, and psychotic disorder, and personality changes. However, the DSM-V leaves uncertain the appropriate use of these diagnoses in clinical practice (Wortzel & Arciniegas, 2014).

Indeed, many neuropsychiatric symptoms after sTBI are similar to those seen in patients with no history of TBI. The differential diagnosis between primary neuropsychiatric conditions and secondary symptoms could be a challenge. For example, it is complicated for a clinician to interpret psychomotor retardation in a patient with TBI as depressive episodes, since it could be outcome result of the TBI itself. On the other hand, patients with a history of TBI may present with positive signs of psychosis, but rarely show negative symptoms that are hallmarks of schizophrenia. Thus, the neuropsychiatric diagnosis in sTBI patients strictly related to DSM criteria is a complex matter (Llauterbach, Notarangelo, Nichols, Llane, & Koliatsos, 2015).

Stéfan et al. (2016) proposed four subgroup of symptoms after sTBI: (a) disruptive primary behaviors *by excess*; (b) disruptive primary behaviors *by default*; (c) *affective* disorders, *anxiety and psychosis*; and (d) *suicide attempts* and *suicide*. In fact, a series of specific neuropsychiatric symptoms can be clearly detectable, as follows:

Agitation

It is one of the main symptoms after coma recovery (being its incidence around 50%) (Nott, Chapparo, & Baguley, 2006). It usually lasts a short time and it is related to LOC, PTA, decreasing as cognitive functions improve (Stéfan et al., 2016). Agitation is usually accompanied by other symptoms, such as aggression, disinhibition, impulsivity, disorganized thinking, and reduced adaptation, limiting rehabilitation gains at discharge. Long lasting agitation is associated with persisting limitations to functional independence (Nott et al., 2006).

Aggression

Present between 25% and 39% of cases (Stéfan et al., 2016), it can appear as verbal or physical aggression against objects, self or other persons, as well as severe irritability, violence, hostility, or assaultive behavior and "episodic dyscontrol" (Fleminger, Greenwood, & Oliver, 2006). It is related to the initial severity of TBI and is more frequent in older males, also in association with language disorders. Depression and anxiety are common in the aggressive patients with sTBI, which also show executive dysfunctioning (Stéfan et al., 2016).

Irritability

It reflects both internal subjective feelings (e.g., annoyance and anger) and commensurate external behavioral responses (Alderman, 2003). Its incidence ranges from 29% to 71% (Stéfan et al., 2016). Risk factors are gender (being male at higher risk), age (younger), unemployment, social isolation or depression. Contrarily than for mild TBI, there is no correlation between cognitive impairment and irritability after sTBI (Stéfan et al., 2016).

Disinhibition

Disinhibition can be considered a synonym of impulsivity (Lezak, 2012) As an executive dysfunction after sTBI it is related with no appropriate behaviors, such as speaking without thinking at the consequences of their words, laughing inappropriately, or engaging in dangerous or costly activities. It is a common after sTBI, and can have many repercussions on the patient's and on their relative's QoL, and on the patient's social and professional outcomes (Kocka & Gagnon, 2014).

Substance abuse disorders

Risky behaviors can be a direct result of substance abuse disorders, and also may lead to subsequent injury or continued use of substances (Ahmed et al., 2017). These patients show an increased preference for small and immediate rewards, than larger but delayed rewards. Increased rates of drug or alcohol use in survivors compared to community controls may be a coping response to the psychosocial stressors of disability or pain, rather than the consequence of proximal neurobiological sequelae of ABI itself (Bjork & Grant, 2009).

Depression and apathy

Depression is typically characterized by episodes of sadness, loss of pleasure, guilt, low self-esteem, hopelessness, and suicidal thoughts. However, current evidence suggests that depressed mood in sTBI is characterized by irritability, anger, and aggression, rather than sadness and tearfulness (Reeves & Panguluri, 2009).

Regarding risk factors, a significant association between major depressive disorder (MDD) and female gender, pre-injury depression, and social and personal circumstances (i.e., unemployment) has been found (Cnossen et al., 2017). A moderately strong association has been also found between patient self-awareness and depressive symptoms, which is consistent with the cognitive–behavioral model of depression (Bivona et al., 2019; Malec, Brown, Moessner, Stump, & Monahan, 2010).

Apathy is particularly common after sTBI, with or without depression. It consists of impaired initiative, diminished activity, lack of concern, and a decrease in the cognitive, behavioral, and emotional components of goal-directed activity. Patients may appear to lack emotional responsivity, motivation, and interest in their environments. Some patients may appear to be experiencing fatigue (Reeves & Panguluri, 2009).

An important issue relates to the possible difference between depression, which can include apathy among its symptoms according to the DSM-V, and apathy itself as a neurological result. The NPI addresses this possible misdiagnosis issue, thanks to an accurate screening allowing discriminating between apathy itself and depression.

Anxiety and post-traumatic stress disorder (PTSD)

The traumatic nature of many circumstances that cause brain injuries confers a significantly greater risk of development anxiety disorders, and especially a PTSD.

The presence of a pre-morbid anxiety is statistically predictive of post-injury anxiety. However, since anxiety is more than doubled post-ABI, it is expected even without pre-morbid problems (Zgaljardic et al., 2015). With a high variability, incidence rates of PTSD post-TBI have been reported around 20%. However, the co-existence of ABI and PTSD has often been considered as improbable because of post-traumatic amnesia (PTA), since the PTA itself is considered as incompatible with one of the relevant risk factors of PTSD, i.e., the memory of the traumatic event. Some researchers have argued that individuals who had unconscious LOC and/or PTA (i.e., those with sTBI) are could be unable to develop PTSD because they would be unable to consciously experience the symptoms of fear, helplessness, and horror associated with the development of PTSD (Sbordone & Liter, 1995). However, in a recent systematic review by Stéfan et al. (2016), it has been stated that if sABI (and subsequent PTA) and PTSD can co-occur due to the presence of "islands of memory" during PTA, unconscious cognitive traces of the trauma, reconstructed memories (pseudo-memory), and/or traumatic experiences related to the event and experienced at a certain distance.

Psychosis, bipolar disorder, obsessive-compulsive symptoms

Psychotic disorders are relatively uncommon in sTBI population as compared with others neuropsychiatric syndromes. Risk factors include gender (i.e., being males), pre-injury neuropsychiatric problems, family history of schizophrenia, greater severity of injury, brain damage location (i.e., involvement of temporal and frontal lobes, or left hemispheric laterality), EEG abnormalities, post-traumatic epilepsy and cognitive impairment (Bhalerao et al., 2013).

Transient psychotic symptoms can certainly appear early in recovery from coma during the post-traumatic delirium. However, it consists of confusion, gross inattention, within a context of extreme cerebral disorganization and the psychotic fragments are incorporated. More clinically problematic is the more persistent, late-onset psychosis with delusions, hallucinations, and other psychotic symptoms that may occur in a delayed fashion following sTBI (Lux, 2007).

In addition, some patients suffer from bipolar disorder, manic episodes or obsessive-compulsive disorder (OCD) after sTBI, but this occurrence is lower as compared to depression, and it is related to dysexecutive syndrome.

Sleep disturbances

They are important and common symptoms, alone or within other neuropsychiatric disorders, disrupting recovery from TBI. Assessment of sleep patterns is essential because there are many different types of sleep disturbances, which requires specialized treatment. Patients usually complain about disturbed sleep patterns, hypersomnia, and difficulty maintaining sleep. Sleep disturbances can be secondary to brain injury or caused by secondary factors related or unrelated to sTBI (e.g., medical comorbidities, pain, medication side effects, etc.) (Rao et al., 2015) (Fig 3).

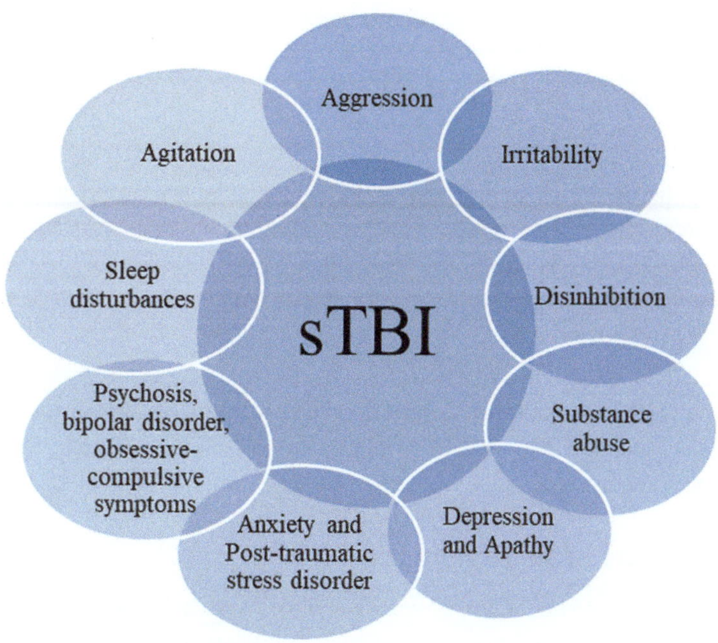

FIG. 3 The most common neuropsychiatric symptoms after sTBI. The figure shows a wide range of possible neuropsychiatric symptoms after sTBI, which are clearly detectable even if their prevalence can be considerably variable.

A brief consideration of possible interventions

Given the relevance of neuropsychiatric symptoms after sTBI, important efforts should be made to diagnose and treat them as best as possible. Indeed, a holistic treatment of sTBI requires the convergence of multiple disciplines, according to an inter-professional team work and within the bio-psycho-social perspective. Treatment is often a complex undertaking, and therapeutic interventions include both pharmacologic and non-pharmacologic approaches to manage its complications associated with sTBI.

Interventions should address the understanding of the event and its neuropsychiatric consequences, describing the recovery course, making appropriate referrals to brain injury support groups, providing support and reassurance, and structuring the treatment approach to accommodate cognitive impairments and difficulties with treatment adherence that are inherent in any neuropsychiatric condition. A strong therapeutic alliance and the clinician expertise in appropriately structuring the therapeutic environment are crucial (Rao et al., 2015).

Patients with ISA deserve special consideration, since improving patients' self-awareness must be one of the preliminary steps in the rehabilitation process (Villalobos et al., 2020). It is also fundamental for the therapist building a therapeutic alliance with patients, by "entering their phenomenological field and reducing their frustration and confusion" (Prigatano & Sherer, 2020). However, even after the patients recover self-awareness, defense mechanisms such as denial can be another obstacle, and the therapist must help patient to make reasonable choices (Prigatano & Sherer, 2020).

In addition to pharmacological interventions (here not addressed for reason of brevity), a variety of other non-pharmacologic therapies have been shown to be beneficial. Interventions based on cognitive behavioral therapy (CBT) include education, cognitive restructuring, self-monitoring, self-talk training and supporting communication. Comprehensive reviews concluded that therapies involving CBT should be effective in patients with TBI with anxiety (Soo & Tate, 2007) and it is practice options for persons with impaired executive functioning and emotional self-regulation (Cattelani, Zettin, & Zoccolotti, 2010). Besides, McCarron et al. (2019) showed that a combined approach using psychotherapy and cognitive rehabilitation targeting multiple layers of social competence seems to be promising.

Other techniques, such a transcranial magnetic stimulation (TMS), enhancing cortical excitability seems reducing depressive symptoms and improving motor, cognitive and behavioral functions after TBI (Demirtas-Tatlidede, Vahabzadeh-Hagh, Bernabeu, Tormos, & Pascual-Leone, 2012).

Of note, intervention programs for caregivers, during patient rehabilitation and also after home re-entry, deserve attention. Families may need psycho-educational support (Bivona et al., 2020) that should be focused on caregiver's emotional state, and on helping them to cope with the problems they will face both within family and the broader social context (Zgaljardic et al., 2015).

Conclusion

Fig. 4 summarizes the main issues that sTBI implies with respect to the diagnosis and treatment of neuropsychiatric disorders.

First, a more accurate and early possible diagnosis is the preliminary step. Indeed, the choice of specific and sensitive assessment tools, based on the levels self-awareness and, more generally, of the cognitive features of survivors', as well as the involvement of their caregivers, are fundamental steps to address, accordingly, a specific and more likely successful treatment. Of note, neuropsychiatric disorders can, in turn, affect the possibility to improve self-awareness (for instance, when the survivor's anger hamper his/her compliance with the neuro-rehabilitation therapist—see the Chapter of Bivona et al. in the present volume, for more details), as well as the caregivers psychological well-being, thus increasing their burden.

Second, based on an accurate diagnosis, the neuropsychiatric disorders can be treated both at a pharmacological and psychotherapeutic levels. For reasons of brevity we could not address here the former, whereas we showed some possible non-pharmacological techniques, which within a holistic and inter-professional perspective represent one of the main challenges in the field of severe acquired brain damage.

Applications to other areas of neuroscience

In this chapter, we have addressed the main neuropsychiatric symptoms after a sTBI, as well as some important issues related to their assessment and a brief consideration of possible non-pharmacological interventions, based both on the therapeutic alliance and on the important role of caregivers.

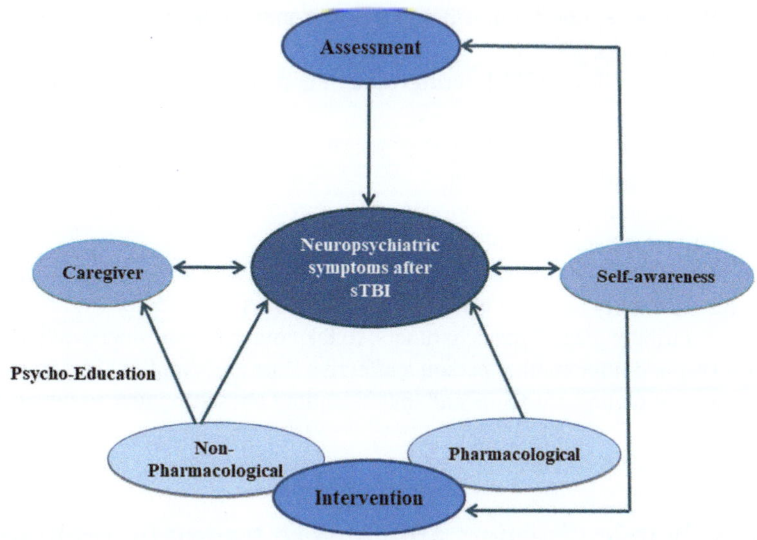

FIG. 4 Assessment and treatment of neuropsychiatric symptoms after sTBI within a holistic perspective. The figure evidences the possible determining variables related to assessment and treatment of neuropsychiatric symptoms after sTBI. As shown, to approach at best the neuropsychiatric sequelae of sTBI, we recommended: (A) to diagnose them as early as possible, through specific and sensitive assessment tools, based on the levels self-awareness and, more generally, of the cognitive features of survivors. We also recommend involving their caregivers; (B) to treat them both at a pharmacological and neuropsychological and psychotherapeutic level.

As descripted above, severe TBI can lead to specific features, which require targeted diagnostic and therapeutic paths (Stéfan et al., 2016). For instance, it has been underlined the importance to avoid misdiagnoses. Indeed, a careful anamnesis is needed, involving also caregivers who, knowing well the survivors, may help clinicians to have a clear picture of premorbid personality, as well as of the current features beyond the clinical setting.

Of note, despite the specificity of neuropathology and phenomenology of sTBI, the considerations made in relation to the neuropsychiatric assessment after sTBI can be applicable also to survivors of severe acquired brain damages of other etiologies. Indeed, the above diagnostic consideration can be considered transversal, being some of the symptoms above mentioned (for instance, fatigue, apathy, or sleep or appetite disturbances) also typical of sABI regardless of its etiology (Kohnen, Lavrijsen, Smals, Gerritsen, & Koopmans, 2019). For instance, since ISA (i.e., anosognosia) is a typical result not only of sTBI but more in general of sABI (Winkens, Van Heugten, Visser-Meily, & Boosman, 2014), the usefulness of involving caregivers in the neuropsychiatric assessment, based on the poor reliability of survivors themselves, can be applicable more in general in the wide range of sABI.

Accordingly, also the treatment of neuropsychiatric disturbances after sABI, as for sTBIs, should be structured to accommodate cognitive impairments and difficulties with treatment adherence that is inherent in any neuropsychiatric condition. Thus, the preliminary and transversal step for clinician treating all severe brain damaged survivors, is aiming to create the crucial therapeutic alliance (Rao et al., 2015). Indeed, especially in the presence of ISA, improving patients' self-awareness should be one of the preliminary steps in the rehabilitation process, due to its mediator role in the survivors' functional outcome (Villalobos et al., 2020).

Secondly, also the mentioned interventions can be utilizable also in case of sABI of non-traumatic etiologies.

Last, but not least, once again we want to underline the fundamental role of caregivers and, more in general, of the whole family system, that in the whole field of severe brain damage represent a fundamental resource in helping clinician to better address the neuropsychiatric assessment, as well as, if opportunely assisted and guided, in sensibly contributing to facilitate the treatment path of such invalidating and often long-lasting disturbances. This issue will be specifically addressed in another chapter of this Handbook.

Mini-dictionary of terms

Acquired brain injury: It is a brain injury occurred after birth, with different etiology (anoxia, bleeding, infection, hypoxia, ischemia, and tumor).

Traumatic brain injury: It is a brain injury caused by a blow to the hand (closed head injury) or by an object that pierces the skull and enters brain tissue (open head injury).

Loss of consciousness: Loss of the perception of our self and the environment; here the term corresponds to period of coma and severe disorder of consciousness (DOC).

Post-traumatic amnesia: The period following TBI during which the patient is not able to memorize everyday events of the last 24 h.

Self-awareness: The ability to perceive the own self, to be aware of one's own thoughts, feelings and mental states, recognizing problems caused by damaged brain function.

Dysexecutive disorders: Cognitive–motivational–emotional and behavior disturbances correlated to lesions of the frontal lobes and anterior brain networks, usually involved in TBI.

Informal caregiver: The primary person who, voluntarily and without receiving any payment, takes care of the patient who is not self-sufficient in his/her family or social network, after an illness or an injury.

Theory of mind: The ability to attribute mental states to others, to take other person's perspective.

Empathy: The ability to observer and infer another person's affective state and emotional responses.

Pre-morbid personality: Patterns of feeling, thinking, and understanding oneself relative to the environment, existing prior to injury.

Coping style: The way to response to stress and change.

Key facts of neuropsychiatric disorders after severe traumatic brain injury: An overview

- The incidence of neuropsychiatric sequelae after TBI is heterogeneously described in the literature, because it depends on the level of TBI severity, the kind of symptoms, and how they are assessed.
- A classification in three main groups, based on different features, has been proposed: disruptive primary behaviors by excess or by default and affective disorders.
- The most important risk factors are anatomo-pathology of the sTBI, environmental variables, pre-morbid person variables and post-injury person variables.
- The neuropsychiatric diagnosis after sTBI is often challenging: it should include phenomenology and specific neuropsychiatric disorders, pre-morbid personality feature, through clinical observation and adapted scales.
- The treatment of the neuropsychiatric sequelae is mandatory, and it should holistic and inter-professional, within the bio-psycho-social perspective, including both pharmacologic and non-pharmacologic approaches and involving all the family.

Summary points

- After sTBI many patients may present neuropsychiatric symptoms.
- They include cognitive, mood, motivational and behavioral symptoms, which have an impact in ADL.
- Brain injury severity is one of the most important factors. Neuropsychiatric symptoms seem to be presented in around 50% of moderate-to-severe patients, 35% in mild, and less than 20 in non-TBI patients.
- A comprehensive assessment, including specific tools for neuropsychiatric symptoms, is crucial for an adequate management.
- ISA has an important impact in neuropsychiatric symptoms manifestation and their approach.
- Agitation, aggression, irritability, disinhibition, substance abuse, depression and apathy, anxiety, post-traumatic stress disorder, psychosis, bipolar disorder, obsessive-compulsive symptoms and sleep disturbances, have been descripted as the most important symptoms.
- Non-pharmacological interventions, based on the therapeutic alliance and considering the important role of caregivers, have been shown to be beneficial.

References

Ahmed, S., Venigalla, H., Mekala, H. M., Dar, S., Hassan, M., & Ayub, S. (2017). Traumatic brain injury and neuropsychiatric complications. *Indian Journal of Psychological Medicine*, *39*(2), 114–121. https://doi.org/10.4103/0253-7176.203129.

Alderman, N. (2003). Contemporary approaches to the management of irritability and aggression following traumatic brain injury. *Neuropsychological Rehabilitation*, *13*(1–2), 211–240. https://doi.org/10.1080/09602010244000327.

American Psychiatric Association DSM-5 Task Force. (2013). *Diagnostic and statistical manual of mental disorders: DSM-5* (5th ed.). American Psychiatric Association.

Arciniegas, D. B., & Wortzel, H. S. (2014). Emotional and behavioral dyscontrol after traumatic brain injury. *Psychiatric Clinics of North America, 37*(1), 31–53. https://doi.org/10.1016/j.psc.2013.12.001.

Baxendale, S., Heaney, D., Rugg-Gunn, F., & Friedland, D. (2019). Neuropsychological outcomes following traumatic brain injury. *Practical Neurology, 19*(6), 476–482. https://doi.org/10.1136/practneurol-2018-002113.

Beck, A., Steer, R., & Brown, G. (1996). *Manual for beck depression inventory-II* (2nd ed.). Psychological Corporation.

Bhalerao, S. U., Geurtjens, C., Thomas, G. R., Kitamura, C. R., Zhou, C., & Marlborough, M. (2013). Understanding the neuropsychiatric consequences associated with significant traumatic brain injury. *Brain Injury, 27*(7–8), 767–774. https://doi.org/10.3109/02699052.2013.793396.

Bird, C. M., Castelli, F., Malik, O., Frith, U., & Husain, M. (2004). The impact of extensive medial frontal lobe damage on "theory of mind" and cognition. *Brain, 127*(4), 914–928. https://doi.org/10.1093/brain/awh108.

Bivona, U., Antonucci, G., Contrada, M., Rizza, F., Leoni, F., Zasler, N. D., et al. (2016). A biopsychosocial analysis of sexuality in adult males and their partners after severe traumatic brain injury. *Brain Injury, 30*(9), 1082–1095. https://doi.org/10.3109/02699052.2016.1165867.

Bivona, U., Costa, A., Contrada, M., Silvestro, D., Azicnuda, E., Aloisi, M., et al. (2019). Depression, apathy and impaired self-awareness following severe traumatic brain injury: A preliminary investigation. *Brain Injury, 33*(9), 1245–1256. https://doi.org/10.1080/02699052.2019.1641225.

Bivona, U., Villalobos, D., De Luca, M., Zilli, F., Ferri, G., Lucatello, S., et al. (2020). Psychological status and role of caregivers in the neuro-rehabilitation of patients with severe acquired brain injury (ABI). *Brain Injury, 34*(13–14), 1714–1722. https://doi.org/10.1080/02699052.2020.1812002.

Bjork, J. M., & Grant, S. J. (2009). Does traumatic brain injury increase risk for substance abuse? *Journal of Neurotrauma, 26*(7), 1077–1082. https://doi.org/10.1089/neu.2008.0849.

Cattelani, R., Zettin, M., & Zoccolotti, P. (2010). Rehabilitation treatments for adults with behavioral and psychosocial disorders following acquired brain injury: A systematic review. *Neuropsychology Review, 20*(1), 52–85. https://doi.org/10.1007/s11065-009-9125-y.

Cattran, C., Oddy, M., & Wood, R. (2011). The development of a measure of emotional regulation following acquired brain injury. *Journal of Clinical and Experimental Neuropsychology, 33*(6), 672–679. https://doi.org/10.1080/13803395.2010.550603.

Cattran, C., Oddy, M., Wood, R., & Moir, J. (2011). Post-injury personality in the prediction of outcome following severe acquired brain injury. *Brain Injury, 25*(11), 1035–1046. https://doi.org/10.3109/02699052.2011.607787.

Ciurli, P., Formisano, R., Bivona, U., Cantagallo, A., & Angelelli, P. (2011). Neuropsychiatric disorders in persons with severe traumatic brain injury: Prevalence, phenomenology, and relationship with demographic, clinical, and functional features. *The Journal of Head Trauma Rehabilitation, 26*(2), 116–126. https://doi.org/10.1097/HTR.0b013e3181dedd0e.

Cnossen, M. C., Scholten, A. C., Lingsma, H. F., Synnot, A., Haagsma, J., Steyerberg, E. W., et al. (2017). Predictors of major depression and posttraumatic stress disorder following traumatic brain injury: A systematic review and meta-analysis. *Journal of Neuropsychiatry and Clinical Neurosciences, 29*(3), 206–224. https://doi.org/10.1176/appi.neuropsych.16090165.

Cummings, J. L., Mega, M., Gray, K., Rosenberg-Thompson, S., Carusi, D. A., & Gornbein, J. (1994). The neuropsychiatric inventory: Comprehensive assessment of psychopathology in dementia. *Neurology, 44*, 2308–2314. http://www.ncbi.nlm.nih.gov/pubmed/12741091.

D'Ippolito, M., Aloisi, M., Azicnuda, E., Silvestro, D., Giustini, M., Verni, F., et al. (2018). Changes in caregivers lifestyle after severe acquired brain injury: A preliminary investigation. *BioMed Research International, 1*, 1–14. https://doi.org/10.1155/2018/2824081.

Deb, S., & Burns, J. (2007). Neuropsychiatric consequences of traumatic brain injury: A comparison between two age groups. *Brain Injury, 21*(3), 301–307. https://doi.org/10.1080/02699050701253137.

Demirtas-Tatlidede, A., Vahabzadeh-Hagh, A. M., Bernabeu, M., Tormos, J. M., & Pascual-Leone, A. (2012). Noninvasive brain stimulation in traumatic brain injury. *The Journal of Head Trauma Rehabilitation, 27*(4), 274–292. https://doi.org/10.1097/HTR.0b013e318217df55.

Fann, J. R., Burington, B., Leonetti, A., Jaffe, K., Katon, W. J., & Thompson, R. S. (2004). Psychiatric illness following traumatic brain injury in an adult health maintenance organization population. *Archives of General Psychiatry, 61*(1), 53–61. https://doi.org/10.1001/archpsyc.61.1.53.

Fleminger, S., Greenwood, R. R., & Oliver, D. L. (2006). Pharmacological management for agitation and aggression in people with acquired brain injury. *Cochrane Database of Systematic Reviews, 4*. https://doi.org/10.1002/14651858.cd003299.pub2.

Formisano, R., Longo, E., Azicnuda, E., Silvestro, D., D'Ippolito, M., Truelle, J.-L., et al. (2017). Quality of life in persons after traumatic brain injury as self-perceived and as perceived by the caregivers. *Journal of the Neurological Sciences, 38*(2), 279–286. https://doi.org/10.1007/s10072-016-2755-y.

Giles, G. M., & Mohr, J. D. (2007). Overview and inter-rater reliability of an incident-based rating scale for aggressive behaviour following traumatic brain injury: The overt aggression scale-modified for neurorehabiltation-extended (OAS-MNR-E). *Brain Injury, 21*(5), 505–511. https://doi.org/10.1080/02699050701311729.

Hamilton, M. (1960). A rating scale for depression. *Journal of Neurology Neurosurgery and Psychiatry, 23*, 56–63.

Jennett, B., & Bond, M. (1975). Assessment of outcome after severe brain damages. A practical scale. *The Lancet, 305*(7905), 480–484. https://doi.org/10.1016/S0140-6736(75)92830-5.

Kocka, A., & Gagnon, J. (2014). Definition of impulsivity and related terms following traumatic brain injury: A review of the different concepts and measures used to assess impulsivity, disinhibition and other related concepts. *Behavioral Sciences, 4*(4), 352–370. https://doi.org/10.3390/bs4040352.

Kohnen, R., Lavrijsen, J., Smals, O., Gerritsen, D., & Koopmans, R. (2019). Prevalence and characteristics of neuropsychiatric symptoms, quality of life and psychotropics in people with acquired brain injury in long-term care. *Journal of Advanced Nursing, 75*(12), 3715–3725. https://doi.org/10.1111/jan.14156.

Lee, B., Gleason, C., & Umucu, E. (2020). Clinical outcome measure: Clinical utility and psychometric properties of the apathy evaluation scale. *Rehabilitation Psychology, 65*(3), 311–312. https://doi.org/10.1037/rep0000356.

Lefebvre, H., & Levert, M. J. (2012). The close relatives of people who have had a traumatic brain injury and their special needs. *Brain Injury, 26*(9), 1084–1097. https://doi.org/10.3109/02699052.2012.666364.

Lezak, M. D. (2012). *Neuropsychological assessment*. Oxford University Press.

Llauterbach, M. D., Notarangelo, P. L., Nichols, S. J., Llane, K. S., & Koliatsos, V. E. (2015). Diagnostic and treatment challenges in traumatic brain injury patients with severe neuropsychiatric symptoms: Insights into psychiatric practice. *Neuropsychiatric Disease and Treatment*, *11*, 1601–1607. https://doi.org/10.2147/NDT.S80457.

Lux, W. E. (2007). A neuropsychiatric perspective on traumatic brain injury. *Journal of Rehabilitation Research and Development*, *44*(7), 951–962. https://doi.org/10.1682/JRRD.2007.01.0009.

Malec, J. F., Brown, A. W., Moessner, A. M., Stump, T. E., & Monahan, P. (2010). A preliminary model for posttraumatic brain injury depression. *Archives of Physical Medicine and Rehabilitation*, *91*(7), 1087–1097. https://doi.org/10.1016/j.apmr.2010.04.002.

Marin, R. S., Biedrzycki, R. C., & Firinciogullari, S. (1991). Reliability and validity of the apathy evaluation scale. *Psychiatry Research*, *38*(2), 143–162. https://doi.org/10.1016/0165-1781(91)90040-V.

McAlister, T. M. (2008). Neurobehavioral sequelae of traumatic brain injury: Evaluation and management. *World Psychiatry*, *7*(1), 3–10. https://doi.org/10.1002/msj.20097.

McCarron, K. K., Dasgupta, M. K., Campbell, C. A., Hull, A. E., Namazi, S., Adams, A. H., et al. (2019). Social rehabilitation for military veterans with traumatic brain injury, psychological trauma, and chronic neuropsychiatric symptoms: Intervention development and initial outcomes. *Psychiatric Rehabilitation Journal*, *42*(3), 296–304. https://doi.org/10.1037/prj0000361.

Nott, M. T., Chapparo, C., & Baguley, I. J. (2006). Agitation following traumatic brain injury: An Australian sample. *Brain Injury*, *20*(11), 1175–1182. https://doi.org/10.1080/02699050601049114.

O'Neil, M. E., Carlson, K., Storzbach, D., Brenner, L., Freeman, M., Quinones, A., et al. (2013). *Complications of mild traumatic brain injury in veterans and military personnel: A systematic review*. Department of Veterans Affairs (US).

Ponsford, J., Janzen, S., McIntyre, A., Bayley, M., Velikonja, D., & Tate, R. (2014). INCOG recommendations for management of cognition following traumatic brain injury, part I: Posttraumatic amnesia/delirium. *The Journal of Head Trauma Rehabilitation*, *29*(4), 307–320. https://doi.org/10.1097/HTR.0000000000000074.

Prigatano, G. P. (2010). *The study of Anosognosia*. Oxford University Press.

Prigatano, G. P., & Klonoff, P. S. (1998). A clinician's rating scale for evaluating impaired self-awareness and denial of disability after brain injury. *The Clinical Neuropsychologist*, *12*(1), 56–67. https://doi.org/10.1076/clin.12.1.56.1721.

Prigatano, G. P., & Sherer, M. (2020). Impaired self-awareness and denial during the postacute phases after moderate to severe traumatic brain injury. *Frontiers in Psychology*, *11*(July), 1–12. https://doi.org/10.3389/fpsyg.2020.01569.

Rao, V., Koliatsos, V., Ahmed, F., Lyketsos, C., & Kortte, K. (2015). Neuropsychiatric disturbances associated with traumatic brain injury: A practical approach to evaluation and management. *Seminars in Neurology*, *35*(1), 64–82. https://doi.org/10.1055/s-0035-1544241.

Reeves, R. R., & Panguluri, R. L. (2009). Neuropsychiatric complications of traumatic brain injury. *Journal of Psychosocial Nursing*, *49*(3), 42–50.

Sandhaug, M., Andelic, N., Berntsen, S. A., Seiler, S., & Mygland, A. (2012). Self and near relative ratings of functional level one year after traumatic brain injury. *Disability and Rehabilitation*, *34*(11), 904–909. https://doi.org/10.3109/09638288.2011.626484.

Sbordone, R. J., & Liter, J. C. (1995). Mild traumatic brain injury does not produce post-traumatic stress disorder. *Brain Injury*, *9*(4), 405–412. https://doi.org/10.3109/02699059509005780.

Sela-Kaufman, M., Rassovsky, Y., Agranov, E., Levi, Y., & Vakil, E. (2013). Premorbid personality characteristics and attachment style moderate the effect of injury severity on occupational outcome in traumatic brain injury: Another aspect of reserve. *Journal of Clinical and Experimental Neuropsychology*, *35*(6), 584–595. https://doi.org/10.1080/13803395.2013.799123.

Soo, C., & Tate, R. (2007). Psychological treatment for anxiety in people with traumatic brain injury. *Cochrane Database of Systematic Reviews*, *18*(3). https://doi.org/10.1002/14651858.CD005239.pub2, CD005239.

Stéfan, A., Mathé, J. F., Dhenain, M., Blanchard, P., Blondet, E., Mathé, J. F., et al. (2016). What are the disruptive symptoms of behavioral disorders after traumatic brain injury? A systematic review leading to recommendations for good practices. *Annals of Physical and Rehabilitation Medicine*, *59*(1), 5–17. https://doi.org/10.1016/j.rehab.2015.11.002.

Stuss, D. T., Binns, M. A., Carruth, F. G., Levine, B., Brandys, C. E., Moulton, R. J., et al. (1999). The acute period of recovery from traumatic brain injury: Posttraumatic amnesia or posttraumatic confusional state? *Journal of Neurosurgery*, *90*(4), 635–643. https://doi.org/10.3171/jns.1999.90.4.0635.

Sumpter, R. E., & McMillan, T. M. (2005). Misdiagnosis of post-traumatic stress disorder following severe traumatic brain injury. *British Journal of Psychiatry*, *186*(May), 423–426. https://doi.org/10.1192/bjp.186.5.423.

Vassallo, J. L., Proctor-Weber, Z., Lebowitz, B. K., Curtiss, G., & Vanderploeg, R. D. (2007). Psychiatric risk factors for traumatic brain injury. *Brain Injury*, *21*(6), 567–573. https://doi.org/10.1080/02699050701426832.

Villalobos, D., Caperos, J. M., Bilbao, Á., Bivona, U., Formisano, R., & Pacios, J. (2020). Self-awareness moderates the association between executive dysfunction and functional independence after acquired brain injury. *Archives of Clinical Neuropsychology*, 1–10. https://doi.org/10.1093/arclin/acaa048.

Walker, W. C., & Pickett, T. C. (2007). Motor impairment after severe traumatic brain injury: A longitudinal multicenter study. *Journal of Rehabilitation Research and Development*, *44*(7). https://doi.org/10.1682/JRRD.2006.12.0158.

Warriner, E. M., & Velikonja, D. (2006). Psychiatric disturbances after traumatic brain injury: Neurobehavioral and personality changes. *Current Psychiatry Reports*, *8*(1), 73–80. https://doi.org/10.1007/s11920-006-0083-2.

Whelan-Goodinson, R., Ponsford, J. L., Schönberger, M., & Johnston, L. (2010). Predictors of psychiatric disorders following traumatic brain injury. *The Journal of Head Trauma Rehabilitation*, *25*(5), 320–329. https://doi.org/10.1097/HTR.0b013e3181c8f8e7.

Winkens, I., Van Heugten, C. M., Visser-Meily, J. M. A., & Boosman, H. (2014). Impaired self-awareness after acquired brain injury: Clinicians' ratings on its assessment and importance for rehabilitation. *The Journal of Head Trauma Rehabilitation*, *29*(2), 153–156. https://doi.org/10.1097/HTR.0b013e31827d1500.

Wortzel, H. S., & Arciniegas, D. B. (2014). The DSM-5 approach to the evaluation of traumatic brain injury and its neuropsychiatric sequelae. *NeuroRehabilitation*, *34*(4), 613–623. https://doi.org/10.3233/NRE-141086.

Yudofsky, S. C., Silver, J. M., Jackson, W., Endicott, J., & Williams, D. (1986). The overt aggression scale for the objective rating of verbal and physical aggression. *American Journal of Psychiatry*, *143*(1), 35–39. https://doi.org/10.1176/ajp.143.1.35.

Zgaljardic, D. J., Seale, G. S., Schaefer, L. A., Temple, R. O., Foreman, J., & Elliott, T. R. (2015). Psychiatric disease and post-acute traumatic brain injury. *Journal of Neurotrauma*, *32*(23), 1911–1925. https://doi.org/10.1089/neu.2014.3569.

Section B

Clinical features of traumatic brain injury

Chapter 7

Cerebral hemorrhages in traumatic brain injury

Andrei Irimia[a,b], Kenneth A. Rostowsky[a], E. Meng Law[c,d], and Helena C. Chui[e]

[a]*Ethel Percy Andrus Gerontology Center, Leonard Davis School of Gerontology, University of Southern California, Los Angeles, CA, United States,* [b]*Corwin D. Denney Research Center, Department of Biomedical Engineering, University of Southern California, Los Angeles, CA, United States,* [c]*iBRAIN Research Laboratory, Departments of Neuroscience, Computer Systems and Electrical Engineering, Monash University, Melbourne, VIC, Australia,* [d]*Department of Radiology, The Alfred Health Hospital, Melbourne, VIC, Australia,* [e]*Department of Neurology, Keck School of Medicine, University of Southern California, Los Angeles, CA, United States*

Abbreviations

AD Alzheimer's disease
ADRD Alzheimer's disease and related dementias
Aβ amyloid β
ApoE apolipoprotein E
APP amyloid precursor protein
BBB blood–brain barrier
BOLD blood oxygenation level-dependent
CAA cerebral amyloid angiopathy
CBF cerebral blood flow
CBV cerebral blood volume
CC corpus callosum
CMB cerebral microbleed
CSF cerebrospinal fluid
DCE dynamic contrast-enhanced
DWI diffusion weighted imaging
FLAIR fluid-attenuated inversion recovery
fMRI functional magnetic resonance imaging
GM gray matter
HC healthy control
ICH intracerebral hemorrhage
ICP intracranial pressure
MCI mild cognitive impairment
MRI magnetic resonance imaging
mTBI mild traumatic brain injury
NFT neurofibrillary tangle
NVC neurovascular coupling
NVD neurovascular disease
NVU neurovascular unit
PD Parkinson's disease
PET positron emission tomography
PSEN presenilin
PVH periventricular hemorrhage
PVS perivascular space
RGB red-green-blue

Diagnosis and Treatment of Traumatic Brain Injury. https://doi.org/10.1016/B978-0-12-823347-4.00030-0

SVD	small vessel disease
SWI	susceptibility-weighted imaging
TAI	traumatic axonal injury
TBI	traumatic brain injury
VCID	vascular cognitive impairment and dementia
VSMC	vascular smooth muscle cell
WM	white matter

Introduction

Neurovascular disease (NVD), Alzheimer's disease (AD), and traumatic brain injury (TBI) have high prevalence in the elderly, where co-morbid NVD and AD may predispose to hemorrhagic brain injury. Because post-traumatic hemorrhages, AD and related dementias (ADRD) are interrelated, there is substantial clinical motivation to study the NVD/AD/TBI interaction. Examining traumatic cerebral hemorrhages and their sequelae is an adequate starting point for investigating this complex topic, and this chapter highlights facts and hypotheses on the vascular dimension of TBI relative to NVD and ADRD. Most clinical efforts to treat NVD focus on large brain vessels, whereas the relationship between NVD, AD, and TBI at the level of arterioles, capillaries, and venules remains understudied. Similarly, TBI-related dysregulation of somatic processes, e.g., via the brain-gut axis, remains underexplored (Irimia & Bradshaw, 2005; Irimia, Richards, & Bradshaw, 2006). Our understanding of small vessel disease (SVD) as a risk factor for hemorrhage after TBI has increased with the advent of susceptibility magnetic resonance imaging (MRI), which identify post-traumatic intracerebral hemorrhages (ICHs) with high sensitivity. This has led to renewed interest in TBI effects upon blood–brain barrier (BBB) dysfunction and neurovascular coupling (NVC) within the neurovascular unit (NVU), a scientific area which overlaps to a surprising degree with the goal of understanding the vascular and neurodegenerative dimensions of ADRD. Here, post-traumatic BBB and NVC dysfunction are discussed and the role of aging in this pathophysiology is highlighted. Strategies for TBI ICH identification are reviewed while considering the differential etiology of traumatic ICHs. (Epi)genetic factors increasing ICH risk are discussed, and their commonality with ADRD risk factors is highlighted.

The BBB and NVC after TBI

The BBB is a complex functional and structural barrier which regulates and preserves the brain's extracellular medium by controlling molecular/ionic gradients and glucose transport, as well as by protecting parenchymal cells (glia and neurons) from toxicity (Buxton, Uludag, Dubowitz, & Liu, 2004). BBB disruption influences the pathogenesis of vascular cognitive impairment and dementia (VCID), AD and Parkinson's disease (PD), where symptomatic onset may be precipitated by TBI or subsequent risk may be heightened (Veksler, Shelef, & Friedman, 2014). In TBI and many neurodegenerative diseases, BBB disruption is associated with abnormally high permeability of blood vessel walls, which can increase cerebrospinal fluid (CSF) levels of albumin and aggravate vasogenic brain edema. In this scenario, fluid diffusion within the parenchyma increases to noxious levels, leading to brain volume increase, blood flow impairment, abnormally high intracranial pressure (ICP), etc. (Veksler et al., 2014). Blood flow dysregulation may be due to the vasculature's inadequate ability to dilate or contract in response to metabolic demands, and/or to blood vessel rupture, as a mechanical effect of injury (Van Horn, Bhattrai, & Irimia, 2016). BBB dysfunction after TBI also leads to abnormal NVC, a biological phenomenon described by the quantitative relationship between neuronal activity and blood flow parameters (which are normally coupled), and by blood flow responses elicited after neuronal activity changes. Typically, such responses occur after a lag of several seconds due to signal transduction delays and to the time required for BBB cells' mechanical parameters to change in response to neuronal signals (Fabiani et al., 2014). NVC has three phases: (A) local neuronal assemblies change their metabolic needs in response to alterations in their neural computation; (B) neurons convey their needs to other neurons and to astrocytes until their messages reach the BBB; and (C) BBB cells react to neuron-initiated glial signals via morphological, structural, and physiological responses which alter the amounts of oxygen and glucose delivered to neurons. NVC occurs within the NVU: an ensemble of neurons, astrocytes, endothelial cells, vascular smooth muscle cells (VSMCs), and capillary pericytes whose integrated, cumulative role is to regulate cerebral blood flow (CBF) and to compound exchange across the BBB.

NVC, TBI, and aging

NVC is affected by cardiovascular fitness and aging. Because ICH risk is also related to these factors, a (nonlinear) relationship between ICH risk and NVC exists. Clinically, ICH often denotes a relatively severe hemorrhage within the telencephalon; here, we emphasize that ICHs include smaller brain hemorrhages like cerebral microbleeds (CMBs), which are ICHs despite not being typically highlighted as such (Goh, Irimia, Vespa, & Van Horn, 2016). Noninvasively, NVC can be quantified indirectly using functional MRI (fMRI) as an increase in the blood oxygenation level-dependent (BOLD) response of oxy- (compared to deoxy-) hemoglobin associated with compensatory vasodilatation (Irimia & Van Horn, 2015b). This measurable response changes nonlinearly as a function of age, physical fitness, and neurons' electrical activity. Thus, the aging brains of physically unfit individuals may have chronically dilated or poorly reactive blood vessels, resulting in sparser capillary beds, abnormally low tissue perfusion, and impaired NVC (Isaacs, Anderson, Alcantara, Black, & Greenough, 1992). Like the BOLD signal, glucose concentrations measured using fluorodeoxyglucose positron emission tomography (PET) change nonlinearly with age and fitness. When neuronal metabolic needs are high, NVC exhibits nonlinear behavior partly due to the vastly different spatial scale of electrical brain activity compared to that of oxygen consumption dynamics and glucose concentrations. In the aging brain, incoming oxygen is utilized relatively faster than in younger adults, leading to lower oxyhemoglobin concentrations and to sluggish blood flow changes. Because ICH risk and NVC are correlated, ICH risk is higher in older and/or physically unfit persons. This is independently reflected by the neurobiology of post-traumatic NVC: after TBI, CMBs are associated with loss of both pericytes and endothelial tight junction molecules, leading to alterations in BBB amyloid β (Aβ) transporter expression (Nelson, Sweeney, Sagare, & Zlokovic, 2016). Normally, CBF within microvessels is regulated via cellular mechanisms which prevent intra-vascular coagulation by preserving a balance of signal transduction between pro- and anti-thrombotic pathways (Wang, Kittaka, Sun, Schreiber, & Zlokovic, 1997). Neuronal activity contributes to this balance through NVC, which is impaired after TBI by VSMC hypercontractility (Roher et al., 2012). As the brain ages, blood vessels stiffen and VSMC contractility becomes additionally impaired. This is reflected in the dysregulation of signal transduction pathways mediating NVC, such that the risks for intra-vascular coagulation, micro-vessel stenosis and ICH increase.

Vascular pathophysiology after TBI

BBB disruption frequently entails increased blood vessel permeability and NVC impairment. These phenomena can lead to vasogenic edema, whereby water diffuses abnormally broadly within the brain parenchyma, potentially leading to brain dilation, high ICP, abnormal ion concentrations and improper neurotransmission. Edema is thus a risk factor for post-traumatic ictogenesis and even epileptogenesis, leading to post-traumatic epilepsy (Irimia et al., 2013; Irimia & Van Horn, 2015a). For example, NVC alterations after TBI aggravate cortical spreading depolarizations, which are electric depolarization waves capable of inducing vasoconstriction and prolonged depolarization oligemia (Tarantini, Tran, Gordon, Ungvari, & Csiszar, 2017). Such electrophysiological phenomena—which are investigated using electro- and magneto-encephalography (Halgren, Sherfey, Irimia, Dale, & Marinkovic, 2011; Irimia, 2005; Lima, Irimia, & Wikswo, 2006)—occur in parallel with BBB disruption after TBI. Both can play a role in SVD pathophysiology (Pantoni, 2010), leading to CMBs and aggravating VCID risk. In dynamic contrast enhanced (DCE) MRI contrast agents improve the ability to explore vascular pathophysiology, such that researchers can map reduced CBF and BBB dysfunction in extra-cortical regions. The magnitudes of these phenomena increase with age (Hanyu et al., 2010; Montagne et al., 2015) and are paralleled by pericyte injury as reflected by CSF levels of platelet-derived growth factor receptor-β expressed by capillary pericytes and by the VSMCs of cerebral arteries and arterioles (Montagne et al., 2015). In some TBI patients, BBB dysfunction persists for months, years and even decades, thereby contributing to the risk for VCID, AD, PD and amyotrophic lateral sclerosis (Wardlaw, Smith, & Dichgans, 2013). Perivascular spaces (PVSs) are pial-lined, fluid-filled structures that accompany vessels entering (i.e., penetrating arteries) or leaving (i.e., draining veins) brain tissue. In place of true brain lymphatic vessels, the PVS-lymphatic system functions to drain away soluble waste proteins and toxic metabolic products such as Aβ protein and τ oligomers. BBB dysfunction, impaired CBF and decrease CSF clearance from PVSs may contribute to CMBs, and then to post-traumatic epileptogenesis, cerebral amyloid angiopathy (CAA) and AD (Sepehrband et al., 2019).

TBI hemorrhage identification and localization

After TBI, hemorrhages are frequent radiological findings (Fig. 1), particularly in older adults (where their incidence exceeds ~40%), males, diabetics, hypertensives and apolipoprotein (ApoE) $\varepsilon4$ allele carriers (Beach et al., 2015). In fact, ~40% of healthy individuals over 80 have CMBs (Werring, 2011), which are vascular neuropathology manifestations of focal BBB disruption and of high BBB permeability (Fig. 2). Because SWI is useful for CMB detection, the clinical use of this technique has increased within the last decade. All major MRI scanner vendors now offer susceptibility-weighted sequences, which use gradient-echo imaging under trade names other than SWI (Siemens), including susceptibility weighted angiography (General Electric), SWI-phase and VenoBOLD (Phillips), blood-sensitive imaging (Hitachi), or flow-sensitive

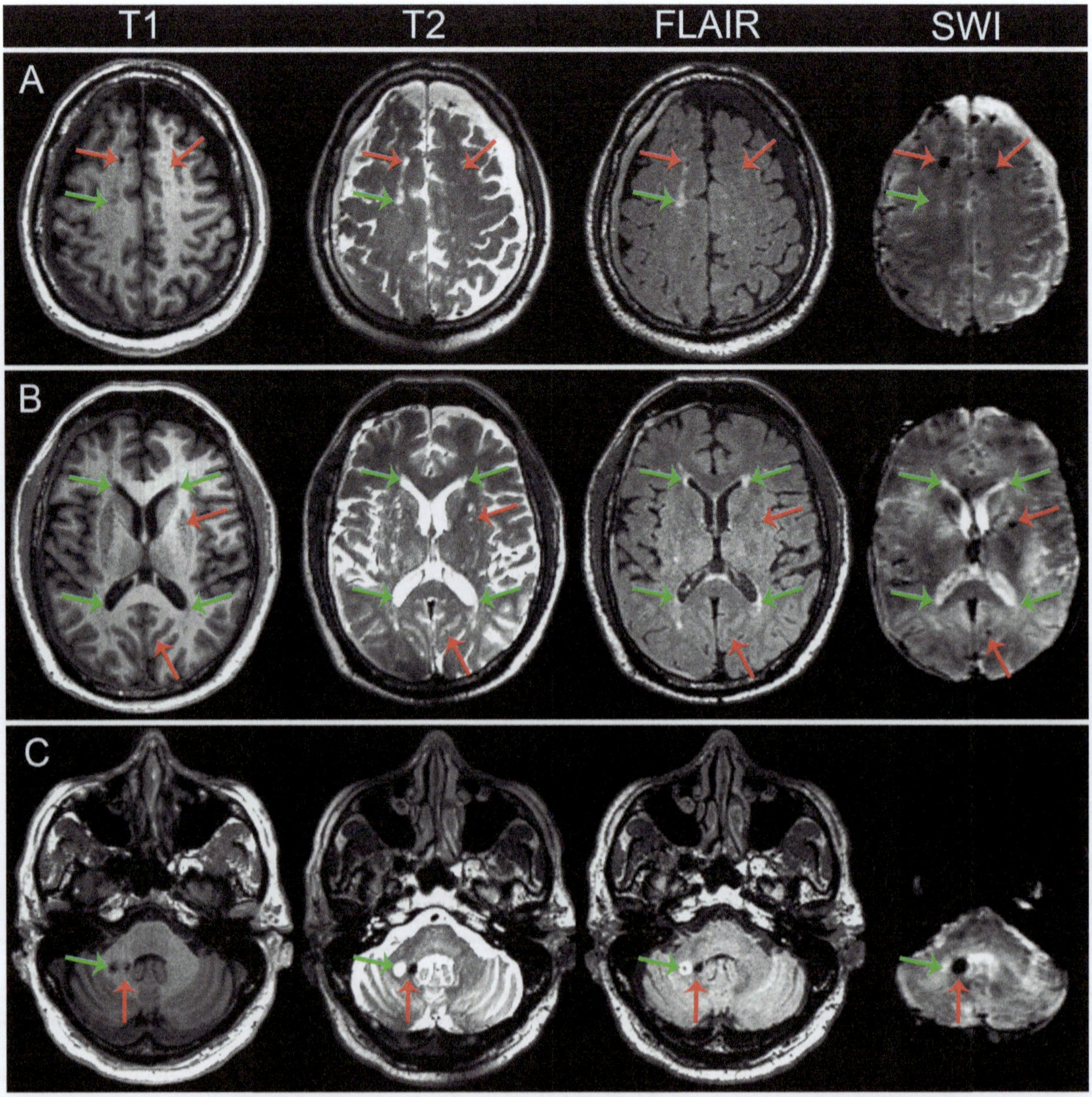

FIG. 1 CMBs across several MRI modalities in a 69-year-old male with acute mTBI. Shown are T_1 and T_2 MRI, FLAIR, and SWI, with red and green arrows indicating the locations of CMBs and WM abnormalities, respectively. Depicted in radiological convention are axial slices at the levels of (A) the superior cerebral cortex, (B) insulae, lateral ventricles and subcortical structures, and (C) cerebellar GM/WM. T_1 MRI only identifies relatively large bleeds (A and C), whereas T_2-weighted MRI can additionally localize a CMB in the putamen (B). FLAIR can assist the localization of WM hyperintensities associated with peri-hemorrhagic edema and likely also with post-traumatic demyelination (A and C) as well as leukoaraiosis (B). SWI sensitivity to paramagnetic blood compounds is highlighted (A–C). Abbreviations: CMB, cerebral microbleed; FLAIR, fluid-attenuated inversion recovery; GM, gray matter; MRI, magnetic resonance imaging; mTBI, mild traumatic brain injury; SWI, susceptibility weighted imaging; WM, white matter.

black blood (Cannon). SWI has been leveraged to demonstrate that ~25% of patients with mild cognitive impairment (MCI) and ~45% of patients with AD have CMBs and abnormal iron complex deposits in the brain. CMB incidence is also higher in MCI/AD patients with a history of neurotrauma (Raven, Lu, Tishler, Heydari, & Bartzokis, 2013). After TBI, ICHs are correlated with Aβ burden, as frequently quantified in vivo using ^{11}C-labeled Pittsburgh compound B PET (Yates et al., 2014). TBI ICHs are also associated with Aβ-related imaging abnormalities; for the latter, vasogenic edema and CMBs occur partially because of inadequate Aβ clearance, especially in areas concomitantly affected by CAA. This latter condition is common in the elderly and involves Aβ accumulation along brain vessel walls, possibly due to improper Aβ clearance (Ramanathan, Nelson, Sagare, & Zlokovic, 2015) or to inadequate Aβ removal by VSMCs (Bell et al., 2009). In ApoE-ε4 carriers and AD, CMBs are commonplace at the boundary between gray matter (GM) and white matter (WM), as well as along leptomeningeal arteries (e.g., superficial siderosis) (Thal, Grinberg, & Attems, 2012). By contrast, in hypertensive SVD, CMB are typically located in the basal ganglia, pons, and cerebellum.

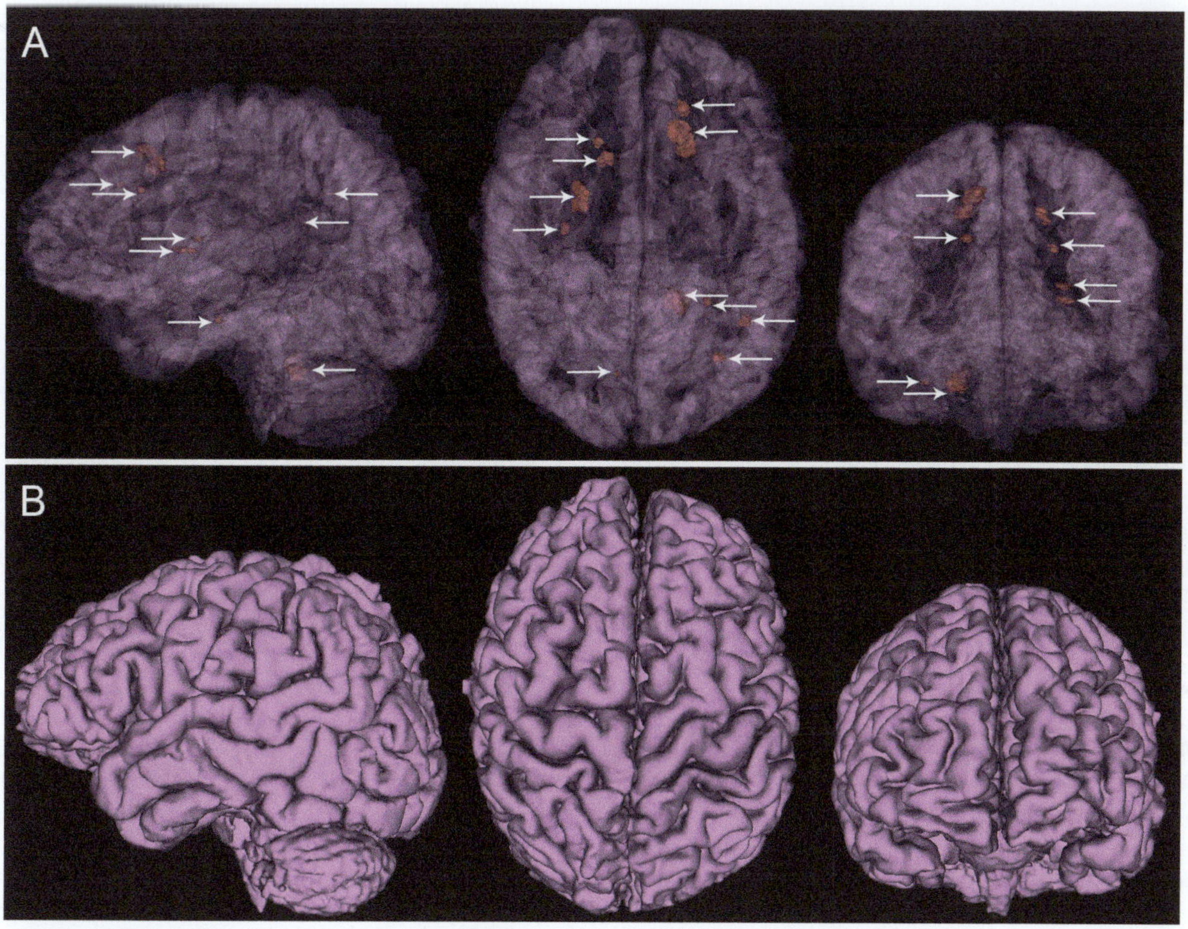

FIG. 2 CMB modeling. Shown are 11 CMBs for the TBI patient in Fig. 1, with lateral (*left*), superior (*center*) and anterior (*right*) views of the brain in radiological convention. (A) Translucent brain rendering allows direct visualization of CMBs relative to neuroanatomic landmarks. In the lateral and anterior views, some CMBs overlap, such that only the superior view allows all of them to be visible simultaneously. (B) The opaque surface rendering for the cerebral and cerebellar cortices suggests only relative moderate pre-traumatic atrophy in this patient. Abbreviations: CMB, cerebral microbleed; TBI, traumatic brain injury.

Differential etiology of TBI-related hemorrhages

Of clinical significance to the neuroradiological interpretation of imaging scans acquired from TBI patients is the task of differential ICH etiology, especially in geriatric patients with mild TBI (mTBI). Brain imaging scans antedating traumatic events are rarely available for most patients, and CMB presence on acute MRIs is often the best surrogate marker for traumatic axonal injury (TAI), especially if mTBI diagnosis is suspected. In young persons, TBI CMBs are typically located at the GM/WM interface. To some extent, TBI CMBs can be distinguished from hypertensive hemorrhages because the latter usually occur in the pons, cerebellum, basal ganglia, and thalamus (Thal, Ghebremedhin, Orantes, & Wiestler, 2003). However, the distributions of TBI and CAA CMBs overlap, with the latter occurring in the territories of meningeal and cortical vessels (Thal et al., 2003). Ascertaining the etiology of hemorrhages identified after TBI is problematic in older adults, where CAA CMBs and TBI CMBs often coexist, and where neuroimaging provides little specificity beyond characteristic deep versus peripheral neuroanatomic localization.

The prevalence of CMBs increases with age, as does hypertension, CAA and leukoaraiosis (Koennecke, 2006). Thus, in elderly patients with mTBI, CMB etiology is particularly difficult to ascertain (Del Ser, Hachinski, Merskey, & Munoz, 2005). Unfortunately, older adults are also at greater risk for neurodegenerative diseases associated with higher CMB prevalence, as reflected by the fact that individuals at high risk for AD exhibit more CMBs than healthy controls (HCs), which complicates etiological classification. For example, some degree of CAA occurs in ~80% of AD patients (Zlokovic, 2011) and at least ~12% of persons with AD have a history of TBI. Thus, it may be particularly difficult to distinguish traumatic from non-traumatic CMBs in patients with co-existing CAA and TBI. Additionally, establishing CMB etiology in vivo remains an important goal for clinical research because (A) CAA pathology often exists in locations affected by atherosclerosis (Yates et al., 2014) and (B) such locations are frequently affected by primary TBI pathologies other than

CMBs. This may be explained by the confounding effects of older age and CAA, although further research is required. Among elderly persons with TBI, CMBs identified at GM/WM junctions are more likely to be due to TBI rather than CAA. Although TBI CMBs occur more frequently at the GM/WM interface and in occipital regions, this is also true of CAA CMB. Thus, CMB location cannot always allow one to distinguish CAA CMBs from TBI CMBs, and new imaging techniques are needed to tell traumatic CMBs from those of other etiologies. Such strategies might involve imaging TBI CMB penumbrae (which are often edematous) and/or monitoring the temporal evolution of CMBs (since CAA CMBs are relatively stable by comparison).

(Epi)genetic risk for hemorrhage after experimental TBI

Although ApoE-$\varepsilon4$ genotype is a key risk factor for ICH, the influence of presenilin (PSEN) 1 and 2 genotypes on TBI CMB risk may be underappreciated. Because the proteins encoded by the PSEN1 and PSEN2 genes help to cleave amyloid precursor protein (APP) to generate Aβ (Tanzi, 2012), PSEN1/PSEN2 genotype can influence hemorrhage risk after experimental TBI. For example, PSEN1 transgenic mice have a notably large number of CMBs, and feature abnormal microvascular morphology with reduced micro-vessel densities in the hippocampus (Gama Sosa et al., 2010). PSEN1/PSEN2 genotype influences microvasculature morphology in animal models, suggesting that post-traumatic CMB risk is partially determined by PSEN-moderated biomechanical abnormalities in blood vessel structure, which increase vasculature vulnerability to TBI shear and strain. The biomechanical consequences of such phenomena are difficult to control and reverse because TAI repair frequently requires axonal guidance across long spatial distances within WM.

One potential strategy for alleviating the detrimental effects of adverse ApoE/PSEN genotype involves improving BBB penetrance using therapeutic antibodies. These could be transported across the barrier using endogenous mechanisms like transferrin receptor-mediated transcytosis, although such techniques remain underdeveloped (Watts & Dennis, 2013). The deposition of vascular amyloid peptides is strongly predicated upon ApoE/PSEN genotype and predicts BBB dysfunction resulting in cerebrovascular lesions and in CAA/CMB risk after TBI (De Reuck, 2012). In APP-PS1 mice (which have no CAA), vascular volume is minimally affected by aging (Do et al., 2014), whereas the BBBs of ApoE-$\varepsilon4^+$ mice and of mutant human τ transgenic mice are disrupted even without Aβ deposition, leading to further post-TBI vulnerability to brain hemorrhage (Bell et al., 2012). By contrast, minimal BBB disruption is observed in mouse models like the PSEN2-APP (a model of age-related Aβ pathology), human τ P301L and P301S (models of late and early tauopathy, respectively), as well as in both the ApoE-knockout and ApoE-$\varepsilon4$-knockin mouse models (both of which exhibit lower brain hemorrhage risk after neurotrauma) (Bien-Ly et al., 2015). Recent research confirmed that TBI results in histone modification and in reductions of deoxyribonucleic acid methylation within brain cells, leading to accelerated brain aging and to stronger epigenetic risk for AD after TBI-related brain hemorrhage (Sivanandam & Thakur, 2012).

Novel approaches for TBI CMB neuroimaging CMBs

TBI is associated not only with ICHs and CMBs but also with increased *microscale* events like amyloidogenesis, leading to vascular dysfunction, reduced CBV, higher Aβ concentration, APP overexpression, chronic mild hypoperfusion and decreased lymphatic/PVS clearance. Importantly, however, TBI triggers cellular processes which also have *macroscale* manifestations, like brain structure alterations which can be interrogated using neuroimaging (Nelson et al., 2016). Such alterations include cortical thinning, WM atrophy and connectome fragmentation (Rostowsky, Chuadhari, Calvillo, Lee, & Irimia, 2019). Periventricular hemorrhage (PVH) is correlated with venulopathies, which are themselves associated with post-traumatic hemorrhage risk (Gorelick, Counts, & Nyenhuis, 2016), and which are MRI-detectable. Specifically, PVHs are identifiable using fluid-attenuated inversion recovery (FLAIR) MRI (Fig. 1) and, because they are very frequent after TBI due to injury biomechanics, their presence makes CMB occurrence more likely after injury. CMB effects on the connectome can be leveraged to assess cognitive decline risk by (A) integrating SWI-based CMB identification with diffusion tensor imaging (DTI) tractography and (B) localizing lesions to structures like the thalamus and/or basal ganglia (Fig. 3). This is otherwise difficult without mapping WM circuitry (Fig. 4) because CMB locations and distributions vary highly across patients (Jellinger, 2013). For this reason, personalized connectome mapping is often needed to identify CMB-affected neural pathways (Fig. 5).

To assess BBB leakage post-injury, neuroimaging techniques can assist patient monitoring throughout the clinical evolution of their TBIs (Fig. 6). These techniques include DCE computed tomography to quantify vascular permeability (Hom et al., 2011), as well as DCE MRI (which offers higher spatial resolution). Because BBB dysfunction is paralleled by chronic cerebral edema, MRI modalities like FLAIR and diffusion weighted imaging (DWI) are frequent adjuvants to DCE MRI (Fig. 1), particularly in cases where the injection of gadolinium-based contrast agents for DCE imaging is contraindicated. Using DCE MRI, van de Haar et al. (2016) found that GM hypoperfusion and relatively low cerebral blood volume

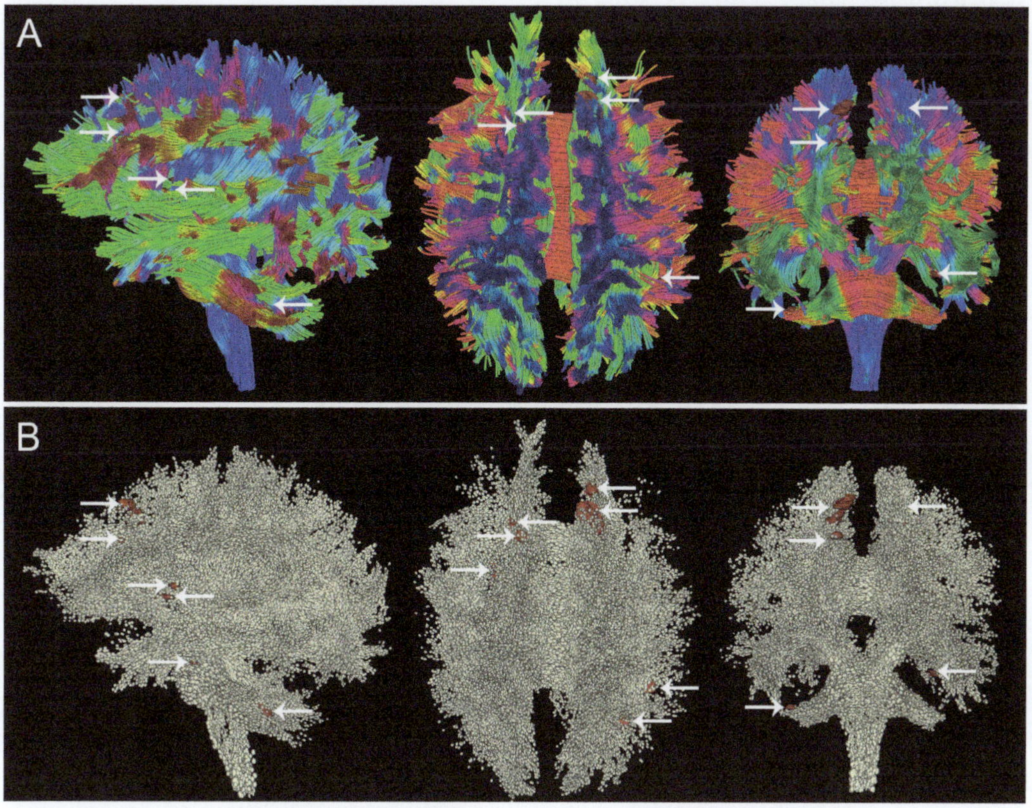

FIG. 3 Rendering of WM DTI tractography streamlines. Reconstructed WM fascicles for the same TBI patient shown in Figs. 1 and 2, viewed from a lateral (*left*), superior (*center*), and anterior (*right*) perspective using radiological convention. (A) Streamline colors encode WM bundle trajectories (*red*: right–left; *green*: anterior–posterior; *blue*: inferior–superior). *Arrows* indicate CMBs, whose visibility is imperfect in this view due to the obstruction caused by 3D WM rendering. (B) Ellipsoidal glyph view of (A), which facilitates CMB inspection relative to DTI tensor representations. Tensor properties are depicted as small ellipsoids located at each imaging voxel, thus providing a more granular appearance. Abbreviations: CMB, cerebral microbleed; DTI, diffusion tensor imaging; WM, white matter.

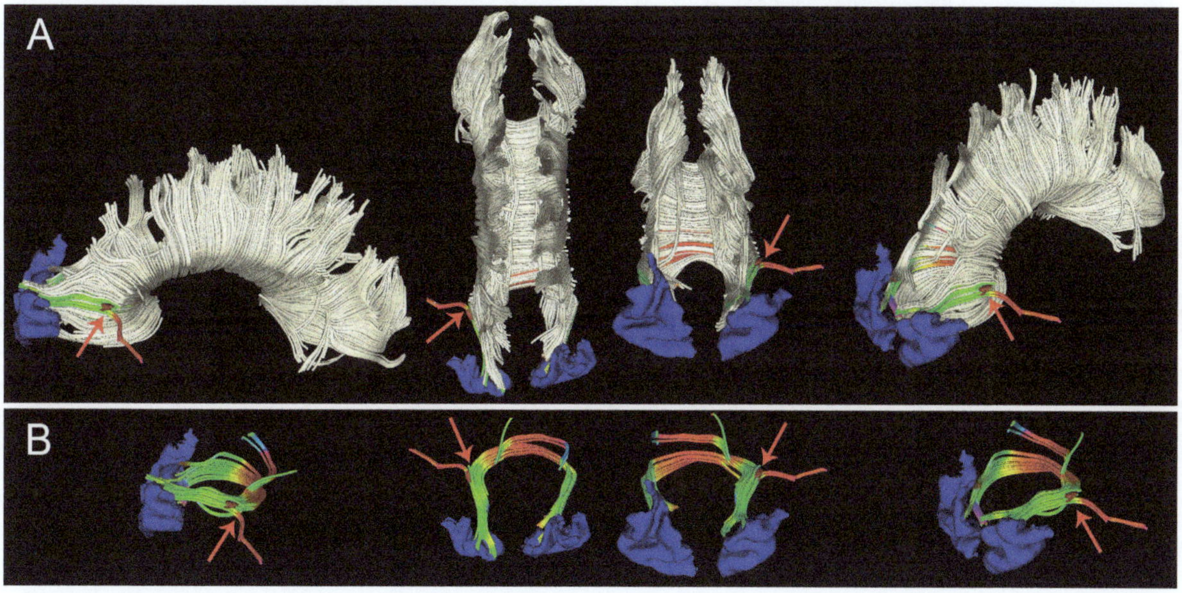

FIG. 4 Visualization of the corpus callosum (CC) and of a CMB affecting its connections. For the subject in Figs. 1–3, a CMB intersecting with streamlines in the genu of the CC is indicated by a red arrow. (A) Callosal WM streamlines (*light beige*) are shown in four distinct views (from left to right: lateral, superior, anterior, and antero-lateral). GM regions innervated by CMB-affected streamlines (i.e., the left and right frontopolar gyri and sulci) are painted blue. (B) CMB-affected DTI streamlines are drawn using the red-green-blue (RGB) convention of DTI (*red*: right–left; *green*: anterior–posterior; *blue*: inferior–superior). The presence of the CMB causes incorrect reconstruction of a few WM streamlines, which are erroneously perceived as branching away from the CC (see arrow location and red-colored WM bundles whose trajectories diverge from the CC). This highlights the potential pitfalls of DTI reconstructions in the presence of CMBs, which alter local water diffusion and may result in neuroanatomically implausible streamline orientations, as illustrated here. Abbreviations: CC, corpus callosum; CMB, cerebral microbleed; DTI, diffusion tensor imaging; GM, gray matter; RGB, red-green-blue; WM, white matter.

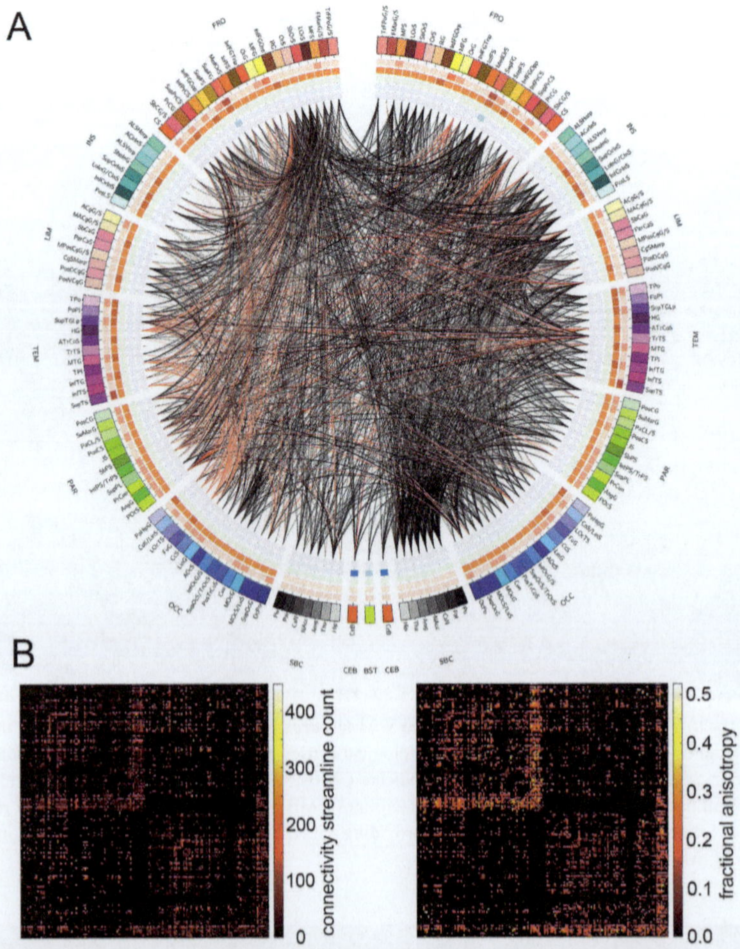

FIG. 5 Connectogram vs conventional representations of the TBI-affected connectome. (A) The connectogram provides a strategy to systematically depict WM connections between brain regions, which are represented as wedges around a circle. Each wedge corresponds to a gyrus or sulcus. The left and right halves of the connectogram correspond to the left and right hemispheres, respectively. Regions within each lobe are grouped together and are separated from other lobes by spacings around the circle. Radially concentric rings encode various structural properties, including GM surface, area, curvature, and WM connectivity density. The colors of links between regions encode the lowest (*blue*), medium (*green*) and highest (*orange*) terciles of values in the histogram of DTI fractional anisotropy (a DWI measure of WM integrity; see also the color scheme for B). The reader is referred elsewhere (Irimia, Chambers, Torgerson, & Van Horn, 2012) for connectogram interpretation. (B) Conventional matrix representations of the connectome provide little insight into its neuroanatomic organization, in contrast to the connectogram. *Left*: color encodes the count of connectivity streamlines between region pairs. *Right*: color encodes mean fractional anisotropy along fascicles connecting region pairs. Abbreviations: DTI, diffusion tensor imaging; GM, gray matter; TBI, traumatic brain injury; WM, white matter.

(CBV) are both associated with high BBB permeability in persons with AD compared to HCs. This is a scenario where the NVU modulates the relationships between CBF, CBV, BBB leakage, and blood perfusion in ways which can be evaluated using noninvasive or minimally invasive macroscale imaging. However, because the BBB may leak relatively slowly after mTBI, DCE MRI has limited sensitivity for detecting BBB disruption after mTBI. One potential exception pertains to the context of dual-time resolution approaches (Jelescu et al., 2011), where fast MRI signal alterations can be captured during initial contrast agent uptake (van de Haar et al., 2016). Nevertheless, because of logistic barriers, DCE imaging is not frequently used in clinical settings.

TBI hemorrhages and AD

TBI is thought to accelerate structural brain aging by about ~5 years (Irimia, Torgerson, Goh, & Van Horn, 2015), the onset of MCI by ~2 years, and to increase dementia risk ~1.6- to ~3-fold, even after adjusting for sex and ApoE-ε4 carrier status (Li, Risacher, McAllister, & Saykin, 2016). Interestingly, mTBI patients with MCI often have WM abnormalities like those noted in AD, including evidence of CMBs and of other cerebral hemorrhagic events (Fakhran, Yaeger, & Alhilali, 2013). In

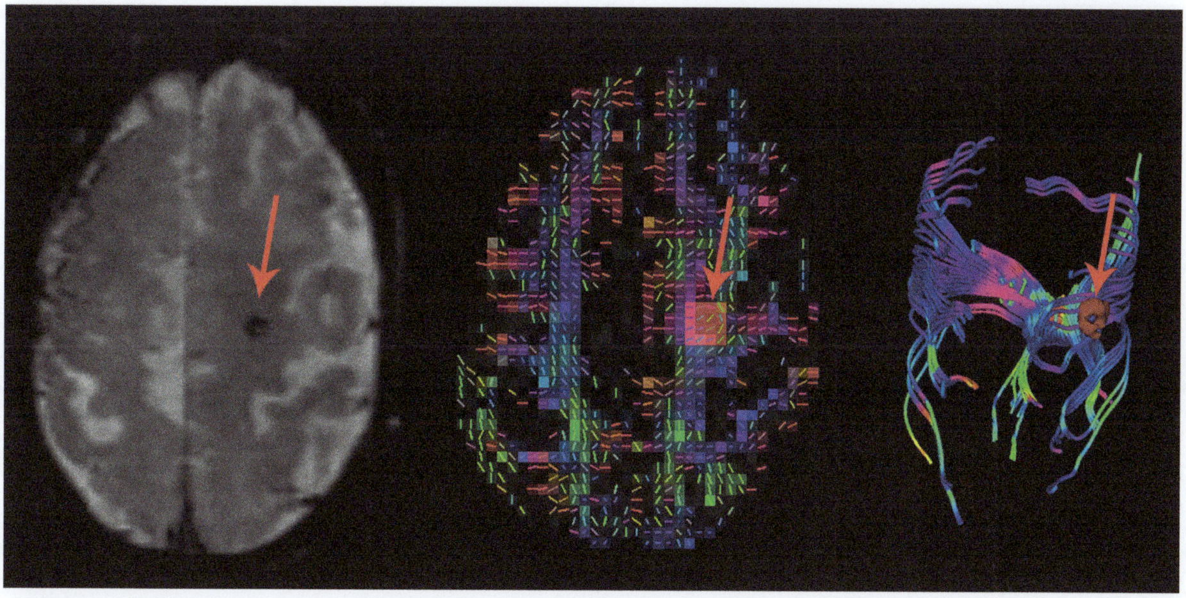

FIG. 6 Multimodality integration for TBI hemorrhage visualization and analysis. For the sample subject in Figs. 1–5, images displayed in radiological convention show (A) a CMB (*red arrow*) and (B) a DWI image with ticks at each voxel to indicate the principal axes of local diffusion tensor orientations. The RGB convention of DTI encodes the orientations of tensors' principal axes. CMB-affected WM voxels are highlighted by a *white square*. (C) 3D rendering of the corona radiata, part of whose streamlines intersect the CMB. The corona is displayed asymmetrically with respect to the anterior-posterior axis because the SWI and DWI volumes were acquired at an angle relative to the plane of the longitudinal fissure. Abbreviations: CMB, cerebral microbleed; DTI, diffusion tensor imaging; DWI, diffusion weighted imaging; RGB, red-green-blue; SWI, susceptibility weighted imaging; TBI, traumatic brain injury; WM, white matter; 3D, three-dimensional.

AD, like in TBI, CMBs are indicative of co-morbid CAA (Charidimou, Shakeshaft, & Werring, 2012), and over 75% of AD patients undergoing *post-mortem* autopsy exhibit CAA CMBs (Arvanitakis et al., 2011). Similarly, severe TBI survivors frequently experience AD-like neuropathology soon after injury, including APP overexpression and Aβ plaques (Roberts et al., 1994). Importantly, TBI impacts MCI severity as a function of both age at injury *and* time between injury and psychometric assessment (Li et al., 2016). Like in AD, *post-mortem* neuropathology examinations of many TBI survivors exhibit both deposits of hyperphosphorylated τ proteins within neurofibrillary tangles (NFTs) as well as Aβ plaques. Vascular pathology, Aβ plaques and NFTs often co-exist in such patients (Venkat, Chopp, & Chen, 2015), suggesting that both post-traumatic neural and vascular degeneration contribute to AD risk after TBI (Kapasi & Schneider, 2016). According to the vascular hypothesis of AD, neurodegeneration is preceded and accelerated by neurovascular injuries, much like those experienced after TBI. BBB permeability is associated with CMBs, CAA and dementia, and increased BBB permeability is related to ApoE genotype (Montagne et al., 2020). This finding can be interpreted as a boost to the vascular hypothesis of AD because it may suggest independent—if coexistent—mechanisms for Aβ plaque formation and AD onset or progression. Recent fMRI evidence indicates that TBI patients with CMBs exhibit AD-like alterations in the functional correlations of their default mode networks (Irimia et al., 2020). Significantly, because such alterations are correlated with both acute and chronic MCI, functional connectomics may be useful for identifying biomarkers of future AD risk in TBI survivors.

Applications to other areas of neuroscience

Because there are many similarities between hemorrhagic pathologies related to traumatic brain injury and Alzheimer's disease-related, post-traumatic hemorrhage risk is likely moderated by Alzheimer's disease risk factors, many of which have macroscale correlates which have been described. Such factors include lifestyle variables (e.g., diet, exercise, smoking, alcohol consumption), an adverse environment (e.g., air pollution), psychiatric pre-morbidities, cardiovascular disease, and metabolic disease. Thus, the evaluation of patients' health profiles and medical histories can serve traumatic brain injury treatment by highlighting hemorrhage risk and the likely extent of brain hemorrhage effects upon cerebral parenchyma. Crucial to this evaluation is the assessment of blood–brain barrier integrity using non-invasive neuroimaging or invasive strategies (in more severe traumatic brain injury cases). Because there appears to be an important relationship between cerebral amyloid angiopathy, blood–brain barrier dysfunction, and cerebral microhemorrhages, the assessment of blood–brain barrier disruption after traumatic brain injury is particularly important when prognosticating the occurrence and severity of subacute/chronic hemorrhages, as well as risk for Alzheimer's disease and Parkinson's disease. Such assessments can assist gauging pharmacological agent bioavailability in the brain and is therefore significant for the development of drugs for use in neurointensive care settings. More broadly, insights from the neurobiology of post-traumatic neurovascular impairment can inform neurodegenerative disease research and our fundamental understanding of brain function under adverse metabolic conditions.

Mini-dictionary of terms

Blood–brain barrier: A structurally intricate functional and structural barrier which regulates and protects the extracellular medium of the brain by modulating glucose transport, balancing the gradients of molecules and ions, and protecting glia and neurons) from toxic compounds.

Brain edema: Abnormal fluid volume increase within living parenchymal tissue. Brain edema can be vasogenic (extracellular) or cytotoxic (intracellular). In the former case, fluid accumulation is due to the disruption of the blood–brain barrier and to the extravasation of blood serum compounds. In the latter case, fluid volume within parenchymal cells increases due to defective cell energetics, oxidative stress, abnormal neurotransmission, and/or improper ion transport across cell membranes.

Cerebral amyloid angiopathy: A neuropathological condition of the brain unrelated to systemic amyloidosis and associated with the deposition of amyloid β predominantly within the tunicae mediae and adventitae of moderately sized brain arteries and, occasionally, of brain veins.

Cerebral microbleed: Neuropathological event involving the deposition of hemorrhagic products within a focal, small, and relatively confined portion of cerebral parenchyma.

Connectomics: The branch of neuroscience whose object is to comprehensively map, classify and investigate the structural and functional connectivity of nervous systems.

Intracerebral hemorrhage: A phenomenon in which blood is released, sometimes profusely, from ruptured blood vessels located within the cerebrum or telencephalon, i.e., from the largest part of the brain containing the cerebral cortex, basal ganglia, hippocampus, and olfactory bulb.

Leukoaraiosis: A type of abnormality in the white matter (from Greek: λευκό (leuko) = white; αραίωση (araiose) = rarefaction, attenuation) presenting as a hyperintensity on fluid-attenuated inversion recovery imaging and consisting of white matter density reduction. Leukoaraiosis can be due to hemorrhage, glial cell proliferation, damage to the blood–brain barrier and/or cerebrospinal fluid-brain barrier, or to axonal demyelination.

Neurovascular coupling: A biological phenomenon whereby neuronal activity is bidirectionally related to blood flow parameters, such that blood flow changes are prompted by changes in neuronal activity, and vice versa.

Neurovascular unit: An ensemble of glial cells (oligodendrocytes, microglia, astrocytes), vascular cells (vascular smooth muscle cells, pericytes, endothelial cells) and neurons which together interact via signal transduction pathways and complex intercellular communication mechanisms to ensure the adequate function of the nervous system.

Susceptibility weighted imaging: A type of magnetic resonance imaging which is sensitive to compounds with paramagnetic (e.g., deoxyhemoglobin, hemosiderin, and ferritin) and diamagnetic (e.g., calcification) properties, thereby allowing the localization and imaging of tissue regions which are relatively rich in compounds with such properties, like hemorrhages.

Key facts of cerebral hemorrhages after traumatic brain injury

- Brain hemorrhages due to traumatic brain injury can have a lasting impact upon brain function and structure.
- Traumatic brain hemorrhages lead to both micro- and macroscale effects, being associated with longer time to recovery.
- Hemorrhages due to brain trauma are more frequent in older adults.
- Hemorrhages due to traumatic brain injury reflect heightened risk for Alzheimer's disease, Parkinson's disease, and other neurodegenerative diseases.
- Magnetic resonance imaging and brain mapping methods are important for the identification of hemorrhages related to traumatic brain injury and for establishing their etiology.

Summary points

- Cerebral hemorrhages related to traumatic brain injury are associated with dysregulation of the blood–brain barrier and may be critically involved in the acceleration of pathogenic processes leading to neurodegenerative diseases, including Parkinson's disease, Alzheimer's disease, and related dementias.
- The risk of intracerebral hemorrhage after traumatic brain injury is higher in older adults and/or physically unfit individuals partly because neurovascular coupling parameters are strongly influenced by cardiovascular fitness and by aging processes.
- Neurovascular coupling disruptions and hemorrhages related to traumatic brain injury are frequently accompanied by vasogenic edema and by other complications including high intracranial pressure, abnormal ion concentrations, improper neurotransmission and post-traumatic ictogenesis.
- Although more frequent in older adults, males, diabetics, hypertensives, and apolipoprotein $\varepsilon4$ allele carriers, hemorrhages due to traumatic brain injury may be difficult to distinguish from hemorrhages due to other causes like cerebral amyloid angiopathy.
- Apolipoprotein $\varepsilon4$ and presenilin 1/2 genotypes modulate hemorrhage risk after traumatic brain injury and can be used together with epigenomic findings to forecast the acceleration of brain aging after injury.
- After traumatic brain injury, hemorrhages are typically identified using susceptibility weighted imaging, which can be integrated with diffusion magnetic resonance imaging to evaluate cerebral microbleed effects on the connectome and to evaluate patients' risk for cognitive decline via patient-tailored mapping of hemorrhage-affected white matter circuitry.
- Hemorrhages due to traumatic brain injury are more frequent in individuals at high risk for neurodegenerative diseases, and the vascular degeneration prompted by such bleeds contributes to Alzheimer's disease risk in survivors of traumatic brain injury.

Acknowledgments

A.I. is supported by the National Institutes of Health (NIH) under award R01 NS 100973, by the US Department of Defense (DoD) under contract W81XWH-18-1-0413, by the James J & Sue Femino Foundation, and by a Hanson-Thorell Family Research Scholarship. H.C.C. is supported by the NIH under award P30 AG 066530. The authors are grateful to Nahian F. Chowdhury, Alyssa Cho, and David J. Robles for their assistance.

References

Arvanitakis, Z., Leurgans, S. E., Wang, Z., Wilson, R. S., Bennett, D. A., & Schneider, J. A. (2011). Cerebral amyloid angiopathy pathology and cognitive domains in older persons. *Annals of Neurology*, *69*(2), 320–327.

Beach, T. G., Adler, C. H., Sue, L. I., Serrano, G., Shill, H. A., Walker, D. G., et al. (2015). Arizona study of aging and neurodegenerative disorders and brain and body donation program. *Neuropathology*, *35*(4), 354–389.

Bell, R. D., Deane, R., Chow, N., Long, X., Sagare, A., Singh, I., et al. (2009). SRF and myocardin regulate LRP-mediated amyloid-beta clearance in brain vascular cells. *Nature Cell Biology*, *11*(2), 143–153.

Bell, R. D., Winkler, E. A., Singh, I., Sagare, A. P., Deane, R., Wu, Z., et al. (2012). Apolipoprotein E controls cerebrovascular integrity via cyclophilin A. *Nature*, *485*(7399), 512–516.

Bien-Ly, N., Boswell, C. A., Jeet, S., Beach, T. G., Hoyte, K., Luk, W., et al. (2015). Lack of widespread BBB disruption in Alzheimer's disease models: Focus on therapeutic antibodies. *Neuron*, *88*(2), 289–297.

Buxton, R. B., Uludag, K., Dubowitz, D. J., & Liu, T. T. (2004). Modeling the hemodynamic response to brain activation. *NeuroImage*, *23*(Suppl 1), S220–S233.

Charidimou, A., Shakeshaft, C., & Werring, D. J. (2012). Cerebral microbleeds on magnetic resonance imaging and anticoagulant-associated intracerebral hemorrhage risk. *Frontiers in Neurology*, *3*, 133.

De Reuck, J. L. (2012). The significance of small cerebral bleeds in neurodegenerative dementia syndromes. *Aging and Disease*, *3*(4), 307–312.

Del Ser, T., Hachinski, V., Merskey, H., & Munoz, D. G. (2005). Alzheimer's disease with and without cerebral infarcts. *Journal of the Neurological Sciences*, *231*(1–2), 3–11.

Do, T. M., Alata, W., Dodacki, A., Traversy, M. T., Chacun, H., Pradier, L., et al. (2014). Altered cerebral vascular volumes and solute transport at the blood-brain barriers of two transgenic mouse models of Alzheimer's disease. *Neuropharmacology*, *81*, 311–317.

Fabiani, M., Gordon, B. A., Maclin, E. L., Pearson, M. A., Brumback-Peltz, C. R., Low, K. A., et al. (2014). Neurovascular coupling in normal aging: A combined optical, ERP and fMRI study. *Neuroimage*, *85*(Pt 1), 592–607.

Fakhran, S., Yaeger, K., & Alhilali, L. (2013). Symptomatic white matter changes in mild traumatic brain injury resemble pathologic features of early Alzheimer dementia. *Radiology*, *269*(1), 249–257.

Gama Sosa, M. A., Gasperi, R. D., Rocher, A. B., Wang, A. C., Janssen, W. G., Flores, T., et al. (2010). Age-related vascular pathology in transgenic mice expressing presenilin 1-associated familial Alzheimer's disease mutations. *The American Journal of Pathology*, *176*(1), 353–368.

Goh, S. Y. M., Irimia, A., Vespa, P. M., & Van Horn, J. D. (2016). Patient-tailored multimodal neuroimaging, visualization and quantification of human intra-cerebral hemorrhage. In *Proceedings of the SPIE conference on medical imaging—PACS and imaging informatics: Next generation and innovations* (p. 9789).

Gorelick, P. B., Counts, S. E., & Nyenhuis, D. (2016). Vascular cognitive impairment and dementia. *Biochimica et Biophysica Acta, 1862*(5), 860–868.

Halgren, E., Sherfey, J., Irimia, A., Dale, A. M., & Marinkovic, K. (2011). Sequential temporo-fronto-temporal activation during monitoring of the auditory environment for temporal patterns. *Human Brain Mapping, 32*(8), 1260–1276. https://doi.org/10.1002/hbm.21106.

Hanyu, H., Sato, T., Hirao, K., Kanetaka, H., Iwamoto, T., & Koizumi, K. (2010). The progression of cognitive deterioration and regional cerebral blood flow patterns in Alzheimer's disease: A longitudinal SPECT study. *Journal of the Neurological Sciences, 290*(1–2), 96–101.

Hom, J., Dankbaar, J. W., Soares, B. P., Schneider, T., Cheng, S. C., Bredno, J., et al. (2011). Blood-brain barrier permeability assessed by perfusion CT predicts symptomatic hemorrhagic transformation and malignant edema in acute ischemic stroke. *AJNR. American Journal of Neuroradiology, 32*(1), 41–48.

Irimia, A. (2005). Electric field and potential calculation for a bioelectric current dipole in an ellipsoid. *Journal of Physics A: Mathematical and General, 38*, 8123–8138.

Irimia, A., & Bradshaw, L. A. (2005). Ellipsoidal electrogastrographic forward modelling. *Physics in Medicine and Biology, 50*(18), 4429–4444. https://doi.org/10.1088/0031-9155/50/18/012.

Irimia, A., Chambers, M. C., Torgerson, C. M., & Van Horn, J. D. (2012). Circular representation of human cortical networks for subject and population-level connectomic visualization. *NeuroImage, 60*(2), 1340–1351. https://doi.org/10.1016/j.neuroimage.2012.01.107.

Irimia, A., Goh, S. Y., Torgerson, C. M., Chambers, M. C., Kikinis, R., & Van Horn, J. D. (2013). Forward and inverse electroencephalographic modeling in health and in acute traumatic brain injury. *Clinical Neurophysiology, 124*(11), 2129–2145. https://doi.org/10.1016/j.clinph.2013.04.336.

Irimia, A., Maher, A. S., Chaudhari, N. N., Chowdhury, N. F., Jacobs, E. B., & Alzheimer's Disease Neuroimaging, I. (2020). Acute cognitive deficits after traumatic brain injury predict Alzheimer's disease-like degradation of the human default mode network. *Geroscience, 42*, 1411–1429.

Irimia, A., Richards, W. O., & Bradshaw, L. A. (2006). Magnetogastrographic detection of gastric electrical response activity in humans. *Physics in Medicine and Biology, 51*(5), 1347–1360. https://doi.org/10.1088/0031-9155/51/5/022.

Irimia, A., Torgerson, C. M., Goh, S. Y., & Van Horn, J. D. (2015). Statistical estimation of physiological brain age as a descriptor of senescence rate during adulthood. *Brain Imaging and Behavior, 9*(4), 678–689. https://doi.org/10.1007/s11682-014-9321-0.

Irimia, A., & Van Horn, J. D. (2015a). Epileptogenic focus localization in treatment-resistant post-traumatic epilepsy. *Journal of Clinical Neuroscience, 22*(4), 627–631. https://doi.org/10.1016/j.jocn.2014.09.019.

Irimia, A., & Van Horn, J. D. (2015b). Functional neuroimaging of traumatic brain injury: Advances and clinical utility. *Neuropsychiatric Disease and Treatment, 11*, 2355–2365. https://doi.org/10.2147/NDT.S79174.

Isaacs, K. R., Anderson, B. J., Alcantara, A. A., Black, J. E., & Greenough, W. T. (1992). Exercise and the brain: Angiogenesis in the adult rat cerebellum after vigorous physical activity and motor skill learning. *Journal of Cerebral Blood Flow and Metabolism, 12*(1), 110–119.

Jelescu, I. O., Leppert, I. R., Narayanan, S., Araujo, D., Arnold, D. L., & Pike, G. B. (2011). Dual-temporal resolution dynamic contrast-enhanced MRI protocol for blood-brain barrier permeability measurement in enhancing multiple sclerosis lesions. *Journal of Magnetic Resonance Imaging, 33*(6), 1291–1300.

Jellinger, K. A. (2013). Pathology and pathogenesis of vascular cognitive impairment—A critical update. *Frontiers in Aging Neuroscience, 5*, 17.

Kapasi, A., & Schneider, J. A. (2016). Vascular contributions to cognitive impairment, clinical Alzheimer's disease, and dementia in older persons. *Biochimica et Biophysica Acta, 1862*(5), 878–886.

Koennecke, H. C. (2006). Cerebral microbleeds on MRI: Prevalence, associations, and potential clinical implications. *Neurology, 66*(2), 165–171. https://doi.org/10.1212/01.wnl.0000194266.55694.1e.

Li, W., Risacher, S. L., McAllister, T. W., & Saykin, A. J. (2016). Traumatic brain injury and age at onset of cognitive impairment in older adults. *Journal of Neurology, 263*(7), 1280–1285.

Lima, E. A., Irimia, A., & Wikswo, J. P. (2006). The magnetic inverse problem. In J. C. A. Braginski (Ed.), *Vol. 2. The SQUID handbook: Applications of SQUIDs and SQUID systems* Wiley-VCH.

Montagne, A., Barnes, S. R., Sweeney, M. D., Halliday, M. R., Sagare, A. P., Zhao, Z., et al. (2015). Blood-brain barrier breakdown in the aging human hippocampus. *Neuron, 85*(2), 296–302.

Montagne, A., Nation, D. A., Sagare, A. P., Barisano, G., Sweeney, M. D., Chakhoyan, A., et al. (2020). APOE4 leads to blood-brain barrier dysfunction predicting cognitive decline. *Nature, 581*(7806), 71–76. https://doi.org/10.1038/s41586-020-2247-3.

Nelson, A. R., Sweeney, M. D., Sagare, A. P., & Zlokovic, B. V. (2016). Neurovascular dysfunction and neurodegeneration in dementia and Alzheimer's disease. *Biochimica et Biophysica Acta, 1862*(5), 887–900.

Pantoni, L. (2010). Cerebral small vessel disease: From pathogenesis and clinical characteristics to therapeutic challenges. *Lancet Neurology, 9*(7), 689–701.

Ramanathan, A., Nelson, A. R., Sagare, A. P., & Zlokovic, B. V. (2015). Impaired vascular-mediated clearance of brain amyloid beta in Alzheimer's disease: The role, regulation and restoration of LRP1. *Frontiers in Aging Neuroscience, 7*, 136.

Raven, E. P., Lu, P. H., Tishler, T. A., Heydari, P., & Bartzokis, G. (2013). Increased iron levels and decreased tissue integrity in hippocampus of Alzheimer's disease detected in vivo with magnetic resonance imaging. *Journal of Alzheimer's Disease, 37*(1), 127–136.

Roberts, G. W., Gentleman, S. M., Lynch, A., Murray, L., Landon, M., & Graham, D. I. (1994). Beta amyloid protein deposition in the brain after severe head injury: Implications for the pathogenesis of Alzheimer's disease. *Journal of Neurology, Neurosurgery, and Psychiatry, 57*(4), 419–425.

Roher, A. E., Debbins, J. P., Malek-Ahmadi, M., Chen, K., Pipe, J. G., Maze, S., et al. (2012). Cerebral blood flow in Alzheimer's disease. *Vascular Health and Risk Management, 8*, 599–611.

Rostowsky, K. A., Chuadhari, N. N., Calvillo, M., Lee, S. K., & Irimia, A. (2019). Geriatric traumatic brain injury and Alzheimer's disease share patterns of white matter decline. *Alzheimer's & Dementia*, *15*(7S), P1407–P1408.

Sepehrband, F., Cabeen, R. P., Choupan, J., Barisano, G., Law, M., Toga, A. W., et al. (2019). Perivascular space fluid contributes to diffusion tensor imaging changes in white matter. *NeuroImage*, *197*, 243–254. https://doi.org/10.1016/j.neuroimage.2019.04.070.

Sivanandam, T. M., & Thakur, M. K. (2012). Traumatic brain injury: A risk factor for Alzheimer's disease. *Neuroscience and Biobehavioral Reviews*, *36*(5), 1376–1381.

Tanzi, R. E. (2012). The genetics of Alzheimer disease. *Cold Spring Harbor Perspectives in Medicine*, *2*(10).

Tarantini, S., Tran, C. H. T., Gordon, G. R., Ungvari, Z., & Csiszar, A. (2017). Impaired neurovascular coupling in aging and Alzheimer's disease: Contribution of astrocyte dysfunction and endothelial impairment to cognitive decline. *Experimental Gerontology*, *94*, 52–58.

Thal, D. R., Ghebremedhin, E., Orantes, M., & Wiestler, O. D. (2003). Vascular pathology in Alzheimer disease: Correlation of cerebral amyloid angiopathy and arteriosclerosis/lipohyalinosis with cognitive decline. *Journal of Neuropathology and Experimental Neurology*, *62*(12), 1287–1301. https://doi.org/10.1093/jnen/62.12.1287.

Thal, D. R., Grinberg, L. T., & Attems, J. (2012). Vascular dementia: Different forms of vessel disorders contribute to the development of dementia in the elderly brain. *Experimental Gerontology*, *47*(11), 816–824.

van de Haar, H. J., Jansen, J. F. A., van Osch, M. J. P., van Buchem, M. A., Muller, M., Wong, S. M., et al. (2016). Neurovascular unit impairment in early Alzheimer's disease measured with magnetic resonance imaging. *Neurobiology of Aging*, *45*, 190–196.

Van Horn, J. D., Bhattrai, A., & Irimia, A. (2016). Multimodal imaging of neurometabolic pathology due to traumatic brain injury. *Trends in Neurosciences*, *40*(1), 39–59.

Veksler, R., Shelef, I., & Friedman, A. (2014). Blood-brain barrier imaging in human neuropathologies. *Archives of Medical Research*, *45*(8), 646–652.

Venkat, P., Chopp, M., & Chen, J. (2015). Models and mechanisms of vascular dementia. *Experimental Neurology*, *272*, 97–108.

Wang, L., Kittaka, M., Sun, N., Schreiber, S. S., & Zlokovic, B. V. (1997). Chronic nicotine treatment enhances focal ischemic brain injury and depletes free pool of brain microvascular tissue plasminogen activator in rats. *Journal of Cerebral Blood Flow and Metabolism*, *17*(2), 136–146.

Wardlaw, J. M., Smith, C., & Dichgans, M. (2013). Mechanisms of sporadic cerebral small vessel disease: Insights from neuroimaging. *Lancet Neurology*, *12*(5), 483–497.

Watts, R. J., & Dennis, M. S. (2013). Bispecific antibodies for delivery into the brain. *Current Opinion in Chemical Biology*, *17*(3), 393–399.

Werring, D. J. (2011). *Cerebral microbleeds: Pathophysiology to clinical practice*. Cambridge University Press.

Yates, P. A., Desmond, P. M., Phal, P. M., Steward, C., Szoeke, C., Salvado, O., et al. (2014). Incidence of cerebral microbleeds in preclinical Alzheimer disease. *Neurology*, *82*(14), 1266–1273.

Yates, P. A., Villemagne, V. L., Ellis, K. A., Desmond, P. M., Masters, C. L., & Rowe, C. C. (2014). Cerebral microbleeds: A review of clinical, genetic, and neuroimaging associations. *Frontiers in Neurology*, *4*, 205. https://doi.org/10.3389/fneur.2013.00205.

Zlokovic, B. V. (2011). Neurovascular pathways to neurodegeneration in Alzheimer's disease and other disorders. *Nature Reviews. Neuroscience*, *12*(12), 723–738.

Chapter 8

Linking fibrinogen, coagulopathy prophylaxis, and traumatic brain injury

Ryuta Nakae and Shoji Yokobori
Department of Emergency and Critical Care Medicine, Nippon Medical School, Tokyo, Japan

List of abbreviations

DVT deep vein thrombosis
FDP fibrinogen/fibrin degradation product
FFP fresh frozen plasma
FbDP fibrin degradation product
FgDP fibrinogen degradation product
IQR interquartile range
ISS injury severity score
MTP massive transfusion protocol
PAI-1 plasminogen activator inhibitor type 1
RBC red blood cell
TBI traumatic brain injury
TF tissue factor
VHA viscoelastic hemostasis assay
VTE venous thromboembolism
tPA tissue plasminogen activator
α_2-PI α_2-plasmin inhibitor

Introduction

Coagulopathy often occurs in patients with acute trauma (Maegele, 2014). It is well known that coagulopathy is particularly serious in traumatic brain injury (TBI) patients (Epstein, Mitra, O'Reilly, Rosenfeld, & Cameron, 2014; Harhangi, Kompanje, Leebeek, & Maas, 2008; Wafaisade et al., 2010). Abnormalities of coagulation and fibrinolysis are risk factors for unfavorable outcomes. In a recent meta-analysis, the occurrence rate of coagulopathy induced by TBI was 32.7%–35.2% (Epstein et al., 2014; Harhangi et al., 2008). Similarly, a large trauma registry in Germany showed that 22.7% of isolated TBI patients presented with coagulopathy, which was associated with an elevated in-hospital mortality rate (Wafaisade et al., 2010).

TBI-induced coagulopathy includes both hyper- and hypocoagulable states that can complicate secondary brain damage with micro-thrombosis and hemorrhage (Laroche, Kutcher, Huang, Cohen, & Manley, 2012; Maegele, 2013). In particular, hypocoagulability in the setting of TBI is considered a risk factor for early hemorrhagic progression leading to critical and fatal complications (Harhangi et al., 2008; Kurland, Hong, Aarabi, Gerzanich, & Simard, 2012). Several reports have found that this hypocoagulability is linked to consumption of coagulation factors accompanied by acceleration of fibrinolytic pathways (Hess et al., 2008; Maegele, Schöchl, & Cohen, 2014).

Fibrinogen is the major plasma protein coagulation factor. It is the first factor to drop to a critically low level during severe bleeding. As fibrinogen is the only precursor of fibrin, fibrin levels are compromised (Mosesson, 2005; Sorensen et al., 2012).

In this chapter, we review the latest findings about the relationship between fibrinogen and post-TBI coagulopathy, and the diagnosis and management steps needed to improve the prognosis.

Diagnosis and Treatment of Traumatic Brain Injury. https://doi.org/10.1016/B978-0-12-823347-4.00013-0

Fibrinogen

Fibrinogen is a glycoprotein that is produced in the liver. Its concentration in plasma is 2.0%–4.0 g/L. It is composed of three polypeptide chains linked by disulfide bridges. Its approximate molecular weight is 340 kDa and has a half-life of 3–5 d. As the precursor of fibrin, which is indispensable for coagulation, it occupies a central role in primary and secondary hemostasis. The plasma fibrinogen concentration of healthy adults increases with age. Hager, Felicetti, Seefried, and Platt (1994) showed that fibrinogen concentration increases by 0.25 g/L per decade; approximately 81% of those $\geq$65 y have fibrinogen levels $\geq$3.2 g/L. Increased fibrinogen concentration is associated with processes such as acute phase reactions, and carries an increased risk for subclinical degeneration of vascular endothelium, or the more common coagulation/fibrinolysis activation, in which interleukin 6 is a potential mediator (Hager et al., 1994; Wilkerson & Sane, 2002).

TBI-induced coagulation/fibrinolysis cascade

Tissue factor (TF) is the primary initiator of the coagulation protease cascade. It is an essential player in hemostasis. It was first reported in 1973 and 1974 that TBI results in the release of TF across the damaged blood–brain barrier, leading to coagulopathy (Goodnight et al., 1974; Keimowitz & Annis, 1973). TF is abundant in the central nervous system as well as in perivascular smooth muscle cells, fibroblasts, and pericytes (Drake, Morrissey, & Edgington, 1989; Eddleston et al., 1993). The leakage of TF into the blood circulation following direct vascular injury or defragmentation from vascular damage is a trigger for the coagulation cascade, of which the TF:Factor VII/VIIa (FVII/FVIIa) complex is the prime initiator. This cascade is called the "extrinsic pathway," because of the need of an exogenous factor for activation. The TF:FVIIa complex is responsible for converting both FIX and FX to their activated forms FIXa and FXa, leading to the formation of small amounts of thrombin, which then activates the cofactors FV and FVIII. The tenase complex (FVIIIa:FIXa) amplifies the clotting cascade with the activation of FX to FXa. The prothrombinase complex (FVa:FXa) and thrombin make up the "common pathway," whereby prothrombin is activated into thrombin, the central protease of the clotting cascade (Fig. 1).

Fibrinolysis begins when thrombin cleaves fibrinogen into a soluble fibrin monomer, which spontaneously polymerizes into a soluble fibrin polymer. Thrombin also activates FXIII, which, in the presence of calcium, crosslinks the fibrin polymer, producing cross-linked fibrin polymer. Cleavage of cross-linked fibrin polymer by plasmin results in fibrin degradation products (FbDPs). D-dimer, a neoantigen produced by the FXIIIa-mediated cross-linking of fibrin, is exposed after the degradation of the cross-linked fibrin polymer. Fibrinogenolysis occurs when the fibrinolytic pathway is activated independent of the activation of coagulation. This results in the pathologic degradation of fibrinogen by plasmin, producing fibrinogen degradation products (FgDPs). "FDP" (fibrinogen/fibrin degradation products) usually refers to FbDP and FgDP combined (Fig. 2). Both fibrinolytic and fibrinogenolytic products are detectable by the measurement of FDP, while only fibrinolytic products are detected by the measurement of D-dimer.

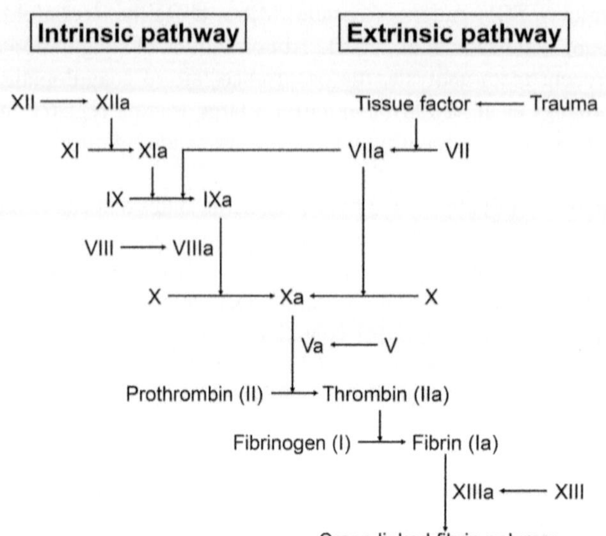

FIG. 1 Traumatic brain injury-induced coagulation cascade. The coagulation cascade is divided into the intrinsic pathway and the extrinsic pathway.

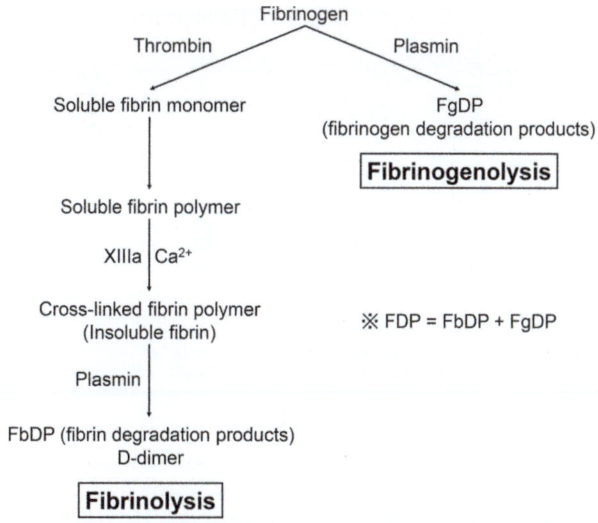

FIG. 2 Traumatic brain injury-induced fibrinolysis cascade. The fibrinolysis cascade is divided into primary fibrinogenolysis and secondary fibrinolysis.

Time course of fibrinogen concentration in the acute phase of TBI

As fibrinogen levels reach a critically low value much earlier than any other coagulation factor during severe bleeding, the drop can induce hemostatic dysfunction (Hiippala, Myllyla, & Vahtera, 1995; Rourke et al., 2012). We investigated coagulation/fibrinolytic parameters' time courses during the acute phase of isolated severe TBI by taking blood samples from severe TBI patients on admission and at 3, 6, and 12h after injury (Nakae et al., 2016), and found a significant decrease in plasma fibrinogen concentrations from admission (median: 2.4g/L, interquartile range [IQR]: 2.0–3.0) to 3h after injury (median: 2.2g/L, IQR: 1.6–2.6). Levels increased insignificantly from 3 to 6h after injury, only to increase significantly again from 6 (median: 2.2g/L, IQR: 1.7–2.6) to 12h after injury (median: 2.6g/L, IQR: 2.0–3.2) (Fig. 3). We also analyzed the relationship between age and time course of fibrinogen concentration in the acute phase of TBI (Nakae et al., 2020), and found that decreases in plasma fibrinogen concentration between admission and 3–6h following injury were more severe in older patients aged ≥75y than younger patients aged 16–74y (Fig. 4). A higher consumption of fibrinogen may be a factor in poorer outcomes in aged TBI patients.

Three possible mechanisms may be involved in the bimodal decrease in fibrinogen: consumption associated with coagulation activation, decomposition associated with increased fibrinolysis, or dilution by fluid/transfusion therapy (Schlimp & Schöchl, 2014).

Bleeding obviously leads to plasma fibrinogen consumption. Hiippala et al. (1995) reported that fibrinogen levels dropped earlier than any other hemostatic abnormality when severe bleeding was treated with colloids, crystalloids, or

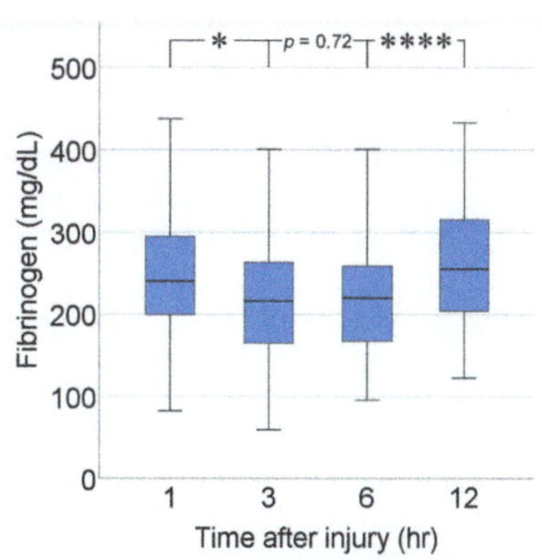

FIG. 3 Admission plasma levels of fibrinogen and levels at 3, 6, and 12h post-traumatic brain injury. Plasma fibrinogen concentration was significantly decreased from admission to 3h after injury, increased insignificantly from 3 to 6h after injury, and continued to increase significantly from 6 to 12h after injury. Grey areas indicate normal ranges for each time. $*P<0.05$, $****P<0.0001$. *Reproduced from Nakae, R., et al. (2016). Time course of coagulation and fibrinolytic parameters in patients with traumatic brain injury.* Journal of Neurotrauma, 33, *688–695, with permission from the publishers.*

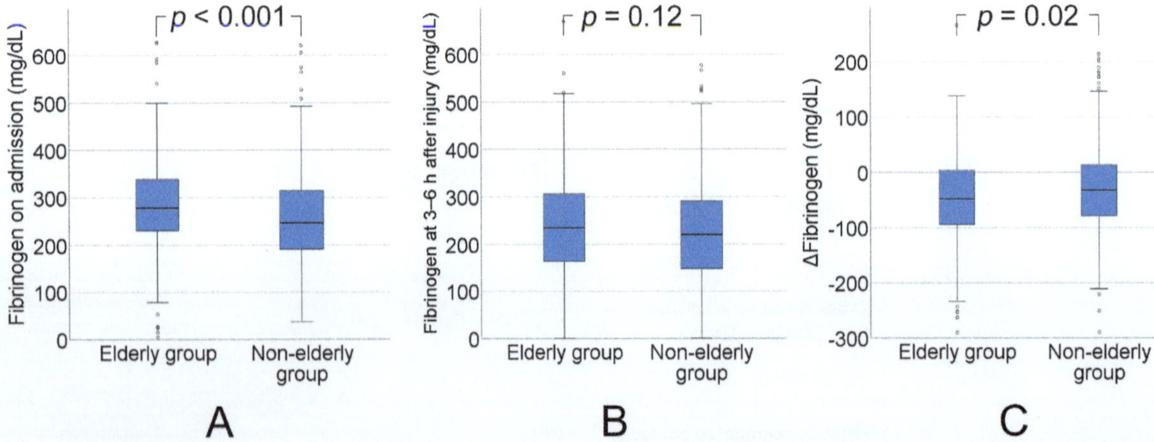

FIG. 4 Admission fibrinogen concentration on admission (A) and at 3–6 h post-traumatic brain injury (B), and the decreases in fibrinogen concentrations between admission and 3–6 h post-traumatic brain injury (C) of older patients (the "elderly group," aged ≥ 75 years) vs younger patients (the "non-elderly group," aged 16–74 years). (A) Admission fibrinogen concentrations were significantly higher in the elderly group than in the non-elderly group. (B) They decreased from admission to 3–6 h after injury in both groups, but there was no significant difference in fibrinogen concentration at 3–6 h after injury between the groups. (C) The reduction in fibrinogen concentration between admission and 3–6 h after injury was significantly higher in the elderly group than in the non-elderly group. ΔFibrinogen = fibrinogen at 3–6 h—fibrinogen on admission. *Reproduced from Nakae, R., et al. (2020). Age-related differences in the time course of coagulation and fibrinolytic parameters in patients with traumatic brain injury.* International Journal of Molecular Sciences, 21, *E5613 under the Creative Commons license CC-BY (http://creativecommons.org/licenses/by/4.0/).*

plasma-depleted red blood cell (RBC) concentrates. After losing 1.5 times the total estimated blood volume, plasma fibrinogen concentrations dropped to 1.0 g/L. McLoughlin, Fontana, Alving, Mongan, and Bunger (1996) measured a fibrinogen concentration of 0.5 g/L followed an exchange of 75% blood volume. However, the cause of low fibrinogen concentrations on admission cannot be caused only by blood loss and fluid resuscitation in the pre-hospital environment. In the acute phase of trauma, especially severe blunt trauma, massive thrombin production occurs, reflecting organ/tissue damage (Engelman, Gabram, Allen, Ens, & Jacobs, 1996; Hayakawa et al., 2011). Dunbar and Chandler (2009) observed that trauma patients with acute coagulopathy show dysregulated hemostasis characterized by the generation of excessive thrombin associated with a combination of circulating pro-coagulants that systematically activate coagulation and, at the same time, reduced levels of coagulation inhibitors. This coagulation cascade acceleration drives the conversion of fibrinogen into fibrin, with fibrinogen undergoing a consumptive decrease. The consumption of fibrinogen following fibrin clot generation significantly increases through the activation of the extrinsic coagulation pathway occurring in the acute phase of TBI.

In the acute phase, fibrinogen degradation (fibrinogenolysis) occurs in addition to fibrin degradation (fibrinolysis), as shown by the fibrinolytic system undergoing increased activation. Primary fibrinogenolysis is independent of coagulation activation in which tissue plasminogen activator (tPA) is either released from injured endothelial cells or by activation of the intrinsic fibrinolytic pathway (Fig. 1). Secondary fibrinolysis is the condition seen in patients with disseminated intravascular coagulation and is pursuant to activation of the coagulation cascade. As a result of hypoperfusion induced by severe trauma and shock, large amounts of tPA are released from the endothelial cells. In addition, thrombomodulin within the endothelium binds to thrombin. Protein C undergoes activation by this thrombin-thrombomodulin complex, and this activated form inactivates FVa and FVIIIa while consuming plasminogen activator inhibitor type 1 (PAI-1), the major tPA antagonist. This results in high concentrations of tPA converting plasminogen into plasmin, which then results in a pro-fibrinolytic state allowing the cleavage of fibrin and fibrinogen. This process also results in the consumption of α_2-plasmin inhibitor (α_2-PI), and in decreased activity of α_2-PI proportional to plasmin activation. Kushimoto, Yamamoto, Shibata, Sato, and Koido (2001) showed that this decrease in α_2-PI activity with a corresponding increase in plasmin was responsible for hyperfibrinolysis and led to a bleeding diathesis from newly created fibrin clot dissolving in the damaged brain.

The major risk of crystalloid resuscitation in these cases is a dilutional coagulopathy. Decreased plasma fibrinogen concentrations are often detected just after admission, when it is important to restore intravascular volume in massively bleeding patients (De Lorenzo, Calatzis, Welsch, & Heindl, 2006; Schlimp, Cadamuro, Solomon, Redl, & Schöchl, 2013). Data from the German trauma registry found that trauma patients received a median of 2.2 L crystalloids in the pre-hospital setting, and that 34% of them had a coagulation abnormality at admission (Levrat et al., 2008). Floccard et al. (2012) obtained blood samples in the pre-hospital environment and found a median plasma fibrinogen concentration of 1.2 g/L in trauma patients with injury severity scores (ISS) > 40. The plasma fibrinogen concentration also tends to decline with hemodilution from subsequent fluid/transfusion therapy (Hiippala, 1998). Although coagulopathy is often perceived as being due to blood loss, extensive hemodilution affects pro-coagulants and pro-fibrinolytic elements, leading to a complex coagulation disorder.

Fibrinogen concentration and outcome in TBI

There is strong evidence of an association between low fibrinogen concentration and poor outcome in trauma patients. McQuilten, Wood, Bailey, Cameron, and Cooper (2017) showed that abnormally decreased fibrinogen concentrations at admission were independently correlated with higher in-hospital mortality rates in their five-year statewide cohort study of 4773 trauma patients. Ishii et al. (2019) observed that an admission plasma fibrinogen level < 1.3 g/L was independently correlated with 28-d mortality in their multicentre retrospective observational study of 666 trauma patients. Rourke et al. (2012) demonstrated that plasma fibrinogen concentrations at admission correlated with the firmness of clots using rotational thromboelastometry (ROTEM®; TEM International GmbH, Munich, Germany) and were independent predictors of early and late mortality in their 517-patient cohort.

Low fibrinogen concentration is also known as a negative prognostic factor in TBI. Lee, Lee, Noh, and Cho (2018) showed that the admission FDP/fibrinogen ratio was a predictor of 28-d mortality in their retrospective observational study of 433 TBI patients. Schöchl, Solomon, et al. (2011) found that the amplitudes of fibrin-based extrinsically activated tests with tissue factor and platelet inhibitor cytochalasin D (FIBTEM) analyses, which is an assay to determine the function of fibrinogen, were significantly lower in non-survivors in their ROTEM® analysis of 88 TBI patients. Wada et al. (2017) observed higher FDP/D-dimer ratios and higher mortality rates in hyper-fibrinolytic TBI patients, reflecting the activation of primary fibrinogenolysis, compared with patients without coagulopathy. These results suggest that early correction of abnormal fibrinogen levels may improve outcomes in TBI.

Therapeutic target level of fibrinogen

The therapeutic target level of fibrinogen is uncertain. American guidelines in 2006 and European guidelines in 2007 following trauma or in the perioperative period recommended a target fibrinogen concentration of 1.0 g/L (Spahn et al., 2007; Tsau, Chen, Tsai, & Sheu, 1991). However, recent guidelines and consensus recommendations from additional prospective, randomized studies (American Society of Anesthesiologists Task Force on Perioperative Blood Management, 2015; Spahn et al., 2019) have resulted in the upward revision of target fibrinogen levels to at least 1.5–2.0 g/L. Clinical data from cardiac surgery, neurosurgery, and gynecologic surgery have demonstrated that perioperative and post-operative hemorrhagic tendencies increase with fibrinogen levels < 1.5 to 2.0 g/L (Blome et al., 2005; Charbit et al., 2007; Gerlach et al., 2002). It is important to recognize that a fibrinogen lower threshold value of 1.5 g/L is the minimum fibrinogen level needed to maintain physiological hemostasis.

Measurements of fibrinogen

The major restriction in using standard laboratory testing for plasma fibrinogen levels is the delay between blood drawing and result availability, which can be > 30 to 60 min. The viscoelastic hemostasis assay (VHA) is an alternative to standard laboratory testing, and is becoming more popular. Two commercially available devices are in widespread use: thromboelastography (TEG®; Hemonetics, Braintree, MA, USA) and ROTEM®. Both devices measure the clotting ability of whole blood under low shear stress, which is hypothesized to mimic in vivo thrombus formation (Winearls et al., 2017). VHA gives information regarding time to clot formation, clot strength, and clot lysis, which allows assessment of different components of the coagulation cascade and their respective contributions to clot kinetics. Assays such as FIBTEM for ROTEM® and Functional Fibrinogen (FF) for TEG® are designed to measure fibrinogen's functional capacity. These assays have been proven to correlate well with standard laboratory testing methodology (Meyer et al., 2015) and can rapidly identify bleeding patients with hypofibrinogenemia (Schöchl, Cotton, et al., 2011; Spahn et al., 2007). Evidence is increasing to support the use of TEG® and ROTEM® to decrease the need for transfusion and to improve outcomes in severe trauma patients. In their retrospective study of 323 severe trauma cases, Schöchl, Cotton, et al. (2011) showed that FIBTEM allows early prediction of the need for massive transfusion. In their single-center randomized controlled trial, Gonzalez et al. (2016) found that utilization of a goal-directed, results-guided massive transfusion protocol (MTP) to resuscitate severely trauma patients greatly improved survival compared to an MTP decision guided by conventional coagulation assays. The TEG-based protocol also utilized fewer fresh frozen plasma (FFP) transfusions during the early phase of resuscitation.

While the administration of the VHA is widely accepted for trauma patients, the advantages for TBI patients are uncertain. In their retrospective study of 88 patients with isolated TBI, Schöchl, Solomon, et al. (2011) showed that plasma fibrinogen concentration was significantly higher in survivors than in non-survivors, and that FIBTEM for ROTEM® had the best predictive factor for mortality. In their prospective, case-control study of 134 patients with isolated TBI requiring craniotomy, Rimaitis et al. (2020) found that ROTEM®-based management led to higher fibrinogen levels and improvement of FIBTEM results in patients with abnormal standard coagulation tests, suggesting better recognition and treatment of fibrinogen deficiency. Further randomized, controlled studies are needed to confirm the benefits of VHA in patients with TBI.

Correlation between fibrinogen and fibrinolytic parameters

We often care for TBI patients with normal plasma fibrinogen concentrations on admission that decline rapidly after admission. We showed that the levels of plasma D-dimer increase abnormally even earlier than plasma fibrinogen levels in the acute phase of TBI. We evaluated the correlation between initial plasma levels of D-dimer and the decreases in fibrinogen levels between admission and 3h following injury (Nakae et al., 2019), and found that admission plasma D-dimer levels significantly negatively correlated with the changes in fibrinogen levels between admission and 3h post-injury (Fig. 5). This suggests that, when the admission plasma D-dimer level at admission is elevated, even if the admission plasma fibrinogen level is within normal limits, supplementation with fibrinogen should be considered in order to maintain high plasma fibrinogen levels. Hayakawa et al. (2016) observed that elevated D-dimer levels on arrival were a strong negative prognostic factor for early death or requirement for massive transfusion in severe trauma patients, even those with high fibrinogen levels. These observations also support our findings.

Treatment by replenishment of fibrinogen

Coagulation factor supplementation is deemed necessary to prevent coagulopathy in TBI. Among the various hemostatic treatments available, replenishment of fibrinogen with FFP, cryoprecipitate, or fibrinogen concentrate have received more attention in the last several years.

In actual clinical practice, FFP is the most frequently used means of coagulation factor replacement. While the immediate administration of FFP, which contains approximately 2.0g/L fibrinogen, is widely accepted as one of the primary treatments for trauma patients, its effects on TBI is uncertain. As for cryoprecipitate and fibrinogen concentrate, there is no evidence of their effectiveness for TBI. Although there remains uncertainty as to the optimal treatment for fibrinogenolysis in TBI, monitoring of plasma fibrinogen concentration by VHA and other methods is understandably important. Prospective, randomized trials of therapeutic alternatives seem warranted.

Applications to other areas of neuroscience

In perioperative brain tumor patients, the plasma fibrinogen concentration has been used as a marker of deep vein thrombosis (DVT) (Guo et al., 2018). Tumor patients are hypercoagulable, with an increased incidence of venous thromboembolism

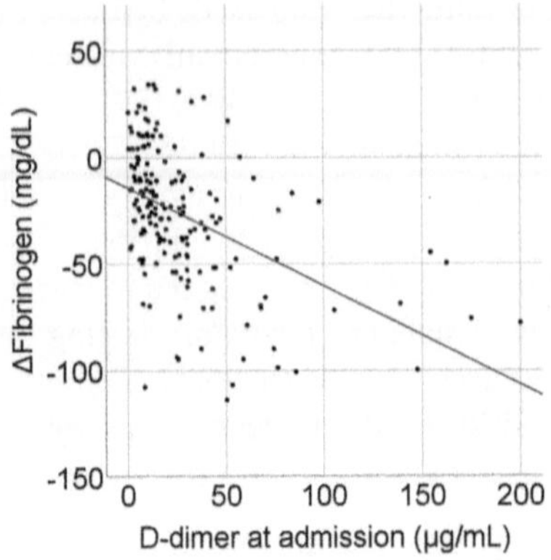

FIG. 5 Correlation between admission D-dimer level and the reduction in fibrinogen level between admission and 3h post-traumatic brain injury. Admission plasma D-dimer levels showed a significantly negative correlation with the changes in fibrinogen levels between admission and 3h following traumatic brain injury. ΔFibrinogen = fibrinogen at 3h—fibrinogen on admission. *Reproduced from Nakae, R., et al. (2019). A retrospective study of the effect of fibrinogen levels during fresh frozen plasma transfusion in patients with traumatic brain injury.* Acta Neurochirurgica, 161, *1943–1953, with permission from the publishers.*

(VTE) (Sorensen, Mellemkjaer, Olsen, & Baron, 2000). The rate of VTE is particularly elevated in patients with brain tumors (Simanek et al., 2007), and this condition is the most frequent complication following craniotomy (Hamilton, Hull, & Pineo, 1994). VTE can be a life-threatening complication when pulmonary thromboembolism forms, so its prediction is vital. The Leiden Thrombophilia Study showed that elevated fibrinogen levels were associated with an increased risk of DVT (Koster et al., 1994).

Fibrinogen is also clinically applied as a hemostatic agent during surgery. The product used most commonly is fibrin glue (Beriplast®, CSL Behring, USA), which contains fibrinogen concentrate/aprotinin and thrombin/calcium chloride solutions. The mechanism is based on separate injections of the highly concentrated adhesive components fibrinogen and thrombin to elicit a hemostatic effect that, when the two components combine in the tissue, allows the formation of a mechanically stable fibrin clot. This is also used for prophylaxis of cerebral spinal fluid dural leakage following craniotomy (Sawamura, Asaoka, Terasaka, Tada, & Uchida, 1999) or trans-sphenoidal microsurgery (Van Velthoven, Clarici, & Auer, 1991), to reinforce aneurysm clips and vascular anastomosis sites (Cho et al., 2017), and for olfactory nerve preservation during surgery with an anterior inter-hemispheric approach (Matano et al., 2016).

Mini-dictionary of terms

- ***Fibrinogen***: A glycoprotein coagulation factor synthesized in the liver. The normal range of plasma fibrinogen is 2.0–4.0 g/L.
- ***Coagulopathy***: A condition characterized by abnormal systemic activation of coagulation and fibrinolysis, potentially causing micro-thrombosis and hemorrhage.
- ***Primary fibrinogenolysis***: Activation of the fibrinogen degradation pathway independently of the activation of coagulation. In the acute phase of traumatic brain injury, primary fibrinogenolysis occurs in addition to secondary fibrinolysis, associated with fibrinolytic system over-activation.
- ***Secondary fibrinolysis***: A fibrin degradation pathway through the activation of coagulation.
- ***FDP***: Abbreviation of fibrinogen/fibrin degradation products. It refers to degradation products of both fibrin and fibrinogen combined.
- ***D-dimer***: Cross-linked fibrin degradation products.

Key facts about fibrinogen

- As far back as the fifth to fourth century BCE, physicians recorded the presence of fibers in circulating blood (Costa-Filho, Hochleitner, Wendt, Teruya, & Spahn, 2016).
- At the end of the 17th century, Malpighi used a single lens microscope to note that whole blood clots have a fibrous and corpuscular component (Douglas, 1999).
- In 1847, the term "fibrinogen" was coined by Virchow, who stated "should one want to give it a name, it could be called fibrinogen" (Virchow, 1847), but Virchow had not discovered fibrinogen (Beck, 1975).
- In 1859, Denis was the first to recognize that plasma contained a clottable substance that differed from fibrin and subsequently attempted to purify and characterize this protein.
- In 1879, Hammarsten was the first to isolate fibrinogen (Rosenfeld, 1982).

Summary points

- This chapter focuses on fibrinogen, which is an important factor in coagulation in patients with traumatic brain injury.
- Fibrinogen is the most reactive coagulation factor, decreasing to a critically low threshold much sooner than any other coagulation factors during severe bleeding.
- Plasma fibrinogen concentrations are often already low at the time of admission after injury, and reach a nadir 3 h after injury.
- There are three possible mechanisms of the decrease in fibrinogen after traumatic brain injury: consumption associated with coagulation activation, decomposition associated with increased fibrinolysis, and dilution by fluid/transfusion therapy.
- A low fibrinogen concentration at admission is a negative prognostic risk factor in patients with traumatic brain injury.
- Admission plasma D-dimer levels have a significantly negative correlation with the changes between admission and 3 h following traumatic brain injury in fibrinogen levels.
- Recent guidelines and consensus recommendations following trauma or for surgery recommend a target fibrinogen concentration of least 1.5–2.0 g/L.

- An alternative to standard laboratory tests, the use of viscoelastic hemostasis assays such as TEG® and ROTEM® to assess the functional capacity of fibrinogen is becoming more popular.
- The effect of fibrinogen supplementation using fresh frozen plasma, cryoprecipitate, or fibrinogen concentrate on traumatic brain injury patients is uncertain.

References

American Society of Anesthesiologists Task Force on Perioperative Blood Management. (2015). Practice guidelines for perioperative blood management: An updated report by the American Society of Anesthesiologists Task Force on Perioperative Blood Management. *Anesthesiology*, *122*, 241–275.

Beck, E. (1975). Historical development of the prothrombin concept. In H. Hemker, & J. Veltkamp (Eds.), *Prothrombin related coagulation factors* (pp. 15–24). Leiden: Leiden University Press.

Blome, M., Isgro, F., Kiessling, A. H., Skuras, J., Haubelt, H., Hellstern, P., et al. (2005). Relationship between factor XIII activity, fibrinogen, haemostasis screening tests and postoperative bleeding in cardiopulmonary bypass surgery. *Thrombosis and Haemostasis*, *93*, 1101–1107.

Charbit, B., Mandelbrot, L., Samain, E., Baron, G., Haddaoui, B., Keita, H., et al. (2007). The decrease of fibrinogen is an early predictor of the severity of postpartum hemorrhage. *Journal of Thrombosis and Haemostasis*, *5*, 266–273.

Cho, A. B., Paulos, R. G., Bersani, G., Iamaguchi, R. B., Torres, L. R., Wei, T. H., et al. (2017). A reinforcement of the sutured microvascular anastomosis with fibrin glue application: A retrospective comparative study with the standard conventional technique. *Microsurgery*, *37*, 218–221.

Costa-Filho, R., Hochleitner, G., Wendt, M., Teruya, A., & Spahn, D. R. (2016). Over 50 years of fibrinogen concentrate. *Clinical and Applied Thrombosis/Hemostasis*, *22*, 109–114.

De Lorenzo, C., Calatzis, A., Welsch, U., & Heindl, B. (2006). Fibrinogen concentrate reverses dilutional coagulopathy induced in vitro by saline but not by hydroxyethyl starch 6%. *Anesthesia and Analgesia*, *102*, 1194–1200.

Douglas, S. (1999). Coagulation history, Oxford 1951–53. *British Journal of Haematology*, *107*, 22–32.

Drake, T. A., Morrissey, J. H., & Edgington, T. S. (1989). Selective cellular expression of tissue factor in human tissues. Implications for disorders of hemostasis and thrombosis. *American Journal of Pathology*, *134*, 1087–1097.

Dunbar, N. M., & Chandler, W. L. (2009). Thrombin generation in trauma patients. *Transfusion*, *49*, 2652–2660.

Eddleston, M., de la Torre, J. C., Oldstone, M. B., Loskutoff, D. J., Edgington, T. S., & Mackman, N. (1993). Astrocytes are the primary source of tissue factor in the murine central nervous system. A role for astrocytes in cerebral hemostasis. *Journal of Clinical Investigation*, *92*, 349–358.

Engelman, D. T., Gabram, S. G., Allen, L., Ens, G. E., & Jacobs, L. M. (1996). Hypercoagulability following multiple trauma. *World Journal of Surgery*, *20*, 5–10.

Epstein, D. S., Mitra, B., O'Reilly, G., Rosenfeld, J. V., & Cameron, P. A. (2014). Acute traumatic coagulopathy in the setting of isolated traumatic brain injury: A systematic review and meta-analysis. *Injury*, *45*, 819–824.

Floccard, B., Rugeri, L., Faure, A., Saint Denis, M., Boyle, E. M., Peguet, O., et al. (2012). Early coagulopathy in trauma patients: An on-scene and hospital admission study. *Injury*, *43*, 26–32.

Gerlach, R., Tolle, F., Raabe, A., Zimmermann, M., Siegemund, A., & Seifert, V. (2002). Increased risk for postoperative hemorrhage after intracranial surgery in patients with decreased factor XIII activity: Implications of a prospective study. *Stroke*, *33*, 1618–1623.

Gonzalez, E., Moore, E. E., Moore, H. B., Chapman, M. P., Chin, T. L., Ghasabyan, A., et al. (2016). Goal-directed hemostatic resuscitation of trauma-induced coagulopathy: A pragmatic randomized clinical trial comparing a viscoelastic assay to conventional coagulation assays. *Annals of Surgery*, *263*, 1051–1059.

Goodnight, S. H., Kenoyer, G., Rapaport, S. I., Patch, M. J., Lee, J. A., & Kurze, T. (1974). Defibrination after brain-tissue destruction: A serious complication of head injury. *New England Journal of Medicine*, *290*, 1043–1047.

Guo, X., Zhang, F., Wu, Y., Gao, L., Wang, Q., Wang, Z., et al. (2018). Coagulation alteration and deep vein thrombosis in brain tumor patients during the perioperative period. *World Neurosurgery*, *114*, e982–e991.

Hager, K., Felicetti, M., Seefried, G., & Platt, D. (1994). Fibrinogen and aging. *Aging*, *6*, 133–138.

Hamilton, M. G., Hull, R. D., & Pineo, G. F. (1994). Venous thromboembolism in neurosurgery and neurology patients: A review. *Neurosurgery*, *34*, 280–296 (discussion 296).

Harhangi, B. S., Kompanje, E. J., Leebeek, F. W., & Maas, A. I. (2008). Coagulation disorders after traumatic brain injury. *Acta Neurochirurgica*, *150*, 165–175. discussion 175.

Hayakawa, M., Maekawa, K., Kushimoto, S., Kato, H., Sasaki, J., Ogura, H., et al. (2016). High D-dimer levels predict a poor outcome in patients with severe trauma, even with high fibrinogen levels on arrival: A multicenter retrospective study. *Shock*, *45*, 308–314.

Hayakawa, M., Sawamura, A., Gando, S., Kubota, N., Uegaki, S., Shimojima, H., et al. (2011). Disseminated intravascular coagulation at an early phase of trauma is associated with consumption coagulopathy and excessive fibrinolysis both by plasmin and neutrophil elastase. *Surgery*, *149*, 221–230.

Hess, J. R., Brohi, K., Dutton, R. P., Hauser, C. J., Holcomb, J. B., Kluger, Y., et al. (2008). The coagulopathy of trauma: A review of mechanisms. *Journal of Trauma*, *65*, 748–754.

Hiippala, S. (1998). Replacement of massive blood loss. *Vox Sanguinis*, *74*(Suppl 2), 399–407.

Hiippala, S. T., Myllyla, G. J., & Vahtera, E. M. (1995). Hemostatic factors and replacement of major blood loss with plasma-poor red cell concentrates. *Anesthesia and Analgesia*, *81*, 360–365.

Ishii, K., Kinoshita, T., Kiridume, K., Watanabe, A., Yamakawa, K., Nakao, S., et al. (2019). Impact of initial coagulation and fibrinolytic markers on mortality in patients with severe blunt trauma: A multicentre retrospective observational study. *Scandinavian Journal of Trauma, Resuscitation and Emergency Medicine*, *27*, 25.

Keimowitz, R. M., & Annis, B. L. (1973). Disseminated intravascular coagulation associated with massive brain injury. *Journal of Neurosurgery*, *39*, 178–180.

Koster, T., Rosendaal, F. R., Reitsma, P. H., van der Velden, P. A., Briet, E., & Vandenbroucke, J. P. (1994). Factor VII and fibrinogen levels as risk factors for venous thrombosis. A case-control study of plasma levels and DNA polymorphisms–the Leiden thrombophilia study (LETS). *Thrombosis and Haemostasis*, *71*, 719–722.

Kurland, D., Hong, C., Aarabi, B., Gerzanich, V., & Simard, J. M. (2012). Hemorrhagic progression of a contusion after traumatic brain injury: A review. *Journal of Neurotrauma*, *29*, 19–31.

Kushimoto, S., Yamamoto, Y., Shibata, Y., Sato, H., & Koido, Y. (2001). Implications of excessive fibrinolysis and alpha(2)-plasmin inhibitor deficiency in patients with severe head injury. *Neurosurgery*, *49*, 1084–1089. discussion 1089–1090.

Laroche, M., Kutcher, M. E., Huang, M. C., Cohen, M. J., & Manley, G. T. (2012). Coagulopathy after traumatic brain injury. *Neurosurgery*, *70*, 1334–1345.

Lee, D. H., Lee, B. K., Noh, S. M., & Cho, Y. S. (2018). High fibrin/fibrinogen degradation product to fibrinogen ratio is associated with 28-day mortality and massive transfusion in severe trauma. *European Journal of Trauma and Emergency Surgery*, *44*, 291–298.

Levrat, A., Gros, A., Rugeri, L., Inaba, K., Floccard, B., Negrier, C., et al. (2008). Evaluation of rotation thrombelastography for the diagnosis of hyperfibrinolysis in trauma patients. *British Journal of Anaesthesia*, *100*, 792–797.

Maegele, M. (2013). Coagulopathy after traumatic brain injury: Incidence, pathogenesis, and treatment options. *Transfusion*, *53*(Suppl 1), 28S–37S.

Maegele, M. (2014). The coagulopathy of trauma. *European Journal of Trauma and Emergency Surgery*, *40*, 113–126.

Maegele, M., Schöchl, H., & Cohen, M. J. (2014). An update on the coagulopathy of trauma. *Shock*, *41*(Suppl 1), 21–25.

Matano, F., Murai, Y., Mizunari, T., Tateyama, K., Kobayashi, S., Adachi, K., et al. (2016). Olfactory preservation during anterior interhemispheric approach for anterior skull base lesions: Technical note. *Neurosurgical Review*, *39*, 63–68. discussion 69.

McLoughlin, T. M., Fontana, J. L., Alving, B., Mongan, P. D., & Bunger, R. (1996). Profound normovolemic hemodilution: Hemostatic effects in patients and in a porcine model. *Anesthesia and Analgesia*, *83*, 459–465.

McQuilten, Z. K., Wood, E. M., Bailey, M., Cameron, P. A., & Cooper, D. J. (2017). Fibrinogen is an independent predictor of mortality in major trauma patients: A five-year statewide cohort study. *Injury*, *48*, 1074–1081.

Meyer, M. A. S., Ostrowski, S. R., Sorensen, A. M., Meyer, A. S. P., Holcomb, J. B., Wade, C. E., et al. (2015). Fibrinogen in trauma, an evaluation of thrombelastography and rotational thromboelastometry fibrinogen assays. *Journal of Surgical Research*, *194*, 581–590.

Mosesson, M. W. (2005). Fibrinogen and fibrin structure and functions. *Journal of Thrombosis and Haemostasis*, *3*, 1894–1904.

Nakae, R., Fujiki, Y., Takayama, Y., Kanaya, T., Igarashi, Y., Suzuki, G., et al. (2020). Age-related differences in the time course of coagulation and fibrinolytic parameters in patients with traumatic brain injury. *International Journal of Molecular Sciences*, *21*, E5613.

Nakae, R., Takayama, Y., Kuwamoto, K., Naoe, Y., Sato, H., & Yokota, H. (2016). Time course of coagulation and fibrinolytic parameters in patients with traumatic brain injury. *Journal of Neurotrauma*, *33*, 688–695.

Nakae, R., Yokobori, S., Takayama, Y., Kanaya, T., Fujiki, Y., Igarashi, Y., et al. (2019). A retrospective study of the effect of fibrinogen levels during fresh frozen plasma transfusion in patients with traumatic brain injury. *Acta Neurochirurgica*, *161*, 1943–1953.

Rimaitis, M., Bilskiene, D., Tamosuitis, T., Vilcinis, R., Rimaitis, K., & Macas, A. (2020). Implementation of thromboelastometry for coagulation management in isolated traumatic brain injury patients undergoing craniotomy. *Medical Science Monitor*, *26*, e922879.

Rosenfeld, L. (1982). *Origins of clinical chemistry: The evolution of protein analysis*. New York: Academic Press. Chapter 18.

Rourke, C., Curry, N., Khan, S., Taylor, R., Raza, I., Davenport, R., et al. (2012). Fibrinogen levels during trauma hemorrhage, response to replacement therapy, and association with patient outcomes. *Journal of Thrombosis and Haemostasis*, *10*, 1342–1351.

Sawamura, Y., Asaoka, K., Terasaka, S., Tada, M., & Uchida, T. (1999). Evaluation of application techniques of fibrin sealant to prevent cerebrospinal fluid leakage: A new device for the application of aerosolized fibrin glue. *Neurosurgery*, *44*, 332–337.

Schlimp, C. J., Cadamuro, J., Solomon, C., Redl, H., & Schöchl, H. (2013). The effect of fibrinogen concentrate and factor XIII on thromboelastometry in 33% diluted blood with albumin, gelatine, hydroxyethyl starch or saline in vitro. *Blood Transfusion. Trasfusione del Sangue*, *11*, 510–517.

Schlimp, C. J., & Schöchl, H. (2014). The role of fibrinogen in trauma-induced coagulopathy. *Hämostaseologie*, *34*, 29–39.

Schöchl, H., Cotton, B., Inaba, K., Nienaber, U., Fischer, H., Voelckel, W., et al. (2011). FIBTEM provides early prediction of massive transfusion in trauma. *Critical Care (London, England)*, *15*, R265.

Schöchl, H., Solomon, C., Traintinger, S., Nienaber, U., Tacacs-Tolnai, A., Windhofer, C., et al. (2011). Thromboelastometric (ROTEM) findings in patients suffering from isolated severe traumatic brain injury. *Journal of Neurotrauma*, *28*, 2033–2041.

Simanek, R., Vormittag, R., Hassler, M., Roessler, K., Schwarz, M., Zielinski, C., et al. (2007). Venous thromboembolism and survival in patients with high-grade glioma. *Neuro-Oncology*, *9*, 89–95.

Sorensen, B., Larsen, O. H., Rea, C. J., Tang, M., Foley, J. H., & Fenger-Eriksen, C. (2012). Fibrinogen as a hemostatic agent. *Seminars in Thrombosis and Hemostasis*, *38*, 268–273.

Sorensen, H. T., Mellemkjaer, L., Olsen, J. H., & Baron, J. A. (2000). Prognosis of cancers associated with venous thromboembolism. *New England Journal of Medicine*, *343*, 1846–1850.

Spahn, D. R., Bouillon, B., Cerny, V., Duranteau, J., Filipescu, D., Hunt, B. J., et al. (2019). The European guideline on management of major bleeding and coagulopathy following trauma: Fifth edition. *Critical Care (London, England)*, *23*, 98.

Spahn, D. R., Cerny, V., Coats, T. J., Duranteau, J., Fernandez-Mondejar, E., Gordini, G., et al. (2007). Management of bleeding following major trauma: A European guideline. *Critical Care (London, England)*, *11*, R17.

Tsau, Y. K., Chen, C. H., Tsai, W. S., & Sheu, J. N. (1991). Complications of nephrotic syndrome in children. *Journal of the Formosan Medical Association, 90*, 555–559.

Van Velthoven, V., Clarici, G., & Auer, L. M. (1991). Fibrin tissue adhesive sealant for the prevention of CSF leakage following transsphenoidal microsurgery. *Acta Neurochirurgica, 109*, 26–29.

Virchow, R. (1847). Faserstoffarten und fibrinogene Substanz. *Virchows Archiv, 1*, 572–583.

Wada, T., Gando, S., Maekaw, K., Katabami, K., Sageshima, H., Hayakawa, M., et al. (2017). Disseminated intravascular coagulation with increased fibrinolysis during the early phase of isolated traumatic brain injury. *Critical Care (London, England), 21*, 219.

Wafaisade, A., Lefering, R., Tjardes, T., Wutzler, S., Simanski, C., Paffrath, T., et al. (2010). Acute coagulopathy in isolated blunt traumatic brain injury. *Neurocritical Care, 12*, 211–219.

Wilkerson, W. R., & Sane, D. C. (2002). Aging and thrombosis. *Seminars in Thrombosis and Hemostasis, 28*, 555–568.

Winearls, J., Campbell, D., Hurn, C., Furyk, J., Ryan, G., Trout, M., et al. (2017). Fibrinogen in traumatic haemorrhage: A narrative review. *Injury, 48*, 230–242.

Chapter 9

Linking traumatic brain injury and nosocomial infections

Marwan Bouras[a,b], Karim Asehnoune[a,b], and Antoine Roquilly[a,b]
[a]*Surgical Intensive Care Unit, Hotel-Dieu, University Hospital of Nantes, Nantes, France,* [b]*Anti-Infectious Therapeutics, Health Research Institute 2, Medical University of Nantes, Nantes, France*

List of abbreviations

ACTH	adreno corticotropic hormone
APCs	antigen-presenting cell
ARDS	acute respiratory distress syndrome
CARS	compensatory anti-inflammatory response syndrome
CIRCI	critical illness-related corticosteroid insufficiency
CRH	corticotropin-releasing hormone
CRP	C-reactive protein
DAMPS	damage-associated molecular patterns
DC	dendritic cell
GMCSF	granulocyte-macrophage colony-stimulating factor
HAP	hospital-acquired pneumonia
HPA	hypothalamic-pituitary-adrenal
ICU	intensive care unit
IL	interleukine
INF	interferon
NK	natural killer
SIRS	systemic inflammatory response syndrome
TBI	traumatic brain injury
TLR	Toll-like receptor
TNF	tumor necrosis factors
VAP	ventilatory-acquired pneumonia

Introduction

Traumatic brain injury (TBI) is a major cause of morbidity and mortality, especially in young people (Lozano et al., 2012). In addition to high initial mortality, the severity of these acute brain injuries also includes significant morbidity and disability (Bombardier et al., 2010). While most of the morbidity and mortality factors are direct consequences of the trauma itself (initial severity of lesions, hemorrhage), about one-third are preventable complications, including nosocomial infectious complications (Myburgh et al., 2008).

Hospital-acquired pneumonia (HAP), especially ventilator-associated pneumonia (VAP), is the most common nosocomial infection in intensive care unit (ICU) patients (de Miguel-Díez et al., 2017). The morbidity and health costs of ventilator-associated pneumonia (VAP) make it one of the most severe care complications (Ohannessian et al., 2018). Between 40% and 60% of patients with severe TBI in ICU will develop VAP, half of which caused by methicillin-sensitive *Staphylococcus aureus* (Asehnoune et al., 2014). During hospitalization in the intensive care unit, a patient with traumatic brain injury suffering from VAP will experience more systemic cerebral aggression (fever, hypotension, hypoxia, hypercapnia); mechanical ventilation will also be prolonged, as well as the stay in ICU. On the one hand, all these factors contribute

Diagnosis and Treatment of Traumatic Brain Injury. https://doi.org/10.1016/B978-0-12-823347-4.00040-3

to the appearance of new ischemic cerebral lesions on hypoxia and, on the other hand, to a delay in post-ICU care (rehabilitation…). The development of a VAP in a patient with TBI is an independent risk factor for poorer long-term neurological outcomes (Bronchard et al., 2004).

The medical consequences of VAP are multiple with prolonged hospitalization, long-term asthenia and depression, and increased risk of death (Melsen et al., 2013). The cost for the society is high; the average cost for each episode of VAP is 40.000 euros (Safdar, Dezfulian, Collard, & Saint, 2005). Moreover, difficult-to-treat multi-resistant bacteria are commonly involved in VAP, making their prevention and treatment more difficult in the next decade (Bassetti et al., 2017). Given the morbidity and mortality associated with pneumonia, their prevention has been the subject of numerous studies, and different drug and non-drug strategies have been proposed. International recommendations for the management of hospital-acquired pneumonia have recently been reviewed: in 2016 for the United States (Kalil et al., 2016) and 2017 for Europe (Torres et al., 2017). The effectiveness of measures to prevent pneumonia remains controversial, and they do not all recommend the same means of prevention. Over the last decade, the incidence of care-associated pneumonia has declined only slightly in the ICU, particularly in traumatic brain-injured patients requiring prolonged mechanical ventilation (Roquilly et al., 2020). A renewed understanding of the pathophysiology and the close link between TBI and nosocomial pneumonia and objective effects that clinicians should expect from effective prevention and treatment gives us hope that we can make good progress in the coming years (Roquilly, Torres, Dickson, Becher, & Asehnoune, et al., 2019).

Post-traumatic immunosuppression

The susceptibility of patients suffering from TBI to nosocomial infections and particularly to VAP is due to the development of an immunodepression that sets in following cerebral aggression (Woiciechowsky et al., 1998). In the aftermath of a severe brain injury, the production of pro-inflammatory cytokines is a necessary physiological phenomenon that favors, on the one hand, the healing of contused tissue and, on the other hand, the defense against the appearance of secondary bacterial infections. However, an exacerbated Systemic Inflammatory Response Syndrome (SIRS) can trigger multiple organ failure syndromes with a very high mortality rate (>50%). To regulate the exacerbated pro-inflammatory response, the central nervous system, in association with the cell regulatory mechanisms of innate immunity, triggers a systemic anti-inflammatory response (Compensatory Anti-inflammatory Response Syndrome [CARS]). The aim of this response is a return to homeostasis. In the brain, after acute aggression, the first cells to suffer are glial cells (Colton, 2009). The extracellular release and accumulation of many normally intracellular components (such as adenosine triphosphate) are then recognized as Damage-Associated Molecular Patterns (DAMPS) and activate the TLR receptors of glial cells, macrophages, dendritic cells, and endothelial cells as well as leukocytes. This activation induces the production of pro-inflammatory cytokines and stimulates antigen delivery from dendritic cells to effector cells (T and NK lymphocytes) (Silver & Miller, 2004). This pro-inflammatory cascade is initially counterbalanced by the release of neurotransmitters, which activate the anti-inflammatory receptors of microglia through a feedback loop and induces a double M1/M2 polarization of cerebral macrophages (Morganti, Riparip, & Rosi, 2016). However, as brain cells are destroyed, neurons enter in apoptosis, and neurotransmitters are depleted, leading to a generalization of this local pro-inflammation state (Kawabori et al., 2015). In severe TBI, permeabilization of the blood-brain barrier also leads to the contact of central nervous system epitopes (such as the basic myelin protein) with cells of innate immunity. The recognition of these antigens as DAMPS could maintain the state of local inflammation (Schwartz, 2001). It is to fight against an overwhelming pro-inflammatory response, potentially causing multiple organ failure syndromes and against subsequent immune aggressions that the CNS induces a systemic anti-inflammatory response leading to an increased susceptibility to infections (Meisel, Schwab, Prass, Meisel, & Dirnagl, 2005) (Fig. 1). This immunosuppressive response mainly involves cells of innate immunity through neurohormonal mechanisms. Several neurohormonal systems are involved in this immunodepression induced by traumatic brain injury:

- The sympathetic nervous system (Engel et al., 2015). All types of acute lesions of the central nervous system can lead to direct lesions of the sympathetic centers, which are very involved in immunomodulation phenomena. These sympathetic centers are situated in the brain and the spinal cord. The sympathetic neurons innervate the adrenal medulla, the spleen, the thymus, and the lymph nodes;
- The parasympathetic nervous system (Wang et al., 2003). For example, cholinergic neurons have been shown to inhibit acute inflammation; the parasympathetic nervous system thus reflexively regulates the inflammatory response in real-time after TBI, just as it controls heart rate and other vital functions;
- The hypothalamic-pituitary axis via the secretion of glucocorticoids (Annane et al., 2017). Impulses from the site of brain injury stimulate the secretion of ACTH from the pituitary gland. Cortisol is thus secreted by the fasciculata area of

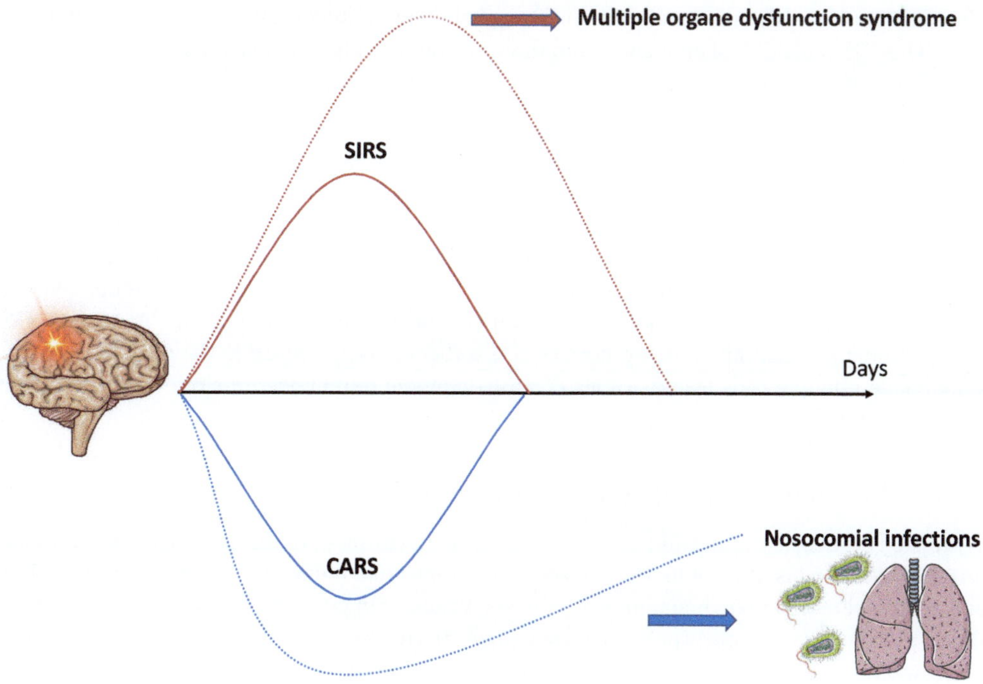

FIG. 1 Post-traumatic inflammatory and anti-inflammatory response in physiological conditions (*solid*) and pathological conditions (*dashed*) leading to post-traumatic immunosuppression: the *red lines* represent the physiological (*solid*) and pathological (*dashed*) inflammatory responses. The *blue lines* represent the physiological (*solid*) anti-inflammatory responses allowing a return to homeostasis and pathological (*dashed*) responses leading to nosocomial infections.

the adrenal cortex, prolonging and deepening this state of immunosuppression. In the hours following the neurological event, these patients have an exuberant cortisol secretion (see below).

There is a close link between the neurohormonal systems damaged by the TBI and the innate immune system.
At the cellular level, this immunodepression develops at several levels:

- An alteration in the antigen presentation capacity of the presenting cells (APCs). These are represented mainly by dendritic cells (DCs) and monocytes. The DCs have a central role in capturing and presenting the antigen and activation of lymphocytes (T and NK) via the secretion of pro-inflammatory cytokines (such as Interleukin-12). These functions, which are essential for developing an inflammatory response, are altered in traumatic brain-injured patients, notably with a decrease in the number and reactivity of DCs (Bouras, Asehnoune, & Roquilly, 2018). For example, our team demonstrated a reduction in the membrane expression of the major histocompatibility complex type II (MHC II) (Roquilly et al., 2015).
- A decrease in the secretion capacity of pro-inflammatory cytokines. Cooperation between APCs and lymphocytes in acute brain injury is essential (Brown et al., 2016). Thus, an alteration in IL-12 production as a consequence of TBI will induce a decrease in the secretion of pro-inflammatory cytokines such as INF-g by lymphoid cells (Spolarics et al., 2003). Furthermore, the sympathetic activation deepens this immunosuppressive state by stimulating the release of anti-inflammatory cytokines (IL-10) (Roquilly et al., 2014). The alteration of this "loop" of the innate immune response causes a decrease in the patient ability to fight nosocomial infection (Deknuydt, Roquilly, Cinotti, Altare, & Asehnoune, 2013). Furthermore, Wong et al. have also shown the critical role of sympathetic activation in immunosuppression after acute brain injury by demonstrating the alteration of the noradrenergic activation of iNKTs (Wong, Jenne, Lee, Leger, & Kubes, 2011);
- T cell exhaustion, a phenomenon also found in post-septic immunosuppression. This exhaustion corresponds to the progressive loss of pro-inflammatory functions of lymphocytes in the presence of a high antigenic load (Bandyopadhyay et al., 2007). T-lymphocytes in traumatic brain-injured patients have a higher expression of the PD1 receptor and the tolerogenic transcription factor BLIMP1 (Yi, Cox, & Zajac, 2010). This immunosuppressive microenvironment is mainly found in the pulmonary parenchyma, where PD1-ligand induces IL-10 overproduction

and lymphocyte apoptosis (Boomer et al., 2011). This phenomenon may partly explain the high incidence of VAP in TBI patients. This "exhaustion" phenomenon and the resulting lymphopenia is a risk factor for mortality (Heffernan et al., 2012). This lymphopenia persists beyond 6 months in most patients.

After a severe TBI, the mechanisms leading to the inflammatory state are very early. Innate immune cells release pro-inflammatory cytokines, the most important of which are IL-1-b, TNF-a, and IL-6, whose half-lives are shorter than 20 min. IL-6 is first detectable in plasma within 1 h after trauma and correlates with its severity (Nast-Kolb et al., 1997). Elevated levels of pro-inflammatory cytokines have also been found in cerebrospinal fluid or damaged nerve tissue, particularly after severe TBI. These pro-inflammatory cytokines activate innate immune cells and inhibit the apoptosis of neutrophils observed after trauma. IL-6 also has anti-inflammatory activity by inducing prostaglandin production that stimulates the release of the potent anti-inflammatory cytokine IL-10 (Köller, Clasbrummel, Kollig, Hahn, & Muhr, 1998).

After a severe traumatic brain injury, neuro-hormonal disorders closely related to the alteration of the immunological capacities of the innate immune cells lead to a state of post-traumatic immunodepression, which favors the occurrence of nosocomial infections in ICU.

Post-traumatic critical illness-related corticosteroid insufficiency

Critical illness-related corticosteroid insufficiency (CIRCI) is the impairment of the hypothalamic-pituitary–adrenal (HPA) axis during critical illness and is thought to occur in acute conditions like trauma (Annane et al., 2017). It occurs in 50% to 75% of patients with severe traumatic brain injury and is associated with poor outcomes (Cohan et al., 2005).

In critically ill patients, reduced cortisol breakdown contributes to abnormal blood cortisol levels (Boonen et al., 2013). This CIRCI corresponds to the impairment of the hypothalamic pituitary axis (HPA) resulting from the inadequate anti-inflammatory activity for the severity of the medical condition (Marik et al., 2008). Three major pathophysiologic events account for CIRCI: dysregulation of the HPA axis, altered cortisol metabolism, and tissue resistance to corticosteroids (Annane et al., 2017). In severe traumatic brain-injured patients, CIRCI occurs frequently and is associated with uncontrolled inflammatory response, more prolonged vasopressor infusions, and poor outcomes (Hoen et al., 2002).

The post-traumatic alteration of the HPA axis is also a consequence of the post-traumatic cytokine storm. For example, interleukin-6 (IL-6), is a pro-inflammatory cytokine known to have HPA-activating activity independent of ACTH. Thus, human IL-6 increases plasma concentrations cortisol in mice and during immune system activation, such as post-traumatic inflammation, IL-6 is a potent CRH independent stimulator of the HPA axis (Bethin, Vogt, & Muglia, 2000). In head-injured children, serum IL-6 and CRP levels are elevated and correlated to the severity of head trauma (Mastorakos, Chrousos, & Weber, 1993) and increased levels of IL-6 in the early phase of severe acute traumatic brain injury is associated with the high inflammatory response such as the development of ARDS. This significant inflammatory state leads to a compensatory anti-inflammatory response (CARS) and, therefore, to a more marked state of immunosuppression, facilitating the occurrence of nosocomial infections. Structural damage to the hypothalamic-pituitary glands could also cause the development of CIRCI in TBI patients. The blood-brain barrier's rupture inducing neuronal death in the hypothalamus is considered a pathological feature of HPA axis dysfunction associated with TBI. Hence, the impairment of HPA axis during critical illness provides the sustainability of the CARS and induces nosocomial infections (Schneider, Kreitschmann-Andermahr, Ghigo, Stalla, & Agha, 2007).

Post-traumatic dysbiosis

Recent microbiome association studies showed that the gut microbiome is involved in the regulation of many vital functions. Many factors can affect microbial organisms residing in the gut, leading individuals to microbial dysbiosis (Hall, Tolonen, & Xavier, 2017). This state, in which the standard structure of the microbial population is disrupted, is a crucial feature of metabolic and immune disorders. Experimental models have been essential for moving research in the human microbiota gut-brain axis forward (Sherwin, Dinan, & Cryan, 2018). The microbiota gut-brain axis comprises the central nervous system, the endocrine and immune systems, the sympathetic and parasympathetic systems, the enteric nervous system, and the intestinal microbiota (Grenham, Clarke, Cryan, & Dinan, 2011). The gut-brain axis can be defined as a physiological model composed of afferent and efferent signals of neural, hormonal, and immunological origins. The dysfunction of this axis can have many consequences, like infectious or immune disorders (Cryan & Dinan, 2012). Sympathetic and parasympathetic innervations modulate the intestinal function and are likely to mediate immune activity in the gastrointestinal tract (Mayer, 2011). TBI can alter the composition and functions of the microbiota by altering intestinal permeability, allowing bacterial antigens to penetrate the epithelium and stimulate an immune response in the mucosa. Mucosal immunity is ubiquitous in the intestinal lumen. Patients with severe TBI suffer from an acute disorder of microbial homeostasis in the gut, such as an increased secretion of interferon γ from colonic mucosal cells, leading to an increase in permeability (Demaude, Salvador-Cartier, Fioramonti, Ferrier, & Bueno, 2006). The overactivation of the sympathetic nervous system linked to TBI leads to an increase in the number and degranulation and activity of mast cells, resulting in local vasodilatation and loss of the impermeability of the intestinal barrier (Rhee, Pothoulakis, & Mayer, 2009). Traumatic brain injury can induce significant damage to the gut structure and impairment of barrier function, which occurs rapidly as early as 3h following brain injury and lasts for more than 7days with marked mucosal atrophy (Hang, Shi, Li, Wu, & Yin, 2003). An experimental trial in mice (Houlden et al., 2016) shows that acute brain injury caused changes in the gut altered the composition of cecal microbiota, correlating with the injury's extent. These effects are mediated by noradrenaline release from the autonomic nervous system with altered cecal mucoprotein production leading to post-traumatic complications like infections (Treangen, Wagner, Burns, & Villapol, 2018). TBI disrupt communication between the intestine and the vagal complex, causing dysautonomia, which is characterized by increasing post-traumatic SIRS (Nicholson et al., 2019) and the release of stress hormones in the hypothalamic-pituitary–adrenal axis and can also result in intestinal changes in permeability, motility, and secretion (Grenham et al., 2011). The intestinal barrier of critically ill patients suffering from TBI is permeable, and the degree of permeability correlates with subsequent risks of organ injury and death (Doig et al., 1998). The physiological changes induced by TBI alters the composition and enrichment of the intestinal microbiome, the behavior and virulence of individual pathogens are also changed, and a bacterial strain that is usually inert and invisible to the host immune system can be transformed by the post-traumatic disorder, gaining virulence and induce systemic inflammation and sepsis (Dickson, 2016). The appearance of nosocomial infections after a TBI secondary to a dysbiosis of the gut microbiota is the result of three different mechanisms:

- The "gut-derived sepsis" by bacterial translocation,
- The "gut-lymph hypothesis" (Deitch, 2012): The lower gastrointestinal tract drains into mesenteric lymph nodes, which in turn empties to the thoracic duct and empties into the sub-clavian vein. Therefore, the lungs are the first body's first capillary system to filter the lymph from the digestive tract, leading to pneumonia development
- The "in situ virulence" via the persistence of local inflammation maintained by sympathetic, parasympathetic, and hypothalamic-hypophyseal neuro-mediators

The gut microbiota is not the only one to be disturbed by traumatic brain injury. The lungs, the organ most affected by nosocomial infections, have a very rich microbiota disturbed by the trauma and favors VAP occurrence. The lungs were considered a sterile organ for decades. New culture-independent techniques have shown that healthy lungs present a dynamic ecosystem of bacteria, helping to understand that dysbiosis of the respiratory microbiome and altered mucosal immunity is a key to understanding pulmonary infection's pathophysiology in ICU (Roquilly et al., 2019). In the lungs of critically ill patients suffering from TBI, the normal microbiome becomes rapidly disrupted. Bacterial diversity is decreased, and commensals might become displaced by potential pathogens, often originating from other ecosystems (gut). In these conditions, the pulmonary microbiome was frequently enriched with gut-associated bacteria, primarily represented by Bacteroidetes and Enterobacteriaceae (Dickson et al., 2016). Recent studies in trauma patients showed that the decrease in bacterial diversity of the lung microbiome leads to nosocomial pulmonary infection (Panzer et al., 2018). Impaired pulmonary clearance of the pathogens due to the post-traumatic immune dysfunction might perpetuate the circle of inflammation and in-situ pulmonary injury. Both immune dysfunction and dysbiosis might persist for extended periods. Patients continue to be susceptible to infections with nosocomial pathogens and are more likely to develop several sepsis episodes (Roquilly et al., 2011) (Figs. 1–3).

Glucocorticoids: A therapeutic avenue in the prevention of nosocomial infections after trauma

To correct the post-traumatic immunosuppression and in prevention for nosocomial infections, many therapies have been tested in recent years. They aimed either to limit the initial SIRS (and thus the CARS), notably by using low-dose glucocorticoids (Asehnoune et al., 2014; Roquilly et al., 2011), or to restore the antigen presentation or cytokine secretion capacities by using glucans, IFN-g, GM-CSF or Interleukin-12 (Roquilly et al., 2017). In a large randomized trial in multiple trauma patients, we found that hydrocortisone therapy prevented the development of hospital-acquired pneumonia by day 28 in patients with CIRCI (defined by a change in baseline cortisol at 60min of $<9\,\mu g/dL$ after ACTH ($250\,\mu g$) administration) (Roquilly et al., 2011) and we found a strong trend in conclusion in TBI patients (Asehnoune et al., 2014).

In TBI patients, corticosteroids seem to be an interesting avenue of investigation. The administration of corticosteroids in this context of relative post-traumatic immunosuppression may certainly seem inappropriate, but the effects of corticosteroids, especially at low doses, are multiple. Indeed, in cases of TBI, corticosteroids can reduce the occurrence of opportunistic infections in two ways:

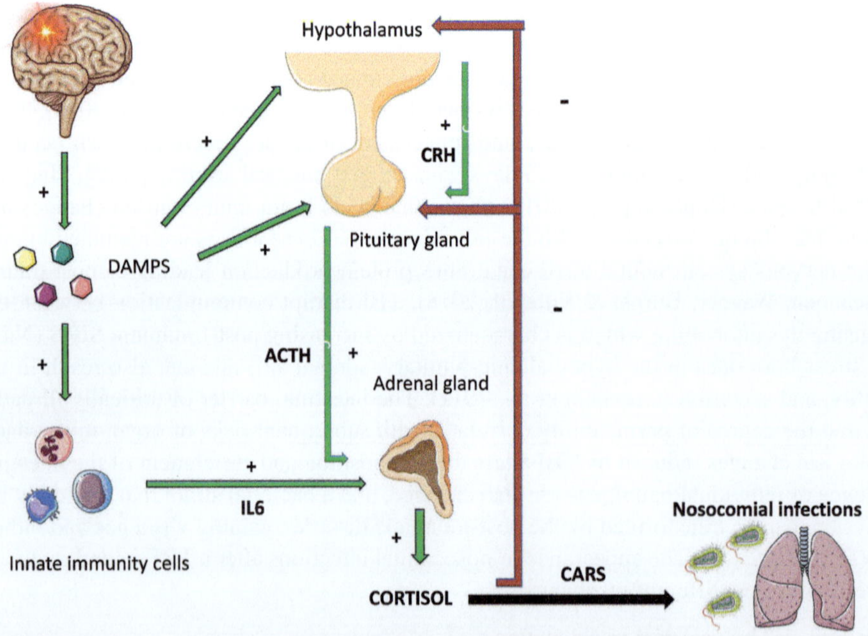

FIG. 2 The post-traumatic critical illness-related corticosteroid insufficiency: in the immediate aftermath of TBI, the HPA axis is stimulated by DAMPS and IL6 leading to an impairment of corticotropic function and favoring the occurrence of nosocomial infections.

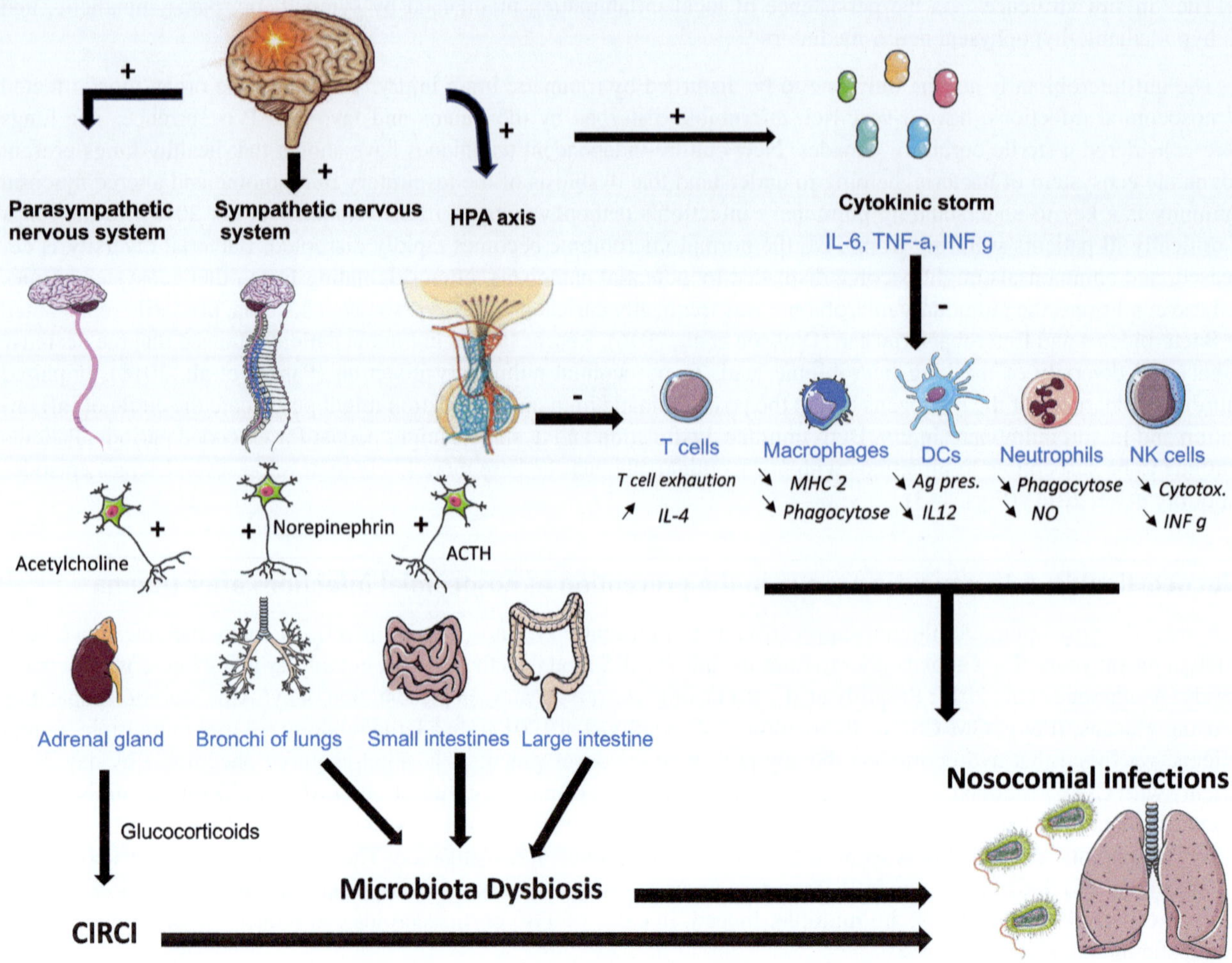

FIG. 3 Cellular and hormonal mechanisms favoring the occurrence of nosocomial infections traumatic brain injured patients: TBI favors the occurrence of nosocomial infection by a inducing state of immunosuppression, by altering intestinal and pulmonary microbiotas and by altering the HPA axis leading to CIRCI.

1. By an anti-inflammatory effect that decreases the immunosuppressive response (CARS), making the host less prone to nosocomial infection by a return to immune homeostasis. The introduction of corticosteroid therapy may, for example, reduce IL-6-dependent stimulation of the hypothalamic-pituitary axis, limit the anti-inflammatory response, and thus the prevalence of VAP in patients with severe inflammation (Fujio et al., 2016). In a recent sub-study of the Corti TC trial, we have shown that hydrocortisone could be useful in preventing nosocomial pneumonia in patients with a high imbalance between the pro- and anti-inflammatory mechanisms (Bouras et al., 2019). Our team has also shown that immunosuppression secondary to head injury is partly due to the elution of dendritic cells by NK cells via an IL-10 dependent mechanism and that glucocorticoids limit this immunosuppressive loop (Roquilly et al., 2014);
2. It is widely shown that glucocorticoids act on innate immune cells during this post-traumatic period by restoring effector capacities affected by TBI. During inflammation, glucocorticoids improve neutrophils' phagocytic abilities and increase the blood concentrations of interferon-g and interleukin-12, pro-inflammatory cytokines involved in the host defense against infections (Keh et al., 2003). Glucocorticoids also modulate dendritic cells during and after inflammation to restore their pivotal role in the immune response (Bouras et al., 2018), reducing trauma-induced tissue damage and susceptibility to bacterial infections.

All these arguments suggest that in patients with severe traumatic brain injury and an exacerbated overcompensated inflammatory response, glucocorticoids may be beneficial in preventing nosocomial infections (Table 1).

TABLE 1 Therapeutic approaches to reduce nosocomial infections in TBI patients.

Limitation of excessive inflammatory response	Molecules/Therapeutic Targets	Potential impact
	Corticosteroids	To inhibit PPRs, PAMPs, cytokines and inflammatory mediators released by neutrophils, NK cells etc.
	IL 1 antagonist receptors	Anti-IL1 antibodies may be beneficial in patients with severe SIRS complicated by lympho-histiocytic syndrome.
	Anti TNFα	Administered at the early stage of trauma could have a benefit in sub-populations of the most inflammatory patients.
Treatment of post-traumatic immunosuppression	**Molecules/therapeutic targets**	**Potential impact**
	Corticosteroids	• Improve neutrophils' phagocytic abilities • Increase the blood concentrations of inf γ and interleukin-12 • Modulate dendritic cells during and after inflammation to restore their pivotal role in the immune response
	IL12	• Restores the inflammation loop between APC and effectors cells • Restores NK cells & T cells capacities to produce Inf γ
	INF γ	• Increases monocytes expression of inflammatory cytokines • Increases HLA-DR expression • - Increases macrophages, monocytes and neutrophiles inflammatory activities
	IL 7	• Induces T cells proliferation and activation • Recovers from T cells exhaustion • - Increases neutrophiles recruitment in damaged tissues
	GMCSF	• Activates inflammatory cells • Increases HLA DR expression on APC • Increases neutrophiles phagocytosis

These therapeutics aim either to reduce the initial inflammatory state (*red*) or to restore failing immune capacities (*blue*).

Application to other areas of neuroscience

In this chapter we have developed the mechanisms underlying the interplay between traumatic brain injury and nosocomial infections which involves the main cells of innate and acquired immunity. The close link between immunology, endocrinology, infectiology, and neurology makes TBI a complex pathology at the interface of many specialties whose knowledge must be combined with the neurosciences in order to achieve optimal management for all patients. For example, strict control of the corticotropic axis in the post-traumatic period, in particular through corticosteroid therapy is an approach favored by our team in the prevention of infectious complications. The impairment of gut and pulmonary microbiota requires a multi-disciplinary approach to the management of these patients in order to develop therapeutics targeted on the cellular microenvironment and on neuro-hormonal regulation. Bidirectional communication between immune cells and the autonomic nervous system implies extending the field of prevention of nosocomial infections in brain injured patients to targeted therapies aimed at limiting the establishment of an immunosuppressed cellular microenvironment and the pathogenic fate of different microbiotas.

All of these post-traumatic disorders can be found in other non-traumatic brain injuries such as subarachnoid hemorrhage or hemorrhagic or ischemic strokes. Each entity of brain injury has its own specificities due to their location and their mechanisms of occurrence, but there is an interconnection, particularly with regard to the risk of nosocomial infections.

Mini-dictionary of terms

Brain-gut axis: The gut-brain axis is a bidirectional communication between the central nervous system and the enteric nervous system, involving the main neuroendocrine functions, immunological mechanisms, and the intestinal bacterial microbiota.

CARS: The compensatory anti-inflammatory response syndrome is a systemic deactivation of the immune system tasked with restoring homeostasis from an overwhelming inflammatory state following trauma or sepsis.

CIRCI: Critical illness-related corticosteroid insufficiency is an adrenal insufficiency in critically ill patients who present a dysregulated systemic inflammation resulting from inadequate intracellular glucocorticoid-mediated anti-inflammatory activity, a decreased glucocorticoid receptor sensitivity and tissue response to corticosteroids.

DAMPS: Damage-associated molecular patterns are endogen cellular molecules that have physiological functions which are released into the extracellular environment after trauma and that activate innate and adaptive immunity.

Dendritic cells: Dendritic cells are the most potent antigen presenting cells and play an essential interface between innate and adaptive immunity by the induction and the regulation of T cell-dependent immune response.

Post-traumatic immunosuppression: In the aftermath of a severe trauma, neuro-hormonal mechanisms and innate and adaptive immunity lead to a state of immune depression that promotes the development of nosocomial infections.

SIRS: A systemic inflammatory response syndrome is defined as a state of activation of inflammatory cascades following an acute injury such as trauma or severe infection. This syndrome involves all cellular immune mechanisms and can lead to a multiple organ failure syndrome when is exacerbated and prolonged.

VAP: Ventilator-associated pneumonia is usually defined as pneumonia occurring more than 48 h after patients have been intubated and received mechanical ventilation. This is the most common nosocomial infection in the intensive care unit setting.

Key facts of ventilatory-associated pneumonia

- Ventilator-associated pneumonia (VAP) is one the most frequent complication and the main cause of antibiotic use in intensive care units (ICU).
- VAP is associated with longer duration of mechanical ventilation, prolonged hospitalization and elevated healthcare cost.
- VAP caused by difficult-to-treat multi-drug–resistant bacteria are increasingly prevalent and results in significant mortality.
- Despite the large number of studies on the prevention of VAP and the implementation of new recommendations, the incidence rate has remained high for several decades at around 25% in 2019 in French ICUs.
- Preventing the occurrence of VAP in patients with traumatic brain injury is a major concern because the occurrence of such infections is a risk factor for poorer neurological outcomes.

Summary points

- There is a close link between traumatic brain injury and nosocomial infections, especially VAP.
- Traumatic brain injury induces a systemic inflammatory response which is compensated by a physiological anti-inflammatory response.
- If this anti-inflammatory response is exacerbated, a state of immunosuppression sets in, leading to nosocomial infections.
- The disturbance of the brain-gut axis and the pulmonary microbiota following the trauma causes a dysbiosis that predispose to nosocomial infections.
- The adrenal insufficiency following traumatic brain injury, called "post-traumatic critical illness-related corticosteroid insufficiency" prolongs this state of immunosuppression and maintains susceptibility to nosocomial infections.
- The early administration of low dose of glucocorticoids in the immediate aftermath of the traumatic brain injury is a therapeutic option in the prevention of nosocomial infections.

References

Annane, D., Pastores, S. M., Arlt, W., Balk, R. A., Beishuizen, A., Briegel, J., et al. (2017). Critical illness-related corticosteroid insufficiency (CIRCI): A narrative review from a Multispecialty Task Force of the Society of Critical Care Medicine (SCCM) and the European Society of Intensive Care Medicine (ESICM). *Intensive Care Medicine*, 1–12. https://doi.org/10.1007/s00134-017-4914-x.

Asehnoune, K., Seguin, P., Allary, J., Feuillet, F., Lasocki, S., Cook, F., et al. (2014). Hydrocortisone and fludrocortisone for prevention of hospital-acquired pneumonia in patients with severe traumatic brain injury (Corti-TC): A double-blind, multicentre phase 3, randomised placebo-controlled trial. *The Lancet Respiratory Medicine*, *2*(9), 706–716. https://doi.org/10.1016/S2213-2600(14)70144-4.

Bandyopadhyay, G., De, A., Laudanski, K., Li, F., Lentz, C., Bankey, P., et al. (2007). Negative signaling contributes to T-cell anergy in trauma patients. *Critical Care Medicine*, *35*(3), 794–801. https://doi.org/10.1097/01.CCM.0000256847.61085.A5.

Bassetti, M., Poulakou, G., Ruppe, E., Bouza, E., Van Hal, S. J., & Brink, A. (2017). Antimicrobial resistance in the next 30 years, humankind, bugs and drugs: A visionary approach. *Intensive Care Medicine*, *43*(10), 1464–1475. https://doi.org/10.1007/s00134-017-4878-x.

Bethin, K. E., Vogt, S. K., & Muglia, L. J. (2000). Interleukin-6 is an essential, corticotropin-releasing hormone-independent stimulator of the adrenal axis during immune system activation. *Proceedings of the National Academy of Sciences of the United States of America*, *97*(16), 9317–9322.

Bombardier, C. H., Fann, J. R., Temkin, N. R., Esselman, P. C., Barber, J., & Dikmen, S. S. (2010). Rates of major depressive disorder and clinical outcomes following traumatic brain injury. *JAMA*, *303*(19), 1938–1945. https://doi.org/10.1001/jama.2010.599.

Boomer, J. S., To, K., Chang, K. C., Takasu, O., Osborne, D. F., Walton, A. H., et al. (2011). Immunosuppression in patients who die of sepsis and multiple organ failure. *JAMA*, *306*(23), 2594–2605. https://doi.org/10.1001/jama.2011.1829.

Boonen, E., Vervenne, H., Meersseman, P., Andrew, R., Mortier, L., Declercq, P. E., et al. (2013). Reduced cortisol metabolism during critical illness. *New England Journal of Medicine*, *368*(16), 1477–1488. https://doi.org/10.1056/NEJMoa1214969.

Bouras, M., Asehnoune, K., & Roquilly, A. (2018). Contribution of dendritic cell responses to sepsis-induced immunosuppression and to susceptibility to secondary pneumonia. *Frontiers in Immunology*, *9*, 2590. https://doi.org/10.3389/fimmu.2018.02590.

Bouras, M., Roquilly, A., Mahé, P.-J., Cinotti, R., Vourc'h, M., Perrot, B., et al. (2019). Cortisol total/CRP ratio for the prediction of hospital-acquired pneumonia and initiation of corticosteroid therapy in traumatic brain-injured patients. *Critical Care*, *23*(1), 394. https://doi.org/10.1186/s13054-019-2680-6.

Bronchard, R., Albaladejo, P., Brezac, G., Geffroy, A., Seince, P.-F., Morris, W., et al. (2004). Early onset pneumonia: Risk factors and consequences in head trauma patients. *Anesthesiology*, *100*(2), 234–239.

Brown, A. S., Yang, C., Fung, K. Y., Bachem, A., Bourges, D., Bedoui, S., et al. (2016). Cooperation between monocyte-derived cells and lymphoid cells in the acute response to a bacterial lung pathogen. *PLoS Pathogens*, *12*(6). https://doi.org/10.1371/journal.ppat.1005691, e1005691.

Cohan, P., Wang, C., McArthur, D. L., Cook, S. W., Dusick, J. R., Armin, B., et al. (2005). Acute secondary adrenal insufficiency after traumatic brain injury: A prospective study. *Critical Care Medicine*, *33*(10), 2358–2366.

Colton, C. A. (2009). Heterogeneity of microglial activation in the innate immune response in the brain. *Journal of Neuroimmune Pharmacology*, *4*(4), 399–418. https://doi.org/10.1007/s11481-009-9164-4.

Cryan, J. F., & Dinan, T. G. (2012). Mind-altering microorganisms: The impact of the gut microbiota on brain and behaviour. *Nature Reviews. Neuroscience*, *13*(10), 701–712. https://doi.org/10.1038/nrn3346.

de Miguel-Díez, J., López-de-Andrés, A., Hernández-Barrera, V., Jiménez-Trujillo, I., Méndez-Bailón, M., de Miguel-Yanes, J. M., et al. (2017). Decreasing incidence and mortality among hospitalized patients suffering a ventilator-associated pneumonia: Analysis of the Spanish national hospital discharge database from 2010 to 2014. *Medicine*, *96*(30). https://doi.org/10.1097/MD.0000000000007625, e7625.

Deitch, E. A. (2012). Gut-origin sepsis: Evolution of a concept. *The Surgeon: Journal of the Royal Colleges of Surgeons of Edinburgh and Ireland*, *10*(6), 350–356. https://doi.org/10.1016/j.surge.2012.03.003.

Deknuydt, F., Roquilly, A., Cinotti, R., Altare, F., & Asehnoune, K. (2013). An in vitro model of mycobacterial granuloma to investigate the immune response in brain-injured patients. *Critical Care Medicine*, *41*(1), 245–254. https://doi.org/10.1097/CCM.0b013e3182676052.

Demaude, J., Salvador-Cartier, C., Fioramonti, J., Ferrier, L., & Bueno, L. (2006). Phenotypic changes in colonocytes following acute stress or activation of mast cells in mice: Implications for delayed epithelial barrier dysfunction. *Gut*, *55*(5), 655–661. https://doi.org/10.1136/gut.2005.078675.

Dickson, R. P. (2016). The microbiome and critical illness. *The Lancet Respiratory Medicine*, *4*(1), 59–72. https://doi.org/10.1016/S2213-2600(15)00427-0.

Dickson, R. P., Singer, B. H., Newstead, M. W., Falkowski, N. R., Erb-Downward, J. R., Standiford, T. J., et al. (2016). Enrichment of the lung microbiome with gut bacteria in sepsis and the acute respiratory distress syndrome. *Nature Microbiology*, *1*(10), 16113. https://doi.org/10.1038/nmicrobiol.2016.113.

Doig, C. J., Sutherland, L. R., Dean Sandham, J., Fick, G. H., Verhoef, M., & Meddings, J. B. (1998). Increased intestinal permeability is associated with the development of multiple organ dysfunction syndrome in critically ill ICU patients. *American Journal of Respiratory and Critical Care Medicine*, *158*(2), 444–451. https://doi.org/10.1164/ajrccm.158.2.9710092.

Engel, O., Akyüz, L., da Costa Goncalves, A. C., Winek, K., Dames, C., Thielke, M., et al. (2015). Cholinergic pathway suppresses pulmonary innate immunity facilitating pneumonia after stroke. *Stroke*, *46*(11), 3232–3240. https://doi.org/10.1161/STROKEAHA.115.008989.

Fujio, N., Masuoka, S., Shikano, K., Kusunoki, N., Nanki, T., & Kawai, S. (2016). Apparent hypothalamic-pituitary-adrenal axis suppression via reduction of Interleukin-6 by glucocorticoid therapy in systemic autoimmune diseases. *PLoS One*, *11*(12). https://doi.org/10.1371/journal.pone.0167854.

Grenham, S., Clarke, G., Cryan, J. F., & Dinan, T. G. (2011). Brain-gut-microbe communication in health and disease. *Frontiers in Physiology*, *2*, 94. https://doi.org/10.3389/fphys.2011.00094.

Hall, A. B., Tolonen, A. C., & Xavier, R. J. (2017). Human genetic variation and the gut microbiome in disease. *Nature Reviews Genetics*, *18*(11), 690–699. https://doi.org/10.1038/nrg.2017.63.

Hang, C.-H., Shi, J.-X., Li, J.-S., Wu, W., & Yin, H.-X. (2003). Alterations of intestinal mucosa structure and barrier function following traumatic brain injury in rats. *World Journal of Gastroenterology*, *9*(12), 2776–2781. https://doi.org/10.3748/wjg.v9.i12.2776.

Heffernan, D. S., Monaghan, S. F., Thakkar, R. K., Machan, J. T., Cioffi, W. G., & Ayala, A. (2012). Failure to normalize lymphopenia following trauma is associated with increased mortality, independent of the leukocytosis pattern. *Critical Care*, *16*(1), R12. https://doi.org/10.1186/cc11157.

Hoen, S., Asehnoune, K., Brailly-Tabard, S., Mazoit, J.-X., Benhamou, D., Moine, P., et al. (2002). Cortisol response to corticotropin stimulation in trauma patients: Influence of hemorrhagic shock. *Anesthesiology*, *97*(4), 807–813.

Houlden, A., Goldrick, M., Brough, D., Vizi, E. S., Lénárt, N., Martinecz, B., et al. (2016). Brain injury induces specific changes in the caecal microbiota of mice via altered autonomic activity and mucoprotein production. *Brain, Behavior, and Immunity*, *57*, 10–20. https://doi.org/10.1016/j.bbi.2016.04.003.

Kalil, A. C., Metersky, M. L., Klompas, M., Muscedere, J., Sweeney, D. A., Palmer, L. B., et al. (2016). Management of Adults With Hospital-acquired and Ventilator-associated Pneumonia: 2016 Clinical Practice Guidelines by the Infectious Diseases Society of America and the American Thoracic Society. *Clinical Infectious Diseases: An Official Publication of the Infectious Diseases Society of America*, *63*(5), e61–e111. https://doi.org/10.1093/cid/ciw353.

Kawabori, M., Kacimi, R., Kauppinen, T., Calosing, C., Kim, J. Y., Hsieh, C. L., et al. (2015). Triggering receptor expressed on myeloid cells 2 (TREM2) deficiency attenuates phagocytic activities of microglia and exacerbates ischemic damage in experimental stroke. *The Journal of Neuroscience: The Official Journal of the Society for Neuroscience*, *35*(8), 3384–3396. https://doi.org/10.1523/JNEUROSCI.2620-14.2015.

Keh, D., Boehnke, T., Weber-Cartens, S., Schulz, C., Ahlers, O., Bercker, S., et al. (2003). Immunologic and hemodynamic effects of "low-dose" hydrocortisone in septic shock: A double-blind, randomized, placebo-controlled, crossover study. *American Journal of Respiratory and Critical Care Medicine*, *167*(4), 512–520. https://doi.org/10.1164/rccm.200205-446OC.

Köller, M., Clasbrummel, B., Kollig, E., Hahn, M. P., & Muhr, G. (1998). Major injury induces increased production of interleukin-10 in human granulocyte fractions. *Langenbeck's Archives of Surgery*, *383*(6), 460–465. https://doi.org/10.1007/s004230050161.

Lozano, R., Naghavi, M., Foreman, K., Lim, S., Shibuya, K., Aboyans, V., et al. (2012). Global and regional mortality from 235 causes of death for 20 age groups in 1990 and 2010: A systematic analysis for the Global Burden of Disease Study 2010. *Lancet*, *380*(9859), 2095–2128. https://doi.org/10.1016/S0140-6736(12)61728-0.

Marik, P. E., Pastores, S. M., Annane, D., Meduri, G. U., Sprung, C. L., Arlt, W., et al. (2008). Recommendations for the diagnosis and management of corticosteroid insufficiency in critically ill adult patients: Consensus statements from an international task force by the American College of Critical Care Medicine. *Critical Care Medicine*, *36*(6), 1937–1949. https://doi.org/10.1097/CCM.0b013e31817603ba.

Mastorakos, G., Chrousos, G. P., & Weber, J. S. (1993). Recombinant interleukin-6 activates the hypothalamic-pituitary-adrenal axis in humans. *The Journal of Clinical Endocrinology and Metabolism*, *77*(6), 1690–1694. https://doi.org/10.1210/jcem.77.6.8263159.

Mayer, E. A. (2011). Gut feelings: The emerging biology of gut-brain communication. *Nature Reviews. Neuroscience*, *12*(8), 453–466. https://doi.org/10.1038/nrn3071.

Meisel, C., Schwab, J. M., Prass, K., Meisel, A., & Dirnagl, U. (2005). Central nervous system injury-induced immune deficiency syndrome. *Nature Reviews Neuroscience*, *6*(10), 775–786. https://doi.org/10.1038/nrn1765.

Melsen, W. G., Rovers, M. M., Groenwold, R. H. H., Bergmans, D. C. J. J., Camus, C., Bauer, T. T., et al. (2013). Attributable mortality of ventilator-associated pneumonia: A meta-analysis of individual patient data from randomised prevention studies. *The Lancet Infectious Diseases*, *13*(8), 665–671. https://doi.org/10.1016/S1473-3099(13)70081-1.

Morganti, J. M., Riparip, L.-K., & Rosi, S. (2016). Call off the dog (ma): M1/M2 polarization is concurrent following traumatic brain injury. *PLoS One*, *11*(1). https://doi.org/10.1371/journal.pone.0148001, e0148001.

Myburgh, J. A., Cooper, D. J., Finfer, S. R., Venkatesh, B., Jones, D., Higgins, A., et al. (2008). Epidemiology and 12-month outcomes from traumatic brain injury in Australia and New Zealand. *The Journal of Trauma*, *64*(4), 854–862. https://doi.org/10.1097/TA.0b013e3180340e77.

Nast-Kolb, D., Waydhas, C., Gippner-Steppert, C., Schneider, I., Trupka, A., Ruchholtz, S., et al. (1997). Indicators of the posttraumatic inflammatory response correlate with organ failure in patients with multiple injuries. *The Journal of Trauma: Injury, Infection, and Critical Care*, *42*(3), 446–455. https://doi.org/10.1097/00005373-199703000-00012.

Nicholson, S. E., Watts, L. T., Burmeister, D. M., Merrill, D., Scroggins, S., Zou, Y., et al. (2019). Moderate traumatic brain injury alters the gastrointestinal microbiome in a time-dependent manner. *Shock (Augusta, Ga.)*, *52*(2), 240–248. https://doi.org/10.1097/SHK.0000000000001211.

Ohannessian, R., Gustin, M.-P., Bénet, T., Gerbier-Colomban, S., Girard, R., Argaud, L., et al. (2018). Estimation of extra length of stay attributable to hospital-acquired infections in adult ICUs using a time-dependent multistate model. *Critical Care Medicine*, *46*(7), 1093–1098. https://doi.org/10.1097/CCM.0000000000003131.

Panzer, A. R., Lynch, S. V., Langelier, C., Christie, J. D., McCauley, K., Nelson, M., et al. (2018). Lung microbiota is related to smoking status and to development of acute respiratory distress syndrome in critically ill trauma patients. *American Journal of Respiratory and Critical Care Medicine*, *197*(5), 621–631. https://doi.org/10.1164/rccm.201702-0441OC.

Rhee, S. H., Pothoulakis, C., & Mayer, E. A. (2009). Principles and clinical implications of the brain-gut-enteric microbiota axis. *Nature Reviews. Gastroenterology & Hepatology*, *6*(5), 306–314. https://doi.org/10.1038/nrgastro.2009.35.

Roquilly, A., Broquet, A., Jacqueline, C., Masson, D., Segain, J. P., Braudeau, C., et al. (2014). Hydrocortisone prevents immunosuppression by interleukin-10+ natural killer cells after trauma-hemorrhage. *Critical Care Medicine*, *42*(12), e752–e761. https://doi.org/10.1097/CCM.0000000000000658.

Roquilly, A., Chanques, G., Lasocki, S., Foucrier, A., Fermier, B., De Courson, H., et al. (2020). Implementation of French recommendations for the prevention and the treatment of hospital-acquired pneumonia: A Cluster-randomized Trial. *Clinical Infectious Diseases*. https://doi.org/10.1093/cid/ciaa1441, ciaa1441.

Roquilly, A., David, G., Cinotti, R., Vourc'h, M., Morin, H., Rozec, B., et al. (2015). Role of IL-12 in overcoming the low responsiveness of NK cells to missing self after traumatic brain injury. *Clinical Immunology*. https://doi.org/10.1016/j.clim.2015.08.006.

Roquilly, A., Mahe, P. J., Seguin, P., Guitton, C., Floch, H., Tellier, A. C., et al. (2011). Hydrocortisone therapy for patients with multiple trauma: The randomized controlled HYPOLYTE study. *JAMA*, *305*(12), 1201–1209. https://doi.org/10.1001/jama.2011.360.

Roquilly, A., McWilliam, H. E. G., Jacqueline, C., Tian, Z., Cinotti, R., Rimbert, M., et al. (2017). Local modulation of antigen-presenting cell development after resolution of pneumonia induces long-term susceptibility to secondary infections. *Immunity*, *47*(1), 135–147.e5. https://doi.org/10.1016/j.immuni.2017.06.021.

Roquilly, A., Torres, A., Villadangos, J. A., Netea, M. G., Dickson, R., Becher, B., et al. (2019). Pathophysiological role of respiratory dysbiosis in hospital-acquired pneumonia. *The Lancet Respiratory Medicine*, *7*(8), 710–720. https://doi.org/10.1016/S2213-2600(19)30140-7.

Safdar, N., Dezfulian, C., Collard, H. R., & Saint, S. (2005). Clinical and economic consequences of ventilator-associated pneumonia: A systematic review. *Critical Care Medicine*, *33*(10), 2184–2193.

Schneider, H. J., Kreitschmann-Andermahr, I., Ghigo, E., Stalla, G. K., & Agha, A. (2007). Hypothalamopituitary dysfunction following traumatic brain injury and aneurysmal subarachnoid hemorrhage: A systematic review. *JAMA*, *298*(12), 1429–1438. https://doi.org/10.1001/jama.298.12.1429.

Schwartz, M. (2001). Harnessing the immune system for neuroprotection: Therapeutic vaccines for acute and chronic neurodegenerative disorders. *Cellular and Molecular Neurobiology*, *21*(6), 617–627.

Sherwin, E., Dinan, T. G., & Cryan, J. F. (2018). Recent developments in understanding the role of the gut microbiota in brain health and disease: The gut microbiota in brain health and disease. *Annals of the New York Academy of Sciences*, *1420*(1), 5–25. https://doi.org/10.1111/nyas.13416.

Silver, J., & Miller, J. H. (2004). Regeneration beyond the glial scar. *Nature Reviews Neuroscience*, *5*(2), 146–156. https://doi.org/10.1038/nrn1326.

Spolarics, Z., Siddiqi, M., Siegel, J. H., Garcia, Z. C., Stein, D. S., Denny, T., et al. (2003). Depressed interleukin-12-producing activity by monocytes correlates with adverse clinical course and a shift toward Th2-type lymphocyte pattern in severely injured male trauma patients. *Critical Care Medicine*, *31*(6), 1722–1729. https://doi.org/10.1097/01.CCM.0000063579.43470.AA.

Torres, A., Niederman, M. S., Chastre, J., Ewig, S., Fernandez-Vandellos, P., Hanberger, H., et al. (2017). International ERS/ESICM/ESCMID/ALAT guidelines for the management of hospital-acquired pneumonia and ventilator-associated pneumonia: Guidelines for the management of hospital-acquired pneumonia (HAP)/ventilator-associated pneumonia (VAP) of the European Respiratory Society (ERS), European Society of Intensive Care Medicine (ESICM), European Society of Clinical Microbiology and Infectious Diseases (ESCMID) and Asociación Latinoamericana del Tórax (ALAT). *European Respiratory Journal*, *50*(3), 1700582. https://doi.org/10.1183/13993003.00582-2017.

Treangen, T. J., Wagner, J., Burns, M. P., & Villapol, S. (2018). Traumatic brain injury in mice induces acute bacterial dysbiosis within the fecal microbiome. *Frontiers in Immunology*, *9*, 2757. https://doi.org/10.3389/fimmu.2018.02757.

Wang, H., Yu, M., Ochani, M., Amella, C. A., Tanovic, M., Susarla, S., et al. (2003). Nicotinic acetylcholine receptor α7 subunit is an essential regulator of inflammation. *Nature*, *421*(6921), 384–388. https://doi.org/10.1038/nature01339.

Woiciechowsky, C., Asadullah, K., Nestler, D., Eberhardt, B., Platzer, C., Schöning, B., et al. (1998). Sympathetic activation triggers systemic interleukin-10 release in immunodepression induced by brain injury. *Nature Medicine*, *4*(7), 808–813. https://doi.org/10.1038/nm0798-808.

Wong, C. H. Y., Jenne, C. N., Lee, W.-Y., Leger, C., & Kubes, P. (2011). Functional innervation of hepatic iNKT cells is immunosuppressive following stroke. *Science*, *334*(6052), 101–105. https://doi.org/10.1126/science.1210301.

Yi, J. S., Cox, M. A., & Zajac, A. J. (2010). T-cell exhaustion: Characteristics, causes and conversion. *Immunology*, *129*(4), 474–481. https://doi.org/10.1111/j.1365-2567.2010.03255.x.

Chapter 10

Long-term sequelae of mild-repetitive and severe traumatic brain injury: Clinical manifestations, neuropathology and diagnosis by tau PET imaging

Keisuke Takahata[a,b], Kenji Tagai[a], Makoto Higuchi[a], and Masaru Mimura[b]

[a]*Department of Functional Brain Imaging Research, National Institute of Radiological Sciences, National Institutes for Quantum and Radiological Science and Technology, Chiba, Japan,* [b]*Department of Neuropsychiatry, Keio University School of Medicine, Tokyo, Japan*

List of abbreviations

AD	Alzheimer's disease
BP_{ND}	binding potential
CTE	chronic traumatic encephalopathy
^{18}**F-FDG**	^{18}F-fluoro-deoxy-glucose
FTLD	frontotemporal lobe degeneration
GM	gray matter
MAO	monoamine oxidase
MRTM$_O$	multi-linear reference tissue model
PBB3	pyridinyl–butadienyl–benzothiazole 3
PDFTBI	psychotic disorder following traumatic brain injury
PET	positron emission tomography
PM-PBB3	propanol modification of PBB3
PSP	progressive supranuclear palsy
SD	standard deviation
SUVR	standard uptake value ratio
TBI	traumatic brain injury
TSPO	18 kDa translocator protein
WM	white matter

Introduction

Recent neuropathological studies have brought a growing awareness of a variety of neurodegenerative diseases that occur long after traumatic brain injury (TBI) (Fig. 1). A typical delayed-onset condition caused by mild-repetitive TBI is boxer's encephalopathy or chronic traumatic encephalopathy (CTE). Long-lasting and progressive neuropsychiatric sequelae caused by mild-repetitive TBI was formerly thought to be a rare condition restricted to retired boxers and other athletes of combat sports (Martland, 1928; Millspaugh, 1937), but recent reports have shown that it can be caused by a wide range of contact sports and occupations (McKee et al., 2013; Omalu et al., 2005). Furthermore, compelling evidence indicates that neurodegenerative diseases following TBI can be initiated regardless of injury types such as mild-repetitive and severe TBI (LoBue et al., 2018; Smith, Johnson, & Stewart, 2013). At present, CTE is diagnosed only with histological assessments of post-mortem brains, and there is no way to diagnose CTE while the patient is alive. Therefore, current efforts in this field have been focused on the development of reliable neuroimaging biomarkers for detection of neuropathological changes due to TBI.

Diagnosis and Treatment of Traumatic Brain Injury. https://doi.org/10.1016/B978-0-12-823347-4.00010-5

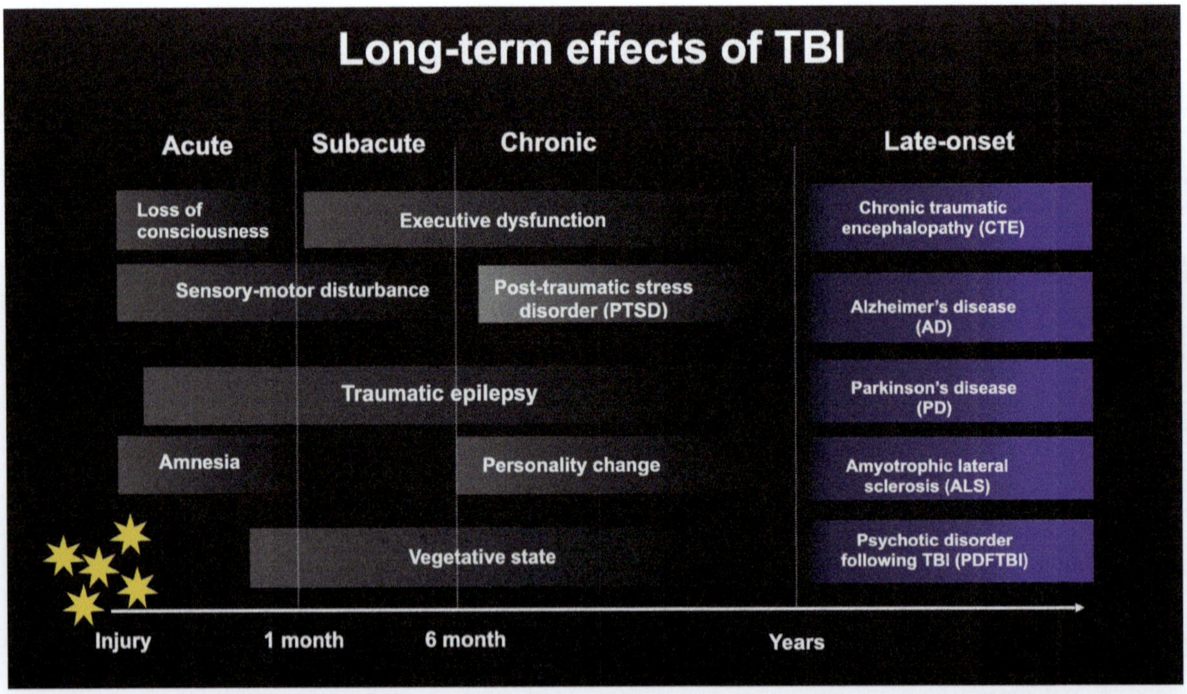

FIG. 1 Long-term effects of TBI. This slide shows that traumatic brain injury (TBI) causes various neurocognitive symptoms depending on time after injury. In the acute phase, loss of consciousness is the most important symptom. In the sub-acute phase, patients show executive dysfunction or attentional deficits. Vegetative states often become apparent in this period. In the chronic phase, several months after injury, patients often develop persistent symptoms such as post-stress traumatic disorder (PTSD) or personality change. Recent research reports and media coverage have drawn significant public attention to the late-onset sequelae that occur several years after injury.

Clinical symptoms and neuropathology of chronic states of TBI

Mild-repetitive TBI: Mild-repetitive TBI can lead to progressive neurocognitive dysfunction such as cognitive impairments, psychiatric symptoms and motor symptoms as years or decades pass (Jordan, 2013). These conditions have been known as "boxer's encephalopathy" since the early 20th century. The first systematic report of a long-term clinical course of mild-repetitive TBI was made in 1927. In this report, Osnato and Giliberti examined more than 100 cases of concussed brains, noting the possibility of slow progressive syndromes occurring following mild-repetitive TBI (Osnato & Giliberti, 1927). One of the most important early literatures of the late-onset sequelae of mild-repetitive TBI was a 1928 report by Martland (1928), in which clinical features and neuroanatomical findings of 23 retired boxers were described in detail. In this report, characteristic cognitive and behavioral problems such as dementia, depression, psychosis, personality changes and motor symptoms were described and termed "punch-drunk syndrome." Ten years later, dementia pugilistica was proposed by Millspaugh, and it was considered a subtype of senile dementia (Millspaugh, 1937). The neuropathological features of boxer's encephalopathy were reported in detail by Brandenburg (Brandenburg and Hallervorden, 1954) and Corsellis (Corsellis, Bruton, & Freeman-Browne, 1973). These early post-mortem studies revealed that repetitive blows to the head cause progressive cognitive impairments, and showed that boxing and other combat sports are a risk for neurodegenerative diseases that can eventually lead to dementia.

Neurodegenerative diseases due to TBI were formerly believed to be caused only by sports that include heavy blows to the head, and little recognition had been given to the dangers of head blows in other sports and occupations. In recent years, however, it has been revealed that a wider range of contact sports and occupations can cause neurodegenerative diseases (Jordan, 2013). The key event that triggered public awareness of the long-term impact of contact sports was the death of a former NFL player named Mike Webster, who was a great center of the Pittsburgh Steelers and was inducted into the Football Hall of Fame in 1997. After his retirement, he began to suffer from addiction to alcohol, financial waste, memory problems, easy rage and depression. Abandoned by his family and eventually becoming homeless, he passed away in 2002 due to a myocardial infarction. His autopsy was performed by Nigerian-African neuropathologist Bennet Omalu. Omalu, who knew from media reports that Webster had been suffering from dementia-like symptoms, speculated that Webster had developed early-onset Alzheimer's disease (AD). However, the gross findings of Webster's brain were different from that of AD. Omalu performed immunostaining for various dementia-related proteins, and found that abundant neurofibrillary tangles appeared in a wide area of the brain. The neurofibrillary tangles were a previously reported pathological hallmark of boxer's encephalopathy, and the topology of neurofibrillary tangles in Webster's brain was different from that of AD. At the time, however, it was believed that boxer's encephalopathy occurs only in athletes of hard contact sports such as boxing and martial arts. Surprised that American football, which was previously believed to be safe, caused neuropathological changes similar to boxer's encephalopathy, Omalu reported the results of Webster's autopsy in the journal *Neurosurgery*. In this report, he revived the forgotten name of the disease, "chronic traumatic encephalopathy (CTE)," to emphasize that boxing is not the only cause of neurodegenerative diseases. Omalu's report sparked fierce debate in the United States, including a rejection by the NFL organization, but similar reports followed, and it was eventually recognized that CTE is caused by a wide range of contact sports, including American football, ice hockey, rugby and wrestling (Goldstein et al., 2012; Maroon et al., 2015; Omalu, Fitzsimmons, Hammers, & Bailes, 2010).

Clinical symptoms of CTE are shown in Fig. 2. The most common symptom of CTE is memory impairment (Mez, Stern, & McKee, 2020). Dysfunction of working memory and attentional functions are also frequently observed. Aggression and violent behavior due to emotional control difficulties are often present. Headache is also a frequent physical symptom. In some cases, symptoms progress slowly, and in others the progression stops at a certain stage (Stern et al., 2013). CTE is characterized by a higher frequency of psychiatric symptoms such as depression and delusions, and a higher risk of suicide.

CTE is a neurodegenerative disease characterized by the deposition of hyperphosphorylated tau (p-tau) protein as neurofibrillary tangles, astrocytic tangles and neurites in striking clusters around small blood vessels of the cortex, typically at the depth of the cortical sulcus (McKee et al., 2013). Neurofibrillary tangles, which have traditionally been described in boxer encephalopathy (Fig. 3), are caused by excessive intracellular phosphorylation of tau protein abnormally deposited intra-cellularly as insoluble aggregates. Biochemical analysis of the tau isoform has shown that 3- and 4-repeat tau proteins appear in CTE brains, similar to AD. Tau lesions in CTE are characterized by irregular and patchy distributions, unlike neurodegenerative diseases such as AD and other non-AD tauopathies (Mckee, Abdolmohammadi, & Stein, 2018). Axonal injury, axonal degeneration, myelin degeneration, and white-matter loss are consistent features of CTE, and they may play a critical role in initiation and propagation of p-tau pathology. A recent brain bank study conducted by Boston University reported that CTE pathology was found in 99% of deceased NFL players (Mez et al., 2017). Therefore, it is assumed that the frequency of neuropathological findings of CTE in individuals with long-term exposure to repetitive mild head trauma is significantly higher than previously believed.

In addition to tau accumulations, TAR DNA-binding protein-43 kDa (TDP-43)-positive inclusions also occur at high rates in CTE (McKee et al., 2010); TDP-43-positive inclusions are the ubiquitin-positive inclusions found in frontotemporal

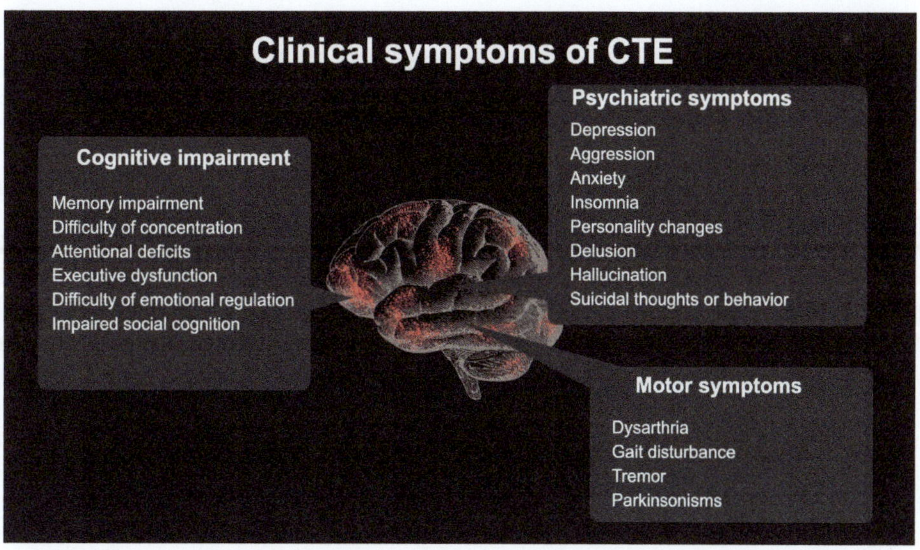

FIG. 2 Clinical symptoms of chronic traumatic encephalopathy. The clinical manifestations of chronic traumatic encephalopathy (CTE) fall into three categories: cognitive impairment, psychiatric symptoms, and motor impairment. Unlike other neurodegenerative diseases, psychiatric symptoms tend to occur at a high frequency. In terms of cognitive dysfunction, impairment of recent memory is the most frequently observed. Parkinsonism is often observed in advanced cases. The difference between the clinical manifestations of boxer's encephalopathy and CTE is that parkinsonism such as gait disturbance, tremor and dysarthria are highly prevalent in the former, whereas rare in the latter. In addition to these neurological symptoms, headache is also frequently observed. The clinical manifestations of CTE are often progressive, eventually leading to dementia in some cases. On the other hand, there are cases in which the progression stops at a certain stage, but the details are not known.

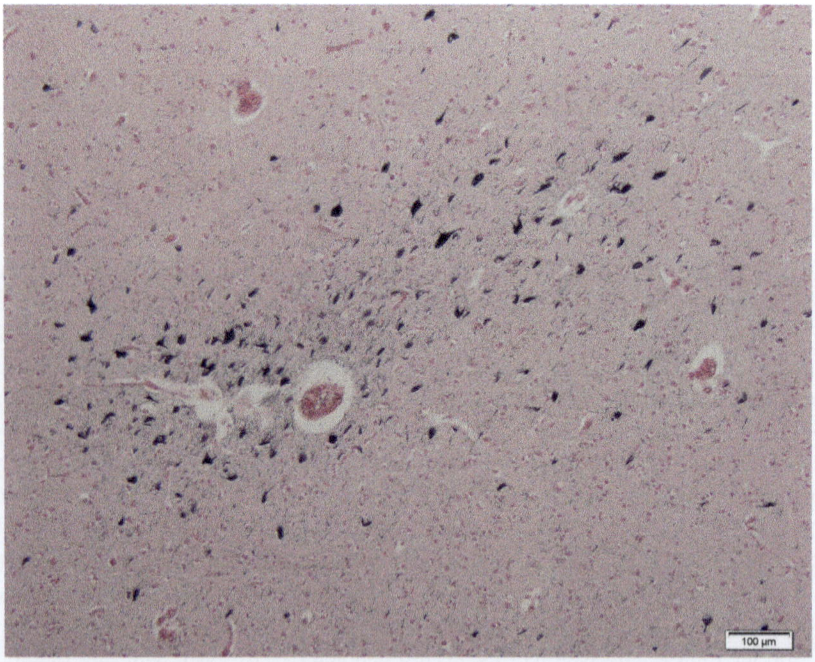

FIG. 3 Neurofibrillary tangles of boxer's encephalopathy. Typical neuropathology of boxer's encephalopathy (courtesy of Haruhiko Akiyama and Hisaomi Suzuki). This brain sample was obtained from a professional boxer who died suddenly after a fight. Gallyas–Braak silver staining visualized neurofibrillary tangles around the small vessel in the depth of cortical sulcus.

lobe degeneration (FTLD) and amyotrophic lateral sclerosis (ALS). Accumulations of α-synuclein and β-amyloid are also seen in the brain of advanced cases (McKee et al. 2013). In other words, neuropathologically, CTE is a multiple protein pathology (poly-proteinopathy). However, the mechanisms by which tau and other proteins aggregate and propagate, leading to changes in brain circuitry, remain unclear.

Severe TBI: Long-term neuropsychiatric consequences of severe TBI share certain clinical features with those of mild-repetitive TBI, despite the fact that clinical manifestations at the acute periods of both injury types are markedly different (Takahata, Tabuchi, & Mimura, 2016). Late-onset neurocognitive symptoms were reported at the chronic stage of severe TBI, and was shown to accelerate age-related cognitive decline (Johnson & Stewart, 2015). Behavioral symptomsare also distinctive features in the chronic stage of severe TBI (Fujii & Ahmed, 1996; Fujii & Fujii, 2012). Difficulties of emotional regulation are frequently observed in long-term survivors of severe TBI.

It has been reported that severe TBI is a major risk factor for various neurodegenerative diseases including Alzheimer's disease and Parkinson's disease (Johnson & Stewart, 2015; Shively, Scher, Perl, & Diaz-Arrastia, 2012). Progressive cognitive impairments and dementia similar to CTE has also been reported to occur after severe TBI. The most significant feature in the late-onset of symptoms due to severe TBI is the high frequency of psychosis as well as mood disorders. In some cases, schizophrenia-like psychotic states emerge late after a single incidence of moderate-to-severe TBI (Achté, Hillbom, & Aalberg, 1969; Fujii & Ahmed, 2014). Historically, post-traumatic psychosis was first reported in the monograph by von Krafft-Ebing (1868), which mentioned several cases of schizophrenic-like psychosis and paranoia that occurred sometime after severe TBI (Achté et al., 1969). A paper published by Adolf Meyer (1904) describes the long-term course of psychotic symptoms following severe TBI and their autopsy results in greater detail (Meyer, 1908). The psychotic symptoms that appear late after severe TBI have recently been conceptualized by Fujii as a psychotic disorder following TBI (PDFTBI). Epidemiological studies have shown that PDFTBI occurs in 3.4%–8.9% of severe TBI and that the average onset of symptoms is approximately 4–6 years after severe TBI (Fujii & Ahmed, 2002; Sachdev, Smith, & Cathcart, 2001). PDFTBI is typically caused by damage to the temporal pole and frontal lobe, but it can also be caused by damage to other areas of the brain (Takahata, Tabuchi, & Mimura, 2016). In some cases, widespread white-matter lesions due to diffuse axonal injury can cause psychosis later in life, even in the absence of focal brain damage. In addition to psychotic symptoms such as delusions and hallucinations, most cases showed impairments of executive dysfunction and memory, as well as attentional deficits. PDFTBI shares a commonality with CTE in that it develops late from an impact to the head, and also a high incidence of psychiatric symptoms. The temporal nature of the delayed onset of PDFTBI strongly suggests that some neurodegenerative process in the brain, triggered by head injury, may be involved in the onset of psychotic symptoms (Takahata, Kato, Mimura, & Shimada, 2016).

Less is known about the neuropathologic findings of severe TBI than of mild repetitive TBI. Amyloid pathologies are perhaps the most extensively researched in the context of potential links with dementia following severe TBI. One of the earliest observations in this field described Aβ plaques in up to 30% of patients who died in the acute phase following a single moderate to severe TBI, and plaque pathology exceeded that seen in age-matched controls (Ikonomovic et al., 2004; Roberts et al., 1994; Roberts, Allsop, & Bruton, 1990). Although there are few reported cases, neurofibrillary tangles have been reported to occur in the brain of severe TBI (Johnson, Stewart, & Smith, 2012). In that study, post-mortem brains from 39 survivors of 1 year or more from a moderate-to-severe TBI demonstrated greater density of neurofibrillary tangles and wider distribution compared with age-matched controls. Furthermore, neurofibrillary tangles were found in the depth of cortical sulci and showed features identical to CTE. Tau pathology similar to CTE were also found in the post-mortem brains of two patients who had undergone pre-frontal leucotomy (Okamura et al., 2019). The neuropathological findings of CTE have also been confirmed in cases of blast injury (Goldstein et al., 2012). At present, it is inconclusive whether CTE is a condition specific to mild-repetitive TBI. Rather, the chronic pathologies following severe and mild-repetitive injuries show similarities (Smith et al., 2013), although comparative studies are lacking at present.

Growing need for neuroimaging biomarkers of chronic pathologies of TBI

There is currently no established treatment or prevention for CTE. As noted above, symptoms of CTE tend to worsen with more blows to the head; therefore, prevention and early detection of deposition of tau aggregates are the highest priority for athletes, soldiers, and others at high risk for CTE. However, the diagnosis of CTE is currently based on autopsy, and there has been no established method to detect tau pathology in the living brain, so that in most cases, the disease is already quite advanced at the time typical symptoms appear. These problems call for neuroimaging biomarkers for detecting the accumulation of tau proteins due in living persons. Reliable neuroimaging biomarkers will be extremely useful as diagnostic marker but also as indicators for early initiation of interventions such as cognitive rehabilitation and social support for individuals with CTE (Lin, Charney, Shenton, & Koerte, 2018). Sensitive and specific biomarkers are also necessary for the development of immunotherapies targeting the accumulation of tau in CTE brains.

One of the most promising methods of detecting CTE pathology is by positron emission tomography (PET). Amyloid PET is widely used as an imaging technique for protein accumulation disease. However, unlike AD, amyloid-β is not a pathological finding consistently seen in CTE. Therefore, amyloid PET is effective only when performed to differentiate AD from CTE.. Neuroinflammation PET using radioligands for 18 kDa translocator protein (TSPO) is also useful for studying the pathophysiology of long-term effects due to TBI (Coughlin et al., 2015), but it is less specific, as an enhancement of neuroinflammation can be found in a variety of neurodegenerative diseases and psychiatric disorders. As mentioned earlier, the most important pathological change in CTE is the accumulation of p-tau, and therefore a neuroimaging technique that can visualize tau in vivo holds the most promise as a biomarker for CTE.

PET imaging of TBI with first-generation tau PET tracers

Tau imaging is a molecular imaging technique to detect pathological changes in neurodegenerative diseases non-invasively using PET with a specific probe for tau aggregates. The properties that a reliable tau PET probe should have as an imaging biomarker of CTE are summarized in Table 1. Since 2013, several first-generation tau PET drugs have been developed, including flortaucipir (^{18}F-AV1451/^{18}F-T-807) (Wooten et al., 2017), ^{18}F-THK5351 (Harada et al., 2016), and ^{11}C-PBB3. These first-generation tau PET drugs have similar ability to detect tau lesions in AD, but differ in their ability to detect tau lesions in non-AD tauopathies. Among these tracers, ^{11}C-PBB3 has successfully visualized tau lesions in various non-AD tauopathies such as progressive supranuclear palsy (PSP) (Endo et al., 2019) and corticobasal degeneration (CBD) (Maruyama et al., 2013). First-generation tau PET drugs often have binding properties to monoamine oxidase (MAO), hampering accurate quantification of tau pathology. Previous studies showed that ^{18}F-AV1451 and ^{18}FTHK5351 have off-target binding to MAO-A and MAO-B (Ng et al., 2017), whereas ^{11}C-PBB3 does not cross-react with MAO-A and MAO-B (Ni et al., 2018), therefore allowing specific assessments of tau pathology.

In an effort to establish an early diagnostic method for neurodegenerative disease due to TBI, we conducted a tau PET study with ^{11}C-PBB3 in subjects with a previous history of mild-repetitive TBI and severe TBI. PET data were acquired from 27 subjects in the chronic stage following mild-repetitive ($n=13$) or severe TBI ($n=14$) and healthy control subjects ($n=15$). All of the mild-repetitive TBI patients had years of multiple exposures to concussions (mean ± SD: 16.5 ± 9.6 years; min–max: 3–29 years), and the average period after the first head injury was 28.3 ± 11.7 years. The severe TBI group consisted of patients who were injured in traffic accidents ($n=12$) and falls ($n=2$). The average period after head injury was 14.5 ± 12.1 years, and the average duration of loss of consciousness (LOC) following the injury was 14.1 ± 12.2 days.

TABLE 1 Characteristics of tau PET probes required as a diagnostic marker of CTE.

Synthetic stability and ease of administration
- No photoreactivity
- Longer half-life of radioactive isotopes (^{18}F-labeled)

Pharmacokinetics
- No or few radiometabolites
- Rapid uptake into the brain
- Fast clearance from the brain: T1/2 for free ligands <30min
- Fast clearance from plasma
- Reaching equilibrium within 120min

Binding capacity to tau isoforms
- Binding to 3+4 R tau in AD and CTE
- Binding to 3R tau (Pick's disease)
- Binding to 4R tau (PSP, CBD)

Binding selectivity
- No off-target binding to MAO-A, MAO-B and iron deposition
- No off-target binding in the choroid plexus
- No non-specific accumulation in the venous sinuses
- No binding to β-Amyloid, α-synuclein, TDP-43

Visualization of tau accumulation and image contrast
- The ratio of specific binding [free + non-specific binding] is sufficiently large
- Have single-case diagnostic capabilities
- Have capability to detect propagation of tau accumulation

Late-onset neuropsychiatric symptoms following TBI were diagnosed based on the modified criteria for traumatic encephalopathy syndrome (TES) (Stern et al., 2013), a clinical counterpart of neuropathologically diagnosed CTE. Representative images are shown in Fig. 4 (Takahata et al., 2019). Parametric images of tracer binding capacity (BP^{*}_{ND}) showed little difference in the tracer retention pattern between normal and TBI patients without TES, whereas both mild-repetitive and severe TBI patients with TES showed an increased retention of tracers suggestive of tau accumulation over a wide area, mainly in the frontal, parietal and temporal lobe. To confirm ^{11}C-PBB3 binding to tau deposits in TBI brains, we conducted in vitro PBB3 fluorescence and phospho-tau antibody immunofluorescence labeling of brain sections of CTE. This in vitro assays demonstrated PBB3-positive tau inclusions at the depth of neocortical sulci, confirming ^{11}C-PBB3 binding to tau lesions.

Challenges for quantification of tau load in TBI by tau PET imaging

Tau lesions in AD and other neurodegenerative diseases are usually quantified by the standard uptake value ratio (SUVR) method, which uses the cerebellar gray matter as a reference region. Since the SUVR method is a relative measure of tau accumulation, clinical studies define a specific region of interest (ROI) and calculate the average SUVR value of all voxels within the target region. However, this quantitative method is not suitable for assessing tau lesions caused by TBI by following reasons.

First, use of the cerebellar cortex as reference region may lead to an underestimation in TBI subjects because tau lesions of CTE can also appear in the cerebellum (McKee et al., 2013) unlike AD, in which tau accumulation is minimum in the cerebellum. Therefore, a reference region should be optimized for detecting tau pathology in CTE brains. Furthermore, tau lesions due to TBI are characterized by small and sparsely-distributed (patchy) lesions, and comparing SUVR values within the ROI may not be able to capture TBI-induced tau accumulations accurately due to partial volume effects. Therefore, a method to estimate the total amount of tau protein, rather than its density, is required. To solve these problems, we proposed a new quantitative method.

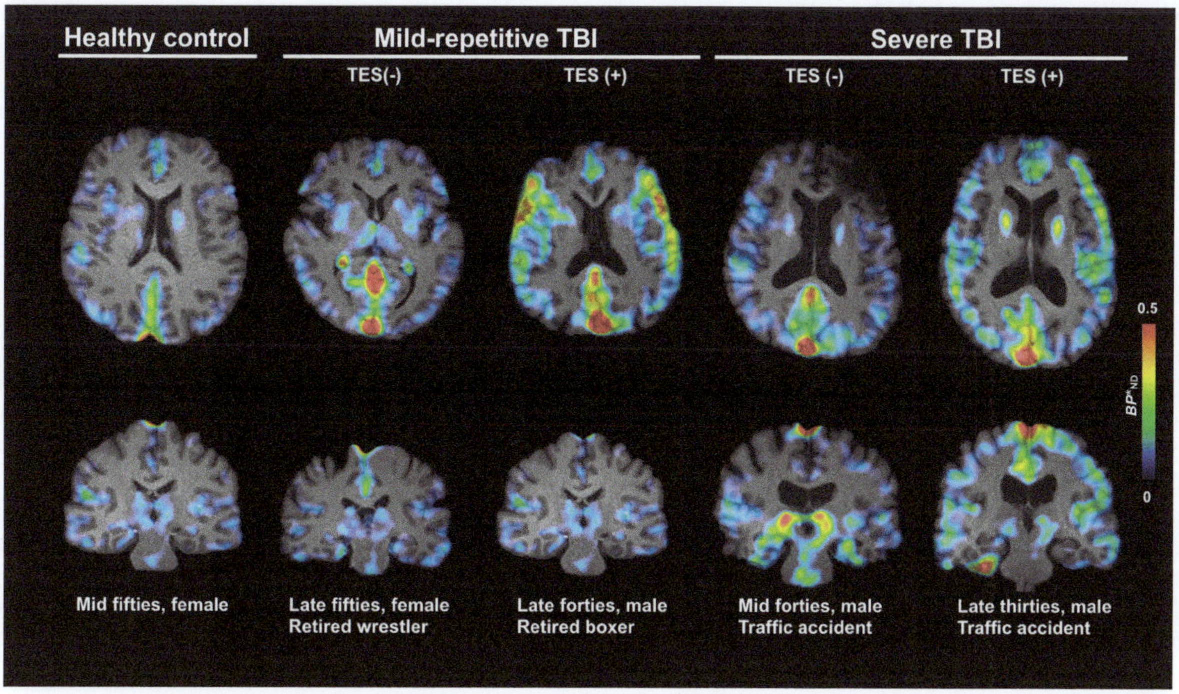

FIG. 4 Representative PET images of first-generation tau tracer in of long-term survivors of mild-repetitive and severe TBI. This figure shows PET (positron emission tomography) images of first-generation tau tracer, ^{11}C-PBB3, in chronic states of mild-repetitive and severe traumatic brain injury (TBI). All patients were long-term survivors from head injuries, more than 10 years after the initial head trauma. BP^*_{ND} (non-displaceable binding potential) parametric images are shown for representative 5 cases. Color bar shows BP^*_{ND} values ranging from 0 to 0.5. Each PET image was overlaid on individual MR T1 images. Healthy controls subjects and TBI patients without traumatic encephalopathy syndrome (TES) showed minimum accumulation of ^{11}C-PBB3 in the brain. TBI patients with TES showed increased retention of ^{11}C-PBB3 in the gray and white matter. There was no obvious difference in retention patterns of ^{11}C-PBB3 between mild-repetitive and severe TBI patients, indicating that late-life tau neuropathology is shared sequelae in both types of TBI, and is associated with late-onset neuropsychiatric symptoms following TBI. Note that all subjects showed non-specific binding of ^{11}C-PBB3 in the choroid plexus.

A. Optimization of a reference region

In consideration of possible tau accumulations in the cerebellar structures, we defined optimized reference voxels for quantification of ^{11}C-PBB3 binding and generated parametric BP^*_{ND} images as follows:

1.	As an initial step, BP^*_{ND} parametric images were generated with an original multi-linear reference tissue model ($MRTM_O$), which employed the cerebellar cortex as reference region.
2.	Next, optimized reference voxels were extracted (Endo et al., 2019; Kimura et al., 2016). Briefly, a range of BP^*_{ND} to cover putative reference voxels was determined as the mean full-width at half-maximum (FWHM) of a control histogram generated by averaging data of age-matched healthy control subjects (mean FWHM: 0.268). Using this optimized reference tissue, we rebuilt parametric BP^*_{ND} images with $MRTM_O$.

B. Quantification of total tau burden

Next, to quantify the tau lesion which is distributed sparsely in the brain, an index which conjunctively reflects the intensity and volume of PET signals in the target region was developed. This technique has been developed to measure the total amount of target proteins with irregular distribution, and has been applied to the estimation of the total cerebral glucose metabolism and accurate tumor size in PET studies with ^{18}F-fluoro-deoxy-glucose (^{18}F-FDG) (Vanhove et al., 2018), and the quantification of pancreatic islet cell signals in ^{11}C-dihydrotetrabenazine-PET (Goland et al., 2009). This technique calculates the mean of PET measures (e.g., BP_{ND}, SUV and SUVR) multiplied by segmented brain volumes as global disease burden, and has also been successfully employed for the analysis of ^{18}F-FDG data in the brains of AD patients (Alavi, Newberg, Souder, & Berlin, 1993). Likewise, we applied this methodological concept to the present study, and measured ^{11}C-PBB3 binding capacity according to the following procedures:

1. As target regions are likely to contain reference voxels with non-displaceable distribution volume below average, BP^*_{ND} might not exceed 0 in such voxels, leading to underestimation of regional mean BP^*_{ND} values. To circumvent this issue,

we defined the threshold of BP^{*}_{ND} separately for the gray matter (GM) and the white matter (WM) segments (in view of the differences in specific versus non-displaceable radioligand binding between these two segments) as one standard deviation away from the mean value in healthy subjects, and picked up voxels with BP^{*}_{ND} above the thresholds.

2. ^{11}C-PBB3 binding capacity in each segment of interest was quantified as a sum of individual voxel BP^{*}_{ND} above threshold multiplied by the voxel volume:

$$^{11}\text{C-PBB3 binding capacity} = \Sigma\left(\text{individual voxel} BP^{*}_{ND} \times \text{voxel volume}\right)$$

The unit is cm^3 (volume) because BP^{*}_{ND} is unitless.

In employing this novel quantitative method, we checked the results using ^{11}C-PBB3 binding capacity against conventional analyses using SUVR and BP^{*}_{ND} values, and showed that they are not only valid as an indicator of tau accumulation, but are also more acutely associated with tau pathology in TBI brains than conventional methods. Our results demonstrated that TBI patients showed increased ^{11}C-PBB3 binding capacity in widespread brain regions compared to age-matched HC subjects (Takahata et al., 2019). To elucidate the association between tau neuropathology and late-onset symptoms following TBI, we separated TBI patients based on the diagnostic criteria of TES. TBI patients with TES showed a higher ^{11}C-PBB3 binding capacity in the WM segment than those without TES, especially in the frontal WM segment. Additional analysis showed that higher ^{11}C-PBB3 binding capacity values were mainly due to increased radiosignals in the surface white matter. Interestingly, such distribution of the tau tracer in the present study is unique to TBI, and is distinct from that of AD. Moreover, ^{11}C-PBB3 binding capacity in the WM segment was positively correlated with the severity of psychosis. These findings indicated that there is a close link between tau neuropathology and late-onset neuropsychiatric symptoms following TBI.

Development of second-generation tau PET tracer

The results described above indicate that tau imaging with the first-generation tau PET tracers can detect tau deposition caused by TBI in living persons. However, first-generation tau PET tracers have several issues that impede practical application, such as non-specific accumulation in the venous sinus, off-target binding to the choroid plexus and basal ganglia, low image contrast due to low signal-to-noise ratio, and metabolic instability. Because of these problems, first-generation tau PET tracers can identify tau pathology following TBI only at the population level, but are difficult to diagnose at the single-case level. To solve these problems, several next-generation tau tracers have been developed, including ^{18}F-PM-PBB3 (^{18}F-APN-1607), ^{18}F-MK-6240 (Walji et al., 2016) and ^{18}F-PI-2620 (Kroth et al., 2019). Compared with first-generation tau PET tracers, second generation tau PET tracers showed less non-specific accumulation, higher image contrast, and metabolic stability. Among these tracers, ^{18}F-PM-PBB3 was shown to capture not only AD but also a wide range of non-AD tauopathies. Notably, the in vivo reactivity of ^{18}F-PM-PBB3 with tau inclusions in non-AD tauopathies was strongly supported by neuropathological examinations of brains derived from Pick's disease, PSP, and CBD patients who underwent PET scans (Tagai et al., 2021) (Fig. 5).

PET imaging of TBI with second-generation tau PET tracer

The second-generation tau PET tracer ^{18}F-PM-PBB3 can visualize tau lesions caused by TBI with higher contrast than first-generation tau PET drugs. Representative images of ^{18}F-PM-PBB3 of 2 cases of former athletes with a long history of concussions are shown in Fig. 6.

Case 1 is an early 60's male, a university professor. He had participated in soccer for 8 years while a teenager, and he had experienced at least one concussion without LOC. At age 15, he began his ice hockey career, retiring at age 55. At age 50, during an international game, he was pushed from behind by an opponent and consequently strongly hit the fence. He lost consciousness for several hours and was diagnosed with a concussion at the hospital. At age 55, he had concussions with LOC again. He remarked to the doctor that he had experienced at least 10 concussions during his ice hockey career. At age 57, he was also involved in a car accident, with his head striking the front windshield. The first indications of psychiatric problems occurred from the mid-50s. At age 56, he began to complain of insomnia, depressive mood, irritability and hypersensitivity. He also suffered from tinnitus, severe headaches and dizziness. Depression persisted for years, and sometimes suicidal ideas occurred. At age 61, he showed a manic episode without any preceding signs. He suddenly stood as a candidate for metropolitan governorship but was unsuccessful. He began to show difficulties in his daily life. His house was overflowing with garbage. Gradually, it became difficult to teach university classes. He underwent PET scans because he suspected the onset of a neurodegenerative disease. Amyloid PET was negative, indicating absence of Aβ pathology.

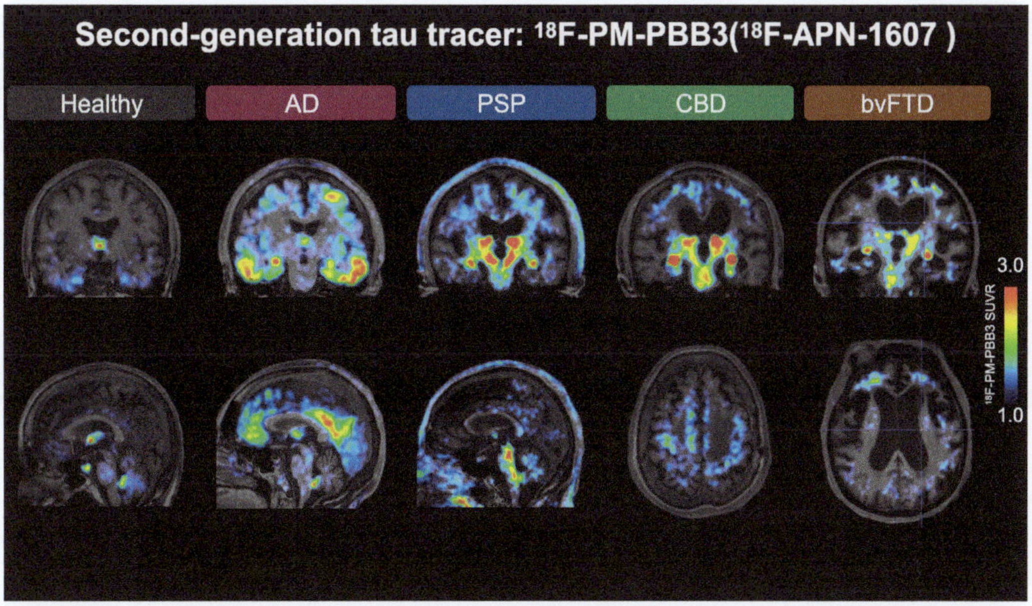

FIG. 5 Representative PET images of second-generation tau tracer in various neurodegenerative disorders. Representative images of PET (positron emission tomography) imaging with ^{18}F-PM-PBB3 (^{18}F-APN-1607) in various tauopathies. In healthy elderly subjects, tracer retention was minimal, but off-target binding was observed in the choroidal plaxus. In Alzheimer's disease, diffuse accumulation was observed in the temporal and parietal lobes, and progressive supranuclear palsy (PSP) showed dense accumulation of the tracer in the midbrain. In patients with corticbasall degeneration (CBD), the cerebral cortex showed accumulation of the tracer in the cerebral cortex with differential topology between left and right hemispheres. bvFTD (behavioral variant of frontotemporal dementia) showed focal accumulation of the tracer in the frontal lobe. In all these diseases, the retention pattern was consistent with known neuropathological findings.

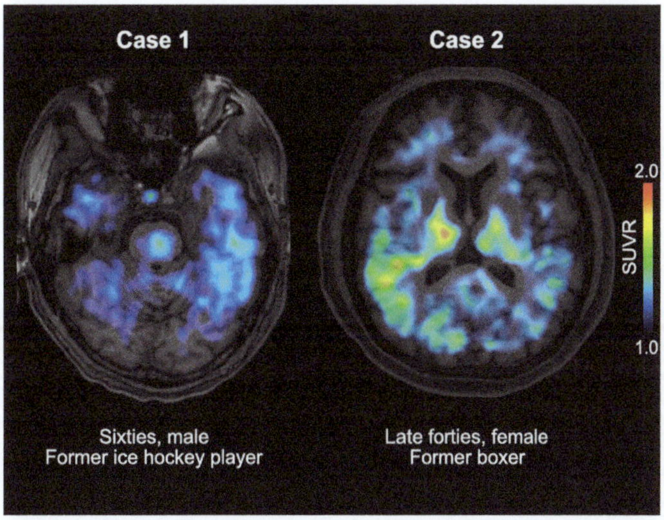

FIG. 6 Representative PET images of second-generation tau tracer in of long-term survivors of mild-repetitive and severe TBI. Representative images of PET (positron emission tomography) imaging with ^{18}F-PM-PBB3 (^{18}F-APN-1607) in mild-repetitive and severe traumatic brain injury (TBI) are shown. Case 1 is a former ice hockey player with long-term history of exposure to concussions. He showed severe cognitive impairment and mood disorder. Focal accumulation of ^{18}F-PM-PBB3 (^{18}F-APN-1607) was observed in the right inferior temporal lobe. Case 2 is a former domestic champion of female boxing. After retiring from boxing, she demonstrated psychotic symptoms such as delusion and auditory hallucination. There was a focal accumulation of ^{18}F-PM-PBB3 in the left occipital and parietal lobe. These cases suggest that second-generation tau PET tracer is superior to first-generation tau PET tracer in terms of image contrast and specificity of binding to tau lesions, allowing for diagnosis at the single-case level.

Tau PET scan with ^{18}F-PM-PBB3 showed focal accumulation of the tracer in the left inferior temporal lobe. The tracer distribution patterns differed from those of AD and other neurodegenerative diseases. Based on these PET findings, he was diagnosed with CTE.

Case 2 is a late 40s female, a former domestic boxing champion. She started boxing at age 23. Since there had been no official federation for female boxers in Japan at that time, she had been sparring with male professional boxers for training. At age 30, she received a strong blow to the right temple, which knocked her down and she hit her head on the concrete wall. Within a few years of her first fight, she won a domestic amateur championship match. In her last fight, she was knocked out by her opponent, and she was taken to a hospital by ambulance. Around age 30, she was diagnosed with depression, and suffered from severe weight loss. She also demonstrated psychotic symptoms such as auditory hallucination, delusional thought, and her diagnosis was changed to schizophrenia. At age 35, she retired from boxing because she felt ataxic and had difficulty walking. She has been hospitalized several times due to recurrence of psychotic symptoms. The psychiatric symptoms were suspected to be due to blows to the head from boxing, and she was referred to our laboratory. Amyloid PET was negative. Tau PET images showed focal accumulation of ^{18}F-PM-PBB3 in the bilateral basal ganglia, right parietal and occipital lobe.

The PET findings of the above two cases indicate that tau PET scans are useful in the diagnosis of neurodegenerative diseases caused by TBI.

Future directions of research

^{18}F-PM-PBB3, a second-generation tau PET tracer, offers the potential for earlier diagnosis of neurodegenerative diseases caused by TBI. It is also expected to allow diagnosis at the single-case level due to its improved image contrast, metabolic stability, and less off-target binding compared to first-generation tau PET tracers. A high degree of accuracy in the visualization of tau lesions in TBI brains may also be useful in the development of new therapies targeting tau lesions caused by head trauma. For this purpose, the process of propagation of tau lesions caused by TBI needs to be captured by PET. Scanning more patients with TBI in longitudinal studies may help to elucidate the pathogenesis of neurodegenerative diseases caused by head trauma. Verification by image-pathological correlation in the same individual is also essential to confirm how accurately we are detecting tau lesions following TBI by PET.

Conclusion

In this chapter, the historical background and pathogenesis of chronic traumatic encephalopathy (CTE) caused by repetitive head trauma are described, and the PET imaging findings using tau tracers are shown. PET imaging with the next-generation tau tracer ^{18}F-PM-PBB3 is a technique that enables the diagnosis of CTE at the single-case level. Early diagnosis of CTE and accurate assessment of tau lesions will enable objective decisions about early interventions, such as cognitive rehabilitation and social support.

Applications to other areas of neuroscience

Tau PET imaging is one of the most promising neuroimaging biomarkers of chronic pathology of traumatic brain injury (TBI). The second-generation tau tracer ^{18}F-PM-PBB3 (^{18}F-APN-1607) has high sensitivity and specificity to tau lesion in the brain and has been indicated to enable the diagnosis of CTE at the single-case level. Tau PET imaging is expected to reveal the process of tau propagation in the brain after head trauma. Tau PET imaging with ^{18}F-PM-PBB3 is expected to reveal the diverse background pathogenesis of neurodegenerative diseases caused by TBI. In addition, ^{18}F-PM-PBB3 has been shown to be a useful imaging biomarker for the assessment of tau accumulation in various tauopathies such as Alzheimer's disease, progressive supranuclear palsy, frontotemporal lobar degeneration and corticobasal degeneration. In order to establish a more accurate diagnostic method, it is also necessary to clarify the relationship between tau PET and blood biomarkers and other brain imaging biomarkers. In vivo evaluation of tau pathologies in the living subjects are also critically important techniques for developing a novel therapy targeting tau aggregates.

Key facts of neurodegenerative diseases following TBI

- Exposure to mild-repetitive TBI causes CTE, a neurodegenerative disease characterized by accumulation of p-tau.
- Severe TBI also cause neurodegenerative diseases such as Alzheimer's disease, Parkinson's disease, amyotrophic lateral sclerosis and CTE-like tauopathy.

- The core pathological feature of CTE is p-tau aggregates in neurons, astrocytes, and cell processes around small vessels in an irregular pattern at the depths of the cortical sulci.
- Tau deposition following TBI can be detected non-invasively by tau PET imaging.

Key facts of tau PET imaging

- First-generation tau PET tracers are useful for approximate assessment of tau lesions, but suffer from lower S/N ratio, off-target binding to MAO-B, non-specific accumulation in various brain areas.
- Second-generation tau PET tracer is a promising tool for detecting brain tau pathology in long-term survivors of TBI at a single-case level.
- PET imaging with second-generation tau PET tracer enable early diagnosis of Alzheimer's disease (AD) and other non-AD tauopathies.
- It is still unclear how early tau PET can detect tau lesions in the brain caused by TBI.

Mini-dictionary of terms

- CTE (chronic traumatic encephalopathy): A neurodegenerative disease which is associated with mild-repetitive TBI. CTE is characterized by accumulation of hyperphosphorylated tau aggregates in the depth of cortical sulci.
- Tau: Protein that has roles primarily in maintaining the stability of microtubules in axons and is abundant in the neurons of the central nervous system.
- TDP-43 (TAR DNA-binding protein-43 kDa): TDP-43 was recently identified as the major pathological protein in sporadic ALS and is the most common pathological subtype of FTLD with ubiquitinated inclusions (FLTD-U).
- β-Amyloid: β-Amyloid is the primary component of senile plaques characteristic of AD.
- PET (positron emission tomography): Positron emission tomography (PET) is a molecular imaging technique that uses radioactive tracers to visualize and measure changes in the density of target proteins, metabolic processes, and in other physiological activities including blood flow, inflammation and neurotransmission.
- ^{18}F-PM-PBB3: A tau imaging probe which can visualize tau deposits in AD and various non-AD tauopathies.

Summary points

- This chapter summarizes clinical and neuropathological feature of CTE and reviewed recent imaging studies targeting late-life tau neuropathology following TBI using tau PET.
- Tau PET imaging is a promising neuroimaging biomarker of CTE and other tauopathies.
- First-generation tau PET drugs can detect tau lesions in the brain caused by head trauma, but technical issues such as lower image contrast and off-target binding make it difficult to diagnose CTE at the single-case level.
- Second-generation tau-PET tracers solved the problems of first-generation tau tracers and are expected to be used for early diagnosis of CTE at the single-case level.

Acknowledgments

PET studies described in this chapter were supported by Japan Society for the Promotion of Science KAKENHI 20K07935, 20H03606, 20H03609, and 21H02805. These studies were also partly supported by AMED under Grant numbers JP19dm0207072 and JP20dm0207072, and a grant from the General Insurance Association of Japan to K.T. (2019-2020). The authors declare no conflict of interest.

References

Achté, K. A., Hillbom, E., & Aalberg, V. (1969). Psychoses following war brain injuries. *Acta Psychiatrica Scandinavica*, *45*(1), 1–18.

Alavi, A., Newberg, A. B., Souder, E., & Berlin, J. A. (1993). Quantitative analysis of PET and MRI data in normal aging and Alzheimer's disease: Atrophy weighted total brain metabolism and absolute whole brain metabolism as reliable discriminators. *Journal of Nuclear Medicine*, *34*(10), 1681–1687.

Brandenburg, W., & Hallervorden, J. (1954). Dementia Pugilistica with anatomical findings. *Virchows Archiv: An International Journal of Pathology*, *325*, 684–709.

Corsellis, J. A., Bruton, C. J., & Freeman-Browne, D. (1973). The aftermath of boxing. *Psychological Medicine*, *3*(3), 270–303.

Coughlin, J. M., Wang, Y., Munro, C. A., Ma, S., Yue, C., Chen, S., Airan, R., et al. (2015). Neuroinflammation and brain atrophy in former NFL players: An *in vivo* multimodal imaging pilot study. *Neurobiology of Disease*, *74*, 58–65. February.

Endo, H., Shimada, H., Sahara, N., Ono, M., Koga, S., Kitamura, S., Niwa, F., et al. (2019). In vivo binding of a tau imaging probe, [^{11}C]PBB3, in patients with progressive supranuclear palsy. *Movement Disorders*, *34*(5), 744–754.

Fujii, D., & Ahmed, I. (1996). Psychosis secondary to traumatic brain injury. *Neuropsychiatry, Neuropsychology, and Behavioral Neurology*, *9*, 133–138.

Fujii, D., & Ahmed, I. (2002). Characteristics of psychotic disorder due to traumatic brain injury: An analysis of case studies in the literature. *The Journal of Neuropsychiatry and Clinical Neurosciences*, *14*(2), 130–140.

Fujii, D. E., & Ahmed, I. (2014). Psychotic disorder caused by traumatic brain injury. *The Psychiatric Clinics of North America*, *37*(1), 113–124.

Fujii, D., & Fujii, D. C. (2012). Psychotic disorder due to traumatic brain injury: Analysis of case studies in the literature. *The Journal of Neuropsychiatry and Clinical Neurosciences*, *24*(3), 278–289.

Goland, R., Freeby, M., Parsey, R., Saisho, Y., Kumar, D., Simpson, N., Hirsch, J., et al. (2009). 11C-Dihydrotetrabenazine PET of the pancreas in subjects with long-standing type 1 diabetes and in healthy controls. *Journal of Nuclear Medicine*, *50*(3), 382–389.

Goldstein, L. E., Fisher, A. M., Tagge, C. A., Zhang, X.-L., Velisek, L., Sullivan, J. A., Upreti, C., et al. (2012). Chronic traumatic encephalopathy in blast-exposed military veterans and a blast neurotrauma mouse model. *Science Translational Medicine*, *4*(134), 134ra60.

Harada, R., Furumoto, S., Tago, T., Furukawa, K., Ishiki, A., Tomita, N., Iwata, R., et al. (2016). Characterization of the radiolabeled metabolite of tau PET tracer ^{18}F-THK5351. *European Journal of Nuclear Medicine and Molecular Imaging*, *43*(12), 2211–2218.

Ikonomovic, M. D., Uryu, K., Abrahamson, E. E., Ciallella, J. R., Trojanowski, J. Q., Lee, V. M.-Y., Clark, R. S., Marion, D. W., Wisniewski, S. R., & DeKosky, S. T. (2004). Alzheimer's pathology in human temporal cortex surgically excised after severe brain injury. *Experimental Neurology*, *190*(1), 192–203.

Johnson, V. E., & Stewart, W. (2015). Traumatic brain injury: Age at injury influences dementia risk after TBI. *Nature Reviews. Neurology*, *11*, 128–130.

Johnson, V. E., Stewart, W., & Smith, D. H. (2012). Widespread τ and amyloid-β pathology many years after a single traumatic brain injury in humans. *Brain Pathology*, *22*(2), 142–149.

Jordan, B. D. (2013). The clinical spectrum of sport-related traumatic brain injury. *Nature Reviews. Neurology*, *9*(4), 222–230.

Kimura, Y., Endo, H., Ichise, M., Shimada, H., Seki, C., Ikoma, Y., Shinotoh, H., et al. (2016). A new method to quantify tau pathologies with ^{11}C-PBB3 PET using reference tissue voxels extracted from brain cortical gray matter. *EJNMMI Research*, *6*(1), 24.

Kroth, H., Oden, F., Molette, J., Schieferstein, H., Capotosti, F., Mueller, A., Berndt, M., et al. (2019). Discovery and preclinical characterization of [^{18}F] PI-2620, a next-generation tau PET tracer for the assessment of tau pathology in Alzheimer's disease and other tauopathies. *European Journal of Nuclear Medicine and Molecular Imaging*, *46*(10), 2178–2189.

Lin, A., Charney, M., Shenton, M. E., & Koerte, I. K. (2018). Chronic traumatic encephalopathy: Neuroimaging biomarkers. *Handbook of Clinical Neurology*, *158*, 309–322.

LoBue, C., Munro Cullum, C., Didehbani, N., Yeatman, K., Jones, B., Kraut, M. A., & Jr, J. H. (2018). Neurodegenerative dementias after traumatic brain injury. *The Journal of Neuropsychiatry and Clinical Neurosciences*, *30*(1), 7–13.

Maroon, J. C., Winkelman, R., Bost, J., Amos, A., Mathyssek, C., & Miele, V. (2015). Chronic traumatic encephalopathy in contact sports: A systematic review of all reported pathological cases. *PLoS One*, *10*(2), e0117338.

Martland, H. S. (1928). Punch drunk. *JAMA*, *91*(15), 1103–1107.

Maruyama, M., Shimada, H., Suhara, T., Shinotoh, H., Ji, B., Maeda, J., Zhang, M.-R., et al. (2013). Imaging of tau pathology in a tauopathy mouse model and in Alzheimer patients compared to normal controls. *Neuron*, *79*(6), 1094–1108.

Mckee, A. C., Abdolmohammadi, B., & Stein, T. D. (2018). The neuropathology of chronic traumatic encephalopathy. *Handbook of Clinical Neurology*, *158*, 297–307.

McKee, A. C., Gavett, B. E., Stern, R. A., Nowinski, C. J., Cantu, R. C., Kowall, N. W., Perl, D. P., et al. (2010). TDP-43 proteinopathy and motor neuron disease in chronic traumatic encephalopathy. *Journal of Neuropathology and Experimental Neurology*, *69*(9), 918–929.

McKee, A. C., Stern, R. A., Nowinski, C. J., Stein, T. D., Alvarez, V. E., Daneshvar, D. H., Lee, H.-S., et al. (2013). The Spectrum of disease in chronic traumatic encephalopathy. *Brain*, *136*(Pt. 1), 43–64.

Meyer, A. (1904). The anatomical facts and clinical varieties of traumatic insanity. *American Journal of Insanity*, *12*(3), 373–441.

Meyer, A. (1908). What do histories of cases of insanity teach us concerning preventive mental hygiene during the years of school life? *The Psychological Clinic*, *2*(4), 89–101.

Mez, J., Daneshvar, D. H., Kiernan, P. T., Abdolmohammadi, B., Alvarez, V. E., Huber, B. R., Alosco, M. L., et al. (2017). Clinicopathological evaluation of chronic traumatic encephalopathy in players of American Football. *JAMA*, *318*(4), 360–370.

Mez, J., Stern, R. A., & McKee, A. C. (2020). Chronic traumatic encephalopathy. *Seminars in Neurology*, *40*(4), 351–352.

Millspaugh, J. A. (1937). Dementia Pugilistica. *Journal of Alzheimer's Disease*, *35*, 297–303.

Ng, K. P., Pascoal, T. A., Mathotaarachchi, S., Therriault, J., Kang, M. S., Shin, M., Guiot, M.-C., et al. (2017). Monoamine oxidase B inhibitor, selegiline, reduces ^{18}F-THK5351 uptake in the human brain. *Alzheimer's Research & Therapy*, *9*(1), 25.

Ni, R., Ji, B., Ono, M., Sahara, N., Zhang, M.-R., Aoki, I., … Higuchi, M. (2018). Comparative in vitro and in vivo quantifications of pathologic tau deposits and their association with neurodegeneration in tauopathy mouse models. *Journal of Nuclear Medicine*, *59*(6), 960–966.

Okamura, Y., Kawakami, I., Watanabe, K., Oshima, K., Niizato, K., Ikeda, K., Akiyama, H., & Hasegawa, M. (2019). Tau progression in single severe frontal traumatic brain injury in human brains. *Journal of the Neurological Sciences*, *407*, 116495. December.

Omalu, B. I., DeKosky, S. T., Minster, R. L., Ilyas Kamboh, M., Hamilton, R. L., & Wecht, C. H. (2005). Chronic traumatic encephalopathy in a National Football League Player. *Neurosurgery*, *57*(1), 128–134.

Omalu, B. I., Fitzsimmons, R. P., Hammers, J., & Bailes, J. (2010). Chronic traumatic encephalopathy in a professional American wrestler. *Journal of Forensic Nursing*, *6*(3), 130–136.

Osnato, M., & Giliberti, V. (1927). Postconcussion neurosis-traumatic encephalitis: A conception of postconcussion phenomena. *Archives of Neurology, 18*, 181–214.

Roberts, G. W., Allsop, D., & Bruton, C. (1990). The occult aftermath of boxing. *Journal of Neurology, Neurosurgery, and Psychiatry, 53*(5), 373–378.

Roberts, G. W., Gentleman, S. M., Lynch, A., Murray, L., Landon, M., & Graham, D. I. (1994). Beta amyloid protein deposition in the brain after severe head injury: Implications for the pathogenesis of Alzheimer's disease. *Journal of Neurology, Neurosurgery, and Psychiatry, 57*(4), 419–425.

Sachdev, P., Smith, J. S., & Cathcart, S. (2001). Schizophrenia-like psychosis following traumatic brain injury: A chart-based descriptive and case–control study. *Psychological Medicine, 31*(2), 231–239.

Shively, S., Scher, A. I., Perl, D. P., & Diaz-Arrastia, R. (2012). Dementia resulting from traumatic brain injury: What is the pathology? *Archives of Neurology, 69*(10), 1245–1251.

Smith, D. H., Johnson, V. E., & Stewart, W. (2013). Chronic Neuropathologies of single and repetitive TBI: Substrates of dementia? *Nature Reviews Neurology, 9*(4), 211–221.

Stern, R. A., Daneshvar, D. H., Baugh, C. M., Seichepine, D. R., Montenigro, P. H., Riley, D. O., Fritts, N. G., et al. (2013). Clinical presentation of chronic traumatic encephalopathy. *Neurology, 81*(13), 1122–1129.

Tagai, K., Ono, M., Kubota, M., Kitamura, S., Takahata, K., Seki, C., Takado, Y., et al. (2021). High-contrast in vivo imaging of tau pathologies in Alzheimer's and non-Alzheimer's disease Tauopathies. *Neuron, 109*(1), 42–58.e8.

Takahata, K., Kato, M., Mimura, M., & Shimada, H. (2016). Late-onset neurocognitive deficits following traumatic brain injury: Chronic traumatic encephalopathy (CTE) and psychotic disorder following TBI (PDF TBI). *Higher Brain Function Research, 35*(3), 276–282.

Takahata, K., Kimura, Y., Sahara, N., Koga, S., Shimada, H., Ichise, M., Saito, F., et al. (2019). PET-detectable tau pathology correlates with long-term neuropsychiatric outcomes in patients with traumatic brain injury. *Brain, 142*(10), 3265–3279.

Takahata, K., Tabuchi, H., & Mimura, M. (2016). Late-onset neurodegenerative diseases following traumatic brain injury: Chronic traumatic encephalopathy (CTE) and Alzheimer's disease secondary to TBI (AD-TBI). *Brain and Nerve, 68*(7), 849–857.

Vanhove, K., Mesotten, L., Heylen, M., Derwael, R., Louis, E., Adriaensens, P., Thomeer, M., & Boellaard, R. (2018). Prognostic value of total lesion glycolysis and metabolic active tumor volume in non-small cell lung cancer. *Cancer Treatment and Research Communications, 15*, 7–12.

von Krafft-Ebing, R. (1868). *Über die durch Gehirnerschütterung und Kopfverletzung hervorgerufenen psychischen Krankheiten: Eine klinisch-forensische Studie*. Erlangen: Verlag von Ferdinand Enke.

Walji, A. M., Hostetler, E. D., Selnick, H., Zeng, Z., Miller, P., Bennacef, I., Salinas, C., et al. (2016). Discovery of 6-(Fluoro-(18)F)-3-(1H-pyrrolo[2,3-C]pyridin-1-Yl)isoquinolin-5-amine ([(^{18})F]-MK-6240): A positron emission tomography (PET) imaging agent for quantification of neurofibrillary tangles (NFTs). *Journal of Medicinal Chemistry, 59*(10), 4778–4789.

Wooten, D. W., Guehl, N. J., Verwer, E. E., Shoup, T. M., Yokell, D. L., Zubcevik, N., Vasdev, N., et al. (2017). Pharmacokinetic evaluation of the tau PET radiotracer ^{18}F-T807 (^{18}F-AV-1451) in human subjects. *Journal of Nuclear Medicine, 58*(3), 484–491.

Chapter 11

Evaluating the integrity of white matter after traumatic brain injury and the utility of diffusion tensor imaging

Eunkyung Kim[a,b], Min-Gu Kang[c], and Byung-Mo Oh[a,d,e,f,g]

[a]*Department of Rehabilitation Medicine, Seoul National University Hospital, Seoul, Korea,* [b]*Biomedical Research Institute, Seoul National University Hospital, Seoul, Korea,* [c]*Department of Physical Medicine and Rehabilitation, Dong-A University College of Medicine, Busan, Korea,* [d]*Department of Rehabilitation Medicine, Seoul National University College of Medicine, Seoul, Korea,* [e]*Institute on Aging, Seoul National University, Seoul, Korea,* [f]*National Traffic Injury Rehabilitation Hospital, Yangpyeong, Korea,* [g]*Neuroscience Research Institute, Seoul National University College of Medicine, Seoul, Korea*

List of abbreviations

AD axial diffusivity
ADC apparent diffusion coefficient
ARAS ascending reticular activating system
CRP corticoreticular pathway
CSF cerebrospinal fluid
CST corticospinal tract
CT computed tomography
DAI diffuse axonal injury
DTI diffusion tensor imaging
DTT diffusion tensor tractography
EPI echo-planar imaging
FA fractional anisotropy
FLAIR fluid-attenuated inversion recovery
GCS Glasgow coma scale
MD mean diffusivity
MEP motor evoked potential
MRC medical research council
MRI magnetic resonance imaging
RD radial diffusivity
ROI region of interest
STT spinothalamic tract
TBI traumatic brain injury

Introduction

Like other neurological diseases such as stroke, imaging tests are essential for the diagnosis and treatment decision-making for traumatic brain injury (TBI). Compared to brain imaging studies on stroke, those for TBI are less sensitive in assessing the extent and severity of neural injury. Cerebrovascular accidents cause rapid and severe damage to a specific region of the brain, whereas TBI causes varying degrees of diffuse brain damage. Therefore, patients can show striking differences between imaging findings and clinical features, e.g., diffuse axonal injury (DAI) without serious intracranial hemorrhage. Typical patients in this category do not show homogeneous abnormalities in specific vascular territories on the initial imaging, but show evidence for cerebral edema and small bleeds throughout the brain (Fig. 1). These patients often experience

Diagnosis and Treatment of Traumatic Brain Injury. https://doi.org/10.1016/B978-0-12-823347-4.00011-7

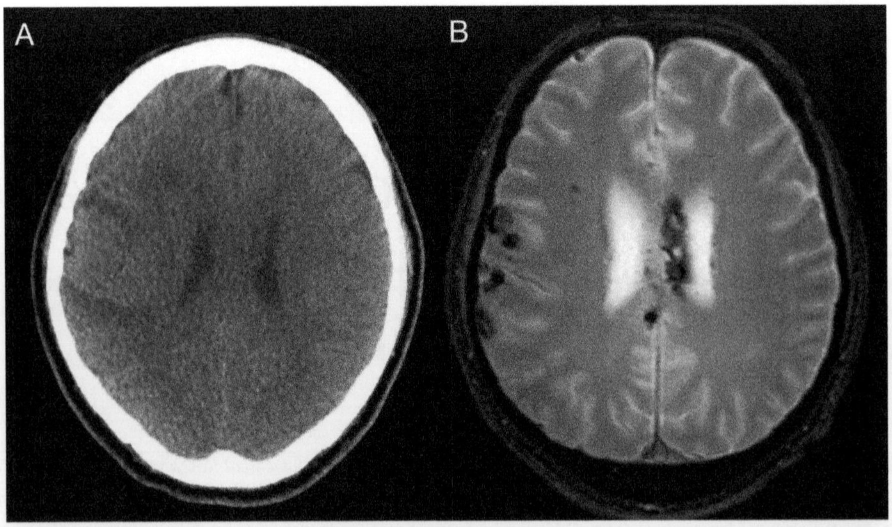

FIG. 1 Computed tomography (A) and magnetic resonance (B) images from a patient with diffuse axonal injury. CT scan shows subtle low attenuation in the right temporal area, while MRI reveals numerous microbleedings in the corpus callosum, corona radiata, and the right inferior frontal to temporal cortex.

a prolonged state of severely disordered consciousness, which is disproportionate to the moderate abnormalities shown by imaging studies. This is partly because white matter injury—a pivotal pathophysiology of TBI—is difficult to visualize sensitively even with modern brain imaging techniques.

Nevertheless, there is a growing body of clinical and preclinical evidence for the vulnerability of brain white matter tracts to traumatic events (Smith, Hicks, & Povlishock, 2013). This particular vulnerability of the white matter to the external force is believed to be accentuated when the primary injury involves a rotational acceleration. For this reason, sensitive and specific brain imaging techniques showing the presence and extent of damage to the white matter are needed.

Here, we review clinical and neuroimaging-based evaluations of white matter integrity with a primary focus on current evidence for the usefulness of diffusion tensor imaging (DTI) in the evaluation of the white matter.

The vulnerable white matter tracts in TBI

Axons in the white matter tracts are selectively more susceptible to head trauma due to their highly anisotropic arrangement (Johnson, Stewart, & Smith, 2013). An autopsy study of 434 cases reported that the parasagittal white matter in the corpus callosum, midbrain, and pons is preferentially affected by diffuse axonal injury (DAI) (Adams et al., 1989).

Neuroimaging studies reported that frequently damaged white matter tracts included the genu and splenium of the corpus callosum, superior and inferior longitudinal fasciculi, anterior corona radiata, uncinate fasciculus, cingulum, fimbria, and corticospinal tract (Bigler & Maxwell, 2012; Herrera et al., 2017). Long white matter tracts are specifically at high risk of injury to mechanical deformation and rotational forces (Fig. 2).

Zhao et al. evaluated injury prediction performances of 50 deep white matter tracts in American National Football players with head impacts (Zhao, Cai, Li, & Ji, 2017). The top four most vulnerable tracts were the genu, body of the corpus callosum, anterior limb of the internal capsule, and superior longitudinal fasciculus.

Clinical and neurophysiological evaluation of injury to white matter

DAIs can be visualized indirectly through hemorrhages on computed tomography (CT) caused by inertial injuries on the blood vessels, or white matter hyperintensities on fluid-attenuated inversion recovery magnetic resonance imaging (FLAIR MRI) (Takaoka et al., 2002). However, conventional CT or MRI is not sufficient to estimate the severity or outcome of DAI (Scheid, Walther, Guthke, Preul, & von Cramon, 2006). Therefore, DAI is still diagnosed in consideration of clinical characteristics and the results of functional evaluation.

The cingulum and fornix, which are the longest limbic tracts, are commonly damaged in DAI. Patients with injuries to these tracts show deficits in attention, working memory, and processing speed (Levin, 1990; Stuss et al., 1985). Recovery of cognitive function following TBI can be evaluated using the Rancho Los Amigos Levels of Cognitive Functioning Scale that describes functioning after TBI based on the interaction with the environment. To accurately assess the severity and

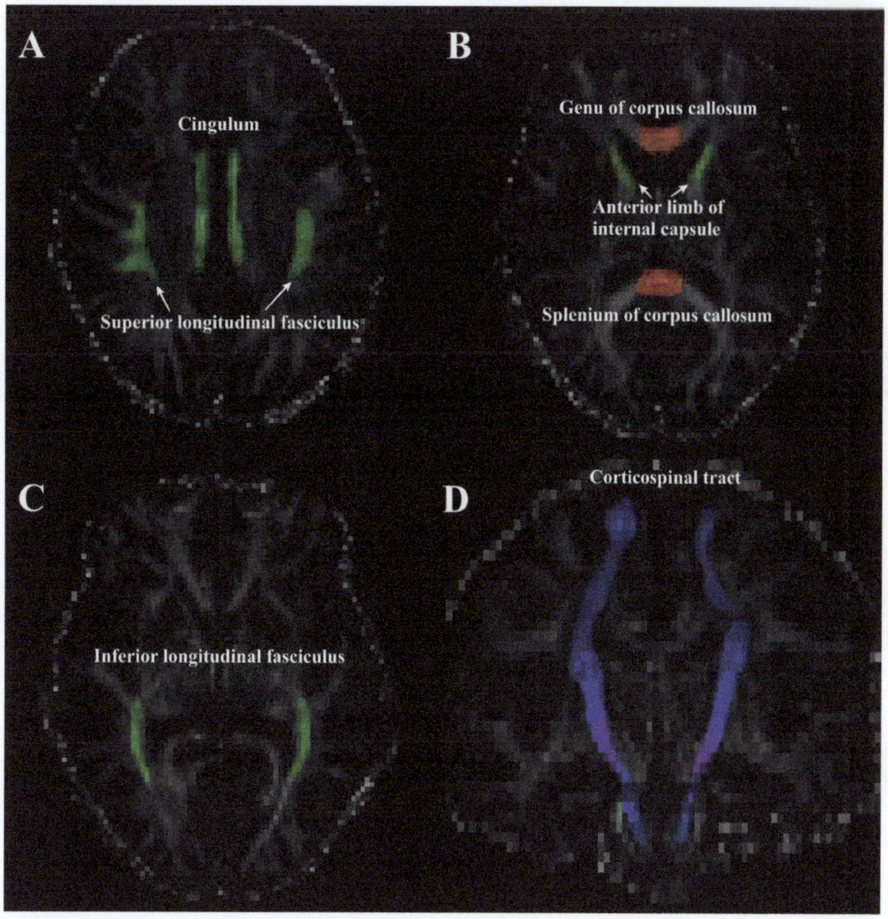

FIG. 2 Vulnerable white matter tracts displayed on a healthy volunteer. Vulnerable tracts are marked with color on the FA maps of a healthy volunteer. (A) Cingulum (*green*) and superior longitudinal fasciculus (*green, arrow*). (B) Genu and splenium of corpus callosum (*red*) and anterior limb of internal capsule (*green, arrow*). (C) Inferior longitudinal fasciculus (*green*). (D) Corticospinal tract (*blue*).

pattern of cognitive impairment and to establish an appropriate treatment plan, comprehensive assessment is necessary in many cases.

Disorders of consciousness are common consequences of TBI. It is considered to be associated with an injury to the ascending reticular activating system (ARAS) in DAI. Since consciousness is a broad spectrum along the continuum, it is not easy to quantify and evaluate it. Glasgow coma scale (GCS) is widely used to assess the disorder of consciousness after TBI. The John F. Kennedy coma recovery scale is also commonly used to evaluate the consciousness in patients with TBI (Kalmar & Giacino, 2005).

The corticospinal tract (CST) and corticoreticular pathway (CRP) are the main white matter pathways responsible for motor deficits. Injury to the CST leads to predominantly distal motor weakness, while CRP injury accounts for proximal weakness. Transcranial magnetic stimulation (TMS) is a neurophysiological tool to examine the motor pathways. While TMS is applied to the primary motor cortex, the motor evoked potential (MEP) response is recorded from the distal limb muscles. However, neurophysiological evaluation is expensive to implement and dependent on the examiner's skill. In addition, it only examines some functional units of the brain and cannot assess the white matter integrity of the entire brain. Therefore, the development of a more objective and quicker method to assess the white matter integrity of the whole brain is needed. A typical assessment that satisfies these requirements is neuroimaging, especially DTI.

Key facts of clinical and neurophysiological assessment after TBI

- Evaluation of the injury to the white matter has largely been based on clinical and functional evaluations.
- To accurately evaluate the cognitive dysfunction after TBI, comprehensive assessment is needed.
- TMS is a neurophysiological tool to investigate the integrity of cardinal motor pathways.
- DTI can be used to assess the white matter integrity of the whole brain.

DTI: A primer

Diffusion-weighted imaging captures the diffusivity of water molecules in biological tissues *in vivo*, particularly for the brain. Previous studies have thoroughly addressed the mechanism of diffusion-weighted imaging in the brain (Mukherjee, Berman, Chung, Hess, & Henry, 2008; Mukherjee, Chung, Berman, Hess, & Henry, 2008). We will briefly address the principles of the diffusion-weighted imaging, which may serve as a primer.

By measuring proton density in brain tissue, water content in the tissue can be relatively measured, with cerebrospinal fluid (CSF) having the highest intensity, followed by gray matter and then white matter. The diffusivity of water occurs in the brain, which is composed of biological structures disrupting free diffusivity, including cell membranes or fibers. Thus, in the brain, water in the CSF diffuses relatively freely compared to water in the gray and white matter. The diffusion coefficient is the moving distance in units of mm^2/s, and the term "apparent diffusion coefficient (ADC)" is used in biological tissue. Taken together, the amount of water content and diffusivity are different per brain tissue type. These characteristics allow researchers to investigate the brain in various ways.

For example, diffusion-weighted imaging uses ADC by applying a pulse magnetic field gradient, which intentionally introduces inhomogeneity in a linear magnetic field in the scanner. When scanning the human brain, there is a main magnetic field that is denoted by B0. Three gradient coils are present in the scanner, along with the x, y, and z directions, modulating the strength of the B0 field. The hydrogen nuclei inside the brain have a spin property and are aligned toward the B0 field when the gradient coils are not applied. When the gradient coils are applied differently for a certain amount of time (δ) and certain amount of strength (G), the frequency of the spin is changed according to the locations (by the gradient coil). Thus, the signals that are in phase (under the B0 field) are de-phased because of the different frequency. After a time Δ from the initial application of the gradient coil, the re-phasing gradient is applied to recover the signal. We eventually obtain the signal loss because of the diffusivity of water, and the amount of signal loss can represent the diffusivity of water. An echo-planar sequence is followed by the re-phasing gradient to obtain a diffusion-weighted image, and the intensity of the signal thereby follows (1) the ADC, (2) gradient factor b (related to the parameters δ, G, and Δ), and (3) original signal intensity *S0* without the diffusion gradients. The equation for estimating the signal intensity is given by

$$S = S_0 \exp(-b\,ADC).$$

A b-zero image represents no diffusion-weighted image, while a b-value higher than zero represents a diffusion-weighted image, which leads to signal attenuation of the tissue depending on its diffusion property. To estimate the ADC map, at least two b-value images are required, one of which is a b-zero image, and other is an image with a b-value higher than zero. A higher intensity is observed in the CSF, while a lower signal intensity is observed in the white matter because of the biological constraint disrupting free diffusivity, i.e., myelination or cell membrane. When the diffusion gradient is applied, the signal intensity of the CSF is lower while the signal intensity of the white matter is greater due to the signal attenuation.

DTI and the white matter integrity

In diffusion-weighted imaging, the diffusion tensor can be used to obtain directional information of the diffusivity in the brain, which is the diffusivity of the ellipsoid. The diffusion tensor matrix (D) contains information of diffusivity along the three gradient coil directions including x, y, and z, which are linearly applied gradients (Dxx, Dyy, Dzz). There are also other directions with other combinations: x and y, x and z, and y and z, represented by Dxy ($=Dyx$), Dxz ($=Dzx$), and Dyz ($=Dzy$). By a linear combination of the three gradient coils, the six tensor components of the diffusion tensor matrix can be estimated. After diagonalization of the diffusion tensor matrix, the three eigenvectors representing the principal diffusion direction and its two orthogonal axes, and the three eigenvalues representing the corresponding diffusivity in that directions can be obtained. Therefore, at least one b-zero image and six diffusion-weighted images are needed for DTI. Practically, more than six diffusion-weighted directions are obtained in a clinical study. For more accurate estimation of tensor orientation, more than 30 directions are recommended (Mukherjee, Chung, et al., 2008).

As described above, the white matter tissue has greater signal intensity in the diffusion-weighted images, and this leads to DTI, which is a promising technique to determine the white matter integrity of the brain. Several measurements from the diffusion tensor matrix have been used to quantify diffusion properties of a given voxel, particularly white matter tissue integrity (Table 1). Volume ratio and volume fraction are other measurements describing diffusion anisotropy but less frequently used.

Among the diffusion parameters in Table 1, the mean diffusivity (MD), fractional anisotropy (FA), axial diffusivity (AD), and radial diffusivity (RD) are representative diffusion property measures in association with the changes in axonal morphology, myelination, and coherence of white matter fiber tracts (Alexander, Lee, Lazar, & Field, 2007; Song et al., 2005), suggesting its role in clinical applications.

TABLE 1 Diffusion tensor parameters describing white matter integrity.

Parameters	Equation	Measured value
Trace (Tr)	$\lambda_1+\lambda_2+\lambda_3$	The amount of diffusivity in the given voxel without containing directional information
Mean diffusivity (MD)	$\frac{\lambda_1+\lambda_2+\lambda_3}{3}$	The average diffusivity in the given voxel
Relative anisotropy (RA)	$\frac{3\sqrt{(\lambda_1-\lambda_2)^2+(\lambda_2-\lambda_3)^2+(\lambda_3-\lambda_1)^2}}{\lambda_1^2+\lambda_2^2+\lambda_3^2}$	The relative ratio of the variance between each three eigenvalues and the average of all of them; ranges from 0 to $\sqrt{2}$
Fractional anisotropy (FA)	$\sqrt{\frac{1}{2}}\frac{\sqrt{(\lambda_1-\lambda_2)^2+(\lambda_2-\lambda_3)^2+(\lambda_3-\lambda_1)^2}}{\sqrt{\lambda_1^2+\lambda_2^2+\lambda_3^2}}$	The relative ratio of the variance between each three eigenvalues and the average of all of them; ranges from 0 to 1
Axial diffusivity (AD)	λ_1	The parallel or longitudinal diffusivity toward principal axis of the diffusivity
Radial diffusivity (RD)	$\frac{\lambda_2+\lambda_3}{2}$	The perpendicular diffusivity along the principal direction

Key facts of DTI

- The water diffusivity is estimated based on the signal loss during an intentional introduction (de-phasing) of the inhomogeneity of a magnetic field in the scanner and the reversal (re-phasing).
- Along the three gradient coil directions (x, y, and z), and the linear combination of the three gradient coils, the diffusion tensor matrix can be estimated.
- Diffusion metrics, which are derived from the diffusion tensor matrix, have been used to quantify tissue integrity.
- Due to the biological tissue compartment, water diffusivity in the brain is restricted, particularly in the white matter tissue.

DTI sequences and analysis tools: Strengths and weaknesses

Echo planar imaging sequence (EPI), which is used for acquiring DTI data, has several strengths and weaknesses. EPI allows for fast acquisition because it requires a minimum acquisition time to cover the full *k*-space, and thus leads to less motion sensitivity. In contrast, the inhomogeneity of the main magnetic field (B0) and the susceptibility effect may contaminate the brain signal, which is recognized as "susceptibility artifacts." To alleviate such disadvantages, a gradient field map can be obtained for post-processing, or more advanced techniques including parallel imaging can be used. For example, the single-shot echo EPI can be used with parallel imaging options such as the generalized auto-calibrating partial parallel acquisition or auto-calibrating reconstruction for Cartesian imaging sequences (pulsed gradient dual echo acquisition protocol).

DTI data can be analyzed via a region-of-interest (ROI) method, voxel-based analysis, or tractography approach. In the ROI analysis, the white matter integrity of given voxels including a specific ROI is averaged to represent the white matter integrity of the ROI. It is a hypothesis-driven approach, which only answers a particular hypothesis and ignores other brain areas than the ROI. Therefore, this method is not appropriate for explorative studies. On the other hand, the voxel-based analysis is a data-driven approach, and generally focuses on the whole brain which has advantage for explorative work. Though this approach confers advantages for investigating white matter integrity over the whole brain, voxel-based analysis can also result in multiple comparison issues. The accurate registration problem should also be considered.

At last, the tractography analysis has been used to reconstruct white matter fiber tracks using diverse mathematical algorithms, including deterministic and probabilistic tractography. The deterministic tractography assumes the principal direction of the fiber in a given voxel according to the principal eigenvector (Chung, Chou, & Chen, 2011). The probabilistic tractography, in contrast, assumes a distribution of fiber orientations in a given voxel. Thus, the reconstructed fiber streamline from the probabilistic tractography is less intuitive but provides more abundant information for the non-dominant tract, which is often ignored in the deterministic approach. Although the tractography results have been used to represent anatomical (or structural) connectivity of the brain or between particular ROIs, the reconstructed tractography should be interpreted with caution. Several factors can also affect to the tractography results, including the crossing-fiber or kissing fiber problems. In recent studies, more advanced techniques to capture the diffusivity of water have been suggested to overcome such limitations, such as diffusion spectrum imaging and high-angular resolution diffusion imaging (HARDI). To use these techniques, a sufficient number of directions and higher *b*-values must be obtained, which requires a long acquisition time to scan. Scanning time and accurate estimation of the diffusivity could be balanced for practical usage of DTI in individuals with TBI.

Moderate-to-severe TBI: DTI Study

Because traumatic axonal injury is a very common pathologic feature, DTI can be used to better understand pathological phenomena and sequelae of the TBI. In a recent meta-analysis study, decreases in FA and increases in MD or ADC were regarded as the pathological features of DTI parameters after TBI (Wallace, Mathias, & Ward, 2018). In particular, decreases in FA and increases in MD or ADC are larger in the moderate-to-severe TBI group than the mild TBI group, although the magnitude and location of the vulnerable white matters vary in individuals (Wallace et al., 2018). Altered white matter integrity after moderate-to-severe TBI prevails in almost all brain areas (Kraus et al., 2007). The reduced FA was observed in the corpus callosum, the brainstem and thalamus (Haberg et al., 2015).

Patients with disorders of consciousness due to severe TBI have also been investigated via DTI (Newcombe et al., 2010). The course of brain recovery could be predicted using the DTI parameters, suggesting its potential prognostic roles in TBI studies. For example, DTI parameters acquired during an early stage of severe TBI were successfully used to predict the outcome at 1 year after the injury (Galanaud et al., 2012). The FA and RD values obtained during the early phase of injury (< 6 weeks) were also associated with neurocognitive sequelae at 5 years after the injury (Dinkel et al., 2014), thereby supporting the usefulness of early-stage DTI parameters for prognostication.

In patients with moderate-to-severe TBI, understanding neurodegenerative changes of the brain is particularly important because it can last through the chronic stage of injury and beyond even after the symptoms are resolved (Corrigan & Hammond, 2013). It has been known that such neurodegenerative changes can develop to the increased rates of dementia (Graham & Sharp, 2019). The DTI parameters were linked to increased amyloid-beta burden, which has been used as an indirect biomarker to represent neurodegenerative changes (Scott et al., 2016).

DTI in special populations: Mild TBI and blast-induced TBI

The prevalence of mild TBI—also known as a concussion—among the TBI population, and the diverse symptoms suffered from individuals with concussion need special attention in clinics as well as research fields. Generally, brain changes after a concussion are considered as an axonal injury represented by the reduced FA and increased diffusivity on ADC image. Because most patients do not show abnormalities in conventional imaging modalities such as CT or structural MRI, DTI is invaluable for diagnostic purposes in individuals with a concussion (Fig. 3).

One of the important factors affecting the changes of DTI parameters is the time post-injury. In the acute phase of injury, TBI results in cerebral hypoxia, edema, or ischemia, which can variously affect DTI parameters (Eierud et al., 2014). In the sub-acute and chronic phases of injury, white matter integrity represented by FA is reduced, reflecting axonal morphological changes or demyelination (Lipton et al., 2012). There have also been inconsistent findings showing decreased white matter integrity during the acute phase after concussion (Dodd, Epstein, Ling, & Mayer, 2014). The discrepancy may be due to the heterogeneous characteristics of a concussion including site of injury, mode of injury, severity, and premorbid mental health information (Dodd et al., 2014). To overcome this heterogeneity, more longitudinal studies are warranted to delineate the trajectories of TBI sequelae or recovery. In a previous study, FA at the chronic stage after injury (6 months later) was reduced compared to the acute stage (< 10 h) (Veeramuthu et al., 2015). At both time points, there was a significant association between integrity of white matter tract and neuropsychological deficits (Veeramuthu et al., 2015). The relationship between the white matter changes after concussion and TBI-associated symptoms is particularly helpful to understand the recovery process of patients. In our previous study, cognitive function and balance control of the patients after mild-to-moderate TBI were associated with altered white matter integrity (Fig. 4) (Kim et al., 2019).

Taken together, the current evidence indicates that the white matter pathophysiology after concussion is likely to take a dynamic course over time, and is under the strong influence of clinical characteristics of the study sample such as symptom severity, premorbid mental health status, age, or time post-injury. Longitudinal quantification of dynamic changes of the

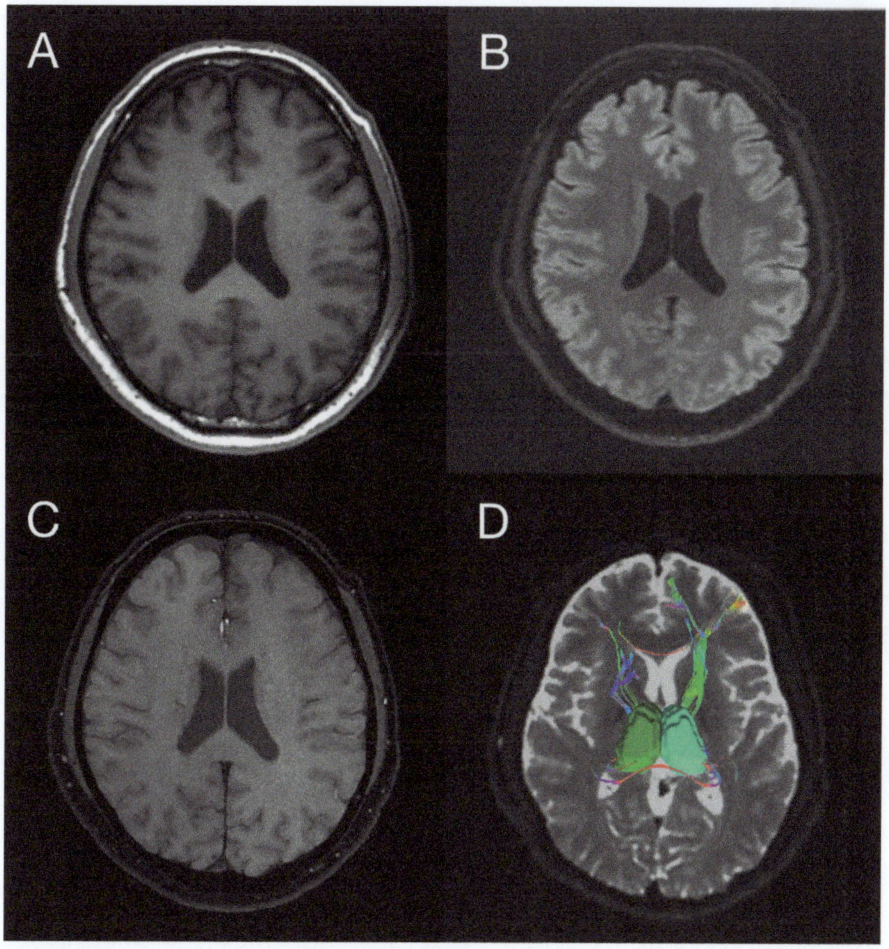

FIG. 3 MR images and diffusion tensor tractography reconstruction for the prefronto-thalamic tract in a patient who sustained personality change after a concussion. The T1-weighted (A), T2 FLAIR (B), and susceptibility-weighted (C) images shows no definite abnormal findings. However, diffusion tensor tractography reveals visible asymmetry of the bilateral prefronto-thalamic tracts (D).

white matter using DTI along with detailed clinical evaluation may provide foundation for the development of diagnostic and/or prognostic biomarkers after concussion.

There has been an increasing interest in the effects of blast-induced TBI on white matter integrity and its association with behavioral changes. Previous studies have reported no direct group difference in brain white matter between patients with blast-induced mild TBI and healthy subjects (McClelland, Fleysher, Mu, Kim, & Lipton, 2018; Sullivan et al., 2019). Nevertheless, axonal injury is also accompanied by blast-induced TBI, showing reduced FA in the brain white matter (Bazarian et al., 2013). Individuals with blast-induced TBI showed a significant association between the FA value and verbal memory scores (Levin et al., 2010). Reduced FA in multiple brain areas lasted even after 5 years after blast-induced injury (Mac Donald et al., 2017).

Diffusion tensor tractography (DTT) in patients with TBI

DTT traces the trajectories of the fiber bundles in the white matter by quantifying the water diffusion (Shenton et al., 2012). The orientation of maximum water diffusion can be acquired at each voxel. By following these orientations, a white matter tract can be reconstructed (Mori & Barker, 1999).

Many studies have been published on the abnormal DTT reconstruction for various white matter tracts in patients with TBI (Table 2). The cingulum is a neural tract that connects the limbic system and brain hemispheres. The anterior part is associated with emotion and behavior, while the posterior part correlates with memory function (Bubb, Metzler-Baddeley, & Aggleton, 2018). DTT studies have demonstrated discontinuation or narrowing of the cingulum in patients with cognitive dysfunction following TBI (Jang, Kim, & Lee, 2019; Kurki et al., 2014).

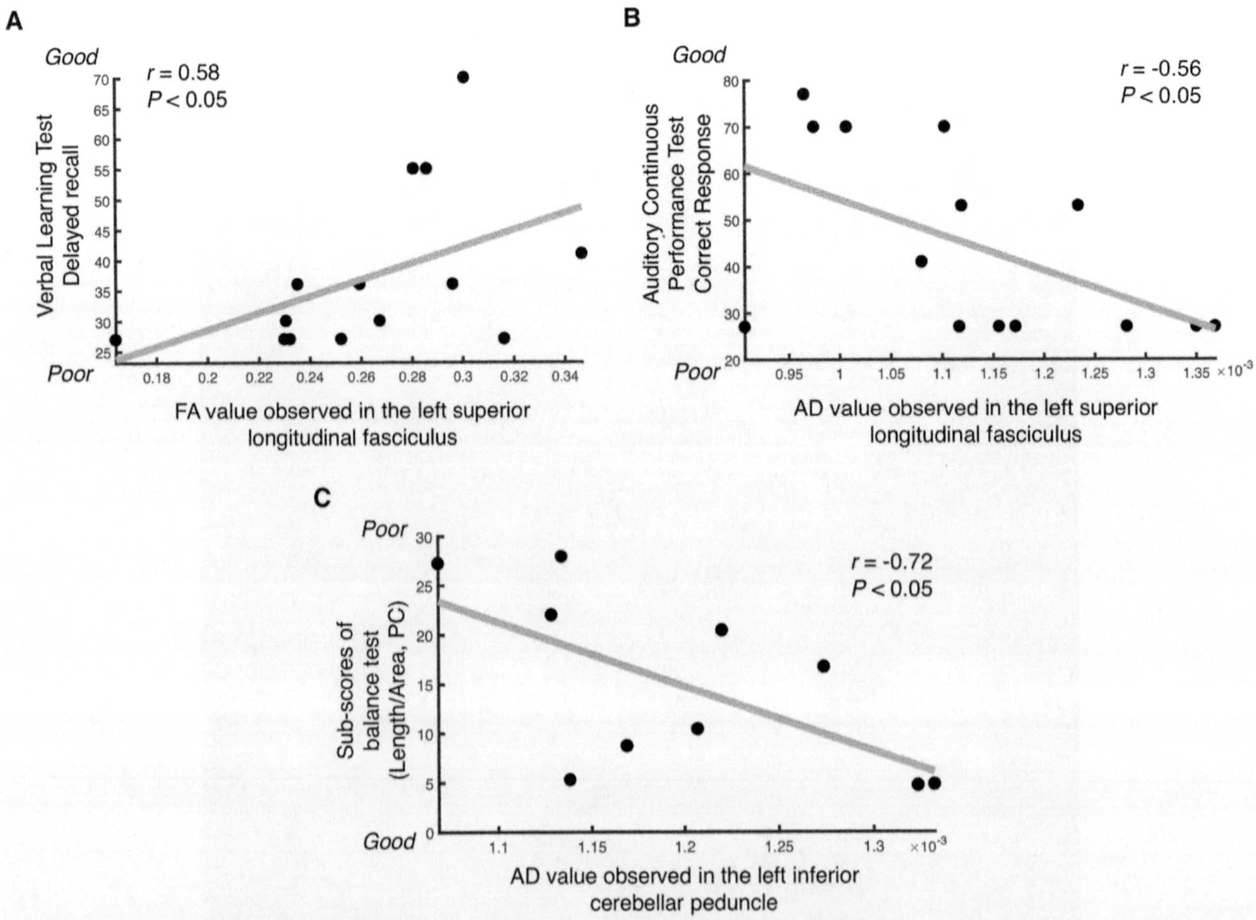

FIG. 4 Correlation between diffusion tensor metrics and clinical assessment results. Significant correlation between FA or AD and the scores of a cognitive test or balance test after mild-to-moderate TBI. FA and AD of the left superior longitudinal fasciculus showed a (A) significantly positive and (B) negative correlation with the sub-score of cognitive test, respectively. (C) AD in the left inferior cerebellar peduncle showed a significantly negative correlation with score of the balance test, indicating decreased AD was associated with poor balance performance after injury (the length-over-area of movement of the center of pressure during eyes closed condition). (The figure was reprinted from Kim, E., Seo, H. G., Lee, H. H., Lee, S. H., Choi, S. H., Cho, W. S., ... Oh, B. M. (2019). Altered white matter integrity after mild to moderate traumatic brain injury. *Journal of Clinical Medicine*, *8*(9), 1318. https://doi.org/10.3390/jcm8091318).

The ARAS is an essential neural component of consciousness (Edlow et al., 2012). It refers to the reticular core and multiple brainstem nuclei mediating arousal, as well as their projections to the hypothalamus, thalamus, basal forebrain, and cortex. Disorders of consciousness following TBI imply injury of the ARAS. DTT studies have shown that the ARAS was narrowed or destroyed in TBI patients with disorders of consciousness (Ordóñez-Rubiano et al., 2017).

The CST is a primary descending motor tract that controls voluntary movements. It runs through the white matter of the precentral gyrus, corona radiata internal capsule, and brainstem. Patients manifesting motor weakness after TBI showed partial discontinuation of CST in studies using DTT (Jang et al., 2009; Seo & Jang, 2015).

The CRP, another key motor pathway, originates from the premotor cortex and descends to the pontomedullary reticular formation. DTT studies have shown that the CRP was narrowed in patients with TBI manifesting proximal weakness and gait disturbance (Kwon & Jang, 2014; Lee & Jang, 2015).

The spinothalamic tract (STT) is a neural tract that carries sensory stimuli from the extremities and body to the somatosensory nuclei of the thalamus. Injury of the STT is believed to be a pathogenetic mechanism to central pain in patients following TBI, which is supported by DTT studies (Jang & Lee, 2019; Kim, Ahn, Cho, Kim, and Jang, 2015).

Limitation of DTI as a clinical tool

Although DTI has been widely used in clinical research, there are certain limitations in practical application that should be noted. Accurate estimation of diffusion parameters, such as FA, AD, and RD, can be affected by various factors. In detail,

TABLE 2 **Diffusion tensor tractography studies in patients following traumatic brain injury.**

Tract	Clinical manifestation	Study population	Outcome measures	Findings of diffusion tensor tractography	References
Cingulum	Cognitive impairment	73 patients with chronic mild TBI	WAIS, MAS	discontinuation of cingulum	Jang, Kim, and Lee (2019)
		80 patients with mild to severe TBI	TMT, WAIS, Stroop test, WMS, BVRT	loss of peripheral fibers and narrowing of the central bundle	Kurki et al. (2014)
		35 patients with chronic TBI	WAIS, MAS	discontinuation of cingulum	Baek et al. (2013)
Ascending reticular activating system (ARAS)	Disorder of consciousness	120 patients with mild to severe TBI	Duration of LOC and PTA, GCS	narrowing of ARAS	Jang, Kim, Kim, and Kim (2019)
		3 patients with severe TBI	GCS, DOCS	destruction of the ARAS	Ordóñez-Rubiano et al. (2017)
Corticospinal tract (CST)	Motor weakness	53 patients with mild TBI	Purdue Pegboard test, grip strength, MRC score	partial tearing at subcortical white matter and narrowing of CST	Jang and Kim (2016)
		5 patients with TBI	Motricity Index	disruption of CST at the subcortical white matter	Jang et al. (2009)
Corticoreticular pathway (CRP)	Proximal muscle weakness	29 patients with mild TBI	MRC score	Narrowing and partial tearing or discontinuation at the subcortical WM or brainstem	Lee and Jang (2015)
Spinothalamic tract (STT)	Neuropathic pain	5 patients with mild TBI	VAS	partial tearing and narrowing of STT	Jang and Lee (2019)
		40 patients with chronic mild TBI	VAS	narrowing of bilateral STTs	Kim, Ahn, Cho, Kim, and Jang (2015)

BCRT, Benton Visual Retention Test; *CLS-DT*, Category Listening Span Dual-Task; *DOCS*, Disorders of Consciousness Scale; *GCS*, Glasgow Coma Scale; *LOC*, loss of consciousness; *MAS*, Memory Assessment Scale; *MRC*, Medical Research Council; *PTA*, post-traumatic amnesia; *TMT*, Trail-Making Test; *VAS*, Visual Analogue Scale; *VSS-DT*, Visuospatial Span Dual-Task; *WAIS*, Wechsler Adult Intelligence Scale; *WMS*, Wechsler Memory Scale.

because diffusion parameters are estimated from molecular displacement of water, subtle head movement can easily contaminate the whole parameter estimation. The sensitivity to motion artifacts provides impetus for technical improvement to reduce the effects by acquiring data with the potential source of motion (Soares, Marques, Alves, & Sousa, 2013). However, this approach is not readily applicable to routine practice because it increases the data acquisition time. In addition, the characteristics of the obtained diffusivity, depending on image resolution and the neighboring voxels, should be considered. The diffusivity is an indirect measurement. If the image resolution is low (i.e., greater voxel size), water diffusivity within a voxel becomes more isotropic. One cubic millimeter of brain tissue contains up to 80,000 neurons and 4.5 million synapses (Insel & Landis, 2013). Considering that the typical DTI voxel size is $2\times2\times2\,mm^3$, subtle microscopic damage affecting a small portion of neuronal groups or axonal bundles may be diluted.

If data are acquired with anisotropic resolution (e.g., $1\times1\times2\,mm^3$), the anisotropy of water diffusivity can be exaggerated inherently. Although an isotropic resolution of small voxel size is recommended (Mukherjee, Chung, et al., 2008), it is often ignored in busy clinical settings. The recent multi-band technique can help to reduce the scanning time while maintaining a good signal-to-noise ratio, but this technique may require costly instrumental upgrades. The partial volume effect in DTI studies is the effect of neighboring voxels to estimate the diffusion tensor matrix, which reduce its accuracy (Alexander, Hasan, Lazar, Tsuruda, & Parker, 2001). It can be more problematic in DTI studies comparing patients with TBI and controls, because the patient group often has more atrophy than the control group, which have different characteristics of neighboring voxels. The image resolution and partial volume effect also affect the tractography results (Chung et al., 2011). In addition, the choice of computational algorithm affects the results of DTT. Generally, in deterministic tractography, the second and third eigenvectors are excluded to track a fiber path, thereby ignoring small fiber bundles that may exist. In probabilistic tractography, the determination of a probabilistic threshold could affect further analysis, associated with identification of spurious tracts versus lost tracts (Chung et al., 2011).

More recently, quality control of inter-site variability between scanners has been considered as an important issue (Palacios et al., 2017). Large white matter tracts showed good reproducibility with <5% coefficient of variation (CV) between scanners, while the CV of the smaller tracts were larger and reaching 10% (Palacios et al., 2017), thereby suggesting the importance of quality control of inter-site variability for large consortium data such as CENTER-TBI (https://www.center-tbi.eu), and TRACK-TBI (https://tracktbi.ucsf.edu).

It should also be mentioned that structural and functional imaging techniques have different strengths and weaknesses. For example, even if axonal continuity of a certain neural circuit is structurally maintained, its neural transmission may not work properly due to regional metabolic pathology. In this case, no specific abnormality may be found in DTI although clinical dysfunction is evident (Bigler, 2019).

Conclusion

Considering the particular vulnerability of white matter in the pathophysiologic cascades after TBI, the need for sensitive and reliable diagnostic tests evaluating the integrity of white matter is more compelling than ever. In this regard, DTI will become even more important in the diagnosis and evaluation of TBI. However, in order for DTI to be useful not only for research purposes but also for clinical usage in the management of TBI, there remains a task of reducing the scanning time to the practical range and improving sensitivity, specificity, and inter-vendor/inter-site reliability.

Applications to other areas of neuroscience

In this chapter, we have reviewed the strengths and weaknesses of DTI in the evaluation of the white matter integrity after TBI. Because TBI is heterogeneous across individuals, the need for personalized biomarkers is much compelling. These imaging biomarkers discovered and validated by various research could also be relevant to other neurological diseases.

The urgent need for more sensitive and specific diagnostic tools is also evident in other diseases such as brain injury of non-traumatic causes, traumatic/non-traumatic spinal cord injury, and stroke. However, there are striking differences in the underlying pathophysiology between stroke and TBI. Stroke causes severe damage and necrosis in localized brain areas of the corresponding vascular supply, whereas TBI causes less severe damage than necrosis throughout the brain. In stroke, diffusion-weighted MRI can be used to visualize acute cytotoxic edema and is commonly used as a gold standard diagnosis. However, there is no sensitive and specific diagnostic tool for TBI comparable to diffusion-weighted MRI for stroke.

Mini-dictionary of terms

Diffuse axonal injury: A type of traumatic diffuse brain injury. The cardinal clinical feature of diffuse axonal injury is a prolonged period of disordered consciousness from the time of injury. This was a post-mortem pathological and clinical term with visualization of microscopic axonal damage involving the whole brain.

Diffusion tensor imaging: A imaging method to capture water diffusivity in the brain. Water diffusivity is interrupted by a biological tissue compartment such as myelination or cell membrane. By assessing disrupted free diffusivity, the integrity of tissue compartment can be measured *in vivo*.

Diffusion metrics: Measured values of tissue integrity based on the diffusion tensor matrix. Fractional anisotropy, axial diffusivity, and radial diffusivity are examples of diffusion metrics.

Diffusion tensor tractography: A imaging technique to reconstruct white matter fiber bundles using a mathematical algorithm.

Corticospinal tract: The primary motor pathway that projects from the cerebral cortex down through the medullary pyramid and into the spinal cord, where they synapse on the spinal motor neurons that control the skeletal muscles.

Ascending reticular activating system: A deep neural system critical for wakefulness and consciousness through controlling the excitability of the thalamus and forebrain.

Mild traumatic brain injury: A special population of traumatic brain injury, which is also known as a concussion. Most patients recover within a short period of time but some of them need special attention with neuropsychological deficits accompanied by functional and/or structural brain changes.

Summary points

- The white matter is especially vulnerable to traumatic brain injury.
- Evaluation of the injury to the white matter has been based on comprehensive clinical and functional evaluations.
- DTI can be used to assess the white matter integrity of the whole brain by calculating diffusion tensor metrics.
- Many DTI studies have demonstrated changes in the white matter of the brain after moderate-to-severe TBI.
- Reports on the changes in diffusion tensor metrics after concussion or blast-induced TBI are less consistent partly because of a dynamic pathophysiology over time, and the heterogeneity of the study samples.

References

Adams, J. H., Doyle, D., Ford, I., Gennarelli, T., Graham, D., & McLellan, D. (1989). Diffuse axonal injury in head injury: Definition, diagnosis and grading. *Histopathology*, *15*(1), 49–59. https://doi.org/10.1111/j.1365-2559.1989.tb03040.x.

Alexander, A. L., Hasan, K. M., Lazar, M., Tsuruda, J. S., & Parker, D. L. (2001). Analysis of partial volume effects in diffusion-tensor MRI. *Magnetic Resonance in Medicine*, *45*(5), 770–780. https://doi.org/10.1002/mrm.1105.

Alexander, A. L., Lee, J. E., Lazar, M., & Field, A. S. (2007). Diffusion tensor imaging of the brain. *Neurotherapeutics*, *4*(3), 316–329. https://doi.org/10.1016/j.nurt.2007.05.011.

Baek, S. O., Kim, O. L., Kim, S. H., Kim, M. S., Son, S. M., Cho, Y. W., … Jang, S. H. (2013). Relation between cingulum injury and cognition in chronic patients with traumatic brain injury; diffusion tensor tractography study. *NeuroRehabilitation*, *33*(3), 465–471. https://doi.org/10.3233/NRE-130979.

Bazarian, J. J., Donnelly, K., Peterson, D. R., Warner, G. C., Zhu, T., & Zhong, J. (2013). The relation between posttraumatic stress disorder and mild traumatic brain injury acquired during operations enduring freedom and Iraqi Freedom. *The Journal of Head Trauma Rehabilitation*, *28*(1), 1–12. https://doi.org/10.1097/HTR.0b013e318256d3d3.

Bigler, E. D. (2019). Pathophysiology. In J. M. Silver, T. W. McAllister, & S. C. Yudofsky (Eds.), *Textbook of traumatic brain injury* (3rd ed., pp. 41–60). Washington, DC: American Psychiatric Publishing.

Bigler, E. D., & Maxwell, W. L. (2012). Neuropathology of mild traumatic brain injury: Relationship to neuroimaging findings. *Brain Imaging and Behavior*, *6*(2), 108–136. https://doi.org/10.1007/s11682-011-9145-0.

Bubb, E. J., Metzler-Baddeley, C., & Aggleton, J. P. (2018). The cingulum bundle: Anatomy, function, and dysfunction. *Neuroscience and Biobehavioral Reviews*, *92*, 104–127. https://doi.org/10.1016/j.neubiorev.2018.05.008.

Chung, H. W., Chou, M. C., & Chen, C. Y. (2011). Principles and limitations of computational algorithms in clinical diffusion tensor MR tractography. *American Journal of Neuroradiology*, *32*(1), 3–13. https://doi.org/10.3174/ajnr.A2041.

Corrigan, J. D., & Hammond, F. M. (2013). Traumatic brain injury as a chronic health condition. *Archives of Physical Medicine and Rehabilitation*, *94*(6), 1199–1201. https://doi.org/10.1016/j.apmr.2013.01.023.

Dinkel, J., Drier, A., Khalilzadeh, O., Perlbarg, V., Czernecki, V., Gupta, R., … NICER (Neuro Imaging for Coma Emergence and Recovery) Consortium. (2014). Long-term white matter changes after severe traumatic brain injury: A 5-year prospective cohort. *American Journal of Neuroradiology*, *35*(1), 23–29. https://doi.org/10.3174/ajnr.A3616.

Dodd, A. B., Epstein, K., Ling, J. M., & Mayer, A. R. (2014). Diffusion tensor imaging findings in semi-acute mild traumatic brain injury. *Journal of Neurotrauma*, *31*(14), 1235–1248. https://doi.org/10.1089/neu.2014.3337.

Edlow, B. L., Takahashi, E., Wu, O., Benner, T., Dai, G., Bu, L., … Folkerth, R. D. (2012). Neuroanatomic connectivity of the human ascending arousal system critical to consciousness and its disorders. *Journal of Neuropathology and Experimental Neurology*, *71*(6), 531–546. https://doi.org/10.1097/NEN.0b013e3182588293.

Eierud, C., Craddock, R. C., Fletcher, S., Aulakh, M., King-Casas, B., Kuehl, D., & LaConte, S. M. (2014). Neuroimaging after mild traumatic brain injury: Review and meta-analysis. *Neuroimage. Clinical*, *4*, 283–294. https://doi.org/10.1016/j.nicl.2013.12.009.

Galanaud, D., Perlbarg, V., Gupta, R., Stevens, R. D., Sanchez, P., Tollard, E., … Recovery, C. (2012). Assessment of white matter injury and outcome in severe brain trauma: A prospective multicenter cohort. *Anesthesiology*, *117*(6), 1300–1310. https://doi.org/10.1097/ALN.0b013e3182755558.

Graham, N. S., & Sharp, D. J. (2019). Understanding neurodegeneration after traumatic brain injury: From mechanisms to clinical trials in dementia. *Journal of Neurology, Neurosurgery, and Psychiatry*, *90*(11), 1221–1233. https://doi.org/10.1136/jnnp-2017-317557.

Haberg, A. K., Olsen, A., Moen, K. G., Schirmer-Mikalsen, K., Visser, E., Finnanger, T. G., … Eikenes, L. (2015). White matter microstructure in chronic moderate-to-severe traumatic brain injury: Impact of acute-phase injury-related variables and associations with outcome measures. *Journal of Neuroscience Research*, *93*(7), 1109–1126. https://doi.org/10.1002/jnr.23534.

Herrera, J. J., Bockhorst, K., Kondraganti, S., Stertz, L., Quevedo, J., & Narayana, P. A. (2017). Acute white matter tract damage after frontal mild traumatic brain injury. *Journal of Neurotrauma*, *34*(2), 291–299. https://doi.org/10.1089/neu.2016.4407.

Insel, T. R., & Landis, S. C. (2013). Twenty-five years of progress: The view from NIMH and NINDS. *Neuron*, *80*(3), 561–567. https://doi.org/10.1016/j.neuron.2013.09.041.

Jang, S. H., & Kim, S. Y. (2016). Injury of the corticospinal tract in patients with mild traumatic brain injury: A diffusion tensor tractography study. *Journal of Neurotrauma*, *33*(19), 1790–1795. https://doi.org/10.1089/neu.2015.4298.

Jang, S. H., Kim, O. L., Kim, S. H., & Kim, J. B. (2019). The relation between loss of consciousness, severity of traumatic brain injury, and injury of ascending reticular activating system in patients with traumatic brain injury. *American Journal of Physical Medicine & Rehabilitation*, *98*(12), 1067–1071. https://doi.org/10.1097/PHM.0000000000001243.

Jang, S. H., Kim, S. H., & Lee, H. D. (2019). Traumatic axonal injury of the cingulum in patients with mild traumatic brain injury: A diffusion tensor tractography study. *Neural Regeneration Research*, *14*(9), 1556. https://doi.org/10.4103/1673-5374.255977.

Jang, S. H., Kim, D.-S., Son, S. M., Cho, Y. W., Kim, S. H., Kim, O. L., & Ahn, S. H. (2009). Clinical application of diffusion tensor tractography for elucidation of the causes of motor weakness in patients with traumatic brain injury. *NeuroRehabilitation*, *24*(3), 273–278. https://doi.org/10.3233/NRE-2009-0478.

Jang, S. H., & Lee, H. D. (2019). Diagnostic approach to traumatic axonal injury of the spinothalamic tract in individual patients with mild traumatic brain injury. *Diagnostics*, *9*(4), 199. https://doi.org/10.3390/diagnostics9040199.

Johnson, V. E., Stewart, W., & Smith, D. H. (2013). Axonal pathology in traumatic brain injury. *Experimental Neurology*, *246*, 35–43. https://doi.org/10.1016/j.expneurol.2012.01.013.

Kim, J. H., Ahn, S. H., Cho, Y. W., Kim, S. H., & Jang, S. H. (2015). The relation between injury of the spinothalamocortical tract and central pain in chronic patients with mild traumatic brain injury. *The Journal of Head Trauma Rehabilitation*, *30*(6), E40–E46.

Kalmar, K., & Giacino, J. T. (2005). The JFK coma recovery scale—Revised. *Neuropsychological Rehabilitation*, *15*(3–4), 454–460. https://doi.org/10.1080/09602010443000425.

Kim, E., Seo, H. G., Lee, H. H., Lee, S. H., Choi, S. H., Cho, W. S., … Oh, B. M. (2019). Altered white matter integrity after mild to moderate traumatic brain injury. *Journal of Clinical Medicine*, *8*(9), 1318. https://doi.org/10.3390/jcm8091318.

Kraus, M. F., Susmaras, T., Caughlin, B. P., Walker, C. J., Sweeney, J. A., & Little, D. M. (2007). White matter integrity and cognition in chronic traumatic brain injury: A diffusion tensor imaging study. *Brain*, *130*(Pt. 10), 2508–2519. https://doi.org/10.1093/brain/awm216.

Kurki, T., Himanen, L., Vuorinen, E., Myllyniemi, A., Saarenketo, A.-R., Kauko, T., … Tenovuo, O. (2014). Diffusion tensor tractography-based analysis of the cingulum: Clinical utility and findings in traumatic brain injury with chronic sequels. *Neuroradiology*, *56*(10), 833–841. https://doi.org/10.1007/s00234-014-1410-7.

Kwon, H. G., & Jang, S. H. (2014). Delayed gait disturbance due to injury of the corticoreticular pathway in a patient with mild traumatic brain injury. *Brain Injury*, *28*(4), 511–514. https://doi.org/10.3109/02699052.2014.887228.

Lee, H. D., & Jang, S. H. (2015). Injury of the corticoreticular pathway in patients with mild traumatic brain injury: A diffusion tensor tractography study. *Brain Injury*, *29*(10), 1219–1222. https://doi.org/10.3109/02699052.2015.1045028.

Levin, H. S. (1990). Memory deficit after closed-head injury. *Journal of Clinical and Experimental Neuropsychology*, *12*(1), 129–153. https://doi.org/10.1080/01688639008400960.

Levin, H. S., Wilde, E., Troyanskaya, M., Petersen, N. J., Scheibel, R., Newsome, M., … Li, X. (2010). Diffusion tensor imaging of mild to moderate blast-related traumatic brain injury and its sequelae. *Journal of Neurotrauma*, *27*(4), 683–694. https://doi.org/10.1089/neu.2009.1073.

Lipton, M. L., Kim, N., Park, Y. K., Hulkower, M. B., Gardin, T. M., Shifteh, K., … Branch, C. A. (2012). Robust detection of traumatic axonal injury in individual mild traumatic brain injury patients: Intersubject variation, change over time and bidirectional changes in anisotropy. *Brain Imaging and Behavior*, *6*(2), 329–342. https://doi.org/10.1007/s11682-012-9175-2.

Mac Donald, C. L., Barber, J., Andre, J., Evans, N., Panks, C., Sun, S., … Temkin, N. (2017). 5-year imaging sequelae of concussive blast injury and relation to early clinical outcome. *Neuroimage. Clinical*, *14*, 371–378. https://doi.org/10.1016/j.nicl.2017.02.005.

McClelland, A. C., Fleysher, R., Mu, W., Kim, N., & Lipton, M. L. (2018). White matter microstructural abnormalities in blast-exposed combat veterans: Accounting for potential pre-injury factors using consanguineous controls. *Neuroradiology*, *60*(10), 1019–1033. https://doi.org/10.1007/s00234-018-2070-9.

Mori, S., & Barker, P. B. (1999). Diffusion magnetic resonance imaging: Its principle and applications. *The Anatomical Record: Advances in Integrative Anatomy and Evolutionary Biology*, *257*(3), 102–109. https://doi.org/10.1002/(SICI)1097-0185(19990615)257:3<102::AID-AR7>3.0.CO;2-6.

Mukherjee, P., Berman, J., Chung, S., Hess, C., & Henry, R. (2008). Diffusion tensor MR imaging and fiber tractography: Theoretic underpinnings. *American Journal of Neuroradiology*, *29*(4), 632–641. https://doi.org/10.3174/ajnr.A1051.

Mukherjee, P., Chung, S. W., Berman, J. I., Hess, C. P., & Henry, R. G. (2008). Diffusion tensor MR imaging and fiber tractography: Technical considerations. *American Journal of Neuroradiology*, *29*(5), 843–852. https://doi.org/10.3174/ajnr.A1052.

Newcombe, V. F., Williams, G. B., Scoffings, D., Cross, J., Carpenter, T. A., Pickard, J. D., & Menon, D. K. (2010). Aetiological differences in neuroanatomy of the vegetative state: Insights from diffusion tensor imaging and functional implications. *Journal of Neurology, Neurosurgery, and Psychiatry*, *81*(5), 552–561. https://doi.org/10.1136/jnnp.2009.196246.

Ordóñez-Rubiano, E. G., Johnson, J., Enciso-Olivera, C. O., Marín-Muñoz, J. H., Cortes-Lozano, W., Baquero-Herrera, P. E., … Cifuentes-Lobelo, H. A. (2017). Reconstruction of the ascending reticular activating system with diffusion tensor tractography in patients with a disorder of consciousness after traumatic brain injury. *Cureus*, *9*(9), e1723. https://doi.org/10.7759/cureus.1723.

Palacios, E. M., Martin, A. J., Boss, M. A., Ezekiel, F., Chang, Y. S., Yuh, E. L., … Crawford, K. L. (2017). Toward precision and reproducibility of diffusion tensor imaging: A multicenter diffusion phantom and traveling volunteer study. *American Journal of Neuroradiology*, *38*(3), 537–545. https://doi.org/10.3174/ajnr.A5025.

Scheid, R., Walther, K., Guthke, T., Preul, C., & von Cramon, D. (2006). Cognitive sequelae of diffuse axonal injury. *Archives of Neurology*, *63*(3), 418–424. https://doi.org/10.1001/archneur.63.3.418.

Scott, G., Ramlackhansingh, A. F., Edison, P., Hellyer, P., Cole, J., Veronese, M., … Sharp, D. J. (2016). Amyloid pathology and axonal injury after brain trauma. *Neurology*, *86*(9), 821–828. https://doi.org/10.1212/wnl.0000000000002413.

Seo, J. P., & Jang, S. H. (2015). Traumatic axonal injury of the corticospinal tract in the subcortical white matter in patients with mild traumatic brain injury. *Brain Injury*, *29*(1), 110–114. https://doi.org/10.3109/02699052.2014.973447.

Shenton, M. E., Hamoda, H., Schneiderman, J., Bouix, S., Pasternak, O., Rathi, Y., … Koerte, I. (2012). A review of magnetic resonance imaging and diffusion tensor imaging findings in mild traumatic brain injury. *Brain Imaging and Behavior*, *6*(2), 137–192. https://doi.org/10.1007/s11682-012-9156-5.

Smith, D. H., Hicks, R., & Povlishock, J. T. (2013). Therapy development for diffuse axonal injury. *Journal of Neurotrauma*, *30*(5), 307–323. https://doi.org/10.1089/neu.2012.2825.

Soares, J., Marques, P., Alves, V., & Sousa, N. (2013). A hitchhiker's guide to diffusion tensor imaging. *Frontiers in Neuroscience*, *7*, 31. https://doi.org/10.3389/fnins.2013.00031.

Song, S. K., Yoshino, J., Le, T. Q., Lin, S. J., Sun, S. W., Cross, A. H., & Armstrong, R. C. (2005). Demyelination increases radial diffusivity in corpus callosum of mouse brain. *NeuroImage*, *26*(1), 132–140. https://doi.org/10.1016/j.neuroimage.2005.01.028.

Stuss, D., Ely, P., Hugenholtz, H., Richard, M., LaRochelle, S., Poirier, C., & Bell, I. (1985). Subtle neuropsychological deficits in patients with good recovery after closed head injury. *Neurosurgery*, *17*(1), 41–47. https://doi.org/10.1227/00006123-198507000-00007.

Sullivan, D. R., Logue, M. W., Wolf, E. J., Hayes, J. P., Salat, D. H., Fortier, C. B., … Miller, M. W. (2019). Close-range blast exposure is associated with altered white matter integrity in apolipoprotein ε4 carriers. *Journal of Neurotrauma*, *36*(23), 3264–3273. https://doi.org/10.1089/neu.2019.6489.

Takaoka, M., Tabuse, H., Kumura, E., Nakajima, S., Tsuzuki, T., Nakamura, K., … Sugimoto, H. (2002). Semiquantitative analysis of corpus callosum injury using magnetic resonance imaging indicates clinical severity in patients with diffuse axonal injury. *Journal of Neurology, Neurosurgery, and Psychiatry*, *73*(3), 289–293. https://doi.org/10.1136/jnnp.73.3.289.

Veeramuthu, V., Narayanan, V., Kuo, T. L., Delano-Wood, L., Chinna, K., Bondi, M. W., … Ramli, N. (2015). Diffusion tensor imaging parameters in mild traumatic brain injury and its correlation with early neuropsychological impairment: A longitudinal study. *Journal of Neurotrauma*, *32*(19), 1497–1509. https://doi.org/10.1089/neu.2014.3750.

Wallace, E. J., Mathias, J. L., & Ward, L. (2018). Diffusion tensor imaging changes following mild, moderate and severe adult traumatic brain injury: A meta-analysis. *Brain Imaging and Behavior*, *12*(6), 1607–1621. https://doi.org/10.1007/s11682-018-9823-2.

Zhao, W., Cai, Y., Li, Z., & Ji, S. (2017). Injury prediction and vulnerability assessment using strain and susceptibility measures of the deep white matter. *Biomechanics and Modeling in Mechanobiology*, *16*(5), 1709–1727. https://doi.org/10.1007/s10237-017-0915-5.

Chapter 12

Cerebral activation of attention and working memory in traumatic brain injury

Athena Stein[a], Kartik K. Iyer[a], and Karen M. Barlow[a,b]

[a]*Acquired Brain Injury in Children Research Program, Child Health Research Centre, Faculty of Medicine, The University of Queensland, South Brisbane, QLD, Australia,* [b]*Dr. Paul Hopkins Chair of Paediatric Rehabilitation in Acquired Brain Injury, Queensland Children's Hospital, South Brisbane, QLD, Australia*

List of abbreviations

A′	A-prime
ACC	anterior cingulate cortex
ADHD	attention-deficit/hyperactivity disorder
aI	anterior insula
BOLD	blood-oxygen-level-dependent
CSTC	cortico-striatal-thalamo-cortical
DAI	diffuse axonal injury
dlPFC	dorsolateral prefrontal cortex
DMN	default mode network
fMRI	functional magnetic resonance imaging
ldlPFC	left dorsolateral prefrontal cortex
mTBI	mild traumatic brain injury
NIBS	non-invasive brain stimulation
Pre-SMA	pre-supplementary motor cortex
rIFG	right inferior frontal gyrus
rs-fMRI	resting state functional magnetic resonance imaging
rTMS	repetitive transcranial magnetic stimulation
SN	salience network
SST	stop signal task
SSRT	stop signal reaction time
STN	sub-thalamic nucleus
TBI	traumatic brain injury
tDCS	transcranial direct current stimulation
TMS	transcranial magnetic stimulation
TPJ	temporo-parietal junction
WM	working memory

Introduction: Executive dysfunction in traumatic brain injury

Traumatic brain injury (TBI) is a leading cause of neurological morbidity and mortality worldwide (Johnson, DeMatt, & Salorio, 2009). Its incidence is greatest in childhood and early adult life (> 640/100,000 children and young people less than 24 years of age each year) (FitzGerald, Tippett, Schuetz, & Pollard, 2008). Following a TBI, significant problems with memory, learning, and attention may arise, leading to life-long impacts on academic achievement, ability to secure employment, socio-emotional outcomes, and quality of life in over 30% of patients (Sigurdardottir, Andelic, Roe, & Schanke, 2009).

Diagnosis and Treatment of Traumatic Brain Injury. https://doi.org/10.1016/B978-0-12-823347-4.00025-7

Traumatic brain injury results in large-scale disruption and impairment of brain networks, leading to significant functional impairments. These brain networks offer an insight into the interconnected, organized areas of the brain, which integrate locally and distally via axons and white matter tracts. This complex system of networks has been termed the "connectome" and underpins the link between brain activity and behavior (van den Heuvel & Sporns, 2011). The hallmark of TBI is axonal injury (focal or diffuse) and white matter tract injury. This injury results in brain network disruption with long-range disconnections and impaired brain network connectivity (Fagerholm, Hellyer, Scott, Leech, & Sharp, 2015; Sharp, Scott, & Leech, 2014). Axonal injury leads to poor outcomes, even where cortical damage is minimal (Sharp et al., 2014). Whereas injury to focal brain regions correlates poorly with outcome, connectome disruption can more accurately predict function across a range of injury severities and pathologies (Fagerholm et al., 2015). The most well studied brain networks are the seven intrinsic connectivity networks including the default mode network (DMN) and salience network (SN) (Bonnelle et al., 2011). To understand the effect of this network impairment it is important to understand how the functions of attention and working memory (WM) are governed.

The neural basis of attention and working memory

Attention and working memory are supported by rich, interconnected neural networks which rapidly mature between early childhood and early adulthood. The neural attention system involves three major components: alerting, orienting, and executive attention. Alerting involves involuntary capture of attention by salient (i.e., "attention-capturing") stimuli and is functionally associated with the fronto-parieto-thalamic network and thalamus. Orienting refers to both the voluntary or involuntary movement of attention to a stimulus (Vossel, Geng, & Fink, 2014). Voluntary, or endogenous orientation of attention, is facilitated by the dorsal attention network (Fig. 1), which involves the frontal eye fields and parietal areas; whereas involuntary, or exogenous movement of attention, engages the ventral attention network, involving the temporo-parietal junction (TPJ) and the inferior frontal gyrus (Vossel et al., 2014). Executive control of attention will be discussed in the section below.

Much of our knowledge about the neurobiological underpinnings of attention is due to advances in modern neuroimaging techniques. Functional neuroimaging provides insight into brain activity and network interaction during attention and memory tasks, allowing acquisition of robust evidence in both clinical and healthy populations. Task-based functional MRI (fMRI) provides insight into changes in cortical activations during cognitive tasks, such as those related to attention and WM. Conversely, resting state fMRI (rs-fMRI) provides a measure of cortical activation when the brain is at rest (e.g., eyes closed). Generally, fMRI is a non-invasive assessment of changes in neural activity which uses the blood-oxygen-level-dependent (BOLD) signal to infer changes in neural activation with excellent spatial resolution. It is also a useful tool for probing dynamic patterns of connectivity (Iyer et al., 2019).

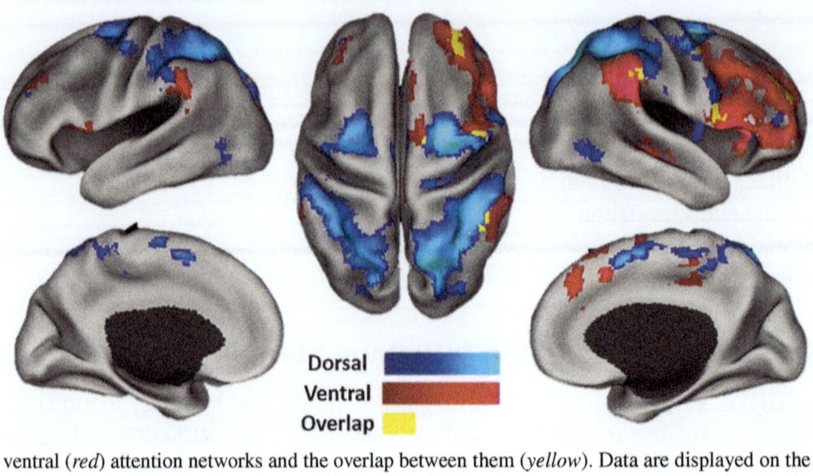

FIG. 1 Dorsal (*blue*) and ventral (*red*) attention networks and the overlap between them (*yellow*). Data are displayed on the lateral and medial surfaces of the left hemisphere (*left*), the dorsal surface (*center*), and the lateral and medial surfaces of the right hemisphere (*right*). Adapted from Fox, M. D., Corbetta, M., Snyder, A. Z., Vincent, J. L., & Raichle, M. E. (2006). Spontaneous neuronal activity distinguishes human dorsal and ventral attention systems. *Proceedings of the National Academy of Sciences of the United States of America, 103 (26)*, 10046–10051. Copyright (2006) National Academy of Sciences, U.S.A.

The dual-network model of attention

As the third major component of attention, executive control of attention refers to the top-down mode of attentional control. The executive network features both cingulo-opercular and fronto-parietal areas (Konrad et al., 2005), which act together in a dual-network architecture. Cingulo-opercular components such as the anterior cingulate cortex and anterior insula (aI) maintain focus during tasks; whereas fronto-parietal components, including the dorsolateral prefrontal cortex (dlPFC), initiate and adjust attentional control (Fig. 2) (Dosenbach, Fair, Cohen, Schlaggar, & Petersen, 2008).

The cingulo-opercular network, also referred to as the salience network, is engaged during behavioral tasks requiring sustained attentional focus and response to salient stimuli. The fronto-parietal network, in contrast, is more strongly engaged during tasks which require a more top-down control of inhibition. The inhibitory component of attention has been found to be a central deficit in TBI (Dosenbach et al., 2008).

It is important to acknowledge another important network which mediates attentional response, known as the default mode network. This network is a task-negative network (or "resting state" network), which is normally activated during internally focused attention (i.e., introspection, mind-wandering) and deactivated during the external focus of attention (e.g., during inhibition and working memory tasks described in this chapter) (Raichle, 2015). The DMN consists primarily of the posterior cingulate cortex, retrosplenial cortex and ventromedial prefrontal cortex (Sharp et al., 2011). Previous evidence has indicated that deactivation of this network is important for mental effort in cognitive tasks (Li, Yan, Bergquist and Sinha, 2007). The DMN will be discussed in more detail in later subsections of this chapter.

Inhibitory control

Inhibitory control relates to the ability to engage in an appropriate response while suppressing a more dominant but less appropriate response. For example, this might involve a person focusing on a task while resisting the urge to check their mobile phone. As a hub in both the cingulo-opercular and fronto-parietal networks, the dlPFC is thought to play an overarching role in the control of inhibition (Chabran et al., 2020; Peters, Dunlop, & Downar, 2016). Inhibition can be broken down into the two domains of *reactive* and *proactive* inhibition:

- *Reactive* inhibition refers to the cue-triggered stopping of a response which has already been initiated. It generally only occurs in close temporal proximity of a presented cue (i.e., it is usually a direct effect of the cue) (Aron, 2011). A common measure of reactive inhibition is the Stop Signal Reaction Time (SSRT) of the Stop Signal Task (SST), where

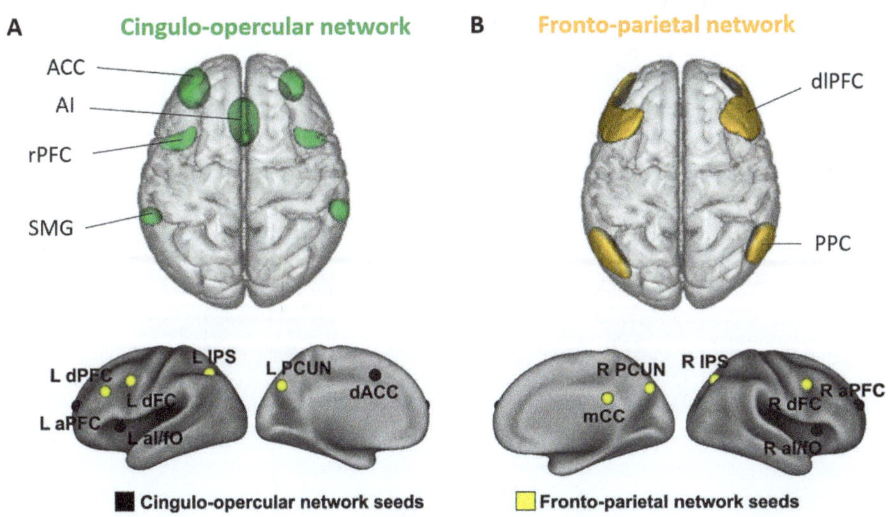

FIG. 2 Diagram representing (A) cingulo-opercular and (B) fronto-parietal networks in healthy participants. These two networks are implicated in the control of attention. L, left; R, right; aPFC, anterior prefrontal cortex; aI, anterior insula; dlPFC, dorsolateral prefrontal cortex; PPC, posterior parietal cortex; IPL, intraparietal lobule; PCUN, precuneus, IPS, intraparietal sulcus, AI, anterior insula; SFG, superior frontal gyrus; MFG, middle frontal gyrus; IFG, inferior frontal gyrus; SMG, supramarginal gyrus; mCC, middle cingulate cortex; fO, frontal operculum. Adapted from Chabran, E., Noblet, V., de Sousa, P. L., Demuynck, C., Philippi, N., Mutter, C., … Blanc, F. (2020). Changes in gray matter volume and functional connectivity in dementia with Lewy bodies compared to Alzheimer's disease and normal aging: Implications for fluctuations. *Alzheimer's Research Therapy, 12 (1)*, 1–13; Han, K., Chapman, S. B., & Krawczyk, D. C. (2018). Neuroplasticity of cognitive control networks following cognitive training for chronic traumatic brain injury. *NeuroImage: Clinical, 18*, 262–278.

a longer reaction time may indicate deficits in reactive inhibition. Reactive inhibition is mediated by *direct* communication of the pre-supplementary motor area (pre-SMA) and right inferior frontal gyrus (rIFG) with the sub-thalamic nucleus (STN) of the basal ganglia (Fig. 3), allowing a quick response.

- *Proactive* inhibition, on the other hand, is a preparatory process which controls the initial instigation of the response. A variety of sensory, cognitive and limbic sources influence proactive inhibition even before presentation of a cue (Aron, 2011). Akin to reactive inhibition processes, proactive inhibition involves communication between the pre-SMA and rIFG. In proactive inhibition, however, these regions then send signals via an *indirect* pathway to the striatum or STN to the output nuclei of the basal ganglia (Fig. 3). Proactive inhibition can be measured using reaction time of tasks involving a "go" trial, such as the "go" component of the Go/No-Go or Stop-Signal tasks (Tan et al., 2019). Delayed response tasks can also be used to measure this type of inhibition, as they create conflicts in decision-making.

Reactive and proactive inhibitory functions improve with age. Young adults typically exhibit faster and more accurate responses on reactive inhibition tasks while displaying greater deactivation of the motor cortex. Proactive inhibition shows a similar developmental pattern, where compared to young children, young adults exhibit better response slowing in anticipation of stopping, alongside increased fronto-striatal connectivity (Vink et al., 2014).

Measuring inhibition

The Stop Signal Task is primarily a reactive inhibition task. It activates the cingulo-opercular network, where the rIFG and pre-SMA areas work in synchrony to inhibit a response, maintaining strong connections with the basal ganglia (Jahfari et al., 2011). The aI is an important node in the cingulo-opercular network, as it is responsible for (1) the maintenance of task-specific attention and (2) responding to salient stimuli (Dosenbach et al., 2007). Increased activation of the left

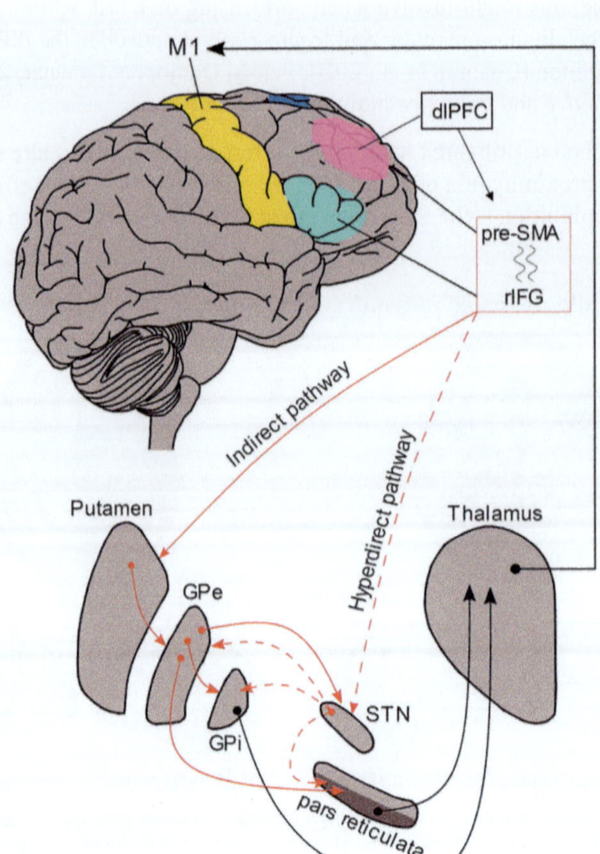

FIG. 3 Diagram representing indirect and direct (hyperdirect) pathway between pre-SMA/rIFG and basal ganglia. These pathways are implicated in reactive and proactive inhibition. M1, primary motor cortex; pre-SMA, pre-supplementary motor cortex; dlPFC, dorsolateral prefrontal cortex; GPe, external globus pallidus; GPi, internal globus pallidus. Adapted from Tan, J., Iyer, K. K., Tang, A. D., Jamil, A., Martins, R. N., Sohrabi, H. R., … Fujiyama, H. (2019). Modulating functional connectivity with non-invasive brain stimulation for the investigation and alleviation of age-associated declines in response inhibition: A narrative review. *NeuroImage, 185*, 490–512. Copyright 2019 Elsevier.

aI has been associated with faster SSRTs (reflecting more efficient stopping) (Swick, Ashley, & Turken, 2011) (Fig. 4). Engagement in the SST is associated with deactivation of the DMN (Li, Yan, Bergquist and Sinha, 2007).

The bilateral dlPFC monitors inputs from interconnected areas such as the rIFG and pre-SMA during the selection of an appropriate response in the SST. The SST is predominantly associated with the activation of the left dlPFC (ldlPFC) compared to the right dlPFC. Here, the ldlPFC is thought to be more involved in analysis, planning, and attention disengagement; whereas the right dlPFC may be more responsible for integration of plans into a strategic sequence (Barbey, Koenigs, & Grafman, 2013).

As previously mentioned, the Go/No-Go task is primarily used to measure proactive inhibition. It is thought to involve neural mechanisms distinct from those activated during the SST (Swick et al., 2011). During the Go/No-Go task, participants must react to a "go" stimulus, and refrain from reacting to a "no-go" stimulus. Inhibition is often measured by the number of errors (or false alarms) a subject makes on "no-go" trials. The Go/No-Go task has been found to activate the pre-SMA, rIFG, bilateral inferior parietal regions, occipital regions, putamen and left premotor cortex (Simmonds, Pekar, & Mostofsky, 2008). These regions are part of the fronto-parietal network, which is involved in adaptive control (Swick et al., 2011). Similarly to the SST, performance of the Go/No-Go task is associated with deactivation of the DMN (Li et al., 2007).

In general, the distinction between the SST and Go/No-Go Task can be thought of as follows: whereas the SST requires participants to "stop," the Go/No-Go task requires them to "not go." In this sense, the SST is thought to more strongly engage the predominantly frontal cingulo-opercular network, and in comparison the Go/No-Go Task engages a wider fronto-parietal network (Swick et al., 2011).

Working memory

Working memory also plays a critical role in executive function and heavily engages prefrontal areas, such as the dlPFC, as well as other task-positive regions such as the premotor, dorsal cingulate, and lateral posterior parietal cortices (Owen, McMillan, Laird, & Bullmore, 2005). WM is a temporary, capacity-limited memory-holding store which is critical for executive function. It is related to attention in that it requires sustained attentional focus on task-relevant mental representations during inhibition of competing distractors. WM plays a crucial role in learning, therefore allowing an individual to manipulate and process information to be stored in long-term memory (Baddeley, 2012). It is commonly investigated using

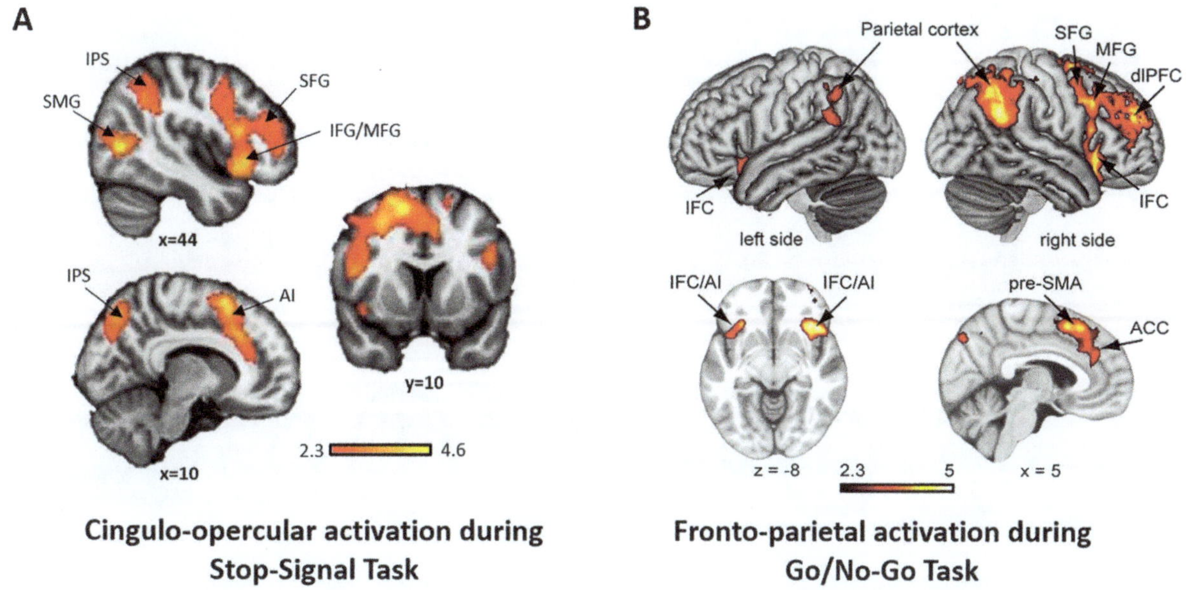

FIG. 4 Diagram representing activation of (A) cingulo-opercular network during Stop Signal Task and (B) fronto-parietal network during Go/No-Go Task in healthy participants. The cingulo-opercular and fronto-parietal networks are implicated differently in tasks requiring proactive and reactive inhibition. L, left; R, right; aPFC, anterior prefrontal cortex; aI, anterior insula; dlPFC, dorsolateral prefrontal cortex; IPL, intraparietal lobule; PCUN, precuneus, IPS, intraparietal sulcus, AI, anterior insula; SFG, superior frontal gyrus; MFG, middle frontal gyrus; IFG, inferior frontal gyrus; SMG, supramarginal gyrus. Adapted from Sharp, D., Bonnelle, V., De Boissezon, X., Beckmann, C., James, S., Patel, M., & Mehta, M. A. (2010). Distinct frontal systems for response inhibition, attentional capture, and error processing. *Proceedings of the National Academy of Sciences of the United States of America, 107 (13)*, 6106–6111; Van Gaal, S., Ridderinkhof, K. R., Scholte, H. S., & Lamme, V. A. (2010). Unconscious activation of the prefrontal no-go network. *Journal of Neuroscience, 30 (11)*, 4143–4150. Copyright 2010 Society for Neuroscience.

the *n*back task, which is associated with activation of the dlPFC (a hub for prefrontal WM-related activation) and deactivation of the DMN (Owen et al., 2005).

Developmental changes in brain networks

Traumatic brain injury is a common diagnosis in children and adolescents, where 691 children per 100,000 are treated for the condition merely in emergency departments each year (Thurman, 2016). Therefore, understanding the developmental changes which occur in the executive network throughout childhood and adolescence is important in the treatment of TBI-related deficits. Beginning in infancy and extending into childhood, there is a rapid anatomical brain volume increase, cortical thickening and surface expansion, formation of new synapses, and myelination (Hagmann et al., 2010). This maturation of attention and WM-related functions is associated with several factors which are summarized below. It is important to consider TBI-related deficits discussed in this Chapter in the context of these developmental changes.

Maturation of attention networks

Maturation of the central executive network from childhood into adulthood is characterized by increasingly focal and anterior activation, including distinct developmental trajectories in attention networks (Fig. 5) (Sherman et al., 2014). From around 4 years of age, the executive attention network gradually differentiates from orienting and alerting systems

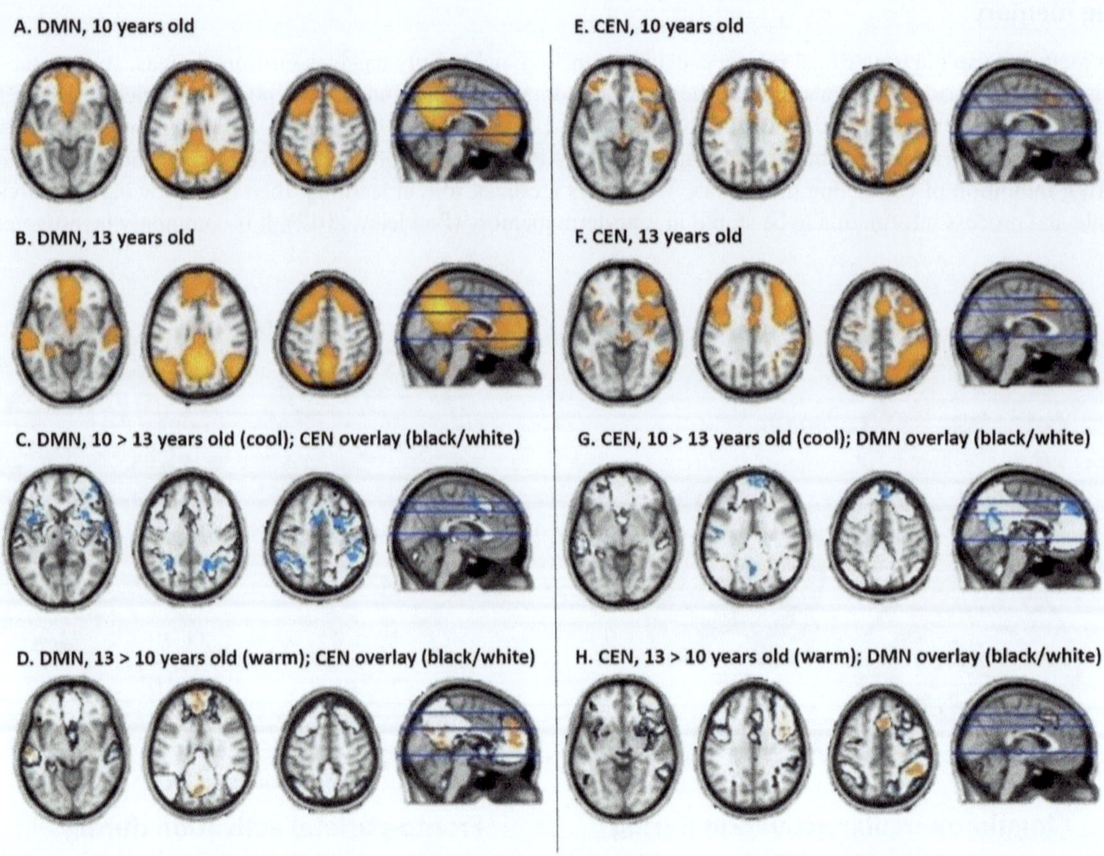

FIG. 5 Default mode network (DMN) and central executive network (CEN) connectivity maps at ages 10 and 13 in healthy participants while listening to a stream of nonsense speech. Overall, children display more focal and anterior activation at 10 years compared to at 13 years of age. Panels represent (A and B) connectivity maps for DMN at 10 and 13 years of age, respectively; (C) DMN-seed connectivity which was higher at 10 than 13 years of age (cool) and (D) DMN-seed connectivity which was higher at 13 than 10 years of age (cool), both overlaid on the 13 year CEN connectivity map (*black and white*); (E and F) connectivity maps for CEN at 10 and 13 years of age, respectively; (G) CEN-seed connectivity which was higher at 10 than 13 years of age (warm) and (H) CEN-seed connectivity which was higher at 13 than 10 years of age (warm), both overlaid on the 13 year DMN connectivity map (*black and white*). Imaged adapted from Sherman, L. E., Rudie, J. D., Pfeifer, J. H., Masten, C. L., McNealy, K., & Dapretto, M. (2014). Development of the default mode and central executive networks across early adolescence: A longitudinal study. *Developmental Cognitive Neuroscience, 10*, 148–159. Copyright 2014 Elsevier.

to become the dominant factor of cognitive control and self-regulation. As a critical component of executive attention, the anterior cingulate cortex (ACC) is thought to be especially implicated in these changes, becoming increasingly differentiated from the orienting network during maturation of sustained attention throughout childhood and into late adolescence (Fair et al., 2009).

Maturation of inhibition

Children are more susceptible to distraction than adults, and yet are also less able to disengage attention (Konrad et al., 2005). The maturation of inhibition in children shows a similar pattern to that of sustained attention; maturing throughout childhood and into late adolescence. This is reflected by changes in intrinsic brain networks. During inhibition tasks, younger children extensively recruit the middle and superior frontal gyri of the prefrontal cortex compared to older children and adolescents. In particular, recruitment of the left inferior frontal gyrus during the Go/No-Go task is decreased in adolescents when compared to adults (Vara, Pang, Vidal, Anagnostou, & Taylor, 2014). Over the course of development, inhibition-related activation becomes increasingly frontally-guided (Levin, Hanten, Zhang, Swank, & Hunter, 2004).

Maturation of working memory

Working memory function exhibits clear maturational changes across puberty, adolescence, and early adulthood. This maturation process is associated with increases in cortical activation in the middle frontal gyrus, and in particular, the dlPFC (Gogtay et al., 2004). In adults, WM-related activation involves several distinct frontal and temporal regions. The dlPFC is widely regarded as a core hub for WM-related processing. In addition to the dlPFC, visual WM tasks involving objects and faces activate areas of the inferior temporal/fusiform gyrus and middle occipital cortex; whereas WM tasks involving spatial and movement judgment activate the inferior parietal/occipital region, superior frontal sulcus and middle temporal region (Ciesielski, Lesnik, Savoy, Grant, & Ahlfors, 2006). Children recruit similar regions to adults during WM tasks, but display more variable and widespread brain activation. In general, the maturation of WM networks throughout childhood and into adulthood involves a shift from diffuse and dorsal activation, to focal and anterior activation (Scherf, Sweeney, & Luna, 2006).

Disruptions following TBI

Brain network disruption following TBI

The underlying neuroanatomical and physiological changes, which occur following TBI, are also important in understanding deficits in executive functioning and brain network activity. TBI results in axonal injury and damage to white matter tracts, thus affecting network connectivity (Fig. 6) (Maria et al., 2019). This axonal injury and white matter tract disruption causes damage to brain networks involved in attention and WM. These changes in network connectivity are particularly important in childhood TBI, where they occur in the context of marked developmental and plasticity changes. As a result, damage acquired at a young age may become ingrained over time during maturation and development, which may result in aberrant structural and functional connectivity throughout childhood and adolescence (Maria et al., 2019).

TBI also causes disruptions in the small-world architecture of the brain. "Small-world" connectivity relates to normally highly connected local modules involved in specialized processing. Following TBI, small-world architecture of the brain is impacted, resulting in overall decreases in functional connectivity and reduced network efficiency. The resultant overall weaker global integration of brain networks in TBI is thought to decrease capacity to integrate information across various brain regions and networks (Caeyenberghs et al., 2014).

Deficits in attention and inhibition following TBI

Attention problems are the most common deficit reported following TBI, with a study finding that 14% of patients experience new-onset attention problems at 1, 4, and 10 months post-TBI (Hooper et al., 2004). Attention is a vital component of learning and memory which allows for the filtering and selection of salient stimuli during inhibition of less important stimuli. Deficits in attentional functions including focused, sustained, divided, alternating and selective attention, as well as response inhibition are common sequelae after TBI (Konrad, Gauggel, Manz, & Schöll, 2000). These deficits are similar to those seen in ADHD, creating an interesting link between the two conditions.

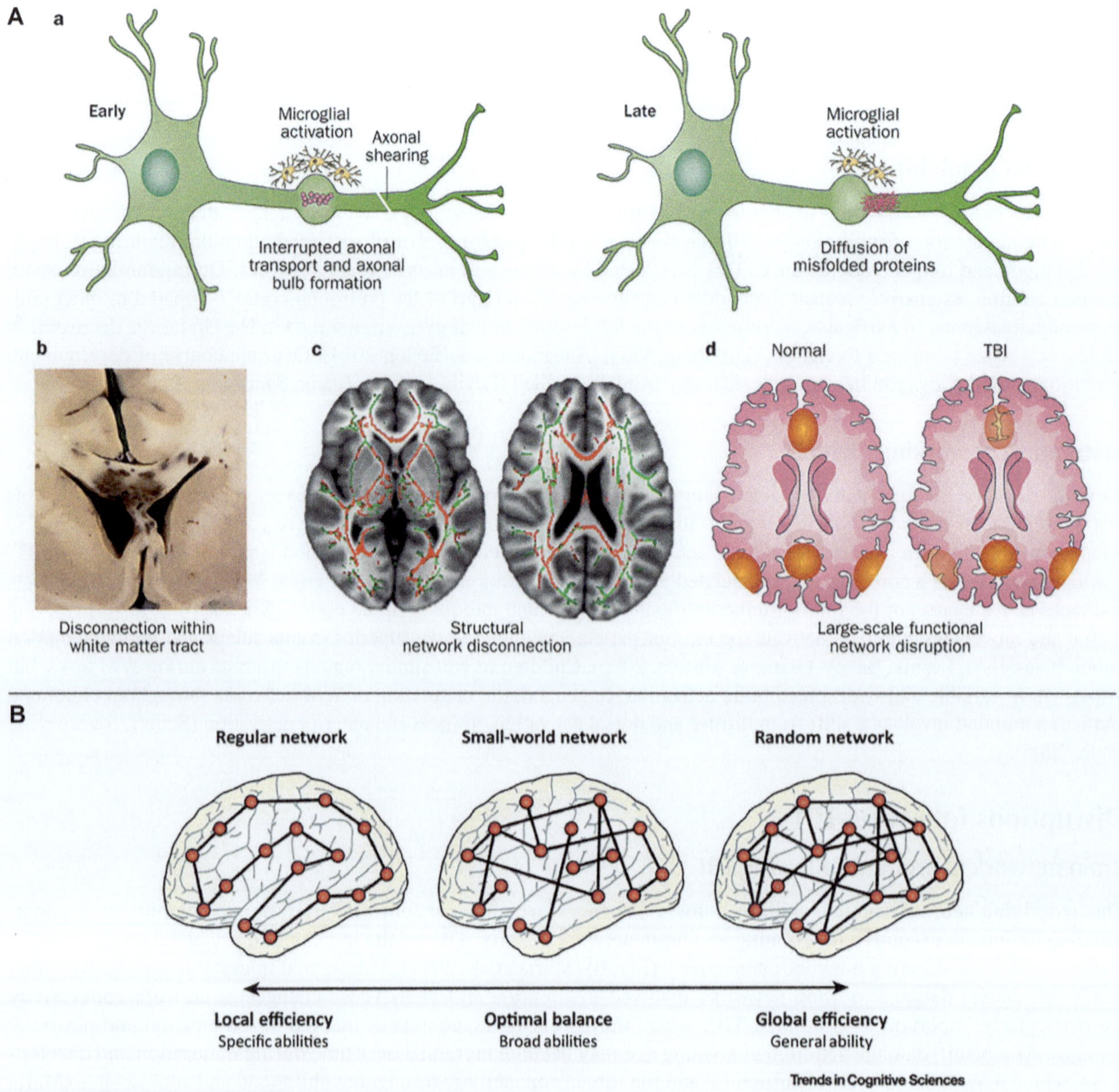

FIG. 6 (A) Traumatic brain injury-related damage across different spatial scales. (a) Diffuse axonal injury can interrupt neuronal processes and trigger neuroinflammation; (b) injuries may be apparent in large-scale white matter tracts, including the corpus callosum (pictured); (c) long distance communication is interrupted following TBI-related damage, often resulting in reduced fractional anisotropy (*red*, white matter skeleton; *green*, intact structure); (d) white matter damage can affect nodes within brain networks (default mode network pictured). (B) Small-world network balance between local and global efficiency. Image reused from Sharp, D. J., Scott, G., & Leech, R. (2014). Network dysfunction after traumatic brain injury. *Nature Reviews Neurology, 10 (3)*, 156. Retrieved from https://www.nature.com/articles/nrneurol.2014.15.pdf, Copyright 2014 Springer Nature and Barbey, A. K. (2018). Network neuroscience theory of human intelligence. Trends in Cognitive Sciences, 22(1), 8–20. Copyright 2018 Elsevier. TBI, traumatic brain injury.

Interestingly, children and adults sustaining a TBI have a 4–5 times greater risk of developing secondary ADHD compared to those without TBI. In fact, secondary ADHD following TBI has been found to occur in 20% of previously unaffected individuals (Schachar, Levin, Max, Purvis, & Chen, 2004). Individuals with ADHD are also more likely to engage in risk-taking behavior where traits of restlessness, inattentiveness and impulsiveness contribute to an increased propensity for accidents, which in turn increases the risk of TBI (Schachar et al., 2004). These functional attention deficits are reflected in structural and functional differences in frontal areas of the brain, which will be discussed later in this chapter.

Inhibition and the default mode network

Patients with TBI often exhibit abnormal default mode network activation during working memory or attention processes. Typically, the DMN exhibits reduced levels of activity during attentionally demanding tasks (Raichle, 2015) in healthy participants. In this sense, the DMN is described to be "anti-correlated" with attention and WM-related regions. Following TBI, abnormalities in DMN activation are common, primarily involving decreased anti-correlation between the DMN and executive areas, where the DMN is hyperactive (Fig. 7). This abnormal activation is termed "DMN interference", due to its interfering effect on normal network operations. Further, damage within the cingulum bundle (an important salience network tract connecting the posterior cingulate and ventromedial prefrontal cortices) causes impairments in sustained attention and inhibition (Sharp et al., 2014). Of note, damage to a SN tract connecting the right aI to the midline pre-supplementary motor area-dorsal ACC strongly and specifically predicts a failure to appropriately deactivate the DMN (Sharp et al., 2014). Abnormal DMN function can be used as a sensitive marker of impairments of attention following TBI (Bonnelle et al., 2011, 2012).

The integrity of the salience network has been found to be an important indicator of inhibitory function in TBI patients and is measured via cognitive tasks such as the SST (Bonnelle et al., 2012). Typically, functional connectivity between the SN and DMN normally increases when motor responses are rapidly inhibited or changed, and the DMN normally shows reduced activity during stopping. However, following TBI, this transient functional connectivity increase is reduced, and the DMN fails to be deactivated (Fig. 8) (Bonnelle et al., 2012; Jilka et al., 2014).

After TBI, damage within the salience network is thought to impair coupling between the right aI and DMN (Jilka et al., 2014). The right aI seems to play an important role in the connectivity between the SN and DMN, where studies have found that the right aI exerts a causal influence on DMN activity across a range of behaviors (Sridharan, Levitin, & Menon, 2008). Disruptions in SN and DMN coupling are thought to affect cognitive control and attentional focus (Jilka et al., 2014). In particular, TBI sustained in childhood can disrupt the maturation of inhibitory networks, leading to long-term deficits (Levin et al., 2004).

Deficits in working memory following TBI

Working memory is important for a broad range of cognitive functions, acting as the interface between perception, long term memory and action (Fig. 9) (Baddeley, 2012). It is closely related to attentional function and is often impacted following TBI. Deficits in WM are seen across the spectrum of TBI severity, where patients find it more difficult to process greater cognitive loads compared to their uninjured counterparts (McAllister et al., 2001). This is thought to be due to disruptions

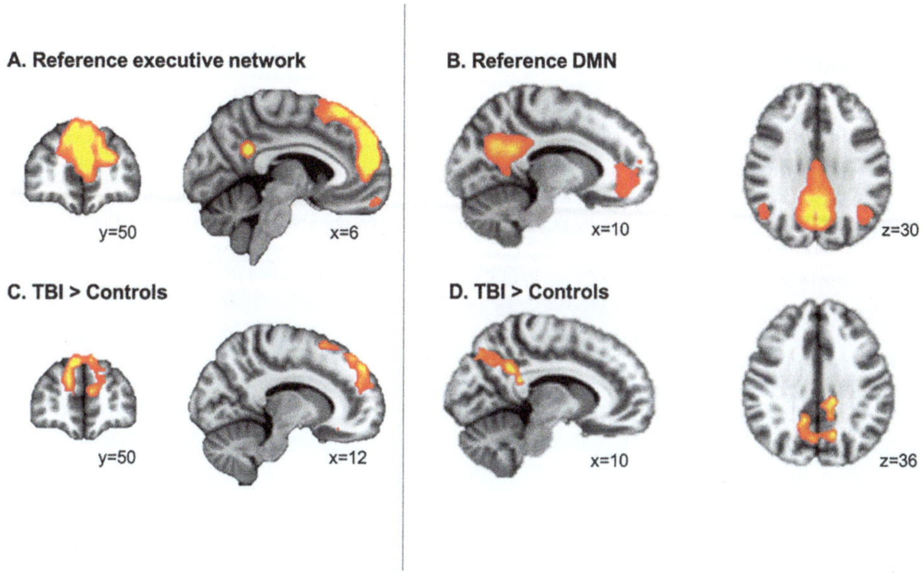

FIG. 7 Hyper-connectivity in TBI patients compared to controls. (A) Reference executive network connectivity; (B) Reference default mode network connectivity; (C) greater activation in executive network of TBI participants compared to controls; (D) greater activation in DMN of TBI patients compared to controls. TBI, traumatic brain injury; DMN, default mode network. Image reused from Sharp, D.J., Beckmann, C.F., Greenwood, R., Kinnunen, K.M., Bonnelle, V., De Boissezon, X., et al. (2011). Default mode network functional and structural connectivity after traumatic brain injury. *Brain, 134* (8), 2233–2247. Copyright 2011 Oxford University Press.

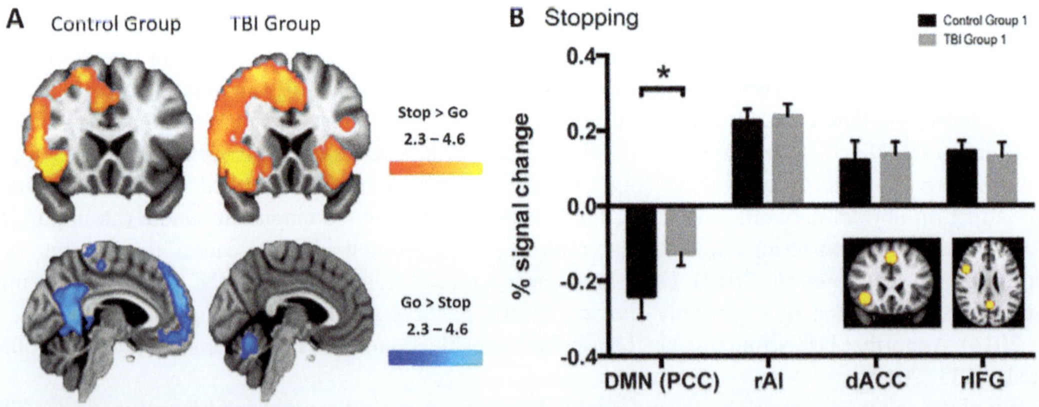

FIG. 8 Stop Signal task activation associated with correct stop vs go trials in control and TBI participants. Patients who have sustained a TBI experience abnormal default mode network connectivity during the stopping phase of the Stop Signal Task. Image adapted from Jilka, S. R., Scott, G., Ham, T., Pickering, A., Bonnelle, V., Braga, R. M., … Sharp, D. J. (2014). Damage to the salience network and interactions with the default mode network. *Journal of Neuroscience, 34 (33)*, 10798–10807. Copyright 2010 Society for Neuroscience.

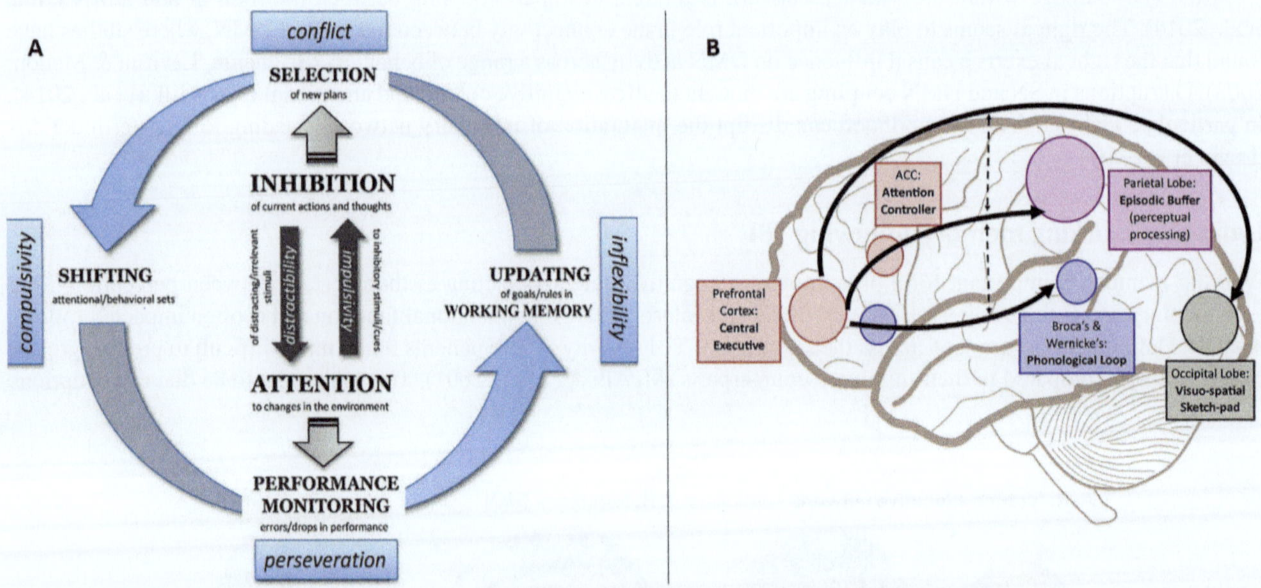

FIG. 9 Inter-relationships between executive functions including attention, inhibition, working memory, and interference control. (A) Schematic representation of the relationship between attention and working memory-related functions and (B) interplay between attention and WM superimposed on a brain template. Images adapted from Bari, A., & Robbins, T. W. (2013). Inhibition and impulsivity: Behavioral and neural basis of response control. *Progress in Neurobiology, 108*, 44–79. Copyright 2013 Elsevier, and Chai, W. J., Abd Hamid, A. I., & Abdullah, J. M. (2018). Working memory from the psychological and neurosciences perspectives: A review. *Frontiers in Psychology, 9*, 401. Retrieved from https://www.ncbi.nlm.nih.gov/pmc/articles/PMC5881171/pdf/fpsyg-09-00401.pdf, Creative Commons CC-BY.

to the central executive network which impacts the brain's ability to allocate processing resources. These deficits are related to comorbid attention problems seen in TBI, where sustained attention and appropriate response inhibition must occur in order for working memory systems to function (McAllister, Flashman, Sparling, & Saykin, 2004).

As a hub for executive functioning, the prefrontal cortex (and in particular, the dlPFC) commonly shows abnormal WM-related activation following TBI. In mild TBI, evidence has shown that the dlPFC is hyperactive during working memory tasks (Hillary, Genova, Chiaravalloti, Rypma, & DeLuca, 2006). This hyper-activation is suggested to occur as a compensatory effort for a loss of processing efficiency due to DAI in various white matter tracts. This pattern has also been noted in moderate to severe adult TBI, where a study found that patients displayed greater bilateral activation while completing higher *n*back task cognitive loads compared to controls. In particular, moderate to severe TBI was related to activation of the aI (Kim et al., 2012). This evidence suggests that hyper-activation of WM-related areas in TBI may reflect network-wide damage.

Following TBI, the DMN often fails to deactivate during working memory tasks, similar to the abnormal coupling observed between areas of the salience and default mode networks during attention tasks (above). Studies have described abnormally increased DMN activation during a WM task following TBI in both children (Stein, Iyer, Khetani, & Barlow, 2021) and adults (Leech, Kamourieh, Beckmann, & Sharp, 2011). Functional connectivity studies have shown abnormally high posterior cingulate cortex connectivity within the DMN during WM tasks to be associated with more efficient response speeds, indicating its important role in cognitive recovery (Sharp et al., 2011). This decreased anti-correlation between WM and default mode-related areas seen in the chronic stage following TBI has been suggested as a compensatory mechanism during cognitive recovery, particularly in studies which find participants to have clinically normal working memory performance (Leech et al., 2011).

Structural and functional changes following TBI

Cortical thickness

In non-injured populations of both children and adults, attention and working memory functions are associated with cortical thickness of frontal and temporal areas, where increased cortical thickness in prefrontal regions is positively correlated with increases in both attention and WM functioning in children and adults (Merkley et al., 2008). Childhood TBI particularly impacts cortical thickness, where injuries can affect the development of brain regions, especially in frontal executive regions (Merkley et al., 2008).

Cortical-striatal loop

Previous evidence has indicated that changes in cortical thickness in TBI also has larger, network-wide effects, including in the cortico-striatal-thalamo-cortical (CSTC) loop. The CSTC loop is thought to be involved in attention-related dysfunctions following TBI. This loop features nodes of the cingulo-opercular (or salience) network including the dorsal ACC and bilateral aI; cortical nodes in the dlPFC and inferior parietal lobule; as well as a set of subcortical nodes in the head of caudate, medio-dorsal nucleus of the thalamus and the dopaminergic brainstem nuclei (Fig. 10). Damage to the CSTC loop has been found to impair cognitive control, including impulsivity, task-switching, and working memory processes (Peters et al., 2016).

Cortico-striatal-thalamo-cortical connectivity is implicated in both TBI and ADHD-related attention and working memory deficits. During the chronic (from approx. 6 months post-injury) stage of adult TBI, increased diffusion of water in the brain (indicative of white matter tract damage) in the bilateral orbitofrontal gyrus, insula and caudate nucleus is associated with impulsivity (Newcombe et al., 2011). This disruption of CSTC white matter networks following TBI results in decreased network efficiency, thus contributing to the clinical deficits seen in TBI (Caeyenberghs et al., 2014). However, there is limited evidence relating to the development of the CSTC loop in younger populations.

Abnormal thalamo-cortical activation has also been noted in ADHD populations. Given the correlation between these conditions, understanding thalamo-cortical activation in ADHD populations may provide insight into deficits following TBI. Stronger thalamo-putamen connectivity has been noted in children diagnosed with ADHD who have poorer working memory function (Mills et al., 2012). Conversely, weaker thalamo-cortical connectivity has been found in adults with persistent and remitted ADHD during a response preparation task (Clerkin et al., 2013). These differences between adult and child studies may indicate altered activation at different stages of development, highlighting the impact of attention deficits on large-scale brain networks. Taken together, this evidence suggests that the CSTC loop provides important insights into the impacts of ADHD and TBI-related attention deficits.

Modulating cortical activations in TBI using non-invasive brain stimulation

Over the past decades, non-invasive brain stimulation (NIBS) has emerged as a neuromodulatory technique which may aid recovery following TBI. The principle of NIBS is to activate or inhibit neural activity in targeted regions of the cortex to improve various functions, such as attention and working memory. NIBS may facilitate clinical recovery by modulating structural and functional interactions within the brain, strengthening synapses, and increasing the number and connections between dendritic spines (Nitsche et al., 2008).

Several studies have reported improvements in attention and WM processes following TBI using NIBS such as transcranial direct current stimulation (tDCS) or repetitive transcranial magnetic stimulation (rTMS) (Fig. 11) (Zaninotto et al., 2019). In particular, tDCS has been demonstrated to modulate both clinical (i.e., behavioral performance on

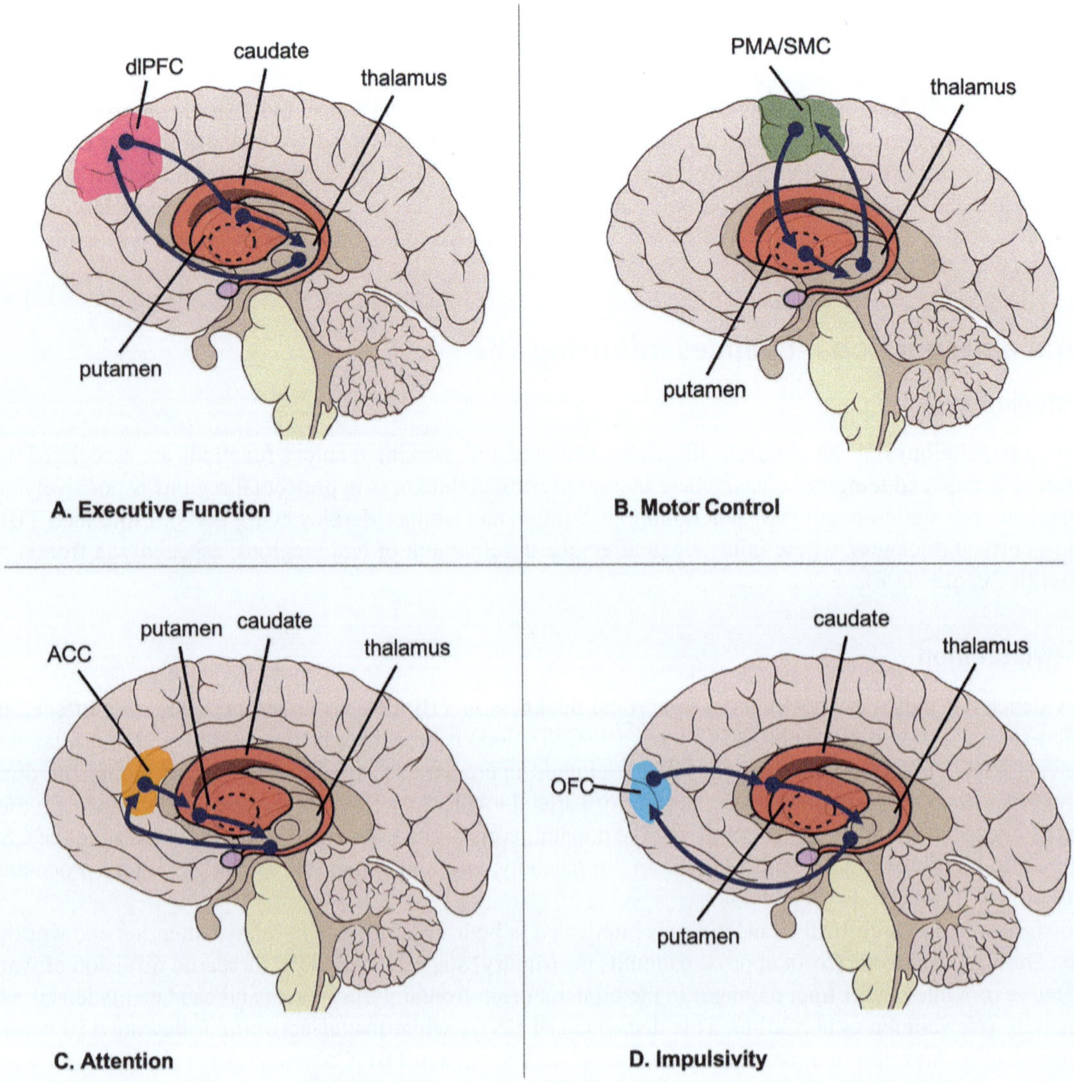

FIG. 10 Various components of the cortico-striatal-thalamo-cortical loop. This loop is implicated in attentional control, involving components related to (A) executive function; (B) motor control; (C) attention; and (D) impulsivity. dlPFC, dorsolateral prefrontal cortex; PMA/SMC, primary motor cortex/supplementary motor cortex; ACC, anterior cingulate cortex; OFC, orbitofrontal cortex. Image is original work with brain template by Patrick J. Lynch; C. Carl Jaffe.

executive tasks) and neurophysiological measures (i.e., fMRI and EEG) when applied during a cognitive task. A potential strength of tDCS in the context of TBI is the indirect nature of current flow in this technique. Given that TBI is a diffuse injury, tDCS may be a useful clinical tool, as the large spatial extent of current flow conferred in this technique may work to address larger areas of diffuse injury in TBI (Fig. 11). Thus, further research is warranted for tDCS in TBI populations. In contrast, rTMS has also been investigated to improve cognitive deficits in TBI. rTMS directly modulates synaptic firing, and therefore can more directly affect brain activity compared to tDCS. Some evidence suggests that rTMS may produce sustained improvement (up to several months) in ongoing symptoms following injury. However, rTMS poses some safety concerns in TBI populations due to its neural excitatory effects which confer a possible seizure risk.

Future neurorehabilitation approaches using NIBS in TBI should address the heterogeneity of injury and resulting variability in patient response to NIBS, for example, through the use of neuroimaging. As such, future research should aim to examine how individual differences in injury affect responses to NIBS, allowing "personalization" of treatment.

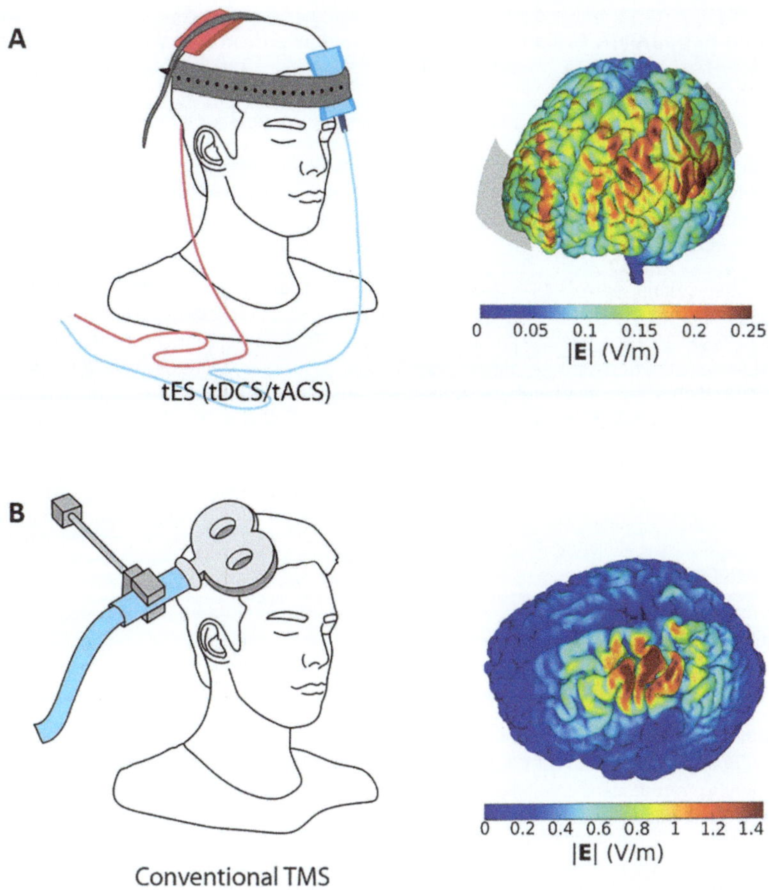

FIG. 11 Schematic representation of (A) tDCS and associated brain current modelling and (B) TMS and associated brain current modelling. tDCS is shown be associated with a more diffuse pattern of current flow, compared to TMS which has a more focal current flow. tDCS, transcranial direct current stimulation; TMS, transcranial magnetic stimulation; tES, transcranial electrical stimulation; V/m, volts per meter (intensity of electric field) Images adapted from Ekhtiari, H., Tavakoli, H., Addolorato, G., Baeken, C., Bonci, A., Campanella, S., … Claus, E. (2019). Transcranial electrical and magnetic stimulation (tES and TMS) for addiction medicine: A consensus paper on the present state of the science and the road ahead. *Neuroscience and Biobehavioral Reviews, 104*, 118–140. Copyright 2019 Elsevier, and Saturnino, G. B., Puonti, O., Nielsen, J. D., Antonenko, D., Madsen, K. H., & Thielscher, A. (2019). SimNIBS 2.1: A comprehensive pipeline for individualized electric field modelling for transcranial brain stimulation. *Brain Human Body Modeling*, 3–25. Creative Commons 4.0.

Applications to other areas of neuroscience

In this chapter, we review attention and working memory related cortical activations in children and adults following traumatic brain injury (TBI). Within this topic, we address the clinical and neurophysiological changes, which occur in the acute and chronic phases of TBI and discuss possible techniques which may improve injury-related deficits.

Executive functioning in the brain depends on a variety of cognitive functions. As such, attention and working memory processes are strongly interrelated with other executive functions including cognitive flexibility, self-regulation, and task-switching. These functions are underpinned by neural networks such as the salience, executive control, and default mode networks. Understanding the mechanisms behind working memory and attention-related dysfunction provides insight into the effect of TBI on various executive functions in the brain. This is particularly important in the context of TBI, which is characterized by diffuse injury to various white matter tracts in the brain.

Traumatic brain injury is associated with various other conditions, which are associated with similar changes in cortical activation. These conditions are presented in Table 1.

Finally, as discussed in the final subsection of this chapter, non-invasive brain stimulation is an evolving and emerging technique which may hold promise in improving attention and working memory-related functions in TBI populations. Understanding the mechanisms underlying non-invasive brain stimulation in TBI may provide greater insight for the application of this technique in other clinical populations.

TABLE 1 Conditions related to traumatic brain injury which impact attention and working memory-related cortical activations.

Condition	Implications for attention and working memory cortical activations
Anxiety disorders	The risk of anxiety disorders is thought to be increased following a traumatic brain injury. Individuals with anxiety disorders may have impaired attention and working memory functions, as these disorders deplete cognitive resources. This is related to abnormal cortical activation in executive areas such as the dorsolateral prefrontal cortex
Attention-deficit/ hyperactivity disorder (ADHD)	Traumatic brain injury is associated with both primary and secondary ADHD. Patients with primary ADHD may have an increased risk of TBI, where the frequency of pre-existing ADHD in moderate to severe cases is between 19% and 48%. TBI is also associated with development of secondary ADHD, where adults and children with TBI have a 4–5 times greater risk of developing secondary ADHD compared to those without TBI. ADHD is associated with difficulties with attention, hyperactive behavior, and impulsiveness, which has been associated with abnormal cortical activation in the right inferior frontal cortex and dorsolateral prefrontal cortex
Depression	Increased rates of depression are one of the more frequently reported neurobehavioral sequelae following TBI. Depression is thought to affect the allocation of attention and elements of working memory. It is associated with decreased activation of the dorsolateral prefrontal cortex, striatum, medial frontal cortex, and medial temporal cortex
Learning difficulties	Traumatic brain injury may result in learning difficulties in children due to problems with executive functions such as attention and working memory. This may impact educational attainment in school. Problems with attention and working memory following TBI are related to abnormal activation of the dorsolateral prefrontal cortex and right inferior frontal gyrus
Post-traumatic stress disorder (PTSD)	Similar to anxiety disorders, the risk of post-traumatic stress disorder is thought to be increased following traumatic brain injury. Post-traumatic stress disorder can impair attention and working memory by depleting cognitive resources. This is related to abnormal cortical activation in executive areas such as the dorsolateral prefrontal cortex
Sleep disturbance	Sleep disturbance is a common symptom following traumatic brain injury. Although complex in etiology, lack of sleep is associated with poorer attention and working memory performance, which may correlate with decreases in activation of the right inferior frontal gyrus

Mini-dictionary of terms

Traumatic brain injury: Commonly abbreviated to TBI. Describes a head injury which may be caused by a blow to the head or shaking of the head and body.

Attention: An executive function in the brain which involves concentration on a certain stimulus while ignoring competing stimuli.

Response inhibition: An executive function in the brain which involves suppressing inappropriate actions in a certain context which hinder a goal-driven behavior.

Working memory: An executive function which allows temporary storage of information to guide decision-making or problem solving.

Dorsolateral prefrontal cortex: An area of the brain located in the frontal region which is important for executive functions such as attention and working memory.

Salience network: A large-scale brain network important in attention which predominantly comprises the anterior and posterior insula and dorsal anterior cingulate cortex.

Default mode network: A large-scale brain network which is active during internally focused attention, such as introspection or meditation

Functional magnetic resonance imaging: Commonly abbreviated to fMRI. A non-invasive imaging technique which is used to measure changes in blood oxygen levels during completion of a task. This is often used to help ascertain brain regions active during certain brain processes.

White matter tracts: Myelinated fibers in the brain which are responsible for conveying information across or within hemispheres, or to areas outside of the brain.

Diffuse axonal injury. An injury that commonly occurs following traumatic brain injury, involving damage to the white matter tracts of the brain due to rapid movement of the brain inside the skull.

Connectivity: This refers to a pattern of structural or functional linkage between areas of the brain. This can be studied from the neuronal to brain region level.

Key facts of working memory and attention

- Attention and working memory are the most commonly affected executive functions following a traumatic brain injury.
- Attention and working memory functions are strongly interrelated, and are controlled by the salience, central executive and default mode networks.
- Diffuse damage sustained following traumatic brain injury can cause a failure to deactivate the default mode network during cognitive tasks, which may cause deficits in working memory and attention
- Working memory has several different components, including the phonological loop, visuo-spatial sketchpad and episodic buffer
- Attention can be captured either voluntarily (i.e., when one chooses to shift their attention to a new stimulus) or involuntarily (i.e., when one's attention is captured by a salient stimulus)
- Attention and working memory undergo rapid development over the course of early childhood to late adolescence

Summary points

- Traumatic brain injury is a leading cause of morbidity and mortality worldwide, with the highest prevalence in children.
- Attention (i.e., inhibition) and working memory are the most commonly affected executive functions following a traumatic brain injury.
- These executive functions are supported by rich, interconnected neural networks such as the salience, central executive and default mode networks.
- The Stop Signal and *n*back tasks are commonly used in both behavioral and neuroimaging settings to measure inhibition and working memory, respectively.
- From childhood to adulthood, executive networks in the brain undergo wide-scale maturational changes, characterized from diffuse to focal activation of cortical regions.
- Traumatic brain injury is a diffuse injury which causes damage to white matter tracts, disrupting structural and functional connectivity within these networks.
- The default mode network is particularly implicated in traumatic brain injury, where abnormal activation of this network is thought to underlie deficits in attention and working memory.
- Traumatic brain injury also results in other structural and functional changes, including changes in cortical thickness and connectivity within the cortico-striatal-thalamo-cortical loop.
- Non-invasive brain stimulation is an emerging technique which has been used to improve attention and working memory function following traumatic brain injury by modulating cortical activity in certain regions of the brain.

References

Aron, A. R. (2011). From reactive to proactive and selective control: Developing a richer model for stopping inappropriate responses. *Biological Psychiatry*, *69*(12), e55–e68. Retrieved from https://www.ncbi.nlm.nih.gov/pmc/articles/PMC3039712/pdf/nihms243856.pdf.

Baddeley, A. (2012). Working memory: Theories, models, and controversies. *Annual Review of Psychology*, *63*, 1–29. Retrieved from https://www.annualreviews.org/doi/pdf/10.1146/annurev-psych-120710-100422.

Barbey, A. K., Koenigs, M., & Grafman, J. (2013). Dorsolateral prefrontal contributions to human working memory. *Cortex*, *49*(5), 1195–1205. Retrieved from https://www.ncbi.nlm.nih.gov/pmc/articles/PMC3495093/pdf/nihms393900.pdf.

Bonnelle, V., Ham, T. E., Leech, R., Kinnunen, K. M., Mehta, M. A., Greenwood, R. J., et al. (2012). Salience network integrity predicts default mode network function after traumatic brain injury. *Proceedings of the National Academy of Sciences of the United States of America*, *109*(12), 4690–4695. Retrieved from https://www.pnas.org/content/pnas/109/12/4690.full.pdf.

Bonnelle, V., Leech, R., Kinnunen, K. M., Ham, T. E., Beckmann, C. F., De Boissezon, X., et al. (2011). Default mode network connectivity predicts sustained attention deficits after traumatic brain injury. *Journal of Neuroscience*, *31*(38), 13442–13451. Retrieved from https://www.ncbi.nlm.nih.gov/pmc/articles/PMC6623308/pdf/zns13442.pdf.

Caeyenberghs, K., Leemans, A., Leunissen, I., Gooijers, J., Michiels, K., Sunaert, S., et al. (2014). Altered structural networks and executive deficits in traumatic brain injury patients. *Brain Structure and Function*, *219*(1), 193–209. Retrieved from https://link.springer.com/content/pdf/10.1007/s00429-012-0494-2.pdf.

Chabran, E., Noblet, V., de Sousa, P. L., Demuynck, C., Philippi, N., Mutter, C., et al. (2020). Changes in gray matter volume and functional connectivity in dementia with Lewy bodies compared to Alzheimer's disease and normal aging: Implications for fluctuations. *Alzheimer's Research Therapy*, *12*(1), 1–13.

Ciesielski, K. T., Lesnik, P. G., Savoy, R. L., Grant, E. P., & Ahlfors, S. P. (2006). Developmental neural networks in children performing a categorical N-back task. *NeuroImage*, *33*(3), 980–990.

Clerkin, S. M., Schulz, K. P., Berwid, O. G., Fan, J., Newcorn, J. H., Tang, C. Y., et al. (2013). Thalamo-cortical activation and connectivity during response preparation in adults with persistent and remitted ADHD. *American Journal of Psychiatry*, *170*(9), 1011–1019.

Dosenbach, N. U., Fair, D. A., Cohen, A. L., Schlaggar, B. L., & Petersen, S. E. (2008). A dual-networks architecture of top-down control. *Trends in Cognitive Sciences*, *12*(3), 99–105. Retrieved from https://www.ncbi.nlm.nih.gov/pmc/articles/PMC3632449/pdf/nihms-452910.pdf.

Dosenbach, N. U., Fair, D. A., Miezin, F. M., Cohen, A. L., Wenger, K. K., Dosenbach, R. A., et al. (2007). Distinct brain networks for adaptive and stable task control in humans. *Proceedings of the National Academy of Sciences of the United States of America*, *104*(26), 11073–11078. Retrieved from https://www.ncbi.nlm.nih.gov/pmc/articles/PMC1904171/pdf/zpq11073.pdf.

Fagerholm, E. D., Hellyer, P. J., Scott, G., Leech, R., & Sharp, D. J. (2015). Disconnection of network hubs and cognitive impairment after traumatic brain injury. *Brain*, *138*(Pt. 6), 1696–1709.

Fair, D. A., Cohen, A. L., Power, J. D., Dosenbach, N. U., Church, J. A., Miezin, F. M., et al. (2009). Functional brain networks develop from a "local to distributed" organization. *PLoS Computational Biology*, *5*(5), e1000381.

FitzGerald, G., Tippett, V., Schuetz, M., & Pollard, C. (2008). The Queensland trauma plan project. *ANZ Journal of Surgery*, *78*(9), 780–783.

Gogtay, N., Giedd, J. N., Lusk, L., Hayashi, K. M., Greenstein, D., Vaituzis, A. C., et al. (2004). Dynamic mapping of human cortical development during childhood through early adulthood. *Proceedings of the National Academy of Sciences of the United States of America*, *101*(21), 8174–8179.

Hagmann, P., Sporns, O., Madan, N., Cammoun, L., Pienaar, R., Wedeen, V. J., et al. (2010). White matter maturation reshapes structural connectivity in the late developing human brain. *Proceedings of the National Academy of Sciences of the United States of America*, *107*(44), 19067–19072.

Hillary, F. G., Genova, H. M., Chiaravalloti, N. D., Rypma, B., & DeLuca, J. (2006). Prefrontal modulation of working memory performance in brain injury and disease. *Human Brain Mapping*, *27*(11), 837–847.

Hooper, S. R., Alexander, J., Moore, D., Sasser, H. C., Laurent, S., King, J., et al. (2004). Caregiver reports of common symptoms in children following a traumatic brain injury. *NeuroRehabilitation*, *19*(3), 175–189. Retrieved from https://content.iospress.com/articles/neurorehabilitation/nre00229.

Iyer, K. K., Barlow, K. M., Brooks, B., Ofoghi, Z., Zalesky, A., & Cocchi, L. (2019). Relating brain connectivity with persistent symptoms in pediatric concussion. *Annals of Clinical Translational Neurology*, *6*(5), 954–961. Retrieved from https://www.ncbi.nlm.nih.gov/pmc/articles/PMC6529928/pdf/ACN3-6-954.pdf.

Jahfari, S., Waldorp, L., van den Wildenberg, W. P., Scholte, H. S., Ridderinkhof, K. R., & Forstmann, B. U. (2011). Effective connectivity reveals important roles for both the hyperdirect (fronto-subthalamic) and the indirect (fronto-striatal-pallidal) fronto-basal ganglia pathways during response inhibition. *Journal of Neuroscience*, *31*(18), 6891–6899. Retrieved from https://www.ncbi.nlm.nih.gov/pmc/articles/PMC6632844/pdf/zns6891.pdf.

Jilka, S. R., Scott, G., Ham, T., Pickering, A., Bonnelle, V., Braga, R. M., et al. (2014). Damage to the salience network and interactions with the default mode network. *Journal of Neuroscience*, *34*(33), 10798–10807.

Johnson, A. R., DeMatt, E., & Salorio, C. F. (2009). Predictors of outcome following acquired brain injury in children. *Developmental Disabilities Research Reviews*, *15*(2), 124–132. Retrieved from https://onlinelibrary.wiley.com/doi/abs/10.1002/ddrr.63.

Kim, J., Whyte, J., Patel, S., Europa, E., Slattery, J., Coslett, H. B., et al. (2012). A perfusion fMRI study of the neural correlates of sustained-attention and working-memory deficits in chronic traumatic brain injury. *Neurorehabilitation and Neural Repair*, *26*(7), 870–880.

Konrad, K., Gauggel, S., Manz, A., & Schöll, M. (2000). Inhibitory control in children with traumatic brain injury (TBI) and children with attention deficit/hyperactivity disorder (ADHD). *Brain Injury*, *14*(10), 859–875. Retrieved from https://www.tandfonline.com/doi/abs/10.1080/026990500445691.

Konrad, K., Neufang, S., Thiel, C. M., Specht, K., Hanisch, C., Fan, J., et al. (2005). Development of attentional networks: An fMRI study with children and adults. *NeuroImage*, *28*(2), 429–439. Retrieved from https://www.sciencedirect.com/science/article/abs/pii/S1053811905004234?via%3Dihub.

Leech, R., Kamourieh, S., Beckmann, C. F., & Sharp, D. J. (2011). Fractionating the default mode network: Distinct contributions of the ventral and dorsal posterior cingulate cortex to cognitive control. *Journal of Neuroscience*, *31*(9), 3217–3224. Retrieved from https://www.ncbi.nlm.nih.gov/pmc/articles/PMC6623935/pdf/zns3217.pdf.

Levin, H. S., Hanten, G., Zhang, L., Swank, P. R., & Hunter, J. (2004). Selective impairment of inhibition after TBI in children. *Journal of Clinical and Experimental Neuropsychology*, *26*(5), 589–597. Retrieved from https://www.tandfonline.com/doi/pdf/10.1080/13803390409609783?needAccess=true.

Li, C.-S. R., Yan, P., Bergquist, K. L., & Sinha, R. (2007). Greater activation of the "default" brain regions predicts stop signal errors. *NeuroImage*, *38*(3), 640–648.

Maria, N. S. S., Sargolzaei, S., Prins, M. L., Dennis, E. L., Asarnow, R. F., Hovda, D. A., et al. (2019). Bridging the gap: Mechanisms of plasticity and repair after pediatric TBI. *Experimental Neurology*, *318*, 78–91.

McAllister, T. W., Flashman, L. A., Sparling, M. B., & Saykin, A. J. (2004). Working memory deficits after traumatic brain injury: Catecholaminergic mechanisms and prospects for treatment—A review. *Brain Injury*, *18*(4), 331–350.

McAllister, T. W., Sparling, M. B., Flashman, L. A., Guerin, S. J., Mamourian, A. C., & Saykin, A. J. (2001). Differential working memory load effects after mild traumatic brain injury. *NeuroImage*, *14*(5), 1004–1012.

Merkley, T. L., Bigler, E. D., Wilde, E. A., McCauley, S. R., Hunter, J. V., & Levin, H. S. (2008). Diffuse changes in cortical thickness in pediatric moderate-to-severe traumatic brain injury. *Journal of Neurotrauma*, *25*(11), 1343–1345.

Mills, K. L., Bathula, D., Costa Dias, T. G., Iyer, S. P., Fenesy, M. C., Musser, E. D., et al. (2012). Altered cortico-striatal–thalamic connectivity in relation to spatial working memory capacity in children with ADHD. *Frontiers in Psychiatry*, *3*, 2.

Newcombe, V. F., Outtrim, J. G., Chatfield, D. A., Manktelow, A., Hutchinson, P. J., Coles, J. P., et al. (2011). Parcellating the neuroanatomical basis of impaired decision-making in traumatic brain injury. *Brain*, *134*(3), 759–768.

Nitsche, M. A., Cohen, L. G., Wassermann, E. M., Priori, A., Lang, N., Antal, A., et al. (2008). Transcranial direct current stimulation: State of the art 2008. *Brain Stimulation*, *1*(3), 206–223. Retrieved from https://www.brainstimjrnl.com/article/S1935-861X(08)00040-5/fulltext.

Owen, A. M., McMillan, K. M., Laird, A. R., & Bullmore, E. (2005). N-back working memory paradigm: A meta-analysis of normative functional neuroimaging studies. *Human Brain Mapping*, *25*(1), 46–59.

Peters, S. K., Dunlop, K., & Downar, J. (2016). Cortico-striatal-thalamic loop circuits of the salience network: A central pathway in psychiatric disease and treatment. *Frontiers in Systems Neuroscience*, *10*, 104.

Raichle, M. E. (2015). The brain's default mode network. *Annual Review of Neuroscience*, *38*, 433–447.

Schachar, R., Levin, H. S., Max, J. E., Purvis, K., & Chen, S. (2004). Attention deficit hyperactivity disorder symptoms and response inhibition after closed head injury in children: Do preinjury behavior and injury severity predict outcome? *Developmental Neuropsychology*, *25*(1–2), 179–198. Retrieved from https://www.tandfonline.com/doi/pdf/10.1080/87565641.2004.9651927?needAccess=true.

Scherf, K. S., Sweeney, J. A., & Luna, B. (2006). Brain basis of developmental change in visuospatial working memory. *Journal of Cognitive Neuroscience*, *18*(7), 1045–1058.

Sharp, D. J., Beckmann, C. F., Greenwood, R., Kinnunen, K. M., Bonnelle, V., De Boissezon, X., et al. (2011). Default mode network functional and structural connectivity after traumatic brain injury. *Brain*, *134*(8), 2233–2247.

Sharp, D. J., Scott, G., & Leech, R. (2014). Network dysfunction after traumatic brain injury. *Nature Reviews Neurology*, *10*(3), 156. Retrieved from https://www.nature.com/articles/nrneurol.2014.15.pdf.

Sherman, L. E., Rudie, J. D., Pfeifer, J. H., Masten, C. L., McNealy, K., & Dapretto, M. (2014). Development of the default mode and central executive networks across early adolescence: A longitudinal study. *Developmental Cognitive Neuroscience*, *10*, 148–159.

Sigurdardottir, S., Andelic, N., Roe, C., & Schanke, A.-K. (2009). Cognitive recovery and predictors of functional outcome 1 year after traumatic brain injury. *Journal of the International Neuropsychological Society*, *15*(5), 740.

Simmonds, D. J., Pekar, J. J., & Mostofsky, S. H. (2008). Meta-analysis of Go/No-go tasks demonstrating that fMRI activation associated with response inhibition is task-dependent. *Neuropsychologia*, *46*(1), 224–232.

Sridharan, D., Levitin, D. J., & Menon, V. (2008). A critical role for the right fronto-insular cortex in switching between central-executive and default-mode networks. *Proceedings of the National Academy of Sciences of the United States of America*, *105*(34), 12569–12574.

Stein, A., Iyer, K. K., Khetani, A. M., & Barlow, K. M. (2021). Changes in working memory-related cortical responses following pediatric mild traumatic brain injury: A longitudinal fMRI study. *Journal of Concussion*, *5*. https://doi.org/10.1177/20597002211006541, 20597002211006541.

Swick, D., Ashley, V., & Turken, U. (2011). Are the neural correlates of stopping and not going identical? Quantitative meta-analysis of two response inhibition tasks. *NeuroImage*, *56*(3), 1655–1665.

Tan, J., Iyer, K. K., Tang, A. D., Jamil, A., Martins, R. N., Sohrabi, H. R., et al. (2019). Modulating functional connectivity with non-invasive brain stimulation for the investigation and alleviation of age-associated declines in response inhibition: A narrative review. *NeuroImage*, *185*, 490–512.

Thurman, D. J. (2016). The epidemiology of traumatic brain injury in children and youths: A review of research since 1990. *Journal of Child Neurology*, *31*(1), 20–27.

van den Heuvel, M. P., & Sporns, O. (2011). Rich-club organization of the human connectome. *The Journal of Neuroscience*, *31*(44), 15775–15786. https://doi.org/10.1523/jneurosci.3539-11.2011.

Vara, A. S., Pang, E. W., Vidal, J., Anagnostou, E., & Taylor, M. J. (2014). Neural mechanisms of inhibitory control continue to mature in adolescence. *Developmental Cognitive Neuroscience*, *10*, 129–139.

Vink, M., Zandbelt, B. B., Gladwin, T., Hillegers, M., Hoogendam, J. M., van den Wildenberg, W. P., et al. (2014). Frontostriatal activity and connectivity increase during proactive inhibition across adolescence and early adulthood. *Human Brain Mapping*, *35*(9), 4415–4427.

Vossel, S., Geng, J. J., & Fink, G. R. (2014). Dorsal and ventral attention systems: Distinct neural circuits but collaborative roles. *The Neuroscientist*, *20*(2), 150–159. Retrieved from https://journals.sagepub.com/doi/pdf/10.1177/1073858413494269.

Zaninotto, A. L., El-Hagrassy, M. M., Green, J. R., Babo, M., Paglioni, V. M., Benute, G. G., et al. (2019). Transcranial direct current stimulation (tDCS) effects on traumatic brain injury (TBI) recovery: A systematic review. *Dementia & Neuropsychologia*, *13*(2), 172–179. Retrieved from https://www.ncbi.nlm.nih.gov/pmc/articles/PMC6601308/pdf/1980-5764-dn-13-02-0172.pdf.

Chapter 13

Traumatic brain injury: Linking intracranial pressure, arterial pressure, and the pressure reactivity index

Alwyn Gomez[a,b], Carleen Batson[b], Logan Froese[c], and Frederick A. Zeiler[a,b,c,d,e]

[a]*Section of Neurosurgery, Department of Surgery, Rady Faculty of Health Sciences, University of Manitoba, Winnipeg, MB, Canada,* [b]*Department of Human Anatomy and Cell Science, Rady Faculty of Health Sciences, University of Manitoba, Winnipeg, MB, Canada,* [c]*Biomedical Engineering, Faculty of Engineering, University of Manitoba, Winnipeg, MB, Canada,* [d]*Centre on Aging, University of Manitoba, Winnipeg, MB, Canada,* [e]*Division of Anaesthesia, Department of Medicine, Addenbrooke's Hospital, University of Cambridge, Cambridge, United Kingdom*

List of abbreviations

ABP	arterial blood pressure
AMP	pulse amplitude of intracranial pressure
AUC	area under the (receiver/operator) curve
CAHR-TBI	Canadian high-resolution traumatic brain injury research collaborative
CBF	cerebral blood flow
CBFV	cerebral blood flow velocity
CBV	cerebral blood volume
CENTER-TBI	collaborative European neurotrauma effectiveness research in traumatic brain injury
CI	confidence interval
CMRO2	cerebral metabolic rate of oxygen
CPP	cerebral perfusion pressure
CPPopt	optimal cerebral perfusion pressure
CT	computer tomography
FV	flow velocity
FVm	mean flow velocity
FVs	systolic flow velocity
GCS	glasgow coma scale
GOS	glasgow outcome scale
GOSE	extended glasgow outcome scale
ICP	intracranial pressure
LLA	lower limit of autoregulation
MAP	mean arterial pressure
MCA	middle cerebral artery
mmHg	millimeters of mercury
mRS	modified Rankin scale
NIRS	near-infrared spectroscopy
PbtO2	brain tissue oxygenation
PET	positron emission tomography
PRx	pressure reactivity index
TBI	traumatic brain injury
TCD	transcranial Doppler
ULA	upper limit of autoregulation

Diagnosis and Treatment of Traumatic Brain Injury. https://doi.org/10.1016/B978-0-12-823347-4.00008-7

Introduction

Dysfunction in the brain's ability to regulate blood flow following traumatic brain injury (TBI) has been described for quite some time (Overgaard & Tweed, 1974). However, the past 25 years has seen a significant expansion to the understanding of cerebral autoregulation following TBI. Much of this can be attributable to developments in the 1990s that allowed for efficient and continuous monitoring of cerebrovascular reactivity following injury (Czosnyka, Smielewski, Kirkpatrick, Menon, & Pickard, 1996). This work primarily utilized transcranial Doppler (TCD) measurements of cerebral blood flow velocities and examined their association with cerebral perfusion pressure (CPP) to gain insight into cerebral autoregulation. CPP is the difference between mean arterial pressure (MAP) and intracranial pressure (ICP). The subsequent development of the pressure reactivity index (PRx), a measure of cerebrovascular reactivity requiring only ICP and MAP measurements (Czosnyka et al., 1997) opened doors to significant exploration of the implications of cerebrovascular reactivity and cerebral autoregulation in the TBI patient.

Cerebral autoregulation

Cerebral autoregulation is the physiologic process by which cerebral blood flow (CBF) is maintained constant over a range of CPPs. First described by Fog in felines (Fog, 1938) and later in humans by Lassen (1959), it is often visualized in a triphasic curve formed by plotting CBF vs MAP or CPP. In the first phase, at CPPs below the lower limit of autoregulation, CBF correlates positively with CPP as autoregulatory mechanisms fail to maintain adequate CBF. As CPP is increased an inflection point is reached which represents the lower limit of autoregulation (LLA). In this second phase, a plateau is found where changes in CPP have little effect on CBF (50–150 mmHg). Over this range, autoregulatory mechanisms maintain CBF constant. This is where normal physiologic conditions are found. Continuing to increase CPP results in a second inflection point which represents the upper limit of autoregulation (ULA). At CPPs beyond the ULA, CBF and CPP once again positively correlate as autoregulatory mechanisms fail to limit CBF to physiologic levels.

The methods of evaluating cerebral autoregulation are generally categorized into two groups: static and dynamic. Static methods, similar to those used by Lassen in his seminal paper, involve measuring CBF in a steady state at a baseline MAP then measuring CBF in a second steady state following perturbation of the MAP. The method to determine cerebral blood flow is characteristically of low temporal resolution but precise. Jugular venous blood sampling of administered radioactive isotopes of inert gasses and positron emission tomography (PET) scans are typically employed as techniques to determine CBF in these methods. While these provide precise evaluation of autoregulation, their poor temporal resolution does not enable assessment of changes in latency of the cerebral autoregulatory response (Armstead, 2016).

This has led to more of a focus on semi-dynamic measurements of cerebral autoregulation. These methods provide information as to the rate at which autoregulatory mechanisms act to stabilize CBF. Aaslid et al. introduced a method of measuring cerebral autoregulation in a dynamic fashion as early as 1989. Their method involved utilizing CBF velocity (CBFV), as measured by transcranial Doppler (TCD), as a surrogate for CBF. A drop in blood pressure was achieved in subjects by rapid deflation of a thigh cuff (Aaslid, Lindegaard, Sorteberg, & Nornes, 1989). This was later confirmed to correlate well with static measurements of cerebral autoregulation by Tiecks, Lam, Aaslid, and Newell (1995).

Early evidence of disrupted cerebral autoregulation in TBI patients came from static measures of autoregulation. In a population of 43 patients with head injury Overgaard & Tweed tested autoregulation utilizing an intra-arterial $Xenon^{133}$ method to determine CBF while MAP was increased using angiotensin. They reported that autoregulation was often impaired in the early time periods following TBI (Overgaard & Tweed, 1974). Utilizing similar methods, Cold and Jensen demonstrated that in a population of 18 unconscious patients with TBI, impaired autoregulation was identified in 83% of assessments with a trend toward normalization of cerebral autoregulation after 5 days post-injury (Cold & Jensen, 1978).

Development of the pressure reactivity index (PRx)

In 1996, Czosnyka et al. reported on a new method of continuously monitoring the state of cerebral autoregulation. These indices relied on TCD measurements of flow velocities (FV) in the middle cerebral artery (MCA). The mean FV (FVm) and the FV during cardiac systole (FVs). The indices were simply the Pearson's correlation coefficients between 36 consecutive 5-s averaged CPP measurements and time matched 5 s averaged FVm (Mx) or FVs (Sx) measurements. The down-sampled FV signals were utilized to focus on the frequency range thought to be associated with cerebral vasomotion. Variation in CPP was not caused by artificial manipulation, but by capturing natural slow fluctuations in MAP. This provided a correlation coefficient value for every 3-min period which was eventually averaged over the entire recording period to obtain an overall value for Mx and Sx (Czosnyka et al., 1996). The correlation coefficient is a standardized value between -1 and $+1$ making interpretation of this normalized index simple with a positive index indicating impaired autoregulation, with zero degree phase-shift between CPP and FV signals. Conversely, a negative index implies intact autoregulation. This association had previously been confirmed in rabbit models (Nelson et al., 1992).

In a population of 82 patients with moderate to severe TBI, Czosnyka et al. reported that Mx and Sx correlated well with admission Glasgow coma scale (GCS) ($r=-0.34$, $p<0.0025$ and $r=-0.38$, $p<0.0008$, respectively) as well as Glasgow outcome score (GOS) at 6 months after injury ($r=0.41$, $p<0.0002$ and $r=0.48$, $p<0.00009$). Once patients were dichotomized into favorable vs unfavorable outcomes, it was noted that those with an unfavorable outcome had Mx >0 (i.e., disrupted cerebral autoregulation) during the first 2 days after injury, and during the 6–8 days following the injury ($p<0.05$) with a period of intact autoregulation in between (Czosnyka et al., 1996).

Czosnyka et al. later reported the development of the PRx the following year. Conceptually the PRx was a measure of phase-shift between a surrogate for pulsatile cerebral blood volume (CBV) (i.e., ICP) and a driver to cerebral blood flow, arterial blood pressure (ABP). If reactivity were intact, an increase in ABP would be met with cerebral vasoconstriction thereby producing a reduction or attenuated response in ICP. If cerebral vasculature were nonreactive then an increase in ABP would be followed by an increase in CBV thereby increasing ICP. As such Czosnyka et al. recognized that ICP could be used as a surrogate for CBV and the relationship between ABP and ICP would provide insight into cerebrovascular reactivity. Initially, the PRx was initially described as the Pearson's correlation coefficient between 40 consecutive 5-s averages of ICP and ABP. Here a positive value indicated passive nonreactive behavior of the cerebrovasculature while negative values are indicative of a normally responsive vascular bed (Czosnyka et al., 1997) (Figs. 1 and 2).

Utilizing the same cohort data as in their TCD based indices Czosnyka et al. found that PRx correlated well with dichotomized (favorable vs unfavorable) global outcomes following TBI ($r=0.484$, $p<0.0000$) and actually outperformed the TCD based index Mx ($r=0.42$, $p<0.0002$). Further analysis revealed that in patients with an unfavorable outcome, PRx was significantly positive in the first 2 days ($p<0.05$). This mirrored the results found by TCD based indices. Interestingly it was noted that all patients with a PRx <-0.2 had a favorable outcome while those with a PRx >0.2 82% had an unfavorable outcome. This was some of the first evidence of a threshold at which outcomes significantly changed and highlighted the possible prognostic implication of PRx. Finally PRx and Mx were also reported to correlate well with one another ($r=0.65$, $p<0.0000$) (Czosnyka et al., 1997).

Cerebrovascular reactivity acts as a surrogate for cerebral autoregulation and the two terms are not necessarily interchangeable. Outside of the upper and lower limits of cerebral autoregulation cerebral blood vessels may still continue to constrict and dilate in response to changes in CPP but fails to maintain a stable CBF (Cold & Jensen, 1978). Despite this PRx has been shown to accurately detect the lower limit of autoregulation, as defined by laser-Doppler flow, in both porcine and rabbit models of physiologic and elevated ICP (Brady et al., 2008; Zeiler, Donnelly, Smielewski, et al., 2018) as well as in porcine models of arterial hypotension (Brady et al., 2008; Zeiler, Lee, Smielewski, Czosnyka, & Brady, 2018). Numerous validation studies in humans have also shown that PRx correlates well with traditional measures of cerebral autoregulation including TCD and PET based methods (Steiner et al., 2003; Steinmeier, Hofmann, Bauhuf, Hübner, & Fahlbusch, 2002). A larger study performed by Budohoski et al. analyzed 486 recordings from 201 TBI patients and found that there was moderate correlation between Mx and PRx ($r=0.58$, $p<0.001$). They noted that discrepancies between Mx and PRx with ICPs exceeding 30 mmHg (Budohoski, Czosnyka, M., de Riva, et al., 2012). These studies all point to a covariance between cerebral autoregulation and cerebrovascular reactivity as measured by PRx.

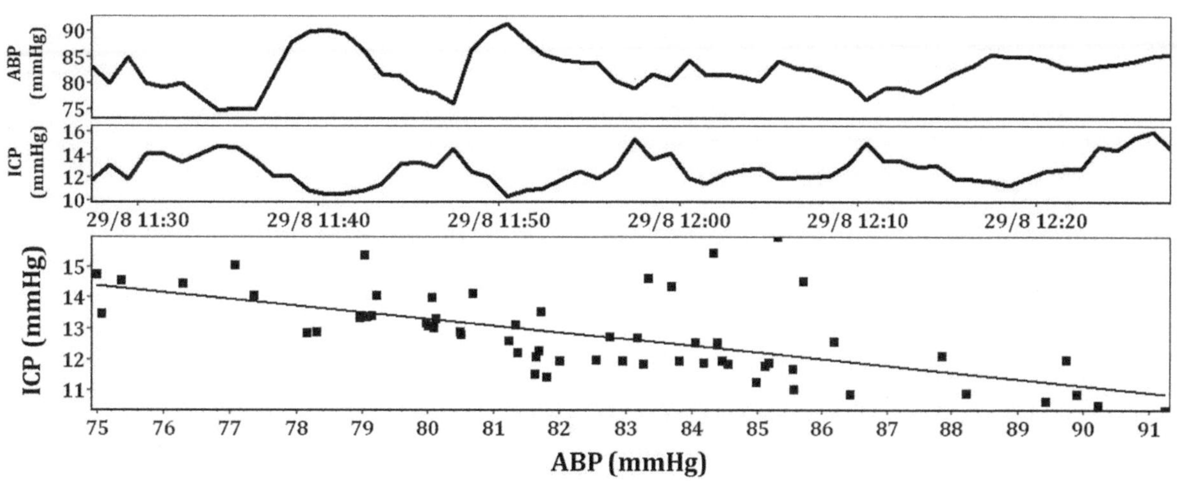

FIG. 1 Derivation of PRx. Arterial blood pressure (ABP) and intracranial pressure (ICP) tracings are in the first two panels, respectively. Note how as ABP increases ICP decreased and vice versa indicating intact cerebral autoregulation. In the third panel we see ICP plotted against ABP with a regression line having an expected negative slope. This is a visual demonstration of the derivation of the pressure reactivity index.

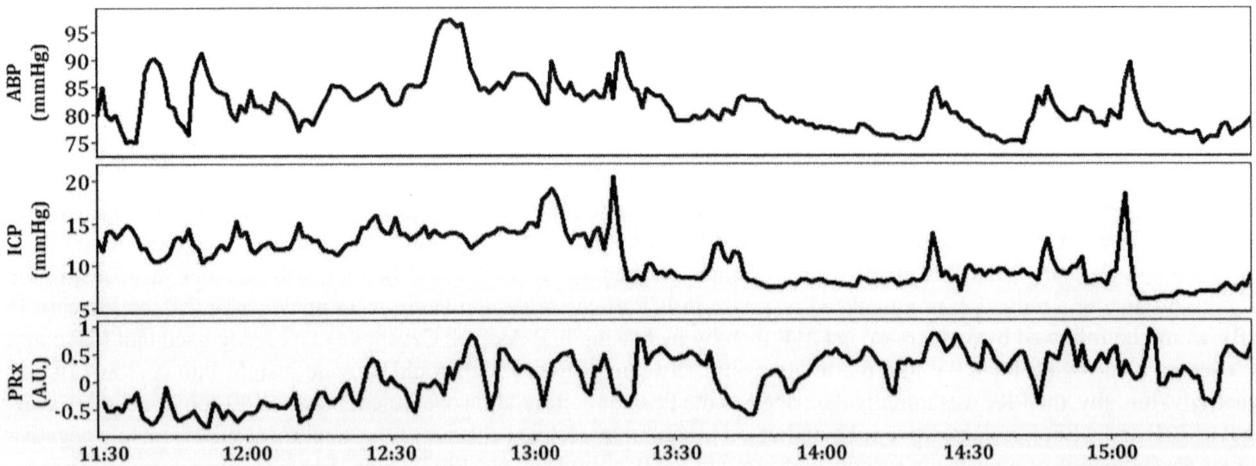

FIG. 2 Pressure reactivity over time. Tracings of arterial blood pressure (ABP), intracranial pressure (ICP) and the pressure reactivity index (PRx) over time in a single patient. Note how the patient goes from a pressure active state (PRx = – 0.5) with ICP negatively correlating with ABP to a pressure passive state (PRx = 0.5) with ABP and ICP positively correlating.

Association of PRx with outcomes

Global outcomes following TBI

Numerous studies have been reported to relate PRx with global outcomes following TBI. In this section we will review studies as they pertain to linking outcomes following TBI and PRx (Table 1). Other outcomes will be addressed late in the chapter.

One of the earliest studies to validate PRx was by Kirkness et al. in a population of 27 TBI patients. This study utilized GCS at the time of discharge as their outcome and did find that elevated PRx was a significant predictor of GCS at discharge when controlled for GCS at admission and age ($r = 0.2515$, $p = 0.029$). Interestingly this study also had a cohort of 25 patients with acute cerebrovascular pathology for which no association with outcome and PRx was found (Kirkness, Mitchell, Burr, & Newell, 2001). Steiner et al. reported more compelling results the following year in a study of 114 TBI patients. They found a significant correlation with PRx and 6-month outcomes as defined by GOS ($r = -0.274$, $p = 0.003$). Notably, GCS on admission did not significantly correlate with GOS at 6 months or PRx. Patient age, however, did correlate with both PRx ($r = 0.358$, $p = 0.0001$) and GOS ($r = -0.207$, $p = 0.027$). In their cohort ICP was also found to be predictive of GOS ($r = -0.252$, $p = 0.007$) but did not significantly correlate with PRx (Steiner et al., 2002).

Several smaller studies have examined association of PRx with outcomes following TBI. Jaeger et al. reported on 27 severe TBI patients in which PRx was found to be associated with GOS at 6 months ($r = -0.52$, $p < 0.01$) (Jaeger, Schuhmann, Soehle, & Meixensberger, 2006). Eide et al. found that in a group of 76 TBI patients that outcome was related to PRx by multiple regression analysis ($p = 0.03$) (Eide, Czosnyka, Sorteberg, Pickard, & Smielewski, 2007). In a group of 18 TBI patients Dias et al. found that a dichotomized PRx was associated with GOS at 6 months and in a group of 28 TBI patients Petkus et al. found a correlation between PRx and GOS at discharge ($r = -0.448$, $p < 0.05$) and at 6 months ($r = -0.402$, $p < 0.05$) (Dias et al., 2015; Petkus et al., 2016). Schmidt et al. also found in their group of 36 patients an association with PRx and outcome, this time defined by a modified Rankin Scale (mRS) at 3 months ($r = 0.38$, $p < 0.05$) (Schmidt et al., 2016). It should be noted that while a GOS increases with better outcome a lower mRS indicates a better outcome. This explains the positive correlation seen in the study by Schmidt et al. These studies are all in concordance that as PRx increases outcomes generally worsen.

One of the first large studies assessing the association of outcome following TBI and PRx was performed by Zweifel et al. In their report 398 TBI patients underwent monitoring of PRx. Outcomes 6 months following injury were correlated with PRx in a near linear fashion. This association was independent of ICP ($r = 0.37$, $p < 0.0001$) indicating that PRx provides prognostic information aside from that encoded by ICP. Age and, GCS on presentation and mean ICP were also found to be independent predictors of outcome (Zweifel et al., 2008). In a study involving 327 TBI patients Aries et al. found a similar association with outcome, however, at ICPs < 15 mmHg, PRx was not found to be able to differentiate between those with fatal and non-fatal outcomes (Aries et al., 2012). This may be secondary to ICP being a poor surrogate for CBV when cerebral compliance is high.

TABLE 1 Major literature associating PRx with global outcomes following TBI.

Study	Number of patients	Key findings about PRx
Steiner et al. (2002)	114	• The correlation with PRx and 6-month GOS was significant ($r=0.27$, $p<0.005$) • The percentage of time with PRx > 0.2 correlated with 6-month GOS significantly ($r=-0.28$, $p<0.005$)
Hiler et al. (2006)	126	• PRx in the first 24 h correlated with 6-month GOS ($r=-0.36$, $p<0.00002$) • Patients with a positive PRx in the first 24 h had a significantly higher mortality rate than those with a negative PRx in the first 24 h (28.6% vs 9.5%, $p<0.001$)
Zweifel et al. (2008)	398	• PRx and Mx was compared in 298 patients and there was a good correlation ($r=0.36$, $p<0.001$) • PRx was associated with the static rate of autoregulation as measured by PET in a sub group of 17 patients ($r=0.31$, p 0.02) • A PRx > 0.25 was associated with a significantly higher mortality than PRx < 0.25 (69% vs < 20%, $p<0.0001$)
Aries et al. (2012)	327	• Mean PRx was worse in patient who died compared to those that survived (0.10 ± 0.21 vs 0.01 ± 0.15 $p=0.001$) • PRx was an independent predictor of mortality by multivariate analysis (OR = 9.8, 95% CI 1.5–64, $p<0.001$)
Budohoski, Czosnyka, de Riva, et al. (2012)	201	• PRx was predictive of mortality (AUC = 0.658, 95% CI 0.566–0.750, $p=0.001$) • PRx significantly correlated with Mx over an average of all recordings ($r=0.58$, $p<0.001$)
Sorrentino et al. (2012)	459	• A PRx threshold of 0.25 was found for mortality ($\chi^2=25.41$, $p<0.001$) • A PRx threshold of 0.02 was found for favorable outcome ($\chi^2=17.47$, $p<0.001$)
Howells, Johnson, McKelvey, and Enblad (2015)	131	• PRx significantly correlated with 6-month GOSE ($r=-0.41$, $p<0.01$)
Adams et al. (2017)	601	• Patients with a fatal outcome due to neurologic cause had a higher mean PRx than those that survived (0.16 vs 0.05, $p<0.001$) • On average those with a fatal outcome due to neurologic cause spent more total monitoring time with a PRx > 0.25 than those that survived (42.9% vs 32.2%, $p<0.001$)
Zeiler et al. (2018)	358	• Univariate logistic regression found that PRx over the entire recording period was able to discriminate favorable vs unfavorable outcomes as determined by 6-month GOS (AUC = 0.573, $p<0.0001$) • Note: All patients in this study did not receive a decompressive craniectomy
Svedung Wettervik, Howells, Enblad, and Lewén (2019)	362	• PRx was found to be a significant predictor of outcomes, as define by 6-month GOS, by a multivariate logistic regression including age, GCS, pupillary reactivity, and deviation from CPPopt (OR = 7.460, 95% CI 1.048–53.076, $p=0.045$)
Zeiler et al. (2019)	204	• Mean PRx was found to be predictive of mortality and favorable outcome (AUC = 0.748, 95% CI 0.662–0.835, $p<0.0001$; AUC = 0.649, 95% CI 0.574–0.724, $p=0.0002$, respectively) • Percentage of time spent over PRx threshold of 0, +0.25, +0.35 was found to be predictive of mortality (AUC = 0.731, 95% CI 0.643–0.818, $p<0.0001$; AUC = 0.736, 95% CI 0.648–0.824, $p<0.0001$; AUC = 0.737, 95% CI 0.648–0.825, $p<0.0001$, respectively) and favorable outcome (AUC = 0.637, 95% CI 0.561–0.712, $p=0.0005$; AUC = 0.647, 95% CI 0.572–0.722, $p=0.0002$; AUC = 0.561 95% CI 0.576–0.726 $p=0.0002$, respectively)

Summary of the major large (> 100 patient) studies that have examined the association between global patient outcomes following TBI.

In a study of 201 TBI patients Budohoski et al. noted that patients who were alive at follow-up had a significantly lower PRx on post-admission day 1 than those that died ($p < 0.01$). This is concordant with the initial findings of Czosnyka et al. that found early dysfunctional cerebrovascular reactivity to herald poor outcomes (Budohoski, Czosnyka, de Riva, et al., 2012; Czosnyka et al., 1997). In a subsequent study of 601 TBI patients, impaired cerebrovascular reactivity in the first 72h following injury was associated with fatal outcome of neurologic cause. This is in contrast to those with non-neurological fatal outcomes that demonstrated impaired cerebrovascular reactivity after 168h post-injury (Adams et al., 2017). These studies point to prognostic implications to the time course of PRx in TBI patients.

The predictive value of PRx in TBI was further studied by Zeiler et al. In 2018 study of 358 TBI without decompressive hemicraniectomy they found that PRx was prognostically useful in differentiating patients with a favorable vs an unfavorable outcome 6months post-injury (AUC=0.573, $p < 0.0001$). Additionally PRx was found to be prognostic of mortality at 6months (AUC=0.651, $p < 0.0001$) (Zeiler, Donnelly, Smielewski, et al., 2018).

The importance of the previously discussed smaller studies becomes relevant when reviewing the literature encompassing larger studies associating PRx with outcome. As it can be seen, there is no paucity of large studies; however, there may be a lack of variety. All the previously mentioned larger cohorts originate from the same single institution's database of TBI patients. The same institution that Czosnyka et al. drew their data from in their 1997 report. The overlap in source patient data does reduce the strength of these findings as they are not altogether independent. Those smaller studies do provide some variety and increase the strength of the literature supporting the utility of PRx in TBI patients.

There are other institutions contributing to the data around the PRx in TBI patients. In a study coming out of patient data from Sweden, Howells et al. found that PRx correlated well with 6-month outcomes, patients as defined by the Extended Glasgow Outcome Scale (GOSE), in a population of 131 TBI patients ($r = -0.41$; Howells et al., 2015). In a follow-up study from the same institution, Svedung et al. reported on outcomes following TBI in 362 patients. Their data showed an interesting trend that differed from previous reports. They noted that PRx was not significantly different in the first 2days post-injury in those with a favorable compared to unfavorable 6-month outcome ($p > 0.05$). They found that 2–5days post-injury was when there was the largest separation in PRx values for those with a favorable vs unfavorable outcome ($p = 0.00003$; Svedung Wettervik et al., 2019). While this points to cerebrovascular reactivity playing more of a prognostic role in the later days following TBI further work is needed prior to any definitive conclusions being drawn.

Recently data coming from large, multi-institutional/multi-national collaborative networks is being reported on in the literature. Zeiler et al. recently published data collected at 21 different institutions around the European Union through the Collaborative European Neuro Trauma Effectiveness Research in TBI (CENTER-TBI) study. In this cohort of 204 TBI patients (159 without decompressive craniectomy) mean PRx was significantly higher in patients that did not survive (0.226 vs 0.0017, $p < 0.0001$) and univariate logistic regression demonstrated a strong predictive value of mean PRx (AUC=0.748, $p < 0.0001$). Similarly, PRx was significantly higher in those with an unfavorable outcome at follow-up (0.121 vs 0.006, $p = 0.001$). Here univariate analysis showed a slightly weaker discriminatory ability (AUC=0.649, $p = 0.0002$; Zeiler, Ercole, et al., 2019). Large multi-center studies such as this one are necessary to gain further insight into the utility of PRx at predicting global outcomes.

PRx thresholds

Given the body of evidence showing the prognostic significance of cerebrovascular reactivity is growing, much work has focused on determining critical prognostic thresholds of PRx. In their initial study Czosnyka et al. noted that 81% of the patients with a PRx > 0.2 experienced an unfavorable outcome while a favorable outcome was found in all patients with a PRx < −0.02 (Czosnyka et al., 1997). This was some of the first evidence of a clinically relevant tipping point. This was further enforced by Steiner et al. in their study of 114 TBI patients where they found that the percentage of monitored time with a PRx > 0.2 correlated significantly with outcomes ($r = -0.28$, $p < 0.05$; Steiner et al., 2002).

As expected, the threshold PRx for predicting mortality appears to be slightly higher. In a study of 193 TBI patients Zweifel et al. noted that mortality rates abruptly increased to > 50% once PRx was > 0.35 (Zweifel et al., 2008). Sorrentino et al. confirmed this discrepancy in a study of 459 TBI patients, with and without decompressive craniectomy, where they compared the Pearson's chi-square values for numerous 2×2 tables constructed at various threshold values of PRx comparing survival and death as well as favorable and unfavorable outcomes. Utilizing this method they determined a survival threshold of PRx=0.25 while a lower threshold was found for favorable outcomes (PRx=0.05; Sorrentino et al., 2012). Utilizing a similar method in a distinct group of 28 TBI patients Petlus et al. also found a similar threshold for mortality at 6months (PRx > 0.24; Petkus et al., 2016).

Contrary to this Zeiler et al. found that in this study of 358 TBI patients without decompressive craniectomy a sequential chi-square method found a single optimal threshold for discerning both favorable vs unfavorable outcome and survival vs death at follow-up (PRx = 0.35; Zeiler, Donnelly, Smielewski, et al., 2018). In follow-up study utilizing 204 patients in the CENTER-TBI study, percent time with PRx > 0.35 was significantly higher in patients with an unfavorable outcome compared to those with a favorable one (30.9 vs 20.6, $p = 0.001$) and in those that died compared to those that survived (41.0 vs 21.4, $p < 0.0001$). However, similar trends were seen with percent time with PRx > 0 and PRx > 0.25. Unfortunately, no analysis to identify the optimal threshold was performed (Zeiler, Ercole, et al., 2019).

These studies have led to the concept of a tiered approach to PRx. A PRx below 0 appears to be ideal with reduced mortality and favorable outcomes while a PRx above 0.3 appears to be indicative of increased mortality and poor outcomes. When PRx is found between 0 and 0.3 mortality may be reduced but outcomes may not be favorable (Czosnyka et al., 2016). Eventually, the time spent at various thresholds (dose of impaired cerebrovascular reactivity), may prove to able to refine prognostic models as has been in the case with CPP and ICP thresholds (Güiza et al., 2015, 2017).

PRx and imaging findings

Evidence linking PRx to imaging findings was described in early studies. Hiler et al. examined a group of 126 TBI patients and found that those with a positive PRx at 24 h post-injury had more midline shift (0.43 vs 0.23 cm, $p = 0.026$) and larger lesion volume (26 vs 14 cm^3, $p = 0.023$) on CT imaging. They were also more likely to have a non-evacuated mass lesion (28.6% vs 12.75, $p = 0.014$; Hiler et al., 2006). In this study it was unclear as to if these imaging findings resulted in disrupted cerebrovascular reactivity or vice versa.

In a single-center study Zeiler et al. examined the association between initial intracranial injury burden as defined by index CT scores (Marshall, Rotterdam, Helsinki, and Stockholm scores) and cerebrovascular reactivity in 358 patients. While none of the CT scores were significantly associated with impaired cerebrovascular reactivity, multivariable logistic regression models for binary PRx threshold outcomes consistently found the presence of subdural hematoma and traumatic subarachnoid hemorrhage predictive of impaired cerebrovascular reactivity as defined by various thresholds (PRx = 0, +0.25, +0.35; Zeiler, Donnelly, Nourallah, et al., 2018). A follow-up study using high-resolution multi-center data from the CENTER-TBI database examined 165 patients with TBI. In general, they found that diffuse patterns of injury as demonstrated on CT were strongly associated with impaired cerebrovascular reactivity (Zeiler, Mathieu, et al., 2020).

Furthermore, in an exploratory analysis Mathieu et al. examined the relationship between progression of pericontusional edema and the extent of cerebrovascular impairment in 23 TBI patients with sequential CT scans. Their CT scans were analyzed using quantitative CT analytics that segmented each scan into various regions of interest. They found that, after controlling initial volume size, timing of scan and coagulopathy, there was a strong correlation with edema progress and percentage of time PRx > 0.25 ($r = 0.69$, $p = 0.002$; Mathieu, Zeiler, Whitehouse, et al., 2020). In a follow-up report of 88 patients in the CENTER-TBI study Mathieu et al. found that under univariate analysis percentage time with a positive PRx was associated with increase in volume of the hemorrhagic contusion core on CT. Notably it was not associated with increase in pericontusional edema and mean PRx was not associated with change in edema or contusion size. Other indices that assess cerebrovascular reactivity were found to be significantly associated with these CT parameters (Mathieu, Zeiler, Ercole, et al., 2020). This hints towards some link radiographic progression of injury and impairment of cerebrovascular reactivity.

PRx and cerebral oxygen metabolism

The primary function of cerebral autoregulation is to ensure that there is an adequate supply of oxygen and nutrients to the brain. The disruption of cerebrovascular reactivity following TBI resulting in poor outcomes is then potentially mediated by changes in oxygen metabolism. Early work by Steiner et al. found that in a cohort of 22 TBI patients a low cerebral metabolic rate of oxygen ($CMRO_2$), as measured by PET scanning, was associated with disrupted cerebrovascular reactivity ($r = 0.21$, $p = 0.018$; Steiner, 2003). Interpretation of these results is difficult as low $CMRO_2$ could be the result of either critical ischemia or a primary metabolic failure seen in this cohort of TBI patients.

Brain tissue oxygen ($PbtO_2$) monitoring has become more popular parameter in the management of critically ill TBI patients and is used as a marker of cerebral oxygenation. In a study by Ang et al. of 40 TBI patients no association was found between changes in PRx and changes in $PbtO_2$. In a subgroup analysis of patients with fatal outcomes increases in PRx were associated with decreases in $PbtO_2$ ($r = -0.685$, $p = 0.042$). Only eight patients were included in this subgroup which calls into question the strength of the relationship (Ang, Wong, Lee, Wang, & Ng, 2006). In a later study of 27 TBI patients Jaeger et al. found conflicting results and noted that in their cohort PRx was negatively associated with $PbtO_2$ ($r = -0.42$, $p = 0.28$). Despite the increased interest in brain tissue oxygenation, in the era of modern BTF based therapy the majority of physiologic insult burden in moderate and severe TBI is related to impaired cerebrovascular reactivity and not low $PbtO_2$ (Zeiler et al., 2020).

Definitive conclusions about the relationship between cerebrovascular reactivity and cerebral oxygenation are difficult to make in the setting of such limited and conflicting evidence. More work is needed before our understanding of this relationship will be complete.

PRx in targeted therapy

In the initial study of PRx, Czosnyka et al. reported that in their cohort of patients plotting PRx vs CPP identified ranges of CPP at which PRx was at a nadir. They speculated that this would be a method of identifying CPP targets in the traumatically injured brain (Czosnyka et al., 1997). This concept was further developed, and it was soon realized that individual targets of CPP could be developed in real time for each patient by continuously plotting PRx vs CPP (Steiner et al., 2002). This sparked a significant research effort attempting to validate the possibility of PRx derived personal targeted therapy. The reader should refer to the Chapter 14 for a full discussion of this topic.

Limitation of PRx

Evidence for the clinical utility of PRx in TBI patients has been building over the past 25 years, but no discussion of PRx would be complete without reviewing some of its limitations. One of the most obvious limitations is that at present, monitoring of PRx requires invasive physiologic monitoring. Continuous ABP, derived from arterial lines, and continuous ICP monitoring, derived from intra-parenchymal probes, require invasive procedures that are not without risk. This has limited monitoring of PRx to TBI patients in the ICU setting. Recently work has been done to develop non-invasive means of deriving continuous measure of cerebrovascular reactivity but these efforts need further validation (Gomez, Dian, Froese, & Zeiler, 2020; Gomez, Dian, & Zeiler, 2020).

The validity of PRx as a measure of cerebrovascular reactivity in all TBI patients is also questionable. Since PRx relies on ICP as a surrogate for CBV, a relatively linear relationship between pressure and volume needs to be present, as is found in a closed calvarium. It is unclear how robust the prognostic ability of PRx is in situations of increased compliance, such as in patients with a decompressive craniectomy (Timofeev et al., 2008; Zeiler, Aries, et al., 2020). Decompressive craniectomy is a commonly utilized means of management of medically refractory ICP in severe TBI patients and so the relevance of PRx in this population is of clinical significance.

As a physiologic parameter PRx is also quite noisy. Since it relies on natural physiologic fluctuations in ABP averaging over long periods of time, at least 30 min, is recommended to provide a true signal. When utilizing PRx to derive patient specific parameters (i.e., CPPopt) the periods of recording time required to generate a meaningful signal is well over 4 h (Czosnyka, Czosnyka, & Smielewski, 2017).

One of the most obvious limitations of PRx has been its limited validation in multiple diverse institutions. As previously mentioned, a large amount of the body of literature supporting PRx as a useful physiologic parameter in TBI patients has been based off the same institutional database used in the original study by Czosnyka et al. Large multi-institutional/multinational collaborative networks such as the ones found in the CENTER-TBI study and CAnadian High-Resolution TBI (CAHR-TBI) Research Collaborative will hopefully help remove this limitation (Bernard et al., 2020).

Future of cerebrovascular reactivity

In this chapter, we focused on PRx as a measure of cerebrovascular reactivity; however, the world of cerebral autoregulation monitoring in TBI is ever expanding. New indices are being developed that involve more computational complexity but are still derived from the same physiologic parameters as PRx. Examples of such include the PAx (correlation between pulse amplitude of ICP (AMP) and MAP), and RAC (correlation between AMP and cerebral perfusion pressure (CPP)). In a recent single-center study by Zeiler et al. both PAx (AUC = 0.606, $p < 0.0001$) and RAC (AUC = 0.655, $p < 0.0001$) were found to have superior prognostic ability than PRx (AUC = 0.573, $p < 0.0001$) when it came to discriminating between TBI patients with favorable vs unfavorable outcomes (Zeiler, Donnelly, Smielewski, et al., 2018). A follow-up report based on patients from the CENTER-TBI study found concordance between PRx, PAx and RAC (Zeiler, Ercole, et al., 2019). The body of evidence for these indices will likely develop quickly as they can use existing databases for validation and comparison to PRx.

The limitations of the invasive nature of PRx have also opened the door for non-invasive measure of cerebrovascular reactivity. Advances in near-infrared spectroscopy (NIRS) and the development of robotic TCD has led to increased interest continuous non-invasive methods of quantifying cerebrovascular reactivity (Gomez, Dian, Froese, et al., 2020; Gomez, Dian, & Zeiler, 2020; Zeiler & Smielewski, 2018; Zeiler, Czosnyka, & Smielewski, 2018; Zeiler, Smielewski, et al., 2019).

The concordance of these non-invasive indices with PRx will be a necessary first step. The prospect of non-invasively monitoring cerebrovascular reactivity opens the door to analysis of autoregulation months after injury as well as in mild TBI and even healthy populations.

Developing an understanding of the impact of cerebrovascular reactivity and the mechanisms by which dysfunction occurs would ideally lead us to molecular targets to modulate autoregulation in the TBI patient. The largest study to date examining PRx in TBI patients reported data collected from 1146 TBI patients over a 25-year period at a single institution. They noted that despite various treatment paradigm shifts over those 25 years PRx values remained relatively unchanged in their cohort (Donnelly et al., 2019). This indicates that current physiologic management of TBI patients has failed to meaningfully alter cerebrovascular reactivity in the TBI patient. A therapeutic agent that would restore cerebral autoregulation following TBI would take us a step further than the "least worst" option presented with PRx based targeted therapy and may finally provide us with a substantial tool in the fight against secondary brain injury (Froese, Batson, Gomez, Dian, & Zeiler, 2020).

Application to other areas of neuroscience

In this chapter, we explored the prognostic utility of the pressure reactivity index (PRx) as a surrogate marker of cerebral autoregulation (CA) in traumatic brain injury (TBI) patients. We also discussed the possibility of future measure of CA that leverage non-invasive technologies such as transcranial Doppler (TCD) and near-infrared spectroscopy (NIRS) based indices. The utility of non-invasive indices of CA has early evidence in the world of aneurysmal subarachnoid hemorrhage. In a study of 98 patients with aneurysmal subarachnoid hemorrhage (SAH) Budohoski et al. demonstrated that disrupted CA within the first 5 days post injury, as determined by TCD and NIRS based indices of CA, significantly increased the risk of delayed cerebral vasospasm (Budohoski, Czosnyka, Smielewski, et al., 2012). This study indicated that determination of CA can both be determined non-invasively and can have prognostic utility outside of the world of TBI, as such, a significant body of literature has developed to determine the link between CA and SAH (Budohoski et al., 2013).

Mini-dictionary of terms

AMP: The pulse amplitude of intracranial pressure (ICP) which is the amplitude of the ICP wave.

Cerebral autoregulation (CA): The ability of the cerebral vasculature to maintain cerebral blood flow (CBF) over a range of cerebral perfusion pressure (CPPs).

Cerebrovascular reactivity: The ability of the cerebrovascular tree to contract and dilate in response to changes in CPP and the metabolic environment.

Lower limit of autoregulation (LLA): This is the CPP below which cerebral autoregulatory mechanisms fail to maintain a constant CBF. Below this level cerebral ischemia is present.

Mx: The continuously updated correlation coefficient between mean flow velocities through the middle cerebral artery as measured with transcranial Doppler and CPP. It acts as semi-continuous index of CA.

PAx: The continuously updated correlation coefficient between the AMP and mean arterial pressure (MAP). It acts as continuous index of cerebrovascular reactivity and can act as a surrogate for CA.

PRx: The continuously updated correlation coefficient between ICP and MAP. It acts as continuous index of cerebrovascular reactivity and can act as a surrogate for CA.

RAC: The continuously updated correlation coefficient between the AMP and CPP. It acts as continuous index of cerebrovascular reactivity and can act as a surrogate for CA.

Upper limit of autoregulation (ULA): This is the CPP above which cerebral autoregulatory mechanisms fail to maintain a constant CBF. Above this level cerebral hyperemia is present.

Key facts of cerebrovascular reactivity in TBI

- Following moderate/severe traumatic brain injury cerebral autoregulation can often be disturbed.
- PRx is a normalized metric of cerebrovascular reactivity ranging from −1 (completely intact cerebrovascular reactivity) to +1 (completely disrupted cerebrovascular reactivity).

- PRx, while being a method of evaluating cerebrovascular reactivity, has been shown to accurately detect the lower limit of autoregulation in animal models.
- Impaired cerebrovascular reactivity, as measured by PRx, has been shown to be associated with outcomes following TBI.
- Different thresholds for mortality and poor outcome have been identified for PRx with mortality having a higher PRx threshold than that for poor outcome.

Summary points

- In this chapter we explored cerebral autoregulation and cerebrovascular reactivity following traumatic brain injury (TBI).
- Cerebral autoregulation is the ability of the brains vasculature to maintain a constant blood flow over a range of cerebral perfusion pressures.
- Following TBI there is significant evidence that CA is disrupted.
- The pressure reactivity index (PRx) was developed as a measure of cerebrovascular reactivity and has been shown to be a good metric of cerebral autoregulation following TBI.
- PRx has been found to be associated with global outcomes, imaging findings and metabolic parameters following TBI.
- There is strong evidence that clinically relevant PRx thresholds exist for mortality and poor outcome.
- While there is a large body of literature supporting the prognostic utility of PRx, data from large multi-center studies is needed to validate these claims.
- In the future, non-invasive means of continuously evaluating cerebral autoregulation following TBI may provide safe and meaningful clinical data.

Acknowledgments

F.A.Z. receives research support from the Manitoba Public Insurance (MPI) Neuroscience/TBI Research Endowment, the Health Sciences Centre Foundation Winnipeg, the United States National Institutes of Health (NIH) through the National Institute of Neurological Disorders and Stroke (NINDS), the Canadian Institutes of Health Research (CIHR), the Canada Foundation for Innovation (CFI), Research Manitoba, the University of Manitoba VPRI Research Investment Fund (RIF), the University of Manitoba Centre on Aging, and the University of Manitoba Rudy Falk Clinician-Scientist Professorship.

L.F. is supported through the University of Manitoba—Department of Surgery GFT Research Grant, and the University of Manitoba Office of Research Services (ORS)—University Research Grant Program (URGP).

C.B. is supported through the Centre on Aging Fellowship at the University of Manitoba.

A.G. is supported through the University of Manitoba Clinician Investigator Program.

References

Aaslid, R., Lindegaard, K. F., Sorteberg, W., & Nornes, H. (1989). Cerebral autoregulation dynamics in humans. *Stroke*, *20*(1), 45–52. https://doi.org/10.1161/01.STR.20.1.45.

Adams, H., Donnelly, J., Czosnyka, M., Kolias, A. G., Helmy, A., Menon, D. K., … Hutchinson, P.J. (2017). Temporal profile of intracranial pressure and cerebrovascular reactivity in severe traumatic brain injury and association with fatal outcome: An observational study. *PLoS Medicine*, *14*(7). https://doi.org/10.1371/journal.pmed.1002353, e1002353.

Ang, B. T., Wong, J., Lee, K. K., Wang, E., & Ng, I. (2006). Temporal changes in cerebral tissue oxygenation with cerebrovascular pressure reactivity in severe traumatic brain injury. *Journal of Neurology, Neurosurgery & Psychiatry*, *78*(3), 298–302. https://doi.org/10.1136/jnnp.2005.082735.

Aries, M. J. H., Czosnyka, M., Budohoski, K. P., Kolias, A. G., Radolovich, D. K., Lavinio, A., … Smielewski, P. (2012). Continuous monitoring of cerebrovascular reactivity using pulse waveform of intracranial pressure. *Neurocritical Care*, *17*(1), 67–76. https://doi.org/10.1007/s12028-012-9687-z.

Armstead, W. M. (2016). Cerebral blood flow autoregulation and dysautoregulation. *Anesthesiology Clinics*, *34*(3), 465–477. https://doi.org/10.1016/j.anclin.2016.04.002.

Bernard, F., Gallagher, C., Griesdale, D., Kramer, A., Sekhon, M., & Zeiler, F. A. (2020). The CAnadian High-Resolution Traumatic Brain Injury (CAHR-TBI) research collaborative. *The Canadian Journal of Neurological Sciences. Le Journal Canadien des Sciences Neurologiques*, *47*(4), 551–556. https://doi.org/10.1017/cjn.2020.54.

Brady, K. M., Lee, J. K., Kibler, K. K., Easley, R. B., Koehler, R. C., & Shaffner, D. H. (2008). Continuous measurement of autoregulation by spontaneous fluctuations in cerebral perfusion pressure: Comparison of 3 methods. *Stroke*, *39*(9), 2531–2537. https://doi.org/10.1161/STROKEAHA.108.514877.

Budohoski, K. P., Czosnyka, M., de Riva, N., Smielewski, P., Pickard, J. D., Menon, D. K., … Lavinio, A. (2012). The relationship between cerebral blood flow autoregulation and cerebrovascular pressure reactivity after traumatic brain injury. *Neurosurgery*, *71*(3), 652–661. https://doi.org/10.1227/NEU.0b013e318260feb1.

Budohoski, K. P., Czosnyka, M., Kirkpatrick, P. J., Smielewski, P., Steiner, L. A., & Pickard, J. D. (2013). Clinical relevance of cerebral autoregulation following subarachnoid haemorrhage. *Nature Reviews Neurology*, *9*(3), 152–163. https://doi.org/10.1038/nrneurol.2013.11.

Budohoski, K. P., Czosnyka, M., Smielewski, P., Kasprowicz, M., Helmy, A., Bulters, D., … Kirkpatrick, P.J. (2012). Impairment of cerebral autoregulation predicts delayed cerebral ischemia after subarachnoid hemorrhage: A prospective observational study. *Stroke*, *43*(12), 3230–3237. https://doi.org/10.1161/STROKEAHA.112.669788.

Cold, G. E., & Jensen, F. T. (1978). Cerebral autoregulation in unconscious patients with brain injury. *Acta Anaesthesiologica Scandinavica*, *22*(3), 270–280. https://doi.org/10.1111/j.1399-6576.1978.tb01301.x.

Czosnyka, M., Aries, M., Weersink, C., Wolf, S., Budohoski, K., Dias, C., … Kordasti, S. (2016). "Solid red line": An observational study on death from refractory intracranial hypertension. *Acta Neurochirurgica. Supplement*, *122*, 113–116. https://doi.org/10.1007/978-3-319-22533-3_22.

Czosnyka, M., Czosnyka, Z., & Smielewski, P. (2017). Pressure reactivity index: Journey through the past 20 years. *Acta Neurochirurgica*, *159*(11), 2063–2065. https://doi.org/10.1007/s00701-017-3310-1.

Czosnyka, M., Smielewski, P., Kirkpatrick, P., Laing, R. J., Menon, D., & Pickard, J. D. (1997). Continuous assessment of the cerebral vasomotor reactivity in head injury. *Neurosurgery*, *41*(1), 11–19. https://doi.org/10.1097/00006123-199707000-00005.

Czosnyka, M., Smielewski, P., Kirkpatrick, P., Menon, D. K., & Pickard, J. D. (1996). Monitoring of cerebral autoregulation in head-injured patients. *Stroke*, *27*(10), 1829–1834. https://doi.org/10.1161/01.STR.27.10.1829.

Dias, C., Silva, M. J., Pereira, E., Monteiro, E., Maia, I., Barbosa, S., … Czosnyka, M. (2015). Optimal cerebral perfusion pressure management at bedside: A single-center pilot study. *Neurocritical Care*, *23*(1), 92–102. https://doi.org/10.1007/s12028-014-0103-8.

Donnelly, J., Czosnyka, M., Adams, H., Cardim, D., Kolias, A. G., Zeiler, F. A., … Budohoski, K.P. (2019). Twenty-five years of intracranial pressure monitoring after severe traumatic brain injury: A retrospective, single-center analysis. *Neurosurgery*, *85*(1), E75–E82. https://doi.org/10.1093/neuros/nyy468.

Eide, P. K., Czosnyka, M., Sorteberg, W., Pickard, J. D., & Smielewski, P. (2007). Association between intracranial, arterial pulse pressure amplitudes and cerebral autoregulation in head injury patients. *Neurological Research*, *29*(6), 578–582. https://doi.org/10.1179/016164107X172167.

Fog, M. (1938). The relationship between the blood pressure and the tonic regulation of the PIal arteries. *Journal of Neurology and Psychiatry*, *1*(3), 187–197. https://doi.org/10.1136/jnnp.1.3.187.

Froese, L., Batson, C., Gomez, A., Dian, J., & Zeiler, F. A. (2020). The limited impact of current therapeutic interventions on cerebrovascular reactivity in traumatic brain injury: A narrative overview. *Neurocritical Care*. https://doi.org/10.1007/s12028-020-01003-4.

Gomez, A., Dian, J., Froese, L., & Zeiler, F. A. (2020). Near-infrared based cerebrovascular reactivity as a means of monitoring cerebral autoregulation and predicting outcome in moderate/severe traumatic brain injury: A pilot study protocol and planned analyses (preprint). *JMIR Research Protocols*. https://doi.org/10.2196/18740.

Gomez, A., Dian, J., & Zeiler, F. A. (2020). Continuous and entirely non-invasive method for cerebrovascular reactivity assessment: Technique and implications. *Journal of Clinical Monitoring and Computing*. https://doi.org/10.1007/s10877-020-00472-4.

Güiza, F., Depreitere, B., Piper, I., Citerio, G., Chambers, I., Jones, P. A., … Meyfroidt, G. (2015). Visualizing the pressure and time burden of intracranial hypertension in adult and paediatric traumatic brain injury. *Intensive Care Medicine*, *41*(6), 1067–1076. https://doi.org/10.1007/s00134-015-3806-1.

Güiza, F., Meyfroidt, G., Piper, I., Citerio, G., Chambers, I., Enblad, P., … Depreitere, B. (2017). Cerebral perfusion pressure insults and associations with outcome in adult traumatic brain injury. *Journal of Neurotrauma*, *34*(16), 2425–2431. https://doi.org/10.1089/neu.2016.4807.

Hiler, M., Czosnyka, M., Hutchinson, P., Balestreri, M., Smielewski, P., Matta, B., & Pickard, J. D. (2006). Predictive value of initial computerized tomography scan, intracranial pressure, and state of autoregulation in patients with traumatic brain injury. *Journal of Neurosurgery*, *104*(5), 731–737. https://doi.org/10.3171/jns.2006.104.5.731.

Howells, T., Johnson, U., McKelvey, T., & Enblad, P. (2015). An optimal frequency range for assessing the pressure reactivity index in patients with traumatic brain injury. *Journal of Clinical Monitoring and Computing*, *29*(1), 97–105. https://doi.org/10.1007/s10877-014-9573-7.

Jaeger, M., Schuhmann, M. U., Soehle, M., & Meixensberger, J. (2006). Continuous assessment of cerebrovascular autoregulation after traumatic brain injury using brain tissue oxygen pressure reactivity. *Critical Care Medicine*, *34*(6), 1783–1788. https://doi.org/10.1097/01.CCM.0000218413.51546.9E.

Kirkness, C. J., Mitchell, P. H., Burr, R. L., & Newell, D. W. (2001). Cerebral autoregulation and outcome in acute brain injury. *Biological Research for Nursing*, *2*(3), 175–185. https://doi.org/10.1177/109980040100200303.

Lassen, N. A. (1959). Cerebral blood flow and oxygen consumption in man. *Physiological Reviews*, *39*(2), 183–238. https://doi.org/10.1152/physrev.1959.39.2.183.

Mathieu, F., Zeiler, F. A., Ercole, A., Monteiro, M., Kamnitsas, K., Glocker, B., … Younsi, A. (2020). Relationship between measures of cerebrovascular reactivity and intracranial lesion progression in acute traumatic brain injury patients: A CENTER-TBI study. *Journal of Neurotrauma*, *37*(13), 1556–1565. https://doi.org/10.1089/neu.2019.6814.

Mathieu, F., Zeiler, F. A., Whitehouse, D. P., Das, T., Ercole, A., Smielewski, P., … Menon, D.K. (2020). Relationship between measures of cerebrovascular reactivity and intracranial lesion progression in acute TBI patients: An exploratory analysis. *Neurocritical Care*, *32*(2), 373–382. https://doi.org/10.1007/s12028-019-00885-3.

Nelson, R. J., Czosnyka, M., Pickard, J. D., Maksymowicz, W., Perry, S., & Lovick, A. H. J. (1992). Experimental aspects of cerebrospinal hemodynamics: The relationship between blood flow velocity waveform and cerebral autoregulation. *Neurosurgery*, *31*(4), 705–710. https://doi.org/10.1227/00006123-199210000-00013.

Overgaard, J., & Tweed, W. A. (1974). Cerebral circulation after head injury: Part 1: Cerebral blood flow and its regulation after closed head injury with emphasis on clinical correlations. *Journal of Neurosurgery*, *41*(5), 531–541. https://doi.org/10.3171/jns.1974.41.5.0531.

Petkus, V., Krakauskaitė, S., Preikšaitis, A., Ročka, S., Chomskis, R., & Ragauskas, A. (2016). Association between the outcome of traumatic brain injury patients and cerebrovascular autoregulation, cerebral perfusion pressure, age, and injury grades. *Medicina*, *52*(1), 46–53. https://doi.org/10.1016/j.medici.2016.01.004.

Schmidt, B., Reinhard, M., Lezaic, V., McLeod, D. D., Weinhold, M., Mattes, H., & Klingelhöfer, J. (2016). Autoregulation monitoring and outcome prediction in neurocritical care patients: Does one index fit all? *Journal of Clinical Monitoring and Computing*, *30*(3), 367–375. https://doi.org/10.1007/s10877-015-9726-3.

Sorrentino, E., Diedler, J., Kasprowicz, M., Budohoski, K. P., Haubrich, C., Smielewski, P., … Czosnyka, M. (2012). Critical thresholds for cerebrovascular reactivity after traumatic brain injury. *Neurocritical Care*, *16*(2), 258–266. https://doi.org/10.1007/s12028-011-9630-8.

Steiner, L. A. (2003). Cerebrovascular pressure reactivity is related to global cerebral oxygen metabolism after head injury. *Journal of Neurology, Neurosurgery & Psychiatry*, *74*(6), 765–770. https://doi.org/10.1136/jnnp.74.6.765.

Steiner, L. A., Coles, J. P., Johnston, A. J., Chatfield, D. A., Smielewski, P., Fryer, T. D., … Czosnyka, M. (2003). Assessment of cerebrovascular autoregulation in head-injured patients: A validation study. *Stroke*, *34*(10), 2404–2409. https://doi.org/10.1161/01.STR.0000089014.59668.04.

Steiner, L. A., Czosnyka, M., Piechnik, S. K., Smielewski, P., Chatfield, D., Menon, D. K., & Pickard, J. D. (2002). Continuous monitoring of cerebrovascular pressure reactivity allows determination of optimal cerebral perfusion pressure in patients with traumatic brain injury. *Critical Care Medicine*, *30*(4), 733–738. https://doi.org/10.1097/00003246-200204000-00002.

Steinmeier, R., Hofmann, R. P., Bauhuf, C., Hübner, U., & Fahlbusch, R. (2002). Continuous cerebral autoregulation monitoring by cross-correlation analysis. *Journal of Neurotrauma*, *19*(10), 1127–1138. https://doi.org/10.1089/08977150260337949.

Svedung Wettervik, T., Howells, T., Enblad, P., & Lewén, A. (2019). Temporal neurophysiological dynamics in traumatic brain injury: Role of pressure reactivity and optimal cerebral perfusion pressure for predicting outcome. *Journal of Neurotrauma*, *36*(11), 1818–1827. https://doi.org/10.1089/neu.2018.6157.

Tiecks, F. P., Lam, A. M., Aaslid, R., & Newell, D. W. (1995). Comparison of static and dynamic cerebral autoregulation measurements. *Stroke*, *26*(6), 1014–1019. https://doi.org/10.1161/01.STR.26.6.1014.

Timofeev, I., Czosnyka, M., Nortje, J., Smielewski, P., Kirkpatrick, P., Gupta, A., & Hutchinson, P. (2008). Effect of decompressive craniectomy on intracranial pressure and cerebrospinal compensation following traumatic brain injury. *Journal of Neurosurgery*, *108*(1), 66–73. https://doi.org/10.3171/JNS/2008/108/01/0066.

Zeiler, F. A., Aries, M., Cabeleira, M., van Essen, T. A., Stocchetti, N., Menon, D. K., … CENTER-TBI High Resolution ICU (HR ICU) Sub-Study Participants and Investigators. (2020). Statistical cerebrovascular reactivity signal properties after secondary decompressive craniectomy in traumatic brain injury: A CENTER-TBI pilot analysis. *Journal of Neurotrauma*, *37*(11), 1306–1314. https://doi.org/10.1089/neu.2019.6726.

Zeiler, F. A., Beqiri, E., Cabeleira, M., Hutchinson, P. J., Stocchetti, N., Menon, D. K., … Collaborative European NeuroTrauma Effectiveness Research in Traumatic Brain Injury (CENTER-TBI) High-Resolution ICU (HR ICU) Sub-Study Participants and Investigators. (2020). Brain tissue oxygen and cerebrovascular reactivity in traumatic brain injury: A collaborative european neurotrauma effectiveness research in traumatic brain injury exploratory analysis of insult burden. *Journal of Neurotrauma*. https://doi.org/10.1089/neu.2020.7024.

Zeiler, F. A., Czosnyka, M., & Smielewski, P. (2018). Optimal cerebral perfusion pressure via transcranial Doppler in TBI: Application of robotic technology. *Acta Neurochirurgica*, *160*(11), 2149–2157. https://doi.org/10.1007/s00701-018-3687-5.

Zeiler, F. A., Donnelly, J., Calviello, L., Lee, J. K., Smielewski, P., Brady, K., … Czosnyka, M. (2018). Validation of pressure reactivity and pulse amplitude indices against the lower limit of autoregulation, part I: Experimental intracranial hypertension. *Journal of Neurotrauma*, *35*(23), 2803–2811. https://doi.org/10.1089/neu.2017.5603.

Zeiler, F. A., Donnelly, J., Nourallah, B., Thelin, E. P., Calviello, L., Smielewski, P., … Menon, D.K. (2018). Intracranial and extracranial Injury burden as drivers of impaired cerebrovascular reactivity in traumatic brain injury. *Journal of Neurotrauma*, *35*(14), 1569–1577. https://doi.org/10.1089/neu.2017.5595.

Zeiler, F. A., Donnelly, J., Smielewski, P., Menon, D. K., Hutchinson, P. J., & Czosnyka, M. (2018). Critical thresholds of intracranial pressure-derived continuous cerebrovascular reactivity indices for outcome prediction in noncraniectomized patients with traumatic brain injury. *Journal of Neurotrauma*, *35*(10), 1107–1115. https://doi.org/10.1089/neu.2017.5472.

Zeiler, F. A., Ercole, A., Cabeleira, M., Zoerle, T., Stocchetti, N., Menon, D. K., … Czosnyka, M. (2019). Univariate comparison of performance of different cerebrovascular reactivity indices for outcome association in adult TBI: A CENTER-TBI study. *Acta Neurochirurgica*, *161*(6), 1217–1227. https://doi.org/10.1007/s00701-019-03844-1.

Zeiler, F. A., Lee, J. K., Smielewski, P., Czosnyka, M., & Brady, K. (2018). Validation of intracranial pressure-derived cerebrovascular reactivity indices against the lower limit of autoregulation, part II: Experimental model of arterial hypotension. *Journal of Neurotrauma*, *35*(23), 2812–2819. https://doi.org/10.1089/neu.2017.5604.

Zeiler, F. A., Mathieu, F., Monteiro, M., Glocker, B., Ercole, A., Beqiri, E., … CENTER-TBI High-Resolution ICU (HR ICU) Sub-Study Participants and Investigators, the CENTER-TBI High-Resolution ICU (HR ICU) Sub-Study Participants and Investigators. (2020). Diffuse intracranial Injury patterns are associated with impaired cerebrovascular reactivity in adult traumatic brain injury: A CENTER-TBI validation study. *Journal of Neurotrauma*, *37*(14), 1597–1608. https://doi.org/10.1089/neu.2019.6959.

Zeiler, F. A., & Smielewski, P. (2018). Application of robotic transcranial Doppler for extended duration recording in moderate/severe traumatic brain injury: First experiences. *Critical Ultrasound Journal*, *10*(1), 16. https://doi.org/10.1186/s13089-018-0097-0.

Zeiler, F. A., Smielewski, P., Stevens, A., Czosnyka, M., Menon, D. K., & Ercole, A. (2019). Non-invasive pressure reactivity index using Doppler systolic flow parameters: A pilot analysis. *Journal of Neurotrauma*, *36*(5), 713–720. https://doi.org/10.1089/neu.2018.5987.

Zweifel, C., Lavinio, A., Steiner, L. A., Radolovich, D., Smielewski, P., Timofeev, I., … Czosnyka, M. (2008). Continuous monitoring of cerebrovascular pressure reactivity in patients with head injury. *Neurosurgical Focus*, *25*(4), E2. https://doi.org/10.3171/FOC.2008.25.10.E2.

Chapter 14

Cerebral perfusion pressure thresholds in traumatic brain injury

Alwyn Gomez[a,b], Carleen Batson[b], Logan Froese[c], and Frederick A. Zeiler[a,b,c,d,e]

[a]*Section of Neurosurgery, Department of Surgery, Rady Faculty of Health Sciences, University of Manitoba, Winnipeg, MB, Canada,* [b]*Department of Human Anatomy and Cell Science, Rady Faculty of Health Sciences, University of Manitoba, Winnipeg, MB, Canada,* [c]*Biomedical Engineering, Faculty of Engineering, University of Manitoba, Winnipeg, MB, Canada,* [d]*Centre on Aging, University of Manitoba, Winnipeg, MB, Canada,* [e]*Division of Anaesthesia, Department of Medicine, Addenbrooke's Hospital, University of Cambridge, Cambridge, United Kingdom*

List of abbreviations

ABP	arterial blood pressure
AR	absolute risk
ARDS	acute respiratory distress syndrome
AUC	area under the (receiver/operator) curve
BTF	brain trauma foundation
CA	cerebral autoregulation
CBF	cerebral blood flow
CENTER-TBI	collaborative European neurotrauma effectiveness research in traumatic brain injury
CI	confidence interval
COGiTATE	optimal cerebral perfusion pressure guided therapy: assessment of target effectiveness
CPP	cerebral perfusion pressure
CPPopt	optimal cerebral perfusion pressure
CSF	cerebrospinal fluid
GCS	glasgow coma scale
GOS	glasgow outcome scale
ICA	internal carotid artery
ICP	intracranial pressure
ICU	intensive care unit
IVH	intra-ventricular hemorrhage
LLA	lower limit of autoregulation
LLR	lower limit of reactivity
MAP	mean arterial pressure
mmHg	millimeters of mercury
NIRS	near infrared spectroscopy
OR	odds ratio
PRx	pressure reactivity index
SCI	spinal cord injury
SIBICC	Seattle International Severe Traumatic Brain Injury Consensus Conference
TBI	traumatic brain injury
TCD	transcranial Doppler
TCDB	traumatic coma data bank
ULA	upper limit of autoregulation
ULR	upper limit of reactivity
χ^2	chi-square

Diagnosis and Treatment of Traumatic Brain Injury. https://doi.org/10.1016/B978-0-12-823347-4.00019-1

Introduction

The principle aim in the management of patients with traumatic brain injury (TBI) is to minimize secondary injury to the brain. As no therapeutic agents have been shown to reliably achieve this in humans, critical care management of these patients primarily focuses on ensuring adequate blood flow and oxygenation of the traumatically injured brain. The brain's limited ability to store metabolic substrates leaves it particularly vulnerable to ischemic damage following TBI.

In this chapter, we will explore the evidence guiding current targets in cerebral perfusion pressure (CPP) and examine the expanding body of literature that will likely guide CPP management in the future. We start by giving the reader an in-depth explanation of CPP and then go on to explore how TBI can impact CPP. We then review the literature that forms current recommendations. In particular, we will discuss the evidence supporting the current clinical guidelines. We will then explore the concept of personalized CPP targets and the data supporting this management paradigm. Finally, we will present some areas for future development in CPP management in TBI.

Defining cerebral perfusion pressure

The concept of perfusion pressure is not one that is unique to cerebral physiology. Perfusion pressure, as defined in other organs as the arithmetic difference between the pressure of the arterial inflow to that organ and the pressure of the venous outflow from that organ. In the setting of the brain, this would be equivalent to the pressure in the internal carotid artery (ICA) less the jugular venous pressure. However, this is not how CPP is clinically derived (Wagner & Traystman, 1986).

In clinical settings, necessitating the monitoring of CPP, the previously described method of deriving a perfusion pressure is both impractical and inaccurate. Accurately measuring pressure in the ICA is overly invasive and comes with a significant risk of stroke. As a result, mean arterial pressure (MAP) as derived from a radial or femoral line is utilized as a surrogate. To correct for the difference in hydrostatic pressure the transducer is often zeroed at the level of the tragus. Additionally, in the setting of brain injury, intracranial pressure (ICP) is usually higher than jugular venous pressure and therefore plays a much more significant role in opposing cerebral perfusion. This leads us to the clinically useful formula for CPP as the difference between MAP and ICP typically in units of millimeters of Mercury (mmHg) (Wagner & Traystman, 1986).

Cerebral autoregulation

Cerebral autoregulation (CA) refers to the ability of the cerebral vasculature to adapt in order to maintain a constant cerebral blood flow (CBF) over a wide range of MAPs or CPPs. While it was first described by Fog et al. during his examination of CBF in felines, Lassen et al. is often credited with describing the process in humans (Fog, 1938; Lassen, 1959). At CPPs below the lower limit of autoregulation (LLA) autoregulatory mechanisms are unable to maintain adequate CBF and ischemia is present. At CPPs above the upper limit of autoregulation (ULA), autoregulatory mechanisms are overwhelmed and hyperemia occurs. However, at CPPs between the LLA and ULA, CBF is adequate and relatively stable (Fig. 1). This concept is necessary to consider when determining thresholds for CPP in TBI patients as it implies that, so long as some

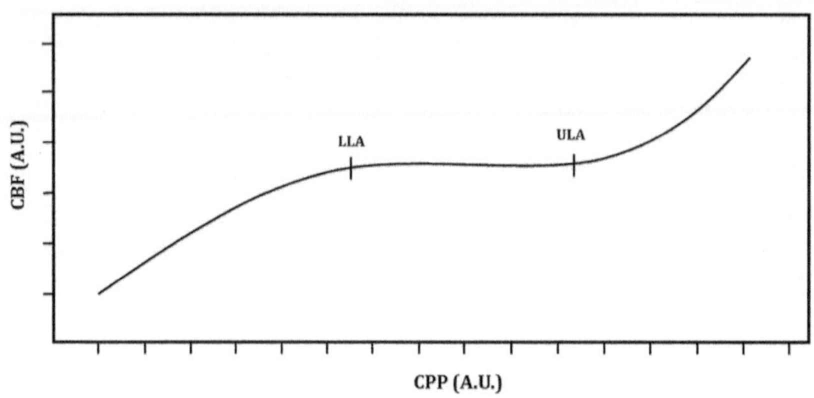

FIG. 1 Lassen curve. Cerebral blood flow (CBF) in arbitrary units (AU) is plotted against cerebral perfusion pressure (CPP) to demonstrate cerebral autoregulation. In the plateau region between the lower limit of autoregulation (LLA) and the upper limit of autoregulation (ULA) CBF is constant as CPP changes. As CPP falls below the LLA, CBF decreases and as CPP increases above the ULA, CBF increases.

autoregulatory mechanisms are functioning, CPP may be sufficient over a range of pressures as opposed to just at a single value. The concept of cerebral autoregulation is more fully discussed in the previous chapter.

CBF variability in TBI

In the setting of TBI, there are three main mechanisms by which CBF can fall resulting in brain ischemia and worsening secondary injury. The first two are mediated by CPP and the third mediated by CA.

During any significant trauma, a patient may become hemodynamically unstable. This is most commonly attributable to hemodynamic shock, however neurogenic and cardiac shock can also be present (ATLS Subcommittee et al., 2013). The net effect of this is a global reduction in MAP which can result in a decrease of CPP to below the LLA, leading to a reduction of CBF and eventual brain ischemia. Manley et al. reported on the prognostic implication of hypotension in TBI. In their study of 170 patients hypotension had a higher association with mortality than hypoxia, age or Glasgow coma scale (GCS) (OR = 3.39, 95% CI 1.34–8.56, $p = 0.009$; Manley et al., 2001).

In the setting of acute TBI, ICP is often increased. This can be as a result of a mass lesion, such as an epidural hematoma or subdural hematoma, or as a result of general brain swelling and edema. While there are compensatory mechanisms that function to normalize ICP in this setting, they can often be overwhelmed in the setting of severe TBI (Kellie, 1824; Monro, 1783). As ICP increases, CPP decreases so long as MAP is constant and eventually falls below the level of LLA. This results in a reduction of CBF, followed by ischemia and further brain swelling. Death can eventually result from brainstem infarction or secondarily through a herniation syndrome. The prognostic implications of increased ICP in TBI are well known. In a study of 459 TBI patients found that an ICP threshold of 22 mmHg had the best predictive value for mortality ($\chi^2 = 58.8$, $p < 0.001$) and favorable outcome ($\chi^2 = 18.15$, $p < 0.001$; Sorrentino et al., 2012). The details of ICP management in TBI are discussed in more detail elsewhere in this textbook.

The final mechanism by which CBF can be jeopardized following TBI is by disruption of CA. It has been known for some time now that following TBI, CA can be disrupted resulting in what was previously referred to as "vasoparalysis" (Cold & Jensen, 1978; Overgaard & Tweed, 1974). Conceptually, this can be thought of as an increase in the LLA and a decrease in ULA resulting in a narrow, and sometimes non-existent, range of CPP at which CBF is stable. The prognostic implications of disrupted CA in the setting of TBI is beyond the scope of this chapter (see Chapter 13).

CPP targets in TBI

Changaris et al. were one of the first to report on the implications of CPP post-injury on long-term outcomes. In their retrospective analysis of 136 severe TBI patients, they noted that those with a CPP ≤ 60 mmHg for more than a third of the hourly measurements on post-admission day, two had a significantly increased mortality (AR = 0.44, 95% CI 0.30–0.55, $p < 0.05$) while those with a CPP ≥ 80 mmHg had an increased likelihood of a favorable GOS (AR = 0.57, 95% CI 0.26–0.74, $p < 0.05$; Changaris et al., 1987). A similar early retrospective study out of the University of Louisville of 221 TBI patients found a CPP of ≥ 80 as a breaking point for reduced mortality and improved outcomes (odds ratios not reported; McGraw, 1989). These early studies helped identify CPP as a clinically useful parameter following TBI and helped identify a range of interest where outcomes seemed to vary.

In the era preceding guideline-based management of TBI, therapeutic modulation of CPP generally fell into one of two camps. The approach of Rosner et al. focused mainly on elevating CPP and placed less emphasis on solely ICP directed therapy. CPP was targeted to be at least 70 mmHg and utilized hyperosmolar agents and cerebrospinal fluid (CSF) drainage to reduce ICP while vasoactive agents were utilized to augment ICP. In their 1995 study of 158 TBI patients, they found that morbidity and mortality improved with their techniques compared to those reported in the Traumatic Coma Data Bank (TCDB) data (Mantel-Haenszel chi-square = 16.26, $p < 0.0001$; Rosner, Rosner, & Johnson, 1995). The excitement surrounding these results was somewhat dampened by a randomized control trial by Robertson et al. dividing 189 TBI patients into two treatment arms, one ICP directed and one CPP directed. The CPP goals in each arm were > 50 mmHg in the ICP arm, and > 70 mmHg in the CPP arm. No difference in neurologic outcomes were found at 6-month follow-up; however, the CPP group had more systemic complications with a five times greater incidence of acute respiratory distress syndrome (ARDS) ($p = 0.007$; Robertson et al., 1999).

The alternative Lund approach came from research out of the University of Lund in Sweden. This protocol focused on reducing vasogenic edema in the brain following TBI. This study analyzed just 10 patients with severe TBI and tolerated CPPs as low at 50 mmHg. At 8 months post injury, 9 members of their cohort survived and 6 had good outcomes (Asgeirsson, Grände, & Nordström, 1994). This work along with a rising concern for the systemic consequences of prolonged elevated CPP opened the door to more clinical analysis of CPP directed therapy following TBI.

Global CPP targeted therapy

A significant body of research had been produced in an attempt to identify global CPP parameters in the era before personalized TBI care. Juul et al. found in their study of 427 TBI patients that those with an initial CPP of less than 60 mmHg was somewhat predictive of later neurologic deterioration (RR = 1.181, 95% CI 0.970–1.437, $p > 0.05$) but initial CPPs greater than 60 mmHg seemed to have little influence on outcome (Juul, Morris, Marshall, & Marshall, 2000). Similarly, Clifton et al. reported that in a retrospective review of 393 patients with TBI a single instance of CPP < 60 mmHg in the first 92 h post-injury increased the odds of a poor outcome (OR = 0.86, 95% CI 0.53–1.41, $p = 0.005$). A similar effect was not seen for a CPP threshold of 70 mmHg (Clifton, Miller, Choi, & Levin, 2002). These seemed to point to a low water mark for CPP of 60 mmHg without increased benefit of elevated CPP. This should, however, be tempered by the fact that these were retrospective studies and only point values of CPP were measured. Additionally, no actual management of CPP was reported.

Prospective data on 81 TBI patients reported by Elf et al. contradicted the idea of a CPP of 60 mmHg as a low water mark. They found that in their cohort, using multi-variable regression including pertinent admission and insult variables, that increased duration of time with a CPP < 60 mmHg increased the odds of a favorable outcome at 6 months (OR = 1.59, 95% CI 1.06–2.39, $p < 0.05$). Interestingly, univariant regression also found time with CPP > 70 mmHg and CPP > 80 mmHg to be associated with decreased odds of a favorable outcome (OR = 0.71, 95% CI 0.51–0.99, $p < 0.05$ OR = 0.69, 95% CI 0.49–0.98, $p < 0.05$, respectively; Elf, Nilsson, Ronne-Engström, Howells, & Enblad, 2005). These results were more in keeping with the Lund approach focused primarily on ICP goals and cautioned against elevated CPPs.

A retrospective cohort study of 213 TBI patients was reported on by Huang et al. that compared outcomes in patients treated with a CPP target of > 70 mmHg to those treated with a target of 60 mmHg. They found that there was no difference in mortality rate (14.3% vs 13.5%, $p = 0.55$) or frequency of unfavorable outcome (22.1% vs 17.3%, $p = 0.38$). They also noted that the rate of pulmonary complications was twice as high in the CPP > 70 mmHg cohort than the CPP > 60 mmHg ($p = 0.02$, no rates given; Huang et al., 2006). This study provided evidence that a CPP guided therapy at a low threshold provided similar results to a high threshold. Unfortunately, no significant data were provided as to the adherence of these thresholds in their cohort.

In an analysis of their 305 severe TBI patients, retrospective cohort study Lin et al. found that those with CPP maintained above 60 mmHg had similar mortality to those that maintained a CPP above 70 mmHg. Subgroup analysis found that when the patient had a more severe injury (GCS = 3–5) maintaining a CPP > 70 mmHg produced a good outcome more frequently than when only maintained above 60 mmHg (25% vs 20%, $p < 0.05$; Lin et al., 2008). This study indicated that the degree of injury may play a role selecting an appropriate target for CPP, however, it failed to provide adequate detail about the frequency of CPP measurements and the time spent at each CPP threshold. Similarly, in a retrospective study Sorrentino et al. found that in their cohort of 459 TBI patients, a threshold CPP of 70 mmHg was best suited to discriminate patients with regards to mortality and favorable outcome ($\chi^2 = 24.78$, $p < 0.001$ and $\chi^2 = 8.79$, $p = 0.003$, respectively; Sorrentino et al., 2012).

CPP-related thresholds below 60 mmHg have been explored. In a retrospective cohort of 1757 severe TBI patients, the effect of time at various CPP thresholds on survival at 14 days post injury was examined. At 14 days post-injury those with maintained CPP > 60 mmHg and those with episodes of CPP falling to between 50 mmHg and 60 mmHg had similar survival rates (84.0% vs 83.6%, respectively, $p = 0.81$). However, those with events of CPP that fell below 50 mmHg had a significantly reduced rate of survival (62.3%, RR = 2.35, 95% CI 1.88–2.95, $p < 0.0001$; Allen, Chiu, Gerber, Ghajar, & Greenfield, 2014). This large retrospective study seemed to indicate that a threshold of CPP > 50 mmHg would substantially increase survival. There are many unanswered questions as a result of the limitations of this study. Selecting a study endpoint of mortality at 14 days post-injury likely allowed for a much larger cohort but significantly reduced the clinical relevance of the study. There is no data presented on long-term survival and more importantly functional outcomes 6 or 12 months post-injury. While 14-day survival was not affected by episodes of CPP between 50 and 60 mmHg, the effect on long-term functional outcomes is not known. There is physiologic data reported by Chang et al. on 27 severe TBI patients that found CPP levels below 60 mmHg were associated with cerebral hypoxia as determined by intra-parenchymal brain tissue oxygen monitoring (RR = 3.01, 95% CI 2.51–3.61, $p < 0.0001$; Chang et al., 2009).

The literature surrounding global thresholds for CPP in TBI is inconsistent and generally of low quality. The most recent Brain Trauma Foundation (BTF) guidelines recommend a CPP target between 60 and 70 mmHg. Recognizing the limitations of the data on which this recommendation is based they classified it only as a Level II B recommendation (Carney et al., 2017). Similar recommendations are made by the Seattle International Severe Traumatic Brain Injury Consensus Conference (SIBICC) in their algorithm for the management of TBI patients with ICP monitoring where a target CPP of between 60 and 70 mmHg is recommended for initial management (Hawryluk et al., 2019). Given the inconsistency of the literature, a move away from global to more personalized targets has begun.

Personalized CPP targets

Traumatic brain injury is a heterogeneous condition and recently some of the inconsistent results of CPP management and targets are being attributed to this heterogeneity. A significant area in which TBI patients vary is in the state of their CA and cerebrovascular relativity with some patients losing this ability to appropriately regulate CBF following injury. The pressure reactivity index (PRx) was first described by Czosnyka et al. as a moving Pearson correlation coefficient between slow fluctuations in MAP and ICP, and is a continuously derived metric of cerebrovascular reactivity. PRx is a normalized metric ranging from $+1.0$ to -1.0, with negative values indicating intact cerebrovascular reactivity and positive values representing disrupted cerebrovascular reactivity (Czosnyka et al., 1997; For a more complete review of the impact of TBI on cerebrovascular reactivity and PRx please see Chapter 13). Even in this seminal paper Czosnyka et al. recognized a relationship between CPP and PRx in the TBI patient with CPP < 60 mmHg being associated with increasing disrupted cerebrovascular reactivity.

This concept was taken further by Steiner et al. (2002) by plotting PRx vs CPP for individual patients in their retrospective review of 114 TBI patients with continuous ICP and MAP monitoring. They noted that in 68 patients a somewhat parabolic distribution was observed with an identifiable CPP at which PRx was at a minimum (cerebrovascular reactivity was most intact). They termed this CPP as the optimal CPP (CPPopt) (Fig. 2). They went on to compare each patient's CPPopt with their mean CPP and found that 38 patients had a mean CPP above their CPPopt and 30 had a mean CPP below. They also noted that for these 68 patients the absolute difference between CPP an CPPopt correlated well with GOS at 6 months post-injury ($r=-0.51$, $p<0.0001$). This indicated a trend that the further a patient's mean CPP was from their CPPopt the worse their outcome was (Steiner et al., 2002). In a follow-up study at the same institution, data from 398 patients were retrospectively assessed. They found that 6-month outcomes were worse in those that had a mean CPP below their CPPopt ($r^2=0.53$, $p<0.001$) and those with a mean CPP above CPPopt ($r^2=-0.40$, $p<0.001$; Zweifel et al., 2008). Sorrentino et al. also noted in their cohort of 459 patients that in patients with disrupted cerebrovascular reactivity, as defined by a PRx > 0.05, those with a CPP less than 70 mmHg were significantly more likely to have an unfavorable outcome (no relative risk given, $p=0.02$; Sorrentino et al., 2012). These studies served to strengthen the relationship between PRx and CPP, and raised interest in the possibility of a targeted CPP threshold.

A CPP close to CPPopt has been found to be associated with outcomes following TBI in a number of additional studies and has even been the subject of systematic review (Needham et al., 2017). In a study of 327 TBI patients Aries et al. found that the difference between CPP and CPPopt was correlated with outcomes ($\chi^2=45$, $p<0.001$; Aries et al., 2012). Depreitere et al. found that in their cohort of 264 TBI patients the proportion of time spent within 5 mmHg of CPPopt was higher in those that survived than those that did not (25.6% vs 19.6%, $p=0.01$; Depreitere et al., 2014). Colton et al. found that poor outcomes were in those with a CPP greater than 10 mmHg away from CPPopt than those within 5 mmHg of CPPopt (RR 2.14, $p<0.05$; Colton, Yang, Hu, Scalea, & Stein, 2014). In a study involving 204 patients form the CENTER-TBI dataset Zeiler et al. showed that average deviation from CPPopt was predictive of mortality (AUC=0.702, 95% CI 0.107–0.795,

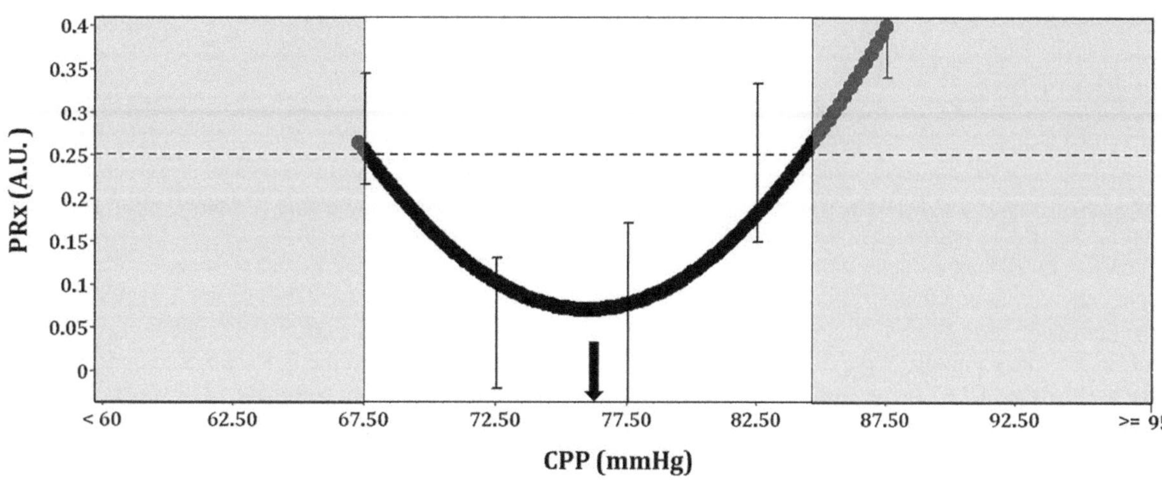

FIG. 2 Derivation of CPPopt. Demonstrated is the derivation of optimal cerebral perfusion pressure (CPPopt). The pressure reactivity index (PRx) in arbitrary units (AU) is plotted against cerebral perfusion pressure (CPP) in millimeters Mercury (mmHg). The CPP corresponding to the minimum of the parabolic curve of best fit (*solid line*) represents the CPPopt (*solid arrow*). The *dashed line* represents a PRx threshold, here set at 0.25. The CPPs in white represent the optimal CPP range. That is the CPPs at which the parabolic curve of best fit is less than the threshold.

$p<0.0001$) and unfavorable outcome (AUC=0.640, 95% CI 0.563–0.718, $p=0.0020$). Additionally, the % time spent more than 5 mmHg and more than 10 mmHg below CPPopt was found to be associated with mortality (AUC=0.692, 95% CI 0.595–0.788, $p<0.0001$; AUC=0.679, 95% CI 0.583–0.776, $p=0.0002$, respectively) and unfavorable outcome (AUC=0.632, 95% CI 0.556–0.710, $p=0.0021$; AUC=0.642, 95% CI 0.571–0.725, $p=0.0035$, respectively; Zeiler, Ercole, et al., 2019; Zeiler et al., 2019). This study provided strong multi-center observational data to support the utility of CPPopt.

The prognostic implications of CPPopt have also been demonstrated in a pediatric population. In a retrospective analysis of 12 pediatric TBI patients, Young et al. found that while no difference was found between the CPP of those that survived and those that did not, time spent with a CPP within 10 mmHg of CPPopt was significantly longer in those that survived compared to those that did not (90.7 ± 12.6% vs 70.6 ± 21.8%, $p=0.02$; Young et al., 2016).

Some physiologic data have also been found to correlate well with CPPopt. In a study of 38 TBI patients Jaeger et al. found the CPP and brain tissue oxygenation correlated well for values below CPPopt ($r=0.51$, $p<0.001$; Jaeger, Schuhmann, Soehle, & Meixensberger, 2006). This implication being that CPPopt may serve as a low water mark for cerebral oxygenation.

Much of the data supporting CPPopt has been retrospective with an emphasis on the prognostic ability of CPPopt. Though, recent multi-center prospective data from the Collaborative European NeuroTrauma Effectiveness Research in TBI (CENTER-TBI) high-resolution sub-study, confirms these strong associations between CPPopt and global outcome (Zeiler, Ercole, et al., 2019). The strength of this data, however, has led to the desire to test CPPopt as a therapeutic target in a randomized fashion. Given that CPPopt can be generated in real time with as little as 4 h of patient data prospective, comparative studies are possible. The CPPopt Guided Therapy: Assessment of Target Effectiveness (COGiTATE) trial is a phase II study that has recently completed data collection. It is a prospective, multi-center, non-blinded randomized, controlled trial that seeks to evaluate the feasibility, safety, and physiologic implications of CPPopt-guided therapy. A control arm will be based on CPP management as described by the BTF guidelines (Beqiri et al., 2019).

ARDS in TBI

As previously mentioned, much of the reluctance to utilize CPP targets that are above 70 mmHg comes from a conserve of an increase in the incidence of ARDS. This was found to be a fivefold increase in a study by Robertson et al. (1999). Subsequent analysis found that the increased risk was mostly attributable to the increased use of vasopressors (Contant, Valadka, Gopinath, Hannay, & Robertson, 2001). Notably this data is over 20 years old and there have been significant advancements in ventilatory management of intensive care unit (ICU) patients over that time. More recent data reported by Thiara et al. found no difference in CPP in those that developed ARDS and those that did not (74 mmHg vs 73 mmHg, $p=0.86$) in their study population of 113 patients with severe TBI (Thiara, Griesdale, Henderson, & Sekhon, 2018). Additionally, Moreira et al. found that in their cohort of 38 patients managed with CPPopt-guided therapy, there was no association with ARDS, and attributed this to lung protective ventilation strategies (Moreira et al., 2018). These more contemporary studies call to question the previously held dogma that elevating CPP above 70 mmHg results in an increased risk of ARDS (Zeiler, 2018). However, large prospective multicenter studies need to be performed before the safety of elevated CPPs can be confirmed.

Future directions

The future of CPP directed management in TBI is full of possibilities. While the era of fixed universal CPP targets has likely come to an end, targeted management of CPP could hold the key to significantly improved clinical outcomes following TBI. The COGiTATE phase II study serves as a stepping off point and based on its results a phase III trial assessing the therapeutic merit of CPPopt-guided management on a large scale is likely not far.

The concept of CPPopt provides a point target for CPP management following TBI. In reality, ranges of CPP will likely prove to be much more practical. Donnelly et al. explored this concept in a population of 729 severe TBI patients. In their study, they utilized a threshold of 0.3 for PRx to identify a CPP range for which the parabolic CPPopt curve was below threshold. Where the parabolic curve intersected the 0.3 PRx threshold were labeled the lower limit and upper CPP limit of reactivity (LLR and ULR, respectively). They reported that % time spent with CPP below the LLR was associated with unfavorable outcome and mortality when controlled for age, initial GCS, and mean ICP (OR=1.04, 95% CI 1.02–1.06, $p<0.001$; OR=1.06, 95% CI 1.04–1.08, $p<0.001$, respectively; Donnelly et al., 2017). This method may be utilized to determine personalized optimal CPP ranges in real time.

While PRx is now the routine parameter used to derive CPPopt new data is being reported on more robust methods of determining CPPopt (Liu et al., 2017). Additionally, new indices are being described that may provide more accurate assessments of cerebrovascular reactivity and these have demonstrated a similar ability to determine a CPPopt (Zeiler et al., 2020; Zeiler, Ercole, et al., 2019). Beyond this, non-invasive methods of determining a patient's state CA, utilizing near infrared spectroscopy (NIRS) and transcranial Doppler (TCD), are being developed and validated (Gomez, Dian, Froese, & Zeiler, 2020; Gomez, Dian, & Zeiler, 2020).

Zeiler et al. have recently reported early evidence of using robotic TCD to monitor CA in a more continuous fashion. Previously TCD based indices were limited to an hour of continuous monitoring by shifting of the probe and subsequent loss of signal. Advances in robotics have led to the development of robotically driven TCDs with automated algorithms for MCA CBFV detection and optimization. This has been shown to allow for much longer periods of uninterrupted data collection (Zeiler & Smielewski, 2018). TCD based indices of CA that leverage robotic TCD technology have been shown to trend closely with PRx and have even demonstrated the ability to identify a CPPopt (Zeiler, Czosnyka, & Smielewski, 2018; Zeiler, Smielewski, et al., 2019).

Outside of the world of TBI, NIRS based indices of CA have been leveraged to guide arterial blood pressure (ABP) management. In situations such as pediatric cardiac surgery and intra-ventricular hemorrhage (IVH) of prematurity ICP monitoring is usually not available and so determination of CPP and PRx is not possible. In this setting, the moving, linear correlation coefficient between slow waves of ABP and cerebral oximetry measured with NIRS has been able to provide targets MAP without the need for invasive intracranial monitoring (Brady et al., 2010; Brady et al., 2010; da Costa et al., 2015; da Costa, Czosnyka, Smielewski, & Austin, 2018). Similar work in TBI may allow for ABP guided therapy and negate the need for invasive ICP monitoring.

The state of CA is already being incorporated into treatment algorithms such as the SIBICC algorithm for the management of TBI patients with ICP monitoring (Hawryluk et al., 2019). Non-invasive methods of determining CA may simplify these algorithms and reduce the need for potentially harmful interventional procedures. Ultimately, a non-invasive method of determining an optimal CPP range utilizing NIRS or robotic TCD may provide the safest means of targeting CPP following TBI.

Applications to other areas of neuroscience

In this chapter, the concept of optimal cerebral perfusion pressure (CPPopt) was discussed as a possible method of determining personalized cerebral perfusion pressure (CPP) targets following traumatic brain injury (TBI). A similar concept has developed in the world of spinal cord injury (SCI). It has been shown that following SCI subdural intra-spinal pressure can be measured safely. A spinal cord perfusion pressure can be calculated and can improve the amplitude of motor-evoked potentials when increased by manipulation of mean arterial pressure (MAP) (Werndle et al., 2014). Follow-up work has shown that the same spectral analysis used to develop the pressure reactivity index (PRx) can be utilized in intra-spinal pressure monitoring and a spinal pressure reactivity index can be derived (Varsos et al., 2015). An analogous optimal spinal cord perfusion pressure has been developed and early observational data has shown improvements in metabolic parameters and clinical outcomes when perfusion pressures were near the calculated optimum (Chen, Smielewski, Czosnyka, Papadopoulos, & Saadoun, 2017). Like with CPPopt a prospective multi-center randomized trial will be needed to validate the therapeutic potential of deriving and targeting an optimal spinal cord perfusion pressure following SCI.

Mini-dictionary of terms

Cerebral autoregulation (CA): The ability of the cerebral vasculature to maintain CBF over a range of CPPs.

Cerebral perfusion pressure (CPP): The driving force of cerebral perfusion often expressed as the difference between the mean arterial pressure and intracranial pressure. It is typically expressed in units of mmHg in the clinical setting.

Cerebrovascular reactivity: The ability of the cerebrovascular tree to contract and dilate in response to changes in CPP and the metabolic environment.

Lower limit of autoregulation (LLA): This is the CPP below which cerebral autoregulatory mechanisms fail to maintain a constant cerebral blood flow. Below this level, cerebral ischemia is present.

Upper limit of autoregulation (ULA): This is the CPP above which cerebral autoregulatory mechanisms fail to maintain a constant cerebral blood flow. Above this level, cerebral hyperemia is present.

Key facts of CPP thresholds in TBI

- Cerebral perfusion pressure (CPP) following traumatic brain injury (TBI) may be compromised both due to decreased mean arterial pressure (MAP) and increased intracranial pressure (ICP).
- CPP has been linked to outcomes following TBI with low CPP associated with increased mortality while the effects of high CPP following TBI are not as clear.
- Current Brain Trauma Foundation guidelines suggest a global target of CPP between 60 and 70 mmHg following TBI.
- Cerebrovascular reactivity as measured by the pressure reactivity index (PRx) appears to be impaired following TBI.
- Personalized CPP targets (CPPopt) can be obtained from monitoring PRx following TBI and has been shown to be associated with patient outcomes when retrospectively analyzed.
- The clinical utility of maintaining personalized CPPopt targets is only now being explored in a prospective fashion.

Summary points

- This chapter focused on the implication of cerebral perfusion pressure (CPP) thresholds following traumatic brain injury (TBI).
- The CPP is the difference between the mean arterial pressure (MAP) and the intracranial pressure (ICP).
- Following TBI, CPP can be reduced due to decreased MAP or increased ICP.
- There is evidence that CPP following TBI has prognostic importance.
- Current guideline-based management of CPP following TBI emphasizes a global range to target with evidence pointing to deleterious effects of both low and high CPP.
- In recent years, work has been done to utilize measures of cerebrovascular reactivity to obtain personalized targets for CPP following TBI but the effects of these targets is only now being tested prospectively.
- In the future, we may see personalized targeting of CPP, and even MAP, based on non-invasive methods of monitoring cerebrovascular reactivity.

Acknowledgments

F.A.Z. receives research support from the Manitoba Public Insurance (MPI) Neuroscience/TBI Research Endowment, the Health Sciences Centre Foundation Winnipeg, the United States National Institutes of Health (NIH) through the National Institute of Neurological Disorders and Stroke (NINDS), the Canadian Institutes of Health Research (CIHR), the Canada Foundation for Innovation (CFI), Research Manitoba, the University of Manitoba VPRI Research Investment Fund (RIF), the University of Manitoba Centre on Aging, and the University of Manitoba Rudy Falk Clinician-Scientist Professorship.

L.F. is supported through the University of Manitoba—Department of Surgery GFT Research Grant, and the University of Manitoba Office of Research Services (ORS)—University Research Grant Program (URGP).

C.B. is supported through the Centre on Aging Fellowship at the University of Manitoba.

A.G. is supported through the University of Manitoba Clinician Investigator Program.

References

Allen, B. B., Chiu, Y., Gerber, L. M., Ghajar, J., & Greenfield, J. P. (2014). Age-specific cerebral perfusion pressure thresholds and survival in children and adolescents with severe traumatic brain injury. *Pediatric Critical Care Medicine*, *15*(1), 62–70. https://doi.org/10.1097/PCC.0b013e3182a556ea.

Aries, M. J. H., Czosnyka, M., Budohoski, K. P., Steiner, L. A., Lavinio, A., Kolias, A. G., et al. (2012). Continuous determination of optimal cerebral perfusion pressure in traumatic brain injury. *Critical Care Medicine*, *40*(8), 2456–2463. https://doi.org/10.1097/CCM.0b013e3182514eb6.

Asgeirsson, B., Grände, P. O., & Nordström, C. H. (1994). A new therapy of post-trauma brain oedema based on haemodynamic principles for brain volume regulation. *Intensive Care Medicine*, *20*(4), 260–267. https://doi.org/10.1007/BF01708961.

ATLS Subcommittee, American College of Surgeons' Committee on Trauma, & International ATLS working group. (2013). Advanced trauma life support (ATLS®): The ninth edition. *The Journal of Trauma and Acute Care Surgery*, *74*(5), 1363–1366. https://doi.org/10.1097/TA.0b013e31828b82f5.

Beqiri, E., Smielewski, P., Robba, C., Czosnyka, M., Cabeleira, M. T., Tas, J., et al. (2019). Feasibility of individualised severe traumatic brain injury management using an automated assessment of optimal cerebral perfusion pressure: The COGiTATE phase II study protocol. *BMJ Open*, *9*(9). https://doi.org/10.1136/bmjopen-2019-030727, e030727.

Brady, K., Joshi, B., Zweifel, C., Smielewski, P., Czosnyka, M., Easley, R. B., et al. (2010). Real-time continuous monitoring of cerebral blood flow autoregulation using near-infrared spectroscopy in patients undergoing cardiopulmonary bypass. *Stroke*, *41*(9), 1951–1956. https://doi.org/10.1161/STROKEAHA.109.575159.

Brady, K. M., Mytar, J. O., Lee, J. K., Cameron, D. E., Vricella, L. A., Thompson, W. R., et al. (2010). Monitoring cerebral blood flow pressure autoregulation in pediatric patients during cardiac surgery. *Stroke*, *41*(9), 1957–1962. https://doi.org/10.1161/STROKEAHA.109.575167.

Carney, N., Totten, A. M., O'Reilly, C., Ullman, J. S., Hawryluk, G. W. J., Bell, M. J., et al. (2017). Guidelines for the management of severe traumatic brain injury, fourth edition. *Neurosurgery*, *80*(1), 6–15. https://doi.org/10.1227/NEU.0000000000001432.

Chang, J. J. J., Youn, T. S., Benson, D., Mattick, H., Andrade, N., Harper, C. R., et al. (2009). Physiologic and functional outcome correlates of brain tissue hypoxia in traumatic brain injury. *Critical Care Medicine*, *37*(1), 283–290. https://doi.org/10.1097/CCM.0b013e318192fbd7.

Changaris, D. G., McGraw, C. P., Richardson, J. D., Garretson, H. D., Arpin, E. J., & Shields, C. B. (1987). Correlation of cerebral perfusion pressure and Glasgow coma scale to outcome. *The Journal of Trauma*, *27*(9), 1007–1013. https://doi.org/10.1097/00005373-198709000-00009.

Chen, S., Smielewski, P., Czosnyka, M., Papadopoulos, M. C., & Saadoun, S. (2017). Continuous monitoring and visualization of optimum spinal cord perfusion pressure in patients with acute cord injury. *Journal of Neurotrauma*, *34*(21), 2941–2949. https://doi.org/10.1089/neu.2017.4982.

Clifton, G. L., Miller, E. R., Choi, S. C., & Levin, H. S. (2002). Fluid thresholds and outcome from severe brain injury. *Critical Care Medicine*, *30*(4), 739–745. https://doi.org/10.1097/00003246-200204000-00003.

Cold, G. E., & Jensen, F. T. (1978). Cerebral autoregulation in unconscious patients with brain injury. *Acta Anaesthesiologica Scandinavica*, *22*(3), 270–280. https://doi.org/10.1111/j.1399-6576.1978.tb01301.x.

Colton, K., Yang, S., Hu, P., Scalea, T., & Stein, D. (2014). Simplifing the calculations of optimal cerebral perfusion pressure without continuous waveforms. *Critical Care Medicine*, *42*, A1493. https://doi.org/10.1097/01.ccm.0000458052.77460.0b.

Contant, C. F., Valadka, A. B., Gopinath, S. P., Hannay, H. J., & Robertson, C. S. (2001). Adult respiratory distress syndrome: A complication of induced hypertension after severe head injury. *Journal of Neurosurgery*, *95*(4), 560–568. https://doi.org/10.3171/jns.2001.95.4.0560.

Czosnyka, M., Smielewski, P., Kirkpatrick, P., Laing, R. J., Menon, D., & Pickard, J. D. (1997). Continuous assessment of the cerebral vasomotor reactivity in head injury. *Neurosurgery*, *41*(1), 11–19. https://doi.org/10.1097/00006123-199707000-00005.

da Costa, C. S., Czosnyka, M., Smielewski, P., & Austin, T. (2018). Optimal mean arterial blood pressure in extremely preterm infants within the first 24 hours of life. *The Journal of Pediatrics*, *203*, 242–248. https://doi.org/10.1016/j.jpeds.2018.07.096.

da Costa, C. S., Czosnyka, M., Smielewski, P., Mitra, S., Stevenson, G. N., & Austin, T. (2015). Monitoring of cerebrovascular reactivity for determination of optimal blood pressure in preterm infants. *The Journal of Pediatrics*, *167*(1), 86–91. https://doi.org/10.1016/j.jpeds.2015.03.041.

Depreitere, B., Güiza, F., Van den Berghe, G., Schuhmann, M. U., Maier, G., Piper, I., et al. (2014). Pressure autoregulation monitoring and cerebral perfusion pressure target recommendation in patients with severe traumatic brain injury based on minute-by-minute monitoring data: Clinical article. *Journal of Neurosurgery*, *120*(6), 1451–1457. https://doi.org/10.3171/2014.3.JNS131500.

Donnelly, J., Czosnyka, M., Adams, H., Robba, C., Steiner, L. A., Cardim, D., et al. (2017). Individualizing thresholds of cerebral perfusion pressure using estimated limits of autoregulation. *Critical Care Medicine*, *45*(9), 1464–1471. https://doi.org/10.1097/CCM.0000000000002575.

Elf, K., Nilsson, P., Ronne-Engström, E., Howells, T., & Enblad, P. (2005). Cerebral perfusion pressure between 50 and 60 mm Hg may be beneficial in head-injured patients: A computerized secondary insult monitoring study. *Neurosurgery*, *56*(5), 962–971 (discussion 962-971).

Fog, M. (1938). The relationship between the blood pressure and the tonic regulation of the plal arteries. *Journal of Neurology and Psychiatry*, *1*(3), 187–197. https://doi.org/10.1136/jnnp.1.3.187.

Gomez, A., Dian, J., Froese, L., & Zeiler, F. A. (2020). Near-infrared based cerebrovascular reactivity as a means of monitoring cerebral autoregulation and predicting outcome in moderate/severe traumatic brain injury: A pilot study protocol and planned analyses (preprint). *JMIR Research Protocols*. https://doi.org/10.2196/18740.

Gomez, A., Dian, J., & Zeiler, F. A. (2020). Continuous and entirely non-invasive method for cerebrovascular reactivity assessment: Technique and implications. *Journal of Clinical Monitoring and Computing*. https://doi.org/10.1007/s10877-020-00472-4.

Hawryluk, G. W. J., Aguilera, S., Buki, A., Bulger, E., Citerio, G., Cooper, D. J., et al. (2019). A management algorithm for patients with intracranial pressure monitoring: The Seattle International Severe Traumatic Brain Injury Consensus Conference (SIBICC). *Intensive Care Medicine*, *45*(12), 1783–1794. https://doi.org/10.1007/s00134-019-05805-9.

Huang, S.-J., Hong, W.-C., Han, Y.-Y., Chen, Y.-S., Wen, C.-S., Tsai, Y.-S., et al. (2006). Clinical outcome of severe head injury using three different ICP and CPP protocol-driven therapies. *Journal of Clinical Neuroscience: Official Journal of the Neurosurgical Society of Australasia*, *13*(8), 818–822. https://doi.org/10.1016/j.jocn.2005.11.034.

Jaeger, M., Schuhmann, M. U., Soehle, M., & Meixensberger, J. (2006). Continuous assessment of cerebrovascular autoregulation after traumatic brain injury using brain tissue oxygen pressure reactivity. *Critical Care Medicine*, *34*(6), 1783–1788. https://doi.org/10.1097/01.CCM.0000218413.51546.9E.

Juul, N., Morris, G. F., Marshall, S. B., & Marshall, L. F. (2000). Intracranial hypertension and cerebral perfusion pressure: Influence on neurological deterioration and outcome in severe head injury. The Executive Committee of the International Selfotel Trial. *Journal of Neurosurgery*, *92*(1), 1–6. https://doi.org/10.3171/jns.2000.92.1.0001.

Kellie, G. (1824). An account of the appearances observed in the dissection of two of three individuals presumed to have perished in the storm of the 3d, and whose bodies were discovered in the vicinity of Leith on the morning of the 4th, November 1821; with some reflections on the pathology of the brain: Part I. *Transactions. Medico-Chirurgical Society of Edinburgh*, *1*, 84–122. Pub med.

Lassen, N. A. (1959). Cerebral blood flow and oxygen consumption in man. *Physiological Reviews*, *39*(2), 183–238. https://doi.org/10.1152/physrev.1959.39.2.183.

Lin, J. W., Tsai, J. T., Lin, C. M., Lee, L. M., Hung, K. S., Huang, S. J., et al. (2008). Evaluation of optimal cerebral perfusion pressure in severe traumatic brain injury. *Acta Neurochirurgica. Supplement*, *101*, 131–136. https://doi.org/10.1007/978-3-211-78205-7_22.

Liu, X., Maurits, N. M., Aries, M. J. H., Czosnyka, M., Ercole, A., Donnelly, J., et al. (2017). Monitoring of optimal cerebral perfusion pressure in traumatic brain injured patients using a multi-window weighting algorithm. *Journal of Neurotrauma*, *34*(22), 3081–3088. https://doi.org/10.1089/neu.2017.5003.

Manley, G., Knudson, M. M., Morabito, D., Damron, S., Erickson, V., & Pitts, L. (2001). Hypotension, hypoxia, and head injury: Frequency, duration, and consequences. *Archives of Surgery*, *136*(10), 1118. https://doi.org/10.1001/archsurg.136.10.1118.

McGraw, C. P. (1989). A cerebral perfusion pressure greater than 80 mm Hg is more beneficial. In J. T. Hoff, & A. L. Betz (Eds.), *Intracranial pressure VII* (pp. 839–841). Berlin Heidelberg: Springer.

Monro, A. (1783). Observations on the structure and functions of the nervous system, illustrated with tables. *The London Medical Journal*, *4*(2), 113–135. PMC.

Moreira, M., Fernandes, D., Pereira, E., Monteiro, E., Pascoa, R., & Dias, C. (2018). Is there a relationship between optimal cerebral perfusion pressure-guided management and PaO2/FiO2 ratio after severe traumatic brain injury? *Acta Neurochirurgica. Supplement*, *126*, 59–62. https://doi.org/10.1007/978-3-319-65798-1_13.

Needham, E., McFadyen, C., Newcombe, V., Synnot, A. J., Czosnyka, M., & Menon, D. (2017). Cerebral perfusion pressure targets individualized to pressure-reactivity index in moderate to severe traumatic brain injury: A systematic review. *Journal of Neurotrauma*, *34*(5), 963–970. https://doi.org/10.1089/neu.2016.4450.

Overgaard, J., & Tweed, W. A. (1974). Cerebral circulation after head injury: Part 1: Cerebral blood flow and its regulation after closed head injury with emphasis on clinical correlations. *Journal of Neurosurgery*, *41*(5), 531–541. https://doi.org/10.3171/jns.1974.41.5.0531.

Robertson, C. S., Valadka, A. B., Hannay, H. J., Contant, C. F., Gopinath, S. P., Cormio, M., et al. (1999). Prevention of secondary ischemic insults after severe head injury. *Critical Care Medicine*, *27*(10), 2086–2095. https://doi.org/10.1097/00003246-199910000-00002.

Rosner, M. J., Rosner, S. D., & Johnson, A. H. (1995). Cerebral perfusion pressure: Management protocol and clinical results. *Journal of Neurosurgery*, *83*(6), 949–962. https://doi.org/10.3171/jns.1995.83.6.0949.

Sorrentino, E., Diedler, J., Kasprowicz, M., Budohoski, K. P., Haubrich, C., Smielewski, P., et al. (2012). Critical thresholds for cerebrovascular reactivity after traumatic brain injury. *Neurocritical Care*, *16*(2), 258–266. https://doi.org/10.1007/s12028-011-9630-8.

Steiner, L. A., Czosnyka, M., Piechnik, S. K., Smielewski, P., Chatfield, D., Menon, D. K., et al. (2002). Continuous monitoring of cerebrovascular pressure reactivity allows determination of optimal cerebral perfusion pressure in patients with traumatic brain injury. *Critical Care Medicine*, *30*(4), 733–738. https://doi.org/10.1097/00003246-200204000-00002.

Thiara, S., Griesdale, D. E., Henderson, W. R., & Sekhon, M. S. (2018). Effect of cerebral perfusion pressure on acute respiratory distress syndrome. *The Canadian Journal of Neurological Sciences. Le Journal Canadien des Sciences Neurologiques*, *45*(3), 313–319. https://doi.org/10.1017/cjn.2017.292.

Varsos, G. V., Werndle, M. C., Czosnyka, Z. H., Smielewski, P., Kolias, A. G., Phang, I., et al. (2015). Intraspinal pressure and spinal cord perfusion pressure after spinal cord injury: An observational study. *Journal of Neurosurgery Spine*, *23*(6), 763–771. https://doi.org/10.3171/2015.3.SPINE14870.

Wagner, E. M., & Traystman, R. J. (1986). Hydrostatic determinants of cerebral perfusion. *Critical Care Medicine*, *14*(5), 484–490. https://doi.org/10.1097/00003246-198605000-00011.

Werndle, M. C., Saadoun, S., Phang, I., Czosnyka, M., Varsos, G. V., Czosnyka, Z. H., et al. (2014). Monitoring of spinal cord perfusion pressure in acute spinal cord injury: Initial findings of the injured spinal cord pressure evaluation study. *Critical Care Medicine*, *42*(3), 646–655. https://doi.org/10.1097/CCM.0000000000000028.

Young, A. M. H., Donnelly, J., Czosnyka, M., Jalloh, I., Liu, X., Aries, M. J., et al. (2016). Continuous multimodality monitoring in children after traumatic brain injury—Preliminary experience. *PLoS One*, *11*(3). https://doi.org/10.1371/journal.pone.0148817, e0148817.

Zeiler, F. A. (2018). Cerebral perfusion pressure targets in traumatic brain injury: The "fuzzy" spots above optimal cerebral perfusion pressure. *The Canadian Journal of Neurological Sciences. Le Journal Canadien des Sciences Neurologiques*, *45*(6), 721–722. https://doi.org/10.1017/cjn.2018.337.

Zeiler, F. A., Czosnyka, M., & Smielewski, P. (2018). Optimal cerebral perfusion pressure via transcranial Doppler in TBI: Application of robotic technology. *Acta Neurochirurgica*, *160*(11), 2149–2157. https://doi.org/10.1007/s00701-018-3687-5.

Zeiler, F. A., Ercole, A., Cabeleira, M., Carbonara, M., Stocchetti, N., Menon, D. K., et al. (2019). Comparison of performance of different optimal cerebral perfusion pressure parameters for outcome prediction in adult traumatic brain injury: A Collaborative European NeuroTrauma Effectiveness Research in Traumatic Brain Injury (CENTER-TBI) study. *Journal of Neurotrauma*, *36*(10), 1505–1517. https://doi.org/10.1089/neu.2018.6182.

Zeiler, F. A., Ercole, A., Czosnyka, M., Smielewski, P., Hawryluk, G., Hutchinson, P. J. A., et al. (2020). Continuous cerebrovascular reactivity monitoring in moderate/severe traumatic brain injury: A narrative review of advances in neurocritical care. *British Journal of Anaesthesia*. https://doi.org/10.1016/j.bja.2019.11.031.

Zeiler, F. A., & Smielewski, P. (2018). Application of robotic transcranial Doppler for extended duration recording in moderate/severe traumatic brain injury: First experiences. *Critical Ultrasound Journal*, *10*(1), 16. https://doi.org/10.1186/s13089-018-0097-0.

Zeiler, F. A., Smielewski, P., Stevens, A., Czosnyka, M., Menon, D. K., & Ercole, A. (2019). Non-invasive pressure reactivity index using Doppler systolic flow parameters: A pilot analysis. *Journal of Neurotrauma*, *36*(5), 713–720. https://doi.org/10.1089/neu.2018.5987.

Zweifel, C., Lavinio, A., Steiner, L. A., Radolovich, D., Smielewski, P., Timofeev, I., et al. (2008). Continuous monitoring of cerebrovascular pressure reactivity in patients with head injury. *Neurosurgical Focus*, *25*(4), E2. https://doi.org/10.3171/FOC.2008.25.10.E2.

Chapter 15

Brain swelling in traumatic brain injury

Silvia De Rosa[a], Lucia Cattin[a], Luca Ricciardi[b], and Alba Scerrati[c]

[a]*Department of Anesthesiology and Intensive Care, San Bortolo Hospital, Vicenza, Italy,* [b]*Dipartimento NESMOS, UOC di Neurochirurgia, AOU Sant'Andrea, Sapienza, Roma, Italy,* [c]*Department of Morphology, Surgery and Experimental Medicine, University Hospital S. Anna, Ferrara, Italy*

List of abbreviations

BBB blood–brain barrier
CE cerebral edema
CSF cerebrospinal fluid
GCS Glasgow Coma Scale
ISF interstitial fluid
TBI traumatic brain injury

Introduction

Several forms of edema have been identified under a single generic name of "cerebral edema." However, brain swelling refers to the increased brain volume, while cerebral edema (CE) refers to abnormal accumulation of water within the brain tissue and it is a prime example of treatable causes of secondary neurological deterioration in patients with brain injury (Shah & Kimberly, 2016). CE can be defined as an increase in brain tissue water (brain volume), included in individual cells and their surrounding interstitial space. The generation of CE after traumatic brain injury (TBI) is a complex heterogeneous process and it may differ based on the non-modifiable nature of primary injury, patient characteristics, and additional clinical insults as hypoxia, hypotension, hyperthermia, seizure. This complex process has a crucial impact on morbidity and mortality following TBI: increment of intracranial pressure, impairment of cerebral perfusion and oxygenation, contributing to additional ischemic injuries (see Chapter 15). This chapter is intended to serve as a foundation and a reference from which researchers and clinicians might have practical understanding of cerebral edema. In addition, we will classify clinical characteristics of brain swelling in TBI patients on the base of type of TBI and GCS.

Historical background

Before the 20th century, severe traumatic brain injury (TBI) was considered fatal. During the American Civil War, gunshot wounds to the head were seen in large numbers. Even if mortality rate was not available, survival was poor because of the high rates of infection (Boake, 1989). Epidural and subdural hematomas were well described and a confederate surgeon named John Chisolm, described the complexities of diffuse axonal injury (Kaufman, 1993). In 1885, a scientist named Paul Ehrlich made an accidental discovery. He was injecting blue dye into animals to study their physiology, when he realized that the dye would not pass into their brains. When he injected dye directly into the animals' brains, it would not pass into the rest of their bodies. He concluded that the brain is protected from the bloodstream with a potential and special barrier that is between the vasculature and the brain, which later became known as the blood–brain barrier. However, the existence of a barrier at the level of cerebral vessels was first postulated by Bield and Kraus (1898) and Lewandowsky (1900) based on the observation that the intravenous injection of cholic acids or sodium ferrocyanide had no pharmacological effects on the CNS, whereas neurological symptoms occurred after intra-ventricular application of the same substances. Successively, Goldmann demonstrates that dye injected into CSF did not circulate in the systemic circulation (Goldmann, 1909). With the development of the electron microscope, it became possible to define the ultrastructural features of blood vessels throughout the body. Van Harrefeld et al. showed that the lack of an extracellular space in the CNS in ultrastructural studies was due to the swelling of tissue components, especially of astrocytic endfeet before fixation (Van Harrefeld, Crowell, & Malhotra, 1966). In 1905, Reichardt introduced the term "brain edema" (Reichardt, 1905). In the 1967, Klatzo using a "cold injury"

Diagnosis and Treatment of Traumatic Brain Injury. https://doi.org/10.1016/B978-0-12-823347-4.00041-5

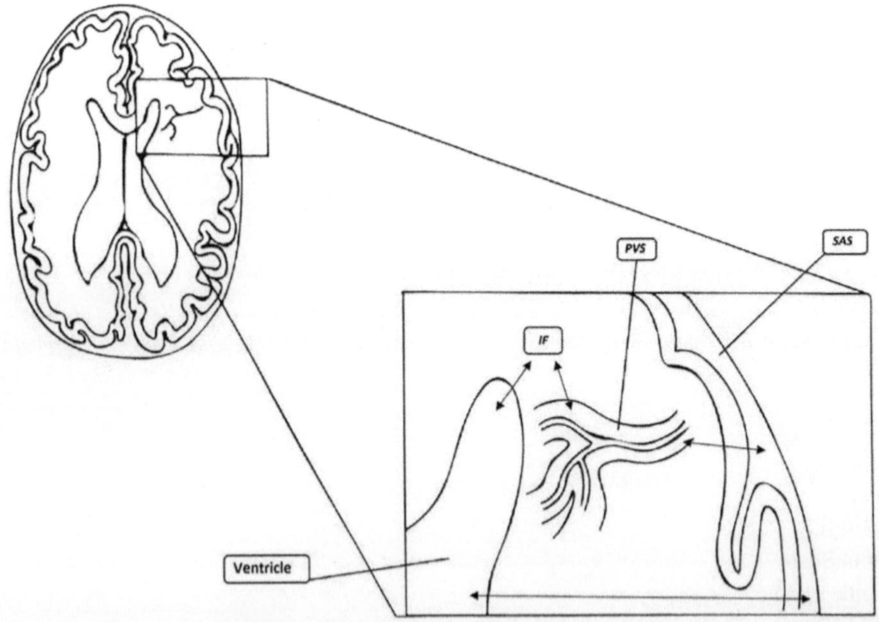

FIG. 1 Fluid omeostasis theory. A new concept suggests a fluid homeostasis between different compartments: parenchyma, interstitial space, subarachnoid space (SAS), perivascular space (PVS), and ventricles. *(Figure obtained with permission from Mantovani, G., Menegatti, M., Scerrati, A., Cavallo, M. A., & De Bonis, P. (2018) Controversies and misconceptions related to cerebrospinal fluid circulation: A review of the literature from the historical pioneers' theories to current models.* BioMed Research International, 26, 2928378.*)*

model described two major types of brain edema, the cytotoxic or vasogenic edema (Klatzo, 1967). In 1975, Fishman added these "interstitial edema" observed in patients with hydrocephalus (Fishman, 1975). Recently a new structural model has been proposed. In this model, the fluids exchange among different intracranial compartments. These exchanges are represented by a dynamic equilibrium among parenchyma, interstitial space, subarachnoid space, vascular compartment, and ventricles (Mantovani, Menegatti, Scerrati, Cavallo, & De Bonis, 2018) (Fig. 1). Several clinical entities, presenting with an imbalance of CSF dynamics, found their physiopathology in this new concept (De Bonis, Menegatti, Cavallo, et al., 2019; Zamboni et al., 2019).

Overview on the blood–brain barrier

The blood–brain barrier (BBB) is a term used to describe the unique properties of the microvasculature of the CNS. The BBB is formed by two types of cells: the endothelial cells and the mural cells. The first are simple squamous epithelial cells that form the walls of the blood vessels, they're extremely thin and are held together by tight junctions that limit the paracellular flux of solutes (Reese & Karnovsky, 1967). Those vessels are non-fenestrated and they provide a tight obstacle against the passage of drugs, pathogens, and injuries. Its importance has been pointed out describing neurological diseases such as multiple sclerosis, brain traumas, and neurodegenerative disorders where the disruption of this barrier plays a major role. Brain endothelial cells reciprocally connect and constitute tight junctions building BBB with astrocytes and pericytes (Fig. 2) (Abbott, Rönnbäck, & Hansson, 2006). Astrocytes secrete a range of factors that can induce aspects of the BBB phenotype while brain endothelium enhances the growth and differentiation of associated astrocytes, inducing a complex system of signaling inside the neurovascular unit (Abbott et al., 2006). The BBB is peculiar in the genesis of brain edema. When cerebral edema results from insults other than closed head trauma, changes in brain tissue specific gravity or water content are associated with increased permeability of the blood–brain barrier (Kuchiwaki et al., 1984).

The pathogenesis of brain edema

CE is a common end result of a variety of neurological and systemic disorders. It is associated with two separate pathophysiological processes with distinct molecular and physiological antecedents: those related to cytotoxic (cellular) edema of neurons and astrocytes, and those related to transcapillary flux of Na+ and other ions, water, and serum macromolecules (Liang, Bhatta, Gerzanich, & Simard, 2007). Swelling of the brain causes a bulking effect that exerts pressure on surrounding tissue,

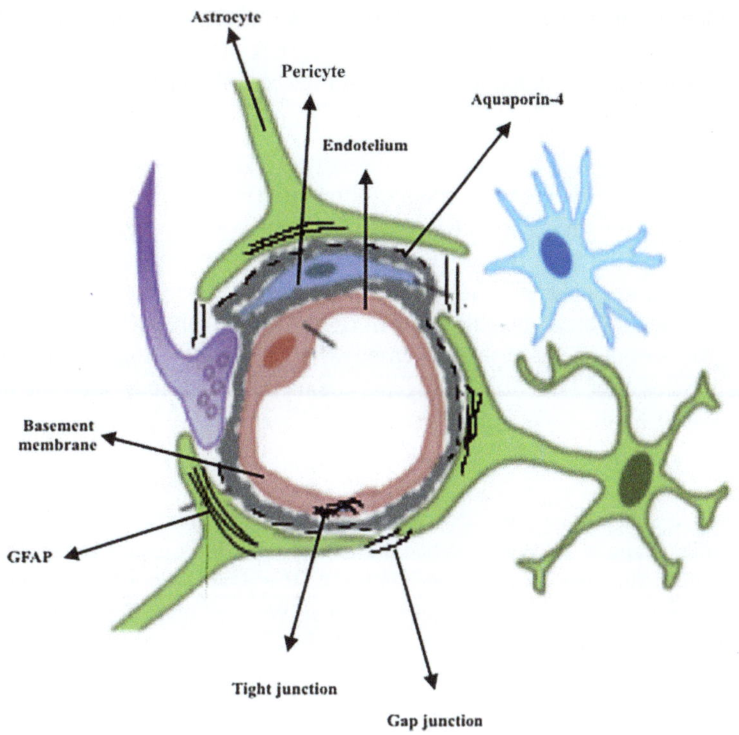

FIG. 2 Blood–brain barrier. BBB in transverse section showing endothelium, basement membrane, pericytes, astrocytes, and tight junctions. In addition, the localization of gap junction, GFAP, and aquaporin-4 are shown.

and which is amplified by the stiffness of the skull, placing an upper limit on the volume at which the brain could expand. Swelling of the brain exerts mechanical forces within the skull, thereby increasing the pressure of the intracranial tissue. When the tissue pressure exceeds the capillary pressure, the capillary lumens collapse. Surrounding shell ischemia triggers further edema formation and further swelling in the subsequent shell (Stokum, Gerzanich, & Simard, 2016).

Most classifications of cerebral edema describe four types (Table 1):

- *Cytotoxic/cellular*, an oncotic cell swelling secondary to sustained intracellular water collection;
- *Vasogenic/extracellular*, which results from vascular leakage through blood–brain barrier (BBB) disruption resulting in extracellular water accumulation and altered concentrations of ions, peptides, and macromolecules in the extracellular space. If the BBB is reconstituted early in the process, the VE can be reversed;

TABLE 1 Characteristics of different traumatic brain edema types.

Type of cerebral edema	Common causes	MR radiological appearance (Ho, Rojas, & Eisenberg, 2012)
Cytotoxic/cellular	Small vessel disease, ischemic stroke	Overall restriction of diffusion of water molecules across the cell membrane and within the cytoplasm resulting in regional high signal on DWI and low signal on ADC images
Vasogenic/extracellular	Tumors, hemorrhages, venous occlusion or thrombosis, arteriovenous shunts	Expanded extracellular fluid yields decreased T1 signal, increased T2 signal, and decreased fractional anisotropy. The white matter is preferentially affected because of its lower density with multiple unconnected parallel axonal tracts
Interstitial	Hydrocephalus	Hypointensity on T1-weighted imaging and periventricular hyperintensity on T2-weighted imaging/FLAIR
Combined	Trauma, infections, inflammations, hypoxic/ischemic encephalopathy	Global loss of gray-white matter differentiation and effacement of the sulci, ventricles, and basal cisterns

- *Osmotic*, caused by osmotic imbalances between blood and tissue. In this case, the brain is hyperosmolar relative to plasma and thus induces water to flow passively across an intact BBB along its concentration gradient.
- *Interstitial/hydrocephalic*, which occurs with transependymal flow of CSF in patients with obstruction of cerebrospinal fluid outflow (hydrocephalus);

Unfortunately, it is difficult to distinguish between all of these types of edema which are often present due to the nature and timing of the underlying disorder. Traditionally, edema following TBI was primarily viewed as being vasogenic, as showed by classical neuropathological findings and CT imaging based on hyperdense hemorrhagic lesions early after injury followed by hypo-dense halo around the hemorrhagic lesion (Unterberg, Stover, Kress, & Kiening, 2004). Later, microscopy and MR/CT imaging using intravascular contrast agents demonstrated massive swelling of astrocytic endfeet and that cytotoxic brain edema is also present in the acute phase of TBI (Foda & Marmarou, 1994).

Particularly, following TBI, various mediators are released which enhance vasogenic and/or cytotoxic brain edema. Cytotoxic edema manifests minutes after TBI, while vasogenic edema manifests hours after the initial insult.

Cytotoxic edema

Cellular survival requires that Na+ be continuously extruded from the intracellular compartment, because this is critical to maintaining normal cell volume (Liang et al., 2007). When an insult to the brain results in ischemia or hypoxia, very little new ATP can be produced and cells quickly use up their reserves of ATP and, unless normoxia is restored, the deranged cellular machinery loses its ability to sustain homeostasis (Sweeney, Yager, Walz, & Juurlink, 1995). Depletion of ATP is accompanied by unchecked influx of sodium that drives Cl− influx via chloride channels, and the resultant increase in intracellular osmolarity drives inflow of water via AQP channels. The molecular basis of water transport in the brain (Amiry-Moghaddam & Ottersen, 2003; Badaut et al., 2001). Thus, this oncotic cell swelling is a premorbid process whereby cells swell due to influx of osmolites (mainly Na^+ and Cl^-) and water from the interstitial spaces into the intracellular compartment, in part due to failure of energy-dependent mechanisms of extrusion. Extracellular water flows into the cell interior, resulting in an increase in intracellular fluid volume. In the initial stages, the blood–brain barrier is intact and largely impermeable to ions and fluids, so extracellular ions and water loss are not replenished. The cytotoxic edema by itself does not generate tissue swelling: fluid movement involved in formation of cytotoxic edema does not lead to any change in total brain volume, despite the observable increase in cell size. However, formation of cytotoxic edema depletes the extracellular space of Na+, Cl−, and water, thereby creating a new gradient for these molecules across the capillary of the blood–brain barrier. It simply represents a rearrangement of parenchymal water mass and does not involve the addition of new water mass to the brain. The new gradient results in driving transcapillary formation of ionic edema. In this context, cellular edema is an important initial step in the formation of cerebral edema and swelling, as it generates the driving force for influx of ionic and vasogenic edema, which do cause swelling. Nitric oxide and pro-inflammatory peptides such as kinins have been indicated to be involved in BBB disruption because synthesis inhibitors and receptor antagonists showed less BBB hyper-permeability and edema formation after cold injury (Oury, Piantadosi, & Crapo, 1993; Raslan et al., 2010). Particularly, in this setting, nitric oxide (NO), that enhances blood flow through vasodilation, exert toxic effects exacerbating cerebral edema in certain circumstances 24–48 h after the initiating insult has passed (Muresanu et al., 2019). When compensatory mechanisms such as ionic pumps in the plasma membrane are exceeded or fail altogether, the swollen cell dies (Fig. 3) (Csaba, 2006).

Vasogenic edema

Vasogenic edema is a landmark of TBI pathophysiology that implies BBB disruption. Precisely, two major phases seems to occur: (1) one phase at 1–3 h post-injury, as a direct consequence of mechanical disruption of microvascular vessels at the site of the impact; (2) second phase at 3–7 days post-injury due to secondary injury events with the presence of pro-permeability factors, called delayed *vasogenic edema* (Li et al., 2016). In regard to the recent advances, it was been proposed a re-organization of the events contributing to brain edema formation as follows: *primary vasogenic edema*, accompanied/followed by ionic/cytotoxic edema, and then d*elayed vasogenic edema* with massive disruption of the BBB. BBB disruption has various devastating consequences including vasogenic edema, hemorrhage, and leukocyte infiltration followed by inflammation (Song et al., 2015). These events result in an increased total brain volume with devastating consequences for the final outcome of TBI.

Primary vasogenic edema

Immediately after the traumatic insult, a sudden mechanical rupture of cerebral blood vessels results in hemorrhage and focal ischemia (Badaut & Bix, 2014). The lack or the severe reduction of perfusion in this area causes edema formation, while

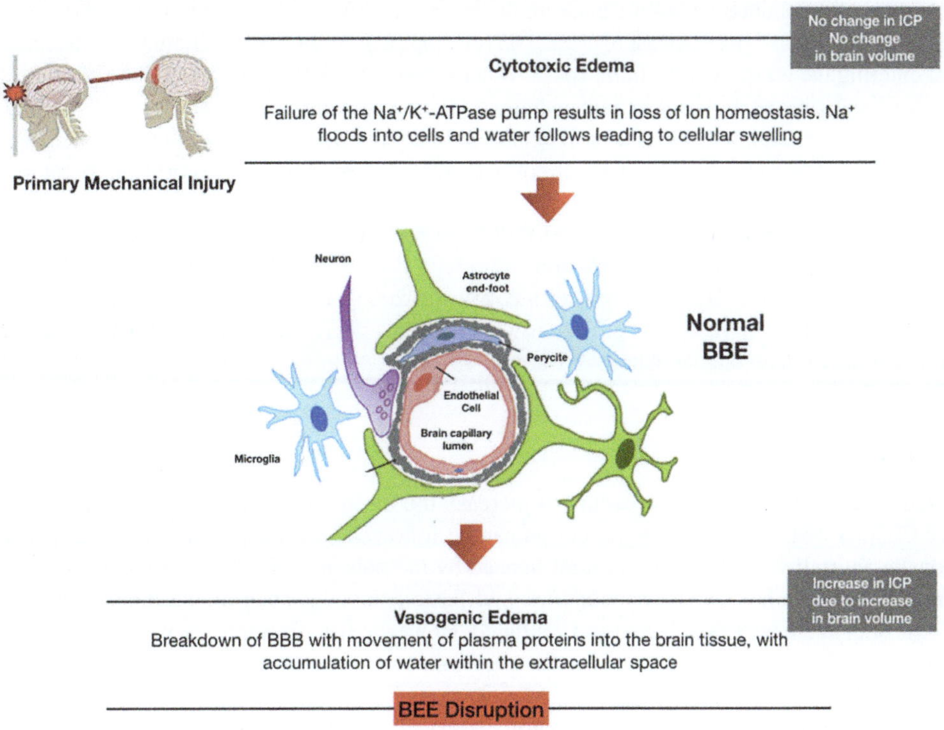

FIG. 3 Cerebral edema. The temporal evolution of the edema process in the context of a contusion type of TBI is very likely starting with an initial breakdown of blood vessels conducting to a vasogenic edema.

away from this area vessels are less damaged and are able to maintain their barrier function. Due to stretch injury, there is an impairment of the BBB and immediate extravasation of a protein-rich plasma transudate occurs into the brain parenchyma. This formation of vasogenic edema contributes to the development and progress of secondary injuries (including delayed vasogenic edema) (Badaut & Bix, 2014).

Delayed vasogenic edema

Alterations of tight junctions contribute to BBB disruption and vasogenic edema formation in TBI (Dore-Duffy et al., 2000). Several molecules released during inflammation process such as endothelin, as well as bradykinin, substance P have direct or indirect effects on matrix metalloproteinases (MMPs), and vascular endothelial growth factor (VEGF) which alter the structure and integrity of tight junction, then, BBB, leading to the influx of protein-rich fluids and leukocytes inside the brain parenchyma. Leukocytes, pericytes, activated astrocytes, and microglia secrete large amounts of vascular permeability factors which promote BBB hyper-permeability and thereby further increase the extravasation of blood-derived protein-rich fluids into the brain parenchyma, all part of the delayed and progressive vasogenic edema post-TBI (Abdul-Muneer, Pfister, Haorah, & Chandra, 2016; Stefano & Herrero, 2016; Yuan et al., 2016).

Clinical translation

Several pathologies of the CNS involve disturbance of blood–brain barrier (BBB) function, and, in many of these, astrocyte–endothelial cooperation is also abnormal (Abbott et al., 2006). In stroke, Astrocytes secrete transforming growth factor-β, which down-regulates brain capillary endothelial expression of fibrinolytic enzyme tissue plasminogen activator and anticoagulant thrombomodulin, there's proteolysis of the vascular basement membrane/matrix (Lo, Dalkara, & Moskowitz, 2003), induction of aquaporin 4 mRNA and protein at BBB disruption.

Losing the anatomical and functional integrity of the BBB, the capillaries act as fenestrated capillary. In this scenario, the hydrostatic pressure, including systemic arterial pressure and intracranial pressure, plays an important role (Simard, Kent, Chen, Tarasov, & Gerzanich, 2007). For instance, systemic arterial pressure should be sufficiently high to ensure brain perfusion but, at the same time, it should be not too high to allow edema formation. The clinical translation of these principles has produced two different therapeutic approaches. The pharmacological approach has promoted the use osmotically active substances such (mannitol) to counteract brain edema. In the brain, the effective pore size of those junctions is 7Å, while in the rest of the body tight junctions' pores measure at least 65Å (Shapira & Artru, 1998). These pores are small enough to prevent sodium from entering the brain's extracellular space, water crosses the BBB rapidly under the osmotic's gradients, even more rapidly than sodium (3 min vs 1–4 h). Albumin and mannitol are completely excluded from the brain by the intact BBB.

There is a growing understanding of the role and regulation of aquaporins, whose modulation is already being tested in animal studies and may lead to novel therapies to alter not both the location of fluid by osmotic gradient and to augment the brains ability to compensate for edema (Okuno et al., 2008).

The surgical approach (decompressive craniotomy) should be performed to lower the tissue pressure Since decompressive craniotomy may help in restoring the reperfusion by reducing intracranial pressure, it should be safely performed early when the BBB is intact (Mori, Nakao, Yamamoto, & Maeda, 2004). On the contrary, if such a surgical procedure is performed later, during the stage of vasogenic edema, it will provide a decrease in tissue pressure, thus worsening the vasogenic edema formation (Cooper, Hagler, Clark, & Barnett, 1979).

Clinical classification

CE is characterized by a biological process leading to increase the brain volume and consequently to raise the intracranial pressure (see Chapter 15) (Fig. 1). These processes can be followed by a reduction in cerebral perfusion, herniation syndromes and death. Mortality in TBI is strongly influenced by the presence of cerebral edema (Tucker et al., 2017). As previously described, physiopathology is very complex and basic principles such as the distinction between vasogenic edema and cytotoxic edema are well established and validated. However, new frontiers are widely opening to research such as the role of the oligodendroglial cell changes in human traumatic edematous cortex (Castejón, 2015), biomarkers of BBB damage (Blyth et al., 2009), neuroinflammation processes (Corrigan, Mander, Leonard, & Vink, 2016), endothelial-derived influences (Lutton et al., 2019), or age-dependent responses (Hazy et al., 2019). All these mechanisms deeply influence the clinical symptoms and signs of brain swelling that are also dependent on:

I. Focal or diffuse swelling
II. Type of TBI (mild, moderate, or severe)
III. Glasgow Coma Scale (GCS) score (Table 2) ≤ 8 or >8

TABLE 2 Glasgow coma scale components.

Component	Points
Best Eye Opening (E)	4 Eyes opening spontaneously 3 Eyes opening to speech 2 Eyes opening in response to pain 1 No eyes opening
Best Verbal Response (V)	5 Oriented 4 Confused 3 Inappropriate words 2 Incomprehensible sounds 1 None
Best Motor Response (M)	6 Obeys commands 5 Localizes to pain 4 Withdraws from pain 3 Flexion in response to pain 2 Extension to pain 1 No motor response

Total: E+V+M 3/15–15/15.
Ranges: normal = 15.
Minimal brain injury = 13–14.
Moderate brain injury = 9–12.
Coma = 8.
Unresponsive = 3.

Following TBI, ischemic conditions of the brain tissue are frequently observed. When cerebral blood flow (CBF) drops below a critical level of $10\,mL/100\,g\,min^{-1}$, ischemia occurs, and water and electrolyte homeostasis are disturbed. During this "low flow status" cytotoxic edema develops. TBI encompasses a variety of intracranial lesions, including contusions, hematomas, subarachnoid hemorrhages, and diffuse axonal injuries. Type, extent, and speed of the development of post-traumatic brain edema are therefore dependent on the expansion of the primary lesion and associated conditions, such as arterial hypotension or hypoxia (Unterberg et al., 2004).

Focal/perifocal traumatic brain swelling

A traumatic cerebral contusion is characterized by a hemorrhagic core with necrotic tissue, that is poorly or no longer perfused, associated to a decrease of perfusion within the surrounding perifocal tissue (see Chapter "Traumatic brain injury and computed tomography assessment of brain swelling"). In the first 3 days after TBI, brain edema progresses from the core incorporating primarily uninjured tissue adjacent to the lesion (Bullock et al., 1990; Unterberg et al., 1997). Focal as well as perifocal traumatic brain edema is predominantly cytotoxic. It seems, that the focal necrosis is surrounded by a progressively growing edematous zone in peri-contusional tissue as observed by magnetic resonance imaging (Kawamata, Katayama, Aoyama, & Mori, 2000; Maeda, Katayama, Kawamata, Koyama, & Sasaki, 2003). In addition, it is suggested that vasogenic edema occurs in the necrotic core of the contusion and that cytotoxic edema is mainly present in the rim of the contusion, which is affected by secondary injury.

Clinical signs are correlated to ICP increase and to local mass effect.

- Seizures development: focal or generalized seizures, depending on the site of the post-traumatic lesion
- Motor deficit development
- Sensory deficit development
- Speech impairment

Diffuse/global traumatic brain swelling

In addition to focal edema, a diffuse brain swelling may occur after TBI. Unfortunately, this condition is often associated with diffuse brain injury following acceleration/deceleration injuries. In the past, it was believed that a "vascular engorgement" was responsible of global brain swelling (Bouma et al., 1992). Recently, studies in literatures showed that the major contributor for the development of diffuse posttraumatic brain swelling is the critically decreased CBF levels (Marmarou et al., 2000).

Clinical signs are mainly correlated to ICP increase.

- Headache
- Deterioration of consciousness: this implies a lowering in GCS and can be associated to vacant stare, delayed motor and verbal responses, difficulty in focusing the attention, unresponsiveness to stimuli, difficulties in staying asleep.
- Nausea and vomiting
- Fatigue
- Photophobia
- Phonophobia
- Incoordination stumbling

Patients with diffuse brain swelling can also develop systemic complications such as hypoxia and hypotension that can, in return, worsen the edema.

Diagnostic tools, in addition to clinical signs, are neuro-radiological (reduction of subarachnoid spaces and cisterns or hypodensity proximal to post-traumatic contusions/hemorrhages) and instrumental (ICP invasive monitoring). Clinical signs depend on the type of cerebral edema (see Chapter "Traumatic brain injury and computed tomography assessment of brain swelling"). MRI is not a recommended in the examination in unstable patient with brain swelling and increased ICP though ultrasound and CT scans are diagnostically useful. These show brain swelling (Fig. 4A: Ultrasound; Fig. 5B: CT scans) and increased ICP (Fig. 4A and B: Ultrasound).

Clinical characteristics of brain swelling in TBI patients should be classified on the base of type of TBI and GCS (Table 3).

Mild/moderate TBI (GCS > 8)

Mild TBI is characterized by an initial aberration of awareness ranging from confusion to brief loss of consciousness, on the order of a few minutes or less. There may even be post-traumatic amnesia (PTA). Neuro-imaging using CT or conventional

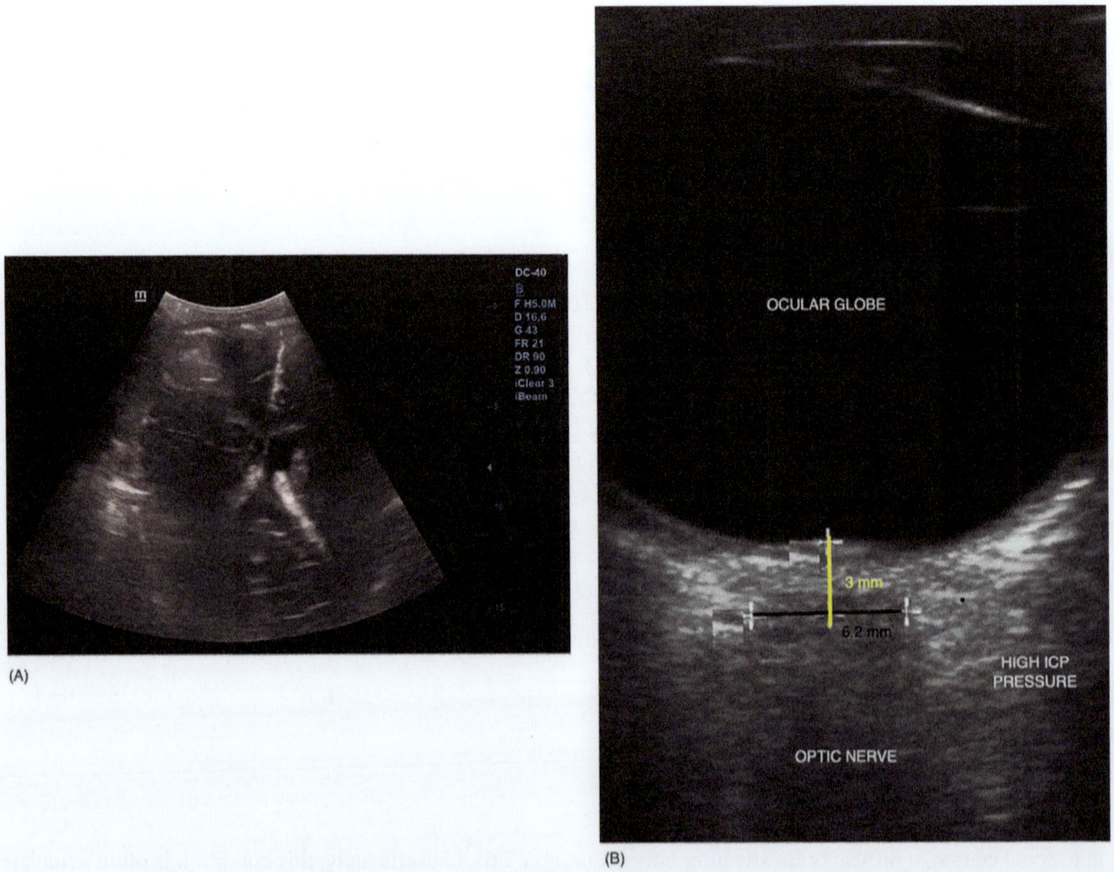

FIG. 4 Axial and coronal US scan of a craniectomized patient showing brain swelling and several contusions after a severe traumatic brain injury (A and B). (B) shows optic nerve sheath diameter (ONSD) obtained from ultrasound imaging of the eye (measurement performed at 3 mm posterior to the globe) and used to calculate intracranial pressure. An increase of ICP > 5 mm is 100% sensitive for elevated ICP.

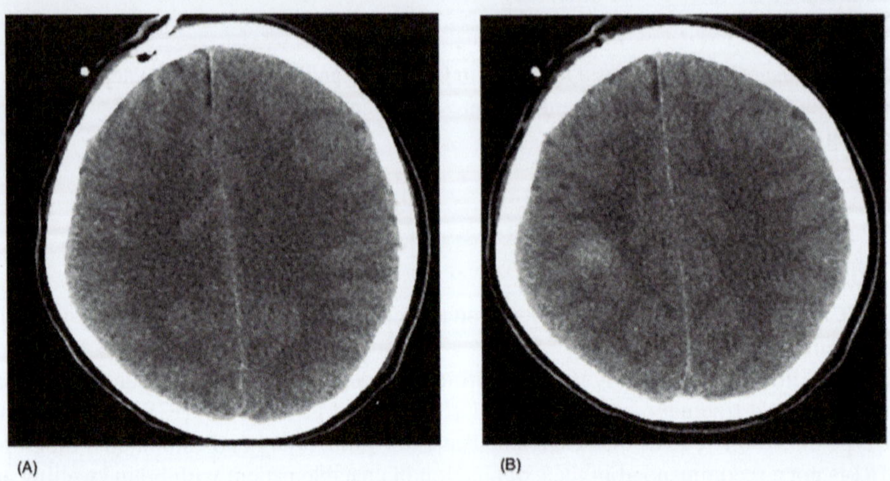

FIG. 5 Diffuse brain swelling. Axial CT images (A and B) showing brain swelling due to severe traumatic brain injury, with no normal representation of subarachnoid spaces.

T1/T2 weighted MRI is typically normal. The patients are usually awake and conscious with different grade of orientation or psychomotor agitation. There are no vital parameters to be monitored which can predict the development of cerebral edema. Moderate brain injuries are characterized by more serious effects, and it is often difficult to be detected. Patients with moderate TBI have great variability in injury severity and acute phase course (Papa et al., 2008). Although in acute phase, both intra and inter-cranial injury may induce secondary brain injury, moderate TBI detection should be done in the

TABLE 3 Clinical classification.

Head Injury severity	Mild	Moderate	Severe
Imaging (CT/MR)	Normal	Normal or abnormal	Normal or abnormal
Loss of consciousness	0–30 min	> 30 min to 24 h	> 24 h
Glasgow coma scale	13–15	9–12	< or 8
Post traumatic amnesia	0–1 day	> 1–7 days	> 7 days
Prognosis	Good Most recover completely	Good	Most impossible to recover completely with physical and/or cognitive disability

shortest time (Maas, Stocchetti, & Bullock, 2008). Moderate TBI (GCS 9–12) is characterized by a loss of consciousness for 1–24 h, and a PTA from 1 to 24 h of the TBI. In addition, it could be characterized by inability to communicate, partial paralysis, and diminished cognitive skills. Patients with moderate brain injuries are more likely to recover than those with a more severe prognosis, and may benefit from rehabilitation therapy. Neuro-imaging using CT or conventional T1/T2-weighted MRI is useful. For patients with mild TBI (GCS < 13 or higher), it is critically important to define a group of patients with TBI without missing a patient with significant TBI that necessitates surgical intervention. However, employing the CT or MRI scan to every patient is expensive and time consuming. Limited resources in the hospitals can also cause the delay of performing CT scan and MRI scan on the patient that poses the risk of moderate TBI.

Severe TBI (GCS ≤ 8)

Severe TBI is characterized by prolonged unconscious state or coma that lasts days, weeks, or months. Symptoms of severe traumatic brain injury can include loss of consciousness, headache, nausea, vomiting, lack of coordination, dizziness, trouble with balance, dilation of one or more pupils, slurred speech, behavioral or mood changes, loss of coordination, restlessness, and agitation. Patients with severe TBI are at high risk for secondary brain injury including hypotension, hypoxemia, and brain swelling (Hukkelhoven et al., 2006). Generally, brain edema is associated with more severe forms of TBI, playing an important role in the outcome and survival of injured patients.

Following severe TBI, there is presumably a pressure gradient where tissue pressure is higher than CSF pressure, leading to movement of interstitial fluid (ISF) to cerebrospinal fluid (CSF). However, following TBI, CSF could be shifted from the cerebral cisterns to the brain, through the para-vascular spaces, leading to a severe brain swelling. It could be assumed that following TBI there is an increased pressure in the subarachnoid compartment, as a consequence of post-traumatic subarachnoid, that, would exacerbate the brain swelling and allow the intra-cerebral pressure to steeply rise. This is an additional phenomenon, "CSF-shift edema," that could be present as additional source of brain swelling in the pathophysiology of post-traumatic edema (Iliff et al., 2012; Yang et al., 2013). By performing cisternostomy, a novel surgical and adjuvant approach, CSF shift is reversed and efflux resistance drops, more fluid can be removed from the cranium, and ICP falls. This approach could replace decompressive craniectomy in severe TBI (see Chapter "Craniotomy for traumatic brain injury") (Grasso, 2015). Large pragmatic, international, multicenter trials are required to evaluate the effectiveness and safety.

The effects of a TBI can vary significantly, depending on the severity. Individuals with a mild TBI generally experience short-term symptoms and feel better within a couple of weeks, whereas individuals with a moderate or severe TBI may have long-term or life-long effects from the injury.

Conclusion(s)

This review discusses history, physiopathology, clinical and radiological characteristics of cerebral edema in general and specifically after traumatic brain injury. Although the horizons of research are still wide and constantly advancing, still several mechanisms are not clearly understood. In this respect, further breakthroughs could lead to acquire and perform better treatments in order to minimally reduce the brain damage and subsequently severe neurological sequelae associated to TBI.

Applications to other areas of neuroscience

In this chapter, we have reviewed the pathogenesis of brain edema and clinical translation of pathologic principles for different therapeutic approaches. We also interlink the role of BBB function in this pathological process. In neurological diseases, disruption of the integrity of the BBB is usually the first pathological change that occurs before clinical symptoms appear. The BBB, as a crucial gate of brain–blood molecular exchange, is involved in the pathogenesis of multiple neurological diseases. Since oxidative stress plays a significant role in the production and maintenance of the BBB, the cerebrovascular system is especially vulnerable to it. In neurological diseases, disruption of the integrity of the BBB is usually the first pathological change that occurs before clinical symptoms appear.

The interconnection between neurological diseases and disruption of the integrity of the BBB is seen in other studies. In neurological diseases, disruption of the integrity of the BBB is usually the first pathological change that occurs before clinical symptoms appear.

Oxidative stress-mediated BBB disruption in other neurological diseases, such as stroke (Kleikers et al., 2012), Alzheimer's disease (Yao, Huang, Chen, Chen, & Li, 2017), and Parkinson's disease (Belarbi, Cuvelier, Destée, Gressier, & Chartier-Harlin, 2017).

Studying the relevance of oxidative stress to the development and outcome of neurological diseases and protecting the BBB in the early stages of diseases will help limit disease progression and improve clinical prognosis.

Mini-dictionary of terms

Brain edema: A number of interconnected processes which result in abnormal shifts of water in various compartments of the brain parenchyma.

Brain swelling: An enlargement of the brain due to excessive fluid collection in the chambers, or ventricles, of the brain or the accumulation of fluid within the brain tissue itself.

Cytotoxic edema: A type of cerebral edema in which extracellular water passes into cells, resulting in their swelling.

Glasgow Coma Scale: A practical method for assessment of impairment of conscious level in response to defined stimuli.

Interstitial edema: A type of cerebral edema caused by an accumulation of cerebral spinal fluid in the brain referred to as hydrocephalus.

Mild TBI: An immediate and transient impairment of neural function with a recovery within 30 min and a GCS of 13–15.

Moderate TBI: An immediate and transient impairment of neural function or 30 min to 24 h, with a GCS of 9–12.

Osmotic edema: A type of cerebral edema caused by osmotic imbalances between blood and tissue. The blood–brain barrier is intact and maintains the osmotic gradient.

Severe TBI: State of coma for more than 24 h will have a GCS of 8 or lower and are at a very high risk for life-long disabilities, a permanent vegetative condition, or death.

Vasogenic edema: An extracellular edema which mainly affects the white matter via leakage of fluid from capillaries (BBB disruption).

Key facts of cerebral edema

Cerebral edema is a major cause of brain damage and contributes significantly to the mortality.

Symptoms includes headaches, nausea, vomiting, seizures, drowsiness, visual disturbances, dizziness, and in severe cases, coma and death.

Diagnosis is based on symptoms and physical examination findings and confirmed by neuroimaging.

Monitoring of the person's airway and intracranial pressure, proper positioning, controlled hyperventilation, medications, fluid management, steroids is recommended.

Extensive cerebral edema can also be treated surgically with a decompressive craniectomy.

Summary points

The blood–brain barrier is peculiar in the genesis of brain edema.

Following traumatic brain injury, various mediators are released which enhance vasogenic and/or cytotoxic brain edema.

Cytotoxic edema manifests minutes after traumatic brain injury.

Vasogenic edema manifests hours after the initial insult.
Focal as well as perifocal traumatic brain edema is predominantly cytotoxic.
The development of diffuse swelling is due to critically decrement of cerebral blood flow levels.
Cerebrospinal fluid could shift from the cerebral cisterns to the brain leading to a severe traumatic brain injury.

Acknowledgments

None.

Conflict of interest

The authors declare no conflict of interest.

References

Abbott, N. J., Rönnbäck, L., & Hansson, E. (2006). Astrocyte-endothelial interactions at the blood-brain barrier. *Nature Reviews. Neuroscience*, *7*(1), 41–53.

Abdul-Muneer, P. M., Pfister, B. J., Haorah, J., & Chandra, N. (2016). Role of matrix metalloproteinases in the pathogenesis of traumatic brain injury. *Molecular Neurobiology*, *53*(9), 6106–6123.

Amiry-Moghaddam, M., & Ottersen, O. P. (2003). The molecular basis of water transport in the brain. *Nature Reviews. Neuroscience*, *4*(12), 991–1001.

Badaut, J., & Bix, G. J. (2014). Vascular neural network phenotypic transformation after traumatic injury: Potential role in long-term sequelae. *Translational Stroke Research*, *5*(3), 394–406.

Badaut, J., Hirt, L., Granziera, C., Bogousslavsky, J., Magistretti, P. J., & Regli, L. (2001). Astrocyte-specific expression of aquaporin-9 in mouse brain is increased after transient focal cerebral ischemia. *Journal of Cerebral Blood Flow and Metabolism*, *21*(5), 477–482.

Belarbi, K., Cuvelier, E., Destée, A., Gressier, B., & Chartier-Harlin, M. C. (2017). NADPH oxidases in Parkinson's disease: A systematic review. *Molecular Neurodegeneration*, *12*(1), 84.

Bield, A., & Kraus, R. (1898). Über eine bisher unbekannte toxische Wirkung der Gallensauren auf das Zentralnervensystem. *Zhl Inn Med*, *19*, 1185–1200.

Blyth, B. J., Farhavar, A., Gee, C., Hawthorn, B., He, H., Nayak, A., et al. (2009). Validation of serum markers for blood-brain barrier disruption in traumatic brain injury. *Journal of Neurotrauma*, *26*(9), 1497–1507.

Boake, C. (1989). A history of cognitive rehabilitation of head-injured patients, 1915 to 1980. *The Journal of Head Trauma Rehabilitation*, *4*(3), 1–8.

Bouma, G. J., Muizelaar, J. P., Stringer, W. A., Choi, S. C., Fatouros, P., & Young, H. F. (1992). Ultra-early evaluation of regional cerebral blood flow in severely head-injured patients using xenon-enhanced computerized tomography. *Journal of Neurosurgery*, *77*, 360–368.

Bullock, R., Statham, P., Patterson, J., Wyper, D., Hadley, D., & Teasdale, E. (1990). The time course of vasogenic edema after focal human head injury—Evidence from SPECT mapping of blood brain barrier defects. *Acta Neurochirurgica. Supplement*, *51*, 286–288.

Castejón, O. J. (2015). Ultrastructural pathology of oligodendroglial cells in traumatic and hydrocephalic human brain edema: A review. *Ultrastructural Pathology*, *39*(6), 359–368.

Cooper, P. R., Hagler, H., Clark, W. K., & Barnett, P. (1979). Enhancement of experimental cerebral edema after decompressive craniectomy: Implications for the management of severe head injuries. *Neurosurgery*, *4*(4), 296–300.

Corrigan, F., Mander, K. A., Leonard, A. V., & Vink, R. (2016). Neurogenic inflammation after traumatic brain injury and its potentiation of classical inflammation. *Journal of Neuroinflammation*, *13*(1), 264.

Csaba, S. (2006). Mechanisms of cell necrosis. *Critical Care Medicine*, *33*(Suppl. 12), S530–S534.

De Bonis, P., Menegatti, E., Cavallo, M. A., et al. (2019). JEDI (jugular entrapment, dilated ventricles, intracranial hypertension) syndrome: A new clinical entity? A case report. *Acta Neurochirurgica*, *161*(7), 1367–1370.

Dore-Duffy, P., Owen, C., Balabanov, R., Murphy, S., Beaumont, T., & Rafols, J. A. (2000). Pericyte migration from the vascular wall in response to traumatic brain injury. *Microvascular Research*, *60*(1), 55–69.

Fishman, R. A. (1975). Brain edema. *The New England Journal of Medicine*, *293*, 706–711.

Foda, M. A., & Marmarou, A. (1994). A new model of diffuse brain injury in rats. Part II: Morphological characterization. *Journal of Neurosurgery*, *80*(2), 301–313.

Goldmann, E. (1909). Die aussere und innere sekretion des gesunden und kranken Organismus im Licht der vitalen Farburg. *Beiträge zur Klinischen Chirurgie*, *64*, 192–265.

Grasso, G. (2015). Surgical treatment for traumatic brain injury: Is it time for reappraisal? *World Neurosurgery*, *84*(2), 594.

Hazy, A., Bochicchio, L., Oliver, A., Xie, E., Geng, S., Brickler, T., et al. (2019). Divergent age-dependent peripheral immune transcriptomic profile following traumatic brain injury. *Scientific Reports*, *9*(1), 8564.

Ho, M. L., Rojas, R., & Eisenberg, R. L. (2012). Cerebral edema. *American Journal of Roentgenology*, *199*(3), W258–W273.

Hukkelhoven, C. W., Rampen, A. J., Maas, A. I., Farace, E., Habbema, J. D., Marmarou, A., et al. (2006). Some prognostic models for traumatic brain injury were not valid. *Journal of Clinical Epidemiology*, *59*(2), 132–143.

Iliff, J. J., Wang, M., Liao, Y., Plogg, B. A., Peng, W., Gundersen, G. A., et al. (2012). A paravascular pathway facilitates CSF flow through the brain parenchyma and the clearance of interstitial solutes, including amyloid β. *Science Translational Medicine*, *4*(147), 147ra111.

Kaufman, H. H. (1993). Treatment of head injuries in the American civil war. *Journal of Neurosurgery*, *78*(5), 838–845.

Kawamata, T., Katayama, Y., Aoyama, N., & Mori, T. (2000). Heterogeneous mechanisms of early edema formation in cerebral contusion: Diffusion MRI and ADC mapping study. *Acta Neurochirurgica. Supplement*, *76*, 9–12.

Klatzo, I. (1967). Presidential address: Neuropathological aspects of brain edema. *Journal of Neuropathology and Experimental Neurology*, *26*, 1–14.

Kleikers, P. W., Wingler, K., Hermans, J. J., Diebold, I., Altenhöfer, S., Radermacher, K. A., et al. (2012). NADPH oxidases as a source of oxidative stress and molecular target in ischemia/reperfusion injury. *Journal of Molecular Medicine (Berlin, Germany)*, *90*(12), 1391–1406.

Kuchiwaki, H., Faruse, M., Gonda, T., Hirai, N., Inao, S., Ikeyama, A., et al. (1984). Studies on changes in tissue water constitution and focal brain tissue pressure in experimental cerebral infarction. In K. G. Go, & A. Baethmann (Eds.), *Vol. 1984. Recent progress in the study and therapy of brain edema* (pp. 459–470). Springer US.

Lewandowsky, M. (1900). Zur Lehre der Zerebrospinalflussigkeit. *Zeitschrift für Klinische Medizin*, *40*, 480–484.

Li, W., Watts, L., Long, J., Zhou, W., Shen, Q., Jiang, Z., et al. (2016). Spatiotemporal changes in blood–brain barrier permeability, cerebral blood flow, T2 and diffusion following mild traumatic brain injury. *Brain Research*, *1646*, 53–61.

Liang, D., Bhatta, S., Gerzanich, V., & Simard, J. M. (2007). Cytotoxic edema: Mechanisms of pathological cell swelling. *Neurosurgical Focus*, *22*(5), E2.

Lo, E. H., Dalkara, T., & Moskowitz, M. A. (2003). Mechanisms, challenges and opportunities in stroke. *Nature Reviews Neuroscience*, *4*(5), 399–414.

Lutton, E. M., Farney, S. K., Andrews, A. M., Shuvaev, V. V., Chuang, G. Y., Muzykantov, V. R., et al. (2019). Endothelial targeted strategies to combat oxidative stress: Improving outcomes in traumatic brain injury. *Frontiers in Neurology*, *10*, 582.

Maas, A. I., Stocchetti, N., & Bullock, R. (2008). Moderate and severe traumatic brain injury in adults. *Lancet Neurology*, *7*(8), 728–741.

Maeda, T., Katayama, Y., Kawamata, T., Koyama, S., & Sasaki, J. (2003). Ultra-early study of edema formation in cerebral contusion using diffusion MRI and ACD mapping. *Acta Neurochirurgica. Supplement*, *89*, 329–331.

Mantovani, G., Menegatti, M., Scerrati, A., Cavallo, M. A., & De Bonis, P. (2018). Controversies and misconceptions related to cerebrospinal fluid circulation: A review of the literature from the historical Pioneers' theories to current models. *BioMed Research International*, *2018*, 2928378.

Marmarou, A., Fatouros, P. P., Barzó, P., Portella, G., Yoshihara, M., Tsuji, O., et al. (2000). Contribution of edema and cerebral blood volume to traumatic brain swelling in head-injured patients. *Journal of Neurosurgery*, *93*, 183–193.

Mori, K., Nakao, Y., Yamamoto, T., & Maeda, M. (2004). Early external decompressive craniectomy with duroplasty improves functional recovery in patients with massive hemispheric embolic infarction: Timing and indication of decompressive surgery for malignant cerebral infarction. *Surgical Neurology*, *62*(5), 420–429.

Muresanu, D. F., Sharma, A., Patnaik, R., Menon, P. K., Mössler, H., & Sharma, H. S. (2019). Exacerbation of blood-brain barrier breakdown, edema formation, nitric oxide synthase upregulation and brain pathology after heat stroke in diabetic and hypertensive rats. Potential neuroprotection with cerebrolysin treatment. *International Review of Neurobiology*, *146*, 83–102.

Okuno, K., Taya, K., Marmarou, C. R., Ozisik, P., Fazzina, G., Kleindienst, A., et al. (2008). The modulation of aquaporin-4 by using PKC-activator (phorbol myristate acetate) and V1a receptor antagonist (SR49059) following middle cerebral artery occlusion/reperfusion in the rat. *Acta Neurochirurgica. Supplement*, *102*, 431–436.

Oury, T. D., Piantadosi, C. A., & Crapo, J. D. (1993). Cold-induced brain edema in mice. Involvement of extracellular superoxide dismutase and nitric oxide. *The Journal of Biological Chemistry*, *268*, 15394–15398.

Papa, L., Robinson, G., Oli, M., Pineda, J., Demery, J., Brophy, G., et al. (2008). Use of biomarkers for diagnosis and management of traumatic brain injury patients. *Expert Opinion on Medical Diagnostics*, *2*(8), 937–945.

Raslan, F., Schwarz, T., Meuth, S. G., Austinat, M., Bader, M., Renné, T., et al. (2010). Inhibition of bradykinin receptor B1 protects mice from focal brain injury by reducing blood-brain barrier leakage and inflammation. *Journal of Cerebral Blood Flow and Metabolism*, *8*, 1477–1486.

Reese, T. S., & Karnovsky, M. J. (1967). Fine structural localization of a blood-brain barrier to exogenous peroxidase. *The Journal of Cell Biology*, *34*(1), 207–217.

Reichardt, M. (1905). Zur Entstehung des Hirndrucks bei Hirngeschwülsten und anderen Hirnkrankheiten und über eine hei diesen zu beobachtende besondere Art der Hirnschwellung. *Deutsche Zeitschrift für Nervenheilkunde*, *28*, 306.

Shah, S., & Kimberly, W. T. (2016). Today's approach to treating brain swelling in the neuro intensive care unit. *Seminars in Neurology*, *36*(6), 502–507.

Shapira, Y., & Artru, A. A. (1998). Cerebral edema in brain ischemia and trauma. In A. Gullo (Ed.), *Anaesthesia, pain, intensive care and emergency medicine—A.P.I.C.E* (pp. 265–275). Springer Milan.

Simard, J. M., Kent, T. A., Chen, M., Tarasov, K. V., & Gerzanich, V. (2007). Brain oedema in focal ischaemia: Molecular pathophysiology and theoretical implications. *The Lancet Neurology*, *6*(3), 258–268.

Song, J., Wu, C., Korpos, E., Zhang, X., Agrawal, S. M., Wang, Y., et al. (2015). Focal MMP-2 and MMP-9 activity at the blood–brain barrier promotes chemokine-induced leukocyte migration. *Cell Reports*, *10*(7), 1040–1054.

Stefano, M. E., & Herrero, M. T. (2016). The multifaceted role of metalloproteinases in physiological and pathological conditions in embryonic and adult brains. *Progress in Neurobiology*, *155*, 36–56.

Stokum, J. A., Gerzanich, V., & Simard, J. M. (2016). Molecular pathophysiology of cerebral edema. *Journal of Cerebral Blood Flow and Metabolism*, *36*(3), 513–538.

Sweeney, M. I., Yager, J. Y., Walz, W., & Juurlink, B. H. (1995). Cellular mechanisms involved in brain ischemia. *Canadian Journal of Physiology and Pharmacology*, *73*(11), 1525–1535.

Tucker, B., Aston, J., Dines, M., Caraman, E., Yacyshyn, M., McCarthy, M., et al. (2017). Early brain edema is a predictor of in-hospital mortality in traumatic brain injury. *The Journal of Emergency Medicine*, *53*(1), 18–29.

Unterberg, A. W., Stover, J., Kress, B., & Kiening, K. L. (2004). Edema and brain trauma. *Neuroscience*, *129*(4), 1021–9.45.

Unterberg, A. W., Stroop, R., Thomale, U. W., Kiening, K. L., Päuser, S., & Vollmann, W. (1997). Characterization of brain edema following controlled cortical impact injury in rats. *Acta Neurochirurgica. Supplement*, *70*, 106–108.

Van Harrefeld, A., Crowell, J., & Malhotra, S. K. (1966). Water, electrolyte and extracellular space in hydrated and dehydrated brains. *The American Journal of Physiology*, *210*, 251–256.

Yang, L., Kress, B. T., Weber, H. J., Thiyagarajan, M., Wang, B., Deane, R., et al. (2013). Evaluating glymphatic pathway function utilizing clinically relevant intrathecal infusion of CSF tracer. *Journal of Translational Medicine*, *11*, 107.

Yao, Y., Huang, J. Z., Chen, L., Chen, Y., & Li, X. (2017). In vivo and in vitro studies on the roles of p38 mitogen-activated protein kinase and NADPH-cytochrome P450 reductase in Alzheimer's disease. *Experimental and Therapeutic Medicine*, *14*(5), 4755–4760.

Yuan, J., Wang, A., He, Y., Si, Z., Xu, S., Zhang, S., et al. (2016). Cordycepin attenuates traumatic brain injury-induced impairments of blood–brain barrier integrity in rats. *Brain Research Bulletin*, *127*, 171–176.

Zamboni, P., Scerrati, A., Menegatti, E., Galeotti, R., Lapparelli, M., Traina, L., et al. (2019). The eagle jugular syndrome. *BMC Neurology*, *19*(1), 333.

Chapter 16

Post-coma syndrome in the context of severe acquired brain injury: Traumatic brain injury and beyond

Rita Formisano, Marta Aloisi, and Marianna Contrada
Post-Coma Unit, IRCCS Santa Lucia Foundation, Rome, Italy

List of abbreviations

AEPs	acoustic evoked potentials
BAERS	brainstem auditory evoked potential
CIP	critical illness polyneuropathy
CRIMYNE	critical illness myoneuropathy
CRS-R	coma recovery scale-revised
DoC	disorders of consciousness
***E*-MCS**	emergence/exit from minimally conscious state
ERPs	event-related potentials
FAST	familiar auditory sensory training
fLIS	functional locked-in syndrome
fMRI	functional magnetic resonance imaging
GOAT	Galveston orientation and amnesia test
GOS	Glasgow Outcome Scale
GOSE	Glasgow Outcome Scale extended
GOSE-R	Glasgow Outcome Scale extended revised
ICU	intensive care unit
ICU-AW	ICU acquired weakness
LCF	level of cognitive functioning
MCS	minimally conscious state
MCS-	minimally conscious state minus
MCS+	minimally conscious state plus
NCS-R	Nociception Coma Scale-revised
PAO	periarticular ossification
PEG	percutaneous endoscopic gastrostomy
PET	positron emission tomography
PSH	paroxysmal sympathetic hyperactivity
PTA	post-traumatic amnesia
sABI	severe acquired brain injury
SMART	sensory modality assessment and rehabilitation technique
SPECT	single-photon emission computed tomography
TBI	traumatic brain injury
tDCS	transcranial direct current stimulation
TMS	transcranial magnetic stimulation
UWS	unresponsive wakefulness syndrome

Diagnosis and Treatment of Traumatic Brain Injury. https://doi.org/10.1016/B978-0-12-823347-4.00029-4

VAS	visual analogue scale
VEPs	visual evoked potentials
VS	vegetative state

Definition issues

Severe acquired brain injury

Severe acquired brain injury (sABI), according to the Italian Guidelines of Rehabilitation, is defined as "a neurological condition due to a severe acquired brain damage with a coma lasting at least 24 hours, according to a Glasgow Coma Scale equal or lower than 8 **and/or** complex and severe neurological disabilities treatable only in high neurorehabilitative specialty wards" (Guidelines Minister of Health, 1998).

Disorders of consciousness

Disorders of consciousness (DoC) after sABI include: coma, vegetative state (VS), recently named unresponsive wakefulness syndrome (UWS) (Laureys et al., 2010) and minimally conscious state (MCS) (Fig. 1).

DoC may have different etiologies, including: traumatic, vascular, anoxic, infective, toxic, and others (Table 1).

Coma

Definition of coma includes the clinical triad of "closed eyes, not obeying simple commands and no comprehensible verbal utterances" (Jennett, 1986). Coma may last from few hours to several weeks and comatose patients generally need mechanical ventilation, parenteral and/or enteral nutrition, support of vital functions and sometimes monitoring of the intracranial pressure. In case of coma persistence, the correct timing of tracheostomy from intubation and percutaneous endoscopic gastrostomy (PEG) for enteral nutrition has to be individually decided, although nasogastric tube should be maintained no longer than 1 month, because of risks of esophageal ulcers and esophago-tracheal fistulae.

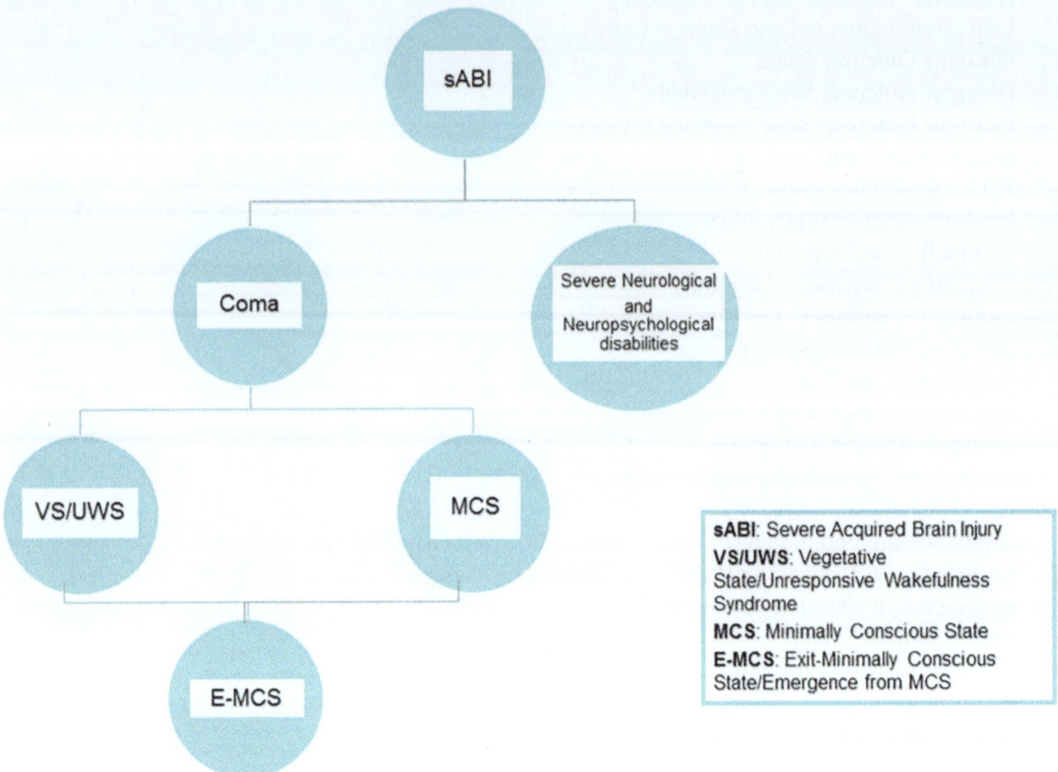

FIG. 1 Disorders of consciousness. The figure shows the consequences of sABI.

TABLE 1 Types of injuries under the umbrella of severe acquired brain injury.

Traumatic	Vascular	Anoxic		Other
Traumatic brain injury	Ischemic or hemorrhagic stroke	Cardiac failure		Cerebral infections
	Subarachnoid hemorrhage	Respiratory failure		Tumors
		Drowning		Sequelae of neurosurgical interventions
		Hanging		
		Drug intoxication		
		Carbon oxide intoxication	⟷	
			⟷	Suicidal attempts
		Anesthesiological accidents	⟷	
			⟷	Opioids overdose
		Epileptic state/status epilepticus	⟷	

The table shows different etiologies of brain injury. Please, note that some etiologies were classified in two categories, since the final effect could be as well as cardio-respiratory failure.

Also, the ideal time to transfer the patient to post-acute neurorehabilitation facilities has to be carefully evaluated, to avoid back transfers to Intensive Care Unit (ICU) (Formisano, Azicnuda, et al., 2017).

Early tracheostomy seems to reduce ICU length of stay (Gabrielli & Layon, 2004), because of the possible weaning from mechanical ventilation within 48 h of tracheostomy.

Vegetative State/Unresponsive Wakefulness Syndrome

Vegetative state (VS), more recently named unresponsive wakefulness syndrome (UWS) is a condition that follows coma, when the patient recovers the vigilance or alertness (eyes opening), but not the awareness of self and surroundings. The patient, in fact, is unable to interact with the environment, in spite of the eyes opening and the recovery of some sleeping-wake cycle. The interval for the definition of "permanent" is considered 1 year for the traumatic cases and 3–6 months for the hypoxic and non-traumatic patients.

More recently the term UWS has been proposed as more neutral and respectful definition in comparison with VS (Formisano, Pistoia, & Sarà, 2011; Zasler, Aloisi, Contrada, & Formisano, 2019).

A possible undiagnosed epileptic activity at admission in ICU may represent the clinical correlate of non-convulsive epileptic state in prolonged DoC (Vespa et al., 2007).

Minimally Conscious State (MCS)

This state may follow either coma or VS/UWS as a transient or permanent condition (Beaumont & Kenealy, 2005) and it should be included as a possible outcome category of the Glasgow Outcome Scale (GOS) (Jennett & Bond, 1975), because of the gap between "VS" and "severe disability" categories. More recently a revised version of the GOS Extended (GOS-E) (McMillan et al., 2016) has been proposed as GOS-E revised (GOSE-R) (Formisano, Aloisi, Ferri, Schiattone, & Contrada, 2018; Formisano, Contrada, Ferri, et al., 2019).

Consciousness level may be affected by several medical, neurological, iatrogenic and neurosurgical complications, such as recurrent infections and sub-continuous fever; convulsive and non-convulsive seizures; normotensive or hypertensive hydrocephalus (Missori et al., 2006; Wang et al., 2020); psychomotor agitation, restlessness, aggressiveness, oppositive behaviors; sedative, antiepileptic and muscle-relaxant drugs; all conditions in which DoC may fluctuate.

Some of these features were taken into account in the proposed Brief Post-Coma Scale (Formisano, Aloisi, Ferri, et al., 2018; Formisano, Aloisi, Iosa, et al., 2018).

As for the diagnosis of akinetic mutism, many MCS patients may be able to show optical fixation, eye pursuing and tracking and may be also mute and akinetic; therefore, the definition of akinetic mutism has been classified as a MCS subcategory (Giacino & Kalmar, 1997).

On the other hand, psychomotor agitation and aggressiveness are frequent behaviors in severe brain injury patients. Aggressiveness might be also the manifestation of physical or emotional discomfort or of mental confusion that the patient is not able to express otherwise. For these reasons agitated patients should also be diagnosed as MCS instead of VS/UWS (Formisano, D'Ippolito, et al., 2011; Formisano, Pistoia, and Sarà, 2011).

MCS is a condition of severe consciousness alteration in which a minimal but not constant awareness of self and surroundings is present (Giacino, 2004).

MCS has been more recently classified in MCS MINUS (MCS −), in which patients show non-reflex responses, such as localizing noxious stimuli or visual pursuing of a moving object/person or emotional reactions to salient stimuli and MCS PLUS (MCS +), in which patients show intelligible verbalization, yes/no responses, or command-following (Bruno, Vanhaudenhuyse, Thibaut, Moonen, & Laureys, 2011).

The emergence from MCS, also defined as Exit–MCS (E-MCS), is diagnosed when reliable and consistent evidence of either functional communication or functional object use can be documented (Giacino, 2004).

Functional Locked-in Syndrome

The term "functional locked-in syndrome" (fLIS) has been proposed for patients with a dissociation between extreme motor dysfunction and preserved higher cortical functions, only identified by functional imaging techniques (para-clinical tests) (Bruno et al., 2011). Recently, advanced electrophysiologic techniques were able to reveal language disorder in some non-communicative patients with DoC (Formisano, Toppi, Risetti, et al., 2019). More recently, Cruse et al. (2011) demonstrated that it is also possible to detect the hidden awareness of patients using an electroencephalography (EEG) technique, by means of mental imagery tasks of motor commands following. Indeed, it has been demonstrated that some patients, who are behaviorally in a VS/UWS, in whom residual cortical processing has been found by means of functional Magnetic Resonance (fMRI) or neurophysiology may also present with willful communication ability (Monti et al., 2019; Owen et al., 2006). Thus, functional neuroimaging and advanced electrophysiology tools could be used, together with the caregivers involvement in the responsiveness assessment, in doubtful cases of DoC, also for nutrition/hydration withdrawal decision making (Formisano & Zasler, 2020).

Covert cognition

Patients with residual cognitive functions (covert cognition) universally are referred to as patients with Cognitive-Motor dissociation (CMD) (Curley, Forgacs, Voss, Conte, & Schiff, 2018), who seem to have better functional outcomes. The Motor Behavior Tool revised (MBT-r) has been suggested to help detect even very subtle signs of motor behavior and thereby to better differentiate patients with CMD from other patients categories (Møller, Højlund, Jensen, Gansonre, & Shtyrov, 2020).

This subgroup of DoC patients might be included in the percentage of misdiagnosis reported in literature as around 40% (Schnakers et al., 2009).

In summary, the condition of covert cognition has been described with several different definitions, such as: fLIS (Formisano, D'Ippolito, & Catani, 2013); non-behavioral MCS (Stender et al., 2014); CMD; but besides the terms, revealing such condition by means of advanced diagnostic techniques (EEG with mental imagery tasks, event-related-evoked potentials, functional MRI, SPECT, PET) is a duty towards the patients and their families.

Prognostic indicators

Clinical predictive factors

Several positive predictive clinical features have been reported as significant, in patients with a duration of unconsciousness of at least 28 days, as in the VS/UWS definition of the American Guidelines (Giacino et al., 2018). Indeed, spontaneous motor activity, sustained optical eye tracking and command following, as well as the interval of safe oral feeding recovery and psychomotor agitation may predict a statistically significant better outcome (Formisano et al., 2005).

Post-traumatic psychomotor agitation represents, according to some authors (Corrigan & Bogner, 1994) a positive indicator for consciousness recovery. Also the Klüver-Bucy syndrome (Formisano, Saltuari, & Gerstenbrand, 1995) has been reported as a possible favorable predictive indicator in prolonged DoC and diagnosed as a transient phase with behavioral disinhibition (Gościński, Kwiatkowski, Polak, & Orlowiejska, 1997).

Pathological postures in decerebration and decortication (Figs. 2 and 3) are rarely described in literature (Dolce & Sazbon, 2002), although they generally represent a negative indicator of the long-term functional outcome.

A dysautonomic syndrome, more recently defined as paroxysmal sympathetic hyperactivity (PSH), has been reported as a negative prognostic indicator of the final outcome (Lucca et al., 2019). Another negative predictive feature for consciousness recovery may be the presence of primitive oral automatisms, such as chewing, sucking, and yawning (Dolce & Sazbon, 2002).

Mioclonic jerks may be secondary to cortical or brainstem lesions or cerebral hypoxic damage and they are also early predictive indicators of poor prognosis for long-lasting disability (Caltagirone, Imbriani, & Piras, 2021).

Evaluation scales

During the first phase following the coma, responsiveness monitoring can be conducted with different scales, as described by Seel et al. (2010); the CRS-R (Coma Recovery Scale-Revised) (Giacino & Trott, 2004) is however considered the gold standard of responsiveness assessment in patients with DoC (Giacino et al., 2018; Kondziella et al., 2020). Another utilized assessment scale is the STAR (Sensory Tool to Assess Responsiveness) (Stokes, Gunn, Schouwenaars, & Badwan, 2018), with an agreement at a good level with the CRS-R.

FIG. 2 Decortication. The figures show patient with post-traumatic pathological postures. www.arco92.org

FIG. 3 Decerebration. The figures show patient with post-traumatic pathological postures. www.arco92.org

Patients with DoC suffer from fluctuations in responsiveness; therefore, the number of assessments by means of CRS-R has a significant effect on the truthfulness of the diagnosis and a minimum of five assessments within a short time interval (e.g., 2 weeks), seems to reduce the risk of misdiagnosis (Wannez, Heine, Thonnard, Gosseries, & Laureys, 2017).

Formisano, Contrada, Iosa, et al. (2019) demonstrate that the inclusion of the caregiver in the CRS-R assessment promotes more reliable signs of responsiveness than the scores obtained with the professional only.

Considering the painful standard stimulus provided by the CRS-R scale, it has been found that patients with no responses to this latter stimulation, may show more responses to personalized painful stimuli in 43% of a sample of patients with DoC (Formisano, Contrada, Aloisi, et al., 2019), thus, the administration of the CRS-R scale may be integrated with the use of the NCS-R (Chatelle, Majerus, Whyte, Laureys, & Schnakers, 2012). The DoC-Feeling Visual Analogue Scale (VAS) also allows to look for information by all professionals and caregivers involved in the care of patients with DoC (Hermann et al., 2019).

Post-traumatic amnesia (PTA), defined as the time interval following TBI during which the patient is not able to memorize everyday events of the last 24 h (Russell & Smith, 1961), is assessed on a prospective daily basis, using validated tools until resolution, such as the Galveston Orientation and Amnesia Test (GOAT) (Levin, O'Donnell, & Grossman, 1979) and other specific scales (Caltagirone, Imbriani, & Piras, 2021). A retrospective assessing of PTA is commonly used, asking the patients to recall their first memory following the TBI or coma onset, using the McMillan "Questionnaire" (1996).

Nakase-Thomson, Sherer, Yablon, Nick, and Trzepacz (2004) and Sherer, Nakase-Thompson, Yablon, and Gontkovsky (2005) also proposed the Confusion Assessment Protocol (CAP), whereas other authors (Snow & Ponsford, 2013) recommended to plan a secure, quiet and consistent environment, to avoid overstimulation and to minimize triggering of agitation.

In conclusion, a holistic and inter-professional evaluation of the patients with sABI, from DoC to PTA resolution, is mandatory, involving also the family in the neuropsychological rehabilitation project (Caltagirone, Imbriani, & Piras, 2021).

Instrumental diagnostic and prognostic indicators

The availability of reliable prognostic markers could help clinicians in identifying patients with higher likelihood for clinical recovery and survival (Estraneo & Trojano, 2018). The European Academy of Neurology guidelines on the diagnosis of coma and other DoC (Kondziella et al., 2020) and a recent international survey (Formisano, Giustini, Aloisi, et al., 2019) showed that multimodal assessment using PET, EEG-based techniques, including evoked potentials, fMRI, together with standardized clinical behavioral scales, provides more robust evaluation of consciousness than routine bedside examination alone.

Neurophysiology

Visual EEG analysis, including reactivity to stimuli and the presence of sleep architecture, may be used for the differentiation between VS/UWS and MCS as a part of a multimodal assessment; in fact, the presence of slow-wave sleep, sleep stage 3 and REM sleep are possibly more accurate for the differentiation between VS/UWS and MCS than sleep spindles (Giubilei et al., 1995; Kang et al., 2014; Valente et al., 2002).

Several studies showed that better EEG background organization at baseline was associated with better outcome (Risetti et al., 2013; Toppi et al., 2016). Moreover, EEG reactivity to eye opening and closing to acoustic stimuli is significantly more frequent in patients who improve their level of consciousness at 6 months after severe brain injury with DoC (Estraneo et al., 2020).

The alterations of the Somatosensory Evoked Potentials (SEPs), Acoustic Evoked Potentials (AEPs)/Brainstem Auditory Evoked Potential (BAERS) and flash Visual Evoked Potentials (VEPs) have a wide predictive value when are considered together (Houlden et al., 2010).

Prognostic value of neurophysiological assessment increases if long-latency auditory event-related potentials (ERPs) are also taken into account (Caltagirone, Imbriani, & Piras, 2021; Toppi et al., 2016). In a meta-analysis the presence of positive component occurring 300 ms after the stimulus (P300), moduled by arousal and attention, appeared to be a reliable predictor of recovery of consciousness in comatose patients following vascular, traumatic, and metabolic encephalopathy (Daltrozzo, Wioland, Mutschler, & Kotchoubey, 2007). The strong relationship between AEPS/BAERS and recovery of consciousness might be indicative for the initial consolidation of neural network underlying communicative ability of patients' interaction with the environment (Formisano, Toppi, Risetti, et al., 2019). Current evidence suggests that repetitive Transcranial Magnetic Stimulation (TMS)-EEG has high sensitivity and specificity for the differentiation between VS/UWs and MCS, as well as motor imagery tasks assessed by EEG were able to reveal some covert cognition (Curley et al., 2018; Massimini, Ferrarelli, Sarasso, & Tononi, 2012).

In summary, multimodal assessment based on advanced EEG and neuroimaging techniques is useful to detect covert consciousness, if present, and to avoid misdiagnosis in patients without command following or other behavioral signs of consciousness at the bedside (Zasler et al., 2019).

Neuroimaging

More recent MRI technologies, such as DTI techniques, perfusion MRI, Turbo-Pepsy (Giugni, Sabatini, Hagberg, Formisano, & Castriota-Scanderbeg, 2005) and functional MRI (fMRI), may provide respectively better information for prognosis and shorter time of acquisition, although a consistent relationship between MRI lesions and clinical or neuropsychological outcomes has been only rarely demonstrated (Ciurli et al., 2010).

SPECT appears to be better than CT or MRI in determining long-term prognosis (Baulieu et al., 2001). Current evidence suggests that resting state PET has a high sensitivity and specificity for the differentiation between VS/UWS and MCS. Neuroimaging techniques analyze structural connectivity (DTI) and functional connectivity in the resting brain (resting state fMRI and cerebral F-fluorodeoxyglucose (PET), investigating the functioning of the neural network related to the awareness state, the so-called default-mode network or intrinsic network (Raichle et al., 2001). This neural network includes precuneus, medial frontal cortex, and the temporal-parietal junction bilaterally and is consistently characterized by a high activation at rest. The study of spontaneous neural activity in "resting state" in DoC patients has demonstrated that preservation of functional connectivity in DMN has been related to the recovery of consciousness in VS/UWS patients after 3 months (Qin et al., 2015).

Passive and active fMRI paradigms also allow identification of a specific and important group of patients, who can follow commands despite appearing completely unresponsive at the bedside (covert cognition) (Monti et al., 2019; Owen et al., 2006). Salient stimuli and/or familiar pictures or voices may increase sensitivity in active and passive fMRI paradigms compared to non-salient stimuli (De Pasquale et al., 2015).

Sequelae of severe Acquired BRAIN Injury (sABI)

Post-traumatic amnesia

As soon as the patient becomes responsive, agitation and restlessness often occur and the recurrence of communication skills reveals a clear state of confusion. This period, called PTA, coincides with level 4 of the Level of Cognitive Functioning (LCF) (Hagen, 2000), while level 5 is characterized by a reduction of agitation, with patients who may be responsive, but confused or inert and apathetic (i.e., akinetic mutism). Because of the presence of all these disturbances, besides memory impairment, the use of the term "Post-Traumatic Confusional State" or "Post Traumatic Delirium" have been proposed, rather than PTA, since all post-coma patients, with different etiologies (not only TBI) may pass through a similar phase, this condition should be named "Post-Coma Confusional State or Post-Coma Amnesia" (PCA) (Caltagirone, Imbriani, & Piras, 2021).

A reliable PTA/PCA assessment is important to be collected, as one of the most significant predictive factors for the final outcome of severe TBI and sABI, as well as for legal considerations and to monitor cognitive–behavioral progress.

Neuropsychological disorders

The resolution of PTA/PCA phase in post-coma patients corresponds to the achieving of level 7 of LCF-R scale (Automatic-Appropriate), in which patients begin to be sufficiently oriented to person, time and place. The variable combination of pathological aspects regarding specific cognitive functions does not usually arise as a simple sum of single deficits, but as a compound and pervasive condition, going from milder to more severe forms involving all cognitive, behavioral, psychological, and relational skills (Caltagirone, Imbriani, & Piras, 2021) (Table 2).

Personality and behavioral changes, also impact on family, social, vocational reintegration and on quality of life (D'Ippolito et al., 2018; Formisano, Silvestro, et al., 2017) (Table 3).

TABLE 2 Cognitive disorders.

Attentional deficits	Pero, Incoccia, Caracciolo, Zoccolotti, and Formisano (2006), Formisano, Bivona, Brunelli, Giustini, and Taggi (2001), and Bivona et al. (2014)
Learning and memory deficits	Carlesimo et al. (1998) and Carlesimo, Formisano, Bivona, Barba, and Caltagirone (2009)
Executive-function deficits	Bivona et al. (2008) and Ciurli et al. (2010)

The table shows the most common neuropsychological disorders.

TABLE 3 Personality and behavioral changes.

Disruptive primary behaviors by excess
Disruptive primary behaviors by default
Affective and psychiatric disorders (depression, anxiety, obsessive-compulsive disorder, psychosis or disorders affecting awareness and social cognition; Bivona et al., 2014; Ciurli, Formisano, Bivona, Cantagallo, & Angelelli, 2011)

The table shows post-traumatic personality and behavioral changes.

Neuromotor disorders

The main aim of the acute phase is the prevention of secondary damage, as in the prolonged bed-rest syndrome, such as bed sores, muscle atrophy and contractures, ICU-Acquired Weakness (ICU-AW), osteoporosis, cardiovascular and respiratory complications, joint ankylosis and peri-articular ossification (PAO) (Caltagirone, Imbriani, & Piras, 2021).

Indeed, recurrent sepsis may cause diffuse muscular atrophy by Critical Illness Polineuropathy (CIP)/Critical Illness Myoneuropathy (CRIMYNE)/ICU-AW (Intiso et al., 2011), or compression neuropathies, due to pathological postures, such as decortication, decerebration or triple flexion at the lower limbs.

The adequate posturing of the patient and the frequent posture changes is fundamental to prevent and counteract the pathological postures and the shortening of the joint capsules and tendons, also by means of anti-decubitus mattresses, soft splinting, reclining chairs, and early verticalization.

During the post-acute phase, the patients with sABI need to be admitted in high specialty neurorehabilitation wards (Formisano, 2020), where a multidisciplinary team is able to propose a patient-centered, personalized and integrated Individual Neurorehabilitation Project (INRP), aimed at obtaining the maximum possible neuromotor and neuropsychological recovery, defining achievable goals at brief, medium and long-term.

The initial functional evaluation investigates the level of vigilance, responsiveness, collaboration, spatial and temporal orientation and the presence of diffuse pain, sensory and motor deficits. The accuracy of the assessment is essential to tailor the INRP (Caltagirone, Imbriani, & Piras, 2021).

Intervention—A multidisciplinary approach

The Sensory Regulation Program could be the reference model for stimulation conducted in the neurorehabilitation of patients with sABI (Caltagirone, Imbriani, & Piras, 2021). According to Wood, Winkowski, Miller, Tierney, and Goldman (1992), the diffuse lesions of white matter, such as in diffuse axonal injury, hamper an adequate reaction to environmental stimuli. The sensory stimulation has to be given softly and slowly, organizing the environment as more reassuring and familiar as possible. This model requires an integrated teamwork and a specific training for the caregiver. A stimulation protocol in which the team, family, and friends of the patient work together is the Sensory Modality Assessment and Rehabilitation Technique (SMART) (Gill-Thwaites & Munday, 1999), a comprehensive program of assessment and intervention (Caltagirone, Imbriani, & Piras, 2021).

Another important aim of the neurorehabilitation team in the first recovery phases is the identification of a communication channel, which needs a prolonged observation of cognitive, sensory, and neuro-motor abilities. Therefore, a holistic approach involving the entire neurorehabilitation team is important, in order to introduce verbal and non-verbal communication responses and some technological compensations to stabilize responses (Caltagirone, Imbriani, & Piras, 2021).

Caregivers need to be guided and educated with their involvement in the neurorehabilitation program (Caltagirone, Imbriani, & Piras, 2021), indeed protocols like the Familiar Auditory Sensory Training (FAST) (Pape et al., 2012) aim to stimulate the patient with structuring methods and family times of relationship and support caregivers in becoming aware of new interaction ways, to reduce their inadequacy and helplessness.

The holistic neurorehabilitation approach to patients with sABI include respiratory and swallowing training, inclusive of more advanced techniques, such as neuromuscular taping and Vitalstim, neuropsychological evaluation and treatment, advanced robotic rehabilitation, telerehabilitation, dual task training and assessment through alternative-augmentative communication devices, including brain-computer interface (BCI) approach, occupational therapy, aquatic therapy, vestibular training and urological rehabilitation (Caltagirone, Imbriani, & Piras, 2021).

Pharmacotherapeutic approach

Pharmacotherapy also accompanies the different recovery phases, according to a symptomatic and syndromic approach (Caltagirone, Imbriani, & Piras, 2021).

Among the nootropic drugs, Piracetam may be able to improve the consciousness level after severe TBI and to promote recovery of language disturbances, memory, and learning abilities (Dimond, 1979; Huber, Willmes, Poeck, Van Vleymen, & Deberdt, 1997). Another interesting effect of piracetam has been reported on myoclonus, with possible improvement of post-hypoxic myoclonus (Guerrini et al., 1996). Mioclonic jerks may be successfully treated also by clonazepam or by levetiracetam (Genton et al., 2006).

Another nootropic drug, L-Acetyl Carnitine, may be useful also for neuroperipheral impairment (ICU-AW); however, all nootropic drugs should be avoided in agitated patients.

As for the treatment of early and late post-traumatic epilepsy (PTE), levetiracetam nowadays replaces phenytoin as one of the drugs of choice in ICU, since phenytoin induces cognitive impairment and hepatoxicity, although also levetiracetam has shown psychiatric and behavioral adverse effects.

Some controlled studies demonstrated the lack of efficacy of prophylactic treatment for late PTE (Formisano et al., 2007). Among the new anti-epileptic agents, oxcarbazepine may substitute carbamazepine, because the former determines less hepatotoxicity, ataxia, dizziness, diplopia and hematological side effects. On the other hand, oxcarbazepine may cause hyponatremia, therefore serial monitoring of electrolytes is indicated during this treatment for epilepsy or agitation (Sommer et al., 2009). Other new anti-epileptic drugs such as lamotrigine, topiramate, and lacosamide may be used for late PTE (Caltagirone, Imbriani, & Piras, 2021). Valproic acid is able to control seizures in patients with hyper-arousal agitation or disturbances in the sleep-wake cycle, while lacosamide may be chosen because of less cognitive impairment (Williamson et al., 2019).

As for severe spasticity and pathological postures, oral anti-spastic drugs may be useful, such as baclofen, dantrolene, tizadine, and diazepam. Baclofen should be avoided in epileptic patients, as well as tizanidine is contraindicated in patients with arterial hypotension; also diazepam should be avoided for its detrimental effect on neural plasticity and cognitive impairment (Bhatnagar, Iaccarino, & Zafonte, 2016; Feeney, Weisend, & Kline, 1993). However, focal spasticity should be treated with botulinum toxin infiltrations, as an early approach, associated with soft splinting or functional compression bandage, to reduce pain and to prevent tendon and capsule retraction and joint ankylosis, but also to avoid the sedation of the oral anti-spastic drugs (Clemenzi et al., 2012).

Diffuse severe spasticity or abnormal pathological postures, such as decortication and decerebration, could be early treated by intrathecal baclofen pump implantation, to improve motor output, to reduce pain and dysautonomic symptoms, as well as to facilitate behavioral responsiveness (Formisano, Aloisi, Contrada, et al., 2019; Sarà, Pistoia, Mura, Onorati, & Govoni, 2009).

Besides diazepam, other drugs with potential deleterious effects on neural sprouting and functional recovery should be avoided in the acute and post-acute phase, such as clonidine, phenobarbital, cortisone and haloperidol (Feeney et al., 1993).

Propranolol is the treatment of choice for controlling agitation (Fleminger, Greenwood, & Olivier, 2006), PSH (Baguley et al., 2014) and cerebellar tremor (Formisano, Saltuari, Sailer, Birbarmer, & Gerstenbrand, 1987).

Psychomotor agitation may be also successfully treated with antiepileptic drugs, such as carbamazepine, oxcarbazepine, and valproic acid, also as mood stabilizers (Gualtieri, 1988).

Among the new neuroleptics, quetiapine seems to be the most useful and tolerated drugs, since its less extrapyramidal side effects. However, during quetiapine treatment, monitoring of glycaemia, cholesterol, triglycerides and electrocardiographic QTc interval is warranted (Weiden, 2005).

Anecdotic reports on the positive effects of Amitriptyline on agitation (Jackson, Corrigan, & Arnett, 1985) have also been reported, although its use should be avoided in epileptic patients, as well as baclofen, haloperidol and some new generation neuroleptics, such as clozapine, for their epileptogenic side effect. Amitriptyline may be also used in patients with food refusal, loss of appetite (Domecq et al., 2015), tremor, and pathologic laughing and weeping (Schiffer, Herndon, & Rudick, 1985).

Neuropathic pain has been widely reported after sTBI and it may be one of the causes of psychomotor agitation during DoC recovery. Given the difficulty to accurately diagnose MCS versus VS/UWS and to demonstrate the lack of pain perception in these conditions, adequate treatment of pain should be provided (Formisano & Zasler, 2020). Indeed, prophylactic treatment with pain killers has been recommended in all patients with DoC (Boly et al., 2007).

Indomethacin may also improve the collaboration and the tolerance of the patient with reducing pain during passive range of motion and decrease the risk of PAO development (Ippolito, Formisano, Caterini, Farsetti, & Penta, 1999a, 1999b; Ippolito, Formisano, Farsetti, Caterini, & Penta, 1999). The progressive and gradual assisted verticalization may be facilitated by elastic stockings at the lower limbs and indomethacin therapy, which may counteract orthostatic hypotension (Kochar & Itskovitz, 1978). Also valproic acid, carbamazepine, oxcarbazepine, gabapentin, pregabalin, clonazepam and duloxetine may be utilized when neurogenic pain of central or peripheral origin (CIP/CRIMYNE/ICU-AW) limits stretching exercises and tolerance of the patient to dynamic splinting (Vargas-Espinosa, Sanmartí-García, Vázquez-Delgado, & Gay-Escoda, 2012). Finally, clonazepam could be useful in patients with diffuse pain, rest or intentional tremor and epilepsy (Caltagirone, Imbriani, & Piras, 2021).

L-Dopa and dopaminergic drugs (Amantadine, Selegiline) may enhance consciousness and verbal communication recovery and may improve cognitive functions and parkinsonian symptoms (Formisano et al., 2009; Formisano & Zasler, 2014; Giacino et al., 2012; Péran et al., 2014).

A paradoxical awakening effect of Zolpidem has been described in some patients with DoC (Clauss & Nel, 2006), as well as modafinil (Provigil) has demonstrated some interesting effects on vigilance (Ferraro et al., 2013).

An anecdotic report of responsiveness improvement in one patient with chronic MCS has been reported after deep brain stimulation (Schiff & Fins, 2007), whereas several studies have demonstrated interesting results with non-invasive neurostimulation techniques, such as transcranial direct current stimulation (tDCS) (Thibaut, Chatelle, Gosseries, Laureys, & Bruno, 2013).

Controlled studies on the possible efficacy of the acetilcholinesterase inhibitors (AchE-I), such as donepezil, rivastigmine, memantine, and galantamine should be of extreme interest, since the efficacy of some AchE-I in the treatment of cognitive and behavioral disturbances after TBI (Tenovuo, 2005).

In conclusion, randomized and multicenter controlled studies on the efficacy of the most commonly used drugs in the treatment of patients with sABI are needed (Caltagirone, Imbriani, & Piras, 2021; Pisa et al., 2015).

Applications to other areas of neuroscience

In this chapter, we have reviewed the post-coma syndrome in the context of severe acquired brain injury. We also interlink the holistic approach to post-coma patients and prolonged disorder of consciousness in the diagnostic and neurorehabilitative approach.

The study of post-coma syndrome and severe acquired brain injury is useful for several applications:

Neurophysiology and neuroimaging

The clinical and instrumental (neurophysiology and neuroimaging) prognostic markers could help clinicians in identifying patients with higher likelihood for clinical recovery and survival (Estraneo & Trojano, 2018). The European Academy of Neurology guidelines on the diagnosis of coma and other disorder of consciousness (Kondziella et al., 2020), on the basis of the review of the existing scientific evidence for the evaluation of coma and other disorder of consciousness, has recently established that multimodal assessment using PET, EEG-based techniques, including evoked potentials, fMRI, together with standardized clinical behavioral scales, provides more robust evaluation of consciousness and higher-order cortical function than routine bedside examination alone. A recent international survey on the use of such advanced diagnostic techniques in patients with disorder of consciousness has also demonstrated some interesting cross-cultural differences (Formisano, Giustini, Aloisi, et al., 2019).

Neuropsychology

The neuropsychological approach to coma survivors in the post-acute phase, is generally based on the Levels of Cognitive Functioning, from disorder of consciousness to post-traumatic amnesia (Russell & Smith, 1961). An individualized neuropsychological treatment to improve functioning in daily life after severe traumatic brain injury is also required, according to cognitive and behavioral features, neuropsychological and functional assessment, and a goal-oriented intervention (Ciurli et al., 2010), psychological adjustment and caregiver education (Silvestro et al., 2017), according to International Classification of Functioning (Erikson, 2002).

Neurorehabilitation

The multidisciplinary take in charge of post-comatose patients pass through the synergic and holistic management of the patient by means of the medical staff, including intensivist, cardiologic, phoniatric, pneumologic, and urologic evaluations, to establish a specific individual neurorehabilitation project, shared by the neuropsychologist, the physical and occupational therapists, speech therapist, orthoptist, nutritionist and social service (Formisano, 2020). A holistic approach to post-coma patients and prolonged disorder of consciousness is fundamental, including nursing care and basal stimulation, with a constant verbalization of all the maneuvers during positioning, hygiene and standard care, respiratory and swallowing management. The cooperation of many professionals during the evolution of the decannulation protocol, with assessment of swallowing, coughing, inspiratory flow and any signs of respiratory distress. Indeed, hydrokinesitherapy, occupational therapy, urological rehabilitation, phoniatric assessment and swallowing and respiratory training are all part of a multiprofessional neurorehabilitation project.

Mini-dictionary of terms

Severe acquired brain injury: A neurological condition due to a brain damage with a coma lasting at least 24h, according to a GCS ≤ 8 **and/or** complex and severe neurological disabilities treatable only in high neurorehabilitative specialty wards.

Coma: Inability to obey simple commands, closed eyes and absence of any verbal communication.

Vegetative State/Unresponsive Wakefulness Syndrome: Condition that follows coma, patients recover the vigilance or alertness, but not the interaction with the environment.

Minimally conscious state: Condition of severe consciousness alteration in which a minimal but not constant awareness of self and surroundings is present.

Exit-minimally conscious state: It is diagnosed when consistent evidence of either functional communication or functional object use can be documented.

Functional Locked-In: Patients with a dissociation between extreme motor dysfunction and preserved higher cortical functions.

Covert cognition: Residual cognitive functions in the absence of any behavioral interaction with the surroundings.

Post-traumatic amnesia: Period during which the patient is not able to memorize everyday events of the last 24 hours.

Paroxysmal sympathetic hyperactivity: Dysautonomic vegetative symptoms associated with disorder of consciousness.

Key facts of post-coma syndrome

The European Academy of Neurology, on the basis of the review of the existing scientific evidence for the evaluation of coma and other disorders of consciousness has recently published guidelines on the multimodal assessment for the diagnosis.

An international survey on the use of advanced diagnostic techniques in patients with disorder of consciousness has also demonstrated some interesting cross-cultural differences.

Tracheostomy is often required in severe acquired brain injury patients until their airways protective reflexes improved. Early tracheostomy seems to reduce intensive care unit length of stay.

Mechanical ventilation needs endotracheal tube with cuff, in order to provide a seal to reduce air leakage, whereas one of the first rehabilitation goals is the utilization of tracheal tube without cuff.

Hydrocephalus represents one of the neurosurgical late complications that could affect the recovery of consciousness.

Summary points

This chapter focuses on post-coma syndrome.

Post-coma syndrome may follow severe acquired brain injury.

Disorders of consciousness may have different origins including: traumatic, vascular, anoxic, infective, toxic and others.

Severe acquired brain injury is defined as "a neurological condition due to a severe acquired brain damage with a coma lasting at least 24 hours, according to a Glasgow Coma Scale equal or ≥ 8 and/or complex and severe neurological disabilities treatable only in high neurorehabilitation specialty wards."

Disorders of consciousness after severe acquired brain injury include: coma, vegetative state, recently named wakefulness syndrome, and minimally conscious state.

A holistic approach to patients with sABI and prolonged disorders of consciousness is fundamental.

References

Baguley, I. J., Perkes, I. E., Fernandez-Ortega, J. F., Rabinstein, A. A., Dolce, G., Hendricks, H. T., & Consensus Working Group. (2014). Paroxysmal sympathetic hyperactivity after acquired brain injury: consensus on conceptual definition, nomenclature, and diagnostic criteria. *Journal of Neurotrauma, 31*, 1515–1520.

Baulieu, F., Fournier, P., Baulieu, J. L., Dalonneau, M., Chiaroni, P., Eder, V., … Legros, B. (2001). Technetium-99m ECD single photon emission computed tomography in brain trauma: Comparison of early scintigraphic findings with long-term neuropsychological outcome. *Journal Neuroimaging, 11*, 112.

Beaumont, J. G., & Kenealy, P. M. (2005). Incidence and prevalence of the vegetative and minimally conscious states. *Neuropsychological Rehabilitation, 15*, 184.

Bhatnagar, S., Iaccarino, M. A., & Zafonte, R. (2016). Pharmacotherapy in rehabilitation of post-acute traumatic brain injury. *Brain Research, 1640*, 164–179.

Bivona, U., Ciurli, P., Barba, C., Onder, G., Azicnuda, E., Silvestro, D., et al. (2008). Executive function and metacognitive self-awareness after severe traumatic brain injury. *Journal of the International Neuropsychological Society, 14*, 862–868.

Bivona, U., Riccio, A., Ciurli, P., Carlesimo, G. A., Delle Donne, V., Pizzonia, E., et al. (2014). Low self-awareness of individuals with severe traumatic brain injury can lead to reduced ability to take another person's perspective. *The Journal of Head Trauma Rehabilitation, 29*, 157–171.

Boly, M., Coleman, M. R., Davis, M. H., Hampshire, A., Bor, D., Moonen, G., … Owen, A. M. (2007). When thoughts become action: an fMRI paradigm to study volitional brain activity in non-communicative brain injured patients. *Neuroimage, 36*, 979–992.

Bruno, M. A., Vanhaudenhuyse, A., Thibaut, A., Moonen, G., & Laureys, S. (2011). From unresponsive wakefulness to minimally conscious PLUS and functional locked-in syndromes: Recent advances in our understanding of disorders of consciousness. *Journal of Neurology, 258*, 1373–1384.

Caltagirone, Imbriani, P., & Piras, F. (2021). *Handbook of neurorehabilitation – Fondazione Santa Lucia*. Giunti (Chapter IX).

Carlesimo, G. A., Formisano, R., Bivona, U., Barba, L., & Caltagirone, C. (2009). Prospective memory in patients with severe closed-head injury: Role of concurrent activity and encoding instructions. *Behavioural Neurology, 22*, 101–110.

Carlesimo, G. A., Sabbadini, M., Bombardi, P., Di Porto, E., Loasses, A., & Caltagirone, C. (1998). Retrograde memory deficits in severe closed-head injury patients. *Cortex, 34*, 1–23.

Chatelle, C., Majerus, S., Whyte, J., Laureys, S., & Schnakers, C. (2012). A sensitive scale to assess nociceptive pain in patients with disorders of consciousness. *Journal of Neurology, Neurosurgery & Psychiatry, 83*, 1233–1237.

Ciurli, P., Bivona, U., Barba, C., Onder, G., Silvestro, D., Azicnuda, E., et al. (2010). Metacognitive unawareness correlates with executive function impairment after severe traumatic brain injury. *Journal of the International Neuropsychological Society, 16*, 360–368.

Ciurli, P., Formisano, R., Bivona, U., Cantagallo, A., & Angelelli, P. (2011). Neuropsychiatric disorders in persons with severe traumatic brain injury: Prevalence, phenomenology, and relationship with demographic, clinical, and functional features. *The Journal of Head Trauma Rehabilitation, 26*, 116–126.

Clauss, R., & Nel, W. (2006). Drug induced arousal from the permanent vegetative state. *NeuroRehabilitation, 21*, 23–28.

Clemenzi, A., Formisano, R., Matteis, M., Gallinacci, L., Cochi, G., Savina, P., et al. (2012). Care management of spasticity with botulinum toxin-A in patients with severe acquired brain injury: A 1-year follow-up prospective study. *Brain Injury, 26*, 979–983.

Dimond, S. (1979). Drugs to improve learning in man: implications and neuropsychological analysis. In R Knight, & O Bakker (Eds.), *The neuropsychology of learning disorders* (p. 367). London: University Park Press.

D'Ippolito, M., Aloisi, M., Azicnuda, E., Silvestro, D., Giustini, M., Verni, F., et al. (2018). Changes in caregivers lifestyle after severe acquired brain injury: A preliminary investigation. *BioMedical Research International, 2018*, 2824081.

Corrigan, J. D., & Bogner, J. A. (1994). Factor structure of the Agitated Behavior Scale. *Journal of Clinical and Experimental Neuropsychology, 16*, 386.

Cruse, D., Chennu, S., Chatelle, C., Bekinschtein, T. A., Fernández-Espejo, D., Pickard, J. D., … Owen, A. M. (2011). Bedside detection of awareness in the vegetative state: A cohort study. *The Lancet, 378*, 2088–2094.

Curley, W. H., Forgacs, P. B., Voss, H. U., Conte, M. M., & Schiff, N. D. (2018). Characterization of EEG signals revealing covert cognition in the injured brain. *Brain, 141*, 1404–1421.

Daltrozzo, J., Wioland, N., Mutschler, V., & Kotchoubey, B. (2007). Predicting coma and other low responsive patients outcome using event related brain potentials: A meta-analysis. *Clinical Neurophysiology, 118*, 606–614.

De Pasquale, F., Caravasso, C. F., Péran, P., Catani, S., Tuovinen, N., Sabatini, U., et al. (2015). Functional magnetic resonance imaging in disorders of consciousness: Preliminary results of an innovative analysis of brain connectivity. *Functional Neurology, 30*, 19.

Dolce, G., & Sazbon, L. (2002). Neurophysiopathology. In *The post-traumatic vegetative state* (pp. 11–17). Thieme.

Domecq, J. P., Prutsky, G., Leppin, A., Sonbol, M. B., Altayar, O., Undavalli, C., … Lababidi, M. H. (2015). Drugs commonly associated with weight change: A systematic review and meta-analysis. *The Journal of Clinical Endocrinology & Metabolism, 100*, 363–370.

Erikson. (2002). *International Classification of Functioning, Disability and Health – trad. it. ICF* (pp. 14–22). Trento: Erikson.

Estraneo, A., Fiorenza, S., Magliacano, A., Formisano, R., Mattia, D., Grippo, A., et al. (2020). Multi-center prospective study on predictors of short-term outcome in disorders of consciousness. *Neurology, 95*, e1488–e1499.

Estraneo, A., & Trojano, L. (2018). Prognosis in disorders of consciousness. In *Coma and disorders of consciousness* (pp. 17–36). Cham: Springer.

Feeney, D. M., Weisend, M. P., & Kline, A. E. (1993). Noradrenergic pharmacotherapy, intracerebral infusion and adrenal transplantation promote functional recovery after cortical damage. *Neural Plasticity*, *4*, 199–213.

Ferraro, L., Antonelli, T., Beggiato, S., Cristina Tomasini, M., Fuxe, K., & Tanganelli, S. (2013). The vigilance promoting drug modafinil modulates serotonin transmission in the rat prefrontal cortex and dorsal raphe nucleus. Possible relevance for its postulated antidepressant activity. *Mini Reviews in Medicinal Chemistry*, *13*, 478–492.

Fleminger, S, Greenwood, R. J., & Olivier, D. L. (2006). Pharmacological management for agitation and aggression in people with acquired brain injury. *Cochrane Database of Systematic Reviews*, *18*, CD003299.

Formisano, R. (2020). Severe acquired brain injury and high specialty neurorehabilitation needs. *Neurological Sciences*, *42*, 1–2.

Formisano, R., Aloisi, M., Contrada, M., Spanedda, F., Schiattone, S., Niedbala, S., et al. (2019). Late recovery of responsiveness after intra-thecal baclofen pump implantation and the role of diffuse pain and severe spasticity: A case report. *Acta Neurochirurgica*, *161*, 1965–1967.

Formisano, R., Aloisi, M., Ferri, G., Schiattone, S., & Contrada, M. (2018). The Glasgow outcome scale extended-revised (GOSE-R) to include minimally conscious state in the vegetative state category. *Journal of Neurological Sciences*, *388*, 22.

Formisano, R., Aloisi, M., Iosa, M., Contrada, M., Rizza, F., Sattin, D., et al. (2018). A new tool to assess responsiveness in disorders of consciousness (DoC): A preliminary study on the brief post-coma scale (BPCS). *Neurological Sciences*, *39*, 1651–1656.

Formisano, R., Azicnuda, E., Sefid, M. K., Zampolini, M., Scarponi, F., & Avesani, R. (2017). Early rehabilitation: Benefits in patients with severe acquired brain injury. *Neurological Sciences*, *38*, 181–184.

Formisano, R., Barba, C., Buzzi, M. G., Newcomb-Fernandez, J., Menniti-Ippolito, F., Zafonte, R., et al. (2007). The impact of prophylactic treatment on post-traumatic epilepsy after severe traumatic brain injury. *Brain Injury*, *21*, 499–504.

Formisano, R., Bivona, U., Brunelli, S., Giustini, M., & Taggi, F. (2001). Driving competence after severe brain injury: A retrospective study. *European Journal of Physical and Rehabilitation Medicine*, *37*, 257.

Formisano, R., Bivona, U., Penta, F., Giustini, M., Buzzi, M. G., Ciurli, P., et al. (2005). Early clinical predictive factors during coma recovery. In *Re-engineering of the damaged brain and spinal cord* (pp. 201–205). Vienna: Springer.

Formisano, R., Cicinelli, P., Buzzi, M. G., Brunelli, S., Zafonte, R., Vinicola, V., et al. (2009). Blink reflex changes in parkinsonism following severe traumatic brain injury correlates with diffuse axonal injury. *Medical Science Monitor*, *15*, 101–106.

Formisano, R., Contrada, M., Aloisi, M., Ferri, G., Schiattone, S., Iosa, M., et al. (2019). Nociception coma scale with personalized painful stimulation versus standard stimulus in non-communicative patients with disorders of consciousness. *Neuropsychological Rehabilitation*, *30*, 1–12.

Formisano, R., Contrada, M., Ferri, G., Schiattone, S., Iosa, M., & Aloisi, M. (2019). The Glasgow outcome scale extended-revised (GOSE-R) to include minimally conscious state in the vegetative state/unresponsive wakefulness syndrome category: A correlation with coma recovery scale-revised (CRS-R). *European Journal of Physical and Rehabilitation Medicine*, *55*, 139–140.

Formisano, R., Contrada, M., Iosa, M., Ferri, G., Schiattone, S., & Aloisi, M. (2019). Coma recovery scale-revised with and without the emotional stimulation of caregivers. *Canadian Journal of Neurological Sciences*, *46*, 607–609.

Formisano, R., D'Ippolito, M., & Catani, S. (2013). Functional locked-in syndrome as recovery phase of vegetative state. *Brain Injury*, *27*, 1332.

Formisano, R., D'Ippolito, M., Risetti, M., Riccio, A., Caravasso, C. F., Catani, S., et al. (2011). Vegetative state, minimally conscious state, akinetic mutism and parkinsonism as a continuum of recovery from disorders of consciousness: An exploratory and preliminary study. *Functional Neurology*, *26*, 15–24.

Formisano, R., Giustini, M., Aloisi, M., Contrada, M., Schnakers, C., Zasler, N., et al. (2019). An international survey on diagnostic and prognostic protocols in patients with disorder of consciousness. *Brain Injury*, *33*, 974–984.

Formisano, R., Pistoia, F., & Sarà, M. (2011). Disorders of consciousness: A taxonomy to be changed? *Brain Injury*, *25*, 638–639.

Formisano, R., Saltuari, L., & Gerstenbrand, F. (1995). Presence of Klüver-Bucy syndrome as a positive prognostic feature for the remission of traumatic prolonged disturbances of consciousness. *Acta Neurologica Scandinavica*, *91*, 54.

Formisano, R., Saltuari, L., Sailer, U., Birbarmer, G., & Gerstenbrand, G. (1987). Post-traumatic cerebellar syndrome. *New Trends Clinical Neuropharmacology*, *1*, 115–118.

Formisano, R., Silvestro, D., Azicnuda, E., Longo, E., Barba, C., Rigon, J., et al. (2017). Quality of life after brain injury (QOLIBRI): Italian validation of the proxy version. *Internal and Emergency Medicine*, *12*, 187–198.

Formisano, R., Toppi, J., Risetti, M., Aloisi, M., Contrada, M., Ciurli, P. M., et al. (2019). Language-related brain potentials in patients with disorders of consciousness: A follow-up study to detect "covert" language disorders. *Neurorehabilitation and Neural Repair*, *33*, 513–522.

Formisano, R., & Zasler, N. D. (2014). Posttraumatic parkinsonism. *The Journal of Head Trauma Rehabilitation*, *29*, 387–390.

Formisano, R., & Zasler, N. (2020). Discontinuation of artificial nutrition and hydration and covert cognition. *Brain Injury*, *24*, 1.

Gabrielli, W. A., & Layon, A. J. (2004). Airway management in the neurointensive care unit. In A. J. Layon, & W. A. Gabrielli (Eds.), *Friedman textbook of neurointensive care* (p. 499). Saunders Elsevier.

Genton, P., Sadzot, B., Fejerman, N., Peltola, J., Despland, P. A., Steinhoff, B., … Van Hammée, G. (2006). Levetiracetam in a broad population of patients with refractory epilepsy: interim results of the international SKATE trial. *Acta Neurologica Scandinavica*, *113*, 387.

Giacino, J. T. (2004). The vegetative and minimally conscious states: Consensus-based criteria for establishing diagnosis and prognosis. *NeuroRehabilitation*, *19*, 293–298.

Giacino, J. T., & Kalmar, K. (1997). The vegetative and minimally conscious states: a comparison of clinical features and functional outcome. *The Journal of Head Trauma Rehabilitation*, *12*, 36–51.

Giacino, J. T., Katz, D. I., Schiff, N. D., Whyte, J., Ashman, E. J., Ashwal, S., et al. (2018). Practice guideline update recommendations summary: Disorders of consciousness: Report of the Guideline Development, Dissemination, and Implementation Subcommittee of the American Academy of Neurology; the American Congress of Rehabilitation Medicine; and the National Institute on Disability, Independent Living, and Rehabilitation Research. *Neurology*, *91*, 450–460.

Giacino, J. T., & Trott, C. T. (2004). Rehabilitative management of patients with disorders of consciousness. *The Journal of Head Trauma Rehabilitation*, *19*, 254.

Giacino, J. T., Whyte, J., Bagiella, E., Kalmar, K., Childs, N., Khademi, A., et al. (2012). Placebo-controlled trial of amantadine for severe traumatic brain injury. *New England Journal of Medicine*, *366*, 819–826.

Gill-Thwaites, H., & Munday, R. (1999). The Sensory Modality Assessment and Rehabilitation Technique (SMART): A comprehensive and integrated assessment and treatment protocol for the vegetative state and minimally responsive patient. *Neuropsychological Rehabilitation*, *9*, 305–320.

Giubilei, F., Formisano, R., Fiorini, M., Vitale, A., Faroni, J., Toni, D., & Santilli, V. (1995). Sleep abnormalities in traumatic apallic syndrome. *Journal of Neurology, Neurosurgery & Psychiatry*, *58*, 84–86.

Giugni, E., Sabatini, U., Hagberg, G. E., Formisano, R., & Castriota-Scanderbeg, A. (2005). Turbo-proton Echo planar spectroscopic imaging (t-PEPSI) MR technique in the detection of diffuse axonal damage in brain injury. Comparison with gradient-recalled Echo (GRE) sequence. *La Radiologia Medica*, *109*, 563–572.

Gościński, I., Kwiatkowski, S., Polak, J., & Orlowiejska, M. (1997). The Kluver-Bucy syndrome. *Acta Neurochirurgica*, *139*, 303–306.

Gualtieri, C. T. (1988). Pharmacotherapy and the neurobehavioural sequelae of traumatic brain injury. *Brain Injury*, *2*, 101–129.

Guerrini, R., Bonanni, P., de Lorey, Serratosa, J. M., Moncla, A., & Suisse, G. (1996). Cortical myoclonus in Angelman syndrome. *Annals of, Neurology*, *40*, 39.

Guideline Minister of Health. (1998). Linee guida del Ministero della Sanità per le attività di riabilitazione-Provvedimento 7 maggio. In *1998 approvato dalla Conferenza Stato Regioni*.

Hagen, C. (2000). *LCF-revised*. Downey, CA: Professional Staff Association of Rancho Los Amigos Hospital.

Hermann, B., Goudard, G., Courcoux, K., Valente, M., Labat, S., Despois, L., … Sitt, J. D. (2019). Wisdom of the caregivers: Pooling individual subjective reports to diagnose states of consciousness in brain-injured patients, a monocentric prospective study. *BMJ Open*, *9*, e026211.

Houlden, D. A., Taylor, A. B., Feinstein, A., Midha, R., Bethune, A. J., Stewart, C. P., & Schwartz, M. L. (2010). Early somatosensory evoked potential grades in comatose traumatic brain injury patients predict cognitive and functional outcome. *Critical Care Medicine*, *38*, 167–174.

Huber, W., Willmes, K., Poeck, K., Van Vleymen, B., & Deberdt, W. (1997). Piracetam as an adjuvant to language therapy for aphasia: A randomized double-blind placebo-controlled pilot study. *Archives of Physical Medicine and Rehabilitation*, *78*(3), 245–250.

Intiso, D., Amoruso, L., Zarrelli, M., Pazienza, L., Basciani, M., Grimaldi, G., et al. (2011). Long term functional outcome and health status of patients with critical illness polyneuromyopathy. *Acta Neurologica Scandinavica*, *123*, 211–219.

Ippolito, E., Formisano, R., Caterini, R., Farsetti, P., & Penta, F. (1999a). Operative treatment of heterotopic hip ossification in patients with coma after brain injury. *Clinical Orthopaedics and Related Research*, *365*, 130–138.

Ippolito, E., Formisano, R., Caterini, R., Farsetti, P., & Penta, F. (1999b). Resection of elbow ossification and continuous passive motion in postcomatose patients. *The Journal of Hand Surgery*, *24*, 546–553.

Ippolito, E., Formisano, R., Farsetti, P., Caterini, R., & Penta, F. (1999). Excision for the treatment of periarticular ossification of the knee in patients who have a traumatic brain injury. *The Journal of Bone and Joint Surgery*, *81*(6), 783–789.

Jackson, R. D., Corrigan, J. D., & Arnett, J. A. (1985). Amitriptyline for agitation in head injury. *Archives of Physical Medicine and Rehabilitation*, *6*, 180–181.

Jennett, B. (1986). Clinical assesment of consciousness. Introduction. *Acta Neurochirurgica*, (Suppl 36), 90.

Jennett, B., & Bond, M. (1975). Assessment of outcome after severe brain damage. *Lancet*, *1*, 480.

Kang, X. G., Li, L., Wei, D., Xu, X. X., Zhao, R., Jing, Y. Y., et al. (2014). Development of a simple score to predict outcome for unresponsive wakefulness syndrome. *Critical Care*, *18*(1), 1–8.

Kochar, M. S, & Itskovitz, H. D. (1978). Treatment of idiopathic orthostatic hypotension. Shy–Dràger syndrome with indomethacin. *Lancet*, *I*,, 1011–1014.

Kondziella, D., Bender, A., Diserens, K., van Erp, W., Estraneo, A., Formisano, R., et al. (2020). European Academy of Neurology guideline on the diagnosis of coma and other disorders of consciousness. *European Journal of Neurology*, *27*(5), 741–756.

Laureys, S., Celesia, G. G., Cohadon, F., Lavrijsen, J., León-Carrión, J., Sannita, W. G., et al. (2010). Unresponsive wakefulness syndrome: A new name for the vegetative state or apallic syndrome. *BMC Medicine*, *8*, 1–4.

Levin, H. S., O'Donnell, V. M., & Grossman, R. G. (1979). The Galveston Orientation and Amnesia Test. A practical scale to assess cognition after head injury. *Journal of Nervous and Mental Disease*, *167*, 675–684.

Lucca, L. F., Pignolo, L., Leto, E., Ursino, M., Rogano, S., & Cerasa, A. (2019). Paroxysmal sympathetic hyperactivity rate in vegetative or minimally conscious state after acquired brain injury evaluated by paroxysmal sympathetic hyperactivity assessment measure. *Journal of Neurotrauma*, *36*, 2430–2434.

Massimini, M., Ferrarelli, F., Sarasso, S., & Tononi, G. (2012). Cortical mechanisms of loss of consciousness: insight from TMS/EEG studies. *Archives Italiennes de Biologie*, *150*, 44–55.

McMillan, T., Wilson, L., Ponsford, J., Levin, H., Teasdale, G., & Bond, M. (2016). The Glasgow Outcome Scale—40 years of application and refinement. *Nature Reviews Neurology*, *12*, 477–485.

Missori, P., Miscusi, M., Formisano, R., Peschillo, S., Polli, F. M., Melone, A., et al. (2006). Magnetic resonance imaging flow void changes after cerebrospinal fluid shunt in post-traumatic hydrocephalus: Clinical correlations and outcome. *Neurosurgical Review*, *29*, 224.

Møller, M. L. H., Højlund, A., Jensen, M., Gansonre, C., & Shtyrov, Y. (2020). Applied potential of task-free event-related paradigms for assessing neurocognitive functions in disorders of consciousness. *Brain Communications*, *2*(2), fcaa087.

Monti, M. M., Vanhaudenhuyse, A., Coleman, M. R., Boly, M., Pickard, J. D., Tshibanda, L., et al. (2019). Willful modulation of brain activity in disorders of consciousness. *New England Journal of Medicine*, *362*, 579–589.

Nakase-Thomson, R., Sherer, M., Yablon, S. A., Nick, T. G., & Trzepacz, P. T. (2004). Acute confusion following TBI. *Brain Injury*, *18*, 131–142.

Owen, A. M., Coleman, M. R., Boly, M., Davis, M. H., Laureys, S., & Pickard, J. D. (2006). Detecting awareness in the vegetative state. *Science*, *313*, 1402.

Pape, T. L. B., Rosenow, J. M., Harton, B., Patil, V., Guernon, A., Parrish, T., et al. (2012). Preliminary framework for familiar auditory sensory training (FAST) provided during coma recovery. *Journal of Rehabilitation Research and Development*, *49*, 1137–1152.

Péran, P., Catani, S., Falletta Caravasso, C., Nemmi, F., Sabatini, U., & Formisano, R. (2014). Supplementary motor area activation is impaired in severe traumatic brain injury parkinsonism. *Journal of Neurotrauma*, *31*, 642–648.

Pero, S., Incoccia, C., Caracciolo, B., Zoccolotti, P., & Formisano, R. (2006). Rehabilitation of attention in two patients with traumatic brain injury by means of 'attention process training'. *Brain Injury*, *20*, 1207–1219.

Pisa, F. E., Cosano, G., Giangreco, M., Giorgini, T., Biasutti, E., Barbone, F., et al. (2015). Prescribing practice and off-label use of psychotropic medications in post-acute brain injury rehabilitation centres: A cross-sectional survey. *Brain Injury*, *29*, 508–516.

Qin, P., Wu, X., Huang, Z., Duncan, N. W., Tang, W., Wolff, A., … Zhang, J. (2015). How are different neural networks related to consciousness? *Annals of Neurology*, *78*, 594–605.

Raichle, M. E., MacLeod, A. M., Snyder, A. Z., Powers, W. J., Gusnard, D. A., & Shulman, G. L. (2001). A default mode of brain function. *Proceedings of the National Academy of Sciences*, *98*(2), 676–682.

Risetti, M., Formisano, R., Toppi, J., Quitadamo, L. R., Bianchi, L., Astolfi, L., et al. (2013). On ERPs detection in disorders of consciousness rehabilitation. *Frontiers in Human Neuroscience*, *7*, 775.

Russell, W. R., & Smith, A. (1961). Posttraumatic amnesia in closed head injury. *Archives of Neurology*, *5*, 16–29.

Sarà, M., Pistoia, F., Mura, E., Onorati, P., & Govoni, S. (2009). Intrathecal baclofen in patients with persistent vegetative state: 2 hypotheses. *Archives of Physical Medicine and Rehabilitation*, *90*, 1245–1249.

Schiff, N. D., & Fins, J. J. (2007). Deep brain stimulation and cognition: moving from animal to patient. *Current Opinion in Neurology*, *20*, 638–642.

Schiffer, R. B., Herndon, R. M., & Rudick, R. A. (1985). Treatment of pathologic laughing and weeping with amitriptyline. *New England Journal of Medicine*, *312*, 1480–1482.

Schnakers, C., Vanhaudenhuyse, A., Giacino, J., Ventura, M., Boly, M., Majerus, S., et al. (2009). Diagnostic accuracy of the vegetative and minimally conscious state: Clinical consensus versus standardized neurobehavioral assessment. *BMC Neurology*, *9*, 35.

Seel, R. T., Sherer, M., Whyte, J., Katz, D. I., Giacino, J. T., Rosenbaum, A. M., et al. (2010). Assessment scales for disorders of consciousness: Evidence-based recommendations for clinical practice and research. *Archives of Physical Medicine and Rehabilitation*, *91*, 1795–1813.

Sherer, M., Nakase-Thompson, R., Yablon, S.A., & Gontkovsky, S.T. (2005). Multidimensional assessment of acute confusion after traumatic brain injury. *Archives of Physical Medicine and Rehabilitation*, *86*(5), 896–904.

Silvestro, D., Mazzetti, M., Melia, C., Stagno, M. T., Carlesimo, G. A., Bivona, U., & Formisano, R. (2017). Educational action in the rehabilitation of severe acquired brain injuries: The role of self-awareness. *Annali dell'Istituto Superiore di Sanità*, *53*(1), 82–85.

Snow, P. C., & Ponsford, J. L. (2013). Assessing and managing impairment of consciousness following TBI. *Traumatic brain injury: Rehabilitation for everyday adaptive living* (pp. 34–65). Psychology Press.

Sommer, O. H., Aga, O., Cvancarova, M., Olsen, I. C., Selbaek, G., & Engedal, K. (2009). Effect of oxcarbazepine in the treatment of agitation and aggression in severe dementia. *Dementia and Geriatric Cognitive Disorders*, *27*, 155–163.

Stender, J., Gosseries, O., Bruno, M. A., Charland-Verville, V., Vanhaudenhuyse, A., Demertzi, A., … Laureys, S. (2014). Diagnostic precision of PET imaging and functional MRI in disorders of consciousness: A clinical validation study. *The Lancet*, *384*, 514–522.

Stokes, V., Gunn, S., Schouwenaars, K., & Badwan, D. (2018). Neurobehavioural assessment and diagnosis in disorders of consciousness: A preliminary study of the Sensory Tool to Assess Responsiveness (STAR). *Neuropsychological Rehabilitation*, *28*, 966–983.

Tenovuo, O. (2005). Central acetylcholinesterase inhibitors in the treatment of chronic traumatic brain injury-clinical experience in 111 patients. *Progress in Neuro-Psychopharmacology and Biological Psychiatry*, *29*, 61–67.

Thibaut, A., Chatelle, C., Gosseries, O., Laureys, S., & Bruno, M. A. (2013). Transcranial direct current stimulation: A new tool for neurostimulation. *Revue Neurologique*, *169*, 108.

Toppi, J., Mattia, D., Risetti, M., Formisano, R., Babiloni, F., & Astolfi, L. (2016). Testing the significance of connectivity networks: Comparison of different assessing procedures. *IEEE Transactions on Biomedical Engineering*, *63*, 2461–2473.

Valente, M., Placidi, F., Oliveira, A. J., Bigagli, A., Morghen, I., Proietti, R., et al. (2002). Sleep organization pattern as a prognostic marker at the subacute stage of post-traumatic coma. *Clinical Neurophysiology*, *113*(11), 1798–1805.

Vargas-Espinosa, M. L., Sanmartí-García, G., Vázquez-Delgado, E., & Gay-Escoda, C. (2012). Antiepileptic drugs for the treatment of neuropathic pain: A systematic review. *Medicina Oral, Patologia Oral y Cirugia Bucal*, *17*, 786.

Vespa, P. M., Miller, C., McArthur, D., Eliseo, M., Etchepare, M., Hirt, D., et al. (2007). Nonconvulsive electrographic seizures after traumatic brain injury result in a delayed, prolonged increase in intracranial pressure and metabolic crisis. *Critical Care Medicine*, *35*, 2830.

Wang, Y., Wen, L., You, W., Zhu, Y., Wang, H., Sun, Y., … Yang, X. (2020). Predicting the outcomes of shunt implantation in patients with post-traumatic hydrocephalus and severe conscious disturbance: a scoring system based on clinical characteristics. *Journal of Integrative Neuroscience*, *19*, 31–37.

Wannez, S., Heine, L., Thonnard, M., Gosseries, O., & Laureys, S. (2017). Coma science group collaborators. The repetition of behavioral assessments in diagnosis of disorders of consciousness. *Annals of Neurology*, *81*, 883–889.

Weiden, P. J. (2005). Switching antipsychotics: an updated review with a focus on quetiapine. *Journal of Psychopharmacology*, *20*, 104–118.

Williamson, D., Frenette, A. J., Burry, L. D., Perreault, M., Charbonney, E., Lamontagne, F., … Bernard, F. (2019). Pharmacological interventions for agitated behaviours in patients with traumatic brain injury: a systematic review. *BMJ Open*, *9*, e029604.

Wood, R. L., Winkowski, T. B., Miller, J. L., Tierney, L., & Goldman, L. (1992). Evaluating sensory regulation as a method to improve awareness in patients with altered states of consciousness: A pilot study. *Brain Injury*, *6*, 411–418.

Zasler, N. D., Aloisi, M., Contrada, M., & Formisano, R. (2019). Disorders of consciousness terminology: History, evolution and future directions. *Brain Injury*, *33*, 1684–1689.

Section C

Diagnosis and evaluation

Chapter 17

Features of the World Health Organization Disability Assessment Schedule 2.0 and its applications for traumatic brain injury assessment

Yu-Hao Lee, Shih-Wei Huang, and Tsan-Hon Liou
Department of Physical Medicine and Rehabilitation, Shuang Ho Hospital, Taipei Medical University, New Taipei City, Taiwan ROC

List of abbreviations

CIQ	Community Integration Questionnaire
DRS	Disability Rating Scale
DSM-5	Diagnostic and Statistical Manual of Mental Disorders, fifth edition
FAM	functional assessment measure
FIM	functional independence measure
GOS	Glasgow Outcome Scale
GOS-E	Glasgow Outcome Scale extended
ICC	Intra-class correlation coefficient
ICF	International Classification of Functioning
PART-O	participation assessment with recombined tools–objective
TBI	traumatic brain injury
WHO	World Health Organization
WHODAS 2.0	World Health Organization Disability Assessment Schedule 2.0

Introduction

Traumatic brain injury (TBI) is a leading cause of disability and socioeconomic problems globally (Maas, Stocchetti, & Bullock, 2008). Patients with TBI experience varied combinations of physical, behavioral, and cognitive problems. According to one study, these problems are related to TBI severity and associated with long-term disability (Langlois, Rutland-Brown, & Wald, 2006). Because of the complexity of the disease presentation, the diagnosis and disability assessment of patients with TBI can be extremely challenging to clinicians. Providing a comprehensive assessment, predicting outcomes, and properly formulating rehabilitation plans typically requires the use of various scales and assessment tools targeting different domains.

More than 700 instruments have been used in assessing patients with TBI (Tate, Godbee, & Sigmundsdottir, 2013). The Functional Independence Measure (FIM; Keith, Granger, Hamilton, & Sherwin, 1987), although widely used in rehabilitation settings for measuring functional improvement and progress, has been criticized for its lack of consideration of cognitive impairment, which deeply affects the overall functioning of patients with brain injuries. The Functional Assessment Measure (FAM), which evaluates behavioral, communication, and community reintegration skills, was developed to compensate for the shortcomings of the FIM. However, according to Linn et al. (1999), the FAM reduces test efficiency and provides minimal additional protection against ceiling effects. The Disability Rating Scale (DRS), which measures general functional status, is also widely used for follow-up from injury onset to recovery and community reintegration (Rappaport, Hall, Hopkins, Belleza, & Cope, 1982; Whyte, Cifu, Dikmen, & Temkin, 2001). However, it fails to assess specific tasks. The Glasgow Outcome Scale (GOS) and the GOS Extended (GOS-E; (Jennett & Bond, 1975) are simple assessment tools

Diagnosis and Treatment of Traumatic Brain Injury. https://doi.org/10.1016/B978-0-12-823347-4.00004-X

used to track the recovery and functional status of patients with TBI. The principal disadvantage of the scales is their inability to detect small but substantial changes (Teasdale, Pettigrew, Wilson, Murray, & Jennett, 1998). The Community Integration Questionnaire (CIQ), another tool widely used in the assessment of community integration for people with TBI, does not assess all of the domains defined by the World Health Organization (WHO), and its items may more closely pertain to activities than to participation (Kuipers, Kendall, Fleming, & Tate, 2004; Salter, Foley, Jutai, Bayley, & Teasell, 2008). The Participation Assessment with Recombined Tools–Objective (PART-O; Whiteneck et al., 2011), is derived from two frequently used instruments in assessing TBI: the CIQ and the Participation Objective, Participation Subjective. Designed to investigate long-term outcomes, the 17-item PART-O assesses activity, productivity, and social relationships and potentially serves as an objective evaluation of intervention effectiveness (Bogner, Bellon, Kolakowsky-Hayner, & Whiteneck, 2013). PART-O has demonstrated promising results, as indicated by its inclusion in the United States TBI Model Systems National Database funded by the National Institute on Disability, Independent Living, and Rehabilitation Research. However, PART-O is limited because it does not consider environmental factors, which contribute crucially to individuals' functionality and personal circumstances. Notably, Hall et al. (1996) reported the presence of ceiling effects in many of the aforementioned tools.

According to Laxe et al. (2012), the majority of items (73%) on the most commonly used assessment tools focus on activity and participation. Body function components and environmental factors were addressed in 24% and a considerably low 3% of the items, respectively. Because every assessment tool has its pros and cons, clinical professionals may need to combine results from several types of assessments to obtain a clear and complete picture of the functional status of a patient with TBI. Moreover, to tailor an optimal rehabilitation plan for each patient, clinicians require a shared language for describing functioning.

Thus, a comprehensive assessment tool is a must in evaluating patients with TBI. The WHO Disability Assessment Schedule 2.0 (WHODAS 2.0), an assessment tool developed under the International Classification of Functioning, Disability and Health (ICF) framework, shows promise in becoming a standard evaluation tool for patients with TBI. This chapter reviews the history, content, and current use of WHODAS 2.0 in the field of TBI care.

ICF framework

ICF, officially endorsed by all 191 WHO Member States in 2001 as the international standard by which health and disability are measured (World Health Organization, 2001), comprehensively defines functioning and disability through the following components: body functions, activities and participation, environmental factors, and body structures. It constitutes an integrative biopsychosocial model and a framework for measuring the effects of health conditions on individuals' functionality and disability. The ICF framework is universally applied by professionals from various fields, including rehabilitation practitioners, policymakers, advocacy groups, and scientists (Stucki, Cieza, & Melvin, 2007). Studies have revealed the utility of ICF in detecting improvement in body functions and activities related to mobility, self-care, mental function, and other activities in patients with TBI following rehabilitation. It provides a holistic view of disability and exerts impacts on service interventions related to functional and societal participation, enabling clinicians to tailor interventions to patients' actual needs (Ptyushkin, Vidmar, Burger, & Marincek, 2010; Svestkova, Angerova, Sladkova, Bickenbach, & Raggi, 2010).

WHODAS 2.0

Developed under the conceptual framework of the ICF model, the WHODAS 2.0 was released in 2010 (Ustun, Chatterji, et al., 2010). It considers an individual's level of functioning in six major life domains: cognition (understanding and communication), mobility (ability to move and get around), self-care (ability to attend to personal hygiene, dress and feed oneself, and live alone), getting along (ability to interact with others), life activities (ability to carry out responsibilities at home, work, and school), and participation in society (ability to engage in community, civil, and recreational activities). The WHODAS 2.0 provides a standardized summary measure of functioning and disability that is reliable and applicable to adults across populations and cultures. It has been translated into 47 languages and dialects, is applied in 27 areas of research, and has been used in hundreds of studies in the past 15 years (Federici, Bracalenti, Meloni, & Luciano, 2017). Furthermore, the *Diagnostic and Statistical Manual of Mental Disorders, Fifth Edition* (DSM-5) Disability Study Group regards the WHODAS 2.0 as the best current measure of disability for routine clinical use and has recommended its inclusion in the DSM-5 since 2013 (Gold, 2014).

WHODAS 2.0 scores are derived from a questionnaire administered to patients (or their healthcare agents if patients cannot answer the questions themselves). The questionnaire comprises 36 items from 6 domains (6, 5, 4, 5, 8, and 8 items from each, respectively): cognition, mobility, self-care, getting along with people, life activities, and participation in society. The items are scored on a five-point Likert scale (1 = *no difficulty*, 2 = *mild difficulty*, 3 = *moderate difficulty*, 4 = *severe difficulty*, and 5 = *extreme difficulty*). Each domain is linked to corresponding ICF codes (Table 1 and Fig. 1). A well-trained interviewer records the results. Missing data (i.e., the unrated items) are substituted using the mean score of the domain. Regarding score computation, according to the WHODAS 2.0 manual, a maximum of 30% of unrated items are allowed in each domain. In addition to the total score, the calculation of standardized domain-specific and summary scores is available on the official WHODAS 2.0 website (https://www.who.int/classifications/icf/whodasii/en/). In the bustle of clinical settings, a short version of the WHODAS 2.0, which comprises a self-administered 12-item questionnaire (Table 2), can be used instead. The short and full versions require 5 and 20 min to complete, respectively (Ustun, Chatterji, et al., 2010).

TABLE 1 World Health Organization Disability Assessment Schedule 2.0 with corresponding international classification of functioning, disability, and health codes.

Domain	Items	ICF code(s)
1. Cognition: In the last 30 days, how much difficulty did you have in:	1.1. Concentrating on a task for 10 min	d160 Focusing attention; b140 Attention functions; d110–d129 Purposeful sensory experiences
	1.2. Remembering to do important things	b144 Memory functions
	1.3. Analyzing and finding solutions to problems in day-to-day life	d175 Solving problems; d130–d159 Basic learning
	1.4. Learning to do something new, such as getting to an unfamiliar location	d1551 Acquiring complex skills
	1.5. Generally understanding what people say	d310 Communicating with—receiving—spoken messages
	1.6. Starting and maintaining a conversation	d3500 Starting a conversation; d3501 Sustaining a conversation
2. Mobility: In the last 30 days how much difficulty did you have in:	2.1. Standing for long periods of time (e.g., 30 min)	d4154 Maintaining a standing position
	2.2. Standing up from sitting down	d4104 Standing
	2.3. Moving around inside your home	d4600 Moving around within the home
	2.4. Leaving your home	d4602 Moving around outside the home and other buildings
	2.5. Walking a long distance, such as 1 km	d4501 Walking long distances
3. Self-care: In the last 30 days, how much difficulty did you have in:	3.1. Washing your entire body	d5101 Washing whole body
	3.2. Getting dressed	d540 Dressing
	3.3. Eating	d550 Eating
	3.4. Being by yourself for a few days	d510–d650 Combination of multiple self-care and domestic life tasks
4. Getting along: In the last 30 days, how much difficulty did you have in:	4.1. Dealing with people you do not know	d730 Relating with strangers
	4.2. Maintaining a friendship	d7500 Informal relationships with friends
	4.3. Getting along with people who are close to you	d760 Family relationships; d770 Intimate relationships; d750 Informal social relationships
	4.4. Making new friends	d7500 Informal relationships with friends; d7200 Forming relationships
	4.5. Sexual activity	d7702 Sexual relationships

Continued

TABLE 1 World Health Organization Disability Assessment Schedule 2.0 with corresponding international classification of functioning, disability, and health codes—cont'd

Domain	Items	ICF code(s)
5. Life activities: In the last 30 days, how much difficulty did you have in:	5.1. Taking care of your household responsibilities	d6 Domestic life
	5.2. Doing most important household tasks well	d640 Doing housework; d210 Undertaking a single task; d220 Undertaking multiple tasks
	5.3. Completing all the housework that you needed to do	d640 Doing housework; d210 Undertaking a single task; d220 Undertaking multiple tasks
	5.4. Completing housework as quickly as needed	d640 Doing housework; d210 Undertaking a single task; d220 Undertaking multiple tasks
	5.5. Your day-to-day work or school activities	d850 Remunerative employment; d830 Higher education; d825 Vocational training; d820 School education
	5.6. Doing your most important work or school tasks well	d850 Remunerative employment; d830 Higher education; d825 Vocational training; d820 School education; d210 Undertaking a single task; d220 Undertaking multiple tasks
	5.7. Completing all the work that you needed to do	d850 Remunerative employment; d830 Higher education; d825 Vocational training; d820 School education; d210 Undertaking a single task; d220 Undertaking multiple tasks
	5.8. Completing your work as quickly as needed	d850 Remunerative employment; d830 Higher education; d825 Vocational training; d820 School education; d210 Undertaking a single task; d220 Undertaking multiple tasks
6. Participation: How much of a problem do you have:	6.1. Joining community activities	d910 Community life
	6.2. Because of barriers or hindrances that the world imposes	d9 Community, social, and civic life
	6.3. Living with dignity	d940 Human rights
	6.4. From time spent on attending to your health condition(s)	Not applicable (impact question)
	6.5. Feeling emotionally affected	b152 Emotional functions
	6.6. Because your health condition(s) is/are a drain on your financial resources	d8700 Personal economic resources
	6.7. With your family facing difficulties because of your health condition(s)	Not applicable (impact question)
	6.8. Doing things by yourself for relaxation or pleasure	d920 Recreation and leisure

Modified from Ustun, T. B., Chatterji, S., Kostanjsek, N., Rehm, J., Kennedy, C., Epping-Jordan, J., ... Project, W. N. J. (2010). Developing the World Health Organization disability assessment schedule 2.0. Bulletin of the World Health Organization, 88 (11), 815–823. doi:10.2471/BLT.09.067231.

Brief introductions of what each domain evaluates are presented as follows (Ustun, Kostanjesek, Chatterji, Rehm, & World Health, 2010):

Domain 1: Cognition

Attention, memory, problem-solving ability, task-learning ability, comprehension, and communication skills.

Domain 2: Mobility

Activities of standing, moving around inside the home, getting out of the home, and walking long distances.

Domain 3: Self-care

Bathing, dressing, eating, and staying by oneself.

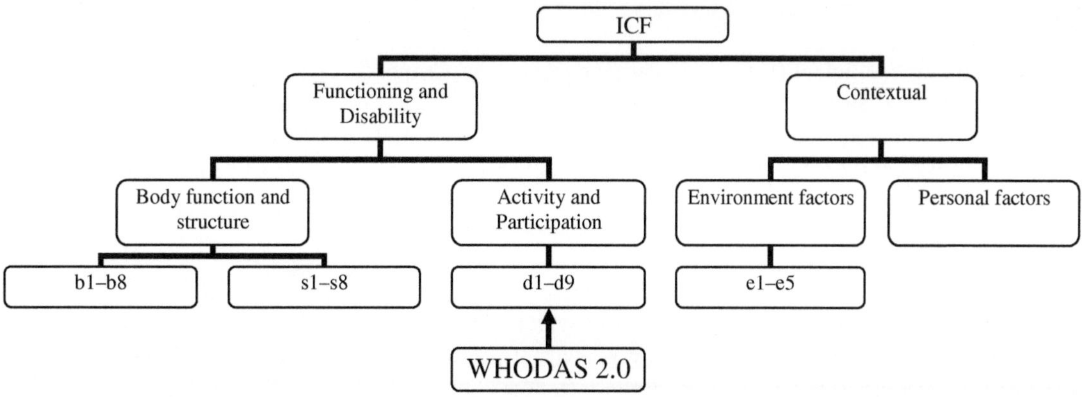

FIG. 1 Focus of the World Health Organization Disability Assessment Schedule 2.0 (WHODAS 2.0) on restrictions to individual activity and participation.

TABLE 2 Short (12-item) version of the WHODAS 2.0, modified from the WHODAS 2.0 manual (Ustun, Chatterji, et al., 2010).

Domain	In the past 30 days, how much difficulty did you have in:
Mobility	1. Standing for long periods (e.g., 30 min)?
Life activities	2. Taking care of your household responsibilities?
Cognition	3. Learning to do something new, such as getting to an unfamiliar location?
Participation	4. Joining community activities (e.g., festivities or religious services) in the same way as anyone else can?
Participation	5. Being emotionally affected by your health problems?
Cognition	6. Concentrating on a task for 10 min?
Mobility	7. Walking a long distance (e.g., 1 km)?
Self-care	8. Washing your whole body?
Self-care	9. Getting dressed?
Getting along	10. Dealing with people you do not know?
Getting along	11. Maintaining a friendship?
Life activities	12. Your day-to-day work or school activities?

WHODAS 2.0, World Health Organization Disability Assessment Schedule 2.0.

Domain 4: Getting along with people

Interacting with people and relevant challenges attributable to one's health conditions. In this context, "people" include both individuals whom the respondent knows well (e.g., family members or close friends) and strangers.

Domain 5: Life activities

Difficulty in performing everyday tasks, including household chores, work, and schoolwork or school activities.

Domain 6: Participation in society

In this domain, which represents a shift from the line of questioning used in the other five, respondents are asked to consider how other people and the world around them make it difficult for them to take part in society. They report not on their activity limitations but rather on the restrictions imposed on them by people, laws, and other features of their world. Respondents must understand that the focus of these questions is on problems encountered because of the society in which they live.

Reliability

The intra-class correlation coefficient (ICC) and internal consistency (Cronbach's α) of the traditional Chinese version of the WHODAS 2.0 are 0.80–0.89 and 0.73–0.99, respectively (Chiu et al., 2014; Yen et al., 2014). The ICCs of WHODAS 2.0 scores in two samples of older adults with schizophrenia were 0.89 and 0.92 (Guilera et al., 2012; McKibbin, Patterson, & Jeste, 2004). A high ICC of 0.88 was observed in the WHODAS 2.0 scores of a group of 64 children in Rwanda (Scorza et al., 2013).

Validity

Studies have reported correlations between scores on the WHODAS 2.0 scores and the abbreviated version of the WHO Quality of Life questionnaire. The level of disability was inversely correlated with that on the WHODAS 2.0, ranging from moderate (0.41–0.58) to strong (0.60–0.70). The correlations noted between assessment scores for the general populations of numerous countries and those of individuals with arthritis, depression, back pain, ankylosing spondylitis, HIV/AIDS, schizophrenia, acquired hearing loss, systemic sclerosis, vestibular disorders, anxiety, posttraumatic stress disorder, acute myocardial infarction, and personality disorders are moderate to strong (Federici et al., 2017).

Current research on the use of the WHODAS 2.0 in TBI care

An increasing number of studies have discussed the utility and reliability of the WHODAS 2.0 in the field of TBI care.

Comparison of disability in spinal cord injury and TBI, as assessed using the WHODAS 2.0

Kuo et al. (2015) were the first to conduct a population-based study ($n=2663$) comparing disability evaluation on the WHODAS 2.0 in patients with TBI and in those with spinal cord injury (SCI) in Taiwan. Compared with the patients with SCI, those with TBI had worse cognition (odds ratio [OR] = 1.988), self-care (OR = 1.073), relationships (OR = 1.429), life activities (OR = 1.029), and participation in society (OR = 1.016). Patients with TBI were also more likely to reside in institutions, which is attributable to their higher WHODAS 2.0 summary scores (OR = 1.178, $p<0.05$).

Evaluation of post-concussion syndrome in patients with mild TBI

In their examination of a group of patients who were recovering slowly after sustaining mild TBI, Snell, Iverson, Panenka, and Silverberg (2017) found the WHODAS 2.0 to have high internal consistency (Cronbach's $\alpha=0.92$) and adequate construct and concurrent validity. Specifically, scores were higher in participants with multiple comorbidities, and those who returned to work successfully had lower scores than those who did not. The WHODAS 2.0 was not only sensitive to post-concussion syndrome but also to conditions concurrent with mild TBI, such as mental health problems and chronic pain.

Association between history of lifetime TBI and disabling functional restrictions

By using the 12-item version of the WHODAS 2.0, Ilie et al. (2018) conducted a large cross-sectional survey ($n=6048$) among adults in Ontario, Canada. Even after adjustments for age, sex, marital status, and household income and education, adults with a history of lifetime TBI had greater odds of global and specific disability compared with adults without a history of TBI. Specifically, these deficits included restricted cognition (moderate-to-severe difficulty in concentrating for 10 min), less mobility (mild-to-extreme difficulty in standing for longer than 30 min and walking long distances), reduced self-care capacity (mild, moderate, or extreme difficulty in washing one's entire body and mild-to-extreme difficulty in getting dressed), difficulties with social relationships (moderate-to-extreme difficulty in dealing with strangers and mild difficulty in maintaining a friendship), fewer life activities (moderate-to-extreme difficulty with day-to-day work and in completing household chores), and reduced participation in society (mild-to-extreme difficulty in joining community activities and remaining emotionally unaffected by one's health problems).

Assessment of chronic TBI with the WHODAS 2.0

Tarvonen-Schroder, Tenovuo, Kaljonen, and Laimi (2018) assessed 112 patients with varying severities (mild, moderate, and severe) of chronic TBI with the short version of the WHODAS 2.0, and a physician performed the ratings for the

WHO minimum data set for health workforce registry (Cieza, Oberhauser, Bickenbach, Chatterji, & Stucki, 2014). The WHODAS 2.0 scores indicated an increased level of impairment in functioning and working ability with increased severity of TBI, especially in standing, walking, and washing and dressing oneself. The results also revealed that the assessments of functioning made by patients, significant others, and treating neurologists were consistent and reliable. Although the WHODAS 2.0 is a general assessment of functioning rather than one designated for TBI, it appears to be a sensible tool for measuring difficulties perceived by patients with different severities of TBI.

Psychometric properties of the WHODAS 2.0

Through Rasch analysis, Snell, Siegert, and Silverberg (2020) found the 12-item version of the WHODAS 2.0 to be a psychometrically sound measure of functional outcomes in adults with mild TBI, adequately covering the spectrum of disability with minimal floor and ceiling effects in the total score. In addition, associations were observed between changes in total Rasch WHODAS 2.0 scores and changes in symptom burden reduction, return to productivity, and global perceptions of symptom improvement. Herrold et al. (2019) reported that although some activities on the 36-item WHODAS 2.0 may be easy for some veterans with mild TBI or behavioral health conditions to perform, the assessment adequately captures the impact of complexity for those with moderate-to-severe disability.

Prediction of institutionalization among patients with TBI

As mentioned, the WHODAS 2.0 provides a comprehensive assessment of patients with TBI. In their examination of a population-based cohort ($n=8630$), Huang, Chang, Escorpizo, Chang, and Liou (2019) found that the WHODAS 2.0 standardized summary score can be used to predict institutionalization in patients who have experienced TBI with acceptable sensitivity (79.6%) and specificity (63.1%) when a cutoff point of 66.85 is applied (area under the receiving operating characteristic curve: 77%).

Conclusion

Because of the characteristics of the disease, functional impairment in patients who have experienced TBI is complex and diverse. The WHODAS 2.0 is a comprehensive assessment tool covering the domains of cognition, mobility, self-care, getting along with people, life activities, and participation in society, with each item linked to ICF codes. It has been proven to be reliable and sensitive to changes in functioning among patients with TBI. Apart from serving as a monitoring tool for rehabilitation progress, the WHODAS 2.0 can potentially be used to predict outcomes of TBI.

Applications to other neuroscience fields

In this chapter, we discussed the assessment of patients with TBI by using the WHODAS 2.0, which was developed under the ICF framework as a comprehensive, reliable, and transcultural assessment tool addressing six domains of functionality. Because of the complexity of neurological diseases, accompanying impairments are diverse and multifaceted. There are more and more studies about the use of the WHODAS 2.0 to assess numerous neurological or neurodegenerative diseases, including SCI, cerebral infarction, dementia, and Parkinson's disease, Huntington disease, epilepsy, multiple sclerosis, has been conducted. Many of these studies have indicated promising results on using the WHODAS 2.0 for functionality assessment, prognostic prediction (e.g., functional recovery, social participation, and return-to-work possibility). We believe that the WHODAS 2.0 will 1 day become a standard tool for evaluating patients with every type of neurological disease.

Mini-dictionary of terms

- ***Cognition***: Mental processes for knowledge acquisition and information retrieval, including thinking, knowing, remembering, judging, and problem solving.
- ***Mobility***: The ability to move freely and easily in one's environment.
- ***Participation***: A means of contributing to broader society by consulting, empowering, and working with communities.
- ***Disability***: Physical or mental impairment that substantially limits one or more of an individual's major life activities.
- ***Impairment***: Partial or complete loss of function of a body structure.

Key facts of the World Health Organization Disability Assessment Schedule 2.0

- Developed by the World Health Organization in 2010 under the conceptual framework of the International Classification of Functioning, Disability, and Health (ICF) model.
- Has been translated into 47 languages and dialects.
- Every item is linked to at least one ICF code concerning functioning in activity and participation.
- The short (12-item) and long (36-item) versions require 5 and 20 min to complete, respectively.
- Numerous studies have confirmed its reliability validity, and utility for disability assessment, follow-up, and prognosis prediction.

Summary points

- Because of the complexity of the disease presentation, the diagnosis and disability assessment of patients with TBI can be extremely challenging to clinicians.
- Every assessment tool has its pros and cons. The majority (73%) of commonly used assessment tools focus on activity and participation. Component body functions and environmental factors are addressed in 24% and a considerably low 3% of the items, respectively.
- The WHODAS 2.0, an assessment tool developed under the ICF framework, shows promise in becoming a standard evaluation tool for patients with TBI.
- The WHODAS 2.0 is sensitive to not only post-concussion syndrome but also conditions concurrent with mild TBI, such as mental health problems and chronic pain.
- The WHODAS 2.0 standardized summary score predicts institutionalization in patients who have experienced TBI with acceptable sensitivity and specificity.
- The WHODAS 2.0 not only serves as a monitoring tool for rehabilitation progress but might also be useful for predicting outcomes in patients with TBI.

References

Bogner, J., Bellon, K., Kolakowsky-Hayner, S. A., & Whiteneck, G. (2013). Participation assessment with recombined tools-objective (PART-O). *The Journal of Head Trauma Rehabilitation*, *28*(4), 337–339. https://doi.org/10.1097/HTR.0b013e31829af969.

Chiu, T. Y., Yen, C. F., Chou, C. H., Lin, J. D., Hwang, A. W., Liao, H. F., et al. (2014). Development of traditional Chinese version of World Health Organization disability assessment schedule 2.0 36—Item (WHODAS 2.0) in Taiwan: Validity and reliability analyses. *Research in Developmental Disabilities*, *35*(11), 2812–2820. https://doi.org/10.1016/j.ridd.2014.07.009.

Cieza, A., Oberhauser, C., Bickenbach, J., Chatterji, S., & Stucki, G. (2014). Towards a minimal generic set of domains of functioning and health. *BMC Public Health*, *14*, 218. https://doi.org/10.1186/1471-2458-14-218.

Federici, S., Bracalenti, M., Meloni, F., & Luciano, J. V. (2017). World Health Organization disability assessment schedule 2.0: An international systematic review. *Disability and Rehabilitation*, *39*(23), 2347–2380. https://doi.org/10.1080/09638288.2016.1223177.

Gold, L. H. (2014). DSM-5 and the assessment of functioning: The World Health Organization disability assessment schedule 2.0 (WHODAS 2.0). *The Journal of the American Academy of Psychiatry and the Law*, *42*(2), 173–181.

Guilera, G., Gómez-Benito, J., Pino, O., Rojo, J. E., Cuesta, M. J., Martínez-Arán, A., et al. (2012). Utility of the World Health Organization disability assessment schedule II in schizophrenia. *Schizophrenia Research*, *138*(2–3), 240–247. https://doi.org/10.1016/j.schres.2012.03.031.

Hall, K. M., Mann, N., High, W. M. J., Wright, J., Kreutzer, J. S., & Wood, D. (1996). Functional measures after traumatic brain injury: Ceiling effects of FIM, FIM+FAM, DRS, and CIQ. *The Journal of Head Trauma Rehabilitation*, *11*(5), 27–39.

Herrold, A. A., Kletzel, S. L., Mallinson, T., Pape, T. L. B., Weaver, J. A., Guernon, A., et al. (2019). Psychometric measurement properties of the world health organization disability assessment schedule 2.0 (WHODAS) evaluated among veterans with mild traumatic brain injury and behavioral health conditions. *Disability and Rehabilitation*, 1–10. https://doi.org/10.1080/09638288.2019.1660914.

Huang, S. W., Chang, K. H., Escorpizo, R., Chang, F. H., & Liou, T. H. (2019). WHODAS 2.0 can predict institutionalization among patients with traumatic brain injury. *International Journal of Environmental Research and Public Health*, *16*(9). https://doi.org/10.3390/ijerph16091484.

Ilie, G., Adlaf, E. M., Mann, R. E., Ialomiteanu, A., Hamilton, H., Rehm, J., et al. (2018). Associations between self-reported lifetime history of traumatic brain injuries and current disability assessment in a population sample of Canadian adults. *PLoS One*, *13*(1). https://doi.org/10.1371/journal.pone.0188908, e0188908.

Jennett, B., & Bond, M. (1975). Assessment of outcome after severe brain damage. *Lancet*, *1*(7905), 480–484. https://doi.org/10.1016/s0140-6736(75)92830-5.

Keith, R. A., Granger, C. V., Hamilton, B. B., & Sherwin, F. S. (1987). The functional independence measure: A new tool for rehabilitation. *Advances in Clinical Rehabilitation*, *1*, 6–18.

Kuipers, P., Kendall, M., Fleming, J., & Tate, R. (2004). Comparison of the Sydney psychosocial reintegration scale (SPRS) with the community integration questionnaire (CIQ): Psychometric properties. *Brain Injury*, *18*(2), 161–177. https://doi.org/10.1080/0269905031000149524.

Kuo, C. Y., Liou, T. H., Chang, K. H., Chi, W. C., Escorpizo, R., Yen, C. F., et al. (2015). Functioning and disability analysis of patients with traumatic brain injury and spinal cord injury by using the World Health Organization Disability Assessment Schedule 2.0. *International Journal of Environmental Research and Public Health*, *12*(4), 4116–4127. https://doi.org/10.3390/ijerph120404116.

Langlois, J. A., Rutland-Brown, W., & Wald, M. M. (2006). The epidemiology and impact of traumatic brain injury: A brief overview. *The Journal of Head Trauma Rehabilitation*, *21*(5), 375–378. https://doi.org/10.1097/00001199-200609000-00001.

Laxe, S., Tschiesner, U., Zasler, N., Lopez-Blazquez, R., Tormos, J. M., & Bernabeu, M. (2012). What domains of the international classification of functioning, disability and Health are covered by the most commonly used measurement instruments in traumatic brain injury research? *Clinical Neurology and Neurosurgery*, *114*(6), 645–650. https://doi.org/10.1016/j.clineuro.2011.12.038.

Linn, R. T., Blair, R. S., Granger, C. V., Harper, D. W., O'Hara, P. A., & Maciura, E. (1999). Does the functional assessment measure (FAM) extend the functional independence measure (FIM) instrument? A rasch analysis of stroke inpatients. *Journal of Outcome Measurement*, *3*(4), 339–359.

Maas, A. I., Stocchetti, N., & Bullock, R. (2008). Moderate and severe traumatic brain injury in adults. *Lancet Neurology*, *7*(8), 728–741. https://doi.org/10.1016/S1474-4422(08)70164-9.

McKibbin, C., Patterson, T. L., & Jeste, D. V. (2004). Assessing disability in older patients with schizophrenia: Results from the WHODAS-II. *The Journal of Nervous and Mental Disease*, *192*(6), 405–413. https://doi.org/10.1097/01.nmd.0000130133.32276.83.

Ptyushkin, P., Vidmar, G., Burger, H., & Marincek, C. (2010). Use of the international classification of functioning, disability and health (ICF) in patients with traumatic brain injury. *Brain Injury*, *24*(13–14), 1519–1527. https://doi.org/10.3109/02699052.2010.523054.

Rappaport, M., Hall, K. M., Hopkins, K., Belleza, T., & Cope, D. N. (1982). Disability rating scale for severe head trauma: Coma to community. *Archives of Physical Medicine and Rehabilitation*, *63*(3), 118–123.

Salter, K., Foley, N., Jutai, J., Bayley, M., & Teasell, R. (2008). Assessment of community integration following traumatic brain injury. *Brain Injury*, *22*(11), 820–835. https://doi.org/10.1080/02699050802425428.

Scorza, P., Stevenson, A., Canino, G., Mushashi, C., Kanyanganzi, F., Munyanah, M., et al. (2013). Validation of the "World Health Organization disability assessment schedule for children, WHODAS-child" in Rwanda. *PLoS One*, *8*(3). https://doi.org/10.1371/journal.pone.0057725, e57725.

Snell, D. L., Iverson, G. L., Panenka, W. J., & Silverberg, N. D. (2017). Preliminary validation of the World Health Organization Disability Assessment Schedule 2.0 for mild traumatic brain injury. *Journal of Neurotrauma*, *34*(23), 3256–3261. https://doi.org/10.1089/neu.2017.5234.

Snell, D. L., Siegert, R. J., & Silverberg, N. D. (2020). Rasch analysis of the World Health Organization disability assessment schedule 2.0 in a mild traumatic brain injury sample. *Brain Injury*, *34*(5), 610–618. https://doi.org/10.1080/02699052.2020.1729417.

Stucki, G., Cieza, A., & Melvin, J. (2007). The international classification of functioning, disability and Health (ICF): A unifying model for the conceptual description of the rehabilitation strategy. *Journal of Rehabilitation Medicine*, *39*(4), 279–285. https://doi.org/10.2340/16501977-0041.

Svestkova, O., Angerova, Y., Sladkova, P., Bickenbach, J. E., & Raggi, A. (2010). Functioning and disability in traumatic brain injury. *Disability and Rehabilitation*, *32*(Suppl 1), S68–S77. https://doi.org/10.3109/09638288.2010.511690.

Tarvonen-Schroder, S., Tenovuo, O., Kaljonen, A., & Laimi, K. (2018). Usability of World Health Organization disability assessment schedule in chronic traumatic brain injury. *Journal of Rehabilitation Medicine*, *50*(6), 514–518. https://doi.org/10.2340/16501977-2345.

Tate, R. L., Godbee, K., & Sigmundsdottir, L. (2013). A systematic review of assessment tools for adults used in traumatic brain injury research and their relationship to the ICF. *NeuroRehabilitation*, *32*(4), 729–750. https://doi.org/10.3233/NRE-130898.

Teasdale, G. M., Pettigrew, L. E., Wilson, J. T., Murray, G., & Jennett, B. (1998). Analyzing outcome of treatment of severe head injury: A review and update on advancing the use of the Glasgow outcome scale. *Journal of Neurotrauma*, *15*(8), 587–597. https://doi.org/10.1089/neu.1998.15.587.

Ustun, T. B., Chatterji, S., Kostanjsek, N., Rehm, J., Kennedy, C., Epping-Jordan, J., et al. (2010). Developing the World Health Organization disability assessment schedule 2.0. *Bulletin of the World Health Organization*, *88*(11), 815–823. https://doi.org/10.2471/BLT.09.067231.

Ustun, T. B., Kostanjesek, N., Chatterji, S., Rehm, J., & World Health, O. (2010). In T. B. Üstün, N. Kostanjsek, S. Chatterji, & J. Rehm (Eds.), *Measuring health and disability: Manual for WHO disability assessment schedule (WHODAS 2.0)*. Geneva: World Health Organization.

Whiteneck, G. G., Dijkers, M. P., Heinemann, A. W., Bogner, J. A., Bushnik, T., Cicerone, K. D., et al. (2011). Development of the participation assessment with recombined tools-objective for use after traumatic brain injury. *Archives of Physical Medicine and Rehabilitation*, *92*(4), 542–551. https://doi.org/10.1016/j.apmr.2010.08.002.

Whyte, J., Cifu, D., Dikmen, S., & Temkin, N. (2001). Prediction of functional outcomes after traumatic brain injury: A comparison of 2 measures of duration of unconsciousness. *Archives of Physical Medicine and Rehabilitation*, *82*(10), 1355–1359. https://doi.org/10.1053/apmr.2001.26091.

World Health Organization. (2001). *The international classification of functioning, disability and health (ICF)*. Geneva, Switzerland: WHO.

Yen, C. F., Hwang, A. W., Liou, T. H., Chiu, T. Y., Hsu, H. Y., Chi, W. C., et al. (2014). Validity and reliability of the functioning disability evaluation scale-adult version based on the WHODAS 2.0—36 items. *Journal of the Formosan Medical Association*, *113*(11), 839–849. https://doi.org/10.1016/j.jfma.2014.08.008.

Chapter 18

Biomarkers in pediatric traumatic injury: The brain and beyond

Dimitrios Metaxas[a], Georgios Alexiou[a], Olga Ygropoulou[b], Dimitrios Rizos[b], and Spyridon Voulgaris[a]
[a]*Department of Neurosurgery, University Hospital of Ioannina, Ioannina, Greece,* [b]*Department of Intensive Care Unit, "Hatzikosta" General Hospital, Ioannina, Greece*

List of abbreviations

AHT	abusive head trauma
CDC	center for disease control
CNS	central nervous system
CSF	cerebrospinal fluid
CT	computed tomography
GCS	Glasgow Coma Scale
GFAP	glial fibrillary acidic protein
ICU	intensive care unit
IL-8	interleukin 8
miR	micro RNA
MRI	magnetic resonance imaging
NSE	neuron specific enolase
TBI	traumatic brain injury
UCH-l1	umbiquitin C-terminal hydrolase-l1

Introduction

Traumatic brain injury (TBI) as defined by CDC most recently in 2015 is a non-degenerative acquired disruption of the normal brain function caused by a bump, jolt, or blow to the head or penetrating head trauma. It can affect any age group but those in greater risk are the elderly and children among which TBI is the leading cause of mortality and morbidity.

Annually there are 837,000 cases of pediatric TBI in the United States alone (CDC, 2019) 3% of which result in hospitalization and a 0.25% result in death. Most cases that require consultation concern children of the 0–4 age group (1618 per 100,000) while hospitalization is more common in adolescents (60 per 100,000). There is a higher prevalence in boys than in girls for both treatment in emergency department and hospitalization. The mechanisms of injury vary according to age groups. In infants, the most usual causes are falls and abusive head trauma (AHT), in older children the leading causes are again falls and also transportation accidents, including motor vehicle and bicycle accidents, and in adolescents sport related injuries and motor vehicle accidents concern most of the cases.

As in adults, TBI in children can present with injuries of the scalp, fractures of the bony compartments of the skull and intra-axial and extra-axial intracranial injuries and it is classified according to Glasgow Coma Scale score in mild (13–15), moderate (9–12), and severe (3–8) TBI. However, the pathophysiology and management of TBI are considerably different in pediatric patients. The developing brain of the children, the age related features that causes different mechanisms of injury and the difficulties that the neurological evaluation in younger ages present to physicians, are the main reasons that TBI in children needs a considerably altered approach than the adult cases. The clinical presentation of a TBI and its severity varies considerably according to the age group of the patient. For example, blood loss in the age group of 0–4 can cause severe disturbances even in small blood volumes comparing to an adult, due to a much smaller total volume of blood. Regarding the actual injuries of the brain, they differ substantially from those in adults as the brain in younger ages has less space in the skull thus moving as one with it and absorbing higher forces than in adults, and also in case of intracranial hemorrhage, either intra-axial or extra-axial, the intracranial pressure becomes critical much faster, presenting with a

Diagnosis and Treatment of Traumatic Brain Injury. https://doi.org/10.1016/B978-0-12-823347-4.00003-8

more rapid onset and progression of neural tissue damage and its accompanying symptoms. Moreover, the not fully formed central nervous system of an infant or a young child is more susceptible to late sequelae due to disturbances in its development even in cases when there is no apparent symptomatology at the time of injury.

Despite the great progress in imaging and laboratory tests, diagnosing and managing TBI in a pediatric patient may be challenging. A MRI during the initial evaluation is the most potent tool in the hands of a physician. However, it is time consuming, it has a high cost, and it is not easily available in emergency departments. A CT scan is usually performed in cases where a intracranial lesion is suspected, however there is not a clear algorithm on when to perform a CT scan or not, and considering the late effects that radiation might be associated, decision making is difficult. Furthermore, early and late management of a child with TBI proves challenging due to difficulties in determining the prognosis and the severity of the trauma. A promising solution for the problems listed above would be the identification of reliable laboratory biomarkers helpful in determining trauma severity, indicating the need for further imaging studies and allowing a more accurate calculation of short-, mid- and late-term prognosis.

Biomarkers in traumatic brain injury

A biomarker can be defined as quantitative or qualitative measurement that provides information helpful in diagnosis, prognosis, and prediction of a disease. The use of several body fluid biomarkers, mainly from blood and CSF, has proven valuable in the diagnosis and management of traumatic brain injury. Decisions such as the need of neuroimaging, can be determined by levels of several biomarkers such as S100B, which outperforms decision making by clinical algorithms and guidelines, such as the New Orleans' criteria (Jones, Harmon, McCann, Gunyan, & Bazarian, 2020) and proved to have a negative predictive value for results of brain CT even in patients under antithrombotic treatment (David et al., 2017), a decision aided also by the levels of several other biomarkers such as glial fibrillary acidic protein (GFAP) (Bazarian et al., 2018), Ubiquitin C-terminal hydrolase-L1 (UCH-L1)) (Bazarian et al., 2018) serum glucose (Alexiou et al., 2019) and others. Prediction of outcome in patients with TBI is also aided by biomarkers such as neuron specific enolase (NSE) which is however outperformed by S100B (Thelin et al., 2016). Furthermore, biomarkers are helpful in the determination of TBI severity as described for example by Schindler et al. (2020) with the detection of miR-423-3p levels in the first hours after an isolated TBI, a high value of which indicates a severe brain trauma. Research is still ongoing and the integration of more biomarkers and their uses in clinical practice can possibly solve many problems and make the management of TBI easier and more cost effective.

Biomarkers in pediatric traumatic brain injury

As described above there is a proven value for several biomarkers in traumatic brain injuries in adults. In contrast on pediatric TBI, research is not as extended and clinical application is far more limited. However, there are a lot of promising biomarkers with possible applications in early diagnosis, decision for neuroimaging, prognosis of TBI, prediction of late sequalae in adult life and other aspects of pediatric TBI. Below we analyze more specifically the possible applications of several biomarkers.

Biomarkers and neuroimaging

Physicians often face the question of how to effectively manage a TBI in children at the emergency department, whether to perform a neuroimaging study and which one. Emergency care rarely provides the solution of an MRI scan as it is time consuming and costly, thus leaving the physicians with the only choice of a brain CT scan. However, the decision to perform a CT scan is not lightly taken especially in mild pediatric TBI because of radiation exposure and CTs, even though they are associated with lower cost in comparison to a MRI, still costs. Several biomarkers under research have shown promising results in identifying patients with positive findings in CTs thus possible enabling a reduction of unnecessary studies if used effectively. Studies have shown that measurement of S100B serum levels at 3 h after head trauma can identify patients with positive results in CT scans (Bouvier et al., 2012; Kelmendi et al., 2018), with a sensitivity of up to 100% with a proposed cutoff threshold of 0.18 μg/L (Bouvier et al., 2012). Saliva levels of S100B are also proposed to predict positive findings in CT scans with an optimal threshold, for highest sensitivity and specificity, of 86.9 pg/mL (Yeung, Bhatia, Bhattarai, & Sinha, 2020). GFAP serum levels obtained up to 6 h after trauma have also been associated with predicting intracranial lesions in imaging studies (Papa et al., 2015) with a threshold of 0.15 ng/mL yielding 94% sensitivity, 47% specificity and 98% negative predictive value, and have also been described as outperforming S100B in this aspect (Papa et al., 2016). Another biomarker with very promising results in this field is UCH-L1 (serum levels obtained up to 6 h after trauma) with

93% sensitivity and 25% specificity in differentiating patients with normal CTs to those with intracranial lesions, at a cut-off threshold of 0.09 ng/mL (Mondello et al., 2016), increasing though its sensitivity to 100%, and yielding a 100% negative predictive value at a cut-off threshold of 0.18 ng/mL (Papa et al., 2017). One study (Langness et al., 2018) also proposed that measurement of D-dimer levels in blood within 6 h after trauma can safely indicate unnecessary CT scans with a value of < 750 pg/mL as negative screen, but further research is needed (Table 1).

Biomarkers as predictors of severity and prognosis of TBI

Traumatic brain injury is a major cause of morbidity and mortality in children. Emergency care physicians face many challenges in determining the severity of TBI in children. Complaints are usually vague; symptomatology varies according to age and development and medical history is not always accurate. Combining all these with the fact that intracranial injuries in children, even those with no initial symptomatology, can rapidly progress and deteriorate, it is made apparent that there is a need for aiding tools in predicting the severity and the outcome of pediatric TBI. Several biomarkers have been proposed for their usefulness in this field. S100B, one of the most studied biomarker, in both adult and pediatric TBI, has been proposed to differentiate between mild and moderate/severe TBI (Babcock, Byczkowski, Mookerjee, & Bazarian, 2012) and having a positive correlation with poor outcome (Thelin et al., 2016). Studies comparing S100b to neuron specific enolase (NSE), though suggesting that the first outperforms the second (Nakhjavan-Shahraki, Yousefifard, Oraii, Sarveazad, & Hosseini, 2017; Thelin et al., 2016), prove that both of them have a significant clinical value in predicting both the severity of a TBI, consistently showing a negative correlation to GCS scores, and the outcome with a positive correlation with the probability of death, prolonged mechanical ventilation, intensive care unit stay and hospitalization (Park & Hwang, 2018; Park, Park, & Hwang, 2019). Glucose levels and more specifically, hyperglycemia in the first 24 h after severe head injury is associated with higher probability of death, prolonged pediatric ICU stay and prolonged mechanical ventilation (Chong et al., 2015). Similar results have been reported for serum lactate, albumin and hemoglobin levels, showing a strong association of high serum lactate (values more than 2 mmol/L) (Fu, Bai, & Liu, 2019), hypo-albuminemia (serum albumin < 30 g/L) and hypo-globulinemia (hemoglobin < 90 g/L) (Luo, Fu, You, Liu, & Xu, 2019), with higher probability of death, reduced ventilation-free days in a 14-day period, reduced ICU free days also in a 14-day period and reduced hospitalization free days in a 28-day period. Isolated studies have also attributed clinical significance in predicting severity and outcome for biomarkers as serum tau (Stukas et al., 2019), serum and CSF UCH-L1 (Metzger et al., 2018), plasma copeptin (Lin, Wang, Shen, & Zhao, 2013), CSF NLRP3 (Wallisch et al., 2017) and autophagy biomarker p62 (Au et al., 2017) but further research is needed.

Biomarkers predicting late sequalae of TBI in children

Traumatic brain injury in children, even in mild to moderate cases, is associated with late sequalae concerning cognitive and psychosocial functioning. However, recognizing cases with higher probability of developing sequalae such as fatigue and attention deficit, remains a problem for physicians and other attending specialists. Recent studies have shown promising results in the application of certain biomarkers in this field. Wilkinson et al. (2017) showed a positive correlation of the levels of neuron specific enolase and negative correlation of the levels of soluble neuron cell adhesion molecule with the probability of attention and executive function deficit. Fatigue resulting from TBI is also possible to be predicted by measuring levels of IL-8 (Crichton et al., 2019) with research results showing that one unit increase in IL-8 levels increases the probability of 12-month post-TBI fatigue by 4%.

TABLE 1 Biomarkers to predict positive CT findings in minor TBI.

Biomarker	Study	Sample	Cut-off value	Sensitivity	Specificity	Negative predictive value
S100	Bouvier (2012)	Blood	0.18 μg/L	100%	33%	n/a
	Yeung (2020)	Saliva	86.9 pg/mL	n/a	n/a	n/a
GFAP	Pappa (2015)	Blood	0.15 ng/mL	94%	47%	98%
UCH-L1	Pappa (2017)	Blood	0.18 ng/mL	100%	47%	100%
D-dimer	Langness (2018)	Blood	< 750 pg/mL	n/a	n/a	100%

Recognizing abusive trauma

Treating children with injuries, especially in younger ages (< 5 years old), frequently presents physicians with the problem of recognizing child abuse and maltreatment. Abusive head trauma, according to Spies and Klevens (2016), results in one third of child maltreatment fatalities. Spies and Klevens (2016) has also published results showing that 35 out of 100,000 infants under 1 year old suffer from abusive head trauma. Recognizing an abusive behavior toward a child might be difficult. Physicians rely in facts such as repetitive emergency department visits, multiple body injuries of different time of infliction and the social background and status of the family. A recent study (Blackwell et al., 2020) indicated a positive correlation of plasma osteopontin levels and higher probability of abusive head trauma thus indicating osteopontin as a possible biomarker for abusive head trauma recognition, however further research is needed.

Applications to other areas of neurosciences

In this chapter, we analyzed the current knowledge and research in the use of biomarkers in the diagnosis and management of traumatic brain injury in pediatric patients. Several of those biomarkers are proven or are under research for their use in any kind of neural tissue damage. Biomarkers such as GFAP and S100B for example are used in the diagnosis and outcome prediction of CNS tumors, neurodegenerative diseases and autoimmune diseases affecting the CNS. Moreover, several of the above-mentioned biomarkers have possible uses in any traumatic injury of neural tissue like spinal cord injury or peripheral nerve injury.

Mini-dictionary of terms

Traumatic brain injury. Brain tissue damage caused by trauma.

Biomarkers. A measurement value indicating the presence, severity, and possible outcome of a disease.

Glasgow Coma Scale. A scale used to measure the level of consciousness by measuring a person's motor and verbal response and the level of stimulation they need to open their eyes.

Glial Fibrilary Acidic Protein. A protein expressed by neural tissue cells, mainly astrocytes, whose elevated values indicate neural tissue damage.

S100B. A calcium binding protein expressed in many cells, including astrocytes that is involved in cellular structure, metabolism, multiplication, and differentiation. Elevated values may indicate neural tissue damage.

Neuron Specific Enolase. An acidic protease present in neurons and neuroendocrine cells. Its levels can indicate the severity of neural tissue damage.

Ubiquitin C-terminal hydrolase-L1. Enzyme abundant in brain tissue, required for axonal integrity. Changes in its levels can indicate dysfunctions of brain tissue.

Osteopontin. Protein present in bone matrix with significant value in bone maintenance. Indicator of bone injury.

Key facts of traumatic brain injury

- Leading cause of disability and death in pediatric patients.
- Classified in mild moderate and severe according to GCS.
- Diagnosis and treatment in children present several difficulties.
- Neuroimaging is crucial in diagnosis, but the choice of study is difficult due to possible late side effects and restriction in availability.
- Outcome difficult to predict with many possible late sequalae.

Key facts of biomarkers

- Medical measured values (biochemical results, biometrics, etc.) that can indicate a dysfunction or disease.
- Several biomarkers have a proven application in recognizing and predicting the outcome of neural tissue damage including traumatic brain injury, but further research is underway.
- GFAP and S100B are two of the most researched biomarkers for neural tissue damage.

- Usefulness in pediatric care is still in early stages of research but with promising results.
- Application of research results in every day clinical practice concerning trauma is limited.
- Further research is needed.

Summary points

- Traumatic brain injury is a major cause of hospitalization, disability, and death in pediatric clinical care.
- Diagnosis and management present physicians with many difficulties, especially in emergency pediatric care.
- Several biomarkers have been proposed as clinical tools in the care of adult trauma but research in children's head injuries is still limited.
- Decision for use of neuroimaging studies can be aided by simple biochemical values.
- Prediction of outcome and possible future disabilities may be simpler by the integration of biomarker research in clinical practice.
- Recognition of child abuse is critical but extremely complicated. Recent research promises use of simple biochemical tests (osteopontin levels) can make the process easier.
- Further research in the field is still necessary.

References

Alexiou, G. A., Sotiropoulos, A., Lianos, G. D., Zigouris, A., Metaxas, D., Nasios, A., et al. (2019). Blood glucose levels may aid the decision for CT scan in minor head trauma. *Disease Markers*, *2019*, 1065254. https://doi.org/10.1155/2019/1065254.

Au, A. K., Aneja, R. K., Bayır, H., Bell, M. J., Janesko-Feldman, K., Kochanek, P. M., et al. (2017). Autophagy Biomarkers Beclin 1 and p62 are increased in cerebrospinal fluid after traumatic brain injury. *Neurocritical Care*, *26*(3), 348–355. https://doi.org/10.1007/s12028-016-0351-x.

Babcock, L., Byczkowski, T., Mookerjee, S., & Bazarian, J. J. (2012). Ability of S100B to predict severity and cranial CT results in children with TBI. *Brain Injury*, *26*(11), 1372–1380. https://doi.org/10.3109/02699052.2012.694565.

Bazarian, J. J., Biberthaler, P., Welch, R. D., Lewis, L. M., Barzo, P., Bogner-Flatz, V., et al. (2018). Serum GFAP and UCH-L1 for prediction of absence of intracranial injuries on head CT (ALERT-TBI): A multicentre observational study. *The Lancet Neurology*, *17*(9), 782–789.

Blackwell, L. S., Martinez, M., Fournier-Goodnight, A., Figueroa, J., Appert, A., Vats, A., et al. (2020). Patterns of osteopontin expression in abusive head trauma compared with other causes of pediatric traumatic brain injury. *The Journal of Pediatrics*, *227*, 170–175. https://doi.org/10.1016/j.jpeds.2020.06.080.

Bouvier, D., Fournier, M., Dauphin, J. B., Amat, F., Ughetto, S., Labbé, A., et al. (2012). Serum S100B determination in the management of pediatric mild traumatic brain injury. *Clinical Chemistry*, *58*(7), 1116–1122. https://doi.org/10.1373/clinchem.2011.180828.

Centers for Disease Control and Prevention. (2019). Surveillance report of traumatic brain injury-related emergency department visits, hospitalizations, and deaths—United States, 2014. In *Centers for Disease Control and Prevention* U.S. Department of Health and Human Services.

Chong, S. L., Harjanto, S., Testoni, D., Ng, Z. M., Low, C. Y., Lee, K. P., et al. (2015). Early hyperglycemia in pediatric traumatic brain injury predicts for mortality, prolonged duration of mechanical ventilation, and intensive care stay. *International Journal of Endocrinology*, *2015*, 719476. https://doi.org/10.1155/2015/719476.

Crichton, A. J., Ignjatovic, V., Babl, F., Oakley, E., Greenham, M., Hearps, S., et al. (2019). Interleukin-8 predicts fatigue at 12 months post-injury in children with traumatic brain injury. *Journal of Neurotrauma*. https://doi.org/10.1089/neu.2018.6083. Advance online publication.

David, A., Mari, C., Vignaud, F., Masson, D., Planche, L., Bord, E., et al. (2017). Evaluation of S100B blood level as a biomarker to avoid computed tomography in patients with mild head trauma under antithrombotic medication. *Diagnostic and Interventional Imaging*, *98*(7–8), 551–556. https://doi.org/10.1016/j.diii.2017.03.010.

Fu, Y. Q., Bai, K., & Liu, C. J. (2019). The impact of admission serum lactate on children with moderate to severe traumatic brain injury. *PLoS One*, *14*(9). https://doi.org/10.1371/journal.pone.0222591, e0222591.

Jones, C., Harmon, C., McCann, M., Gunyan, H., & Bazarian, J. J. (2020). S100B outperforms clinical decision rules for the identification of intracranial injury on head CT scan after mild traumatic brain injury. *Brain Injury*, *34*(3), 407–414. https://doi.org/10.1080/02699052.2020.1725123.

Kelmendi, F. M., Morina, A. A., Mekaj, A. Y., Blyta, A., Alimehmeti, R., Dragusha, S., et al. (2018 April 23). Serum S100B levels can predict computed tomography findings in paediatric patients with mild head injury. *BioMed Research International*, *2018*. https://doi.org/10.1155/2018/6954045, 6954045. 29850551. PMC5937551.

Langness, S., Ward, E., Halbach, J., Lizardo, R., Davenport, K., Bickler, S., et al. (2018 Apr). Plasma D-dimer safely reduces unnecessary CT scans obtained in the evaluation of pediatric head trauma. *Journal of Pediatric Surgery*, *53*(4), 752–757. https://doi.org/10.1016/j.jpedsurg.2017.08.017. Epub 2017 Sep 1 29555324.

Lin, C., Wang, N., Shen, Z. P., & Zhao, Z. Y. (2013). Plasma copeptin concentration and outcome after pediatric traumatic brain injury. *Peptides*, *42*, 43–47. https://doi.org/10.1016/j.peptides.2013.01.015.

Luo, H. C., Fu, Y. Q., You, C. Y., Liu, C. J., & Xu, F. (2019). Comparison of admission serum albumin and hemoglobin as predictors of outcome in children with moderate to severe traumatic brain injury: A retrospective study. *Medicine*, *98*(44). https://doi.org/10.1097/MD.0000000000017806, e17806.

Metzger, R. R., Sheng, X., Niedzwecki, C. M., Bennett, K. S., Morita, D. C., Zielinski, B., et al. (2018). Temporal response profiles of serum ubiquitin C-terminal hydrolase-L1 and the 145-kDa alpha II-spectrin breakdown product after severe traumatic brain injury in children. *Journal of Neurosurgery. Pediatrics*, *22*(4), 369–374. https://doi.org/10.3171/2018.4.PEDS17593.

Mondello, S., Kobeissy, F., Vestri, A., Hayes, R. L., Kochanek, P. M., & Berger, R. P. (2016). Serum concentrations of ubiquitin C-terminal hydrolase-L1 and glial fibrillary acidic protein after pediatric traumatic brain injury. *Scientific Reports*, *6*, 28203. https://doi.org/10.1038/srep28203.

Nakhjavan-Shahraki, B., Yousefifard, M., Oraii, A., Sarveazad, A., & Hosseini, M. (2017). Meta-analysis of neuron specific enolase in predicting pediatric brain injury outcomes. *EXCLI Journal*, *16*, 995–1008. https://doi.org/10.17179/excli2017-405.

Papa, L., Mittal, M. K., Ramirez, J., Ramia, M., Kirby, S., Silvestri, S., et al. (2016). In children and youth with mild and moderate traumatic brain injury, glial fibrillary acidic protein out-performs S100β in detecting traumatic intracranial lesions on computed tomography. *Journal of Neurotrauma*, *33*(1), 58–64. https://doi.org/10.1089/neu.2015.3869.

Papa, L., Mittal, M. K., Ramirez, J., Silvestri, S., Giordano, P., Braga, C. F., et al. (2017). Neuronal biomarker ubiquitin C-terminal hydrolase detects traumatic intracranial lesions on computed tomography in children and youth with mild traumatic brain injury. *Journal of Neurotrauma*, *34*(13), 2132–2140. https://doi.org/10.1089/neu.2016.4806.

Papa, L., Zonfrillo, M. R., Ramirez, J., Silvestri, S., Giordano, P., Braga, C. F., et al. (2015). Performance of glial fibrillary acidic protein in detecting traumatic intracranial lesions on computed tomography in children and youth with mild head trauma. *Academic Emergency Medicine: Official Journal of the Society for Academic Emergency Medicine*, *22*(11), 1274–1282. https://doi.org/10.1111/acem.12795.

Park, S. H., & Hwang, S. K. (2018). Prognostic value of serum levels of S100 calcium-binding protein B, neuron-specific enolase, and Interleukin-6 in pediatric patients with traumatic brain injury. *World Neurosurgery*, *118*, e534–e542. https://doi.org/10.1016/j.wneu.2018.06.234.

Park, D. W., Park, S. H., & Hwang, S. K. (2019). Serial measurement of S100B and NSE in pediatric traumatic brain injury. *Child's Nervous System*, *35*(2), 343–348. https://doi.org/10.1007/s00381-018-3955-y.

Schindler, C. R., Woschek, M., Vollrath, J. T., Kontradowitz, K., Lustenberger, T., Störmann, P., et al. (2020). miR-142-3p expression is predictive for severe traumatic brain injury (TBI) in trauma patients. *International Journal of Molecular Sciences*, *21*(15), 5381. https://doi.org/10.3390/ijms21155381.

Spies, E. L., & Klevens, J. (2016). *Fatal abusive head trauma among children aged <5 years – United States, 1999–2014. MMWR Morbidity and Mortality Weekly Report*, *65*(20), 505–509.

Stukas, S., Higgins, V., Frndova, H., Gill, J., Hubara, E., Guerguerian, A. M., et al. (2019). Characterisation of serum total tau following paediatric traumatic brain injury: A case-control study. *The Lancet Child & Adolescent Health*, *3*(8), 558–567. https://doi.org/10.1016/S2352-4642(19)30194-4.

Thelin, E. P., Jeppsson, E., Frostell, A., Svensson, M., Mondello, S., Bellander, B. M., et al. (2016). Utility of neuron-specific enolase in traumatic brain injury; relations to S100B levels, outcome, and extracranial injury severity. *Critical Care (London, England)*, *20*, 285. https://doi.org/10.1186/s13054-016-1450-y.

Wallisch, J. S., Simon, D. W., Bayır, H., Bell, M. J., Kochanek, P. M., & Clark, R. (2017). Cerebrospinal fluid NLRP3 is increased after severe traumatic brain injury in infants and children. *Neurocritical Care*, *27*(1), 44–50. https://doi.org/10.1007/s12028-017-0378-7.

Wilkinson, A. A., Dennis, M., Simic, N., Taylor, M. J., Morgan, B. R., Frndova, H., et al. (2017). Brain biomarkers and pre-injury cognition are associated with long-term cognitive outcome in children with traumatic brain injury. *BMC Pediatrics*, *17*(1), 173. https://doi.org/10.1186/s12887-017-0925-6.

Yeung, C., Bhatia, R., Bhattarai, B., & Sinha, M. (2020). Role of salivary Biomarkers in predicting significant traumatic brain injury: An exploratory study. *Pediatric Emergency Care*. https://doi.org/10.1097/PEC.0000000000002050. Advance online publication.

Chapter 19

Brain injury biomarkers: Proteins and autoantibodies interplay

Samar Abdelhady[a], Hawraa Issa[b], Ohanes Ashekyan[c], Muhammad Ali Haidar[c], Oumaima Outani[d], Yasmine Samir[e], Eslam Belal[a], Zaynab Shakkour[c], Stefania Mondello[f], and Firas H. Kobeissy[c,g]

[a]*Faculty of Medicine, Alexandria University, Alexandria, Egypt,* [b]*College of Public Health, Phoenicia University, Zahrani, Lebanon,* [c]*Department of Biochemistry and Molecular Genetics, Faculty of Medicine, American University of Beirut, Beirut, Lebanon,* [d]*Faculty of Medicine and Pharmacy of Rabat, Muhammad-V University, Rabat, Morocco,* [e]*Departments of Emergency Medicine, Psychiatry, Neuroscience and Chemistry, Faculty of Science, Alexandria University, Alexandria, Egypt,* [f]*Department of Biomedical and Dental Sciences and Morphofunctional Imaging, University of Messina, Messina, Italy,* [g]*Program for Neurotrauma, Neuroproteomics & Biomarkers Research, Departments of Emergency Medicine, Psychiatry, Neuroscience and Chemistry, University of Florida, Gainesville, FL, United States*

List of abbreviations

AchR	acetylcholine Receptor
AMPA	α-amino-3-hydroxy-5-methyl-4-isoxazolepropionic acid
APRIL	a proliferation inducing ligand
AQP4	aquaporin 4
ATP	adenosine triphosphate
BAFF	B-cell activation factor
BBB	blood–brain barrier
CNS	central nervous system
CSF	cerebrospinal fluid
CT	computed tomography
DAMPs	damage-associated molecular patterns
GFAP-BDP	glial fibrillary acidic protein—breakdown products
GS	glymphatic system
HMGB-1	high mobility group protein B1
IL	interleukin
ISF	interstitial fluid
IV-IG	intravenous immunoglobulin
MBP	myelin basic protein
MRI	magnetic resonance imaging
NMDA	*N*-methyl-D-aspartic acid
PRDX6	peroxiredoxin 6
ROS	reactive oxygen species
SCI	spinal cord injury
TBI	traumatic brain injury
TGF-β	transforming growth factor-beta
Th-1	T helper cell 1
TNF-α	tumor necrosis factor-alpha
Tregs	regulatory T-cells

Diagnosis and Treatment of Traumatic Brain Injury. https://doi.org/10.1016/B978-0-12-823347-4.00036-1

Introduction

Traumatic brain injury (TBI), the most recognized subtype of central nervous system (CNS) injuries, is a major cause of death and morbidity worldwide, with millions of people afflicted every year. Approximately 1.7 million individuals are reported to suffer from TBI in the USA alone each year (Coronado et al., 2011; Faul & Coronado, 2015; McKee & Lukens, 2016). Thereby, TBI draws a global concern due to its debilitating implication and serious socioeconomic burden, especially among military personnel, athletes, and the elderly populations. Moreover, the TBI-induced cascade of inflammatory and pathophysiological events has been of significant concern since scientific literature data have confirmed its role as a precursor to the development of severe neurological and neuropsychiatric disorders such as Alzheimer's disease (AD), chronic traumatic encephalopathy (CTE), Parkinson's disease (PD), amyotrophic lateral sclerosis (ALS), and post-traumatic stress disorder (PTSD), to name a few (McKee & Lukens, 2016).

According to the Centers for Disease Control and Prevention (CDC, 2019), TBI is defined as "the trauma induced by a bump, blow, or jolt to the head leading to disruption of the brain structure and homeostasis." Following the Glasgow Coma Scale (GCS) guidelines, TBI can be classified into (a) mild; (b) moderate; and (c) severe according to the severity of cognitive, behavioral, and physical outcomes (McKee & Lukens, 2016; Teasdale & Jennett, 1974). TBI is a multi-faceted disease process that triggers structural and functional deficits, inconsequential, from primary to secondary injury mechanisms (Nizamutdinov & Shapiro, 2017). The immediate mechanical damage caused by the direct force to the brain induces the primary injury insults. This includes neuronal and glial apoptosis, hemorrhage and Blood–brain barrier (BBB) disruption, leakage of immune cells, and axonal shearing injury, in the minutes to hours following the traumatic impact. In accordance, the secondary injury mechanisms develop on top of the primary insult. They are the results of biochemical alterations, mitochondrial dysfunction, oxidative stress, glutamate excitotoxicity, and diffuse axonal injury (DAI) that evolve from minutes to days, months, and even years following the injury. Unless timely rescued, the end result could be an irreversible brain cell atrophy and apoptosis (Nizamutdinov & Shapiro, 2017). Recent advances in diagnostics research could match the previously mentioned biomolecular events to distinct signature biomarkers detected in body biofluids (i.e., Cerebrospinal fluid (CSF), Serum, urine, and saliva). The Food and Drug Administration (FDA, 2016) defines biomarkers as a "characteristic that is measured as an indicator of normal biological processes, pathogenic processes, or responses to an exposure or intervention, including therapeutic interventions." The crucial challenge of TBI biomarkers, on the one hand, is to substitute the need to and reduce the cost and risk of radiation exposure to unnecessary imaging modalities (Slavoaca et al., 2020). On the other hand, improving the speed of detecting the impact of the repeated mild injuries with long-lasting minor insults, that could otherwise be uncorrelated, could improve the prognostic clinical outcomes.

Herein we give an overview on the current challenges in tracking the immune reactions following TBI both in animal models and clinical observations, highlight the role of autoantibodies as theragnostic biomarkers, each corresponding to a distinct pathophysiological mechanism and a specific temporal signature, and give an update on the recent theories in TBI therapies using a personalized approach.

The immunological cascade of events post-TBI

Innate immune response in TBI: A tug of war between pro- and anti-inflammatory mediators

TBI leads to meningeal and neuronal tear or contusion, axonal shearing injuries, and damage to blood vessels (hemorrhage) in the brain and surrounding meninges. Thus, it exposes the immune system to damage-associated molecular patterns (DAMPs) molecules such as ATP, matricryptins, cold-inducible RNA-binding protein, high mobility group protein B1 (HMGB-1), histones, and dysfunctional mitochondrial DNA (Huber-Lang, Lambris, & Ward, 2018). When exposed to DAMPs, microglia are rapidly recruited and transformed to their phagocytic form to clear debris, seal defective barriers, produce neurotrophic factors, and release pro-inflammatory cytokines such as IL-1 β, IL-18, IL-6, and TNF-α (Huber-Lang et al., 2018).

Interleukin-1β has a pro-inflammatory phase lasting hours or days followed by a reparative phase in the ensuing days and months. After exposure to IL-1β, astrocytes rapidly generate immune signals resulting in further recruitment of neutrophils and systemic release of cytokines. The resulting inflammatory response is associated with the additional release of reactive oxygen species (ROS) and excitatory neurotransmitters, inducing further cell-to-cell signaling and microcirculatory disturbance (Needham et al., 2019). This results in more progressive damage known as secondary TBI. These pro-inflammatory cascades are usually confronted with an anti-inflammatory counterbalance embodied in the release of cytokines such as IL-10 and transforming growth factor-beta (TGF-β) (Jassam, Izzy, Whalen, McGavern, & El Khoury, 2017).

There is considerable evidence suggesting that the balance between the IL-1 cytokine family members, particularly between IL-1β and its endogenous inhibitor IL-1Ra, is an essential determinant of the degree of inflammation as opposed to the absolute concentration of IL-1β. In contrast to IL-1β, IL-6 has a predominant neuroprotective effect. In accordance, raised CSF levels of IL-6 are correlated with improved immunological outcomes. IL-18, a member of the neutrophil chemotactic factors family, promotes neutrophil infiltration and consequently increases BBB permeability. Moreover, intrathecal levels of soluble membrane attack complex (sC5b-9) have been reported to correlate with BBB dysfunction (Helmy, De Simoni, Guilfoyle, Carpenter, & Hutchinson, 2011).

It should be noted that IL-10, considered as an archetypal "anti-inflammatory" cytokine due to its suppressive effect on the Th-1 axis, is a potent driver of B-cell activation, maturation, and antibody synthesis, and can subsequently drive detrimental inflammation via B-cell–mediated pathways (Needham et al., 2019).

Astrocyte-derived immunomodulatory mechanisms post-TBI

Astrocytes, also called astroglia, are the most abundant type of glial cells in the brain. Under physiological conditions, these cells are implicated in structural support, internal homeostasis, and BBB maintenance. Pathological mechanisms associated with TBI, such as BBB disruption, oxidative stress, inflammation, apoptosis, edema, and glutamate excitotoxicity, are correlated with the release of danger signals promoting astrocytic activation. This process, known as reactive astrogliosis, is characterized by structural and functional modifications of astrocytes. It is highlighted by cellular hypertrophy, processes elongation, modified gene expression, altered molecular signaling, and increased proliferation. Activated astrocytes migrate to the injury site and form a physical barrier called glial scar to wall off the damaged and inflamed area. They contain the inflammation and exert anti-oxidative and anti-apoptotic effects to trigger functional recovery. Moreover, reactive astrocytes produce specific factors to enhance neurogenesis, synaptogenesis, synaptic plasticity, angiogenesis, and BBB repair after TBI (Burda, Bernstein, & Sofroniew, 2016; Cheng et al., 2019; Zhou et al., 2020).

Persistent inflammation and excitotoxicity associated with severe TBI might stimulate dysfunctional astrogliosis. In such case, astrocytes become maladaptive and mediate adverse effects through inhibition of brain reconstruction and neuronal plasticity. In fact, astroglia can drive inflammatory damage and cerebral edema via continuous recruitment of immune cells to the site of injury and upregulation of aquaporin 4, respectively. Furthermore, the scar tissue can secrete growth inhibitors to limit axonal regeneration and neuronal connectivity. Also, extracellular matrix remodeling of perineuronal nets structures found around fast-spiking inhibitory interneurons, alters synaptic plasticity and stimulates excitation/inhibition imbalance and post-traumatic epilepsy during the course of chronic astrogliosis.

On the other hand, heterogeneous astrogliosis is correlated with different TBI subtypes, stages, and damaged areas. For instance, acute injury is associated with reversible hypertrophy and limited astrogliosis, while severe lesions are linked to highly reactive astrogliosis and glial scar formation. Therefore, serum biomarkers can be used to assess astrogliosis to determine injury severity and progression during TBI diagnosis. It seems that anti-astrogliosis effects rely on the nature and/or the concentration of the astrocytes-derived factors such as inflammatory cytokines and extracellular matrix components. Hence, research is warranted to unravel the main neuroprotective and neurotoxic astrocytic mediators released under specific disease state. Moreover, therapeutic interventions focusing on these molecules can be promising for TBI patients with minimal adverse effects (Cheng et al., 2019; Kim, Porter, Friedman, & Kaufer, 2016; Zhou et al., 2020).

Adaptive immune response in TBI: The role of T- and B-lymphocytes

During TBI progression, the adaptive immune system is engaged to promote enhanced recovery through antibody production and expansion of T-cells according to the type of injury. B- and T-cells are rapidly recruited to the injury site by the release of chemokines and the upregulation of vascular adhesion molecules. B-lymphocytes differentiate into mature plasma cells producing antibodies to support CNS repair. However, BBB compromised integrity can promote auto-inflammation through increased leakage of brain-specific proteins and breakdown products to the blood, thus triggering autoantibody production. Circulating blood contents can also get to the brain and stimulate the activation and survival of dormant memory B-cells. It has been proposed that some autoantibodies hold neuro-therapeutic potentials by favoring molecular pathways leading to cellular protection and repair. Autoantibodies can also involve impaired recovery while inducing systemic complications and more damage to the CNS. Exogenous antibodies and drugs neutralizing autoantibodies or inhibiting B-cell synthesis, growth, activation, and differentiation are recommended as a source of therapy. Notably, highly sensitive and specific autoantibodies can be used as potential biomarkers for TBI diagnosis and prognosis to ensure early therapeutic intervention with minimal adverse effects (Ankeny & Popovich, 2010; Filiano, Gadani, & Kipnis, 2017; Nasser et al., 2016; Raad et al., 2014).

B-cells, as well as apoptotic and necrotic cells, can serve as antigen-presenting cells stimulating T-cells activation. Multiple types of T-lymphocytes can be detected at the injury site such as cytotoxic T-cells and subsets of T helper cells, including TH1, TH2, TH17, and regulatory T-cells (Tregs). Generally, Tregs suppress the immune response by releasing anti-inflammatory mediators that block other types of T-cells. It seems that an appropriate number of Tregs is essential to contain the damage since low levels of these cells are associated with excessive inflammation, while high levels are linked to neuroprotection suppression. A protective Tregs function appears to be correlated with the production of autoreactive T-cells and with a shift toward TH2 phenotype delivering neurotrophic factors that maintain neuronal survival. Harmful effects of Tregs can be associated with the generation of destructive autoreactive T-cells (Ankeny & Popovich, 2010).

Moreover, it is highly probable that the shift toward TH1 and TH17 phenotypes induce detrimental cytokine production and neuron destruction. Tregs can also modulate myeloid cells and B-cell activation. Thus strategies targeting T-cell phenotype regulation might be promising (Filiano et al., 2017; Raad et al., 2014).

The glymphatic system disruption: A source of biomarkers in TBI

The tightness and selectiveness of the BBB, proofed by the low hydraulic conductivity of brain vessels, presumed that the brain lacks a conventional lymphatic system (Abbott, 2004). Although the brain possesses a high metabolic rate, which quickly builds up waste products, it was not until 2013 when the term "Glymphatic System (GS)" was first coined by Iliff and Nedergaard (2013) to describe a para-vascular pathway for CSF distribution and interstitial solute clearance.

The GS relies on three serial constituents: (1) The para-arterial CSF influx, which describes the distribution of CSF among brain parenchyma, originating from the subarachnoid space and the perivascular spaces of penetrating leptomeningeal arteries. From this point, the CSF flows across the glial basement membrane and is connected to the astrocytic end-feet, which express the water channel aquaporin 4 that, in turn, allows the flow of CSF and CSF-ISF exchange. (2) The trans-parenchymal pathway which describes the movement of interstitial fluids (ISF) within the brain parenchyma from penetrating arteries and empties along the veins. (3) The para-venous CSF-ISF clearance route that depends on the drainage of fluids into veins by astrocytic end-feet AQP4, as illustrated in Fig. 1 (Abbott, 2004; Iliff & Nedergaard, 2013; Jessen, Munk, Lundgaard, & Nedergaard, 2015; Rasmussen, Mestre, & Nedergaard, 2018).

The GS activity is functional during sleep and is mostly disengaged during wakefulness. This suggests the importance of sleep to allow the clearance of neurotoxic waste products. Otherwise, it would be accumulated, causing neurodegeneration or masking neurological assessment after brain disruption cases such as TBI (Jessen et al., 2015; Plog et al., 2015). It is worth mentioning that after TBI, glymphatic pathway function was reduced by 60%, with this impairment persisting for at least 1-month post-injury (Iliff et al., 2014). Among the common clinically used biomarkers in the assessment of TBI are the cytosolic proteins S100β and GFAP that leak from glial or neuronal cells after brain insult to body fluids, e.g., blood, CSF, or saliva (Plog et al., 2015).

The necessity to link the relationship between the function of GS and the transport of TBI biomarkers in peripheral blood is vital to determine and predict the severity of the injury and also to investigate whether GS suppression could affect

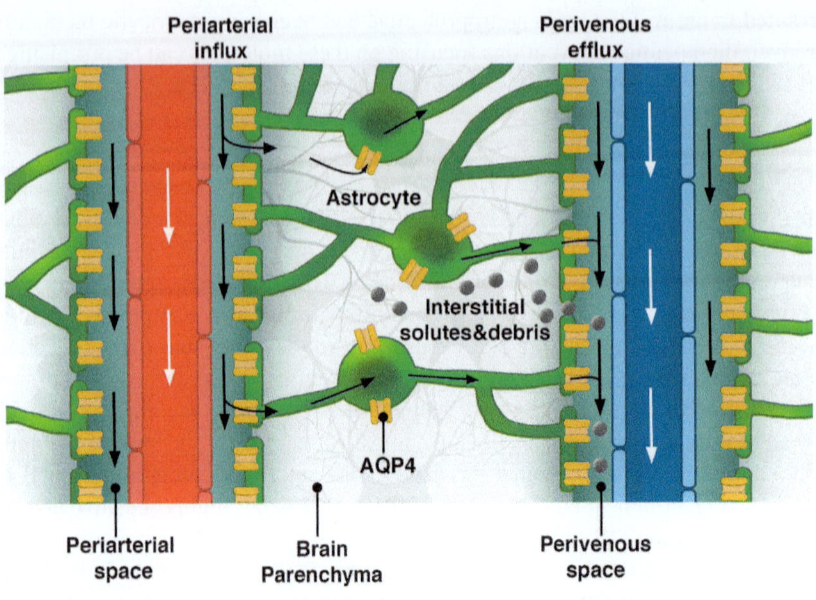

FIG. 1 A Schematic representation of the Glymphatic System (GS) of the brain. CSF flows through the brain parenchyma via the periarterial influx. The aquaporin-4 (AQP4) channels located on the astrocytes end feet processes mediates the wash out of solutes and debris, from the interstitial space, along the CSF flow into the perivenous space.

the delivery of TBI biomarker to the serum. In 2015, Plog et al. tackled these questions through several methods. Firstly, they applied four different manipulations to affect the influx pathway of GS, resulting in blocking solute and metabolites clearance, which led to protein aggregates and neurodegeneration. Secondly, they tried to identify the best candidate biomarker as a theragnostic tool (Plog et al., 2015).

Although different fluctuating factors affect the level of TBI biomarkers in serum (Dadas, Washington, Diaz-Arrastia, & Janigro, 2018), biomarker levels measured 18 h after experimental TBI induction showed that there was a significant elevation relative to the baseline of S100β and GFAP in wild-type mice, but not with AQP4KO (Plog et al., 2015). The latter findings support the hypothesis that GS functionality is a vital contributor when evaluating TBI biomarkers.

Autoantibodies as biomarkers in TBI: An overview

Recent studies have highlighted the crucial role of the immune system in TBI, especially the autoimmune events following the injury, and precisely following the disruption of the BBB. Through a plethora of mechanisms, this disruption leads to the leakage of immune cells to the brain and brain-derived proteins into the bloodstream, which subsequently leads to the activation of B-cells and the formation of autoantibodies against CNS antigens (Raad et al., 2014). Another mechanism that could explain the formation of these autoantibodies in TBI is the transportation of brain antigens to cervical lymph nodes through the newly discovered brain glymphatic system (Absinta et al., 2017).

These autoantibodies are increasingly thought to be promising long-term biomarkers to detect and diagnose TBI, predict the patient's neurological outcome and injury severity, and can open new horizons for therapeutic targets.

Putative biomarkers that have been documented in experimental TBI and then in patients post-TBI cover many processes including injury to neuronal cell bodies, demyelination, astroglial injury, and axonal injury (Wang et al., 2018). Some of these identified markers include Glia Calcium-Binding Protein (S100B), Myelin Basic Protein (MBP), and neurofilament proteins. All these can be used for diagnostic purposes alongside the currently available Imaging techniques (Zhang et al., 2014). The levels of these autoantibodies in accessible biofluids is a promising diagnostic and monitoring tool that can serve as a prognostic indicator of long-term outcomes and neurodegenerative consequences. It may also help detect TBI in polytraumatized and unstable patients on whom neuroimaging modalities such as MRI and CT cannot be performed.

Nevertheless, caution should be made since these autoantibodies are not specific to brain injury. Therefore, we need more investigations on these overlapping specificities, especially with other neurodegenerative diseases like Alzheimer's Disease, and on their true prevalence.

A list of identified autoantibodies post-TBI

Myelin basic protein (MBP)

MBP, a hydrophilic protein involved in the structural organization of the myelin sheath and used as a marker for demyelination, is one of the first identified autoantigens (Lisianyi, Cheren'ko, & Terletskaia Ia, 1987). Increased anti-MBP was reported in human TBI with a concentration of CSF levels, on the first-day post-insult, being correlated with the severity of injury (Ngankam, Kazantseva, & Gerasimova, 2011). Other studies on spinal cord injury (SCI) have displayed elevated levels of antibodies against MBP by immunohistochemistry (Arevalo-Martin et al., 2018).

S100B protein

Specific to astrocytes, S100B is a calcium-binding protein used as a biomarker for CNS injury, with its varying levels being correlated with the severity of TBI (Kobeissy et al., 2020). S100B autoantibodies have been detected in repeated mild TBI (rmTBI) and single severe TBI with chronic effects associated with prolonged post-concussive symptoms, early cognitive decline, and chronic traumatic encephalopathy (Kobeissy et al., 2020; Marchi et al., 2013).

Acetylcholine receptor (AChR)

The Acetylcholine receptor is involved in neuronal plasticity and anti-inflammatory effects by suppressing the production of specific cytokines, e.g., TNF-α (Kobeissy et al., 2020). Its autoantigenic properties are related to the α7 subunit (α7 AChR), as revealed in a study by Sorokina et al., where levels of serum auto anti-α7 AChR seemed proportional to the severity of the insult (Sorokina et al., 2011).

Glial fibrillary acidic protein (GFAP)

GFAP is an astrocyte marker that has been notably identified with its break down products (BDPs) in the peripheral circulation after damage to BBB (Kobeissy et al., 2020). A study by Zhang et al. showed a 3.77-fold increase of anti-GFAP autoantibodies from early to late time points after TBI, as depicted in Fig. 2 (Zhang et al., 2014). These were also revealed, in the same study, to enter astroglia in culture and affect their survival. Additionally, anti-GFAP autoantibodies exhibited a prominent increase in the plasma of patients with acute SCI and correlated positively with SCI-associated neuropathic pain (Hergenroeder et al., 2018).

Peroxiredoxin 6

Peroxiredoxin 6 is another astrocyte-specific protein that functions as an anti-oxidative enzyme and was identified as a TBI biomarker through autoimmune profiling (Buonora et al., 2015). Research on this protein revealed that it is highly expressed in the cortex post-injury, with circulating levels increasing up to four times 4 to 24 h after a mild to moderate insult (Buonora et al., 2015; Kobeissy et al., 2020).

Glutamate receptor

Glutamate receptors play a role in inducing excitatory signals in neuronal and non-neuronal cells. Several studies on TBI and SCI reported the increased production of autoantibodies against these receptors (Kobeissy et al., 2020). One study examining 60 children with chronic post-traumatic headache between the ages of 7 and 16 years exhibited an increase of autoantibodies against AMPA glutamate receptor (GluR1 subunit) and NMDA glutamate receptor (NR2A) (Goryunova et al., 2007). The research team rationalized the resulting neurological symptoms by suggesting that hypoxic brain lesions, coupled with evidence for hyper-stimulation of NMDA glutamate receptors, increased BBB permeability and over-activated an autoimmune response (Goryunova et al., 2007). An increase in autoantibodies against neuronal glutamate receptors was also notable in mice with mid-thoracic spinal contusion injury (Ankeny, Lucin, Sanders, McGaughy, & Popovich, 2006).

- The following Table 1 and Fig. 3 depict on the characteristics and specificity, respectively, of the aforementioned autoantibodies and antigens as Biomarkers in different models of TBI.

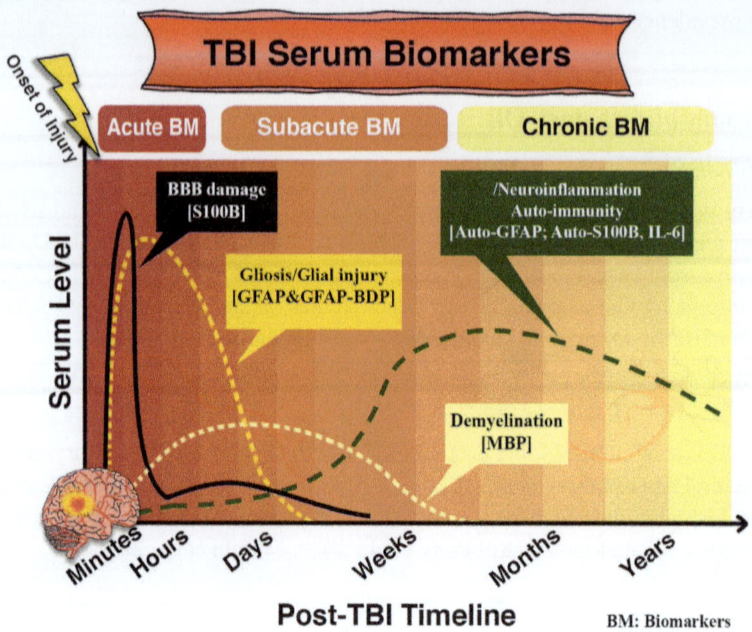

FIG. 2 Blood-based TBI biomarkers' time course. A distinct temporal profile of proteases, released into the bloodstream, in conjunction with the pathophysiological mechanisms triggered by primary and secondary TBI cascade of events at multiple acute and chronic timepoints, respectively. BBB damage marker, S100BBM; gliosis/glial injury markers, GFAP & GFAP-BDP (glial fibrillary acidic protein-breakdown products); demyelination marker, MBP (myelin-based protein); neuroinflammation/autoimmunity markers, auto-GFAP, auto-S100B, IL-6 (interleukin-6). *BM*, biomarker; *BBB*, blood–brain barrier.

TABLE 1 A list of autoantibody biomarkers observed in traumatic brain injuries.

Autoantibody	Study type	Cell type specificity	Biofluid	Timepoint identification	Outcome	Reference
MBP	Human study	Myelin	CSF	1, 10, 21 days after injury	Levels of MBP antibodies were correlated with the severity of the injury (day 1) and the degree of recovery (day 21)	Ngankam et al. (2011)
S100B	Human Study	Astrocytes	Serum	24h before any football-related activity 1 h and 24h after every game for five consecutive games	Anti-S100B elevation was correlated with repeated concussive events characterized by a disruption of the BBB	Marchi et al. (2013)
AChR	Human Study	Neurons	Serum	7 days after brain injury	Increased anti-α7 AChR was correlated with the severity of the insult	Sorokina et al. (2011)
GFAP	Human Study	Astrocytes	Serum	1–10days after brain injury	Anti-GFAP antibodies increased by 3.77 folds up to 10days post-injury and negatively correlated with the outcome at 6months	Zhang et al. (2014)
	Human Study	Astrocytes	Serum	16±7days post-SCI	Higher levels of concentrations of anti-GFAP were present in patients that developed neuropathic pain as opposed to those who did not	Hergenroeder et al. (2018)
PRDX6	Animal Study (Rats)	Astrocytes	Serum	4–24h and 2–7 days post-TBI	Anti-PRDX6 were elevated fourfold within the first 24h post-injury.	Buonora et al. (2015)
GluR	Human Study	Neuron	Serum	6months and 1 year after skull injury	Elevation of autoantibodies against the NMDA subunit of GluR was observed up to a year post-injury and is linked to hypoxic processes in the brain	Goryunova et al. (2007)

Characterization of potential biomarkers of TBI according to the brain cell-type specificity, biofluid, surging time-point profile post-injury, and the translational pathological outcomes. *AChR*, acetylcholine receptor; *GFAP*, glial fibrillary acidic protein; *GluR*, glutamate receptor; *MBP*, myelin-based protein; *PRDX6*, peroxiredoxin-6.

Evaluation of the prospective power of autoantibodies as prognostic markers in TBI

Autoantibodies demonstrate key diagnostic potential as TBI biomarkers because of their prolonged existence in serum as opposed to their respective antigens (Raad et al., 2014). Preliminary data have displayed that human post-TBI serum can maintain anti-GFAP autoantibodies for as long as 6months. Therefore, the amount of GFAP autoantibodies in serum could possibly serve as the foundation of a test for TBI (Zhang et al., 2014). Moreover, it was reported by Marchi et al. (2013) that in football players who were subject to repetitive sub-concussive headshots, elevated amounts of S100B were noticed in the wake of individual matches. Players with the most substantial increases of S100B had the greatest titer of serum S100B autoantibodies at the end of a season. S100B escaping into serum could, therefore, have played an essential role in the development of autoantibodies that could consequently serve as diagnostic biomarkers. Furthermore, Zhang et al. (2014) have also demonstrated an immunodominant autoantibody reaction to GFAP and its cleaved fragments in a group of subacute and chronic patients with TBI.

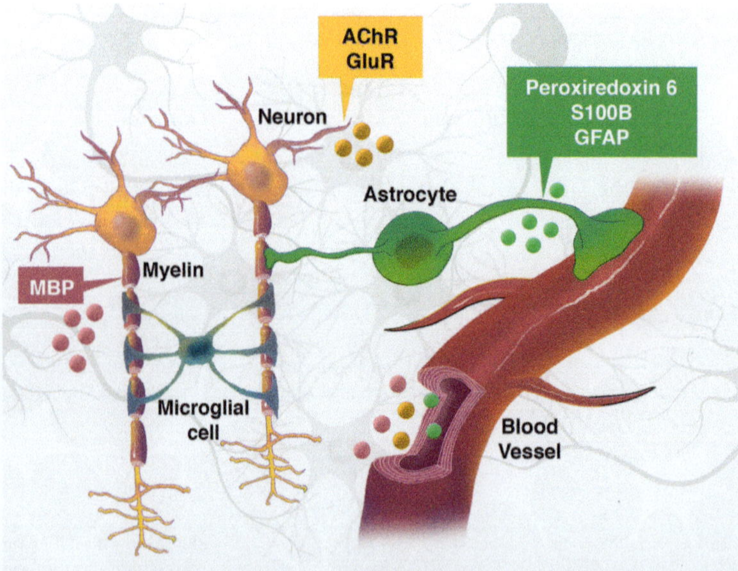

FIG. 3 Neuroanatomical localization of the Autoantibodies implicated in Traumatic Brain Injury. A schematic representation of the anatomical origin of different proteases and breakdown products, released into biofluids; CSF and blood, which are reputed as potential biomarkers of TBI pathological insults. *AChR*, acetylcholine receptor; GFAP, Glial Fibrillary acidic protein; *GluR*, glutamate receptor; *MBP*, myelin-based protein.

The prognostic power of autoantibodies is well documented too. An autoantibody, explicitly targeting a key brain protein GFAP, might initiate a continued autoimmune insult to the CNS and may adversely affect TBI patients' long-term outcomes (Wang et al., 2018). Also (Zhang et al., 2014) observed a negative correlation among fold change in TBI autoantibody concentrations and GOS-E (Glasgow Outcome Score Extended) at discharge (6 months after injury), suggesting that elevated anti-GFAP autoantibody concentrations may imply a more serious injury. Similarly, S100B protein, along with its autoantibodies, has been assumed to be a manifestation of an autoimmune response that persists over time and may lead to various adverse outcomes, negatively affecting disease prognosis (Raad et al., 2014).

B-lymphocytes as a therapeutic target post-TBI

More data are suggesting how determinant B Lymphocytes are in the post-traumatic immune responses following TBI. From the release of pro-inflammatory cytokines and protective cytokines to the production of pathogenic antibodies, overactivation of B-cells exacerbates the aftermath of TBI. The pathogenic role of B lymphocytes was suggested after mice, lacking this cell type, showed an important improvement of locomotor function (Ankeny & Popovich, 2010). Therefore, targeting B-cell activation in patients who sustained TBI is a hypothesis worth exploring.

Therapies with the potential of inhibiting de-novo B-cells or eliminating pathogenic B Lymphocytes such as Intravenous Immunoglobulin (IV-Ig) or plasmapheresis or immunomodulatory biologics should be considered (Ankeny & Popovich, 2010). However, when considering these therapeutic targets, we should not lose sight of the complications that might follow the inhibition of B-cells.

Before moving these hypotheses to pre-clinical trials, it is of interest to do proteomics analysis to identify highly specific autoantibodies. It is also worth mentioning that when delivering these neuromodulator drugs, intracranial delivery pathways should be further investigated.

Two key molecules could be targeted that play an essential role in B-cell differentiation and activation: B-cell Activation factor (BAFF) and A Proliferation Inducing Ligand (APRIL) (Raad et al., 2014).

While there is a lack of reconciled data in the field of Neuroimmune Pharmacology, it still holds great promise for the recovery and the prevention of chronic complications of TBI.

Conclusion

The heterogeneity and multi-faceted nature of TBI pathophysiological mechanisms necessitate the need to thoroughly develop experimental paradigms and clinical research modalities to fully characterize the immunological dynamics fueling the primary and secondary insults for months and even years post-injury.

The growing interest in autoantibodies as promising long-term biomarkers to diagnose TBI, predict the patient's clinical outcome, and injury severity, can open new prospects for therapeutic targets. Given the dynamic surging profile of the previously discussed TBI biomarkers, their usefulness would be attained when assigning each to a different stage post-TBI according to the corresponding pathogenic mechanism.

However, many challenges remain before these biomarkers can be translated from experimental applicability to bedside clinical practice. Accordingly, whether biomarkers could play the deterministic role as TBI theragnostic tools, independently of the appliance of neuroimaging modalities, is still unknown. In parallel, therapeutic interventions that inhibit B-cells immunomodulation could be of therapeutic potential.

Applications to other areas of neuroscience: The efficacy of GFAP in assessing the diagnostic power of different neuroimaging modalities

About 29% of TBI patients with negative computed tomography (CT-) have positive magnetic resonance imaging (MRI+) results. Recently, it has been suggested that the use of serum biomarkers with MRI has the potential to help in the diagnosis of CT-occult TBI (Yue, Upadhyayula, Avalos, Deng, & Wang, 2020). The Discovery of intracranial injuries often depends on head CT, which is overused and resource-intensive. Blood-based brain biomarkers can predict a lack of intracranial injury and consequently lead to a decrease in needless head CT scanning (Bazarian et al., 2018).

Glial fibrillary acidic protein and its breakdown products (GFAP-BDP) displayed excellent predictive potential and showed substantial discrimination of injury severity. The use of GFAP-BDP has been shown to come up with a clear advantage over the usage of clinical examination alone and a clear drop (12%–30%) in unnecessary scans (McMahon et al., 2015). Also, the GFAP-BDP amount was the most precise anticipator of the occurrence of intracranial injury identified by radiographic imaging compared with other accepted clinical predictors of intracranial injury. Thus, GFAP-BDP has appeared as a feasible early indicator of intracranial injury and as a beneficial assistant to existing diagnostic methods for TBI (McMahon et al., 2015). Furthermore, (Yue et al., 2019) reported that median GFAP level was higher in patients with negative CT and positive MRI findings than in those with negative CT and negative MRI finding, further favoring the predictive ability of the usage of GFAP as an MRI-complementary biomarker to detect TBI.

Both glial fibrillary acidic protein (GFAP) and S100β are located in glial cells. They are discharged into serum following a traumatic brain injury (TBI). Still, the clinical significance of S100β as a biomarker has been doubted because it is released from bones (Papa et al., 2014). It has been demonstrated that the level of S100β was considerably greater in patients with fractures, compared to those without fractures whether mild-to-moderate (mmTBI) was present or not. However, GFAP levels were not significantly altered by the presence of fractures. Expectedly, in a general trauma population, GFAP performed considerably better than S100β in identifying intracranial CT lesions, especially in the case of extracranial fractures (Papa et al., 2014).

Mini-dictionary of terms

- **Biomarker:** Breakdown products released into the body biofluids such as, blood, CSF, urine, saliva, etc., in correspondence to and indicative of distinct internal pathogenic mechanisms.
- **Cytokines:** A group of cell-signaling molecules modulating cell-to-cell communication in response to inflammatory signals.
- **Excitotoxicity:** A pathogenic process induced by the overactivation of receptors of the excitatory neurotransmitter glutamate, such as NMDA and AMPA receptors.
- **Glymphatic System:** The newly discovered brain clearance system that washes out cellular solutes and debris that, otherwise, their accumulation would induce a state of oxidative stress and neurodegeneration.
- **Proteomics:** A field of study concerned with the characterization of the entire set of proteins produced or modified by a living organism or system.
- **Reactive Oxygen Species (ROS):** A group of free radicals of molecular oxygen originates as byproducts of oxidoreductase enzymes reactions and mitochondrial electron transport cascades. ROS play a pivotal role in cell signaling mechanisms, such as: apoptosis and gene expression.
- **TBI Secondary Injury:** The damage implicated to the brain parenchyma that is not directly caused by the causative injury tool. It may appear hours to days after the primary injury and involves a cascade of biochemical changes that eventually end in neuronal cell death.

Key-facts of TBI biomarkers

- Each TBI Biomarker has a distinct surging profile corresponding to specific timepoint of pathogenic mechanisms.
- Myelin-Based Protein (MBP) is one of the first identified autoantigens as biomarkers of acute TBI, back then in 1987.
- GFAP, a distinct astrocyte marker, has been an indicative of BBB disruption at very early time points post-injury.
- The application of GFAP-BDP as a diagnostic TBI biomarker has shown clear drop (12%–30%) in unnecessary imaging scans.
- GFAP has a better discriminative power than S100β in identifying intracranial CT lesions.

Key facts of the glymphatic system (GS)

- The "glymphatic system" was first coined in 2013 to describe the newly discovered, back then, brain clearance system.
- The GS activity is functional during sleep and mostly disengaged during wakefulness.
- Aquaporin channels (AQP4) at the astrocytes end feet processes are an important component of the GS mediating the CSF-ISF exchange.
- Recent discoveries suggest the role of the GS in mediating the delivery of TBI biomarkers to the serum.
- Any disruption to the GS, induced by TBI, leads to the accumulation of debris and breakdown products which in turn induces a state of excitotoxicity and neurodegeneration.

Summary points

- TBI is a leading cause of death and disability, worldwide, with approximately 1.7 million victims, reported per year, in the USA.
- TBI-induced cascade of inflammatory and pathophysiological events could be a precursor to the development of serious neurological and neuropsychiatric disorders such as, Alzheimer's disease (AD), chronic traumatic encephalopathy (CTE), Parkinson's disease (PD), Amyotrophic lateral sclerosis (ALS), and post-traumatic stress disorder (PTSD).
- As a multi-dimensional disease process, TBI induces two stages of injury; primary and secondary, that run both parallel and inconsequential and result in neuronal and glial apoptosis, hemorrhage, BBB disruption, oxidative stress, and leakage of immune cells into brain parenchyma.
- Recent discoveries suggest the role of the glymphatic system in mediating the delivery of TBI biomarkers to the serum.
- The released autoantibodies post-injury could be a potential biomarker for TBI diagnosis with a distinct surging profile in correspondence to specific pathogenic mechanisms.

References

Abbott, N. J. (2004). Evidence for bulk flow of brain interstitial fluid: Significance for physiology and pathology. *Neurochemistry International*, *45*(4), 545–552.

Absinta, M., Ha, S.-K., Nair, G., Sati, P., Luciano, N. J., Palisoc, M., et al. (2017). Human and nonhuman primate meninges harbor lymphatic vessels that can be visualized noninvasively by MRI. *eLife*, *6*, 1–15.

Ankeny, D. P., Lucin, K. M., Sanders, V. M., McGaughy, V. M., & Popovich, P. G. (2006). Spinal cord injury triggers systemic autoimmunity: Evidence for chronic B lymphocyte activation and lupus-like autoantibody synthesis. *Journal of Neurochemistry*, *99*(4), 1073–1087.

Ankeny, D. P., & Popovich, P. G. (2010). B cells and autoantibodies: Complex roles in CNS injury. *Trends in Immunology*, *31*(9), 332–338.

Arevalo-Martin, A., Grassner, L., Garcia-Ovejero, D., Paniagua-Torija, B., Barroso-Garcia, G., Arandilla, A. G., et al. (2018). Elevated autoantibodies in subacute human spinal cord injury are naturally occurring antibodies. *Frontiers in Immunology*, *9*, 2365.

Bazarian, J. J., Biberthaler, P., Welch, R. D., Lewis, L. M., Barzo, P., Bogner-Flatz, V., et al. (2018). Serum GFAP and UCH-L1 for prediction of absence of intracranial injuries on head CT (ALERT-TBI): A multicentre observational study. *Lancet Neurology*, *17*(9), 782–789.

Buonora, J. E., Mousseau, M., Jacobowitz, D. M., Lazarus, R. C., Yarnell, A. M., Olsen, C. H., et al. (2015). Autoimmune profiling reveals Peroxiredoxin 6 as a candidate traumatic brain injury biomarker. *Journal of Neurotrauma*, *32*(22), 1805–1814.

Burda, J. E., Bernstein, A. M., & Sofroniew, M. V. (2016). Astrocyte roles in traumatic brain injury. *Experimental Neurology*, *275*(Pt 3), 305–315.

Centers for Disease Control and Prevention. (2019). Surveillance report of traumatic brain injury-related emergency department visits, hospitalizations, and deaths—United States, 2014. In *Centers for Disease Control and Prevention* U.S. Department of Health and Human Services.

Cheng, X., Wang, J., Sun, X., Shao, L., Guo, Z., & Li, Y. (2019). Morphological and functional alterations of astrocytes responding to traumatic brain injury. *Journal of Integrative Neuroscience*, *18*(2), 203–215.

Coronado, V. G., Xu, L., Basavaraju, S. V., McGuire, L. C., Wald, M. M., Faul, M. D., et al. (2011). Surveillance for traumatic brain injury-related deaths—United States, 1997–2007. *Morbidity and Mortality Weekly Report. Surveillance Summaries*, *60*(5), 1–32.

Dadas, A., Washington, J., Diaz-Arrastia, R., & Janigro, D. (2018). Biomarkers in traumatic brain injury (TBI): A review. *Neuropsychiatric Disease and Treatment, 14*, 2989–3000.

Faul, M., & Coronado, V. (2015). Epidemiology of traumatic brain injury. *Handbook of Clinical Neurology, 127*, 3–13.

FDA-NIH Biomarker Working Group. (2016). *BEST (Biomarkers, EndpointS, and other Tools) Resource [Internet]*. Silver Spring, MD: Food and Drug Administration (US).

Filiano, A. J., Gadani, S. P., & Kipnis, J. (2017). How and why do T cells and their derived cytokines affect the injured and healthy brain? *Nature Reviews. Neuroscience, 18*(6), 375–384.

Goryunova, A. V., Bazarnaya, N. A., Sorokina, E. G., Semenova, N. Y., Globa, O. V., Semenova Zh, B., et al. (2007). Glutamate receptor autoantibody concentrations in children with chronic post-traumatic headache. *Neuroscience and Behavioral Physiology, 37*(8), 761–764.

Helmy, A., De Simoni, M. G., Guilfoyle, M. R., Carpenter, K. L. H., & Hutchinson, P. J. (2011). Cytokines and innate inflammation in the pathogenesis of human traumatic brain injury. *Progress in Neurobiology, 95*(3), 352–372.

Hergenroeder, G. W., Redell, J. B., Choi, H. A., Schmitt, L., Donovan, W., Francisco, G. E., et al. (2018). Increased levels of circulating glial fibrillary acidic protein and collapsin response mediator protein-2 autoantibodies in the acute stage of spinal cord injury predict the subsequent development of neuropathic pain. *Journal of Neurotrauma, 35*(21), 2530–2539.

Huber-Lang, M., Lambris, J. D., & Ward, P. A. (2018). Innate immune responses to trauma review-article. *Nature Immunology, 19*(4), 327–341.

Iliff, J. J., Chen, M. J., Plog, B. A., Zeppenfeld, D. M., Soltero, M., Yang, L., et al. (2014). Impairment of glymphatic pathway function promotes tau pathology after traumatic brain injury. *The Journal of Neuroscience: The Official Journal of the Society for Neuroscience, 34*(49), 16180–16193.

Iliff, J. J., & Nedergaard, M. (2013). Is there a cerebral lymphatic system? *Stroke, 44*(Suppl. 1), 93–95.

Jassam, Y. N., Izzy, S., Whalen, M., McGavern, D. B., & El Khoury, J. (2017). Neuroimmunology of traumatic brain injury: Time for a paradigm shift. *Neuron, 95*(6), 1246–1265.

Jessen, N. A., Munk, A. S., Lundgaard, I., & Nedergaard, M. (2015). The glymphatic system: A beginner's guide. *Neurochemical Research, 40*(12), 2583–2599.

Kim, S. Y., Porter, B. E., Friedman, A., & Kaufer, D. (2016). A potential role for glia-derived extracellular matrix remodeling in postinjury epilepsy. *Journal of Neuroscience Research, 94*(9), 794–803.

Kobeissy, F. H., Ahmad, F., Shaito, A., Hasan, H., Hady, S. A., Nasrallah, L., et al. (2020). Autoantibodies in central nervous system trauma: New frontiers for diagnosis and prognosis biomarkers. In A. H. B. Wu, & W. F. Peacock (Eds.), *Biomarkers for traumatic brain injury* (pp. 431–451). Academic Press (Chapter 29).

Lisianyi, N. I., Cheren'ko, T. M., & Terletskaia Ia, T. (1987). Detection of antibodies to myelin basic proteins in patients with closed cranio-cerebral trauma. *Vrachebnoe Delo*, (10), 101–104.

Marchi, N., Bazarian, J. J., Puvenna, V., Janigro, M., Ghosh, C., Zhong, J., et al. (2013). Consequences of repeated blood-brain barrier disruption in football players. *PLoS One, 8*(3), e56805.

McKee, C. A., & Lukens, J. R. (2016). Emerging roles for the immune system in traumatic brain injury. *Frontiers in Immunology, 7*(December), 1–17.

McMahon, P. J., Panczykowski, D. M., Yue, J. K., Puccio, A. M., Inoue, T., Sorani, M. D., et al. (2015). Measurement of the glial fibrillary acidic protein and its breakdown products GFAP-BDP biomarker for the detection of traumatic brain injury compared to computed tomography and magnetic resonance imaging. *Journal of Neurotrauma, 32*(8), 527–533. https://doi.org/10.1089/neu.2014.3635.

Nasser, M., Bejjani, F., Raad, M., Abou-El-Hassan, H., Mantash, S., Nokkari, A., et al. (2016). Traumatic brain injury and blood-brain barrier cross-talk. *CNS & Neurological Disorders Drug Targets, 15*(9), 1030–1044.

Needham, E. J., Helmy, A., Zanier, E. R., Jones, J. L., Coles, A. J., & Menon, D. K. (2019). The immunological response to traumatic brain injury. *Journal of Neuroimmunology, 332*(April), 112–125.

Ngankam, L., Kazantseva, N. V., & Gerasimova, M. M. (2011). Immunological markers of severity and outcome of traumatic brain injury. *Zhurnal Nevrologii i Psikhiatrii Imeni S.S. Korsakova, 111*(7), 61–65.

Nizamutdinov, D., & Shapiro, L. A. (2017). Overview of traumatic brain injury: An immunological context. *Brain Sciences, 7*(1).

Papa, L., Silvestri, S., Brophy, G. M., Giordano, P., Falk, J. L., Braga, C. F., et al. (2014). GFAP out-performs S100beta in detecting traumatic intracranial lesions on computed tomography in trauma patients with mild traumatic brain injury and those with extracranial lesions. *Journal of Neurotrauma, 31*(22), 1815–1822. https://doi.org/10.1089/neu.2013.3245.

Plog, B. A., Dashnaw, M. L., Hitomi, E., Peng, W., Liao, Y., Lou, N., et al. (2015). Biomarkers of traumatic injury are transported from brain to blood via the glymphatic system. *The Journal of Neuroscience: The Official Journal of the Society for Neuroscience, 35*(2), 518–526.

Raad, M., Nohra, E., Chams, N., Itani, M., Talih, F., Mondello, S., et al. (2014). Autoantibodies in traumatic brain injury and central nervous system trauma. *Neuroscience, 281*, 16–23.

Rasmussen, M. K., Mestre, H., & Nedergaard, M. (2018). The glymphatic pathway in neurological disorders. *Lancet Neurology, 17*(11), 1016–1024.

Slavoaca, D., Muresanu, D., Birle, C., Rosu, O. V., Chirila, I., Dobra, I., et al. (2020). Biomarkers in traumatic brain injury: New concepts. *Neurological Sciences, 41*(8), 2033–2044. https://doi.org/10.1007/s10072-019-04238-y.

Sorokina, E. G., Vol'pina, O. M., Semenova Zh, B., Karaseva, O. V., Koroev, D. O., Kamynina, A. V., et al. (2011). Autoantibodies to alpha7-subunit of neuronal acetylcholine receptor in children with traumatic brain injury. *Zhurnal Nevrologii i Psikhiatrii Imeni S.S. Korsakova, 111*(4), 56–60.

Teasdale, G., & Jennett, B. (1974). Assessment of coma and impaired consciousness. A practical scale. *Lancet, 2*, 81–84.

Wang, K. K., Yang, Z., Zhu, T., Shi, Y., Rubenstein, R., Tyndall, J. A., et al. (2018). An update on diagnostic and prognostic biomarkers for traumatic brain injury. *Expert Review of Molecular Diagnostics, 18*(2), 165–180.

Yue, J. K., Upadhyayula, P. S., Avalos, L. N., Deng, H., & Wang, K. K. W. (2020). The role of blood biomarkers for magnetic resonance imaging diagnosis of traumatic brain injury. *Medicina (Kaunas, Lithuania), 56*(2). https://doi.org/10.3390/medicina56020087.

Yue, J. K., Yuh, E. L., Korley, F. K., Winkler, E. A., Sun, X., Puffer, R. C., et al. (2019). Association between plasma GFAP concentrations and MRI abnormalities in patients with CT-negative traumatic brain injury in the TRACK-TBI cohort: A prospective multicentre study. *Lancet Neurology*, *18*(10), 953–961.

Zhang, Z., Zoltewicz, J. S., Mondello, S., Newsom, K. J., Yang, Z., Yang, B., et al. (2014). Human traumatic brain injury induces autoantibody response against glial fibrillary acidic protein and its breakdown products. *PLoS One*, *9*, e92698.

Zhou, Y., Shao, A., Yao, Y., Tu, S., Deng, Y., & Zhang, J. (2020). Dual roles of astrocytes in plasticity and reconstruction after traumatic brain injury. *Cell Communication and Signaling: CCS*, *18*(1), 62.

Chapter 20

Chitinase-3-like protein 1: Features and applications as a biomarker in traumatic brain injury

C.S. Carabias[a,b] and A. Lagares[a,b,c]

[a]*Neurotraumatology and Subarachnoid Hemorrhage Research Unit, Area 8: Neuroscience and Mental Health, Research Institute Hospital 12 de Octubre (i+12), Madrid, Spain,* [b]*Department of Neurosurgery, Hospital Universitario 12 de Octubre, Madrid, Spain,* [c]*Department of Surgery, Faculty of Medicine, Complutense University of Madrid (UCM), Madrid, Spain*

List of abbreviations

AIS	acute ischemic stroke
APP	amyloid precursor protein
aSAH	aneurysmal subarachnoid hemorrhage
AUC	area under the receiver operating characteristic curve
bFGF	basic fibroblast growth factor
BOP	blast over pressure
BRP-39	breast regression protein 39/Chitinase-3-like protein 1
CCI	controlled cortical impact
CCL2	C-C motif chemokine ligand 2/Monocyte chemoattractant protein-1
CI	confidence interval
COX-2	cyclooxygenase 2
CRP	C-reactive protein
CSF	cerebrospinal fluid
CT	computed tomography
CTE	chronic traumatic encephalopathy
CXCL1–3	C-X-C motif chemokine ligand 1
DPI	days post-injury
DWI	diffusion-weighted imaging
ECM	extracellular matrix
EDH	epidural hematoma
ERK1/2	extracellular signal-regulated kinases
FLAIR	fluid-attenuated inversion recovery
GCS	Glasgow Coma Scale
GFAP	glial fibrillary acidic protein
GOS	Glasgow Outcome Scale
HR	hazard ratio
Iba-1	ionized calcium-binding adapter molecule 1/Allograft inflammatory factor 1
ICC	immunocytochemistry
ICH	intracerebral hemorrhage
ICP	intracranial pressure
IFNγ	interferon gamma
IL13Rα2	interleukin-13 receptor subunit alpha-2
IL-1β	interleukin 1β
IL-4Rα	interleukin-4 receptor subunit alpha

Diagnosis and Treatment of Traumatic Brain Injury. https://doi.org/10.1016/B978-0-12-823347-4.00009-9

iNOS	inducible nitric oxide synthase
ISS	injury severity score
IVH	intra-ventricular hemorrhage
J	Youden index
JAK	Janus kinase
JNK	Jun N-terminal kinase/stress-activated protein kinase
KO	knockout
MAPK	mitogen-activated protein kinase
MCAO	middle cerebral artery occlusion
MMP	metalloproteinases
MRI	magnetic resonance imaging
NF-L	neurofilament light chain
NF-κB	nuclear factor kappa B
NIHSS	National Institutes of Health Stroke Scale
NSE	neuron-specific enolase
OR	odds ratio
PCS	post-concussion syndrome
PCT	procalcitonin
pNF-H	phosphorylated neurofilament heavy chain
RAGE	receptor for advanced glycation end-products
rcTBI	repetitive concussive traumatic brain injury
RPQ	Rivermead post-concussion symptoms questionnaire
S100β	calcium-binding protein B
SAA1	serum amyloid A1
SDH	subdural hematoma
STAT	signal transducer and activator of transcription
TBI	traumatic brain injury
TCDB	traumatic coma data bank
TGFβ	transforming growth factor beta
TNF-α	tumor necrosis factor alpha
tSAH	traumatic subarachnoid hemorrhage
T-tau	total tau
WT	wild type
YKL-40	chitinase-3-like protein 1.

Introduction

Chitinase 3-like protein 1 (CHI3L1, also known as YKL-40) was discovered in 1989 in the culture supernatant of the human osteosarcoma cell line MG63. This secreted glycoprotein was termed YKL-40 by Johansen, Williamson, Rice, and Price (1992) based on the one-letter code for its three N-terminal amino acids (tyrosine [Y], lysine [K] and leucine [L]) and its molecular mass of 40 kDa. However, in the last few years, this protein has received various nomenclatures: human cartilage glycoprotein-39 (HC gp39), breast regression protein 39 (BRP-39), 38-kDa heparin-binding glycoprotein (gp38k), chondrex, and 40 kDa mammary gland protein (MGP-40). YKL-40 belongs to the 18-glycosyl-hydrolase family of proteins but lacks chitinase/hydrolase activity (Johansen, 2006). Crystallographic analysis of the native protein revealed that CHI3L1 is composed of a (β/α)8-barrel fold with a β + α domain insertion responsible for its functions in both physiological and pathological conditions (Fusetti, Pijning, Kalk, Bos, & Dijkstra, 2003). Nevertheless, the physiological role of CHI3L1 remains unclear. This chitinase-like protein has been hypothesized to be involved in antigen-induced and oxidant-induced injury responses, inflammation, cellular migration and proliferation, and other tissue-remodeling responses (Zhao, Su, Li, Zhang, & You, 2020). Furthermore, circulating and/or tissue YKL-40 levels are increased in a variety of inflammatory and remodeling conditions (Lee et al., 2011). Recent clinical studies found that increased expression of YKL-40 in the cerebrospinal fluid (CSF) and serum of patients with traumatic brain injury (TBI), as well as post-traumatic levels, were associated with clinical and radiological lesion burden and unfavorable outcome in these patients (Bonneh-Barkay et al., 2010; Carabias et al., 2020; Shahim et al., 2017). Elegant transgenic mouse models have been used to elucidate the potential role of this molecule in neuropathological outcomes after TBI (Wiley et al., 2016, 2015). Evidence documenting the diametrically opposing actions of YKL-40 in the pathophysiology of head injury is reviewed in the following sections.

Preclinical studies of YKL-40 in TBI

Evaluation of YKL-40 expression in pre-clinical TBI has been determined predominantly in rats and chiefly in models of focal cortical impact, with only a single study in diffuse blast-induced injury (Table 1).

Pericontusional astrocytic YKL-40 transcription is driven by cytokines released from macrophages infiltrating the necrotic area

In a pivotal study, Bonneh-Barkay, Zagadailov, et al. (2010) characterized temporal changes in YKL-40 expression after parasagittal controlled cortical impact (CCI) (6mm impactor, 2.8mm depth, 4.0m/s impact velocity) over the parietal cortex. In situ hybridization and immunohistochemical analysis revealed evidence of YKL-40 expression in astrocytes restricted to the pericontusional area and surrounding the site of injury (Fig. 1) (Bonneh-Barkay, Zagadailov, et al., 2010). Likewise, a correlation between CHI3L1 transcription and glial fibrillary acidic protein (GFAP) staining was demonstrated in a variety of acute and chronic human neurological conditions. A previous study identified increased tissue localization of YKL-40 in different stages of the infarcted cortical brain tissue of patients following ischemic stroke. Again, focal and temporal expression of YKL-40 was noted in astrocytes proximal to the lesion (Bonneh-Barkay, Wang, Starkey, Hamilton, & Wiley, 2010). Furthermore, YKL-40-induced transcription in astrocytes was related to morphological changes attributable to a reactive phenotype and enhanced migratory capacity (Bonneh-Barkay et al., 2012). Astrocytic YKL-40 transcription in the traumatic penumbra was noted as early as 24h post-CCI, remained increased for several days, and decreased by day 12. Moreover, the time course of YKL-40 cortical astrocytic proliferation was coincident with the neuroinflammatory infiltrate, evaluated by interleukin 1β (IL-1β) mRNA. As inflammation resolved, YKL-40 and IL-1β expression diminished, indicating temporal cytokine-driven expression of YKL-40 in astrocytes (Bonneh-Barkay, Zagadailov, et al., 2010). These findings are in agreement with other studies showing that IL-1β and tumor necrosis factor alpha (TNF-α) regulate YKL-40 expression by a RelB-dependent mechanism (Bhardwaj et al., 2015; Bonneh-Barkay et al., 2012; Recklies, Ling, White, & Bernier, 2005). Moreover, it has been proposed that the cytokine-driven expression of YKL-40 in astrocytes requires binding of both signal transducer and activator of transcription 3 (STAT3) and RelB/p50 complexes to the YKL-40 promoter and is independent of the p65 subunit of nuclear factor kappa B (NF-κB) (Bhardwaj et al., 2015). However, in the rat contusion-TBI model, differences in TNF-α transcription levels between the ipsilateral and contralateral cortices were undetectable (Bonneh-Barkay, Zagadailov, et al., 2010). Consequently, although TNF-α was capable of inducing astrocytic YKL-40 transcription in vitro (Bonneh-Barkay et al., 2012), a TNFα-induced YKL-40 expression course was not established in vivo. Nevertheless, it has been postulated that additional factors other than IL-1β and TNF-α are responsible for YKL-40-induced transcription in astrocytes through NF-κB cell signaling activation. Substantial YKL-40 transcription has been observed in macrophages in vitro, induced by classical (M1) activation and inhibited by alternative (M2) activation (Bonneh-Barkay et al., 2012). Paradoxically, macrophages infiltrating the necrotic area did not hybridize to YKL-40 transcripts in vivo. Therefore, the authors concluded that macrophages infiltrating the necrotic contusional core secreted the cytokines responsible for local astrocytic YKL-40 transcription surrounding the traumatic penumbra (Bonneh-Barkay, Zagadailov, et al., 2010).

CHI3L1: Friend or foe following TBI?

Further study in focal contusion (1.5mm depth, 6m/s impact velocity) parsed out temporal and cellular differences in neuropathological outcomes in wild type (WT) and *chi3l1* knockout (KO) mice to determine the role of YKL-40 in response to CCI (Wiley et al., 2015). Wiley et al. (2015) used mice in which the murine YKL-40 homologue, the *chi3l1/brp-39 gene*, had been homozygously disrupted, eliminating exons 1–6 and part of the BRP-39 promoter. A deformation of 1.5mm in the left parietal–temporal cortex generated consistent meningeal inflammatory infiltrate and cortical tissue damage (Table 1). Interestingly, the extent of neuritic damage was similar in WT and *chi3l1* KO mice 24h after CCI. Therefore, it has been suggested that CHI3L1 is not involved in post-traumatic axonal degeneration (Wiley et al., 2015), since this model of TBI generates a great amount of axonal injury as early as 6h in both mice and rats (Hall, Bryant, Cho, & Sullivan, 2008). Seven days post-injury, astrogliosis and microgliosis were comparable in WT and *chi3l1* KO mice after a severe degree of injury. Although *chi3l1* homozygous disruption appeared to be unrelated to gross tissue loss, CHI3L1 influenced the subacute astrocytic and microglial responses after TBI. On days 14–19 after CCI, spatial memory acquisition and retention were assessed in these animals. Surprisingly, no significant differences were observed between groups with respect to the Morris water maze variables, suggesting that the absence of CHI3L1 does not condition learning capacity. Twenty-one days post-CCI, homozygous *chi3l1* KO mice showed more noticeable histopathological changes than WT mice (Fig. 2). Although astrocytosis associated with TBI has been proposed to exhibit diametrically opposed functions on surviving neural tissue

TABLE 1 Focal and temporal YKL-40 expression in experimental models of TBI.

Reference	Experimental procedure, region, and severity	Animals, age	Methods	Time course evaluated	Histopathological effects of TBI injury model	YKL-40 involvement in the neuropathology of TBI
Bonneh-Barkay, Zagadailov, et al. (2010)	Parasagittal CCI, right parietal cortex, 6 mm diameter, 2.8 mm depth, 4 m/s impact velocity	Sprague–Dawley rats, age not specified	ISH, IHC, IF (GFAP, CD68 staining), real-time PCR	1, 2, 3, 4, 6, 8 and 12 DPI	Not specified	3 DPI: YKL-40 transcription was related to reactive astrocytes restricted to the traumatic penumbra. YKL-40 co-localized with GFAP staining. Macrophages infiltrating the necrotic area did not hybridize to YKL-40 transcripts. YKL-40 transcription began at 1 DPI, remained increased for 8 days, and decreased by 12 DPI. YKL-40 transcription after CCI mirrors the time course of the acute neuroinflammatory response, determined by IL-1β mRNA
Wiley et al. (2015)	CCI, left parietal–temporal cortex, 1.5 mm depth, 6 m/s impact velocity	Adult female C57BL/6 WT and *chi3l1* KO (both after craniotomy alone and craniotomy with CCI), 9–12 weeks	ISH, IHC (GFAP, Iba-1, Nissel staining, de Olmo's amino-cupric-silver method). Morris water maze.	1, 7 and 21 DPI	Consistent meningeal inflammatory infiltrate and cortical tissue damage underlying the craniotomy with or without CCI. Astrogliosis was evidenced in the ipsilateral hemisphere with extension into the underlying hippocampus, thalamus, striatum, corpus callosum and fornix. Certain degree of astrogliosis was observed in the contralateral cortex. Similar pattern of immunostaining was observed for markers of microglia/macrophage and degenerating axonal and nerve terminals.	1 DPI: No significant differences were found between WT and *chi3l1* KO regarding the extent of neuritic damage. 7 DPI: Comparable astrogliosis in the ipsilateral and contralateral cortices of WT and *chi3l1* KO after CCI. 14–19 DPI: No significant differences were observed between groups with respect to spatial memory acquisition and retention. 21 DPI: Gross tissue loss was not statistically different between WT and *chi3l1* KO. *Chi3l1* KO displayed greater astrocytosis and microglial/macrophage presence in the ipsilateral (42% and 31%) and contralateral hemisphere (55% and 230%) than WT mice.

Wiley et al. (2016)	CCI, left parietal–temporal cortex, 1.5 mm depth, 6 m/s impact velocity	Adult female C57BL/6 WT and *chi3l1* KO mice, 9–12 weeks	IHC (de Olmo's amino-cupric-silver, APP, and *Wisteria floribunda* agglutinin staining)	1, 7 and 21 DPI	Craniotomy alone was associated with some level of cortical pathology. CCI-induced lesions ranged from cortical contusion to brain herniation with tissue loss. Most prominent silver-staining was noted in the neuritic process arising from contusion cortical surface, which descended into thalamic or striatal regions and traversed the corpus callosum, penetrating the contralateral neocortex. The loss of perineural nets in the ipsilateral neocortex persisted up to 21 DPI.	7 DPI: *Chi3l1* KO exhibited 160% and 200% increase in severity in the ipsilateral and contralateral hemisphere compared to WT. 21 DPI: *Chi3l1* KO mice showed slight histopathological differences compared to WT animals in the lesioned and contralateral hemispheres (6% and 16%, respectively). WT and *chi3l1* KO animals did not show subjective differences on APP staining. The ultrastructural characteristics of the traumatic lesions were consistent in both groups.
Szmydynger-Chodobska, Shan, Thomasian, and Chodobski (2016)	CCI, right cortex 3 mm posterior to bregma and 2.5 mm lateral to midline, 2.5 mm diameter, 0.5, 1.0 or 1.5 mm depth, 5 m/s impact velocity	Adult male Long-Evans rats, age not specified	Real-time PCR, IHC, ELISA	6 and 24 h post-mild TBI	Cerebral cortex revealed no evidence of tissue injury at 4 weeks after 1 mm deformation. 6 h: Unlike brain parenchyma, large number of neutrophils crossed the pial microvessels and invaded the subarachnoid space and the cistern of the velum inter-positum. 24 h: Small number of monocytes appeared into the cerebral microvasculature.	The magnitude of changes in CHI3L1 expression was comparable in the ipsilateral hippocampus and cortex in response to a brain deformation of 1.5 mm. In presence of lower injury severity, CHI3L1 cortical expression levels were greater than those found in the hippocampus. Conversely to other pro-inflammatory mediators evaluated, CHI3L1 mRNA levels followed a time-dependent increased.

Continued

TABLE 1 Focal and temporal YKL-40 expression in experimental models of TBI—cont'd

Reference	Experimental procedure, region, and severity	Animals, age	Methods	Time course evaluated	Histopathological effects of TBI injury model	YKL-40 involvement in the neuropathology of TBI
					They entered mostly the subarachnoid space and the cistern, with a limited invasion of brain parenchyma. Some neutrophils and monocytes leaved the CSF compartments, moving along the perivascular spaces to invade the brain parenchyma. Neutrophil and monocyte chemoattracts were not detected in the cerebrovascular endothelium in the injured parenchyma.	
Malik, Motamedi, Royo, LeBold, and Watson (2011)	Single LFP, left cortex midway between lambda and bregma and midway between the sagittal suture and the temporal ridge, 2.65–2.75 atm, 20 ms + brain infusion cannula above the subcortical white matter (above the injured hippocampus)	Adult male Sprague–Dawley rats, age not specified	Microarray analysis	After 30 h of infusion	Not specified	*Chi3l1* gene was transcriptionally upregulated 2.3 folds by recombinant human neurotrophin-4/5 infusion in hippocampal CA3 neurons relative to normal brains.

Pun et al. (2011)	Single sub-lethal BOP exposure of 48.9 kPa, positive duration of 14.5 msec, distance of 30 m from the blast source	Adult male Sprague–Dawley rats, age not specified	Microarray analysis	1, 4 and 7 DPI	Gross examination did not reveal extra-axial hemorrhages. Morphological analysis provided evidence of degenerative processes in cortical neurons, which presented a darkened and shrunken appearance and a narrowed vasculature. Changes in vascular profiles pointed to the occurrence of vasospasm. The DNA fragmentation in the oligodendrocytes and astrocytes of the white matter with corresponding accumulation of APP suggested axonal pathology. Low BOP exposure did not result in blood–brain-barrier permeability changes.	*Chi3l1* gene expression was significantly altered following a BOP of 48.9 kPa.

BOP, blast over pressure; *CA3*, cornu ammonis subfield 3; *CCI*, controlled cortical impact; *DPI*, days post-injury; *GFAP*, glial fibrillary acidic protein; *Iba-1*, ionized calcium-binding adapter molecule 1/allograft inflammatory factor 1; *IF*, immunofluorescence; *IHC*, immunohistochemistry; *ISH*; in situ hybridization; *KO*, knockout; *LFP*, lateral fluid percussion; *TBI*, traumatic brain injury; *WT*, wild type; *YKL-40*, chitinase-3-like protein 1.

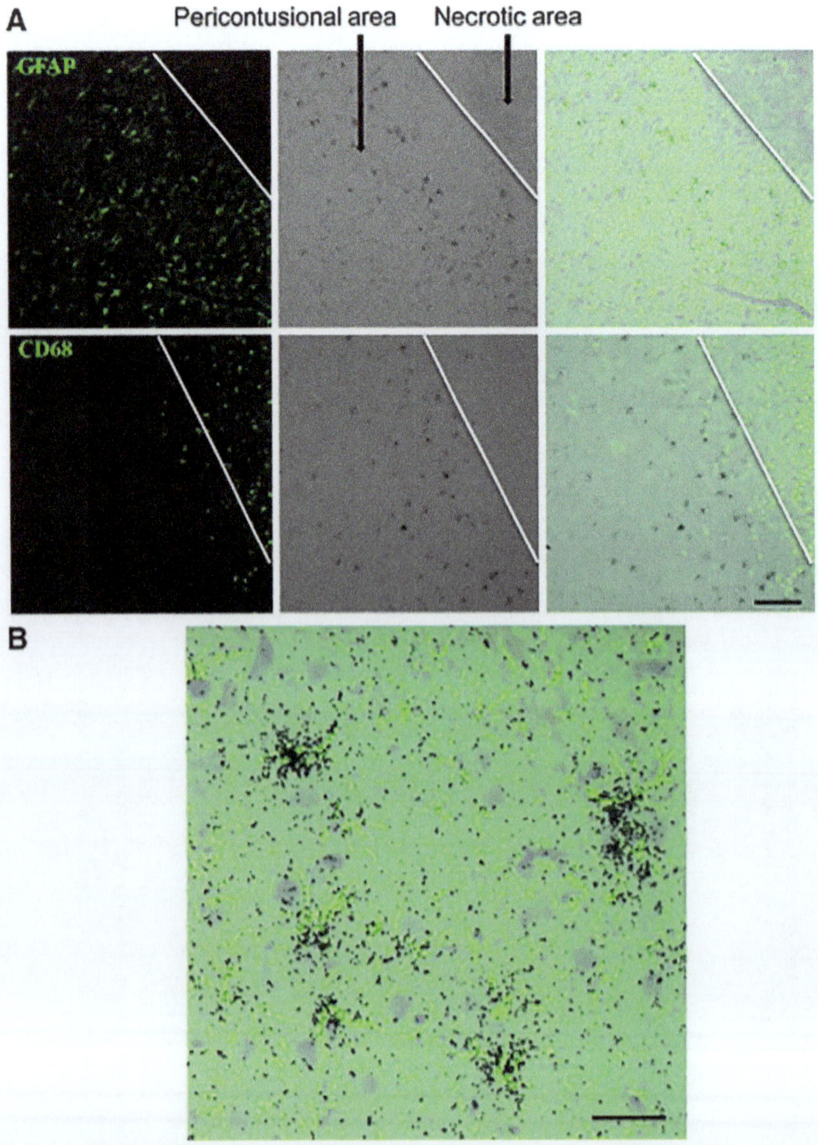

FIG. 1 Characterization of focal and temporal changes in YKL-40 expression after parasagittal CCI. (A, middle panels) Paraffin-embedded cortical tissue sections from rats after TBI hybridized for ^{35}S-labeled RNA probes for YKL-40. (A, left panels) Formalin-fixed, paraffin-embedded sections with GFAP (astrocytes) or CD68 (macrophages) immunostaining. (A, right panels) Combined images of the ISH and immunofluorescence. (B) YKL-40 induced expression in astrocytes was restricted to the pericontusional area on day 3 post-CCI as demonstrated by YKL-40 co-localization with GFAP staining. Micron bar in A = 0 μm; B = 5 μm. CCI, controlled cortical impact; GFAP, glial fibrillary acidic protein; ISH, in situ hybridization; TBI, traumatic brain injury; YKL-40, chitinase-3-like protein 1. *(Adapted with permission from Bonneh-Barkay, D., Zagadailov, P., Zou, H., Niyonkuru, C., Figley, M., Starkey, A., ... Wagner, A. K. (2010). YKL-40 expression in traumatic brain injury: An initial analysis. Journal of Neurotrauma, 27(7), 1215–1223. 10.1089/neu.2010.1310.)*

(Laird, Vender, & Dhandapani, 2008), CHI3L1 astrocytic expression was suggested by Wiley et al. to mediate downregulation of neuroinflammation, restricting the extent of both astrocytic and microglial activation. Consistent with these results, Im et al. (2020) observed that middle cerebral artery occlusion (MCAO)-injured *chi3l1* KO mice exhibited increased infarct volume and greater neurological deficits 24 h after ischemia/reperfusion. MCAO-injured KO mice exhibited higher expression of inflammatory proteins, such as inducible nitric oxide synthase (iNOS), cyclooxygenase 2 (COX-2), GFAP and ionized calcium-binding adapter molecule 1 (Iba-1), IL-1β and IL-6 levels, and decreased IL-4 and IL-10, with corresponding apoptotic neuronal cell death in the brain. In addition, *chi3l1* KO mice exhibited decreased expression levels of interleukin-4 receptor subunit alpha (IL-4Rα), Janus kinase 1 (p-JAK1), p-JAK3, and p-STAT6, suggesting that the absence of *chi3l1* promotes stroke development by increasing M1-polarized microglia and decreasing STAT6-dependent M2 microglia activation (Im et al., 2020). Furthermore, the *chi3l1* gene was upregulated in experimental intracerebral

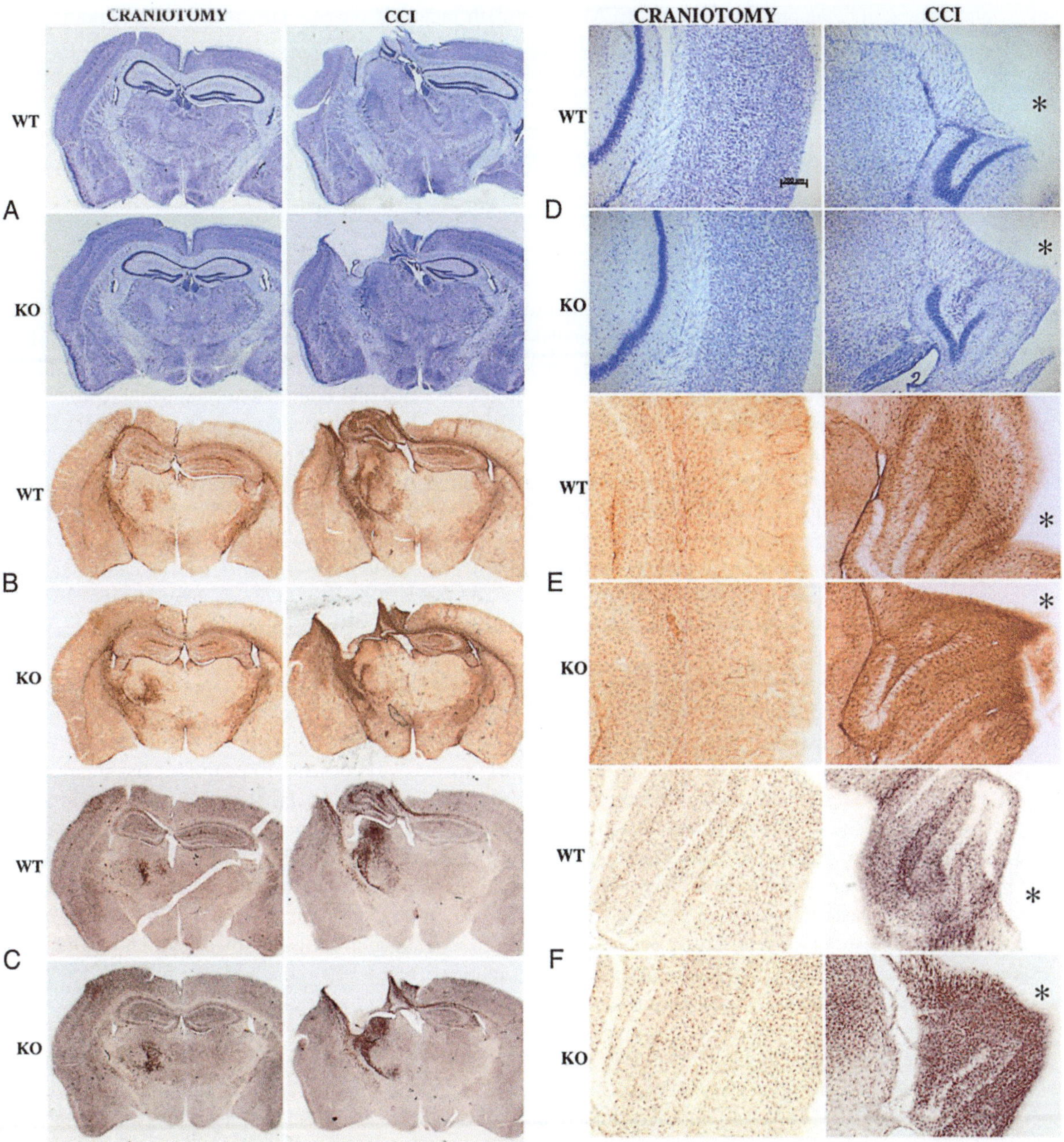

FIG. 2 Histopathological changes observed in WT and chi3l1 KO mice with left parietal–temporal craniotomy alone or with craniotomy and controlled cortical impact at 21 days post-injury. (A) Detail of coronal sections from paraformaldehyde-fixed, cryoprotected, and Nissl-stained mice brains. (B) Immunohistochemical staining for GFAP (*brown*) revealed ipsilateral astrogliosis with expansion into the underlying hippocampus, thalamus, corpus callosum and contralateral hippocampus. (C) Staining for microglia/macrophages showed widespread presence of Iba-1 (*brown*) at the site of induced-injury (right) and in the ipsilateral thalamus of both CCI and craniotomy alone (left). (D, E and F) Higher-power micrographs of hippocampus and overlying cortex. *Asterisks* indicate the region of cortical tissue loss in the animals with CCI. Micron bar in D=200 μm. *CCI*, controlled cortical impact; *GFAP*, glial fibrillary acidic protein; *Iba-1*, ionized calcium-binding adapter molecule 1/allograft inflammatory factor 1; *KO*, knockout; *WT*, wild type. *(Adapted with permission from Wiley, C. A., Bonneh-Barkay, D., Dixon, C. E., Lesniak, A., Wang, G., Bissel, S. J., & Kochanek, P. M. (2015). Role for mammalian chitinase 3-like protein 1 in traumatic brain injury. Neuropathology, 35(2), 95–106. 10.1111/neup.12158.)*

hemorrhage (ICH) induced in the anterior striatum (caudate and putamen) and was consistent with alternative macrophage/microglia activation, M2 or acquired deactivation (Lively & Schlichter, 2012). It should be noted that M1-like phenotypes cause neuronal damage, whereas M2-like phenotypes may result in neuroprotection (Im et al., 2020). Moreover, Bonneh-Barkay et al. (2008) demonstrated in vitro that YKL-40 displaces extracellular matrix-bound basic fibroblast growth factor (bFGF) from the heparan sulfate proteoglycan extracellular matrix (ECM) and inhibits bFGF-induced axonal branching in hippocampal neurons (Bonneh-Barkay et al., 2008). These findings indicate that YKL-40 modulates neurotrophic factor-associated interactions among neurons, the ECM and growth factors. Consequently, it has been hypothesized that cytokine-driven expression of YKL-40 in astrocytes inhibits survival cues from the brain ECM to invading leukocytes, and the abolition of astrocytic YKL-40 secretion may increase leukocyte survival in the intracranial compartment, resulting in exacerbation of the neuroinflammatory response observed in mice deficient in *chi3l1* (Wiley et al., 2015). These interesting data suggest that YKL-40 may exert both protective and detrimental roles following TBI and emphasize the importance of understanding the temporal regulation of reactive astrocytic factors, such as YKL-40, in the pathogenesis of brain injury (Laird et al., 2008).

CHI3L1 is not involved in the ultrastructural characteristics of TBI lesions

Subsequently, Wiley et al. (2016) characterized the ultrastructural changes associated with diaschisis lesions after the same model of severe focal impact (1.5 mm deformation over the dural surface) using homozygous deficient *chi3l1* KO and WT mice. Histopathological effects of the CCI injury model were evidenced by amino-cupric-silver staining (Table 1). Quantitative analysis of the amount of silver staining regions in the lesioned and contralateral hemispheres revealed differences between *chi3l1* KO and WT mice. Seven days post-CCI, ipsilateral and contralateral lesions in chi3l1 KO brains showed a substantial increase in severity compared to WT mice (160% and 200%, respectively). Regarding silver-positive lesions observed 21 days post-injury, *chi3l1* KO animals displayed a modest increase in the severity of ipsilateral (6%) and contralateral (16%) lesions compared to WT animals. Amyloid precursor protein (APP) immunohistochemical staining did not reveal congested neuritic profiles or axonal retraction bulbs in *chi3l1* KO or WT brains. Homozygously disrupted animals exhibited a more severe neuroinflammatory lesion than WT mice. Nonetheless, additional histochemical staining revealed that the ultrastructural characteristics of the CCI lesions were equivalent in both mouse lines (Wiley et al., 2016).

YKL-40 may play a role in immune cell trafficking into the injured brain

In a focal mild CCI model (2.5 mm impactor, 0.5, 1.0, or 1.5 mm depth, 5.0 m/s impact velocity) that resembles human complicated mild TBI or concussion, Szmydynger-Chodobska et al. (2016) examined the role of neuroinflammation in the pathophysiology of TBI. Real-time PCR analysis revealed an early increase in cortical production of the chemokine synthesis inductors TNF-α and IL-1β within 6 h post-injury. Similar changes in the expression of neutrophil and monocyte chemoattractants (CXCL1–3 and CCL2, respectively) were observed in an injury severity-dependent manner. As early as 6 h after mTBI, cortical and hippocampal expression of CXCL1–3 was coincident with a robust neutrophil influx into the subarachnoid space and the cistern of the velum interposum near the injury site. Analogous temporal changes in the production of CCL2 were observed in astrocytes and neutrophils. However, 24 h after CCI, synthesis of these chemokines in the injured brain had decreased to pre-injury levels, a time point at which cerebral neutrophil infiltration was evident. Moreover, synthesis of YKL-40 in both regions experienced a time-dependent increase, and levels remained upregulated for a sustained period. Unlike other pro-inflammatory mediators evaluated, the magnitude of changes in YKL-40 expression was similar in the ipsilateral hippocampus and cortex at the highest depth of brain deformation investigated (Szmydynger-Chodobska et al., 2016). Recent evidence has highlighted specific YKL-40 reactivity in the subpial marginal zone, particularly in the hippocampus, within the end feet layer and in surrounding parenchymal blood vessels of the human brain cortex (Bjørnbak, Brøchner, Larsen, Johansen, & Møllgård, 2014; Brøchner, Holst, & Møllgård, 2015). Noticeable discrepancies between the time-course changes in post-injury synthesis of neutrophil and monocyte chemoattractants and the trafficking of inflammatory cells indicate that additional chemokines should appear based on neutrophil-driven recruitment of monocytes to the injured brain. Interestingly, Szmydynger-Chodobska et al. suggested that CHI3L1 might be implicated in monocyte/macrophage recruitment into the damaged parenchyma after TBI. In addition, in a study by Kazakova et al. (2018), immunocytochemical examination of CSF smears from TBI patients revealed strong diffuse cytoplasmic expression of YKL-40 with perinuclear accentuation in polymorphonuclear leukocytes. Consistent with a study from Szmydynger-Chodobska et al., it has been shown a significant association between CHI3L1 expression and the number of infiltrated macrophages and micro-vessel density in other pathological conditions. In fact, the macrophage chemotaxis induced by CHI3L1 overexpression is mediated by activation of the mitogen-activated protein kinase (MAPK) signaling pathway. In particular, increased expression of CHI3L1 upregulated extracellular signal-regulated kinases 1/2 (ERK1/2) and jun N-terminal kinase (JNK) phosphorylation without affecting p38 phosphorylation. Moreover, Kawada et al. (2012) observed involvement of MAPK in CHI3L1-induced IL-8 and CCL-2 secretion and enhancement of human macrophage and endothelial cell migration in response to CHI3L1 overexpression. This hypothesis is in agreement with results from studies conducted by Bjørnbak et al. (2014) and Brøchner et al. (2015). They examined the morphological features and functional aspects of the human and rat outer brain barriers and observed that YKL-40 immunoreactivity was precisely detected at sites of brain barriers and, astonishingly, at entry points for microglia (Brøchner et al., 2015). Moreover, the spatiotemporal distribution of membrane-associated immunoreactivity for YKL-40 suggests its secretion from choroid plexus epithelium, leptomeningeal cells and pericytes and implicates it in controlling access of peripheral cells (microglia and conceivably macrophages) to the brain (Bjørnbak et al., 2014; Brøchner et al., 2015). This was further supported by the correlation obtained between YKL-40 and metalloproteinases (MMPs) (Ku et al., 2011; Michelsen et al., 2010) and the involvement of MMP-2 and MMP-9 in immune cell transmigration across the glia limitans, suggesting that YKL-40 may play a role in immune cell trafficking at these anatomic interfaces (Brøchner et al., 2015). Nevertheless, whether YKL-40 immunoreactivity in the meninges is elevated in response to TBI-induced neuroinflammation warrants further investigation.

CHI3L1 is a potential pathological fingerprint of blast-induced neurotrauma

Pun et al. (2011) evaluated the effects of a single sub-lethal blast over pressure (BOP) exposure in rat brain to profile acute histopathological and gene expression changes. They showed that even an acute low-level blast-induced neurotrauma of 48.9 kPa resulted in changes in both cortical cerebrovasculature and white matter tracts (Table 1). Microarray analysis revealed significant gene expression alterations in mild TBI, identifying 676 established genes with well-defined functions. Curiously, *chi3l1* gene expression was significantly altered following low BOP exposure. Nevertheless, it is unclear whether the *chi3l1* gene or its protein product decisively affect injury outcome in blast-exposed victims or whether it represents a reliable pathological fingerprint of blast-induced mild TBI. Even though there is a pressing urgency for accurate and reliable diagnostic, prognostic and therapeutic approaches to blast injuries, further studies are necessary to confirm whether the fingerprints identified in this study are consistent and reproducible across the spectrum of blast-injury severities (Pun et al., 2011).

Clinical studies of YKL-40 in TBI

The magnitude of CHI3L1 elevation has been found to be associated with neurologic deficits, symptom severity, volume and location of intracranial findings and unfavorable outcomes in head injury patients (Table 2).

Release patterns of CSF YKL-40 are related to the trauma-induced pro-inflammatory response that negatively impacts patient clinical outcomes

Bonneh-Barkay, Zagadailov, et al. (2010) reported for the first time that YKL-40 protein levels were elevated in the CSF of TBI patients who required an extra-ventricular drainage catheter for intracranial pressure (ICP) monitoring and management. In this prospective study, they assessed the relationship between the temporal profile of YKL-40 levels and injury severity (Glasgow Coma Scale [GCS] score) as well as the long-term survival rate in a population of 20 patients with severe TBI. CSF YKL-40 levels in patients with a mixture of diffuse and focal injuries were significantly elevated compared to healthy subjects after the first day, reaching a peak 4 days post-injury, followed by a decrease 5 and 6 days post-injury. In acute ischemic stroke patients, however, Park et al. (2012) observed that the temporal dynamics of circulating YKL-40 levels were quite different, with peak concentration observed within 18–24 h of symptom onset. Different release patterns of CSF YKL-40 were not observed between admission GCS score subgroups (Bonneh-Barkay, Zagadailov, et al., 2010). They noticed that patients who died during their intensive care unit stay tended to present higher CSF YKL-40 concentrations than patients who survived their injuries. Therefore, an inverse but non-significant relationship between 4-day YKL-40 levels and Glasgow Outcome Scale (GOS) scores 6 months after injury was observed. The evidence from this study supports the hypothesis that the mechanism of TBI-induced YKL-40 expression is associated with the level of neuroinflammation mediated by cytokines. In this clinical cohort of severe TBI, several candidate cytokines were evaluated for their overall association with YKL-40 expression. There was a significant relationship between YKL-40 concentrations and IL-1β, TNF-α and C-reactive protein (CRP) levels; nevertheless, the associations with IL-6 and IL-10 levels were not significant. Given the greater YKL-40 levels in patients with unfavorable outcomes and the correlations between the CSF patterns of this protein and cytokines, the researchers determined that YKL-40 elevations are related to the trauma-induced pro-inflammatory response that negatively impacts patient clinical outcomes. In line with this claim, Mygind et al. (2013) showed that YKL-40 levels within the first 24 h after hospital admission identified patients with a high risk of short- and long-term mortality independent of discharge diagnosis. Patients with YKL-40 $\geq$ 289 ng/mL had a hazard ratio (HR) of 7.1 for all-cause mortality during the first year post-injury and 3.4 in the 12-year study period (Mygind et al., 2013). In patients with carotid, vertebral or basilar artery stenosis, Chen et al. (2017) reported that high YKL-40 levels predicted poor functional outcome (odds ratio [OR] of 6.47). Additionally, the combination of YKL-40 determinations with the National

TABLE 2 Clinical studies of YKL-40 as a potential biomarker in TBI.

Reference	Type of study	Patients, clinical severity (median, range), controls	Radiological features (*n*, %)	Biofluid, time interval between extraction and TBI (median, range), method	Outcome (*n*, %)	Main results
Bonneh-Barkay, Zagadailov, et al. (2010)	Prospective observational analytic cohort study	20 Patients, severe TBI (GCS 6, 3–8), 6 healthy controls	*Type of injury* EDH: 4, 20% SDH: 12, 60% tSAH: 15, 75% Contusion: 6, 30% ICH: 9, 45% IVH: 4, 20% DAI: 5, 25% Coalescence of multiple intracranial findings	CSF, collection every 12 h for up 7 DPI, ELISA	*6-month GOS* GOS 4/5: 7, 35% GOS 2/3: 8, 40% GOS 1: 1, 5%	YKL-40 levels were significantly increased in the CSF of severe TBI patients compared to controls after 1 DPI, levels peaked 4 DPI, and decreased 5–6 DPI. Different release patterns of CSF YKL-40 were not observed between patients with different admission GCS scores. Patients who died during their hospitalization tended to present higher CSF YKL-40 levels than surviving patients. 4-day YKL-40 levels were inversely associated with 6-month GOS scores (trended toward statistical significance). Significant relationships were found between YKL-40 and other CSF inflammatory mediators (IL-1β, TNF-α and CRP, but not with IL-6 or IL-10).
Shahim et al. (2017)	Multi-center cross-sectional study	28 Professional athletes (ice hockey and soccer players) with PCS > 3 months following rcTBI (RPQ score 13, 6–35), 19 controls without a history of concussion	No evidence of structural damage on conventional MRI (T1/T2 and FLAIR)	CSF, 1.5 years after the last concussion (0.25–12), ELISA	Not specified	CSF YKL-40 concentrations were elevated in athletes with persistent PCS due to rcTBI compared to controls. YKL-40 levels were not significantly different between subgroups (PCS ≤ 1 year and PCS > 1 year). CSF YKL-40 was associated with RPQ scores and the lifetime number of concussions, but not with the time since the last concussion. YKL-40 levels reflected similar changes as GFAP, and levels were correlated with CSF NF-L.

Kazakova et al. (2018)		19 Patients (GCS not specified), 15 age- and gender-matched controls	Not specified	CSF and plasma, 24 and 96 h post-injury, ELISA and ICC (indirect immunoperoxidase method)	Not specified	No differences in 1-day YKL-40 levels were observed between patients and controls. Immunocytochemical identification of YKL-40 in the cytoplasm of polymorphonuclear leukocytes (strong diffuse expression).
Carabias et al. (2020)	Prospective observational analytic study of a hospital cohort	115 Patients GCS 3–8: 26, 23% GCS 8–12: 3, 3% GCS 13–15: 86, 75% *Pupil responsiveness* Normal: 109, 94% Unilateral or bilateral mydriasis: 6, 6% *ISS:* 9, 4–26 15 healthy controls	*Admission Marshall CT Classification* DI I 49, 42% DI II 58, 50% DI III 2, 2% DI IV 2, 2% Evacuated mass lesion 3, 3% Non-evacuated mass lesion 1, 1% *Type of intracranial hemorrhage* EDH 7, 6% SDH 18, 16% tSAH 47, 41% Contusion 11, 10% IVH 10, 9% Coalescence of multiple intracranial findings	Serum, first 24 h (9.2, 4–15), ELISA	Not specified	Serum YKL-40 levels were significantly correlated with the degree of neurologic deficit assessed by total GCS and motor subcomponent at various evaluation time points. Serum YKL-40 concentrations were associated with computer-assisted volumetric measurements of various types of intracranial disease (SDH, tSAH, ICH, IVH and total mass) as well as Marshall CT classification. Serum YKL-40 levels exhibited an outstanding capacity for identifying bleeding in the several intracranial compartments. A relationship between YKL-40 and other neurotrauma biomarkers was observed.

CRP, C-reactive protein; *CSF*, cerebrospinal fluid; *CT*, computed tomography; *DAI*, diffuse axonal injury; *DI*, diffuse injury; *DPI*, days post-injury; *EDH*, epidural hematoma; *ELISA*, enzyme-linked immunosorbent assay; *FLAIR*; fluid-attenuated inversion recovery; *GCS*, Glasgow Coma Scale; *GFAP*, glial fibrillary acidic protein; *GOS*, Glasgow Outcome Scale; *ICC*, immunocytochemistry; *ICH*, intracerebral hemorrhage; *ISS*, Injury Severity Score; *IVH*, intra-ventricular hemorrhage; *MRI*, magnetic resonance imaging; *NF-L*, neurofilament light chain; *PCS*, Post-concussion syndrome; *rcTBI*, repetitive concussive traumatic brain injury; *RPQ*, Rivermead Post-Concussion Symptoms Questionnaire; *SDH*, subdural hematoma; *TBI*, traumatic brain injury; *tSAH*, traumatic subarachnoid hemorrhage; *YKL-40*, chitinase-3-like protein 1.

Institutes of Health Stroke Scale (NIHSS) score slightly improved the prognostic accuracy of the NIHSS score (Chen et al., 2017). Likewise, Jiang, Zhang, Wang, and Yang (2014) identified YKL-40 as an independent predictive factor of 3-month mortality and 3-month unfavorable outcome in patients with ICH. Additionally, YKL-40 levels predicted severe neurologic defects during the acute phase of ICH, suggesting that YKL-40 may serve as an additional valuable tool for risk stratification and decision-making during the acute stage of cerebral pathology (Jiang et al., 2014). Unfortunately, Bonneh-Barkay, Zagadailov, et al. did not indicate a possible association between YKL-40 concentrations and computed tomography (CT) severity, neurological deterioration or improvement during the early post-traumatic period. This study also failed to determine the ability of YKL-40 to predict an increase in ICP. We should keep in mind that the type of management of such injuries (conservative or surgical) could condition the release patterns of post-traumatic YKL-40 protein. Pre-clinical studies of severe TBI have shown that sham animals undergoing a parieto-temporal craniotomy without CCI presented some level of astrogliosis in the ipsilateral cortex. Moreover, quantitative analysis of the area immunostained for GFAP showed differences between WT mice and homozygously deficient *chi3l1* mice (Wiley et al., 2016, 2015). Consequently, the possible alteration of YKL-40 levels in patients who underwent craniotomy for mass evacuation or ventriculostomy placement should be taken into consideration. Despite the limited results, Bonneh-Barkay, Zagadailov, et al. laid the foundations of this research field and concluded that YKL-40 is a promising marker for assessing the pathophysiology and prognosis of severe head injury patients.

CSF YKL-40 levels as a reflection of post-concussion syndrome severity

In a multi-center cross-sectional study, Shahim et al. (2017) evaluated whether post-concussion syndrome (PCS) secondary to repetitive concussive TBI (rcTBI) was related to biomarker evidence of astroglial activation and amyloid dysmetabolism. Twenty-eight professional athletes (ice hockey and soccer players) with PCS for more than 3 months following rcTBI and 19 controls without a history of known concussion were selected. The individuals exposed to rcTBI and persistent PCS exhibited higher concentrations of CSF YKL-40 compared to neurologically healthy controls. However, differences did not remain significant after correcting for multiple comparisons and age. Concentrations of YKL-40 protein were not conditioned by the time since the last sport-related concussion. Despite this, CSF YKL-40 levels reflected symptom severity assessed by the Rivermead Post-Concussion Symptoms Questionnaire (RPQ). Remarkably, there were no associations between any of the other CSF biomarkers and RQP scores. In addition, YKL-40 levels in athletes with PCS were related to the number of lifetime concussion events. This study of concussion-like injuries revealed that patients with no evidence of structural damage on conventional magnetic resonance imaging sequences exhibited CSF astroglial activation markers (Shahim et al., 2017). Because patients with concussion usually do not show visible intracranial pathology on brain imaging studies, there is a potential opportunity for CSF- and blood-based biomarkers to provide objective and reliable information regarding the existence of concussion and the underlying pathology (Gan et al., 2019). The operating characteristics of YKL-40 for documenting concussions might be relevant for further exploration from diagnostic and therapeutic standpoints. Given the relationship between YKL-40 levels and PCS symptoms due to rcTBI and the proportion of individuals with PCS developing progressive neurodegeneration or chronic traumatic encephalopathy (CTE), it might be reasonable to examine this marker regarding its accuracy for predicting CTE in rcTBI (Shahim et al., 2017).

YKL-40 concentration as an objective and complementary assessment of impairment of consciousness levels and radiologic lesion burden in head injury patients

Assessment of neurologic status

Recently, our group assessed the release patterns of a select blood-based biomarker panel and their association with volumetric measurement of traumatic lesions on admission CT. In a cohort of 115 head injury patients (GCS score 3–15), we attempted to determine whether the amount and distribution of intracranial hemorrhagic lesions could be diagnosed or classified by quantitative determination of several biomarkers during the acute stage. Initial serum YKL-40 levels were not significantly related to patient age or gender. Both patients with mild and moderate to severe head injury presented significantly increased YKL-40 serum concentrations compared to the values of age-matched controls. The receiver operating characteristics of admission YKL-40 levels for distinguishing between controls and TBI patients were excellent (area under the curve [AUC] = 0.908, 95% confidence interval [CI] 0.848–0.969) (Fig. 3) (Carabias et al., 2020). Similar diagnostic accuracies were observed for discriminating acute ischemic stroke (AIS) patients from controls in the study of Park et al. (2012), regardless of the presence or absence of several modifiable vascular risk factors. YKL-40 presented one of the

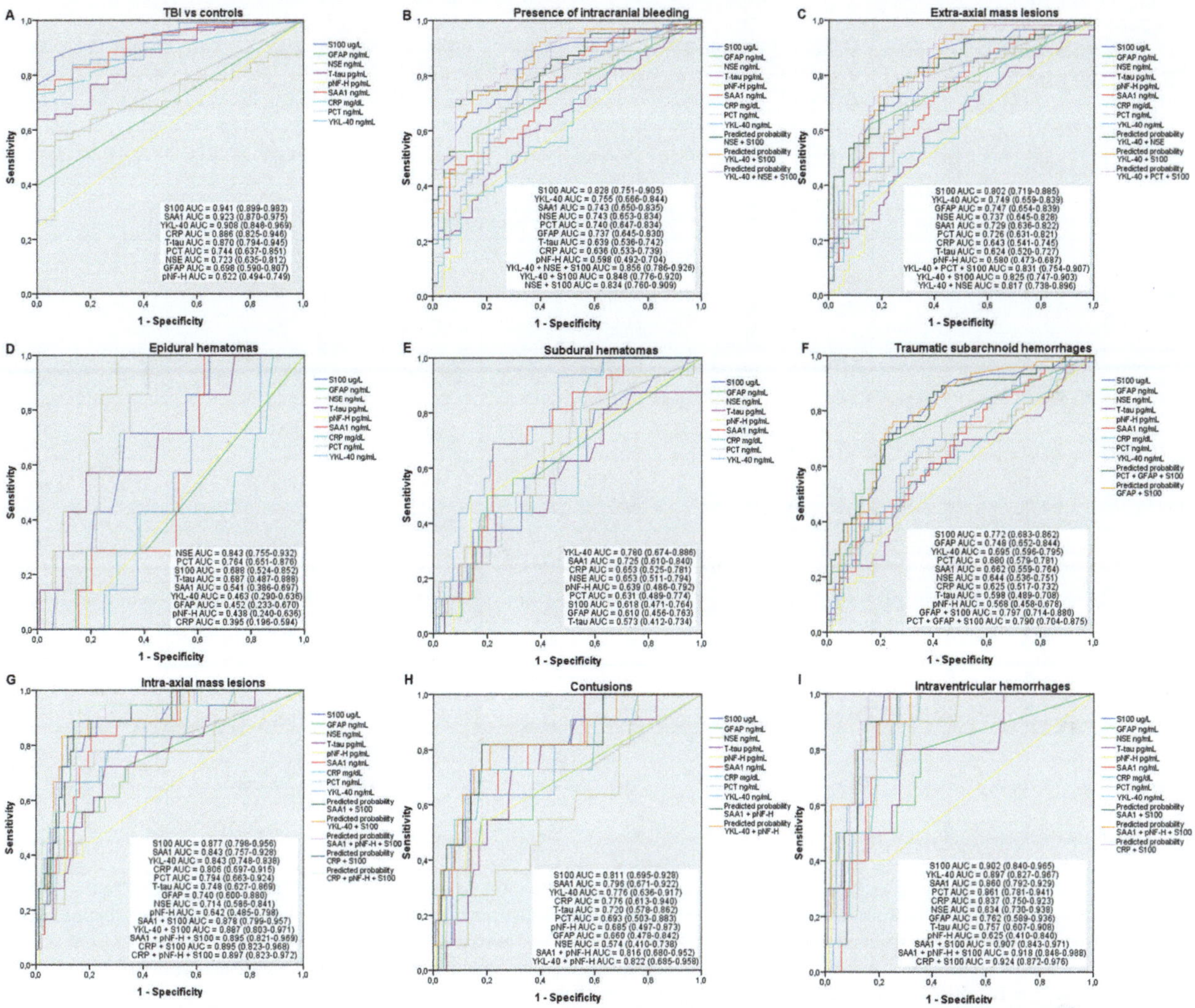

FIG. 3 Operating characteristics of YKL-40 and its combination with other TBI biomarkers after logistic regression for identification of bleeding in the several intracranial compartments. (A) Traumatic brain injury versus age-matched controls. (B) Presence of intracranial bleeding. (C) Extra-axial mass lesions. (D) Epidural hematomas. (E) Subdural hematomas. (F) Traumatic subarachnoid hemorrhages. (G) Intra-axial mass lesions. (H) Contusions. (I) Intra-ventricular hemorrhages. *AUC*, area under the curve; *CRP*, C-reactive protein; *GFAP*, glial fibrillary acidic protein; *NSE*, neuron-specific enolase; *PCT*, procalcitonin; *pNF-H*, phosphorylated neurofilament heavy chain; *S100β*, calcium-binding protein B; *SAA1*, serum amyloid A1; ; *T-tau*, total tau; *YKL-40*, chitinase-3-like protein 1. *(Adapted from Carabias, C. S., Gomez, P. A., Panero, I., Eiriz, C., Castaño-León, A. M., Egea, J., ... i+12 Neurotraumatology Group Collaborators. (2020). Chitinase-3-like protein 1, serum amyloid A1, C-reactive protein, and procalcitonin are promising biomarkers for intracranial severity assessment of traumatic brain injury: Relationship with Glasgow coma scale and computed tomography volumetry. World Neurosurgery, 134, e120–e143. https://doi.org/10.1016/j.wneu.2019.09.143.)*

strongest correlations with level of consciousness, both with total GCS and motor subcomponent, at initial and admission after non-surgical resuscitation (Fig. 4), indicating an association between the severity of neurologic deficit and the underlying process of reactive gliosis. A shift toward a weaker association between YKL-40 levels and GCS motor subcomponent was apparent from the pre-hospital stage to 6h post-admission. In fact, the GCS motor score assessed in the field exhibited the highest correlation with marker levels (Carabias et al., 2020), which is supported by the fact that aggressive pre-hospital stabilization maneuvers obscure the neurological assessment that can become problematic in such cases (Carabias et al., 2020; Majdan et al., 2015). Moreover, YKL-40 maintained the ability to reflect the clinical characteristics of the patients when analyzing a subsample of 79 TBIs with a non-head injury severity score (ISS) ≤ 4, suggesting that concomitant extracranial injury in our series did not condition the results obtained (Carabias et al., 2020). In line with our results, Jiang et al. (2014) observed that serum YKL-40 levels were highly associated with the severity of stroke-related neurologic deficits in a series of 172 patients with spontaneous basal ganglia hemorrhage. Moreover, in the presence of dynamic factors, such as

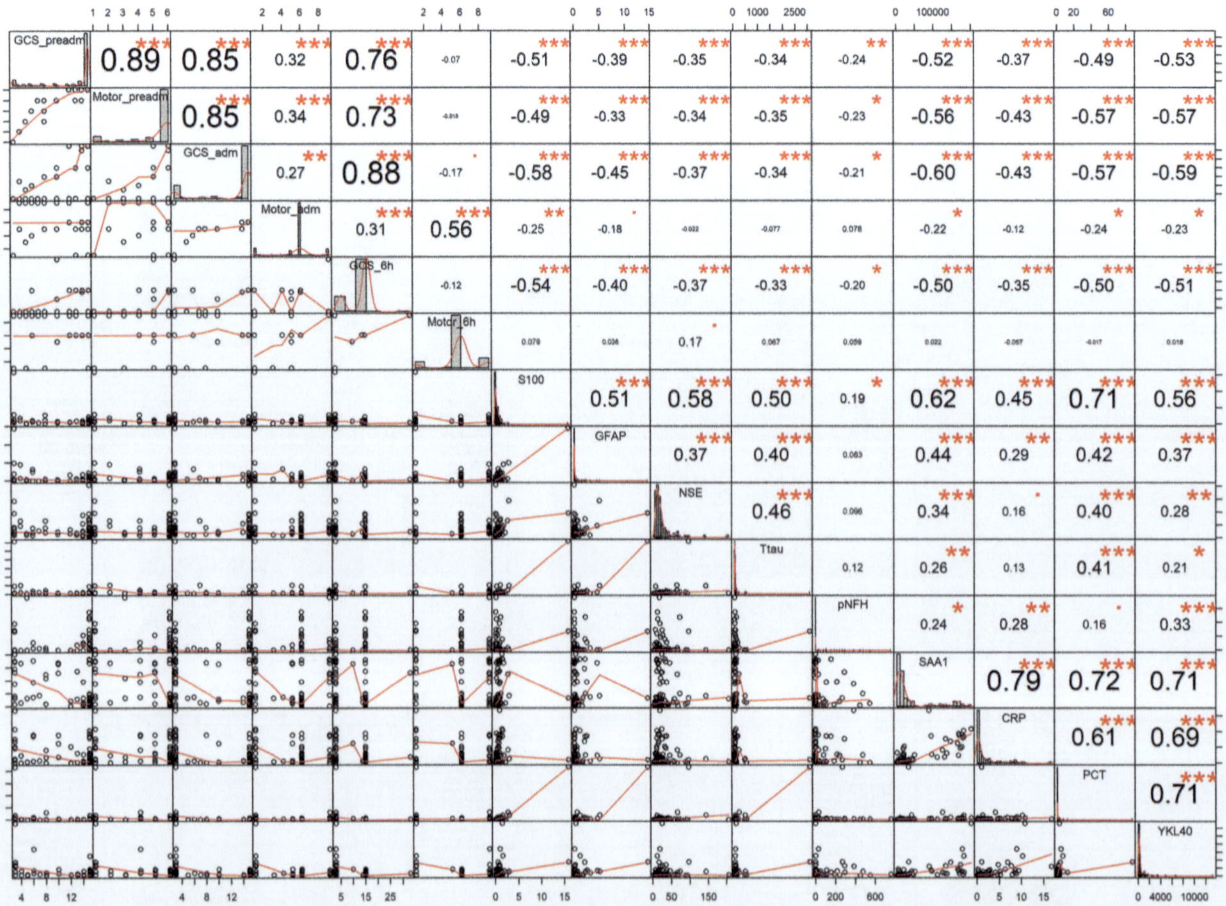

FIG. 4 Chart correlation matrices between serum biomarkers and post-resuscitation GCS score at various evaluation time points. The distribution of each variable shown as histograms is exposed on the diagonal line of the graph. Above the diagonal the value of the correlation plus significance level from the Spearman rho correlation test are represented as $***q<0.001$, $**q<0.01$, $*q<0.05$, $\cdot\ q<0.1$. The greater the strength of the correlation, the larger the font size. Below the diagonal, bivariate scatter plots with a fitted line for each variable are shown. *CRP*, C-reactive protein (mg/dL); *GCS_adm*, Glasgow Coma Scale at hospital admission; *GCS_preadm*, Glasgow Coma Scale at pre-hospital admission or initial assessment; *GCS_6h*, Glasgow Coma Scale at 6h after admission; *GFAP*, glial fibrillary acidic protein (ng/mL); *Motor_6h*, motor component at 6h after admission; *Motor_adm*, motor component at hospital admission; *Motor_preadm*, motor component at pre-hospital admission or initial assessment; *NSE*, neuron-specific enolase (ng/mL); *PCT*, procalcitonin (ng/mL); *pNF-H*, phosphorylated neurofilament heavy chain (pg/mL); *S100β*, calcium-binding protein B (μg/L); *SAA1*, serum amyloid A1 (ng/mL); *T-tau*, total tau (pg/mL); *YKL-40*, chitinase-3-like protein 1 (ng/mL). *(Adapted from Carabias, C. S., Gomez, P. A., Panero, I., Eiriz, C., Castaño-León, A. M., Egea, J., … i+12 Neurotraumatology Group Collaborators. (2020). Chitinase-3-like protein 1, serum amyloid A1, C-reactive protein, and procalcitonin are promising biomarkers for intracranial severity assessment of traumatic brain injury: Relationship with Glasgow coma scale and computed tomography volumetry. World Neurosurgery. 134, e120–e143. https://doi.org/10.1016/j.wneu.2019.09.143.)*

hematoma volume expansion or oedema formation, YKL-40 demonstrated a high predictive value for identifying patients with severe neurologic defects defined as NIHSS > 11 (AUC = 0.830, 0.765–0.882, cut-off value of 242.8 ng/mL, sensitivity 95%, specificity 58%) (Jiang et al., 2014).

Identification of CT-positive findings

Surprisingly, YKL-40 was among the highest-performing biomarkers for detecting bleeding in several intracranial compartments (Fig. 3, Table 3). While YKL-40 alone showed the second-best discriminative ability for identifying CT-positive findings (regardless of the anatomic compartment in which the hemorrhage was found) (AUC = 0.755, 0.666–0.844), its combination with neuron-specific enolase (NSE) and calcium-binding protein B (S100β) yielded the best receiver operating characteristic (AUC = 0.856, 0.786–0.926). With respect to the Traumatic Coma Data Bank (TCDB) classification of TBI, a reasonably strong correlation was observed in our study between initial YKL-40 serum levels and CT Marshall classification. Accordingly, YKL-40 levels strongly represented the total hemorrhagic lesion burden as estimated from quantitative CT (Fig. 5) (Carabias et al., 2020). This is consistent with a previous study showing significant associations between YKL-40 levels and clinical characteristics of AIS, including NIHSS, and diffusion-weighted imaging (DWI) infarct

TABLE 3 The AUC, cut-off points, sensitivity, and specificity of the biomarkers for the identification of bleeding in the several intracranial compartment.

Biomarker	AUC (95% CI)	Cut-off points (J)	Sensitivity	Specificity
CT-positive findings				
S100β	**0.828** (0.751–0.905) $q<0.001$	0.425 μg/L	66.7	87.5
YKL-40	**0.755** (0.666–0.844) $q<0.001$	171.62 ng/mL	69.8	68.8
SAA1	**0.743** (0.650–0.835) $q<0.001$	25,733.37 ng/mL	46.0	93.8
NSE	0.743 (0.653–0.834) $q<0.001$	16.78 ng/mL	65.1	79.2
PCT	0.740 (0.647–0.834) $q<0.001$	0.0252 ng/mL	76.2	68.8
GFAP	0.737 (0.645–0.830) $q<0.001$	0.0123 ng/mL	58.7	85.4
T-tau	0.639 (0.536–0.742) $q<0.05$	50.68 pg/mL	58.7	66.7
CRP	0.636 (0.533–0.739) $q<0.05$	0.655 mg/dL	50.8	70.8
pNF-H	0.598 (0.492–0.704) $q=0.078$	0.588 pg/mL	33.3	87.5
NSE+S100β	**0.834** (0.760–0.909) $q<0.001$	–		
YKL-40+S100β	**0.848** (0.776–0.920) $q<0.001$	–		
YKL-40+NSE+S100β	**0.856** (0.786–0.926) $q<0.001$	–		
Extra-axial mass lesions				
S100β	**0.802** (0.719–0.885) $q<0.001$	0.425 μg/L	67.2	83.0
YKL-40	**0.749** (0.659–0.839) $q<0.001$	171.616 ng/mL	72.4	67.9
GFAP	**0.747** (0.654–0.839) $q<0.001$	0.0123 ng/mL	62.1	84.9
NSE	0.737 (0.645–0.828) $q<0.001$	16.78 ng/mL	65.5	75.5
SAA1	0.729 (0.636–0.822) $q<0.001$	25,733.37 ng/mL	46.6	90.6
PCT	0.726 (0.631–0.821) $q<0.001$	0.0252 ng/mL	77.6	66.0
CRP	0.643 (0.541–0.745) $q<0.01$	0.770 mg/dL	50.0	73.6
T-tau	0.624 (0.520–0.727) $q<0.05$	50.668 pg/mL	58.6	64.2
pNF-H	0.580 (0.473–0.687) $q=0.146$	0.588 pg/mL	32.8	84.9
YKL-40+NSE	**0.817** (0.738–0.896) $q<0.001$	–		
YKL-40+S100β	**0.825** (0.747–0.903) $q<0.001$	–		
YKL-40+PCT+S100β	**0.831** (0.754–0.907) $q<0.001$	–		
Epidural hematomas				
NSE	**0.843** (0.755–0.932) $q<0.01$	21.27 ng/mL	100	70.2
PCT	**0.764** (0.651–0.876) $q<0.05$	0.0604 ng/mL	100	58.7
S100β	**0.688** (0.524–0.852) $q=0.097$	0.525 μg/L	71.4	67.3
T-tau	0.687 (0.487–0.888) $q=0.098$	70.87 pg/mL	57.1	81.7
SAA1	0.541 (0.386–0.697) $q=0.716$	6688.83 ng/mL	100	37.5
YKL-40	0.463 (0.290–0.636) $q=0.743$	53.790 ng/mL	100	16.3
GFAP	0.452 (0.233–0.670) $q=0.671$	0.3903 ng/mL	28.6	81.7
pNF-H	0.438 (0.240–0.636) $q=0.585$	27.415 pg/mL	14.3	81.7
CRP	0.395 (0.196–0.594) $q=0.353$	0.070 mg/dL	100	11.5

Continued

TABLE 3 The AUC, cut-off points, sensitivity, and specificity of the biomarkers for the identification of bleeding in the several intracranial compartment—cont'd

Biomarker	AUC (95% CI)	Cut-off points (J)	Sensitivity	Specificity
Subdural hematomas				
YKL-40	**0.780** (0.674–0.886) $q<0.001$	171.62 ng/mL	93.8	53.7
SAA1	**0.725** (0.610–0.840) $q<0.01$	25,733.37 ng/mL	68.8	77.9
CRP	**0.653** (0.525–0.781) $q=0.0504$	0.205 mg/dL	100	36.8
NSE	0.653 (0.511–0.794) $q=0.051$	16.62 ng/mL	75.0	55.8
pNF-H	0.639 (0.486–0.792) $q=0.076$	26.00 pg/mL	50.0	86.3
PCT	0.631 (0.489–0.774) $q=0.094$	0.0272 ng/mL	81.3	48.4
S100β	0.618 (0.471–0.764) $q=0.133$	0.435 μg/L	62.5	61.1
GFAP	0.610 (0.456–0.763) $q=0.161$	0.2256 ng/mL	50.0	80.0
T-tau	0.573 (0.412–0.734) $q=0.351$	43.19 pg/mL	81.3	36.8
Traumatic subarachnoid hemorrhage				
S100β	**0.772** (0.683–0.862) $q<0.001$	0.425 μg/L	71.7	76.9
GFAP	**0.748** (0.652–0.844) $q<0.001$	0.0123 ng/mL	67.4	80.0
YKL-40	**0.695** (0.596–0.795) $q<0.001$	198.15 ng/mL	67.4	66.2
PCT	0.680 (0.579–0.781) $q<0.01$	0.0119 ng/mL	78.3	55.4
SAA1	0.662 (0.559–0.764) $q<0.01$	32,039.04 ng/mL	41.3	87.7
NSE	0.644 (0.536–0.751) $q=0.0101$	16.78 ng/mL	63.0	66.2
CRP	0.625 (0.517–0.732) $q<0.05$	0.850 mg/dL	52.2	72.3
T-tau	0.598 (0.489–0.708) $q=0.079$	50.67 pg/mL	60.9	61.5
pNF-H	0.568 (0.458–0.678) $q=0.224$	0.588 pg/mL	32.6	81.5
S100β + GFAP + PCT	**0.790** (0.704–0.875) $q<0.001$	–		
S100β + GFAP	**0.797** (0.714–0.880) $q<0.001$	–		
Intra-axial mass lesions				
S100β	**0.877** (0.798–0.956) $q<0.001$	0.570 μg/L	88.9	77.4
SAA1	**0.843** (0.757–0.928) $q<0.001$	20,834.42 ng/mL	83.3	79.6
YKL-40	**0.843** (0.748–0.938) $q<0.001$	953.19 ng/mL	66.7	92.5
CRP	0.806 (0.697–0.915) $q<0.001$	2.360 mg/dL	66.7	84.9
PCT	0.794 (0.663–0.924) $q<0.001$	0.872 ng/mL	66.7	91.4
T-tau	0.748 (0.627–0.869) $q<0.001$	58.289 pg/mL	72.2	74.2
GFAP	0.740 (0.600–0.880) $q<0.01$	0.2637 ng/mL	55.6	84.9
NSE	0.714 (0.586–0.841) $q<0.01$	18.70 ng/mL	72.2	65.6
pNF-H	0.642 (0.485–0.798) $q=0.058$	59.481 pg/mL	38.9	89.2
SAA1 + S100β	**0.878** (0.799–0.957) $q<0.001$	–		
YKL-40 + S100β	**0.887** (0.803–0.971) $q<0.001$	–		
CRP + S100β	**0.895** (0.823–0.968) $q<0.001$	–		
CRP + pNF-H + S100β	**0.897** (0.823–0.972) $q<0.001$	–		
SAA1 + pNF-H + S100β	**0.895** (0.821–0.969) $q<0.001$	–		

TABLE 3 The AUC, cut-off points, sensitivity, and specificity of the biomarkers for the identification of bleeding in the several intracranial compartment—cont'd

Biomarker	AUC (95% CI)	Cut-off points (J)	Sensitivity	Specificity
Intra-parenchymal hemorrhage / Contusions				
S100β	**0.811** (0.695–0.928) $q<0.001$	0.770 μg/L	72.7	86.0
SAA1	**0.796** (0.671–0.922) $q<0.01$	47,165.74 ng/mL	63.6	86.0
YKL-40	**0.776** (0.636–0.917) $q<0.01$	953.19 ng/mL	63.6	88.0
CRP	**0.776** (0.613–0.940) $q<0.01$	2.360 mg/dL	72.7	82.0
T-tau	0.720 (0.578–0.862) $q<0.05$	58.289 pg/mL	72.7	71.0
PCT	0.693 (0.503–0.883) $q<0.05$	0.872 ng/mL	63.6	87.0
pNF-H	0.685 (0.497–0.873) $q<0.05$	5.284	54.5	82.0
GFAP	0.660 (0.478–0.842) $q=0.082$	0.2637 ng/mL	54.5	82.0
NSE	0.574 (0.410–0.738) $q=0.421$	9.920 ng/mL	100	24.0
SAA1 + pNF-H	**0.816** (0.680–0.952) $q<0.001$	–		
YKL-40 + pNF-H	**0.822** (0.685–0.958) $q<0.001$	–		
Intra-ventricular hemorrhages				
S100β	**0.902** (0.840–0.965) $q<0.001$	0.630 μg/L	100	78.2
YKL-40	**0.897** (0.827–0.967) $q<0.001$	293.49 ng/mL	100	72.3
SAA1	**0.860** (0.792–0.929) $q<0.001$	20,834.42 ng/mL	100	76.2
PCT	0.861 (0.781–0.941) $q<0.001$	0.0760 ng/mL	100	65.3
CRP	0.837 (0.750–0.923) $q<0.001$	0.6950 mg/dL	100	64.4
NSE	0.834 (0.730–0.938) $q<0.001$	19.670 ng/mL	90.0	68.3
GFAP	0.762 (0.589–0.936) $q<0.01$	0.0123 ng/mL	80	64.4
T-tau	0.757 (0.607–0.908) $q<0.01$	58.288 pg/mL	80.0	71.3
pNF-H	0.625 (0.410–0.840) $q=0.194$	254.137 pg/mL	40.0	98.0
SAA1 + S100β	**0.907** (0.843–0.971) $q<0.001$	–		
SAA1 + pNF-H + S100β	**0.918** (0.848–0.988) $q<0.001$	–		
CRP + S100β	**0.924** (0.872–0.976) $q<0.001$	–		

The three best AUC of each biomarker and the combination of several after logistic regression are highlighted in bold. *AUC*, area under the receiver operating characteristic curve; *CI*, confidence interval; *CRP*, C-reactive protein (mg/dL); *CT*, computed tomography; *EDH*, epidural hematoma; *GFAP*, glial fibrillary acidic protein (ng/mL); *ICH*, contusion; *IVH*, intra-ventricular hemorrhage; *J*, Youden Index; *NSE*, neuron-specific enolase (ng/mL); *PCT*, procalcitonin (ng/mL); *pNF-H*, phosphorylated neurofilament heavy chain (pg/mL); *S100β*, calcium-binding protein B (μg/L); *SDH*, subdural hematoma; *tSAH*, traumatic subarachnoid hemorrhage; *T-tau*, total tau (pg/mL); *SAA1*, serum amyloid A1 (ng/mL); *YKL-40*, chitinase-3-like protein 1 (ng/mL).
(Adapted from Carabias, C. S., Gomez, P. A., Panero, I., Eiriz, C., Castaño-León, A. M., Egea, J., … i+12 Neurotraumatology Group Collaborators. (2020). Chitinase-3-like protein 1, serum amyloid A1, C-reactive protein, and procalcitonin are promising biomarkers for intracranial severity assessment of traumatic brain injury: Relationship with Glasgow coma scale and computed tomography volumetry. World Neurosurgery, 134, e120–e143. https://doi.org/10.1016/j.wneu.2019.09.143.)

volume within 18–24 h of symptom onset. In addition, YKL-40 levels reflected not only cerebral infarct extension but also the stroke subtype. Park et al. (2012) found that YKL-40 concentrations tended to be greater in patients with large artery atherosclerosis or atherothrombosis than in patients with small-vessel occlusion or lacunar infarcts. The authors attributed the increased YKL-40 to acute ischemic injury or peri-infarction inflammation rather than to chronic stable atherosclerotic lesions. Remarkably, their study demonstrated that YKL-40 measurements were even capable of detecting the existence of a tiny infarction (Park et al., 2012). Nonetheless, an association between CSF YKL-40 concentrations and the degree and distribution of white matter lesions on CT scans according to the age-related white matter changes (ARWMC) scale was not identified by Hjalmarsson et al. (2014) in patients with cortico-subcortical infarctions. Nevertheless, it should be taken

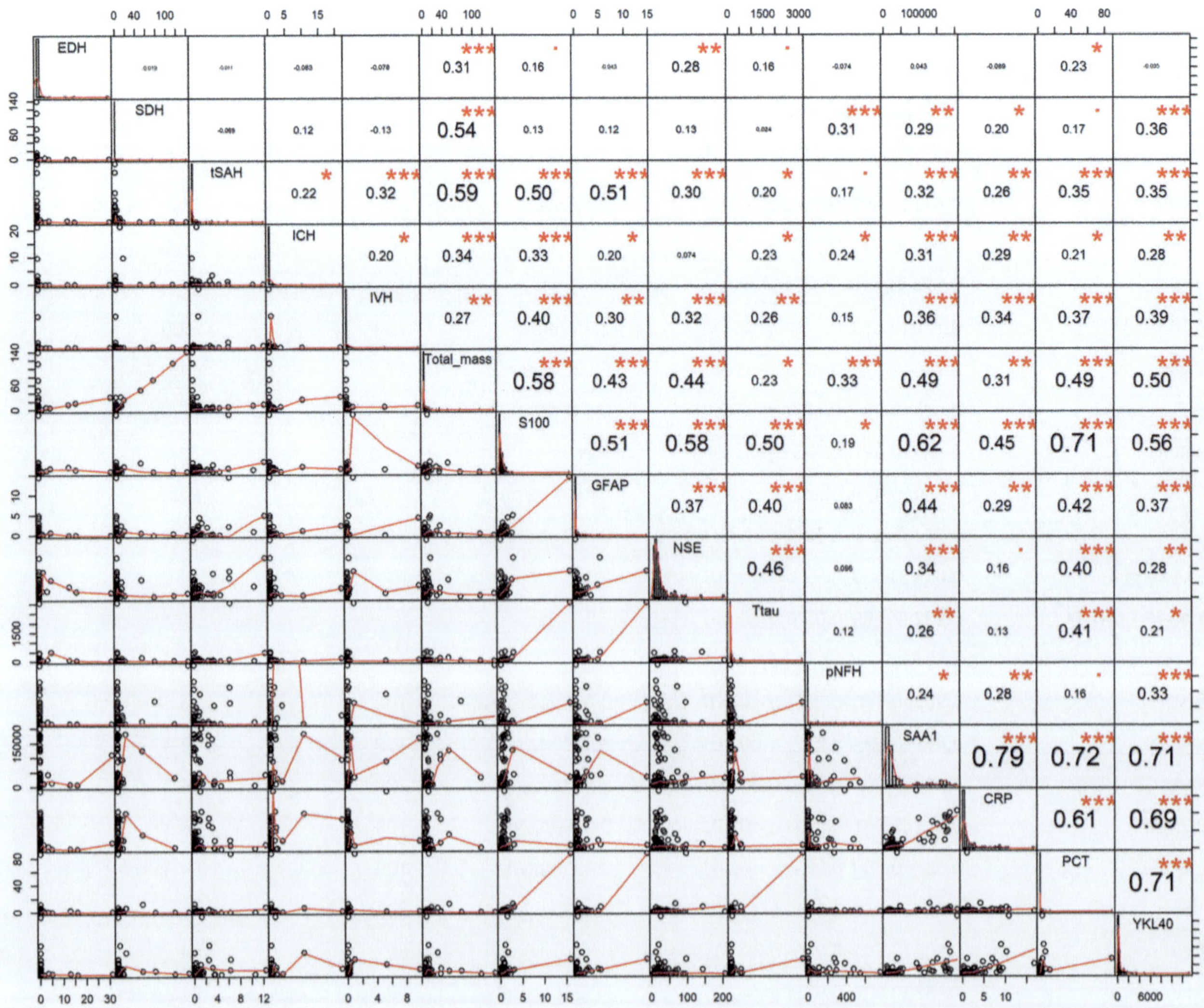

FIG. 5 Chart correlation matrices between serum biomarkers and computer-assisted volumetric measurement of traumatic intracranial hemorrhages on admission CT. The distribution of each variable shown as histograms is exposed on the diagonal line of the graph. Above the diagonal the value of the correlation plus significance level from the Spearman rho correlation test are represented as $***q<0.001$, $**q<0.01$, $*q<0.05$, $\cdot q<0.1$. The greater the strength of the correlation, the larger the font size. Below the diagonal, bivariate scatter plots with a fitted line for each variable are shown. *CRP*, C-reactive protein (mg/dL); *EDH*, epidural hematoma; *GFAP*, glial fibrillary acidic protein (ng/mL); *ICH*, contusion; *IVH*, intra-ventricular hemorrhage; *NSE*, neuron-specific enolase (ng/mL); *PCT*, procalcitonin (ng/mL); *pNF-H*, phosphorylated neurofilament heavy chain (pg/mL); *S100β*, calcium-binding protein B (μg/L); *SDH*, subdural hematoma; *tSAH*, traumatic subarachnoid hemorrhage; *T-tau*, total tau (pg/mL); *SAA1*, serum amyloid A1 (ng/mL); *YKL-40*, chitinase-3-like protein 1 (ng/mL). *(Adapted from Carabias, C. S., Gomez, P. A., Panero, I., Eiriz, C., Castaño-León, A. M., Egea, J., ... i+12 Neurotraumatology Group Collaborators. (2020). Chitinase-3-like protein 1, serum amyloid A1, C-reactive protein, and procalcitonin are promising biomarkers for intracranial severity assessment of traumatic brain injury: Relationship with Glasgow coma scale and computed tomography volumetry. World Neurosurgery, 134, e120–e143. https://doi.org/10.1016/j.wneu.2019.09.143.)*

into account that the receiver operating characteristics for detecting intracranial pathology in CT scans in our series were assessed over a broad spectrum of severities, which could overestimate the discriminatory ability of this marker. Hence, its receiver operating characteristics should be confirmed in an isolated cohort of GCS 13–15 patients.

Classification of extra-axial mass lesions

In our quantitative volumetric study, YKL-40 was the second-best marker for suggesting the presence of extra-axial mass lesions (AUC = 0.749, 0.659–0.839) (Fig. 3, Table 3). In patients with mild and moderate head injury, detection of hematomas, particularly extradural hematoma, is extremely crucial; however, no correlation was observed between the volume of epidural bleeding and YKL-40 levels (Fig. 5). Consequently, the receiver operating characteristics for documenting epidural hematomas were poor (AUC = 0.463, 0.290–0.636). Nevertheless, YKL-40 was the top-performing biomarker for identifying subdural blood collection (AUC = 0.780, 0.674–0.886), outperforming more commonly studied biomarkers, such as S100β, NSE and GFAP. Furthermore, post-traumatic YKL-40 levels were significantly correlated with the volume of blood localized in the subarachnoid space. In fact, YKL-40 exhibited the third-best ability to differentiate the subarachnoid blood clot load (AUC=0.695, 0.596–0.795) (Carabias et al., 2020). Kaynar et al. (2005) were the first to demonstrate that patients with aneurysmal subarachnoid hemorrhage (aSAH) exhibited significantly higher YKL-40 concentrations in both serum and CSF samples than controls with normotensive hydrocephalus or hydrocephalus secondary to aqueductal stenosis. However, Kacira et al. (2008) noticed that symptomatic vasospasm or neurologic outcome had no effect on either CSF or serum YKL-40 levels in 22 patients with an aneurysm rupture. Therefore, they concluded that YKL-40 was not related to the severity of aneurysmal SAH-induced inflammation (Kacira et al., 2008). In light of these findings, the authors suggested that YKL-40 might play a role in a different manner in the intrathecal compartment following aSAH (Kacira et al., 2008; Kaynar et al., 2005). However, a relationship with the amount and distribution of CT-visualized subarachnoid blood clots was not investigated in these studies. Moreover, Bjørnbak et al. (2014) and Brøchner et al. (2015) showed that YKL-40 mRNA expression was particularly prominent in pia-arachnoid tissue, and dense immunoreactivity was observed from the emergence of meningeal tissues, suggesting that YKL-40 is produced by leptomeninges and secreted into the subarachnoid space due to IL13Rα2 receptor-mediated uptake (Brøchner et al., 2015; He et al., 2013; Lee et al., 2016). These findings are quite remarkable, especially in the context of the correlations observed in our study between YKL-40 levels and bleeding located in the extra-axial compartment (Carabias et al., 2020), although how these types of bleeding directly or indirectly affect YKL-40 release is a question that requires additional studies.

Discrimination of intra-axial mass lesions

In contrast, YKL-40 had excellent receiver operating characteristics for identifying intra-axial mass lesions (AUC=0.843, 0.748–0.938) (Fig. 3, Table 3), and combining the YKL-40 Youden index (*J*) (953.19 ng/mL) with *J* phosphorylated neurofilament heavy chain (pNF-H) (59.48 pg/mL) and *J* S100β (0.570 µg/L) yielded the best predictive probability (AUC=0.918, 0.834–1). Consequently, a significant correlation was demonstrated between serum YKL-40 levels and the volume of parenchymal contusion and intra-ventricular extension visible on CT scans (Fig. 5). For classifying intracerebral hematomas, YKL-40 exhibited the same detection ability as CRP (AUC=0.776, 0.636–0.917). The combination of YKL-40 and pNF-H levels further increased the discriminatory capacity for the detection of contusion, showing the best performance after logistic regression (AUC=0.822, 0.685–0.958). However, CRP levels showed mild superiority to YKL-40 for contusion detection after dichotomizing *J* (Carabias et al., 2020). Our findings are in agreement with pre-clinical evidence that YKL-40 expression is increased within pericontusional lesions after experimental brain contusion (Bonneh-Barkay, Zagadailov, et al., 2010). Nonetheless, one of the most important and devastating clinical features of TBI is the spatiotemporal progression or "blossoming" of a hemorrhagic contusion. Therefore, it would be especially relevant to evaluate the ability of YKL-40 for predicting progressive secondary hemorrhage in further studies. Similarly, YKL-40 presented the second-best ability for reflecting intra-ventricular hemorrhage (AUC=0.897, 0.827–0.967), and the combination of *J* YKL-40 (293.49 ng/mL) with *J* S100β (0.630 µg/L) yielded the highest predictive probability (AUC=0.955, 0.919–0.992) (Carabias et al., 2020). In this study, we identified YKL-40 as one of the leading candidate biomarkers with potential utility for clinical decision-making in TBI, as demonstrated by the high AUC. Nonetheless, further research is warranted to confirm the results obtained in our study and to determine whether YKL-40 remains a top-performing biomarker for reflecting the multi-faceted pathophysiology associated with brain injury.

The relationship between YKL-40 and other neurotrauma biomarkers

Additionally, we observed that levels of YKL-40 were mostly associated with levels of acute-phase proteins and S100β, followed by GFAP, pNF-H, NSE and total tau (T-tau) (Carabias et al., 2020) (Figs. 4 and 5). Interestingly, inflammation-associated cytokines are the major stimuli for inducing acute phase changes. Members of the IL-6 family, IL-1, TNF-α, transforming growth factor beta (TGFβ), and interferon gamma (IFNγ), influenced by circulating modulators, such as IL-1Rα, have received the most attention as inducers of acute-phase proteins. Furthermore, C/EBP family members, STAT proteins, the Rel family (typically NF-κB) or Sp1 usually mediates enhanced expression of acute-phase genes in response to these cytokines (Samols, Agrawal, & Kushner, 2001). In addition to the other molecular mechanisms noted above (Bhardwaj et al., 2015; Kawada et al., 2012), it has been pointed out that the transcription factor Sp1 regulates expression of the *CHI3L1* gene (Lee et al., 2011). Previous studies suggested that S100β secretion involves the MAP kinase-mediated pathway and may involve NF-κB. The activation of receptor for advanced glycation end-products (RAGE)-signaling mediates pro-inflammatory cytokine production, leading to the migration of neutrophils, monocytes, and macrophages to the site of injury (Xia, Braunstein, Toomey, Zhong, & Rao, 2018). Furthermore, exposure of primary cultures of cortical neurons to CHI3L1 resulted in significant neurite retraction and reduced neuronal survival, indicating an early effect of CHI3L1 on neuronal functionality and degeneration (Matute-Blanch et al., 2020). Thus, the association patterns observed in our series between YKL-40 and neuroaxonal markers may mirror complex neuronal-glial interactions rather than simply reflecting a correlate of the severity of TBI.

Future directions

Despite the increased evidence on the promising features of CHI3L1, there are still doubts regarding the reliability of any CSF or peripheral blood biomarker as an indicator of brain injury severity. The putative association of common variations in the *CHI3L1* locus with inter-individual YKL-40 levels in biofluids and the prevalence and severity of TBI should be considered in future studies. To date, there are no studies that have characterized the cellular source of CHI3L1 in human post-traumatic brain contusions. Although the brain tissue extraction process could add some degree of unpredictable inflammatory reaction, there is a great need to acquire knowledge on YKL-40 overexpression in surgically evacuated human contusion specimens to determine whether CHI3L1 upregulation is related to time post-injury. The precise roles and mechanisms of CHI3L1 in the pathogenesis of TBI remain controversial and largely unexplored. Nevertheless, the opposing functions currently attributed to this protein open the possibility of a fruitful area for further investigation. Research focused on elucidating the inflammatory pathways regulated by CHI3L1 could yield potential therapeutic strategies for modulating neuroinflammation in this devastating neurological condition.

Applications to other areas of neuroscience

In this book chapter, we discuss the potential contributions of this chitinase-like protein to the pathogenesis of brain trauma. We also summarize the emerging evidence suggesting that CHI3L1 influences the response of the brain in other pathological scenarios, such as large-artery atherosclerotic stroke, small-vessel occlusion or lacunar infarct, aneurysmal subarachnoid hemorrhage and spontaneous basal ganglia hemorrhage (Chen et al., 2017; Hjalmarsson et al., 2014; Jiang et al., 2014; Kacira et al., 2008; Kaynar et al., 2005; Kjaergaard, Bojesen, Johansen, & Nordestgaard, 2010; Park et al., 2012; Rathcke, Thomsen, Linneberg, & Vestergaard, 2012; Xu et al., 2014). Moreover, we discuss the effects of *chi3l1* gene deletion on cerebral ischemia development through pre-clinical models of middle cerebral artery occlusion (Im et al., 2020) or collagenase-induced intracerebral hemorrhage (Lively & Schlichter, 2012). Although stroke and TBI arises from different types of primary insults, similar pathomechanisms participate in the structural and functional abnormalities associated with these cerebral injuries. Mechanisms of injury include excitotoxicity, cortical spreading depression, glial swelling, vascular perturbations, white matter vulnerability, inflammation and cell death mechanisms (Bramlett & Dietrich, 2004). Nevertheless, the inter-connection between neurological injury and CHI3L1 expression is not limited to acute neurological diseases. Previous studies have reported altered expression of CHI3L1 in Alzheimer's disease, multiple sclerosis, amyotrophic lateral sclerosis, lentiviral encephalitis, Pick's disease and schizophrenia, indicating that its expression is a nonspecific phenomenon of neuroinflammation-induced reactive gliosis. Consequently, CHI3L1 could potentially be explored as a biomarker for the neuroinflammatory process in a broad spectrum of diseases (Bonneh-Barkay, Wang, et al., 2010).

Mini-dictionary of terms

Contusion: traumatic lesions on the surface of the brain, more frequently located in the temporal and frontal lobes, which are caused by the brain striking to the skull or a fold of the dura mater. Pathologically they are composed of scattered areas of bleeding in the surface or undersurface of the brain. These areas of bleeding normally are considered nonviable tissue and called contusion core.

Pericontusional area: Area of viable tissue surrounding the contusion core. It could be subjected to progressive damage due to inflammation, oedema or loss of normal brain perfusion.

Concussion: Traumatic brain injury that temporarily affects brain functioning and that could be identified by symptoms such as loss of consciousness, memory loss, impaired thinking or concentration, sleep and mood changes.

Post-concussion syndrome: Sometimes all the different symptoms and signs related to concussion continue for weeks, months or even a year after the traumatic event. This persistence of symptoms constitutes the post-concussion syndrome.

Rivermead Post-Concussion Symptoms Questionnaire: A self-report questionnaire consisting of a compendium of symptoms related to concussion and their severity. It has been designed to assess the severity of post-concussion syndrome.

Youden index: Statistical index that informs on the performance of a diagnostic test. It is defined as Specificity + Sensitivity—1. Its maximal value is used for selecting cut-off points for diagnostic tests.

Extra-axial mass lesions: Lesions in the intracranial compartment but outside the brain, mainly epidural and subdural hematomas.

Intra-axial mass lesions: Intra-parenchymal brain lesions, mainly contusions.

Controlled cortical impact: Experimental traumatic brain injury model that resembles human brain contusions through a pneumatic or electromagnetic device impacting on the surgically exposed cortical dural surface. In cases of more severe injury the subcortical structures are affected by the impact. A broad spectrum of severities can be reproduced by controlling the injury parameters, such as depth of cortical deformation, dwell time, impact velocity or tip size.

Key facts of CHI3L1: Features and applications as a biomarker in TBI

- YKL-40 is a promising biomarker related to the inflammatory response, reactive to different pathological processes such as TBI or different types of stroke.
- Both experimental and clinical evidence suggests that this biomarker has a relation to inflammation in the central nervous system triggered by different pathological mechanisms.
- Though its expression in astrocytes is increased after trauma it may have an immunomodulatory effect as experimental studies show increase lesion burden in knockout models.
- As a clinical biomarker, YKL-40 levels have been related to the clinical and pathological severity of TBI, as well as prognosis in terms of both post-concussion syndrome and long-term outcome in all severities of TBI.
- There is still need for clinical and experimental studies that could increase the knowledge on the role of CHI3L1 in TBI, including its clinical applicability, its role in lesion progression and possible therapeutic interventions related to its role on inflammation.

Summary points

- This chapter focuses on CHI3L1/YKL-40 which is a secreted glycoprotein that belongs to the 18-glycosyl-hydrolase family of proteins but lacks chitinase/hydrolase activity.
- CHI3L1 transcription is associated with reactive astrocytes restricted to the pericontusional region after experimental TBI and is related to acute neuroinflammation process.
- Cytokines released from macrophages that infiltrate the contusional core drive the focal YKL-40 transcription.
- *Chi3l1* homozygous disruption increases the neuroinflammatory response evidenced by the prominent astrocytosis and microgliosis after experimental TBI without changing the ultrastructural character of brain injury lesions.
- CHI3L1 could play a critical role in disease-involved immune cells transmigration across brain barriers.
- Cerebrospinal fluid YKL-40 levels are associated with clinical outcome of severe head injury patients as a reflection of the ongoing acute neuroinflammatory process.
- YKL-40 is a neurochemical aftermath of repetitive mild TBI, and levels are associated with the number of lifetime concussion events and post-concussion syndrome symptom severity.
- The magnitude of CHI3L1 elevation is related to neurologic status and radiologic lesion burden in patients with TBI.

References

Bhardwaj, R., Yester, J. W., Singh, S. K., Biswas, D. D., Surace, M. J., Waters, M. R., … Kordula, T. (2015). RelB/p50 complexes regulate cytokine-induced YKL-40 expression. *The Journal of Immunology*, *194*(6), 2862–2870. https://doi.org/10.4049/jimmunol.1400874.

Bjørnbak, C., Brøchner, C. B., Larsen, L. A., Johansen, J. S., & Møllgård, K. (2014). Brain barriers and a subpopulation of astroglial progenitors of developing human forebrain are immunostained for the glycoprotein YKL-40. *Journal of Histochemistry and Cytochemistry*, *62*(5), 369–388. https://doi.org/10.1369/0022155414528514.

Bonneh-Barkay, D., Bissel, S. J., Kofler, J., Starkey, A., Wang, G., & Wiley, C. A. (2012). Astrocyte and macrophage regulation of YKL-40 expression and cellular response in neuroinflammation. *Brain Pathology*, *22*(4), 530–546. https://doi.org/10.1111/j.1750-3639.2011.00550.x.

Bonneh-Barkay, D., Bissel, S. J., Wang, G., Fish, K. N., Nicholl, G. C. B., Darko, S. W., … Wiley, C. A. (2008). YKL-40, a marker of simian immunodeficiency virus encephalitis, modulates the biological activity of basic fibroblast growth factor. *The American Journal of Pathology*, *173*(1), 130–143. https://doi.org/10.2353/ajpath.2008.080045.

Bonneh-Barkay, D., Wang, G., Starkey, A., Hamilton, R. L., & Wiley, C. A. (2010). In vivo CHI3L1 (YKL-40) expression in astrocytes in acute and chronic neurological diseases. *Journal of Neuroinflammation*, *7*(1), 34. https://doi.org/10.1186/1742-2094-7-34.

Bonneh-Barkay, D., Zagadailov, P., Zou, H., Niyonkuru, C., Figley, M., Starkey, A., … Wagner, A. K. (2010). YKL-40 expression in traumatic brain injury: An initial analysis. *Journal of Neurotrauma*, *27*(7), 1215–1223. https://doi.org/10.1089/neu.2010.1310.

Bramlett, H. M., & Dietrich, W. D. (2004). Pathophysiology of cerebral ischemia and brain trauma: Similarities and differences. *Journal of Cerebral Blood Flow & Metabolism*, *24*(2), 133–150. https://doi.org/10.1097/01.WCB.0000111614.19196.04.

Brøchner, C. B., Holst, C. B., & Møllgård, K. (2015). Outer brain barriers in rat and human development. *Frontiers in Neuroscience*, *9*(75). https://doi.org/10.3389/fnins.2015.00075.

Carabias, C. S., Gomez, P. A., Panero, I., Eiriz, C., Castaño-León, A. M., Egea, J., … i+12 Neurotraumatology Group Collaborators. (2020). Chitinase-3-like protein 1, serum amyloid A1, C-reactive protein, and procalcitonin are promising biomarkers for intracranial severity assessment of traumatic brain injury: Relationship with Glasgow coma scale and computed tomography volumetry. *World Neurosurgery*, *134*, e120–e143. https://doi.org/10.1016/j.wneu.2019.09.143.

Chen, X.-L., Li, Q., Huang, W.-S., Lin, Y.-S., Xue, J., Wang, B., … Shao, B. (2017). Serum YKL-40, a prognostic marker in patients with large-artery atherosclerotic stroke. *Acta Neurologica Scandinavica*, *136*(2), 97–102. https://doi.org/10.1111/ane.12688.

Fusetti, F., Pijning, T., Kalk, K. H., Bos, E., & Dijkstra, B. W. (2003). Crystal structure and carbohydrate-binding properties of the human cartilage glycoprotein-39. *Journal of Biological Chemistry*, *278*(39), 37753–37760. https://doi.org/10.1074/jbc.M303137200.

Gan, Z. S., Stein, S. C., Swanson, R., Guan, S., Garcia, L., Mehta, D., & Smith, D. H. (2019). Blood biomarkers for traumatic brain injury: A quantitative assessment of diagnostic and prognostic accuracy. *Frontiers in Neurology*, *10*, 446. https://doi.org/10.3389/fneur.2019.00446.

Hall, E. D., Bryant, Y. D., Cho, W., & Sullivan, P. G. (2008). Evolution of post-traumatic neurodegeneration after controlled cortical impact traumatic brain injury in mice and rats as assessed by the De Olmos silver and Fluorojade staining methods. *Journal of Neurotrauma*, *25*(3), 235–247. https://doi.org/10.1089/neu.2007.0383.

He, C. H., Lee, C. G., Dela Cruz, C. S., Lee, C.-M., Zhou, Y., Ahangari, F., … Elias, J. A. (2013). Chitinase 3-like 1 regulates cellular and tissue responses via IL-13 receptor α2. *Cell Reports*, *4*(4), 830–841. https://doi.org/10.1016/j.celrep.2013.07.032.

Hjalmarsson, C., Bjerke, M., Andersson, B., Blennow, K., Zetterberg, H., Åberg, N. D., … Wallin, A. (2014). Neuronal and glia-related biomarkers in cerebrospinal fluid of patients with acute ischemic stroke. *Journal of Central Nervous System Disease*, *6*, 51–58. https://doi.org/10.4137/JCNSD.S13821.

Im, J. H., Yeo, I. J., Park, P. H., Choi, D. Y., Han, S.-B., Yun, J., & Hong, J. T. (2020). Deletion of Chitinase-3-like 1 accelerates stroke development through enhancement of neuroinflammation by STAT6-dependent M2 microglial inactivation in Chitinase-3-like 1 knockout mice. *Experimental Neurology*, *323*. https://doi.org/10.1016/j.expneurol.2019.113082, 113082.

Jiang, Y. X., Zhang, G. H., Wang, Z. M., & Yang, H. (2014). Serum YKL-40 levels as a prognostic factor in patients with intracerebral hemorrhage. *Clinical Biochemistry*, *47*(18), 302–306. https://doi.org/10.1016/j.clinbiochem.2014.09.015.

Johansen, J. S. (2006). Studies on serum YKL-40 as a biomarker in diseases with inflammation, tissue remodelling, fibroses and cancer. *Danish Medical Bulletin*, *53*(2), 172–209. Retrieved from http://www.ncbi.nlm.nih.gov/pubmed/17087877.

Johansen, J. S., Williamson, M. K., Rice, J. S., & Price, P. A. (1992). Identification of proteins secreted by human osteoblastic cells in culture. *Journal of Bone and Mineral Research*, *7*(5), 501–512. https://doi.org/10.1002/jbmr.5650070506.

Kacira, T., Hanimoglu, H., Kucur, M., Sanus, G. Z., Kafadar, A. M., Tanriverdi, T., & Kaynar, M. Y. (2008). Elevated cerebrospinal fluid and serum YKL-40 levels are not associated with symptomatic vasospasm in patients with aneurysmal subarachnoid haemorrhage. *Journal of Clinical Neuroscience*, *15*(9), 1011–1016. https://doi.org/10.1016/j.jocn.2006.11.014.

Kawada, M., Seno, H., Kanda, K., Nakanishi, Y., Akitake, R., Komekado, H., … Chiba, T. (2012). Chitinase 3-like 1 promotes macrophage recruitment and angiogenesis in colorectal cancer. *Oncogene*, *31*(26), 3111–3123. https://doi.org/10.1038/onc.2011.498.

Kaynar, M. Y., Tanriverdi, T., Kafadar, A. M., Kacira, T., Yurdakul, F., Uzun, H., & Gumustas, K. (2005). YKL-40 levels in the cerebrospinal fluid and serum of patients with aneurysmal subarachnoid hemorrhage: Preliminary results. *Journal of Clinical Neuroscience*, *12*(7), 754–757. https://doi.org/10.1016/j.jocn.2005.06.001.

Kazakova, M., Pavlov, G., Dikov, D., Simitchiev, K., Dichev, V., Stefanov, C., & Sarafian, V. (2018). Protein YKL-40 in cerebrospinal fluid in traumatic brain injury. *Acta Morphologica et Anthropologica*, *25*, 61–66.

Kjaergaard, A. D., Bojesen, S. E., Johansen, J. S., & Nordestgaard, B. G. (2010). Elevated plasma YKL-40 levels and ischemic stroke in the general population. *Annals of Neurology*, *68*(5), 672–680. https://doi.org/10.1002/ana.22220.

Ku, B. M., Lee, Y. K., Ryu, J., Jeong, J. Y., Choi, J., Eun, K. M., … Kang, S. S. (2011). CHI3L1 (YKL-40) is expressed in human gliomas and regulates the invasion, growth and survival of glioma cells. *International Journal of Cancer*, *128*(6), 1316–1326. https://doi.org/10.1002/ijc.25466.

Laird, M. D., Vender, J. R., & Dhandapani, K. M. (2008). Opposing roles for reactive astrocytes following traumatic brain injury. *Neurosignals*, *16*(2–3), 154–164. https://doi.org/10.1159/000111560.

Lee, C. G., Da Silva, C. A., Dela Cruz, C. S., Ahangari, F., Ma, B., Kang, M.-J., … Elias, J. A. (2011). Role of chitin and chitinase/chitinase-like proteins in inflammation, tissue remodeling, and injury. *Annual Review of Physiology*, *73*(1), 479–501. https://doi.org/10.1146/annurev-physiol-012110-142250.

Lee, C.-M., He, C. H., Nour, A. M., Zhou, Y., Ma, B., Park, J. W., … Elias, J. A. (2016). IL-13Rα2 uses TMEM219 in chitinase 3-like-1-induced signalling and effector responses. *Nature Communications*, *7*(1), 12752. https://doi.org/10.1038/ncomms12752.

Lively, S., & Schlichter, L. C. (2012). Age-related comparisons of evolution of the inflammatory response after intracerebral hemorrhage in rats. *Translational Stroke Research*, *3*(S1), 132–146. https://doi.org/10.1007/s12975-012-0151-3.

Majdan, M., Steyerberg, E. W., Nieboer, D., Mauritz, W., Rusnak, M., & Lingsma, H. F. (2015). Glasgow coma scale motor score and pupillary reaction to predict six-month mortality in patients with traumatic brain injury: Comparison of field and admission assessment. *Journal of Neurotrauma*, *32*(2), 101–108. https://doi.org/10.1089/neu.2014.3438.

Malik, S. Z., Motamedi, S., Royo, N. C., LeBold, D., & Watson, D. J. (2011). Identification of potentially neuroprotective genes upregulated by neurotrophin treatment of CA3 neurons in the injured brain. *Journal of Neurotrauma*, *28*(3), 415–430. https://doi.org/10.1089/neu.2010.1487.

Matute-Blanch, C., Calvo-Barreiro, L., Carballo-Carbajal, I., Gonzalo, R., Sanchez, A., Vila, M., … Comabella, M. (2020). Chitinase 3-like 1 is neurotoxic in primary cultured neurons. *Scientific Reports*, *10*(1), 7118. https://doi.org/10.1038/s41598-020-64093-2.

Michelsen, A. E., Rathcke, C. N., Skjelland, M., Holm, S., Ranheim, T., Krohg-Sørensen, K., … Halvorsen, B. (2010). Increased YKL-40 expression in patients with carotid atherosclerosis. *Atherosclerosis*, *211*(2), 589–595. https://doi.org/10.1016/j.atherosclerosis.2010.02.035.

Mygind, N. D., Iversen, K., Køber, L., Goetze, J. P., Nielsen, H., Boesgaard, S., … Kastrup, J. (2013). The inflammatory biomarker YKL-40 at admission is a strong predictor of overall mortality. *Journal of Internal Medicine*, *273*(2), 205–216. https://doi.org/10.1111/joim.12006.

Park, H. Y., Jun, C. D., Jeon, S. J., Choi, S. S., Kim, H. R., Choi, D. B., … Park, D. S. (2012). Serum YKL-40 levels correlate with infarct volume, stroke severity, and functional outcome in acute ischemic stroke patients. *PLoS One*, *7*(12). https://doi.org/10.1371/journal.pone.0051722, e51722.

Pun, P. B. L., Kan, E. M., Salim, A., Li, Z., Ng, K. C., Moochhala, S. M., … Lu, J. (2011). Low level primary blast injury in rodent brain. *Frontiers in Neurology*, *2*, 19. https://doi.org/10.3389/fneur.2011.00019.

Rathcke, C. N., Thomsen, S. B., Linneberg, A., & Vestergaard, H. (2012). Variations of CHI3L1, levels of the encoded glycoprotein YKL-40 and prediction of fatal and non-fatal ischemic stroke. *PLoS One*, *7*(8). https://doi.org/10.1371/journal.pone.0043498, e43498.

Recklies, A. D., Ling, H., White, C., & Bernier, S. M. (2005). Inflammatory cytokines induce production of CHI3L1 by articular chondrocytes. *Journal of Biological Chemistry*, *280*(50), 41213–41221. https://doi.org/10.1074/jbc.M510146200.

Samols, D., Agrawal, A., & Kushner, I. (2001). Acute-phase proteins. In *Encyclopedia of life sciences*. Chichester, UK: John Wiley & Sons, Ltd. https://doi.org/10.1006/rwcy.2002.0213.

Shahim, P., Tegner, Y., Marklund, N., Höglund, K., Portelius, E., Brody, D. L., … Zetterberg, H. (2017). Astroglial activation and altered amyloid metabolism in human repetitive concussion. *Neurology*, *88*(15), 1400–1407. https://doi.org/10.1212/WNL.0000000000003816.

Szmydynger-Chodobska, J., Shan, R., Thomasian, N., & Chodobski, A. (2016). The involvement of pial microvessels in leukocyte invasion after mild traumatic brain injury. *PLoS One*, *11*(12). https://doi.org/10.1371/journal.pone.0167677, e0167677.

Wiley, C. A., Bissel, S. J., Lesniak, A., Dixon, C. E., Franks, J., Beer Stolz, D., … Murdoch, G. (2016). Ultrastructure of diaschisis lesions after traumatic brain injury. *Journal of Neurotrauma*, *33*(20), 1866–1882. https://doi.org/10.1089/neu.2015.4272.

Wiley, C. A., Bonneh-Barkay, D., Dixon, C. E., Lesniak, A., Wang, G., Bissel, S. J., & Kochanek, P. M. (2015). Role for mammalian chitinase 3-like protein 1 in traumatic brain injury. *Neuropathology*, *35*(2), 95–106. https://doi.org/10.1111/neup.12158.

Xia, C., Braunstein, Z., Toomey, A. C., Zhong, J., & Rao, X. (2018). S100 proteins as an important regulator of macrophage inflammation. *Frontiers in Immunology*, *8*, 1908. https://doi.org/10.3389/fimmu.2017.01908.

Xu, X., Ma, H., Xu, J., Huang, H., Wu, X., Xiong, Y., & Huang, F. (2014). Elevation in circulating YKL-40 concentration in patients with cerebrovascular disease. *Bosnian Journal of Basic Medical Sciences*, *14*(3), 120. https://doi.org/10.17305/bjbms.2014.3.42.

Zhao, T., Su, Z., Li, Y., Zhang, X., & You, Q. (2020). Chitinase-3 like-protein-1 function and its role in diseases. *Signal Transduction and Targeted Therapy*, *5*(1), 201. https://doi.org/10.1038/s41392-020-00303-7.

Chapter 21

microRNA biomarkers in traumatic brain injury: A new narrative

Steven D. Hicks
Department of Pediatrics, Penn State College of Medicine, Hershey, PA, United States

List of abbreviations

CTE chronic traumatic encephalopathy
miRISC miRNA-induced silencing complex
miRNA micro-ribonucleic acid
mTBI mild traumatic brain injury
ncRNA non-coding ribonucleic acid
PPCS persistent post-concussion symptoms
qPCR quantitative polymerase chain reaction
RNAseq RNA sequencing
TBI traumatic brain injury

Introduction

Rates of traumatic brain injury (TBI) have increased steadily in recent decades, and public awareness surrounding long-term effects of TBI have risen (Coronado, McGuire, Faul, Sugerman, & Pearson, 2012). Both mild TBI (mTBI) and severe TBI can lead to prolonged symptoms and functional deficits. Nearly 25% of individuals with TBI suffer from persistent post-concussion symptoms (PPCS) (Zemek et al., 2016). Together, the acute and chronic effects of TBI have a staggering economic impact –$76.5 billion in lifetime medical costs (Coronado et al., 2012). There is evidence that early, accurate diagnosis, and prognosis can improve outcomes for individuals with TBI (Levin & Diaz-Arrastia, 2015). Unfortunately, many clinicians report that the clinical tools available for identifying mTBI, or predicting the long-term functional impacts, are insufficient (Zonfrillo et al., 2012). Studies have shown that a significant proportion of mTBI diagnoses are missed (Rowe et al., 2018). This may result because neurocognitive assessments are under-utilized, or because they can be manipulated by individuals seeking to delay/expedite return to activities (Higgins, Denney, & Maerlender, 2017). Recent mTBI guidelines note the lack of tools for providing accurate, individualized prognoses following injury (Lumba-Brown et al., 2018). The addition of a rapid, objective, biologic marker for TBI could profoundly impact the way clinicians diagnose and manage TBI.

Characteristics of the ideal biomarker

The ideal TBI biomarker should be sensitive enough to detect mTBI, but capable of discerning mTBI from sub-concussive head impacts. It should convey information about the severity of injury, ideally displaying a direct relationship with subjective symptoms or functional impairment (Mannix et al., 2020). A biomarker capable of providing prognostic information about functional status (severe TBI) or symptom duration (mTBI) would also fill a crucial gap in care (Jeter et al., 2013). Currently, clinicians have no accurate tools to predict recovery trajectory for patients and their families (Lumba-Brown et al., 2018). During the recovery period, there is also a paucity of assessments that can guide treatment decisions. A biomarker that tracks brain recovery after TBI might allow clinicians to quickly intervene when patients deviate from the expected course. Such a biomarker could support the development and testing of novel therapeutic approaches. For example, a marker that accurately predicts symptom duration would allow researchers to stratify patients into groups, and accurately assess the efficacy of interventions aimed at expediting recovery (Jeromin, Maas, Buki, & Mondello, 2015). A biomarker

Diagnosis and Treatment of Traumatic Brain Injury. https://doi.org/10.1016/B978-0-12-823347-4.00018-X

that "normalizes" with brain recovery would provide an objective tool aiding safe return to activities. This could reduce sandbagging (wherein athletes manipulate functional assessments to expedite return-to-play), and malingering (wherein individuals exaggerate subjective symptom reports to delay return-to-work or school) (Higgins et al., 2017).

For any biomarker to obtain widespread clinical utility, it must be rapidly assessed. On the side-line, or in the emergency department, clinicians require answers within minutes. A biomarker is unlikely to be clinically adopted if it takes hours or days to measure (Mannix et al., 2020). Further, biomarkers that can be rapidly collected, but rely on specialist interpretation may not be suitable outside of hospital settings. On the playing field, or the battlefield, biomarker technology should be portable. Medics and athletic trainers are most likely to adopt new technology that is accessible at the point of injury and provides real-time answers. Ensuring that biomarker technology is cost-effective will allow adoption in resource-poor countries that experience high rates of TBI (Jeter et al., 2013).

Unique properties of micro-ribonucleic acids (miRNAs)

Micro-ribonucleic acids (miRNAs) may provide a novel, molecular window into the pathophysiology of TBI (Atif & Hicks, 2019). miRNAs are a class of non-coding RNA (ncRNA). They do not code for proteins. They block the translation of messenger RNAs (mRNAs). Precursor miRNAs are double-stranded hairpin structures that arise from an individual's DNA (Mott & Mohr, 2015). They are exported outside the nucleus by Exportin 5, and then cleaved into single-stranded molecules by Dicer. The resulting mature miRNA is capable of binding specific mRNAs within the miRNA-induced silencing complex. This complex prevents gene expression and promotes mRNA degradation. Thus, miRNAs effectively act as "off" switches (Fig. 1). The mRNA targets of each miRNA are determined by a seed-sequence—a set of 19 to 22 base-pairs that anneal the miRNA to complementary mRNA codes. A single miRNA may bind hundreds of different mRNAs, depending on the distinctiveness of its seed sequence.

Because miRNAs are rapidly transcribed, they allow humans to respond to sudden changes in the external environment, such as a TBI (Redell, Liu, & Dash, 2009). miRNAs are functionally conserved. Many miRNAs found in human cells have been found to target similar physiologic processes in rodents, or even invertebrates (Mott & Mohr, 2015). miRNAs are found in cell types throughout the human body; however, they are highly concentrated within the central nervous system. Studies have demonstrated that miRNA expression is critical for brain development and plasticity. Inhibition of miRNA function dysregulates neuronal growth and repair, and disrupts synaptic transmission (Kapsimali et al., 2007). Perhaps the most unique feature of miRNAs that lends itself to biomarker development is that miRNAs can be used as extracellular signaling molecules (Fig. 2). Neurons can encapsulate miRNAs within protective vesicles (e.g., exosomes), allowing them to travel through the extra-cellular space, dock at distant cells, and impact mRNA expression (Shi, Yang, Malichewe, Li,

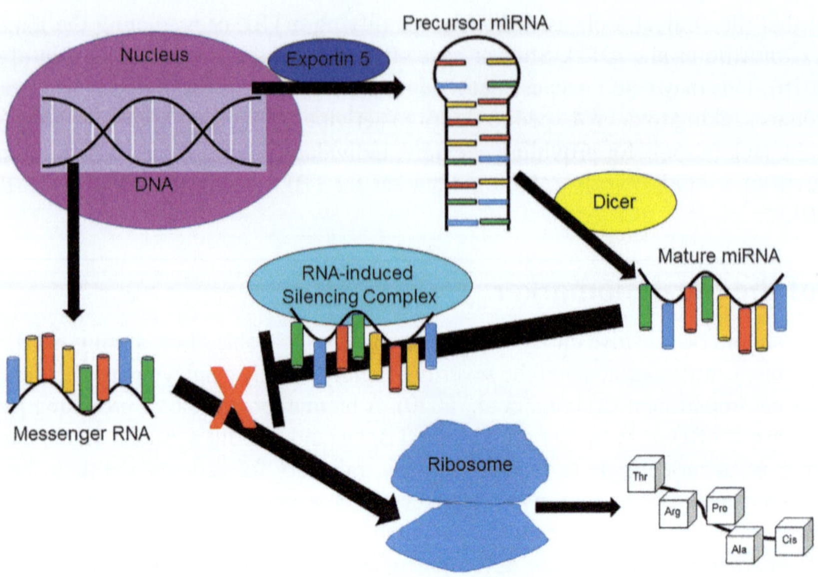

FIG. 1 Micro-ribonucleic acid physiology. Like messenger RNA, precursor micro-ribonucleic acids (miRNAs) are transcribed from DNA within the nucleus. Exportin 5 delivers precursor miRNA outside the nucleus, where it is processed into a mature, single-stranded form by Dicer. Mature miRNA can bind complementary messenger RNA within the RNA-induced silencing complex, inhibiting ribosome-mediated translation into functional proteins.

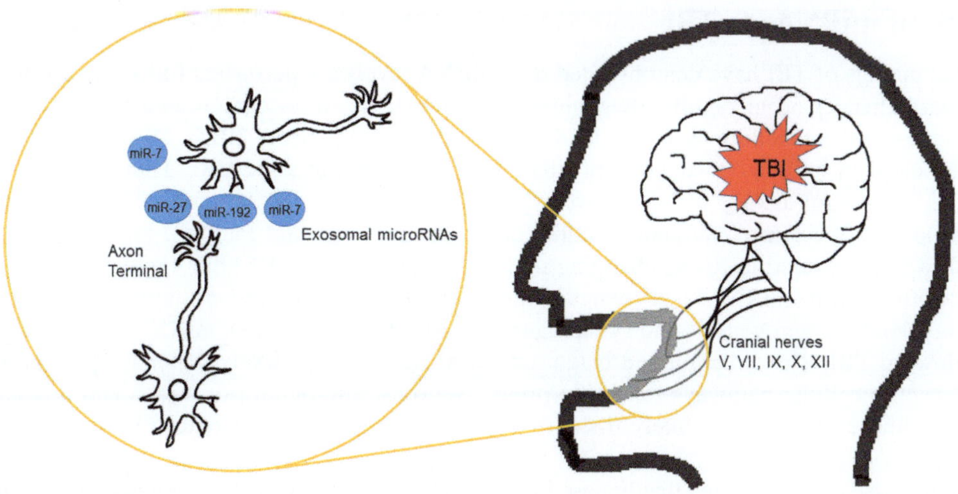

FIG. 2 Exosomal micro-ribonucleic acid signaling in traumatic brain injury. Following traumatic brain injury (TBI), neurons and glia up- or down-regulate micro-ribonucleic acid (miRNA) expression to swiftly regulate translation of proteins critical to oxidative stress, cell death, or structural repair. Neurons can export miRNAs within exosomes to coordinate intracellular gene expression. Measurement of exosomal miRNA levels in saliva provides a non-invasive technique for measuring neuronal miRNA expression following TBI.

& Guo, 2020). Therefore, measurement of miRNAs within extracellular fluids may provide insight into central nervous system injury, even in the absence of cell death or blood–brain barrier disruption (Jin, Wu, Wang, & Li, 2013).

Measurement of miRNAs

The small size and vesicular protection of extracellular miRNAs allows them to evade enzymatic degradation, and confers concentration stability (Cheng, Sharples, Scicluna, & Hill, 2014; Shi et al., 2020). Measurement of miRNAs can be achieved through a number of modalities (Liu, Tong, Li, Wang, & Chen, 2020). High throughput RNA sequencing (RNAseq) is the most comprehensive approach, allowing quantification of all known miRNAs, as well as other families of ncRNAs (e.g., small nucleolar RNAs, piwi-interacting RNAs) within the same experiment. With adequate sequencing depth, this approach allows interrogation of molecules present in pico-molar concentrations. However, the cost of RNAseq can be prohibitive in large studies, and the time required for sample processing impedes clinical translation. Variability in library preparation techniques can also make it difficult to replicate RNAseq results from site-to-site. For these reasons, RNAseq is perhaps best reserved for initial biomarker exploration (Ari & Arikan, 2016).

Quantitative polymerase chain reaction (qPCR) is often used to validate biomarkers identified through initial high throughput experiments. qPCR provides a wide dynamic range with low quantification limits, and generally demonstrates good test–retest reliability (de Ronde, Ruijter, Moerland, Creemers, & Pinto-Sietsma, 2018). When analyzing <30 miRNAs, this approach is more cost-effective than RNAseq. However, qPCR has limited portability, and still requires several hours to go from "sample in" to "answer out." In addition, it requires comparison against a stabile reference gene—which can be difficult to define across the spectrum of TBI. An emerging technology for miRNA quantification is immunoplex assay (Godoy et al., 2019). This approach uses sequence-specific probes to capture miRNAs of interest, followed by enzymatic ligation and biotin labeling, prior to addition of a fluorescent reporter and measurement with flow cytometry. As many as 65 miRNAs can be quantified in one experiment. Immunoplex assays do not reduce the time required to quantify miRNAs (relative to qPCR), but they do allow direct miRNA measurement from small quantities of crude biofluids. They also deliver comparable reliability, albeit at a slightly higher price point.

There are several novel approaches that allow exploration of the cellular origin of miRNA biomarkers. Such approaches can validate the premise that miRNAs measured in extracellular fluids provide physiologic information about the brain (Jin et al., 2013). Single-cell RNAseq can be used to characterize the entire transcriptome of cells within a sample, and then compare patterns in transcriptional profiles across cell types (Faridani et al., 2016). This approach ensures that miRNAs highly expressed within epithelial cells, or white blood cells are not selected as biomarker candidates. Similarly, isolation and characterization of cell-specific exosomes can be used to focus on cargo originating from neurons, glia, or other cells relevant to brain plasticity and repair (Yoo et al., 2018). This is accomplished by targeting neuron- or glia-specific antigens on the exosome surface, followed by high throughput RNAseq. Although single-cell RNAseq and exosome-specific sequencing are costly and time-consuming, they can provide biologic rationale for the selection of specific miRNAs within a biomarker panel.

The response of miRNAs to TBI

Studies in animal models of TBI have demonstrated that miRNA levels are perturbed following cortical impacts (Ko et al., 2019). These changes occur rapidly after injury and can extend across numerous brain regions, including the cortex, hippocampus, and cerebellum (Schober, Ondruschka, Dreßler, & Abend, 2015). Dysregulated miRNA expression has been reported in both mTBI and severe TBI (Hicks, Johnson, et al., 2018). Though the specific miRNAs implicated in the TBI response vary widely between studies (based upon the severity of impact, timing of sample collection, method of RNA extraction/quantification, and other experimental factors), the function of TBI-related miRNAs is relatively consistent; miRNAs changing immediately following TBI often regulate pathways involved in ischemia and apoptosis, while miRNAs that change in the sub-acute or chronic phases of recovery have been shown to target mRNAs involved in neurogenesis and synaptogenesis (Atif & Hicks, 2019). miRNA biomarkers may increase or decrease following TBI. The direction and magnitude of miRNA change likely depends upon the need to up- or down-regulate specific cellular pathways following injury. Neurons, microglia, astrocytes, and oligodendrocytes all produce miRNAs, and each cell type likely displays a unique response to TBI (Ziu, Fletcher, Rana, Jimenez, & Digicaylioglu, 2011).

Studies in humans have demonstrated that brain-related miRNAs can be detected in peripheral biofluids (Lafourcade, Ramírez, Luarte, Fernández, & Wyneken, 2016). Further, changes in peripheral miRNA profiles mirror changes seen in cerebrospinal fluid following severe TBI (Hicks, Johnson, et al., 2018). In both mild and severe TBI, peripheral miRNA changes can be seen within hours of injury. The miRNA response to TBI can be detected in both blood and saliva; however, the salivary miRNA response occurs more rapidly (LaRocca et al., 2019). In humans, peripheral miRNA perturbations evolve across the acute, sub-acute, and chronic post-injury periods (LaRocca et al., 2019; Yang et al., 2016). Altered levels of miRNAs are reliably detected up to 2 weeks following mTBI (Johnson et al., 2018), and have been reported in blood as long as 5 years after severe TBI (Taheri et al., 2016). The evolution of miRNA expression following TBI may parallel functional improvements in neurocognitive performance (Di Pietro et al., 2018; Papa et al., 2019) and symptom severity (Hicks, Johnson, et al., 2018; Johnson et al., 2018). It remains unknown how longitudinal miRNA patterns may relate to structural abnormalities on neuroimaging in patients with severe TBI.

There are clear physiologic differences between mTBI and severe TBI. However, a subset of miRNAs display consistent changes across the spectrum of severity (Bhomia, Balakathiresan, Wang, Papa, & Maheshwari, 2016; Hicks, Johnson, et al., 2018). These miRNAs may target physiologic pathways that are common to all TBI, such as mechanical micro-trauma or oxidative stress. The ability of miRNAs to regulate a multitude of pathways makes them a dynamic class of biomarkers (Cheng et al., 2014; Jin et al., 2013). For example, one set of miRNAs may orchestrate the initial response to injury and cellular stress, while another set of miRNAs may influence mRNAs involved in plasticity and repair (Atif & Hicks, 2019). Coupling critical miRNAs from the acute and chronic phases could provide biologic support for the diagnosis, prognosis, and clinical management of TBI.

Use of miRNAs for mTBI diagnosis

A number of studies have explored the utility of miRNA levels for aiding detection of TBI (Table 1). This clinical application is most useful in mTBI, where structural changes are difficult to detect on neuroimaging, and the diagnosis relies predominantly on subjective symptom reports. In mTBI, miRNA levels may eventually serve as an adjunct to semi-objective measures such as neurocognitive testing (Hicks, Onks, et al., 2020). Levels of miRNAs in both blood and saliva have demonstrated utility for mTBI diagnosis. Several studies have shown that the diagnostic accuracy of saliva miRNAs compares favorably with blood-based protein biomarkers (Di Pietro et al., 2018; LaRocca et al., 2019).

The largest diagnostic study of miRNA biomarkers to date, compared saliva miRNA levels between 251 individuals with mTBI and 287 controls with mTBI-like symptomology (i.e., exercise-related fatigue, orthopedic injury, attention difficulty, anxiety/depression). In a naïve test set, a panel of nine ncRNAs identified mTBI-status with similar accuracy to balance and neurocognitive testing, and increased diagnostic accuracy when added to standardized symptom scales (Hicks, Onks, et al., 2020). Despite these promising results, important questions remain. Can miRNA levels discern mTBI from sub-concussive head impacts? Can miRNA profiles delineate the mechanism of TBI—such as abusive head trauma? Do specific miRNAs reflect the underlying pathology of TBI, or are they bi-products of the resulting symptomology (e.g., headache, dizziness, confusion)? Large prospective studies designed to answer these important questions will be necessary to advance miRNA technology from bench to bedside.

TABLE 1 Clinical applications for micro-ribonucleic acid biomarkers.

Clinical application	Current tools	Studies assessing miRNA utility (sample size)
Diagnosis	Validated symptom scales +/− neurocognitive testing	Hicks, Onks, et al. (2020) (*n*=538) Di Pietro et al. (2017) (*n*=135) Hicks, Johnson, et al. (2018) (*n*=89) Davies et al. (2019) (*n*=56) Redell, Moore, Ward III, Hergenroeder, and Dash (2010) (*n*=47) Di Pietro et al. (2018) (*n*=42) Bhomia et al. (2016) (*n*=31) Zhangjie et al. (2017) (*n*=23)
Prognosis	5P prediction tool (Zemek et al., 2016)	Hicks, Olympia et al. (2020) (*n*=341) Johnson et al. (2018) (*n*=52) Ko et al. (2019) (*n*=36) Mitra et al. (2017) (*n*=35)
Clinical decision Making (i.e., return to activities, when to image, etc.)	Validated symptom scales +/− neurocognitive and balance testing	Yang et al. (2016) (*n*=114) Papa et al. (2019) (*n*=53) LaRocca et al. (2019) (*n*=50) You et al. (2016) (*n*=47) Taheri et al. (2016) (*n*=38) Polito et al. (2020) (*n*=20)

The table describes the clinical scenarios where specific micro-ribonucleic acids may be used to aid traumatic brain injury assessment.

Predicting TBI outcomes with miRNAs

Perhaps a greater unmet clinical need, which miRNAs are poised to address, is the lack of tools for predicting medical outcomes for patients with TBI. Individuals with severe TBI face significant morbidity and mortality (Rondina et al., 2005). An objective biologic tool capable of predicting functional outcomes could provide reassurance for these patients and their families, or at the very least, set realistic expectations regarding the duration and course of therapy (Bond, Draeger, Mandleco, & Donnelly, 2003). To date, few studies have investigated the potential of miRNAs to fill this gap (Taheri et al., 2016; Yang et al., 2016). However, studies which describe longitudinal "recovery" of miRNA concentrations toward baseline levels in parallel with brain function (Di Pietro et al., 2018; Hicks, Johnson, et al., 2018; Polito et al., 2020), suggest that tracking miRNA dynamics could portend functional outcomes in severe TBI.

Even in mTBI, a surprising proportion of individuals are at risk for PPCS (Zemek et al., 2016). The ability to gauge PPCS risk at the time of initial diagnosis, would provide an opportunity for earlier pharmacologic intervention, and open the door for clinical trials aimed at reducing symptom duration. Several studies have examined the relationship between miRNA expression and PPCS-risk (Johnson et al., 2018; Mitra et al., 2017). These pilot studies suggest that a miRNA-based approach may perform favorably, when compared to initial symptom burden, or a clinical risk score incorporating clinical variables. Serial measurement of miRNAs across the acute and sub-acute periods may provide more granular information about the trajectory of individual recovery (LaRocca et al., 2019). It will be important that future work in this area account for the social and demographic features that may influence concussion recovery (Zemek et al., 2016). It is likely that a combination of miRNA measures and clinical variables will yield the most accurate toolset.

Emerging evidence suggests that recurrent episodes of mTBI increase risk for chronic effects, such as postural instability, neurocognitive difficulty, or chronic traumatic encephalopathy (CTE). Currently, there are no biologic tests to confirm the diagnosis of CTE until the post-mortem period, and there are no tests to predict which concussed individuals are at risk of CTE (Saulle & Greenwald, 2012). The epi-transcriptional nature of miRNAs makes them well-suited as a diagnostic or prognostic marker for CTE. miRNAs regulate interactions between the genetics that confer CTE susceptibility (You et al., 2016), and the environmental impacts that lead to chronic effects of mTBI (Devoto et al., 2020; Ghai et al., 2020). Several miRNAs target transcription of tau-related proteins that characterize CTE pathology (Hébert, Sergeant, & Buée, 2012). Other miRNAs are critical to the neurometabolic cascade that characterize CTE. Because miRNAs respond to sub-concussive head impacts (LaRocca et al., 2019), they are capable of reflecting sub-concussive trauma that may raise CTE risk in competitive athletes (Alvia, 2020; Papa et al., 2019). Alterations in peripheral miRNA levels have been described in former professional football athletes with increased CTE risk. Examination of these same miRNAs among 310 young adult athletes, identified two miRNAs that were associated with the number of previous concussions (Hicks, Olympia, et al., 2020).

Clinical management of mTBI

Current guidelines recommend that youths with mTBI restrict physical and cognitive activity during the first several days after mTBI, before gradual resumption of non-contact activities (Lumba-Brown et al., 2018). Return-to-school protocols should be customized based on the severity and duration of symptoms. Unfortunately, these guidelines rely on accurate report of subjective symptoms. Such an approach can be derailed by adolescents wishing to expedite return to athletics (Higgins et al., 2017). Subjective reports from young children can also be inaccurate (Patsimas et al., 2020). A biomarker that responds acutely to mTBI and returns to baseline alongside subjective and functional symptoms could aid clinical management (Jeter et al., 2013; Davies et al., 2019). Longitudinal studies of peripheral miRNA levels have identified several promising candidates. These studies suggest that miRNAs respond dynamically in the days and weeks following mTBI, and that miRNA levels may reflect subjective symptom severity (Hicks, Johnson, et al., 2018; Zhangjie et al., 2017), as well as reaction time performance (Di Pietro et al., 2018; LaRocca et al., 2019; Papa et al., 2019).

The possibility that specific miRNAs may correspond with specific TBI symptoms (e.g., headache, dizziness, fatigue) holds substantial promise for biological phenotyping (Hicks, Onks, et al., 2020). Currently, individuals with mTBI receive similar recommendations and treatment, regardless of symptomology. The ability to differentiate mTBI subtypes based on biologic patterns could open the doors for personalized treatment, and the development of new pharmacologic targets (Jeromin et al., 2015). Studies of youths with mTBI have identified levels of salivary miRNAs correlated with headache, fatigue, attention and confusion (Hicks, Johnson, et al., 2018; Hicks, Onks, et al., 2020).

Top miRNA candidates

The specific miRNAs identified in individual studies are highly dependent upon the methods of sample collection, RNA extraction, and miRNA quantification (Sourvinou, Markou, & Lianidou, 2013). It is clear that miRNA patterns may vary considerably depending upon the timing of sample collection (Hicks, Khurana, et al., 2018; LaRocca et al., 2019), the severity of injury (Bhomia et al., 2016), or even the history of previous TBIs (Hicks, Olympia, et al., 2020). Nonetheless, literature reviews (Atif & Hicks, 2019) have identified several miRNAs that appear across multiple studies, regardless of biofluid, injury severity, or miRNA quantification technique (Table 2).

Decreases in miR-27 levels have been described in the blood (Bhomia et al., 2016; Di Pietro et al., 2017), saliva (Di Pietro et al., 2018; Hicks, Onks, et al., 2020), and cerebrospinal fluid (Hicks, Johnson, et al., 2018; Taheri et al., 2016) of individuals with TBI. Perturbations in miR-27 occur in both mild and severe TBI. Down-regulation of miR-27 may be detected within hours of injury, and persists in samples nearly two weeks after mTBI (Hicks, Onks, et al., 2020).

TABLE 2 Promising micro-ribonucleic acid candidates for traumatic brain injury.

	miR-7	miR-27	miR-192
Biofluids	Blood, saliva, cerebrospinal fluid	Blood, saliva, cerebrospinal fluid	Blood, saliva
Timing	Acute and chronic	Acute and chronic	Acute and chronic
Predominant direction of change	Increased	Decreased	Decreased
Severity	Mild and severe	Mild and severe	Mild
Studies	Hicks, Johnson, et al. (2018) Taheri et al. (2016) Hicks, Onks, et al. (2020)	Bhomia et al. (2016) Di Pietro et al. (2017) Di Pietro et al. (2018) Hicks, Onks, et al. (2020) Taheri et al. (2016) Hicks, Johnson, et al. (2018)	Johnson et al. (2018) Hicks, Onks, et al. (2020) Taheri et al. (2016) Di Pietro et al. (2017)

The table summarizes evidence for three micro-ribonucleic acids with potential utility in traumatic brain injury assessment.

Concentrations of miR-27 have demonstrated utility for mTBI diagnosis. Given the important role of miR-27 in blood–brain barrier maintenance and repair (Xi et al., 2018), serial miR-27 measurement may also serve to guide TBI prognosis.

Levels of miR-192 are perturbed in blood (Taheri et al., 2016; Di Pietro et al., 2017), and saliva (Hicks, Onks, et al., 2020; Johnson et al., 2018) following TBI. Individuals with mTBI display alterations in miR-192 expression. These changes have been reported within hours and persist for weeks following injury. The majority of studies (3/4) have reported decreased levels of miR-192 following TBI. This change shows potential utility for differentiating youths at risk for PCSS, and identifying athletes with sports-related concussion. Such a change is consistent with functional studies demonstrating that miR-192 inhibition promotes neuronal repair and prevents apoptosis in a murine model of peripheral nerve injury (Liu, Cui, Guan, Dong, & Zhang, 2020).

Another biomarker candidate that has been described in both mild and severe TBI is miR-7. Dysregulated levels of miR-7 occur in blood (Taheri et al., 2016), saliva (Hicks, Onks, et al., 2020) and cerebrospinal fluid (Hicks, Johnson, et al., 2018) and persist through the sub-acute post-injury period. Concentrations of miR-7 are increased within saliva and cerebrospinal fluid of individuals with TBI. This change is consistent with a neuroprotective effect reported in cellular models of Parkinson's, wherein elevated miR-7 prevents apoptosis and alpha-synuclein accumulation (Li et al., 2016).

Future directions for miRNA research in TBI

Pilot studies have laid the groundwork for future investigations of miRNA in TBI (Atif & Hicks, 2019). Their results demonstrate that miRNA levels respond to TBI, and that measurement of miRNAs may provide important tools for the diagnosis, prognosis, and clinical management of TBI (Meehan & Mannix, 2018). However, important questions remain (Pinchi et al., 2020). What are the optimal time points for biofluid collection and measurement of TBI-related miRNAs? Can miRNA testing overcome inter-individual variability in miRNA expression? Will the addition of miRNA testing improve the diagnostic/prognostic accuracy of neurocognitive and balance testing? Answering these questions will require large-scale, prospective studies of individuals with strictly defined TBI (Table 3). Methods for sample collection, sample storage, RNA isolation, and RNA quantification should be described in detail and widely available (Jeter et al., 2013). Moreover, the test–retest reliability and external validity of these techniques must be validated through repeated sampling and naïve

TABLE 3 Recommendations for future biomarker studies involving micro-ribonucleic acids.

Experimental elements	Recommendations
Study design	Large scale (sample sizes > 1000 may be necessary to overcome inter-individual variability in miRNA expression) Prospective (old samples = questionable RNA quality) Employ separate training and naïve hold out sets Longitudinal (time points selected for physiologic or clinical relevance)
Participant selection	TBI participants diagnosed by expert clinicians using up-to-date clinical guidelines Focused on a narrow range of TBI severity Control participants selected based on clinical application (e.g., diagnostic biomarkers of sports related concussion will require control participants with sub-concussive head impacts, recent exercise, and neuropsychologic symptoms mimicking TBI)
Measures	Validated, age-appropriate symptom scales should be used to assess subjective symptoms alongside miRNA levels Clinical utility of miRNA biomarkers should be compared with other adjuncts for TBI assessment (e.g., neurocognitive testing, balance testing) Performance of miRNA biomarkers can be compared with other emerging biomarker modalities (e.g., blood-based protein biomarkers, neuroimaging, neuroelectrophysiology)
RNA quantification	RNA sequencing may be used for initial selection of candidate miRNAs, while allowing exploration of other non-coding RNA families Migration to qPCR or immunoassay-based approaches will likely be required for downstream clinical application and finalization of predictive algorithms using miRNA measurement Development of novel technology that permits rapid, affordable measurement of miRNAs from crude biofluids at the point-of-care will be crucial for widespread clinical adoption as TBI biomarkers

The table identifies the qualities of future studies necessary for building upon the existing literature regarding micro-ribonucleic acid perturbations in traumatic brain injury.

hold-out sets. Diagnostic studies should use appropriate control participants, including individuals with sub-concussive head impacts, orthopedic injuries, recent exercise, and symptoms mimicking concussion (e.g., headache, attention difficulty, depression, anxiety, sleep difficulty). Numerous medical and demographic variables may impact miRNA levels and TBI physiology. It is crucial that studies collect and control for factors such as sex, age, race/ethnicity, injury severity, number of previous concussions, duration of past concussions, timing of sample collection, physical activity, medication use, and baseline psychological characteristics. Comparison with other objective or semi-objective measures (e.g., neurocognition, balance, serum proteins, neuroimaging, electrophysiology, eye tracking) will help establish the relative utility of miRNAs, and may lead to multi-disciplinary toolsets with superior accuracy. Finally, standardized measurement of primary medical outcomes (e.g., post-injury function, symptom duration, symptom character) will help to define the clinical application for which individual miRNA features are best suited.

Applications to other areas of neuroscience

The ability to measure molecules in peripheral biofluids that convey information about neurologic inflammation and injury holds great potential for neuroscience. Like mTBI, there are many neurologic conditions for which neuroimaging holds limited utility, and for which few objective tests exist. One example is autism spectrum disorder (Barbaro & Dissanayake, 2009). Autism has seen a dramatic increase in prevalence over the past several decades. Autism arises from an interaction between a child's underlying genetics and the environment, yet genetic testing cannot reliably be used for autism diagnosis (Roesser, 2011). Instead, autism diagnoses are based entirely on subjective observations of child behavior. Measurement of miRNA levels have demonstrated potential for differentiating children with autism from peers with non-autistic developmental delay (Hicks, Carpenter, et al., 2020; Hicks & Middleton, 2016). miRNA levels may also inform prognoses for children with autism, by providing information about the response to therapy, or the severity of symptoms. Similarly, growing rates of Alzheimer's disease necessitate the development of non-invasive biomarkers to aid identification and track the biologic response to emerging therapies. Here too, miRNAs are poised to support diagnosis and clinical management (Hébert et al., 2012). miRNAs hold potential as a non-invasive biomarker in a wide variety of neurologic and neuropsychiatric conditions (Ignacio et al., 2015; Jin et al., 2013) (Fig. 3). Their ability to cross the blood–brain barrier allows them to convey information about neuronal function, even in the absence of major cell death, and obviates the need for invasive cerebrospinal fluid collection. These same characteristics may also allow us to 1 day deliver them as targeted, individualized therapeutic interventions for a wide range of brain disorders (Shi et al., 2020).

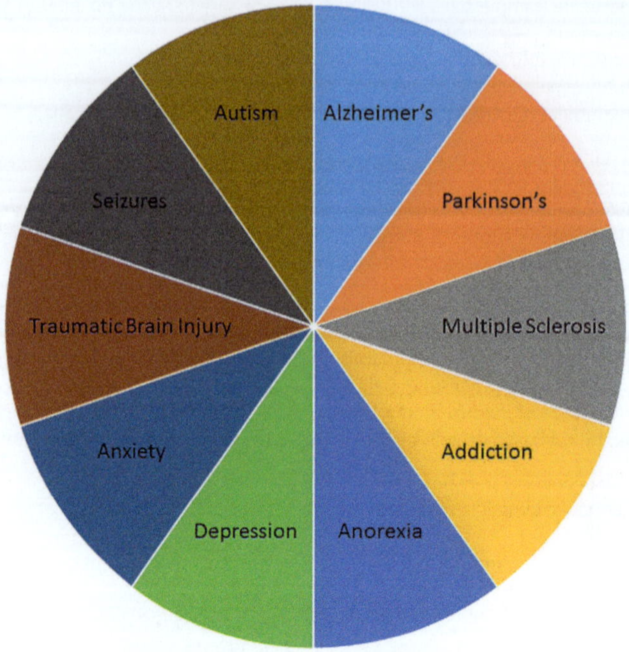

FIG. 3 Neurologic conditions amenable to micro-ribonucleic acid biomarker development. The figure displays neurologic and neuropsychologic conditions with growing evidence for micro-ribonucleic acid involvement.

Mini-dictionary of terms

Micro-ribonucleic acids: Class of short, non-coding nucleic acids that are transcribed from the genome and processed into mature single-stranded molecules capable of selectively binding messenger RNAs within the RNA-induced silencing complex, and preventing protein translation via transcript degradation.

Exosome: Small (30-100 nm) capsules found in the extracellular space of eukaryotic organisms that may contain nucleic acids or proteins within a protective outer membrane.

Biomarker: A biological characteristic that can be objectively measured to define a specific physiologic state.

Transcription: The initial step in gene expression, in which RNA polymerase copies a strand of DNA into its complementary RNA intermediary.

Translation: The final step in gene expression, in which ribosomes convert strands of messenger RNA into amino acid chains that form the building blocks for proteins.

Epi-transcriptome: Biochemical features (such as micro-ribonucleic acids or small nucleolar RNA) capable of altering the process of translation through modifications to messenger RNA.

Quantitative polymerase chain reaction: A laboratory method that measures the amount of a specific nucleic acid sequence (RNA or DNA) within a biologic sample by monitoring longitudinal change in fluorescently tagged sequences that occur during targeted enzymatic amplification.

Chronic traumatic encephalopathy: Constellation of symptoms, including cognitive impairment, depression, impulsive behavior, and emotional instability, believed to arise from repeated head trauma, and characterized by the accumulation of phosphorylated tau protein within the central nervous system that are detected on autopsy.

Biofluid: A liquid (such as saliva, urine, serum, plasma, or cerebrospinal fluid) that is naturally formed by a living organism, and contains organic compounds that can be measured in a basic science or clinical laboratory.

Non-coding ribonucleic acids: Classes of RNA that arises from DNA, but cannot be translated into proteins. These molecules (e.g., long non-coding RNA, small nucleolar RNA, piwi-interacting RNA, and micro-ribonucleic acids) typically regulate the processes of transcription and translation.

Key facts of micro-ribonucleic acids (miRNAs)

- Micro-ribonucleic acids are small, non-coding nucleic acids that regulate protein transcription throughout the human body.
- Micro-ribonucleic acids are particularly concentrated within the central nervous system, where they may be extruded within exosomes as inter-cellular signaling tools.
- Micro-ribonucleic acids can be easily measured in biofluids such as blood or saliva using quantitative polymerase chain reaction, RNA sequencing, or immunoplex assays.
- The methods used to collect, extract, and quantify micro-ribonucleic acids from human biofluids impact the specific micro-ribonucleic acids that are detected within a sample.
- Levels of micro-ribonucleic acids within blood or saliva are perturbed in individuals who suffer traumatic brain injury.
- Perturbations in micro-ribonucleic acid levels may distinguish individuals with traumatic brain injury from individuals without head trauma.
- The extent of micro-ribonucleic acid perturbation may reflect the severity of brain injury and telegraph the duration of concussion symptoms.
- Additional research is required to determine whether micro-ribonucleic acid measurement can be used to aid the diagnosis, prognosis, or clinical management of traumatic brain injury.

Summary points

- Micro-ribonucleic acids are an emerging class of biomarkers for traumatic brain injury.
- Micro-ribonucleic acids are small, non-coding nucleic acids that regulate the translation of messenger RNA into proteins.
- Neurons and glia release micro-ribonucleic acids into the extracellular space, where they can dock at neighboring cells and coordinate gene expression after traumatic brain injury.
- Because micro-ribonucleic acids are small (17–22 nucleotides), and often packaged within exosomes, they resist enzymatic degradation, and are easily measured within blood or saliva.

- Levels of specific micro-ribonucleic acids are altered in the brain and extracellular fluid of individuals who suffer traumatic brain injury.
- Perturbations of peripheral micro-ribonucleic acid levels occur within minutes and persist for weeks after traumatic brain injury.
- Preliminary studies suggest that measurement of micro-ribonucleic acids may aid the diagnosis or prognosis of traumatic brain injury, which currently rely on subjective symptom reports.
- Concentrations of micro-ribonucleic acids correlated with symptom character and duration have potential utility for guiding clinical management of patients with traumatic brain injury.
- A number of micro-ribonucleic acid candidates have been identified as biomarkers of traumatic brain injury, including miR-7, miR-27, and miR-192, which demonstrate consistent dysregulation across various time points, biofluids, and injury severities.
- Additional large-scale, longitudinal studies are needed to define the precise relationships between micro-ribonucleic acids and traumatic brain injury physiology, and confirm their value as clinical biomarkers.

Conflicts of interest

SDH is a paid consultant and advisory board member for Quadrant Biosciences, Inc. He is named as a co-inventor on intellectual property involving microRNA biomarkers for neurological diseases, which is owned by the Penn State College of Medicine.

References

Alvia, M. (2020). *MicroRNA alterations in chronic traumatic encephalopathy and amyotrophic lateral sclerosis*. Doctoral dissertation Boston University.

Ari, Ş., & Arikan, M. (2016). Next-generation sequencing: Advantages, disadvantages, and future. In *Plant omics: Trends and applications* (pp. 109–135). Cham: Springer.

Atif, H., & Hicks, S. D. (2019). A review of microRNA biomarkers in traumatic brain injury. *Journal of Experimental Neuroscience*, *13*, 1179069519832286.

Barbaro, J., & Dissanayake, C. (2009). Autism spectrum disorders in infancy and toddlerhood: A review of the evidence on early signs, early identification tools, and early diagnosis. *Journal of Developmental & Behavioral Pediatrics*, *30*(5), 447–459.

Bhomia, M., Balakathiresan, N. S., Wang, K. K., Papa, L., & Maheshwari, R. K. (2016). A panel of serum MiRNA biomarkers for the diagnosis of severe to mild traumatic brain injury in humans. *Scientific Reports*, *6*(1), 1–12.

Bond, A. E., Draeger, C. R. L., Mandleco, B., & Donnelly, M. (2003). Needs of family members of patients with severe traumatic brain injury: Implications for evidence-based practice. *Critical Care Nurse*, *23*(4), 63–72.

Cheng, L., Sharples, R. A., Scicluna, B. J., & Hill, A. F. (2014). Exosomes provide a protective and enriched source of miRNA for biomarker profiling compared to intracellular and cell-free blood. *Journal of Extracellular Vesicles*, *3*(1), 23743.

Coronado, V. G., McGuire, L. C., Faul, M., Sugerman, D., & Pearson, W. (2012). The epidemiology and prevention of TBI. In *Brain injury medicine. New York, NY: Demos* (pp. 45–56).

Davies, D., Yakoub, K. M., Scarpa, U., Bentley, C., Grey, M., Hammond, D., et al. (2019). Serum miR-502: A potential biomarker in the diagnosis of concussion in a pilot study of patients with normal structural brain imaging. *Journal of Concussion*, *3*, 2059700219886190.

de Ronde, M. W., Ruijter, J. M., Moerland, P. D., Creemers, E. E., & Pinto-Sietsma, S. J. (2018). Study design and qPCR data analysis guidelines for reliable circulating miRNA biomarker experiments: A review. *Clinical Chemistry*, *64*(9), 1308–1318.

Devoto, C., Lai, C., Qu, B. X., Guedes, V. A., Leete, J., Wilde, E. A., et al. (2020). Exosomal microRNAs in military persons with mild traumatic brain injury: Preliminary results from a chronic effects of neurotrauma consortium (CENC) biomarker discovery project. *Journal of Neurotrauma*, *37*, 2482–2492.

Di Pietro, V., Porto, E., Ragusa, M., Barbagallo, C., Davies, D., Forcione, M., et al. (2018). Salivary microRNAs: Diagnostic markers of mild traumatic brain injury in contact-sport. *Frontiers in Molecular Neuroscience*, *11*, 290.

Di Pietro, V., Ragusa, M., Davies, D., Su, Z., Hazeldine, J., Lazzarino, G., et al. (2017). MicroRNAs as novel biomarkers for the diagnosis and prognosis of mild and severe traumatic brain injury. *Journal of Neurotrauma*, *34*(11), 1948–1956.

Faridani, O. R., Abdullayev, I., Hagemann-Jensen, M., Schell, J. P., Lanner, F., & Sandberg, R. (2016). Single-cell sequencing of the small-RNA transcriptome. *Nature Biotechnology*, *34*(12), 1264.

Ghai, V., Fallen, S., Baxter, D., Scherler, K., Kim, T. K., Zhou, Y., et al. (2020). Alterations in plasma microRNA and protein levels in War Veterans with chronic mild traumatic brain injury. *Journal of Neurotrauma*, *37*(12), 1418–1430.

Godoy, P. M., Barczak, A. J., DeHoff, P., Srinivasan, S., Etheridge, A., Galas, D., et al. (2019). Comparison of reproducibility, accuracy, sensitivity, and specificity of miRNA quantification platforms. *Cell Reports*, *29*(12), 4212–4222.

Hébert, S. S., Sergeant, N., & Buée, L. (2012). MicroRNAs and the regulation of tau metabolism. *International Journal of Alzheimer's Disease*, *2012*, 406561. https://doi.org/10.1155/2012/406561.

Hicks, S. D., Carpenter, R. L., Wagner, K. E., Pauley, R., Barros, M., Tierney-Aves, C., et al. (2020). Saliva microRNA differentiates children with autism from peers with typical and atypical development. *Journal of the American Academy of Child & Adolescent Psychiatry*, *59*(2), 296–308.

Hicks, S. D., Johnson, J., Carney, M. C., Bramley, H., Olympia, R. P., Loeffert, A. C., et al. (2018). Overlapping microRNA expression in saliva and cerebrospinal fluid accurately identifies pediatric traumatic brain injury. *Journal of Neurotrauma*, *35*(1), 64–72.

Hicks, S. D., Khurana, N., Williams, J., Dowd Greene, C., Uhlig, R., & Middleton, F. A. (2018). Diurnal oscillations in human salivary microRNA and microbial transcription: Implications for human health and disease. *PLoS One*, *13*(7), e0198288.

Hicks, S. D., & Middleton, F. A. (2016). A comparative review of microRNA expression patterns in autism spectrum disorder. *Frontiers in Psychiatry*, *7*, 176.

Hicks, S. D., Olympia, R. P., Onks, C., Kim, R. Y., Zhen, K. J., Fedorchak, G., et al. (2020). Saliva microRNA biomarkers of cumulative concussion. *International Journal of Molecular Sciences*, *21*(20), 7758. https://doi.org/10.3390/ijms21207758.

Hicks, S. D., Onks, C., Kim, R. Y., Zhen, K. J., Loeffert, J., Loeffert, A. C., et al. (2020). Diagnosing mild traumatic brain injury using saliva RNA compared to cognitive and balance testing. *Clinical and Translational Medicine*, *10*(6), e197.

Higgins, K. L., Denney, R. L., & Maerlender, A. (2017). Sandbagging on the immediate post-concussion assessment and cognitive testing (ImPACT) in a high school athlete population. *Archives of Clinical Neuropsychology*, *32*(3), 259–266.

Ignacio, C., Hicks, S. D., Burke, P., Lewis, L., Szombathyne-Meszaros, Z., & Middleton, F. A. (2015). Alterations in serum microRNA in humans with alcohol use disorders impact cell proliferation and cell death pathways and predict structural and functional changes in brain. *BMC Neuroscience*, *16*(1), 55.

Jeromin, A., Maas, A. I., Buki, A., & Mondello, S. (2015). Developing a molecular taxonomy for traumatic brain injury: A perspective to enable the development of diagnostics and therapeutics. *Biomarkers in Medicine*, *9*(7), 619–621.

Jeter, C. B., Hergenroeder, G. W., Hylin, M. J., Redell, J. B., Moore, A. N., & Dash, P. K. (2013). Biomarkers for the diagnosis and prognosis of mild traumatic brain injury/concussion. *Journal of Neurotrauma*, *30*(8), 657–670.

Jin, X. F., Wu, N., Wang, L., & Li, J. (2013). Circulating microRNAs: A novel class of potential biomarkers for diagnosing and prognosing central nervous system diseases. *Cellular and Molecular Neurobiology*, *33*(5), 601–613.

Johnson, J. J., Loeffert, A. C., Stokes, J., Olympia, R. P., Bramley, H., & Hicks, S. D. (2018). Association of salivary microRNA changes with prolonged concussion symptoms. *JAMA Pediatrics*, *172*(1), 65–73.

Kapsimali, M., Kloosterman, W. P., de Bruijn, E., Rosa, F., Plasterk, R. H., & Wilson, S. W. (2007). MicroRNAs show a wide diversity of expression profiles in the developing and mature central nervous system. *Genome Biology*, *8*(8), R173.

Ko, J., Hemphill, M., Yang, Z., Beard, K., Sewell, E., Shallcross, J., et al. (2019). Multi-dimensional mapping of brain-derived extracellular vesicle microRNA biomarker for traumatic brain injury diagnostics. *Journal of Neurotrauma*, *37*, 2424–2434.

Lafourcade, C., Ramírez, J. P., Luarte, A., Fernández, A., & Wyneken, U. (2016). MIRNAS in astrocyte-derived exosomes as possible mediators of neuronal plasticity: Supplementary issue: Brain plasticity and repair. *Journal of Experimental Neuroscience*, *10*, JEN-S39916.

LaRocca, D., Barns, S., Hicks, S. D., Brindle, A., Williams, J., Uhlig, R., et al. (2019). Comparison of serum and saliva miRNAs for identification and characterization of mTBI in adult mixed martial arts fighters. *PLoS One*, *14*(1), e0207785.

Levin, H. S., & Diaz-Arrastia, R. R. (2015). Diagnosis, prognosis, and clinical management of mild traumatic brain injury. *The Lancet Neurology*, *14*(5), 506–517.

Li, S., Lv, X., Zhai, K., Xu, R., Zhang, Y., Zhao, S., et al. (2016). MicroRNA-7 inhibits neuronal apoptosis in a cellular Parkinson's disease model by targeting Bax and Sirt2. *American Journal of Translational Research*, *8*(2), 993.

Liu, X., Cui, X., Guan, G., Dong, Y., & Zhang, Z. (2020). microRNA-192-5p is involved in nerve repair in rats with peripheral nerve injury by regulating XIAP. *Cell Cycle*, *19*(3), 326–338.

Liu, K., Tong, H., Li, T., Wang, X., & Chen, Y. (2020). Research progress in molecular biology related quantitative methods of MicroRNA. *American Journal of Translational Research*, *12*(7), 3198.

Lumba-Brown, A., Yeates, K. O., Sarmiento, K., Breiding, M. J., Haegerich, T. M., Gioia, G. A., et al. (2018). Centers for Disease Control and Prevention guideline on the diagnosis and management of mild traumatic brain injury among children. *JAMA Pediatrics*, *172*(11), e182853.

Mannix, R., Levy, R., Zemek, R., Yeates, K. O., Arbogast, K. B., Meehan, W., et al. (2020). Fluid biomarkers of pediatric mild traumatic brain injury: A systematic review. *Journal of Neurotrauma*, *37*, 2029–2044.

Meehan, W. P., & Mannix, R. (2018). Promise of salivary MicroRNA for assessing concussion. *JAMA Pediatrics*, *172*(1), 14–15.

Mitra, B., Rau, T. F., Surendran, N., Brennan, J. H., Thaveenthiran, P., Sorich, E., et al. (2017). Plasma micro-RNA biomarkers for diagnosis and prognosis after traumatic brain injury: A pilot study. *Journal of Clinical Neuroscience*, *38*, 37–42.

Mott, J. L., & Mohr, A. M. (2015, February). Overview of microRNA biology. In *Vol. 35. Seminars in liver disease* (p. 3). NIH Public Access. no. 1.

Papa, L., Slobounov, S. M., Breiter, H. C., Walter, A., Bream, T., Seidenberg, P., et al. (2019). Elevations in microRNA biomarkers in serum are associated with measures of concussion, neurocognitive function, and subconcussive trauma over a single national collegiate athletic association division I season in collegiate football players. *Journal of Neurotrauma*, *36*(8), 1343–1351.

Patsimas, T., Howell, D. R., Potter, M. N., Provance, A. J., Kirkwood, M. W., & Wilson, J. C. (2020). Concussion-symptom rating correlation between pediatric patients and their parents. *Journal of Athletic Training*, *55*, 1020–1026.

Pinchi, E., Luigi, C., Paola, S., Gianpietro, V., Raoul, T., Mauro, A., et al. (2020). MicroRNAs: the new challenge for traumatic brain injury diagnosis. *Current Neuropharmacology*, *18*(4), 319–331.

Polito, F., Famà, F., Oteri, R., Raffa, G., Vita, G., Conti, A., et al. (2020). Circulating miRNAs expression as potential biomarkers of mild traumatic brain injury. *Molecular Biology Reports*, *47*(4), 2941–2949. https://doi.org/10.1007/s11033-020-05386-7.

Redell, J. B., Liu, Y., & Dash, P. K. (2009). Traumatic brain injury alters expression of hippocampal microRNAs: Potential regulators of multiple pathophysiological processes. *Journal of Neuroscience Research*, *87*(6), 1435–1448.

Redell, J. B., Moore, A. N., Ward, N. H., III, Hergenroeder, G. W., & Dash, P. K. (2010). Human traumatic brain injury alters plasma microRNA levels. *Journal of Neurotrauma*, *27*(12), 2147–2156.

Roesser, J. (2011). Diagnostic yield of genetic testing in children diagnosed with autism spectrum disorders at a regional referral center. *Clinical Pediatrics*, *50*(9), 834–843.

Rondina, C., Videtta, W., Petroni, G., Lujan, S., Schoon, P., Mori, L. B., et al. (2005). Mortality and morbidity from moderate to severe traumatic brain injury in Argentina. *The Journal of Head Trauma Rehabilitation*, *20*(4), 368–376.

Rowe, B. H., Eliyahu, L., Lowes, J., Gaudet, L. A., Beach, J., Mrazik, M., et al. (2018). Concussion diagnoses among adults presenting to three Canadian emergency departments: Missed opportunities. *The American Journal of Emergency Medicine*, *36*(12), 2144–2151.

Saulle, M., & Greenwald, B. D. (2012). Chronic traumatic encephalopathy: A review. *Rehabilitation Research and Practice*, *2012*, 816069.

Schober, K., Ondruschka, B., Dreßler, J., & Abend, M. (2015). Detection of hypoxia markers in the cerebellum after a traumatic frontal cortex injury: A human postmortem gene expression analysis. *International Journal of Legal Medicine*, *129*(4), 701–707.

Shi, Z. Y., Yang, X. X., Malichewe, C., Li, Y. S., & Guo, X. L. (2020). Exosomal microRNAs-mediated intercellular communication and exosome-based cancer treatment. *International Journal of Biological Macromolecules*, *158*, 530–541.

Sourvinou, I. S., Markou, A., & Lianidou, E. S. (2013). Quantification of circulating miRNAs in plasma: Effect of preanalytical and analytical parameters on their isolation and stability. *The Journal of Molecular Diagnostics*, *15*(6), 827–834.

Taheri, S., Tanriverdi, F., Zararsiz, G., Elbuken, G., Ulutabanca, H., Karaca, Z., et al. (2016). Circulating microRNAs as potential biomarkers for traumatic brain injury-induced hypopituitarism. *Journal of Neurotrauma*, *33*(20), 1818–1825.

Xi, T., Jin, F., Zhu, Y., Wang, J., Tang, L., Wang, Y., et al. (2018). miR-27a-3p protects against blood–brain barrier disruption and brain injury after intracerebral hemorrhage by targeting endothelial aquaporin-11. *Journal of Biological Chemistry*, *293*(52), 20041–20050.

Yang, T., Song, J., Bu, X., Wang, C., Wu, J., Cai, J., et al. (2016). Elevated serum miR-93, miR-191, and miR-499 are noninvasive biomarkers for the presence and progression of traumatic brain injury. *Journal of Neurochemistry*, *137*(1), 122–129.

Yoo, Y. K., Lee, J., Kim, H., Hwang, K. S., Yoon, D. S., & Lee, J. H. (2018). Toward exosome-based neuronal diagnostic devices. *Micromachines*, *9*(12), 634.

You, W. D., Tang, Q. L., Wang, L., Lei, J., Feng, J. F., Mao, Q., et al. (2016). Alteration of microRNA expression in cerebrospinal fluid of unconscious patients after traumatic brain injury and a bioinformatic analysis of related single nucleotide polymorphisms. *Chinese Journal of Traumatology*, *19*(1), 11–15.

Zemek, R., Barrowman, N., Freedman, S. B., Gravel, J., Gagnon, I., McGahern, C., et al. (2016). Clinical risk score for persistent postconcussion symptoms among children with acute concussion in the ED. *JAMA*, *315*(10), 1014–1025.

Zhangjie, S. U., Davies, D., Wang, M., Evans, S., Hill, L., Grey, M., et al. (2017). Circulating microRNA as novel early biomarker of concussion in elite athletes. *British Journal of Sports Medicine*, *51*(11), A2–A3.

Ziu, M., Fletcher, L., Rana, S., Jimenez, D. F., & Digicaylioglu, M. (2011). Temporal differences in microRNA expression patterns in astrocytes and neurons after ischemic injury. *PLoS One*, *6*(2), e14724.

Zonfrillo, M. R., Master, C. L., Grady, M. F., Winston, F. K., Callahan, J. M., & Arbogast, K. B. (2012). Pediatric providers' self-reported knowledge, practices, and attitudes about concussion. *Pediatrics*, *130*(6), 1120–1125.

Chapter 22

Pediatric minor head injury imaging

Anastasios Nasios, Spyridon Voulgaris, and George Alexiou
Department of Neurosurgery, University Hospital of Ioannina, Ioannina, Greece

List of abbreviations

Cho	choline
Cr	creatine
CT	computed tomography
DWI	diffusion-weighted imaging
fMRI	functional magnetic resonance imaging
Lac	lactate
MRI	magnetic resonance imaging
MRS	magnetic resonance spectroscopy
mTBI	minor traumatic brain injury
NAA	*N*-acetylaspartate
NAT	non-accidental trauma
PET	positron emission tomography
pTBI	pediatric traumatic brain injury
rMRI	rapid magnetic resonance imaging
SPECT	single-photon emission computed tomography
SRC	sports-related concussion
SWI	susceptibility-weighted imaging
TBI	traumatic brain injury

Introduction

Epidemiology of pediatric traumatic brain injury (pTBI)

Traumatic brain injury is a leading cause of morbidity and mortality in children and young adults. The annual incidence of pTBI in the USA is approximately estimated as 475,000 new cases per year in the age group of 0–14 years of age. Studies reveal that children in the age group of 0–4 years old are more often admitted to emergency departments for consultation in comparison to those in the age group of 5–14. Hospitalization for pTBI is more common in the age group 5–14 and more common in boys than girls, but death rates seem to be higher in children less than 4 years old (Centers for Disease Control and Prevention WISCARS, 2015). Falls are the most common cause of pTBI followed by transportation related injuries (motor vehicle accidents, bicycle associated accidents) and non-accidental injuries (Langlois, Rutland-Brown, & Thomas, 2004). Abusive head trauma is more often encountered in infants (Keenan et al., 2003).

Types of injury in the pediatric population

Significant differences exist in the pathophysiology of TBI in children and adults. A wide variety of structural and functional impairment is encountered in pTBI that depends on the age group of the patients, the special features of the developing brain and skull, as well as the mechanism of injury (Goldsmith & Plunkett, 2004). The craniofacial properties of infants differ compared to those of the adults. The developing scalp is thin and more vulnerable to injury with less likelihood to cushion against forces. Infant skulls tend to be less rigid and more mobile due to the presence of opened sutures and fontanelles resulting in a different pattern of external forces absorption (Ommaya, Goldsmith, & Thibault, 2002). The head

Diagnosis and Treatment of Traumatic Brain Injury. https://doi.org/10.1016/B978-0-12-823347-4.00016-6

to body ratio is at its highest rate with progressive depression over time making a head injury more possible during a fall, accident etc. Moreover, facial structures and paranasal sinuses are under development and cannot protect the underlying brain sufficiently during direct injury of the face or frontal area. Furthermore, white matter fibers are not yet heavily myelinated and their concentration in water is high, making fibers less prone to shearing but more fragile to direct injury due to the softness of the brain parenchyma. All these features listed above are responsible for the types of injuries encountered in pTBI. Cephalic hematoma (which can rarely become calcified) and sub-galeal hematoma, which are usually seen during head injury, are caused due to the micro-vascular breakdown and water retention properties of the thin scalp. Skull fractures of all types can occur, such as linear cranial vault fractures, depressed skull fractures, basal skull fractures (less common in pTBI) and growing skull fractures (AKA post-traumatic leptomeningeal cysts) with of variable degree of dural tearing at the same time (Orman et al., 2015). Acute epidural hematomas are not so common in infants due to the tight adhesion of the dura matter to the inner skull surface. They are usually of venous origin and are caused by injury to the emissary veins or dural venous sinuses and are less commonly arterial (Leggate, Lopez-Ramos, Genitori, Lena, & Choux, 1989). Acute subdural hematomas can occur both from direct external forces applied to the head and from sudden deceleration type of injuries. They are usually coexistent with cerebral contusions and are thinner compared to those seen in adults (Loh, Lin, Kwan, & Howng, 2002). Multiple subdural hematomas can sometimes occur and if there is evidence that they were sustained at different time periods, then an increased level of suspicion for abusive head trauma (oftenly associated with shaken baby syndrome) should be raised (Duhaime, Christian, Rorke, & Zimmerman, 1998). Cerebral contusions are common in pediatric patients especially those seen directly beneath the injury site—coup contusions, whereas contrecoup contusions are less common (Matsumoto et al., 1988). They can rapidly progress and cause secondary brain edema, because the autoregulatory mechanism of cerebral blood flow is under development in infants and profound post-traumatic cerebral hyperemia can be encountered (Lang, Teasdale, Macpherson, & Lawrence, 1994). They can also cause cerebral ischemia due to increased intracranial pressure, as well as traumatic subarachnoid hemorrhage or intra-ventricular hemorrhage due to direct penetration of the subarachnoid spaces or ventricles respectively from the adjacent contusion/ions (Hochstadter, Stewart, Alharfi, Ranger, & Fraser, 2014). Diffuse axonal injury is caused by shearing forces to the nerve fibers and blood vessels and is strongly associated with high cognitive function impairment, and therefore should be closely monitored and managed (Abu Hamdeh et al., 2017).

Minor pTBI features

Minor traumatic brain injury in the pediatric population is one of the most common reasons of presentation to emergency departments and primary care settings. It includes a wide range of clinical features and the majority of these type of injuries are referred as concussions.

The pathophysiology of concussion includes a metabolic cascade of events following TBI. Sheer and rotational forces cause neuronal and axonal injury in the cellular environment, ultimately resulting in a metabolic mismatch between metabolic demands and supply to the injured areas of brain tissue (Barkhoudarian, Hovda, & Giza, 2016). Clinical presentation features include physical symptoms such as nausea, vomiting, headache, visual impairment, gait disturbances, altered level of consciousness, but also include cognitive symptoms such as memory disturbances, attention disorders, mental slowness and emotional changes of irritability or depression (Blinman, Houseknecht, Snyder, Wiebe, & Nance, 2009). Post-traumatic migraine is a distinct presentation including headache and photophobia and phonophobia, symptoms that can last for weeks or months after injury (Mihalik et al., 2013). Physical symptoms show a tendency of earlier presentation than cognitive ones, but also resolve sooner in most cases (Eisenberg, Meehan 3rd, & Mannix, 2014).

The diagnostic approach of these patients includes a mixture of physical examination, imaging and laboratory studies that determine appropriate management. Physical examination includes routine neurologic examination (level of consciousness, mental examination, cranial nerve examination, assessment of motor and sensory function, gait and coordination) and some special physical examination tests that mainly focus on vestibular and oculomotor function evaluation, which are usually showing evidence of dysfunction in concussion patients, plus are of predictive value for learning difficulties in school aged patients who suffer from mTBI (Mucha et al., 2014).

Even if concussion is usually diagnosed based only on history and clinical examination, imaging is a useful adjunct in mTBI patient evaluation. CT scan remains the gold standard imaging modality in evaluating patients with suspected skull fractures and intracranial hemorrhages in the acute setting. Several models have been constructed in the recent years, in order to aid clinicians during clinical decision making to rule out possible severe intracranial complications of pTBI and to predict when CT scans should be obtained or not (Zonfrillo, Kim, & Arbogast, 2015). MRI is of limited value in diagnosing concussion, especially during the acute phase, because in most cases no abnormalities related to injury are depicted.

MRI scans are mostly used in the evaluation of patients with prolonged concussive symptoms and during post-TBI follow-up. Newer techniques based on MRI expand the limitations of standard MRI protocols and aid in the detection of subtle findings related to brain trauma. Laboratory testing used to be of limited predictive value in TBI. Nowadays certain serum biomarkers have come up, many of which have been thoroughly studied and show important correlation with imaging findings in TBI such as GFAP or S-100 calcium binding protein B (Mannix, Eisenberg, Berry, Meehan 3rd, & Hayes, 2014). Neurocognitive testing models through computerized software are available and can help clinicians mainly in the follow-up of patients suffering from concussions, through monitoring of performance scores during the recovery phase, so that a more individualized plan of returning to normal physical and mental activity can be scheduled (Hang, Babcock, Hornung, Ho, & Pomerantz, 2015).

The foundation of pediatric mTBI management is based on its pathophysiological nature that involves biochemical metabolic changes and neuronal activity impairment. The standard management includes rest, both physical and cognitive. Duration of rest and timing of return to normal activity is controversial in medical literature. A suggested management plan proposes 24–48 h of rest followed by gradual return to previous normal activity (Thomas, Apps, Hoffmann, McCrea, & Hammeke, 2015). Medication is used mainly to alleviate pain during the acute phase through analgesics administration. Melatonin has been used in the management of sleep disturbances following concussion. For patients suffering from prolonged post-traumatic headaches, amitriptyline has been shown to be beneficial (Bramley et al., 2015).

Imaging

Imaging techniques available in pTBI

A wide variety of imaging techniques is available to clinicians who come across managing pTBI. These techniques can evaluate pTBI in every phase of its natural progression including the acute, post-acute and chronic phase of this entity. This makes imaging a very useful tool that allows clinicians to estimate the impact of pediatric brain injury more thoroughly. Both conventional and advanced imaging modalities are available, each one of them covering different aspects of pTBI evaluation and diagnosis. Conventional techniques, which clinicians are more familiar with and are broadly used mainly in the acute setting, are skull x-rays, computed tomography scans (CT) and structural magnetic resonance imaging (MRI). Advanced techniques are less commonly used in the acute setting, but offer valuable information mainly during the post-acute phase following TBI. These include diffusion-weighted imaging (DWI), susceptibility-weighted imaging (SWI), diffusion tensor imaging (DTI), magnetic resonance spectroscopy (MRS), functional magnetic resonance imaging (fMRI), single-photon emission computed tomography (SPECT), and positron emission tomography (PET). Combination of these techniques based on every patient individualized profile offers a spherical approach to pTBI diagnosis and consequent management (Figs. 1 and 2).

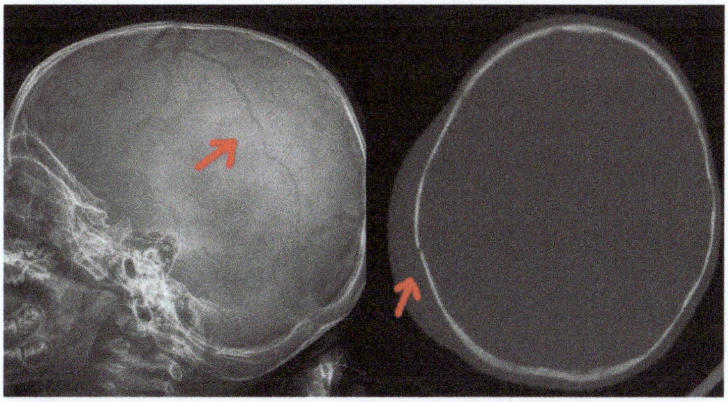

FIG. 1 Calvarial fracture. This figure shows a skull fracture of a 1-year-old boy after a fall. The left image is a lateral skull x-ray projection that shows the fracture line along the calvarium of the skull (*red arrow*). Notice that the fracture is limited by the bony sutures. The right image shows the same fracture as seen in an axial CT scan section (University Hospital of Ioannina).

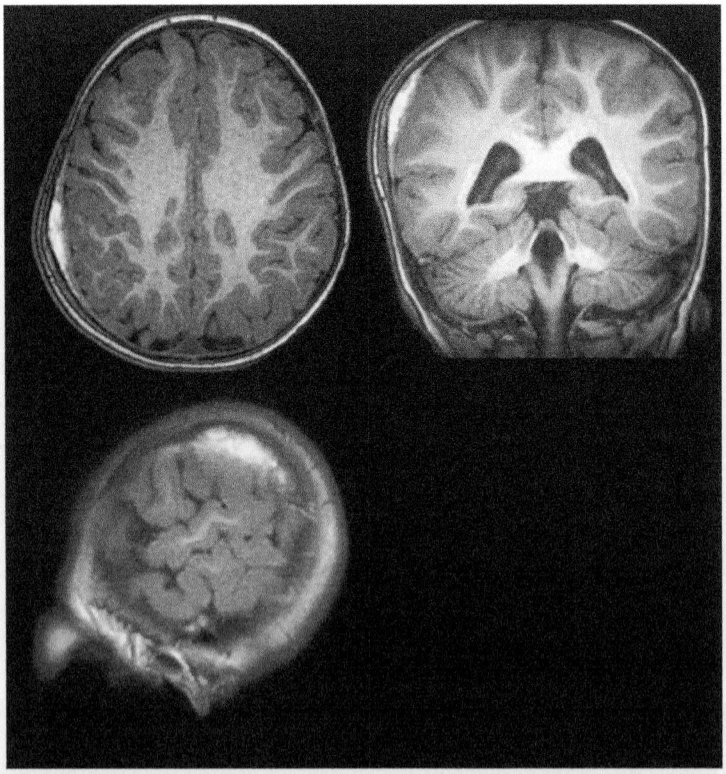

FIG. 2 Intracranial hematoma. This figure shows the MRI scan obtained in the same patient a few days after admission. T1W images without contrast administration are shown in all three planes (transverse: top left, coronal: top right, sagittal: bottom left). Notice the hyper-intense signal present on the right hemisphere indicating a mixed epidural and subdural blood collection just underneath the skull fracture (University Hospital of Ioannina).

Conventional imaging techniques

X-rays

Skull radiographic evaluation is an indispensable technique used in the diagnosis and management of pTBI. Even if CT scan is the gold standard imaging modality in TBI, skull x-rays are used as a preliminary tool in patients presenting with head trauma. A skull fracture detected on x-ray is associated with an increased risk of serious intracranial pathology. Skull x-ray sensitivity is low in depicting skull fractures. The detection of a fracture is dependent on fracture width and direction. Temporal and skull base fractures are very difficult to delineate through x-rays. All the above suggest that x-rays are of limited value in evaluating TBI and should not be obtained if CT scan is feasible to be conducted, because they can only delay the diagnosis of possibly severe intracranial pathology (Chawla et al., 2015).

Computed tomography (CT scan)

CT scan remains the gold standard examination in the evaluation of the acute stage of head injury in pediatric patients presenting in the emergency departments (Klora et al., 2019). CT scan is a fast, safe and easy to obtain survey in the acute phase following pTBI. It is reliable in depicting intracranial abnormalities related to brain injury and offers valuable information to clinicians, especially regarding patients who present with symptomatic TBI and an evolving intracranial lesion is clinically suspected which might need urgent intervention. Its sensitivity allows early detection of severe traumatic pathology and offers certainty in clinical decision making whenever history and clinical examination alone cannot safely evaluate the severity of pTBI. It reduces the amount of unnecessary hospital admissions for pTBI observation and reduces awaiting in the ED, alleviating the stressful feeling of parents related to their children acute medical situation and prognosis. There is a significant awareness regarding its use due to ionizing radiation exposure and subsequent increased risk for developing cancer (Brenner, 2002). CT scans can depict a whole range of TBI related pathology such as, scalp hematomas, skull fractures, extra-axial hematomas (epidural and subdural hematomas), cerebral contusions, traumatic subarachnoid hemorrhage, brain edema, pneumocephalus and hydrocephalus, thus evaluating pTBI from its most mild to its most severe form (Toyama et al., 2005). Even though CT can detect larger volume intracranial hemorrhagic lesions, it is less sensitive in

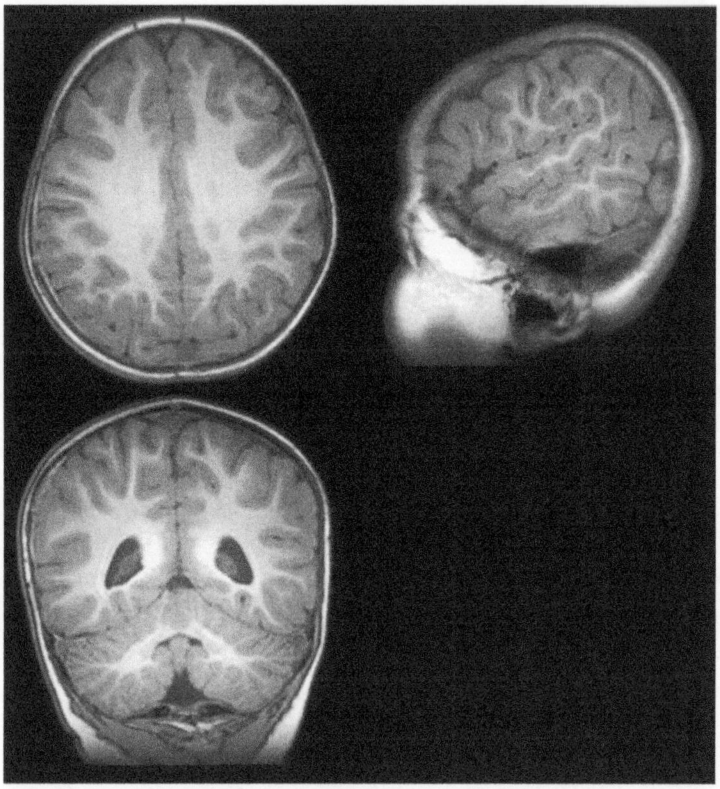

FIG. 3 Resolution of intracranial hematoma. This figure shows the MRI scan of the same patient 1 month after injury. T1W images with no contrast administration are shown in all planes (transverse: top left, sagittal: top right, coronal: bottom left). In these images the hematoma is significantly reduced in size indicating its gradual resolution (University Hospital of Ioannina).

detecting smaller abnormalities related to injury and cannot appreciate the extend of brain injury in patients presenting with prolonged TBI symptoms but who show no abnormalities on routine CT scans (Lee et al., 2008). CT scan is less valuable in evaluating diffuse axonal injury (DAI) in patients who suffer from concussions and cannot safely correlate clinical findings to imaging in this subgroup of patients (Fig. 3).

Magnetic resonance imaging (MRI scan)

MRI is an alternative to CT scan in evaluating pTBI patients. The most significant advantage of MRI in comparison to CT scan is that patients are not exposed to radiation and its future complications. Standard MRI protocols are time consuming and bare the risk of motion artifacts making them inappropriate for the evaluation of pTBI in the acute phase. Modified MRI protocols have been used in the follow-up of patients with hydrocephalus which are less time consuming and are adequate tools in neuroimaging evaluation (O'Neill et al., 2013). Rapid MRI (rMRI) protocols can be obtained in less than 5 min and show validity in pTBI evaluation, or even 2–3 min in some cases (Mehta et al., 2016). In a lot of patients, sedation is needed thus creating some amount of concern in its use (Malviya et al., 2000). The most rapid sequence that can be obtained in 5 s is the single-shot T2 fast filed echo EPI (Missios et al., 2008). More specifically it has been shown that rMRI can detect traumatic abnormalities in a similar fashion to CT scans, especially in minor pTBI patients. It is particularly valuable in depicting extra-axial blood collections and especially intra-parenchymal contusions with a higher sensitivity than CT scan for smaller hemorrhagic contusions near the convexity. Even if skull fractures are better evaluated with CT scans rather than MRI, rMRI can detect skull fractures especially depressed ones or those in need of neurosurgical intervention. MRI shows greater imaging properties in evaluating DAI in pTBI which is a reflexion of its ability to depict smaller hemorrhagic contusions in more detail (Mittl et al., 1994). CT scan is still the gold standard in the acute phase especially in patients that are critically ill and possible neurosurgical candidates, but rMRI can be successfully used in the post-acute phase as a follow-up tool reducing radiation exposure, especially in minor pTBI patients (Yuh et al., 2013). Problems that lead to the underuse of MRI in pTBI include cost, lack of emergency access to MRI facilities and lack of staff during night shifts and weekends (Fig. 4).

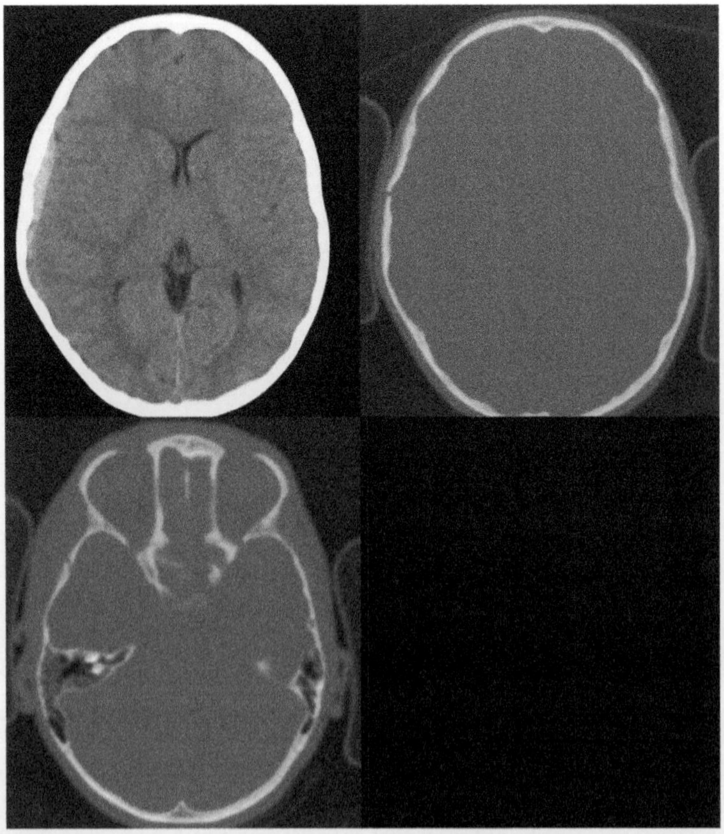

FIG. 4 Acute epidural hematoma. This figure shows the CT scan of a 6-year-old patient who sustained head injury after a fall. The scan was obtained urgently, a few moments after impact. Top left: This image shows a transverse CT scan section (brain window), which depicts an extra-axial hyperdense blood collection suggesting an acute epidural hematoma. Top right and bottom left: These images show transverse sections of CT scans (bone window) which depict the skull fracture of the right temporal and parietal bone (University Hospital of Ioannina).

Advanced imaging techniques

Advanced imaging techniques show significant potential in aiding in the evaluation and diagnosis of pTBI. Careful interpretation of these modalities can depict injuries that could not be detected by CT or standard MRI techniques and allows further evaluation of TBI of the "normal appearing brain" in conventional studies. Susceptibility-weighted imaging (SWI) is a sequence that utilizes the paramagnetic properties of deoxyhemoglobin. Thus, it can be used to detect hemorrhagic lesions that could not be detected in standard MRI (Sehgal et al., 2005). One of its main advantages lies in detecting very small hemorrhages that appear in patients suffering from DAI (Ashwal et al., 2006). DAI is related to widespread injuries to the brain parenchyma especially in areas such as the corpus callosum, the subcortical parasagittal white matter and the gray-white matter junction and can lead to devastating physical and cognitive disability (Adams et al., 1989). This shows the importance of early detection of these types of injury that are usually undiagnosed. Magnetic resonance spectroscopy (MRS) is a useful neuroimaging technique that allows analysis of the metabolic and neurochemical activity of various brain regions (Kreis, Ernst, & Ross, 1993). Metabolites studied include *N*-acetylaspartate (NAA), creatine, choline, lactate and glutamate. Different brain regions express different amounts of substances during various stages of neuronal development (Frahm et al., 1989). Alterations in the neurochemical composition of brain regions during TBI can be studied through MRS. Application of this technique is also found in the study of DAI in which increased focal levels of choline (Cho) and reduced levels of NAA are depicted, that result from the breakdown of myelin and the cellular membranes resulting in neuronal and axonal injury (Ross et al., 1998). Also increased levels of Cho and creatine (Cr) have been shown in subtle subarachnoid hemorrhage without the presence of lactate (Lac) which is indicative of neuronal ischemia (Macmillan et al., 2002). These findings increase the examiners ability to detect DAI abnormalities more thoroughly (Holshouser, Ashwal, & Tong, 2005). Combination of information derived from standard T2W and FLAIR sequences with SWI and MRS allows clinicians to evaluate DAI patients in detail and tailor further management and outline prognosis. Diffusion-weighted imaging (DWI) is a rapid imaging sequence that represents the random thermal movement of molecules. With DWI we can differentiate lesions of the brain that show evidence of increased or decreased diffusion when compared to normal tissue

(Huisman, 2003). Restricted diffusion is believed to be related to cytotoxic edema, while increased diffusion correlates to vasogenic edema (Barzó, Marmarou, Fatouros, Hayasaki, & Corwin, 1997). The main use of this technique lies in the detection of acute cerebral ischemia but it shows properties in the evaluation of TBI as well. DWI is particularly important in depicting shearing injuries and less valuable in detecting hemorrhagic lesions compared to other techniques. Studies show that DWI is a sensitive indicator of brain injury in pTBI patients following non-accidental trauma (NAT). Reports suggest that hypoxia and ischemia are common findings in TBI near injured areas, possibly due to focal reactive vasospasm, thus making DWI a more sensitive technique for early TBI evaluation in comparison to standard routine MRI, especially in DAI patients (Hergan, Schaefer, Sorensen, Gonzalez, & Huisman, 2002). Diffusion tensor imaging (DTI) is a more complex form of DWI sequences that allows us to study the positioning and direction of the white matter fiber tracts in the brain. It has used in the study of the myelination process during the development of the nervous system (Huppi et al., 1998). Since DAI affects white matter fibers it is suggested that DTI can detect injury by depicting changes in axonal structure (Arfanakis et al., 2002). Cytoskeletal network disruption is thought to be an important neuropathologic abnormality in DAI. DTI uses the phenomena of diffusion and anisotropy as well as the functional impairment of the axons throughout their whole length to conduct conclusions in DAI (Neil et al., 1998). Functional magnetic resonance imaging (fMRI) is a technique that is based on the blood oxygen levels and measures the ratio between oxygenated and deoxygenated blood, thus giving a picture of brain function during rest and during activity of different brain regions (Ogawa, Lee, Kay, & Tank, 1990). It has been mainly used in the study of post-traumatic neurocognitive disorders, memory disturbances, attention difficulties and emotional changes by interpreting regional brain activity and functional connectivity of various sensory and motor control brain regions. Only a few studies have used single photon emission computed tomography (SPECT) (Chiu Wong et al., 2006) and positron emission tomography (PET) in the evaluation of pTBI (Amyot et al., 2015). The idea behind this is that these modalities interpret the metabolic activity of different brain regions which is thought to be disturbed in TBI (Figs. 5 and 6).

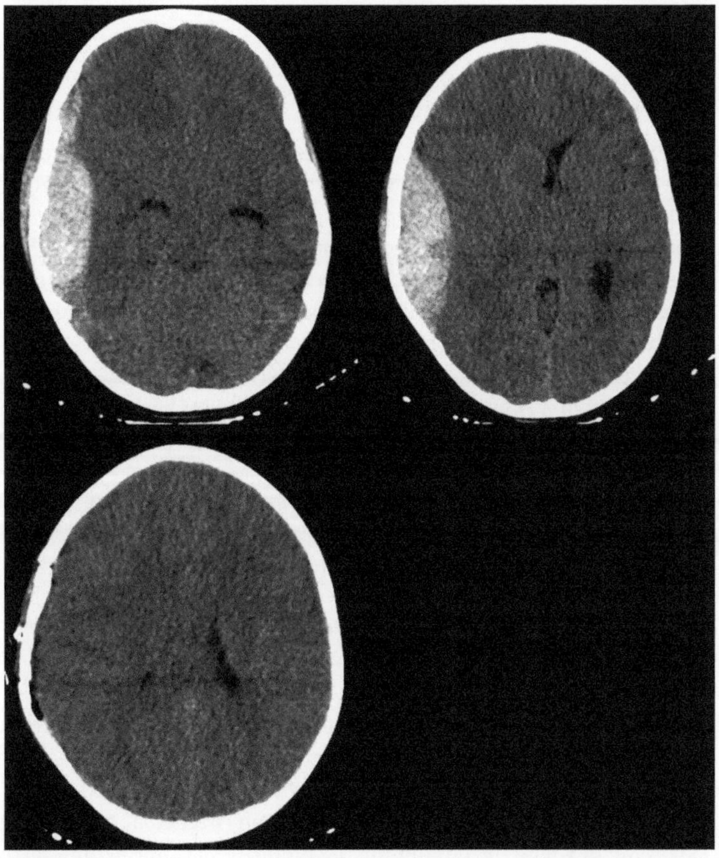

FIG. 5 Progression of the acute epidural hematoma. This figure shows the CT scan (transverse sections) of the 6-year-old patient approximately 6h after impact. The top images show the progression of the acute epidural hematoma. The volume of the hematoma has been significantly increased causing compression of the brain parenchyma and midline shift. The bottom image shows the post-operative result after hematoma evacuation. Notice the restoration of the midline after the evacuation (University Hospital of Ioannina).

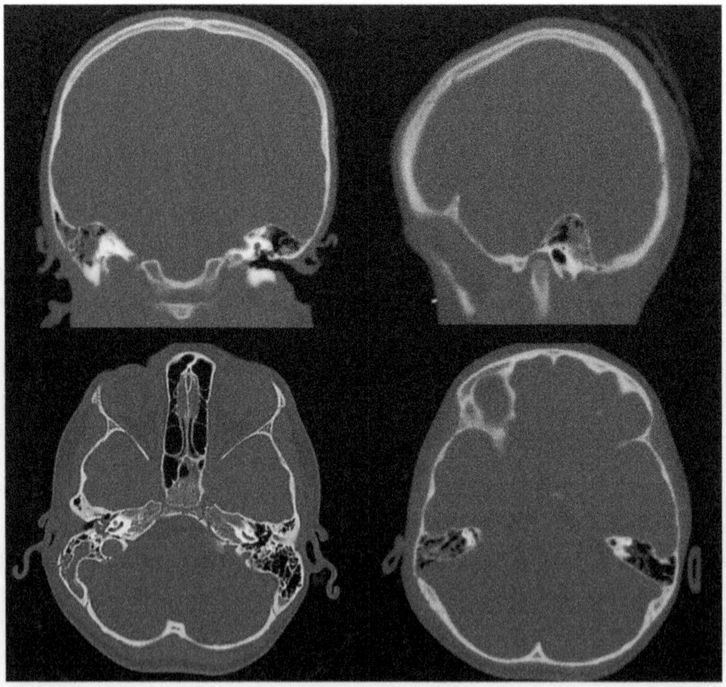

FIG. 6 Petrous bone fracture. This figure shows the CT scan of a 9-year-old girl after a fall. Top left (coronal) and top right (sagittal) images: Note the hyperdensity seen within the mastoid air cells of the right temporal bone. These suggest the presence of acute blood within the air sinuses. Bottom left (transverse): This image shows a fracture of the petrous part of the right temporal bone. Bottom right (transverse): Note the presence of air bubbles within the cranial cavity posterior to the petrous bone on the right side that suggest a skull base fracture (University Hospital of Ioannina).

Applications to other areas of neuroscience

In this chapter, we have reviewed the imaging modalities mainly used in the evaluation of pediatric head injury and their special features regarding brain trauma diagnosis. As mentioned before, most of the patients in the acute setting sustain minor injuries that resolve spontaneously and require no further management. A special subgroup of patients includes those young patients that sustain head injury in the form of concussions in whom symptoms last longer than expected and are sometimes underestimated in severity and underdiagnosed. Many young patients show signs of neurocognitive disorders during their psychosocial development, even years after TBI suggesting a possible interconnection between these two entities. Since conventional imaging modalities cannot further evaluate pTBI in these cases, a strong suggestion to proceed with further examination through advanced imaging techniques should be advised. In some cases, DAI is 'hidden' behind those findings and can only be depicted through advanced modalities. Children who have a medical history of minor pTBI and show signs of neurocognitive disorders should be referred to specialists such as pediatric neurologists, pediatric psychologists etc. for further evaluation of their condition and management.

Mini-dictionary of terms

Diffuse axonal injury: A form of primary brain injury caused by rapid acceleration and deceleration of the brain. It results in shearing of the white matter fibers. It is classified into three distinct types.

Neurocognitive disorders: A term describing the situation in which the mental function of an individual is impaired due to medical causes.

Magnetic resonance spectroscopy: An imaging technique that studies the chemical composition of brain tissue. Its usefulness is primarily seen in tumors but also in other areas of neuroscience as well.

Diffusion-weighted imaging: An imaging modality based on MRI which studies the diffusion of molecules throughout the brain allowing the mapping of diffusion phenomena seen in either physiological or pathological processes.

Hydrocephalus: A clinical syndrome in which the circulation of cerebrospinal fluid within the brain is disturbed resulting in accumulation of fluid intracranially. Common classification includes, congenital and acquired hydrocephalus, as well as obstructive (non-communicating) or non-obstructive (communicating) types.

Vasospasm: A condition in which cerebral vasculature is constricted resulting in reduced blood supply to the brain and ultimately ischemia. It can be focal or diffuse. Usually encountered in subarachnoid hemorrhage.

Vasogenic edema: A form of extracellular brain swelling caused by disruption of the blood brain barrier. Commonly seen in the white matter surrounding intra-parenchymal brain tumors.

Contusions: A common type of intra-parenchymal brain injury. They are a form of primary brain injury. Two types are seen, coup (underneath the site of head impact) and contra-coup (at the opposite site of head impact).

Post-traumatic leptomeningeal cyst: Also known as growing skull fracture. A rare post-traumatic complication seen in children in which a calvarial fracture results in non-union and leptomeningeal or brain parenchyma herniation through it.

Deoxyhemoglobin: Hemoglobin is a protein found in abundance in red blood cells. Its function is to carry oxygen from the lungs to peripheral tissues. The form in which hemoglobin does not form a bond with oxygen is called deoxyhemoglobin and is responsible for the purple-blue color of venous blood.

Key facts of concussions

- Concussion is an injury induced by biomechanical forces on the brain.
- Concussions can be caused through direct head impact or by the impact of some force on a different body part and get transmitted with an impulsive way to the head.
- Concussions result in rapid onset transient impairment of neurological function.
- In some patients signs and symptoms evolve over a number of minutes to hours and can last for longer periods of time.
- Their acute clinical signs and symptoms suggest more of a functional disturbance rather than a structural type of injury.
- No structural abnormality is usually seen on conventional neuroimaging studies.

Summary points

- Traumatic brain injury is a leading cause of morbidity and mortality in children and young adults.
- Pediatric minor head injury is one of the most common causes of admission to emergency department.
- Diffuse axonal injury in the pediatric population can lead to future neurocognitive disorders and can sometimes be missed in the acute setting through conventional imaging.
- Conventional imaging studies still remain the modalities of first choice in the acute evaluation of pediatric head trauma.
- Advanced imaging techniques can help further evaluate head injury, especially in cases of "normal appearing brain" on conventional studies.

References

Abu Hamdeh, S., Marklund, N., Lannsjö, M., Howells, T., Raininko, R., Wikström, J., et al. (2017). Extended anatomical grading in diffuse axonal injury using MRI: Hemorrhagic lesions in the substantia nigra and mesencephalic tegmentum indicate poor long-term outcome. *Journal of Neurotrauma*, *34*(2), 341–352. https://doi.org/10.1089/neu.2016.4426.

Adams, J. H., Doyle, D., Ford, I., Gennarelli, T. A., Graham, D. I., & McLellan, D. R. (1989). Diffuse axonal injury in head injury: Definition, diagnosis and grading. *Histopathology*, *15*(1), 49–59. https://doi.org/10.1111/j.1365-2559.1989.tb03040.x.

Amyot, F., Arciniegas, D. B., Brazaitis, M. P., Curley, K. C., Diaz-Arrastia, R., Gandjbakhche, A., et al. (2015). A review of the effectiveness of neuroimaging modalities for the detection of traumatic brain injury. *Journal of Neurotrauma*, *32*(22), 1693–1721. https://doi.org/10.1089/neu.2013.3306.

Arfanakis, K., Haughton, V. M., Carew, J. D., Rogers, B. P., Dempsey, R. J., & Meyerand, M. E. (2002). Diffusion tensor MR imaging in diffuse axonal injury. *AJNR. American Journal of Neuroradiology*, *23*(5), 794–802.

Ashwal, S., Babikian, T., Gardner-Nichols, J., Freier, M.-C., Tong, K. A., & Holshouser, B. A. (2006). Susceptibility-weighted imaging and proton magnetic resonance spectroscopy in assessment of outcome after pediatric traumatic brain injury. *Archives of Physical Medicine and Rehabilitation*, *87*(Suppl. 12), 50–58. https://doi.org/10.1016/j.apmr.2006.07.275.

Barkhoudarian, G., Hovda, D. A., & Giza, C. C. (2016). The molecular pathophysiology of concussive brain injury—An update. *Physical Medicine and Rehabilitation Clinics of North America*, *27*(2), 373–393. https://doi.org/10.1016/j.pmr.2016.01.003.

Barzó, P., Marmarou, A., Fatouros, P., Hayasaki, K., & Corwin, F. (1997). Contribution of vasogenic and cellular edema to traumatic brain swelling measured by diffusion-weighted imaging. *Journal of Neurosurgery*, *87*(6), 900–907. https://doi.org/10.3171/jns.1997.87.6.0900.

Blinman, T. A., Houseknecht, E., Snyder, C., Wiebe, D. J., & Nance, M. L. (2009). Postconcussive symptoms in hospitalized pediatric patients after mild traumatic brain injury. *Journal of Pediatric Surgery*, *44*(6), 1223–1228. https://doi.org/10.1016/j.jpedsurg.2009.02.027.

Bramley, H., Heverley, S., Lewis, M. M., Kong, L., Rivera, R., & Silvis, M. (2015). Demographics and treatment of adolescent posttraumatic headache in a regional concussion clinic. *Pediatric Neurology*, *52*(5), 493–498. https://doi.org/10.1016/j.pediatrneurol.2015.01.008.

Brenner, D. J. (2002). Estimating cancer risks from pediatric CT: Going from the qualitative to the quantitative. *Pediatric Radiology*, *32*(4), 228–244. https://doi.org/10.1007/s00247-002-0671-1.

Centers for Disease Control and Prevention, & WISCARS. (2015). *Leading causes of death reports, national and regional, 1999–2013*. http://webappa.cdc.gov/cgi-bin/broker.exe. Access 2015.9.22.

Chawla, H., Malhotra, R., Yadav, R. K., Griwan, M. S., Paliwal, P. K., & Aggarwal, A. D. (2015). Diagnostic utility of conventional radiography in head injury. *Journal of Clinical and Diagnostic Research: JCDR*, *9*(6), TC13–TC15. https://doi.org/10.7860/JCDR/2015/13842.6133.

Chiu Wong, S. B., Chapman, S. B., Cook, L. G., Anand, R., Gamino, J. F., & Devous, M. D., Sr. (2006). A SPECT study of language and brain reorganization three years after pediatric brain injury. *Progress in Brain Research*, *157*, 173–185. https://doi.org/10.1016/s0079-6123(06)57011-6.

Duhaime, A. C., Christian, C. W., Rorke, L. B., & Zimmerman, R. A. (1998). Nonaccidental head injury in infants—The "shaken-baby syndrome". *The New England Journal of Medicine*, *338*, 1822–1829.

Eisenberg, M. A., Meehan, W. P., 3rd, & Mannix, R. (2014). Duration and course of post-concussive symptoms. *Pediatrics*, *133*(6), 999–1006. https://doi.org/10.1542/peds.2014-0158.

Frahm, J., Bruhn, H., Gyngell, M. L., Merboldt, K. D., Hänicke, W., & Sauter, R. (1989). Localized proton NMR spectroscopy in different regions of the human brain in vivo. Relaxation times and concentrations of cerebral metabolites. *Magnetic Resonance in Medicine*, *11*(1), 47–63. https://doi.org/10.1002/mrm.1910110105.

Goldsmith, W., & Plunkett, J. (2004). A biomechanical analysis of the causes of traumatic brain injury in infants and children. *The American Journal of Forensic Medicine and Pathology*, *25*(2), 89–100. https://doi.org/10.1097/01.paf.0000127407.28071.63.

Hang, B., Babcock, L., Hornung, R., Ho, M., & Pomerantz, W. J. (2015). Can computerized neuropsychological testing in the emergency department predict recovery for young athletes with concussions? *Pediatric Emergency Care*, *31*(10), 688–693. https://doi.org/10.1097/PEC.0000000000000438.

Hergan, K., Schaefer, P. W., Sorensen, A. G., Gonzalez, R. G., & Huisman, T. A. (2002). Diffusion-weighted MRI in diffuse axonal injury of the brain. *European Radiology*, *12*(10), 2536–2541. https://doi.org/10.1007/s00330-002-1333-2.

Hochstadter, E., Stewart, T. C., Alharfi, I. M., Ranger, A., & Fraser, D. D. (2014). Subarachnoid hemorrhage prevalence and its association with short-term outcome in pediatric severe traumatic brain injury. *Neurocritical Care*, *21*(3), 505–513. https://doi.org/10.1007/s12028-014-9986-7.

Holshouser, B. A., Ashwal, S., & Tong, K. A. (2005). Proton MR spectroscopic imaging depicts diffuse axonal injury in children with traumatic brain injury. *AJNR. American Journal of Neuroradiology*, *26*, 1276–1285.

Huisman, T. A. (2003). Diffusion-weighted imaging: Basic concepts and application in cerebral stroke and head trauma. *European Radiology*, *13*, 2283–2297.

Huppi, P. S., Maier, S. E., Peled, S., Zientara, G. P., Barnes, P. D., & Jolesz, F. A. (1998). Microstructural development of human newborn cerebral white matter assessed in vivo by diffusion tensor magnetic resonance imaging. *Pediatric Research*, *44*, 584–590.

Keenan, H. T., Runyan, D. K., Marshall, S. W., Nocera, M. A., Merten, D. F., & Sinal, S. H. (2003). A population-based study of inflicted traumatic brain injury in young children. *JAMA*, *290*(5), 621–626. https://doi.org/10.1001/jama.290.5.621.

Klora, M., Zeidler, J., Bassler, S., Hirsch, F. W., Gosemann, J. H., Lacher, M., et al. (2019). Frequency of neuroimaging for pediatric minor brain injury is determined by the primary treating medical department. *Medicine*, *98*(28). https://doi.org/10.1097/MD.0000000000016320, e16320.

Kreis, R., Ernst, T., & Ross, B. D. (1993). Development of the human brain: In vivo quantification of metabolite and water content with proton magnetic resonance spectroscopy. *Magnetic Resonance in Medicine*, *30*, 424–437.

Lang, D. A., Teasdale, G. M., Macpherson, P., & Lawrence, A. (1994). Diffuse brain swelling after head injury: More often malignant in adults than children? *Journal of Neurosurgery*, *80*(4), 675–680. https://doi.org/10.3171/jns.1994.80.4.0675.

Langlois, J., Rutland-Brown, W., & Thomas, K. (2004). *Traumatic brain injury in the United States: Emergency department visits, hospitalizations, and deaths*. Atlanta: Centers for Disease Control and Prevention, National Center for Injury Prevention and Control. https://www.cdc.gov/traumaticbraininjury/tbi_ed.html. Access 2015.9.22.

Lee, H., Wintermark, M., Gean, A. D., Ghajar, J., Manley, G. T., & Mukherjee, P. (2008). Focal lesions in acute mild traumatic brain injury and neurocognitive outcome: CT versus 3T MRI. *Journal of Neurotrauma*, *25*(9), 1049–1056. https://doi.org/10.1089/neu.2008.0566.

Leggate, J. R., Lopez-Ramos, N., Genitori, L., Lena, G., & Choux, M. (1989). Extradural haematoma in infants. *British Journal of Neurosurgery*, *3*, 533–539.

Loh, J. K., Lin, C. L., Kwan, A. L., & Howng, S. L. (2002). Acute subdural hematoma in infancy. *Surgical Neurology*, *58*(3–4), 218–224. https://doi.org/10.1016/s0090-3019(02)00830-3.

Macmillan, C. S., Wild, J. M., Wardlaw, J. M., Andrews, P. J., Marshall, I., & Easton, V. J. (2002). Traumatic brain injury and subarachnoid hemorrhage: In vivo occult pathology demonstrated by magnetic resonance spectroscopy may not be "ischaemic". A primary study and review of the literature. *Acta Neurochirurgica*, *144*(9), 853–862. https://doi.org/10.1007/s00701-002-0966-x.

Malviya, S., Voepel-Lewis, T., Eldevik, O. P., Rockwell, D. T., Wong, J. H., & Tait, A. R. (2000). Sedation and general anaesthesia in children undergoing MRI and CT: Adverse events and outcomes. *British Journal of Anaesthesia*, *84*(6), 743–748. https://doi.org/10.1093/oxfordjournals.bja.a013586.

Mannix, R., Eisenberg, M., Berry, M., Meehan, W. P., 3rd, & Hayes, R. L. (2014). Serum biomarkers predict acute symptom burden in children after concussion: A preliminary study. *Journal of Neurotrauma*, *31*(11), 1072–1075. https://doi.org/10.1089/neu.2013.3265.

Matsumoto, M., Sanpei, K., Nishikawa, H., Seki, T., Shibata, I., & Terao, H. (1988). Characteristics of traumatic intracerebral hematomas in children. *Neurologia Medico-Chirurgica (Tokyo)*, *28*, 1081–1088.

Mehta, H., Acharya, J., Mohan, A. L., Tobias, M. E., LeCompte, L., & Jeevan, D. (2016). Minimizing radiation exposure in evaluation of pediatric head trauma: Use of rapid MR imaging. *AJNR. American Journal of Neuroradiology*, *37*(1), 11–18. https://doi.org/10.3174/ajnr.A4464.

Mihalik, J. P., Register-Mihalik, J., Kerr, Z. Y., Marshall, S. W., McCrea, M. C., & Guskiewicz, K. M. (2013). Recovery of posttraumatic migraine characteristics in patients after mild traumatic brain injury. *The American Journal of Sports Medicine*, *41*(7), 1490–1496. https://doi.org/10.1177/0363546513487982.

Missios, S., Quebada, P. B., Forero, J. A., Durham, S. R., Pekala, J. S., Eskey, C. J., et al. (2008). Quick-brain magnetic resonance imaging for nonhydrocephalus indications. Journal of neurosurgery. *Pediatrics*, *2*(6), 438–444. https://doi.org/10.3171/PED.2008.2.12.438.

Mittl, R. L., Grossman, R. I., Hiehle, J. F., Hurst, R. W., Kauder, D. R., Gennarelli, T. A., et al. (1994). Prevalence of MR evidence of diffuse axonal injury in patients with mild head injury and normal head CT findings. *AJNR. American Journal of Neuroradiology*, *15*(8), 1583–1589.

Mucha, A., Collins, M. W., Elbin, R. J., Furman, J. M., Troutman-Enseki, C., DeWolf, R. M., et al. (2014). A brief vestibular/ocular motor screening (VOMS) assessment to evaluate concussions: Preliminary findings. *The American Journal of Sports Medicine*, *42*(10), 2479–2486. https://doi.org/10.1177/0363546514543775.

Neil, J. J., Shiran, S. I., McKinstry, R. C., Schefft, G. L., Snyder, A. Z., Almli, C. R., et al. (1998). Normal brain in human newborns: Apparent diffusion coefficient and diffusion anisotropy measured by using diffusion tensor MR imaging. *Radiology*, *209*(1), 57–66. https://doi.org/10.1148/radiology.209.1.9769812.

Ogawa, S., Lee, T. M., Kay, A. R., & Tank, D. W. (1990). Brain magnetic resonance imaging with contrast dependent on blood oxygenation. *Proceedings of the National Academy of Sciences of the United States of America*, *87*(24), 9868–9872. https://doi.org/10.1073/pnas.87.24.9868.

Ommaya, A. K., Goldsmith, W., & Thibault, L. (2002). Biomechanics and neuropathology of adult and paediatric head injury. *British Journal of Neurosurgery*, *16*(3), 220–242. https://doi.org/10.1080/02688690220148824.

O'Neill, B. R., Pruthi, S., Bains, H., Robison, R., Weir, K., Ojemann, J., et al. (2013). Rapid sequence magnetic resonance imaging in the assessment of children with hydrocephalus. *World Neurosurgery*, *80*(6), e307–e312. https://doi.org/10.1016/j.wneu.2012.10.066.

Orman, G., Wagner, M. W., Seeburg, D., Zamora, C. A., Oshmyansky, A., Tekes, A., et al. (2015). Pediatric skull fracture diagnosis: Should 3D CT reconstructions be added as routine imaging? *Journal of Neurosurgery. Pediatrics*, *16*(4), 426–431. https://doi.org/10.3171/2015.3.PEDS1553.

Ross, B. D., Ernst, T., Kreis, R., Haseler, L. J., Bayer, S., Danielsen, E., et al. (1998). 1H MRS in acute traumatic brain injury. *Journal of Magnetic Resonance Imaging: JMRI*, *8*(4), 829–840. https://doi.org/10.1002/jmri.1880080412.

Sehgal, V., Delproposto, Z., Haacke, E. M., Tong, K. A., Wycliffe, N., Kido, D. K., et al. (2005). Clinical applications of neuroimaging with susceptibility-weighted imaging. *Journal of Magnetic Resonance Imaging: JMRI*, *22*(4), 439–450. https://doi.org/10.1002/jmri.20404.

Thomas, D. G., Apps, J. N., Hoffmann, R. G., McCrea, M., & Hammeke, T. (2015). Benefits of strict rest after acute concussion: A randomized controlled trial. *Pediatrics*, *135*(2), 213–223. https://doi.org/10.1542/peds.2014-0966.

Toyama, Y., Kobayashi, T., Nishiyama, Y., Satoh, K., Ohkawa, M., & Seki, K. (2005). CT for acute stage of closed head injury. *Radiation Medicine*, *23*(5), 309–316.

Yuh, E. L., Mukherjee, P., Lingsma, H. F., Yue, J. K., Ferguson, A. R., Gordon, W. A., et al. (2013). Magnetic resonance imaging improves 3-month outcome prediction in mild traumatic brain injury. *Annals of Neurology*, *73*(2), 224–235. https://doi.org/10.1002/ana.23783.

Zonfrillo, M. R., Kim, K. H., & Arbogast, K. B. (2015). Emergency department visits and head computed tomography utilization for concussion patients from 2006 to 2011. *Academic Emergency Medicine: Official Journal of the Society for Academic Emergency Medicine*, *22*(7), 872–877. https://doi.org/10.1111/acem.12696.

Chapter 23

Transcranial Doppler ultrasound in acute traumatic brain injury

Alexey O. Trofimov[a,b], Anatoly Y. Sheludyakov[b], Andrew Y. Abashkin[c], Darya I. Agarkova[d], Michael Dobrzeniecki[a], Zhanna B. Semenova[e], and Denis E. Bragin[f,g]

[a]*Department of Neurologic Diseases, Privolzhsky Research Medical University, Nizhny Novgorod, Russia,* [b]*Department of Neurosurgery, Trauma Center Level I, City Clinical Hospital, Nizhny Novgorod, Russia,* [c]*Department of Neurotrauma, Trauma Center Level II, Urgent Care Hospital, Dzerzhinsk, Russian Federation,* [d]*Department of Ophthalmology, Trauma Center Level I, Regional Clinical Hospital named after NA Semashko, Nizhny Novgorod, Russian Federation,* [e]*Department of Neurosurgery, Children's Clinical and Research Institute of Emergency Surgery and Trauma, Moscow, Russian Federation,* [f]*Lovelace Biomedical Research Institute, Albuquerque, NM, United States,* [g]*University of New Mexico, School of Medicine, MSC 105615 Neurosurgery, Albuquerque, NM, United States*

List of abbreviations

ACA	anterior cerebral artery
BA	basilar artery
BP	blood pressure
CBF	cerebral blood flow
CCP	critical closing pressure
CPP	cerebral perfusion pressure
FV	flow velocity
GOS	Glasgow outcome score
ICA	internal carotid artery
ICH	intracerebral hemorrhage
ICP	intracranial pressure
LR	Lindegaard ratio
MAP	mean arterial pressure
MCA	middle cerebral artery
PCA	posterior cerebral artery
PI	pulsatility index
RI	resistance index
SAH	subarachnoid hemorrhage
TBI	traumatic brain injury
TCD	transcranial Doppler
VA	vertebral artery

Introduction

Transcranial Doppler (TCD) ultrasound is a non-invasive, informative and unsophisticated technique used for the assessment of cerebral hemodynamics by ultrasonography of the main cerebral arteries and convenient for the work in the monitoring mode.

This method was initially used to diagnose vasospasm resulting from ruptured cerebral aneurysms, which complicated the course of this pathology. It is currently widely used in neurological and neurosurgical clinics to study various physiological and pathological states of cerebral hemodynamics. The information obtained through its application makes it possible to clarify the causes of cerebrovascular disorders (ruptured aneurysms, vasospasm, thrombosis), to diagnose intracranial

Diagnosis and Treatment of Traumatic Brain Injury. https://doi.org/10.1016/B978-0-12-823347-4.00039-7

complications in the case of TBI and diseases, in particular, intracranial hypertension and cerebral ischemia. The term "cerebral macrocirculation" is associated with this technique. It means the blood flow in the main cerebral vessels. The term "main cerebral blood flow" is also used to characterize the TCD object.

It is generally accepted (Al-Mufti et al., 2018; Niu et al., 2020) that TCD in TBI is most often used to diagnose hypoperfusion, hyperemia, vasospasm, and cessation of cerebral blood flow in the brain death.

As shown by Calviello (2019), TCD can serve as a source of secondary, "derivative" data, on the basis of which the parameters of the state of the microvasculature can be calculated. For example, critical closure pressure, time constant, arterial compliance, etc.

It is generally accepted (Martin et al., 1997) that TCD in TBI is most often used to diagnose hypoperfusion, hyperemia, vasospasm and cessation of cerebral blood flow in brain death.

However, as shown by Schmidt (2016), Zeiler et al. (2019), and Nag (2019), and many other authors, TCD can serve as a source of secondary "derivative" data, based on which the parameters of the state of the microvasculature can be calculated. For example, critical closure pressure, time constant, arterial compliance, etc.

As noted by Robba (2018), monitoring of such parameters, carried out with the help of TCD, allows a new look at cerebral blood flow and its regulation.

Physical basis for transcranial Doppler ultrasound

According to Aaslid et al. (1982), in the human body, the ultrasound is a wave-like propagating oscillation of tissue density within the frequency range from 16,000 Hz and more. The frequencies ranging from 2 to 10 MHz are used to insonate the vessels. The higher the ultrasound frequency (i.e., the shorter the wavelength), the higher is the device resolution. The crystal-based piezoelectric transducers are used to generate megahertz waves. When investigating blood, flow such a property of the ultrasound is used as the reflection from moving objects, i.e., formed blood elements. The Doppler effect manifests itself in the shift of the reflected signal frequency, formed red blood elements therein. An approaching flow creates a positive shift and increases a frequency, while a receding one forms a negative shift or decrease in the signal frequency. Thus, the direction of blood flow relative to the Doppler sensor may be determined based on frequency shift direction. An ultrasonic sensor operating a pulsed mode is used to examine cerebral vessels. The sensor crystal emits at different periods ultrasonic waves, which are reflected from moving red blood cells with relevant frequency changes (the Doppler effect), and perceives a reflected signal. The best-distinct signal is obtained when the sensor and the vascular plane form a 45-degree angle. The ultrasound signal frequency identifies the ultrasound penetration depth and the lower threshold for the blood flow velocity determination. The higher frequency, the less the ultrasound penetration depth, and the lower blood flow velocities can be located.

According to Aaslid et al. (1982), the information during ultrasonic Doppler examination is provided for analysis as follows: in audio form as characteristic pulsating noise of the artery being located and in visual form as a curve on the monitor screen or a *Doppler image*. The Doppler image visible on the monitor screen is a graphical reflection of the Doppler shift of all frequencies resulting from contact with moving blood flow in real time. Dominant frequencies are thereat reflected more clearly. The Doppler image clarity depends significantly on the transparency of the "window" used and the vessel locating angle. There are available low frequency "noises" caused by oscillations of the blood vessel walls, the summation of arterial and venous signals and places of the vessel bifurcation.

The TCD method records the velocity spectrum in a vessel. It is known that in a constant laminar blood flow, the highest velocity is in the middle of the bloodstream, which is visualized in the envelope curve mode Fig. 1.

Nowadays, quite a significant number of the equipment has been developed for transcranial Doppler sonography, which differs by the level of computerization, the transducer sensitivity, the quantity and operating modes of a spectrum analyzer, the number of channels for Doppler imaging, the operating frequencies of sensors, etc. To perform the transcranial Doppler sonography, a 2 MHz transducer (PW) with power up to 100 MW/cm must be available. Reference values of Doppler parameters of cerebral blood flow in intracranial arteries are shown in Table 1.

Technical features and ultrasound neurovascular anatomy

The conclusion about the cerebral macrocirculation state is made based on computer processing of obtained Doppler image parameters.

According to Akbik (2016), a linear relationship is described between CBF and FV assuming a stable vascular cross-sectional area and constant angle of insonation:

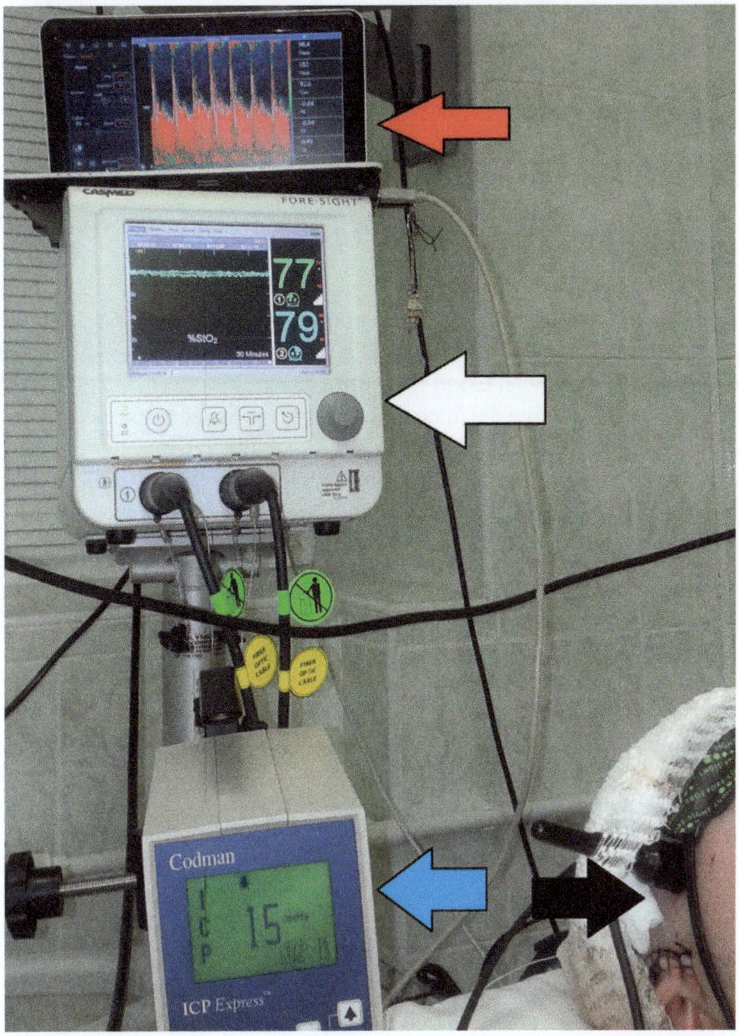

FIG. 1 The investigation system appearance including TCD. A *red arrow* indicates a TCD, a *white arrow* shows a cerebral oximeter, a *blue arrow* shows ICP monitor and a *black arrow* marks a TCD probe.

TABLE 1 Reference values of Doppler parameters of cerebral blood flow in intracranial arteries.

Artery/depth (mm)	Age	Maximum systolic cerebral blood flow velocity (cm/s)	Minimum diastolic cerebral blood flow velocity (cm/s)	Average cerebral blood flow velocity (cm/s)	PI
MCA (50–60)	≤ 40	102.2 ± 9.7	49.1 ± 3.2	64.3 ± 2.8	0.82 ± 0.07
	41–59	89.3 ± 6.5	42.0 ± 3.9	56.4 ± 2.1	0.84 ± 0.08
	≥ 60	75.5 ± 5.4	33.4 ± 3.1	48.3 ± 2.8	0.86 ± 0.08
ACA (65–70)	≤ 40	76.7 ± 6.8	38.1 ± 3.3	45.8 ± 3.2	0.84 ± 0.08
	41–59	74.6 ± 4.9	36.8 ± 3.5	43.9 ± 3.7	0.86 ± 0.08
	≥ 60	60.7 ± 3.8	27.3 ± 2.8	37.0 ± 2.3	0.9 ± 0.09
PCA (60–75)	≤ 40	60.5 ± 5.7	29.5 ± 2.9	38.1 ± 3.1	0.82 ± 0.09
	41–59	55.6 ± 4.3	26.1 ± 2.2	37.0 ± 3.7	0.80 ± 0.08
	≥ 60	60.5 ± 8.4	27.0 ± 2.9	37.3 ± 2.8	0.90 ± 0.09

$$\text{CBF} = \text{FV} \times \text{S area of the insonated vessel} \times \text{Cosine of the angle of the insonation}$$

Flow velocity characteristics include:

- the systolic flow velocity (FVs),
- diastolic flow velocity (FVd)
- and mean flow velocity (FVm) measured in cm/s

The FV depends on the vessel diameter, CO_2 concentration, blood pressure, peripheral cerebrovascular resistance as well on vascular elastance (compliance).

The latter's level depends on the blood viscosity, the intracranial pressure and the state of the microvasculature or the pial-capillary vascular network, calculated from the sFV, dFV, and mFV ratios. The most sensitive in relation to changes in the peripheral resistance level is the pulsation index or the Gosling index (PI) = the ratio of the difference in systolic (S) and end-diastolic (D) blood flow velocity to the mean blood flow velocity.

The second important indicator characterizing the microvasculature state is the resistance index (RI) or the Purselot index.

As distinct from the blood flow velocity, PI and RI do not depend on the vessel's angle. The available close correlation between the Purselot and Gosling indices on the one hand and intracranial pressure (ICP) and cerebral perfusion pressure (CPP) on the other, RI = (S–D)/D, is also of great importance.

Other indices are shown in Table 2.

For the evaluation of cerebral autoregulation, the Cambridge group «Brain Physics» proposed the following TCD indices (Zeiler et al., 2019):

- mean flow index [Mx]: correlation between FVm and CPP;
- Mx_a: correlation between FVm and MAP;

TABLE 2 Different Doppler parameters and explanation.

Abbreviation	Term	Calculation formula
RI	Resistance index (Pourcelot)	Vs – Vd/Vs
PI	Pulsatility index (Gosling)	Vs – Vd/MV
mFV	Mean velocity	Mean velocity calculated over a series of cardiac cycles
sFV	Peak-systolic velocity	Highest velocity in the cardiac cycle
dFV	End-systolic velocity	Velocity at the end of the systolic phase of cardiac cycle
CPP	Cerebral perfusion pressure	eCPP = MAP × Vd/Vm + 14
CBFF	Cerebral blood flow fluctuation	Interquartile range of velocity
Min–max MV ratio	Ratio of minimum and maximum MV	Minimal MV/maximal MV
Vmean ratio	Ratio of VM at different times	MV in first 12 h/MV at 6–168 h
TAV	Time averaged velocity	Vmax/2
CV%	Coefficient of variability	Coefficient of variation of AUVC values of 20 consecutive cardiac cycles
BFV	Blood flow velocity	
CPP	Cerebrovascular perfusion pressure	Mean BFV/(mean BFV – diastolic BFV) × (mean BP – diastolic BP)
CPP	Critical closing pressure	$\text{CPP} = \text{APsys} - \frac{\text{APsys} - \text{APdia}}{\text{Vsys} - \text{Via}} \times \text{Vsys}$
CVR	Cerebrovascular resistance	CVR = CPP/CBFV
CAC	Cerebral arterial compliance	cAC = ΔCBV ÷ ABPamp
CTC	Cerebrovascular time constant	CTC = CAC × CVR
SI	Soustiel index	maximum MFV BA/extracranial VA MFV

- Sx: correlation between FVs and CPP;
- Sx_a: correlation between FVs and MAP;
- Dx: correlation between FVd and CPP;
- Dx_a: correlation between FVd and MAP.

When examining a patient with TBI, the first task is to obtain a clear Doppler signal. It is not possible in all patients because of the lack of the window "transparency." There are four acoustic windows: the temporal, occipital (foraminal), orbital, and submandibular (retromandibular). To get the complete information during the TCD, one must adhere to the following rules:

1. Enough special gel for ultrasound examination should be applied to the patient's skin under the sensor for reliable contact.
2. Prior to performing the examination, the findings of other investigation methods should be taken into account, especially those about any possible displacement of arteries in case of intracranial hematomas and brain contusions.
3. Spectrograms should be recorded when receiving the maximum clear and stable ultrasonic signal.
4. In any doubtful and unclear cases, the examination is to be repeated.
5. The examination is to be conducted (if possible) prior to drug administration.
6. The examination is to be postponed in case of the patient's condition deterioration (Fig. 1).

The insonation of the main cerebral arteries is performed using the main ultrasound windows described below. The temporal window is used to locate the middle cerebral artery, the supraclinoid segment of the internal carotid artery, the anterior cerebral artery and the posterior cerebral artery. The temporal window is most frequently used in practice. It appears to be "non-transparent" because of age-related changes in 25%–30% of examinees aged over 60 and in 80% of them only on one side. In these cases, the insonation of the arteries of the large cerebral arterial circle is carried out from the side of the "transparent" window through the midline. In typical cases, the sensor is placed at a point located 1 cm above the zygomatic arch along the hair growth border and in the position perpendicular to the skin at the 50 mm depth a stable signal is obtained from the M1 segment of the internal carotid artery.

The middle cerebral artery (MCA)—the search shall be carried out from the point in the temporal fossa located opposite the auditory canal starting from the insonation depth of 50–55 mm in the frontal plane.

If no signal is found, the sensor axis perpendicular to its plane is to be shifted towards the opposite eye; the search for the vessel is performed by small movements of the sensor end. If it fails, the sensor is to be moved slightly to the frontal direction. The correctness of the signal from the middle cerebral artery is achieved by shifting the sensor somewhat in different directions with searching for a clear image of the signal. Upon obtaining thereof, it is necessary to go to a depth of 45–40 mm and to record a signal from the distal part of the MCA. The average blood flow velocity therein is 60 ± 10 cm/s.

The anterior cerebral artery (ACA) is at the initial location depth of 60–70 mm in the frontal-basal direction; the blood flow therein is directed to the opposite side; therefore, the study is carried out with reversing.

Sometimes significant difficulties occur in distinguishing the ACA from the siphon of the internal carotid artery. For this purpose, several tests can be used. The ipsilateral compression of the carotid artery in the neck should reverse the ACA's direction (if there is no communicating artery dysplasia). Conversely, the contralateral compression of the common carotid artery increases the ACA's signal clarity without changing the direction. The average blood flow velocity therein is 50 ± 12 cm/s (Fig. 2).

After the insonation of the M2 segment and the proximal parts of the M1 segment at the depths of 30–45 and 55 mm, respectively, the control volume is immersed to 60–65 mm, where under ideal conditions, a bi-directional blood flow is recorded from the M1 section, the supraclinoid segment of the internal carotid artery and the A1 segment of the ipsilateral anterior cerebral artery, the detailed study thereof is carried out at the 70 mm depth. Upon examination completion of the anterior parts of Willis circle, the sensor is shifted by 1 cm to the ear and it is given a caudal direction. After that, the P1 segment of the ipsilateral artery is probed at a depth of 65–75 mm and the contralateral posterior cerebral artery at a depth of 75–85 mm. After tilting the sensor in relation to the skin by 60–500 at a depth of 60–70 mm, the P2–3 segments of the posterior cerebral artery were probed. The posterior cerebral artery is seldom examined due to the difficulty of its clear identification. Moreover, the insolation angle exceeds 500, thus affecting the reliability of findings. The search for the artery is carried out at a depth of 60–75 mm in the posterior direction from the middle cerebral artery. As a reference-point here may serve in some cases the posterior communicating artery with the blood flow therein directed to the opposite side. Upon its clear location, the distance and retrodeviation angle from the temporal window is increased. Sometimes P1 segments of the posterior cerebral arteries are located at the maximum depth. Having obtained the reverse blood flow, the posterior cerebral artery is identified. The blood flow velocity therein varies within 42 ± 10 cm/s (Fig. 3).

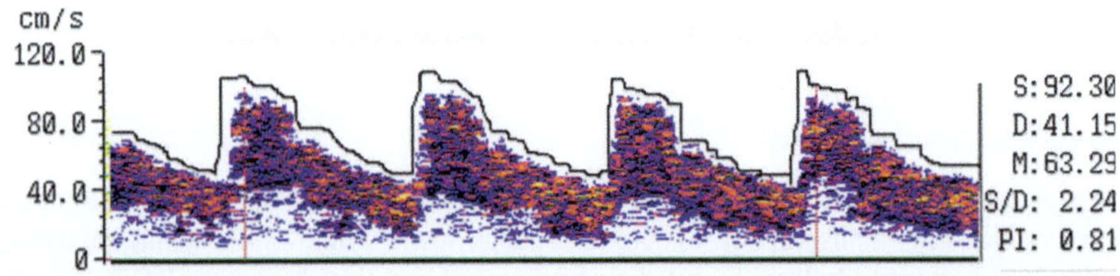

FIG. 2 Transcranial Doppler sonography identifies the linear velocity of blood flow in systolic and diastolic phases. The direction and nature of the blood flow, gradients in different parts of the vascular bed, may be determined by the shape and structure of Doppler images. With consideration for additional indices, the state and elasticity of the vascular wall may be characterized.

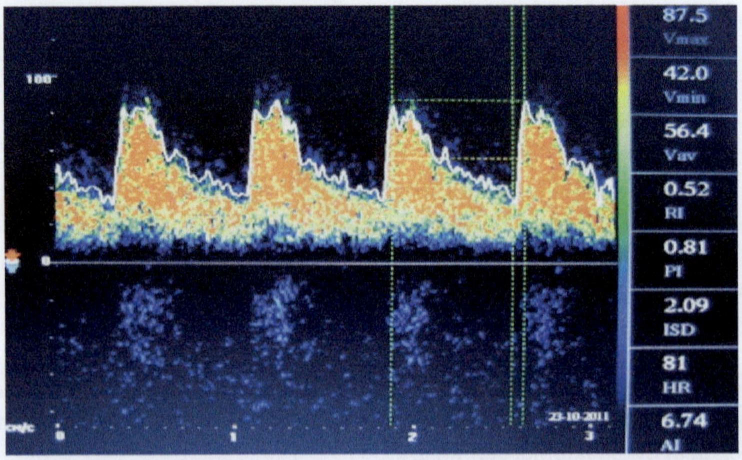

FIG. 3 Transcranial Doppler sonography of anterior cerebral artery.

The diagnostic criteria for a vessel as the Doppler shift source are the perception of the stereotopic structure of the Willis circle and the blood flow direction in a vessel under study.

The orbital window is used to insonate the proximal segments of the ophthalmic artery, the siphon of the internal carotid artery and the posterior communicating artery. The sensor is placed on the upper eyelid in the orbit's upper-lateral corner. The beam is directed to the point of the projection of its apex, the optic nerve canal, where the orbital artery is probed at a depth of 35–45mm. After the "immersion" to 55–70mm, signals are recorded from the "lead-in" and "lead-off" necks of the siphon and, occasionally, from the posterior communicating artery. The radiation power in the transorbital mode of operation is 10–25MW/cm.

The ophthalmic artery is examined through the eyeball under low ultrasound power. Usually, the 4 or 8MHz sensors are used, in some cases a 2MHz sensor at a minimum distance of 20–30mm.

The sub-occipital window is used to insonate the distal segments of the vertebral and basilar arteries. The sensor is placed in the occipital region at the level of the spinous process of the first cervical vertebra. The beam is directed to the point corresponding to the projection of the middle of the frontal region, where vertebral arteries are probed at a depth of 35–45mm with a slight lateral rotation. After the "immersion" to 55–70mm, signals are recorded from the basilar artery, which may be in some cases traced over the length of 10–20mm.

The basilar artery is located in the midline. The sensor is placed at the level of the spinous processes of the C1–C2 vertebrae at the location depth of 60–90mm; the location direction is declivous; the ultrasound signal thereat penetrates through the great occipital foramen. If there are no severe osteochondrosis and gross meningeal syndrome, the basilar artery's location is usually uncomplicated, and the artery can be examined throughout. The mean blood flow velocity therein varies within 42 ± 10cm/s (Fig. 3).

Typical blood flow values in vessels depending on the type of the main arterial vessel and the patient's age are presented in Table 1.

Cerebrovascular disorders in severe traumatic brain injury

A traumatic impact on the brain causes early cerebrovascular disorders at the hemispheric, regional and local levels.

Classic works of Zurynski (1995), Martin et al. (1997) and Steiger (1994) have shown three phases of CBF response after TBI:

1. Hypoperfusion (day 0)—a low CBF and relatively normal MCA velocity.
2. Hyperperfusion (days 1–3)—a raised CBF, high MCA velocity, and normal LR.
3. Vasospasm (days 4–14)—a lower CBF, high MCA velocity, and high LR.

The importance of ischemic brain damages in TBI was first established over 40 years ago by Graham et al. (1978). They revealed it on sectional material in 91% of patients who had died from brain injuries. Ischemic foci were most frequently found in the hippocampus, basal ganglia as well as in cerebral cortex and cerebellum. In the survivors, such lesions were found depending on the severity of the injury in 27%–43% of cases. In many respects, based on these findings, the brain compression treatment main principles were formulated, which consisted in the immediate elimination of extra- and intracranial causes of ischemia: hypoxia, hypotension and compression of the brain by intracranial hematomas. Such tactics resulted in a decrease in the total post TBI mortality, but the ischemia frequency remains nearly the same.

The mechanisms of cerebral tissue damage in ischemic and traumatic strokes are currently recognized as similar. The leading causes of unfavorable outcomes in such cases are secondary pathological processes initiated by the primary insult and promoting the medulla destruction zone's expansion (Graham et al., 1978).

As shown by van Santbrink (2002), inadequate CPP is deemed to play one of the leading roles in the post-TBI ischemia development. High intracranial pressure with associated changes in CPP is regarded among the main mechanisms of its development. The CPP is calculated by the equation:

$$\text{CPP} = (1.1 \times \text{CPP estimated}) - 5 = 1.1 \times \text{BPsys} \times \text{FVm} / \text{FVs}.$$

It is noted by Zeiler et al. (2020), reducing CPP lower than 50 mmHg results in global cerebral ischemia, 90% of such TBI sufferers die. Hence, the importance of systemic hemodynamics indicators, in particular, a drop in systolic pressure below 90 mmHg and PaO_2 below 8 kPa causes 100% mortality.

In summary, it may be stated that the mechanisms of vascular disorders in severe TBI comprise the following three main factors: cerebral hypoperfusion due to intracranial hypertension, vasospasm, and cerebral hyperemia. These pathological processes may be diagnosed with the transcranial Doppler sonography aid based on characteristic changes in Doppler patterns or the so-called TCD patterns (Fig. 4).

Hypoperfusion

Studies by Alexandrov (2013) have shown that the registration of this pattern (which is also called "hindered perfusion") in patients with TBI has the greatest clinical significance since it is highly likely to indicate high intracranial pressure as a result of cerebral edema or intracranial hematoma.

As shown by Lewis (2012), the mechanisms for changing the velocity of blood flow in the arteries can be as follows. After the loss of compensatory capabilities to withstand high intracranial pressure, the smooth muscle wall of the artery is compressed, primarily in the diastole phase, when the hydrostatic arterial pressure in the vessel lumen is minimal. This phenomenon is called by Kondakov (2002) "diastolic compression" and can be diagnosed on TCD by a significant decrease

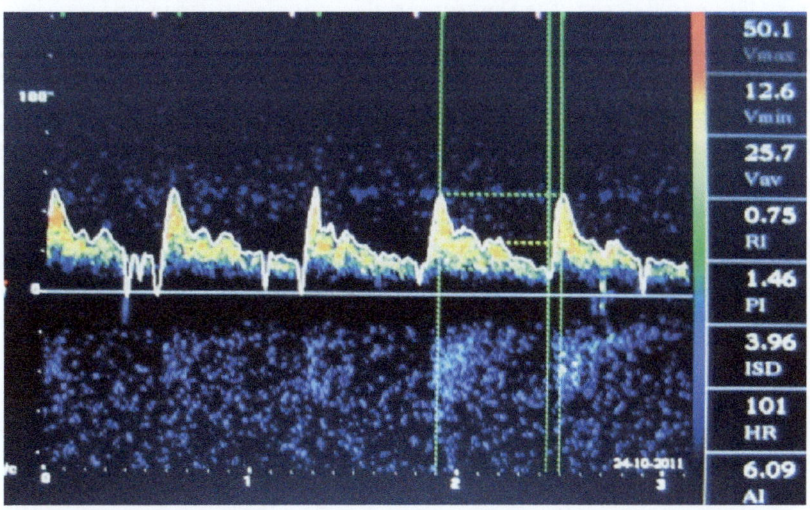

FIG. 4 Transcranial Doppler sonography of basilar artery.

in dFV. The increasing difference between systolic and diastolic FV is accompanied by an increase in the pulsation index. PI values most closely reflect the degree of impaired perfusion and therefore ICH. The PI also increases, reflecting the increase in CVR. The triad described above (decrease in dFV, increase in PI and RI) shows a decrease in CPP as a key indicator in assessing the brain blood supply adequacy.

The recording of this pattern in patients with TBI is of the greatest clinical significance since it proves with great probability high intracranial pressure as a result of cerebral edema or intracranial hematoma (Abecasis et al., 2020).

Studies by Varsos (2014), Kaczmarska et al. (2018), and Robba et al. (2018) have shown that diastolic blood flow velocity is most sensitive to changes in ICP. Its significant decrease is observed with ICP equal to half the diastolic blood pressure. On the other hand, diastolic velocity approaches zero when ICP approaches diastolic blood pressure (51 ± 12 mmHg).

The pressure at which the collapse of the cerebral vascular bed and cerebral blood flow stops is called the critical closing pressure or zero flow pressure. The CCP-model was proposed by Burton (1951) and developed by Dewey (1974) and Permutt and Riley (1963).

It has been shown by Varsos (2014) that with moderate intracranial hypertension in cases of traumatic brain injury, the CCP value is on the average 40.36 ± 17.44 mmHg but at the "plateau-wave" development, it increases by almost 40% to 54.71 ± 17.66 mmHg.

In a broad sense, the CCP plays a key role in limiting excessive cerebral perfusion, which damages the structure of the blood–brain barrier and perhaps the glymphatic system (Plog et al., 2015).

Thus, TCD can be used not only to diagnose an increase in ICP but also to determine the parameters of the state of the microcirculatory bed.

TCD - diagnostics of intracranial hypertension in TBI:

1. signs of hypoperfusion (hindered perfusion) in all the main arteries,
2. a progressive homogeneous decrease in the mean blood flow velocity,
3. increase indices of peripheral resistance in all basins.

To solve tactical problems in traumatic brain injury, monitoring the flow velocity in the middle cerebral artery is most practical. As intracranial hypertension grows and the critical ICP level approaches close to diastolic blood pressure, the cerebral blood flow becomes ineffective, namely intermittent, monophasic, or reverberating and this was shown previously by Eder et al. (2012) and Vicenzini (2010).

Vasospasm

According to different authors' findings, a post-traumatic arterial spasm is observed in 30% to 40% of TBI sufferers. The after-effects of a micro-vascular VS (VS of arteries less than 250 μm and arterioles) are the most dangerous. They result in a significant reduction of cerebral blood flow in large areas of the brain and may cause an infarction (Turek et al., 2013).

According to Aaslid et al. (1982), the connection with the introduction of TCD into clinical practice in the eighties of the 20th century, the problem of post-traumatic vasospasm caught a "second wind."

As Martin et al. (1997) showed, FV is inversely proportional to the square of the radius of the studied vessel: FV increases due to narrowing of the arterial lumen. The Doppler signal of vascular spasm in TBI appears after a latent period on the 3–5th day and exists on average for 4–5 days.

Thus, based on the determination of CBFV in the main arteries of the brain, it became possible to assess: (1) the degree of vasospasm as a parameter associated with arterial narrowing, (2) its prevalence, as well as the dynamics of its development.

Sloan (1996) developed diagnostic criteria for vasospasm (Table 3).

TABLE 3 TCD diagnostic criteria of cerebral vasospasm.

Artery	Possible VS (cm/s)	Probable VS (cm/s)	Definitive VS (cm/s)
ICA	>80	>110	>130
ACA	>90	>110	>120
PCA	>60	>80	>90
BA	>70	>90	>100
VA	>60	>80	>90

According to Alexandrov (2013), the vasospasm is divided into the following three degrees:

Degree I—80–140 cm/s,
Degree II—141–200 cm/s,
Degree III—>200 cm/s.

How to differentiate traumatic vasospasm from cerebral hyperemia with the aid of transcranial Doppler sonography? In case of traumatic brain injury, the blood flow velocity in the homolateral internal carotid artery will be lower because of the high cerebrovascular resistance, while in later cases, there may be hyperemia because of the low vascular resistance. Thus, the ratio of the blood flow velocity in the internal carotid and middle cerebral arteries helps to distinguish these two limit states; for this purpose, the Lindegaard index is used. If the coefficient is higher than 3, the vasospasm of the medial cerebral artery is diagnosed (the norm is $1.7 + 0.4$).

TCD – diagnostics of vasospasm:

1. increasing linear blood flow velocity in the MCA,
2. a slight decrease in pulsation index (PI) values caused by compensatory peripheral vasodilatation or an increase thereof in case of concomitant ICH,
3. in the case of gross spasm, the Doppler spectrum violation (with the formation of the acoustic "murmur" phenomenon),
4. the ratio of the blood flow velocity in extra—and intracranial vessels (the ratio of the linear blood flow velocity in the MCA and ICA on the side of the spasm) exceeds 3.0.

Hyperemia

Among other post-TBI cerebrovascular disorders, hyperemia is of certain importance. The increased blood supply to brain tissue either at the hemispheric or regional level results in increased volume exacerbating vasogenic cerebral edema.

According to Nemoto (2014), another phenomenon of impaired cerebral blood flow in TBI is "shunt blood flow." They observed a progressive increase in micro-vascular shunting with increasing ICP and suggested that the decrease in CPP is associated with micro-vascular shunting, resulting in high CBF with low blood CPP.

Probably, a similar mechanism underlies the development of luxurious perfusion syndrome (luxurious perfusion)—increased blood filling of the brain with vasodilatation of micro-vessels, but the lumen of large vessels remains unchanged. This phenomenon is characterized by a significant increase (mainly due to the pre-diastolic) linear blood flow velocity, a decrease in peripheral resistance due to dilatation of the pial-capillary bed, which determines a low level of hydrodynamic resistance in the system (Lassen, 1966).

According to Bor-Seng-Shu (2019), one of the leading reasons for the development of pattern hyperemiain TBI is "rapid" decompression of the brain (wide craniotomy/craniectomy) against the background of gross violations of cerebral autoregulation.

The main causes of cerebral hyperemia are: anemia, hypercapnia, arterial hypertension, hypervolemia, cerebral metabolic crisis and micro-vascular dilation.

Diagnosis of the hyperemia in TBI.

1. increase in the linear velocity of blood flow (mainly due to the diastolic component),
2. decrease in peripheral resistance,
3. decrease in cerebrovascular reactivity due to a decrease in the vasoconstrictor reserve,
4. MCA mean blood velocity > 100 cm/s along with LR < 3

TCD peculiarities in certain clinical forms of traumatic brain injury

TCD-patterns at cerebral contusions

In this group of sufferers, most often, patterns of difficult perfusion, hyperemia, vasospasm and norm are recorded (in descending order).

In general, the quantitative changes in Doppler images in case of some mild TBI are off the border-line nature as they correspond to the initial degree of one or another TCD-pattern.

The duration of recording FV changes is of clinical importance. The so-called "smooth" normalization of the blood flow velocity parameters within 7–9 days is followed by the gradual improvement in the patients' condition.

More persistent and longer (up to 2–3 weeks) deviations of the FV are wave-like with periodically increasing velocity indicators by over 15–20 cm/s.

Some pathological deviations of Doppler parameters are observed in 90% of patients with moderate brain contusions. When performing the initial TCD, the hindered perfusion pattern dominates. Often, some signs of "obstruction" for the blood flow are recorded in the basilar artery basin. It proved higher intracranial pressure. MLCS varied from 24.7 cm/s to 47.9 cm/s; PI was from 1.0 to 1.45.

The above classification more often corresponds to a "soft" one (mFV of 80–120 cm/s), the PI is from 0.87 to 1.28.

The TCD-pattern of hyperemia is recorded less frequently. mFV is from 77.42 to 110.09 cm/s, the PI is from 0.63 to 0.77.

The dynamics of indicators for the main cerebral blood flow are of significant clinical importance. The so-called "peak-like" dynamics of the increasing FV with the development of VS signs on the 4th–5th day after injury are accompanied by the aggravation of focal symptoms, in particular, speech disorders.

In severe brain contusions, the various macrocirculation disorders are recorded in each and all sufferers. According to our findings, vasospasm development is most often noted in this group.

A distinctive feature is its earlier recording and involvement of both the medial cerebral arteries and the basilar artery. This version of the macrocirculation disorder development is defined as acute diffuse or widespread vasospasm. The clinical course in this version was characterized by profound disorders of consciousness at the admission to the hospital, the availability of gross symptoms of damage to the brainstem parts of the brain with vital conditions. According to our classification, early spasm in severe TBI is regarded as "prominent" (mFV is 120–200 cm/s). The velocity indicators vary from 130.15 cm/s to 185.18 cm/s; the PI is from 0.87 to 1.28.

The Doppler pattern of the "hindered" perfusion in severe TBI is more often identified in the basal artery than in the carotid basins. It indicates the availability of intracranial hypertension and brainstem dislocation. It should be noted that such macro-circulation disorders can develop even on the first day after injury. The FV varies from 22.43 cm/s to 45.87 cm/s, the PI is from 0.97 to 1.56. The hindered perfusion pattern's duration may exceed 3 weeks and directly prove persistent cerebral edema and intracranial hypertension.

Algorithm for the assessment of TCD findings in cerebral contusions

- Type of TCD for the primary examination.
- Type of TCD for secondary measurement (on the 2nd–4th day after injury) or in case of the clinical deterioration of the patient's condition.
- VS-signs—vasodilator therapy, control CT to exclude the formation of secondary ischemic foci.
- Hypoperfusion/"hindered perfusion"—cerebral edema treatment, CT scan to exclude the formation of intracranial hematomas.
- Hyperemia—the prevention and treatment of vasogenic cerebral edema.

TCD-patterns at traumatic ICH

As mentioned above, the following three types of Doppler pattern changes are identified at TBI: low mFV and high PI (hypoperfusion/hindered perfusion pattern); high mFV and normal or higher PI denote a pattern of the vasospasm; high mLBF and normal or lower PI denote a pattern of hyperemia.

According to Kondakov (2002), the TCD-pattern of hypoperfusion/hindered perfusion (55.2%) has the largest specific weight; the cause thereof is the ICH and the diastolic compression of the great vessels caused thereby.

Such a type of Doppler pattern change is most typical for severe TBI. It occurs according to our findings in 85.7% of cases. DFV in these cases is especially noticeably reduced and is 26.6 ± 16.6 cm/s; the PI is increased to (1.41 ± 0.53). Blood flow disorders occur both on the side of the brain compression and also on the opposite side. The pattern of macrocirculation disorders in the cerebral hemispheres is complemented by the data of FV in the basilar artery, where signs of diastolic compression are also evident (1.67 ± 0.43) and reflect the brain dislocation phenomena. This type of blood flow disorders in severe TBI correlates with a higher incidence of adverse outcomes in this group.

As shown by Bor-Seng-Shu (2019), it may be challenging to diagnose vasospasm using TCD if there are intracranial hematomas available because typical TCD signs of VS (mFV is higher 80–90 cm/s) may be concealed by concomitant intracranial hypertension.

Our study has shown that the hyperemia-like changes in cerebral perfusion are only observed in mild TBI patients with sub-acute and chronic subdural hematomas (Trofimov et al., 2016).

Thus, the type of changes in the blood flow of the main cerebral vessels with intracranial hematomas depends on the severity of the injury and the degree of concomitant ICH, cerebral edema, tSAH. In severe TBI, this is "hindered perfusion," in moderate TBI, there are combined signs of hindered perfusion and VS; in mild TBI, there is hyperemia.

TCD-patterns after surgical removal of ICH

As shown by Sarı (2020), the dynamics of TCD findings after the removal of intracranial hematomas are of clinical relevance. The blood flow dynamics in the great cerebral vessels are of different nature depending on the type of initial disorders.

The reperfusion after eliminating the brain compression with regression of diastolic compression and a significant increase in the FV is a favorable prognosis. The phenomenon of "reperfusion" in the large cerebral vessels reflects the creation of the brain decompression and the decreasing intracranial pressure (Bonow et al., 2019).

Nevertheless, a significant increase in FV (hyperemia) is potentially dangerous in terms of the development of postoperative vasogenic cerebral edema and hemorrhagic complications.

According to Kondakov (2002), the most negative event is the decrease in the FV after the hematoma removal; the dropping blood flow in the main artery is exceedingly disastrous. It is followed by general enhancing instability hemodynamics, a deepening degree of coma and death within the first days after the operation.

The lack of dynamics in FV parameters after surgery most often indicates the non-eliminated compression of the brain. Postoperative changes in the FV with VS are greatly variable; surgery could contribute to both its regression and aggravation.

Doppler pattern parameters in the MCA on the side of the operation may be normalized and this is a positive sign, but it does not always mean the FV normalization in the other hemisphere or at the sub-tentorial level. The delayed post-surgery development of VS signs is especially negative. Clinically it results in the deepening of consciousness disorders to deep com, the aggravation of focal and dislocation symptoms.

As noted by Fatima (2019) that targeted intensive therapy in TBI should consider the type of hemodynamic disorders wherein the TCD distinguishes the pole states of cerebral hemodynamics: the hypoperfusion (or "hindered perfusion") on the one hand and the hyperemia on the other.

According to Kondakov (2002), by assessing both "fast" (developing within several tens of minutes) and "slow" vascular reactions and determining a dominant pathological link (cerebral hyperemia, intracranial hypertension, cerebral vasospasm), the nature of cerebral circulation disorders is specified, and an individual therapeutic and tactical program is determined.

The widespread use of the transcranial Doppler ultrasound technique has enabled us to ensure of its high efficiency. However, in our opinion, in the TBI patient examination protocol, the dynamic TCD of cerebral vessels is more expedient. It allows to identify not only the initial pattern of cerebral macro-circulation disorders but also to assess any tendencies for its changes during targeted treatment. The dynamic TCD is of particular importance in neuroresuscitation practice in the protocol of multimodal neuro-monitoring of the condition of patients with TBI.

Applications to other areas of neuroscience

In this chapter, we presented the practice of using TCD in TBI to diagnose hypoperfusion, vasospasm, and hyperemia. At the same time, the possibilities of using Doppler ultrasonography are much broader than a simple assessment of cerebral circulation.

For example, in the case of steno-occlusive carotid diseases, TCD allows assessing the degree of the stenosis and determining the indications for the carotid stenting or endarteriectomy.

Moreover, TCD is an indispensable tool for assessing the restoration of CBF during reperfusion therapy in ischemic strokes.

In addition, TCD makes it possible to determine the early signs and assess the dynamics of cerebral vasospasm in the SAH due to the ruptured aneurysms, as well as to define signs of brain death due to a critical rise of ICP.

TCD makes possible a non-invasive assessment of not only the arterial blood flow but also venous outflow (for example, in cerebral sinus thrombosis or in the diagnosis of arteriovenous malformations).

Although an increase in the frequency of ultrasound decreases the insonation depth, in the presence of a skull defect (for example, after a craniotomy), it becomes possible to examine the blood flow in the micro-vascular bed.

We would like to point out that our review is only a small piece of the vast canvas of the use of Doppler ultrasound in biomedical research and clinical practice.

Mini-dictionary of terms

Laminar parabolic blood flow is when there is a gradient of flow velocity between the central (maximum velocity) and near-wall (minimum) layers. The difference between these velocities is maximal in systole and minimal in diastole. If layers do not mix with each other, the flow is laminar. It is typical for large vessels of the brain.

Laminar plug-like (parallel) blood flow is when flow velocities of all blood layers, both central and parietal, are nearly equal, which is not typical for cerebral vessels.

Turbulent blood flow is when the layers begin to mix because of the non-homogeneity of the vascular wall and the high velocity of blood flow. In that case, the normal distribution of velocities in the flow is disrupted and red blood cells make chaotic movements.

"Luxury perfusion" syndrome is a phenomenon of excessive cerebral blood flow that exceeds the metabolic and oxygen demands of the brain tissue. It develops in the area of previous ischemia and reflects the impairments of cerebral autoregulation.

Brain death is the irreversible stop of the brain function due to insufficient cerebral perfusion, severe cerebral ischemia and hypoxia, brain edema, and cerebral circulation stop.

Key facts of transcranial Doppler ultrasound

- Doppler's method is associated with the works of C. Doppler (1803–1854).
- Doppler discovered the following phenomenon—then an ultrasonic signal is reflected from a moving object, the radiation frequency shift (the Doppler effect) occurs directly proportional to the velocity of the said objects. While performing TCD, such moving objects are red blood cells—the erythrocytes in cerebral vessels.
- The recording of CBFV in intracranial vessels through the temporal bone was first applied and described by Aaslid et al. (1982). The first methodological publication was issued in 1986. Further one appeared in 1992, co-authored with Newell.
- Approximately 10%–20% of patients having acoustically non-transparent temporal windows.
- The TCD is an operator-dependent technique.
- The TCD demonstrates a 100% agreement with the "gold standard" of the angiography for the confirmation of brain death.

Summary points

- TCD is a portable, repeatable, non-invasive technique in the bedside assessment and monitoring of cerebral blood flow with high temporal resolution.
- Moderate-to-severe TBI may lead to hypoperfusion (day 0), hyperemia (days 1–3), and vasospasm (days 4–15).
- TCD can identify TBI-related cerebral hemodynamic changes, which can be used as early biomarkers of outcome according to GOS.
- Non-invasive TCD estimates of ICP, CCP and cerebral autoregulation require further validation in randomizing clinical trials for the evaluation of the benefit of CPP-targeting and autoregulation-oriented therapy in TBI.
- TCD can complement cerebral CT/MRI to more accurately screen the patients at risk of neurological deterioration after TBI.

References

Aaslid, R., Markwalder, T. M., & Nornes, H. (1982). Noninvasive transcranial Doppler ultrasound recording of flow velocity in basal cerebral arteries. *Journal of Neurosurgery*, *57*(6), 769–774.

Abecasis, F., Cardim, D., Czosnyka, M., Robba, C., & Agrawal, S. (2020). Transcranial Doppler as a non-invasive method to estimate cerebral perfusion pressure in children with severe traumatic brain injury. *Child's Nervous System: ChNS: Official Journal of the International Society for Pediatric Neurosurgery*, *36*(1), 125–131.

Akbik, O. S., Carlson, A. P., Krasberg, M., & Yonas, H. (2016). The utility of cerebral blood flow assessment in TBI. *Current Neurology and Neuroscience Reports*, *16*(8), 72.

Alexandrov, A. V. (2013). *Neurovascular ultrasound examination*. Wiley.

Al-Mufti, F., Amuluru, K., Lander, M., Mathew, M., El-Ghanem, M., Nuoman, R., et al. (2018). Low Glasgow coma score in traumatic intracranial hemorrhage predicts development of cerebral vasospasm. *World Neurosurgery*, *120*, e68–e71.

Bonow, R. H., Young, C. C., Bass, D. I., Moore, A., & Levitt, M. R. (2019). Transcranial Doppler ultrasonography in neurological surgery and neurocritical care. *Neurosurgical Focus*, *47*(6). https://doi.org/10.3171/2019.9.FOCUS19611, E2.

Bor-Seng-Shu, E., de Lima-Oliveira, M., Nogueira, R. C., Almeida, K. J., Paschoal, E., & Paschoal, F. M., Jr. (2019). Decompressive Craniectomy for traumatic brain injury: Postoperative TCD cerebral hemodynamic evaluation. *Frontiers in Neurology*, *10*, 354.

Burton, A. C. (1951). On the physical equilibrium of small blood vessels. *The American Journal of Physiology*, *164*(2), 319–329.

Calviello, L. A., Zeiler, F. A., Donnelly, J., Uryga, A., de Riva, N., Smielewski, P., et al. (2019). Estimation of pulsatile cerebral arterial blood volume based on transcranial doppler signals. *Medical Engineering & Physics*, *74*, 23–32.

Dewey, R. C., Pieper, H. P., & Hunt, W. E. (1974). Experimental cerebral hemodynamics. Vasomotor tone, critical closing pressure, and vascular bed resistance. *Journal of Neurosurgery*, *41*(5), 597–606.

Eder, K. E., Haussen, D. C., Searls, D. E., & Henninger, N. (2012). Reverberating TCD flow pattern in brain death. *Neurology*, *79*(9), e79.

Fatima, N., Shuaib, A., Chughtai, T. S., Ayyad, A., & Saqqur, M. (2019). The role of transcranial Doppler in traumatic brain injury: A systemic review and Meta-analysis. *Asian Journal of Neurosurgery*, *14*(3), 626–633.

Graham, D. I., Adams, J. H., & Doyle, D. (1978). Ischaemic brain damage in fatal non-missile head injuries. *Journal of the Neurological Sciences*, *39*(2–3), 213–234.

Kaczmarska, K., Kasprowicz, M., Grzanka, A., Zabołotny, W., Smielewski, P., Lalou, D. A., et al. (2018). Critical closing pressure during a controlled increase in intracranial pressure. *Acta Neurochirurgica. Supplement*, *126*, 133–137.

Kondakov, E. N. (2002). *Severe traumatic brain injury Sankt-Peterburg, Albi. Vol. 2002* (p. 146). Albi (in Russian).

Lassen, N. A. (1966). The luxury-perfusion syndrome and its possible relation to acute metabolic acidosis localised within the brain. *Lancet (London, England)*, *2*(7473), 1113–1115.

Lewis, P. M., Smielewski, P., Rosenfeld, J. V., Pickard, J. D., & Czosnyka, M. (2012). Monitoring of the association between cerebral blood flow velocity and intracranial pressure. *Acta Neurochirurgica. Supplement*, *114*, 147–151.

Martin, N. A., Patwardhan, R. V., Alexander, M. J., Africk, C. Z., Lee, J. H., Shalmon, E., et al. (1997). Characterization of cerebral hemodynamic phases following severe head trauma: Hypoperfusion, hyperemia, and vasospasm. *Journal of Neurosurgery*, *87*(1), 9–19.

Nag, D. S., Sahu, S., Swain, A., & Kant, S. (2019). Intracranial pressure monitoring: Gold standard and recent innovations. *World Journal of Clinical Cases*, *7*(13), 1535–1553.

Nemoto, E. M., Bragin, D. E., Statom, G., Krasberg, M., Pappu, S., Sena, B., et al. (2014). Role of microvascular shunts in the loss of cerebral blood flow autoregulation. *Advances in Experimental Medicine and Biology*, *812*, 43–49.

Niu, N., Tang, Y., Hao, X., & Wang, J. (2020). Non-invasive evaluation of brain death caused by traumatic brain injury by ultrasound imaging. *Frontiers in Neuroinformatics*, *14*, 607365. https://doi.org/10.3389/fninf.2020.607365.

Permutt, S., & Riley, R. L. (1963). Hemodynamics of collapsible vessels with tone: The vascular waterfall. *Journal of Applied Physiology*, *18*, 924–932.

Plog, B. A., Dashnaw, M. L., Hitomi, E., Peng, W., Liao, Y., Lou, N., et al. (2015). Biomarkers of traumatic injury are transported from brain to blood via the glymphatic system. *The Journal of Neuroscience: The Official Journal of the Society for Neuroscience*, *35*(2), 518–526.

Robba, C., Cardim, D., Sekhon, M., Budohoski, K., & Czosnyka, M. (2018). Transcranial Doppler: A stethoscope for the brain-neurocritical care use. *Journal of Neuroscience Research*, *96*(4), 720–730.

Sarı, R., Bolukbasi, F. H., Kahraman Özlü, E. B., Isik, N., Güra Çelik, M., & Elmacı, İ. (2020). Decompressive craniectomy in traumatic brain injury: Transcranial Doppler sonography used as a guide. Travmatik beyin hasarında dekompresif kraniektomi: Transkraniyal Doppler ultrasonografi yol gösterici olabilir mi? *Ulusal travma ve acil cerrahi dergisi = Turkish journal of trauma & emergency surgery: TJTES*, *26*(3), 418–424.

Schmidt, B., Czosnyka, M., Smielewski, P., Plontke, R., Schwarze, J. J., Klingelhöfer, J., et al. (2016). Noninvasive assessment of ICP: Evaluation of new TBI data. *Acta Neurochirurgica. Supplement*, *122*, 69–73.

Sloan, M. A. (1996). Transcranial Doppler monitoring of vasospasm after subarachnoid hemorrhage. *Neurosonology* (pp. 156–171). St. Louis: Mosby.

Steiger, H. J., Aaslid, R., Stooss, R., & Seiler, R. W. (1994). Transcranial Doppler monitoring in head injury: Relations between type of injury, flow velocities, vasoreactivity, and outcome. *Neurosurgery*, *34*(1), 79–86.

Trofimov, A. O., Kalentiev, G., Voennov, O., Yuriev, M., Agarkova, D., Trofimova, S., et al. (2016). Comparison of two algorithms for analysis of perfusion computed tomography data for evaluation of cerebral microcirculation in chronic subdural hematoma. *Advances in Experimental Medicine and Biology*, *923*, 407–412.

Turek, R., Linzer, P., Filip, M., Sámal, F., & Jurek, P. (2013). Application of transcranial color-coded sonography in severe brain injury. *Acta Neurochirurgica. Supplement*, *118*, 265–267. https://doi.org/10.1007/978-3-7091-1434-6_50.

van Santbrink, H., Schouten, J. W., Steyerberg, E. W., Avezaat, C. J., & Maas, A. I. (2002). Serial transcranial Doppler measurements in traumatic brain injury with special focus on the early posttraumatic period. *Acta Neurochirurgica*, *144*(11), 1141–1149.

Varsos, G. V., Richards, H. K., Kasprowicz, M., Reinhard, M., Smielewski, P., Brady, K. M., et al. (2014). Cessation of diastolic cerebral blood flow velocity: The role of critical closing pressure. *Neurocritical Care*, *20*(1), 40–48.

Vicenzini, E., Pulitano, P., Cicchetti, R., Randi, F., Papov, P., Spadetta, G., et al. (2010). Transcranial Doppler for brain death in infants: The role of the fontanelles. *European Neurology*, *63*(3), 164–169.

Zeiler, F. A., Cabeleira, M., Hutchinson, P. J., Stocchetti, N., Czosnyka, M., Smielewski, P., et al. (2020). Evaluation of the relationship between slow-waves of intracranial pressure, mean arterial pressure and brain tissue oxygen in TBI: A CENTER-TBI exploratory analysis. *Journal of Clinical Monitoring and Computing*. https://doi.org/10.1007/s10877-020-00527-6.

Zeiler, F. A., Smielewski, P., Stevens, A., Czosnyka, M., Menon, D. K., & Ercole, A. (2019). Non-invasive pressure reactivity index using Doppler systolic flow parameters: A pilot analysis. *Journal of Neurotrauma*, *36*(5), 713–720.

Zurynski, Y. A., Dorsch, N. W., & Fearnside, M. R. (1995). Incidence and effects of increased cerebral blood flow velocity after severe head injury: A transcranial Doppler ultrasound study II. Effect of vasospasm and hyperemia on outcome. *Journal of the Neurological Sciences*, *134*(1–2), 41–46.

Chapter 24

Computed tomography assessment of brain swelling

Matheus Fernandes de Oliveira[a,b], Anselmo Alves Boa Sorte, Jr[a], and Gustavo Fernandes Leobas[a]
[a]*Neurosurgery Department, Public Servant State Hospital, São Paulo, Brazil,* [b]*Neurosurgery Department, DFV Neuro, São Paulo, Brazil*

List of abbreviations

CT	computed tomography
MR	magnetic resonance
TBI	traumatic brain injury
ADC	apparent diffusion coefficient
SAH	subarachnoid hemorrhage
SAHt	traumatic subarachnoid hemorrhage
CE	cerebral edema
ICP	intracranial pressure
ATLS	advanced trauma life support
GCS	Glasgow Coma Scale
ZI	Zumkeller index
MS	midline shift
HT	hematoma thickness
IVH	intra-ventricular hemorrhage

Introduction

Traumatic brain injury is a relevant epidemiological health problem. It is the main cause of death in children and adults worldwide and brings elevated costs for public and private healthcare systems (Adelson, Bratton, Carney, et al., 2003; Ahmed, Biswas, & Paul, 2019; Amorim, Morais, Pereira, et al., 2019; Badri, Chen, Barber, et al., 2012; Bobinski, Olivecrona, & Koskinen, 2012; Carney, Totten, O'Reilly, et al., 2017; Chesnut, Temkin, Carney, et al., 2012; Greenberg, 1984; Hukkelhoven, Steyerberg, Habbema, et al., 2005). According to the Centers for Disease Control and Prevention (CDC) (2019), in 2014 in the USA traumatic brain injury was responsible for 2.87 million visits to emergency services, hospitalizations and deaths, and it is estimated that 150 Americans die every day due to TBI-related injuries. In Europe, traumatic brain injuries have a significant socioeconomic and health impact, with a progressively higher incidence in recent decades, which varies from 262 to 326 individuals per 100,000 inhabitants/year (Peeters et al., 2015). They mainly result from falls suffered by children and the elderly, while in young adults transport accidents predominate.

Primary brain injury occurs at the time of trauma, related to direct injury (blunt or penetrating impact), in addition to acceleration, deceleration and consequent shear forces. Meanwhile, secondary brain injury refers to the set of inflammatory, cellular and ischemic activities, which also begin at the time of the trauma, but can continue for hours or days (Jha, Kochanek, & Simard, 2019; Kim, 2011; Kim, Dong, Yoon, et al., 2014; Lobato et al., 2005; Lobato, Gomez, Alday, et al., 1997; Maas, Hukkelhoven, Marshall, & Steyerberg, 2005; Marmarou, Signoretti, Fatouros, et al., 2006; Marshall, Marshall, Klauber, et al., 1992; Metting, Rödiger, Stewart, et al., 2009).

In this pathophysiological context, cerebral edema (CE) is one of the main determinants related to intracranial hypertension (ICP), a factor that in turn is directly involved in the mechanisms of secondary brain injury, leading to cerebral hypoperfusion and ischemia, which is associated with more edema, within a process of continuous self-feedback, which can lead to brain death. Shortly after the trauma, a variety of disorders in the tissue microstructure contributes to the development of edema (Nujaimin et al., 2009; Pinto et al., 2015; Pinto, Capone-Neto, Prist, Silva, & Poli-de-Figueiredo, 2006; Rahman,

Diagnosis and Treatment of Traumatic Brain Injury. https://doi.org/10.1016/B978-0-12-823347-4.00017-8

Al Den, & Maull, 2010; Stiell, Wells, Vandemheen, et al., 2001; Talari, Hamidian, Moussavi, et al., 2019; Toutant, Klauber, Marshall, et al., 1984; Tucker, Aston, Dines, et al., 2017; Williamson, Morgan, & Klein, 2017; Wintermark et al., 2006).

Initially, it was believed that the main responsible for the highest concentration of water in the brain tissue was vascular engorgement associated with the breakdown of the blood-brain barrier, leakage of proteins and water through the vessel wall and consequent vasogenic edema (Bobinski et al., 2012; Jha et al., 2019; Marmarou et al., 2006; Pinto et al., 2006, 2015). However, recent analyzes of the apparent diffusion coefficient (ADC) in images generated by magnetic resonance have identified that the cerebral edema observed in patients suffering from focal or diffuse trauma is predominantly of cellular origin (cytotoxic edema) and contributes to about 75% of the increase total brain volume in these cases. This, in turn, probably results from the failure of the energy-dependent sodium-potassium pump, leading to sodium influx and water accumulation in the intracellular compartment. Despite this classic division between the mechanisms for the installation of cerebral edema in cytotoxic and vasogenic, new concepts go beyond this fragmented interpretation and increasingly defend that these processes seem to be interrelated (Bobinski et al., 2012; Jha et al., 2019; Marmarou et al., 2006; Pinto et al., 2006, 2015).

Brain swelling follows the increase of intracranial pressure (ICP), occurring in 16% of all TBI victims and in 28% of pediatric head-injured patients (Adelson et al., 2003; Ahmed et al., 2019;Amorim et al., 2019; Badri et al., 2012).

One of the main evaluation parameters is the monitoring of intracranial pressure (ICP). Increases in ICP are directly associated with higher mortality and worse outcomes (Adelson et al., 2003). However, it is not a widely available tool. A randomized clinical trial demonstrated equivalence in relation to the prognosis among patients submitted to intracranial pressure monitoring, when compared to serial evaluation by tomography (Ahmed et al., 2019). Thus, for better intensive management of these critically ill patients, serial brain imaging is an important tool, allowing the assessment of cerebral edema, and the response to adopted therapeutic measures.

Identifying radiological parameters to assess CE, its extension and signs related to clinical-neurological worsening is fundamental in the management of patients suffering from severe traumatic brain injury (TBI) (Amorim et al., 2019; Badri et al., 2012). One of the major challenges in this regard is to determine which parameters related to CE should be used to change therapeutic measures (Amorim et al., 2019; Badri et al., 2012).

Skull X-ray was routinely used in the pre-CT era; however, it has several limitations and too low sensitivity and specificity to perform TBI radiological evaluation (Nujaimin et al., 2009; Pinto et al., 2006, 2015; Rahman et al., 2010; Stiell et al., 2001; Talari et al., 2019; Toutant et al., 1984; Tucker et al., 2017; Williamson et al., 2017; Wintermark et al., 2006). Its application today is solely reserved to anecdotal situations. Magnetic resonance (MR) presents high sensitivity and good resolution in the evaluation of TBI and especially CE, superior to computed tomography (CT). Other nuclear medicine image modalities, such as positron emission tomography and single-photon emission tomography may also bring more information than simple skull CT, remarking inflammatory processes and tissue metabolism. However, such parameters are not correlated with changes in the management of TBI (Amorim et al., 2019; Badri et al., 2012). Besides, MR and nuclear medicine neuroimage use in trauma is decreased due to prolonged time demanded to perform exam and should be chosen according to tailored and individual features of each case. Thus, despite the advances presented by MR and other modalities of neuroimage, CT remains as the exam of choice, both in the evaluation of the acute phase and in the follow-up of patients who are victims of brain trauma. CT advantages include its wide availability, accessibility, rapid image acquisition, especially when considering hemodynamic unstable patients.

CT should be performed early on patients suffering from TBI, as soon as there is clinical stability, within the ATLS (Advanced Trauma Life Support) recommended protocol. Non-contrasted CT examination has high accuracy in the diagnosis of cranial fractures, focal intracranial injuries, hydrocephalus, herniation and cerebral edema. According to Canadian CT Rule criteria (Stiell et al., 2001), every patient with Glasgow Coma Scale (GCS) < 14 or risk factors for tomographic changes, must undergo a tomographic examination (Bobinski et al., 2012), with the initial objective of evaluating injuries that require immediate neurosurgical intervention (such as extra-axial hematomas with midline shift and herniation signs). In addition, it is able to provide information to guide therapy in conservative patients, or in post-operative management, and to provide prognostic parameters.

In this chapter, we present the main aspects of CT use in cerebral edema following TBI.

Parameters to evaluate brain swelling in computed tomography

While evaluating CT scan of a patient suffering from TBI, some signs are typical of intracranial hypertension, potentially associated with risk of life, such as midline shift greater than 5 mm or acute expansive focal lesions with deviation. However, the difficulty in evaluating CE lies in critically ill patients, without these clearly surgical lesions. In these cases, careful evaluation and determination of radiological signs potentially associated with increased intracranial pressure are essential.

High convexity sulci (Fig. 1) pattern should be addressed. The absence of clear delimitation suggests areas of CE and brain swelling. However, there can be great variability between patients with different age groups, notably in the elderly, who have diffuse atrophy, compared to younger patients.

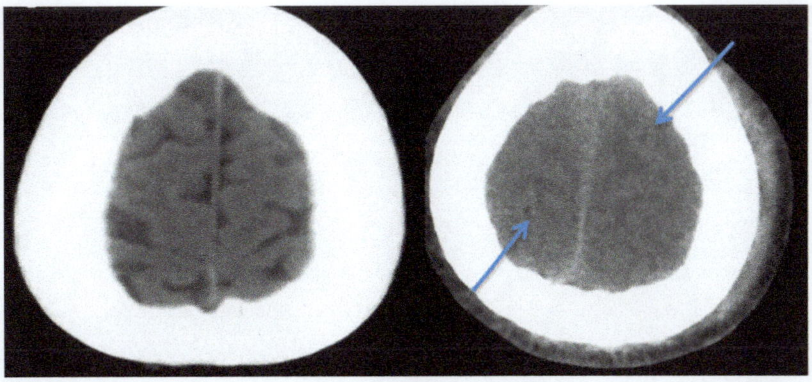

FIG. 1 Evaluation of high convexity sulci. At *left*, normal Skull CT displaying convexity sulci in a healthy patient. At *right*, obliterated sulci in a severe TBI patient with CE.

Basal cisterns status (Fig. 2) is another extremely important and reproducible factor. Relevant cisterns are: interpeduncular, crural, ambiens, and quadrigeminal. Obliteration and disappearance of these structures infers an increase in intracranial pressure, and is directly related to a worse prognosis. This is one of the main parameters evaluated, being also present in the main CT classifications, as Rotterdam and Marshall scales.

The status of **Sylvian fissure** (Fig. 3) may also be clarified. Similarly to high convexity sulci and basal cistern status it has been suggested that Sylvian fissure obliteration correlates with the degree of CE.

Several studies have pointed to the presence of **traumatic Subarachnoid Hemorrhage (SAH)** as an early predictor of brain damage and prognosis after TBI. However, it is a finding widely identified in CT scans of a patient with mild TBI, with good clinical and neurological evolution. Therefore, it must be evaluated in together with other findings, especially in critically ill patients.

Another interesting skull CT finding in cerebral edema following TBI is the "**white cerebellum**" sign (Fig. 4). This happens because due to increased edema in supratentorial brain it turns swollen and remarkedly grayish, compared to the relatively spared cerebellum, which in turn seems more white than supratentorial brain. This sign indicates severe cerebral edema and loss of autoregulation principles and this difference of Hounsfield units between supratentorial and infratentorial brain is only possible due to the existence of cerebellar tent, which is an almost inelastic dural protection allowing different degrees of pressure simultaneously in infratentorial and supratentorial compartments.

Radiological signs of herniation are other indispensable aspects in radiological evaluation. Notably, the presence of uncal and sub-falcine herniation. The evaluation of these findings has different values depending on the moment of evaluation:

(1) At arrival CT, signs of cerebral herniation are usually associated with focal intra- or extra-axial lesions, with a mass effect and suggest the need for a surgical approach.
(2) At follow-up CT, especially in critically ill patients, with diffuse cerebral edema, suggests greater severity, and the need to adopt measures (conservative or surgical).

Uncal herniation is directly related to the obliteration of basal cisterns.

Midline shift, especially when greater than 5 mm is also another important factor for assessing CE and increased intracranial pressure. In the presence of a focal lesion, midline shift is usually proportional to the size of the lesion. However, when midline shift is greater than the extent of the injury, this difference is associated with trauma-related brain edema and focal injury.

This is the principle that guides the use of **Zumkeller Index (ZI)** (Zumkeller, Behrmann, Heissler, & Dietz, 1996) in decision making and assessment of severity in trauma. ZI is calculated with the difference between the midline shift (MS) and the thickness of the lesion (HT—Hematoma Thickness) in the same CT image.

$$ZI = MS - HT$$

This difference in thickness is credited by the authors to "brain swelling" (brain swelling factor). When this difference exceeds 3 mm, the survival rate is around 50%, while deviations exceeding 5 mm from the thickness of the focal lesion are associated with 25% survival.

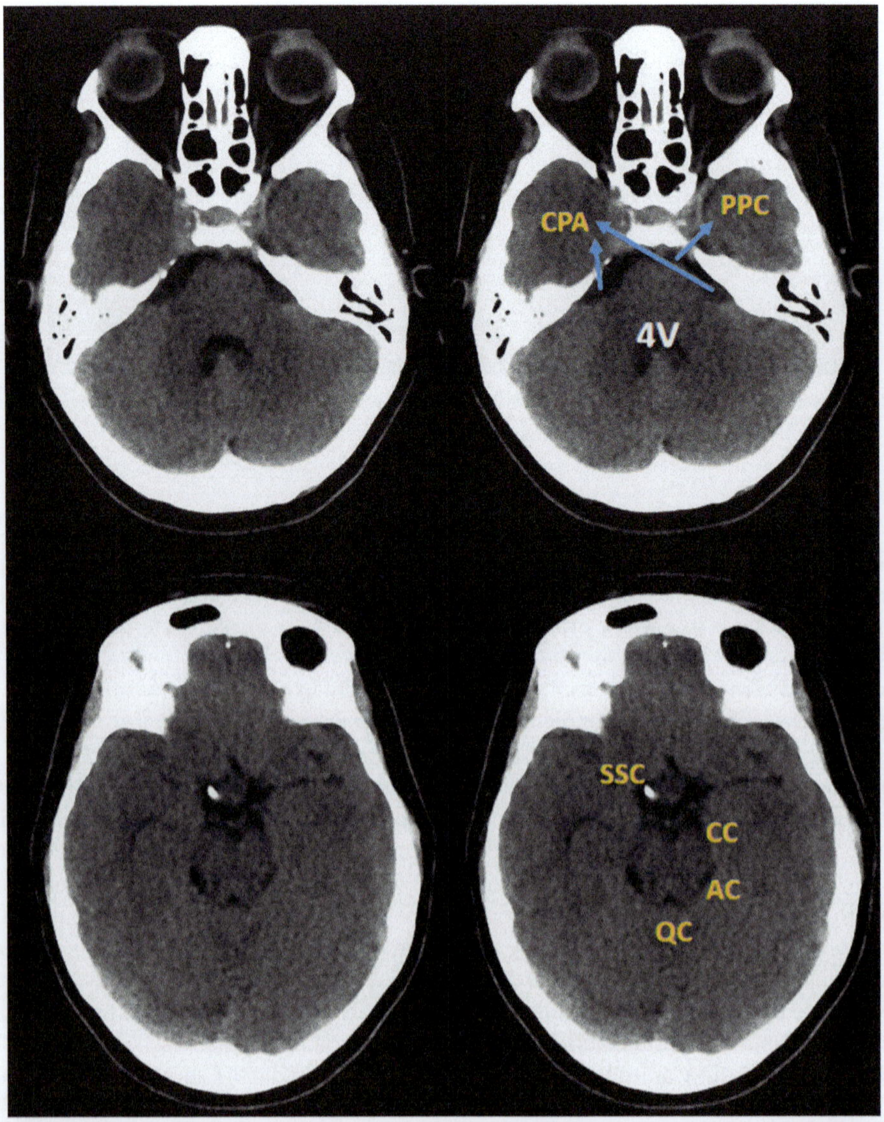

FIG. 2 Evaluation of basal cisterns. Normal Skull CT displaying basal cisterns which should be evaluated in routine TBI evaluation and may be obliterated when cerebral edema occurs. *CPA*, cerebellopontine angle; *PPC*, prepontine cistern; *SSC*, suprasellar cistern; *CC*, crural cistern; *AC*, ambiens cistern; *QC*, quadrigeminal cistern.

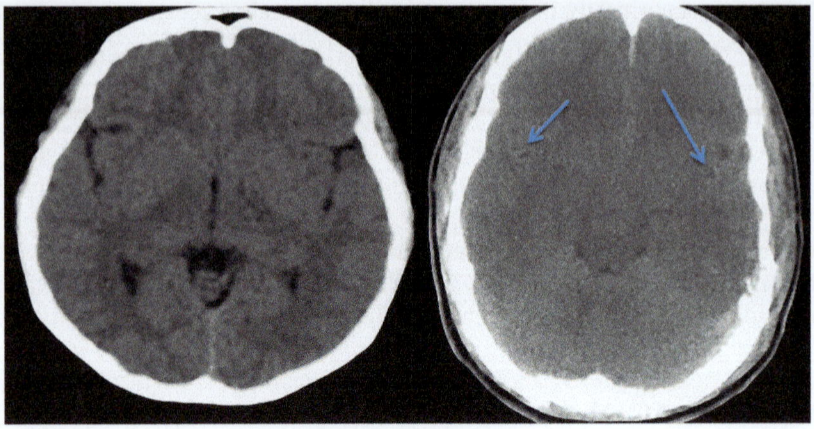

FIG. 3 Evaluation of Sylvian fissure. At *left*, normal skull CT displaying Sylvian fissure in a healthy patient. At *right*, obliterated Sylvian fissure in a severe TBI patient with CE.

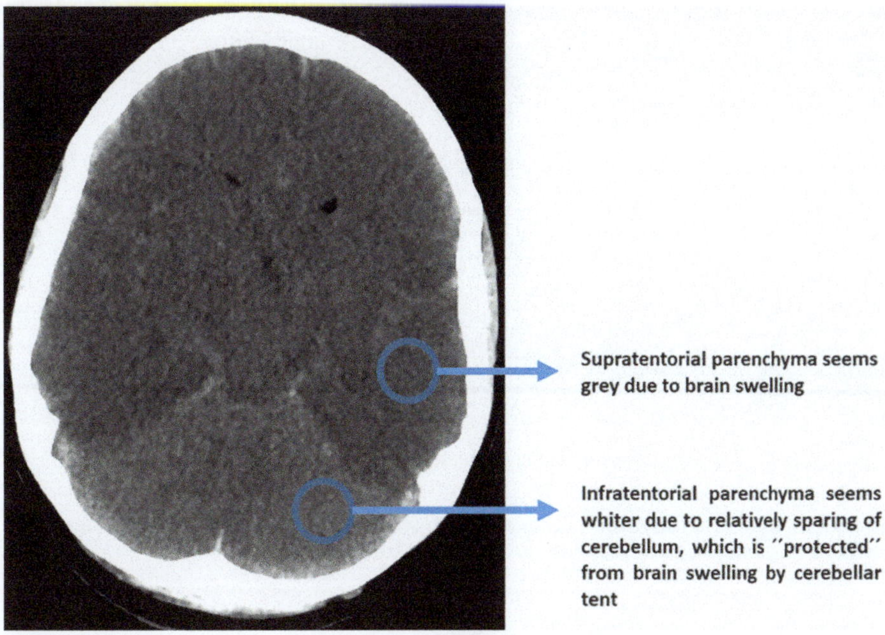

FIG. 4 "White cerebellum" sign, revealing a whiter infratentorial content compared to darker supratentorial content.

Marshall scale

Marshall Scale came up with the proposal to identify radiological characteristics on tomography that could be correlated with the increase in intracranial pressure. Proposed in 1991 by Marshall et al., it was based on the Traumatic Coma Data Bank (United States), classifying the TBI in 06 different descriptive categories (Table 1 and Fig. 5). Until then, traumatic brain injuries were traditionally classified as focal or diffuse. However, especially in relation to the latter, there was a perception of the existence of heterogeneous tomographic patterns grouped under the term "diffuse" and that were not useful in predicting the prognosis or did not help in defining the treatment. Marshall et al. in turn, evaluated and established parameters that allowed identifying among the diffuse lesions those indicative of a higher risk of intracranial hypertension and, therefore, of a worse neurological result. For this, they emphasized the cerebral volumetric status and the effect produced by it on the appearance of the mesencephalic cisterns and on the degree of deviation from the midline on cranial CT, thus creating a scale that provided essential information for determining the risk of death and morbidity in these patients.

Marshall's diagnostic category I (Diffuse Injury I) refers to patients with absolutely normal skull CT for a certain age and pre-traumatic pattern. Therefore, the midline is centered and the mesencephalic cisterns are not extinguished. Marshall's category II (Diffuse Injury II), in turn, describes a pattern with preservation of mesencephalic cisterns, but in which there may be a small deviation from the midline and/or the presence of a variety of injuries resulting from trauma (bruises, typical petechial hemorrhages suggestive of diffuse axonal injury, hemorrhages in the brain stem, among others). In the Diffuse Injury III category, compressed or absent mesencephalic cisterns are described, and here, too, there may or may not be a small deviation from the associated midline as well as the presence of traumatic injuries similar to those already mentioned in category II. Category IV is distinguished by the presence of significant deviation from the midline. The other categories describe the image of any surgically evacuated lesion (category V) or the presence of lesions with a mass effect not surgically addressed (category VI). In these last two categories there are criticisms of Marshall's original classification, since one or the other will exist based on a clinical decision and do not represent a purely tomographic parameter.

It is important to highlight that the patient who is a victim of TBI and undergoing serial imaging tests must have his classification reassessed in the face of new findings compatible with a diagnostic category different from that initially defined.

Although the scale was not designed with the objective of defining prognosis, several subsequent studies demonstrated the prognostic value of the Marshall scale, especially with a worse prognosis in patients with diffuse type IV lesions (up to 44%). However, the scale has some flaws that impact its wide use in TBI treatment centers, such as the lack of reproducibility in patients with multiple types of injuries and the lack of linearity in the prognostic evaluation.

TABLE 1 Marshall CT (computed tomography) classification of traumatic brain injury.

Category	Definition	Mortality at discharge
Diffuse injury I (no visible pathology)	No visible intracranial pathology seen on CT scan	10%
Diffuse injury II	Cisterns are present with midline shift of 0–5 mm and/or lesions densities present; no high or mixed density lesion > 25 cm^3 may include bone fragments and foreign bodies	14%
Diffuse injury III (Swelling or global edema)	Cisterns compressed or absent with midline shift 0–5 mm; no high or mixed density lesion > 25 cm^3	34%
Diffuse injury IV (Hemispheric edema)	Midline shift > 5 mm; no high or mixed density lesion > 25 cm^3	56%
Evacuated mass lesion V	Any lesion surgically evacuated	39%
Non-evacuated mass lesion VI	High or mixed density lesion > 25 cm^3; not surgically evacuated	53%

Reproduced from Adelson, P.D., Bratton, S.L., Carney, N.A., Chesnut, R.M., du Coudray, H.E., Goldstein, B., et al., 2003. Guidelines for the acute medical management of severe traumatic brain injury in infants, children, and adolescents. Chapter 1: Introduction. Pediatric Critical Care Medicine, 4, S2–S4.

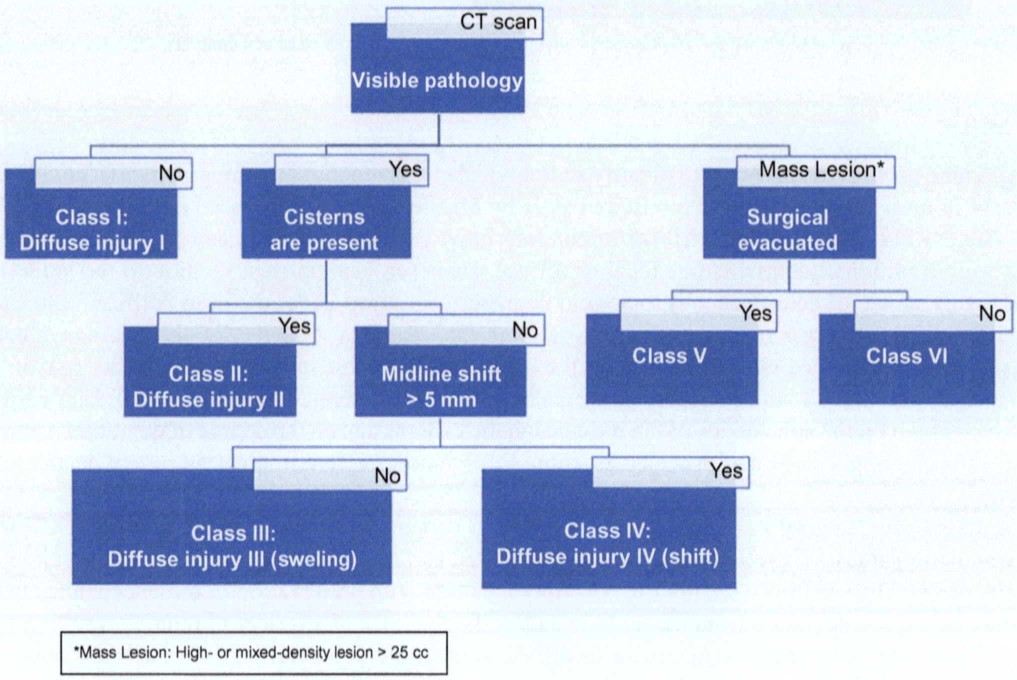

FIG. 5 Marshall CT classification presented in a tree structure. (Reproduced from Mass, A.I., Hukkelhoven, C.W., Marshall, L.F., & Steyerberg, E.W. (2005). Prediction of outcome in traumatic brain injury With computed tomographic characteristics: a comparison between the computed tomographic classification and combinations of computed tomographic predictors. *Neurosurgery*, *57*(6), 1173–1182.)

Rotterdam scale

Published in 2005, the Rotterdam scale emerged from a multivariate analysis of main parameters evaluated in the CT scan of patients suffering from TBI, relating to mortality in 6 months. The study was conducted with combined data from two bases, International and North American Tirilazad trials (n = 2269), evaluating patients between 15 and 65 years of age, victims of severe Brain trauma (GCS 3–8) and moderate brain trauma (GCS 9–12). At admission CT of these patients, more than 90% had abnormalities. In this context, the status of basal cisterns, the presence of midline shift, the presence of epidural hematoma, intra-ventricular hemorrhage and subarachnoid hemorrhage were associated with increased mortality. Based on these findings, a prognostic scale based on CT was created, with higher values indicating a worse prognosis (score from 1 to 6) (Maas et al., 2005).

In assessing the parameters of increased ICP, the highest mortality was observed in patients with changes in basal cisterns, associated with a deviation from the midline greater than 5 mm (DLM > 5 mm). Such parameters had already been related to the prognosis in severe TBI by the Marshall Scale. However, other points not evaluated by MS have shown to have a prognostic value in tomographic evaluation:

- Subarachnoid hemorrhage (SAHt): predictor of both neurological prognosis and mortality in patients with TBI;
- Intra-ventricular hemorrhage (IVH): associated with a worse neurological prognosis;
- Differentiation of the expansive lesion (Epidural x others): The Marshall scale does not divide the types of focal lesions. The literature has shown better evolution in patients with Extra dural Hematoma (HEDA), when compared and Acute Subdural Hematoma and Intra-cerebral Hemorrhage. In the sample evaluated, the mortality of patients with HEDA was zero;
- Differentiation of changes in basal cisterns (Normal × Compressed × Absent)

Rotterdam scale showed a more robust correlation with prognosis compared to the Marshall scale, also showing a direct correlation with intracranial pressure (ICP), especially in the first 24 h. In addition, Rotterdam is useful in the evaluation of serial CT scans, and not only in the analysis of incoming CT, and can be used as a follow-up assessment tool (Tables 2 and 3).

Criteria to guide practical management

Currently, the fundamental point in intensive management of patients suffering from TBI is to avoid secondary injuries maintaining adequate ICP and ensuring a good cerebral perfusion pressure.

TABLE 2 Rotterdam CT (computed tomography) classification of traumatic brain injury.

Predictor value		Score
Basal cisterns	Normal	0
	Compressed	1
	Absent	2
Midline shift	No shift or shift ≤ 5 mm	0
	Shift > 5 mm	1
Epidural mass lesion	Absent	0
	Present	1
Intra-ventricular blood or subarachnoid hemorrhage	Absent	0
	Present	1
Sum score:		Total + 1

TABLE 3 Rotterdam CT (computed tomography) score versus mortality.

Score	Mortality
1	0
2	6.8%
3	16%
4	26%
5	53%
6	61%

Thus, the main parameter used ends up being invasive ICP monitoring. Brain Trauma Foundation Guidelines (Carney et al., 2017) indicate the introduction of a PIC monitor in patients suffering from severe TBI (GCS < 8):

1. Who have any abnormalities on tomography (defined by hematoma, bruises, cerebral edema, herniation or compression of the cisterns).
2. Who have normal skull CT, and 02 or more of the following criteria: Age > 40 years; change in unilateral or bilateral motor posture; systolic blood pressure (SBP) < 90 mmHg.

In a randomized clinical study, Chesnut et al. (2012) demonstrated the validity and feasibility of using a protocol guided by clinical and radiological evaluation (ICE—Imaging and Clinical Examination). Two groups of patients were compared, one submitted to invasive monitoring of intracranial pressure and the other using the ICE protocol. Patients suffering from severe TBI in the imaging protocol were submitted to CT scans in this pattern:

- Arrival/48 h/5 to 7 days/if neurological worsening. Those with surgical lesions were submitted to surgery. Meanwhile, conservative patients were assessed for the presence of cerebral edema, based on 3 parameters
  (a) Compression of peri-mesencephalic cisterns;
  (b) Deviation from the midline;
  (c) Compression or erasure of the cortical grooves.

Based on this assessment, therapy for traumatic intracranial hypertension was adopted. Finally, the study displayed no difference in mortality between groups with invasive ICP monitoring, versus those guided by radiological clinical evaluation. This suggests that in low-income countries or in situations with limited availability of intracranial hypertension monitoring resources, neurological monitoring only with serial CTs is feasible and correlates accurately with outcomes.

Special conditions

Until now we discussed main aspects concerning CT evaluation in brain swelling, pointing pathophysiology, epidemiology and main radiological aspects. There are two other brief points which deserve some discussion. At first, the role of CT in pediatric TBI; then, the application of different CT modalities to evaluate TBI.

As presented above, TBI is the main cause of children deaths in USA and Europe. Pediatric TBI has some details and differences from adults. The most expressive difference is process of skull and brain growing in children. In turn, the ratio brain/skull is bigger in children than in adults and elderly. In practical neuroimage it is possible to see children brains tight inside skulls, with narrowed sulci and cisterns. Somehow it already simulates a "brain swelling".

Thus, at neuroradiological investigations in children, their brains already seem with some degree of "swelling" and this phenomenon makes clinical evaluation more difficult. After TBI, cerebral edema may be more common in children (up to 28% of cases) and also impact prognosis. Up to date there is no specific score to address CT findings of TBI on pediatric population. Thus, findings described in adults are usually applied but should be cautiously interpreted. Routine CT evaluation is useful and should be applied. Both Marshall and Rotterdam Scores correlates with prognosis, together with Glasgow Coma Scale (GCS).

Another important point that should be presented is the use of contrasted CT to allow vascular investigation and performing also perfusion study. With the adjunct of both techniques, it is possible to increase diagnostic accuracy of CT. The use of angiotomography or perfusion CT may delay examination time and their indication should be analyzed case by case. They are usually applied in the scenario of severe head trauma with suspicion of traumatic vascular disorders (Nujaimin et al., 2009; Williamson et al., 2017; Wintermark et al., 2004, 2006; Yamamura, Morioka, Yamamoto, & Mizobata, 2016).

Angiotomography is able to identify arterial and venous macro-vasculature disturbances like acute ongoing hemorrhages, traumatic aneurysms and vascular occlusions and thrombosis. Its application specifically in brain swelling is limited. However, it may be necessary to assess suspected vascular lesions in major brain trauma (Nujaimin et al., 2009; Williamson et al., 2017; Wintermark et al., 2004, 2006; Yamamura et al., 2016).

Perfusion CT is able to assess microvasculature disclosing tissue relationship with distal blood supply. Perfusion studies may be useful to evaluate cerebral edema once they disclose more informations than just anatomical details. It is possible to identify hyperfusional or hypoperfusional regions which might behave differently during follow-up and may tailor intensive care management. Hypoperfusional regions may point towards the risk of oliguemia and ischemia, demanding improvement of blood supply, while hyperperfusional regions indicate hyperemia and may provoke bleeding, dynamic contusions and should be managed to decrease blood support to that area. Some studies found association of Perfusion CT findings with late outcomes in patients with severe head trauma (Nujaimin et al., 2009; Williamson et al., 2017; Wintermark et al., 2006; Wintermark et al., 2004; Yamamura et al., 2016).

Applications to other areas of neuroscience

Skull computed tomography is widely used in experimental, animal and human neuroscience studies, especially those concerning evaluation of traumatic brain injury. On the other hand, neuroscience of traumatic brain injury is complex and aim of several studies worldwide. Computed tomography provides fast, trustworthy and practical brain anatomical evaluation and has high accuracy of diagnosing and monitoring consequences of traumatic brain injury like fractures, focal lesions and diffuse lesions, like brain edema. Although tomography has limited accuracy in diagnosing and detailing diffuse axonal injury, it also provides some suggestive signs which are especially used in the emergency room at the acute phase of trauma.

Traumatic brain injury is a prevalent health problem potentially reaching all human beings. There is still much to be discovered in order to provide diagnostic and prognostic information and improve clinical management. This chapter has specific application to radiological evaluation of traumatic brain injury but our summarized concepts may be applied in different scientific and practical study, diagnosis and treatment of trauma.

Computed tomography may be applied in any area of neuroscience concerning neuroimage. It is cheaper, faster and may be improved by combining vascular study (angiotomography) or perfusion studies (CT perfusion), increasing accuracy of identifying, detailing and monitoring any kind of neuroscience application demanding anatomical, metabolism and physiological investigation.

Key facts of brain swelling and computed tomography

- Brain swelling follows the increase of intracranial pressure (ICP), occurring in 16% of all TBI victims and in 28% of pediatric head-injured patients;
- Despite the advances presented by MR and other modalities of neuroimage, CT remains as the exam of choice, both in the evaluation of the acute phase and in the follow-up of patients who are victims of brain trauma. CT advantages include its wide availability, accessibility, rapid image acquisition, especially when considering hemodynamic unstable patients;
- Useful parameters include evaluation of high convexity sulci, basal cisterns status, Sylvian fissure status, presence of traumatic sub-arachnoid hemorrhage (SAH) and radiological signs of herniation.
- Marshall and Rotterdam tomographic scales are widely used in clinical practice. They describe tomographic findings and correlate with prognosis.
- Skull CT remains the gold standard for diagnosing, monitoring and treating traumatic brain injury.

Mini-dictionary of terms

- **Traumatic brain swelling**: a post-traumatic state of cell engorgement due to water imbalance after injury;
- **Canadian CT Rule**: a practical guideline to propose the use or not of neuroimage in cases of trauma;
- **Zumkeller index**: a mathematical relationship between degree of midline shift and size of clot, indicating the presence of cerebral edema;
- **Convexity sulci**: sulci which are situated peripherically inside brain;
- **Subarachnoid hemorrhage**: presence of blood clot within arachnoidal membrane;
- **Basal cisterns**: anatomical cavities delimited by arachnoidal membrane and filled with cerebrospinal fluid;
- **Sylvian fissure**: main sulcus in lateral brain projection, also called lateral sulcus.
- **Brain herniation**: when cerebral tissue is displaced from its anatomical location due to edema or compressive forces, thus occupying other locations. In this process, the ectopic tissue may displaced other brain tissues.

Summary points

- This chapter focuses on the relevance of CT image to evaluate TBI and especially brain edema and swelling.
- Severe traumatic brain injury correlates with increased intracranial pressure (ICP) and brain swelling;
- Controlling ICP is one of the main factors to improve prognosis;
- CT remains as the exam of choice to evaluate brain swelling in TBI;
- Marshall and Rotterdam tomographic scales are widely used in clinical practice. They describe tomographic findings and correlate with prognosis.

References

Adelson, P. D., Bratton, S. L., Carney, N. A., et al. (2003). Introduction. *Pediatric Critical Care Medicine*, *4*(Suppl 3), 4–6. https://doi.org/10.1097/01.CCM.0000066600.71233.01.

Ahmed, S., Biswas, T., & Paul, S. P. (2019). The white cerebellum sign: An under-recognized red flag finding with grave prognosis. *Journal of the College of Physicians and Surgeons–Pakistan*, *29*(11), 1123–1124.

Amorim, R. L., Morais, B. A., Pereira, F. O. C., et al. (2019). Evaluation of head computed tomography assessment of brain swelling after acute traumatic brain injury: A pilot study. *Arquivos Brasileiros de Neurocirurgia Brazilian Neurosurgery*, *38*(04), 257–262. https://doi.org/10.1055/s-0039-1698785.

Badri, S., Chen, J., Barber, J., et al. (2012). Mortality and long-term functional outcome associated with intracranial pressure after traumatic brain injury. *Intensive Care Medicine*, *38*(11), 1800–1809. https://doi.org/10.1007/s00134-012-2655-4.

Bobinski, L., Olivecrona, M., & Koskinen, L. O. D. (2012). Dynamics of brain tissue changes induced by traumatic brain injury assessed with the Marshall, Morris-Marshall, and the Rotterdam classifications and its impact on outcome in a prostacyclin placebo-controlled study. *Acta Neurochirurgica*, *154*(6), 1069–1079. https://doi.org/10.1007/s00701-012-1345-x.

Carney, N., Totten, A. M., O'Reilly, C., et al. (2017). Guidelines for the Management of Severe Traumatic Brain Injury, Fourth Edition. *Neurosurgery*, *80*(1), 6–15. https://doi.org/10.1227/NEU.0000000000001432.

Centers for Disease Control and Prevention. (2019). *Surveillance report of traumatic brain injury-related emergency department visits, hospitalizations, and deaths—United States, 2014*. Centers for Disease Control and Prevention, U.S. Department of Health and Human Services. https://www.cdc.gov/traumaticbraininjury/pdf/TBI-Surveillance-Report-FINAL_508.pdf.

Chesnut, R. M., Temkin, N., Carney, N., et al. (2012). A trial of intracranial-pressure monitoring in traumatic brain injury. *The New England Journal of Medicine*, *367*(26), 2471–2481. https://doi.org/10.1056/NEJMoa1207363.

Greenberg, J. O. (1984). Neuroimaging in brain swelling. *Neurologic Clinics*, *2*(4), 677–694.

Hukkelhoven, C. W. P. M., Steyerberg, E. W., Habbema, J. D. F., et al. (2005). Predicting outcome after traumatic brain injury: Development and validation of a prognostic score based on admission characteristics. *Journal of Neurotrauma*, *22*(10), 1025–1039. https://doi.org/10.1089/neu.2005.22.1025.

Jha, R. M., Kochanek, P. M., & Simard, J. M. (2019). Pathophysiology and treatment of cerebral edema in traumatic brain injury. *Neuropharmacology*, *145*(Pt B), 230–246.

Kim, H., Dong, K. G., Yoon, B. C., et al. (2014). Quantitative analysis of computed tomography images and early det CEtion of cerebral edema for pediatric traumatic brain injury patients: RetrospCEtive study. *BMC Medicine*, *12*(1), 1–16. https://doi.org/10.1186/s12916-014-0186-2.

Kim, Y. J. (2011). A systematic review of factors contributing to outcomes in patients with traumatic brain injury. *Journal of Clinical Nursing*, *20*(11–12), 1518–1532. https://doi.org/10.1111/j.1365-2702.2010.03618.x.

Lobato, R. D., Alen, J. F., Perez-Nuñez, A., Alday, R., Gómez, P. A., Pascual, B., et al. (2005). Value of serial CT scanning and intracranial pressure monitoring for detecting new intracranial mass effect in severe head injury patients showing lesions type I-II in the initial CT scan. *Neurocirugía (Asturias, Spain)*, *16*(3), 217–234.

Lobato, R. D., Gomez, P. A., Alday, R., et al. (1997). Sequential computerized tomography changes and related final outcome in severe head injury patients. *Acta Neurochirurgica*, *139*(5), 385–391. https://doi.org/10.1007/BF01808871.

Maas, A. I. R., Hukkelhoven, C. W. P. M., Marshall, L. F., & Steyerberg, E. W. (2005). Prediction of outcome in traumatic brain injury with computed tomographic characteristics: A comparison between the computed tomographic classification and combinations of computed tomographic predictors. *Neurosurgery*, *57*(6), 1173–1181. https://doi.org/10.1227/01.NEU.0000186013.63046.6B.

Marmarou, A., Signoretti, S., Fatouros, P. P., et al. (2006). Predominance of cellular edema in traumatic brain swelling in patients with severe head injuries. *Journal of Neurosurgery*, *104*, 720–730.

Marshall, L. F., Marshall, S. B., Klauber, M. R., et al. (1992). The diagnosis of head injury requires a classification based on computed axial tomography. *Journal of Neurotrauma*, *9*(Suppl 1), S287–S292.

Metting, Z., Rödiger, L. A., Stewart, R. E., et al. (2009). Perfusion computed tomography in the acute phase of mild head injury: Regional dysfunction and prognostic value. *Annals of Neurology*, *66*(6), 809–816.

Nujaimin, U., Saufi, A., Rahman, A. G., Badrisyah, I., Sani, S., Zamzuri, I., et al. (2009). Post traumatic cerebral oedema in severe head injury is related to intracranial pressure and cerebral perfusion pressure but not to cerebral compliance. *Asian Journal of Surgery*, *32*(3), 157–162.

Peeters, W., van den Brande, R., Polinder, S., Brazinova, A., Steyeberg, E. W., Lingsma, H. F., et al. (2015). Epidemiology of traumatic brain injury in Europe. *Acta Neurochirurgica*, *157*(10), 1683–1696.

Pinto, F. C., Capone-Neto, A., Prist, R., Silva, E. M. R., & Poli-de-Figueiredo, L. F. (2006). Volume replacement with lactated Ringer's or 3% hypertonic saline solution during combined experimental hemorrhagic shock and traumatic brain injury. *The Journal of Trauma*, *60*(4), 758–763 (discussion 763-4).

Pinto, F. C., Oliveira, M. F., Prist, R., Silva, M. R., Silva, L. F., & Capone, N. A. (2015). Effect of volume replacement during combined experimental hemorrhagic shock and traumatic brain injury in prostanoids, brain pathology and pupil status. *Arquivos de Neuro-Psiquiatria*, *73*(6), 499–505.

Rahman, Y. S. A., Al Den, A. S. S., & Maull, K. I. (2010). Prospective study of validity of neurologic signs in predicting positive cranial computed tomography following minor head trauma. *Prehospital and Disaster Medicine*, *25*(1), 59–62.

Stiell, I. G., Wells, G. A., Vandemheen, K., et al. (2001). The Canadian CT Head Rule for patients with minor head injury. *Lancet*, *357*, 1391–1396.

Talari, H. R., Hamidian, Y., Moussavi, N., et al. (2019 May). The prognostic value of Rotterdam computed tomography score in predicting early outcomes among children with traumatic brain injury. *World Neurosurgery*, *125*, e139–e145.

Toutant, S. M., Klauber, M. R., Marshall, L. F., et al. (1984). Absent or compressed basal cisterns on first CT scan: Ominous predictors of outcome in severe head injury. *Journal of Neurosurgery*, *61*(4), 691–694. https://doi.org/10.3171/jns.1984.61.4.0691.

Tucker, B., Aston, J., Dines, M., et al. (2017). Early brain edema is a predictor of in-hospital mortality in traumatic brain injury. *The Journal of Emergency Medicine*, *53*(1), 18–29.

Williamson, C., Morgan, L., & Klein, J. P. (2017). Imaging in neurocritical care practice. *Seminars in Respiratory and Critical Care Medicine*, *38*(6), 840–852.

Wintermark, M., Chiolero, R., Van Melle, G., Revelly, J. P., Porchet, F., Regli, L., et al. (2006). Cerebral vascular autoregulation assessed by perfusion-CT in severe head trauma patients. *Journal of Neuroradiology*, *33*(1), 27–37.

Wintermark, M., van Melle, G., Schnyder, P., Revelly, J. P., Porchet, F., Regli, L., et al. (2004). Admission perfusion CT: Prognostic value in patients with severe head trauma. *Radiology*, *232*(1), 211–220.

Yamamura, H., Morioka, T., Yamamoto, T., & Mizobata, Y. (2016). Head computed tomographic measurement as a predictor of outcome in patients with subdural hematoma with cerebral edema. *Scandinavian Journal of Trauma, Resuscitation and Emergency Medicine*, *24*, 83.

Zumkeller, M., Behrmann, R., Heissler, H. E., & Dietz, H. (1996). Computed tomographic criteria and survival rate for patients with acute subdural hematoma. *Neurosurgery*, *39*(4), 708–713. https://doi.org/10.1097/00006123-199610000-00011.

Chapter 25

Machine learning and prediction of traumatic brain injury mortality

João Gustavo Rocha Peixoto dos Santos[a] and Wellingson Silva Paiva[b]
[a]*Neurosurgery Department, Santa Casa de Maceio, Maceio, Brazil,* [b]*Neurology Department, Hospital das Clínicas, University of Sao Paulo, Sao Paulo, Brazil*

List of abbreviations

AIS	abbreviated injury score
CI	confidence interval
CT	computed tomography
EEG	eletroencephalography
GCS	Glasgow Coma Scale score
ICU	intensive care unit
ORs	odds ratios
SSEPs	somatosensory evoked potentials
TBI	traumatic brain injury
tSAH	traumatic subarachnoid hemorrhage

Introduction

Traumatic brain injury (TBI) is a pathology that accounts for significant healthcare, socioeconomic, and psychological issues in developing countries. Disability of affected subjects reduces quality of life, work capacity, and life expectancy. Etiology varies according to country and region of the same country, but motor vehicle accidents are the main cause, followed by height falls and urban violence. Young adults (age group 20–40 years old) are more frequently affected (Taylor, Bell, Breiding, & Xu, 2017).

Ability to estimate prognosis at admission is an important factor for decision making and to define the intensity of care. Searching the medical literature, it is possible to find some prediction models and calculators (Majdan et al., 2016; Stevens & Sutter, 2013) created based on data obtained from large scale multi-centric studies (Maas, Hukkelhoven, Marshall, & Steyerberg, 2005; MRC CRASH Trial Collaborators et al., 2008).

Some variables are found repeatedly in the models. The most common variable is Glasgow coma scale score (GCS). Glasgow coma scale score is one of the most used for TBI evaluation due to its practicality, objectivity and universal understanding. Many papers studied its relation to mortality (Moore et al., 2006) and 6-month outcomes. The prognostic evaluation of GCS is combined with other variables. One of them integrates age and brain stem reflexes (Prasad, 1996). Others combine GCS and Abbreviated Injury Score for outcome measures in 12 months after initial injury. Recently, GCS was associated with pupillary response to light to determinate mortality and unfavorable outcome in 6 months (Brennan, Murray, & Teasdale, 2018).

Independently, no prognostic model fits to all populations and some systematic reviews demonstrated this (Hukkelhoven et al., 2005; Perel, Edwards, Wentz, & Roberts, 2006). Some factors implicated to justify this are the differences among populations and samples on socioeconomic pattern and healthcare infrastructure. Also, most of these studies have been performed in developed countries.

Diagnosis and Treatment of Traumatic Brain Injury. https://doi.org/10.1016/B978-0-12-823347-4.00015-4

TBI mortality

Considering absolute numbers, TBI is the most frequent cause of death and morbidity in Brazil and worldwide nowadays. TBI is responsible for at least 50% of deaths from traumatic injuries (Dutton et al., 2010; Settervall, Sousa, & Silva, 2011). There is a complex physiopathology which includes molecular and cellular, as well as brain hemodynamics changes, leading to brain swelling and other secondary lesions. Cytotoxic and vasogenic edema seem to be proportional to the severity of trauma and reach their maximum level in approximately 72h (Settervall et al., 2011).

In the United States, the incidence of TBI is estimated at 538 per 100 thousand inhabitants, which represents at least 1.7 million new cases per year since 2003 (Faul, Xu, Wald, & Coronado, 2010). In Europe the incidence is 235 per 100 thousand individuals (Cassidy et al., 2004). In Brazil and other Latin American countries, studies on the topic are scarce. Records carried out between 2001 and 2007 showed a total of 440,000 hospitalizations due to TBI, with an average of 68,200 admissions per year, which reflects an incidence of 37 per 100 thousand inhabitants in the studied age group (14–69 years). In addition, there was a 50% increase in mortality in the registered period, that is, a rate of 4.5 per 100 thousand inhabitants in 2001 and 6.7 per 100 thousand inhabitants in 2007. DATASUS data from 2008 to 2012 revealed about 125,500 hospital admissions per year associated with TBI in Brazil, with an incidence of 65.7 hospital admissions per 100 thousand inhabitants per year, with 9715 deaths, corresponding to a mortality rate of about 5.1 per 100 thousand inhabitants per year (Roozenbeek et al., 2012).

International Mission for Prognosis and Analysis of Clinical Trials (IMPACT) and TBI-trac databases (King Jr, Carlier, & Marion, 2005) (over a 10-year period), estimate that 23% of the patients had died 14 days after severe TBI. Originally, IMPACT analyzed outcome 6 months after TBI. Some authors argue that 14-day mortality is more relevant, because in the Corticosteroid randomization after significant head injury (CRASH) trial, for example, 85% of deaths occurred in the first 2 weeks.

Moderate to severe TBI patients must be hospitalized and treated in an Intensive Care Unit (ICU). Studies showed that mortality rate of these ICU patients is about 30% (Andriessen et al., 2011; Myburgh et al., 2008). ICU objective for TBI patients is to reduce secondary lesions by: (1) controlling intracranial pressure; (2) maintaining cerebral homeostasis and cerebral perfusion pressure (Stocchetti et al., 2017).

Predicting TBI mortality

During evaluation and monitoring of the patient, at the emergency department, ICU or yards, hundreds of thousands of data can be generated but goes without appropriate interpretation. The human mind is not able to process such quantities of information or use them in decision making. The construction of TBI models helps us clarify prognosis, reduce iatrogenic measures and strengthen good patient care.

Over time, different strategies were developed to try to predict TBI prognosis. Most of them use data from patient admission, which is not ideal. At the beginning, patients are at the emergency department, not stable and subjected to many biases. Considering these problems, other models defend that prediction must be done in an ICU environment (Matsuo et al., 2020).

Since both diagnostic and therapeutic efforts seek to enhance a patient's prognosis and outcome, prognosis is the foundation of clinical medicine. Because of developments in statistical modeling and the availability of vast databases, diagnosis and prognosis may now be thought of as probabilities rather than uncertain predictions. Clinical decision making can benefit from probability estimates (Lingsma, Roozenbeek, Steyerberg, Murray, & Maas, 2010).

Predictors of outcome

Different papers published tried to determine single factors of prognosis in TBI patients. For instance, univariate regression may be used to analyze the relationship between a single indicator and an outcome of interest. Since it lacks the position of other confounding variables that may clarify (part of) the observed correlation, such an interpretation is of limited benefit (Steyerberg, 2009).

To account for confounders in the estimation of relative risks, statistical analysis such as logistic regression are needed. The strength of prognostic effects is generally expressed using odds ratios (ORs). If the OR's 95% confidence interval (CI) does not contain the value 10, the relationship is important. The OR, on the other hand, does not take into account the prevalence of a predictor (Lingsma et al., 2010).

There are models which define prognosis based on *admission characteristics* and others based on *clinical course*.

Admission characteristics

The typical approach of most studies is to analyze the clinical severity on admission to define prognosis. One of the most used scores, focused on TBI, is the GCS. Many studies have shown a correlation between a low GCS score and a worse result. Since eye and vocal responses are often absent in patients with more serious injuries, the motor portion of the GCS has the greatest predictive benefit (Stocchetti et al., 2004). Medical sedation, paralysis, or toxicity can obscure an accurate evaluation of the GCS at the acute setting. Ideal GCS evaluation should be after primary respiratory and hemodynamic stabilization on admission (Murray et al., 1999).

Another important parameter that can be measured and correlated to prognosis on admission is *pupillary abnormalities*. They indicate brainstem compression or damage, being thus strongly associated with poorer outcomes. Even though pupillary reactions may be influenced by sedation and hemodynamic states, they are less influenced by these than GCS. So, they can be considered a more reliable factor for analysis.

Considering the systemic influence of polytraumatized patients in TBI prognosis, the *abbreviated injury score* (AIS) attempts to estimate overall severity of extracranial injuries. The occurrence of both a mild TBI and an extracranial injury is related to a high risk of mortality and morbidity (Gennarelli, Champion, Copes, & Sacco, 1994).

In TBI, *age* is one of the most effective predictors of death and functional outcome; several papers have found that older patients (>60 years old) tend to present a poorer outcome (Brain Trauma Foundation, American Association of Neurological Surgeons, Congress of Neurological Surgeons, Joint Section on Neurotrauma and Critical Care, AANS/CNS, et al., 2007; Demetriades et al., 2004; Ono, Yamaura, Kubota, Okimura, & Isobe, 2001). Other demographic factors were also evaluated in studies, with controversial results. *Secondary insults* (i.e., hypotension, hypoxia) are common after TBI, especially in pre-hospital scenarios and are associated with worse prognosis (Chesnut et al., 1993).

Computed tomographic alterations are also well-known and correlated with prognosis. The Marshall classification itself has prognostic value (Marshall et al., 1991) (Table 1). Status of basal cisterns, midline shift, presence of mixed density lesion $>25\,cm^2$ are the main factors observed by this score.

Clinical course

Changes in admission variables are important prognostic factors. For example, worsening of GCS score is a high predictor of poorer outcome because this usually reflects progressive brain damage (Lannoo et al., 2000). Secondary insults follow

TABLE 1 Marshall computed tomography classification.

Category	Definition
Diffuse injury I (normal CT)	No intracranial pathology visible on CT.
Diffuse injury II	– Cisterns are present. – Lesion absent or < 25 mL – Midline shift 0–5 mm;
Diffuse injury III	– Cisterns compressed or absent – Lesion absent or < 25 mL – Midline shift 0–5 mm
Diffuse injury IV	– Cisterns presents or absents – Lesion absent or < 25 mL – Midline shift > 5 mm
Diffuse injury V	Any lesion > 25 mL evacuated
Diffuse injury VI	Any lesion > 25 mL not evacuated

A prognostic score based on image finding for severe and moderate TBI patients.
Source: Modified from Marshall, L.F., Marshall, S.B., Klauber, M.R., Clark, M.V.B., Eisenberg, H.M., Jane, J.A., Luerssen, T.G., Marmarou, A., & Foulkes, M.A. (1991). A new classification of head injury based on computerized tomography. Journal of Neurosurgery, 75(Supplement), S14–S20. Retrieved from https://thejns.org/view/journals/j-neurosurg/75/Supplement/article-pS14.xml (Accessed 16 May 2021).

the same pattern. They may occur at any time during a patient's care, especially during transport to the hospital, despite clinicians' attempts to avoid them.

High intracranial pressure, decreased brain oxygen tension and low cerebral perfusion pressure are associated with worse outcome in many studies (Huang et al., 2007; Vik et al., 2008). TBI pathophysiology emphasizes that intracranial and blood arterial pressure control are needed to avoid hypoperfusion and other secondary damages.

Neurophysiological exams have been studied as an attempt to predict prognosis in TBI. Electroencephalography (EEG) results are limited due to the quantity of subcortical axonal injury in TBI in comparison to cortical. In a chronic TBI phase, EEG results can be correlated to clinical scores to differentiate between a vegetative and a minimally conscious state (Schnakers et al., 2008). Somatosensory evoked potentials (SSEPs) are also studied for prognosis in TBI patients. Bilateral absence of cortically recorded median nerve SSEPs within 8 days of severe TBI was strongly predictive (had a 98.5% positive likelihood ratio) of unfavorable outcome (death or persistent vegetative state) (Carter & Butt, 2001). However, absence of bilateral SSEPs may occur with focal lesions as well (subdural effusions, and after recent decompressive craniectomy).

Outcomes

Mortality is the most important outcome evaluated in most prognostic models (4, 5). Another relevant outcome is disability in 6 months. Disability impacts directly in quality of life, work capacity, healthcare policies and costs. The ideal time varied between published studies, but in general, three to 6 months post TBI were implicated as effective measures of one-year TBI outcome (Lingsma et al., 2010).

Health-related quality of life and Neuropsychological assessment are also important outcomes to measure, which cause impact on different aspects of a patient's social life. Work problems, self-care issues, interpersonal relationships and many other points may be affected, and their evaluation helps in understanding the real dimension of TBI in patients' lives.

Prognostic models

Building a prognostic model

The construction of a predicting model should follow a pre-established protocol. An effective strategy is the *Transparent reporting of a multivariable prediction model for individual prognosis or diagnosis* (TRIPOD) statement recommendations for the reporting of prognostic models (Collins, Reitsma, Altman, & Moons, 2015). Based on a literature review, an exhaustive list of items was compiled, which was narrowed after a web-based survey and updated at a 3-day conference in June 2011 with methodologists, health care experts, and journal editors. Several meetings of the steering committee and e-mail exchanges with the larger community of TRIPOD contributors helped fine-tune the list. In this statement, it was established a set of reporting guidelines for studies that create, validate, or update a prediction model, whether for diagnostic or prognostic purposes.

In the TRIPOD checklist, six sections/topics are evaluated (*Title and abstract, Introduction, Methods, Results, Discussion* and *Other information.*) Each topic is evaluated as validated or developed and there is a description for each section which the article that presents the model must follow precisely.

The *Title* must specify the target population and the result to be expected after developing and/or validating a multivariable prediction model.

The *Abstract* has to include a description of the study's goals, architecture, context, participants, sample size, predictors, effect, statistical analysis, findings, and conclusions.

In the *Introduction*, the *Background and objectives* subsections need to explain the medical sense (including whether the multivariable prediction model is diagnostic or prognostic) and reasoning for creating or validating the multivariable prediction model, including comparisons with current models. Objectives must be clear, even specifying if the study model is for validation or development.

In the Methods section, there are 8 subsections required: Source of data, Participants, Outcome, Predictors, Sample size, Missing data, Statistical analysis methods, Risk groups, Development vs. validation.

The *Results* section requires the description of *Participants*, *model development*, *model specification*, its *performance* and *updating*. The Discussion asks for *limitations*, *interpretation* and *implications*. Other information such as study protocol, Web calculator, data sets and funding should also be specified.

Every component on the checklist is illustrated, along with examples of positive reporting from published publications. The TRIPOD authors also advocate that researchers outline the qualities of good (and the shortcomings of less good) research, since many of these studies are methodologically poor. To endorse and justify the model, a detailed database from previous systematic analyses of prediction models was used as well (Collins et al., 2015).

Important prognostic models

The IMPACT (Maas et al., 2005) models were built on large data, appropriated methodology, including internal and external validation processes. Eight admission characteristics were chosen based on previous studies (Bullock et al., 1996) which correlated them to prognosis:

(1) Age
(2) Gender
(3) Cause of injury (traffic accidents, falls, and other causes).
(4) Motor score (dichotomized—it combined the category of "no response" with "extensor response" and category of "localizing" with "obeying commands")
(5) Pre-admission hypotension and hypoxia (systolic blood pressure below 90 mmHg or a pO2 below 60 mmHg)
(6) Pupillary reactivity
(7) Computed Tomography (CT) classification (Marshall) (Marshall et al., 1983)
(8) Presence of traumatic Subarachnoid hemorrhage (tSAH).

CT classification and tSAH were evaluated by CT scans obtained within the first 4 h after the injury. CT classification is detailed in Table 1. Whenever any of the predictors' values were absent (4.8% of the required values), their values were calculated statistically using regression models that included the other predictors. Authors developed a score chart to estimate the probability of mortality and unfavorable outcome (Table 2).

TABLE 2 Prognostic score chart for the risk of unfavorable outcome and mortality.

Predictor	Value	Mortality	Unfavorable outcome
Age (years)	15–39 40–54 55–64 ≥65	0 1 2 3	0 1 2 3
Motor score	None/extensor Abnormal flexion Withdraws Localizes/Obeys	3 2 1 0	3 2 1 0
Pupillary reactivity	Both react One reacts None reacts	0 1 2	0 1 2
Hypoxia	No Yes	0 1	0 1
Hypotension	No Yes	0 2	0 1
CT classification	I or II III IV V or VI	0 2 4 2	0 1 1 1
tHSA	No Yes	0 2	0 1
Score: 0–2 3 4–5 6–7 ≥8		 7% 15% 30% 47% 68%	 14% 34% 55% 76% 89%

Prognostic score chart for the risk of unfavorable outcome and mortality is a scale for patients with severe or moderate traumatic brain injury and it is calculated by the sum of points of each variable. The final result defines prognosis.
Source: Modified from Hukkelhoven, C.W., Steyerberg, E.W., Habbema, J.D., Farace, E., Marmarou, A., Murray, G.D., Marshall, L.F., & Maas, A.I. (2005). Predicting outcome after traumatic brain injury: Development and validation of a prognostic score based on admission characteristics. Journal of Neurotrauma, 22(10), 1025–1039. doi:10.1089/neu.2005.22.1025.

The continuous study of other variables arose the knowledge that the presence of epidural mass could be a protective factor (good prognosis with surgery). Other variables were also included based on their influence on prognosis such as glucose and hemoglobin levels (Steyerberg et al., 2008). Fig. 1 illustrates the site interface and the characteristics required for calculation. This calculator is available online at the link:

http://tbi-impact.org/?p=impact/calc

The head injury prognosis by MRC CRASH trial collaborators was designed with two different perspectives: (1) a basic model with clinical and demographic variables only and (2) CT model with these perspectives associated with CT features.

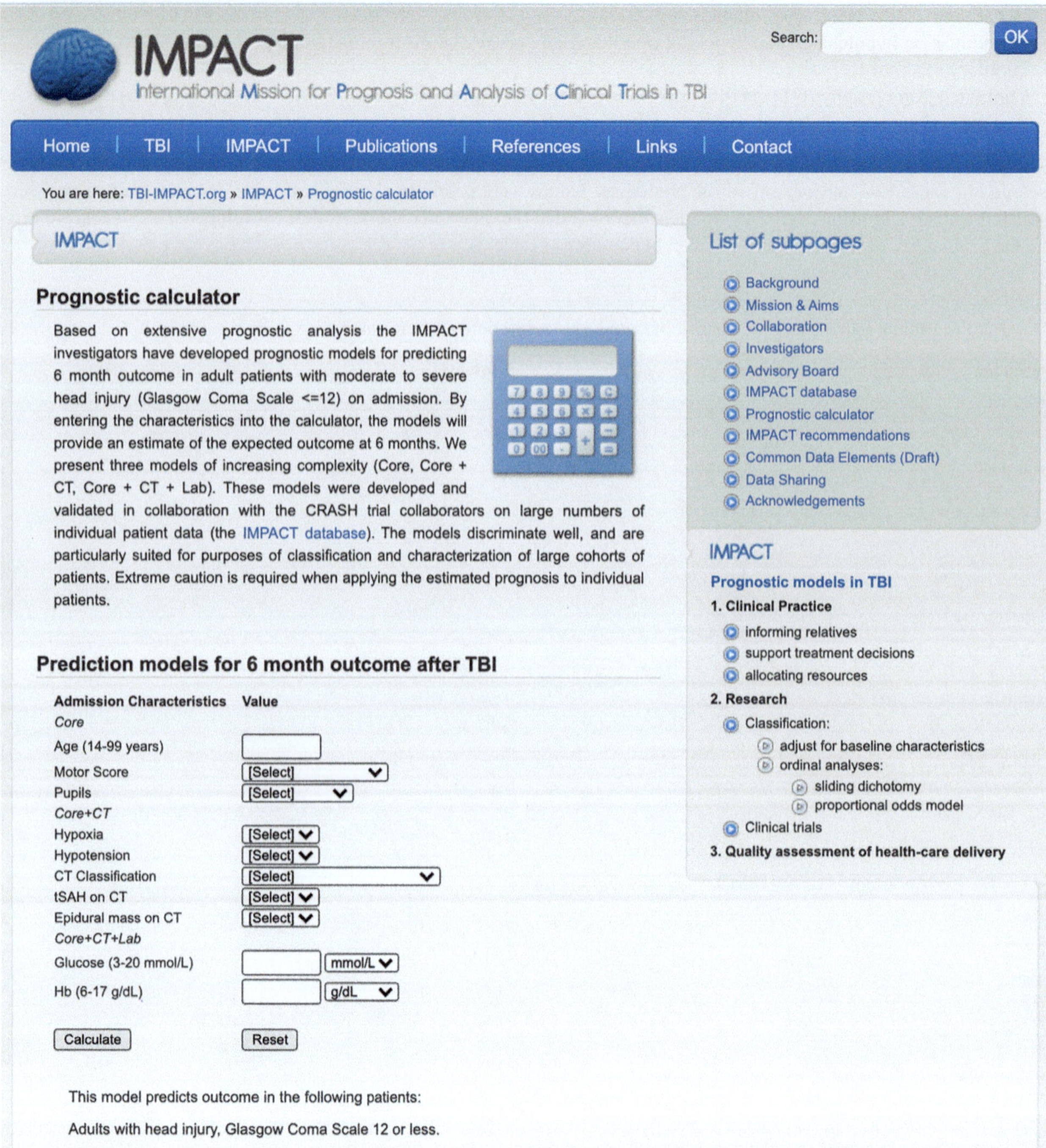

FIG. 1 IMPACT site interface. The calculator requires some characteristics which involves clinical, radiological e and laboratorial features.

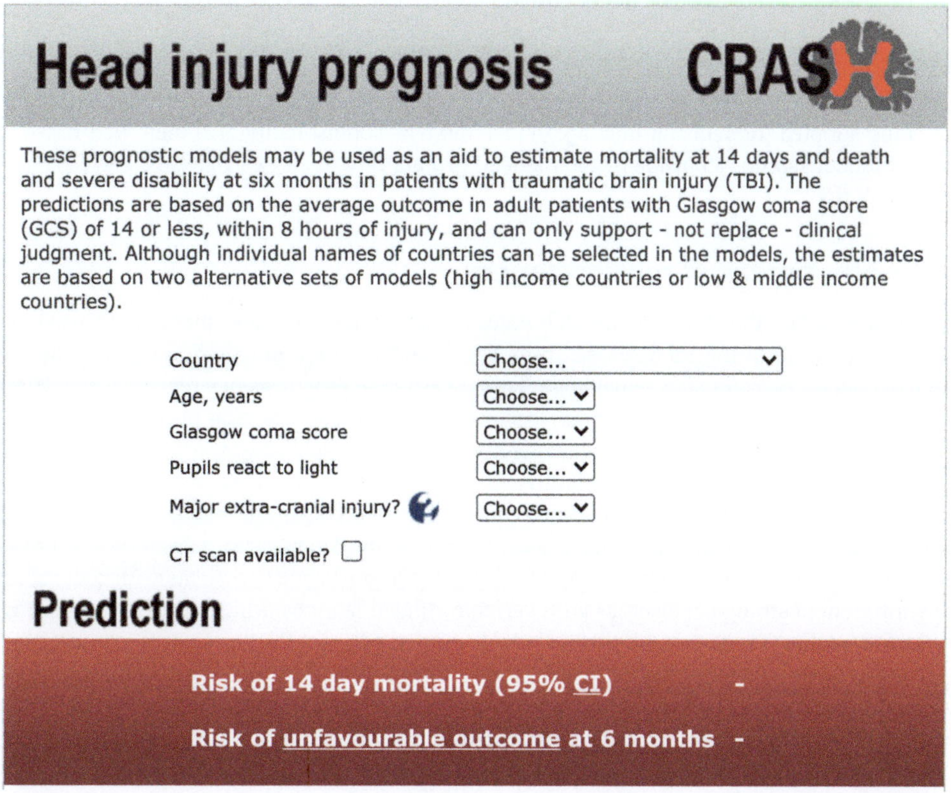

FIG. 2 MRC CRASH prognosis chart. This calculator allows prediction of mortality and unfavorable outcome at 6 months after TBI.

Authors used multivariable logistic regression to choose variables which were independently associated with prognosis. Four variables were found independently associated with prognosis in the basic model (age, Glasgow coma scale, pupil reactivity, and the presence of major extracranial injury). The CT model evaluated the following features: Presence of petechial hemorrhages, obliteration of the third ventricle or basal cisterns, subarachnoid bleeding, midline shift, and non-evacuated hematoma. The required points and the home page view are exposed in Fig. 2. The calculator is available in the following link: http://www.crash.lshtm.ac.uk/Risk%20calculator/index.html

However, these two models available were created based on studies for developed countries, and TBI is a pathology which happens predominantly in developing countries. A prognostic predictor was created and presented interesting results (Amorim et al., 2020).

Machine learning

After building the model, automatization is necessary. The creation of algorithms with calculation and computer programming to teach the machine how to process and interpret the combination of predictors is the next step. Different devices could be used to calculate these informations. First, computer programs are easier to elaborate. Even modern models can be found for smartphones in apps with multi calculators to speed up decision making.

Regularized least squares and linear regression are usually used as machine learning classification methods for categorical variable prediction. Random forest, neural network, decision tree, boosting, generalized linear model, partial least squares, and multivariate adaptive regression splines were among the machine learning regression models that are usually aimed at numeric outcomes (Amorim et al., 2020).

Random forest, discriminant analysis, Bayesian approaches, neural network, decision tree, boosting, generalized linear model, partial least squares, and multivariate adaptive regression splines are among the regression models commonly used to classify numeric variables. The region under the curve, sensitivity, and specificity, kappa values and positive and negative predictive values are tools that can be used to compare models (Amorim et al., 2020).

Conclusions

Despite the pre-established protocol (TRIPOD), rigid statistical methodology and appropriated variables, the construction of an ideal prediction model for TBI victims seems to not be possible. Some factors could be hypothesized for the difficulty of current models being adopted for decision-making: (1) the models' sophistication, (2) their web-based architecture, (3) the lack of a coherent model summary outside of the models' originally written explanations, and (4) questions about the models' predictions' validity.

Applications to other areas of neuroscience

Machine learning for traumatic brain injury is an elaborate, methodological process that can be used as an example to build similar models for other neurological diseases. In moderate and severe traumatic brain injury, the goal is to analyze admission and clinical course characteristics which together may estimate prognosis in a few days and determine mortality.

This is an important tool for public health policies because it helps focus attention on patients with better prognosis, especially when the resources are scarce. In some degenerative or untreatable diseases, it could also be possible to inform the family what the likely clinical course of that patient is.

The construction of the model should follow the same one as for traumatic brain injury. Papers with large samples can show which features could be specifically studied. Besides, time is an important factor in neurological diseases. Thus, automatization is necessary in our digital, globalized world to expedite the assistance to patients with acute neurological issues.

There are many different neurological diseases with variable clinical features. Machine-learning for prognosis may not be applicable to all of them. On the other hand, the statistical understanding of the creation of a model may be derived for other interesting tasks, such as a diagnostic tool, a therapeutic approach or even an image alteration probability strategy. This technology may help doctors to preview and to check neuroimages. Statistical analyses would collaborate to certify the quality, efficiency and effectiveness of diagnostic and therapeutic interventions.

Mini-dictionary of terms

- **Glasgow coma scale**: It is a score to determine the level of consciousness. It varies from 3 to 15 and evaluates three different parts of the neurological examination: Eye opening, motor response and verbal response.
- **Brain stem reflexes**: They are reflexes obtained to evaluate the brainstem function. The arc reflex happens by the function integration of specific cranial nerves and their nuclei. The corneo palpebral reflex is an example, where the afferent information gets to the brainstem through V1 division of trigeminal nerve, and the "blink" response (efferent) is executed by the facial nerve.
- **Brain swelling**: Expression used for the process of swelling of the brain after a traumatic injury. It is a cascade of inflammatory and metabolic changes which leads to secondary insults.
- **Cytotoxic and vasogenic edema**: They are the two more common forms of edema which happen in the brain. In the first, the cause is the death cell and apoptosis and the second one, the overflow of substances from the intracellular for the extracellular environment.
- **Impact**: Acronym for International Mission for Prognosis and Analysis of Clinical Trials.
- **Crash**: Acronym for Corticosteroid randomization after significant head injury
- **Datasus**: Informatic system which stores data about health issues in Brazil.
- **Odds ratios**: It's a metric that measures how strong the connection between two cases, A and B, is. It is the ratio of A's odds in the presence of B to A's odds in the absence of B, or (due to symmetry) the ratio of B's odds in the presence of A to B's odds in the absence of A. If and only if the OR equals 1, two events are separate.
- **Confidence interval**: Interval which expresses the probability of an event to repeat itself at the same time and in the same sample studied.
- **Abbreviated injury score**: It is an anatomically based injury severity score that presents each injury by body region on a 6-point scale.
- **Secondary insults**: Injuries caused by associated trauma events like hypotension and hypoxia.
- **Somatosensory evoked potentials**: Cerebral response after peripheral sensory electrical stimulation.
- **Persistent vegetative state**: Chronic state of altered level of consciousness, with the presence of a sleep-wake cycle but without interaction with the environment.
- **Neuropsychological assessment**: Battery of tests with the objective to obtain a complete cognitive evaluation of the patient.

Key facts of quality of life of patients after TBI

- The quality of life is hard to measure due to patients' impressions, which are individual.
- Quality of Life after Brain Injury (QOLIBRI) and Abbreviated World Health Organization Quality of Life (WHOQOL-BREF) questionnaires are two important questionnaires which try to measure the quality of life of TBI victims.
- The total QOLIBRI score will reach a limit of 100. A score of less than 60 suggests a worse result, with a higher chance of developing an affective and/or anxiety disorder.
- WHOQOL-BREF is made up of 26 questions about quality of life, two of which are universal and 24 of which are divided into four categories: physical wellbeing, psychological, social interactions, and climate.
- In terms of general QoL queries, the majority of TBI victims ranked their quality of life as "good or very good" (65.9%) and said they were "satisfied or very satisfied" with their wellbeing (66%).
- The average overall QOLIBRI score was 65.5 (±22.6), suggesting good HRQoL. Autonomy in everyday life and perception are factors that contributed to a higher degree of happiness.

Key facts of Glasgow coma scale score

- It is one of the most important scales for the evaluation of level of consciousness.
- Its importance is mainly due to its capacity of simple and fast quantification of the level of consciousness.
- It evaluates three different behaviors: eye opening, verbal and motor responses.
- It can be used for medical quick decision for intubation (patient too drowsy cannot protect airways from bronchoaspiration).
- There were attempts to correlate the final sum score to prognosis but its value changes too much depending on clinical condition.

Key facts of pupillary abnormalities

- An important predictor of prognosis in many severe and moderate TBI studies.
- Usually divided in: Both pupil reaction, one pupil reaction and no pupil reaction.
- The relation with severity of TBI is based on the likelihood of uncal herniation: Both pupil reaction—no uncal herniation; One pupil reaction—ipsilateral uncal herniation; No pupil reaction—bilateral uncal herniation.
- A possible confounding factor in TBI patients is that direct trauma to the III cranial nerve—may mimic the same response.
- There may be a limitation for evaluation of patients with ocular diseases such as cataract and or ciliary muscle affections.

Key facts of abbreviated injury score (AIS)

- It was created by the Association for the Advancement of Automotive Medicine to classify and describe the severity of injuries.
- Rather than a thorough evaluation of the seriousness of the injuries, it reflects the danger to life associated with the injury.
- The score uses seven numbers to reflect three facets of the accident: type, localization and severity.
- The group of findings are compiled on a scale of 1 to 6 (Abbreviated Injury Score-Code): 1—Minor; 2—Moderate; 3—Serious; 4—Severe; 5—Critical; 6—Maximum.
- The probability of death of each score: 1 = 0%; 2 = 1%–2%; 3 = 8%–10%; 4 = 5%–50%; 6 = 100%.

Key facts of CT Marshall classification

- The observations from a major cohort of head injury cases described in 1990 were used to develop this scheme, which was first published in 1992.
- The Marshal scheme categorizes patients into one of six severity levels (I–VI) based on results from a non-contrast computed tomography scan of the brain.
- Two parameters are evaluated: (1) midline shift and/or compression of basal cisterns presence and (2) size of contusions/hemorrhages referred to "high or mixed density lesions" (< or > 25 mL).

- Each category has a prognostic value for mortality or unfavorable outcome (dead, vegetative state, severe disability).
- I—38%; II—65%; III—84%; IV—94%; V—77%; VI—89%.

Summary points

- Traumatic brain injury is a pathology that accounts for significant healthcare, socioeconomic and psychological issues all over the world.
- The capacity to estimate prognosis is necessary for the decision-making process and to define the intensity of care. Some prediction models and calculators are available in the literature.
- There are models which define prognosis based on *admission characteristics* and others based on *clinical course*.
- Mortality is the most important outcome evaluated in most prognostic models. Another relevant outcome is disability in 6 months.
- The construction of a predicting model should follow a pre-established protocol. An effective strategy is transparent reporting of a multivariable prediction model for individual prognosis or diagnosis (TRIPOD) statement recommendations for the reporting of prognostic models.
- In the TRIPOD checklist, six sections/topics are evaluated: *Title and abstract*, *Introduction*, *Methods*, *Results*, *Discussion*, and *Other information*.)
- Two important prognostic models are International Mission for Prognosis and Analysis of Clinical Trials (IMPACT) and Corticosteroid randomization after significant head injury (CRASH).
- The IMPACT model was built on large data, appropriated methodology, including internal and external validation processes.
- The head injury prognosis by MRC CRASH trial collaborators was designed in two different perspectives: A basic model with clinical and demographic variables only and computed tomography (CT) model with these perspectives associated with CT features.
- After building the model, automation is necessary. Creation of algorithms with calculation and computer programming to teach the machine how to process and interpret the combination of predictors is the next step.

References

Amorim, R. L., Oliveira, L. M., Malbouisson, L. M., Nagumo, M. M., Simoes, M., Miranda, L., … Paiva, W. S. (2020). Prediction of early TBI mortality using a machine learning approach in a LMIC population. *Frontiers in Neurology*, *10*, 1366. https://doi.org/10.3389/fneur.2019.01366.

Andriessen, T. M., Horn, J., Franschman, G., van der Naalt, J., Haitsma, I., Jacobs, B., … Vos, P. E. (2011). Epidemiology, severity classification, and outcome of moderate and severe traumatic brain injury: A prospective multicenter study. *Journal of Neurotrauma*, *28*(10), 2019–2031. https://doi.org/10.1089/neu.2011.2034.

Brain Trauma Foundation, American Association of Neurological Surgeons, Congress of Neurological Surgeons, Joint Section on Neurotrauma and Critical Care, AANS/CNS, Bratton, S. L., Chestnut, R. M., Ghajar, J., McConnell Hammond, F. F., Harris, O. A., Hartl, R., Manley, G. T., Nemecek, A., Newell, D. W., Rosenthal, G., Schouten, J., Shutter, L., Timmons, S. D., Ullman, J. S., Videtta, W., Wilberger, J. E., … Wright, D. W. (2007). Guidelines for the management of severe traumatic brain injury. VIII. Intracranial pressure thresholds. *Journal of Neurotrauma*, *24*(Suppl 1), S55–S58. https://doi.org/10.1089/neu.2007.9988.

Brennan, P. M., Murray, G. D., & Teasdale, G. M. (2018). Simplifying the use of prognostic information in traumatic brain injury. Part 1: The GCS-pupils score: An extended index of clinical severity. *Journal of Neurosurgery*, *128*(6), 1612–1620. https://doi.org/10.3171/2017.12.JNS172780.

Bullock, R., Chesnut, R. M., Clifton, G., Ghajar, J., Marion, D. W., Narayan, R. K., … Wilberger, J. W. (1996). Guidelines for the management of severe head injury. Brain Trauma Foundation. *European Journal of Emergency Medicine: Official Journal of the European Society for Emergency Medicine*, *3*(2), 109–127. https://doi.org/10.1097/00063110-199606000-00010.

Carter, B. G., & Butt, W. (2001). Review of the use of somatosensory evoked potentials in the prediction of outcome after severe brain injury. *Critical Care Medicine*, *29*(1), 178–186. https://doi.org/10.1097/00003246-200101000-00036.

Cassidy, J. D., Carroll, L. J., Peloso, P. M., Borg, J., von Holst, H., Holm, L., … WHO Collaborating Centre Task Force on Mild Traumatic Brain Injury. (2004). Incidence, risk factors and prevention of mild traumatic brain injury: results of the WHO Collaborating Centre Task Force on Mild Traumatic Brain Injury. *Journal of Rehabilitation Medicine*, (43 Suppl), 28–60. https://doi.org/10.1080/16501960410023732.

Chesnut, R. M., Marshall, L. F., Klauber, M. R., Blunt, B. A., Baldwin, N., Eisenberg, H. M., … Foulkes, M. A. (1993). The role of secondary brain injury in determining outcome from severe head injury. *The Journal of Trauma*, *34*(2), 216–222. https://doi.org/10.1097/00005373-199302000-00006.

Collins, G. S., Reitsma, J. B., Altman, D. G., & Moons, K. G. (2015). Transparent reporting of a multivariable prediction model for individual prognosis or diagnosis (TRIPOD): The TRIPOD statement. *Journal of Clinical Epidemiology*, *68*(2), 134–143. https://doi.org/10.1016/j.jclinepi.2014.11.010.

Demetriades, D., Murray, J., Martin, M., Velmahos, G., Salim, A., Alo, K., & Rhee, P. (2004). Pedestrians injured by automobiles: Relationship of age to injury type and severity. *Journal of the American College of Surgeons*, *199*(3), 382–387. https://doi.org/10.1016/j.jamcollsurg.2004.03.027.

Dutton, R. P., Stansbury, L. G., Leone, S., Kramer, E., Hess, J. R., & Scalea, T. M. (2010). Trauma mortality in mature trauma systems: Are we doing better? An analysis of trauma mortality patterns, 1997-2008. *The Journal of Trauma*, *69*(3), 620–626. https://doi.org/10.1097/TA.0b013e3181bbfe2a.

Faul, M., Xu, L., Wald, M. M., & Coronado, V. G. (2010). *Traumatic brain injury in the United States: Emergency department visits, hospitalizations and deaths 2002–2006*. Atlanta: National Center for Injury Prevention and Control, Centers for Disease Control and Prevention.

Gennarelli, T. A., Champion, H. R., Copes, W. S., & Sacco, W. J. (1994). Comparison of mortality, morbidity, and severity of 59,713 head injured patients with 114,447 patients with extracranial injuries. *The Journal of Trauma*, *37*(6), 962–968. https://doi.org/10.1097/00005373-199412000-00016.

Huang, S. J., Hong, W. C., Han, Y. Y., Chen, Y. S., Wen, C. S., Tsan, Y. S., & Tu, Y. K. (2007). Clinical outcome of severe head injury in different protocol-driven therapies. *Journal of Clinical Neuroscience: Official Journal of the Neurosurgical Society of Australasia*, *14*(5), 449–454. https://doi.org/10.1016/j.jocn.2005.12.012.

Hukkelhoven, C. W., Steyerberg, E. W., Habbema, J. D., Farace, E., Marmarou, A., Murray, G. D., … Maas, A. I. (2005). Predicting outcome after traumatic brain injury: Development and validation of a prognostic score based on admission characteristics. *Journal of Neurotrauma*, *22*(10), 1025–1039. https://doi.org/10.1089/neu.2005.22.1025.

King, J. T., Jr., Carlier, P. M., & Marion, D. W. (2005). Early Glasgow outcome scale scores predict long-term functional outcome in patients with severe traumatic brain injury. *Journal of Neurotrauma*, *22*(9), 947–954. https://doi.org/10.1089/neu.2005.22.947.

Lannoo, E., Van Rietvelde, F., Colardyn, F., Lemmerling, M., Vandekerckhove, T., Jannes, C., & De Soete, G. (2000). Early predictors of mortality and morbidity after severe closed head injury. *Journal of Neurotrauma*, *17*(5), 403–414. https://doi.org/10.1089/neu.2000.17.403.

Lingsma, H. F., Roozenbeek, B., Steyerberg, E. W., Murray, G. D., & Maas, A. I. (2010). Early prognosis in traumatic brain injury: From prophecies to predictions. *The Lancet. Neurology*, *9*(5), 543–554. https://doi.org/10.1016/S1474-4422(10)70065-X.

Maas, A. I., Hukkelhoven, C. W., Marshall, L. F., & Steyerberg, E. W. (2005). Prediction of outcome in traumatic brain injury with computed tomographic characteristics: A comparison between the computed tomographic classification and combinations of computed tomographic predictors. *Neurosurgery*, *57*(6), 1173–1182. https://doi.org/10.1227/01.neu.0000186013.63046.6b.

Majdan, M., Plancikova, D., Brazinova, A., Rusnak, M., Nieboer, D., Feigin, V., & Maas, A. (2016). Epidemiology of traumatic brain injuries in Europe: A cross-sectional analysis. *The Lancet Public Health*, *1*(2), e76–e83. https://doi.org/10.1016/s2468-2667(16)30017-2.

Marshall, L. F., Becker, D. P., Bowers, S. A., Cayard, C., Eisenberg, H., Gross, C. R., … Warren, J. (1983). The national traumatic coma data bank. Part 1: Design, purpose, goals, and results. *Journal of Neurosurgery*, *59*(2), 276–284. https://doi.org/10.3171/jns.1983.59.2.0276.

Marshall, L. F., Marshall, S. B., Klauber, M. R., Clark, M. V. B., Eisenberg, H. M., Jane, J. A., … Foulkes, M. A. (1991). A new classification of head injury based on computerized tomography. *Journal of Neurosurgery*, *75*(Supplement), S14–S20. Retrieved from https://thejns.org/view/journals/j-neurosurg/75/Supplement/article-pS14.xml. (Accessed 16 May 2021).

Matsuo, K., Aihara, H., Nakai, T., Morishita, A., Tohma, Y., & Kohmura, E. (2020). Machine learning to predict in-hospital morbidity and mortality after traumatic brain injury. *Journal of Neurotrauma*, *37*(1), 202–210. https://doi.org/10.1089/neu.2018.6276.

Moore, L., Lavoie, A., Camden, S., Le Sage, N., Sampalis, J. S., Bergeron, E., & Abdous, B. (2006). Statistical validation of the Glasgow Coma Score. *The Journal of Trauma*, *60*(6), 1238–1244. https://doi.org/10.1097/01.ta.0000195593.60245.80.

MRC CRASH Trial Collaborators, Perel, P., Arango, M., Clayton, T., Edwards, P., Komolafe, E., … Yutthakasemsunt, S. (2008). Predicting outcome after traumatic brain injury: Practical prognostic models based on large cohort of international patients. *BMJ (Clinical research ed.)*, *336*(7641), 425–429. https://doi.org/10.1136/bmj.39461.643438.25.

Murray, G. D., Teasdale, G. M., Braakman, R., Cohadon, F., Dearden, M., Iannotti, F., … Unterberg, A. (1999). The European Brain Injury Consortium survey of head injuries. *Acta Neurochirurgica*, *141*(3), 223–236. https://doi.org/10.1007/s007010050292.

Myburgh, J. A., Cooper, D. J., Finfer, S. R., Venkatesh, B., Jones, D., Higgins, A., … Australasian Traumatic Brain Injury Study (ATBIS) Investigators for the Australian, & New Zealand Intensive Care Society Clinical Trials Group. (2008). Epidemiology and 12-month outcomes from traumatic brain injury in Australia and New Zealand. *The Journal of Trauma*, *64*(4), 854–862. https://doi.org/10.1097/TA.0b013e3180340e77.

Ono, J., Yamaura, A., Kubota, M., Okimura, Y., & Isobe, K. (2001). Outcome prediction in severe head injury: Analyses of clinical prognostic factors. *Journal of Clinical Neuroscience: Official Journal of the Neurosurgical Society of Australasia*, *8*(2), 120–123. https://doi.org/10.1054/jocn.2000.0732.

Perel, P., Edwards, P., Wentz, R., & Roberts, I. (2006). Systematic review of prognostic models in traumatic brain injury. *BMC Medical Informatics and Decision Making*, *6*, 38. https://doi.org/10.1186/1472-6947-6-38.

Prasad, K. (1996). The Glasgow Coma Scale: A critical appraisal of its clinimetric properties. *Journal of Clinical Epidemiology*, *49*(7), 755–763. https://doi.org/10.1016/0895-4356(96)00013-3.

Roozenbeek, B., Chiu, Y. L., Lingsma, H. F., Gerber, L. M., Steyerberg, E. W., Ghajar, J., & Maas, A. I. (2012). Predicting 14-day mortality after severe traumatic brain injury: Application of the IMPACT models in the brain trauma foundation TBI-trac® New York State database. *Journal of Neurotrauma*, *29*(7), 1306–1312. https://doi.org/10.1089/neu.2011.1988.

Schnakers, C., Ledoux, D., Majerus, S., Damas, P., Damas, F., Lambermont, B., … Laureys, S. (2008). Diagnostic and prognostic use of bispectral index in coma, vegetative state and related disorders. *Brain Injury*, *22*(12), 926–931. https://doi.org/10.1080/02699050802530565.

Settervall, C. H. C., Sousa, R. M. C., & Silva, S. C. F. (2011). In-hospital mortality and the Glasgow Coma Scale in the first 72 hours after traumatic brain injury. *Revista Latino-Americana de Enfermagem*, *19*(6), 1337–1343. https://doi.org/10.1590/S0104-11692011000600009.

Stevens, R. D., & Sutter, R. (2013). Prognosis in severe brain injury. *Critical Care Medicine*, *41*(4), 1104–1123. https://doi.org/10.1097/CCM.0b013e318287ee79.

Steyerberg, E. (2009). Clinical prediction models. In *Statistics for biology and health*. https://doi.org/10.1007/978-0-387-77244-8.

Steyerberg, E. W., Mushkudiani, N., Perel, P., Butcher, I., Lu, J., McHugh, G. S., … Maas, A. I. (2008). Predicting outcome after traumatic brain injury: Development and international validation of prognostic scores based on admission characteristics. *PLoS Medicine*, *5*(8). https://doi.org/10.1371/journal.pmed.0050165, e165.

Stocchetti, N., Carbonara, M., Citerio, G., Ercole, A., Skrifvars, M. B., Smielewski, P., … Menon, D. K. (2017). Severe traumatic brain injury: Targeted management in the intensive care unit. *The Lancet. Neurology*, *16*(6), 452–464. https://doi.org/10.1016/S1474-4422(17)30118-7.

Stocchetti, N., Pagan, F., Calappi, E., Canavesi, K., Beretta, L., Citerio, G., … Colombo, A. (2004). Inaccurate early assessment of neurological severity in head injury. *Journal of Neurotrauma*, *21*(9), 1131–1140. https://doi.org/10.1089/neu.2004.21.1131.

Taylor, C. A., Bell, J. M., Breiding, M. J., & Xu, L. (2017). Traumatic brain injury-related emergency department visits, hospitalizations, and deaths–United States, 2007 and 2013. *Morbidity and mortality weekly report. Surveillance summaries (Washington, D.C. 2002)*, *66*(9), 1–16. https://doi.org/10.15585/mmwr.ss6609a1.

Vik, A., Nag, T., Fredriksli, O. A., Skandsen, T., Moen, K. G., Schirmer-Mikalsen, K., & Manley, G. T. (2008). Relationship of "dose" of intracranial hypertension to outcome in severe traumatic brain injury. *Journal of Neurosurgery*, *109*(4), 678–684. https://doi.org/10.3171/JNS/2008/109/10/0678.

Chapter 26

Pituitary dysfunction after traumatic brain injury: A focus on screening, diagnosis, and treatment

Aysa Hacioglu[a], Zuleyha Karaca[a], and Fahrettin Kelestimur[b]
[a]*Department of Endocrinology, Erciyes University Medical School, Kayseri, Turkey,* [b]*Department of Endocrinology, Faculty of Medicine, Yeditepe University, Kosuyolu Hospital, Istanbul, Turkey*

List of abbreviations

ACTH	adrenocorticotropic hormone
ADH	anti-diuretic hormone
AHA	anti-hypothalamus antibody
APA	anti-pituitary antibody
ApoE	apolipoprotein E
BMI	body mass index
CBG	cortisol binding globulin
DI	diabetes insipidus
FSH	follicle stimulating hormone
fT3	free triiodothyronine
fT4	free thyroxine
GCS	Glasgow coma scale
GH	Growth hormone
GHRH	growth hormone releasing hormone
GST	glucagon stimulation test
HPA axis	hypothalamic-pituitary-adrenal axis
HPG axis	hypothalamic-pituitary-gonadal axis
ICU	intensive care unit
IGF-1	insulin-like growth factor-1
ITT	insulin tolerance test
LH	luteinizing hormone
MRI	magnetic resonance imaging
QoL	quality of life
TBI	traumatic brain injury
TSH	thyroid stimulating hormone

Introduction

Traumatic brain injury (TBI) is a major health problem that causes a huge burden on health care. There have been 2.5 million emergency department visits and 282,000 hospitalizations related to TBI in the United States in 2013 (Taylor, Bell, Breiding, & Xu, 2017). TBI may cause life-long disabilities and unnoticed pituitary dysfunction may be a serious contributor. It was in 1918 that the association between TBI and pituitary dysfunction was reported for the first time in the literature (Cryan, 1918). Accumulating reports of new cases and new evidence obtained through time increased awareness during last decades.

Diagnosis and Treatment of Traumatic Brain Injury. https://doi.org/10.1016/B978-0-12-823347-4.00026-9

The prevalence of pituitary dysfunction has been reported in a range of 5% to 76% following TBI (Glynn & Agha, 2019). The prevalence was 15% in a meta-analysis that included only studies performing confirmatory tests (Tanriverdi et al., 2015). Diversity between studies in terms of patient demographics, severity of TBI, time interval of hormone evaluation and different tests used for diagnosis are the main reasons for the wide range. Pituitary dysfunction that results in deficiencies of cortisol, thyroid hormones or anti-diuretic hormone (ADH) may be life-threatening and insufficiencies of growth hormone (GH) and gonadal hormones are closely related to decreased quality of life and sense of well-being. On the other hand, signs and symptoms of pituitary dysfunction may be insidious and non-specific, requiring a high index of suspicion for the diagnosis. High prevalence of pituitary dysfunction following TBI as well as its debilitating and possibly life-threatening effects that may be successfully treated made development of screening strategies essential.

Obviously not every patient who had TBI should be screened for hormone deficiencies and predictive factors have been proposed to detect patients likely to develop pituitary dysfunction. Screening algorithms have been based on these risk factors though there are still no definite factors that can perfectly predict patients who will suffer from neuroendocrine consequences post-TBI.

There are considerable differences in terms of patient characteristics, diagnostic tests, screening intervals and duration of follow-up between screening algorithms proposed so far. Clinicians that deal with TBI patients should be aware of these algorithms as well as diagnostic tests and replacement therapies of pituitary hormone deficiencies especially which could have life-threatening consequences.

In this section, we will focus on predictive factors, screening algorithms, diagnostic strategies and replacement therapies of pituitary dysfunction post-TBI.

Screening of pituitary dysfunction following traumatic brain injuries

Predictive factors of post-TBI pituitary dysfunction

One of the main queries during development of screening algorithms is the characteristics of the patients that should be offered routine follow-up for pituitary dysfunction post-TBI. Screening all patients will neither be cost-effective nor necessary and false positive results can cause unnecessary treatments and burden on health care. For these reasons predictive factors have been proposed for patient selection and screening algorithms have been based on these factors (Ghigo et al., 2005; Tan et al., 2017; Tanriverdi et al., 2015; Tritos, Yuen, Kelly, Neuroendocrine, & Pituitary Scientific, 2015). The factors that have been reported to be associated with post-TBI pituitary dysfunction are presented in Table 1. However, there are inconsistent results between studies and none of these factors are proved to be precise.

The most commonly used predictive factor as the basis for screening algorithms is the severity of TBI. Glasgow coma scale (GCS) score is generally used to define severity. Patients with low GCS scores develop pituitary dysfunction more frequently (Hari Kumar et al., 2016; Klose et al., 2007). However, some authors stated that GCS may not be discriminative enough and that secondary injuries that cannot be graded by initial GCS (brain edema, bleeding etc.) contribute to severity (Schneider et al., 2006). Another group of experts used loss of consciousness, length of post-traumatic amnesia and being discharged from emergency unit in addition to GCS to define severe and mild TBI and integrated the term of 'complicated mild TBI' in their screening algorithm (Table 1) (Tanriverdi et al., 2015).

Hypocortisolemia, central diabetes insipidus (DI) or low IGF-1 levels during acute phase were associated with long term pituitary deficiencies (Hannon et al., 2013; Tanriverdi et al., 2006) though this association was not confirmed by all studies (Klose et al., 2007). Dysfunctions of multiple axes during acute phase were predictive of long term deficiencies (Hari Kumar et al., 2016) and panhypopituitarism detected at 3rd month post-TBI was reported to persist thereafter (Aimaretti et al., 2005).

Patients with radiological abnormalities during acute phase were more likely to develop long term pituitary dysfunctions. Bavisetty et al. (2008) reported the severity of brain parenchymal injury on CT imaging during acute phase as the strongest predictor of pituitary dysfunction and diffuse brain swelling, multiple contusions and evacuated intracranial hematoma were associated with increased risk for hypopituitarism. Basal skull fractures and diffuse axonal injuries were also associated with hormonal impairments (Schneider et al., 2008).

Old age, female gender and high body mass index (BMI) were associated with pituitary dysfunction (Frendl et al., 2017; Lauzier et al., 2014; Schneider et al., 2008). However, a meta-analysis reported that the association of age was lost if studies with high risk of bias were excluded and that gender was not associated with pituitary dysfunction (Lauzier et al., 2014). Considering the inconsistencies in the literature, these demographic features cannot be regarded as strong predictors.

TABLE 1 Factors reported to be associated with development of pituitary deficiency following TBI.

Factors reported in the literature	References (author, year of publication)
Severity of TBI Glasgow Coma Scale Modified Rankin Scale	Hari Kumar, Swamy, and Khan (2016); Kelly et al. (2000); Klose et al. (2007) Schneider et al. (2006)
Hormonal changes during acute phase Adrenal insufficiency Diabetes insipidus Low Insulin-like growth factor-1 Multiple hormonal deficiencies Panhypopituitarism	Hannon et al. (2013) Hannon et al. (2013) Tanriverdi et al. (2006) Hari Kumar et al. (2016) Aimaretti et al. (2005)
Radiological changes during acute phase Diffuse brain swelling Basal skull fracture Diffuse axonal injury Multiple contusions	Bavisetty et al. (2008); Kelly et al. (2000) Lauzier et al. (2014); Schneider et al. (2008) Schneider et al. (2008) Bavisetty et al. (2008)
Potential biomarkers microRNA-126-3p, microRNA-3610 Anti-pituitary antibody, anti-hypothalamus antibody Plasminogen activator inhibitor-1	Taheri et al. (2016) Tanriverdi et al. (2013) Frendl et al. (2017)
Genetic polymorphism Apolipoprotein E	Tanriverdi et al. (2008))
Patient demographics Older age Female gender High body mass index	Schneider et al. (2008); Tanriverdi et al. (2008) Frendl et al. (2017) Bavisetty et al. (2008); Tanriverdi et al. (2006)
Miscellaneous Hypotensive or hypoxic insults Neurosurgical intervention	Kelly et al. (2000) Nemes et al. (2016)

Predictive factors of pituitary dysfunction following TBI and relevant studies are presented.

Development of biomarkers that will accurately predict pituitary dysfunction post-TBI is a novel area of research. Tanriverdi et al. (2013) reported that positivity of anti-pituitary (APA) and anti-hypothalamus antibodies (AHA) are associated with pituitary dysfunction post-TBI. These results pointed to a possible autoimmune mechanism that take role in the development of hormonal disturbances following TBI. Taheri et al. (2016) studied the role of microRNAs (miRNA) and reported significant association between pituitary dysfunction post-TBI and altered expression of miR-126-3p and miR-3610. Frendl et al. (2017) reported that low levels of plasminogen activator inhibitor type-1 during acute phase were predictive of long term pituitary deficiencies and it was suggested as a biomarker.

Apolipoprotein E (apoE) is known to be involved in neural branching, growth and injury repair and genetic polymorphisms of apoE have been reported to be associated with various diseases such as Alzheimer's disease, cerebral infarction, subarachnoid hemorrhage as well as prognosis and outcome following TBI (Maiti et al., 2015). Tanriverdi et al. reported that pituitary dysfunction post-TBI was associated with genetic polymorphisms of apoE, occurring significantly less in patients with APOE3/E3 genotype (Tanriverdi, Taheri, et al., 2008). However, these studies were all single-centered and proposed biomarkers are not eligible for clinical utility unless the results are confirmed on larger patient groups.

Screening algorithms of pituitary dysfunction after traumatic brain injuries

Taking into consideration the high prevalence of pituitary dysfunction post-TBI and its detrimental effects, many authors agreed on the necessity of routine screening programs (Bavisetty et al., 2008; Lauzier et al., 2014; Schneider et al., 2008). Systematic long-term screening algorithms developed by expert groups and authors are presented in Table 2. These algorithms have been mainly based on the literature data and clinical experiences and they vary in terms of criteria of patient selection, screening intervals and duration of screening.

TABLE 2 Long-term screening algorithms for pituitary dysfunction post-TBI.

<table>
<tr><th>First author, year of publication</th><th>Inclusion criteria in the screening strategy</th><th>Exclusion criteria from the screening strategy</th><th>Screening intervals and tests used to evaluate pituitary axes</th></tr>
<tr><td>Ghigo et al. (2005)</td><td>Acute phase post-TBI:
TBI patients who are hospitalized and who have symptoms (e.g., hyponatremia or hypotension)
Prospective screening:
All TBI patients
Retrospective screening:
Symptomatic patients with a history of a moderate or severe TBI before >12 months</td><td>- Being in a vegetative state
- Functioning at a very low level
- Being institutionalized due to a TBI
(These patients are only to be evaluated and replaced for; central hypocortisolemia, central hypothyroidism, central diabetes insipidus and inappropriate ADH syndrome.)</td><td>Acute phase:
- Basal hormonal evaluation*
- Patients with adrenal insufficiency, diabetes insipidus or those symptomatic for hypopituitarism should be evaluated for other pituitary hormone deficiencies without waiting for 3rd-month control.
Prospective screening:
- Basal hormonal evaluation at 3rd and 12th months following TBI or discharge from the hospital or intensive care unit*¥
- Patients who have signs or symptoms of pituitary dysfunction should be evaluated whenever medically indicated.
Retrospective screening:
- Immediate basal hormonal evaluation when signs or symptoms of pituitary dysfunction are detected*¥
- Testing is suggested once in these patients.
*(morning cortisol, urinary free cortisol, fT3, fT4, TSH, IGF-1, FSH, LH, Testosterone (men), Estradiol (women), prolactin, in cases of polyuria: urine density, serum sodium and osmolality)
¥Stimulation tests are recommended if baseline hormonal evaluation is not diagnostic.</td></tr>
<tr><td>Tanriverdi et al. (2015)</td><td>- Complicated mild TBI;
- Radiological changes on initial imaging by CT or MRI
- Acute central adrenal insufficiency and/or central DI
- Hospitalization for >24h
- ICU monitoring
-Neurosurgical intervention
- APA and AHA positivity
- Moderate TBI
- Glasgow Coma Score of 9–12
- Severe TBI;
- Glasgow Coma Score of 3–8
or
- Loss of consciousness, and/or confusion, disorientation >30min or post-traumatic amnesia >24h
- Independent of severity of TBI;
- Hospitalization for ≥24h
- Need for ICU monitoring</td><td>- Low life expectancy
- Being in a vegetative state
- Being severely disabled
-Having mild TBI and
- being discharged from emergency unit and/or
- having no loss of consciousness and/or
- having post-traumatic amnesia <30min</td><td>Acute phase (first 10–14 days following TBI):
- Basal cortisol measurement on 1st–4th day post-TBI
- If clinical suspicion exists; basal cortisol measurement on 5th–10th day post-TBI
- If basal cortisol is not diagnostic, stimulation test (insulin tolerance test, glucagon stimulation test) is suggested when the patient is stable before discharge
6th month
- Basal hormone levels
- Stimulation tests (ACTH stimulation test, insulin tolerance test, glucagon stimulation test) only for central adrenal insufficiency
12th month
- Basal hormone levels
- Stimulation tests for central adrenal insufficiency and GH deficiencies
Follow-up beyond 12th month
- Patients with complicated mild TBI;
- Yearly re-evaluation until 5 years post-TBI
- Patients with moderate or severe TBI;
- Stop follow-up if no hormone deficiencies are detected at 12th month
- Follow-up as for other causes of hypopituitarism for those with ≥1 hormone deficiencies</td></tr>
</table>

Continued

TABLE 2 Long-term screening algorithms for pituitary dysfunction post-TBI—cont'd

First author, year of publication	Inclusion criteria in the screening strategy	Exclusion criteria from the screening strategy	Screening intervals and tests used to evaluate pituitary axes
Tritos et al. (2015)	- Patients with mild TBI* who are symptomatic and have decrease in quality of life - All patients with moderate and severe TBI* *Veterans Administration/ Department of Defense Guideline was used for evaluation of the severity of TBI (Management of Concussion/m, 2009)	Not specified	*Acute phase (first 2 weeks following TBI):* - Evaluation for central adrenal insufficiency and central diabetes insipidus - Basal cortisol levels are to be used for evaluation and not stimulation tests - Central hypothyroidism is suggested to be evaluated ≥4–6 weeks after TBI *Chronic phase (>3 months):* - Basal hormone levels - Stimulation tests for evaluation of central adrenal insufficiency and GH deficiency if indicated - Re-evaluation following >12 months after TBI is not recommended due to insufficiency of data
Tan et al. (2017)	- All patients who are admitted for >48 h following TBI - Symptomatic patients who are not admitted or admitted for less than 48 h	Not specified	*Acute phase:* - No routine evaluation during acute phase after TBI - Clinician should consider central diabetes insipidus/inappropriate ADH syndrome in cases of sodium imbalance *Following discharge from hospital:* - Patients admitted for >48 h should be evaluated at 3–6 months post-TBI with basal hormone levels - If test results at 3–6 months are abnormal, patients should be fully evaluated with stimulation tests - Screening for GH insufficiency is not suggested during first year following TBI *> 1 year:* - TBI patients with symptoms attributable to pituitary dysfunction should be evaluated for hormone deficiencies by an endocrinologist
Gilis-Januszewska, Kluczynski, and Hubalewska-Dydejczyk (2020)	- Patients with mild TBI* who have; - Old age, - Repetitive injuries, - Skull fractures, - Long duration of coma, - Polyuria, polydipsia - Patients with moderate and severe TBI* *Severity is defined according to GCS	- Mild TBI patients without risk factors - Severe TBI patients with low life expectation	*Acute phase (1–7 days following TBI):* - Evaluation for central adrenal insufficiency *Chronic phase (3–6, 12 months following TBI):* - Re-evaluation for central adrenal insufficiency and central DI in patients with previous diagnoses - Evaluation for central hypothyroidism, central hypogonadism, GH deficiency *Annual screening after 12 months:* - Re-evaluation for central adrenal insufficiency and central DI in patients with previous diagnoses - Evaluation for new insufficiencies

Screening algorithms for pituitary dysfunction following TBI that are developed by expert groups and suggested to be used on a routine basis. *ACTH*, adrenocorticotropic hormone: *ADH*, anti-diuretic hormone; *AHA*, anti-hypothalamus antibody; *APA*, anti-pituitary antibody; *CT*, computed tomography; *DI*, diabetes insipidus; *FSH*, follicle stimulating hormone; *fT3*, free triiodothyronine; *fT4*, free thyroxine; *GCS*, Glasgow come scale; *GH*, growth hormone; *ICU*, intensive care unit; *IGF-1*, insulin-like growth factor-1; *LH*, luteinizing hormone; *MRI*, magnetic resonance imaging; *TBI*, traumatic brain injury; *TSH*, thyroid stimulating hormone.

The severity of TBI was commonly used for patient selection and GCS was generally the main determinant for severity. Tritos et al. (2015) used the classification proposed by Veterans Administration/Department of Defense Guideline. Alteration and loss of consciousness and post-traumatic amnesia were other parameters used besides GCS in this classification. Tanriverdi et al. (2015) used GCS for severity of TBI but they included a subgroup of mild TBI patients (complicated mild TBI) that was defined by predictive factors based on previous literature data. Tan et al. (2017) based patient selection on duration of stay in the hospital post-TBI (Table 2).

Although inclusion criteria vary across guidance, authors agree that patients in a vegetative state or those that are severely disabled should be excluded from long-term screening. These patients should be tested and replaced for central adrenal insufficiency, central hypothyroidism and posterior pituitary dysfunctions. Routine evaluation of gonadal functions and GH deficiency is not recommended as no benefits are expected.

Acute phase is generally defined as first 14 days and chronic phase as a period of ≥3 months since TBI (Tanriverdi et al., 2015; Tritos et al., 2015). Authors are in consensus that patients should be evaluated only for development of central adrenal insufficiency and central DI during acute phase (Table 2). Central hypothyroidism may be confused with non-thyroidal illness syndrome and evaluation is suggested to be deferred until 4th week post-TBI (Tritos et al., 2015). Testing for dysfunctions of gonadal and GH axes should be performed during chronic phase because of suppressive adaptive response during acute phase and possibility of spontaneous improvement during follow-up.

Pituitary functions tend to be dynamic following TBI. Hormone deficiencies detected during early phase may improve or new impairments may develop over time. Tanriverdi et al. (2013) observed that improvement of pituitary functions was more likely in patients with mild TBI rather than severe TBI. The dynamic alterations generally occur during first 6 months but may be seen even years following trauma (Tanriverdi et al., 2015). For this reason long-term screening is recommended. Some authors suggested that routine screening should be continued until 5 years post-TBI while other groups limited it to one year due to lack of evidence (Table 2).

Diagnosis of pituitary dysfunction following traumatic brain injuries

The diagnosis of pituitary dysfunction is mainly based on clinical evaluation and laboratory analysis. Imaging studies of the pituitary during acute phase may reveal basal skull fractures, brain edema or hemorrhages of pituitary region. However, these radiological abnormalities are not diagnostic and patients may still have normal hormonal profile. Decrease in pituitary volume was detected during chronic phase in some patients with pituitary dysfunction (Tanriverdi et al., 2008) but completely normal pituitary imaging may be observed too. Radiologic analysis is required for exclusion of other pituitary pathologies such as tumors in patients with hormonal disturbances.

Central adrenal insufficiency

Central adrenal insufficiency may cause non-specific complaints such as fatigue, loss of appetite, nausea, weight loss but may have life-threatening consequences if left unnoticed. Hypotension resistant to fluid resuscitation and vasopressors, electrolyte imbalances, hypoglycemia in critically ill patients should alert the clinician and appropriate therapy should be started immediately.

The diagnosis of central adrenal insufficiency during acute phase is based on basal cortisol concentrations. Basal cortisol should be obtained at 8 to 9 am and levels below 3 μg/dL (80 nmol/L) are highly diagnostic (Fleseriu et al., 2016). The levels higher than 15 μg/dL indicate a sufficient response but some centers suggest a cut-off of 18 μg/dL (Arlt & Allolio, 2003; Fleseriu et al., 2016). Diagnostic cut-offs are different for critically ill patients followed in ICU (Annane et al., 2017).

Stimulation tests are required for patients with basal cortisol values between 3 μg/dL and 15 μg/dL (Fleseriu et al., 2016). The tests used and their diagnostic cut-off values are presented in Table 3. ACTH stimulation test should be deferred until at least 3 months following TBI as adrenals may still have normal stimulatory response to ACTH and the results may be false negative (Dorin, Qualls, & Crapo, 2003). Insulin tolerance test (ITT) is the gold standard diagnostic test but caution should be practiced as the test is contraindicated in patients with seizures. Glucagon stimulation test (GST) may be an option in such cases. GST gives the opportunity of evaluation of both GH and HPA axes but it is time consuming, dependent on BMI and may be disturbing to patients. There is no single test to correctly diagnose all patients and clinical evaluation together with test results is critical for correct diagnosis.

Growth hormone deficiency

Growth hormone deficiency causes non-specific signs and symptoms such as fatigue, weakness, increase in adiposity and decline in cognition that may require a high index of suspicion for diagnosis. Serum IGF-1 is the main screening test and results below the age- and sex-specific reference range indicate GH deficiency but stimulation tests are required for definitive diagnosis. Patients with three or more pituitary hormone deficiencies and low IGF-1 levels are the exception and the diagnosis is straightforward in these cases (Fleseriu et al., 2016). Symptomatic TBI patients with low IGF-1

TABLE 3 Stimulation tests used for the evaluation of hypothalamic-pituitary-adrenal axis (Fleseriu et al., 2016; Karaca et al., 2011; Tanriverdi et al., 2007).

Stimulation tests	Test protocol	Interpretation of the results
ACTH stimulation test Standard dose (250 μg) Low dose (1 μg)	- ACTH 1–24 applied intravenously at 0.min - Blood samples obtained for cortisol at 0.-30.-60.min *(Some centers obtain samples at 90. and 120. min also)*	*Sufficient response (250 μg and 1 μg):* Peak cortisol: - > 18.1 μg/dL *Cut-offs used by some centers:* *Sufficient response:* Peak cortisol: - 1 μg ACTH: > 12.5 μg/dL - 250 μg ACTH: > 20 μg/dL
Insulin tolerance test	- Insulin (0.05–0.15 U/kg) applied as intravenous bolus - Glucose measured < 40 mg/dL at 0.min - Blood samples obtained for cortisol at -30.-0.-30.-60.-120.min	*Sufficient response*: Peak cortisol: - > 18.1 μg/dL
Glucagon stimulation test	- Glucagon (1 mg; if patient is > 90 kg 1.5 mg) applied intramuscularly/ subcutaneously at 0.min - Blood samples obtained for cortisol at 0.-30.-60.-90.-120.-150.-180.-210.-240.min	*Sufficient response*: Peak cortisol: - > 18 μg/dL *Cut-offs used by some centers:* *Sufficient response:* Peak cortisol: - > 10.74 μg/dL

Stimulation tests, test protocols and diagnostic cut-offs used for the diagnosis of central adrenal insufficiency are presented.

levels and otherwise normal pituitary functions are to be diagnosed only after insufficient responses to two stimulation tests (Molitch et al., 2011). Table 4 displays the stimulation tests and various cut-offs used by different centers. ITT is the gold standard test but it may not be convenient for some TBI patients. GHRH (growth hormone releasing hormone)-arginine test may cause false positivity in patients with hypothalamic injury and GHRH is not available in many countries. The results of GST are reported to be affected by BMI and lower cut-offs have been offered for obese patients (Diri et al., 2015).

The diagnostic tests for GH deficiency should only be pursued in patients who are intended to be treated. All guidelines are in consensus that the evaluation of GH deficiency should be deferred until at least 12 months following TBI as there are no evidences that the patients will benefit from the replacement during early phase.

Central hypogonadism

The evaluation of hypothalamo-pituitary-gonadal (HPG) axis is suggested to be performed during chronic phase (Table 2) as the axis may be suppressed during acute phase as an adaptive response to stress (Vanhorebeek, Langouche, & Van den Berghe, 2006).

Low testosterone with inappropriately low FSH and LH levels obtained in the morning during fasting are indicative of the diagnosis in male patients but other causes such as hyperprolactinemia should be excluded first (Fleseriu et al., 2016).

The HPG axis is considered intact in women who are menstruating on a regular basis and routine testing is not required. In cases of oligo-anovulation, low baseline estradiol with low gonadotropin levels are diagnostic but other etiologic factors that may cause ovulatory dysfunction should be excluded. Post-menopausal women with inappropriately low FSH and LH levels are diagnosed with central hypogonadism. There are no stimulation tests to evaluate HPG axis (Fleseriu et al., 2016).

Central hypothyroidism

The diagnosis of central hypothyroidism is straightforward in patients with low free thyroxine (fT4) and inappropriately low, normal or mildly elevated thyrotropin (TSH) levels during chronic phase (Fleseriu et al., 2016). Stimulation tests are not required.

TABLE 4 Stimulation tests used for the evaluation of hypothalamic-pituitary-somatotropic axis (Fleseriu et al., 2016; Tanriverdi, Unluhizarci, Coksevim, et al., 2007).

Stimulation tests	Test protocol	Interpretation of the results
Insulin tolerance test	- Insulin (0.05–0.15 U/kg) applied as intravenous bolus - Glucose measured < 40 mg/dL at 0.min - Blood samples obtained for GH at -30.-0.-30.-60.-120.min	*Sufficient response*: Peak GH: - > 3 μg/L
Glucagon stimulation test	- Glucagon (1 mg; if patient is > 90 kg 1.5 mg) applied intramuscularly at 0.min - Blood samples obtained for GH at 0.-30.-60.-90.-120.-150.-180.-210.-240.min	*Sufficient response*: Peak GH: - > 3 μg/L *Cut-offs used by some centers:* *Sufficient response:* Peak GH: - > 1.18 μg/L
GHRH + arginine test	- GHRH (1 μg/kg) applied intravenously and then arginine infusion (0.5 g/kg) administered over 30 min - Blood samples obtained for GH at 0.-30.-45.-60.-75.-90.-105. and 120 min	*Insufficient response*: Peak GH: - ≤ 11.5 μg/L (BMI (kg/m^2) < 25) - ≤ 8.0 μg/L (BMI (kg/m^2) 25–30) - ≤ 4.2 μg/L (BMI (kg/m^2) ≥ 30)

Stimulation tests, test protocols and diagnostic cut-offs used for the diagnosis of growth hormone deficiency are presented. *BMI*, body mass index, *GH*: growth hormone, *GHRH*, growth hormone releasing hormone.

The evaluation should be performed 4 to 6 weeks following TBI because fT4 has a long half-life and non-thyroidal illness syndrome may confound early phase thyroid functions (Tritos et al., 2015). Free triiodothyronine (fT3) levels are usually first to be decreased in non-thyroidal illness while it is generally within the normal range in patients with central hypothyroidism. However, severe stressful events or trauma may cause decrease in both fT3 and fT4 levels making it difficult to differentiate from central hypothyroidism.

Central diabetes insipidus

Traumatic brain injury that causes hypothalamic, posterior pituitary or stalk damage may lead to decrease in ADH secretion by posterior pituitary. Insufficient secretion of ADH in response to increased osmolality causes central DI. Central DI usually develops during acute phase but may occur during chronic phase too. Conscious patients with intact sense of thirst will complain of polydipsia and polyuria. Patients who lost their sense of thirst or unconscious patients will present with increased output of dilute urine, hypernatremia and increased serum osmolality (Seckl & Dunger, 1989).

Simultaneous urine and serum osmolality and serum sodium levels are required for the diagnosis. The diagnostic criteria are presented in Table 5. The differential diagnosis may be challenging in some cases and water deprivation test may be required. Measurement of serum ADH level is not routinely practiced. Loss of the bright spot of posterior pituitary on magnetic resonance imaging (MRI) supports the diagnosis of DI (Fleseriu et al., 2016).

Contrary to central DI, increased secretion of ADH independently from feed-back mechanisms of serum osmolality and volume status causes hyponatremia and increase in urine osmolality leading to syndrome of inappropriate ADH secretion. Readers are referred to relevant articles for detailed information (Spasovski et al., 2014).

Treatment of pituitary dysfunction following traumatic brain injuries

Central adrenal insufficiency

Adrenal insufficiency may have life-threatening consequences and treatment should be started immediately without waiting for test results. If clinical suspicion exists in critically ill patients, a random sample for cortisol should be obtained and immediate parenteral glucocorticoid replacement should be started (Table 6). Clinically stable patients with non-specific symptoms should be tested as noted in 'Diagnosis' section and patients with confirmed diagnosis should be treated as suggested in relevant guidelines (Bornstein et al., 2016; Fleseriu et al., 2016) (Table 6).

TABLE 5 Diagnostic criteria of central diabetes insipidus (Seckl & Dunger, 1989).

Polyuria	> 2 mL/kg/hour > 300 mL/hour (during two consecutive hours)
Serum osmolality	> 300 mOsm/kg
Urine osmolality	< 300 mOsm/kg

Urine volume, serum and urine osmolalities required for the diagnosis of central DI are presented.

TABLE 6 Replacement modalities in patients with pituitary dysfunction (Bornstein et al., 2016; Fleseriu et al., 2016; Molitch et al., 2011).

Pituitary hormone deficiency	Treatment strategy	Recommendations during follow-up and target values
Central adrenal insufficiency	- Hydrocortisone 15–25 mg/day - Two or three divided doses - Highest dose taken in the morning - The lowest tolerable dose should be prescribed *In critically ill patients with suspected adrenal crisis (hypotension, electrolyte imbalance, hypoglycemia):* - Immediate treatment with hydrocortisone 100 mg as intravenous bolus and then 200 mg/24 h as infusion - The decision of the dose, route and duration of therapy during follow-up should be based on clinical evaluation.	- Monitoring of the therapy is based on clinical evaluation and not on serum cortisol levels; - Hypertension, edema, weight-gain, Cushingoid appearance, osteoporosis may be observed in overtreated patients - Insufficient treatment causes ongoing symptoms related to adrenal insufficiency - Patients should be recommended to increase the dose in stress-related situations
Growth hormone deficiency	- Initial GH doses: - > 60 years: 0.1–0.2 mg/day - 30–60 years: 0.2–0.3 mg/day - < 30 years: 0.4–0.5 mg/day - The dose may be increased 0.1–0.2 mg after 1–2 months	- Monitoring of clinical response, adverse effects, IGF-1 is suggested every 1–2 months during titration and every 6 months thereafter - The upper half of age-adjusted IGF-1 reference range is targeted - Patients on estrogen replacement require higher daily doses of GH - GH therapy may require dose adjustments of thyroid and steroid replacement - GH is contraindicated in patients with active malignancy - The most frequent adverse effects are arthralgia, myalgia, fluid retention, edema, increase in blood pressure and insulin resistance
Central hypogonadism	*Male patients;* - Testosterone replacement should be started in a formulation (oral, transdermal, intramuscular) convenient to the patient. *Female patients;* - Pre-menopausal women with intact uterus should be prescribed estrogen and progesterone replacement. - Hysterectomized patients should be started only on estrogen. - Patients who seek for fertility should be prescribed gonadotropin therapy.	*Male patients;* - Follow-up is suggested at 3–6 months after initiation of the therapy and then annually. - Serum testosterone level of mid-normal reference range is targeted *Female patients;* - Clinical evaluation and assessment of adverse effects during follow-up is suggested. - Serum estrogen level is not used during follow-up.
Central hypothyroidism	- Usual daily levothyroxine dose is 1.6 µg/kg. - Initial dose should be based on patient's fT4 level, age and comorbid diseases.	- The dose should be titrated to target the upper half of fT4 reference range - Serum TSH has no value during follow up. - Adrenal insufficiency should be excluded and if present should be replaced properly before levothyroxine replacement.

Continued

TABLE 6 Replacement modalities in patients with pituitary dysfunction (Bornstein et al., 2016; Fleseriu et al., 2016; Molitch et al., 2011)—cont'd

Pituitary hormone deficiency	Treatment strategy	Recommendations during follow-up and target values
Central diabetes insipidus	- Appropriate doses of desmopressin should be started initially and increased as needed during follow-up - Various desmopression formulations (oral, sublingual, parenteral) can be preferred according to patient's clinical status	- Urine output, fluid balance, serum and urine osmolalities and serum sodium levels are monitored during dose titration - The aim is to control polyuria and restore a neutral fluid balance with no electrolyte abnormalities

Hormone replacement strategies in pituitary dysfunction and important tips during follow-up are summarized. *GH*, growth hormone; *IGF-1*, insulin-like growth factor-1; *fT4*, free thyroxine, *TSH*, thyroid stimulating hormone.

Growth hormone deficiency

Growth hormone deficiency may have various detrimental consequences such as decrease in muscle mass, increase in adiposity, disruption of bone structure, changes in lipid profile, fatigue, decline in cognitive functions and quality of life (QoL) (Kelly et al., 2006; Molitch et al., 2011). Replacement of GH has been reported to reverse these effects to some extent (Molitch et al., 2011). However, studies investigating the effects of GH replacement mostly included non-TBI patients and the data on TBI patients is scarce.

Growth hormone has anabolic effects on muscle tissue increasing lean body mass, though improvement in muscle strength was not observed meanwhile (Widdowson & Gibney, 2010). Lipolytic effects of GH cause reduction in body fat; however, inconsistent results have been reported on lipid profile (Chihara et al., 2006; Kreitschmann-Andermahr et al., 2008; Tanriverdi, Unluhizarci, Karaca, Casanueva, & Kelestimur, 2010). Improvement in some cardiovascular parameters such as endothelial function and carotid intima-media thickness was observed but the impact on cardiovascular mortality is not known yet (Chihara et al., 2006; Gibney et al., 1999; Molitch et al., 2011).

Replacement of GH has been reported to improve cognitive functions in GH deficient TBI patients (Reimunde et al., 2011). These effects are not surprising as GH and IGF-1 have been reported to have stimulatory effects on cellular proliferation, renewal and neurogenesis as well as on angiogenesis, cerebral blood flow, glucose metabolism, neurotransmitter systems and neuroprotection (Aberg, Brywe, & Isgaard, 2006). Wright et al. (2020) reported that GH replacement caused alterations in the brain structure of GH deficient mild TBI patients. They observed increase in connectivity between brain networks and increase in gray matter volume and reported that these changes were correlated with improvements in some symptoms such as fatigue and sleep quality.

Increase in QoL was reported following GH therapy in GH deficient TBI patients (Kreitschmann-Andermahr et al., 2008; Moreau et al., 2013). Interestingly, TBI patients have been reported to gain more improvement in terms of QoL following GH replacement compared to non-TBI patients and that this benefit was long-term (Gardner et al., 2015).

The replacement of GH is recommended to be started one year after TBI (Table 2). GH secretion may recover during early phase and there is no evidence on beneficial effects of replacement during this period. The initial dose should be based on the age of patient and dose titration during follow-up should be individualized according to serum IGF-1, clinical outcome and adverse effects (Table 6).

There is no evidence based data on proper duration of GH replacement in adults. Discontinuation of the treatment leads to deterioration in parameters that had improved with GH replacement (Filipsson Nystrom et al., 2012; Maric et al., 2010). The therapy should be stopped if adverse effects are observed or if there are no beneficial effects after at least one year.

Central hypogonadism

Hypogonadism may cause multiple systemic detrimental effects. Male patients may experience decrease in muscle mass and strength, anemia, osteoporosis, fatigue, sexual dysfunction, infertility, loss of body hair and decreased QoL while female patients may have menstrual irregularities, osteoporosis and increased risk of cardiovascular diseases (Bhasin et al., 2010; Fleseriu et al., 2016).

Treatment of hypogonadism should be initiated during chronic phase after eliminating the possibility of adaptive response. Restoration of a normal serum testosterone level improves muscle strength, bone mineral density, anemia, sexual function and sense of well-being in men (Fleseriu et al., 2016). Regular evaluation for polycythemia, sleep apnea and prostate pathologies is suggested (Bhasin et al., 2010).

Hormone replacement in pre-menopausal women restores normal menstruation, prevents osteoporosis and decreases cardiovascular diseases and mortality. The therapy is suggested to be continued until the age of menopause (Fleseriu et al., 2016; Webber, Anderson, Davies, Janse, & Vermeulen, 2017). Clinicians should be aware of the contraindications and adverse effects.

The possibility of spontaneous recovery of the axis should not be overlooked (Tanriverdi, Unluhizarci, Selcuklu, Casanueva, & Kelestimur, 2007). The replacement therapy may be interrupted and hormonal evaluation repeated if clinical suspicion of recovery occurs.

Central hypothyroidism

Central hypothyroidism may cause non-specific symptoms such as decreased energy status, fatigue, weight gain, cognitive dysfunction and signs such as electrolyte imbalance, insulin resistance, bradycardia and hypertension (Fleseriu et al., 2016). Thyroid hormones are vital and must be replaced properly. Treatment of central hypothyroidism in TBI patients is the same as in non-TBI patients (Table 6).

Central diabetes insipidus

The patient with central DI should be started on desmopressin to decrease free water clearance. The therapy is started with low doses initially and individualized according to urine output. As most cases of central DI are transient and resolve spontaneously during early phase post-TBI, initial desmopressin therapy should be given as required and not on a routine basis (Tudor & Thompson, 2019). In cases of spontaneous improvement of central DI, probable development of adrenal insufficiency should be kept in mind (Fleseriu et al., 2016).

Serum sodium and urine output should be monitored during follow-up. As free water clearance will be inhibited by desmopressin, hyponatremia may abruptly develop unless fluid replacement is dynamically adjusted according to decreased urine output. The therapy is more complicated in patients with hypothalamic damage resulting in loss of sense of thirst (adipsic DI). These patients should be managed with individualized desmopressin dose and fixed fluid intake with frequent weighing and serum sodium monitoring (Tudor & Thompson, 2019).

Applications to other areas of neuroscience

Growth hormone (GH) replacement is known to improve neurocognitive functions in patients who developed GH deficiency after traumatic brain injury (TBI) (Moreau et al., 2013). This effect may be explained by the regulatory role of GH/IGF-1 (insulin-like growth factor-1) on cellular renewal, neurogenesis and increase in cerebral blood flow (Aberg et al., 2006). However, GH and IGF-1 were also reported to take role in repair of brain injury secondary to TBI.

Studies indicated that GH/IGF-1 has effects on neural progenitors and neurons in terms of induction of proliferation, differentiation and migration, thus regulating the neural repair that leads to improvement of both motor and cognitive neurological sequel following head trauma (Devesa, Reimunde, Devesa, Barbera, & Arce, 2013). A patient with a completely destroyed right cerebral hemisphere and severe neurological disabilities following TBI was reported to strikingly recover after GH therapy and rehabilitation. Patient did not have GH deficiency and authors commented that GH promoted compensation of brain functions through neural plasticity in intact brain areas (Devesa et al., 2015). Devesa et al. (2013) administered GH along with rehabilitation therapy in patients with neurologic sequel following brain injury that did not improve with solely rehabilitation previously. Both cognitive and motor improvements were observed even if GH was started years following TBI regardless of GH insufficiency. However, authors hypothesized that early administration of GH following TBI may cause better results. Rat studies supported this hypothesis and Heredia et al. (2013) reported that GH treatment improved motor functions through induction of neurogenesis of perilesional area and increased brain plasticity in contralateral brain area during early phase following neural injury in rats. It is not known yet if GH replacement can amplify neuronal repair and enhance neuroregeneration during acute phase post-TBI. Further studies are required to investigate beneficial effects of GH during acute phase in terms of neuronal repair.

Another area of research is the association between TBI, GH deficiency and gut microbiota. A novel mechanism that has been hypothesized to contribute to neurocognitive dysfunction post-TBI is gut dysbiosis (Yuen et al., 2020). Traumatic brain injury and subsequent GH deficiency may cause alteration in the composition of gut microbiota that might further deteriorate cognitive functions. Replacement of GH is assumed to reverse these changes through modulation of gut microbiota (Yuen et al., 2020). Investigations are currently ongoing on this interesting topic.

Mini-dictionary of terms

- **Anti-pituitary antibody**: An antibody against human pituitary cells that can be detected in the serum of patients who have hypophysitis or some other autoimmune endocrine disorders.
- **Apolipoprotein E**: A protein that mainly takes role in lipid metabolism but that is also found in the brain and involved in neural growth and repair.
- **Central diabetes insipidus**: A clinical condition characterized with increased excretion of free water and polyuria due to insufficient anti-diuretic hormone secretion from posterior pituitary.
- **Desmopressin:** A drug that inhibits aquaresis and free water loss and that is used for the treatment of central diabetes insipidus.
- **Empty sella:** A radiological term that refer to significant loss of pituitary volume and the appearance of emptiness given by cerebrospinal fluid filling the sella.
- **Syndrome of inappropriate anti-diuretic hormone (ADH) secretion:** A clinical condition characterized with euvolemic hyponatremia resulting from excess secretion of ADH and inability to excrete free water.
- **Insulin-like growth factor-1 (IGF-1):** A protein secreted by liver in response to growth hormone which is responsible for its peripheral effects.
- **Microbiota:** All micro-organisms that reside in a habitat.
- **microRNA:** Small RNA particles that do not code proteins but have effects on gene expression.
- **Panhypopituitarism:** Impairment of secretion of all anterior pituitary hormones.

Key facts of screening strategies, diagnostic and treatment modalities of pituitary dysfunction after traumatic brain injury

- Pituitary gland secretes hormones that are vital such as cortisol and thyroid hormones as well as those that are crucial for health and general well-being such as gonadal hormones, prolactin and growth hormone.
- Traumatic brain injury may directly or indirectly damage pituitary causing impairment in its functionality.
- Pituitary dysfunction following TBI may significantly affect quality of life and even may be life-threatening depending on the deficient hormone.
- Symptoms of pituitary dysfunction are generally not specific and diagnosis requires a high index of suspicion. For this reason screening algorithms have been developed by expert groups.
- Severity of head trauma increases risk of pituitary dysfunction and it is used for patient inclusion in screening algorithms.
- Patients with low life expectancy are not candidates for routine screening algorithms and they should be tested and replaced only for vital hormones.
- Cortisol, thyroid and anti-diuretic hormone deficiencies should be evaluated during early phase post-TBI and screening of gonadal hormones and growth hormone is recommended to be delayed until 6 to 12 months following TBI.
- Basal hormone measurements are first-line diagnostic tests and stimulation tests are recommended if indicated.
- Replacement therapies reverse detrimental effects caused by hormone deficiencies and improve quality of life of TBI patients.

Summary points

- Traumatic brain injury (TBI) is a common health problem that causes pituitary dysfunction in 15% of patients.
- Pituitary dysfunction may cause a broad range of symptoms from insidious and non-specific complaints to acute and potentially life-threatening conditions. Thus, routine screening of the patients who carry risk factors for post-TBI pituitary dysfunction is required for timely diagnosis and treatment.
- Severity of traumatic brain injury, patient demographics, radiologic abnormalities and hormone deficiencies during acute phase are reported as risk factors of post-TBI pituitary dysfunction but there are inconsistent results between studies.
- Screening algorithms are usually based on severity of TBI as patients with moderate to severe TBI are more prone to develop post-TBI pituitary dysfunction.

- Patients with low life-expectancy or those in a vegetative state should only be tested and replaced for central adrenal insufficiency, central hypothyroidism and diabetes insipidus.
- Screening of central adrenal insufficiency and close follow-up for posterior pituitary dysfunction are recommended during acute phase and evaluation of central hypogonadism and growth hormone deficiency should be performed during chronic-phase.
- Recommendations of relevant guidelines developed for treatment of pituitary dysfunction in non-TBI patients are recommended for TBI patients too.

References

Aberg, N. D., Brywe, K. G., & Isgaard, J. (2006). Aspects of growth hormone and insulin-like growth factor-I related to neuroprotection, regeneration, and functional plasticity in the adult brain. *ScientificWorldJournal*, *6*, 53–80.

Aimaretti, G., Ambrosio, M. R., Di Somma, C., Gasperi, M., Cannavo, S., Scaroni, C., … Ghigo, E. (2005). Residual pituitary function after brain injury-induced hypopituitarism: A prospective 12-month study. *The Journal of Clinical Endocrinology and Metabolism*, *90*(11), 6085–6092.

Annane, D., Pastores, S. M., Rochwerg, B., Arlt, W., Balk, R. A., Beishuizen, A., … Van den Berghe, G. (2017). Guidelines for the diagnosis and management of critical illness-related corticosteroid insufficiency (CIRCI) in critically ill patients (part I): Society of critical care medicine (SCCM) and European society of intensive care medicine (ESICM) 2017. *Critical Care Medicine*, *45*(12), 2078–2088.

Arlt, W., & Allolio, B. (2003). Adrenal insufficiency. *Lancet*, *361*(9372), 1881–1893.

Bavisetty, S., Bavisetty, S., McArthur, D. L., Dusick, J. R., Wang, C., Cohan, P., … Kelly, D. F. (2008). Chronic hypopituitarism after traumatic brain injury: Risk assessment and relationship to outcome. *Neurosurgery*, *62*(5), 1080–1093. discussion 1093–1084.

Bhasin, S., Cunningham, G. R., Hayes, F. J., Matsumoto, A. M., Snyder, P. J., Swerdloff, R. S., … Task Force, E. S. (2010). Testosterone therapy in men with androgen deficiency syndromes: An Endocrine Society clinical practice guideline. *The Journal of Clinical Endocrinology and Metabolism*, *95*(6), 2536–2559.

Bornstein, S. R., Allolio, B., Arlt, W., Barthel, A., Don-Wauchope, A., Hammer, G. D., … Torpy, D. J. (2016). Diagnosis and treatment of primary adrenal insufficiency: An endocrine society clinical practice guideline. *The Journal of Clinical Endocrinology and Metabolism*, *101*(2), 364–389.

Chihara, K., Kato, Y., Kohno, H., Takano, K., Tanaka, T., Teramoto, A., & Shimatsu, A. (2006). Efficacy and safety of growth hormone (GH) in the treatment of adult Japanese patients with GH deficiency: A randomised, placebo-controlled study. *Growth Hormone & IGF Research*, *16*(2), 132–142.

Cryan, E. (1918). Pituitary damage due to skull base fracture. *Deutsche Medizinische Wochenschrift*, *44*, 1261.

Devesa, J., Diaz-Getino, G., Rey, P., Garcia-Cancela, J., Loures, I., Nogueiras, S., … Devesa, P. (2015). Brain recovery after a plane crash: Treatment with growth hormone (GH) and neurorehabilitation: A case report. *International Journal of Molecular Sciences*, *16*(12), 30470–30482.

Devesa, J., Reimunde, P., Devesa, P., Barbera, M., & Arce, V. (2013). Growth hormone (GH) and brain trauma. *Hormones and Behavior*, *63*(2), 331–344.

Diri, H., Karaca, Z., Simsek, Y., Tanriverdi, F., Unluhizarci, K., Selcuklu, A., & Kelestimur, F. (2015). Can a glucagon stimulation test characterized by lower GH cut-off value be used for the diagnosis of growth hormone deficiency in adults? *Pituitary*, *18*(6), 884–892.

Dorin, R. I., Qualls, C. R., & Crapo, L. M. (2003). Diagnosis of adrenal insufficiency. *Annals of Internal Medicine*, *139*(3), 194–204. https://doi.org/10.7326/0003-4819-139-3-200308050-00009.

Filipsson Nystrom, H., Barbosa, E. J., Nilsson, A. G., Norrman, L. L., Ragnarsson, O., & Johannsson, G. (2012). Discontinuing long-term GH replacement therapy—a randomized, placebo-controlled crossover trial in adult GH deficiency. *The Journal of Clinical Endocrinology and Metabolism*, *97*(9), 3185–3195.

Fleseriu, M., Hashim, I. A., Karavitaki, N., Melmed, S., Murad, M. H., Salvatori, R., & Samuels, M. H. (2016). Hormonal replacement in hypopituitarism in adults: An endocrine society clinical practice guideline. *The Journal of Clinical Endocrinology and Metabolism*, *101*(11), 3888–3921.

Frendl, I., Katko, M., Galgoczi, E., Boda, J., Zsiros, N., Nemeti, Z., … Nagy, E. V. (2017). Plasminogen activator inhibitor type 1: A possible novel biomarker of late pituitary dysfunction after mild traumatic brain injury. *Journal of Neurotrauma*, *34*(23), 3238–3244.

Gardner, C. J., Mattsson, A. F., Daousi, C., Korbonits, M., Koltowska-Haggstrom, M., & Cuthbertson, D. J. (2015). GH deficiency after traumatic brain injury: Improvement in quality of life with GH therapy: Analysis of the KIMS database. *European Journal of Endocrinology*, *172*(4), 371–381.

Ghigo, E., Masel, B., Aimaretti, G., Leon-Carrion, J., Casanueva, F. F., Dominguez-Morales, M. R., … Urban, R. (2005). Consensus guidelines on screening for hypopituitarism following traumatic brain injury. *Brain Injury*, *19*(9), 711–724.

Gibney, J., Wallace, J. D., Spinks, T., Schnorr, L., Ranicar, A., Cuneo, R. C., … Russell-Jones, D. (1999). The effects of 10 years of recombinant human growth hormone (GH) in adult GH-deficient patients. *The Journal of Clinical Endocrinology and Metabolism*, *84*(8), 2596–2602.

Gilis-Januszewska, A., Kluczynski, L., & Hubalewska-Dydejczyk, A. (2020). Traumatic brain injuries induced pituitary dysfunction: A call for algorithms. *Endocrine Connections*, *9*(5), R112–R123.

Glynn, N., & Agha, A. (2019). The frequency and the diagnosis of pituitary dysfunction after traumatic brain injury. *Pituitary*, *22*(3), 249–260.

Hannon, M. J., Crowley, R. K., Behan, L. A., O'Sullivan, E. P., O'Brien, M. M., Sherlock, M., … Thompson, C. J. (2013). Acute glucocorticoid deficiency and diabetes insipidus are common after acute traumatic brain injury and predict mortality. *The Journal of Clinical Endocrinology and Metabolism*, *98*(8), 3229–3237.

Hari Kumar, K. V., Swamy, M. N., & Khan, M. A. (2016). Prevalence of hypothalamo pituitary dysfunction in patients of traumatic brain injury. *Indian Journal of Endocrinology and Metabolism*, *20*(6), 772–778.

Heredia, M., Fuente, A., Criado, J., Yajeya, J., Devesa, J., & Riolobos, A. S. (2013). Early growth hormone (GH) treatment promotes relevant motor functional improvement after severe frontal cortex lesion in adult rats. *Behavioural Brain Research*, *247*, 48–58.

Karaca, Z., Lale, A., Tanriverdi, F., Kula, M., Unluhizarci, K., & Kelestimur, F. (2011). The comparison of low and standard dose ACTH and glucagon stimulation tests in the evaluation of hypothalamo-pituitary-adrenal axis in healthy adults. *Pituitary*, *14*(2), 134–140.

Kelly, D. F., Gonzalo, I. T., Cohan, P., Berman, N., Swerdloff, R., & Wang, C. (2000). Hypopituitarism following traumatic brain injury and aneurysmal subarachnoid hemorrhage: A preliminary report. *Journal of Neurosurgery*, *93*(5), 743–752.

Kelly, D. F., McArthur, D. L., Levin, H., Swimmer, S., Dusick, J. R., Cohan, P., … Swerdloff, R. (2006). Neurobehavioral and quality of life changes associated with growth hormone insufficiency after complicated mild, moderate, or severe traumatic brain injury. *Journal of Neurotrauma*, *23*(6), 928–942.

Klose, M., Juul, A., Struck, J., Morgenthaler, N. G., Kosteljanetz, M., & Feldt-Rasmussen, U. (2007). Acute and long-term pituitary insufficiency in traumatic brain injury: A prospective single-centre study. *Clinical Endocrinology*, *67*(4), 598–606.

Kreitschmann-Andermahr, I., Poll, E. M., Reineke, A., Gilsbach, J. M., Brabant, G., Buchfelder, M., … Wallaschofski, H. (2008). Growth hormone deficient patients after traumatic brain injury—Baseline characteristics and benefits after growth hormone replacement—An analysis of the German KIMS database. *Growth Hormone & IGF Research*, *18*(6), 472–478.

Lauzier, F., Turgeon, A. F., Boutin, A., Shemilt, M., Cote, I., Lachance, O., … Cook, D. (2014). Clinical outcomes, predictors, and prevalence of anterior pituitary disorders following traumatic brain injury: A systematic review. *Critical Care Medicine*, *42*(3), 712–721.

Maiti, T. K., Konar, S., Bir, S., Kalakoti, P., Bollam, P., & Nanda, A. (2015). Role of apolipoprotein E polymorphism as a prognostic marker in traumatic brain injury and neurodegenerative disease: A critical review. *Neurosurgical Focus*, *39*(5), E3.

Management of Concussion/m, T. B. I. W. G. (2009). VA/DoD clinical practice guideline for management of concussion/mild traumatic brain injury. *Journal of Rehabilitation Research and Development*, *46*(6), CP1–68.

Maric, N. P., Doknic, M., Pavlovic, D., Pekic, S., Stojanovic, M., Jasovic-Gasic, M., & Popovic, V. (2010). Psychiatric and neuropsychological changes in growth hormone-deficient patients after traumatic brain injury in response to growth hormone therapy. *Journal of Endocrinological Investigation*, *33*(11), 770–775.

Molitch, M. E., Clemmons, D. R., Malozowski, S., Merriam, G. R., Vance, M. L., & Endocrine, S. (2011). Evaluation and treatment of adult growth hormone deficiency: An endocrine society clinical practice guideline. *The Journal of Clinical Endocrinology and Metabolism*, *96*(6), 1587–1609.

Moreau, O. K., Cortet-Rudelli, C., Yollin, E., Merlen, E., Daveluy, W., & Rousseaux, M. (2013). Growth hormone replacement therapy in patients with traumatic brain injury. *Journal of Neurotrauma*, *30*(11), 998–1006.

Nemes, O., Kovacs, N., Szujo, S., Bodis, B., Bajnok, L., Buki, A., … Mezosi, E. (2016). Can early clinical parameters predict post-traumatic pituitary dysfunction in severe traumatic brain injury? *Acta Neurochirurgica*, *158*(12), 2347–2353. https://doi.org/10.1007/s00701-016-2995-x.

Reimunde, P., Quintana, A., Castanon, B., Casteleiro, N., Vilarnovo, Z., Otero, A., … Devesa, J. (2011). Effects of growth hormone (GH) replacement and cognitive rehabilitation in patients with cognitive disorders after traumatic brain injury. *Brain Injury*, *25*(1), 65–73.

Schneider, H. J., Schneider, M., Saller, B., Petersenn, S., Uhr, M., Husemann, B., … Stalla, G. K. (2006). Prevalence of anterior pituitary insufficiency 3 and 12 months after traumatic brain injury. *European Journal of Endocrinology*, *154*(2), 259–265.

Schneider, M., Schneider, H. J., Yassouridis, A., Saller, B., von Rosen, F., & Stalla, G. K. (2008). Predictors of anterior pituitary insufficiency after traumatic brain injury. *Clinical Endocrinology*, *68*(2), 206–212.

Seckl, J., & Dunger, D. (1989). Postoperative diabetes insipidus. *BMJ*, *298*(6665), 2–3.

Spasovski, G., Vanholder, R., Allolio, B., Annane, D., Ball, S., Bichet, D., … Hyponatraemia Guideline Development, G. (2014). Clinical practice guideline on diagnosis and treatment of hyponatraemia. *European Journal of Endocrinology*, *170*(3), G1–47.

Taheri, S., Tanriverdi, F., Zararsiz, G., Elbuken, G., Ulutabanca, H., Karaca, Z., … Kelestimur, F. (2016). Circulating MicroRNAs as potential biomarkers for traumatic brain injury-induced hypopituitarism. *Journal of Neurotrauma*, *33*(20), 1818–1825.

Tan, C. L., Alavi, S. A., Baldeweg, S. E., Belli, A., Carson, A., Feeney, C., … Hutchinson, P. J. (2017). The screening and management of pituitary dysfunction following traumatic brain injury in adults: British neurotrauma group guidance. *Journal of Neurology, Neurosurgery, and Psychiatry*, *88*(11), 971–981.

Tanriverdi, F., De Bellis, A., Ulutabanca, H., Bizzarro, A., Sinisi, A. A., Bellastella, G., … Kelestimur, F. (2013). A five year prospective investigation of anterior pituitary function after traumatic brain injury: Is hypopituitarism long-term after head trauma associated with autoimmunity? *Journal of Neurotrauma*, *30*(16), 1426–1433.

Tanriverdi, F., Schneider, H. J., Aimaretti, G., Masel, B. E., Casanueva, F. F., & Kelestimur, F. (2015). Pituitary dysfunction after traumatic brain injury: A clinical and pathophysiological approach. *Endocrine Reviews*, *36*(3), 305–342.

Tanriverdi, F., Senyurek, H., Unluhizarci, K., Selcuklu, A., Casanueva, F. F., & Kelestimur, F. (2006). High risk of hypopituitarism after traumatic brain injury: A prospective investigation of anterior pituitary function in the acute phase and 12 months after trauma. *The Journal of Clinical Endocrinology and Metabolism*, *91*(6), 2105–2111.

Tanriverdi, F., Taheri, S., Ulutabanca, H., Caglayan, A. O., Ozkul, Y., Dundar, M., … Kelestimur, F. (2008). Apolipoprotein E3/E3 genotype decreases the risk of pituitary dysfunction after traumatic brain injury due to various causes: Preliminary data. *Journal of Neurotrauma*, *25*(9), 1071–1077.

Tanriverdi, F., Ulutabanca, H., Unluhizarci, K., Selcuklu, A., Casanueva, F. F., & Kelestimur, F. (2008). Three years prospective investigation of anterior pituitary function after traumatic brain injury: A pilot study. *Clinical Endocrinology*, *68*(4), 573–579.

Tanriverdi, F., Unluhizarci, K., Coksevim, B., Selcuklu, A., Casanueva, F. F., & Kelestimur, F. (2007). Kickboxing sport as a new cause of traumatic brain injury-mediated hypopituitarism. *Clinical Endocrinology*, *66*(3), 360–366.

Tanriverdi, F., Unluhizarci, K., Karaca, Z., Casanueva, F. F., & Kelestimur, F. (2010). Hypopituitarism due to sports related head trauma and the effects of growth hormone replacement in retired amateur boxers. *Pituitary*, *13*(2), 111–114.

Tanriverdi, F., Unluhizarci, K., Kocyigit, I., Tuna, I. S., Karaca, Z., Durak, A. C., … Kelestimur, F. (2008). Brief communication: Pituitary volume and function in competing and retired male boxers. *Annals of Internal Medicine*, *148*(11), 827–831.

Tanriverdi, F., Unluhizarci, K., Selcuklu, A., Casanueva, F. F., & Kelestimur, F. (2007). Transient hypogonadotropic hypogonadism in an amateur kick-boxer after head trauma. *Journal of Endocrinological Investigation*, *30*(2), 150–152.

Taylor, C. A., Bell, J. M., Breiding, M. J., & Xu, L. (2017). Traumatic brain injury-related emergency department visits, hospitalizations, and deaths—United States, 2007 and 2013. *MMWR Surveillance Summaries*, *66*(9), 1–16.

Tritos, N. A., Yuen, K. C., Kelly, D. F., Neuroendocrine, A., & Pituitary Scientific, C. (2015). American association of clinical endocrinologists and american college of endocrinology disease state clinical review: A neuroendocrine approach to patients with traumatic brain injury. *Endocrine Practice*, *21*(7), 823–831.

Tudor, R. M., & Thompson, C. J. (2019). Posterior pituitary dysfunction following traumatic brain injury: Review. *Pituitary*, *22*(3), 296–304.

Vanhorebeek, I., Langouche, L., & Van den Berghe, G. (2006). Endocrine aspects of acute and prolonged critical illness. *Nature Clinical Practice. Endocrinology & Metabolism*, *2*(1), 20–31.

Webber, L., Anderson, R. A., Davies, M., Janse, F., & Vermeulen, N. (2017). HRT for women with premature ovarian insufficiency: A comprehensive review. *Human Reproduction Open*, *2017*(2), hox007.

Widdowson, W. M., & Gibney, J. (2010). The effect of growth hormone (GH) replacement on muscle strength in patients with GH-deficiency: A meta-analysis. *Clinical Endocrinology*, *72*(6), 787–792.

Wright, T., Urban, R., Durham, W., Dillon, E. L., Randolph, K. M., Danesi, C., … Sheffield-Moore, M. (2020). Growth hormone alters brain morphometry, connectivity, and behavior in subjects with fatigue after mild traumatic brain injury. *Journal of Neurotrauma*, *37*(8), 1052–1066.

Yuen, K. C. J., Masel, B. E., Reifschneider, K. L., Sheffield-Moore, M., Urban, R. J., & Pyles, R. B. (2020). Alterations of the GH/IGF-I Axis and gut microbiome after traumatic brain injury: A new clinical syndrome? *The Journal of Clinical Endocrinology and Metabolism*, *105*(9). https://doi.org/10.1210/clinem/dgaa398, dgaa398.

Section D

Treatments: Experimental and clinical

Chapter 27

European aspects of guidelines used in traumatic brain injury

Alvaro Cordoba[a,b], Bernardina Frache[c], and Jose Maria Dominguez Roldan[d]
[a]*Neurosurgery and Intensive Care Unit ASESP, Montevideo, Uruguay,* [b]*Medical University of Vienna, Vienna, Austria,* [c]*Intensive Care Unit, Sanatorio Americano, Montevideo, Uruguay,* [d]*Intensive Care Unit Hospital Virgen del Rocio, Sevilla, Spain*

List of abbreviations

BTF	brain trauma foundation
EBIC	European brain injury consortium
GCS	Glasgow Coma Scale
ICP	intracranial pressure
LP	Lund protocol
RCT	randomized control trials
S-TBi	severe traumatic brain injury

Introduction

Cranioencephalic trauma is a nosological entity that includes a wide variety of clinical processes derived from a common element the transmission of kinetic energy to the patient's skull and brain. Variability in the type of aggression, and the organic response to it, means that head trauma includes multiple pathological situations that require unique management at all times. Unlike defined diseases, there is no unique specific treatment for brain trauma, but it will depend on the type of brain aggression, and the brain's response to it. That is why it has been necessary to establish patient management guides with brain trauma that facilitate and homogeneity care for patients with traumatic brain injuries.

Clinical guidelines are systematically developed statements to help medical professionals make decisions that facilitate patient care in a given clinical context. Guidelines increase the quality of clinical care provided and have advantages for patients, physicians, and healthcare systems (Woolf, Grol, Hutchinson, Eccles, & Grimshaw, 1999). In the case of head trauma, guidelines should include diagnostic and therapeutic as well as preventive aspects, as the prevention of head trauma is the basis for minimizing the aftermath of this epidemic.

The use of guides is based on the fact that patients with similar clinical problems should receive similar care worldwide, which is facilitated by guidelines. For medical professionals, the guides provide explicit recommendations on how to act and make their work easier, especially for less experienced physicians. The use of guides prevents variability in clinical practice and facilitates the study of cases both retrospectively and prospectively.

The use of clinical practice guidelines may also have some drawbacks including hindering the continuous progress of knowledge when used as an unquestionable dogma by doctors. They can also be an obstacle when, in certain circumstances, clinical processes require flexible approach. Guides should always be interpreted through local adaptations to clinical problems and facilities. Guides should never be considered rigid, inflexible schemes, nor should they be assumed to encompass the absolute truth about patient management.

Head injury is the most frequent cause of mortality and morbidity in those aged up to 45 years. Outcome depends on the nature and severity of the primary lesion, and the consequence of secondary brain damage (extra- and intracranial). Ischemia is the most important mechanism of cerebral damage in cerebral injury resulting from intracranial hypertension caused by traumatic brain swelling, intracranial hemorrhage and systemic complications. The knowledge of the treatment is limited in principle to general symptomatic measures; so improvements in patient management are required on a comprehensive approach and testing the results. In this clinical and experimental context, coordinate efforts are being made all over the world, not only to assess the current efficacy of primary care of severe head trauma, but also towards this

Diagnosis and Treatment of Traumatic Brain Injury. https://doi.org/10.1016/B978-0-12-823347-4.00028-2

concepts the development of a consensus in order to optimize the management and treatment procedures in intensive care units. Secondary ischemia of the brain are forever at the recent experimental and clinical pathophysiological studies focus on the measurement of the cerebral blood flow (CBF), brain autoregulation processes including post-traumatic vasospasm (Baethmann et al., 1998) concerned with the role of cytokines, leucocyte-endothelial interactions and molecular genetics in severe head injury (apolipoprotein E gene).

The foundation of international organizations as the American and European Brain Injury Consortium (EBIC) were the natural consequence of simultaneous efforts to improve outcomes in severe head trauma.

The head injury guidelines (Narayan, 1995) are in general evidenced-based, about published data rather than on expert opinion. The quality of data has been carefully analyzed and graded. The level of the recommendations is based upon the strength of the data available. A summary of the terminology and methodology is used to develop guidelines. After the introduction global aspects and arguments for and against practice guidelines must be checked and usefulness analyzed.

Interpretation and application of guidelines requires a comprehension of the reasons to adapt the concepts and conclusions and simultaneously the understanding of the methodological processes on which they are supported. The Lund protocol (Grande, 2017), The North American guidelines and the EBIC guidelines, were the main first international exponents of guidelines in head injury (Maas, 2002). As the gold standards concepts over decades the evidence underlying these guidelines needs to be updated and critically discussed and the conclusions put in the field of a perspective for next years. The European guidelines were formulated like a common approach to basic therapy in centers comparing clinical trials. Based on a more realistic understanding of the pathophysiology, it contains a lot of aspects not analyzed in the North American guidelines. The recommendations of the Guidelines should be considered a very important topic in clinical practice, but not in a dogmatic way. The adherence and the lack of evidence underlying many aspects of management in head injury are the results of years of experience and clinical application and should form an incentive for further scientific studies, mainly to consider the real impact in outcome. Clinical experience and an understanding of the pathophysiology are the basic first steps; then for developing more targeted approaches which can then be subjected to scientific analysis we need a new impulse on the field of neuropharmacology and restorative neurosurgery. New Guidelines should be considered combining these fields in neural recovery like our next new generation of neurocritical therapies.

The treatment of patients suffering from severe head injury is normally restricted to general procedures, and specific pharmacological agents of neuroprotection including hypothermia have not been found to improve the outcome in clinical trials. The preclinical rescue of patients (i.e. stabilization or reestablishment of the circulatory and respiratory system) or of the early clinical care prompt diagnosis and treatment of an intracranial space occupying mass constitute the current treatment based on the efforts of the health care systems. The efficacy or remaining deficiencies are continuously analyzed worldwide.

In 1997, Grande, Asgeirsson, & Nordstrom (1997b) developed a novel pathological concept of head trauma, different from that used until then by other medical schools. These principles underpin that in patients with head trauma there are modifications of brain blood flow, in tissue with disruption of the blood brain barrier, and that it is an important factor for the increase of intracranial pressure and the commitment of microcirculation. Lund therapy, aimed at the treatment of patients with head trauma, has as a relevant pathological element the compression of large intracranial veins against the skull, which makes it difficult to get blood out of the brain and consequently also infusion. Therapeutic strategies under this therapy include: maintaining plasma osmolarity, reducing systemic blood pressure and a modern conception of the resistance of capillary vessels with low doses of thiopental and dihydroergotamine (Grande, Asgeirsson, & Nordstrom, 1997a). This therapeutic strategy can be considered more as a pathological and therapeutic approach without as a clinical practice guide.

Projects in Europe to develop guidelines for the management of patients with head trauma have been multiple (Table 1). In Europe, the first treatment guides for patients with head trauma emerge at the level of some local scientific groups, mainly Italy. In 1996, the Study Group on Head Injury of the Italian Society for Neurosurgery developed the Guidelines for minor head injured patients' management in adult age (Guidelines for minor head injured patients' management in adult age. The Study Group on Head Injury of the Italian Society for Neurosurgery, 1996). These guidelines were aimed at treating adult patients with moderate early-stage head trauma. These guidelines, endorsed by the Italian Society for Neurosurgery, were a useful tool for addressing the management of patients with brain trauma and the protocolized manner. From a methodological point of view, however, they lacked sufficient levels of evidence to achieve a high degree of recommendation.

In 1997, the European Brain Injury Consortium developed the first European guides involving various specialists from different countries on the continent. The objective was to present a pragmatic document, based on consensus and expert opinion, covering the management of patients with head trauma from the site of the accident to their treatment in intensive care units. A first part of the paper discusses the importance of primary intracranial lesions, as well as the importance of secondary both systemic and intracranial insult. The guides mentioned previously published guides such as those for primary

TABLE 1

Guideline title, year, and country	Main characteristics of guideline
The Study Group on Head Injury of the Italian Society for Neurosurgery: Guidelines for minor head injured patients' management in adult age. Italy 1996	Early management. Adults. Mild TBI
EBIC-guidelines for management of severe head injury in adults. 1997. Europe	Pre-hospital, early management, and ICU. Adults. Severe TBI care
Guidelines for the pre-hospital care of patients with severe head injuries. P Working Group for Neurosurgical Intensive Care of the ESICM. 1998. Europe	Pre-hospital management. No defined population. Severe TBI
Guidelines for the treatment of adults with severe head trauma. Initial assessment; evaluation and pre-hospital treatment; current criteria for hospital admission; systemic and cerebral monitoring. 2000. Italy	Early management of severe TBI. Adults
National Institute for Health and Clinical Excellence. Head injury. Triage, assessment, investigation and early management of head injury in infants, children, and adult: 2007. UK	Early management. Pediatrics and Adults population. All TBI
Treatment of minor and severe traumatic brain injury. National reference guidelines. 2008. Italy	Early Management. Adults population. All TBI
Scandinavian guidelines for initial management of minimal, mild and moderate head injuries in adults: an evidence and consensus-based update. 2013. Scandinavia	Early management. Adults. All TBI

assessment and traumatized patient resuscitation (ATLS), and those made by the section on Neurotrauma and Critical Care of the American Association of Neurological Surgeons and the Congress of Neurological Surgeons, supported by the Brain Trauma Foundation. The recommendations included concerned the management of adult patients in both the hospital area and the intensive care unit. Recommendations for pre-hospital management essentially pursued the patient's physicians recording data useful for patient evaluation. Emphasis was also placed on the early onset of resuscitation measures. An important aspect of these guidelines in the importance attached to the transfer of patients with severe head trauma to neurosurgical centers with permanent availability of neurosurgeons and CT scans. Although they were useful and novel guides in those years, so overall quality was moderate as the evidence used was low.

In 1998, the Working Group for Neurosurgical Intensive Care of the ESICM (European Society of Intensive Care Medicine) developed the Guidelines for the pre-hospital care of patients with severe head injuries. These guides, published in Intensive Care Medicine (Piek, 1998) were more a brief summary of the advisable actions, and based on variable levels of evidence than a guide itself.

In 2000, a Group of Italian Neurocritical doctors and Neurosurgeons belonging to the Italian Societies of Neurosurgery (SINch) and Anesthesiology & Intensive Care (SIAARTI) developed for the evaluation, treatment, and monitoring of patients with severe head trauma. The main objectives of these were: (1) Standardize the methods of neurological evaluation of these patients, (2) Prevent secondary brain damage through careful management of systemic homeostasis, (3) Establish guidelines for monitoring intracranial phenomena associated with brain trauma, (4) Establish organizational recommendations of care for these patients and (5) Guide to medical and surgical treatment. The recommendations were developed after a thorough analysis of the scientific evidence existing at the time, and were of great use and impact after publication (Procaccio et al., 2000a, 2000b).

The UK National Institute for Health and Clinical Excellence published in 2007 the guides entitled "Head injury, Triage, assessment, investigation, and early management of head injury in infants, children, and adult." They were published in an extensive document, with detailed content, and based on excellent levels of evidence. From a methodological point of view, excellent guides can be considered and applicable to all patients with early stage head trauma (whatever their age) (Head Injury Triage & NBK53036.pdf.).

In 2013, the Scandinavian Guidelines for Initial Management of Minimal, Mild and Moderate head injuries in adults were published. These guidelines focused primarily on the initial management of the adult patient with mild and moderate head trauma and were developed by medical teams from Norway, Sweden, and Finland. The guidelines sought to establish what the initial management guidelines for these patients should be, the criteria for performing CT scans and the patient's entry and discharge from the hospital.

In 2014 The Neurocritical Care Society in collaboration with the European Society of Intensive Care Medicine, the Society for Critical Care Medicine, and the Latin America Brain Injury Consortium organized an international, multidisciplinary consensus conference to begin to address these needs. International experts from neurosurgery, neurocritical care, neurology, critical care, neuro anesthesiology, nursing, pharmacy, and informatics were recruited on the basis of their research, publication record, and expertise. They undertook a systematic literature review to develop recommendations about specific topics on physiologic processes important to the care of patients with disorders that require neurocritical care. This review does not make recommendations about treatment, imaging, and intraoperative monitoring. A multidisciplinary jury, selected for their expertise in clinical investigation and development of practice guidelines, guided this process. The GRADE system was used to develop recommendations based on literature review, discussion, integrating the literature with the participants' collective experience, and critical review by an impartial jury. Recommendations should be based on data quality and translation into clinical practice providing pragmatic guidelines and recommendations in neuromonitoring (Le Roux et al., 2014).

Current data update and analysis

Deficits are commonly associated with delays in pre-clinical rescue procedures until intubation of the patient or administration of fluids. Delays also occur during hospital admission delay with the diagnostic of intracranial lesions requiring prompt neurosurgical intervention. By support of the Federal Ministry of Education and Research and under the auspices of the Forschungsverbund Neurotraumatologie, University of Munich, a prospective system analysis was carried out on major aspects of the pre- and early clinical management at a population based level in patients with traumatic brain injury. Documentation of pertinent data was made from August 1998 to July 1999 covering a catchment area of Southern Bavaria (5.6 mio inhabitants). Altogether 528 cases identified to suffer from severe head injury (GCS < or = 8 or deteriorating to that level within 48 h) were enrolled following admission to the hospital and establishment of the diagnosis. Further, patients dying on the scene or during transport to the hospital were also documented, particularly as to the frequency of severe head injury as underlying cause of mortality. The analysis included also cases with additional peripheral trauma (polytrauma). The efficacy of the logistics and organization of the management was studied by documentation of prognosis-relevant time intervals, as for example until arrival of the rescue squad at the scene of an accident, until intubation and administration of fluid, or upon hospital admission until establishment of the CT-diagnosis and commencement of surgery or transfer to the intensive care unit, respectively. The severity of cases studied in the present analysis is evident from a mortality of far above 40% of cases admitted to the hospital, which was increased by about 20% when including pre-hospital mortality. The outcome data notwithstanding, the emerging results demonstrate a high efficacy of the pre- and early clinical management, as indicated by a prompt arrival of the rescue squad at the scene, a competent pre-hospital and early clinical management and care, indicative of a low rate of avoidable complications. It is tentatively concluded on the basis of these findings that the patient prognosis is increasingly determined by the manifestations of primary brain damage vs. the development of secondary complications (Wirth et al., 2004). In Europe there are great differences between centers in the frequency of CPP monitoring and ventilatory support applied to head-injured patients. ICP measurement disclosed a high rate of intracranial hypertension, which was not suspected in patients evaluated on a clinical basis alone. ICP monitoring was associated with a low rate of complications. Cases with severe neurological impairment, and with the worse outcome, were treated and monitored more intensively.

The first publication of the EBIC was in 1995 and was recruited over a 3-month period. A three-page questionnaire comprising 60 items of information has been compiled by 67 centers in 12 European countries. Sixty items of information on demography, clinical features, investigations, management and early complications were captured on a simple, two-page questionnaire and, information on outcome at 6 months on a third page. The median age of the subjects was 38 years, 74% were male and 51% injured in road traffic accidents; 57% of patients were transferred to the "neuro" center from another hospital. Assessment of clinical responsiveness was limited by the use of sedation and intubation and information from four early time points (pre-hospital, arrival at the Accident and Emergency department, post-resuscitation, and arrival at the "neuro" unit) was combined to stratify the subjects as severe (58%), moderate (17%) or intermediate (19%). In 48% of patients classified the CT scan showed features of a "mass lesion" and in 40% showed a subarachnoid hemorrhage. Fifty-five centers provided the data on outcome for 94% of the cases recruited in these centers 6 months after injury. Thirty-one percent died, 3% were vegetative, 16% severely disabled, 20% moderately disabled, and 31% had made a good recovery. Information was collected prospectively regarding all severe and moderate head injuries in adults (> 16 years) admitted to neurosurgery within 24 h of injury. A total of 1005 adult head injury cases were enrolled in the study from February 1, 1995 to April 30, 1995. The Glasgow Outcome Scale was administered at 6 months.

Early surgery was performed in 346 cases (35%); arterial pressure was monitored invasively in 631 (68%), ICP in 346 (37%), and jugular bulb saturation in 173 (18%). Artificial ventilation was provided to 736 patients (78%). Intracranial hypertension was noted in 55% of patients in whom ICP was recorded, while it was suspected in only 12% of cases without ICP measurement. There were great differences in the use of ventilation and CPP monitoring among the centers. Mortality at 6 months was 31%.

There was an association between an increased frequency of monitoring and intervention and an increased severity of injury; correspondingly, patients who more frequently underwent monitoring and ventilation had a less favorable outcome.

Comparison of the data from different parts of Europe showed differences in the frequency of secondary transfer, cause of injury, occurrence of major extracranial injury, CT scan findings, intracranial operation, clinical severity of injury, and utilization of the components of intensive care and the occurrence of a favorable outcome, although the latter difference was not statistically significant when variations in the initial severity of injury were taken into account. The findings in the present survey are compared with newly analyzed information for three previous large series: the International Data Bank involving the UK, the Netherlands and the USA, the North American Traumatic Coma Data Bank, and data from four centers in the UK. The comparisons showed substantial similarities and also differences that may reflect variations in policy for admission of the head injury to "neuro" units, and evolution in methods of assessment, investigation and management. The effects of these differences on outcome require further, rigorous prospective study (Murray et al., 1999).

In 2001, 24 brain centers participated of a survey about brain monitoring and therapies on guidelines (two respondents from 18 sites). The average proportion of agreement between duplicate respondents was 0.79 (range 0.44–1.00). All Brain IT centers monitored ICP. The reported order of treatment for intracranial hypertension was: evacuation of mass-lesions and head elevation (1), increase of sedation and Mannitol Scheme (2), hyperventilation (3), ventricular drainage (4), craniectomy and pentothal coma (5), and decompressive lobectomy (6). The respondents were less prone to evacuate expansive contusions in relation to extra cerebral hematomas. The most common suggested interventions (alone or in combination) for treatment of intracranial hypertension without mass lesions was the Mannitol scheme (included in 71% of the suggestions), CSF drainage (included in 56%), hyperventilation (included in 32%), and pentothal coma (included in 22%). The suggested management of TBI was mainly in accordance with published guidelines, although a minor proportion of the answers deviated to some extent. The suggested order and combinations of different treatment interventions varied. Variation of treatment within the range of prescribed standards provides optimal conditions for an interesting future analysis of treatment and monitoring data as collected prospectively in a Brain IT database (Enblad et al., 2004).

In 2008, the Austrian study were design as a prospective multi-center cohort study in order to elucidate the reasons why patients did or did not receive intracranial pressure (ICP) monitoring and to describe factors influencing hospital mortality after severe traumatic brain injury (TBI). About 88,274 patients were consecutively admitted to 32 medical, surgical and mixed Austrian ICUs between 1998 and 2004. No interventions were performed. Measurements and results: 1856 patients (2.1% of all ICU admissions) exhibited severe TBI (GCS $<$9); of these, 1031 (56%) had ICP monitoring. The "worst" and the "best" cases were both less likely to receive ICP monitoring. Younger patients, female patients, and patients with isolated TBI were more likely to receive ICP monitoring. Compared with large centers ICP was monitored more frequently [odds ratio (OR) 3.09, CI 2.42–3.94] in medium-sized centers. The 20% of patients with the highest likelihood to receive ICP monitoring were monitored in 91% of cases, and had the lowest hospital mortality (31%, OR 0.78, CI 0.37–1.64). Multivariate analysis revealed that severity of illness, TBI severity, isolated TBI, and the number of cases treated per year was associated with hospital outcome. Compared with the large centers, ORs for hospital mortality were 1.85 (CI 1.42–2.40) for patients from medium-sized centers and 1.91 (CI 1.24–2.93) for patients from small centers. Conclusions showed that ICP monitoring may possibly have some beneficial effects, but this needs further evaluation. Patients with severe TBI should be admitted to experienced centers with high patient volumes since this might improve hospital mortality rates (Mauritz, Steltzer, Bauer, Dolanski-Aghamanoukjan, & Metnitz, 2008).

European Guidelines for the management of severe head injury in adults are updated and discussed since 2017. Prevention and treatment of secondary insults is emphasized and the treatment reviewed. Guidelines presented are pragmatic, mainly based on consensus and expert opinion, covering the treatment in intensive care unit. Other approaches to the development of guidelines, such as evidence based analysis are considered.

Although the basic investigations regarding TBI have allowed a greater knowledge of the mechanisms involved, those aspects that are related to the improvement of the clinical management of patients have not followed the same course. This has generated differences in the control, treatment, and management of patients within the different centers, both nationally and internationally. Management guidelines for TBI are available and available to professionals, but the evidence supporting these recommendations is weak. TBI management guidelines generally come from observational studies, meta-analysis, guideline development, since RCTs have strict inclusion criteria and usually focus on subpopulations of patients, making them difficult to extrapolate in general. Those studies or actions aimed at improving the characterization of the disease, with overwhelming results regarding clinical management recommendations and impact on the results, will lead to a homogenization of the care and management of patients with TBI. Comparative effectiveness research (CER) allows the benefits and harms of different alternatives for prevention, diagnosis, treatment and care improvements to be compared. They are useful tools that give health professionals and researchers more information to make decisions and define actions that can improve the quality of care and results, both individually and generally. The CENTER-TBI was a project designed to allow evaluating the heterogeneity in the management of patients with traumatic brain injury, to compare these differences and thus to be able to identify effective interventions in this regard. A database compiled from different European centers was generated, in order to achieve a cohort large enough to allow CER analysis of differences in clinical care and management pathways in TBI. The CENTER-TBI data are a good resource, accessible, and very well characterized to be used for long-term monitoring.

CENTER-TBI is a prospective longitudinal nonrandomized observational study across the severity spectrum of TBI in up to 80 sites from 22 countries for 18 months (Fig. 1). Center was characterized with regard to their structural profile in order to explore effects of organizational aspects. The study consisted of 2 parts: CENTER-TBI core data study (n=5400) and CENTER-TBI registry (n=15,000–25,000). The inclusion criteria were: for core study: Clinical diagnosis of TBI, clinical indication for CT scan, presentation within 24 h of injury, informed consent obtained according to local and national requirements. Exclusion criteria: Severe preexisting neurological disorder that would confound outcome assessments. For registry study: clinical diagnosis of TBI and Clinical indication for CT scan. This part didn't have exclusion criteria. Recruitment was initiated across the participating study centers between November 2014 and January 2015. The index follow-up for outcome assessment will be at 6 months.

Three clinical groups differentiated by clinical care path: Emergency room (ER) stratum: patients evaluated in the ER and discharged (n=1800). Admission stratum: patients admitted to the hospital but not to the intensive care unit (ICU; n=1800). ICU stratum: patients admitted directly from ER or other hospital to the ICU (n=1800) (Maas et al., 2015).

The CENTER-TBI registry is important for two purposes: (1) assessing representativeness of the CENTER-TBI core study and (2) providing opportunities for comparative effectiveness analysis of organization of care.

Blood extraction for biomarker study and genetic study was performed on the patients, and skull tomography performed acutely, and at least 1 performed in the following 2 to 7 days of the event, will be uploaded to the neuroimaging database of the CENTER-TBI.

A pre-specified neuropsychological evaluation was planned to be performed in all strata at 6 months after injury and longitudinally at various time points in the 3 strata up to 24 months after injury.

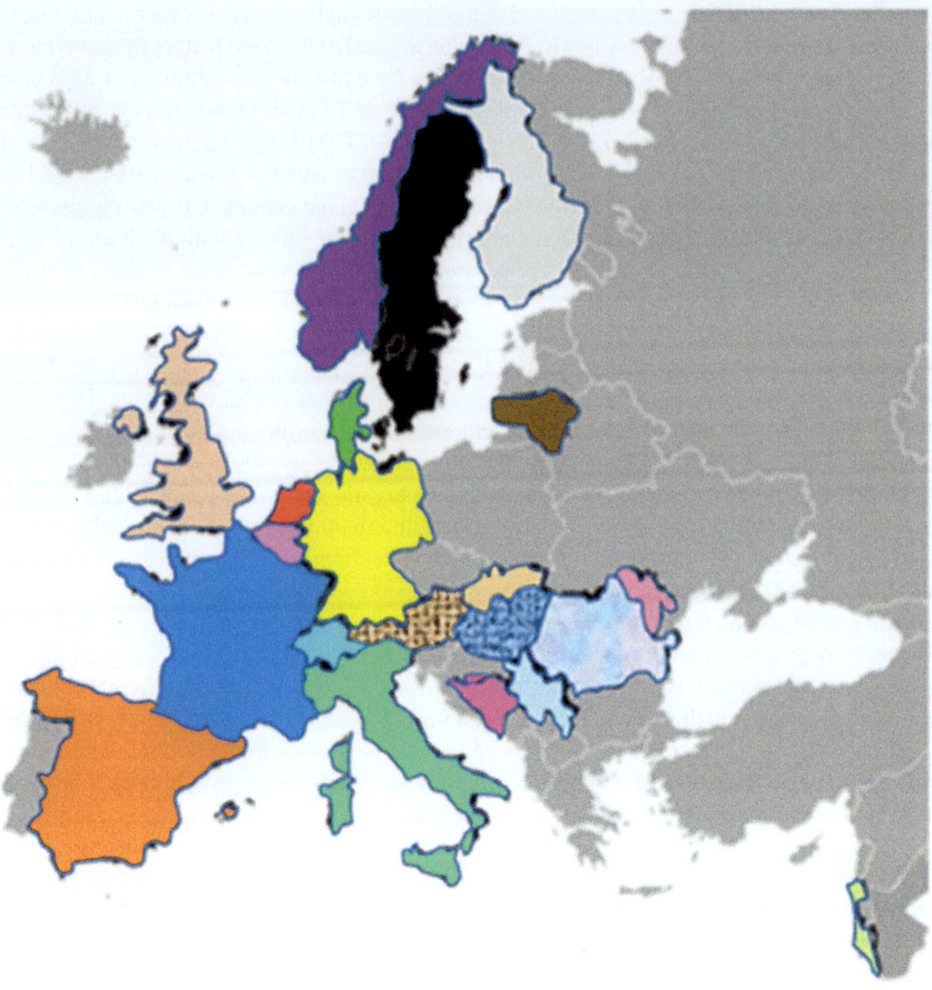

FIG. 1 Countries invited to participate in the CENTER-TBI study.

In this way CENTER-TBI becomes a useful tool, which contains a large amount of data on patients with TBI and policies on it in different centers in Europe, and can be the starting point for different studies and projects related to head trauma.

Since the last 5 years, up today, all the data registered in the CENTER-TBI have contributed to the design of new studies, using the large database that is available from all the participating centers.

These studies have allowed evaluating the variability in structure, design, practical tools, among others, of the care centers for patients with head trauma, as well as the difference in the management policies of each center, and the application and monitoring of the TBI management guidelines, among other not less important data.

In 2016, Cnossen et al. (2016) published a study to assess variations in the conditions in which patients are cared for in health centers, and the process of care for those who suffer a trauma of the skull, and thus obtain a profile of the European health centers. Scientists wanted to quantify variation in care structure and processes. The study was carried out through a prospective, observational, longitudinal design, in which 11 questionnaires with more than 300 questions were formulated, which were distributed in 71 centers in 20 countries participating in the CENTER-TBI, obtaining a rate of 100% response.

The results obtained were the following: 92% of the centers were academic hospitals, more than 60% were designated as a level I trauma center and almost all of the centers were located in urban areas. A little more than half of the centers had a specific neurointensive unit; 51% had a rehabilitation unit in the hospital; 79% of the centers admitted to using indication guidelines with respect to CT scans for patients with TBI. In addition, a wide variation was found in the care processes, such as the ICU admission policy and the intracranial pressure monitoring policy between the centers 11.

What is remarkable about this study is that it allows us to know an overview of the care centers in the BIT and thus be able to assess the different needs for improvement in terms of quality of care. It is not possible to extrapolate the results universally since it presented some limitations that may condition the results.

Another problem that has been present for many decades, which generates an increase in the mortality and morbidity of patients with head trauma, permanently studied and with frequent updates of recommendations in this regard, is the management and treatment of ICP in TBI. This has been and is a matter of controversy for all professionals who treat patients with brain injury resulting from head trauma. The evidence is not conclusive so that all those professionals who face severe head trauma and endocraneal hypertension apply the same methods of control and treatment.

In 2017, a study was published that aimed to examine the variation in policies for monitoring and treating endocranial hypertension in patients with TBI. Again Cnossen et al. (2016) designed a prospective, longitudinal study, based on questionnaires that were delivered to centers participating in the CENTER-TBI study, of which 66 of them fully answered the questions.

The results showed that the Brain Trauma Foundation guidelines were used in 74% of the centers, 90% of those who participated in the study indicated placing an ICP monitor in patients with severe traumatic brain injury and abnormalities on CT. A great variation was found in the use and application of first and second level treatments for endocranial hypertension; there was a relatively equitable distribution regarding the consideration of centers as aggressive or conservative regarding monitoring and treatment.

In a prospective study conducted in Spain over 5 years, with the participation of 52 hospitals caring for patients with severe head trauma, including 1463 patients with severe head trauma (Llompart-Pou et al., 2020), it was found that recommendations for monitoring intracranial pressure according to BTF guidelines were followed in 635 patients (45.1%). In contrast to this data, a survey-type study conducted in hospitals caring for patients with severe brain trauma under the auspices of the Neurotrauma Group of the Spanish Society of Neurosurgery (Joanes, 2017) in 29 Spanish neurotraumatological centers it concluded that almost 90% of these centers monitored intracranial pressure in more than 75% of patients with severe head trauma. In that study, neurosurgeons responded to the survey claimed that 86.2% used brain Trauma Foundation protocols as a guide, while none of the respondents recognized using EBIC guidelines.

These results and literature related to the subject reveal the lack of consensus to be able to unify methods of monitoring and treating this pathology that has high mortality and morbidity, becoming a major public and social health problem, for which the weight of scientific evidence has not reached a sufficient level to facilitate its management.

The management guidelines of those patients with brain injuries resulting from a skull trauma help to decrease the variability in the care of these patients, as well as to decrease the variations in the results.

A study published in 2019, carried out by Volovici et al. (2019), was based on the study of variations in the implementation of the TBI Management Guidelines, adherence to them and the causes of non-adherence to them. Again, the questionnaire technique was used, which was answered by 65 participating CENTER-TBI centers. Of these, 75% reported using the Trauma Brain Foundation guidelines for the medical treatment of traumatic brain injury or related institutional protocols, 17% did not use guidelines, and 8% used other guidelines. The most frequent reasons for non-compliance were "each patient is unique" and the existence of extracranial injuries (Fig. 2).

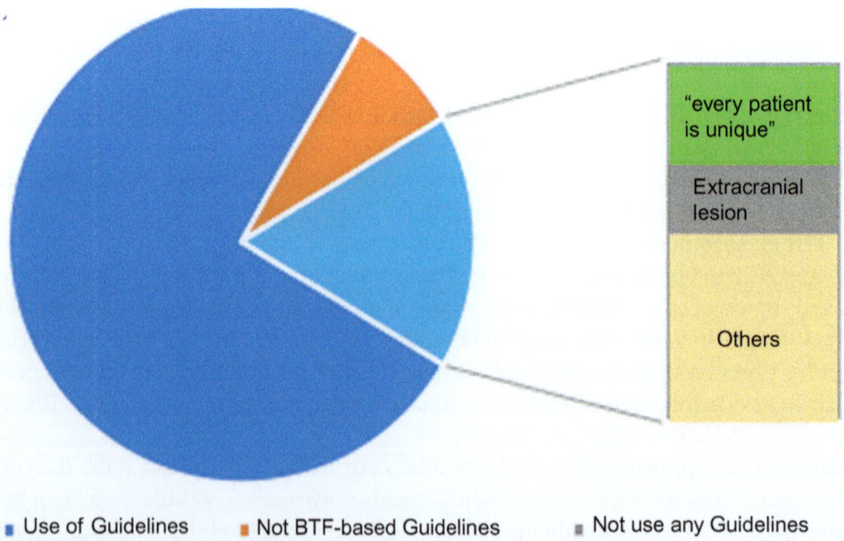

FIG. 2 Results of adherence or not to the guidelines. 75% use Guidelines; 8% not use guidelines; 17% not use any guidelines, in this group, the reason for non-adherence were "every patients is unique" 39%; extracranial lesion 16%; other 55%.

This study shows the existence of great variability regarding the use of guidelines, and shows the complexity of managing TBI, and may be different for each patient, depending on the clinical presentation, the injuries generated, age, comorbidities, among others, which may explain the changes between different centers regarding adherence to established guidelines.

Other areas of neuroscience

It is necessary to continue searching for measures and actions that improve the quality of care, facilitate the management of these patients for health professionals, as well as achieve a positive impact on the results of this pathology (Cordoba, 2019; Steyerberg et al., 2019).

In European neurological intensive care units there is a tendency to use protocols and clinical practice guidelines for the management of patients with head trauma. Thus, in the survey conducted by Suarez et al. (2020) within the PRINCE study (Point PRevalence In Neurocritical CarE was found to be 70.45% of intensive care units in Europe using the Brain Trauma Foundation Guidelines for the management of patients with head trauma, placing that percentage above the employment rate of these guidelines at U.S. UCIs. Similarly, standardized protocols were used for the use of osmolar therapy up to 75% of all units, up to 77.27% or the use of standardized protocols for the treatment of intracranial hypertension of the neurocritic patient. Overall, and according to the results of this survey, in European intensive care units there is a high rate of use of protocols for the management of the neurocritical patient, extending beyond the patient with cranioencephalic trauma to other diseases such as intracerebral hemorrhage, spontaneous subarachnoid hemorrhage, or ischemic stroke.

Discussion

Clinical practice guidelines applied to patients with head trauma have been shown useful for acquiring knowledge from non-expert physicians in the care of patients with traumatic brain injury, allowing the dissemination of pathological, diagnostic, and therapeutic bases of head trauma.

In Europe, there have been multiple attempts by different groups to develop clinical practice guidelines aimed at managing traumatic brain injury in the different phases (pre-hospital, in areas of hospital emergencies, in intensive care, etc.); as well as for the different severity levels of these patients (mild, moderate, or severe head trauma).

However, it should be noted that severe head trauma is not a disease, but a syndromic complex in which in all patients there is only one common element, is the transmission of kinetic energy to the skull and brain, either as an active subject or as a taxable person. However, the primary damage caused by trauma, or the patient's organic response to traumatic brain aggression is often diverse. That is why there is sometimes a variability in the results obtained despite strict application of the recommendations of the clinical practice guidelines.

It is also important to note that there are a number of factors influencing the implementation of brain trauma protocols at European level. The main one of these factors is the existence of very different patient care systems when comparing different countries. These differences are manifested, for example, in the variability of health financing systems (public, private, mixed), and also in the variability in the level of coordination that exists between pre-hospital and hospital health systems. The level of leadership and collaboration between the different specialties involved (neurosurgery, radiology, intensive care, etc.), not always homogeneous in the different countries of Europe, can also influence this variability. In fact, the existence of intensive care units of neurocritical patients remains an exceptionality within the models of critical patient care in Europe.

Mini dictionary terms

- Lund's Protocol: First Swedish protocol on head trauma treatment.
- Head Trauma Guidelines: Recommendations about the management and treatment of head injury.
- Adherence to Guidelines: Grade of application of the recommendations.
- Severe Head Injury: Encephalic lesion with GCS minor to 8.
- CENTER-TBI is a prospective longitudinal non-randomized observational study across the severity spectrum of TBI.

Key facts of adherence of guidelines for treatment of severe head injury

- Clinical guidelines are systematically developed statements to help medical professionals make decisions that facilitate patient care in a given clinical context.
- The use of clinical practice guidelines may also have some drawbacks including hindering the continuous progress of knowledge when used as an unquestionable dogma by doctors.
- The foundation of international organizations as the American and European Brain Injury Consortium (EBIC) were the natural consequence of simultaneous efforts to improve outcomes in severe head trauma.
- European Guidelines for the management of severe head injury in adults are updated and discussed since 2017. Prevention and treatment of secondary insults is emphasized and the treatment reviewed.
- The results showed that the Brain Trauma Foundation guidelines were used in 74% of the centers, 90% of those who participated in the study indicated placing an ICP monitor in patients with severe traumatic brain injury and abnormalities on CT. A great variation was found in the use and application of first and second level treatments for endocranial hypertension; there was a relatively equitable distribution regarding the consideration of centers as aggressive or conservative regarding monitoring and treatment.

Summary points

- This chapter is focus on the current European Guidelines in head trauma and his application.
- Severe head trauma is a crucial problem in public health assistance.
- Guidelines are a pool of wide recommendations on the treatment on severe head trauma.
- Adherence to guidelines are analyzed in order to discuss the results and the outcomes.
- Prospective trials for further evaluation of the definitively outcomes are needed.
- Restorative neurosurgery is a new possibility in the treatment of head trauma.[i]

References

Baethmann, A., Eriskat, J., Stoffel, M., Chapuis, D., Wirth, A., & Plesnila, N. (1998). Special aspects of severe head injury: Recent developments. *Current Opinion in Anaesthesiology*, *11*(2), 193–200. https://doi.org/10.1097/00001503-199804000-00013.

Cnossen, M. C., Polinder, S., Lingsma, H. F., Maas, A. I., Menon, D., Steyerberg, E. W., … Participants. (2016). Variation in structure and process of care in traumatic brain injury: Provider profiles of European neurotrauma centers participating in the CENTER-TBI study. *PLoS One*, *11*(8). https://doi.org/10.1371/journal.pone.0161367, e0161367.

Cordoba, A. (2019). Perspective: Variation in guideline implementation and adherence regarding severe traumatic Brain injury treatment: A CENTER-TBI survey study in Europe. *World Neurosurgery*, *126*, 414. https://doi.org/10.1016/j.wneu.2019.02.235.

Enblad, P., Nilsson, P., Chambers, I., Citerio, G., Fiddes, H., Howells, T., … Brain, I. T. G. (2004). R3-survey of traumatic brain injury management in European Brain IT centres year 2001. *Intensive Care Medicine*, *30*(6), 1058–1065. https://doi.org/10.1007/s00134-004-2206-8.

Grande, P. O. (2017). Critical evaluation of the Lund concept for treatment of severe traumatic head injury, 25 years after its introduction. *Frontiers in Neurology*, *8*, 315. https://doi.org/10.3389/fneur.2017.00315.

Grande, P. O., Asgeirsson, B., & Nordstrom, C. (1997a). Aspects on the cerebral perfusion pressure during therapy of a traumatic head injury. *Acta Anaesthesiologica Scandinavica. Supplementum, 110*, 36–40. https://doi.org/10.1111/j.1399-6576.1997.tb05493.x.

Grande, P. O., Asgeirsson, B., & Nordstrom, C. H. (1997b). Physiologic principles for volume regulation of a tissue enclosed in a rigid shell with application to the injured brain. *The Journal of Trauma, 42*(5 Suppl), S23–S31. https://doi.org/10.1097/00005373-199705001-00005.

Guidelines for minor head injured patients' management in adult age. The Study Group on Head Injury of the Italian Society for Neurosurgery. (1996). *Journal of Neurosurgical Sciences, 40*(1), 11–15.

Joanes, V. (2017). Organisation of care and initial management of severe head injury in Spain: Results of a national survey. *Neurocirugía (Asturias, Spain), 28*(4), 167–175. https://doi.org/10.1016/j.neucir.2017.01.001.

Le Roux, P., Menon, D. K., Citerio, G., Vespa, P., Bader, M. K., Brophy, G. M., … Taccone, F. (2014). Consensus summary statement of the international multidisciplinary consensus conference on multimodality monitoring in neurocritical care: A statement for healthcare professionals from the neurocritical care society and the European Society of Intensive Care Medicine. *Neurocritical Care, 21*(Suppl 2), S1–26. https://doi.org/10.1007/s12028-014-0041-5.

Llompart-Pou, J. A., Barea-Mendoza, J. A., Sanchez-Casado, M., Gonzalez-Robledo, J., Mayor-Garcia, D. M., Montserrat-Ortiz, N., … en representacion del Grupo de Trabajo de Neurointensivismo y Trauma de la, S. (2020). Neuromonitoring in the severe traumatic brain injury. Spanish Trauma ICU Registry (RETRAUCI). *Neurocirugía (Asturias, Spain), 31*(1), 1–6. https://doi.org/10.1016/j.neucir.2019.05.005.

Maas, A. I. (2002). Guidelines for head injury: Their use and limitations. *Neurological Research, 24*(1), 19–23. https://doi.org/10.1179/016164102101199495.

Maas, A. I., Menon, D. K., Steyerberg, E. W., Citerio, G., Lecky, F., Manley, G. T., … Investigators. (2015). Collaborative European NeuroTrauma Effectiveness Research in Traumatic Brain Injury (CENTER-TBI): A prospective longitudinal observational study. *Neurosurgery, 76*(1), 67–80. https://doi.org/10.1227/NEU.0000000000000575.

Mauritz, W., Steltzer, H., Bauer, P., Dolanski-Aghamanoukjan, L., & Metnitz, P. (2008). Monitoring of intracranial pressure in patients with severe traumatic brain injury: An Austrian prospective multicenter study. *Intensive Care Medicine, 34*(7), 1208–1215. https://doi.org/10.1007/s00134-008-1079-7.

Murray, G. D., Teasdale, G. M., Braakman, R., Cohadon, F., Dearden, M., Iannotti, F., … Unterberg, A. (1999). The European Brain injury consortium survey of head injuries. *Acta Neurochirurgica, 141*(3), 223–236. https://doi.org/10.1007/s007010050292.

Narayan, R. K. (1995). Development of guidelines for the management of severe head injury. *Journal of Neurotrauma, 12*(5), 907–912. https://doi.org/10.1089/neu.1995.12.907.

Piek, J. (1998). Guidelines for the pre-hospital care of patients with severe head injuries. Working Group for Neurosurgical Intensive Care of the European Society of Intensive Care Medicine. *Intensive Care Medicine, 24*(11), 1221–1225. https://doi.org/10.1007/s001340050748.

Procaccio, F., Stocchetti, N., Citerio, G., Berardino, M., Beretta, L., Della Corte, F., … Tomei, G. (2000a). Guidelines for the treatment of adults with severe head trauma (part I). Initial assessment; evaluation and pre-hospital treatment; current criteria for hospital admission; systemic and cerebral monitoring. *Journal of Neurosurgical Sciences, 44*(1), 1–10.

Procaccio, F., Stocchetti, N., Citerio, G., Berardino, M., Beretta, L., Della Corte, F., … Tomei, G. (2000b). Guidelines for the treatment of adults with severe head trauma (part II). Criteria for medical treatment. *Journal of Neurosurgical Sciences, 44*(1), 11–18.

Steyerberg, E. W., Wiegers, E., Sewalt, C., Buki, A., Citerio, G., De Keyser, V., … Investigators. (2019). Case-mix, care pathways, and outcomes in patients with traumatic brain injury in CENTER-TBI: A European prospective, multicentre, longitudinal, cohort study. *Lancet Neurology, 18*(10), 923–934. https://doi.org/10.1016/S1474-4422(19)30232-7.

Suarez, J. I., Martin, R. H., Bauza, C., Georgiadis, A., Venkatasubba Rao, C. P., Calvillo, E., … Investigators, P. S. (2020). Worldwide organization of neurocritical care: Results from the PRINCE study part 1. *Neurocritical Care, 32*(1), 172–179. https://doi.org/10.1007/s12028-019-00750-3.

Volovici, V., Ercole, A., Citerio, G., Stocchetti, N., Haitsma, I. K., Huijben, J. A., … collaborators, C.-T. (2019). Variation in guideline implementation and adherence regarding severe traumatic brain injury treatment: A CENTER-TBI survey study in Europe. *World Neurosurgery, 125*, e515–e520. https://doi.org/10.1016/j.wneu.2019.01.116.

Wirth, A., Baethmann, A., Schlesinger-Raab, A., Assal, J., Aydemir, S., Bayeff-Filloff, M., … Neuropsychological, R. (2004). Prospective documentation and analysis of the pre- and early clinical management in severe head injury in southern Bavaria at a population based level. *Acta Neurochirurgica. Supplement, 89*, 119–123. https://doi.org/10.1007/978-3-7091-0603-7_17.

Woolf, S. H., Grol, R., Hutchinson, A., Eccles, M., & Grimshaw, J. (1999). Clinical guidelines: Potential benefits, limitations, and harms of clinical guidelines. *BMJ, 318*(7182), 527–530. https://doi.org/10.1136/bmj.318.7182.527.

Further reading

Head injury: Triage, assessment, investigation and early management of head injury in infants, children and adults (1st ed.). London: National Institute for Health and Clinical Excellence. Guidance [Internet] http://www.ncbi.nlm.nih.gov/books/NBK53036/pdf/Bookshelf_NBK53036.pdf.

Chapter 28

Anesthesia in traumatic brain injury

Siddharth Chavali[a] and Girija Prasad Rath[b]

[a]*Institute of Neurosciences, AIG Hospitals, Hyderabad, India,* [b]*Department of Neuroanaesthesiology and Critical Care, Neurosciences Centre, All India Institute of Medical Sciences (AIIMS), New Delhi, India*

List of abbreviations

AMPA	alpha amino-3-hydroxy-5-methyl-4-isoxazoleproprionic acid
ARDS	acute respiratory distress syndrome
ATP	adenosine triphosphate
CBF	cerebral blood flow
$CMRO_2$	cerebral metabolic rate of oxygen or cerebral oxygen consumption
CPP	cerebral perfusion pressure
EEG	electroencephalogram
GABA	gamma amino butyric acid
GCS	Glasgow Coma Scale
ICP	intracranial pressure
MAC	minimum alveolar concentration
MAP	mean arterial pressure
MILS	manual in-line stabilization
NMDA	*N*-methyl D-aspartate
PEEP	positive end expiratory pressure
TBI	traumatic brain injury
TCD	transcranial Doppler

Introduction

Traumatic brain injury (TBI) is a modern epidemic that contributes significantly to mortality and morbidity in young adults worldwide. Approximately a tenth of these patients suffer moderate to severe injuries, requiring surgical intervention (Yates, Williams, Harris, Round, & Jenkins, 2006). Postoperative cognitive and neurological deficits leading to disability are dreaded complications and constitute a significant socioeconomic and healthcare burden. The primary goal of perioperative management of these patients is to minimize the extent of brain injury through various modalities available in the clinician's arsenal. For the purpose of this chapter, the perioperative management goals for the patient arriving with TBI will be discussed through the course of their clinical presentations.

Preoperative management

Pathophysiology

TBI remains a heterogeneous disease process with regards to etiology, presenting symptoms, severity, and outcomes. Clinical course and outcomes are broadly associated with the presenting Glasgow Coma Scale (GCS) score following appropriate resuscitation. The GCS is formed by the sum of the motor, eye, and verbal scales, and TBI is classified according to the total score as mild (13–15), moderate (9–13), and severe TBI (<8) (Teasdale & Jennett, 1974). TBI is further divided into primary and secondary injuries sustained by the brain. Primary brain injury is the direct consequence of the mechanical trauma sustained due to the transfer of kinetic energy, and its pattern depends on the nature of the impact. Typical examples of primary brain injury are calvarial fractures, hematomas, and diffuse axonal injuries. This primary injury then leads to the

Diagnosis and Treatment of Traumatic Brain Injury. https://doi.org/10.1016/B978-0-12-823347-4.00044-0

initiation of local as well as systemic inflammatory processes that result in edema, increased intracranial pressure (ICP), and reduced cerebral perfusion pressure (CPP).

Secondary injury results from myriad systemic physiological insults following the primary impact, which leads to additional neurological damage. Hypotension and hypoxemia are seen frequently following trauma, and are the major causes of secondary injury in TBI, resulting in increased morbidity and an approximately twofold increase in mortality (Spaite et al., 2017). Such data have led to an interest in securing definitive airways in patients with impaired consciousness in the prehospital setting; however, suboptimal intubation and ventilation have worsened outcomes. Patients with moderate to severe TBI should be managed in a designated center with expertise in trauma management.

Airway management

Endotracheal intubation and mechanical ventilation remain the standard management for patients presenting with a GCS < 8. Full stomach, facial injuries, or cervical cord trauma may complicate airway management in the emergency setting, making adequate pre-oxygenation essential. Anesthetic induction agents are used to attenuating the hemodynamic response to laryngoscopy/intubation, and muscle relaxants may be used to avoid surges in ICP. Adequate sedation and muscle relaxation also reduce cerebral metabolic rate of oxygen ($CMRO_2$) requirement and prevent increases in ICP associated with coughing. Rapid sequence induction with cricoid pressure is the standard technique for airway management.

The cervical spine may be injured in a significant fraction (5%–6%) of TBI cases, which may be explained by its high degree of mobility (Holly et al., 2002). During acceleration-deceleration injuries, this range of motion of the cervical spine may be exceeded, leading to various injuries in bones and soft tissues, most frequently between the occiput and C3 vertebra. High cervical injuries may lead to loss of diaphragmatic innervation leading to respiratory inadequacy, and neurogenic shock may result as a loss of efferent sympathetic nervous system activity. Accordingly, all patients with severe TBI should be presumed to have associated cervical spine trauma, and measures to limit cervical spine motion such as hard collar application, use of video laryngoscopy, and manual in-line stabilization (MILS) during laryngoscopy should be performed until the injury is ruled out via imaging.

Systemic complications of brain injury

TBI is associated with a massive surge of catecholamines and a neuro-inflammatory reaction, which may manifest in extracranial organ systems (Zygun, Kortbeek, Fick, Laupland, & Doig, 2005). This surge in catecholamines may be due to stimulation of brainstem centers such as the nucleus of the tractus solitarius, hypothalamus and medulla secondary to increased ICP (Šedý, Kuneš, & Zicha, 2015). Apart from electrocardiographic (ECG) changes such as QT-interval prolongation, ST-segment, and T-wave abnormalities, this surge is associated with vasoconstriction, leading to hypertension, tachycardia, increased myocardial oxygen demand and subsequent myocardial injury. Subendocardial ischemia and hemorrhage have been reported in up to 50% of patients who die due to TBI (Clifton, McCormick, & Grossman, 1981). Other complications that occur due to the sympathetic surge are neurogenic stunned myocardium, regional wall motion anomalies, and Takotsubo cardiomyopathy (Komamura, Fukui, Iwasaku, Hirotani, & Masuyama, 2014). The management goals for cardiac dysfunction following TBI are treating the primary cause and reducing fluctuations in ICP. Continuous ECG monitoring, cardiac biomarkers trends, and invasive cardiac monitoring may be used to optimize fluid and inotropic support.

Respiratory complications are prevalent after TBI. The incidence of lung injury is linked to the severity of TBI, e.g., nonoperated lesions and significant diffuse axonal injuries are linked with an increased incidence of lung injury. The most significant respiratory anomalies encountered following TBI are apnea/bradypnea leading to hypoxic-ischemic encephalopathies (Atkinson, Anderson, & Murray, 1998). Impaired cough/gag reflexes predispose patients to aspiration pneumonia, lower respiratory tract infections, etc. Neurogenic pulmonary edema, characterized by intraalveolar accumulation of protein-rich fluid and hemorrhage is another dreaded complication within the first 2 weeks following severe TBI, with an observed incidence of 20% (Davison, Terek, & Chawla, 2012). The priority in managing respiratory failure in TBI should be securing the airway, followed by aggressive management of increased ICP to limit sympathetic dysfunction.

Endocrine dysfunction manifesting as hypothalamic-pituitary-adrenal axis disruption following severe TBI may be due to primary trauma or secondary insults such as hypoxia, hypotension, and pyrexia. Injury to anterior pituitary may lead to cortisol insufficiency, with approximately 15% of patients requiring vasopressor support (Dimopoulou et al., 2004). One of the significant conditions in severe TBI is posterior pituitary dysfunction leading to altered antidiuretic hormone (ADH) secretion, causing hypo-osmolar euvolemic hyponatremia and increased urinary osmolarity. Significantly raised ICP may manifest as central diabetes insipidus (DI) characterized by hypernatremia, hypovolemic and polyuria (Tsagarakis, Tzanela, &

Dimopoulou, 2005). Disorders of sodium homeostasis following TBI can be diagnosed via clinical examination and paired blood and urine osmolalities.

Hematological complications are seen in up to a quarter of patients with severe TBI, and severity is associated with TBI severity (May, Young, Butler, Bassam, & Brady, 1997; Piek et al., 1992). This is thought to be caused by high concentrations of tissue thromboplastin within the brain, which is released into the systemic circulation following severe TBI and disruption of the blood-brain barrier (Astrup, 1965). Coagulopathy following TBI shows an initial hypercoagulable condition leading to microthrombi formation. As physiological hemostatic mechanisms become overwhelmed by excessive coagulation, increased fibrinolysis, and consumptive coagulopathy develop (Zhang et al., 2012). Correction of coagulopathy is vital to prevent intracranial hemorrhage, and various options such as administration of blood components, recombinant factor VII, and hemostatic agents have been studied. The recent CRASH-3 trial suggested that early (<3 h) administration of tranexamic acid was associated with reduced in-hospital mortality due to head injury at 28 days, and given its low risk of harm, its early use may be considered (The CRASH-3 trial collaborators, 2019).

Anesthetic management

Anesthetic management of TBI patients, inside an operating room (Fig. 1), is a continuation of the resuscitation initiated in the prehospital setting, with similar management goals. The focus of these patients' perioperative management is a limitation of secondary injury, understanding that hemodynamic alterations caused by the surgical procedure or anesthetic agents can exacerbate injury sustained by the brain due to potentially reversible causes such as hypoxia, hypercapnia, hypoglycemia, hyperglycemia, hypotension, and hyperthermia. Essentially, the goals of anesthetic management of patients with TBI are:

- Maintenance of CPP
- Management of raised ICP
- Avoidance of secondary causes of brain injury
- Providing good operative conditions

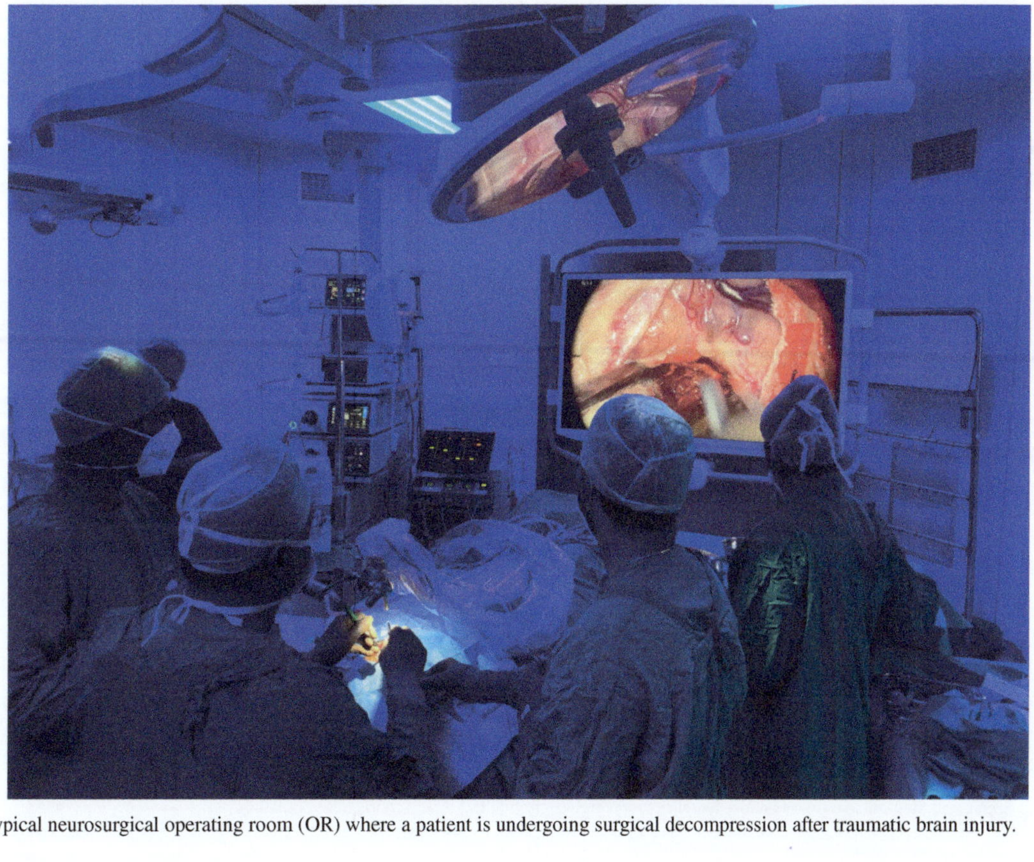

FIG. 1 A typical neurosurgical operating room (OR) where a patient is undergoing surgical decompression after traumatic brain injury.

Neuropharmacology of commonly used anesthetic agents

Several agents may be used to induce and maintain anesthesia in patients with TBI (Table 1). Anesthetic agents form an important part of TBI therapy and limit secondary brain injury by permitting manipulation of CO_2 levels via ventilation and reducing cerebral oxygen consumption ($CMRO_2$). $CMRO_2$ and cerebral blood flow (CBF) are usually coupled. Anesthetic agents attenuate secondary injury by reducing $CMRO_2$, improving ischemia tolerance, and reducing demand/supply mismatch in autoregulation impairment areas. Reduction of $CMRO_2$ with anesthetic agents is dose-dependent to the point of isoelectricity on electroencephalography (EEG), following which there is no further reduction.

CBF reduction is an adaptive phenomenon linked to a decrease in $CMRO_2$. Although anesthetic agents reduce CBF, peripheral vasodilation, and myocardial depression induced by anesthesia may reduce mean arterial pressure (MAP), leading to reduced CPP in impaired autoregulation areas (Table 2). Even in intact autoregulation areas, a reduction in MAP may lead to increased ICP due to physiological vasodilation, making a careful assessment of volume and cardiac status critical to minimize hypotensive episodes. Literature to advocate using one anesthetic agent over another is scarce (Roberts, Haroon, & Hall, 2012). Consensus guidelines also do not recommend particular agents, with barbiturate use for refractory ICP being the exception.

Propofol, a phenol derivative, is formulated as a lipid emulsion with a rapid onset of action due to its high lipid solubility. It acts via the potentiation of $GABA_A$ receptors and the blockade of sodium channels. Due to fast clearance, it is well-suited for early neurological assessment even after prolonged infusions. It has been shown to reduce CBF by 50%–60%,

TABLE 1 Pathophysiology of traumatic brain injury.

- Outcomes in TBI are a function of the etiology of injury and avoidance of subsequent systemic insults
- Secondary factors such as hypoxemia, hypoglycemia, hypotension, hypercarbia, hyperthermia, etc. are associated with increased morbidity and mortality
- A significant proportion of patients with TBI may present with concurrent cervical spine trauma—experts should perform airway management
- The catecholamine surge associated with TBI can manifest across multiple organ systems
- Neurogenic stunned myocardium and Takutsubo cardiomyopathy are examples of cardiological sequelae of TBI
- Aspiration pneumonitis due to impaired airway reflexes and neurogenic pulmonary edema is the most significant respiratory sequelae of TBI
- TBI is associated with abnormalities of coagulation and may manifest like disseminated intravascular coagulation

Key facts on the pathophysiology of TBI and its other systemic manifestations.

TABLE 2 Anesthetic agents and cerebral vascular dynamics.

- Anesthetic agents are categorized as inhalational and intravenous agents
- They limit secondary brain injury by reducing $CMRO_2$ and subsequent reduction of CBF and ICP
- The effects of anesthetic agents on $CMRO_2$ are dose-dependent unto the point of burst suppression on EEG
- Intravenous agents typically reduce CBF and $CMRO_2$
- Ketamine is the exception which increases CBF without increasing $CMRO_2$
- Volatile agents are potent cerebral vasodilators; they reduce $CMRO_2$ with increased CBF
- Cerebral autoregulation is the mechanism by which CBF is maintained at a MAP range of 50–150 mmHg
- Intravenous agents preserve both the static and dynamic components of cerebral autoregulation
- Volatile agents cause loss of dynamic autoregulation at 0.5 MAC and loss of static autoregulation at 1.5 MAC, making perfusion pressure-dependent.
- CO_2 reactivity is typically preserved with the use of both intravenous and inhalational anesthetics

Lists anesthetic agents and their mechanism of action, the effects on cerebral blood flow, and cerebral autoregulation.

particularly in areas regulating arousal and autonomic control (Schlünzen, Juul, Hansen, & Cold, 2012). This drug is also associated with significant peripheral vasodilation, and care should be taken for MAP and CPP maintenance (Illievich et al., 1993). Reactivity of cerebral vasculature to CO_2 is preserved, ensuring that hyperventilation remains an option to reduce ICP during propofol anesthesia. Prolonged infusions of propofol (> 48 h) are associated with the development of propofol infusion syndrome (PRIS), which presents as lactic acidosis, rhabdomyolysis, renal dysfunction, arrhythmias, and cardiac dysfunction. This condition is thought to be due to a demand/supply imbalance at the mitochondrial level or dysfunction of normal lipid metabolism pathways (Wong, 2010).

Barbiturates such as thiopentone are prototypical anesthetic agents in the management of TBI patients. These drugs stimulate GABA receptors and inhibit AMPA receptors, producing sedation and anesthesia. Their high lipid solubility ensures rapid equilibration between blood and CNS, with dose-dependent reductions in CBF and $CMRO_2$ seen until isoelectricity is observed, leading to a reduction of ICP. Barbiturates may have neuroprotective effects unrelated to EEG effects, such as free radical scavenging, reduced excitotoxicity, etc. (Warner et al., 1996). However, thiopental administration in patients undergoing aneurysmal clipping was not associated with significant neuroprotection (Hindman et al., 2010). Due to the cumulative nature of barbiturates and slow metabolism, use of these agents has largely been reduced and is typically used as rescue therapy for refractory increases in ICP and refractory status epilepticus.

Ketamine is an NMDA receptor antagonist which has typically been avoided due to concerns about increased ICP and possible epileptogenesis (Gibbs, 1972). Ketamine increases CBF by 60% without significantly changing global $CMRO_2$ (Takeshita, Okuda, & Sari, 1972). It has been seen to raise ICP; however, this can be prevented by hyperventilation or by concurrent administration of barbiturates or propofol. In contrast to other intravenous induction agents, ketamine administration is not associated with significant hemodynamic fluctuations, making it a safe drug for TBI patients and preserving CPP (Filanovsky, Miller, & Kao, 2010). The epileptogenic potential of ketamine has also been questioned, with research suggesting that NMDA receptor antagonism leading to reduced calcium influx may in fact, lead to reduced excitotoxicity and increased seizure thresholds. Ketamine is often used as an adjunct in the management of refractory status epilepticus.

Etomidate, a carboxylated imidazole compound, is used as an induction agent in patients with cardiovascular compromise due to minimal effects on systemic hemodynamics. Induction doses of etomidate cause a 30%–50% reduction in $CMRO_2$ and CBF, with preserved reactivity to CO_2. CBF is reduced in a dose-dependent manner until an isoelectric EEG is observed. Due to minimal reduction of MAP, etomidate may decrease ICP without significantly affecting CPP. Despite these beneficial effects, questions have been raised regarding the safety of etomidate, primarily due to its side effect of adrenocortical suppression via inhibition of the 11-beta-hydroxylase enzyme, a key enzyme for cortisol synthesis. This may lead to reduced cortisol levels and aldosterone for up to 24 h following induction and may be associated with hypotension (Malerba et al., 2005). Injection of etomidate is also associated with myoclonic movements, nausea, and epileptogenesis (Forman & Warner, 2011).

Volatile anesthetics are potent cerebral vasodilators and are capable of increasing ICP in a dose-dependent manner. Volatile agents produce a decrease in $CMRO_2$, with an increase in CBF caused by direct cerebral vasodilation. The rise in ICP with these agents is parallel to the increase in CBF, and may be associated with reduced CPP, especially in the setting of systemic hypotension. This increase in ICP may be attenuated by hyperventilation, or by concurrent use of intravenous anesthetic agents. In patients with hemorrhagic and ischemic stroke with moderately elevated baseline ICP in the ICU setting, 0.5–0.8 MAC of isoflurane was not associated with significant increases in ICP (Bösel et al., 2012; Villa et al., 2012). Due to profound reductions in $CMRO_2$ seen with these agents, volatile agents were predicted to have significant neuroprotective properties. However, there is no conclusive evidence in support of this notion. Isoflurane and sevoflurane have also been studied as agents to induce ischemic preconditioning or postconditioning (Kapinya et al., 2002; Xiong et al., 2003). In severe TBI, although volatile anesthetic agents may be used, it may be prudent to use propofol in patients when rigid control of ICP is desired.

Opioid agents act on the mu, kappa, epsilon, and delta-opioid receptors at the spinal as well as the supraspinal level and have variable effects on CBF, $CMRO_2$, and ICP, possibly due to variability in the usage of background anesthetics. Opioid agents act as cerebral vasoconstrictors when volatile agents are being used and have minimal effect with intravenous agents (Hänel, Werner, von Knobelsdorff, & Schulte am Esch, 1997). Large doses of opioids such as fentanyl have been seen to induce seizure activity and increase global CBF. Large doses may be associated with a transient spike in ICP, possibly caused by increased CO_2 levels due to hypoventilation. Histamine-release associated with large doses may result in vasodilation and reduced CPP. Newer agents such as remifentanil have increased lipid solubility and rapid metabolism, leading to a more predictable offset after the termination of infusion. This makes them attractive choices in patients with neurological injuries, facilitating shorter extubation time, and neurological assessment.

Autoregulation and anesthetic agents

Autoregulation is the ability of cerebral vasculature to ensure perfusion to the central nervous system (CNS) across a wide variety of perfusing pressures. The autoregulatory range of cerebral circulation is between 50 and 150 mmHg of MAP, which means that CPP remains relatively constant across these pressures due to cerebral vasoconstriction or vasodilation. This autoregulatory response is characterized by a fast component (dynamic autoregulation) and a steady-state component (static autoregulation). Intravenous anesthetic agents are known to preserve autoregulation, and volatile agents may disrupt it, with dynamic autoregulation being more affected by anesthetic agents. Propofol is an attractive choice for anesthesia in neurologically injured patients, since it has been shown to preserve both dynamic and static autoregulation across a range of doses. In contrast, with volatile anesthetic agents, dynamic autoregulation is typically lost at 0.5 MAC and static autoregulation is impaired at 1.5 MAC, leading to exaggerated CBF fluctuations with abrupt changes in MAP (Strebel et al., 1995). Such hemodynamic changes may exacerbate the secondary injury, necessitating tight control of blood pressure in patients with intracranial space-occupying lesions.

Anesthetic agents also affect the cerebrovascular reactivity to CO_2. At typically used concentrations of both intravenous and inhalational anesthetic agents, cerebrovascular reactivity to changes in CO_2 is well preserved. High concentrations of volatile agents may attenuate the CO_2 reactivity of the cerebral vasculature.

Hemodynamic management

The goal in the management of TBI patients is to prevent the progression of secondary brain injury (Table 3). Implementation of protocol-based management of these patients has led to a significant reduction in mortality. Factors contributing to secondary neurological injuries include hypotension, hypercapnia, hypoxemia, and variations in plasma glucose concentrations. Of these factors, hypotension, hypoxemia, and pyrexia have been seen to be associated with the worst outcomes in severe TBI (Jones et al., 1994).

Even a single **hypotension** episode, defined as a systolic BP <90 mmHg, is associated with higher mortality and morbidity in patients with severe TBI (Chesnut et al., 1993). The duration of time spent below this threshold BP also correlates with neurological outcomes; hence, the goal in these patients is to target an SBP >90 mmHg. Hypotension in this patient population is most frequently induced by bleeding from extracranial injuries, administration of osmotic diuretics, and attempts to restrict fluid intake. Some patients with significantly raised ICP may present with low-normal blood pressures and tachycardia, owing to the hemodynamic response to increased ICP. Such patients could be hypovolemic, and may decompensate when the brain is surgically decompressed unless judicious fluid therapy is administered. Monitoring tools such as pulse pressure variability, stroke volume variation, capillary refill, inferior vena caval diameter, etc. can be used to monitor these patients' volume status.

Fluid management in these patients is undertaken via the use of isotonic crystalloids. The blood-brain barrier can sustain damage during TBI, allowing unregulated flux of fluids across this normally impermeable barrier. Excessive volume administration may lead to vasogenic cerebral edema and complications such as acute respiratory distress syndrome (ARDS). Underfilling may result in hemodynamic compromise, leading to a reduction in CPP. Water movement across the

TABLE 3 Secondary brain injury after trauma.

- Primary brain injury is a function of energy transference during trauma and cannot be reversed
- Secondary brain injury is caused by systemic insults following the primary insult and represents a potentially reversible/avoidable neurological injury
- Causes of secondary neurological injury may include hypotension, hypercapnia, hypoxemia, and variations in plasma glucose
- Hypotension is an independent predictor of mortality, and even a single episode of systolic blood pressure less than 90 mmHg can be associated with worse outcomes
- Hypoglycemia and hyperglycemia have been seen to be associated with worse outcomes, and a target blood glucose range of 140–180 mmHg has been suggested
- With oxygen saturations of $<90\%$, hypoxemia is also a strong predictor of poor outcomes in TBI
- The arterial CO_2 level is a powerful tool to regulate CBF and ICP, with a linear correlation between a $PaCO_2$ of 20–60 mmHg

Lists key facts regarding primary and secondary brain injury and its potential causes.

blood-brain barrier is primarily governed by the osmotic pressure gradient between plasma and the brain. There is renewed interest in the use of hypertonic saline as a resuscitation fluid in patients with TBI, since it leads to expansion of circulating volume and acts as an osmotic agent in reducing ICP without the disadvantage of excessive diuresis. Hypotonic fluids and dextrose-containing fluids are avoided during the management of TBI patients since they may exacerbate the neuronal injury. Glucose administration has been shown to increase infarct sizes and augment secondary neurological injuries in animal models. Frequent monitoring of plasma glucose levels should be done, targeting a value between 140 and 180 mg/dL (de Courten-Myers, Myers, & Schoolfield, 1988).

Hypoxemia and **CO2** fluctuations are also significant contributors of secondary injuries. Hypoxemia and period spent below an oxygen saturation of 90% are independent predictors of mortality and morbidity. The minimum goal in the management of these patients should be a partial pressure of arterial oxygenation (PaO_2) of more than 60 mmHg. Administration of positive end-expiratory pressure (PEEP) has been a subject of controversy since it was thought that increased intrathoracic pressure might reduce venous outflow from the brain, resulting in increased ICP. However, studies have found that up to 10 cmH_2O of PEEP may be safely used in severe TBI without significantly affecting the ICP (Cooper, Boswell, & Choi, 1985). Arterial CO_2 levels are among the most powerful physiological variables affecting CBF and, by extension, ICP (Harp & Wollman, 1973). There is a linear association between CBF and $PaCO_2$ levels at the $PaCO_2$ range of 20–60 mmHg. Hypercapnia leads to increased CBF due to vasodilation caused by nitric oxide-mediated relaxation of pial arterioles due to alteration of extracellular pH. Hypocapnia leads to decreased CBF due to vasoconstriction, and excessive hyperventilation may aggravate ischemic insult.

Management of intraoperative intracranial hypertension (Fig. 2)

Head-up positioning, with the neck in a neutral position, allows for adequate venous drainage. Tight fixation of the endotracheal tube or excessively turning the head may drastically increase ICP due to impaired venous return from the cranium.

Hyperventilation is frequently used as a tool in the management of intracranial hypertension. As discussed previously, although hypocapnia can produce vasoconstriction mediated reductions in ICP and CBF, it may exaggerate neuronal ischemia. Since the injured brain sustains a dramatic decrease in CBF of approximately 50% in the first 24 h after injury, it may be prudent to avoid hyperventilation below $PaCO_2$ levels of 25 mmHg in the first 24 h following TBI (Sioutos et al., 1995). Hypocapnia may also reduce venous return and blood pressure, reduce cardiac output, and reduce oxygen delivery to tissues due to a leftward shift in the oxygen-hemoglobin dissociation curve. Due to these concerns, the clinical target in the management of patients with severe TBI should be normocapnia.

Hyperosmolar agents such as mannitol and hypertonic saline are used to increase plasma osmolality and create a gradient across the plasma and the blood-brain barrier, allowing a decrease in brain water content and ICP. The reduction in ICP is a function of the osmotic gradient generated and the structural integrity of the blood-brain barrier. Mannitol, typically administered at a dose of 0.25–2 g/kg to target a serum osmolality of up to 320 mOsm/L, acts via various means to reduce brain injury. Mannitol exerts a triphasic hemodynamic response; transient hypotension, followed by increased intravascular volume due to the osmotic gradient generated leading to slightly increased ICP values, and finally, its peak diuretic

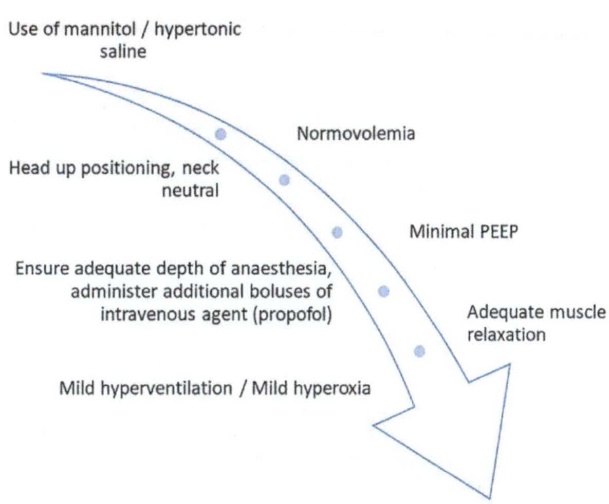

FIG. 2 Management of intraoperative intracranial hypertension. Tiered-approach to the management of intraoperative intracranial hypertension.

effect at 30 min leading to reduced ICP. Mannitol also reduces plasma viscosity and increased red blood cell distensibility, improving oxygen delivery within the brain microcirculation. Many clinicians increasingly prefer hypertonic saline due to "low-volume resuscitation," i.e., increased plasma volume, reduced ICP, and improved systemic blood pressure. However, studies had not shown any significant outcome benefit when hypertonic saline was used as the initial resuscitation fluid in severe TBI (Cooper et al., 2004).

Temperature control is vital in patients with severe TBI because temperatures of > 38°C have been associated with worse outcomes. Patients with pyrexia should be aggressively managed with antipyretic therapy or external and internal cooling measures. Conversely, hypothermia has been studied as a therapeutic modality in patients with severe TBI due to potential benefits associated with a reduction of $CMRO_2$. Several studies have looked at the induction of hypothermia as a neuroprotective modality in the prehospital as well as an intraoperative setting; however, no significant outcome benefits have been seen (Andrews et al., 2015; Cooper et al., 2018; Todd, Hindman, Clarke, Torner, & Intraoperative Hypothermia for Aneurysm Surgery Trial (IHAST) Investigators, 2005). Induction of hypothermia may increase the risk of bleeding, arrhythmias, cardiac dysfunction, and infective complications such as pneumonia.

Intraoperative monitoring

In addition to the standard American Society of Anesthesiology (ASA) monitoring devices, arterial catheterization should be performed in patients requiring decompressive surgery for continuous pressure monitoring, blood gas analysis, and glucose sampling. ICP monitoring is usually instituted in all patients with severe TBI (GCS < 9) and an abnormal computed tomographic (CT) scan, and in patients with severe TBI with a normal CT scan if greater than two of the following features are seen: age > 40, motor posturing, or SBP < 90 mmHg. Therapy may be instituted for raised ICP, and ICP values of < 20 mmHg may be targeted.

Multimodal monitoring for perioperative management of patients with TBI is gaining popularity, and monitoring cerebral oxygen delivery, CBF, and local brain metabolites may help provide tailored, patient-specific targets for therapy. Jugular venous oximetry reflects the balance between global cerebral oxygen supply and demand, and values < 50% may indicate the need to optimize ventilation or institute ICP lowering measures. If hyperventilation is to be used to reduce ICP, the use of jugular oximetry to guide ventilation may be associated with a reduced incidence of cerebral oligemia. Unlike other modalities which assess global changes in cerebral oxygenation, insertion of brain tissue oxygenation monitors may identify focal areas of oxygen demand/supply mismatch, with values of < 15 mmHg suggestive of ischemia. Management of severe TBI guided by a combination of ICP and $PbtO_2$ was seen to trend toward more favorable outcomes than management guided by ICP alone (Okonkwo et al., 2017). Near-Infrared Spectroscopy (NIRS) offers the capacity to monitor cerebral oxygenation noninvasively and may have a predictive role in raised ICP(Davies et al., 2015). Extravascular blood such as hematomas may significantly affect NIRS values, and this has been used as a technique to localize pathologies. Transcranial Doppler (TCD) is a noninvasive bedside tool that can provide real-time information regarding changes in cerebral blood flow velocity, vasospasm, and assessment of autoregulation. TCD measurements correlate with ICP and have been found to have prognostic value in patients with mild to moderate TBI (Bouzat et al., 2016).

Postoperative sequelae

Anesthetic agents and neuronal injury

Animal studies have shown that anesthetic agents exhibit cytotoxic/neurotoxic properties in the developing brain, leading to long-term cognitive dysfunction, behavioral changes, and learning disabilities (Jevtovic-Todorovic et al., 2003). In primates, early life exposure to anesthetic agents such as ketamine, sevoflurane, and isoflurane led to persistent cognitive decline and increased anxiety (Alvarado, Murphy, & Baxter, 2017). The total duration of exposure likely influences these anesthetic agents' impact on cognition and behavior, the number of exposures, and the animal's developmental stage. However, these findings have not been replicated in humans (Sun et al., 2016). Molecular mechanisms behind anesthesia-induced neurotoxicity in the developing brain have been studied extensively, with mitochondrial apoptosis being a consistently reported factor. Exposure to isoflurane significantly increased apoptosis, increased reactive oxygen species, promoted cytochrome-C release, and activated caspase-3 (Zhang et al., 2010). This leads to excessive Ca^{2+} release and mitochondrial calcium overload, causing failure of ATP synthesis and caspase activation.

Anesthetic agents have also been shown to be neuroprotective in the setting of global insults such as hypoxic-ischemic brain injury. During hypoxia, cellular metabolism switches from aerobic to anaerobic metabolism, which is insufficient to meet demand. This leads to depletion of ATP reserves and failure cellular pumps such as the Na^+/K^+-ATPase, causing Na^+

and Ca^{2+} influx, followed by cytotoxic edema and apoptosis. Ischemic insult also causes significant release of glutamate, and calcium influx, via the opening of AMPA and NMDA channels. Accumulation of cytosolic Ca^{2+} leads to free radical synthesis, cell membrane damage, phospholipase-C activation, and apoptosis, ultimately leading to cell death. Anesthetic agents seem to antagonize NMDA-induced excitotoxicity, which may have protective effects against global cerebral insults (Kudo et al., 2001). Besides, GABA receptor agonism may also inhibit the depolarization-induced release of glutamate, further reducing neurotoxicity (Bickler, Warner, Stratmann, & Schuyler, 2003).

Applications to other areas of neuroscience

In this chapter, we aimed to address the major considerations for anesthesia TBI patients. The TBI can present with concomitant injuries and myriad extracranial manifestations, which require specific management. The primary goal of anesthetic management is to prevent secondary brain injuries due to systemic insults. The anesthetic agents, both intravenous and inhalational, have been studied for their effects on cerebral vascular dynamics, ICP, and neuroprotective effects (Table 4). General anesthesia causes a reversible alteration in the level of consciousness, and the understanding of the exact neural mechanisms of consciousness remains incomplete. Acute brain injury results in alterations of cerebral blood flow, autoregulation, neurotransmitter levels, integrity of the blood-brain barrier, and release of inflammatory cytokines. Surgery and anesthesia are typically required in the acute phase following injury, a period when the brain is most vulnerable to secondary injuries. The development of newer multimodal monitoring devices may help evaluate cerebral metabolism, perfusion, and oxygenation during anesthesia. It may help provide therapeutic targets that are tailored to individual patients' physiologies. Proper management of these patients during the perioperative setting may improve outcomes and reduce the subsequent burden on healthcare systems and families.

Mini-dictionary of terms

Aspiration pneumonitis: Lung inflammation caused by inhalation of oropharyngeal or gastric contents into the pulmonary system.
Burst suppression: EEG pattern characterized by periods of electrical silence alternating with periods of high-voltage electrical activity, seen in states of brain inactivation.
Colloid: A solution with large molecules unable to pass through capillary membranes, leading to increased osmotic pressure in the blood.
Hypertonic: Solutions with osmolarity higher than that of plasma (typically > 300 mOsm/L).
Isotonic: Solutions with osmolarity similar to that of plasma (280–300 mOsm/L).

TABLE 4 Neuropharmacology of commonly used anesthetic agents.

Agent	Advantages	Disadvantages
Propofol	Reduced CBF, cerebral blood volume and ICP	Hypotension may lead to reduced CPP; prolonged infusions may cause propofol infusion syndrome
Ketamine	NMDA receptor antagonism may be neuroprotective, minimal hemodynamic fluctuation, antiepileptic activity	Early case reports suggestive of increased ICP, increased CBF
Barbiturates	Reduced CBF, CBV, and ICP, neuroprotective effect in addition to EEG activity, used for management of refractory ICP	Hypotension may lead to reduced CPP; prolonged infusions may lead to accumulation and prolonged duration of action
Midazolam	Lack of serious intracranial adverse effects, hemodynamic stability, antiepileptic activity	Tachyphylaxis
Etomidate	Hemodynamic stability, reduced $CMRO_2$, and reduced CBF	Epileptogenesis, nausea, adrenocortical suppression, reduced brain tissue oxygenation
Volatile anesthetic agents	Reduced $CMRO_2$, reduced ICP at low doses, may have preconditioning effect	Cerebral vasodilation, flow-metabolism uncoupling at higher MAC, disruption of autoregulation

Advantages and disadvantages of commonly used anesthetics in the setting of TBI.

Minimum Alveolar Concentration (MAC): The concentration of a volatile anesthetic agent within the alveoli at which 50% of people will not move in response to a surgical stimulus.
Multimodal Monitoring: The use of various tools to monitor cerebral metabolism, perfusion, and oxygenation aimed at the detection of injury at an early stage.
Neurogenic Pulmonary Edema: A rare form of pulmonary edema, caused by increased pulmonary interstitial and alveolar fluid that develops rapidly following acute CNS injury.
Neuroprotection: The relative reduction of neuronal structural and/or functional loss over time.
Positive End-Expiratory Pressure (PEEP): In ventilated patients, PEEP is the pressure that exists at the end of expiration, preventing alveolar collapse.
Pulse Pressure Variability (PPV): Quantification of changes in arterial pressure during mechanical ventilation, one of the dynamic indices of fluid responsiveness.

Key facts of anesthetics for injured brain

- Both inhalational and intravenous anesthetic agents limit secondary brain injury by reducing $CMRO_2$, and subsequent reduction of CBF and ICP.
- Intravenous agents are cerebral vasoconstrictors and reduce CBF and $CMRO_2$; ketamine is the exception to this rule and increases CBF.
- Volatile anesthetics are potent cerebral vasodilators and cause reduced $CMRO_2$ with increased CBF.
- With the mechanism of cerebral autoregulation, the CBF is maintained across a MAP range of 50–150 mmHg.
- Intravenous anesthetic agents preserve cerebral autoregulation, whereas the volatile agents may cause loss of autoregulation, making the perfusion pressure-dependent.
- CO_2 reactivity is typically preserved with the use of both types of anesthetics.

Summary points

- This chapter focuses on the effects of general anesthetics on normal neurophysiology and the injured brain.
- Secondary brain injuries are independent predictors of mortality and morbidity in TBI patients, and management focuses on their avoidance.
- Patients with TBI may also have concomitant cervical spine trauma, and airway management should be performed with adequate precautions.
- TBI may manifest with multisystemic complications involving the heart, lungs, endocrine, and coagulation systems.
- Anesthetic agents reduce the functional brain's oxygen consumption, leading to reduced CBF and lower ICP.
- Anesthetic agents influence cerebral vascular dynamics and autoregulation in a dose-dependent manner.

References

Alvarado, M. C., Murphy, K. L., & Baxter, M. G. (2017). Visual recognition memory is impaired in rhesus monkeys repeatedly exposed to sevoflurane in infancy. *British Journal of Anaesthesia*, *119*(3), 517–523.

Andrews, P. J. D., Sinclair, H. L., Rodriguez, A., Harris, B. A., Battison, C. G., Rhodes, J. K. J., & Murray, G. D. (2015). Hypothermia for intracranial hypertension after traumatic brain injury. *New England Journal of Medicine*, *373*(25), 2403–2412.

Astrup, T. (1965). Assay and content of tissue thromboplastin in different organs. *Thrombosis et Diathesis Haemorrhagica*, *14*(3–4), 401–416.

Atkinson, J. L., Anderson, R. E., & Murray, M. J. (1998). The early critical phase of severe head injury: Importance of apnea and dysfunctional respiration. *The Journal of Trauma*, *45*(5), 941–945.

Bickler, P. E., Warner, D. S., Stratmann, G., & Schuyler, J. A. (2003). Gamma-Aminobutyric acid-A receptors contribute to isoflurane neuroprotection in organotypic hippocampal cultures. *Anesthesia and Analgesia*, *97*(2), 564–571.

Bösel, J., Purrucker, J. C., Nowak, F., Renzland, J., Schiller, P., Pérez, E. B., … Steiner, T. (2012). Volatile isoflurane sedation in cerebrovascular intensive care patients using AnaConDa(®): Effects on cerebral oxygenation, circulation, and pressure. *Intensive Care Medicine*, *38*(12), 1955–1964.

Bouzat, P., Almeras, L., Manhes, P., Sanders, L., Levrat, A., David, J.-S., … Payen, J.-F. (2016). Transcranial doppler to predict neurologic outcome after mild to moderate traumatic brain injury. *Anesthesiology*, *125*(2), 346–354.

Chesnut, R. M., Marshall, L. F., Klauber, M. R., Blunt, B. A., Baldwin, N., Eisenberg, H. M., … Foulkes, M. A. (1993). The role of secondary brain injury in determining outcome from severe head injury. *The Journal of Trauma*, *34*(2), 216–222.

Clifton, G. L., McCormick, W. F., & Grossman, R. G. (1981). Neuropathology of early and late deaths after head injury. *Neurosurgery*, *8*(3), 309–314.

Cooper, D. J., Myles, P. S., McDermott, F. T., Murray, L. J., Laidlaw, J., Cooper, G., … HTS Study Investigators. (2004). Prehospital hypertonic saline resuscitation of patients with hypotension and severe traumatic brain injury: A randomized controlled trial. *JAMA*, *291*(11), 1350–1357.

Cooper, D. J., Nichol, A. D., Bailey, M., Bernard, S., Cameron, P. A., Pili-Floury, S., … McArthur, C. (2018). Effect of early sustained prophylactic hypothermia on neurologic outcomes among patients with severe traumatic brain injury: The POLAR randomized clinical trial. *JAMA*, *320*(21), 2211–2220.

Cooper, K. R., Boswell, P. A., & Choi, S. C. (1985). Safe use of PEEP in patients with severe head injury. *Journal of Neurosurgery*, *63*(4), 552–555.

Davies, D. J., Su, Z., Clancy, M. T., Lucas, S. J. E., Dehghani, H., Logan, A., & Belli, A. (2015). Near-infrared spectroscopy in the monitoring of adult traumatic brain injury: A review. *Journal of Neurotrauma*, *32*(13), 933–941.

Davison, D. L., Terek, M., & Chawla, L. S. (2012). Neurogenic pulmonary edema. *Critical Care*, *16*(2), 212.

de Courten-Myers, G., Myers, R. E., & Schoolfield, L. (1988). Hyperglycemia enlarges infarct size in cerebrovascular occlusion in cats. *Stroke*, *19*(5), 623–630.

Dimopoulou, I., Tsagarakis, S., Kouyialis, A. T., Roussou, P., Assithianakis, G., Christoforaki, M., … Roussos, C. (2004). Hypothalamic-pituitary-adrenal axis dysfunction in critically ill patients with traumatic brain injury: Incidence, pathophysiology, and relationship to vasopressor dependence and peripheral interleukin-6 levels. *Critical Care Medicine*, *32*(2), 404–408.

Filanovsky, Y., Miller, P., & Kao, J. (2010). Myth: Ketamine should not be used as an induction agent for intubation in patients with head injury. *Canadian Journal of Emergency Medicine*, *12*(2), 154–157.

Forman, S. A., & Warner, D. S. (2011). Clinical and molecular pharmacology of etomidate. *Anesthesiology*, *114*(3), 695–707.

Gibbs, J. M. (1972). The effect of intravenous ketamine on cerebrospinal fluid pressure. *British Journal of Anaesthesia*, *44*(12), 1298–1302.

Hänel, F., Werner, C., von Knobelsdorff, G., & Schulte am Esch, J. (1997). The effects of fentanyl and sufentanil on cerebral hemodynamics. *Journal of Neurosurgical Anesthesiology*, *9*(3), 223–227.

Harp, J. R., & Wollman, H. (1973). Cerebral metabolic effects of hyperventilation and deliberate hypotension. *British Journal of Anaesthesia*, *45*(3), 256–262.

Hindman, B. J., Bayman, E. O., Pfisterer, W. K., Torner, J. C., Todd, M. M., & Investigators, I. H. A. S. T. (2010). No association between intraoperative hypothermia or supplemental protective drug and neurologic outcomes in patients undergoing temporary clipping during cerebral aneurysm surgery: Findings from the Intraoperative Hypothermia for Aneurysm Surgery Trial. *Anesthesiology*, *112*(1), 86–101.

Holly, L. T., Kelly, D. F., Counelis, G. J., Blinman, T., McArthur, D. L., & Cryer, H. G. (2002). Cervical spine trauma associated with moderate and severe head injury: Incidence, risk factors, and injury characteristics. *Journal of Neurosurgery: Spine*, *96*(3), 285–291.

Illievich, U. M., Petricek, W., Schramm, W., Weindlmayr-Goettel, M., Czech, T., & Spiss, C. K. (1993). Electroencephalographic burst suppression by propofol infusion in humans: Hemodynamic consequences. *Anesthesia & Analgesia*, *77*(1), 155–160.

Jevtovic-Todorovic, V., Hartman, R. E., Izumi, Y., Benshoff, N. D., Dikranian, K., Zorumski, C. F., … Wozniak, D. F. (2003). Early exposure to common anesthetic agents causes widespread neurodegeneration in the developing rat brain and persistent learning deficits. *The Journal of Neuroscience: The Official Journal of the Society for Neuroscience*, *23*(3), 876–882.

Jones, P. A., Andrews, P. J., Midgley, S., Anderson, S. I., Piper, I. R., Tocher, J. L., … Dearden, N. M. (1994). Measuring the burden of secondary insults in head-injured patients during intensive care. *Journal of Neurosurgical Anesthesiology*, *6*(1), 4–14.

Kapinya, K. J., Diana, L., Carsten, F., Martin, M., Waschke, K. F., Isaev, N. K., & Ulrich, D. (2002). Tolerance against ischemic neuronal injury can be induced by volatile anesthetics and is inducible NO synthase dependent. *Stroke*, *33*(7), 1889–1898.

Komamura, K., Fukui, M., Iwasaku, T., Hirotani, S., & Masuyama, T. (2014). Takotsubo cardiomyopathy: Pathophysiology, diagnosis and treatment. *World Journal of Cardiology*, *6*(7), 602–609.

Kudo, M., Aono, M., Lee, Y., Massey, G., Pearlstein, R. D., & Warner, D. S. (2001). Effects of volatile anesthetics on N-methyl-D-aspartate excitotoxicity in primary rat neuronal-glial cultures. *Anesthesiology*, *95*(3), 756–765.

Malerba, G., Romano-Girard, F., Cravoisy, A., Dousset, B., Nace, L., Lévy, B., & Bollaert, P.-E. (2005). Risk factors of relative adrenocortical deficiency in intensive care patients needing mechanical ventilation. *Intensive Care Medicine*, *31*(3), 388–392.

May, A. K., Young, J. S., Butler, K., Bassam, D., & Brady, W. (1997). Coagulopathy in severe closed head injury: Is empiric therapy warranted? *The American Surgeon*, *63*(3), 233–236.

Okonkwo, D. O., Shutter, L. A., Moore, C., Temkin, N. R., Puccio, A. M., Madden, C. J., … Diaz-Arrastia, R. R. (2017). Brain tissue oxygen monitoring and management in severe traumatic brain injury (BOOST-II): A phase II randomized trial. *Critical Care Medicine*, *45*(11), 1907–1914.

Piek, J., Chesnut, R. M., Marshall, L. F., van Berkum-Clark, M., Klauber, M. R., Blunt, B. A., … Foulkes, M. A. (1992). Extracranial complications of severe head injury. *Journal of Neurosurgery*, *77*(6), 901–907.

Roberts, D. J., Haroon, B., & Hall, R. I. (2012). Sedation for critically ill or injured adults in the intensive care unit. *Drugs*, *72*(14), 1881–1916.

Schlünzen, L., Juul, N., Hansen, K. V., & Cold, G. E. (2012). Regional cerebral blood flow and glucose metabolism during propofol anaesthesia in healthy subjects studied with positron emission tomography. *Acta Anaesthesiologica Scandinavica*, *56*(2), 248–255.

Šedý, J., Kuneš, J., & Zicha, J. (2015). Pathogenetic mechanisms of neurogenic pulmonary edema. *Journal of Neurotrauma*, *32*(15), 1135–1145.

Sioutos, P. J., Orozco, J. A., Carter, L. P., Weinand, M. E., Hamilton, A. J., & Williams, F. C. (1995). Continuous regional cerebral cortical blood flow monitoring in head-injured patients. *Neurosurgery*, *36*(5), 943–949.

Spaite, D. W., Hu, C., Bobrow, B. J., Chikani, V., Barnhart, B., Gaither, J. B., … Sherrill, D. (2017). The effect of combined out-of-hospital hypotension and hypoxia on mortality in major traumatic brain injury. *Annals of Emergency Medicine*, *69*(1), 62–72.

Strebel, S., Lam, A. M., Matta, B., Mayberg, T. S., Aaslid, R., & Newell, D. W. (1995). Dynamic and static cerebral autoregulation during isoflurane, desflurane, and propofol anesthesia. *Anesthesiology*, *83*(1), 66–76.

Sun, L. S., Li, G., Miller, T. L. K., Salorio, C., Byrne, M. W., Bellinger, D. C., … McGowan, F. X. (2016). Association between a single general anesthesia exposure before age 36 months and neurocognitive outcomes in later childhood. *JAMA*, *315*(21), 2312–2320.

Takeshita, H., Okuda, Y., & Sari, A. (1972). The effects of ketamine on cerebral circulation and metabolism in man. *Anesthesiology*, *36*(1), 69–75.

Teasdale, G., & Jennett, B. (1974). Assessment of coma and impaired consciousness. A practical scale. *Lancet*, *2*(7872), 81–84.

The CRASH-3 trial collaborators. (2019). Effects of tranexamic acid on death, disability, vascular occlusive events and other morbidities in patients with acute traumatic brain injury (CRASH-3): A randomised, placebo-controlled trial. *The Lancet*, *394*(10210), 1713–1723.

Todd, M. M., Hindman, B. J., Clarke, W. R., Torner, J. C., & Intraoperative Hypothermia for Aneurysm Surgery Trial (IHAST) Investigators. (2005). Mild intraoperative hypothermia during surgery for intracranial aneurysm. *The New England Journal of Medicine*, *352*(2), 135–145.

Tsagarakis, S., Tzanela, M., & Dimopoulou, I. (2005). Diabetes insipidus, secondary hypoadrenalism and hypothyroidism after traumatic brain injury: Clinical implications. *Pituitary*, *8*(3–4), 251–254.

Villa, F., Iacca, C., Molinari, A. F., Giussani, C., Aletti, G., Pesenti, A., & Citerio, G. (2012). Inhalation versus endovenous sedation in subarachnoid hemorrhage patients: Effects on regional cerebral blood flow. *Critical Care Medicine*, *40*(10), 2797–2804.

Warner, D. S., Takaoka, S., Wu, B., Ludwig, P. S., Pearlstein, R. D., Brinkhous, A. D., & Dexter, F. (1996). Electroencephalographic burst suppression is not required to elicit maximal neuroprotection from pentobarbital in a rat model of focal cerebral ischemia. *Anesthesiology*, *84*(6), 1475–1484.

Wong, J. M. (2010). Propofol infusion syndrome. *American Journal of Therapeutics*, *17*(5), 487–491.

Xiong, L., Zheng, Y., Wu, M., Hou, L., Zhu, Z., Zhang, X., & Lu, Z. (2003). Preconditioning with isoflurane produces dose-dependent neuroprotection via activation of adenosine triphosphate-regulated potassium channels after focal cerebral ischemia in rats. *Anesthesia and Analgesia*, *96*(1), 233–237.

Yates, P. J., Williams, W. H., Harris, A., Round, A., & Jenkins, R. (2006). An epidemiological study of head injuries in a UK population attending an emergency department. *Journal of Neurology, Neurosurgery, and Psychiatry*, *77*(5), 699–701.

Zhang, J., Jiang, R., Liu, L., Watkins, T., Zhang, F., & Dong, J. (2012). Traumatic brain injury-associated coagulopathy. *Journal of Neurotrauma*, *29*(17), 2597–2605.

Zhang, Y., Dong, Y., Wu, X., Lu, Y., Xu, Z., Knapp, A., … Xie, Z. (2010). The mitochondrial pathway of anesthetic isoflurane-induced apoptosis. *The Journal of Biological Chemistry*, *285*(6), 4025–4037.

Zygun, D. A., Kortbeek, J. B., Fick, G. H., Laupland, K. B., & Doig, C. J. (2005). Non-neurologic organ dysfunction in severe traumatic brain injury. *Critical Care Medicine*, *33*(3), 654–660.

Chapter 29

Treatment of raised intracranial pressure in traumatic head injury

Alvaro Cordoba[a] and Bernardina Frache[b]
[a]*ASESP Neurosurgical Department ICU Unit, Montevideo, Uruguay,* [b]*Intensive Care Unit, Sanatorio Americano, Montevideo, Uruguay*

List of abbreviations

BBB blood–brain barrier
BTF Brain Trauma Foundation
CBF cerebral blood flow
CBV cerebral blood volume
CPP cerebral perfusion pressure
DC decompressive craniotomy
GCS Glasgow coma scale
GOS Glasgow outcome scale
ICP intracranial pressure
TBI traumatic brain injury
TCD transcranial Doppler

Introduction

Rosner (Rosner, Rosner, & Johnson, 1995) advocated a theory for de management and treatment of intracranial pressure (ICP) based on the modification of the cerebral perfusion pressure (CPP). This model can be called "the complex vasodilatory/vasoconstriction cascade." Stabilizing CPP at high levels, the ICP could be better controlled without cerebral ischemia.

CPP is a physiological parameter linked with ICP, and is a determinant of cerebral hemodynamic responses and effects, and it can be manipulated. The Rosner theory stated that the minimum level of CPP in Traumatic Brain Injury (TBI) is greater than 70 mmHg and frequently higher, defined by individual circumstances.

Between 1992 and 1994, at the Lund University Hospital in Sweden was developed "The Lund Concept" (Grände, 2006) and was introduced for treatment of severe head trauma. Its main characteristics are based on hypotheses originating from basic physiological principles regarding control of brain volume and cerebral perfusion.

The concept is based in a therapy with two main objectives: reduce or prevent an increase in ICP (ICP-targeted goal) and improve perfusion and oxygenation around contusions (perfusion-targeted goal). The Lund therapy considers the consequences of a disrupted blood–brain barrier. This therapy could be applied to all patients with severe head injury, and should be started early to avoid secondary injuries and to antagonize increase in ICP.

In 1996, BTF (Brain Trauma Foundation) (Bullock, Chesnut, Clifton, et al., 1996) a new approach in traumatic head injury was adopted on the basis of the three levels of treatment focused on the registering of the optimized ICP. Each step is considered on the basis of the response to treatment.

Analysis of the treatment during the so called "brain decade"

During 1984, Rosner et al. (Rosner & Becker, 1984) described the incidence of plateau waves in the ICP monitoring; furthermore, in 1990 (Rosner & Daughton, 1990), CPP limits were described about 70 to 88 mmHg or more in some cases. The concept of brain autoregulation was established. To obtain an adequate CBF, an increase up the normal range of the CPP is necessary.

Euvolemia and vasopressors are mandatory and also to avoid secondary injury.

Diagnosis and Treatment of Traumatic Brain Injury. https://doi.org/10.1016/B978-0-12-823347-4.00023-3

In 1995 (Rosner et al., 1995), 158 patients with GCS 7 or less were managed using vascular volume expansion CSF fluid drainage via ventriculostomy, systemic vasopressors, and Mannitol to maintain a minimum CPP of at least 70 mmHg.

Neither barbiturates hypothermia nor hyperventilation was used. Nearly 100 variables (prospective and observational study) were recorded.

The authors concluded that CPP is the primary goal in the treatment of traumatic intracranial hypertension and substantially improved mortality and morbidity after TBI.

The range of the CPP must be always considered above 70 mmHg mainly in 85 mmHg (Rosner, 1995).

Between 1992 and 1994, a protocol was designed in Sweden (Lund University) like an approach for the treatment of severe brain trauma (Asgeirsson, Grände, & Nordström, 1994).

The authors considered that the opening of the BBB (Blood–Brain Barrier) during severe brain lead to an impairment of the ICP due to an increase of the extracellular edema. That means that we must reduce the hydrostatic capillary pressure and maintain a normal osmotic pressure. Results indicate that the therapy should focus on extracellular compartment rather than intracellular edema. Also they concluded that ischemia is not the main cause of the edema formation.

In 2004 and 2006, Grände et al. published a revision of physiological principles and the clinical application of the protocol. They recommend how to deal with different parameters such as ventilation, nutrition, blood pressure, sedation, volume substitution, and body temperature. Disruption or not of the BBB is the most important point of the treatment.

Head elevation promotes a decrease of ICP due to a reduction of the CBV (Cerebral blood Volume) and the use of PEEP could be used safely to prevent atelectasis after head trauma (Table 2).

Decompressive craniotomy was also considered like a theoretical possibility to reduce brain edema and degree of herniation in a mass expansive lesion. Also the protective effect of the surgery evacuation under the concept of the clearance of neurotoxic factors.

Hyperosmotic therapy was used to treat edema and the efficacy is evident but in a short period. Renal effects are also seen during the treatment.

Vascular resistance reduction in injured areas is more important than to promote high CPP every time.

As the same effect in hypovolemia on the baroreceptors the use of vasoconstrictors such as norepinephrine phenylephrine etc. may lead a compromise on the brain perfusion.

Barbiturates reduce the ICP through a vasoconstrictor and sedative effect.

High dose barbiturate therapy has renal and electrolyte adverse effects also with fever and pulmonary and cardiovascular complications.

Sedatives and analgesia promote reduction on stress status decreasing the affectation in cerebral microcirculation of the pericontusional areas.

Prevention of the temperature increase is beneficial and improves the outcome.

The use of Paracetamol and one bolus dose of steroid like methylprednisolone is recommended.

Hyperventilation is also a matter of discussion and the protocol reveals that must be avoided except in brain stem herniation.

Related to fluid administration, the volume can be replaced by colloids or crystalloids; colloids have a much greater volume expanding effect and they last more than crystalloids.

Lund protocol recommends the use of albumin at the first level of treatment.

HCT (Hematocrit) must be maintained in normal range.

Related to CPP the range for adults in the authors experience is 60 to 70 mmHg;

They concluded favorable outcomes and results (Eker, Asgeirsson, Grande, Schalen, & Nordstrom, 1998; Wahlström, Olivecrona, Koskinen, Rydenhag, & Naredi, 2005) (Table 1).

TABLE 1 Comparison between Lund protocol and BTF concepts.

Lund concept	BTF (1996)	BTF 2007–2016
CPP 50–70 mmHg	CPP above 70 mmHg	CPP 50–70 mmHg
Avoidance of osmotherapy	Osmotherapy IPC-reducing therapy	Osmotherapy with caution
Avoidance of vasopressors	High doses of vasopressors accepted CPP above 70 mmHg	Vasopressor can be used, but Avoid pulmonary distress
Active cooling is not used	Active cooling is used	Active cooling is not used
Albumin recommended as plasma volume expander	Albumin recommended	Albumin not recommended

In 1996, the BTF (Bullock et al., 1996) made the guidelines that address key issues relating to the management of severe head injury in adult patients with Glasgow Coma Scale score of 3 to 8. The guidelines comprised different aspects about head injury like system trauma, pre-hospital resuscitation, monitoring and treatment of intracranial hypertension and intensive care. The value of therapies or interventions is classified into one of three categories according to the following criteria: Class I evidence: prospective randomized controlled trials (PRCT) the gold standard of clinical trials. Class II evidence: clinical studies in which the data were collected prospectively, and retrospective analyses which were based on clearly reliable data. Class III evidence: most studies based on retrospectively collected data. For the technology, the devices were evaluated in terms of their reliability, accuracy, cost effectiveness and therapeutic potential (Table 2).

Recommendations

The first recommendations are about the Trauma System and Neurosurgeon, they propose that all regions in the United States should have an organized trauma care system. Since 1970 (Campbell, Watkins, & Kreis, 1989; Kreis, Plasencia, Augenstein, et al., 1986; West, Trunkey, & Lim, 1979), studies generally attempted to show that in regions without organized trauma care occurred excessive trauma deaths. The centers that treat larger volumes of trauma patients, have better outcomes than centers with fewer admissions (Smith, Frateschi, Sloan, et al., 1990). The mortality is reduced after major trauma in patients treated in the trauma system.

For the integration of brain-specific treatment into the initial resuscitation of the severe head injury patient, they said there is not sufficient data to support a treatment guideline for this topic, but they recommend some aspects. Is very important a rapid physiologic resuscitation. The efficacy of cardiopulmonary resuscitation in improving survival from trauma in general is well accepted. In addition, it is very important to know the negative influence of secondary insults.

The hyperventilation should be established because it can reduce the ICP by causing cerebral vasoconstriction and subsequent reduction in cerebral blood flow (CBF). But hyperventilation has a potential effect exacerbated in intracranial ischemia; therefore, aggressive hyperventilation should be avoided (Obrist, Langfitt, Jaggi, et al., 1984).

The hypotension (systolic blood pressure of less than 90 mmHg) or hypoxia, (apnea or cyanosis in the field, or a $PaO_2 < 60$ mmHg by arterial blood gas analysis), mainly early post-injury are associated with increased mortality and morbidity (Chesnut, Marshall, Klauber, et al., 1993; Fearnside, Cook, McDougall, et al., 1993; Pigula, Wald, Shackford, et al., 1993). Avoid Hypoxia and Hypotension or corrected immediately are other recommendations. The mean arterial pressures should be maintained above 90 mmHg to maintain CPP about 70 mmHg.

TABLE 2 Venous pathology causing intracranial hypertension.

• Types of lesions	• Location resistance	• Examples
• Focal extramural venous sinus compression	• External compression of a significant venous sinus at a focal point	• Depressed skull fracture, periosteal hematoma, tumor
• Focal intermural venous sinus stenosis	• A focal narrowing within the sinus wall	• Idiopathic intracranial hypertension
• Focal intramural venous sinus obstruction	• Obstruction within a significant venous sinus	• Sagittal or transverse sinus thrombosis
• Diffuse venous compression	• Throughout the venous tree	• Brain swelling e.g. hypoxia, cerebral edema, contusions
• Extracranial venous hypertension—cervical	• Within the neck	• Cervical collars, hanging
• Extracranial venous hypertension—thoracic	• Within the thorax	• increased intra-thoracic pressure—Chest infection, adult respiratory distress syndrome, mechanical ventilation
• Extracranial venous hypertension—abdominal	• Within the abdomen	• Increased abdominal pressure—obesity, obstruction
• Orthostatic/gravity	• No resistance	• Visual impairment and raised ICP/space obstruction syndrome

Use of Mannitol could be desirable but the conditions of the patient must be adequate volume resuscitation due to its volume-depleting diuretic effect.

Sedation a Neuromuscular blockade could be used. Pharmacologic relaxations, in the absence of evidence of herniation, should be used in situations where sedation alone is not sufficient to optimize safe and efficient patient transport and resuscitation.

Presence of signs of hypertension intracranial elevated should be treated to lower ICP. These signs include unilateral or bilateral pupillary dilatation, asymmetric pupillary reactivity, motor posturing, or other evidence of deterioration of the neurologic examination. Without signs of ICP, the treatment should be conducted to resuscitations and to avoid the secondary injury.

Intracranial pressure monitoring

Current data update and analysis

Intracranial pressure (ICP) monitoring has been strongly recommended in selected patients with a severe traumatic brain injury (TBI) with the rationale of early identification and treatment of intracranial hypertension to maintain adequate cerebral blood flow and oxygenation (Figs. 1–3). This recommendation was not based on Level 1 evidence, but derived from the retrospective

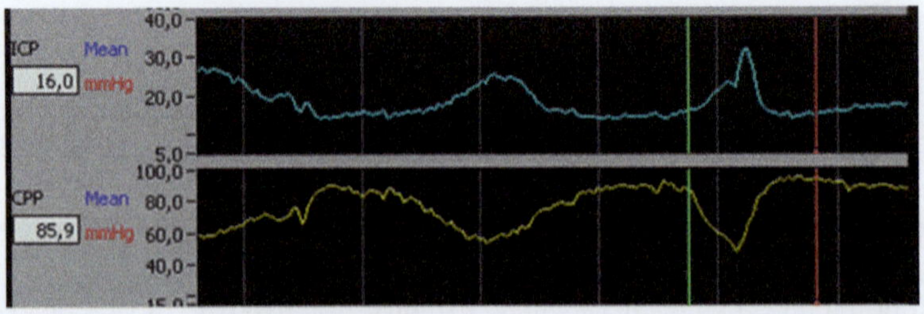

FIG. 1 Monitor curves shown impairment of the CPP while ICP increase.

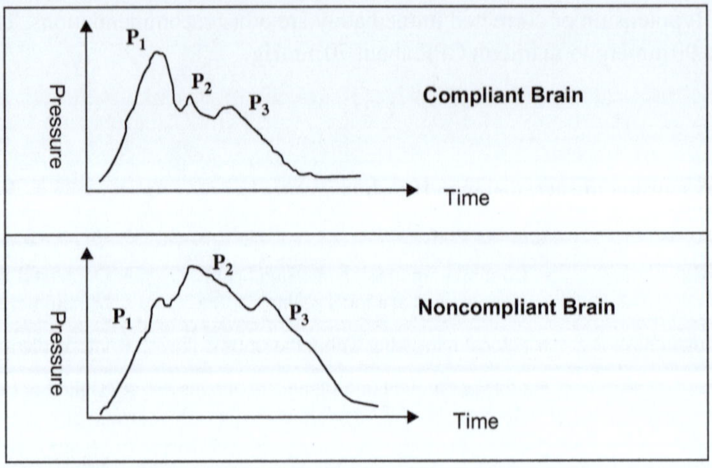

FIG. 2 Waves of compliant and noncompliant brain.

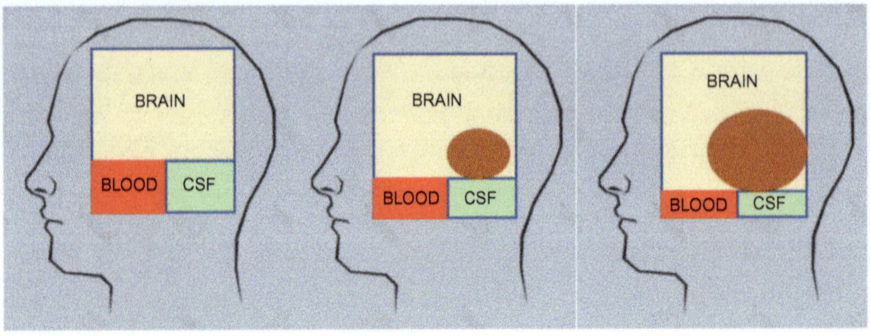

FIG. 3 Brain compartments and fluid displacement due an intracranial mass.

and limited prospective observational studies. The recommendation on ICP monitoring for severe TBI dates from the first edition of the Brain Trauma Foundation (BTF) guidelines. There was no difference in the recommendation between the first and second edition of the BTF guidelines. Both editions mentioned that there is a lack of standard for recommendation on ICP monitoring. However, ICP monitoring is appropriate in patients with a severe TBI with an abnormal computed tomography (CT) scan. ICP monitoring is not routinely indicated in patients with mild or moderate TBI. However, a physician may choose to monitor ICP with traumatic mass lesions. The third edition of the BTF guidelines was more prescriptive. Although there was a lack of Level 1 evidence supporting routine monitoring of ICP, based on Level 2 evidence, the BTF recommended that "ICP should be monitored in all patients with severe TBI with an abnormal CT scan." The Level 3 recommendation was "ICP monitoring is indicated in patients with a severe TBI with a normal CT scan if ≥2 of the following: age > 40 years, unilateral or bilateral motor posturing, and systolic blood pressure < 90 mmHg. After the publication of a randomized trial of ICP monitoring, the latest edition of the BTF guidelines mentioned that there was insufficient evidence to support recommendation for ICP monitoring. However, the management of severe TBI patients using information from ICP monitoring is recommended to reduce in-hospital and 2-week post-injury mortality. The BTF guidelines are US based, and not practiced universally.

In the Japan Neurotrauma Data Bank analysis of 1091 cases from 2009 to 2011, it was found that the rate of ICP monitoring as per the Japanese guidelines for the management of severe TBI was 28%, and earlier it was 55%. The indications for ICP monitoring were not based on the guidelines, but the ICP monitored was performed in younger patients, diffuse injury III-IV, evacuated mass lesions, perimesencephalic cistern compression, and midline shift of ≥5 mm. The patients in ICP-monitoring group received more hyperventilation, hyperosmolar diuretics, sedatives, anticonvulsants, surgical treatment, and temperature management. Besides maximum ICP value, age, Glasgow Coma Scale (GCS) score, papillary abnormalities, and perimesencephalic cisterns were significant determinants of outcome. The mortality rate was 35.2% in the ICP-monitoring group and 45% in the non-monitored group. The favorable outcome rates for these two groups were 29.2% and 30%, respectively (Carney et al., 2017). The Japan Neurotrauma Data Bank analysis showed that the rate of ICP monitoring in the treatment of severe TBI came down. Therapies were performed aggressively in the ICP-monitoring group. However, no increase in the favorable outcome rate was found with ICP monitoring.

In an Indian study analyzing the adherence of BTF guidelines in a single center, ICP monitors were placed in 126 (63%) of 200 patients. Patients with ICP monitor placement experienced lower in-hospital mortality (adjusted relative risk: 0.50 [0.29, 0.87]) than patients without ICP monitoring. However, there was no benefit at 3, 6, and 12 months. The absence of cerebral edema and intra-ventricular hemorrhage was associated with reduced unfavorable outcomes (Suehiro, Fujiyama, Koizumi, & Suzuki, 2017).

The Benchmark Evidence from South American Trials: Treatment of Intracranial Pressure Trial (BEST-TRIP) compared the two treatment protocols: clinical monitoring including imaging and ICP monitoring in addition to the clinical monitoring. There was no difference in early and late mortality rates between the two groups ($P=0.19$ and 0.60, respectively) (CHIRAG Study, 2018). This trial was performed by North Americans, and the results were not acceptable to them; hence, they released a consensus-based interpretation of the BEST-TRIP (Chesnut et al., 2012). The strongest statement was on lack of generalizability of the results. The maximum burden of neurotrauma is in developing countries; hence, these results are definitely generalizable globally.

Even in America, the ICP monitoring is not practiced uniformly. In the American College of Surgeons (ACS)-Trauma Quality Improvement Program (TQIP) database queried for a 2-year time range (2013–2014) for the patients who met the criteria for ICP monitoring, only 10.8% of patients underwent ICP monitoring in Level 1 centers. Logistic regression was used to compare the outcome. The ICP monitoring placement was found to be an independent risk factor for mortality, mechanical ventilation, intensive care unit (ICU) length of stay, systemic complications, and decreased functional independence at discharge (Chesnut et al., 2015).

In the present study, the authors investigated the association of ICP monitoring and clinical outcome in patients with an isolated severe blunt TBI from the ACS-TQIP database during 2014. The ICP was monitored in 18.3% of patients who met the criteria for ICP monitoring. Unlike the previous TQIP database analysis that used logistic regression, the current study used propensity score matching, which is a superior technique for comparing two cohorts. A propensity score is the probability that a unit with certain characteristics will be assigned to the intervention group (as opposed to the control group). The scores can be used to reduce or eliminate selection bias in observational studies by balancing covariates between intervention and control groups. This method approximates a random experiment, eliminating many of the problems that come with observational data analysis. The propensity score matching mimics some of the particular characteristics of a randomized controlled trial. The current study showed that the ICP monitoring was associated with an increased risk of systemic complications such as deep venous thrombosis, acute respiratory distress syndrome, and sepsis. The ICP-monitored group had significantly increased length of stay both in ICU and hospital. The ICP-monitored group had an increased mortality risk with odds ratio of 1.6 (95% confidence interval: 1.1–2.5, $P=0.038$) (Piccinini et al., 2017).

Based on the previous observation and current study, it can be concluded that ICP monitoring cannot be considered as a standard for the management of severe TBI. The question should not be whether to monitor ICP, but in whom to monitor ICP. Unless a lot of studies that globally replicate the results of studies showing benefit of ICP monitoring in improving outcomes in severe TBI, the ICP monitoring will not be incorporated in the management of severe TBI protocols, particularly in resource-limited settings (Ahl, Sarani, Sjolin, & Mohseni, 2018; Austin, 2011).

Concerning developing countries, in March 2017, a consensus conference for the development of the protocol for comprehensive care of adults with TBI was held in Bogota, Colombia. Twenty experts in pre-hospital care, emergency medicine, neurosurgery, and intensive care attended, accompanied by five methodologists. Six participants attended through videoconferencing and 19 in person. The process took 3 days; 2 meeting days with 1 final consensus day during which participants discussed possible answers to the formulated questions.

It is estimated that 80% of the global population live in developing economies where access to resources required for optimum treatment is limited. There is limitation for applications of CPGs recommendations in areas where there is low availability or absence of resources for integral care. Development of mixed methods consensus, including evidence review and expertise points of good clinical practices can fill gaps in application of CPGs. BOOTStraP (Beyond One Option for Treatment of Traumatic Brain Injury: A Stratified Protocol) is intended to be a practical handbook for care providers to use to treat TBI patients with whatever resources are available (Rubiano, Vera, & Montenegro, 2020).

Role of the cranial decompression

The most recent search was run on 8 December 2019 (Sahuquillo & Dennis, 2019). They searched the Cochrane Injuries Group's Specialized Register, CENTRAL (Cochrane Library), Ovid MEDLINE(R), Ovid MEDLINE(R) In-Process & Other Non-Indexed Citations, Ovid MEDLINE(R) Daily and Ovid OLDMEDLINE(R), Embase Classic + Embase (OvidSP) and ISI Web of Science (SCI-EXPANDED & CPCI-S). We also searched trials registries and contacted experts.

The search included randomized studies assessing patients over the age of 12 months with severe TBI who either underwent decompressive craniectomy (DC) to control ICP refractory to conventional medical treatments or received standard care. They included three trials (590 participants).

Decompressive craniectomy holds promise of reduced mortality, but the effects of long-term neurological outcome remain controversial, and involve an examination of the priorities of participants and their families. Future research should focus on identifying clinical and neuroimaging characteristics to identify those patients who would survive with an acceptable quality of life.

The use of secondary DC in refractory elevated ICP after severe TBI was explored in the DECRA trial (recruitment 2002–2010), which offered prophylactic DC (within 72 h of TBI) in cases of diffuse (on CT) or severe (on GCS scoring) TBI. 155 cases were randomized to surgical and control arms, where ICP was raised ≥ 20 mmHg for > 15 min within an hour period—despite first medical treatment (i.e., excluding hypothermia and barbiturates use). Importantly, patients who had intracranial hematoma evacuation without primary DC, patients who had a hemiDC were excluded and patients with bilaterally unreactive pupils were included (forming 27% of surgical arm patients vs 12% of the control group). Mortality rates at 6 months were similar at 19% in the surgery group and 18% medical group. Disability at 6 months in terms of RESCUE-ICP defined unfavorable outcomes were 37% in the surgical group vs 23% in the medical arm, whereas favorable were 44% in the surgical arm compared to 59% in the control cohort. After post hoc adjustment for baseline pupil reactivity, functional differences between the two arms were found to be no longer statistically significant. Similar to the POLAR and Eurotherm3235 studies, it appears that unstratified prophylactic use of this intervention does not confer benefit to the patients (BTF Guidelines, 2016; Khellaf, Khan, & Helmy, 2019).

RESCUE-ICP (recruitment 2004–2014) was an international prospective RCT comparing DC (bifrontal or large unilateral) plus medical management with medical management alone as therapy for TBI patients with severe, sustained and refractory intracranial hypertension. About 408 patients were included and randomized when all medical management, other than barbiturate coma, were exhausted and ICP remained elevated > 25 mmHg for 1–12 h. Importantly, patients with fixed bilateral pupils, unsurvivable injury, bleeding diathesis or those treated with primary DC were excluded. At 6 months after randomization, secondary DC resulted in lower mortality rates (26.9% vs 48.9% in the medical group). Functional outcomes at 6 months were measured via GOS-E Outcomes graded as upper severe disability or better were categorized as favorable and were seen in 42.8% of surgical cases vs 34.6% of medical cases. Unfavorable outcomes (lower severe disability and vegetative state) were 30.4% of surgical cases vs 16.5% of medical cases (Andrews et al., 2018). Of note, incidence of persistent vegetative state was 8.5% in the surgical group compared to 2.1% in the medical group. RESCUE-ICP demonstrated that as last tier therapy for raised ICP, DC can reduce mortality, though at a cost of increased severe disability and persistent vegetative status. There is no simple answer as to when DC should be used in severe TBI with medically refractory ICP; however, RESCUE-ICP provided some clear data on the expected outcomes in this context informing the, often complex, discussions with family when this intervention is considered. This study was published after release of the 4th edition of the BTF guidelines.

One consequence of DC that is undergoing increasing interest, is the need for cranioplasty (replacement of the bony defect with an artificial plate). There is no consensus on the timing of cranioplasty, the cranioplasty material (e.g., titanium, polyethylene, methyl methacrylate, and hydroxyapatite) or the use of autologous bone. It is hoped that initiatives such as the NIHR Global Health Research Group on Neurotrauma and the Collaborative European NeuroTrauma Effectiveness Research in Traumatic Brain Injury (CENTER-TBI) shed light on these questions across resource settings (Cooper et al., 2018).

In August 2020, Gregory W. J. Hawryluk, et al. published the update to the decompressive craniectomy with the finding in RESCUE icp study and the recently published 12-mo outcome data from the DECRA (Decompressive Craniectomy in Patients With Severe TBI) trial. Both trials studied the secondary DC for the treatment of refractory ICP elevation, but the difference in their protocols is that they were designed to investigate conditions of early refractory ICP elevations (DECRA) and late refractory ICP elevations (RESCUE icp).

The new update recommendations after addition of these studies are: (1) Secondary DC performed for late refractory ICP elevation is recommended to improve mortality and favorable outcomes. (2) Secondary DC performed for early refractory ICP elevation is not recommended to improve mortality and favorable outcomes.

In the TBI field, DC remains a controversial topic. DECRA and RESCUEicp, are complex trials, and they, together or separately, have not provided definitive evidence about the performance of DC. The most important conclusion of both studies is that choosing to perform a DC, is a not simple decision and is very important that the benefits should be balanced against the complications and outcomes on a case-by-case basis.

Despite several decades of successful pre-clinical studies that have developed promising neuroprotective therapies, none have widely translated into the clinical scenario, because there is still no proven curative pharmacotherapy for moderate-to-severe TBI nor pharmacotherapy with confirmed benefit in functional outcomes (Evans, Butler, Schofield-Robinson, & Alderson, 2017).

Corticosteroids

The first large scale pharmacological trial in TBI was the MRC-funded (Medical Research Council, United Kingdom) Corticosteroid Randomization after Significant Head Injury study (MRC CRASH, 2004), which randomized moderate-to-severe TBI patients to a 48-h infusion of high dose corticosteroids (methylprednisolone) or placebo. About 10,008 patients were recruited from 1999 to 2004 from 239 hospitals in 49 countries. The trial primary outcome was 2-week mortality, which was higher in the treatment group (21.1%) than in the placebo group (17.9%). At 6 months of follow-up, there were 173 excess deaths in the treatment arm (1248 vs 1075). The complications of corticosteroids such as hyperglycemia and immunosuppression were not considered in this study (Roberts et al., 2004).

Progesterone

With the failure of corticosteroids in TBI, preclinical animal studies focus on the early administration of progesterone, a potent steroid synthesized in the central nervous system, and showed that it reduced neuronal loss, cerebral edema, and behavioral deficits after experimental TBI. Progesterone has then been investigated by large double-blind placebo-controlled phase III multi-center RCTs (SYNAPSE and PROTECT III published in 2014) without demonstrating benefit in patient mortality and functional outcomes, halting enthusiasm generated by two precedent single-center clinical studies (Wright et al., 2007).

Erythropoietin

Erythropoietin (EPO) is a glycoprotein regulating hematopoiesis in the bone marrow which is naturally produced in the kidneys following hypoxic stimulation. Animal studies have demonstrated that EPO can neutralize the neuronal apoptotic program, reduce the inflammatory response and act as a neurotrophic factor, thus it is hypothesized to alleviate the impact of secondary brain injury in TBI. A meta-analysis of 6 RCTs with 1041 patients (up to January 2017) looking at outcomes of EPO-treated patients vs untreated patients following acute (moderate-to-severe) TBI showed that EPO significantly reduced mortality but did not reduce rates of poor functional outcome. There were no significant differences in complication rates including deep venous thrombosis between the treatment groups (Aloizos et al., 2015).

Amantadine

Amantadine hydrochloride works as an *N*-methyl-D-aspartate (NMDA) antagonist and indirect dopamine agonist. A prospective multi-center double-blind RCT examined amantadine administration vs placebo over 4 weeks followed by 2 weeks in which treatment was discontinued in 184 adult patients who were in a vegetative or minimally conscious state 4 to 16 weeks after TBI and who were receiving inpatient rehabilitation. The study concluded that amantadine accelerated the pace of functional recovery as measured by the Disability Rating Scale (DRS) during the active 4-week treatment without significant difference in the incidence of serious adverse events in patients with post-traumatic disorders of consciousness. Another small single-center double-blind RCT evaluating amantadine vs placebo given for 6 weeks in 40 severe TBI patients did not show any significant effect on patient mortality and functional outcomes at 6 months. A multi-center double-blind RCT evaluated the efficacy of amantadine vs placebo administered for 60 days on cognitive function in 119 patients with chronic TBI (> 6-month post-injury). This study failed to show benefits from amantadine administration in chronic TBI and even suggested that amantadine may hinder cognitive processing within first 28 days of use (Ghalaenovi et al., 2018; Hammond et al., 2018).

Tranexamic acid

Tranexamic acid (TXA), a synthetic derivative of the amino acid lysine, is an anti-fibrinolytic agent used to reduce active bleeding. It works by reversible blockade of the lysine sites on plasminogen. In about 30% of TBI patients, laboratory markers for coagulation are altered; frequently peri-hemorrhage edema is associated with worse neurological outcomes.

In the trauma setting TXA administration has been associated with decreased mortality particularly if used early, elucidated by the CRASH-2 trial. This trial randomized 20,211 adult trauma patients with, or at risk of, significant bleeding randomized within 8 h of injury to either TXA or matching placebo.

A nested study within CRASH-2 examining the subgroup of trauma patients with extracranial bleeding and TBI with abnormal CT brain findings found a non-significant trend towards reduction in hemorrhage growth, ischemic lesions, and mortality with TXA administration.

These findings have formed the foundation for the Tranexamic Acid for Significant Traumatic Brain Injury (CRASH-3) trial, a RCT aimed at evaluating whether the benefits of TXA are transferable to TBI patients with traumatic intracranial bleeding, currently ongoing (Dewan et al., 2012).

Citicoline

Citicoline is a cholinergic agent which can increase formation of ATP—hypothesized to promote the functioning of cell-membrane ATP-dependent pumps and thus increase cell-membrane integrity and decrease cellular edema It was considered a potential therapeutic agent against parts of secondary brain injury process. One multicenter double-blind randomized phase III Citicoline Brain Injury Treatment Trial failed to demonstrate improvement in functional and cognitive status at 90-day post-injury from the use of citicoline vs placebo (Tan, Wasiak, Rosenfeld, O'Donohoe, & Gruen, 2014).

Anti-inflammatory therapies

Recombinant interleukin-1 receptor antagonist (rIL1ra) has demonstrated putative benefit in a range of neuronal pathologies by inhibition of the IL1 receptor mediated inflammatory cascade. In TBI it has been shown to be safe and modify the acute neuroinflammatory response in a phase II single-center RCT. A dose ranging study to optimize both the dose and timing of administration is currently in progress (Helmy, 2017).

Role of the ultrasound. Transcranial Doppler TCD

Brain tissue injury and cerebral hemodynamic disturbance with elevated ICP are important contributors to poor outcome in patients with TBI. According to changes in cerebral flow velocity in basal brain arteries, TCD monitoring of patients with TBI may reveal numerous pathologic hemodynamic changes such as the presence of cerebral hyperemia, vasospasm, and ischemia. Therefore, the purpose of our research was to compare the clinical outcomes of routine ICP monitoring with ICP combined with TCD monitoring in patients with TBI after DC operation. The results showed that TCD could be helpful for the serial monitoring of cerebral hemodynamic changes after DC for TBI, which could be beneficial for neurologic outcome improvement. Unconditional multivariable logistic regression results showed that no other factor, including clinic characteristics, was independently associated with GOS-E score at 6 months.

Optimizing CPP is essential to maintain optimal CBF. Guidelines of BTF recommend CPP values of 50–70 mmHg and ICP < 20 mmHg for the management of TBI. Although an ICP < 20 mmHg is acceptable after DC, it remains compromised in cerebrospinal compensation, peripheral cerebrovascular resistance, and micro-vascular perfusion. Adequate individual targets are still difficult to define because patients have different perfusion thresholds. ICP and TCD measurements can be combined to identify patients at risk for secondary neurologic deterioration post-DC to improve their prognosis.

The results showed that TCD and Duplex mode monitoring after DC could assess the cerebral hemodynamics and treat the abnormal cerebral hemodynamics, which may help reduce the extent of secondary ischemic injuries in patients with TBI and elevated ICP.

Despite the use of TCD measurements in the management of patients with TBI remains controversial. TCD allows a better understanding of the ICP and CBF and is practical for serially monitoring the progression of cerebral hemodynamics after DC. It could be a useful bedside tool for TBI management after DC.

Hypothermia

Hypothermia can mitigate the effects of TBI including reducing ICP, reducing the innate inflammatory response and reducing the cerebral metabolic rate.

Two large phase 3 randomized trials have attempted to show the putative benefit of therapeutic hypothermia. The multicenter non-blinded RCT Eurotherm3235 (2015) is the largest trial on hypothermia for patients with intracranial hypertension (> 20 mmHg) after TBI. This study showed that therapeutic moderate hypothermia (32–35 °C) plus standard care to reduce ICP led to slightly increased mortality rates and unfavorable functional outcomes compared to those with standard care alone. Adverse association between hypothermia and worsening multiple organ failure was confirmed. A Cochrane Review on March 2017 on mild hypothermia in severe brain injury, which included 37 studies with 3110 participants, demonstrated no high-quality decrease of mortality and morbidity in patients with severe TBI Despite the evidence, hypothermia is routinely used in many units with two staged therapeutic targets, 35 °C and 33 °C.

More recently, the multi-center Prophylactic Hypothermia Trial to Lessen Traumatic Brain Injury—Randomized Clinical Trial (POLAR-RCT, 2018) (xxxiv) evaluated outcomes of early, prophylactic, sustained hypothermia (33–35 °C), for at minimum 72 h and up to 7 days, in severe TBI patients.

This study has been criticized for the use of severe hypothermia in patients prophylactically, even in those without raised ICP. In this group, the risks of hypothermia may exceed the putative benefits.

Discussion

The new knowledge on TBI due in a large part by the widespread establishment of specialized neurointensive care focused all the effort in improving outcomes since 1990. Continuous updating treatments under the new experience give us advanced management and research in severe TBI. Improving care protocols and new therapies in parallel with pro evidence for decompressive craniectomy and drugs therapies were the first steps and objectives since the last 20 years.

Despite these efforts and exciting approaches in brain monitoring and treatments during the 90s, the experience in the last years shows that the need of outcomes improving in terms of functional recovery is longer to be satisfied.

We learned that brain monitoring is only a tool and new approaches focused on restorative therapy are mandatory. So all of the efforts must go on this direction. Protocols must be updated again including this concept on the basis of the evidence.

Applications in other areas of neuroscience

Monitoring TBI with modern neuroimaging technologies, including functional magnetic resonance imaging (fMRI), diffusion tensor imaging (DTI), positron emission tomography (PET), and high definition fiber tracking (HDFT) increasing sensitivity and specificity, are the basis of the new approach of this pathology.

Physiological monitoring, together with newly established cerebral micro-dialysis devices and techniques, are also beneficial for TBI. Molecular biomarkers under the knowledge of genomic and proteomic changes following TBI, have proven to be useful. Specific biomarkers in head trauma are being developed with the combination of systems related biology approaches.

Biomarkers are currently not performed routinely outside of clinical research contexts. The most important biomarkers in TBI include glia-related biomarkers (GFAP, S100B), neuron/axon-related biomarkers (neuron-specific enolase [NSE], neurofilament light polypeptide [NFL], ubiquitin carboxy-terminal hydrolase [UCH-L1], tau, amyloid β, αII-Spectrin breakdown products among others) and other inflammation-related biomarkers (high mobility group box protein 1 [HMGB1], various cytokines and autoantibodies). Up to now, only S100B is part of a consensus guideline pathway (by the Scandinavian Neurotrauma Committee) for stratification of mild TBI patients at presentation for neuroimaging (Undén, Calcagnile, Undén, Reinstrup, & Bazarian, 2015).

Protein biomarkers with a shorter serum half-life (t1/2), e.g., S100B (t1/2–24 h) are likely more useful than proteins with a longer serum half-life, e.g., NSE (t1/2–48–72 h). A longer half-life offers a longer post-injury window for the detection of secondary neurological insults in severe TBI.

Advances in restoratives treatments also include stem cells and nanotechnology like combining methods; supported by new physical and pharmaceutical approaches and also the use of new drugs which all present favorable neuroprotection and restoring in TBI.

In other field neuromodulation with transcranial magnetic stimulation or direct current stimulation are promising but are still at a relatively early stage (Clayton, Kinley-Cooper, Weber, & Adkins, 2016; O'Brien et al., 2018). It is likely to be some years before sufficiently robust studies are available for TBI-specific evidence-based clinical recommendations.

Pharmacological interventions promoting neuroplasticity are of intense current interest for stroke patients more than trauma. The effects of drugs developed for other purposes are attractive. One example is identification of positive effects of Fluoxetine on motor recovery in non-depressed patients after ischemic stroke, with current studies extending recruitment to also include hemorrhagic stroke. It is unclear whether similar clinical benefits occur after TBI; several of the important mechanisms are certainly relevant after TBI and the simplicity of such an intervention is attractive (Mead et al., 2015). In addition to local neuroinflammatory processes within the brain after TBI, systemic immune and inflammatory processes are also impacted and have recently been reviewed. Peripheral immune cells are mobilized from the bone marrow, thymus, and spleen early after the TBI. While there is a brief recovery in thymocytes and classical monocytes in thymus and spleen within the first 2 weeks post injury, over time these cell populations decline again and may be responsible for the post-TBI immune suppression that is observed in patients. Also implicated in the peripheral immune response and long-term immunosuppression is both the activation of the hypothalamus-pituitary–adrenal axis, through the release of glucocorticoids and the sympathetic nervous system, through release of catecholamines.

Ideally, inflammation will be a self-limiting process. There are endogenously formed products of arachidonic acid metabolism that actively inhibit pro-inflammatory responses. These include the lipoxins, resolvins, protectins, and maresins, which decrease pro-inflammatory cytokine secretion, alter migratory signals for peripheral immune cells, and stimulate neuroprotective and tissue regeneration processes. There has been very little research on these resolving mediators in TBI, but administering the lipoxin, LXA4, into the lateral ventricles 10 min after a TBI induced by weight drop, reduced pro-inflammation, BBB disruption, and lesion size. Further, its receptor, ALX/FPR2, is up-regulated in astrocytes (Hazeldine, Lord, & Belli, 2015).

Unfortunately, in many instances inflammation may not resolve and become chronic. In the aged brain, there is increased recruitment of peripheral macrophage into the TBI brain In addition to cellular infiltration of pro-inflammatory immune cells after TBI, sustained complement C3 activation leads to chronic inflammation by activating microglia and astrocytes in the region around the initial lesion and contributes to further neuronal loss 30 days post-TBI. Using a combination of magnetic resonance imaging, magnetic resonance spectroscopy, and positron emission tomography in a rat lateral fluid percussion model, inflammation was shown to be present still 6 months post-TBI (Nizamutdinov & Shapiro, 2017). Stem cells therapy is widely investigated and also the clinical applications are mostly in Phase 1 level. Cells and their environment with a lot of cytokines Il1–10 chemokine (C–C motif) ligand 20 (CCL20) are the focus of the experimental models. Investigators administered a calpain inhibitor 30 min after TBI and then transplanted the MSCs at 24 h post-TBI.134 There were significant decreases in pro-inflammatory cytokines around the lesion, increased survival of the MSCs.

The first major cell therapy trials for TBI using BMMNCs (bone marrow mononuclear cells) were conducted on children by Cox and colleagues in 2011 (Bisicchia, Sasso, Catanzaro, et al., 2018). In the pre-clinical studies and Phase 1 clinical trial, treatment reduced BBB permeability and neuroinflammation after TBI. In a second pediatric study from the same research group, the TBI-induced increase in intracranial pressure was reduced in the cell treatment group (Cox, Hetz, Liao, et al., 2017; Hu, Chen, Huang, et al., 2019) (Fig. 4).

Study Title	Identifier	Phase	Target Enrollment	Ages	Condition	Primary Outcome	Results	References
Safety of Autologous Stem Cell Treatment for Traumatic Brain Injury in Children	NCT00254722	1	10	5 to 14 y/o	Patients with TBI within 24 h after lesion	Safety of BMPCs autologous Transplant.	-	-
A Double-Blind, Controlled Phase 2 Study of the Safety and Efficacy of Modified Stem Cells (SB623) in Patients with Chronic Motor Deficit From Traumatic Brain Injury (TBI)	NCT02416492	2	61	18 to 75 y/o	Patients with motor deficits from TBI at least 12 months	It has been assessed the motor functions in patients with chronic TBI after stereotactic intracranial implantation of allogeneic SB623	The study results showed a significant improvement in motor function after 24 weeks of intracranial administration of SB623 cells. Furthermore, 5 severe side events were reported in patients treated with SB623 and 3 in the control patients. By contrast, there were no differences in the rate of adverse events emerging from the treatment between the two groups of patient	https://clinicaltrials.gov/
Treatment of Severe Adult Traumatic Brain Injury Using Bone Marrow Mononuclear Cells	NCT01575470	1/2	25	18 to 55 years	Patients with severe TBI within 36 h after injury	It was shown the safety of BMMNCs autologous transplantation after TBI	None of the study participants have shown severe side events after collection as well as the administration of BMMNCs	https://clinicaltrials.gov/

FIG. 4 Current protocols on stem cells therapy in brain injury. https://clinicaltrials.gov/2020.

(Continued)

							transplantation. Likewise, cell therapy has demonstrated the structural conservation of the critical regions of interest brain tissue. Moreover, the authors have reported a decrease in the inflammatory response, as well as a statistically significant reduction in IL-1β, IFN-γ, and TNF-α following autologous BMMNCs transplantation.	
Treatment of Adult Severe Traumatic Brain Injury Using Autologous Bone Marrow Mononuclear Cells	NCT02525432	2	55	18 to 55 years	Patients with TBI within 24 h after injury	Will be evaluated both the macrostructural and microstructural properties of gray matter, white matter as well as the integrity of the regions in the corpus callosum	-	
Neurologic Bone Marrow Derived Stem Cell Treatment Study	NCT02795052	-	300	Over 18 years	Patients with damage to the central or peripheral nervous system	Will be assessed the efficacy of intravenous and intranasal administration of autologous BM-MSCs by measuring the Activities of Daily Living at 3,6 and 12 months	-	-

FIG. 4, CONT'D

						following the procedure		
A Phase 2 Multicenter Trial of Pediatric Autologous Bone Marrow Mononuclear Cells (BMMNCs) for Severe Traumatic Brain Injury (TBI)	NCT01851083	1/2	50	5 to 17 years	Patients with severe TBI within 24 h after injury	Will be evaluated by using diffusion tensor MRI the conservation of both white and grey matter in the groups of patients treated and untreated after the injury	-	-
A Clinical Trial to Determine the Safety and Efficacy of Hope Biosciences Autologous Mesenchymal Stem Cell Therapy for the Treatment of Traumatic Brain Injury and Hypoxic-Ischemic Encephalopathy	NCT04063215	1/2	24	18 to 55 years	Patients with neurological injury at least 6 months	Will be evaluated the safety of autologous AD-MSCs transplantation in brain regions associated with specific neurocognitive deficits of patients with both subacute and chronic neurological injury	-	-

FIG. 4, CONT'D

(Continued)

Use of Adipose-Derived Cellular Stromal Vascular Fraction (AD-cSVF) Parenterally in Post-Concussion Injuries and Traumatic Brain Injuries (TBI)	NCT02959294	1/2	200	16 to 70 years	Patients with mild TBI or concussion syndrome at least 1 months	The safety of AD-cSVF in patients with TBI and concussion syndrome will be assessed by measuring the side events at baseline and 6 months. Moreover, it will be assessed the clinical symptoms associated with TBI and concussion such as recurrent headaches, amnesia, behavioral change, cognitive impairment, as well as sleep disturbances.	-	-
Non-Randomized, Open-Labeled, Interventional, Single Group, Proof of Concept Study with Multimodality Approach in Cases of Brain Death Due to Traumatic Brain Injury Having Diffuse Axonal Injury	**NCT02742857**	**1**	**20**	**15 to 65 years**	**Patients who present brain death caused by TBI with axonal injury shown by MRI**	**Will be demonstrated the reversal of brain death documented by clinical examination or electroencephalogram in the 15 days' frame time**	-	-

BMPCs: bone marrow progenitor cells; SB623: modified bone marrow-derived mesenchymal stem cells; BMMNCs: bone marrow mononuclear cells; IL-1β: interleukin-1beta; IFN-γ: interferon-γ; TNF-α: tumor necrosis factor-α; BM-MSCs: bone marrow mesenchymal stem cells; MRI: magnetic resonance imaging; AD-MSCs: adipose tissue-derived MSCs; AD-cSVF: adipose-derived stromal vascular fraction.

FIG. 4, CONT'D

Mini dictionary terms

- Lund Protocol: Standardized treatment on raised ICP focused on brain volume and perfusion.
- The complex vasodilatory/vasoconstriction cascade: vasoreactivity response to head injury.
- ICP-targeted goal: treatment of brain injury focus on raised ICP.
- perfusion -targeted goal: treatment focus on brain oxygenation.
- Recombinant interleukin-1 receptor antagonist (rIL1ra): Anti-inflammatory therapies by inhibition of the IL1 receptor mediated inflammatory cascade.
- TBI glia-related biomarkers: (GFAP, S100B), neuron/axon-related biomarkers (neuron-specific enolase [NSE], neurofilament light polypeptide [NFL], ubiquitin carboxy-terminal hydrolase [UCH-L1], tau, amyloid β, αII-Spectrin breakdown products among others) and other inflammation-related biomarkers (high mobility group box protein 1 [HMGB1], various cytokines and autoantibodies.

Key facts

- Traumatic brain injury.
- Brain monitoring.
- Intracranial pressure.
- Cerebral perfusion Brain.
- Brain Trauma Foundation.

Summary points

- Head trauma is a very prevalent pathology worldwide, which has implications at the health and can generate severe neurological disabilities, limiting the life of the individual.
- Intracranial hypertension is a determining factor in neurological damage; therefore, studies have been carried out and continue to be carried out to improve the monitoring, control and treatment.
- Between 1992 and 1994, at the Lund University Hospital in Sweden was developed "The Lund Concept" and was introduced for treatment of severe head trauma. Its main characteristics are based on hypotheses originating from basic physiological principles regarding control of brain volume and cerebral perfusion.
- In 1996, the BTF made guidelines that address key issues relating to the management of severe head injury in adult patients with Glasgow Coma Scale score of 3 to 8. The guidelines comprised different aspects about head injury like system trauma, pre-hospital resuscitation, monitoring and treatment of intracranial hypertension and intensive care.
- After the publication of a randomized trial of ICP monitoring, the latest edition of the BTF guidelines mentioned that there was insufficient evidence to support recommendation for ICP monitoring. However, the management of severe TBI patients using information from ICP monitoring is recommended to reduce in-hospital and 2-week post injury mortality. The BTF guidelines are US based, and not practiced universally.
- Despite this efforts and exciting approaches in brain monitoring and treatments during the 90s, the experience in the last years shows that the need of outcomes improving in terms on functional recovery is longer to be satisfied.

References

Ahl, R., Sarani, B., Sjolin, G., & Mohseni, S. (2018). The association of intracranial pressure monitoring and mortality: A propensity score-matched cohort of isolated severe blunt traumatic brain injury. *Journal of Emergencies, Trauma, and Shock*, *12*, 18–22 (Google Scholar).

Aloizos, S., Evodia, E., Gourgiotis, S., Isaia, E. C., Seretis, C., & Baltopoulos, G. J. (2015). Neuroprotective effects of erythropoietin in patients with severe closed brain injury. *Turkish Neurosurgery*, *25*(4), 552–558.

Andrews, P. J., Sinclair, H. L., Rodriguez, A., Harris, B., Rhodes, J., Watson, H., et al. (2018). Therapeutic hypothermia to reduce intracranial pressure after traumatic brain injury: The Eurotherm3235 RCT. *Health Technology Assessment (Winchester, England)*, *22*(45), 1–134.

Asgeirsson, B., Grände, P. O., & Nordström, C. H. (1994). A new therapy of post-trauma brain oedema based on haemodynamic principles for brain volume regulation. *Intensive Care Medicine*, *20*, 260–267.

Austin, P. C. (2011). An introduction to propensity score methods for reducing the effects of confounding in observational studies. *Multivariate Behavioral Research*, *46*, 399–424 (PMC free article) [PubMed] [Google Scholar).

Bisicchia, E., Sasso, V., Catanzaro, G., et al. (2018). Resolvin D1 halts remote neuroinflammation and improves functional recovery after focal brain damage via ALX/FPR2 receptor-regulated microRNAs. *Molecular Neurobiology*, *55*(8), 6894–6905.

Brain Trauma Foundation. (2016). *Guidelines for the management of severe traumatic brain injury*. New York: Mary Ann Liebert Inc Publications.

Bullock, R., Chesnut, R. M., Clifton, G., et al. (1996). Guidelines for the management of severe head injury. Brain trauma foundation. *European Journal of Emergency Medicine*, *3*(2), 109–127.

Campbell, S., Watkins, G., & Kreis, D. (1989). Preventable deaths in self-designated trauma system. *The American Surgeon*, *55*, 478–480.

Carney, N., Totten, A. M., O'Reilly, C., Ullman, J. S., Hawryluk, G. W., Bell, M. J., et al. (2017). Guidelines for the management of severe traumatic brain injury, fourth edition. *Neurosurgery*, *80*, 6–15 (PubMed) [Google Scholar).

Chesnut, R. M., Bleck, T. P., Citerio, G., Classen, J., Cooper, D. J., Coplin, W. M., et al. (2015). A consensus-based interpretation of the benchmark evidence from South American trials: Treatment of intracranial pressure trial. *Journal of Neurotrauma*, *32*, 1722–1724 (PubMed] [Google Scholar).

Chesnut, R. M., Marshall, L. F., Klauber, M. R., et al. (1993). The role of secondary brain injury in determining outcome from severe head injury. *The Journal of Trauma*, *34*, 216–222.

Chesnut, R. M., Temkin, N., Carney, N., Dikmen, S., Rondina, C., Videtta, W., et al. (2012). A trial of intracranial-pressure monitoring in traumatic brain injury. *The New England Journal of Medicine*, *367*, 2471–2481 (PMC free article] [PubMed] [Google Scholar).

CHIRAG (Collaborative Head Injury and Guidelines) Study. (2018). Patients who benefit from intracranial pressure monitoring without cerebrospinal fluid drainage after severe traumatic brain injury. *Neurosurgery* (Epub ahead of print] [PubMed] [Google Scholar).

Clayton, E., Kinley-Cooper, S. K., Weber, R. A., & Adkins, D. L. (2016). Brain stimulation: Neuromodulation as a potential treatment for motor recovery following traumatic brain injury. *Brain Research*, *1640*, 130–138.

Cooper, D. J., Nichol, A. D., Bailey, M., Bernard, S., Cameron, P. A., Pili-Floury, S., et al. (2018). Effect of early sustained prophylactic hypothermia on neurologic outcomes among patients with severe traumatic brain injury: The POLAR randomized clinical trial. *Journal of the American Medical Association*, *320*(21), 2211–2220.

Cox, C. S., Hetz, R. A., Liao, G. P., et al. (2017). Treatment of severe adult traumatic brain injury using bone marrow mononuclear cells. *Stem Cells*, *35*(4), 1065–1079.

Dewan, Y., Komolafe, E. O., Mejía-Mantilla, J. H., Perel, P., Roberts, I., & Shakur, H. (2012). CRASH-3-tranexamic acid for the treatment of significant traumatic brain injury: Study protocol for an international randomized, double-blind, placebo-controlled trial. *Trials*, *13*(1), 87.

Eker, C., Asgeirsson, B., Grande, P. O., Schalen, W., & Nordstrom, C. H. (1998). Improved outcome after severe head injury with a new therapy based on principles for brain volume regulation and preserved microcirculation. *Critical Care Medicine*, *26*(11), 1881–1886.

Evans, D. J., Butler, A. R., Schofield-Robinson, O. J., & Alderson, P. (2017). Hypothermia for traumatic brain injury. *The Cochrane Database of Systematic Reviews*, *9*(9), CD001048.

Fearnside, M. R., Cook, R. J., McDougall, P., et al. (1993). The Westmead head injury project outcome in severe head injury. Comparative analysis of pre-hospital, clinical and CT variables. *British Journal of Neurosurgery*, *7*, 267–279.

Ghalaenovi, H., Fattahi, A., Koohpayehzadeh, J., Khodadost, M., Fatahi, N., Taheri, M., et al. (2018). The effects of amantadine on traumatic brain injury outcome: A double-blind, randomized, controlled, clinical trial. *Brain Injury*, *32*(8), 1050–1055.

Grände, P. (2006). The "Lund concept" for the treatment of severe head trauma—Physiological principles and clinical application. *Intensive Care Medicine*, *32*, 1475–1484.

Hammond, F. M., Sherer, M., Malec, J. F., Zafonte, R. D., Dikmen, S., Bogner, J., et al. (2018). Amantadine did not positively impact cognition in chronic traumatic brain injury: A multi-site, randomized, controlled trial. *Journal of Neurotrauma*, *35*(19), 2298–2305.

Hazeldine, J., Lord, J. M., & Belli, A. (2015). Traumatic brain injury and peripheral immune suppression: Primer and prospectus. *Frontiers in Neurology*, *6*, 235.

Helmy, A. (2017). *A randomised double blind placebo controlled dose-range study using placebo, 1.5 g and 3.0 g of intravenous recombinant interleukin-1 receptor antagonist (Anakinra) for patients with moderate-to-severe TBI*. US National Library of Medicine, ClinicalTrials.gov.

Hu, J., Chen, L., Huang, X., et al. (2019). Calpain inhibitor MDL28170 improves the transplantation-mediated therapeutic effect of bone marrow-derived mesenchymal stem cells following traumatic brain injury. *Stem Cell Research & Therapy*, *10*(1), 96.

Khellaf, A., Khan, D. Z., & Helmy, A. (2019). Recent advances in traumatic brain injury. *Journal of Neurology*, *266*(11), 2878–2889.

Kreis, D., Jr., Plasencia, G., Augenstein, D., et al. (1986). Prevent able trauma deaths: Dade County, Florida. *The Journal of Trauma*, *26*, 649–654.

Mead, G., Hackett, M. L., Lundstrom, E., Murray, V., Hankey, G. J., & Dennis, M. (2015). The FOCUS, AFFINITY and EFFECTS trials studying the effect(s) of fluoxetine in patients with a recent stroke: A study protocol for three multicentre randomised controlled trials. *Trials*, *16*, 369.

Nizamutdinov, D., & Shapiro, L. A. (2017). Overview of traumatic brain injury: An immunological context. *Brain Sciences*, *7*(1), 11. https://doi.org/10.3390/brainsci7010011.

O'Brien, A. T., Bertolucci, F., Torrealba-Acosta, G., Huerta, R., Fregni, F., & Thibaut, A. (2018). Non-invasive brain stimulation for fine motor improvement after stroke: A meta-analysis. *European Journal of Neurology*, *25*, 1017–1026.

Obrist, W. D., Langfitt, T. W., Jaggi, J. L., et al. (1984). Cerebral blood flow and metabolism in comatose patients with acute head injury. Relationship to intracranial hypertension. *Journal of Neurosurgery*, *61*, 241–253.

Piccinini, A., Lewis, M., Benjamin, E., Aiolfi, A., Inaba, K., Demetriades, D., et al. (2017). Intracranial pressure monitoring in severe traumatic brain injuries: A closer look at level 1 trauma centers in the United States. *Injury*, *48*, 1944–1950 (PubMed] [Google Scholar).

Pigula, F. A., Wald, S. L., Shackford, S. R., et al. (1993). The effect of hypotension and hypoxia on children with severe head injuries. *Journal of Pediatric Surgery*, *28*, 310–314 (discussion 315-16).

Roberts, I., Yates, D., Sandercock, P., Farrell, B., Wasserberg, J., Lomas, G., et al. (2004). Effect of intravenous corticosteroids on death within 14 days in 10,008 adults with clinically significant head injury (MRC CRASH trial): Randomised placebo-controlled trial. *Lancet (London, England)*, *364*(9442), 1321–1328.

Rosner, M. J. (1995). Introduction to cerebral perfusion pressure management. *Neurosurgery Clinics of North America*, *6*(4), 761–773.

Rosner, M. J., & Becker, D. P. (1984). Origin and evolution of plateau waves. *Journal of Neurosurgery*, *60*(2), 312–324.

Rosner, M. J., & Daughton, S. (1990). Cerebral perfusion pressure management in head injury. *The Journal of Trauma*, *30*(8), 933–941.

Rosner, M. J., Rosner, S. D., & Johnson, A. H. (1995). Cerebral perfusion pressure: Management protocol and clinical results. *Journal of Neurosurgery*, *83*(6), 949–962.

Rubiano, A. M., Vera, D. S., & Montenegro, J. H. (2020). Recommendations of the Colombian consensus committee for the management of traumatic brain injury in prehospital, emergency department, surgery, and intensive care (beyond one option for treatment of traumatic brain injury, a stratified protocol [BOOTStraP]). *Journal of Neurosciences in Rural Practice*, *11*(1), 7–22.

Sahuquillo, J., & Dennis, J. A. (2019). Decompressive craniectomy for the treatment of high intracranial pressure in closed traumatic brain injury. *The Cochrane Database of Systematic Reviews*, *12*(12). Dec 31, CD003983.

Smith, R., Frateschi, L., Sloan, E., et al. (1990). The impact of volume on outcome in seriously injured trauma patients: Two years' experience of the Chicago Trauma System (see comments). *The Journal of Trauma*, *30*, 1066–1076.

Suehiro, E. M., Fujiyama, Y., Koizumi, H., & Suzuki, M. (2017). Directions for use of intracranial pressure monitoring in treatment of severe traumatic brain injury using data from the Japan Neurotrauma Data Bank. *Journal of Neurotrauma*, *34*, 2230–2234.

Tan, H. B., Wasiak, J., Rosenfeld, J. V., O'Donohoe, T. J., & Gruen, R. L. (2014). Citicoline (CDP-choline) for traumatic brain injury. *Cochrane Database of Systematic Reviews*, *8*, CD011217.

Undén, L., Calcagnile, O., Undén, J., Reinstrup, P., & Bazarian, J. (2015). Validation of the Scandinavian guidelines for initial management of minimal, mild and moderate traumatic brain injury in adults. *BMC Medicine*, *13*, 292.

Wahlström, M. R., Olivecrona, M., Koskinen, L. O. D., Rydenhag, B., & Naredi, S. (2005). Severe traumatic brain injury in pediatric patients: Treatment and outcome using an intracranial pressure targeted therapy—The Lund concept. *Intensive Care Medicine*, *31*(6), 832–839.

West, J., Trunkey, D., & Lim, R. (1979). Systems of trauma care. Study of two counties. *Archives of Surgery*, *114*, 455–460.

Wright, D. W., Kellermann, A. L., Hertzberg, V. S., Clark, P. L., Frankel, M., Goldstein, F. C., et al. (2007). ProTECT: A randomized clinical trial of progesterone for acute traumatic brain injury. *Annals of Emergency Medicine*, *49*(4), 391–402.

Further reading

NIHR Global Health Research Group on Neurotrauma. (2021). http://neurotrauma.world/. Accessed 1 July 2019 Xxxv.

Chapter 30

Seizures after traumatic brain injury and their treatment

Luiz Fernando Freire Royes[a], Michele Rechia Fighera[b], Ana Flavia Furian[c], M.S. Oliveira[c], Leonardo Magno Rambo[d], and Gabriel Corrêa Lima[a]

[a]*Exercise Biochemistry Laboratory, Center of Physical Education and Sports, Federal University of Santa Maria—UFSM, Santa Maria, RS, Brazil,* [b]*Department of Neuropsychiatry, Laboratory of Experimental and Clinical Neuropsychiatry, Health Sciences Center, Medical School, Federal University of Santa Maria, Santa Maria, RS, Brazil,* [c]*Neurotoxicity and Psychopharmacology Laboratory, Health Sciences Center, Federal University of Santa Maria, Santa Maria, RS, Brazil,* [d]*Exercise Biochemistry and Physiology Laboratory, Biochemistry Graduate Program, Federal University of Pampa, Uruguaiana, RS, Brazil*

List of abbreviations

AED antiepileptic drug
ALK5 activin receptor-like kinase 5
Apaf-1 apoptotic Protease Activating Factor
APOE apolipoprotein E
AQP4 aquaporin-4
BBB blood–brain barrier
CCI cortical impact model
DAI diffuse axonal injury
DG dentate gyrus
FPI fluid-percussion injury
GAD glutamate decarboxylase
IL-1β interleukin-1 beta
ILAE international league against epilepsy
PARV$^+$ parvalbumin positive cells
PBBI penetrating ballistic brain injury
PTE post-traumatic epilepsy
PTS post-traumatic seizure
RNS reactive nitrogen species
ROS reactive oxygen species
TBI traumatic brain injury
TGF-β transforming growth factor-β
TNF-α tumor necrosis factor alpha
FPI fluid percussion injury
PBBI penetrating ballistic brain injury
CCI controlled cortical impact

Introduction

Traumatic brain injury (TBI) is a devastating condition that usually initiates a sequence of neurological disorders that affect people's quality of life. This neuronal injury represents 20% of all symptomatic epilepsies in people aged 15 to 24 years, and surprisingly, there is no evidence of effective pharmacological treatment in preventing or treating symptomatic seizures (Piccenna, Shears, & O'Brien, 2017). These epidemiological data corroborate the idea that seizure onset is associated with

Diagnosis and Treatment of Traumatic Brain Injury. https://doi.org/10.1016/B978-0-12-823347-4.00020-8

post-traumatic epilepsy (PTE) development and suggest that early TBI-induced molecular and cellular changes may be important markers of future epileptogenesis.

The International League against Epilepsy (ILAE) emphasizes that an isolated and unprovoked late seizure associated with TBI meets the criteria for epilepsy (Fisher et al., 2017), although not all post-traumatic seizures (PTSs) occur according to the old definition of epilepsy (two unprovoked epileptic seizures at least 24 h apart). Hence, the complexity of PTSs resulting from the heterogeneous nature of TBI and variable seizure onset makes identifying mechanisms of epileptogenesis a challenging process (Diaz-Arrastia, Agostini, Madden, & Van Ness, 2009). Furthermore, this scenario is even more alarming considering PTSs increase the risk of PTE in the first six months, a risk that remains high for up to 10 years (Fisher et al., 2005). Given these obstacles, it is essential to identify potential acute clinical biomarkers of epileptogenesis by longitudinally evaluating a robust and well-fed sample of patients with moderate–severe TBI.

Specific injury types and admission injury characteristics related to injury severity have been linked to increased seizure susceptibility after brain injury. For instance, the pathophysiology of PTE is closely related to the location, type, and severity of the cranial trauma. Because the electrical activity is more frequent in areas such as the posterior gyrus of the parietal lobe of the cerebral cortex and hippocampus, damage to these structures may interrupt specific neural networks after TBI and lower the threshold for epileptic seizures compared to other areas (Liu & Woodson, 2019). When compared to closed trauma, open cranial trauma is also associated with early PTSs. Thus, the abnormal cortical neuronal discharges evidenced in severe brain damage accidents are much more likely to be associated with epilepsy than milder accidents (Salinsky et al., 2019).

In the last decade, information about the involvement of inflammatory and neurodegenerative pathways in experimental models of PTE has been useful in developing promising drugs and procedures in clinical and preclinical trials. In this context, antiepileptic drugs (AEDs) administered to suppress early seizures and counteract the delayed epileptogenesis have been tested in several TBI models (see Table 1). Despite interrupting the epileptogenic process in TBI animal models, there is still a lack of antiepileptic or disease-modifying therapies that can prevent epilepsy and the consequences of TBI and PTE in humans. In human studies, clinical trial planning is usually performed to control seizures symptomatically, and their mechanisms of action may differ from those associated with the development of PTE and be ineffective in epilepsy treatment (Piccenna et al., 2017). In this sense, the animal models need to be clinically validated, and treatments that show efficacy in one model must be equally effective in the others. Considering that one of the main challenges that limit improved PTE diagnosis and treatment is that the pathophysiology is relatively unknown in humans, we reviewed how changes in neuronal metabolism are associated with the pathophysiology of the epileptic focus post-TBI. Additionally, we discussed the integrative actions of procedures and pharmacological interventions on the post-traumatic process of epileptogenesis.

TABLE 1 Descriptive characteristics of the use of AED in patients with traumatic brain injury.

Study (year)	Subjects	Purpose	Methods/Measurements	Main results
Zimmermann, Diaz-Arrastia, and Vespa (2016)	Patients with moderate and severe TBI	Role of (AED) prophylaxis in patients with severe traumatic brain injury (TBI).	Studies that prospectively compared the rates of post-traumatic seizures in patients who received prophylaxis with AED versus controls.	Early treatment with phenytoin reduced the incidence of seizures in the first 7 days after injury, but did not reduce the risk of late posttraumatic seizures or PTE.
Thompson, Pohlmann-Eden, Campbell, and Abel (2015)	Patients with any severity after TBI	To compare the efficacy of antiepileptic drugs and neuroprotective agents in preventing post-traumatic epilepsy.	Included 10 RCTs (reported in 12 articles) consisting of 2326 people diagnosed with any severity of traumatic brain injury.	There was insufficient evidence about the efficacy or safety of neuroprotective agents compared to placebo or phenytoin.
Huijben et al. (2018)	Patients with any severity after TBI	To assess and quantify variation in perceptions on intensive care unit (ICU) management of patients with TBI in European neurotrauma centers.	It was analyzed 23 questions focused on: (1) circulatory and respiratory management; (2) fever control; (3) use of corticosteroids; (4) nutrition and glucose management; and (5) seizure prophylaxis and treatment.	Although the variation in prophylactic indications for seizures, the levetiracetam was the best choice for treatment of seizures.

TABLE 1 Descriptive characteristics of the use of AED in patients with traumatic brain injury—cont'd

Study (year)	Subjects	Purpose	Methods/Measurements	Main results
Yang, Zheng, Xu, and Wang (2016)	Patients with any severity after TBI	To assess the efficacy and safety of levetiracetam versus phenytoin for seizure prophylaxis following TBI	It was conducted a search of the MEDLINE, EMBASE, and Cochrane library databases to March 2016, and screened original research that included patients with TBI who received levetiracetam.	Levetiracetam does not appear to be superior to phenytoin in efficacy or safety with regard to early or late seizure prophylaxis following TBI. However, no class I evidence was identified.
Bakr and Belli (2018)	Patients with any severity after TBI	The systematic review comparing the efficacy of levetiracetam and phenytoin in reducing the incidence of late PTS	Trials and observational studies comparing the efficacy of phenytoin and levetiracetam in the prevention of late PTS	Long-term use of levetiracetam is more effective than phenytoin.
Wat et al. (2019)	Patients with severe TBI	Association between the use of prophylactic AEDs, vs placebo or no treatment, and risk of early seizures after TBI.	Included 3 randomized controlled trials (RCTs) (750 patients) and 6 observational studies (3362 patients).	Protective association between AEDs: phenytoin, levetiracetam, and valproate (mostly phenytoin) and early PTS.
Zangbar et al. (2016)	Retrospective cohort Patients after severe TBI	To evaluate if the treatment with levetiracetam is effective in preventing early post-traumatic seizures	5551 trauma patients suffered a TBI during the study period.	Levetiracetam, phenytoin, or fosphenytoin prophylaxis was ineffective in preventing seizures.
Khan et al. (2016)	Patients diagnosed with severe TBI	Systematic review and meta-analysis to compare levetiracetam with phenytoin for seizure prophylaxis	Search of several electronic databases and bibliographies	Patients treated with either levetiracetam or phenytoin have similar incidences of early seizures after TBI
Harris, Hateley, Tsang, Wilson, and Seemungal (2020)	Patients with any severity after TBI	To compare length of hospital stay in TBI patients receiving the commonest AEDs: Phenytoin and Levetiracetam	Retrospective observational study	Levetiracetam is preferable to Phenytoin for early seizure prophylaxis in TBI
DJohn, Ibrahim, Patel, DeHoff, and Kolbe (2020)	Patients with mild, moderate, and severe TBI	To evaluate the appropriateness of Levetiracetam use in patients with mild, moderate, and severe TBI.	Retrospective cohort study	Patients with mild and moderate TBI probably do not need Levetiracetam prophylaxis. If there is no documented seizure activity during hospitalization, the duration of Levetiracetam therapy should not exceed 7 days.

Pathophysiology of TBI-induced epileptogenesis

Epidemiological and clinical presentation

Epilepsy is a common debilitating consequence of brain injury and contributes to progressive neurodegeneration after TBI. Despite being treated as a "silent epidemic," PTE represents 20% of symptomatic epilepsy in the general population in which approximately 80% of individuals who develop PTE have their first seizure within the first 12 months post-injury, and over 90% by the end of the second year (Roozenbeek, Maas, & Menon, 2013). After the first late-onset seizure (> 1 week. from injury), 86% of patients have reported a second seizure within 2 years (Christensen, 2015).

In order to clarify this point, the PTSs were classified either as early (occurring within 7 days post-injury) or late (occurring after 7 days) events, which are relatively common in TBI patients (Gupta et al., 2014) and sufficient to diagnose PTE. Furthermore, the location of the lesion is related to a high incidence of initial seizures and the longitudinal development of PTE. Despite the injury severity characteristics being compatible, patients who develop PTE have more significant chronic atrophy of the temporal lobe than those who do not develop PTE (Tubi et al., 2019) suggesting that the disruption of normal neural processing may lead to epileptogenic network in the temporal lobe and its underlying structures.

Another critical determinant for PTE is the severity of TBI. In a population-based clinical study of 4541 TBI cases, the cumulative five-year probability of unprovoked seizures was 0.7% in patients with mild TBI, 1.2% for moderate TBI, and 10.0% for severe TBI (Annegers, Hauser, Coan, & Rocca, 1998). Regarding the lesion characteristics with a delayed post-traumatic crisis, it is important to note that some injuries, such as subarachnoid or intra-ventricular hemorrhages that do not require evacuation, do not seem to put individuals at risk of PTE development (de Vries et al., 2016). On the other hand, injuries including dural penetration, depressed cranial fracture, intracranial hematoma, and loss of consciousness or amnesia for more than 1 day (Piccenna et al., 2017) have important prognostic implications for PTE development.

It has long been recognized that early seizures caused by hemorrhage are events associated with a local immune response characterized by the activation of microglia and astrocytes and the release of pro-inflammatory cytokines and chemokines. The hypothesis that neuroinflammation is associated with seizure activity after TBI arose from clinical evidence of immune activation evident in the epileptic foci of brain tissue in epilepsy patients (de Vries et al., 2016). In this scenario, PTSs trigger rapid activation of nearby glial cells, which respond by increasing the inflammatory mediator production to promote an inflammatory immune cascade. After TBI, blood–brain barrier (BBB) dysfunction (endothelial rupture of cells and infiltration of red blood cells) and infiltration of leukocytes derived from blood to the brain are events that influence PTE development throughout the following phases. The first phase is the initial insult and associated molecular and cellular mechanisms within the next minutes to days. During the second phase (days to years after the initial injury), the injury-induced neurogenesis and neuroplasticity contribute to post-injury recovery while ongoing secondary injury processes promote neurodegeneration (Sharma et al., 2019).

Recent clinical evidence has suggested that certain genetic variants may also increase the risk of PTE. In general, the genes involved in TBI can be roughly categorized into those that influence the extent of the injury (e.g., pro- and anti-inflammatory cytokines) and those that affect repair and plasticity (e.g., neurotrophic genes). Although genetic studies on TBI-induced pathophysiology are still in their early stages, clinical data have shown that early PTSs are associated with low plasma concentrations of valproic acid (a first-line antiepileptic drug used after TBI) and UGT1A6/CYP2C9 genetic polymorphisms in patients with severe traumatic brain injury (Sun et al., 2017). In patients with moderate or severe TBI, the inheritance of the apolipoprotein E (APOE) epsilon 4 allele is associated with increased risk of late PTSs suggesting that therapeutic manipulation of lipid or lipoprotein metabolisms in the brain may be useful as anti-epileptogenic therapy (Diaz-Arrastia et al., 2003).

Translational considerations of PTE mechanisms

Regarding the TBI-induced physiopathology, this neurological injury is an inherently heterogeneous condition that leads to debilitating secondary injuries. Therefore, TBI animal models are designed to produce a relatively homogeneous lesion and numerous TBI experimental models have been employed to investigate PTE (see Table 1). Some of these models include the weight-drop model (also referred to as the impact-acceleration model), controlled cortical impact model (CCI; involving focal damage to the cortical surface), fluid-percussion injury (FPI); causing both focal damage and diffuse axonal injury), and blast injury (highly relevant for military personnel). The reproducibility and robustness of these parameters facilitate the study of biological mechanisms and the investigation of new therapies. For example, the laceration/intracranial hemorrhage resulting from contact injury, diffuse axonal injury (DAI), and cerebral edema after diffuse brain injury due to the acceleration/deceleration of the brain injury (Flierl et al., 2009; Golarai, Greenwood, Feeney, & Connor, 2001) are events associated with primary brain damage that activate secondary pathophysiological cascades that result in neuronal hyperexcitability (Pitkänen et al., 2016).

After neuronal injury, the cellular excitotoxicity that primarily occurs as results of up-regulated *N*-methyl-D-aspartate (NMDA) and α-amino-3-hydroxy-5-methyl-4-isoxazolepropionic acid (AMPA) receptors at high concentrations of glutamate in the extracellular space contribute to spreading depolarization waves and the influx of calcium into the intracellular medium (Liu & Woodson, 2019). A high intracellular concentration of calcium activates an array of catabolic enzymes that damage cell cytoskeleton and endonucleases that cause DNA fragmentation (resulting in apoptosis and necrosis). The elevated calcium load in the intracellular environments also affects mitochondrial function since mitochondrial permeability transition pores, which are calcium-dependent, remain active long after TBI (Tubi et al., 2019).

The significant changes over time in mitochondrial phosphorylation capacity are reactions influenced by TBI severity or amount of post-injury time (Tavazzi et al., 2007). Furthermore, the mitochondrial morphological changes in the central zone near the lesion that are not in the peripheral zone (penumbra) observed in TBI patients (Balan et al., 2013) reinforce the idea that mitochondrial architectural profile alterations after TBI can be instrumental in both cell failure and recovery. The reduced ability of cells to maintain ionic gradients due to energy failure caused by post-injury mitochondrial dysfunction leads to cell depolarization, hyperexcitability, and oxidative damage (de Castro et al., 2017). Furthermore, the oxidative stress characterized by an imbalance between reactive oxygen species (ROS) and antioxidants plays a significant role in secondary damage post-TBI. The most common ROS in TBI is superoxide (O_2^-), which propagates ROS and reactive nitrogen species (RNS) production and is primarily produced by mitochondrial dysfunction (Tavazzi et al., 2007) followed by activation of caspase 3, an enzyme considered the central apoptotic effector responsible for many biochemical and morphological characteristics of apoptosis (Hiebert, Shen, Thimmesch, & Pierce, 2015). As neuronal reorganizations and synaptogenesis are well-documented consequences of TBI, it is plausible to propose that intrinsic mitochondrial apoptosis elicited by free radical generation precede neuronal cell death and contributes to the progression of TBI-induced epilepsy.

Although any cellular constituent can target free radical damage, a failure of the selected target increases cell excitability post-TBI. After TBI, the inhibition of Na^+, K^+-ATPase activity and increased levels of oxidative stress markers (de Castro et al., 2017) reinforce the idea that failure of some selected targets, such as Na^+, K^+-ATPase, results in hyperexcitability and eventually leads to the occurrence of first recurrent spontaneous seizures after TBI. This hypothesis has been upheld in experimental models of PTE where the progression of neuronal damage during the first 1 to 2 months post-injury corresponds to a period in which most electrophysiological changes imply an increase in the excitability of the injured tissue, thus engendering the appearance of seizures (Pitkänen et al., 2016).

Regarding the role of ROS in TBI-induced epileptogenes is free radicals generated directly and indirectly by inflammatory cells increase neuronal damage through the continuous cycle of oxidative stress and inflammation (O'Connell & Littleton-Kearney, 2013). For instance: the exposure of brain tissue to extravasated serum albumin caused by BBB rupture induces activation of the transforming growth factor-β/activin receptor-like kinase 5 (TGF-β/ALK5) signaling in astrocytes that produce interleukin-1beta (IL-1β) and tumor necrosis factor-alpha (TNF-α). By acting on glial cells, IL-1β can also mediate astrocyte and microglia activation by forming glial scars and releasing a series of neurotoxic mediators that promote the loss of neuronal cells associated with epileptic foci in the brain (Xu et al., 2019) (Fig. 1).

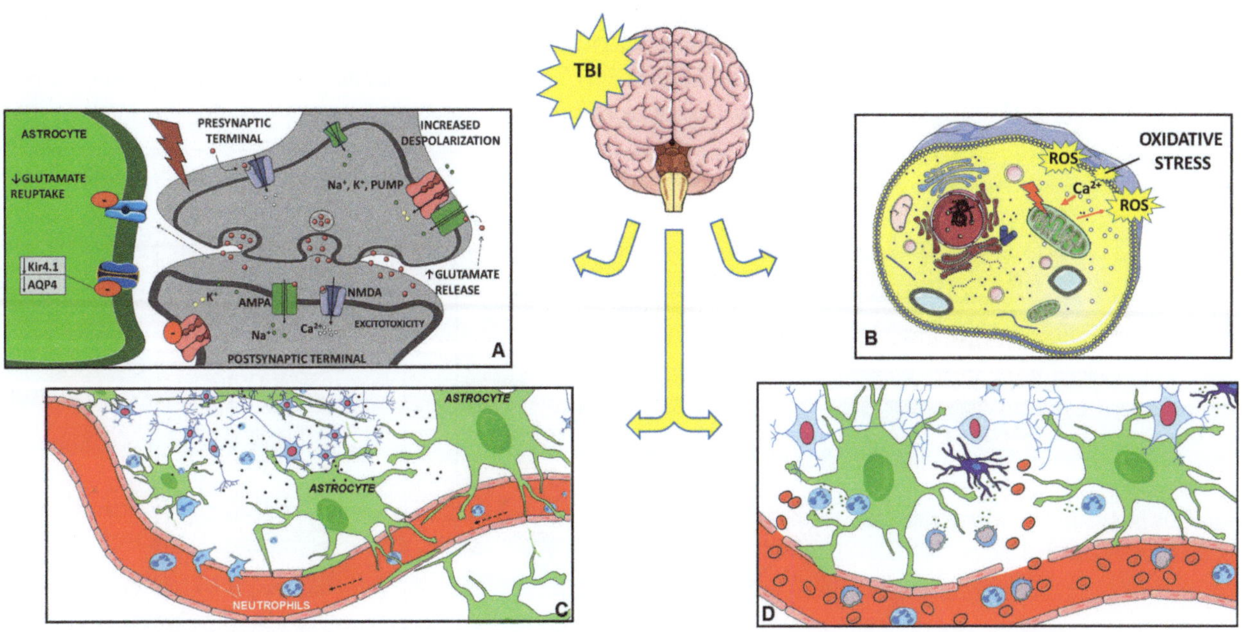

FIG. 1 Pathophysiology of TBI-induced early seizures. The early seizures caused by TBI are cellular events characterized by up-regulated NMDA and AMPA receptors at high concentrations of glutamate in the extracellular space (A) followed by blood–brain barrier (BBB) dysfunction (endothelial rupture of cells and infiltration of red blood cells), infiltration of leukocytes derived from blood to the brain (B) that alter mitochondrial architectural profile, inducing ROS generation and reducing the ability of cells to maintain ionic gradients (C). Regarding the role of ROS in TBI-induced epileptogenes, free radicals generated directly and indirectly by inflammatory cells increase neuronal damage through the continuous cycle of oxidative stress and inflammation in the first days after neuronal injury (D).

Besides promoting the loss of neuronal cells associated with epileptic foci, the accumulation of these cytokines by astrocytes decreases the expression of Kir channels and weakens their extracellular K^+ clearance. Decreased expression of Kir channels and dislocation of Aquaporin-4 (AQP4) channels in astrocytes post-TBI contribute to impairing K^+ buffering, leading to the appearance of seizures (D'Ambrosio, Maris, Grady, Winn, & Janigro, 1999). The reactive astrocytes also dramatically transform their morphology and function, causing a deranged glutamate uptake and decreased GABA release (Xu et al., 2019). In this sense, previous results from our laboratory revealed that the excitability shown post-injury by FPIis associated with significant changes in GABAergic function, including neuronal loss of parvalbumin-positive cells ($PARV^+$), decreased levels and activity of the limiting enzyme in GABA synthesis, glutamate decarboxylase (GAD), and benzodiazepine binding sites part of the $GABA_A$ receptor (binding of [^{3}H]flunitrazepam) in the hippocampus (Gerbatin et al., 2019). However, TBI-induced astrocytes mediate deficits in postsynaptic inhibitory hyperexcitability triggered by currents in circuits in specific areas, such as the hippocampus, without changing intrinsic properties. The inhibitory postsynaptic current dysfunction elicited by TBI may result from a failed astrocytic glutamate-glutamine cycle (Xu et al., 2019).

Although the epileptogenesis elicited by TBI is multifactorial, it is clear that several specific functions of astrocytes such as cytokine regulation, water regulation, neurotransmitters, ion homeostasis, and the BBB are events connected to PTE. Regarding the progression of inflammatory factors involved in epileptogenesis after TBI, the release of the early cytokines that persists for days after injury are events that coincide with the activation of the ion channel and post-translational changes in receptors associated with neuronal excitation and inhibition. Neuroinflammation, which concurs with generalized neuronal loss and the mossy fiber sprouting in the hippocampus, contributes to an increasingly excitable neuronal environment post-TBI (Tubi et al., 2019). Although the contribution of mossy fibers to seizure susceptibility after TBI is controversial, the hyperexcitability of granular cells facilitates the breakdown of intrinsic seizure resistance in these cells in the hippocampus. The consequent loss of this standard "filtering" in the dentate gyrus (DG) can be particularly epileptogenic in combination with other TBI-induced changes, including loss of interneurons (Gerbatin et al., 2019), glial dysfunction (D'Ambrosio et al., 1999), and development of seizure foci in other susceptible circuits, such as among neocortical neurons near the injury site (Golarai et al., 2001) (Fig. 2).

Pre-clinical studies in animal models of PTE

Although the exact mechanisms that lead to spontaneous seizures and their comorbidities after TBI remain unknown, the understanding the molecular bases responsible for the development of PTE are of fundamental importance to design effective therapeutic strategies to treat PTE. In this context, the potential of several strategies has been assessed in PTE animal models with a variable degree of success, such as classic AEDs (Ethosuximide, Phenytoin, Gabapentin, Valproate, and Levetiracetam), natural plant products (Piperine and Glycyrrhizin), ergogenic compounds (*N*-acetylcysteine and Creatine), and drugs targeting a broad range of enzymes and receptors (Table 1) Moreover, non-pharmacological interventions,

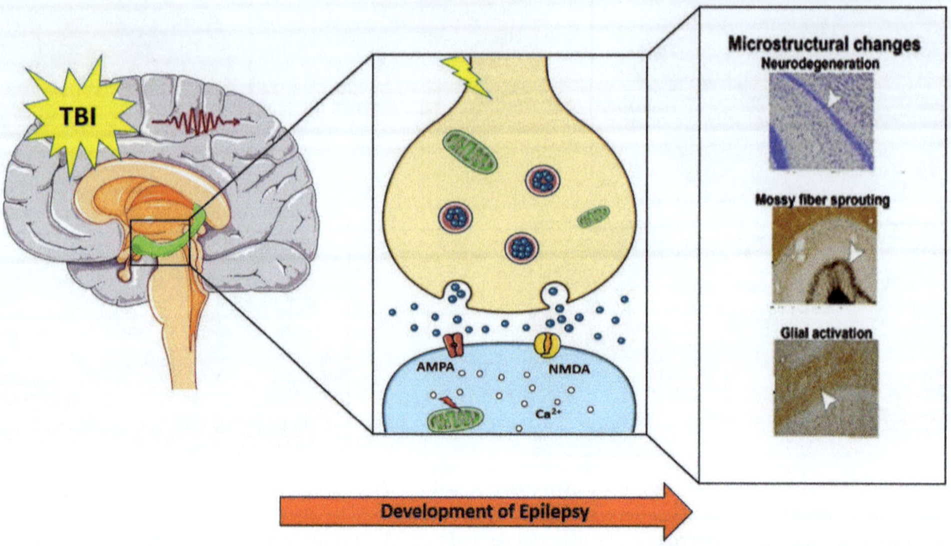

FIG. 2 Pathophysiology of TBI-induced epileptogenesis. The initial acute injury (hours-days) after TBI induces the excitability of glutamate. The remaining surviving neurons participate in structural adaptations in which neuroinflammation, which coincides with generalized neuronal loss and the mossy fiber sprouting in the hippocampus, contributes to an increasingly excitable neuronal environment after TBI, thus generating late-onset crises.

including genetic manipulation (aquaporin-4 knockout and matrix metalloproteinase-9 knockout or overexpression) and treadmill exercise, have been evaluated in experimental models of PTE. An important conclusion from the studies presented in Table 1 is that PTE attenuation and post-traumatic seizure susceptibility may be achievable. Moreover, another point to consider is that most experimental studies on early prevention of PTE, including those by our research group, have used seizure susceptibility as the endpoint to assess protective strategies. Of course, there is no guarantee that protection against increased seizure susceptibility automatically translates to protection against spontaneous PTSs in a chronic individual. Therefore, the studies investigating seizure susceptibility after PTE are more than definitive pre-clinical tests and may serve as a less time- and resource-consuming alternative to long-term 24/7 EEG monitoring that may accelerate the discovery of new strategies to prevent PTE (Fig. 3).

Early intervention after TBI to prevent PTS and PTE

As described above, one possible consequence of TBI is an immediate symptomatic seizure, and it may arise due to several factors involved in head trauma such as injury site, type, and severity (Liu & Woodson, 2019). It is highly recommended that the EAD intervention be started as soon as possible to avoid early seizures, as an early seizure can cause a cascade of excitotoxic events that can aggravate the pre-established damage to the brain and further favor the development of epilepsy (see Table 2). In this sense, several randomized controlled trials (RCTs) have studied the use of anticonvulsant drugs for prophylaxis or therapy in TBI. For instance, meta-analysis studies revealed that phenytoin and carbamazepine prevent the occurrence of early seizures when administered in the first week after TBI, although the same effect was not observed after 1 year of treatment and at 2 years of follow-up (Piccenna et al., 2017).

Randomized clinical trials comparing the effectiveness of phenytoin and levetiracetam also found no differences in the occurrence of initial seizures after treatments, although patients who received levetiracetam reported fewer side effects. After 3 and 6 months of follow-up, people treated with levetiracetam versus phenytoin performed better on the Disability Rating Scale (DRS) and Glasgow Results Scale (GOS); however, the neurological status was worse (Temkin, 2009). In addition to phenytoin and carbamazepine, phenobarbital is commonly used to treat seizures post-TBI. In a randomized,

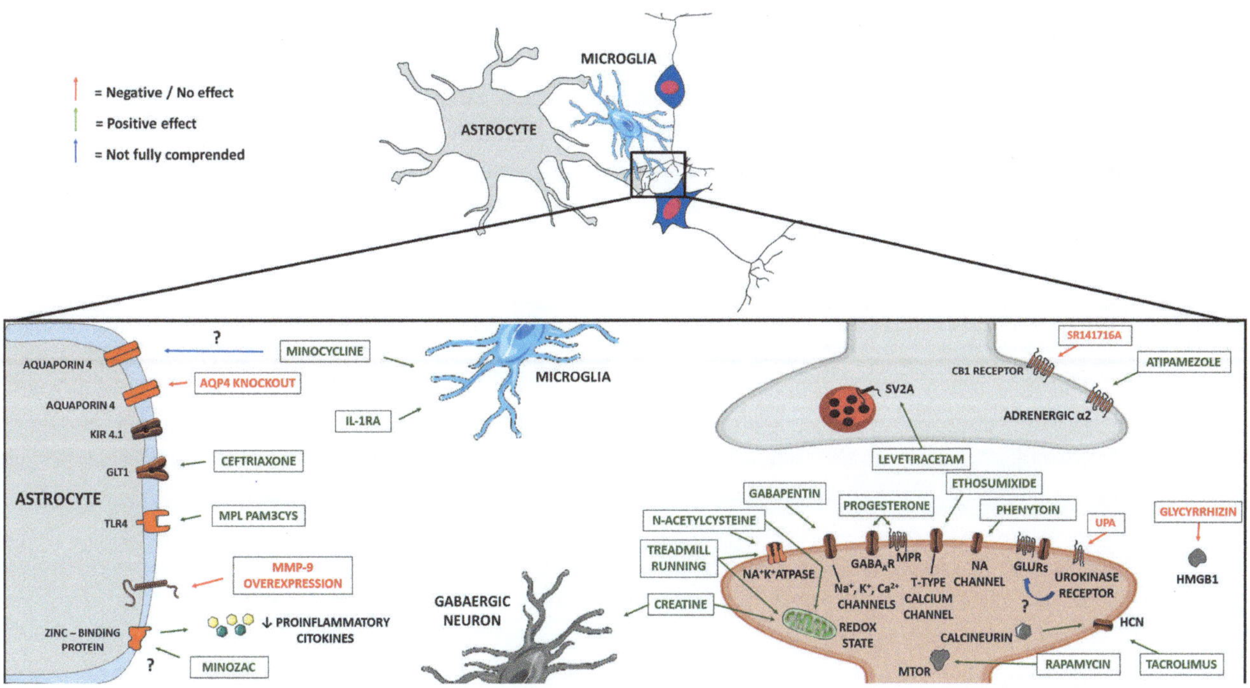

FIG. 3 Effects of different therapeutic interventions on posttraumatic epileptogenesis. In experimental studies assessing the effects of different therapeutic interventions on posttraumatic epileptogenesis in animal models of PTE (Table 1) two major outcomes have been evaluated: occurrence of spontaneous posttraumatic seizures and hypersensitivity to low dose of classical convulsants such as pentylenetetrazole. Here we summarize data from Table 1, presenting the therapeutic interventions and their outcomes, linking them to respective cellular and molecular targets. Interventions shown in green indicate beneficial effects on PTE, whereas interventions shown in red indicate lack of effects on PTE. Blue arrows indicate incompletely understood mechanisms. For more detailed information on the therapeutic interventions and their outcomes in animal models please refer to Table 1.

TABLE 2 The table describes the action of various anticonvulsant strategies in animal models of PTE.

Therapeutic intervention	Dosage and schedule	TBI Model	Animal	Effect on seizures	Study
Piperine	10–40 mg/kg (p.o.), for 7 days starting after TBI	Weight drop	Male ICR mice	Piperine prevented the TBI-induced increase in EEG total power and TBI-induced hypersensitivity to low dose of pentylenetetrazole. These effects were comparable to 200 mg/kg sodium valproate.	Song et al. (2020)
Aquaporin 4 knockout	Not applicable	CCI	Male mice	Aquaporin 4 knockout mice presented longer spontaneous posttraumatic seizures. TBI-induced hypersensitivity to low dose of pentylenetetrazole was decreased in aquaporin 4 knockout mice.	Szu, Patel, Chaturvedi, Lovelace, and Binder (2020) and Lu, Zador, Yao, Fazlollahi, and Manley (2011)
Progesterone (4-Pregnene-3, 20-dione)	16 or 32 mg/kg (i.p) for 15 days starting after TBI	Weight drop	Male Wistar rats	Progesterone prevented the TBI-induced hypersensitivity to low dose of pentylenetetrazole.	Ghadiri, Vakilzadeh, Hajali, and Khodagholi (2019)
Glycyrrhizin	50 mg/kg at 1 h pre-injury; 1 h and 6 h post-injury; once daily for 7 days	CCI	Male C57Bl/6 mice	Glycyrrhizin (High mobility group box protein-1 inhibitor) treatment did not alter the TBI-induced hypersensitivity to low dose of pentylenetetrazole or the rate of animals displaying spontaneous posttraumatic seizures.	Webster et al. (2019)
Monophosphoryl lipid A (MPL) Pam3Cys	Single administration of 1 μg (intracerebroventricular) 5 days before TBI	CCI	Male Wistar rats	MPL or Pam3Cys (Toll-like receptor agonists) prevented TBI-induced decrease in seizure threshold (amygdala kindling)	Hesam et al. (2018)
Genetic manipulation of matrix metalloproteinase-9 (MMP-9)	MMP-9 knockout or overexpression	CCI	Male C57BL/6J mice	MMP-9 overexpression increased the number of mice with spontaneous post-traumatic seizures and TBI-induced hypersensitivity to low dose of pentylenetetrazole. MMP-9 knockout mice presented with the opposite set of responses.	Pijet et al. (2018)
Rapamycin	Rapamycin (3 mg/kg; i.p.) beginning 20–30 min after TBI and continued once daily for 8–12 weeks	CCI	Male GIN[1] mice	Rapamycin (mTOR inhibitor) treatment reduced, TBI-induced hyperexcitability of surviving $eGFP^+$ hilar interneurons.	Butler, Boychuk, and Smith (2017)
Atipamezole	100 μg/kg/h (subcutaneous osmotic minipump) beginning 30 min or 7 day post-TBI and maintained for 9 weeks	FPI	Male Sprague–Dawley rats	Atipamezole (α_2-adrenoceptor antagonist) beginning at day 7 post-TBI reduced seizure susceptibility to low dose of pentylenetetrazol but did not prevent the development of posttraumatic epilepsy.	Nissinen et al. (2017)
SR141716A	10 mg/kg (i.p.) once daily for 9 weeks beginning 2 or 30 min post-TBI	FPI	Male Sprague–Dawley rats	SR141716A (cannabinoid receptor subtype 1 antagonist) did not alter TBI-induced hypersensitivity to low dose of pentylenetetrazole or occurrence of posttraumatic seizures.	Nissinen et al. (2017)

Interleukin-1 receptor antagonist (IL-1Ra)	100 mg/kg (s.c.) at 2, 8, and 24 h post-TBI (acute treatment) Bolus of 100 mg/kg (s.c.) at 2 h post-TBI plus 100 mg/kg/day (subcutaneous osmotic minipump) for 7 days (chronic treatment)	CCI	C57BL/6J mice	Acute or chronic IL-1Ra treatment decreased TBI-induced hypersensitivity to low dose of pentylenetetrazole. Chronic IL-1Ra treatment did not alter occurrence of spontaneous post-traumatic seizures[2].	Semple et al. (2017)
Rapamycin	Rapamycin (3 or 10 mg/kg; i.p.) beginning 20–30 min after TBI and continued once daily for 8–13 weeks	CCI	CD-1 mice	Rapamycin treatment did not alter prevalence of spontaneous posttraumatic seizures[3]. Rapamycin treatment decreased TBI-induced dentate granule cell network hyperexcitability.	Butler, Boychuk, and Smith (2015)
Urokinase-type plasminogen activator (uPA) or uPA receptor (uPAR) knockout	Not applicable	CCI	Male C57BL/6 mice	uPA or uPAR knockout did not alter TBI-induced hypersensitivity to low dose of pentylenetetrazole or occurrence of posttraumatic seizures.	Bolkvadze, Rantala, Puhakka, Andrade, and Pitkänen (2015) and Bolkvadze, Puhakka, and Pitkänen (2016)
Tacrolimus	Single injection (5 mg/kg; i.p.) 1 h after TBI	FPI	Male Sprague–Dawley rats	Tacrolimus (Calcineurin inhibitor)-treated animals showed less non-convulsive post-traumatic seizures	Churn, Campbell, Gandhi, and Singh (2014)
Ethosuximide Phenytoin	Ethosuximide (25–187.5 mg/kg; i.v.) or Phenytoin (5–60 mg/kg; i.v.) twice daily for 3 days beginning 30 min after TBI	PBBI	Male Sprague–Dawley rats	Ethosuximide and Phenytoin attenuated non-convulsive seizures in a dose-dependent manner.	Mountney et al. (2013)
Ceftriaxone	200 mg/kg (i.p.) once daily for 7 days beginning 30 min after TBI	FPI	Male Long-Evans rats	Ceftriaxone reduced the cumulative duration and frequency of spontaneous post-traumatic seizures.	Goodrich et al. (2013)
Caspase 8 knockout	Not applicable	CCI	FVB/N mice		Lu et al. (2011)
Minocycline	45 mg/kg (i.p.) 12 h before TBI and then twice daily for 1 month beginning 30 min after TBI	CCI	Male mice	Minocycline reduced TBI-induced hypersensitivity to low dose of pentylenetetrazole.	Lu et al. (2011)
Minozac	5 mg/kg (i.p.) at 3 h and 6 h after TBI	CCI	Male CD1 mice	Minozac (a suppressor of pro-inflammatory cytokine up-regulation) decreased TBI-induced hypersensitivity to electroconvulsive shock.	Chrzaszcz, Venkatesan, Dragisic, Watterson, and Wainwright (2010)
N-acetylcysteine (NAC) Levetiracetam (LEV) Gabapentin (GBP)	NAC (100 mg/kg), LEV (50 mg/kg), GBP (100 mg/kg), NAC + LEV and NAC + GBP intra-peritoneally for 14 days beginning 7 days after TBI	Weight drop	Male Sprague–Dawley rats	LEV, GBP or NAC + GBP decreased TBI-induced hypersensitivity to low dose of pentylenetetrazole.	Efendioglu et al. (2020)
Creatine	300 mg/kg (p.o.) once daily for 4 weeks beginning 7 days after TBI	FPI	Male Wistar rats	Creatine decreased TBI-induced hypersensitivity to low dose of pentylenetetrazole.	Gerbatin et al. (2019)
Creatine	300 mg/kg (p.o.) once daily for 3 or 7 days beginning 30 min after TBI	FPI	Male Wistar rats	Creatine did not alter TBI-induced hypersensitivity to low dose of pentylenetetrazole.	Saraiva et al. (2012)
N-acetylcysteine	100 mg/kg (p.o.) once daily for 5 weeks beginning immediately after TBI	FPI	Male Wistar rats	*N*-acetylcysteine decreased TBI-induced hypersensitivity to low dose of pentylenetetrazole.	Silva et al. (2011)
Treadmill exercise	60 min/day (at 24–26 m/min) for 4 weeks beginning 7 days after TBI	FPI	Male Wistar rats	Treadmill exercise decreased TBI-induced hypersensitivity to low dose of pentylenetetrazole.	Silva et al. (2013)

double-blind study divided into groups receiving placebo or phenobarbital, Manaka (1992) concluded that there was no reduction in the incidence of late seizures after TBI in patients who received the study drug (Manaka, 1992).

Valproic acid is also the objective of studies regarding its usefulness as an anticonvulsant after TBI. Compared with other anticonvulsants drugs, valproic acid showed no difference in seizure occurrence regardless of the amount of time; however, a trend toward higher mortality was observed after treatment with it. Notably, except for the combination of drugs such as phenytoin/phenobarbital, all RCT PTE prevention has used a single agent. In addition, the phenytoin/phenobarbital combination shows an estimate below a 25% reduction in PTSs, which is a prognosis of less than 50% reduction in the projected. These clinical trials reinforce the assumption that most studies identified thus far use pharmacological treatments on their own, without revealing clear evidence. Although antiepileptic procedures and drugs effectively reduce the frequency of acute post-traumatic seizures, there are caveats to be considered when interpreting the lack of effects on TBI-induced epileptogenesis progression (Table 2).

Conclusions

Traumatic brain injury is the most significant cause of symptomatic epilepsy where the latent or silent period of at least 6 months, and sometimes up to 20 years, between the causing injury and onset seizures provides a potential time window for intervention. However, lack of coordination between preclinical and clinical development and the lack of validated and uniformly accepted in vivo PTE animal models compromises the development of more effective treatments in this neurological pathology. In this scenario, a better understanding of the process of epileptogenesis, testing treatments that demonstrate anti-epileptogenic effects in the laboratory, and performing thorough preclinical and phase II evaluations before attempting definitive trials should significantly improve the chances of identifying ways to prevent post-traumatic epilepsy.

Applications to other areas of neuroscience

Traumatic brain injury (TBI) is a devastating disease frequently followed by significant behavioral disabilities and long-term medical complications including post traumatic epilepsy (PTE). For definition, PTE is characterized by recurrent and unprovoked Post traumatic seizures (PTS) that occur at least 1 week after TBI. Seizures during the first week after TBI are considered provoked an acute complication from head injury, while seizures occurring 1 week after TBI are considered a manifestation of PTE and if only a single seizure occurs it is known as late PTS. Although the exact molecular signaling pathways have not yet been defined, it is known that the occurrence of the epileptic seizure is directly proportional to the severity of the TBI. In addition, it has been demonstrated that the first year after neuronal injury is the most likely critical period for an epileptic seizure and use of anticonvulsant agents does not show benefits with regard to prophylaxis of crises that can occur in a period longer than a week after TBI. In this scenario, it is important to note that the complexity of events resulting from the heterogeneous nature of TBI and the variable onset of seizure makes it difficult to identify mechanisms of epileptogenesis. In order to complete this picture, we reviewed how changes in neuronal metabolism may be associated with the pathophysiology of the epileptic focus after TBI. We also discuss the integrative actions of procedures and pharmacological interventions on post-traumatic process of epileptogenesis.

Mini-dictionary of terms

Secondary Damage: Events that occur hours and days after the primary injury and that play a large role in brain damage and death resulting from TBI.

Epileptogenesis: events during which occur in CNS after TBI that results in the development of recurrent seizures and epilepsy.

Oxidative damage: imbalance between high production oxidative compounds and persistent antioxidant defenses that lead to tissue damage.

Post-translational modifications: Functional diversity of the proteome by covalent addition of functional groups or proteins, proteolytic cleavage of regulatory subunits or degradation of proteins.

Genetic polymorphisms: Event characterized by the occurrence of multiple alleles in a locus, where at least two alleles occur with a frequency greater than 1%.

Fluid-percussion injury: Experimental model of TBI that produces an injury through a craniectomy by applying a momentary fluid pressure pulse on to the exposed dura.

Controlled cortical impact model: Experimental TBI model used that uses a pneumatically or electromagnetically controlled piston to induce reproducible and well-controlled injuries.

Weight-drop model: Experimental TBI model that uses the gravitational force of free-falling weight to generate focal or diffuse brain damage.

Diffuse axonal injury: Traumatic event that occurs when the brain moves back and forth rapidly within the skull as a result of acceleration and deceleration.

Kay facts of post-traumatic epilepsy (PTE)

- Epilepsy is a disorder of the functioning of the brain due to an abnormal discharge from a population of neurons with sudden, unpredictable and generally short (duration of seconds up to 15 min).
- Post-traumatic epilepsy (PTE) is a repeated epileptic seizure in people who have suffered TBI that occurs days and/or years after neuronal injury.
- Although somewhat arbitrarily classified, the post-traumatic seizures (PTS) are characterized by immediate (at the moment of impact), early (within 7 days after TBI) or late (after 7 days) seizures after neuronal injury.
- TBI is one of the most common reasons why patients develop epilepsy. PTE represents 20% of symptomatic epilepsy in the general population.
- According to the International League Against Epilepsy (ILAE), an unprovoked seizure associated with a known traumatic brain injury meets the criteria for epilepsy.

Summary points

- Epileptogenesis occurs as part of the secondary events caused by TBI.
- Interruption of normal neural processing leads to the maturation of an epileptogenic network in hippocampus after TBI.
- The lack of antiepileptic therapies that modify the TBI-induced physiopathology compromises the treatment of EPT.
- Oscillation of Na^+, K^+-ATPase activity in different times causes drastic neurochemical changes after TBI.
- The capacity of mitochondrial phosphorylation is influenced by the severity of TBI.

References

Annegers, J. F., Hauser, W. A., Coan, S. P., & Rocca, W. A. (1998). A population-based study of seizures after traumatic brain injuries. *New England Journal of Medicine*, *338*(1), 20–24.

Balan, I. S., Saladino, A. J., Aarabi, B., Castellani, R. J., Wade, C., Stein, D. M., et al. (2013). Cellular alterations in human traumatic brain injury: Changes in mitochondrial morphology reflect regional levels of injury severity. *Journal of Neurotrauma*, *30*(5), 367–381.

Bakr, A., & Belli, A. (2018). A systematic review of levetiracetam versus phenytoin in the prevention of late post-traumatic seizures and survey of UK neurosurgical prescribing practice of antiepileptic medication in acute traumatic brain injury. *British Journal of Neurosurgery*, *32*(3), 237–244.

Bolkvadze, T., Puhakka, N., & Pitkänen, A. (2016). Epileptogenesis after traumatic brain injury in Plaur-deficient mice. *Epilepsy and Behavior*, *60*, 187–196.

Bolkvadze, T., Rantala, J., Puhakka, N., Andrade, P., & Pitkänen, A. (2015). Epileptogenesis after traumatic brain injury in Plau-deficient mice. *Epilepsy and Behavior*, *51*, 19–27.

Butler, C. R., Boychuk, J. A., & Smith, B. N. (2015). Effects of rapamycin treatment on neurogenesis and synaptic reorganization in the dentate gyrus after controlled cortical impact injury in mice. *Frontiers in Systems Neuroscience*, *9*(27), 1–14. https://doi.org/10.3389/fnsys.2015.00163. 163.

Butler, C. R., Boychuk, J. A., & Smith, B. N. (2017). Brain injury-induced synaptic reorganization in hilar inhibitory neurons is differentially suppressed by rapamycin. *ENeuro*, *4*(5), 134–151.

Christensen, J. (2015). The epidemiology of posttraumatic epilepsy. *Seminars in Neurology*, *35*(3), 218–222.

Chrzaszcz, M., Venkatesan, C., Dragisic, T., Watterson, D. M., & Wainwright, M. S. (2010). Minozac treatment prevents increased seizure susceptibility in a mouse "two-hit" model of closed skull traumatic brain injury and electroconvulsive shock-induced seizures. *Journal of Neurotrauma*, *27*(7), 1283–1295.

Churn, S., Campbell, J. N., Gandhi, A., & Singh, B. (2014). Traumatic brain injury causes a tacrolimus-sensitive increase in non-convulsive seizures in a rat model of post-traumatic epilepsy. *International Journal of Neurology and Brain Disorders*, *1*(1), 1–11.

D'Ambrosio, R., Maris, D. O., Grady, M. S., Winn, H. R., & Janigro, D. (1999). Impaired K+ homeostasis and altered electrophysiological properties of post-traumatic hippocampal gila. *Journal of Neuroscience*, *19*(18), 8152–8162.

de Castro, M. R. T., de Ferreira, A. P. O., Busanello, G. L., da Silva, L. R. H., da Silveira Junior, M. E. P., da Fiorin, F. S., et al. (2017). Previous physical exercise alters the hepatic profile of oxidative-inflammatory status and limits the secondary brain damage induced by severe traumatic brain injury in rats. *Journal of Physiology*, *595*(17), 6023–6044.

de Vries, E. E., van den Munckhof, B., Braun, K. P. J., van Royen-Kerkhof, A., de Jager, W., & Jansen, F. E. (2016). Inflammatory mediators in human epilepsy: A systematic review and meta-analysis. *Neuroscience and Biobehavioral Reviews*, *63*, 177–190. Elsevier Ltd.

Diaz-Arrastia, R., Agostini, M. A., Madden, C. J., & Van Ness, P. C. (2009). Posttraumatic epilepsy: The endophenotypes of a human model of epileptogenesis. *Epilepsia*, *50*(SUPPL. 2), 14–20.

Diaz-Arrastia, R., Gong, Y., Fair, S., Scott, K. D., Garcia, M. C., Carlile, M. C., et al. (2003). Increased risk of late posttraumatic seizures associated with inheritance of APOE ε4 allele. *Archives of Neurology*, *60*(6), 818–822.

DJohn, J., Ibrahim, R., Patel, P., DeHoff, K., & Kolbe, N. (2020). Administration of Levetiracetam in traumatic brain injury: Is it warranted? *Cureus*, *12*(7). https://doi.org/10.7759/cureus.9117.

Efendioglu, M., Basaran, R., Akca, M., Ceman, D., Demirtas, C., & Yildirim, M. (2020). Combination therapy of gabapentin and N-acetylcysteine against posttraumatic epilepsy in rats. *Neurochemical Research*, *45*(8), 1802–1812.

Fisher, R. S., Cross, J. H., French, J. A., Higurashi, N., Hirsch, E., Jansen, F. E., et al. (2017). Operational classification of seizure types by the international league against epilepsy: Position paper of the ILAE Commission for Classification and Terminology. *Epilepsia*, *58*(4), 522–530.

Fisher, R. S., Van Emde Boas, W., Blume, W., Elger, C., Genton, P., Lee, P., et al. (2005). Epileptic seizures and epilepsy: Definitions proposed by the International League Against Epilepsy (ILAE) and the International Bureau for Epilepsy (IBE). *Epilepsia*, *46*(4), 470–472. Epilepsia.

Flierl, M. A., Stahel, P. F., Beauchamp, K. M., Morgan, S. J., Smith, W. R., & Shohami, E. (2009). Mouse closed head injury model induced by a weight-drop device. *Nature Protocols*, *4*(9), 1328–1337.

Gerbatin, R. R., Silva, L. F. A., Hoffmann, M. S., Della-Pace, I. D., do Nascimento, P. S., Kegler, A., et al. (2019). Delayed creatine supplementation counteracts reduction of GABAergic function and protects against seizures susceptibility after traumatic brain injury in rats. *Progress in Neuro-Psychopharmacology and Biological Psychiatry*, *92*, 328–338.

Ghadiri, T., Vakilzadeh, G., Hajali, V., & Khodagholi, F. (2019). Progesterone modulates post-traumatic epileptogenesis through regulation of BDNF-TrkB signaling and cell survival-related pathways in the rat hippocampus. *Neuroscience Letters*, *709*(14). https://doi.org/10.1016/j.neulet.2019.134384.

Golarai, G., Greenwood, A. C., Feeney, D. M., & Connor, J. A. (2001). Physiological and structural evidence for hippocampal involvement in persistent seizure susceptibility after traumatic brain injury. *Journal of Neuroscience*, *21*(21), 8523–8537.

Goodrich, G. S., Kabakov, A. Y., Hameed, M. Q., Dhamne, S. C., Rosenberg, P. A., & Rotenberg, A. (2013). Ceftriaxone treatment after traumatic brain injury restores expression of the glutamate transporter, GLT-1, reduces regional gliosis, and reduces post-traumatic seizures in the rat. *Journal of Neurotrauma*, *30*(16), 1434–1441.

Gupta, A., Ip, I. K., Raja, A. S., Andruchow, J. E., Sodickson, A., & Khorasani, R. (2014). Effect of clinical decision support on documented guideline adherence for head CT in emergency department patients with mild traumatic brain injury. *Journal of the American Medical Informatics Association: JAMIA*, *21*(e2), 347–351. https://doi.org/10.1136/amiajnl-2013-002536.

Harris, L., Hateley, S., Tsang, K. T., Wilson, M., & Seemungal, B. M. (2020). Impact of anti-epileptic drug choice on discharge in acute traumatic brain injury patients. *Journal of Neurology*, *267*(6), 1774–1779.

Hesam, S., Khoshkholgh-Sima, B., Pourbadie, H. G., Babapour, V., Zendedel, M., & Sayyah, M. (2018). Monophosphoryl lipid A and Pam3Cys prevent the increase in seizure susceptibility and epileptogenesis in rats undergoing traumatic brain injury. *Neurochemical Research*, *43*(10), 1978–1985.

Hiebert, J. B., Shen, Q., Thimmesch, A. R., & Pierce, J. D. (2015). Traumatic brain injury and mitochondrial dysfunction. *American Journal of the Medical Sciences*, *350*(2), 132–138. Lippincott Williams and Wilkins.

Huijben, J. A., Volovici, V., Cnossen, M. C., Haitsma, I. K., Stocchetti, N., Maas, A. I. R., et al. (2018). Variation in general supportive and preventive intensive care management of traumatic brain injury: A survey in 66 neurotrauma centers participating in the Collaborative European NeuroTrauma Effectiveness Research in Traumatic Brain Injury (CENTER-TBI) study. *Critical Care*, *22*(1), 1–9. https://doi.org/10.1186/s13054-018-2000-6.

Khan, N. R., Vanlandingham, M. A., Fierst, T. M., Hymel, C., Hoes, K., Evans, L. T., et al. (2016). Should Levetiracetam or phenytoin be used for posttraumatic seizure prophylaxis? A systematic review of the literature and Meta-analysis. *Neurosurgery*, *79*(6), 775–781. Lippincott Williams and Wilkins.

Liu, J., & Woodson, B. (2019). Deep learning classification for epilepsy detection using a single channel electroencephalography (EEG). In *ACM international conference proceeding series* (pp. 23–26).

Lu, D. C., Zador, Z., Yao, J., Fazlollahi, F., & Manley, G. T. (2021). Aquaporin-4 reduces post-traumatic seizure susceptibility by promoting astrocytic glial scar formation in mice. *Journal of Neurotrauma*, *38*(8), 1193–1201. https://doi.org/10.1089/neu.2011.2114.

Manaka, S. (1992). Cooperative prospective study on posttraumatic epilepsy: Risk factors and the effect of prophylactic anticonvulsant. *Psychiatry and Clinical Neurosciences*, *46*(2), 311–315.

Mountney, A., Shear, D. A., Potter, B., Marcsisin, S. R., Sousa, J., Melendez, V., et al. (2013). Ethosuximide and phenytoin dose-dependently attenuate acute nonconvulsive seizures after traumatic brain injury in rats. *Journal of Neurotrauma*, *30*(23), 1973–1982.

Nissinen, J., Andrade, P., Natunen, T., Hiltunen, M., Malm, T., Kanninen, K., et al. (2017). Disease-modifying effect of atipamezole in a model of post-traumatic epilepsy. *Epilepsy Research*, *136*, 18–34.

O'Connell, K. M., & Littleton-Kearney, M. T. (2013). The role of free radicals in traumatic brain injury. *Biological Research for Nursing*, *15*(3), 253–263.

Piccenna, L., Shears, G., & O'Brien, T. J. (2017). Management of post-traumatic epilepsy: An evidence review over the last 5 years and future directions. *Epilepsia Open*, *2*(2), 123–144. Wiley-Blackwell Publishing Ltd.

Pijet, B., Stefaniuk, M., Kostrzewska-Ksiezyk, A., Tsilibary, P. E., Tzinia, A., & Kaczmarek, L. (2018). Elevation of MMP-9 levels promotes epileptogenesis after traumatic brain injury. *Molecular Neurobiology*, *55*(12), 9294–9306.

Pitkänen, A., Löscher, W., Vezzani, A., Becker, A. J., Simonato, M., Lukasiuk, K., et al. (2016). Advances in the development of biomarkers for epilepsy. *The Lancet Neurology*, *15*(8), 843–856. Lancet Publishing Group.

Roozenbeek, B., Maas, A. I. R., & Menon, D. K. (2013). Changing patterns in the epidemiology of traumatic brain injury. *Nature Reviews Neurology*, *9*(4), 231–236.

Salinsky, M., Rutecki, P., Parko, K., Goy, E., Storzbach, D., Markwardt, S., et al. (2019). Health-related quality of life in veterans with epileptic and psychogenic nonepileptic seizures. *Epilepsy and Behavior*, *94*, 72–77.

Saraiva, A. L. L., Ferreira, A. P. O., Silva, L. F. A., Hoffmann, M. S., Dutra, F. D., Furian, A. F., et al. (2012). Creatine reduces oxidative stress markers but does not protect against seizure susceptibility after severe traumatic brain injury. *Brain Research Bulletin*, *87*(2–3), 180–186.

Semple, B. D., O'Brien, T. J., Gimlin, K., Wright, D. K., Kim, S. E., Casillas-Espinosa, P. M., et al. (2017). Interleukin-1 receptor in seizure susceptibility after traumatic injury to the pediatric brain. *Journal of Neuroscience*, *37*(33), 7864–7877.

Sharma, R., Shultz, S. R., Robinson, M. J., Belli, A., Hibbs, M. L., O'Brien, T. J., et al. (2019). Infections after a traumatic brain injury: The complex interplay between the immune and neurological systems. *Brain, Behavior, and Immunity*, *79*, 63–74. Academic Press Inc.

Silva, L. F. A., Hoffmann, M. S., Gerbatin, R. D. R., Fiorin, F. D. S., Dobrachinski, F., Mota, B. C., et al. (2013). Treadmill exercise protects against pentylenetetrazol induced seizures and oxidative stress after traumatic brain injury. *Journal of Neurotrauma*, *30*(14), 1278–1287.

Silva, L. F. A., Hoffmann, M. S., Rambo, L. M., Ribeiro, L. R., Lima, F. D., Furian, A. F., et al. (2011). The involvement of Na +, K +-ATPase activity and free radical generation in the susceptibility to pentylenetetrazol-induced seizures after experimental traumatic brain injury. *Journal of the Neurological Sciences*, *308*(1–2), 35–40.

Song, Y., Cao, C., Xu, Q., Gu, S., Wang, F., Huang, X., et al. (2020). Piperine attenuates TBI-induced seizures via inhibiting cytokine-activated reactive astrogliosis. *Frontiers in Neurology*, *11*, 431. https://doi.org/10.3389/fneur.2020.00431.

Sun, Y., Yu, J., Yuan, Q., Wu, X., Wu, X., & Hu, J. (2017). Early post-traumatic seizures are associated with valproic acid plasma concentrations and UGT1A6/CYP2C9 genetic polymorphisms in patients with severe traumatic brain injury. *Scandinavian Journal of Trauma, Resuscitation and Emergency Medicine*, *25*(1), 85.

Szu, J. I., Patel, D. D., Chaturvedi, S., Lovelace, J. W., & Binder, D. K. (2020). Modulation of posttraumatic epileptogenesis in aquaporin-4 knockout mice. *Epilepsia*, *61*(7), 1503–1514.

Tavazzi, B., Vagnozzi, R., Signoretti, S., Amorini, A. M., Finocchiaro, A., Cimatti, M., et al. (2007). Temporal window of metabolic brain vulnerability to concussions: Oxidative and nitrosative stresses—Part II. *Neurosurgery*, *61*(2), 390–396.

Temkin, N. R. (2009). Preventing and treating posttraumatic seizures: The human experience. *Epilepsia*, *50*(SUPPL. 2), 10–13.

Thompson, K., Pohlmann-Eden, B., Campbell, L. A., & Abel, H. (2015). Pharmacological treatments for preventing epilepsy following traumatic head injury. *Cochrane Database of Systematic Reviews*, *2015*(8). https://doi.org/10.1002/14651858.CD009900.pub2.

Tubi, M. A., Lutkenhoff, E., Blanco, M. B., McArthur, D., Villablanca, P., Ellingson, B., et al. (2019). Early seizures and temporal lobe trauma predict post-traumatic epilepsy: A longitudinal study. *Neurobiology of Disease*, *123*, 115–121.

Wat, R., Mammi, M., Paredes, J., Haines, J., Alasmari, M., Liew, A., et al. (2019). The effectiveness of antiepileptic medications as prophylaxis of early seizure in patients with traumatic brain injury compared with placebo or no treatment: A systematic review and meta-analysis. *World Neurosurgery*, *122*, 433–440. Elsevier Inc.

Webster, K. M., Shultz, S. R., Ozturk, E., Dill, L. K., Sun, M., Casillas-Espinosa, P., et al. (2019). Targeting high-mobility group box protein 1 (HMGB1) in pediatric traumatic brain injury: Chronic neuroinflammatory, behavioral, and epileptogenic consequences. *Experimental Neurology*, *320*. https://doi.org/10.1016/j.expneurol.2019.112979.

Xu, S., Sun, Q., Fan, J., Jiang, Y., Yang, W., Cui, Y., et al. (2019). Role of astrocytes in post-traumatic epilepsy. *Frontiers in Neurology*, *10*, 1149. Frontiers Media S.A.

Yang, Y., Zheng, F., Xu, X., & Wang, X. (2016). Levetiracetam versus phenytoin for seizure prophylaxis following traumatic brain injury: A systematic review and meta-analysis. *CNS Drugs*, *30*(8), 677–688. Springer International Publishing.

Zangbar, B., Khalil, M., Gruessner, A., Joseph, B., Friese, R., Kulvatunyou, N., et al. (2016). Levetiracetam prophylaxis for post-traumatic brain injury seizures is ineffective: A propensity score analysis. *World Journal of Surgery*, *40*(11), 2667–2672.

Zimmermann, L. L., Diaz-Arrastia, R., & Vespa, P. M. (2016). Seizures and the role of anticonvulsants after traumatic brain injury. *Neurosurgery Clinics of North America*, *27*(4), 499–508. W.B. Saunders.

Chapter 31

Neurosurgical treatment of critical brain damage: The early post-surgical period and algorithms

Alexey O. Trofimov[a,b], Anatoly Y. Sheludyakov[b], Andrew Y. Abashkin[c], Darya I. Agarkova[d], Artem A. Kopylov[e], Svyatoslav B. Korolev[f], Michael Dobrzeniecki[a], Xenia A. Trofimova[a], Denis E. Bragin[g,h], and Dmitry S. Martynov[i]

[a]*Department of Neurologic Diseases, Privolzhsky Research Medical University, Nizhny Novgorod, Russia,* [b]*Department of Neurosurgery, Trauma Center Level I, City Clinical Hospital, Nizhny Novgorod, Russia,* [c]*Department of Neurotrauma, Trauma Center Level II, Urgent Care Hospital, Dzerzhinsk, Russian Federation,* [d]*Department of Ophthalmology, Trauma Center Level I, Regional Clinical Hospital named after NA Semashko, Nizhny Novgorod, Russian Federation,* [e]*Department of Neurosurgery, Trauma Center Level I, Regional Clinical Hospital named after NA Semashko, Nizhny Novgorod, Russian Federation,* [f]*Department of Traumatology and Neurosurgery, Privolzhsky Research Medical University, Nizhny Novgorod, Russia,* [g]*Lovelace Biomedical Research Institute, Albuquerque, NM, United States,* [h]*University of New Mexico, School of Medicine, MSC 105615 Neurosurgery, Albuquerque, NM, United States,* [i]*Nizhny Novgorod State Technical University n.a. R.E. Alekseev, Nizhny Novgorod, Russian Federation*

List of abbreviations

ABp arterial brain pressure
AIS abbreviated injury scale
CPP cerebral perfusion pressure
CT computed tomography
DCNS damage control neurosurgery
GCS glasgow coma score
ICH intracranial hematoma
ISS injury severity score
TBI traumatic brain injury

Introduction

According to Hutchinson et al. (2019), current stage of the neurotraumatology development is characterized by the widespread introduction into the practice neuroimaging methods, the emergence and use of minimally aggressive treatment methods, the principles of "evidence-based medicine" in the form of "standards" of care for the patients.

As shown by Maas (2016) on the other hand, it is characterized by a progressive increase in the severity of injuries, the frequency of multiple and associated injuries, accompanied by the massive destruction of the human body's internal structures, bleeding and shock.

According Duchesne, Inaba, and Khan (2018), the resuscitation, intensive care and surgery's success have reduced the mortality rate in severe TBI to 25–35%. However, the mortality rate in critically ill patients with unstable hemodynamics and respiratory disorders still remains more than 80%.

Studies by Stocchetti et al. (2017) were found that the surgical treatment of such patients, according to the existing recommendations (the extensive craniotomy, final stop of the bleeding, removal of brain destruction foci), in the overwhelming majority of the cases ends in death on the operating table or in the first hour's postoperative period.

Outside a large hospital, this problem might not have a solution: the patient is either recognized as inoperable or dies at the craniotomy stage and this was shown previously by Rossini et al. (2019).

Diagnosis and Treatment of Traumatic Brain Injury. https://doi.org/10.1016/B978-0-12-823347-4.00027-0

Critical damage: Physiological features and surgical treatment nuances

In the literature of recent years (this group of suffers was called: "borderline patient" (Bates, Parker, McFadyen, & Pallister, 2016) or "patient in extremis". Similar clusters of the patients are described in emergency traumatology and in general surgery (Sagraves, Toschlog, & Rotondo, 2006).

Further works by Friedland, Porter, and Daryanani (1996) have shown that its development is based on the impairment of humoral immunity with the avalanche-liked changes in the concentration of pro-inflammatory cytokines (IL-6, IL-2, TNFα, IL-10). This is accompanied by the rapid development of a systemic inflammatory response.

Studies by Garcia et al. (2017) and Chio, Lin, and Chang (2015) were found that in the critical condition patients immunity disorders developed within the next hours after injury and significantly associated with the changes in IL-2, IL-6, IL-10 serum concentrations.

According to Cotton et al. (2011) and Mizobata (2017) and one-staged surgery for the critical injuries in the patient with vital parameters disorders is reputed to be to the poor outcome and aggravating the injury that may cause a death.

Stone, Mackrell, and Brandstater (1955) were the first to propose a new approach to treat patients in the hemorrhagic shock due to critical blood loss. The indications for its implementation were signs of hypothermia (less than $34^{\circ C}$ and more than 4 h), the presence of metabolic acidosis and coagulopathy.

The technique included carrying out only "small" operations for the shortest possible time, followed by intensive care. The intensive surgical intervention was carried out only after the stabilization of the patient's condition (Boukerrouche, 2019).

Over the next decade, the works of American researchers: both the civilians (Rotondo M. and others) and military (Datzmann et al., 2019; Neal et al., 2018; Turner, Stockinger, Bell, & Gurney, 2018; Zouris, Walker, Dye, et al., 2006); the scientists group from the University of Hanover (Duchesne et al., 2018; Hofman et al., 2015) have significantly supplemented and expanded the indications for the use of this technique not only in abdominal surgery but also in emergency orthopedics, urgent gynecology, etc. The technique itself is called "Damage control."

According to Duchesne et al. (2018), DCS is characterized by the following features:

(1) phasing of the treatment;
(2) carrying out the resuscitation measures during the entire time between treatment stages;
(3) programmability of the reoperations.

According to Sagraves et al. (2006), the goal of the first stage of surgical treatment is to interrupt the supercritical stimuli with the highest possible speed and minimum surgical trauma. It should be especially noted that restoring normal anatomical relationships between damaged organs is not included in the tasks of the first stage of a surgical treatment since it takes time and is accompanied by the infliction of additional surgical trauma.

During the first stage and in the next 24 to 48 h, intensive therapy is carried out (in fact, resuscitation measures), on the one hand, aimed at restoring vital functions, and on the other, at conservative stopping bleeding using modern hemostatics (such as rFVIIa "Novoseven"), infusion and blood transfusion therapy (Stawicki, Brooks, & Bilski, 2008).

After vital stabilization, one or several reoperations are performed with the purpose of restoring the anatomical relationships and treat the damage delayed at the first stage.

Studies by Helgason and Jobe (2004) have shown that in DCS there are still no precise algorithms for the sequence of surgical interventions, especially for the multiple and combined injuries.

This is due to the impossibility of performing large randomized, placebo-controlled trials on this issue due to both ethical and technical features (Carney et al., 2017). The recommendations are made using multiple retrospective studies that correspond to class II studies, as well as peer review based on consensus (Freeman, Winston, Byers, & Beauchamp, 2015; Rosenfeld, 2004).

According to the above authors of many years work data, the "Damage control" technique has reduced mortality in critical traumatic injuries to 12–67%, averaging about 50%.

As have been shown by Cotton et al. (2011) and the clear benefits of DCS on survival in polytraumazed patients who have been managed with DCR compared to "early total care" method.

As noted by Hsu and Pham (2011) that in these conditions, post-operative pro-inflammatory complications came out on top among the causes of mortality, ahead of decompensated blood clotting disorders and ongoing bleeding. The frequency of such complications is, on average, 30%. And according to the Association of Military Surgeons of the United States, it reaches 100% with penetrating trauma.

"Damage control": A path between the Scylla and the Charybdis

However, the use of the principles of "Damage control" in emergency neurosurgery seems to be quite reasonable, especially from the standpoint of the currently dominant concept of "secondary stroke."

For example, Sokolov (2006) defines the components of "Damage control" in emergency neurosurgery as early prevention of the increase in ICP, decrease in CPP, prevention of secondary brain injury due to hypoxia, arterial hypotension and hyperthermia.

In our opinion, such recommendations are correct and scientifically confirmed in terms of identifying the priorities of intensive care in critically injured patients. At the same time, they do not fully reflect the issues of determining the circle of patients indicated for such treatment, the features of the surgical nuances and the technical aspects of performing surgery from the position of its stages.

Perhaps, the aspects of the use of DCS principles in the neurosurgery are much broader than the simple implementation of the recommendations for intensive care of severe TBI and this was shown previously (Duchesne et al., 2018).

We believe that the elements of "Damage control" are fully justified and applicable in the treatment of patients with cerebral compression associated with multiple injuries to the abdominal cavity, chest, pelvis and long bones. As a rule, such patients arrived at an emergency department in a deep shock, the presence of which practically excludes the possibility of performing a "standard" intracranial surgery. These patients either die during an attempt by a neurosurgeon to perform a craniotomy against a profound vital functions disorder of; or from a dislocation of the brain stem (Fig. 1).

According to Chuck et al. (2021) at the same time, the rapid transportation of a patient with acute hematoma from the injury site to the emergency room and the external drainage of the hematoma cavity during the so-called "golden hour" in some cases can significantly reduce its volume and improve outcomes.

We have previously shown (Trofimov & Voennov, 2013) that trephination can be performed directly in the intensive care room without full preparation of the scalp (release from the scalp at the trephination site). Performing such a minimally aggressive intervention in certain cases leads to a significant decrease not only in the volume of the hematoma but also in the degree of transverse dislocation (Figs. 2 and 3).

As have been shown by Sokolov (2006), the surgical tactics according to DCS principle might be used in the case of the brain compression by envelope hematoma (subdural and/or epidural), combined with the long bones fractures, severe shock, massive blood loss.

As noted above by Hofman et al. (2015), "Damage control" neurosurgery includes two stages: the first stage is the brain compression letup. During the second stage of intensive therapy, blood loss is reimbursed, cerebral edema, circulatory and respiratory disorders are controlled. Under these conditions, an osteosynthesis is postponed for 2 to 3 days or up to 21 days, depending on the suffer's condition.

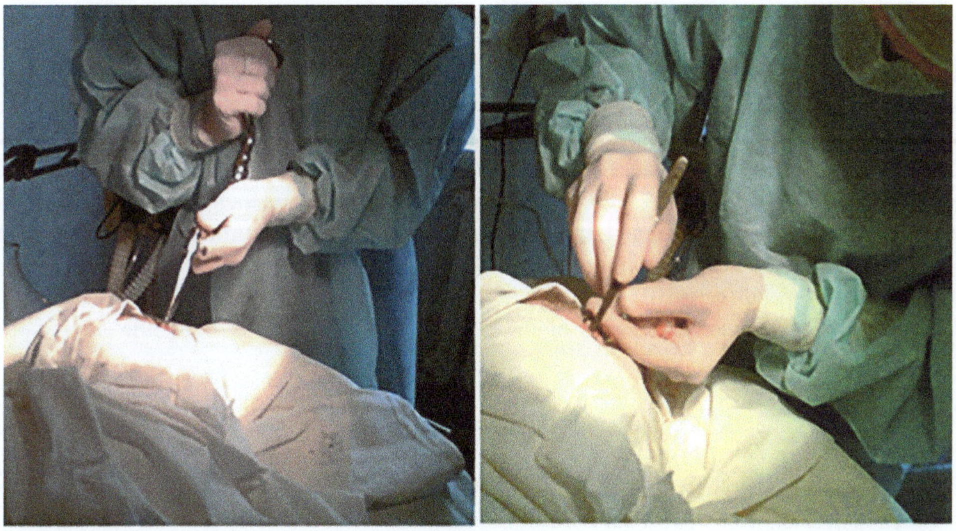

FIG. 1 The first stage of DCNS—a partial removal of an a hyperacute subdural hematoma in emergency room. The neurosurgeon performs the trephination over the area of its maximum thickness.

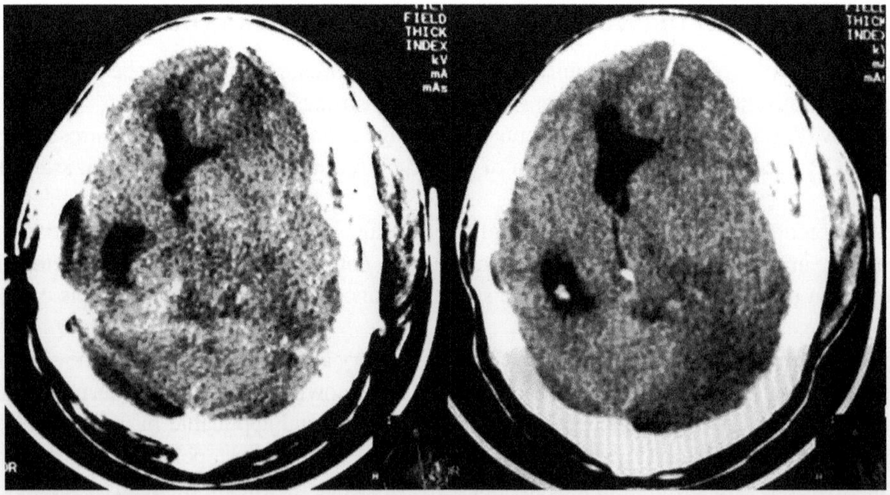

FIG. 2 Patient L. CT, axial view; huge acute subdural hematoma of the left fronto-temporo-parietal region is evident. The volume of the hematoma more than 120 mL. The midline shift is 14 mm.

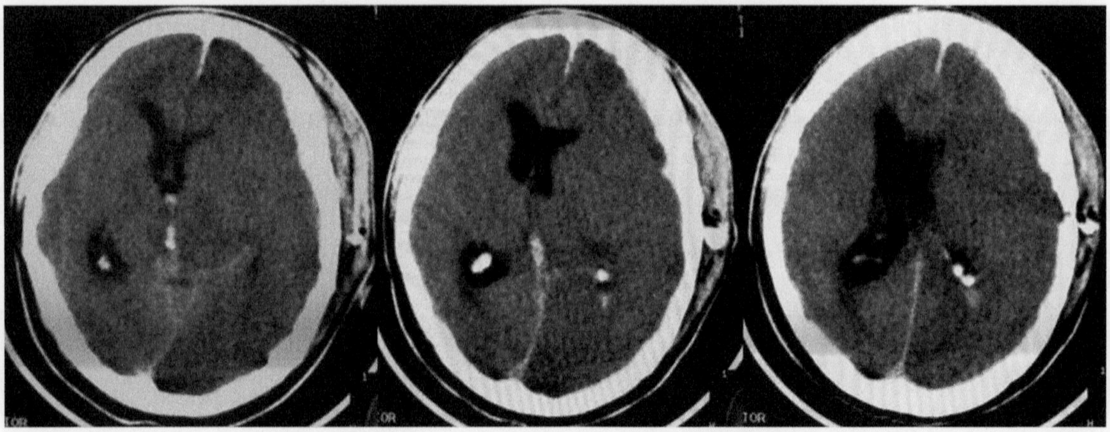

FIG. 3 Patient L. CT, axial view; 30 min after the operation of external drainage of the hematoma through the milling hole. The volume of the hematoma is about 50 mL. The midline shift is 8 mm.

Germanos et al. (2008) has been shown that in most polytrauma cases, an adequate intracranial surgery is possible only after whole-body CT (including the brain) and identification of all brain compression substrates with their exact localization. At the same time, by the end of this study, most such patients are hemodynamically unstable, which imposes significant restrictions on the volume of surgical intervention.

We have previously shown that the surgical intervention begins with the imposition of a diagnostic milling hole and aspiration reduction of the hematoma volume. A drop in systemic blood pressure is the basis for a temporary suspension of the operation and intensification of intensive (and, in fact, resuscitation) therapy (Trofimov & Voennov, 2013).

As shown by Czosnyka (2016) the autoregulation of cerebral blood flow in such patients is severely impaired, and cerebral blood flow directly depends on ABP. The temporary suspension of the operation is a prerequisite, as it helps to prevent critical cerebral ischemia. The carrying on of the operation is possible only after the stabilization of systemic hemodynamics.

It should be noted that the presence of a large (more than 1 cm) midline shift significantly complicates the choice of surgical tactics in critically ill patients. In some cases, performing a wide craniectomy is accompanied by the formation in the postoperative period of so-called "delayed" intracerebral hematomas.

According Yokobori et al. (2018), they are foci of mixed stroke of hemispheric localization, the development of which is associated with the reperfusion in the perifocal zone of the removed hematoma. In this case, intraoperative diagnostics are of particular importance. The surgical intervention begins with the imposition of a diagnostic milling hole and aspiration reduction of the hematoma volume. A drop in systemic blood pressure is the basis for a temporary suspension of the operation and intensification of intensive (and, in fact, resuscitation) therapy.

The autoregulation of cerebral blood flow in such patients is severely impaired, and cerebral blood flow directly depends on systemic blood pressure and this was shown previously by Donnelly et al. (2019).

Temporary suspension of the operation is a prerequisite, as it helps to prevent critical cerebral ischemia. Continuation of the operation is possible only after stabilization of systemic hemodynamics.

It should be noted that the presence of a large (more than 1 cm) transverse displacement of the midline structures significantly complicates the choice of surgical treatment tactics in critically ill patients. In some cases, the implementation of a wide decompression of the brain is accompanied by the formation in the postoperative period of the so-called "delayed" intracerebral hematomas. According to Datzmann et al. (2019), they are foci of mixed stroke of hemispheric localization, the development of which is associated with reperfusion in the perifocal zone after removal of the hematoma (Figs. 4 and 5).

As shown by Ragel et al. (2010), DCNS tactics (a reducing of hematoma volume from a minimal-sized approach) are quite justified in this category of patients in order to prevent the development of reactive cerebral hyperemia with the subsequent formation of parenchymal hemorrhage foci.

At the same time, we believe that the DCNS is currently of little implement for the treatment of critically ill patients with cerebral compression by intracerebral hematomas, which, due to their peculiarities, do not lend themselves to such effective drainage as subdural hematomas.

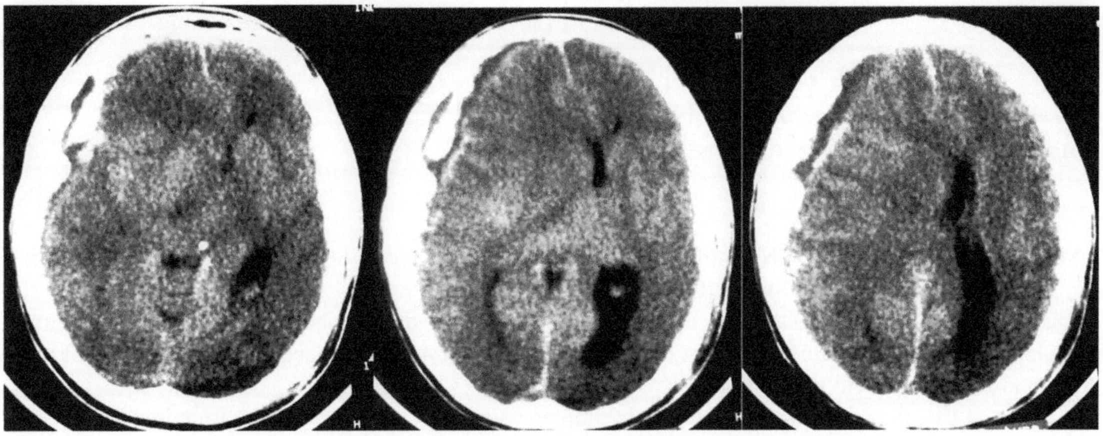

FIG. 4 Patient B. CT, axial view; right subdural hematoma of the right fronto-temporal region is evident. The volume of the hematoma is 118 mL. The midline shift is 14 mm. The hematoma is "crescentic" in shape and layer the entire brain surface.

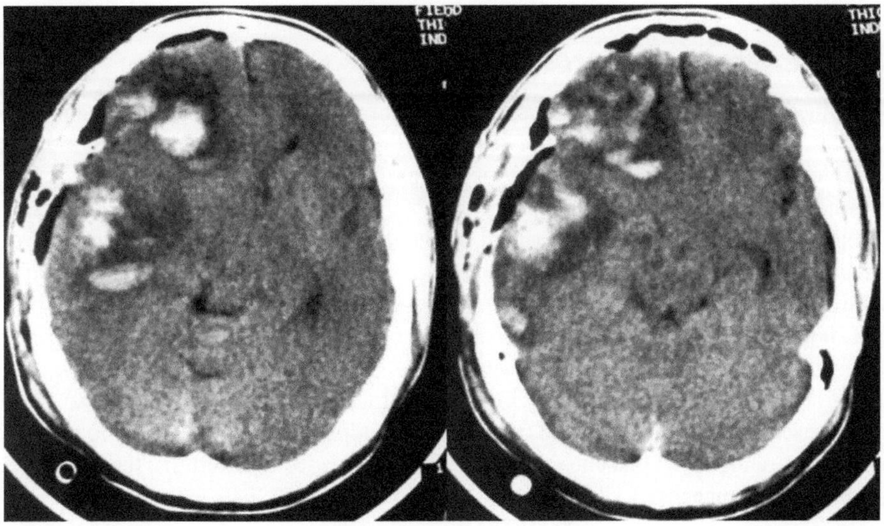

FIG. 5 Patient B. CT, axial view. 16 h after the right-side cranioectomy and the removal of the subdural hematoma. There is the formation of parenchymal heterogeneous density foci in the right fronto-temporal lobe. The midline shift is 9 mm. The total volume of the intracranial substrate is about 70 mL.

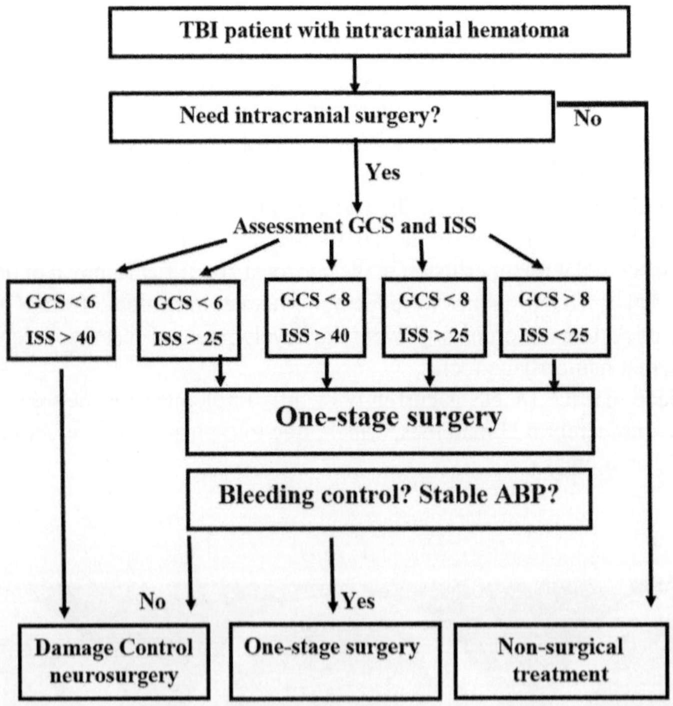

FIG. 6 The algorithm of Surgical Treatment of Polytraumazed Patient with TBI.

In addition, although the use of DCNS might reduce mortality in the group of "borderline" TBI patients, it does not significantly change their severe disability (Teff, 2010).

According Tao and Li-hong (2015) and Xu, Dong, Wang, et al. (2015), the application of DCNS in the treatment of severe TBI patients and take appropriate multi-stage operative strategy intervention can reduce the poor outcome rate, prevent complications and promote patient's recovery.

It should be noted that the "Damage control" can be implemented only in multidisciplinary clinics, where not only is there a neurosurgical service equipped with everything necessary for any urgent surgery, but there is also a round-the-clock trauma team and neuroimaging.

In conclusion, we present an approximate algorithm for staging surgical care for concomitant brain and limb trauma (Fig. 6).

Applications to other areas of neuroscience

In this chapter, we examined the main aspects of the staged tactics of surgical treatment of critical brain damage.

As we presented above, the one-stage treatment of severe TBI combined with severe extracranial injuries in the most cases leads to unfavorable outcomes already at the stage of the surgical approach.

For example, performing a craniotomy in such patients against the background of arterial hypotension, after the restoration of adequate cerebral perfusion, may be accompanied by the development of malignant cerebral edema and cerebral dislocation and herniation. However, as shown above, such a sequence of the events is characteristic in general for all critical brain injuries of various etiologies.

In a broad sense, it should be noted that the DCNS tactics is the surgical implementation of the principle of maintaining the constancy of the cerebral microcirculatory bed perfusion.

At present, the use of DCNS tactics is justified only in the neurotrauma. However, it is likely that the tactics of the staged surgical treatment can be used in the cases of critical conditions development at a non-traumatic etiology.

In the future, multi-center randomized placebo-controlled trials are needed to determine the most effective patient selection protocol and assess surgical nuances of DCNS.

Mini-dictionary of terms

Abbreviated Injury Scale (AIS) is a global severity scoring system that classifies each damage in every body area according to its relative severity on a six-point ordinal scale: minor (1 point), moderate (2 points), serious (3 points), severe (4 points), critical (5 points), maximal-untreatable (6 points).

Cerebral autoregulation is the key mechanism for maintaining a constant cerebral blood flow over a wide range of arterial blood pressure, which protects the brain from ischemia or hyperperfusion.

Glasgow Coma Scale (GCS) is a medical scale used to measure the level of a consciousness after brain injury of a various etiologies. It was proposed by Teasdale and Graham in 1974. The GCS assesses a suffer's consciousness based on an analysis of three elements: the eye movements and gaze direction, the speech and motor response. The GCS score ranges from 3 (no response at all) to 15 (full consciousness). This indicator is widely used not only to assess consciousness at TBI, but also at other conditions accompanied by the disorders of consciousness. As previously shown, lower GCS scores correlate with a higher risk of the mortality.

Hyperacute hematoma is an intracranial hematoma in the first hours of its formation, usually consists of liquid blood with a minimum of clots.

Injury Severity Score (ISS) is a scale for assessing the severity of associated injuries. To calculate ISS value for suffer, the body is divided into six body parts: head/neck, face, chest, abdomen, limbs/pelvis. The sum of the squares of the AIS severity of the three most damaged areas is used to calculate the ISS level. If any of the three points is 6, the severity is set to 75 points (no survival). If any of the three points is 6, the severity is set to 75 points (no survival).

Traumatic brain injury can be defined as damage to the brain parenchyma as a result of external mechanical impact due to acceleration/deceleration, collision, explosion or penetration of a wounding projectile.

Key facts of the surgical treatment of polytraumazed patient with TBI

- TBI with a GCS of 8 point or below is severe, 9 to 12 is moderate and 13 or above is mild.
- Clinical criteria for the "borderline" group are follow: 1) deep coma (GCS 3–4), 2) unstable hemodynamics (systolic blood pressure less than 70 mmHg), 3) ineffective external respiration or apnea, 4) severe concomitant or multiple injuries leading to the shock.
- «Damage control»—the term was borrowed from the maritime code and meant "the fight for the survivability of a damaged vessel, no matter by what means." (Keep the ship afloat no matter what ...—cited by John Keegan).
- Although the three-pin Mayfield skull fixation system facilitates intraoperative correction of the head position during DCNS, it can limit the manipulation of the anesthesiologists and is therefore not necessary.
- Gavriil Abramovich Ilizarov (1921–1992) was a Soviet—Russian physician, known as the inventor of the extrafocal fixation apparatus for lengthening bones and for the treatment of bone fracture named after him.

Summary points

- The trefination, as an initial stage of DCNS, should be performed not in the area of maximum ICH thickness, but stepping back 1–1.5 cm downward for the possibility of its spontaneous emptying under the influence of the gravity.
- In the process of one surgical intervention carried out by two teams with some breaks in time: removing of the ICH, stopping bleeding, and before suturing—an osteosynthesis of limb fractures, then suturing the wounds by two teams simultaneously.
- If the first option is contraindicated for the patient's condition—removal of the ICH, temporary immobilization of the fracture (wire-rod apparatus, Ilizarov apparatus etc.) during the first 3 days or on days 8 to 14—an osteosynthesis.
- In case of traumatic amputations of the extremities and brain compression by ICH—stop bleeding from the stump—tourniquet, clamp in the wound, removing of the ICH, then an amputation.
- Current and future studies will continue to identify the most appropriate suffers that may benefit from neurosurgical damage control.

References

Bates, P., Parker, P., McFadyen, I., & Pallister, I. (2016). Demystifying damage control in musculoskeletal trauma. *Annals of the Royal College of Surgeons of England*, *98*(5), 291–294.

Boukerrouche, A. (2019). Damage control surgery Concept. *Open Journal of Trauma*, *3*(1), 006–009. https://doi.org/10.17352/ojt.000019.

Carney, N., Totten, A. M., O'Reilly, C., Ullman, J. S., Hawryluk, G. W., Bell, M. J., et al. (2017). Guidelines for the management of severe traumatic brain injury, fourth edition. *Neurosurgery*, *80*(1), 6–15.

Chio, C. C., Lin, M. T., & Chang, C. P. (2015). Microglial activation as a compelling target for treating acute traumatic brain injury. *Current Medicinal Chemistry*, *22*(6), 759–770.

Chuck, C. C., Martin, T. J., Kalagara, R., Shaaya, E., Kheirbek, T., & Cielo, D. (2021). Emergency medical services protocols for traumatic brain injury in the United States: A call for standardization. *Injury*. https://doi.org/10.1016/j.injury.2021.01.008. S0020-1383 (21)00024-3. Advance online publication.

Cotton, B. A., Reddy, N., Hatch, Q. M., Le Febvre, E., Wade, C. E., et al. (2011). Damage control resuscitation is associated with a reduction in resuscitation volumes and improvement in survival in 390 damage control laparotomy patients. *Annals of Surgery*, *254*, 598–605.

Czosnyka, M., Aries, M., Weersink, C., Wolf, S., Budohoski, K., Dias, C., et al. (2016). "Solid Red Line": An observational study on death from refractory intracranial hypertension. *Acta Neurochirurgica. Supplement*, *122*, 113–116.

Datzmann, T., Kapapa, T., Scheuerle, A., McCook, O., Merz, T., Unmuth, S., et al. (2019). In-depth characterization of a long-term, resuscitated model of acute subdural hematoma-induced brain injury. *Journal of Neurosurgery*, 1–12. Advance online publication https://doi.org/10.3171/2019.9.JNS191789.

Donnelly, J., Czosnyka, M., Adams, H., Cardim, D., Kolias, A. G., Zeiler, F. A., et al. (2019). Twenty-five years of intracranial pressure monitoring after severe traumatic brain injury: A retrospective, single-center analysis. *Neurosurgery*, *85*(1), E75–E82. https://doi.org/10.1093/neuros/nyy468.

Duchesne, J., Inaba, K., & Khan, M. A. (2018). *Damage control in trauma care: An evolving comprehensive team approach*. Springer International Publishing.

Freeman, J. L., Winston, K. R., Byers, J. T., & Beauchamp, K. (2015). Damage-control neurosurgery: Packing to halt relentless intracranial bleeding. *The Journal of Trauma and Acute Care Surgery*, *79*(5), 865–869.

Friedland, J. S., Porter, J. C., Daryanani, S., et al. (1996). Plasma proinflammatory cytokine concentrations, Acute Physiology and Chronic Health Evaluation (APACHE) III scores and survival in patients in an intensive care unit. *Critical Care Medicine*, *24*(11), 1775–1781. https://doi.org/10.1097/00003246-199611000-00003.

Garcia, J. M., Stillings, S. A., Leclerc, J. L., Phillips, H., Edwards, N. J., Robicsek, S. A., et al. (2017). Role of Interleukin-10 in acute brain injuries. *Frontiers in Neurology*, *8*, 244.

Germanos, S., Gourgiotis, S., Villias, C., Bertucci, M., Dimopoulos, N., & Salemis, N. (2008). Damage control surgery in the abdomen: An approach for the management of severe injured patients. *International Journal of Surgery (London, England)*, *6*(3), 246–252.

Helgason, C. M., & Jobe, T. H. (2004). Fuzzy logic and causal reasoning with an 'n' of 1 for diagnosis and treatment of the stroke patient. *Expert Review of Neurotherapeutics*, *4*(2), 249–254.

Hofman, M., Koopmans, G., Kobbe, P., Poeze, M., Andruszkow, H., Brink, P. R., et al. (2015). Improved fracture healing in patients with concomitant traumatic brain injury: Proven or not? *Mediators of Inflammation*, *2015*, 204842. https://doi.org/10.1155/2015/204842.

Hsu, J. M., Pham, T., et al. (2011). Damage control in the injured patient. *International Journal of Critical Illness and Injury Science*, *1*, 66–72. https://doi.org/10.4103/2229-5151.79285.

Hutchinson, P. J., Kolias, A. G., Tajsic, T., Adeleye, A., Aklilu, A. T., Apriawan, T., et al. (2019). Consensus statement from the international consensus meeting on the role of decompressive craniectomy in the management of traumatic brain injury: Consensus statement. *Acta Neurochirurgica*, *161*(7), 1261–1274.

Maas, A. (2016). Traumatic brain injury: Changing concepts and approaches. *Chinese Journal of Traumatology*, *19*(1), 3–6.

Mizobata, Y. (2017). Damage control resuscitation: A practical approach for severely hemorrhagic patients and its effects on trauma surgery. *Journal of Intensive Care*, *5*, 4.

Neal, C. J., Bell, R. S., Carmichael, J. J., DuBose, J. J., Grabo, D. J., Oh, J. S., et al. (2018). Catastrophic non-survivable brain injury care-role 2/3. *Military Medicine*, *183*(suppl_2), 73–77.

Ragel, B. T., Klimo, P., Jr., Martin, J. E., Teff, R. J., Bakken, H. E., & Armonda, R. A. (2010). Wartime decompressive craniectomy: Technique and lessons learned. *Neurosurgical Focus*, *28*(5), E2.

Rosenfeld, J. V. (2004). Damage control neurosurgery. *Injury*, *35*(7), 655–660.

Rossini, Z., Nicolosi, F., Kolias, A. G., Hutchinson, P. J., De Sanctis, P., & Servadei, F. (2019). The history of decompressive craniectomy in traumatic brain injury. *Frontiers in Neurology*, *10*, 458.

Sagraves, S. G., Toschlog, E. A., & Rotondo, M. F. (2006). Damage control surgery—The intensivist's role. *Journal of Intensive Care Medicine*, *21*(1), 5–16.

Sokolov, V. A. (2006). *Multiple and combined injuries* (p. 245). Moscow: GEOTAR-Media (in Russian).

Stawicki, S. P., Brooks, A., & Bilski, T. (2008). The concept of damage control: Extending the paradigm to emergency general surgery. *Injury*, *39*, 93–101.

Stocchetti, N., Carbonara, M., Citerio, G., Ercole, A., Skrifvars, M. B., Smielewski, P., et al. (2017). Severe traumatic brain injury: Targeted management in the intensive care unit. *The Lancet. Neurology*, *16*(6), 452–464.

Stone, H., Mackrell, T., Brandstater, B., et al. (1955). The effect of induced hemorrhagic shock on the cerebral circulation and metabolism of man. *Surgical Forum*, *5*, 789–794.

Tao, C., & Li-hong, L. (2015). Surgical concept of damage control and its application in treatment of traumatic brain injury. *Journal of Medical Recapitulate*, *21*(15), 2758–2760.

Teff, R. J. (2010). Use of neurosurgical decision-making and damage control neurosurgery courses in the Iraq and Afghanistan conflicts: A surgeon's experience. *Neurosurgical Focus*, *28*(5), E9.

Trofimov, A., & Voennov, O. (2013). «damage control» at severe polytrauma: Neurosurgical aspects. *Ukrainian Neurosurgical Journal*, *3*, 49–53. https://doi.org/10.25305/unj.54149.

Turner, C. A., Stockinger, Z. T., Bell, R. S., & Gurney, J. M. (2018). Neurosurgical workload during US combat operations: 2002 to 2016. *The Journal of Trauma and Acute Care Surgery*, *85*(1), 140–147.

Xu, Q., Dong, J., Wang, Y., et al. (2015). Damage control surgery for bilateral frontal cerebral contusion and intracerebral hematoma. *Chinese Journal of Minimally Invasive Neurosurgery*, *20*(9), 402–404.

Yokobori, S., Nakae, R., Yokota, H., Spurlock, M. S., Mondello, S., Gajavelli, S., et al. (2018). Subdural hematoma decompression model: A model of traumatic brain injury with ischemic-reperfusional pathophysiology: A review of the literature. *Behavioural Brain Research*, *340*, 23–28.

Zouris, J. M., Walker, G. J., Dye, J., et al. (2006). Wounding patterns for U.S. marines and sailors during operation Iraqi freedom, major combat phase. *Military Medicine*, *171*(3), 246–252.

Chapter 32

Hypertonic saline usage in traumatic brain injury: A focus on pediatrics

Andrew Wu[a], Tina Slusher[b,c], and Andrew Kiragu[b,c,d]

[a]*Division of Critical Care Medicine, Department of Anesthesiology, Critical Care, and Pain Medicine, Boston Children's Hospital, Boston, MA, United States,* [b]*Division of Critical Care, Department of Pediatrics, University of Minnesota, Minneapolis, MN, United States,* [c]*Division of Critical Care, Department of Pediatrics, Hennepin Healthcare, Minneapolis, MN, United States,* [d]*Children's Minnesota, Minneapolis, MN, United States*

List of abbreviations

AKI acute kidney injury
BBB blood brain barrier
CNS central nervous system
CSF cerebrospinal fluid
GCS glasgow coma score
HTS hypertonic saline
ICP intracranial pressure
TBI traumatic brain injury

Introduction

Traumatic brain injury (TBI) remains a leading cause of death worldwide (Ogden, Mayer, & Connolly Jr., 2005). In the United States alone, TBI annually results in almost 3 million emergency department visits including over 800,000 in children. There are over 56,000 TBI-related deaths including over 2500 in children (Kochanek et al., 2019; Ogden et al., 2005; Surveillance Report of Traumatic Brain Injury-related Emergency Department Visits, Hospitalizations, and Deaths—United States, 2014, 2019). TBI broadly is classified as primary or secondary. Primary injuries consist of the primary insult that results in the initial brain injury-impact, penetrating, rotational, acceleration-deceleration forces or anoxia. Secondary injuries can exacerbate the initial insult and include ischemia, excitotoxicity, inflammation, mitochondrial dysfunction, cerebral edema, apoptosis, and necrosis (Rosenfeld et al., 2012). One of the main goals in the management of traumatic brain injury is to reduce the impact of these secondary injuries. This includes attenuating and treating cerebral edema (Ogden et al., 2005). Uncontrolled, cerebral edema can lead to intracranial hypertension, brain herniation, poor outcomes in survivors, or death (Ogden et al., 2005). The bony vault that is the cranium is a fixed space (Witherspoon & Ashby, 2017). The Monroe-Kellie hypothesis postulates that any increase in the volume of one of the intracranial contents which include brain, blood and cerebrospinal fluid, must be accompanied by a reduction in one or both of the other contents so as to avoid the resultant increase in intracranial pressure (Witherspoon & Ashby, 2017). Cerebral edema increases the volume of the brain. The degree of cerebral edema secondary to TBI correlates with outcome, and early treatment is key to attenuating neurologic deterioration and decreasing mortality (Ogden et al., 2005). Medical management of cerebral edema, and intracranial hypertension, are critical components of neurocritical care practice and the non-operative management of severe TBI (Khellaf, Khan, & Helmy, 2019; Kochanek et al., 2019; Lovett, O'Brien, & Leonard, 2019; Rosenfeld et al., 2012). A full list of available therapies is shown in Table 1. In this chapter, we will specifically review hyperosmolar therapy.

Diagnosis and Treatment of Traumatic Brain Injury. https://doi.org/10.1016/B978-0-12-823347-4.00042-7

TABLE 1 Full list of management strategies for intracranial hypertension in the setting of traumatic brain injury. Management strategies are categorized by system.

Other Treatments for Traumatic Brain Injury Beyond Hypertonic Saline	
Neurologic	Elevation of head of bed
	Keeping head midline
	Monitoring ICP
	Decompressive Craniectomy
	CSF drainage
	Intravenous mannitol administration
	Anti-epileptic therapy
	Sedation
	Neuromuscular blockade
	Narcotics
Respiratory	General ventilation and oxygenation strategies
	Hyperventilation and hypocarbia (disputed)
Cardiovascular	General blood pressure management
	Prevention of hypotension and use of vasoactive medications
Other	Nutritional support
	Targeting normothermia/cooling (disputed)

Of note, hyperventilation and hypothermia are not currently recommended as therapies unless in emergency situations.

Overview of traumatic brain injury and hyperosmolar therapy

Cerebral edema

Cerebral edema is defined as an abnormal accumulation of water in the brain parenchyma (Jha, Kochanek, & Simard, 2019; Michinaga & Koyama, 2015; Ogden et al., 2005). There are three main types of cerebral edema; vasogenic, cytotoxic and hydrostatic (Koenig, 2018; Ogden et al., 2005). Vasogenic edema is a consequence of the disruption and increased permeability of the blood–brain barrier which results in plasma components such as proteins, water, and electrolytes extravasating into the extracellular compartment of brain parenchyma. Common causes of vasogenic edema include intracerebral and extracerebral brain tumors, trauma, and intracerebral hemorrhage (ICH), meningitis and brain abscesses (Bhardwaj & Ulatowski, 1999; Koenig, 2018). The main impact of vasogenic edema is on the subcortical white matter with sparing of the cerebral cortex and subcortical gray matter (Bhardwaj & Ulatowski, 1999; Koenig, 2018).

Cytotoxic edema is intracellular fluid accumulation due to the disruption of intraparenchymal cellular membranes that results in water shifts from the extracellular to the intracellular compartment (Bhardwaj & Ulatowski, 1999;Koenig, 2018 ; Ogden et al., 2005). It results from a disruption of homeostatic ion channels and pumps leading to intracellular oncotic shifts and failed ionic gradients leading to cellular swelling predominantly in astrocytes and glial cells (Winkler, Minter, Yue, & Manley, 2016).

Cytotoxic edema affects both gray and white matter structures, resulting in loss of gray -white differentiation on imaging studies. Ischemic stroke is the most common cause of cytotoxic edema. Other causes include TBI, ICH, hepatic encephalopathy, and Reye syndrome. Since TBI can result in both types of edema, there are potential implications for effective treatment depending on the predominant type of edema contributing to brain swelling. The third type of edema is hydrostatic edema which is a consequence of the transependymal displacement of cerebrospinal fluid from the intra-ventricular space into brain tissue usually due to obstructive hydrocephalus (Koenig, 2018; Ogden et al., 2005).

Traumatic brain injury-associated cerebral edema

As noted, advances in research have led to a better understanding of the pathophysiology of cerebral edema including the processes that result in disruption of the blood brain barrier and the regulation of cell volume by a variety of ionic pumps, osmotic gradients and inflammation. It has become increasingly clear that cerebral edema following TBI is due to a combination of vasogenic and cytotoxic edema (Jha et al., 2019). Vasogenic edema is biphasic with an initial peak 1 to 3h after the injury followed by increased microglial activation 5days after the injury (Winkler et al., 2016). Animal studies show that TBI induces the expression of nicotinamide adenine dinucleotide phosphate (NADPH) oxidase 1 and inducible nitric oxide synthetase (iNOS) that generate free radicals which cause microvascular injury and endothelial tight junction disruption and apoptosis (Winkler et al., 2016). A combination of post-injury generated pro-inflammatory cytokines and oxidative stress also contribute to BBB disruption through increased expression of matrix metalloproteinases, bradykinins and substance P which increase BBB permeability (Winkler et al., 2016). This BBB disruption, increased endothelial inflammatory cell adhesion molecules, and increased inflammatory cytokine production result in additional cascades of inflammation and reactive oxidative species generation leading to additional BBB disruption (Winkler et al., 2016).

Cytotoxic edema is intracellular swelling secondary to ionic and osmotic flux dysfunction (Jha et al., 2019; Winkler et al., 2016) (Fig. 1). Ischemia or hypoxia leads to an inability to maintain cell membrane electrochemical gradients. Depletion of cellular ATP stores due to impaired oxygen and glucose delivery results in the impaired function of Na1/K1 ATPase. This in turn results in intracellular sodium accumulation. The function of other cellular cation channels that include acid-sensing ion channels (ASIC'S), the sulfonylurea receptor 1 (SUR1)-regulated $NC_{Ca\text{-}ATP}$ channel, and the NKCC 1 and the *N*-methyl-D-aspartate (NMDA)channels is impacted by TBI and also leads to an intracellular influx of ions and consequent influx of water across the resultant concentration gradient (Winkler et al., 2016). Movement of water in the central nervous system is regulated by the aquaporin (AQP) family of transmembrane channels. The main AQP's in the CNS

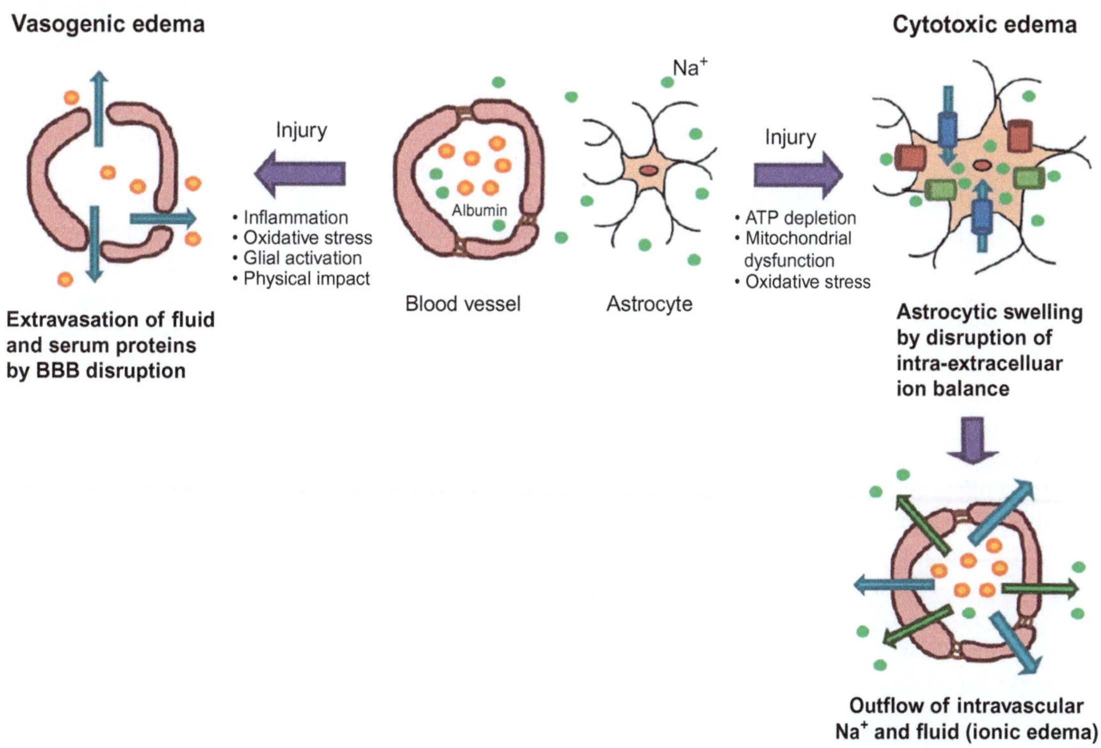

FIG. 1 Pathophysiology of vasogenic and cytotoxic edema. Vasogenic edema: TBI results in endothelial tight junction disruption by inflammation and oxidative stress. The resultant increased permeability of the BBB eventually leads to extracellular accumulation of fluid into the cerebral parenchyma. Cytotoxic edema: TBI results in intracellular ATP depletion, leading to mitochondrial dysfunction and oxidative stress and disturbs intra-extracellular ion balance. This results in excessive inflows of extracellular fluid and Na+ into cells leading to cell swelling. Since extracellular Na+ contents are decreased by this excessive inflow into cells, the outflow of Na+ and fluid from blood vessels is consequently enhanced. The intravascular Na+ outflow then results in extracellular fluid accumulation in the cerebral parenchyma. Blue arrows represent flow of water, the green arrows represent the flow of Na+, orange spheres represent albumin, green spheres represent Na+. The blue columns are water channels, green columns are ion transporters and red columns are the ion channel. *From: Michinaga, S., & Koyama, Y. (2015). Pathogenesis of brain edema and investigation into anti-edema drugs. International Journal of Molecular Sciences 16, 9949–9975 (Open Access).*

are AQP-1 and AQP-4. The latter functions as a water-potassium transport complex in astrocytes (Winkler et al., 2016). Following TBI, AQP-4 may facilitate the movement of water into astrocytes resulting in cellular edema (Jha et al., 2019; Winkler et al., 2016).

Use of hyperosmolar therapy in TBI

The reduction of cerebral water content has long been theorized as an effective way of the management of intracranial hypertension (Kochanek et al., 2019). Initially, this principally involved fluid restriction and diuresis (Kochanek et al., 2019). As it became clear that establishing a concentration gradient across the blood brain barrier was not dependent on systemic dehydration, hyperosmolar therapy, principally mannitol, became the primary means to treat cerebral edema. Hyperosmolar therapy has two principal mechanisms of action: decreasing brain water content and decreasing the production of cerebrospinal fluid. Hyperosmolar agents also alter blood rheology through reduction of blood viscosity and increased oxygen delivery leading to cerebral vasoconstriction and a decrease in cerebral blood flow and consequent blood volume similar to what is seen with hyperventilation (Kochanek et al., 2019). Mannitol was for many years the mainstay of hyperosmolar therapy. The most recent guidelines recommend that hypertonic saline be the principal hyperosmolar agent (Kochanek et al., 2019; Rosenfeld et al., 2012).

History and evolution of the use of HTS for TBI

Treatment of intracranial hypertension with hypertonic saline has been described for almost a century, with the large majority of the experiments being done in adults (Bhardwaj & Ulatowski, 2004; Khanna et al., 2000; Marmarou, Saad, Aygok, & Rigsbee, 2005; Ogden et al., 2005; Raslan & Bhardwaj, 2007; Rockswold et al., 2009; Suarez et al., 1998; Vialet et al., 2003; Ware et al., 2005; White, Cook, & Venkatesh, 2006; Ziai, Toung, & Bhardwaj, 2007). In 1919, Weed and McKibben showed that saline of varying concentrations effectively attenuated cerebral edema in cats (Lutters, Koehler, & Wijdicks, 2020; Weed & McKibben, 1919). Haden in 1919 and Sachs and Belcher in 1920 were successful in using hypertonic solutions clinically; the former with a 25% dextrose solution and the latter with a high concentration salt solution (Lutters et al., 2020). The neurosurgeons Javid and Settlage began the use of urea to treat intracranial hypertension in the 1950's. Urea was thereafter widely used in the management of TBI and brain tumors until it was displaced by mannitol which was introduced by Burton Wise and Norman Chater in 1961. Mannitol was easier to prepare, store and administer (Lutters et al., 2020).

The clinical use of hypertonic saline did not gain acceptance until Worthley et al. reported the use of 29% saline to reduce refractory intracranial pressure (Worthley, Cooper, & Jones, 1988). Concrete and theoretical benefits of hypertonic saline have since been better established (Bhardwaj & Ulatowski, 2004; Harutjunyan et al., 2005; Khanna et al., 2000; Marmarou et al., 2005; Oddo et al., 2009; Ogden et al., 2005; Qureshi & Suarez, 2000; Raslan & Bhardwaj, 2007; Rockswold et al., 2009; Suarez, 2004; Suarez et al., 1998; Vialet et al., 2003; Ware et al., 2005; White et al., 2006; Ziai et al., 2007). Harutjunyan, et al. compared mannitol and hypertonic saline bolus therapy and found that adult patients in the mannitol group experienced more frequent intracranial hypertensive episodes than patients in the hypertonic saline group (Harutjunyan et al., 2005). However, there was no difference in outcome between the two groups. Oddo et al. noted that in adult patients with intracranial hypertension refractory to mannitol, treatment with hypertonic saline significantly improved cerebral oxygenation as well as cerebral and systemic hemodynamics (Oddo et al., 2009). More evidence exists showing that hypertonic saline indeed has beneficial effects on patients' hemodynamic status, specifically in the pre-hospital resuscitation of hemorrhagic shock (Ducey, Mozingo, Lamiell, Okerburg, & Gueller, 1989; Vassar, Perry, Gannaway, & Holcroft, 1991).

Current literature now supports the role of HTS as the first-line drug in the management of intracranial hypertension given its hemodynamic profile and faster improvement of elevated ICP (Niknam, Mistry, & Lumba-Brown, 2019) There is data supporting it's superiority to mannitol, although there is also data showing no significant difference in mean ICP reduction (Battison, Andrews, Graham, & Petty, 2005; Berger, Schürer, Härtl, Messmer, & Baethmann, 1995; Mirski, Denchev, Schnitzer, & Hanley, 2000; Niknam et al., 2019). However, while both agents may temporarily improve cardiac output by drawing fluid into the intravascular space, the osmotic diuresis that follows the administration of mannitol can have a deleterious hemodynamic impact (Ducey et al., 1989).

Pathophysiology/mechanism of HTS use for TBI

Hypertonic saline has a number of physiologic and pharmacologic benefits that are theoretically helpful for the treatment of intracranial hypertension and there are a number of mechanisms that have been proposed for the use of HTS in the treatment of increased ICP. HTS causes a biphasic reduction in intracranial hypertension through its creation of an osmolar gradient between the intracellular and intravascular space and its effects on blood flow (Surani, Lockwood, Macias, Guntupalli, & Varon, 2015).

Mechanistically, the rapid infusion of hypertonic saline leads to an osmotic gradient that draws free water intravascularly leading to hemodilution, endothelial shrinkage and improved microvascular flow (Oddo et al., 2009; Rockswold et al., 2009; Surani et al., 2015). Cardiac output has been shown to increase with hypertonic saline resuscitation which, combined with reduced microcirculatory resistance, improves blood flow through capillary beds with resultant metabolic improvement (Oddo et al., 2009; Rockswold et al., 2009; Ware et al., 2005). Improved oxygen delivery to the brain as documented by brain tissue oxygen monitoring is the result (Oddo et al., 2009; Rockswold et al., 2009). The improved microvascular flow also potentially causes autoregulatory vasoconstriction that reduces cerebral blood volume and consequently a reduction in ICP (Surani et al., 2015).

Sodium chloride has a reflection coefficient of 1.0, an ideal value that denotes effective restriction of transport across the blood–brain barrier (Zornow, 1996). The sodium particle stays in the intravascular space and produces a significant osmotic gradient, pulling water away from injured brain tissue. Other theoretical benefits of saline include maintenance of a cell's resting membrane potential and cell volume (Ogden et al., 2005; Suarez et al., 1998; Zornow, 1996).

A study by Simma, et al. described 32 children with traumatic brain injury that were randomized to receive either Lactated Ringers or 2% saline (Simma, Burger, Falk, Sacher, & Fanconi, 1998). In addition to routine measures such as hyperventilation, sedation, mannitol, and barbiturates, the 2% saline group showed decreased morbidity and mortality. This finding led the authors to recommend that the resuscitation of pediatric patients with traumatic brain injury be done with hypertonic saline solutions. In another study, Peterson et al. showed prolonged infusions of hypertonic saline were effective in decreasing intracranial pressure in pediatric patients and noted no adverse effects of the associated supraphysiologic plasma hyperosmolality (Peterson, Khanna, Fisher, & Marshall, 2000).

HTS has also been reported to decrease or inhibit inflammation (Raslan & Bhardwaj, 2007; Surani et al., 2015). The exact mechanism for this modulation of the immune system is unclear but is felt to be secondary to cell shrinkage caused by HTS which results in altered cellular signaling (Surani et al., 2015). This causes a reduction in macrophage and neutrophil activation and the consequent reduction in adhesion molecule and inflammatory cytokine expression. The subsequent decreased inflammation results in a decrease in secondary brain injury the expansion of plasma volume, reduction of vasospasm, and the diminishment of the post-traumatic inflammatory response (Rockswold et al., 2009; Surani et al., 2015; Winkler et al., 2016).

It is clear that hypertonic saline solutions offer benefits for the treatment of intracranial hypertension, and most studies that describe these benefits using 3% and 7.5% saline. As noted earlier, only three studies in the medical literature detail saline solutions with higher concentrations, and these studies were performed in adults (Rockswold et al., 2009; Suarez et al., 1998; Ware et al., 2005).

Current use of HTS in the pediatric clinical setting

The guidelines detailed below are according to the US Brain Trauma Foundation Guidelines for Pediatric severe TBI, Third Edition in 2019 unless otherwise noted (Kochanek et al., 2019). Recommendations are categorized as level I, level II, or level III depending on if the evidence is of high-quality, moderate-quality, or low-quality, respectively.

Indications

Use of HTS is generally reserved for patients with TBI categorized as "severe," which is commonly defined as a GCS of less than 9 (Guideline for the Management of Severe Traumatic Brain Injury during Retrieval, 2009; Kochanek et al., 2019). However, specific definitions of "severe TBI" may vary by institution, and administration of HTS should be based on individual institutional indications. Using a bolus of 3% HTS to control ICP in pediatric patients with intracranial hypertension should be considered (a level II recommendation). Clinicians may also consider using a continuous infusion of HTS (3% is suggested) and for refractory intracranial hypertension, using a bolus of 23.4% HTS (i.e., level III recommendations). As of this writing, there are no level I recommendations regarding use of HTS in children with intracranial hypertension. These recommendations are congruent with international guidelines also recommending 3% saline for increased ICP due to head injury (Guideline for the Management of Severe Traumatic Brain Injury during Retrieval, 2009; Guideline to make up hypertonic saline (3% NaCl), 2018).

Contraindications

There are no established contraindications to use of HTS in children with TBI. However, current literature has demonstrated adverse events that seem to be linked to sustained, high sodium levels. A sodium level sustained for longer than 72 h above 170 mEq/L should be avoided to minimize risk of thrombocytopenia and anemia (Gonda et al., 2013). A sustained level above 160 mEq/L should be avoided to minimize complications of deep vein thromboses. Additionally, the medical provider must be judicious in fluid administration if the patient has pathology that predisposes him or her to fluid overload. For example, some may argue that congestive heart failure and renal failure are relative contraindications to hyperosmolar therapy in general (Latorre & Greer, 2009).

Dosing

Table 2 shows the various doses for four commonly used concentrations of HTS.

Monitoring

Monitoring ICP is routinely employed to guide the use of HTS in the setting of increased ICP given that sustained elevated ICP can lead to ischemic injury and brain herniation syndromes. However, despite this pathophysiologic rationale, the Brain Trauma Foundation Guidelines deem monitoring ICP a level III recommendation as no controlled trials evaluating ICP monitoring in the setting of TBI management exist (Kochanek et al., 2019). Furthermore, though some studies have shown an association between increased ICP monitoring and positive outcomes in pediatric patients with TBI requiring ICU care (Kochanek et al., 2019; Tilford et al., 2005; Tilford et al., 2001), others have shown either inconsistencies or no association (Chesnut et al., 2012; Keenan, Nocera, & Bratton, 2005). Therefore, though monitoring the ICP intuitively should be done in order to direct therapies intended to decrease ICP, there simply is a lack of high-quality evidence to support this notion. With all that said, measuring ICP in patients receiving HTS therapy is currently recommended given the pathophysiologic rationale; that studies have demonstrated outcomes associated with decreased ICP (Alberico, Ward, Choi, Marmarou, & Young, 1987; Jagannathan et al., 2008; Kochanek et al., 2019); and that there exists no other accurate measure of intracranial hypertension.

The goal of administration of HTS is to achieve a "permissive hypernatremia" and subsequently lower ICP. However, not all hospitals have neurosurgical services (and thus, ICP monitoring) and therefore, sodium levels are often used as a more accessible objective proxy. The goal sodium during hyperosmolar therapy for increased ICP in patients with TBI is generally 145 to 160 meq/L (Guideline for the Management of Severe Traumatic Brain Injury during Retrieval, 2009; Kochanek et al., 2019). However, reaching levels too quickly or beyond this range can lead to complications. Indeed, adverse effects such as coagulopathies and AKI are thought to be attributable to HTS due to inducing supra-therapeutic sodium levels (Gonda et al., 2013; Kochanek et al., 2019). In addition to these complications, hypernatremia has long been associated with other adverse effects including but not limited to central pontine myelinolysis, seizures, and cerebral edema (Schwade, Haftel, Rühe, & Endmann, 2019). When monitoring sodium in this context, levels are generally checked no more than 4 h apart.

Volume overload has been frequently cited as a major risk factor of morbidity and mortality for critically ill children (Arikan et al., 2012; Chen et al., 2016). Therefore, when administering large amounts of fluid, it is critical to balance the benefits of HTS therapy with the risk of volume overload, especially given that these patients are often receiving other fluid

TABLE 2 Dosing and administration for concentrations of HTS.

Formulation	Bolus Dose (mL/kg)	Bolus Infusion Time (min)	Continuous Infusion dose (mL/kg/h)	Route
2%	3–5	30–60	0.1–1	Peripheral or Central
3%	2–5	10–20	0.1–1	Central
7% or 7.5%[a]	2	30	N/A	Central
23.4%	0.5 (max 30 mL)	10	N/A	Central

[a] *Adult doses. Pediatric doses have not been established for this concentration.*

therapies. Of note, the higher the saline concentration, the lower the volume. Therefore, 23.4% saline might be an attractive option for children with severe TBI who are at risk of fluid overload.

Potential adverse effects

Multiple studies cite concerns regarding associations between hyperosmolar therapy and adverse effects including coagulopathies, AKI, and hypernatremia (Khanna et al., 2000; Schell, Applegate 2nd, & Cole, 1996; Wisner, Schuster, & Quinn, 1990).

Hypernatremia is a salient concern when using HTS, as increasing sodium is generally the actual intention when attempting to lower elevated ICP. However, over-correction or too rapid of a correction should be avoided to prevent known potential complications of relatively sudden changes in sodium such as seizures, cerebral edema, and central pontine myelinolysis. An inappropriate increase in sodium leads to an inappropriate increase in blood osmolarity, which in turn causes a large efflux of water from intracellular compartments to the blood, shrinkage of these cells, and apoptosis (Crismale et al., 2017). Indeed, a 2009 retrospective study of adults with TBI found that hypernatremia is associated with increased mortality, adjusted for baseline risk (Maggiore et al., 2009). This association has been linked to mortality rates as high as 15–20% in children (Gossman, Naganathan, & Al-Dhahir, 2020). Taken together, it is crucial that sodium levels be closely monitored during HTS therapy such that levels do not exceed goal levels or goal rate. There is little literature on the appropriate rate of increase in patients with TBI, but in patients with symptomatic hyponatremia, goals are generally to increase sodium no more than 0.5 meq/L per hour to a maximum of 6 meq/L in the first 24 h.

Coagulopathies have been one of more common concerns of HTS use. A 2013 study in children by Gonda, et al. found that continuous infusions of HTS leading to sustained sodium levels above 170 meq/L was associated with higher frequencies of thrombocytopenia, neutropenia, and anemia compared to those whose sodium level was below this threshold (Gonda et al., 2013). Notably, the platelets rebounded to the normal range after the sodium levels were corrected in this study. Other studies have also suggested coagulopathies in patients receiving high volumes of HTS and achieving higher sodium levels (generally > 160 meq/L), though some studies suggest an increase in clotting whereas others suggest an increase in bleeding (Coleman et al., 2020; Webster, Fei, Falcone, & Kaplan, 2015). Potential mechanisms such as ionic interference with enzymatic reactions in the coagulation cascade by sodium have been explored in vitro studies (Coleman et al., 2020; Tan, Tan, Ng, & Loh, 2002), but extrapolation to the clinical setting is not straightforward. For example, severe TBI has been associated with coagulopathies independent of the therapies received (Zhang, Zhang, & Dong, 2018). In contrast, other studies have found no such association. Taken together, the role of HTS in coagulopathies in the context of TBI is still under investigation. At this point, it would be prudent for clinicians to monitor for both uncontrolled bleeding and unwanted thromboses (i.e., deep vein thrombi, cerebral venous sinus thrombi) in patients with severe TBI and in particular those receiving HTS therapy.

The incidence of AKI in children receiving hyperosmolar therapy for increased ICP has ranged from 0% to 53% (Gonda et al., 2013). This variation is likely due to the varying definitions of AKI in children. However, this data is confounded by the incidence of AKI in children with elevated ICP regardless of what therapy is received, which appears to be approximately 8% to 23% (Bailey et al., 2007; Chiwane & Chauhan, 2016; Guardenier, Peterson, Hilfiker, & Shellington, 2015; Prodhan et al., 2012). The potential mechanism underlying AKI in the setting of HTS use this is unknown, though animal studies have shown that acute hypernatremia can lead to reduced renal blood flow (Gerber, Branch, Nies, Hollifield, & Gerkens, 1979). Despite the ambiguity of the relationship between AKI and HTS therapy in the literature, it is common practice to closely follow indices of renal function during HTS therapy.

Literature review

The following tables detail the current existing literature that focuses on the use of hypertonic saline in children. Only clinical trials were included (other literature reviews, surveys, and guideline articles were not included) (Tables 3–6).

Applications to other areas of neuroscience

In this chapter, we review the role of HTS in children with TBI. The application of the information presented here overlaps with other areas of study relevant to TBI that are not necessarily covered in this chapter. This includes studies evaluating other forms of hyperosmolar therapy used in brain injury, namely mannitol which has been studied extensively in this context (Battison, Andrews, Graham, & Petty, 2005; Oddo et al., 2009; Vialet et al., 2003). The proposed mechanism by which mannitol affects the brain during injury is similar, though not identical, to the proposed mechanism for HTS. The

TABLE 3 Studies comparing HTS to other hyperosmolar therapies.

Citation	Study Design 1 = retrospective, Chart review 2 = prospective 3 = RCT 4 = Crossover 5 = Cross-sectional 6 = Case controlled 7 = Case Series 8 = Descriptive 9 = Cohort 10 = Observational NA = not applicable	Ages	Aim/Intervention(s)	Number of enrolled patients	Authors conclusions
Battison et al. (2005)	2,3,4	> 16 yo	Compare effects of equimolar HTS + dextran with mannitol	9	Rapid IV infusion 7.5%saline/6% dextran reduces ICP more effectively than mannitol
Bourdeaux & Brown (2011)	2,3	> 16 yo	Compare equiosmolar 5% HTS with Sodium bicarbonate 8.4% given over 30 min	11	Sodium bicarbonate is as effective as HTS in lowering ICP
Kumar, Devi, Reddy, Shukla et al. (2019)	2,3	≤ 16 yo	Effects of equiosmolar doses of mannitol and HTS in reducing ICP in children with TBI	30	Both mannitol and HTS are equally effective for treatment of increased ICP in children with severe TBI
Lumba-Brown et al. (2014)	2,3	7–16 yo	Investigate use of 3%HS in acute management of pain associated with concussion in children	44	3% HTS is more effective than NS in acutely reducing concussion pain in children
Mangat et al. (2020)	2,6	> 16 yo	Determine difference in effects of HTS and mannitol in the combined burden of elevated ICP and low CPP in patients with severe TBI	502	HS bolus appears to be superior to mannitol in reduction of the combined burden of elevated ICP and associated hypoperfusion in severe TBI
Roumeliotis et al. (2016)	1	1 mo–18 yo	Describe use of hyperosmolar therapy in pediatric TBI and examine its effect on ICP and CPP	16	Both mannitol and HTS are used in pediatric TBI, but dose and indications vary without clear indications for one versus the other, controlled studies needed
Shein, et al., (2016)	2,10	< 18 yo	To describe acute cerebral hemodynamic effects of medications used to treat ICP in children with TBI	25	ICP decreased after multiple drugs but HTS may warrant consideration as 1st line for treating ICP as it was associated with most favorable cerebral hemodynamics and fastest resolution of increased ICP
Simma et al. (1998)	2,3	1–14 yo	Compare Ringer's lactate vs HTS in severe TBI	32	Treatment of severe head injury with HTS is superior to treatment with lactated Ringer's. An increase in serum sodium concentrations significantly correlates with lower ICP and higher CPP. Children treated with hypertonic saline require fewer interventions, fewer complications, and shorter ICU stay.

TABLE 4 Studies detailing potential adverse effects of HTS.

Citation	Study Design 1 = retrospective, Chart review 2 = prospective 3 = RCT 4 = Crossover 5 = Cross-sectional 6 = Case controlled 7 = Case Series 8 = Descriptive 9 = Cohort NA = not applicable	Ages	Aim/Intervention(s)	Number of enrolled patients	Authors conclusions
Brenkert et al. (2013)	1, 5, 8	0–18 yo	Describe use, dosing, administration of HTS in ED	56	3% safely given without central line; dose 3–5 mL/kg
Gonda et al. (2013)	1	3 wks–19 yo	Determine safe upper limits for sodium and look for complications with prolonged HTS infusions	90	Children treated with HTS for ICP with sodium levels above 165–170 experienced significantly more acute renal failure, CBC changes, and ARDS
Luu et al. (2011)	1	2 mo–17 yo	Describe administration of 3% HTS during pediatric critical care transport	101	3% HTS can be administered safely during transport via a PIV
Mesghali et al. (2019)	1	3–51 yo	Compare incidence of extravasation injury with HTS or Mannitol was administered by peripheral IV	192 (85 HS, 107 mannitol)	No incidence of extravasation among patients given 3% HTS or mannitol
Stulce et al. (2020)	1	2–11 yo	Reports prevalence of fluid overload in children treated with 3% HTS and the effect on outcome	138	Patients with severe TBI did have higher rates of fluid overload but this did not contribute to mortality, length of ventilation, rates of AKI or increased PICU stay
Webster et al. (2015)	1,9	children	Evaluate the association between HTS and DVT in pediatric patients with severe TBI	58	In children with severe TBI the total bolus dose of 3% HTS and sustained levels of sodium > 160 mmoles/l are independently associated with DVT

Table is arranged by study citation, the type of study, the age of participants, the number of enrolled participants, and the author's conclusions.

TABLE 5 Studies evaluating the clinical effects of HTS in children with elevated ICP.

Citation	Study Design 1 = retrospective, Chart review 2 = prospective 3 = RCT 4 = Crossover 5 = Cross-sectional 6 = Case controlled 7 = Case Series 8 = Descriptive 9 = Cohort 10 = Observational NA = not applicable	Ages	Aim/Intervention(s)	Number of enrolled patients	Authors conclusions
Berger et al. (2002)	NA	12 yo, 11 yo	Case report(s)	2	12 yo had excellent outcome; 11 yo good outcome
Cummings et al. (2020)	1	1–18 yo	Describe initial experience with HTS 23.4% at 10, 20, 30 mL	16	All 3 volumes increased sodium and decreased ICP. Standard-volume based dosing may simplify weight based dosing, storage, and administration for pediatric emergencies but optimum dose and safety in children remains unknown
Khanna et al. (2000)	1	4 mo–12 yo	Evaluate effect of prolonged infusion of 3% HTS and sustained hypernatremia on refractory ICP in TBI	10	Increased sodium levels by using HTS decreases ICP, increases CPP; sustained hypernatremia and hyperosmolarity are safely tolerated in children
Nakagawa et al. (2012)	NA	23 mo/4 yo	Present 2 children with severe TBI treated with 23.4% HS	2	Able to safely and effectively use 23.4% HTS in these 2 children without apparent complications
Paredes-Andrade et al. (2012)	2	17–67	Evaluate effectiveness of HTS in reducing ICP in presence of wide range of serum and CSF osmolalities	18	23.4% HTS bolus is effective for reduction of ICP in patients with severe TBI even in presence of high serum and CSF osmolalities
Peterson et al. (2000)	1	Children-Average age 8	Determine effects of continuous infusion of 3% HTS on ICP and describe physiologic effects of HS	78	HS to children with closed head injury appears promising to control cerebral edema. More study needed to determine optimal duration of treatment.
Piper & Harrigan (2015)	1,9	8 mo–17 yo	Review 9 years' experience with 23.4% HTS in pediatric TBI	32	The majority of pediatric patients respond to 23.4% HTS administered via bolus with a mean drop in ICP of 10.2 mmHg: Glasgow coma outcome score suggested that 74% of their patients would likely return to an independent existence compared to ≤40% in other studies; consider maintenance of sodium with a lower concentration of HS

TABLE 5 Studies evaluating the clinical effects of HTS in children with elevated ICP—cont'd

Citation	Study Design 1 = retrospective, Chart review 2 = prospective 3 = RCT 4 = Crossover 5 = Cross-sectional 6 = Case controlled 7 = Case Series 8 = Descriptive 9 = Cohort 10 = Observational NA = not applicable	Ages	Aim/Intervention(s)	Number of enrolled patients	Authors conclusions
Rallis et al. (2017)	1,9	≤14 yo	Evaluate effect of 7.5% HTS on ICP and CPP in children with severe TBI	29	7.5% HTS as second-tier therapy was associated with significant reduction in ICP and increase in CPP in children with severe TBI
Rockswold et al. (2009)	1	16–64	Determine effect on HTS on ICP, CPP, and brain tissues oxygenation in severe TBI	25	HTS as single osmotic agent decreased ICP while improving CPP and brain oxygenation in severe TBI; Patients with higher baseline ICP and lower CPP responded to HTS more significantly
Sabers et al. (2020)	1	31 day–18 yo	Describe outcomes with bolus and continuous infusions of HTS in children with TBI	45	Equimolar bolus variable concentrations of HTS were associated with significant and sustained decreased ICP. Reduction in ICP greatest in HTS 12% group as was least of pulmonary edema; children in 6% group had best fluid balance and highest protein and calorie intake Higher concentrations of HTS may improve fluid balance, less pulmonary edema and greater protein and energy intake
Siddiqui et al. (2019)	7	1–16 yo	Determine clinical outcome and length of stay in pediatric patients receiving 3% HTS in the emergency room	105	Pediatric patients with TBI receiving 3% HTS improved GCS and decreased length of stay
Wu et al. (2019)	1	< 18 yo	Explore effect of 23.4% HTS for management of elevated ICP in children	26	Use of 23.4% HTS in children admitted for severe TBI is associated with a significant decrease in ICP within 1 h of use

Table is arranged by study citation, the type of study, the age of participants, the number of enrolled participants, and the author's conclusions.

TABLE 6 Studies evaluating the pathophysiologic effects of HTS in children with acute TBI.

Citation	Study Design 1 = retrospective, Chart review 2 = prospective 3 = RCT 4 = Crossover 5 = Cross-sectional 6 = Case controlled 7 = Case Series 8 = Descriptive 9 = Cohort 10 = Observational NA = not applicable	Ages	Aim/Intervention(s)	Number of enrolled patients	Authors conclusions
Junger et al. (2013)	2,3 (a priori subgroup)	≥15 yo	Does hypertonic fluid resuscitation reduce PMN activation in TBI patients	83 with severe TBI/ 20 healthy controls	Pre-hospital resuscitation with HTS can partially restore normal PMN activity and apoptotic behavior or PMN's while HTS + dextran was largely ineffective
Pascual et al. (2008)	2	14–71	Could HTS be safely standardized in patients with severe TBI and lower ICP while improving CPP and brain oxygenation	12	HS infusion in hypotensive severe TBI reduces ICP and raises CPP. Brain tissue oxygenation tends to improve after HTS infusion
Rhodes et al. (2016)	2	16–65 yo	Report changes in brain oxygen tension over first 24 h in patients with severe TBI;	25	Trends in brain oxygen tension over first 24 h in cohort as a whole but individual trends related to outcome with improved outcome in those with improving brain oxygenation

Table is arranged by study citation, the type of study, the age of participants, the number of enrolled participants, and the author's conclusions.

information in this chapter may also apply to fields of study evaluating the role of sodium in brain injury, namely how the osmotic shifts between intravascular and intracellular spaces during injury change as sodium changes and leads to either beneficial or harmful outcomes (Crismale et al., 2017; Maggiore et al., 2009). In this chapter, we also discuss coagulopathies as a potential adverse effect of HTS. Though this effect has not been well-defined in the literature as of yet, those who are interested in the pathophysiology of TBI as a whole or its systemic effects on the body may find the interplay between changes in sodium levels and coagulopathies of interest.

Mini-dictionary of terms

Traumatic brain injury: The condition in which an insult to the structure or function of the brain (i.e. bleed, deformation, etc.) is caused by a physical force (i.e. motor vehicle accident, projectile, fall, etc.)

Hypertonic saline: Formulations of sodium chloride that are higher than the concentration in normal saline, which is 0.9% sodium chloride (or 0.9 g of sodium chloride in 100 mL of water).

Hyperosmolar therapy: A form of clinical management for TBI in which solutions with an osmolarity higher than that of human plasma are infused into the bloodstream in an effort to increase the osmotic pressure intravascularly, thereby drawing fluid from the tissues into the vessels and subsequently reducing tissue edema.

Edema: A clinical symptom in which fluid leaves the intravascular space and enters the extracellular space of tissues. Commonly and colloquially referred to as "swelling."

Intracranial pressure: The pressure inside the skull that is created by the fluids inside and around the brain, namely cerebrospinal fluid (CSF).

Cerebral perfusion pressure: The pressure that the brain tissues experience as a result from incoming arterial blood flow minus the ICP.
Hypernatremia: Sodium levels higher than the physiologic normal range, which is usually defined as 135 to 145 meq/L.
Cerebrospinal fluid: A clear watery fluid generated inside the brain and that covers the brain and spinal cord.
Acute kidney injury: A clinical condition in which one or both kidneys have suffered a physiologic or structural insult such as decreased perfusion or infection.
Glasgow Coma Score: A scale from 3 to 15 commonly used to measure the degree of neurological impairment in an individual.

Key facts of TBI in children

- TBI contributes to about a third of all pediatric deaths from injury in the U.S.
- Children are the most commonly presenting age group in the emergency department for TBI
- Adolescents are the second most common age group to die from TBI
- Children are more susceptible to TBI compared to adults due to their large brain-to-skull size ratio, larger head-to-body size ratio, and thinner bones.
- Many recommendations for TBI management in children are based on reviews of adult guidelines by expert consensus

Summary points

- The cerebral edema that follows TBI involves vasogenic and cytotoxic edema that altogether precipitates a complex inflammatory cascade and microvascular injury.
- The primary mechanism by which HTS reduces ICP is by the creation of an osmolar gradient between the intracellular and extravascular space.
- A bolus of 3% HTS should be considered in a pediatric patient with suspected or known increased ICP
- Sodium levels above 160 meq/L should be avoided
- Monitoring ICP and sodium levels frequently is key to promoting positive outcomes while minimizing risk of adverse effects

References

Alberico, A. M., Ward, J. D., Choi, S. C., Marmarou, A., & Young, H. F. (1987). Outcome after severe head injury. Relationship to mass lesions, diffuse injury, and ICP course in pediatric and adult patients. *Journal of Neurosurgery*, *67*(5), 648–656. https://doi.org/10.3171/jns.1987.67.5.0648.

Arikan, A. A., Zappitelli, M., Goldstein, S. L., Naipaul, A., Jefferson, L. S., & Loftis, L. L. (2012). Fluid overload is associated with impaired oxygenation and morbidity in critically ill children*. *Pediatric Critical Care Medicine*, *13*(3), 253–258. https://doi.org/10.1097/PCC.0b013e31822882a3.

Bailey, D., Phan, V., Litalien, C., Ducruet, T., Mérouani, A., Lacroix, J., & Gauvin, F. (2007). Risk factors of acute renal failure in critically ill children: A prospective descriptive epidemiological study*. *Pediatric Critical Care Medicine*, *8*(1), 29–35. https://doi.org/10.1097/01.pcc.0000256612.40265.67.

Battison, C., Andrews, P. J., Graham, C., & Petty, T. (2005). Randomized, controlled trial on the effect of a 20% mannitol solution and a 7.5% saline/6% dextran solution on increased intracranial pressure after brain injury. *Critical Care Medicine*, *33*(1), 196–202. discussion 257-198 https://doi.org/10.1097/01.ccm.0000150269.65485.a6.

Berger, S., Schürer, L., Härtl, R., Messmer, K., & Baethmann, A. (1995). Reduction of post-traumatic intracranial hypertension by hypertonic/hyperoncotic saline/dextran and hypertonic mannitol. *Neurosurgery*, *37*(1), 98–107. discussion 107-108 https://doi.org/10.1227/00006123-199507000-00015.

Berger, S., Schwarz, M., & Huth, R. (2002). Hypertonic saline solution and decompressive craniectomy for treatment of intracranial hypertension in pediatric severe traumatic brain injury. *Journal of Trauma and Acute Care Surgery*, *53*(3), 558–563.

Bhardwaj, A., & Ulatowski, J. A. (1999). Cerebral edema: Hypertonic saline solutions. *Current Treatment Options in Neurology*, *1*(3), 179–188. https://doi.org/10.1007/s11940-999-0002-z.

Bhardwaj, A., & Ulatowski, J. A. (2004). Hypertonic saline solutions in brain injury. *Current Opinion in Critical Care*, *10*(2), 126–131. https://doi.org/10.1097/00075198-200404000-00009.

Bourdeaux, C. P., & Brown, J. M. (2011). Randomized controlled trial comparing the effect of 8.4% sodium bicarbonate and 5% sodium chloride on raised intracranial pressure after traumatic brain injury. *Neurocritical Care.*, *15*(1), 42–45.

Brenkert, T. E., Estrada, C. M., McMorrow, S. P., & Abramo, T. J. (2013). Intravenous hypertonic saline use in the pediatric emergency department. *Pediatric Emergency Care.*, *29*(1), 71–73.

Chen, J., Li, X., Bai, Z., Fang, F., Hua, J., Li, Y., … Li, Y. (2016). Association of Fluid Accumulation with clinical outcomes in critically ill children with severe Sepsis. *PLoS One*, *11*(7). https://doi.org/10.1371/journal.pone.0160093, e0160093.

Chesnut, R. M., Temkin, N., Carney, N., Dikmen, S., Rondina, C., Videtta, W., … Hendrix, T. (2012). A trial of intracranial-pressure monitoring in traumatic brain injury. *The New England Journal of Medicine*, *367*(26), 2471–2481. https://doi.org/10.1056/NEJMoa1207363.

Chiwane, S., & Chauhan, M. (2016). 1319: Incidence, risk factors and outcomes of acute kidney injury in children with acute neurologic injury. *Critical Care Medicine*, *44*(12), 405. https://doi.org/10.1097/01.ccm.0000509993.01771.e3.

Coleman, J. R., Moore, E. E., Silliman, C. C., Stettler, G. R., Nunns, G. R., Samuels, J. M., … Sauaia, A. (2020). Examining the effect of hypertonic saline administered for reduction of intracranial hypertension on coagulation. *Journal of the American College of Surgeons*, *230*(3), 322–330.e322. https://doi.org/10.1016/j.jamcollsurg.2019.11.011.

Crismale, J. F., Meliambro, K. A., DeMaria, S., Jr., Bronster, D. B., Florman, S., & Schiano, T. D. (2017). Prevention of the osmotic demyelination syndrome after liver transplantation: A multidisciplinary perspective. *American Journal of Transplantation*, *17*(10), 2537–2545. https://doi.org/10.1111/ajt.14317.

Cummings, B. M., Fernandes, N. D., Parker, L. F., Murphy, S. A., & Yager, P. H. (2020). Standardized volume dosing protocol of 23.4% hypertonic saline for pediatric critical care: Initial experience. *Annals of Pharmacotherapy*, 1060028020907997.

Ducey, J. P., Mozingo, D. W., Lamiell, J. M., Okerburg, C., & Gueller, G. E. (1989). A comparison of the cerebral and cardiovascular effects of complete resuscitation with isotonic and hypertonic saline, hetastarch, and whole blood following hemorrhage. *The Journal of Trauma*, *29*(11), 1510–1518. https://doi.org/10.1097/00005373-198911000-00010.

Gerber, J. G., Branch, R. A., Nies, A. S., Hollifield, J. W., & Gerkens, J. F. (1979). Influence of hypertonic saline on canine renal blood flow and renin release. *The American Journal of Physiology*, *237*(6), F441–F446. https://doi.org/10.1152/ajprenal.1979.237.6.F441.

Gonda, D. D., Meltzer, H. S., Crawford, J. R., Hilfiker, M. L., Shellington, D. K., Peterson, B. M., & Levy, M. L. (2013). Complications associated with prolonged hypertonic saline therapy in children with elevated intracranial pressure. *Pediatric Critical Care Medicine*, *14*(6), 610–620. https://doi.org/10.1097/PCC.0b013e318291772b.

Gossman, W., Naganathan, S., & Al-Dhahir, M. A. (2020). *Hypernatremia*. Treasure Island, FL: StatPearls Publishing.

Guardenier, A., Peterson, B., Hilfiker, M., & Shellington, D. (2015). 1170: Acute renal failure in pediatric patients after moderate to severe traumatic BRAIN injury. *Critical Care Medicine*, *43*(12), 294. https://doi.org/10.1097/01.ccm.0000475001.16167.bc.

Guideline for the Management of Severe Traumatic Brain Injury during Retrieval. (2009). *National Health Service* (p. 9).

Guideline to Make Up Hypertonic Saline (3% NaCl). (2018). *N. H. Service* (p. 1).

Harutjunyan, L., Holz, C., Rieger, A., Menzel, M., Grond, S., & Soukup, J. (2005). Efficiency of 7.2% hypertonic saline hydroxyethyl starch 200/0.5 versus mannitol 15% in the treatment of increased intracranial pressure in neurosurgical patients–a randomized clinical trial [ISRCTN62699180]. *Critical Care (London, England)*, *9*(5), R530–R540. https://doi.org/10.1186/cc3767.

Jagannathan, J., Okonkwo, D. O., Yeoh, H. K., Dumont, A. S., Saulle, D., Haizlip, J., … Jane, J. A., Jr. (2008). Long-term outcomes and prognostic factors in pediatric patients with severe traumatic brain injury and elevated intracranial pressure. *Journal of Neurosurgery. Pediatrics*, *2*(4), 240–249. https://doi.org/10.3171/ped.2008.2.10.240.

Jha, R. M., Kochanek, P. M., & Simard, J. M. (2019). Pathophysiology and treatment of cerebral edema in traumatic brain injury. *Neuropharmacology*, *145*, 230–246. https://doi.org/10.1016/j.neuropharm.2018.08.004.

Junger, W. G., Rhind, S. G., Rizoli, S. B., Cuschieri, J., Baker, A. J., Shek, P. N., … Bulger, E. M. (2013). Pre-hospital hypertonic saline resuscitation attenuates the activation and promotes apoptosis of neutrophils in patients with severe traumatic brain injury. *Shock (Augusta, Ga.)*, *40*(5), 366.

Keenan, H. T., Nocera, M., & Bratton, S. L. (2005). Frequency of intracranial pressure monitoring in infants and young toddlers with traumatic brain injury. *Pediatric Critical Care Medicine*, *6*(5), 537–541. https://doi.org/10.1097/01.pcc.0000164638.44600.67.

Khanna, S., Davis, D., Peterson, B., Fisher, B., Tung, H., O'Quigley, J., & Deutsch, R. (2000). Use of hypertonic saline in the treatment of severe refractory posttraumatic intracranial hypertension in pediatric traumatic brain injury. *Critical Care Medicine*, *28*(4), 1144–1151.

Khellaf, A., Khan, D. Z., & Helmy, A. (2019). Recent advances in traumatic brain injury. *Journal of Neurology*, *266*(11), 2878–2889. https://doi.org/10.1007/s00415-019-09541-4.

Kochanek, P. M., Tasker, R. C., Carney, N., Totten, A. M., Adelson, P. D., Selden, N. R., … Wainwright, M. S. (2019). Guidelines for the management of pediatric severe traumatic brain Injury, third edition: Update of the brain trauma foundation guidelines. *Pediatric Critical Care Medicine*, *20*(3S Suppl 1), S1–s82. https://doi.org/10.1097/pcc.0000000000001735.

Koenig, M. A. (2018). Cerebral edema and elevated intracranial pressure. *Continuum (Minneap Minn)*, *24*(6), 1588–1602. https://doi.org/10.1212/con.0000000000000665.

Kumar, S. A., Devi, B. I., Reddy, M., & Shukla, D. (2019). Comparison of equiosmolar dose of hyperosmolar agents in reducing intracranial pressure—a randomized control study in pediatric traumatic brain injury. *Child's Nervous System.*, *35*(6), 999–1005.

Latorre, J. G., & Greer, D. M. (2009). Management of acute intracranial hypertension: A review. *The Neurologist*, *15*(4), 193–207. https://doi.org/10.1097/NRL.0b013e31819f956a.

Lovett, M. E., O'Brien, N. F., & Leonard, J. R. (2019). Children with severe traumatic brain injury, intracranial pressure, cerebral perfusion pressure, what does it mean? A review of the literature. *Pediatric Neurology*, *94*, 3–20. https://doi.org/10.1016/j.pediatrneurol.2018.12.003.

Lumba-Brown, A., Harley, J., Lucio, S., Vaida, F., & Hilfiker, M. (2014). Hypertonic saline as a therapy for pediatric concussive pain: a randomized controlled trial of symptom treatment in the emergency department. *Pediatric Emergency Care.*, *30*(3), 139.

Lutters, B., Koehler, P. J., & Wijdicks, E. F. (2020). Worth their salt: One hundred years of hyperosmolar therapy. *European Neurology*, 1–6. https://doi.org/10.1159/000510183.

Luu, J. L., Wendtland, C. L., Gross, M. F., Mirza, F., Zouros, A., Zimmerman, G. J., … Abd-Allah, S. A. (2011). Three percent saline administration during pediatric critical care transport. *Pediatric Emergency Care.*, *27*(12), 1113–1117.

Maggiore, U., Picetti, E., Antonucci, E., Parenti, E., Regolisti, G., Mergoni, M., … Fiaccadori, E. (2009). The relation between the incidence of hypernatremia and mortality in patients with severe traumatic brain injury. *Critical Care (London, England)*, *13*(4), R110. https://doi.org/10.1186/cc7953.

Mangat, H. S., Wu, X., Gerber, L. M., Schwarz, J. T., Fakhar, M., Murthy, S. B., … Härtl, R. (2020). Hypertonic saline is superior to mannitol for the combined effect on intracranial pressure and cerebral perfusion pressure burdens in patients with severe traumatic brain injury. *Neurosurgery.*, *86*(2), 221–230.

Marmarou, A., Saad, A., Aygok, G., & Rigsbee, M. (2005). Contribution of raised ICP and hypotension to CPP reduction in severe brain injury: Correlation to outcome. *Acta Neurochirurgica. Supplement*, *95*, 277–280. https://doi.org/10.1007/3-211-32318-x_57.

Mesghali, E., Fitter, S., Bahjri, K., & Moussavi, K. (2019). Safety of peripheral line administration of 3% hypertonic saline and mannitol in the emergency department. *The Journal of Emergency Medicine.*, *56*(4), 431–436.

Michinaga, S., & Koyama, Y. (2015). Pathogenesis of brain edema and investigation into anti-edema drugs. *International Journal of Molecular Sciences*, *16*(5), 9949–9975. https://doi.org/10.3390/ijms16059949.

Mirski, A. M., Denchev, I. D., Schnitzer, S. M., & Hanley, F. D. (2000). Comparison between hypertonic saline and mannitol in the reduction of elevated intracranial pressure in a rodent model of acute cerebral injury. *Journal of Neurosurgical Anesthesiology*, *12*(4), 334–344. https://doi.org/10.1097/00008506-200010000-00006.

Nakagawa, K., Chang, C. W., Koenig, M. A., Yu, M., & Tokumaru, S. (2012). Treatment of refractory intracranial hypertension with 23.4% saline in children with severe traumatic brain injury. *Journal of Clinical Anesthesia.*, *24*(4), 318–323.

Niknam, K., Mistry, R., & Lumba-Brown, A. (2019). Pediatric hypertonic saline use in emergency departments. *The American Journal of Emergency Medicine*, *37*(5), 981–983. https://doi.org/10.1016/j.ajem.2018.09.040.

Oddo, M., Levine, J. M., Frangos, S., Carrera, E., Maloney-Wilensky, E., Pascual, J. L., … LeRoux, P. D. (2009). Effect of mannitol and hypertonic saline on cerebral oxygenation in patients with severe traumatic brain injury and refractory intracranial hypertension. *Journal of Neurology, Neurosurgery, and Psychiatry*, *80*(8), 916–920. https://doi.org/10.1136/jnnp.2008.156596.

Ogden, A. T., Mayer, S. A., & Connolly, E. S., Jr. (2005). Hyperosmolar agents in neurosurgical practice: The evolving role of hypertonic saline. *Neurosurgery*, *57*(2), 207–215. discussion 207-215 https://doi.org/10.1227/01.neu.0000166533.79031.d8.

Paredes-Andrade, E., Solid, C. A., Rockswold, S. B., Odland, R. M., & Rockswold, G. L. (2012). Hypertonic saline reduces intracranial hypertension in the presence of high serum and cerebrospinal fluid osmolalities. *Neurocritical Care.*, *17*(2), 204–210.

Pascual, J. L., Maloney-Wilensky, E., Reilly, P. M., Sicoutris, C., Keutmann, M. K., Stein, S. C., … Gracias, V. H. (2008). Resuscitation of hypotensive head-injured patients: Is hypertonic saline the answer? *The American Surgeon.*, *74*(3), 253–259.

Peterson, B., Khanna, S., Fisher, B., & Marshall, L. (2000). Prolonged hypernatremia controls elevated intracranial pressure in head-injured pediatric patients. *Critical Care Medicine*, *28*(4), 1136–1143. https://doi.org/10.1097/00003246-200004000-00037.

Piper, B. J., & Harrigan, P. W. (2015). Hypertonic saline in paediatric traumatic brain injury: A review of nine years' experience with 23.4% hypertonic saline as standard hyperosmolar therapy. *Anaesthesia and Intensive Care.*, *43*(2), 204–210.

Prodhan, P., McCage, L. S., Stroud, M. H., Gossett, J., Garcia, X., Bhutta, A. T., … Blaszak, R. T. (2012). Acute kidney injury is associated with increased in-hospital mortality in mechanically ventilated children with trauma. *Journal of Trauma and Acute Care Surgery*, *73*(4), 832–837. https://doi.org/10.1097/TA.0b013e31825ab14f.

Qureshi, A. I., & Suarez, J. I. (2000). Use of hypertonic saline solutions in treatment of cerebral edema and intracranial hypertension. *Critical Care Medicine*, *28*(9), 3301–3313. https://doi.org/10.1097/00003246-200009000-00032.

Rallis, D., Poulos, P., Kazantzi, M., Chalkias, A., & Kalampalikis, P. (2017). Effectiveness of 7.5% hypertonic saline in children with severe traumatic brain injury. *Journal of Critical Care.*, *38*, 52–56.

Raslan, A., & Bhardwaj, A. (2007). Medical management of cerebral edema. *Neurosurgical Focus*, *22*(5). https://doi.org/10.3171/foc.2007.22.5.13, E12.

Rhodes, J. K., Chandrasekaran, S., & Andrews, P. J. (2016). Early changes in brain oxygen tension may predict outcome following severe traumatic brain injury. In *InIntracranial Pressure and Brain Monitoring XV* (pp. 9–16). Cham: Springer.

Rockswold, G. L., Solid, C. A., Paredes-Andrade, E., Rockswold, S. B., Jancik, J. T., & Quickel, R. R. (2009). Hypertonic saline and its effect on intracranial pressure, cerebral perfusion pressure, and brain tissue oxygen. *Neurosurgery*, *65*(6), 1035–1041. discussion 1041-1032 https://doi.org/10.1227/01.Neu.0000359533.16214.04.

Rosenfeld, J. V., Maas, A. I., Bragge, P., Morganti-Kossmann, M. C., Manley, G. T., & Gruen, R. L. (2012). Early management of severe traumatic brain injury. *Lancet*, *380*(9847), 1088–1098. https://doi.org/10.1016/s0140-6736(12)60864-2.

Roumeliotis, N., Dong, C., Pettersen, G., Crevier, L., & Emeriaud, G. (2016). Hyperosmolar therapy in pediatric traumatic brain injury: a retrospective study. *Child's Nervous System.*, *32*(12), 2363–2368.

Sabers, E. J., Reiter, P. D., Skillman, H. E., & DeMasellis, G. (2020). Concentrated hypertonic saline in severe pediatric traumatic brain injury. *Brain Injury.*, *34*(6), 828–833.

Schell, R. M., Applegate, R. L., 2nd, & Cole, D. J. (1996). Salt, starch, and water on the brain. *Journal of Neurosurgical Anesthesiology*, *8*(2), 178–182.

Schwade, J.-N., Haftel, L., Rühe, L., & Endmann, M. (2019). Extrapontine Myelinolysis following extreme hypernatremia and hyperosmolarity. *Case Reports in Pediatrics*, *2019*, 7381597. https://doi.org/10.1155/2019/7381597.

Shein, S. L., Ferguson, N. M., Kochanek, P. M., Bayir, H., Clark, R. S., Fink, E. L., … Bell, M. J. (2016). Effectiveness of pharmacological therapies for intracranial hypertension in children with severe traumatic brain injury–results from an automated data collection system time-synched to drug administration. *Pediatric critical care medicine: A journal of the Society of Critical Care Medicine and the World Federation of Pediatric Intensive and Critical Care Societies*, *17*(3), 236.

Siddiqui, E. U., Waheed, S., Perveen, F., Daniyal, M., Khan, M. A., Siddiqui, S., & Siddiqui, Z. (2019). Clinical outcome of paediatric patients with traumatic brain injury (TBI) receiving 3% hypertonic saline (HTS) in the emergency room of a tertiary care hospital. *JPMA.*.

Simma, B., Burger, R., Falk, M., Sacher, P., & Fanconi, S. (1998). A prospective, randomized, and controlled study of fluid management in children with severe head injury: Lactated Ringer's solution versus hypertonic saline. *Critical Care Medicine*, *26*(7), 1265–1270. https://doi.org/10.1097/00003246-199807000-00032.

Stulce, C., Reisner, A., Kane, J. M., Shin, H. S., McCracken, C., Williamson, J., … Paden, M. (2020). Fluid overload in pediatric severe traumatic brain injury. Pediatric Critical Care Medicinel Society of. *Critical Care Medicine.*, *21*(2), 164–169.

Suarez, J. I. (2004). Hypertonic saline for cerebral edema and elevated intracranial pressure. *Cleveland Clinic Journal of Medicine*, *71*(Suppl 1), S9–13. https://doi.org/10.3949/ccjm.71.suppl_1.s9.

Suarez, J. I., Qureshi, A. I., Bhardwaj, A., Williams, M. A., Schnitzer, M. S., Mirski, M., … Ulatowski, J. A. (1998). Treatment of refractory intracranial hypertension with 23.4% saline. *Critical Care Medicine*, *26*(6), 1118–1122. https://doi.org/10.1097/00003246-199806000-00038.

Surani, S., Lockwood, G., Macias, M. Y., Guntupalli, B., & Varon, J. (2015). Hypertonic saline in elevated intracranial pressure: Past, present, and future. *Journal of Intensive Care Medicine*, *30*(1), 8–12. https://doi.org/10.1177/0885066613487151.

Surveillance Report of Traumatic Brain Injury-related Emergency Department Visits, Hospitalizations, and Deaths—United States, 2014. (2019). Retrieved from https://www.cdc.gov/traumaticbraininjury/pdf/TBI-Surveillance-Report-508.pdf.

Tan, T. S., Tan, K. H., Ng, H. P., & Loh, M. W. (2002). The effects of hypertonic saline solution (7.5%) on coagulation and fibrinolysis: An in vitro assessment using thromboelastography. *Anaesthesia*, *57*(7), 644–648. https://doi.org/10.1046/j.1365-2044.2002.02603.x.

Tilford, J. M., Aitken, M. E., Anand, K. J. S., Green, J. W., Goodman, A. C., Parker, J. G., … Adelson, P. D. (2005). Hospitalizations for critically ill children with traumatic brain injuries: A longitudinal analysis. *Critical Care Medicine*, *33*(9), 2074–2081. https://doi.org/10.1097/01.ccm.0000171839.65687.f5.

Tilford, J. M., Simpson, P. M., Yeh, T. S., Lensing, S., Aitken, M. E., Green, J. W., … Fiser, D. H. (2001). Variation in therapy and outcome for pediatric head trauma patients. *Critical Care Medicine*, *29*(5), 1056–1061. https://doi.org/10.1097/00003246-200105000-00037.

Vassar, M. J., Perry, C. A., Gannaway, W. L., & Holcroft, J. W. (1991). 7.5% sodium chloride/dextran for resuscitation of trauma patients undergoing helicopter transport. *Archives of Surgery*, *126*(9), 1065–1072. https://doi.org/10.1001/archsurg.1991.01410330019002.

Vialet, R., Albanèse, J., Thomachot, L., Antonini, F., Bourgouin, A., Alliez, B., & Martin, C. (2003). Isovolume hypertonic solutes (sodium chloride or mannitol) in the treatment of refractory posttraumatic intracranial hypertension: 2 mL/kg 7.5% saline is more effective than 2 mL/kg 20% mannitol. *Critical Care Medicine*, *31*(6), 1683–1687. https://doi.org/10.1097/01.Ccm.0000063268.91710.Df.

Ware, M. L., Nemani, V. M., Meeker, M., Lee, C., Morabito, D. J., & Manley, G. T. (2005). Effects of 23.4% sodium chloride solution in reducing intracranial pressure in patients with traumatic brain injury: A preliminary study. *Neurosurgery*, *57*(4), 727–736 (discussion 727-736).

Webster, D. L., Fei, L., Falcone, R. A., & Kaplan, J. M. (2015). Higher-volume hypertonic saline and increased thrombotic risk in pediatric traumatic brain injury. *Journal of Critical Care*, *30*(6), 1267–1271. https://doi.org/10.1016/j.jcrc.2015.07.022.

Weed, L. H., & McKibben, P. S. (1919). Experimental alteration of brain bulk. *The American Journal of Physiology*, *48*, 531–555.

White, H., Cook, D., & Venkatesh, B. (2006). The use of hypertonic saline for treating intracranial hypertension after traumatic brain injury. *Anesthesia and Analgesia*, *102*(6), 1836–1846. https://doi.org/10.1213/01.ane.0000217208.51017.56.

Winkler, E. A., Minter, D., Yue, J. K., & Manley, G. T. (2016). Cerebral edema in traumatic brain injury: Pathophysiology and prospective therapeutic targets. *Neurosurgery Clinics of North America*, *27*(4), 473–488. https://doi.org/10.1016/j.nec.2016.05.008.

Wisner, D. H., Schuster, L., & Quinn, C. (1990). Hypertonic saline resuscitation of head injury: Effects on cerebral water content. *The Journal of Trauma*, *30*(1), 75–78.

Witherspoon, B., & Ashby, N. E. (2017). The use of mannitol and hypertonic saline therapies in patients with elevated intracranial pressure: A review of the evidence. *The Nursing Clinics of North America*, *52*(2), 249–260. https://doi.org/10.1016/j.cnur.2017.01.002.

Worthley, L. I., Cooper, D. J., & Jones, N. (1988). Treatment of resistant intracranial hypertension with hypertonic saline. Report of two cases. *Journal of Neurosurgery*, *68*(3), 478–481. https://doi.org/10.3171/jns.1988.68.3.0478.

Wu, A. G., Samadani, U., Slusher, T. M., Zhang, L., & Kiragu, A. W. (2019). 23.4% hypertonic saline and intracranial pressure in severe traumatic brain injury among children: a 10-Year retrospective analysis. *Pediatric Critical Care Medicine.*, *20*(5), 466–473.

Zhang, J., Zhang, F., & Dong, J.-F. (2018). Coagulopathy induced by traumatic brain injury: Systemic manifestation of a localized injury. *Blood*, *131*(18), 2001–2006. https://doi.org/10.1182/blood-2017-11-784108.

Ziai, W. C., Toung, T. J., & Bhardwaj, A. (2007). Hypertonic saline: First-line therapy for cerebral edema? *Journal of the Neurological Sciences*, *261*(1–2), 157–166. https://doi.org/10.1016/j.jns.2007.04.048.

Zornow, M. H. (1996). Hypertonic saline as a safe and efficacious treatment of intracranial hypertension. *Journal of Neurosurgical Anesthesiology*, *8*(2), 175–177. https://doi.org/10.1097/00008506-199604000-00021.

Chapter 33

Treatment of vascular lesions in traumatic brain injury

Alvaro Cordoba[a,b]
[a]Neurosurgery and Intensive Care Unit ASESP, Montevideo, Uruguay, [b]Medical University of Vienna, Vienna, Austria

List of abbreviations

BCVI blunt cerebrovascular injury
CCF carotid cavernous fistula
CTA computed tomography angiography
DSA digital subtraction angiography
PTCAD post-traumatic carotid artery dissection
TBI traumatic brain injury
TCVI traumatic cerebrovascular injury
VA vertebral artery
VAD vertebral arterial dissection
VV vasa vasorum

Arterial lesions: Dissections and pseudoaneurysms

The Denver criteria are a set of screening criteria for blunt cerebrovascular injury (BCVI) in trauma used to reduce the need for CT angiography and its associated radiation exposure.

The screening protocol criteria 1,3 for BCVI are divided into signs/symptoms and risk factors:

Signs and symptoms,
arterial hemorrhage,
cervical bruit in patients < 50 years of age,
expanding cervical hematoma,
focal neurologic deficit,
neurologic exam incongruous with head CT scan findings,
stroke on secondary CT scan,
risk factors,
high-energy transfer mechanism with,
Le Fort fracture: type 2 or 3,
base of skull fractures involving the carotid canal,
cervical spine fractures patterns,
fractures involving C1, C2, or C3,
fractures involving the transverse foramen,
subluxation,
near hanging with hypoxic-ischemic (anoxic) brain injury, and
diffuse axonal injury with a Glasgow coma scale < 6,

If a trauma patient meets one or many of these criteria, the diagnosis of BCVI should be excluded with CTA or DSA.

Wide variation exists in the literature regarding the effectiveness of the criteria. In a small retrospective study, a negative Denver criteria screening excluded BCVI in 99.8% of patients. Whereas a large systematic review and meta-analysis of more than 120,000 trauma patients found BCVI present in 20%–22% of patients not meeting traditional screening criteria.

Diagnosis and Treatment of Traumatic Brain Injury. https://doi.org/10.1016/B978-0-12-823347-4.00005-1

The modified Memphis criteria are a set of screening criteria for BCVI (BCVI) in trauma. The presence of one or more of these criteria makes necessary a complementary CTA or DSA study to exclude a BCVI.

The screening protocol criteria for BCVI are:

base of skull fracture with involvement of the carotid canal,
base of skull fracture with involvement of petrous temporal bone,
cervical spine fracture,
neurological exam findings not explained by neuroimaging,
Horner syndrome,
Le Fort II or III fracture pattern, and
neck soft tissue injury (e.g. seatbelt sign, hanging, hematoma).

Ciapetti et al. (2010) showed that a modification of the original Memphis criteria could increase the sensitivity of BCVI 1. If a trauma patient meets one or many of these criteria, the diagnosis of BCVI should be excluded with CTA or DSA. DSA is far more sensitive than 32 CT 2 but less available. However, a study retrospectively applying the criteria has recently concluded that CTA should be performed on all multi-trauma patients for 37.5% of injuries would have been missed.

Non-hemorrhagic spontaneous cerebral dissections had better functional outcomes after treatment (Urasyanandana et al., 2018). Post-traumatic carotid artery dissection (PTCAD) may result in acute arterial ischemic stroke (AIS). Stroke after blunt carotid injury is associated with mortality rates that range from 23% to 28%, with 48% to 58% of survivors suffering permanent severe neurologic deficits. Pediatric PTCAD is rarely diagnosed prior to onset of neurological symptoms PTCAD may result from neck and head trauma. To avoid secondary AIS, radiologists should be familiar with neuroimaging findings in children, especially as acute PTCAD may initially be clinically silent. Consequently, pediatric neuroradiologists should actively exclude PTCAD in children with head and neck trauma. Endovascular and surgical management were effective treatments by parent vessel occlusion or reconstructions.

Vertebral arterial dissection (VAD) is often difficult to diagnose and can be a potentially devastating cause of ischemic stroke (Kumar, Ludwig, & Patel, 2018).

VAD in head trauma is associated with higher rates of stroke and longer length of ICU stay. Early diagnosis and endovascular treatment may be an alternative option to reduce the rate of in-hospital stroke in these patients (Majidi, Hassan, Adil, Jadhav, & Qureshi, 2014).

Bow hunter's syndrome is a rare cause of vertebrobasilar insufficiency arising from mechanical compression of the vertebral artery (VA) during rotation of the head. Surgical treatment usually involves direct decompression of the VA at the site of compression. The PICA-to-PICA in situ bypass is a useful option in the treatment of Bow hunter's-type ischemia in the absence of focal structural compression of the VA or VA stenosis (Kan, Yashar, Langer, Siddiqui, & Levy, 2012).

Surgical procedures included clipping or trapping for traumatic aneurysms, superficial temporal artery—middle cerebral artery bypass, carotid endarterectomy, and direct suture of the injured vessels. Traumatic aneurysms occur in 10% of extracranial blunt traumatic cerebrovascular injuries (TCVI). Optimal management of traumatic aneurysms needs to be defined (Krings, Geibprasert and Lasjaunias, 2008).

Endovascular treatment of traumatic aneurysms

Aside from a risk of thromboembolic stroke due to disruption of the intima, extracranial carotid and VA aneurysms are only rarely symptomatic. Despite the majority of traumatic aneurysms resolve spontaneously or remain stable on follow-up imaging, it seems reasonable to limit endovascular treatment to larger (> 15 mm) aneurysms or aneurysms with enlargement on follow-up. Covered stenting, coiling, stent-assisted coiling, and flow diverter (FD) placement techniques demonstrated a high rate of success. Anti-agregant therapy depends on the endovascular treatment strategy. Long-term studies of the endovascular treatment are also needed.

Therapeutic arterial occlusion

Complete thrombotic arterial occlusion carries not only a risk of hemodynamic failure but also of migration of the intraluminal thrombus into the intracranial circulation. The high clot burden contributes to the higher risk of ischemic stroke. A Manipulation of the cervical spine during surgery to repair a fracture may cause migration of an intraluminal VA clot and ischemic stroke (Fig. 1). An analysis of 52 cases of traumatic VA occlusions actually found that spine surgery was associated with a lower risk of stroke. Coil embolization of the ICA and of the VA proximal to the TCVI lesion prior to cervical spine surgery to reduce the theoretical risk of embolization has been reported. Comparative studies are needed to determine if this strategy is useful.

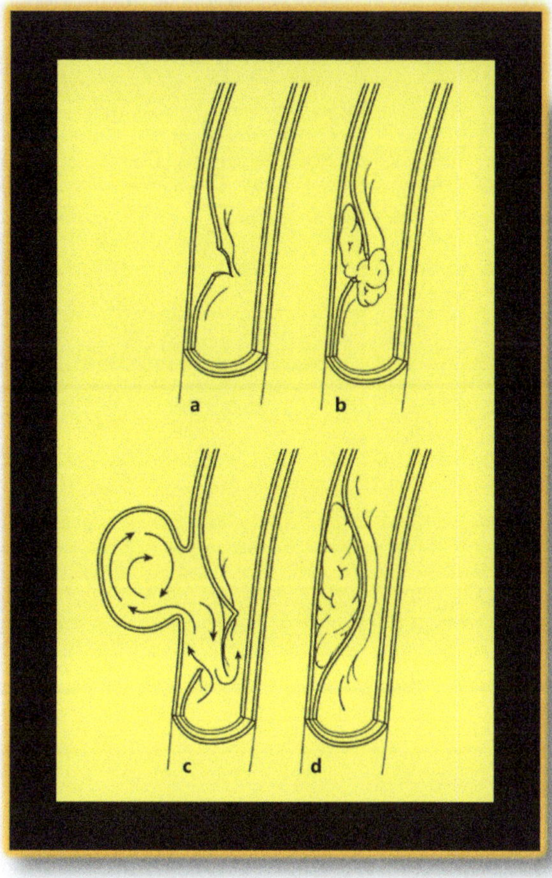

FIG. 1 Formation of a dissecting aneurysm.

Endovascular options include stenting of dissections, intra-arterial thrombolysis for acute ischemic stroke caused by trauma, and embolization of traumatic aneurysms.

The role of catheter angiography in the diagnosis and management of TCVI has evolved rapidly with advances in CT and MR angiography and continued development of endovascular techniques.

Distribution of lesions by location was 42% intracranial, 39% cervical, and 19% extracranial. The most common injuries were pseudoaneurysm (38%), fistula (29%), and dissection (19%). In total, 41% of lesions underwent endovascular treatment, with trends favoring treatment of non-acute, penetrating, non-cervical, and high grade lesions (Fig. 2). Therapy involved coil embolization for 89% of treated lesions. There were a total of 2% immediate neurovascular complications and one delayed neurovascular complication (Kansagra et al., 2014).

While the majority of injuries will resolve with medical management, a proportion will require further intervention in order to reduce the risk of subsequent stroke.

Injury type	Description	Proportion of all blunt traumatic carotid artery injuries [24], %	Proportion of all blunt traumatic vertebral artery injuries [25], %
1	Luminal irregularity or dissection with <25% luminal narrowing	61	53
2	Dissection or intramural hematoma with ≥25% luminal narrowing	17	19
3	Traumatic aneurysm	15	6
4	Occlusion	5	21
5	Transection	4	0

[1] By digital subtraction angiography.

FIG. 2 Four grades of arterial lesions.

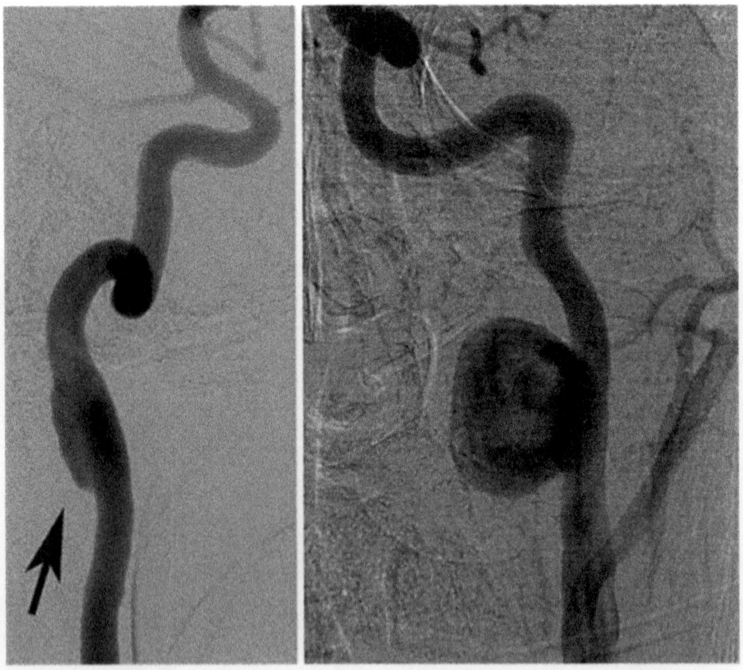

FIG. 3 Arterial dissection and pseudoaneurysm.

13 patients with 26 traumatic aneurysms. All patients were treated with 325 mg aspirin daily and underwent clinical and imaging follow-up. Endovascular treatment was reserved for aneurysms demonstrating significant enlargement on follow-up imaging (Fig. 3).

The majority of traumatic aneurysms can be managed with an anti-platelet regimen daily and serial imaging. Saccular aneurysms have a greater tendency to enlarge when compared with fusiform aneurysms (Fusco & Harrigan, 2011).

Biffle et al. report a total of 76 patients, diagnosed with 109 BCI. Two-thirds of mild intimal injuries (grade I) healed, regardless of therapy. Dissections or hematomas with luminal stenosis (grade II) progressed, despite heparin therapy in 70% of cases. Only 8% of pseudoaneurysms (grade III) healed with heparin, but 89% resolved after endovascular stent placement. Occlusions (grade IV) did not recanalize in the early post-injury period. Grade V injuries (transections) were lethal and refractory to intervention (Fig. 4). Stroke risk increased with injury grade. Severe head injuries (Glasgow Coma Scale score < or = 6) were found in 46% of patients and confounded evaluation of neurologic outcomes (Biffl et al., 1999).

Mechanical thrombectomy

Endovascular clot removal is an attractive option for TCVI patients with acute ischemic stroke, particularly as most trauma patients are not candidates for intravenous alteplase due to other injuries (Fig. 5). However, effective use of mechanical thrombectomy in this setting depends on prompt recognition of symptoms or imaging evidence of large vessel occlusion.

Stenting

Stent placement for TCVI, usually for the treatment of carotid artery dissections, has become increasingly common in recent years. Despite technical success rate is high (> 99%), there is no consensus about the indications for the use of stenting and a significant percentage of stent procedures are done for asymptomatic traumatic carotid artery dissections. Prospective studies of TCVI stenting to assess efficacy are lacking. Importantly, intra-arterial stent placement requires anti-platelet medication, usually with dual anti-platelet coverage. In a series of TCVI carotid stent cases in which anti-platelet agents were not used routinely, acute thrombosis occurred in 45% (Markus et al., 2015). Anti-platelet agents can be problematic in trauma patients, it seems prudent to limit stenting to symptomatic lesions.

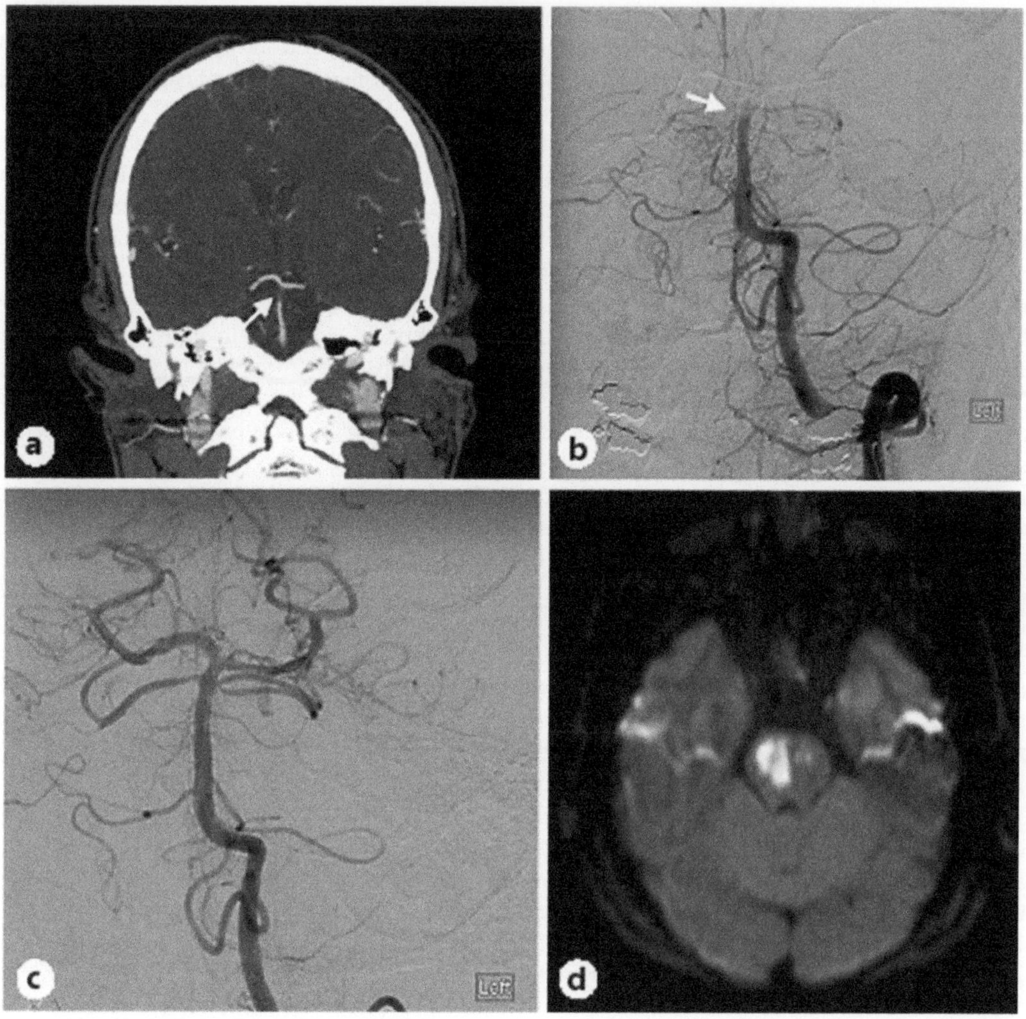

FIG. 4 Proportion of blunt injuries in carotid and vertebral arteries.

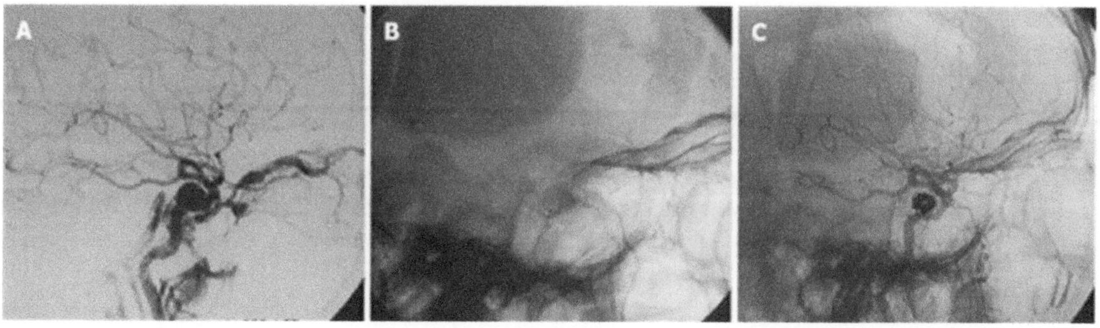

FIG. 5 Carotid cavernous aneurysm.

Carotid-cavernous fistula

Carotid-cavernous sinus fistulas are abnormal communications between the carotid system and the cavernous sinus. Several classification schemes have described carotid cavernous sinus fistulas according to etiology, hemodynamic features, or the angiographic arterial architecture. Increased pressure within the cavernous sinus appears to be the main factor in pathophysiology. Important features are related to size, exact location, and duration of the fistula, venous drainage and the presence of arterial/venous collaterals. Non-invasive imaging (computed tomography, magnetic resonance, computed tomography angiography, magnetic resonance angiography, Doppler) is often used in the initial work-up of a possible carotid cavernous sinus fistulas. Cerebral angiography is the gold standard for the definitive diagnosis, classification, and planning of treatment for these lesions.

Several classification schemes have categorized CCFs according to etiology (traumatic or spontaneous), hemodynamic features (high versus low flow), or the angiographic arterial architecture (direct or indirect). The angiographic classification defines the angioarchitecture of the lesion on which a therapeutic strategy can be planned. According to the angiographic findings, Barrow et al. (1985) provided a detailed anatomical classification which categorizes CCFs into four distinct types based on their arterial supply. Type A fistulas are direct communications between the internal carotid artery (ICA) and the cavernous sinus, usually associated with high flow rates. Indirect fistulas (types B, C, and D) are dural arteriovenous fistulas fed by the meningeal arteries of the ICA, the external carotid artery (ECA), or both. Type B fistulas have dural ICA branches to the cavernous sinus, which are relatively uncommon. Type C fistulas are supplied solely by the dural branches of the ECA. The most prevalent form of indirect CCF is the type D fistula that has dural ICA and ECA branches to the cavernous sinus. Tomsick sub-classified type D CCFs into type D1 or D2 depending on the presence of a unilateral or bilateral arterial supply (Ernst & Tomsick, 1997).

Traumatic CCFs demonstrate a single direct communication between the ICA and the cavernous sinus and are almost always found to be type A direct fistulas. Spontaneous fistulas usually have multiple dural feeders and numerous microfistulas within the cavernous sinus wall (Coskun et al., 2000). Spontaneous CCFs may fall into any of the four angiographic categories defined by Barrow et al., because a type A shunt with high-flow characteristics can develop following spontaneous rupture of an intracavernous ICA aneurysm.

Most direct CCFs occur at the proximal horizontal intracavernous segment (Ringer, Salud, & Tomsick, 2005) of the ICA in the vicinity of the inferolateral trunk. With decreasing frequency, CCFs are found to occur at the junction of the horizontal and intracavernous ascending segments, posterior ascending segment, junction of the anterior ascending and horizontal intracavernous segments, and the anterior ascending segment.

Traumatic disruption of the vessel wall is the most common etiological factor for direct CCFs. Blunt and penetrating head trauma as well as iatrogenic damage (trans-sphenoidal surgery, glycerol rhizotomy, Fogarty catheter manipulation for carotid angioplasty, etc.) may lead to direct CCFs (Connors & Wojak, 1999).

These lesions are usually unilateral although bilateral traumatic CCFs occur in approximately 1%–2% of patients with traumatic CCFs. These unusual bilateral traumatic CCFs are generally associated with more severe head trauma, are more commonly fatal, and are therefore less frequent. Additionally, unilateral fistulas may occur with bilateral or contralateral orbital symptoms, depending on the venous drainage route, via intercavernous communication (Gemmete, Ansari, & Gandhi, 2009). Direct fistulae are less likely to resolve spontaneously and may require intervention if symptomatic. Traumatic indirect CCFs differ from the spontaneous type because these lesions are usually single-hole fistulas in which the accessory meningeal artery and middle meningeal artery are found to be the most common feeders (Luo, Teng, Chang, & Chang, 2009).

CT angiography can be used as a first-line diagnostic tool in evaluating the presence of a CCF it has some limitations. CT angiography rarely depicts small feeding arteries in dural CCFs or the exact site of fistulous communication in direct CCFs. This technique cannot provide information about the blood-flow characteristics within fistulas.

Color Doppler imaging can assist in diagnosis and follow-up of patients with CCFs. Increased velocity with reversal of the direction of blood flow, arterial pulsations are characteristic findings.

Cerebral angiography is the gold standard for the definitive diagnosis, classification, and planning of endovascular intervention in CCFs. The initial angiographic evaluation can be used to obtain the following information: size and location of the fistula, differentiation of direct from indirect lesions, presence of any associated cavernous carotid aneurysm, presence of complete or partial steal phenomena, assessment of the global cortical arterial circulation and collateral flow through the circle of Willis, identification of high-risk features (e.g., cortical venous drainage, pseudoaneurysm, cavernous sinus varix), venous drainage patterns, determination of therapeutic route, associated vascular injuries (e.g., traumatic pseudoaneurysm, arterial dissection), identification of any dangerous collateral pathways and evaluation of carotid bifurcation before compression therapy (Debrun, 1995; Meyers et al., 2002).

The exact location of fistulous communication in the cavernous ICA can be challenging because of the high flow-related washout of intra-arterial contrast and instantaneous opacification of the cavernous sinus.

Tolerance for ICA occlusion should also be assessed to identify the appropriate therapeutic choices before embarking on a therapeutic intervention. Balloon test occlusion (BTO) is the currently accepted technique for evaluation. After confirmation of ICA occlusion, detailed testing of mental status, speech, visual fields, facial animation, and motor power in all four extremities are performed. In the absence of significant deficit, the patient is observed for 15–20 min and re-examined. If the patient tolerates occlusion at normal blood pressure, nitroprusside infusion is initiated and titrated to achieve a mean arterial pressure two thirds of the patient's baseline level. The patient is examined again and observed for 15–20 min. Single proton emission computed tomography (SPECT) may be used to rule out significant asymmetry in perfusion during BTO. The SPECT evaluation can uncover the risk of suffering a major stroke after permanent carotid occlusion even in patients who seem to tolerate BTO during relative hypotensive challenge test.

Treatment

Conservative

Spontaneous resolution of dural fistulas can occasionally occur within days to months after symptomatic presentation secondary to further thrombosis of the involved segment of the cavernous sinus. First attempt accepted in practice is to treat the patient's ocular symptoms medically with prism therapy or patching for diplopia, topical β-adrenergic blockers and acetazolamide for elevated intraocular pressure, lubrication for proptosis-related keratopathy, and/or systemic corticosteroids if needed.

Furthermore, manual external carotid-jugular compression therapy may be initiated as a non-invasive treatment for indirect CCFs. The patient is instructed to compress the carotid artery and jugular vein with the contralateral hand for a period of 10 s while sitting or lying down, four to six times each hour. The aim of the compression therapy is the transient reduction of arteriovenous shunting by decreasing arterial inflow while simultaneously increasing the outlet venous pressure, thereby promoting spontaneous thrombosis within the fistula. Use of the contralateral hand ensures that if ischemia develops, the symptomatic arm will fall away from the neck, thus allowing cortical revascularization immediately. Intermittent self-administered manual carotid-jugular compression alone can result in a cure in 30% of patients with spontaneous CCFs (McConnell et al., 2009). However, this treatment is usually ineffective in the high-flow CCFs, which usually require endovascular intervention.

Surgical management

Early treatments for CCF consisted of various surgical approaches such as ligation of the CCA, surgical trapping of the fistula, and surgical transvenous packing. Although surgical management can be useful for both direct and indirect CCFs, its role is limited because of associated morbidity from cranial nerve deficits and residual fistulous communications. Indications for surgical repair include compromised proximal arterial access that prevents endovascular repair or causes it to fail. Surgical management remains a consideration for salvage of failed endovascular treatments. Complete angiographic documentation of the fistula and BTO should be performed during preoperative evaluation. The appearance and condition of the superficial temporal artery should also be noted because extracranial-to-intracranial bypass can be required to augment blood flow when surgical sacrifice is unavoidable.

Radiosurgery

Stereotactic radiosurgery has emerged as an alternative treatment option and has been investigated for the treatment of CCFs in several institutions. Gamma knife radiosurgery can be used either alone or as an adjunct therapy before/after endovascular intervention (Koebbe et al., 2005; Link, Coffey, Nichols, & Gorman, 1996). Although preliminary data show that radiosurgery is a safe and effective alternative treatment for indirect CCFs, the 22-mo average lag between treatment and complete symptom relief is a significant risk. Furthermore, an inability to manage emergencies and traumatic fistulae inhibit the usage of radiosurgery as a first line treatment.

Endovascular management

Recent advances in endovascular technology have made available a number of different treatment options for CCFs. As a result of these advances, the endovascular approach has evolved as the primary treatment option in clinical emergencies and following the failure of conservative therapy. Although the clinical manifestations of direct and indirect fistulas may overlap, their natural history and method of endovascular treatment are often significantly different. The treatment choice is made according to the type, exact anatomy of the fistula, size of the arterial defect, and operator/institutional preferences.

Direct fistulas occur from a tear in the cavernous segment of the ICA or, less commonly, from the intracavernous rupture of an ICA aneurysm. The goal of treatment in direct CCFs is to occlude the tear between the ICA and the cavernous sinus while preserving the patency of the ICA. This goal can be accomplished by either transarterial obliteration of the fistula with a detachable balloon, transarterial or transvenous obliteration of the ipsilateral cavernous sinus with coils or other embolic material, or deployment of a covered stent across the fistula. Rarely, if the defect is large and cannot be repaired, the ICA may need to be sacrificed or trapped.

Indirect fistulas consist of small dural arteriovenous shunts between the meningeal branches of the ICA, the ECA, or both, and the cavernous sinus. The goal of treatment in this condition is to interrupt fistulous communications and decrease pressure in the cavernous sinus. This can be accomplished by occluding the arterial branches supplying the fistula (transarterial embolization) or, more commonly, by occluding the cavernous sinus that harbors the fistulous communications (transvenous embolization). The following sections provide a brief overview of the various endovascular options for the treatment of CCFs.

The technique for detachable silicone balloon occlusion of a CCF involves transfemoral access to the proximal CCA with a 7-French guide catheter or long 6-French sheath. Next, the uninflated balloon is advanced to the distal end of the guide catheter; at this point, roadmap imaging is used for further balloon positioning. The balloon offers the advantage of being able to be flow-directed through the fistula and into the cavernous sinus. The balloon is inflated to a volume larger than the orifice of the fistula to prevent its retrograde prolapse into the ICA and then is detached. Following successful balloon deployment, cerebral angiography is repeated to ensure closure of the fistula and patency of the ICA. A single silicone balloon is usually sufficient for most CCFs. However, multiple balloons may be required in the setting of a large tear in the ICA. The advantage of balloon occlusion of a CCF is the ability to occlude the fistula rapidly with preservation of the ICA. To provide easier navigation of the balloon into the cavernous sinus and prevent protrusion of the inflated balloon through the fistula site to narrow the adjacent ICA lumen, Teng et al. (2000) developed a double-balloon technique.

Transarterial embolization with coils or other embolic material now is the mainstay of endovascular treatment for high-flow direct CCFs, given the limited availability of detachable balloons. Transarterial CCF embolization can be performed with the same technique as aneurysmal embolization. Embolization can be achieved with detachable platinum coils, silk and liquid embolic agents such as n-butyl cyanoacrylate (n-BCA), and ethylene-vinyl alcohol copolymer (EVOH). The standard transarterial approach consists of placing a guiding catheter in the cervical ICA. Next, a micro-catheter is super-selectively advanced into the cavernous segment of the ICA and through the tear into the cavernous sinus. Through this micro-catheter, embolic material is placed into the cavernous sinus.

Detachable platinum coils are preferred because of their reliable and controlled deployment. The coils can be adjusted easily or removed if the placement is not optimal. Dense coil packing is performed using the same principles as aneurysm coiling. Thrombogenic, non-retrievable fibered micro-coils can also be used but these are associated with the risk of coil deposition into the ICA if the micro-catheter recoils into the parent artery during placement. The advantages of coil occlusion of CCFs, when compared with balloon embolization, include ease of access and availability of a variety of sizes of the embolic device. Potential disadvantages include slower gradual occlusion of the fistula, which increases procedure time, and the risk of incomplete fistula occlusion with loss of transarterial access; a loss which would then require a second transvenous approach. The transvenous approach is discussed more thoroughly under the heading "transvenous embolization" (Fig. 1).

Polyfluorotetraethylene-covered stents have created alternatives to ICA sacrifice in traumatic arterial damage, especially in unsuccessful BTO study. Covered stent grafts can be extremely useful for the immediate obliteration of a direct CCF, while preserving arterial patency.

If a decision to perform endovascular arterial sacrifice is made, assessment of the collateral flow and patient's ability to tolerate ICA occlusion is paramount. In cases of complete steal presenting without any ischemic symptom, the quality of collateral flow through circle of Willis is confirmed and a test occlusion may be unnecessary. If anterior and posterior communicating artery collaterals are found to be patent, the safety of parent artery occlusion is also high. Otherwise, balloon occlusion test is recommended to ensure distal perfusion from collateral circulation before permanent occlusion. Recently, the Hydrocoil Embolization System (HES) has been introduced, which can achieve high rates of volumetric occlusion compared with that in platinum coils. The HES device consists of a carrier platinum coil coupled with an expandable hydrogel material that increases in volume on contact with blood and facilitates vessel sacrifice, decreasing the procedure and fluoroscopy times. Hydrogel-coated (Fig. 6) detachable coils can be utilized to achieve rapid, precise, targeted parent artery occlusion procedures, even in short segments of vessel (Kallmes & Cloft, 2004).

Although transvenous embolization is an alternative technique in direct CCFs that cannot be treated by transarterial route, it is the preferred treatment for indirect CCFs (Fig. 7). For indirect CCFs, transvenous techniques have precedence over transarterial methods because of their simplicity, lower ischemic risk, higher success rates and capability to cure the fistula in a single session. The aim of the transvenous approach is to catheterize the abnormal cavernous sinus superselectively and to occlude the fistula without rerouting venous drainage to cortical structures. Current micro-catheter technology

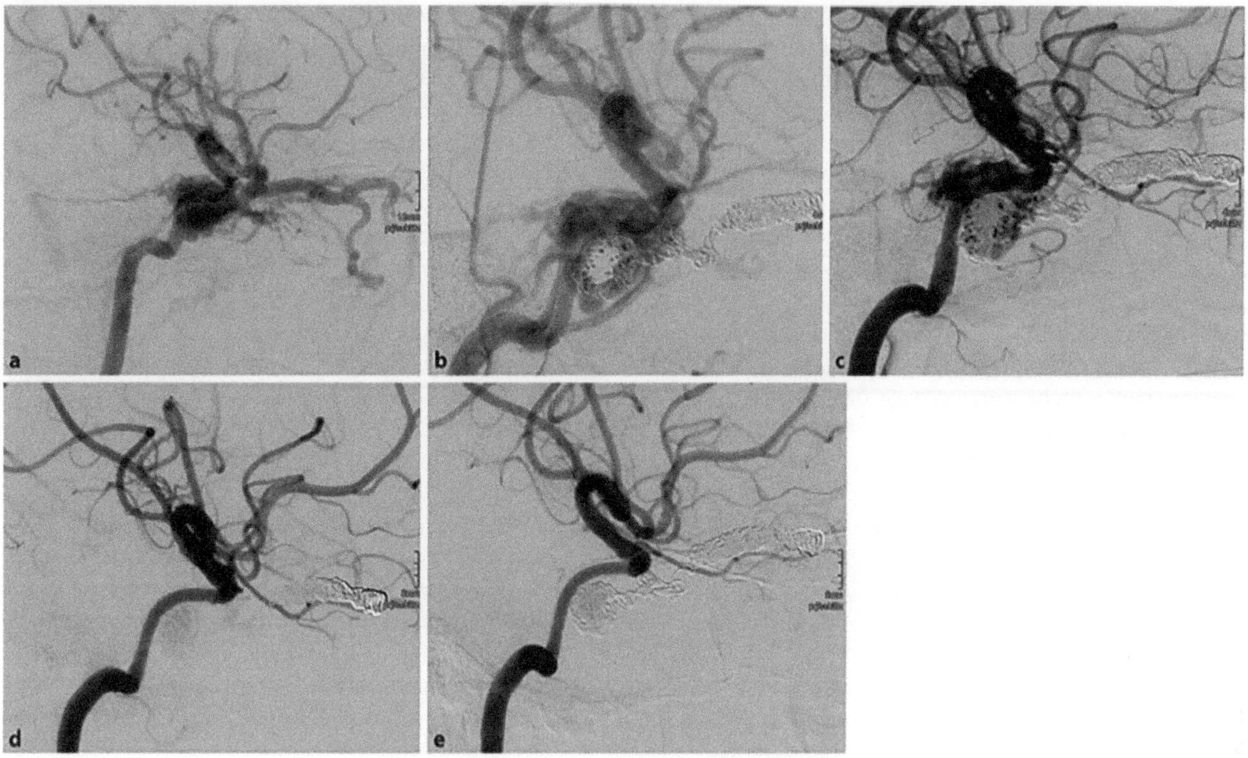

FIG. 6 Coiling of a CCF.

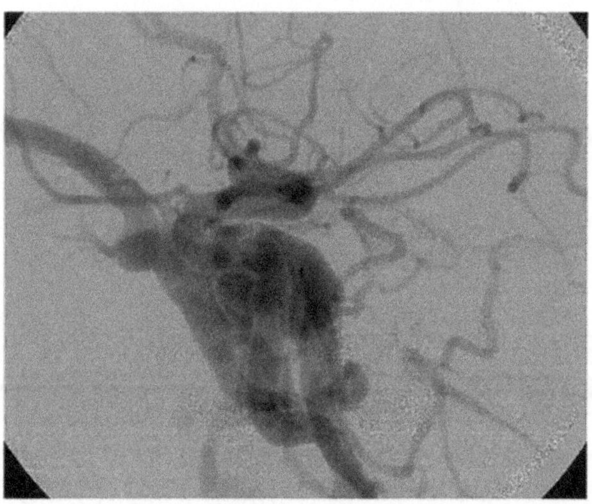

FIG. 7 Anterior venous route for CCF.

permits access to the cavernous sinus via multiple routes. The most commonly used venous pathway for cannulation of the cavernous sinuses is via the IPS. This transvenous route is usually used from a posterior direction through the internal jugular vein and the IPS up to the pathologic shunts of the cavernous sinus (Berkmen, Troffkin, & Wakhloo, 2003).

Transarterial techniques involve distal catheterization of the small meningeal branches supplying the fistula. Ideally, placement of the superselective micro-catheter is performed with the micro-catheter tip as close as possible to the point of fistulous communication. Once a satisfactory micro-catheter position is achieved, liquid embolic agents (n-BCA, EVOH) are injected under fluoroscopic control with the goal of occluding the fistulous connections and penetrating the cavernous sinus. Although coils and particulate agents have been used, these agents used alone cannot provide permanent occlusion of the fistula (Chi, Nguyen, Duc, Chau, & Son, 2014; Korkmazer et al., 2013).

Initially, flow diverting implants were designed for the extrasaccular treatment of complex cerebral aneurysms, but techniques and indications for the use of FD continue to evolve. Recently published case reports showed promising results for the treatment of CCF with FD (Wendl et al., 2017). The dense coverage of FD not only provides protection of the ICA during transvenous coil insertion, but also enables endothelial overgrowth of the lacerated arterial segment.

Applications to other areas of neuroscience

Injury-repair response of vessel wall to hypoxia may differ as time lapses. In the circumstance of chronic occlusion, vessel adventitia can act as a biological central processing unit in vessel wall function, likely from harboring a wide variety of components such as fibroblasts, inflammatory cells, stem/progenitor cells, and vasa vasorum (VV). Therefore, oxygen and nutrients to the inner wall of a vessel in the occluded segment may be completely supplied by adventitial VV. In response to injury, VV conduits proliferate in adventitia and extend to intima by interacting with vascular resident stem cells (VSCs) housed in the VV niche. Angiogenesis of VSCs may be beneficial for vasculature and plaque stabilization, when VSCs supplied and transported by VV enter into intima and differentiate into endothelial cells, pericytes, and fibroblasts and when pericytes interact with endothelial cells enough. However, neovessels of VV are immature with a characteristically thin wall with less pericyte encirclement. The fragile structure may lead to micro-vascular cleft that is responsible for leakage of red blood cells (intramural hemorrhage) and accumulation of inflammatory cells (chronic inflammation). Additionally, VSCs can also differentiate into vascular smooth muscle cells associated with acceleration of plaque growth and reduction of vascular stability. The above-described unfavorable responses may increase vulnerability of the intimal layer of non-acute ICAO to mechanical injury, especially when operating in atherosclerotic vessels. Normalization of the impaired vasa vasorum would be an attractive therapeutic strategy; however, this is not yet available in clinical practice. Of note, the cavernous ICA exit site for the false channel dissection could be partially explained by the concurrent location of the atretic atrophy of the primitive trigeminal artery, potentially resulting in vessel wall weakening (Kawabe & Hasebe, 2014).

Mini-dictionary of terms

- PTCAD: Lesion of the inner wall of the artery after trauma resulting in stenosis of the lumen and impairment of blood circulation.
- Denver criteria: The Denver criteria are a set of screening criteria for BCVI in trauma used to reduce the need for CT angiography and its associated radiation exposure.
- Memphis criteria: The modified Memphis criteria are a set of screening criteria for BCVI in trauma. The presence of one or more of these criteria makes necessary a complementary CTA or DSA study to exclude a BCVI.
- PICA-PICA bypass: The PICA-to-PICA (Posterior inferior cerebellar artery) in situ bypass is a useful option in the treatment of Bow hunter's-type ischemia in the absence of focal structural compression of the VA or VA stenosis.
- Mechanical thrombectomy: Endovascular clot removal option for TCVI patients with acute ischemic stroke, particularly as most trauma patients are not candidates for intravenous alteplase due to other injuries.
- Traumatic Carotid cavernous fistula: Carotid cavernous sinus fistulas are abnormal communications between the carotid system and the cavernous sinus after trauma.
- Direct fistulas: occur from a tear in the cavernous segment of the ICA or, less commonly, from the intracavernous rupture of an ICA aneurysm.

Key facts: Treatment of vascular lesions in traumatic head injury

- Traumatic vascular injuries of head and the neck are rare, but serious complications of head trauma.
- Approximately 93% of blunt cervical trauma involves carotid lesions and 72% in penetrating lesions.
- Arterial and venous site of the vasculature can be involved, the latter one being often overlooked.
- The Denver criteria are a set of screening criteria for BCVI in trauma.
- The modified Memphis criteria are a set of screening criteria for BCVI in trauma.
- Endovascular clot removal is an attractive option for TCVI patients with acute ischemic stroke, particularly as most trauma patients are not candidates for intravenous alteplase due to other injuries.
- Stent placement for TCVI, usually for the treatment of carotid artery dissections, has become increasingly common in recent years.
- Traumatic disruption of the vessel wall is the most common etiological factor for direct CCFs.

- Cerebral angiography is the gold standard for the definitive diagnosis, classification, and planning of endovascular intervention in CCFs.

Summary points

- Traumatic vascular injuries of the head and the neck are rare 1%, but serious complications of head trauma due mainly for the delayed diagnosis.
- Arterial and venous site of the vasculature can be involved, the latter one being often overlooked.
- Traumatic aneurysms occur in 10% of extracranial blunt TCVI.
- Stroke after blunt carotid injury is associated with mortality rates that range from 23% to 28%, with 48% to 58% of survivors suffering permanent severe neurologic deficits.
- Traumatic CCFs demonstrate a single direct communication between the ICA and the cavernous sinus
- Direct fistulas occur from a tear in the cavernous segment of the ICA or, less commonly, from the intracavernous rupture of an ICA aneurysm.
- Indirect fistulas consist of small dural arteriovenous shunts between the meningeal branches of the ICA, the ECA, or both, and the cavernous sinus.
- Surgical management remains a consideration for salvage of failed endovascular treatments.
- Covered stent grafts can be extremely useful for the immediate obliteration of a direct CCF, while preserving arterial patency.
- Flow diverting implants were designed for the extrasaccular treatment of complex cerebral aneurysms, but techniques and indications for the use of FD continue to evolve. Recently published case reports showed promising results for the treatment of CCF with FD.
- In response to injury, VV conduits proliferate in adventitia and extend to intima by interacting with VSCs housed in the VV niche.

References

Barrow, D. L., Spector, R. H., Braun, I. F., Landman, J. A., Tindall, S. C., & Tindall, G. T. (1985). Classification and treatment of spontaneous carotid-cavernous sinus fistulas. *Journal of Neurosurgery*, *62*(2), 248–256.

Berkmen, T., Troffkin, N. A., & Wakhloo, A. K. (2003). Transvenous sonographically guided percutaneous access for treatment of an indirect carotid cavernous fistula.AJNR. *American Journal of Neuroradiology*, *24*(8), 1548–1551.

Biffl, W. L., Moore, E. E., Offner, P. J., Brega, K. E., Franciose, R. J., & Burch, J. M. (1999). SO blunt carotid arterial injuries: Implications of a new grading scale. *The Journal of Trauma*, *47*(5), 845.

Chi, C. T., Nguyen, D., Duc, V. T., Chau, H. H., & Son, V. T. (2014). Direct traumatic carotid cavernous fistula: Angiographic classification and treatment strategies. Study of 172 cases. *Interventional Neuroradiology*, *20*(4), 461–475.

Ciapetti, M., Circelli, A., Zagli, G., et al. (2010). Diagnosis of carotid arterial injury in major trauma using a modification of Memphis criteria. *Scandinavian Journal of Trauma, Resuscitation and Emergency Medicine*, *18*(1), 61.

Connors, J. J., & Wojak, J. C. (1999). *Interventional neuroradiology: Strategies and practical techniques* (pp. 215–226). Philadelphia: WB Saunders Co.

Coskun, O., Hamon, M., Catroux, G., Gosme, L., Courthéoux, P., & Théron, J. (2000). Carotid-cavernous fistulas: Diagnosis with spiral CT angiography. *AJNR. American Journal of Neuroradiology*, *21*(4), 712–716.

Debrun, G. M. (1995). Review angiographic workup of a carotid cavernous sinus fistula (CCF) or what information does the interventionalist need for treatment. *Surgical Neurology*, *44*(1), 75–79.

Ernst, R. J., & Tomsick, T. A. (1997). Classification and angiography of carotid cavernous fistulas. In T. A. Tomsick (Ed.), *Carotid cavernous sinus fistula* (pp. 13–22). Cincinnati: Digital Education Publishing.

Fusco, M. R., & Harrigan, M. R. (2011). Cerebrovascular dissections: A review. Part II: Blunt cerebrovascular injury. *Neurosurgery*, *68*(2), 517–530.

Gemmete, J. J., Ansari, S. A., & Gandhi, D. M. (2009). Review endovascular techniques for treatment of carotid-cavernous fistula. *Journal of Neuro-Ophthalmology*, *29*(1), 62–71.

Kallmes, D. F., & Cloft, H. J. (2004). The use of hydrocoil for parent artery occlusion. *AJNR. American Journal of Neuroradiology*, *25*(8), 1409–1410.

Kan, P., Yashar, P., Langer, D. J., Siddiqui, A. H., & Levy, E. I. (2012). Posterior inferior cerebellar artery to posterior inferior cerebellar artery in situ bypass for the treatment of Bow hunter's-type dynamic ischemia in holovertebral dissection. *World Neurosurgery*, *78*(5). https://doi.org/10.1016/j.wneu.2011.09.050, 553.e15-7 (Epub 2011 Nov 1).

Kansagra, A. P., Cooke, D. L., English, J. D., Sincic, R. M., Amans, M. R., Dowd, C. F., et al. (2014). Current trends in endovascular management of traumatic cerebrovascular injury. *Journal of Neurointerventional Surgery*, *6*(1), 47–50. https://doi.org/10.1136/neurintsurg-2012-010605 (Epub 2013 Jan 15).

Kawabe, J., & Hasebe, N. (2014). Role of the vasa vasorum and vascular resident stem cells in atherosclerosis. *BioMed Research International*, *2014*, 70.

Koebbe, C. J., Singhal, D., Sheehan, J., Flickinger, J. C., Horowitz, M., Kondziolka, D., et al. (2005). Radiosurgery for dural arteriovenous fistulas. *Surgical Neurology*, *64*(5), 392–398 (discussion 398-9).

Korkmazer, B., Kocak, B., Tureci, E., Islak, C., Kocer, N., & Kizilkilic, O. (2013). Endovascular treatment of carotid cavernous sinus fistula: A systematic review. *World Journal of Radiology*, *5*(4), 143–155.

Krings, T., Geibprasert, S., & Lasjaunias, P. L. (2008). Cerebrovascular trauma. *European Radiology*, *18*(8), 1531–1545. https://doi.org/10.1007/s00330-008-0915-z (Epub 2008 Apr 8).

Kumar, G., Ludwig, B., & Patel, V. V. (2018). Vertebral artery dissection masquerading as concussion in an adolescent. *Pediatric Emergency Care*, *34*(5), e97–e99.

Link, M. J., Coffey, R. J., Nichols, D. A., & Gorman, D. A. (1996). Radiosurgery for dural arteriovenous fistulas. *Journal of Neurosurgery*, *84*(5), 804–809.

Luo, C. B., Teng, M. M., Chang, F. C., & Chang, C. Y. (2009). Traumatic indirect carotid cavernous fistulas: Angioarchitectures and results of transarterial embolization by liquid adhesives in 11 patients. *Surgical Neurology*, *71*(2), 216–222.

Majidi, S., Hassan, A. E., Adil, M. M., Jadhav, V., & Qureshi, A. I. (2014). Incidence and outcome of vertebral artery dissection in trauma setting: Analysis of national trauma data base. *Neurocritical Care*, *21*(2), 253–258.

Markus, H. S., Hayter, E., Levi, C., Feldman, A., Venables, G., & Norris, J. (2015). Antiplatelet treatment compared with anticoagulation treatment for cervical artery dissection: A randomised trial. *Lancet Neurology*, *14*, 361–367.

McConnell, K. A., Tjoumakaris, S. I., Allen, J., Shapiro, M., Bescke, T., Jabbour, P. M., et al. (2009). Neuroendovascular management of dural arteriovenous malformations. *Neurosurgery Clinics of North America*, *20*, 431–439.

Meyers, P. M., Halbach, V. V., Dowd, C. F., Lempert, T. E., Malek, A. M., Phatouros, C. C., et al. (2002). Dural carotid cavernous fistula: Definitive endovascular management and long-term follow-up. *American Journal of Ophthalmology*, *134*(1), 85–92.

Ringer, A. J., Salud, L., & Tomsick, T. A. (2005). Review carotid cavernous fistulas: Anatomy, classification, and treatment. *Neuroimaging Clinics of North America*, *16*(2), 279–295.

Teng, M. M., Chang, C. Y., Chiang, J. H., Lirng, J. F., Luo, C. B., Chen, S. S., et al. (2000). Double-balloon technique for embolization of carotid cavernous fistulas. *AJNR. American Journal of Neuroradiology*, *21*(9), 1753–1756.

Urasyanandana, K., Songsang, D., Aurboonyawat, T., Chankaew, E., Withayasuk, P., & Churojana, A. (2018). Treatment outcomes in cerebral artery dissection and literature review. *Interventional Neuroradiology*, *24*(3), 254–262.

Wendl, C. M., Henkes, H., Martinez Moreno, R., Ganslandt, O., Bäzner, H., & Aguilar Pérez, M. (2017). Direct carotid cavernous sinus fistulae: Vessel reconstruction using flow-diverting implants. *Clinical Neuroradiology*, *27*(4), 493–501.

Further reading

Link, M. J., Coffey, R. J., Nichols, D. A., & Gorman, D. A. (2004). The role of radiosurgery and particulate embolization in the treatment of dural arteriovenous fistulas. *Neuroimaging Clinics of North America*, *20*(4), 431–439.

Chapter 34

Airway pressure release ventilation (APRV) in traumatic brain injury

Christopher S. Thomas[a] and Evert A. Eriksson[b]
[a]*Department of Surgery Education Office, Medical University of South Carolina, Charleston, SC, United States,* [b]*Medical University of South Carolina, Charleston, SC, United States*

List of abbreviations

APRV airway pressure release ventilation
CO cardiac output
CPP cerebral perfusion pressure
I:E ratio inspiratory time to expiratory time ratio
ICP intracranial pressure
PEEP positive end expiratory pressure
TBI traumatic brain injury

Introduction

Traumatic brain injury (TBI) patients frequently require mechanical ventilation in the acute setting due to severe injury of the neurological system. Multi-system trauma also can contribute to the need for mechanical ventilation in these critically injured patients. As part of this multi-system trauma, lung injury can occur as a primary injury during the traumatic event, or as a secondary injury due to disease processes such as prolonged mechanical ventilation, aspiration, hospital acquired pneumonia, neurogenic pulmonary edema, etc. The ventilatory management of these critically injured patients varies widely across different hospitals and medical centers, with no clearly defined targets or algorithms for management (Picetti et al., 2020). The use of high positive end expiratory pressure (PEEP) in TBI patients has been well studied, and is still controversial (Vargas, Sutherasan, Gregoretti, & Pelosi, 2014). It has previously been shown that an increase in PEEP decreases cardiac output (CO) in traumatically injured patients (Luecke & Pelosi, 2005), which has been attributed to a decreased left ventricular function secondary to decreased left ventricular stroke volume in patients with high PEEP (Luecke et al., 2004). Despite this, little research has been done in regards to airway pressure release ventilation (APRV) in TBI patients (Davies et al., 2015). The aim of this chapter is to provide a brief overview of APRV, the physiology of positive pressure ventilation on the neurologic system, and examine the use of APRV in TBI patients.

Airway pressure release ventilation

APRV is a pressure controlled, intermittent mandatory ventilation method that has been historically used as a rescue mode of mechanical ventilation although originally created as a weaning mode of ventilation. Within this method of mechanical ventilation the patient is able to spontaneously breath without restriction. The proposed benefits of APRV include increased alveolar recruitment, improved oxygenation, and preservation of spontaneous breathing (Daoud, Farag, & Chatburn, 2012). APRV has traditionally been described as having an "inverse ratio" of inspiration time to expiration time, with APRV typically having an I:E ratio of around 4:1 (compared to 1:2–3 in traditional modes of mechanical ventilation). This higher I:E ratio can lead to high mean airway pressures (Stock, Downs, & Frolicher, 1987).

The settings for APRV include the "P_{high}," "P_{low}," "T_{high}," and "T_{low}." "P_{high}" is the pressure of the mandatory breath, which is delivered for a time period denoted by "T_{high}." "P_{low}" is the expiratory pressure that is maintained for a time denoted by "T_{low}" (see Fig. 1).

Diagnosis and Treatment of Traumatic Brain Injury. https://doi.org/10.1016/B978-0-12-823347-4.00033-6

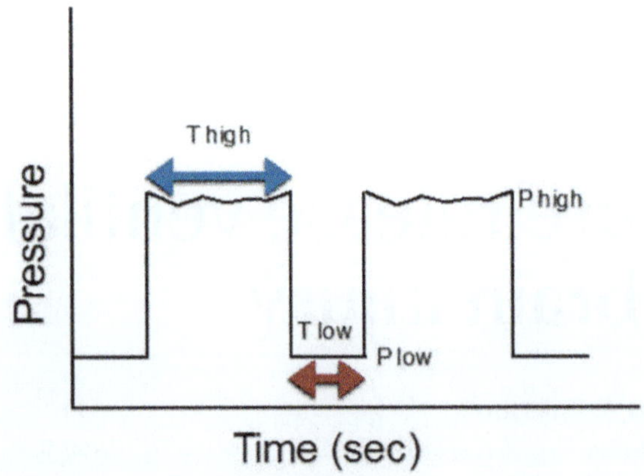

FIG. 1 Pressure vs. time curve for airway pressure release ventilation (Iowa, 2020).

When changing a patient to APRV as a "rescue mode," the plateau pressure is measured to serve as a surrogate for alveolar volume and lung compliance. The "P_{high}" setting should not exceed this plateau pressure in order to avoid volutrauma and barotrauma to the alveoli. The "T_{low}" is typically so short of a time period, due to the inverse ratio of APRV, that an airway pressure of zero is rarely reached (Mireles-Cabodevila & Kacmarek, 2016).

Several studies, as well as systematic reviews and meta-analyses, have shown that APRV may have several benefits when used in critically injured patients. One recent study conducted a systematic review and meta-analysis of seven randomized control trials, which included over 400 patients, to examine the safety and efficacy of APRV in critically ill patients who were intubated for hypoxic respiratory failure (Lim & Litton, 2019). It was determined by these authors that APRV is associated with a lower mortality in these patients. APRV was also found to be associated with greater synchronization with the ventilator, as well as decreased sedation requirements (Lim & Litton, 2019).

Kollisch-Singule et al. found that APRV with particular settings (higher "P_{low}" and lower "T_{low}") reduced secondary lung injury associated with prolonged mechanical ventilation (Kollisch-Singule et al., 2014). Despite extensive use of APRV as a rescue mode of ventilation, and even some use as a standard mode of ventilation, there has been little research into APRV use in TBI patients. This is likely because the increase in positive pressure across the thorax (due to increased I:E ratio and "P_{high}" values) can result in decreased venous return to the heart due to external pressure on the superior and inferior vena cava.

Positive end expiratory pressure and cerebral perfusion pressure

Several studies have examined the effects of positive pressure ventilation and cerebral perfusion pressure (CPP). Early studies hypothesized and showed that high PEEP decreased cardiac output due to increased right ventricular afterload and decreased right ventricular contractility (Biondi et al., 1988; Luecke & Pelosi, 2005). This is hypothesized to be due to the increase in intra-thoracic pressure that is associated by using positive pressure ventilation. This increase in pressure is transduced into the low pressure components of the cardiovascular system and great vessels, thereby reducing cardiac output.

Continued research showed that these changes seen with increased PEEP, although measured statistically, may not be clinically significant in patients with brain injury (Boone et al., 2017). Other studies have shown that while CPP may be effected by PEEP, intracranial pressure is not significantly changed in patients with normal lung compliance (Caricato et al., 2005). It is critical to use comprehensive, real time, monitoring of these values in patients with TBI and multi-system trauma, as the use of close monitoring has been shown to lead to improved neurological outcomes (Lou, Xue, Chen, Xue, & Wang, 2012; Young, Blow, Turrentine, Claridge, & Schulman, 2003; Zhang, Yang, Wang, & Fan, 2011). With close monitoring of hemodynamic parameters and ICP, titration to higher peep values may be safely performed.

APRV in traumatic brain injury

The first published case report that showed the safe use of APRV in a patient with brain injury was published by Marik et al. This documented a 68-year-old female with spontaneous rupture of an intracranial aneurysm and large subarachnoid hemorrhage who underwent surgical treatment and post-operatively developed respiratory failure and prolonged mechanical ventilation. As her respiratory status worsened, she was transitioned to APRV as a rescue mode of mechanical ventilation. The patient's cerebral hemodynamics were recorded as she already had invasive neurologic monitoring in place. It was observed that the cerebral blood flow increased by almost 20% once transitioned to APRV, and the respiratory status improved in as little as 30min. The patient was extubated after 24h on APRV and found to have a stable neurologic examination. This case report lead to increased interest in the use of APRV in brain injury patients (Marik, Young, Sibole, & Levitov, 2012).

Davies et al. were the first group to study the use of APRV in concomitant brain/lung injured animal models. They examined the physiologic parameter changes, as well as the histological changes in lung tissue over time after placing injured swine on APRV. They found that peak inspiratory pressure, mean airway pressure, and P/F ratio increased in the APRV group. However, no histological changes in lung tissue suggestive of increased injury were noted in the swine randomized to APRV. This study called for the increased research in APRV in TBI patients, as more data was needed (Davies et al., 2015).

The first reported study of APRV in human TBI patients was performed by Fletcher et al. This group performed a retrospective cohort study of 21 patients which examined the change in therapeutic intensity level, CPP, and intracranial pressure in patients during the six hours immediately following the change to APRV as a mode of mechanical ventilation. They found that while there were some statistically significant changes in the cerebral hemodynamics during APRV use in TBI patients, these changes were not clinically significant. This research group called for the continued research of APRV use in TBI patients, as well as urged for advanced neuromonitoring in these patients (Fletcher, Wilson, Rajajee, Davidson, & Walsh, 2018).

The most recent study of APRV in TBI patients was performed by Edgerton et al. This study was a retrospective review examining the intracranial pressure in patients following a transition from traditional modes of mechanical ventilation to APRV in patients with poor lung compliance. These authors did show a statistically significant increase in central venous pressure, arterial oxygenation, and P:F ratio. However, there was no statistically significant difference in intracranial pressure, mean arterial pressure, and CPP after transition to APRV. In a subset of patients who did have an increased intracranial pressure, the rate of elevation was similar between APRV and other modes of ventilation. This was the first study to show that despite an increase in central venous pressure, the intracranial pressure and hemodynamics of TBI patients was not affected by transition to APRV (Fig. 2) (Edgerton et al., 2019).

Conclusion

While multiple studies have been done examining the hemodynamic and physiologic changes of TBI patients with conventional modes of mechanical ventilation, there is limited data regarding the effects of APRV use in this population. It had previously been hypothesized that APRV had a negative hemodynamic on the neurological system of the TBI patient due to increased thoracic pressure transduced into the low-pressure cardiovascular system, which would therefore reduce cerebral venous outflow and potentially reduce cerebral vascular inflow. Initial studies in animal models and human case reports showed that APRV may be safe in TBI patients. Subsequent research showed that APRV does not increase intracranial

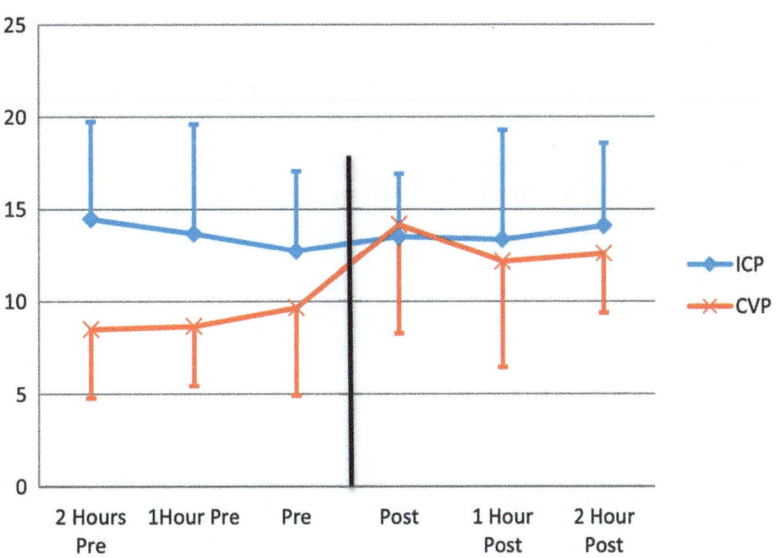

FIG. 2 ICP and CVP before and after transition to APRV (black line) (Edgerton et al., 2019).

pressure or effect hemodynamics of TBI patients, and what changes do occur are not clinically significant. This suggests that APRV is a safe mode of mechanical ventilation in some TBI patients. When using elevated PEEP or APRV to ventilate neurologically injured patients, ICP and CPP should be closely monitored when changing ventilatory setting to avoid adversely effecting perfusion pressures.

Applications to other areas of neuroscience

TBI patients are unfortunately encountered on a daily basis. These patients experience high rates of morbidity and mortality due to their severe injury patterns. Frequently these injuries are associated with multi-system trauma, and require intensive care and mechanical ventilation. APRV may be safely used in a TBI patient as long as advanced neurological and hemodynamic monitoring is in place. This is particularly applicable to other areas of Neuroscience as many clinicians and practitioners are involved in the care of these injured patients.

Due to the frequent multi-disciplinary approach, other fields involving Neuroscience are frequently directly involved in the care of these TBI patients. This may include advanced neurological testing, electroencephalogram monitoring, acute care of neurological injury, and disease, as well as more chronic care such as neurologic rehabilitation of TBI patients. It is important for clinicians and providers to be aware of the methods of care for these patients.

Mini-dictionary of terms

- **Traumatic brain injury**—disruption in brain function due to injury to the head.
- **Airway pressure release ventilation**—a mode of mechanical ventilation which uses an inverse ratio of inspiration to expiration, is pressure controlled, and allows for spontaneous breathing by the patient with intermittent mandatory breaths delivered by the ventilator.
- **Mechanical ventilation**—the use of a machine to assist in ventilation, and oxygenation, in critically ill patients.
- **Positive end expiratory pressure (PEEP)**—the positive pressure that remains in the airway after exhalation.
- **Cerebral Perfusion Pressure (CPP)**—the net pressure gradient (mean arterial pressure minus intracranial pressure) that allows blood flow to the brain.

Key facts of APRV in traumatic brain injury

- APRV is a mode of mechanical ventilation that is typically used as a rescue mode for patients with worsening respiratory status. It is an inverse ratio, pressure controlled, intermittent mandatory ventilation with unrestricted spontaneous breathing.
- Potential advantages to APRV include increased lung recruitment, improved oxygenation, and the allowing of spontaneous breathing.
- Disadvantages of APRV include risk of barotrauma, increased energy expenditure, and increased work of breathing.
- The effect of APRV on intracranial pressure and CPP has been hypothesized and studied; however, no definitive conclusion has been reached regarding clinically significant changes in CPP with the use of APRV.
- There have been few studies examining the effects of APRV on TBI patients; however, recent studies suggest that this method of mechanical ventilation may be safe for TBI patients.

Summary points

- APRV is a mode of mechanical ventilation which is frequently used as a rescue mode for patients with severe respiratory status.
- APRV uses increased positive pressure in order to recruit alveoli for improved oxygenation.
- Increased thoracic pressure was thought to transduce to the great vessels and cerebrovascular system; however, this has been shown to not be clinically significant.
- APRV may be safe for TBI patients with advanced neurological and hemodynamic monitoring.

References

Biondi, J. W., Schulman, D. S., Soufer, R., Matthay, R. A., Hines, R. L., Kay, H. R., et al. (1988). The effect of incremental positive end-expiratory pressure on right ventricular hemodynamics and ejection fraction. *Anesthesia and Analgesia*, *67*(2), 144–151.

Boone, M. D., Jinadasa, S. P., Mueller, A., Shaefi, S., Kasper, E. M., Hanafy, K. A., et al. (2017). The effect of positive end-expiratory pressure on intracranial pressure and cerebral hemodynamics. *Neurocritical Care*, *26*(2), 174–181. https://doi.org/10.1007/s12028-016-0328-9.

Caricato, A., Conti, G., Della Corte, F., Mancino, A., Santilli, F., Sandroni, C., et al. (2005). Effects of PEEP on the intracranial system of patients with head injury and subarachnoid hemorrhage: The role of respiratory system compliance. *Journal of Trauma*, *58*(3), 571–576. https://doi.org/10.1097/01.ta.0000152806.19198.db.

Daoud, E. G., Farag, H. L., & Chatburn, R. L. (2012). Airway pressure release ventilation: What do we know? *Respiratory Care*, *57*(2), 282–292. https://doi.org/10.4187/respcare.01238.

Davies, S. W., Leonard, K. L., Falls, R. K., Jr., Mageau, R. P., Efird, J. T., Hollowell, J. P., et al. (2015). Lung protective ventilation (ARDSNet) versus airway pressure release ventilation: Ventilatory management in a combined model of acute lung and brain injury. *Journal of Trauma and Acute Care Surgery*, *78*(2), 240–249. discussion 249-251 https://doi.org/10.1097/ta.0000000000000518.

Edgerton, C. A., Leon, S. M., Hite, M. A., Kalhorn, S. P., Scott, L. A., & Eriksson, E. A. (2019). Airway pressure release ventilation does not increase intracranial pressure in patients with traumatic brain injury with poor lung compliance. *Journal of Critical Care*, *50*, 118–121. https://doi.org/10.1016/j.jcrc.2018.11.034.

Fletcher, J. J., Wilson, T. J., Rajajee, V., Davidson, S. B., & Walsh, J. C. (2018). Changes in therapeutic intensity level following airway pressure release ventilation in severe traumatic brain injury. *Journal of Intensive Care Medicine*, *33*(3), 196–202. https://doi.org/10.1177/0885066616669315.

Iowa, U.o. (2020). Airway pressure release ventilation (APRV). In *PICU handbook*. Retrieved from: https://uichildrens.org/health-library/airway-pressure-release-ventilation-aprv.

Kollisch-Singule, M., Emr, B., Smith, B., Roy, S., Jain, S., Satalin, J., et al. (2014). Mechanical breath profile of airway pressure release ventilation: The effect on alveolar recruitment and microstrain in acute lung injury. *JAMA Surgery*, *149*(11), 1138–1145. https://doi.org/10.1001/jamasurg.2014.1829.

Lim, J., & Litton, E. (2019). Airway pressure release ventilation in adult patients with acute hypoxemic respiratory failure: A systematic review and Meta-analysis. *Critical Care Medicine*, *47*(12), 1794–1799. https://doi.org/10.1097/ccm.0000000000003972.

Lou, M., Xue, F., Chen, L., Xue, Y., & Wang, K. (2012). Is high PEEP ventilation strategy safe for acute respiratory distress syndrome after severe traumatic brain injury? *Brain Injury*, *26*(6), 887–890. https://doi.org/10.3109/02699052.2012.660514.

Luecke, T., & Pelosi, P. (2005). Clinical review: Positive end-expiratory pressure and cardiac output. *Critical Care (London, England)*, *9*(6), 607–621. https://doi.org/10.1186/cc3877.

Luecke, T., Roth, H., Herrmann, P., Joachim, A., Weisser, G., Pelosi, P., et al. (2004). Assessment of cardiac preload and left ventricular function under increasing levels of positive end-expiratory pressure. *Intensive Care Medicine*, *30*(1), 119–126. https://doi.org/10.1007/s00134-003-1993-7.

Marik, P. E., Young, A., Sibole, S., & Levitov, A. (2012). The effect of APRV ventilation on ICP and cerebral hemodynamics. *Neurocritical Care*, *17*(2), 219–223. https://doi.org/10.1007/s12028-012-9739-4.

Mireles-Cabodevila, E., & Kacmarek, R. M. (2016). Should airway pressure release ventilation be the primary mode in ARDS? *Respiratory Care*, *61*(6), 761–773. https://doi.org/10.4187/respcare.04653.

Picetti, E., Pelosi, P., Taccone, F. S., Citerio, G., Mancebo, J., & Robba, C. (2020). VENTILatOry strategies in patients with severe traumatic brain injury: The VENTILO survey of the European Society of Intensive Care Medicine (ESICM). *Critical Care (London, England)*, *24*(1), 158. https://doi.org/10.1186/s13054-020-02875-w.

Stock, M. C., Downs, J. B., & Frolicher, D. A. (1987). Airway pressure release ventilation. *Critical Care Medicine*, *15*(5), 462–466. https://doi.org/10.1097/00003246-198705000-00002.

Vargas, M., Sutherasan, Y., Gregoretti, C., & Pelosi, P. (2014). PEEP role in ICU and operating room: From pathophysiology to clinical practice. *Scientific World Journal*, *2014*, 852356. https://doi.org/10.1155/2014/852356.

Young, J. S., Blow, O., Turrentine, F., Claridge, J. A., & Schulman, A. (2003). Is there an upper limit of intracranial pressure in patients with severe head injury if cerebral perfusion pressure is maintained? *Neurosurgical Focus*, *15*(6). https://doi.org/10.3171/foc.2003.15.6.2, E2.

Zhang, X. Y., Yang, Z. J., Wang, Q. X., & Fan, H. R. (2011). Impact of positive end-expiratory pressure on cerebral injury patients with hypoxemia. *American Journal of Emergency Medicine*, *29*(7), 699–703. https://doi.org/10.1016/j.ajem.2010.01.042.

Chapter 35

Linking death, the paranasal sinuses, and traumatic head trauma

Ayhan Kanat[a] and Elena Romana Gasenzer[b]
[a]*Department of Neurosurgery, Medical Faculty, Recep Tayyip Erdogan University, Rize, Turkey,* [b]*Department Physician Assistance, Carl Remigius Medical School, Frankfurt a.M., Germany*

Abbreviations

CPP cerebral perfusion pressure
GCS Glasgow coma scale
ICP intracranial pressure
TBI traumatic brain injury

Key points

The paranasal sinuses affect the mortality of patients with a head injury.
The significance of the paranasal sinuses for human health is still a topic of debate.

Introduction

The presence of multiple cranial important structures makes the skull a critical surgical region (Kanat & Aydin, 1999b). There are three paranasal sinuses which include sphenoid, frontal, ethmoid sinuses. These sinuses are located along with the anterior and middle skull bases. Important intracranial structures such as the optic nerves and internal carotid arteries are found near the sphenoid sinus (Lee, Kellman, & Darling, 2014). Blunt cranial traumas are the main cause of craniofacial fractures and traumatic cerebral injury. Cranial trauma is not new in human life; it also has historical importance, e.g., there is some evidence that Mozart had suffered cranial trauma at an early age of his life. His post-mortem cranium shows a fracture of the left temporoparietal bone (Kanat, Romana Gasenzer, & Neugebauer, 2019).

The significance of the paranasal sinuses

The hypothesis of "Crumple Zone" is that the paranasal sinuses protect surrounding vital organs from the impact of trauma (Perry, 2009). The paranasal sinuses air-filled cavities cranial bones (Celiker et al., 2020) (Fig. 1), and their function is a topic of debate (Yu, Branstetter 4th, & Snyderman, 2014). Hypotheses benefits of the paranasal sinuses include: (1) amplification of an individual's voice, (2) humidification of inspired air, (3) strengthening of the host's defense mechanism, (4) thermal insulation of underlying organs, (5) weight reduction of the skull, and (6) flotation (Cai et al., 2017). An external mechanical trauma may lead to brain injury with paranasal sinus fracture. It can affect orchestrating mechanisms of CN with neurologic deficit. It has been proposed that the paranasal sinuses may function as "crumple zones." This zone may be a shock absorber that can protect the surrounding vital neurovascular structures from traumatic injury (Celiker et al., 2020; Cohn, Iezzi, Licata, Othman, & Zwillenberg, 2020).

Anatomy of paranasal sinuses

Frontal sinus

The frontal sinuses are paired air-filled cavities (Kanat, Yazar, Ozdemir, Kazdal, & Balik, 2017). They start to develop in childhood (Celiker et al., 2020). They are radiographically identifiable at age 8 years and matures to full size at the

Diagnosis and Treatment of Traumatic Brain Injury. https://doi.org/10.1016/B978-0-12-823347-4.00031-2

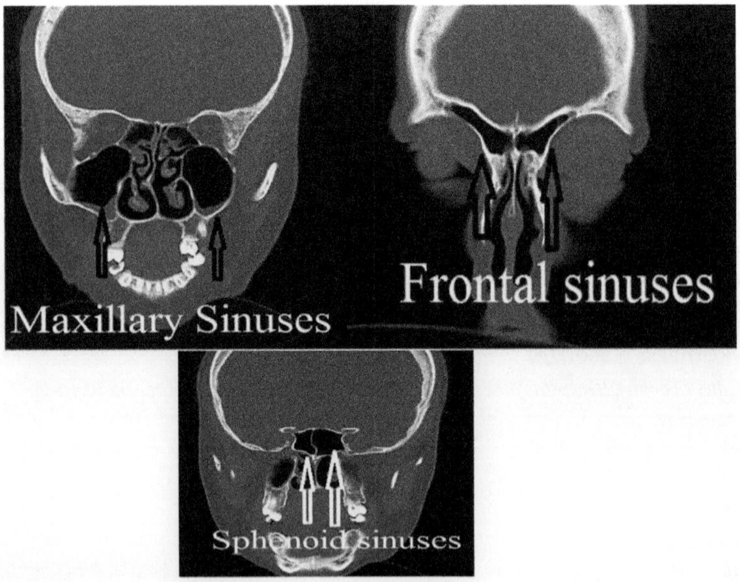

FIG. 1 The paranasal sinuses are seen in a cranial CT of the patient without trauma.

approximate age 15 years (Doonquah, Brown, & Mullings, 2012). The final shape is asymmetric nature. Brain development can show a different course for different individuals (Kanat et al., 2017), producing the size of the frontal sinus may vary anatomically. The frontal bone surrounds the sinuses superiorly and laterally (Kim, Boahene, & Byrne, 2017) and high impact trauma may lead to frontal sinus fractures, which are found in about 5%–12% of all patients with maxillofacial traumas (Schultz et al., 2019). The mortality risk of frontal sinus trauma is 25%, including a 42% risk of permanent neurological deficit (Banks, Grayson, Cho, & Woodworth, 2020). In the frontal fractures, surgeries are performed to resolve the aesthetic deficit as well as prevention of severe complications (Buller, Kreppel, Maus, Zirk, & Zöller, 2019). CSF leaks can also occur. Celiker et al. reported that the size of the frontal sinus after severe head trauma has a role in the outcome of patients (Celiker et al., 2020).

Management: A change in surgical paradigm

The management of frontal sinus fractures has changed over time (Schultz et al., 2019). Advances in endoscopic surgical techniques have altered the treatment algorithms for frontal sinus fractures (Dedhia, Morisada, Tollefson, & Strong, 2019). Traditionally, the surgery of traumatic paranasal fracture has been made using open incisions. In 1978, the method of cranialization was described. In this procedure, the surgeon the posterior table is removed (Schultz et al., 2019). Cranialization procedures have been recommended for the patient with posterior table defects greater than 25% and in general (Schultz et al., 2019).

Sphenoid sinus

The sphenoid sinus is located in the center of the cranial base (Findik et al., 2020). It varies considerably in size (Patel, Fernandez-Miranda, Wang, & Wang, 2016). They are divided by septum (Jaworek-Troć et al., 2019). Right and left SS (s) are paired irregular pneumatic spaces (Jaworek-Troć et al., 2019). These sinuses are classified the three types as conchal, presellar, and sellar (Pirinc et al., 2019).

Maxillary sinus

The maxillae are paired bones. Most cases with maxillary sinus fractures are managed conservatively (Cohn et al., 2020). The classification in maxillary bone fracture, Le Fort classification to is still applied. Le Fort I fracture: The maxilla is separated from the facial skull in a horizontal plane above the teeth apices and the hard palate (Kühnel & Reichert, 2015). Le Fort II fracture is characterized by a dissociation of the maxilla, the nasal bones, and the nasal septum from the cranial skull and the lateral midface (Kühnel & Reichert, 2015). In Le Fort III fracture, the facial skeleton is separated from the cranial skull (Kühnel & Reichert, 2015).

Cribiform plate fractures

The cribriform plate is the portion of the ethmoid bone that forms the roof of the nasal cavity (Gomez & Pickup, 2020). The cribriform plate is the thinnest portion of the base of the skull and is therefore susceptible to fracture in cases of facial trauma (Gomez & Pickup, 2020).

History and physical examination

Non-specific symptoms such as nausea, vomiting, headaches, visual changes, dizziness, confusion can be seen in paranasal sinus fractures (Lizzo & Waseem, 2020). The presence of vomiting and post-traumatic seizure may occur. The timely detection of neurological deterioration is an important issue (Aydin et al., 2019) by performing serial bedside neurological examinations (Kazdal et al., 2017). In neurological examination, Glasgow Coma Scale (GCS) is still used in the evaluation of the severity of the TBI. The Glasgow Coma Scale (GCS) was originally developed, in 1974, to investigate patients with impaired consciousness or coma following head trauma (Kanat & Aydin, 2000). Verbal response (V), motor response (M), and eye-opening (E) are evaluated in this scale.

The severity in paranasal sinus trauma and TBI is determined according to GCS, with minor TBI having a score of 13–15, moderate having a score of 9–12, and severe having a score of <8. We think that the GCS evaluation of a patient with head trauma is important, GCS evaluation after a head injury has repeatedly been criticized for its several failures to reflect verbal reaction in intubated patients of intensive care unit (Kazdal et al., 2016). After a severe paranasal sinus trauma, traumatic brain injury, impairments in physical, cognitive, psychological, behavioral functioning, and early and late complications can be seen (Kanat et al., 2019). As a late complication, cranial trauma may lead to subdural hematoma (Kanat, Kayaci, Yazar, Kazdal, & Terzi, 2010), the inclusion of epidermal cells into the diploe of the skull, and maybe a reason for intradiploic epidermoid cyst (Turk et al., 2018). For that reason, close neurological and radiological observation along with the medical treatment (Kanat, Yazar, & Kazdal, 2009) is necessary for patients with paranasal sinus involvement following head trauma. Measuring and comparing pupil diameter with reactivity to light can help diagnose cranial nerve injury after severe cranial trauma. Initial vitals are important to review. Both parasympathetic and sympathetic nervous system disorders appear to be important factors in pupillary diameter changes (Aydin et al., 2019; Ozturk et al., 2018). The bilateral dilated pupil may indicate a poor outcome. Laboratory testing: currently, there are no standard recommendations for the use of laboratory testing.

Imaging of paranasal sinus trauma

Acute epidural or subdural hematomas, intraparenchymal hemorrhage may occur. Acute subdural hematomas may contiguous through contused, lacerated brain tissue with intracerebral hemorrhage (Kanat, 1998). If the brain shrinks due to brain atrophy and decreased intracranial pressure, the fluid collection such as in hygroma may develop by a passive effusion (Kanat et al., 2009). Paranasal sinus trauma is a strong indication for radiological investigation CT head (Kanat, 1998) and MRI are used to evaluate the radiological changes of paranasal sinus trauma. Computed tomography (CT) has revolutionized the assessment of frontal sinus fractures. MRI is sometimes used in the assessment of soft tissue changes of the frontal sinus. These include fracture, hemorrhage, and edema. Fig. 2 shows anterior and posterior wall fracture of the frontal sinus which is associated with a cerebral contusion. CT scan for acute traumatic intracranial hemorrhage and fractures is still the imaging modality of choice (Konar et al., 2020). Fig. 3 shows retro-orbital fracture of a patient. Magnetic resonance imaging has several advantages, including the absence of ionizing radiations, multi-planar capability, and excellent soft-tissue contrast: its main limitation is the difficulty of imaging the cortical bone (Salvolini, 2002).

Pathogenesis

In paranasal sinus trauma, direct impact and accelerative and decelerative forces may be seen. Primary and secondary injury brain injury may occur after paranasal sinus trauma. Primary injury entails the direct brain damage that occurs immediately after the trauma. In the primary injury, extra parenchymal hemorrhages (epidural hematoma, subdural hematoma, subarachnoid hemorrhage, and intra-ventricular hemorrhage); focal contusions, and intraparenchymal hemorrhages; traumatic axonal (focal or diffuse) injury may be seen. The hematoma volume is an important issue (Yilmaz et al., 2011). Inside of cranium, large hemorrhage volumes often result in poor outcomes (Yilmaz et al., 2011). Secondary injury mechanisms are also initiated at the moment of the traumatic incident. The blood–brain barrier (BBB)-disturbance, excite-toxicity, mitochondrial dysfunction, oxidative stress, inflammation, and cell loss are the principal identified mechanisms orchestrating secondary injury mechanisms.

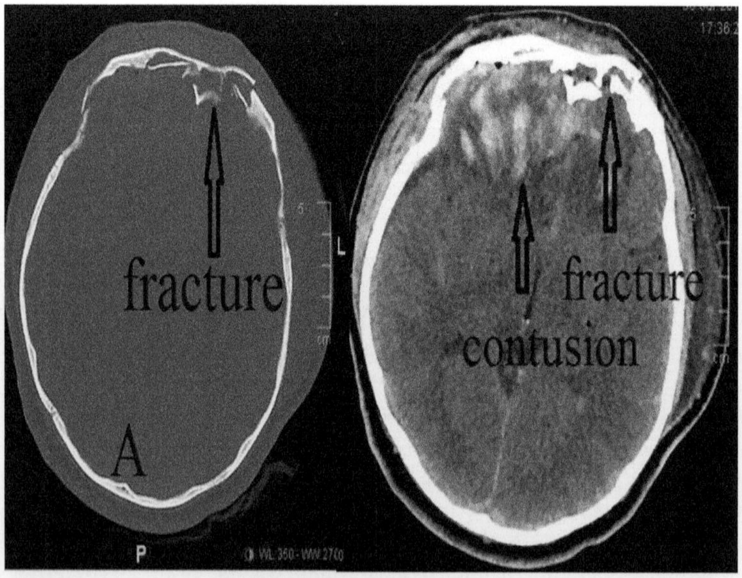

FIG. 2 Anterior and posterior wall fracture of frontal sinus which associated with cerebral contusion is seen.

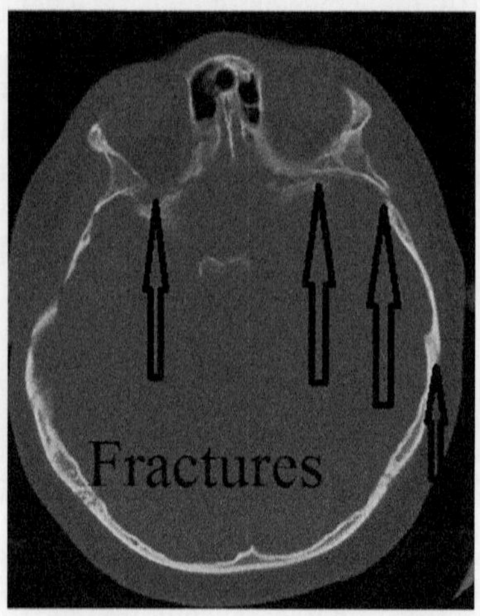

FIG. 3 Retro-orbital fracture of a patient.

Cerebral perfusion pressure

Cerebral perfusion pressure (CPP) is the net pressure gradient that drives oxygen delivery to cerebral tissue (Mount & Das, 2020). CPP is the difference between the mean arterial pressure (MAP) and the ICP (CPP = MAP − ICP) (Shaikh & Waseem, 2020). Target CPP is 55–60 mmHg (Shaikh & Waseem, 2020). MAP and ICP must be measured simultaneously (Mount & Das, 2020). An increase in ICP can decrease the CPP (Shaikh & Waseem, 2020). Lowering MAP in a hypotensive patient may have a detrimental effect on the outcome of patients. A minimum CPP should be maintained to avoid cerebral insult. A sustained ICP > 20 mmHg is associated with increased morbidity and mortality (Ginsburg & Huff, 2020). A head injury can lead to an increased ICP. If ICP approaches the MAP, CPP falls to dangerously low levels (Ginsburg & Huff, 2020). The body responds by increasing the MAP and dilating cerebral blood vessels, which further raises ICP, thus continuing the vicious cycle that can lower CPP and lead to brain ischemia and long-term injury (Ginsburg & Huff, 2020). Maintaining adequate CPP will decrease the risk of poor outcome (Mount & Das, 2020). Cerebral herniation may occur after head trauma with an associated paranasal fracture. There are main four types of herniations: (1) uncal transtentorial, (2) central transtentorial, (3) cerebellar tonsillary, and (4) upward posterior fossa/cerebellar herniation (Shaikh & Waseem, 2020).

Oxidative stress

Oxidative damage may also occur following para nasal and head trauma. The brain is very sensitive to oxidative damage. Oxidative stress represents a state where oxygen levels along with oxygen-derived free radicals overwhelm the scavenging antioxidant system (Ismail et al., 2020). To date, no treatment has been effective in eradicating the consequences of injury. Instead, therapeutics have focused on alleviating the impact of secondary injury and managing its biochemical contributors (Ismail et al., 2020). Antioxidants are being considered as potential therapeutics (Ismail et al., 2020).

Blood–brain barrier disruption

The blood–brain barrier (BBB) has a value concerning brain function (Kanat, 2003). Histopathology only plays a role in post-mortem diagnostics (Bruggeman, Haitsma, Dirven, & Volovici, 2020). It is not used in a clinical setting (Bruggeman et al., 2020). For that reason, animal studies are important (Ozdemir, Kanat, Erturk, et al., 2020). Midline shift, ventricle, and basal cistern compression are often seen in patients with increased ICP following head trauma. In these patients, increased ICP is one of the important causes of the worse outcome (Celiker et al., 2020). Intracranial changes following head injury may lead to hemodynamic changes. These hemodynamic changes are important because the proper functioning of brain cells relies on an abundant and continuous supply of oxygen (Aydin et al., 2020; Kanat, 2003).

The debate about the protective effect of the paranasal sinuses

Severe traumatic paranasal sinus injuries can be very complex because many vital structures are confined within a small volume in the human cranium (Ozdemir et al., 2018). The function and the importance of the paranasal sinuses is a topic of debate (Yu et al., 2014). A hypothesis was outlined by Le Fort in his study of the upper jaw in 1901 (Yu et al., 2014). He theorized that the sinuses may serve to protect the surrounding contents of the head from impact (Yu et al., 2014). Their structure may function to distribute and absorb energy by fracturing like a crumple zone on a car to protect the brain, eyes, and other critical structures (Yu et al., 2014). However, the about role of this zone after severe traumatic paranasal injury and head trauma is still not well-known. Celiker et al. reported that the size of the frontal sinus after severe head trauma has led to poor outcome (Celiker et al., 2020).

Music and paranasal sinus

The paranasal sinuses are complex anatomical structures of unknown significance (Cai et al., 2017). This complexity is likely influenced by the morphology and traumatic injury of the paranasal cavities. Nasal and paranasal cavities are supposed to contribute substantially to the vocal tract resonator properties (Havel et al., 2017). However, their acoustical effect (s) of head and paranasal sinuses are not known. Many studies have been published in patients with traumatic brain and paranasal injury. Several different neural mechanisms of the embodiment are likely to be activated during the perception of human action (Gasenzer, Kanat, & Neugebauer, 2018). Problem is studies about music interventions in different medical indications have a high risk of bias because the condition of patients is very different and methods, music styles, or instruments of the interventions are very different, too. Cultural imprint, music education, and music culture are very different, too. In the acute and late-stage after brain injury, the neuronal growth of the synaptic connections and the possibilities of a neuronal reorganization are not well known. The long-term music training and the associated sensorimotor skill learning can be a strong stimulus for neuroplastic changes of musicians (Gasenzer, Kanat, Ozdemir, Rakici, & Neugebauer, 2020). The neuronal plasticity of a musician's brain may be different from that of a non-musician (Gasenzer, Kanat, & Neugebauer, 2017b). The brain damage caused by the head and paranasal sinus injury of musicians disrupts the functional music processing system, creating a unique opportunity to investigate the effect of this damage, the brain plasticity of a musician may affect his/her recovery rate.

Conclusion

At present, neurosurgical practice is confronted by an explosion of technology (Yolas et al., 2016), neurosurgery and neuroscience have gone through moments of great renewal (Gasenzer, Kanat, Ozdemir, & Neugebauer, 2018), but, traumatic head injuries impose a greater burden on modern society than do other diseases (Kanat & Aydin, 1999a; Kazdal et al., 2016), and paranasal sinus involvement in the head injuries are a major public health problem with high morbidity and mortality. There is a link to death, and traumatic head trauma involved the paranasal sinuses. The thick clivus protects the structure of the brainstem and posterior cranial fossa, but there is no thick structure such as a clivus in the anterior part of the cranium (Ozdemir, Kanat, Erturk, et al., 2020). In paranasal sinus trauma, direct impact and accelerative and decelerative forces may be seen. If we consider the nervous system as a great orchestra that can express a complete range of rhythms and melodies and the most complex harmonic combinations (Gasenzer, Kanat, & Neugebauer, 2017a), we will find it easier to understand how any traumatic paranasal injury with head trauma may be translated into an alteration of the rhythmic systems that synchronize the brain.

Applications to other areas of neuroscience

The paranasal sinuses are generally affected by trauma to the anterior skull base. The cranial trauma can be seen in any age group in the community (Ozdemir, Kanat, & Kazdal, 2020). Paranasal sinus fractures increase mortality and morbidity rates of patients with cranial trauma. These traumas may also lead to some fatal complications such as CSF leakage. Understanding the consequences of paranasal sinus trauma may provide future guidance in the management of patients.

Mini-dictionary of terms

Paranasal sinuses: Air-filled cavity in the bones around the nasal cavity.
Fracture: broken bone.

Key facts of paranasal sinuses

1. The paranasal sinuses are air-filled cavities; the names of sinuses are frontal, maxillary, ethmoid, and sphenoid sinuses.
2. An external mechanical trauma may lead to brain injury with a paranasal sinus fracture.
3. In traumatic injury, direct impact, and accelerative and decelerative forces may be seen.
4. Primary and secondary injury brain injury may occur.
5. It is reported that paranasal sinuses prevent impact energy from being transferred to the brain as a created a "crumple zone".

Key facts of crumple zone

- There are many debates on the function of the paranasal sinuses.
- Crumple zone hypothesizes that the paranasal sinuses protect surrounding neurovascular structures.
- The distinct properties of the "crumple zone" contribute to its protective function but it is likely that the properties of the paranasal sinus bones that build a "crumple zone" may play an equally important role. It can work to decrease the impact effect of energy on areas around the maxilla-facial region on the neurovascular structures.

Summary points

- The paranasal sinuses are generally affected by trauma to the anterior skull base.
- Paranasal sinus fractures increase mortality and morbidity rates of patients with cranial trauma.
- The traumas of paranasal sinuses cause significant complications.
- The hypothesis of "Crumple Zone" is that the paranasal sinuses protect surrounding vital organs from the impact of trauma.
- Crumple zones function to decrease impact energy.

References

Aydin, M. D., Kanat, A., Aydin, A., Aydin, A., Demirci, T., & Ozmen, S. (2020). Estimating basiler artery upper rupture limit by dangerous alarming diameter of arteries (DADA) following bilateral common carotid artery ligation; a new theorem. *The International Journal of Neuroscience*, 1–10. https://doi.org/10.1080/00207454.2020.1803303.

Aydin, M. D., Kanat, A., Yolas, C., Soyalp, C., Onen, M. R., Yilmaz, I., et al. (2019). Spinal subarachnoid hemorrhage induced intractable miotic pupil. A reminder of ciliospinal sympathetic center ischemia based miosis: An experimental study. *Turkish Neurosurgery*, *29*(3), 434–439. https://doi.org/10.5137/1019-5149.JTN.24446-18.1.

Banks, C., Grayson, J., Cho, D. Y., & Woodworth, B. A. (2020). Frontal sinus fractures and cerebrospinal fluid leaks: A change in surgical paradigm. *Current Opinion in Otolaryngology & Head and Neck Surgery*, *28*(1), 52–60. https://doi.org/10.1097/MOO.0000000000000602.

Bruggeman, G. F., Haitsma, I. K., Dirven, C. M. F., & Volovici, V. (2020). Traumatic axonal injury (TAI): Definitions, pathophysiology and imaging-a narrative review. *Acta Neurochirurgica*. https://doi.org/10.1007/s00701-020-04594-1.

Buller, J., Kreppel, M., Maus, V., Zirk, M., & Zöller, J. E. (2019). Risk of frontal sinus anterior table fractures after craniofacial trauma and the role of anatomic variations in frontal sinus size: A retrospective case-control study. *Journal of Cranio-Maxillo-Facial Surgery: Official Publication of the European Association for Cranio-Maxillo-Facial Surgery*, *47*(4), 611–615. https://doi.org/10.1016/j.jcms.2019.01.018.

Cai, S. S., Mossop, C., Diaconu, S. C., Hersh, D. S., AlFadil, S., Rasko, Y. M., et al. (2017). The "crumple zone" hypothesis: Association of frontal sinus volume and cerebral injury after craniofacial trauma. *Journal of Cranio-Maxillo-Facial Surgery: Official Publication of the European Association for Cranio-Maxillo-Facial Surgery*, *45*(7), 1094–1098. https://doi.org/10.1016/j.jcms.2017.04.005.

Celiker, M., Kanat, A., Ozdemir, A., Celiker, F. B., Kazdal, H., Ozdemir, B., et al. (2020). Controversy about the protective role of volume in the frontal sinus after severe head trauma: Larger sinus equates with higher risk of death. *British Journal of Oral and Maxillofacial Surgery*, *58*(3), 314–318. https://doi.org/10.1016/j.bjoms.2019.12.008.

Cohn, J. E., Iezzi, Z., Licata, J. J., Othman, S., & Zwillenberg, S. (2020). An update on maxillary fractures: A heterogenous group. *The Journal of Craniofacial Surgery*, *31*(7), 1920–1924. https://doi.org/10.1097/SCS.0000000000006675.

Dedhia, R. D., Morisada, M. V., Tollefson, T. T., & Strong, E. B. (2019). Contemporary management of frontal sinus fractures. *Current Opinion in Otolaryngology & Head and Neck Surgery*, *27*(4), 253–260. https://doi.org/10.1097/MOO.0000000000000546.

Doonquah, L., Brown, P., & Mullings, W. (2012). Management of frontal sinus fractures. *Oral and Maxillofacial Surgery Clinics of North America*, *24*(2), 265–274. ix https://doi.org/10.1016/j.coms.2012.01.008.

Findik, H., Kanat, A., Celiker, F. B., Celiker, M., Beyazal, M., Ozdemir, B., et al. (2020). What are we missing from asymmetric relationship between the retinal nerve fiber layer thickness profiles and sphenoid sinus volume? *The Journal of Craniofacial Surgery*, *31*(2), e210–e214. https://doi.org/10.1097/SCS.0000000000005909.

Gasenzer, E. R., Kanat, A., & Neugebauer, E. (2017a). Neurosurgery and music; effect of Wolfgang Amadeus Mozart. *World Neurosurgery*, *102*, 313–319. https://doi.org/10.1016/j.wneu.2017.02.081.

Gasenzer, E. R., Kanat, A., & Neugebauer, E. (2017b). The unforgettable neurosurgical operations of musicians in the last century. *World Neurosurgery*, *101*(5), 444–450. https://doi.org/10.1016/j.wneu.2016.11.144.

Gasenzer, E. R., Kanat, A., & Neugebauer, E. (2018). In reply to "hypothesis on the origin of dance: Mirror neuron system and imitation from movement disorders". *World Neurosurgery*, *120*, 605. https://doi.org/10.1016/j.wneu.2018.09.162.

Gasenzer, E. R., Kanat, A., Ozdemir, V., & Neugebauer, E. (2018). Analyzing of dark past and bright present of neurosurgical history with a picture of musicians. *British Journal of Neurosurgery*, *33*(3), 303–304. https://doi.org/10.1080/02688697.2018.1467000.

Gasenzer, E. R., Kanat, A., Ozdemir, V., Rakici, S. Y., & Neugebauer, E. (2020). Interesting different survival status of musicians with malignant cerebral tumors. *British Journal of Neurosurgery*, *34*(3), 264–270. https://doi.org/10.1080/02688697.2019.1701629.

Ginsburg, J., & Huff, J. S. (2020). *Closed head trauma*. Treasure Island, FL: StatPearls Publishing.

Gomez, J., & Pickup, S. (2020). *Cribiform plate fractures*. Treasure Island, FL: StatPearls Publishing.

Havel, M., Becker, S., Schuster, M., Johnson, T., Maier, A., & Sundberg, J. (2017). Effects of functional endoscopic sinus surgery on the acoustics of the sinonasal tract. *Rhinology*, *55*(1), 81–89. https://doi.org/10.4193/Rhin16.229.

Ismail, H., Shakkour, Z., Tabet, M., Abdelhady, S., Kobaisi, A., Abedi, R., et al. (2020). Traumatic brain injury: Oxidative stress and novel anti-oxidants such as mitoquinone and edaravone. *Antioxidants (Basel, Switzerland)*, *9*(10). https://doi.org/10.3390/antiox9100943.

Jaworek-Troć, J., Zarzecki, M., Bonczar, A., Kaythampillai, L. N., Rutowicz, B., Mazur, M., et al. (2019). Sphenoid bone and its sinus—Anatomo-clinical review of the literature including application to FESS. *Folia Medica Cracoviensia*, *59*(2), 45–59.

Kanat, A. (1998). Acute posttraumatic subdural hematomas: "Intradural" computed tomographic appearance as a favorable prognostic factor. *Neurosurgery*, *43*(3), 685–686.

Kanat, A. (2003). Brain oxygenation and energy metabolism: Part I—Biological function and pathophysiology. *Neurosurgery*, *52*(6), 1508–1509.

Kanat, A., & Aydin, Y. (1999a). Postcontrast magnetic resonance imaging to predict progression of traumatic epidural and subdural hematomas in the acute stage. *Neurosurgery*, *44*(3), 685–686.

Kanat, A., & Aydin, Y. (1999b). Recurrent meningiomas. *Journal of Neurosurgery*, *91*(4), 720–721.

Kanat, A., & Aydin, Y. (2000). Prognostic value and determinants of ultraearly angiographic vasospasm after aneurysmal subarachnoid hemorrhage. *Neurosurgery*, *46*(2), 505–507.

Kanat, A., Kayaci, S., Yazar, U., Kazdal, H., & Terzi, Y. (2010). Chronic subdural hematoma in adults: Why does it occur more often in males than females? Influence of patient's sexual gender on occurrence. *Journal of Neurosurgical Sciences*, *54*(3), 99–103.

Kanat, A., Romana Gasenzer, E., & Neugebauer, E. (2019). A different aspect of the unexpected death of Mozart at the age of 35 years. *CNS Spectrums*, *24*(6), 628–631. https://doi.org/10.1017/S1092852918001736.

Kanat, A., Yazar, U., & Kazdal, H. (2009). Chronic subdural hygroma with thrombocythemia: First case report. *Journal of Neurosurgical Sciences*, *53*(4), 165–167.

Kanat, A., Yazar, U., Ozdemir, B., Kazdal, H., & Balik, M. S. (2017). Neglected knowledge: Asymmetric features of lumbar disc disease. *Asian Journal of Neurosurgery*, *12*(2), 199–202. https://doi.org/10.4103/1793-5482.145573.

Kazdal, H., Kanat, A., Aydin, M. D., Yazar, U., Guvercin, A. R., Calik, M., et al. (2017). Sudden death and cervical spine: A new contribution to pathogenesis for sudden death in critical care unit from subarachnoid hemorrhage; first report—An experimental study. *Journal of Craniovertebral Junction and Spine*, *8*(1), 33.

Kazdal, H., Kanat, A., Findik, H., Sen, A., Ozdemir, B., Batcik, O. E., et al. (2016). Transorbital ultrasonographic measurement of optic nerve sheath diameter for intracranial midline shift in patients with head trauma. *World Neurosurgery, 85*. https://doi.org/10.1016/j.wneu.2015.10.015.

Kim, I. A., Boahene, K. D., & Byrne, P. J. (2017). Trauma in facial plastic surgery: Frontal sinus fractures. *Facial Plastic Surgery Clinics of North America, 25*(4), 503–511. https://doi.org/10.1016/j.fsc.2017.06.004.

Konar, S., Pavlov, O., Durango-Espinosa, Y., Garcia-Ballestas, E., Joaquim, A. F., Ghosh, A., et al. (2020). Critical appraisal of traumatic brain injury and its management. *Indian Journal of Neurotrauma, 17*(2), 79–84.

Kühnel, T. S., & Reichert, T. E. (2015). Trauma of the midface. *GMS Current Topics in Otorhinolaryngology—Head and Neck Surgery, 14*. https://doi.org/10.3205/cto000121, Doc06.

Lee, T. S., Kellman, R., & Darling, A. (2014). Crumple zone effect of nasal cavity and paranasal sinuses on posterior cranial fossa. *The Laryngoscope, 124*(10), 2241–2246. https://doi.org/10.1002/lary.24644.

Lizzo, J. M., & Waseem, M. (2020). *Brain trauma*. Treasure Island, FL: StatPearls Publishing.

Mount, C. A., & Das, J. M. (2020). *Cerebral perfusion pressure*. Treasure Island, FL: StatPearls Publishing.

Ozdemir, B., Kanat, A., Batcik, S., Batcik, O. E., Celiker, M., & Kayayurt, K. (2018). Unilateral isolated hypoglossal nerve palsy caused by gunshot injury. *The Journal of Craniofacial Surgery*, 1. https://doi.org/10.1097/SCS.0000000000004126.

Ozdemir, B., Kanat, A., Erturk, C., Batcik, O. E., Celiker, F. B., Celiker, M., et al. (2020). Important aspect of hypoglossal nerve injury following gunshot wound; can the clivus has a role? A case report. *British Journal of Neurosurgery*, 1–2. https://doi.org/10.1080/02688697.2020.1777261.

Ozdemir, B., Kanat, A., & Kazdal, H. (2020). Deneysel beyin yaralanma modelleri experimental cerebral injury models. *Turk Norosirurji Dergisi, 30*(2), 308–311.

Ozturk, C., Ozdemir, N. G., Kanat, A., Aydin, M. D., Findik, H., Aydin, N., et al. (2018). How reliable is pupillary evaluation following subarachnoid hemorrhage? Effect of oculomotor nerve degeneration secondary to posterior communicating artery vasospasm: First experimental study. *Journal of Neurological Surgery, Part A: Central European Neurosurgery, 79*(4), 302–308. https://doi.org/10.1055/s-0037-1608841.

Patel, C. R., Fernandez-Miranda, J. C., Wang, W.-H., & Wang, E. W. (2016). Skull base anatomy. *Otolaryngologic Clinics of North America, 49*(1), 9–20. https://doi.org/10.1016/j.otc.2015.09.001.

Perry, M. (2009). Maxillofacial trauma—Developments, innovations and controversies. *Injury, 40*(12), 1252–1259. https://doi.org/10.1016/j.injury.2008.12.015.

Pirinc, B., Fazliogullari, Z., Guler, I., Unver Dogan, N., Uysal, I. I., & Karabulut, A. K. (2019). Classification and volumetric study of the sphenoid sinus on MDCT images. *European Archives of Oto-Rhino-Laryngology: Official Journal of the European Federation of Oto-Rhino-Laryngological Societies (EUFOS): Affiliated with the German Society for Oto-Rhino-Laryngology—Head and Neck Surgery, 276*(10), 2887–2894. https://doi.org/10.1007/s00405-019-05549-8.

Salvolini, U. (2002). Traumatic injuries: Imaging of facial injuries. *European Radiology, 12*(6), 1253–1261. https://doi.org/10.1007/s00330-002-1445-8.

Schultz, J. J., Chen, J., Sabharwal, S., Halsey, J. N., Hoppe, I. C., Lee, E. S., et al. (2019). Management of frontal bone fractures. *The Journal of Craniofacial Surgery, 30*(7), 2026–2029. https://doi.org/10.1097/SCS.0000000000005720.

Shaikh, F., & Waseem, M. (2020). *Head trauma*. Treasure Island, FL: StatPearls Publishing.

Turk, O., Ozdemir, N. G., Demirel, N., Atci, I. B., Kanat, A., & Yolas, C. (2018). Nontraumatic intradiploic epidermoid cyst and older age: Association or causality? *The Journal of Craniofacial Surgery, 29*(2), e143–e146. https://doi.org/10.1097/SCS.0000000000003897.

Yilmaz, A., Musluman, A. M., Kanat, A., Cavusoglu, H., Terzi, Y., & Aydin, Y. (2011). Rüptüre posterior fossa arteriovenöz malformasyonlarda hematom volümü ile klinik sonuç arasindaki ilişki ve hematomun cerrahi tedavisinin önemi. *Turkish Neurosurgery, 21*(2), 152–159. https://doi.org/10.5137/1019-5149.JTN.3401-10.0.

Yolas, C., Kanat, A., Aydin, M. D., Altas, E., Kanat, I. F., Kazdal, H., et al. (2016). Unraveling of the effect of Nodose ganglion degeneration on the coronary artery vasospasm after subarachnoid hemorrhage: An experimental study. *World Neurosurgery, 86*, 79–87. https://doi.org/10.1016/j.wneu.2015.09.004.

Yu, J. L., Branstetter, B. F., 4th, & Snyderman, C. H. (2014). Frontal sinus volume predicts incidence of brain contusion in patients with head trauma. *The Journal of Trauma and Acute Care Surgery, 76*(2), 488–492. https://doi.org/10.1097/TA.0b013e3182aaa4bd.

Chapter 36

Cannabidiol in traumatic brain injury

Enza Palazzo, Monica Iannotta, Carmela Belardo, Rosmara Infantino, Flavia Ricciardi, Serena Boccella, Francesca Guida, Livio Luongo, and Sabatino Maione

Department of Experimental Medicine, Pharmacology Division, University of Campania "L. Vanvitelli", Naples, Italy

List of abbreviations

AEA	anandamide
2-AG	2-arachidonoylglycerol
AMT	anandamide membrane transporter
BDNF	brain derived neurotrophic factor
CB_1	cannabinoid receptors 1
CB_2	cannabinoid receptor 2
CBCA	cannabichromene acid
CBD	cannabidiol
CBDA	cannabidiol acid
CBDB	cannabidibutol
CBDP	cannabidiphorol
CBDVA	cannabidivarinic acid
CBG	cannabigerol
CBGA	cannabigerolic acid
CBDVA	cannabidivarinic acid
CBGVA	cannabigerovarinic acid
CBM	cannabimovone
CNS	central nervous system
CYP	cytochrome P450
$DAGL_{\alpha/\beta}$	diacylglycerol lipase α and β
FAAH	fatty acid amide hydrolase
GPR55	G protein-coupled orphan receptor 55
$5\text{-}HT_{1A}$	serotonin 1A
MAGL	monoacylglycerol lipase
MGBA	microbiota-gut-brain axis
mPFC	medial prefrontal cortex
NAPE-PLD	N-acylphosphatidylethanolamine phospholipase D
NMDA	N-methyl-D-aspartate
PPARγ	peroxisome proliferator-activated receptor gamma
PTSD	post-traumatic stress disorder
TBI	traumatic brain injury
THC	Δ^9-tetrahydrocannabinol
THCA	acidic Δ^9-tetrahydrocannabinol
THCB	tetrahydrocannabutol
THCP	(−)-trans-Δ^9-tetrahydrocannabiphorol
THCV	Δ^9-tetrahydrocannabidivarine
THCVA	Δ^9-tetrahydrocannabidivarinic acid

Diagnosis and Treatment of Traumatic Brain Injury. https://doi.org/10.1016/B978-0-12-823347-4.00032-4

$TRPA_1$	transient receptor potential ankyrin 1
$TRPV_1$	transient receptor potential vanilloid type 1
$TRPV_2$	transient receptor potential vanilloid type 2
UGT	uridine 5′-diphospho-glucuronosyl transferase

Introduction

Cannabis sativa plant, commonly known as marijuana, contains over 100 phytocannabinoids, of which the Δ^9-tetrahydrocannabinol (THC) and cannabidiol (CBD) are the most abundant and investigated. The therapeutic and side effects of cannabis extracts are due to the interaction with the endocannabinoid system which includes Gi/o-coupled cannabinoid receptors 1 and 2 (CB_1 and CB_2), the endocannabinoids anandamide (AEA) and 2-arachidonoylglycerol (2-AG), the anandamide membrane transporter (AMT), the anabolic N-acylphosphatidylethanolamine phospholipase D (NAPE-PLD) and the catabolic fatty acid amide hydrolase (FAAH) enzymes for AEA and the anabolic diacylglycerol lipase α and β ($DAGL_{\alpha/\beta}$) and catabolic monoacylglycerol lipase (MAGL) for 2-AG (Fig. 1). CB_1 receptors are mainly expressed in the central nervous system (CNS) on the axon and synaptic terminals of the neurons, while CB_2 receptors are mainly expressed in peripheral immune cells and brain microglia. CB_1 receptor activation is associated with the main central effects of cannabinoids including the psychosis associated with THC and synthetic cannabinoids, while CB_2 receptor activation is associated with inflammation attenuation by the inhibition of microglial release of pro-inflammatory cytokines and reactive oxygen species. 2-AG, whose levels are three times higher than those of AEA in the brain, is the full agonist on CB_1 and CB_2 receptors, acts also on $GABA_A$ receptor and serves as the major precursor for arachidonic acid. AEA acts as a partial agonist on CB_1 and CB_2 receptors and also activates the transient receptor potential vanilloid type 1($TRPV_1$). AEA and 2-AG, being lipid messengers, are not stored in synaptic vesicles and released by exocytosis as the other neurotransmitters, but are synthesized "on-demand" by a membrane phospholipid precursor following stimuli such as pain, psychological or physical stress, and pathological conditions. In experimental models of neuropsychiatric disorders and in the post-mortem brain areas of humans affected by these diseases the levels and signaling mechanisms of endocannabinoids are dysregulated, suggesting their involvement and, above all, a target on which to direct effective therapeutic strategies (Di Marzo, 2009).

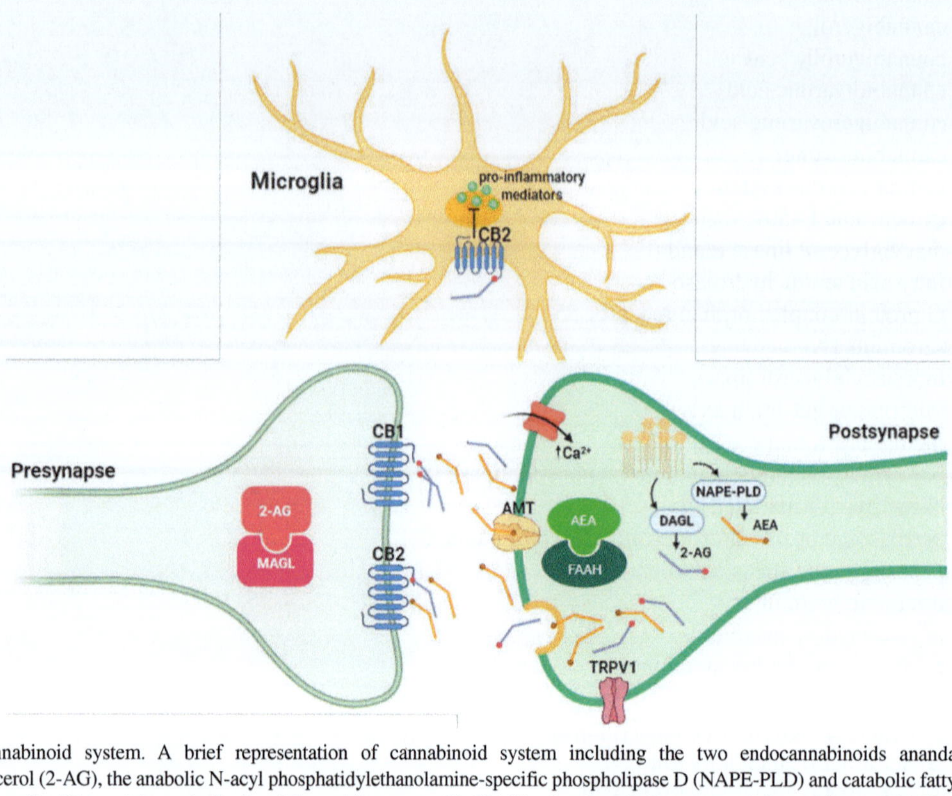

FIG. 1 Endocannabinoid system. A brief representation of cannabinoid system including the two endocannabinoids anandamide (AEA) and 2-arachidonoylglycerol (2-AG), the anabolic N-acyl phosphatidylethanolamine-specific phospholipase D (NAPE-PLD) and catabolic fatty acid amide hydrolase (FAAH) enzymes for AEA and the anabolic diacylglycerol lipase α and β, ($DAGL_{\alpha/\beta}$) and catabolic monoacylglycerol lipase (MAGL) enzymes for 2-AG, the cannabinoid receptor 1 and 2 (CB_1 and CB_2) and transient receptor potential vanilloid type 1 ($TRPV_1$) and the anandamide membrane transporter (AMT).

Traumatic brain injury (TBI) despite the widespread and promising preclinical studies is devoid of effective pharmacological therapies. In severe cases is surgically treated by decompressive craniotomy, often associated with complications such as hematoma, subdural hygroma, and hydrocephalus. Following brain neuronal damage and consequent *N*-methyl-D-aspartate (NMDA) excitotoxicity endocannabinoid levels in certain CNS sites increase suggesting their involvement in the activation of compensatory repair mechanisms (Cristino, Bisogno, & Di Marzo, 2020; Palazzo et al., 2006). CB_1 and CB_2 receptor manipulation, or the blockade of the FAAH and MAGL, inhibit the cellular and molecular processes underlying the TBI such as neuroinflammation, excitotoxicity, cell death, disruption of the blood–brain barrier and cell structures and improve TBI-induced behavioral consequences such as cognitive impairment, anxiety, depression, irritability, sleep disturbance, post-traumatic headaches, epilepsy and motor disorders (Schurman & Lichtman, 2017). The main limits of natural and synthetic cannabinoids are associated with psychotomimetic effects. Unlike THC, CBD is not psychoactive since it does not activate CB_1 receptors, is well-tolerated, and exhibits a broad spectrum of therapeutic properties (Fernández-Ruiz et al., 2013). Among these, the neuroprotective action, the enhancement of neurogenesis, the reduction of inflammation, the recovery of cerebral blood flow, and the protection against reactive oxygen species are all effects underlying the efficacy of CBD in TBI (Singh & Neary, 2020). Moreover, when CBD is administered alongside THC there is an elevation of the therapeutic properties and an attenuation of the side effects of THC.

Cannabidiol targets

CBD is the second main phytocannabinoid, whose presence in the extracts from *Cannabis sativa* can reach 40%. CBD has negligible activity on CB_1 and CB_2 receptors and shows a completely different pharmacological profile from THC. CBD acts as a negative allosteric modulator on the CB_1 receptor, explaining its capability to antagonize some THC effects, such as anxiety, hunger, sedation, and tachycardia, and a partial agonist activity on the CB_2 receptor. Thus, therapeutic doses of CBD do not cause psychotomimetic effects, appetite and anxiety increase as the agonists do, nor determine weight loss and depression, as the antagonists do. CBD inhibits FAAH and AMT activity increasing the AEA levels and the consequent activation of CB_1, CB_2, and $TRPV_1$ receptors. CBD by itself is also a weak agonist on transient receptor potential ankyrin 1 ($TRPA_1$) (EC_{50} = 110 nM), $TRPV_1$ (EC_{50} = 3.5 μM), and $TRPV_2$ (EC_{50} = 1.25 μM) channels (Ibeas Bih et al., 2015). In line with the activation and desensitization of $TRPV_1$, a pain transducer of noxious temperature and chemical stimuli, CBD shows analgesic property against chronic pain (Belardo et al., 2019). Additionally, CBD binds to the serotonin 5-HT_{1A} receptor at 16 μM (Russo, Burnett, Hall, & Parker, 2005), a mechanism behind the rapid antidepressant, anxiolytic, and anti-headache effect. Peroxisome proliferator-activated receptor γ ($PPAR_\gamma$) is also a target of CBD. CBD has proven to restore the increased permeability of the blood–brain barrier associated with oxygen–glucose deprivation, reduce β-amyloid expression, induce apoptosis in lung cancer cells, inhibit the expression of inflammatory cytokines including necrosis factor kappa B, nitric oxide, tumor necrosis factor alpha, and interleukin 1 beta and promote neurogenesis in the hippocampus by $PPAR_\gamma$ activation. Noteworthy, all these neuroprotective effects were completely abolished by a $PPAR_\gamma$ antagonist (Esposito et al., 2011). CBD activates also the adenosine 2A receptor (A2A) inhibiting by this mechanism the release of dopamine and glutamate. Activation of the A2A receptor also reduces microglial activation and levels of pro-inflammatory mediators in a model of multiple sclerosis (Mecha et al., 2013). CBD can also inhibit adenosine reuptake, even if the contribution of this mechanism on the in vivo effects of CBD requires further study. CBD also antagonizes the G protein-coupled orphan receptor 55 (GPR55) with functional consequences in the hippocampus. CBD potently inhibits drug resistance-associated protein 1, a transporter expressed in various tissues with broad substrate specificity including several drugs. This mechanism could underlie the enhancement of the THC effect exerted by CBD or other phytocannabinoids when administered together (entourage effect). In addition to those described here, the literature includes over 65 molecular targets for CBD. Most of these targets are activated at high concentrations that are difficult to reach in vivo, given the low bioavailability of CBD. However, the voltage-gated anion channel 1 and the calcium channel CaV3.x, which regulate responses to changes in intracellular calcium, appear to be relevant. They may plausibly be involved in the beneficial effects of CBD in neurological diseases, given that increased Ca^{2+} is the vehicle for cytotoxicity in neurological and neurodegenerative diseases. A summary of CBD targets is shown in Fig. 2.

Phytocannabinoids

The *Cannabis sativa* plant contains thousands of compounds. Among them, more than a hundred have been defined phytocannabinoids. The plant synthesis of the phytocannabinoids starts from the olivetolic acid that forms a precursor, already identified as a phytocannabinoid, the cannabigerolic acid (CBGA). From the CBGA initiates the synthesis of other compounds such as acidic cannabichromene (CBCA), acidic cannabidiol (CBDA), and acidic Δ^9-tetrahydrocannabinol

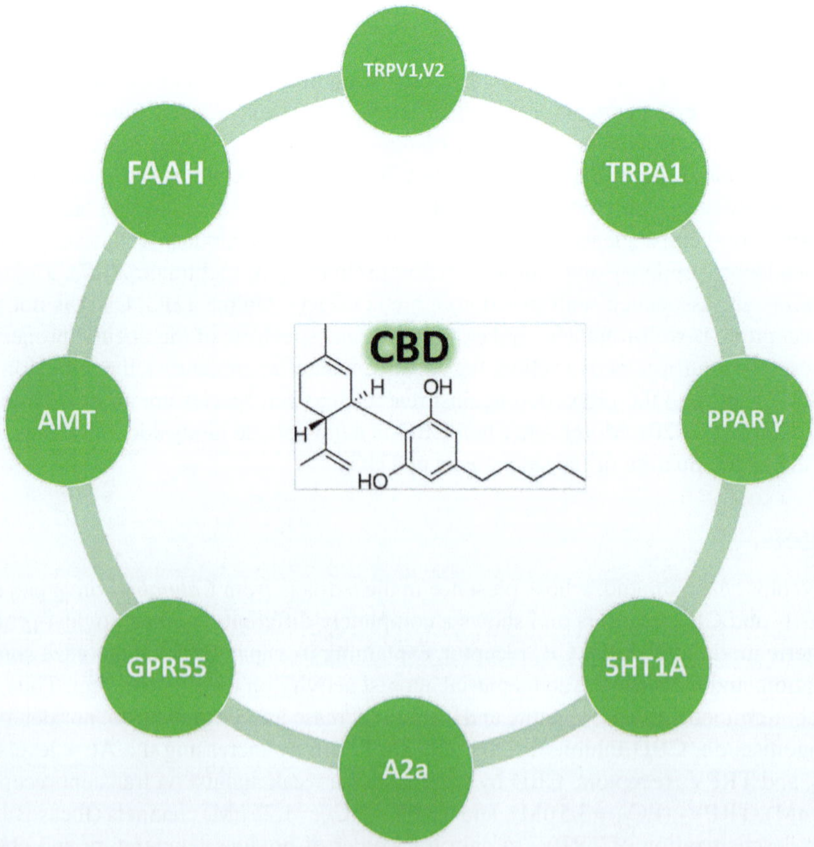

FIG. 2 Cannabidiol targets. The main cannabidiol (CBD) targets including fatty acid amide hydrolase (FAAH) enzyme, serotonin 5-HT1A receptor, adenosine A2A receptor, transient receptor potential vanilloid type 1 and 2 ($TRPV_1$ and $TRPV_2$), transient receptor potential ankyrin 1 ($TRPA_1$), anandamide membrane transporter (AMT), G-protein receptor 55, (GPR55) and peroxisome proliferator-activated receptor γ ($PPAR_γ$).

(THCA). All compounds in the raw plant are in their acidic form and are later on decarboxylated through high temperature and/or drying. Besides the compounds deriving from the cannabigerol (CBG), Cannabis also produces isopropyl derivatives from the cannabigerovarinic acid (CBGVA), producing the corresponding cannabidivarinic acid (CBDVA) and Δ^9-tetrahydrocannabidivarinic acid (THCVA). Also for the isopropyl derivatives, the raw plant contains the acidic compounds that become neutral after decarboxylation (Fig. 3). The phytocannabinoids do not exert pharmacological effects by the same mechanism of action. Indeed, only THC and partly THCV are capable to bind and activate the CB_1 receptor inducing the classical "high" behavior associated with the CB_1 stimulation. The other compounds own several mechanisms of action, often hybrid and engaging several targets such as catabolic or metabolic enzymes, transporters, metabotropic, and ionotropic receptors. The most studied phytocannabinoids are the THC and CBD. The THC has been largely investigated, especially for its potential pattern of abuse and also in terms of toxicology. However, also the beneficial pharmacological effects of the THC have been widely described. Nowadays, the CBD is deeply investigated in both basic (translational) and clinical research, because of its pleiotropic action and the possibility to use this compound in several pathologies. The mechanisms by which CBD exerts its pharmacological effect have been discussed above and in Fig. 2. Recently, four new potential phytocannabinoids have been identified in a specific variety of cannabis. The first two have been the Δ^9-tetrahydrocannabutol (THCB) and the cannabidibutol (CBDB), which have been shown to exert anti-inflammatory effect (Linciano et al., 2020). The other two compounds isolated from the same variety of cannabis in a very small concentration were the (−)-trans-Δ^9-tetrahydrocannabiphorol (THCP) and the cannabidiphorol (CBDP). Interestingly, the THCP is a highly selective and very potent activator for the CB_1 receptors (Citti et al., 2019). More recently, another phytocannabinoid was discovered in the Cannabis plant, the cannabimovone (CBM) (Fig. 3). This compound has been shown to act at the nuclear receptor $PPAR_γ$, thus, potentially useful for the treatment of insulin resistance-related disorders (Iannotti et al., 2020).

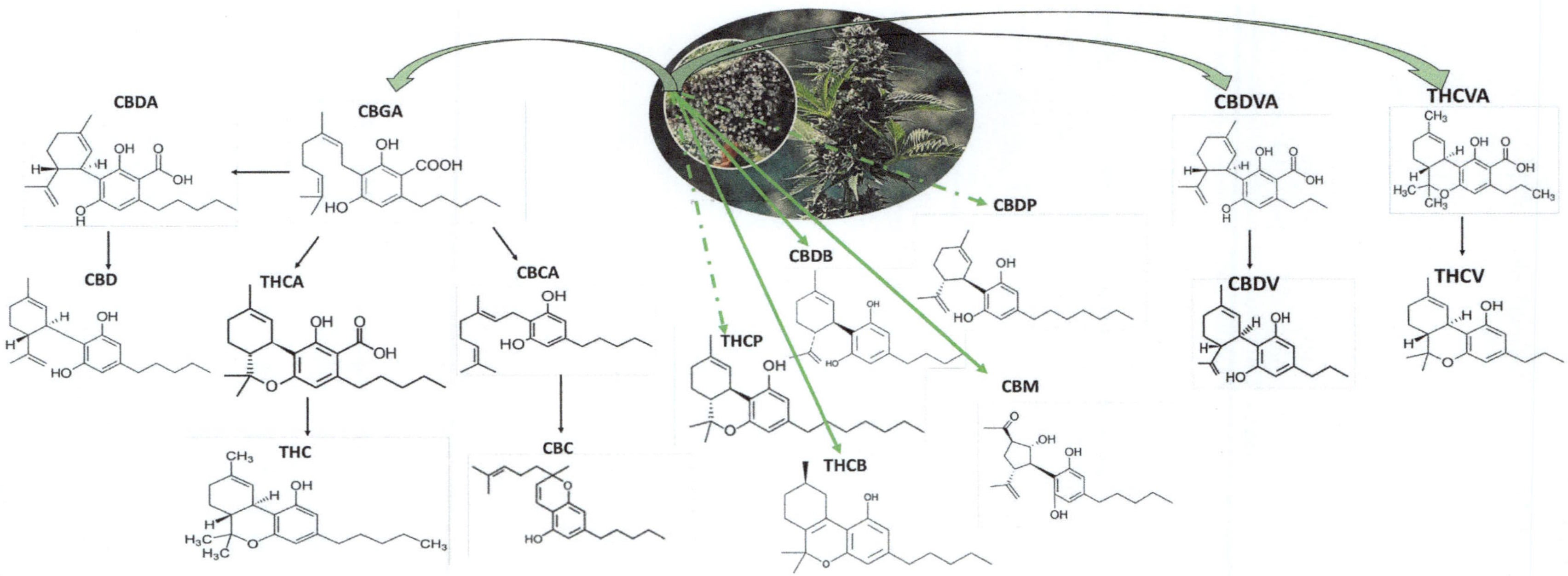

FIG. 3 Phytocannabinoids. Chemical structure of main phytocannabinoids extracted from the *Cannabis sativa* plant. From CBGA, cannabigerolic acid, derive CBCA, acidic cannabichromene, CBDA, acidic cannabidiol and THCA, acidic Δ^9-tetrahydrocannabinol that are decarboxylated in CBC, cannabichromene, CBD, cannabidiol, and THC, Δ^9-tetrahydrocannabinol. Isopropyl derivatives include the CBDVA, cannabidivarinic acid, and THCVA, Δ^9-tetrahydrocannabidivarinic acid that are decarboxylated in CBDV, cannabidivarine, and THCV, Δ^9-tetrahydrocannabidivarine. Other phytocannabinoids recently identified are THCB, Δ^9-tetrahydrocannabutol, CBDB, cannabidibutol, THCP, trans-Δ^9-tetrahydrocannabiphorol, and CBDP, cannabidiphorol, the latter two in low concentrations (dashed arrows).

Neuroprotective role of cannabidiol

Neuroinflammation and cell death are events implicated in the initial phase of TBI (Arciniegas, 2011). Peripheral immune cells, mast cells, and T lymphocytes, involved in the inflammatory process, trigger neurovascular, electrical, chemical, and energy disorders. These processes are characterized by a dynamic interaction involving brain micro-vascular endothelia, peripheral immune cells, and glia, the latter playing a critical role in both, the development and maintenance of neuroinflammation (Luongo, Maione, & Di Marzo, 2014). Evidence from preclinical studies has already shown the beneficial properties of CBD, including neuroprotective effects in CNS disorders (De Gregorio et al., 2019) (Table 1). The mechanisms and the targets underlying the neuroprotective effects of CBD are, however, still unclear. The enhancement of AEA levels by the inhibition of FAAH and AMT induces neuroprotection, whereas the stimulation of TRPV1, which controls vascular responses, can protect against TBI-induced cerebrovascular alterations. The stimulation of A2A receptor and the inhibition of adenosine transporter may be involved in the CBD anti-inflammatory effects while the stimulation of 5-HT_{1A} receptors may decrease excitotoxicity, inflammation, and oxidative stress. Finally, is possible that CBD can help to regulate cerebral blood flow by interacting with the G-protein-coupled receptor 18, which plays a neuroprotective effect by regulating the cell migration (Penumarti & Abdel-Rahman, 2014). A representation of the neuroprotective action of CBD on TBI is shown in Fig. 4. TBI induces neurological, sensorial, and psychiatric symptoms including depression, apathy, anxiety, insomnia, and aggressiveness. Sensory anomalies, especially of sight, but also of hearing, smell, taste, and touch are associated with the lesion of the related brain areas. Injury to the tactile control area may cause persistent tingling,

TABLE 1 Neuroprotective effects of cannabidiol.

Model	Pathological process	Species	Outcome	Dose of CBD	Receptor	References
Carrageenan (1% w/v intra paw)	Acute inflammation	Rats	Antihyperalgesic effect	10 mg/kg, os	TRPV1	Costa, Giagnoni, Franke, Trovato, and Colleoni (2004)
Hypoxic ischemic brain injury	Hypoxic-ischemic encephalopathy	Newborn pigs	Reduction of excitotoxicity, oxidative stress and inflammation	1 mg/kg, i.v.	5-HT1A	Pazos et al. (2013)
Focal cerebral ischemia	Occlusion to the middle cerebral artery	Mice	Increase in cerebral blood flow	0.1– 10 mg/kg, i.p.		Mishima et al. (2005)
Brain microvascular endothelial cells and astrocytes		Human	Reduction of the increased blood brain barrier permeability	100 nM, 10 μM, 1 μM	PPARγ	Hind, England, and O'Sullivan (2016)
Alzheimer's disease	Neuronal damage	Rats	Promotion of hippocampal neurogenesis and reduction in inflammatory biomarkers	10 mg/kg, i.p. for 15 days		Esposito et al. (2011)
Viral multiple sclerosis	Inflammation	Female mice	Downregulation of inflammatory biomarkers and attenuation of microglia activation during initial stages	5 mg/kg, i.p. from 1 to 7 or 10 days post-infection	A2A	Mecha et al. (2013)
In vitro model of newborn hypoxic–ischemic brain injury	Cell death, excitotoxicity, inflammation	7-to-10-days old mice	Reduction of inflammatory biomarkers and glutamate in hypoxic–ischemia	From 0.1 to 1000 μM		Castillo, Tolon, Fernandez-Ruiz, Romero, and Martinez-Orgado (2010)

The table shows the neuroprotective effects of cannabidiol, reporting for each study the experimental model, the pathological process induced, the species used, the effects obtained, the dose and the duration of treatment, and the reference. *i.v.*, intravenous; *os*, oral; *i.p.*, intra-peritoneal.

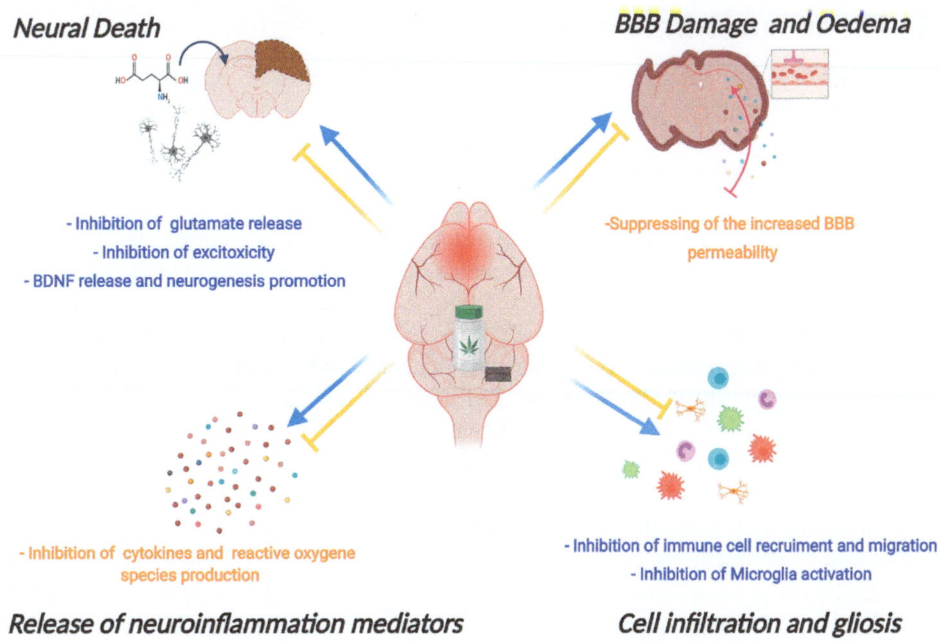

FIG. 4 Cannabidiol and molecular and cellular effects underlying neuroinflammation. The illustration shows the neuroprotective effects of cannabidiol on the main processes underlying neuroinflammation such as neuronal death, blood brain barrier (BBB) damage and edema, release of neuroinflammatory mediators, cell immune recruitment/migration and microglia activation. BDNF: brain-derived neurotrophic factor.

itching, or pain in patients. Although rare, these conditions are difficult to treat. Mild TBI (mTBI) determined abnormal pain responses after innocuous stimuli in mice probably due to the overall inflammatory condition (Feliciano et al., 2014). Interestingly, CBD restored the abnormal pain responses produced by mTBI that lasted for weeks (Belardo et al., 2019). mTBI proved also to be associated with impaired neuronal activity and increased glutamate release in the medial prefrontal cortex (mPFC), an area that regulates the interaction between chronic pain and affective states. CBD normalized glutamate levels and neuronal activity in mPFC, thereby counteracting the maladaptive plastic changes and remodeling that can occur after trauma. These latest data are in line with previous ones showing that dysregulation of the glutamatergic system may underlie several aspects of neuropsychiatric diseases and that the protective effects of cannabinoids on such maladaptive plastic changes correlate with the amelioration of the symptoms (Guida et al., 2017; Palazzo et al., 2015).

Cannabidiol in TBI-related anxiety and depression

Depression and anxiety are common psychiatric consequences of TBI. Anxiety is often the most marked manifestation just after trauma, due to the stress of the injurious event itself as well as pathogenic changes in inhibitory circuits. Depression is the most common post-traumatic mood disorder, affecting 25%–50% of individuals and appearing within the first year after TBI (Jorge & Arciniegas, 2014). Experimental models of TBI can mimic mild, moderate, or severe brain injury based on the degree of head impacts and the following pathophysiological consequences. Several studies describe a complex neuropsychiatric behavioral profile after TBI in both, animal models and humans. Accordingly, TBI induces biochemical and molecular processes that result in neuroinflammation, neuronal disruption, and excitotoxicity (Bullock et al., 1998). Recent preclinical evidence shows that endocannabinoid levels are elevated during and after TBI, which suggests that the endocannabinoid system plays an important neuroprotective role after trauma (Shohami, Cohen-Yeshurun, Magid, Algali, & Mechoulam, 2011). Clinical and pre-clinical studies have highlighted the ability of CBD to treat chronic pain that is also associated with comorbidities such as anxiety and depression. In this context CBD proved to be antiallodynic and anxiolytic, the latter effect through the activation of 5-HT$_{1A}$ receptor in neuropathic rats (De Gregorio et al., 2019). Biphasic development of several neuropsychiatric symptoms has been recently demonstrated after mTBI in the mouse. In particular, mice showed a depressive-like behavior associated with cortical neuronal plasticity 60 days post-TBI (Guida et al., 2017). Indeed, depression-like behavior following mTBI was strictly correlated with the biphasic firing activity of pyramidal neurons in the medial prefrontal cortex (mPFC), a key area regulating chronic pain (Luongo et al., 2013) and the correlated negative affective states, such as anxiety and depression. Thus, the inhibition of depressive-like behavior in TBI mice, induced by repeated oral CBD treatment, may be probably due to its direct modulation of altered synaptic activity in mPFC (Belardo et al., 2019).

Cannabidiol in TBI and post-traumatic stress disorder-related aggression

TBI is often associated with aggression that commonly resolves within 3 months or can persist over a three-month period (Dean, O'Neill, & Sterr, 2012). Commonly, the pharmacological approach to attenuate aggressive behavior is based on antipsychotics such as risperidone, and antidepressant/anxiolytic drugs, such as the 5-HT_{1A} receptor partial agonists buspirone and ipsapirone, or the selective serotonin reuptake inhibitor fluoxetine. Unfortunately, the use of these drugs is limited, due to a number of side effects such as inhibition of locomotor activity, sedation, and reduction of social interaction. Thus, CBD could be an useful compound, since it has been demonstrated to induce an anti-aggressive-like response without causing motor side effects (Hartmann et al., 2019). Clinical studies indicate that CBD, in addition to the high therapeutic potential in patients with mental disorders associated with aggressive traits, such as schizophrenia, anxiety, autism, and others has a favorable safety profile (Bergamaschi, Queiroz, Zuardi, & Crippa, 2011). The only preclinical study that addresses the efficacy of CBD in aggressive behavior related to TBI, reports that repeated oral treatment with CBD reduces the aggressive-like behavior in mice (Belardo et al., 2019). Furthermore, TBI itself is a traumatic event and therefore can be included in post-traumatic stress disorders (PTSD). CBD, which has shown efficacy in TBI-induced psychiatric disorders, could also have a beneficial effect on PTSD. Long-term isolation such as that occurred during the COVID-19 pandemic is associated with a high incidence of PTSD and aggression. It is noteworthy in this context that CBD reduced aggression in mice subjected to social isolation and then tested in the resident-intruder paradigm (Hartmann et al., 2019). The CBD-mediated anti-aggressive effect appears to be mediated by the stimulation of the $5HT_{1A}$ receptors, and indirect facilitation of CB_1 receptor stimulation. Consistent with these results, the administration of a CB_1 receptor agonist reduced aggressive behavior and CB_1 receptor knockout mice exhibit aggressive behavior (Rodriguez-Arias et al., 2013).

Cannabidiol in TBI-related gut microbiota dysfunctions

The microbiota-gut-brain axis (MGBA) is the neural substrate responsible for the bidirectional interaction between the intestine and brain, in both physiological and pathological conditions. In the gut, the perturbation of bacteria composition produces a local, and subsequently, systemic inflammation, which is believed to trigger neuroinflammation and alterations in brain functioning and behavior. Conversely, brain injuries may induce modifications in the gut metabolism with direct regulation of intestinal permeability, inflammation, and microbiota composition. On these bases, the gut microbiota is believed to contribute to the modulation of brain molecular processes relevant for TBI pathophysiology including microglial activation, neuroinflammation, and blood–brain barrier permeability (Rice, Pandya, & Shear, 2019). Indeed, an increase in α-diversity and different overall microbial compositions was recently detected in mild TBI patients (Bai et al., 2020). Therefore, modifications of gut microbiota setting or related metabolites may represent potential brain damage biomarkers.

The endocannabinoid system is implicated in the control of gastrointestinal functions, including motility, food intake, immune cell activity, and dynamic interactions with gut microbiota. The first evidence of crosstalk between gut microbiota and the endocannabinoid system has been proposed by Everard et al. (2013) that isolated the gut bacterium *Akkermansia muciniphila* able to alleviate metabolic syndrome in obese mice, by increasing endocannabinoids levels. More recently, Guida et al. (2018, 2020) correlated modifications of endocannabinoid levels in the CNS, mediated by microbiota perturbation, with sensorial and psychiatric disorders in mice.

The beneficial and immunomodulatory effects of CBD have been widely evidenced in experimental animal models of inflammatory bowel diseases (Esposito et al., 2013), suggesting a possible protective role on gut microbiota. However, the impact of CBD exposure on gut microbiota setting and functioning needs to be further investigated. Interestingly, a recent study by Al-Ghezi, Busbee, Alghetaa, Nagarkatti, and Nagarkatti (2019) indicated that the combination of THC and CBD is able to attenuate neuroinflammatory processes in multiple sclerosis, by inducing substantial modifications in the gut microbiome (e.g., reductions in *A. muciniphila*). Indeed, we cannot exclude that the beneficial effects of CBD in TBI-induced neuroinflammation, could occur, at least in part, across the MGBA. Although the limited existing evidence in this area, the recent discoveries open the possibility of microbiota modulating approaches for the treatment of psychiatric diseases and other CNS disorders, including TBI. Future research is needed to directly address the role of microbiota in the effects exerted by CBD in the TBI pathophysiology and symptomatology by paving the way for clinical translation.

Cannabidiol and clinical trials

In the last decades, several clinical trials have been conducted to test the efficacy of CBD in the management of heterogeneous neuropsychiatric conditions. Indeed, due to the plethora of actions and complex pharmacology, CBD has been suggested to have potential effects in a wide range of disorders (Crippa, Guimarães, Campos, & Zuardi, 2018). Specifically, studies on its anticonvulsant property have led to its use in pediatric drug-resistant epilepsy, refractory epileptic encephalopathy, refractory infantile spasms, Lennox-Gasteaux and Dravet syndromes (Devinsky et al., 2014; Gaston & Szaflarski, 2018). CBD efficacy in treatment-resistant seizures paved the way to ongoing trials testing its effects on neurodevelopmental and neurobehavioral disorders such as Rett syndrome, autism spectrum disorders, and Tourette syndrome in childhood, as well as on neurodegenerative diseases such as Alzheimer's and Parkinson's disease in adults and elderly population (Maroon & Bost, 2018; Pretzsch et al., 2019). Moreover, CBD has also been tested in psychiatric pathologies, in particular anxiety and depressive disorders, addiction (alcohol, cocaine, opioids, marijuana use disorders), psychosis, and phobias (Bonaccorso, Ricciardi, Zangani, Chiappini, & Schifano, 2019). The other important translational target of CBD is represented by pathologies that share the neuroinflammatory etiopathological substrate, ranging from chronic pain to spinal or brain injury, also including TBI and its psychopathological related PTSD (Khorasanizadeh et al., 2019). This paragraph provides an overview of ongoing studies recorded on http://clinicaltrial.gov, by classifying them into recruiting and completed trials (Table 2). The effectiveness of CBD in PTSD has been investigated in 5 clinical studies. Although only one has been completed so far, its results are not yet available because of the approval of a temporal extension. This study tested the use of a *Cannabis sativa* extract containing both CBD and THC at different concentrations. The studies that investigate the use of CBD alone are mostly conducted on PTSD resulting from psychopathological trauma. Among these, a single study, the most recent one, will be focused on the association of PTSD with the TBI medical condition. Specifically, this will be a randomized placebo-controlled parallel dose-finding study. The purpose of this trial is to test the efficacy of CBD on symptoms of PTSD including neurocognitive impairment. Male and female subjects (N=120) suffering from PTSD with, or without comorbid mild/moderate TBI will be enrolled and divided into 60 patients per group. The study will include different treatments: (1) 40 subjects assigned to oral CBD 400 mg daily; (2) 40 subjects assigned to CBD 600 mg daily and (3) 40 subjects assigned to placebo daily. Treatment will have a 8-week duration. The primary outcome will be represented by a reduction in PTSD symptoms expressed as a percentage of subjects with a reduction in PCL-5, a 20-item self-report measure that assesses the DSM-5 symptoms of PTSD, greater than 30% compared to the baseline in participants using CBD. The secondary outcome will consider the reduction of anxiety measured by the self-reported General Anxiety Disorder 7 item scale questionnaire and depression measured by the Patient Health Questionnaire. Regarding the remaining two

TABLE 2 Clinical trials on cannabidiol, post-traumatic stress disorder and traumatic brain injury.

Status	Study title	Conditions	Interventions	Locations	Period
Recruiting	Use of CBD oil in the treatment of posttraumatic stress disorder	PTSD	(1) Cannabidiol (2) Placebo	Laboratory for the Study of Anxiety Disorders, University of Texas, Austin, United States	January 2020–May 2024
Recruiting	Cannabidiol as a treatment for AUD comorbid with PTSD	AUD PTSD	(1) Cannabidiol (2) Placebo	School of Medicine, New York University, New York, United States	September 2019–August 2021
Recruiting	Burnout and distress prevention with cannabidiol in front-line health care workers dealing with COVID-19	Covid19 burnout PTSD	(1) Cannabidiol	Hospital das Clínicas da Faculdade de Medicina de Ribeirão Preto, Brazil	June 2020–September 2020
Not yet recruiting	Cannabidiol for PTSD and TBT	PTSD TBT	(1) Cannabidiol (2) Placebo	Langone Health, New York University , New York, United States	December 2020–March 2023
Completed	Study of four different potencies of smoked Marijuana in 76 Veterans with PTSD	PTSD	(1) High THC Marijuana (2) High CBD Marijuana (3) High THC/high CBD marijuana (4) Placebo marijuana	Scottsdale Research Institute Phoenix, Arizona, United States	January 2017–January 2019 Temporal extension approved in June 2020

The table shows the clinical trials on cannabidiol. For each trial is reported the title, pathology on which the effect of cannabidiol is tested, treatment, location where the trial is carried out and start and presumed end of the trial.

studies, one has been investigating the condition associated with PTSD and alcohol use disorder while, interestingly, the last study has been testing the use of CBD in PTSD associated with Burnout Syndrome in frontline health care workers during the COVID-19 pandemic. Burnout is a pathology relatively limited to some specific clinical areas but given the widespread of the COVID-19 epidemic, the demonstration of CBD efficacy could lead to extensive use of the drug. Preclinical and clinical data reported that CBD seems to be very well tolerated but cannot be considered risk-free. Clinical CBD studies for epilepsy and psychiatric disorders reported hepatic abnormalities, diarrhea, fatigue, vomiting, and somnolence. Moreover, CBD is metabolized by cytochrome P450 (CYP) CYP2C19 and CYP3A4, and by uridine 5′-diphospho-glucuronosyl transferase (UGT), UGT1A7, UGT1A9, and UGT2B7 isoforms, mainly producing hydroxylated and carboxylated metabolites. The involvement of these enzymes exposes to a high probability of interactions with other drugs. Until now, CBD-induced drug–drug interactions have been reported in co-treatments with antiepileptics such as clobazam, valproate, lamotrigine, levetiracetam (Huestis et al., 2019). However, further investigations on pharmacokinetic interaction with other concomitant treatments are required.

Applications to other areas of neuroscience

The chapter discusses the effect of cannabidiol on traumatic brain injury, the molecular mechanisms underlying it and the correlated behavioral symptoms. Cannabidiol is a phytocannabinoid with a pharmacodynamic profile different from Δ^9-tetrahydrocannabinol and a multi-target action that broadens its spectrum of action. Cannabidiol has been shown to inhibit the production of inflammatory cytokines, the activation of microglia, the proliferation of immune cells, the vascular changes resulting from ischemic events, excitotoxicity, apoptosis, oxidative stress, and inflammation while increases brain-derived neurotrophic factor levels and neurogenesis in the hippocampus (Fig. 4). These molecular and cellular effects suggest that cannabidiol counteracts neuroinflammation showing a neuroprotective effect. Neuroinflammation is the pathological mechanism underlying most neurological, psychiatric, and neurodegenerative diseases. Consistently, cannabidiol has shown analgesic, anxiolytic, antidepressant, antipsychotic, antiepileptic, anti-parkinsonian, mood-stabilizing effects, beneficial effects in addictions, sleep disorders, Alzheimer's disease, and post-traumatic stress disorder, and preservation of cerebral circulation during ischemic events and encephalitis. The neuroprotective effect of cannabidiol in neuroinflammatory conditions has enormous potential in traumatic brain injury, not only in the immediate mechanisms following the injury but also in the later sensory, affective and cognitive consequences associated with it. Another consequence associated with traumatic brain injury is the alteration of microbiota with consequences that afflict the gut-brain axis. The efficacy of cannabidiol on traumatic brain injury and on the consequences related to traumatic brain injury described in this chapter can be applied to all pathologies with neuroinflammatory origin involving neurology, geriatrics, psychiatry and gastroenterology. Cannabidiol with a broad spectrum of action combines good tolerability and the lack of psychotomimetic effects. Moreover, most of the neuroprotective effects of cannabidiol are associated with the stimulation of peroxisome proliferator-activated receptor γ, which suggests the use of selective agonists for pathologies with neuroinflammatory pathophysiological mechanisms.

Mini-dictionary of terms

Brain-derived neurotrophic factor: neurotrophin supporting the survival of existing neurons and promoting synaptogenesis.
Cannabinoid system: neurotransmitter system activated by endocannabinoids and *Cannabis sativa* plant-derived cannabinoids such as Δ^9-tetrahydrocannabinol and cannabidiol. It includes endogenous neurotransmitters, anabolic and catabolic enzymes, receptors, and membrane transporters.
Entourage effect: synergetic effect occurring when cannabis plant compounds are assumed together.
Microbiota: gut microorganisms including bacteria, fungi, viruses, protozoa, and archaea that vary for each individual.
Neuroinflammation: the inflammatory response in the central nervous system that involves the activation of microglia and astroglia, the release of pro-inflammatory cytokines, and chemokines.
Peroxisome proliferator-activated receptor: nuclear receptors functioning as transcription factors which regulate energy homeostasis and metabolic function.
Post-traumatic stress disorder: Mental disorder triggered by a traumatizing event characterized by flashbacks, nightmares, severe anxiety, aggression, and disproportionate reactions.
Phytocannabinoids: Plant-derived compounds capable of interacting with cannabinoid system.
Psychotomimetic: a drug activity producing symptoms similar to those of psychosis.
Synaptogenesis: the process underlying formation, maintenance, and activity-dependent plasticity of the synapses.

Key facts of cannabidiol

- Cannabidiol is the most abundant phytocannabinoid after the Δ^9-tetrahydrocannabinol in *Cannabis sativa*.
- Cannabidiol is not psychoactive as Δ^9-tetrahydrocannabinol.
- Cannabidiol targets indirectly cannabinoid system through the blockade of anandamide membrane transporters and anandamide degrading enzyme.
- Other targets of cannabidiol are the transient receptor vanilloid type 1 and 2, transient receptor potential ankyrin 1, serotonin 1A receptor, adenosine 2A receptor, and the peroxisome proliferator-activated receptor γ.
- Cannabidiol shows neuroprotective effects and efficacy in all the pathologies with neuroinflammation origin.
- The most neuroprotective effects of cannabidiol are mediated by peroxisome proliferator-activated receptor γ.

Summary points

- This chapter discusses the main preclinical and clinical evidence on the effects of cannabidiol in traumatic brain injury.
- Cannabidiol is a *Cannabis sativa* extract like Δ^9-tetrahydrocannabinol, but unlike Δ^9-tetrahydrocannabinol has fewer side effects and is not psychoactive.
- The beneficial effects of cannabidiol are only indirectly mediated by the interaction with the cannabinoid system.
- Cannabidiol is a multi-target phytocannabinoid interacting with different receptors, ion channels, and enzymes that increase its spectrum of action.
- Cannabidiol exerts a neuroprotective effect based on the inhibition of excitotoxicity, apoptosis, activation of immune cells and microglia, the release of free radicals and inflammatory cytokines.
- Cannabidiol increases brain-derived neurotrophic factor release and synaptogenesis.
- The broad spectrum of action of cannabidiol includes analgesic, anxiolytic, antidepressant, antipsychotic, antiepileptic, mood stabilizer, beneficial effects in addictions, and sleep disturbances, symptoms that are often associated with TBI.
- Cannabidiol has a beneficial action on the intestinal inflammatory states and microbiota so the efficacy of cannabidiol on the symptoms associated with traumatic brain could partially involve the microbiota gut-brain axis.
- Most clinical trials of cannabidiol on neuroinflammation-based pathologies including traumatic brain injury and its related psychopathological post-traumatic stress disorder are still ongoing.

References

Al-Ghezi, Z. Z., Busbee, P. B., Alghetaa, H., Nagarkatti, P. S., & Nagarkatti, M. (2019). Combination of cannabinoids, delta-9-tetrahydrocannabinol (THC) and cannabidiol (CBD), mitigates experimental autoimmune encephalomyelitis (EAE) by altering the gut microbiome. *Brain, Behavior, and Immunity*, *82*, 25–35.

Arciniegas, D. B. (2011). Clinical electrophysiologic assessments and mild traumatic brain injury: State-of-the-science and implications for clinical practice. *International Journal of Psychophysiology*, *82*, 41–52.

Bai, L., Li, T., Zhang, M., Wang, S., Gan, S., Jia, X., et al. (2020). Association of gut microbiota with cerebral cortex and cerebrovascular abnormality in human mild traumatic brain injury. *bioRxiv*, *07*(19), 211227.

Belardo, C., Iannotta, M., Boccella, S., Rubino, R. C., Ricciardi, F., Infantino, R., et al. (2019). Oral cannabidiol prevents allodynia and neurological dysfunctions in a mouse model of mild traumatic brain injury. *Frontiers in Pharmacology*, *10*, 352.

Bergamaschi, M. M., Queiroz, R. H., Zuardi, A. W., & Crippa, J. A. (2011). Safety and side effects of cannabidiol, a Cannabis sativa constituent. *Current Drug Safety*, *6*, 237–249.

Bonaccorso, S., Ricciardi, A., Zangani, C., Chiappini, S., & Schifano, F. (2019). Cannabidiol (CBD) use in psychiatric disorders: A systematic review. *Neurotoxicology*, *74*, 282–298.

Bullock, R., Zauner, A., Woodward, J. J., Myseros, J., Choi, S. C., Ward, J. D., et al. (1998). Factors affecting excitatory amino acid release following severe human head injury. *Journal of Neurosurgery*, *89*, 507–518.

Castillo, A., Tolon, M. R., Fernandez-Ruiz, J., Romero, J., & Martinez-Orgado, J. (2010). The neuroprotective effect of cannabidiol in an in vitro model of newborn hypoxic-ischemic brain damage in mice is mediated by CB(2) and adenosine receptors. *Neurobiology of Disease*, *37*, 434–440.

Citti, C., Linciano, P., Russo, F., Luongo, L., Iannotta, M., Maione, S., et al. (2019). A novel phytocannabinoid isolated from Cannabis sativa L. with an in vivo cannabimimetic activity higher than Δ9-tetrahydrocannabinol: Δ9-Tetrahydrocannabiphorol. *Scientific Reports*, *9*, 20335.

Costa, B., Giagnoni, G., Franke, C., Trovato, A. E., & Colleoni, M. (2004). Vanilloid TRPV1 receptor mediates the antihyperalgesic effect of the nonpsychoactive cannabinoid, cannabidiol, in a rat model of acute inflammation. *British Journal of Pharmacology*, *143*, 247–250.

Crippa, J. A., Guimarães, F. S., Campos, A. C., & Zuardi, A. W. (2018). Translational investigation of the therapeutic potential of Cannabidiol (CBD): Toward a new age. *Frontiers in Immunology*, *21*, 2009.

Cristino, L., Bisogno, T., & Di Marzo, V. (2020). Cannabinoids and the expanded endocannabinoid system in neurological disorders. *Nature Reviews. Neurology*, *16*, 9–29.

De Gregorio, D., McLaughlin, R. J., Posa, L., Ochoa-Sanchez, R., Enns, J., Lopez-Canul, M., et al. (2019). Cannabidiol modulates serotonergic transmission and reverses both allodynia and anxiety-like behavior in a model of neuropathic pain. *Pain*, *160*, 136–150.

Dean, P. J., O'Neill, D., & Sterr, A. (2012). Post-concussion syndrome: Prevalence after mild traumatic brain injury in comparison with a sample without head injury. *Brain Injury*, *26*, 14–26.

Devinsky, O., Cilio, M. R., Cross, H., Fernandez-Ruiz, J., French, J., Hill, C., et al. (2014). Cannabidiol: Pharmacology and potential therapeutic role in epilepsy and other neuropsychiatric disorders. *Epilepsia*, *55*, 791–802.

Di Marzo, V. (2009). The endocannabinoid system: Its general strategy of action, tools for its pharmacological manipulation and potential therapeutic exploitation. *Pharmacological Research*, *60*, 77–84.

Esposito, G., De Filippis, D., Cirillo, C., Iuvone, T., Capoccia, E., Scuderi, C., et al. (2013). Cannabidiol in inflammatory bowel diseases: A brief overview. *Phytotherapy Research*, *27*, 633–636.

Esposito, G., Scuderi, C., Valenza, M., Togna, G. I., Latina, V., De Filippis, D., et al. (2011). Cannabidiol reduces Aβ-induced neuroinflammation and promotes hippocampal neurogenesis through PPARγ involvement. *PLoS One*, *6*, e28668.

Everard, A., Belzer, C., Geurts, L., Ouwerkerk, J. P., Druart, C., Bindels, L. B., et al. (2013). Cross-talk between *Akkermansia muciniphila* and intestinal epithelium controls diet-induced obesity. *Proceedings of the National Academy of Sciences of the United States of America*, *110*, 9066–9071.

Feliciano, D. P., Sahbaie, P., Shi, X., Klukinov, M., Clark, J. D., & Yeomans, D. C. (2014). Nociceptive sensitization and BDNF up-regulation in a rat model of traumatic brain injury. *Neuroscience Letters*, *583*, 55–59.

Fernández-Ruiz, J., Sagredo, O., Pazos, M. R., García, C., Pertwee, R., Mechoulam, R., et al. (2013). Cannabidiol for neurodegenerative disorders: Important new clinical applications for this phytocannabinoid? *British Journal of Clinical Pharmacology*, *75*, 323–333.

Gaston, T. E., & Szaflarski, J. P. (2018). Cannabis for the treatment of epilepsy: An update. *Current Neurology and Neuroscience Reports*, *18*, 73.

Guida, F., Boccella, S., Belardo, C., Iannotta, M., Piscitelli, F., De Filippis, F., et al. (2020). Altered gut microbiota and endocannabinoid system tone in vitamin D deficiency-mediated chronic pain. *Brain, Behavior, and Immunity*, *85*, 128–141.

Guida, F., Boccella, S., Iannotta, M., De Gregorio, D., Giordano, C., Belardo, C., et al. (2017). Palmitoylethanolamide reduces neuropsychiatric Behaviors by restoring cortical electrophysiological activity in a mouse model of mild traumatic brain injury. *Frontiers in Pharmacology*, *8*, 95.

Guida, F., Turco, F., Iannotta, M., De Gregorio, D., Palumbo, I., Sarnelli, G., et al. (2018). Antibiotic-induced microbiota perturbation causes gut endocannabinoidome changes, hippocampal neuroglial reorganization and depression in mice. *Brain, Behavior, and Immunity*, *67*, 230–245.

Hartmann, A., Lisboa, S. F., Sonego, A. B., Coutinho, D., Gomes, F. V., & Guimarães, F. S. (2019). Cannabidiol attenuates aggressive behavior induced by social isolation in mice: Involvement of 5-HT1A and CB1 receptors. *Progress in Neuropsychopharmacology and Biological Psychiatry*, *94*, 109637.

Hind, W. H., England, T. J., & O'Sullivan, S. E. (2016). Cannabidiol protects an in vitro model of the blood-brain barrier from oxygen-glucose deprivation via PPARγ and 5-HT1A receptors. *British Journal of Pharmacology*, *173*, 815–825.

Huestis, M. A., Solimini, R., Pichini, S., Pacifici, R., Carlier, J., & Busardò, F. P. (2019). Cannabidiol adverse effects and toxicity. *Current Neuropharmacology*, *17*, 974–989.

Iannotti, F. A., De Maio, F., Panza, E., Appendino, G., Taglialatela-Scafati, O., De Petrocellis, L., et al. (2020). Identification and characterization of Cannabimovone, a cannabinoid from Cannabis sativa, as a novel PPARγ agonist via a combined computational and functional study. *Molecules*, *25*, 1119.

Ibeas Bih, C., Chen, T., Nunn, A. V., Bazelot, M., Dallas, M., & Whalley, B. J. (2015). Molecular targets of Cannabidiol in neurological disorders. *Neurotherapeutics*, *12*, 699–730.

Jorge, R. E., & Arciniegas, D. B. (2014). Mood disorders after TBI. *The Psychiatric Clinics of North America*, *37*, 13–29.

Khorasanizadeh, M., Yousefifard, M., Eskian, M., Lu, Y., Chalangari, M., Harrop, J. S., et al. (2019). Neurological recovery following traumatic spinal cord injury: A systematic review and meta-analysis. *Journal of Neurosurgery. Spine*, *15*, 1–17.

Linciano, P., Citti, C., Luongo, L., Belardo, C., Maione, S., Vandelli, M. A., et al. (2020). Isolation of a high-affinity cannabinoid for the human CB1 receptor from a medicinal Cannabis sativa variety: Δ9-Tetrahydrocannabutol, the butyl homologue of Δ9-tetrahydrocannabinol. *Journal of Natural Products*, *83*, 88–98.

Luongo, L., de Novellis, V., Gatta, L., Palazzo, E., Vita, D., Guida, F., et al. (2013). Role of metabotropic glutamate receptor 1 in the basolateral amygdala-driven prefrontal cortical deactivation in inflammatory pain in the rat. *Neuropharmacology*, *66*, 317–329.

Luongo, L., Maione, S., & Di Marzo, V. (2014). Endocannabinoids and neuropathic pain: Focus on neuron-glia and endocannabinoid-neurotrophin interactions. *European Journal of Neuroscience*, *39*, 401–408.

Maroon, J., & Bost, J. (2018). Review of the neurological benefits of phytocannabinoids. *Surgery Neurology International*, *26*, 9–91.

Mecha, M., Feliú, A., Iñigo, P. M., Mestre, L., Carrillo-Salinas, F. J., & Guaza, C. (2013). Cannabidiol provides long-lasting protection against the deleterious effects of inflammation in a viral model of multiple sclerosis: A role for A2A receptors. *Neurobiology of Disease*, *59*, 141–150.

Mishima, K., Hayakawa, K., Abe, K., Ikeda, T., Egashira, N., Iwasaki, K., et al. (2005). Cannabidiol prevents cerebral infarction via a serotonergic 5-hydroxytryptamine1A receptor-dependent mechanism. *Stroke*, *36*, 1077–1082.

Palazzo, E., de Novellis, V., Petrosino, S., Marabese, I., Vita, D., Giordano, C., et al. (2006). Neuropathic pain and the endocannabinoid system in the dorsal raphe: Pharmacological treatment and interactions with the serotonergic system. *European Journal of Neuroscience*, *24*, 2011–2020.

Palazzo, E., Romano, R., Luongo, L., Boccella, S., De Gregorio, D., Giordano, M. E., et al. (2015). MMPIP, an mGluR7-selective negative allosteric modulator, alleviates pain and normalizes affective and cognitive behavior in neuropathic mice. *Pain*, *156*, 1060–1073.

Pazos, M. R., Mohammed, N., Lafuente, H., Santos, M., Martínez-Pinilla, E., Moreno, E., et al. (2013). Mechanisms of cannabidiolneuroprotection in hypoxic-ischemic newborn pigs: Role of 5HT(1A) and CB2 receptors. *Neuropharmacology*, *71*, 282–291.

Penumarti, A., & Abdel-Rahman, A. A. (2014). The novel endocannabinoid receptor GPR18 is expressed in the rostral ventrolateral medulla and exerts tonic restraining influence on blood pressure. *Journal of Pharmacology and Experimental Therapeutics*, *349*, 29–38.

Pretzsch, C. M., Voinescu, B., Mendez, M. A., Wichers, R., Ajram, L., Ivin, G., et al. (2019). The effect of cannabidiol (CBD) on low-frequency activity and functional connectivity in the brain of adults with and without autism spectrum disorder (ASD). *Journal of Psychopharmacology*, *33*, 1141–1148.

Rice, M. W., Pandya, J. D., & Shear, D. A. (2019). Gut microbiota as a therapeutic target to ameliorate the biochemical, neuroanatomical, and Behavioral effects of traumatic brain injuries. *Frontiers in Neurology*, *10*, 875.

Rodriguez-Arias, M., Navarrete, F., Daza-Losada, M., Navarro, D., Aguilar, M. A., Berbel, P., et al. (2013). CB1 cannabinoid receptor-mediated aggressive behavior. *Neuropharmacology*, *75*, 172–180.

Russo, E. B., Burnett, A., Hall, B., & Parker, K. K. (2005). Agonistic properties of cannabidiol at 5-HT1a receptors. *Neurochemical Research*, *30*, 1037–1043.

Schurman, L. D., & Lichtman, A. H. (2017). Endocannabinoids: A promising impact for traumatic brain injury. *Frontiers in Pharmacology*, *8*, 69.

Shohami, E., Cohen-Yeshurun, A., Magid, L., Algali, M., & Mechoulam, R. (2011). Endocannabinoids and traumatic brain injury. *British Journal of Pharmacology*, *163*, 1402–1410.

Singh, J., & Neary, J. P. (2020). Neuroprotection following concussion: The potential role for Cannabidiol. *The Canadian Journal of Neurological Sciences*, *47*, 289–300.

Chapter 37

Valproic acid: A new narrative on its features and effects in traumatic brain injury

Glenn Wakam[a,*], Michael Kemp[a,*], and Hasan Alam[b]

[a]*Department of Surgery, University of Michigan, Ann Arbor, MI, United States,* [b]*Department of Surgery, Feinberg School of Medicine, Northwestern Memorial Hospital, Chicago, IL, United States*

Abbreviations

A_β	amyloid-beta
BDNF	brain-derived neurotrophic factor
CDC	Centers for Disease Control
FDA	Food and Drug Administration
GFAP	glial fibrillary acidic protein
HATs	histone acetyl transferases
HDACIs	histone deacetylase inhibitors
HDACs	histone deacetylases
Iba1	ionized calcium binding activated molecule-1
PECAM-1	platelet endothelial cell adhesion molecule 1
p-Tau	phosphorylated-Tau protein
TBI	traumatic brain injury
TUNEL	terminal deoxynucleotidyl transferase dUTP nick end labeling
VPA	valproic acid

Background

Traumatic brain injury (TBI) is a major cause of morbidity and mortality worldwide. The United States Centers for Disease Control (CDC) estimates that TBI-related emergency department visits, and deaths accounted for 2.8 million cases in 2014 (Center for Disease Control and Prevention, 2020). Throughout the world, others have estimated that 5.48 million people suffer from severe TBI (World Health Organization, 2020). After having an acute episode of TBI, it is estimated that 43% of patients will develop long term disability, an outcome that has significant loss of productivity and increased costs. In fact, it has been previously demonstrated that TBI survivors have difficulty with "activities of daily living, instrumental activities of daily living, social integration, and financial independence" (Ma, Chan, & Carruthers, 2014). A study from 2009 demonstrated that TBI survivors were more likely to be receiving some form of financial disability support (Silver, Kramer, Greenwald, & Weissman, 2001). This same study also found that survivors suffer from worse physical health, memory problems, and higher rates of several psychiatric disorders that are associated with an increased likelihood of suicide attempt. Depending on the population, the reported rates of return to work range from 12% to 70%, further evidence of the significant loss in productivity associated with TBI events (Ma et al., 2014). The costs associated with TBI and long-term care are also quite substantial. In 2013, a structured review indicated that the total annual direct and indirect costs associated with TBI ranged from \$62–\$76 billion in the United States (Ma et al., 2014).

*Designates co-primary authors.

Diagnosis and Treatment of Traumatic Brain Injury. https://doi.org/10.1016/B978-0-12-823347-4.00007-5

TBI can be categorized in a number of ways. For example, one method categorizes the injury into phases: primary, secondary, and chronic. The primary phase injury refers to the direct mechanical injury to the brain and is typically considered irreversible. On the other hand, secondary injury refers to the subsequent damage to the brain parenchyma, due to multiple molecular processes that are set in motion by additional insults (Ladak, Enam, & Ibrahim, 2019). Prevention/attenuation of secondary injury remains the main focus of our management strategies. Finally, the chronic phase refers to the resulting chronic long-term sequalae of primary and secondary TBI. Despite great advances in healthcare the mainstay of treatment following TBI is supportive, with the emphasis on reducing the secondary cellular damage by preventing hypotension, hypoxemia, osmotic changes, or fever. Currently, there is no treatment for TBI that confers pro-survival benefits and enhances short-term and long-term recovery. This gap in care for such a large proportion of trauma victims has been the driver behind the recent efforts to develop novel therapeutic strategies. For more than a decade our lab has sought to refine pharmacologic interventions that would alter the cellular response to create a pro-survival phenotype, which would result in better clinical outcomes following TBI. We have found that Valproic Acid (VPA), a non-selective histone deacetylase inhibitor, can alter gene expression, induce post-translational protein modifications, modulate cellular pathways, and improve outcomes following TBI. These promising results suggest that VPA may change the current paradigm of TBI management by providing a novel therapeutic option. A broad overview depicting the described effects of VPA following TBI is presented in Fig. 1.

Epigenetic changes following TBI

Traumatic injury manifests variably in patients. Following trauma there are strikingly different responses even when patients are matched for injury severity, age, gender, and comorbidities. Cellular regulation of gene transcription and protein functions has been implicated for the wide spectrum of phenotypic expression by different people and by different tissues in the same individual. One form of regulation, epigenetics, is the phenomena of changes that affect a cell's phenotype without affecting the genotype (Candiani, Franceschi, Chignola, et al., 1992). Acetylation, phosphorylation, methylation, and ubiquitylation are all examples of epigenetic processes that act through modification of chromatin, the complex of DNA and proteins within a cell's nucleus. Our lab has shown that following trauma and resuscitation approximately 7% of genes (and their downstream pathways) change their expression in an organ-specific manner (Chen et al., 2006; Lin, Chen, Koustova, et al., 2007). Data has shown, in response to TBI, brain tissue has increased expression of proteins related to neurodegeneration and decreased expression of those related to antioxidant defense (Abu Hamdeh et al., 2018).

Our lab has focused on the post-translational acetylation of proteins (histone and non-histone). Histone proteins in the chromatin modulate gene transcription. When chromatin is condensed, it is inaccessible to transcription factors and therefore considered inactive. Protein acetylation promotes chromatin relaxation which allows transcription factors to bind and activate the transcription of the target genes (Hebbes, Thorne, & Crane-Robinson, 1988). Two families of enzymes regulate the process—histone acetyl transferases (HATs) increase acetylation whereas histone deacetylases (HDACs) tilt the balance in favor of deacetylation. This is an active and reversible process (acetylation homeostasis) that goes on constantly in the cells. The degree of acetylation and the various sites of acetylation on the histone protein can precisely regulate the transcription process by creating highly regulated structural changes in the chromatin. On the other hand, balance in favor of HDAC promotes deacetylation of histones and chromatin "silencing" (Halaweish, Nikolian, Georgoff, Li, & Alam, 2015).

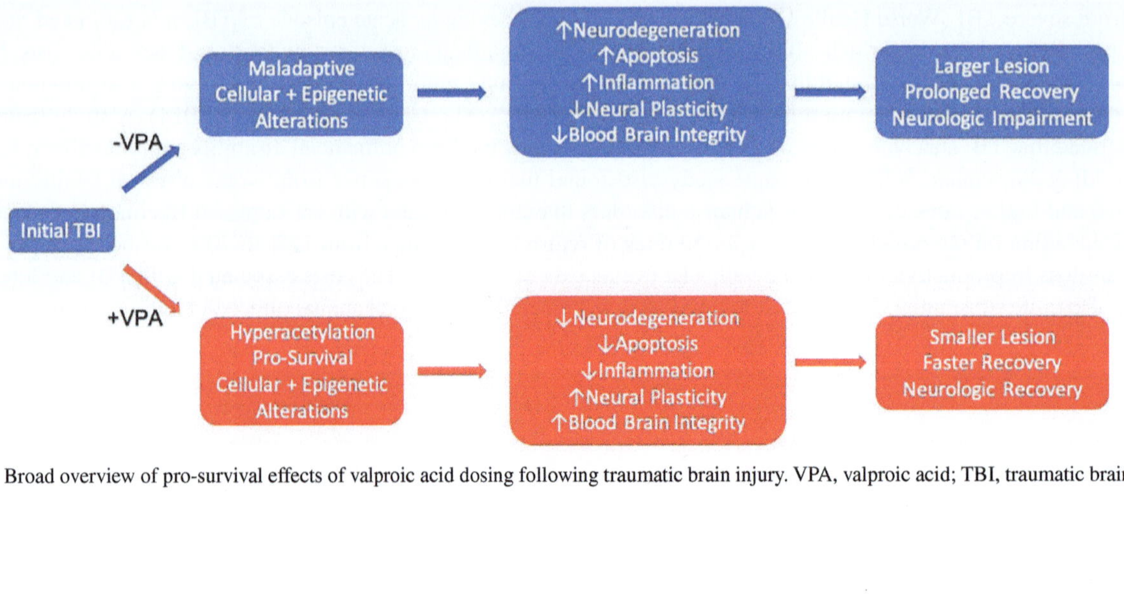

FIG. 1 Broad overview of pro-survival effects of valproic acid dosing following traumatic brain injury. VPA, valproic acid; TBI, traumatic brain injury.

Therapies that can alter the balance between acetylation and deacetylation can promote or inhibit specific gene transcription. Although histones were considered the main target of HDAC, we now know that non-histone proteins involved in key cellular functions such as cell cycle, stress response, cytoskeleton dynamics, signaling, repair/healing/remodeling, communication, and proliferation are equally involved (Glozak, Sengupta, Zhang, & Seto, 2005). At least 50 such non-histone proteins have been well-characterized (Kim & Bae, 2011), but this list is rapidly growing. In one study, 3600 acetylation sites (in 1750 human proteins) were identified, which regulated nearly all nuclear and many key cytoplasmic processes (Choudhary, Kumar, Gnad, et al., 2009). This makes acetylation a regulatory mechanism that is as prevalent and important as phosphorylation (Chang, Weykamp, Dennahy, et al., 2018; Kouzarides, 2000; Norris, Lee, & Yao, 2009), which theoretically can influence nearly all cellular processes. Acetylation of non-histone proteins is actually even more important than histones in the setting of severe shock and/or traumatic brain injury. While it takes some time to produce phenotype changes through modulation of gene transcription, direct (non-transcriptional) acetylation of regulatory proteins is extremely fast and thus much more relevant for rapidly lethal conditions such as shock or TBI. In short, there are many features that make acetylation an attractive therapeutic target: it is *rapid*, completely *reversible*, and can be altered by drugs that are *already in clinical use*.

Valproic acid: A histone deacetylase inhibitor

Valproic acid (VPA) is a well-known drug that has traditionally been utilized for the treatment of epilepsy or seizures. VPA belongs to a class of drugs called histone deacetylase inhibitors (HDACI). In humans, there are at least 18 HDAC isoforms that have been identified, which are categorized into 1 of 4 classes. VPA is considered a non-selective HDACI but has been shown to predominantly inhibit class I (ubiquitous) and class IIa (tissue-specific) (Chang et al., 2018). Since VPA inhibits histone deacetylation, the balance tilts in favor of acetylation of histone and non-histone proteins that affects numerous nuclear and cytoplasmic processes to alter the cellular phenotype (Butt, Sailhamer, Li, et al., 2009). Understanding the resulting cascading cellular responses that are associated with this hyperacetylation has been a primary focus of our lab's work. Treatment with VPA has shown promising results with regard to modulation of several pro-survival, anti-inflammatory, and anti-apoptotic pathways in pre-clinical models, and is being considered as a potential treatment for TBI beyond the standard supportive therapies (Russo, Kemp, Bhatti, et al., 2020).

Effects of valproic acid in pre-clinical models of traumatic brain injury

Following a TBI, the magnitude of brain parenchyma damaged depends on both the severity of the initial mechanical force and the extent of secondary injury caused by hypoxia and hypotension. Hypoxia and hypotension propagate injury by contributing to cellular edema, raising intracranial pressure and furthering cellular destruction; both are independently correlated with significant increase in morbidity and mortality following TBI (Chesnut, Marshall, Klauber, et al., 1993; Rhind, Crnko, Baker, et al., 2010). Given this knowledge of pathophysiology, our lab has utilized models of TBI and hemorrhagic shock of varying degrees to create worst case scenarios following TBI in order to test the efficacy of VPA treatment. Early work with large animal (swine) models of TBI and hemorrhagic shock demonstrated that the addition of high dose VPA (300mg/kg) to traditional resuscitation reduced brain lesion size within 6h post-injury (Jin, Duggan, Imam, et al., 2012). Further work demonstrated that lower doses of VPA not only still improved outcomes but showed long-term benefit as well. In a pre-clinical model of TBI and 40% hemorrhagic shock, the swine that received a single dose of 150mg/kg of VPA had brain lesions that were one-third in volume compared to the control animals that received traditional fluid resuscitation (Fig. 2), and the VPA treated animals showed significantly faster neurologic recovery over a 30-day observation period (Halaweish, Bambakidis, et al., 2015 (Fig. 3). The same benefits of decreased brain lesion size and improved neurologic recovery were also demonstrated in more severe models (Nikolian, Georgoff, Pai, et al., 2017).

Underlying mechanisms of VPA-neuroprotection

We have done extensive work to better understand the underlying cellular mechanisms that lead to the attenuated neurologic impairment, faster neurologic recovery and decreased brain lesion sizes associated with VPA administration. We have discovered a multitude of possible explanations. One possible mechanism behind the neuroprotection is a decrease in neural apoptosis, inflammation, and degenerative changes, in addition to increased neural plasticity. The tissues from swine that were given a single dose of VPA after TBI and hemorrhagic shock were harvested 30days after injury and underwent extensive immunohistochemistry analysis (Chang, Williams, Bhatti, et al., 2019). To assess long-term neural apoptosis, a terminal deoxynucleotidyl transferase dUTP nick end labeling (TUNEL) assay and expression of c-caspase 3 were used.

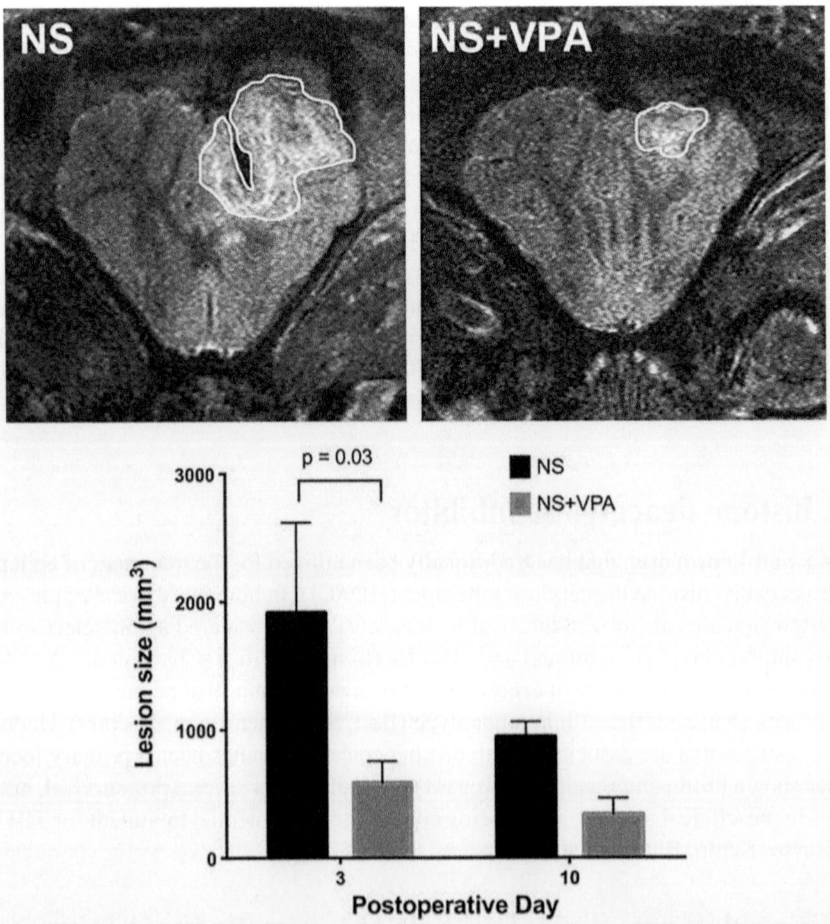

FIG. 2 Brain lesion size following treatment with valproic acid in a pre-clinical model. In a pre-clinical model of TBI and hemorrhagic shock, swine that received a single dose of 150mg/kg of valproic acid (VPA) and normal saline (NS) resuscitation had smaller brain lesions as compared to swine that only received NS resuscitation. The top panels show magnetic resonance images of swine brain with outlining of the lesions and the associated surrounding edema. The bottom panel provides MRI volume measurements of brain lesion sizes. *Adapted from Halaweish, I., Bambakidis, T., Chang, Z., et al. (2015). Addition of low-dose valproic acid to saline resuscitation provides neuroprotection and improves long-term outcomes in a large animal model of combined traumatic brain injury and hemorrhagic shock.* Journal of Trauma and Acute Care Surgery, 79 *(6), 911–919, discussion 919. Copyright Wolters Kluwer (2015).*

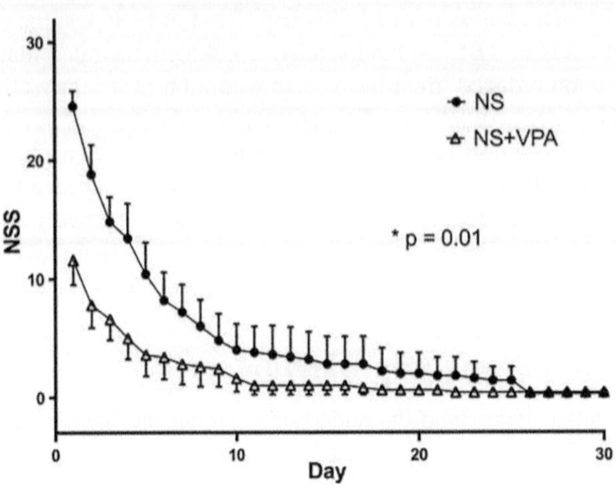

FIG. 3 Neurologic recovery following treatment with valproic acid in a pre-clinical model. In a pre-clinical model of TBI and hemorrhagic shock, swine that received a single dose of 150mg/kg of valproic acid (VPA) and normal saline (NS) resuscitation had significantly lower neurologic impairment and faster neurologic recovery over a 30-day observation period. In the above graph, a higher neurologic severity score (NSS) is demonstrative of more severe injury and a score of 0 is indicative of no neurologic impairment. *Adapted from Halaweish, I., Bambakidis, T., Chang, Z., et al. (2015). Addition of low-dose valproic acid to saline resuscitation provides neuroprotection and improves long-term outcomes in a large animal model of combined traumatic brain injury and hemorrhagic shock.* Journal of Trauma and Acute Care Surgery, 79 *(6), 911–919, discussion 919. Copyright Wolters Kluwer (2015).*

The VPA-treated group demonstrated decreased TUNEL-positive cells and c-caspase 3 expression compared to control groups, consistent with decreased neural apoptosis. VPA-treated animals also had decreased expression of glial fibrillary acidic protein (GFAP) and ionized calcium binding activated molecule-1 (Iba1), both markers of neuroinflammation, in addition to amyloid-beta (Aβ) and phosphorylated-Tau protein (p-Tau), markers of neurodegeneration. Finally, evaluation of brain tissue at 30days following VPA administration in a model of TBI showed increased brain-derived neurotrophic factor (BDNF) expression, a key participant in promoting neural plasticity and survival, consistent with pro-survival changes (Chang et al., 2019).

Another proposed mechanism for improved outcomes with VPA administration is drug-induced structural changes to the blood brain barrier integrity. For example, in combined TBI and hemorrhagic shock models, animals treated with VPA demonstrated less perivascular albumin extravasation suggesting greater integrity of the barrier (Nikolian, Dekker, Bambakidis, et al., 2018). Furthermore, this integrity is associated with higher expression of Claudin-5, a key protein involved in maintaining the barrier, and laminin, an extracellular matrix protein involved in repair and remodeling (Nikolian et al., 2018).

Our group has also analyzed the role that VPA treatment may have in platelet function after TBI. "Exhausted platelet syndrome" is a term used to describe the increased platelet activation but decreased function that occurs following TBI (Jacoby et al., 2001). The underlying mechanism is believed to be initial platelet hyperactivation and concomitant depletion of intracellular mediators, leading to platelet unresponsiveness to stimulation. A large animal model of TBI and hemorrhagic shock demonstrated significant platelet dysfunction that was improved in the brain and serum after administration of combined VPA and fresh frozen plasma as part of the resuscitation (Dekker, Sillesen, Bambakidis, et al., 2014). Further investigation revealed VPA-induced higher sCD40 ligand levels (marker of platelet activation) and lower levels of brain platelet endothelial cell adhesion molecule 1 (PECAM-1), a surface glycoprotein receptor that limits growth of platelet thrombi. VPA also improved blood clot strength and rate of formation in animals with HS and TBI. This work was notable for the fact that VPA did not alter platelet function or coagulation in healthy controls.

Alterations to a number of cellular pathways have been implicated in creating the pro-survival phenotype associated with VPA administration in pre-clinical models involving TBI (Higgins, Georgoff, Nikolian, et al., 2017). Computational biology tools are revealing numerous additional mechanisms and possibilities. For example, network reconstruction has revealed that valproic acid regulates neurogenic transcriptional programs in adult brain following traumatic injury (Higgins, Georgoff, et al., 2017) (Fig. 4), and many of these transcriptional networks can also serve as targets for development of newer generation drugs (Higgins, Williams, Ade, Alam, & Athey, 2019). Our group is also more recently focusing on cellular proteomic and metabolomic network changes to further describe the broad scale effects that VPA has on cellular pathways (Bhatti, Williams, Kathawate, et al., 2019). These types of analyses however also offer the opportunity to explore other pathways not currently appreciated in the literature and identify new targets for therapeutic interventions (Higgins, Allyn-Feuer, et al., 2017).

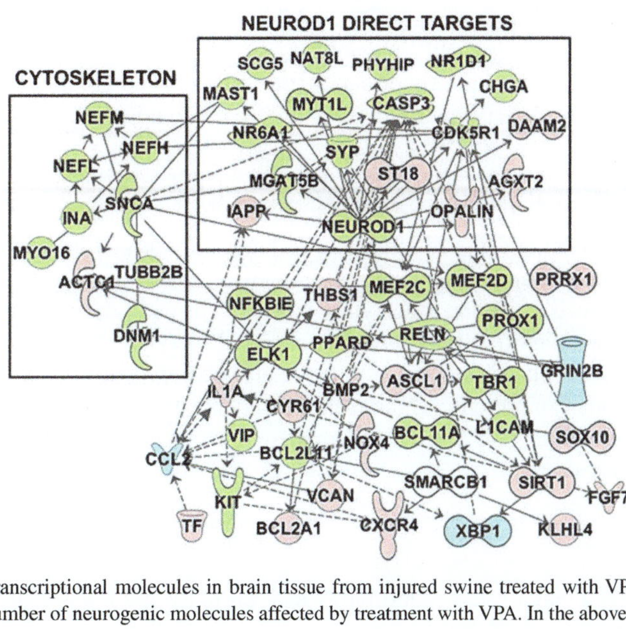

FIG. 4 Regulation of neurogenic transcriptional molecules in brain tissue from injured swine treated with VPA. In injured brain tissue from swine, network reconstruction identified a number of neurogenic molecules affected by treatment with VPA. In the above image, known VPA pharmacogenomic genes are represented in blue. Symbols in green indicate up-regulation and symbols in red indicate down-regulation. *This is adapted from Higgins, G.A., Georgoff, P., Nikolian, V., et al. (2017). Network reconstruction reveals that valproic acid activates neurogenic transcriptional programs in adult brain following traumatic injury.* Pharmaceutical Research, 34, *1658–1672. Copyright Springer (2017).*

Dosing, safety, and future work

VPA has been approved by the Food and Drug Administration (FDA) since 1978 for the routine treatment of epilepsy or seizures (Food and Drug Administration, 2016). Intravenous VPA is currently approved for the treatment of epilepsy with a maximum recommended dose of 60mg/kg/day. Known contraindications to VPA include hepatic disease or significant dysfunction, mitochondrial or related disorders, known hypersensitivity, and urea cycle disorders. Life threatening adverse reactions include hepatotoxicity, fetal risks during pregnancy, and pancreatitis. Although these adverse events are extremely rare and reported after prolonged use of VPA (as opposed to a single dose), liver function enzymes and platelets should be measured and monitored in patients who are on VPA.

Currently, valproic acid is undergoing clinical trials for the treatment of trauma and hemorrhagic shock, and we have already completed a phase 1 clinical trial. In this phase 1 trial, the maximum tolerated dose of intravenous VPA in healthy individuals was 140mg/kg with nausea and headache being key dose-limiting toxicities (Georgoff, Nikolian, Bonham, et al., 2018). Importantly, none of the serious adverse events listed above were noted. The information obtained from this initial phase 1 trial supports the notion that higher doses of VPA above the current recommended dose warrant consideration in trauma, including TBI. Although our lab has shown effectiveness with these higher doses of VPA, ongoing preclinical models are assessing the effectiveness of VPA at lower doses in improving TBI outcomes to determine if lower doses in the current FDA-approved range will improve outcomes in a similar manner.

Investigation is also ongoing to evaluate efficacy of other HDACIs given the potential significant toxicities associated with treatment with the non-selective HDACI, VPA. In fact, some data suggests that treatment with an isoform and tissue-selective HDACI would confer similar therapeutic benefits without the associated toxicities (Chang et al., 2018).

Applications to other areas of neuroscience

VPA is a commonly used neuroleptic which is being evaluated for efficacy in other neurologic disease processes such as TBI. Given its previously demonstrated protective effects, VPA may have promising effects in several other areas of neurologic injury. For example, its therapeutic benefit is also being evaluated in other forms of neurologic injury such as subarachnoid hemorrhage and low perfusion states such as shock or cardiac arrest (Chang et al., 2019; Lee et al., 2015; Tso, Lass, Ai, & Loch, 2015).

Mini-dictionary of terms

- ***Traumatic brain injury (TBI)***: A disease associated with significant morbidity and mortality.
- ***Primary TBI***: The direct mechanical injury to brain parenchyma.
- ***Secondary TBI***: The indirect injury to brain parenchyma caused by molecular processes.
- ***Valproic acid (VPA)***: A non-selective histone deacetylase inhibitor that is commonly used for treatment of epilepsy.
- ***Epigenetics***: A cellular process by which a cell's phenotype is altered through chromatic modification (no genotypic change) and such processes as acetylation, phosphorylation, methylation, and ubiquitylation.
- ***Apoptosis***: Programmed cellular death that can occur secondary to tissue injury.

Key facts of traumatic brain injury and valproic acid

- Traumatic brain injury (TBI) is a major cause of morbidity and mortality primarily treated with supportive measures.
- TBI is associated with tissue-specific epigenetic changes.
- Valproic acid (VPA) is a non-selective histone deacetylase inhibitor currently approved for treatment of epilepsy.
- VPA can modify cellular epigenetic profiles by promoting increased acetylation.
- VPA, through increased acetylation, has been shown to modulate pro-survival, inflammatory, and apoptotic pathways and improves neurologic outcomes in pre-clinical large animal models of TBI.
- Ongoing work is needed to determine if VPA promotes improved outcomes at lower doses that are currently approved by the Food and Drug Administration.

Summary points

- Traumatic brain injury (TBI) is a major cause of morbidity and mortality and is associated with tissue-specific epigenetic changes.
- Valproic acid (VPA) is a non-selective histone deacetylase inhibitor currently approved for treatment of epilepsy.

- VPA modifies cellular epigenetic profiles by promoting increased acetylation.
- VPA modulates pro-survival, inflammatory, and apoptotic pathways and improves neurologic outcomes in pre-clinical large animal models of TBI.
- Ongoing work is needed to determine if the therapeutic benefits of VPA on TBI-associated outcomes occur at lower doses currently approved by the Food and Drug Administration.
- VPA may confer therapeutic benefits in a wide array of clinical scenarios involving various forms of neurologic injury.

References

Abu Hamdeh, S., Shevchenko, G., Mi, J., Musunuri, S., Bergquist, J., & Marklund, N. (2018). Proteomic differences between focal and diffuse traumatic brain injury in human brain tissue. *Scientific Reports*, *8*(1), 6807.

Bhatti, U. F., Williams, A. M., Kathawate, R. G., et al. (2019). Comparative analysis of isoform-specific and non-selective histone deacetylase inhibitors in attenuating the intestinal damage after hemorrhagic shock. *Trauma Surgery & Acute Care Open*, *4*(1), e000321.

Butt, M. U., Sailhamer, E. A., Li, Y., et al. (2009). Pharmacologic resuscitation: Cell protective mechanisms of histone deacetylase inhibition in lethal hemorrhagic shock. *The Journal of Surgical Research*, *156*(2), 290–296.

Candiani, C., Franceschi, A., Chignola, R., et al. (1992). Blocking effect of human serum but not of cerebrospinal fluid on ricin a chain immunotoxin potentiation by monensin or carrier protein-monensin conjugates. *Cancer Research*, *52*(3), 623–630.

Center for Disease Control and Prevention. (2020). Available https://www.cdc.gov/traumaticbraininjury/data/tbi-edhd.html. Last accessed on September 1, 2020.

Chang, P., Weykamp, M., Dennahy, I. S., et al. (2018). Histone deacetylase inhibitors: Isoform selectivity improves survival in a hemorrhagic shock model. *Journal of Trauma and Acute Care Surgery*, *84*(5), 795–801.

Chang, P., Williams, A. M., Bhatti, U. F., et al. (2019). Valproic acid and neural apoptosis, inflammation, and degeneration 30 days after traumatic brain injury, hemorrhagic shock, and polytrauma in a swine model. *Journal of the American College of Surgeons*, *228*(3), 265–275.

Chen, H., Alam, H. B., Querol, R. I., Rhee, P., Li, Y., & Koustova, E. (2006). Identification of expression patterns associated with hemorrhage and resuscitation: Integrated approach to data analysis. *The Journal of Trauma*, *60*(4), 701–723 (discussion 704–723).

Chesnut, R. M., Marshall, L. F., Klauber, M. R., et al. (1993). The role of secondary brain injury in determining outcome from severe head injury. *The Journal of Trauma*, *34*(2), 216–222.

Choudhary, C., Kumar, C., Gnad, F., et al. (2009). Lysine acetylation targets protein complexes and co-regulates major cellular functions. *Science*, *325*(5942), 834–840.

Dekker, S. E., Sillesen, M., Bambakidis, T., et al. (2014). Treatment with a histone deacetylase inhibitor, valproic acid, is associated with increased platelet activation in a large animal model of traumatic brain injury and hemorrhagic shock. *The Journal of Surgical Research*, *190*(1), 312–318.

Food and Drug Administration, 2016. Depakene valproic acid. https://www.accessdata.fda.gov/drugsatfda_docs/label/2016/018081s065_018082s048lbl.pdf. (Accessed 14 November 2021).

Georgoff, P. E., Nikolian, V. C., Bonham, T., et al. (2018). Safety and tolerability of intravenous valproic acid in healthy subjects: A phase I dose-escalation trial. *Clinical Pharmacokinetics*, *57*(2), 209–219.

Glozak, M. A., Sengupta, N., Zhang, X., & Seto, E. (2005). Acetylation and deacetylation of non-histone proteins. *Gene*, *363*, 15–23.

Halaweish, I., Bambakidis, T., Chang, Z., et al. (2015). Addition of low-dose valproic acid to saline resuscitation provides neuroprotection and improves long-term outcomes in a large animal model of combined traumatic brain injury and hemorrhagic shock. *Journal of Trauma and Acute Care Surgery*, *79*(6), 911–919. discussion 919.

Halaweish, I., Nikolian, V., Georgoff, P., Li, Y., & Alam, H. (2015). Creating a "pro-survival phenotype" through histone deacetylase inhibition: Past, present, and future. *Shock*.

Hebbes, T. R., Thorne, A. W., & Crane-Robinson, C. (1988). A direct link between core histone acetylation and transcriptionally active chromatin. *The EMBO Journal*, *7*(5), 1395–1402.

Higgins, G. A., Allyn-Feuer, A., Georgoff, P., Nikolian, V., Alam, H. B., & Athey, B. D. (2017). Mining the topography and dynamics of the 4D nucleome to identify novel CNS drug pathways. *Methods*, *123*, 102–118.

Higgins, G. A., Georgoff, P., Nikolian, V., et al. (2017). Network reconstruction reveals that valproic acid activates neurogenic transcriptional programs in adult brain following traumatic injury. *Pharmaceutical Research*, *34*(8), 1658–1672.

Higgins, G. A., Williams, A. M., Ade, A. S., Alam, H. B., & Athey, B. D. (2019). Druggable transcriptional networks in the human neurogenic epigenome. *Pharmacological Reviews*, *71*(4), 520–538.

Jacoby, R. C., Owings, J. T., Holmes, J., Battistella, F. D., Gosselin, R. C., & Paglieroni, T. G. (2001). Platelet activation and function after trauma. *The Journal of Trauma*, *51*(4), 639–647.

Jin, G., Duggan, M., Imam, A., et al. (2012). Pharmacologic resuscitation for hemorrhagic shock combined with traumatic brain injury. *Journal of Trauma and Acute Care Surgery*, *73*(6), 1461–1470.

Kim, H. J., & Bae, S. C. (2011). Histone deacetylase inhibitors: Molecular mechanisms of action and clinical trials as anti-cancer drugs. *American Journal of Translational Research*, *3*(2), 166–179.

Kouzarides, T. (2000). Acetylation: A regulatory modification to rival phosphorylation? *The EMBO Journal*, *19*(6), 1176–1179.

Ladak, A. A., Enam, S. A., & Ibrahim, M. T. (2019). A review of the molecular mechanisms of traumatic brain injury. *World Neurosurgery*, *131*, 126–132.

Lee, J. H., Kim, K., Jo, Y. H., Lee, M. J., Hwang, J. E., & Kim, M. A. (2015). Effect of valproic acid combined with therapeutic hypothermia on neurologic outcome in asphyxial cardiac arrest model of rats. *The American Journal of Emergency Medicine*, *33*(12), 1773–1779.

Lin, T., Chen, H., Koustova, E., et al. (2007). Histone deacetylase as therapeutic target in a rodent model of hemorrhagic shock: Effect of different resuscitation strategies on lung and liver. *Surgery*, *141*(6), 784–794.

Ma, V. Y., Chan, L., & Carruthers, K. J. (2014). Incidence, prevalence, costs, and impact on disability of common conditions requiring rehabilitation in the United States: Stroke, spinal cord injury, traumatic brain injury, multiple sclerosis, osteoarthritis, rheumatoid arthritis, limb loss, and back pain. *Archives of Physical Medicine and Rehabilitation*, *95*(5), 986–995.e981.

Nikolian, V. C., Dekker, S. E., Bambakidis, T., et al. (2018). Improvement of blood-brain barrier integrity in traumatic brain injury and hemorrhagic shock following treatment with valproic acid and fresh frozen plasma. *Critical Care Medicine*, *46*(1), e59–e66.

Nikolian, V. C., Georgoff, P. E., Pai, M. P., et al. (2017). Valproic acid decreases brain lesion size and improves neurologic recovery in swine subjected to traumatic brain injury, hemorrhagic shock, and polytrauma. *Journal of Trauma and Acute Care Surgery*, *83*(6), 1066–1073.

Norris, K. L., Lee, J. Y., & Yao, T. P. (2009). Acetylation goes global: The emergence of acetylation biology. *Science Signaling*, *2*(97), pe76.

Rhind, S. G., Crnko, N. T., Baker, A. J., et al. (2010). Prehospital resuscitation with hypertonic saline-dextran modulates inflammatory, coagulation and endothelial activation marker profiles in severe traumatic brain injured patients. *Journal of Neuroinflammation*, *7*, 5.

Russo, R., Kemp, M., Bhatti, U. F., et al. (2020). Life on the battlefield: Valproic acid for combat applications. *Journal of Trauma and Acute Care Surgery*, *89*(2S Suppl. 2), S69–s76.

Silver, J. M., Kramer, R., Greenwald, S., & Weissman, M. (2001). The association between head injuries and psychiatric disorders: Findings from the New Haven NIMH epidemiologic catchment area study. *Brain Injury*, *15*(11), 935–945.

Tso, M. K., Lass, E., Ai, J., & Loch, M. R. (2015). Valproic acid treatment after experimental subarachnoid hemorrhage. *Acta Neurochirurgica. Supplement*, *120*, 81–85.

World Health Organization. (2020). Available at: http://www.who.int/violence_injury_prevention/key_facts/VIP_key_fact_1.pdf. Last accessed on June 30, 2020.

Chapter 38

Dietary supplementation for traumatic brain injury

Muthu Kumaradoss MohanMarugaRaja[a], Agilandeswari Devarajan[b], and Vipin V. Dhote[c]
[a]*Department of Pharmacognosy and Phytochemistry, Parul Institute of Pharmacy & Research, Parul University, Vadodara, Gujarat, India,* [b]*Research and Development Department, Synapsesai India Private Limited, Indore, Madhya Pradesh, India,* [c]*VNS Group of Institutions, Neelbud, Bhopal, Madhya Pradesh, India*

List of abbreviations

AD	Alzheimer's disease
ALA	alpha linolenic acid
AMPA	α-amino-3-hydroxy-5-methyl-4-isoxazolepropionic acid
ATP	adenosine triphosphate
BDNF	brain-derived neurotrophic factor
cAMP	cyclic adenosine monophosphate
COBRIT	Citicoline Brain Injury Treatment
CTE	chronic traumatic encephalopathy
DHA	docosahexaenoic acid
EPA	eicosapentaenoic acid
GABA	γ-aminobutyric acid
GCS	Glasgow Coma Scale
IL	interleukin
NADH	nicotinamide adenine dinucleotide
NFL	National Football League
NIH	National Institute of Health
NMDA	*N*-methyl-D-aspartate
PD	Parkinson's disease
ROS	reactive oxygen species
TBI	traumatic brain injury
TNF	tumor necrosis factor
USFDA	United States Food and Drug Administration

Introduction

The National Institute of Health (NIH) estimates that almost 2.5–6.5 million individuals in the United States experience disabilities due to acute and chronic traumatic brain injury (TBI). The annual recorded incidences of TBI are as high as 2 million and almost 15% of the patients required hospitalization. Numerous studies have provided evidence linking TBI with long-term progressive neurodegeneration. Goldman et al. (2006) established an association between mild to moderate closed head injury and increased risk of Parkinson's disease in later years. Epidemiological studies further established the association of TBI and neurodegenerative diseases (Gardner & Yaffe, 2015). Concussion is inevitable in contact-collision sports and combat situations. Physical impact to the head results in transient head injuries and triggers a wide array of neuropathological events like ionic imbalance, excitatory neurotransmitter release, neuronal depolarization, a metabolic shift in neurons, and axonal impairment (Giza & Hovda, 2001). Traumatic brain injury is also associated with long-term health disorders like post-traumatic stress disorder, Parkinsonism, Alzheimer's disease, psychosocial health problems, epilepsy, neurocognitive deficits, pain, and other alterations in behavior or personality. Such possibility of long-term disability and

Diagnosis and Treatment of Traumatic Brain Injury. https://doi.org/10.1016/B978-0-12-823347-4.00038-5

susceptibility to secondary injury poses a public health problem with a substantial financial burden on the society. The absence of an effective treatment has forced the focus on to the benefits of nutrition, health supplements, and alternative therapies for the treatment of TBI and its subsequent symptoms. A collecting body of research now suggests that consumption of certain diets or their associated physiologically active constituents may be linked to disease risk reduction. The new inclination is of reconsidering nutrition and their adequate intakes as the first line of protection against abnormal modalities. The science of nutrition is at a new frontier of transforming from the concept of "adequate nutrition," which is required for survival to that of "optimal nutrition," which is required for improving health. Later in 2011, a consensus was established that, advanced nutritional planning should be started 24–48 h after TBI (Cook, Peppard, & Magnuson, 2008; Scrimgeour & Condlin, 2014). Another study found that patients receiving enteral nutrition within 48 h of traumatic brain injury in acute stage had better survival rates (Chiang et al., 2012).

The role of dietary supplements in traumatic brain injury is an evolving research area. Many dietary supplements have shown beneficial outcomes in pre-clinical studies. Certain studies suggest the use of diet supplements to reduce the injury and treat TBI (Ashbaugh & McGrew, 2016; Oliver, Anzalone, & Turner, 2018), while some studies recommend that dietary supplements when used for longer duration alone can improve TBI (Erdman, Maria, & Pillsbury, 2011). However, there are also reports pointing out the lack of evidence (Trojian, Wang, & Leddy, 2017) about dietary supplements in the recovery of patients following TBI. This chapter reviews some of the extensively explored dietary supplements and their clinical interventions for the treatment of TBI.

Omega-3 fatty acids

Omega-3 fatty acids are polyunsaturated fatty acids that have a double bond at the third atom from the terminal methyl group. Omega-3 fatty acids are essential for maintaining membrane fluidity (Valentine & Valentine, 2004) which in turn impacts neuronal cell adhesion, axon guidance, synapse maintenance, dendritic formation, and the speed of neurotransmission (Hering, Lin, & Sheng, 2003; Innis, 2008; Tsui-Pierchala, Encinas, Milbrandt, & Johnson, 2002). Omega-3 fatty acids are readily taken up by the brain and target the multiple components of the secondary TBI injury cascade. Therapeutic effect of omega-3 fatty acids on different mechanisms of injury in TBI are improvement of cell survival, recovery, and functional outcome after traumatic insult via pleiotropic effects on a variety of pathways. Omega-3 fatty acids mitigate the consequences of several key pathological pathways in TBI, such as mitochondrial malfunction, apoptotic cell death, glutamate-triggered excitotoxicity, and injury-induced oxidative stress and inflammation (Table 1) (Hasadsri et al., 2013).

The three main omega-3 fatty acids are alpha linolenic acid (ALA), eicosapentaenoic acid (EPA), and docosahexaenoic acid (DHA). ALA is found mainly in plant oils whereas, DHA and EPA are found in fish and other seafoods. ALA is an essential fatty acid and only a minute quantity can be converted by human body into EPA and then to DHA. Therefore, EPA and DHA have to be administered as dietary supplements to meet the requirement of omega-3 fatty acids in the body. The USFDA has confirmed the overall safety of fish oil, and both DHA and EPA at levels up to 3 g/day as safe (Davidson

TABLE 1 Effects of omega-3 fatty acids on different mechanisms of injury in TBI.

Injury mechanism	Effects
Axonal loss	Increases the outgrowth of new fibers Increases axonal regeneration Increases survival levels of neurons and oligodendrocytes
Inflammation	Produces neuroprotectins and resolvins Inhibits inflammation cell migration and their activation Increases expression of receptors related to tissue repair Decreases inflammatory markers
Excessive glutamate activity	Decreases glutamate cytotoxicity Decreases calcium influx induced dysfunction and cell death Decreased expression of AMPA receptor
Oxidative stress	Decreases lipid peroxidation and its byproducts Increases antioxidant activity Increases antiapoptotic cascades

AMPA, α-amino-3-hydroxy-5-methyl-4-isoxazolepropionic acid.

et al., 2007; USFDA, 1997; Wohl et al., 2005). DHA is a primary structural component of the mammalian cerebral cortex, and comprises 50% of neuronal membrane phospholipids (Singh, 2005). It is known to protect neurons by regulating signal transduction and gene expression. DHA and EPA (Fig. 1) have been explored more than ALA for TBI. Fish oil contains both DHA and EPA. DHA has been the most promising omega-3 fatty acids for neuroprotective and/or restorative capacities in TBI as per many pre-clinical studies. A randomized, double-blind, placebo-controlled, parallel design showed the neuroprotective effect of DHA supplementation (2–6 g/day) as prophylactics in American football athletes. DHA attenuated the serum neurofilament light (a biomarker of axonal injury) and exhibited neuroprotective effect (Oliver et al., 2016) (Table 2). One clinical study reported that arachidonic acid and DHA intake for 90 days (240 mg/day) improves cognitive dysfunction in patients with mild cognitive impairment. DHA also improved the membrane function and regional cerebral blood flow (Table 2) (Kotani et al., 2006). Despite the accumulation of evidence that DHA has protective antioxidant effects on cells, only two studies have examined markers of oxidative stress in patients supplemented with 4 and 6 g/day of DHA for 6 and 8 weeks, respectively, and found that lipid peroxidation was either unaffected or significantly lower in these individuals compared to placebo-treated controls (Mori et al., 2003; Wu, Lu, Wang, Jou, & Wang, 2006). Due to the anti-inflammatory effect of DHA, doses more than 4.9 g/day can suppress T-cell activation in humans and inhibit natural killer cell activity as well as decrease the number of total circulating white blood cells (Kelley et al., 1999; Kelley, Taylor, Nelson, & Mackey, 1998; Kew et al., 2004).

Sydenham, Dangour, and Lim (2012) reported mild gastrointestinal discomfort after fish oil supplementation. Omega-3 fatty acids specifically DHA may provide benefits in brain injuries but there isn't sufficient convincing clinical evidence

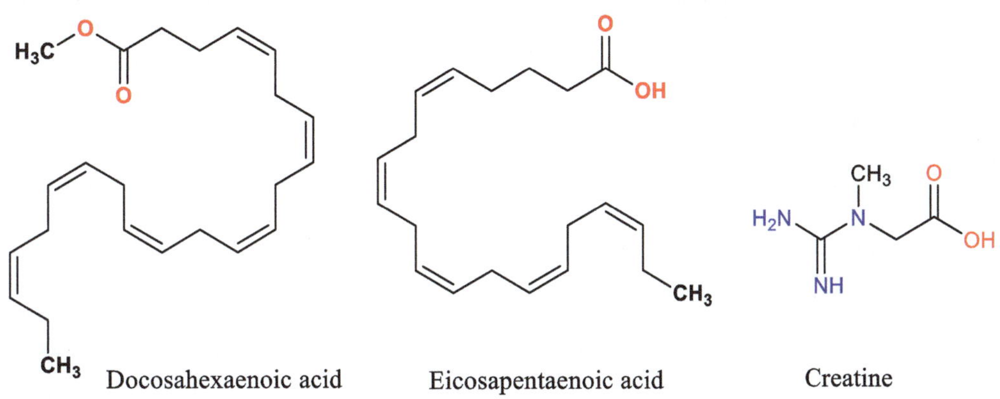

FIG. 1 Chemical structures of DHA, EPA, and creatine. *DHA*, docosahexaenoic acid; *EPA*, eicosapentaenoic acid.

TABLE 2 Clinical outcomes of omega-3 fatty acids and creatine in traumatic brain injury.

Dietary supplement	Treatment	Outcomes	References
Omega-3 fatty acids	90 days of DHA (240 mg/day)	DHA improved the membrane function, regional cerebral blood flow and cognitive dysfunction in patients with mild cognitive impairment	Kotani et al. (2006)
	189 days of DHA (2–6 g/day)	DHA attenuated the serum neurofilament light and exhibited neuroprotective effect	Oliver et al. (2016)
Creatine	06 months of creatine (0.4 g/kg/day)	Reduced the post-traumatic amnesia, headaches, dizziness and fatigue Improved cognitive function and recovery	Sakellaris et al. (2006), Sakellaris et al. (2008), Sakellaris et al. (2012)
	24 weeks of creatine (4 × 5 g/day for 5 days followed by 5 g/day)	No change in cognition nor depression in elderly women between 60 and 80 years	Alves et al. (2013)
	05 days of creatine (20 g/day)	Improved performance on memory task	Benton and Donohoe (2011)

DHA, docosahexaenoic acid.

that may prove as a functional outcome in patients with TBI (Hasadsri et al., 2013). Many pre-clinical studies have reported acute and chronic neuroprotective effects of omega-3 fatty acids in TBI but, further clinical studies are warranted to examine closely both the potential benefits and adverse effects of omega-3 fatty acids in human TBI (Petraglia, Winkler, & Bailes, 2011).

Creatine

Creatine (Fig. 1) is a methyl guanidine acetic acid, a naturally occurring nitrogenous organic acid. It is stored predominately in skeletal muscles but also present in the liver, kidney, testes and brain of vertebrates. The predominant function of creatine is to maintain energy homeostasis by keeping cellular ATP levels constant in muscles and the brain. Mild TBI induces cellular energy crisis causeing an extremely large demand on ATP in an environment of reduced blood flow and hypoxia. The neuropathology of mild TBI overlaps with the role of creatine (Table 3).

Creatine is the most preferred nutritional ergogenic aids for athletes (Knapik et al., 2016), the military (Havenetidis, 2016) and in clinical movement disorders (Gualano et al., 2012; Riesberg et al., 2016) to enhance muscle function. Numerous pre-clinical studies have demonstrated that creatine supplementation after TBI causes alterations in biological markers related to its neuropathology. Several TBI studies demonstrate that creatine supplementation can have neuroprotective effects in animal models (e.g., protect mitochondrial function, antioxidant effects, and increased ATP). Creatine supplementation also improves cognition in non-clinical populations, outcome in anxiety and depression-related mood disorders, and has a protective effect in situations that impair cell energy status (Dean, Arikan, Opitz, & Sterr, 2017). Fewer clinical studies have been carried out on the effects of creatine supplementation after TBI (Table 2). These studies investigated the effects of creatine supplementation for 6 months after TBI in children and adolescents. The administration of creatine reduced post-traumatic amnesia and improved recovery, cognitive function, headaches, dizziness and fatigue, as well as language articulation and understanding (Sakellaris et al., 2006; Sakellaris et al., 2008; Sakellaris et al., 2012). It also should be noted that these studies were conducted in children, after severe to moderate TBI, in open-label trials.

Dietary creatine supplement acted as neuroprotective in humans when cellular energy provision was compromised by hypoxia intervention. Moreover, creatine supplementation augments neural creatine, increases corticomotor excitability, and restricts the decline in attention that occurs during severe oxygen deficit (Turner, Byblow, & Gant, 2015). There is sufficient theoretical evidence that creatine supplementation limits damage from TBI, improves neurodegeneration, and reduces the severity of the injury (Freire Royes & Cassol, 2016; Kreider et al., 2017). There is a large gap in the literature on creatine supplementation and its usage in human TBI. Hence, further clinical studies are essential to investigate the potential use of creatine in TBI along with dosing regimen and safety recommendations using double-blind interventions.

TABLE 3 Neuropathology of traumatic brain injury and the functions of creatine.

Neuropathology of TBI	Functions of creatine	References
Membrane depolarization	Maintains/restores membrane	Andres, Ducray, Schlattner, Wallimann, and Widmer (2008)
Calcium influx	Decreases influx of calcium	Gualano, Roschel, Lancha-Jr, Brightbill, and Rawson (2012)
Impaired mitochondrial function	Stimulates respiration Stabilizes mitochondrial permeability transition pores and restricts pore opening	Gualano et al. (2012), Rae (2014)
Production of ROS and free radicals	Directly scavenges free radicals	Gualano et al. (2012)
Glutamate excitotoxicity	Aids neurotransmission and reuptake	Andres et al. (2008)
Inflammation	Putative anti-inflammatory role	Riesberg, Weed, Mcdonald, Eckerson, and Drescher (2016), Briones, Woods, and Rogozinska (2013)
Altered neurotransmission balance	Putative role as neurotransmitter for GABA and NMDA receptors	Rae and Broer (2015), Rae (2014)

ROS, reactive oxygen species; *GABA*, γ-aminobutyric acid; *NMDA*, *N*-methyl-D-aspartate.

Vitamins

Vitamin supplementation improved the outcomes in pre-clinical models of mild TBI and in the treatment of concussions (Maroon, Lepere, Blaylock, & Bost, 2012). Specifically, vitamin D, E, and B_3 have reduced the markers of inflammation, improved cognition and provided neuroprotection in different rodent models. While vitamin B_9 showed no treatment effect with folic acid administration in rodents following mild TBI. These different pre-clinical studies indicate that vitamin supplementation may have species-specific effects.

Vitamins C and E supplementation to humans who have sustained acute TBI showed decreased mortality and improved Glasgow Outcome Scores. Vitamin E decreased edema while vitamin C reduced lesion size (Razmkon et al., 2011) (Table 4). Aminmansour et al. (2012) reported that the recovery rate in patients with severe brain trauma in the group receiving progesterone and vitamin D together was significantly higher than that of progesterone group, which was in turn higher than that of placebo group (Table 4). Scrimgeour and Condlin (2014) consider that the combination of progesterone and vitamin D should is a novel and compelling approach to TBI treatment. Buell et al. (2010) and Witham, Dove, Sugden, Doney, and Struthers (2012) concluded that vitamin D insufficiency and deficiency is associated with all-cause dementia, Alzheimer's disease and stroke (with and without dementia symptoms). These findings suggest the potential of vitamin D as neuroprotective.

Zinc

Young et al. (1996) tested the effectiveness of using intravenous zinc supplementation in 68 patients following TBI in an attempt to improve neurological outcomes. Zinc supplementation resulted in better Glasgow Coma Scale (GCS) scores in patients with zinc supplemented than control patients. Lai et al. (2012) reported that there are no data on the use of supplemental zinc to reduce or check depressive symptoms in patients with TBI associated depression, but the use of zinc improved mood effectively among the patient population.

Magnesium

Magnesium is commonly depleted following TBI likely through interaction with transient receptor potential melastatin, which leads to neuronal cell death. Magnesium is a key nutrient that correlated with improved somatic scores when supplemented following mild TBI in humans (Wahls, Rubenstein, Hall, & Snetselaar, 2014). Sen and Gulati (2010) found that co-administration of magnesium with mannitol increases brain bioavailability in humans.

Other dietary supplements for TBI

TBI patients have shown decreased melatonin concentrations in traumatic brain injury (Seifman et al., 2014). Sublingual melatonin (3 or 10 mg) for 28 days decreased post-concussion syndrome symptoms compared to the placebo group (Barlow et al., 2014). Melatonin scavenges free radicals, limits cytokine release to induce a neuroprotective effect in Alzheimer's disease, Huntington's disease, and amyotrophic lateral sclerosis (Pandi-Perumal et al., 2013) (Table 5). Citicoline, a rich source of choline showed enhanced cerebral blood flow (Leon-Carrion et al., 2000) and had neuroprotective effect (Başkaya, Doğan, Rao, & Dempsey, 2000) in TBI patients. However, in a follow-up randomized, double-blinded, multicenter COBRIT trial, it showed no difference from placebo (Zafonte et al., 2009) (Table 5). Enzogenol is a flavonoid-rich extract from *Pinus radiata* bark, it reduced self-perceived cognitive failures observed in patients with post-mild TBI after

TABLE 4 Clinical outcomes of vitamins in traumatic brain injury.

Vitamins	Treatment	Outcomes	References
Vitamin C and vitamin E	07 days of vitamin C (500 mg to 10 g/day); 07 days of vitamin E (400 IU/day)	Diffused axonal injury, decreased mortality, decreased edema and lesion size in short TBI patients	Razmkon et al. (2011)
Vitamin D	05 days of vitamin D (5 µg/kg/day)	Neuroprotective and restorative benefits in TBI patients	Aminmansour et al. (2012), Buell et al. (2010)

TBI, traumatic brain injury.

TABLE 5 Other dietary supplements clinically studied for traumatic brain injury.

Dietary supplement	Treatment	Outcomes	References
Melatonin	28 days of melatonin (3–10 mg/day)	Decreased post-concussion syndrome symptoms	Barlow et al. (2014)
Citicoline	03 months of citicoline (1 g/day)	Enhanced cerebral blood flow Neuroprotective effect	Leon-Carrion, Dominguez-Roldan, Murillo-Cabezas, del Rosario Dominguez-Morales, and Munoz-Sanchez (2000)
Enzogenol	12 weeks of enzogenol (1000 mg/day)	Reduced self-perceived cognitive failures in mild TBI patients	Theadom et al. (2013)
Cerebrolysin	05 days of cerebrolysin (30 mL/day)	Improved the cognitive function	Chen, Wei, Tsaia, Chen, and Cho (2013)

TBI, traumatic brain injury.

3–4 months of injury (Theadom et al., 2013) (Table 5). Cerebrolysin, a mixture of peptides purified from pig brains improved the cognitive function of the mild TBI in patients at 3rd month after injury, especially for long-term memory and drawing function (Chen et al., 2013) (Table 5). *S*-Adenosylmethionine, a nutritional supplement, had a 77% improvement in clinical scores of post-concussive symptoms of patients (Bacci-Ballerini, López Anguera, Alcaraz, & Hernández Reyes, 1983). The supplementation of *Ginkgo biloba*, vinpocetine, acetyl-L-carnitine, huperzine, and alpha-lipoic acid (for 6 months) led to cognitive and cerebral blood flow improvements in National Football League (NFL) players (Amen, Wu, Taylor, & Willeumier, 2011).

Promising dietary supplements for TBI

Curcumin, resveratrol, green tea, caffeine, lipoic acid, acetyl-L-carnitine, rosmarinic acid, roots of *Scutellaria baicalensis*, blueberry supplementation and huperzine-A have exhibited improved cognition function, neuroprotective activity and oxidative stress attenuation in various in vivo rodent models and in vitro cell line studies. However, no clinical evidence is available for their benefits in TBI treatment.

Conclusion

It is too early to firmly establish the use of dietary supplements to limit or control the tissue damage due to traumatic injury or treat brain injury. The lack of translation of the promising outcomes of pre-clinical studies to clinical practices warrants further research with systematic clinical trials designs for dietary supplements in traumatic brain injury. Moreover, the complex outcomes of TBI, especially neurobehavioral, may take long-term treatment and follow-up with specific clinical milestone to recommend dietary supplements for the cognitive and behavioral recovery following TBI. On the brighter side, there appear to be no adverse effects to any of these dietary supplements, when used appropriately. The major challenges of the future are performing well-designed clinical studies in patients to confirm anticipated effects, to define the optimal doses, to compare single vs multiple dietary components, and to assess possible side effects.

Applications to other areas of neuroscience

We have discussed the role of dietary supplements on the recovery of various outcomes of TBI. Multiple facets of severe TBI ranging from musculoskeletal to neurobehavioral disturbances can be mitigated by nutrients provided within a specific duration or throughout the recovery period after injury. Omega-3 fatty acids specifically DHA are also extensively studied for stroke patients (Blondeau et al., 2015), moreover the implication of neuroinflammation in both TBI and stroke emphasizes the link between cellular events during both brain injuries. Neuroinflammation contributes to immune excitotoxicity and microglial hyperactivity induced by the repetitive mild head injury (Blaylock & Maroon, 2011; Davalos et al., 2005). The initiation of inflammatory reactivity triggers the release of pro-inflammatory cytokines (interleukin (IL)-6, IL-1β, and TNF-α) and generate free radicals which are inhibited by omega-3 fatty acids (Mori & Beilin, 2004).

The neurodegenerative disorders like Parkinson's disease (PD) and Alzheimer's disease (AD) do have beneficial effects with the administration of dietary supplements (vitamins and melatonin). Vitamin E decreases lipid peroxidation in Alzheimer patients (Arlt, Müller-Thomsen, Beisiegel, & Kontush, 2012), whereas Melatonin demonstrated a neuroprotective protective effect during AD. Melatonin is recognized to scavenge free radicals it exhibits anti-amyloidogenic properties (Rosales-Corral et al., 2012). Nicotinamide (vitamin B3) demonstrates antioxidant effects over various biological stresses and decreases dopaminergic neurotoxicity (Williams & Ramsden, 2005). NADH is a crucial cofactor that can hydroxylate tyrosine and produce dopamine. Similarly, patients of PD exhibit NADH deficiency (Kotani et al., 2006), treatment with nicotinamide improves locomotor activity while it also inhibited the activity of lactate dehydrogenase and nitric oxide synthase (Lu et al., 2014).

Despite the challenges, like the complexity of neurobehavioral alteration, the divergent phenotype of TBI, and lack of translational potential, the constant understanding of the clinical efficacy of dietary supplements provides a hope to develop an effective neurotherapeutic strategy to usher the recovery after TBI. The insight into the impact of various factors, especially metabolic demands, neuroinflammation and ionic flux in brain could become cornerstones of the approach of employing metabolic switching for the development of a safe and effective therapeutic intervention to mitigate the damage after TBI and other neurodegenerative disorders.

Mini-dictionary of terms

1. **Traumatic brain injury (TBI):** The brain damage caused by biomechanical forces which may result in impairment of motor, behavior and learning activity of an individual.
2. **Chronic traumatic encephalopathy (CTE):** It is a condition where long-term behavioral and cognitive disturbances could arise after a single or repetitive TBI. Microstructural damage in vital region of the brain is the hallmark of CTE.
3. **Dietary supplements:** Fortification regular nutritional requirements with the biological agents proven to have beneficial effects to improve performance or recovery from disease or injury. These agents have high nutritional value and can be incorporated in diet.
4. **Mitochondrial malfunction:** Mitochondria are crucial to the anatomical and physiological wellbeing of every cell in the living system. However, an imbalance in any of the myriad functions of mitochondria induces long-term cellular and functional discrepancies termed mitochondrial malfunction.
5. **Glutamate-triggered excitotoxicity:** Glutamate (excitatory neurotransmitter) release also coincides with this change in ionic balance brought by K^+ efflux and Na^+ influx. The lack of an efficient compensatory mechanism to limit the K^+ flux immediately after traumatic injury reflects the indiscriminate release of glutamate, activation of glutamate receptors, and precipitation of excitotoxicity leading to neuronal cell death.
6. **A randomized, double-blind, placebo-controlled, parallel design of clinical trial:** A clinical study where human subjects (with selection criteria) are divided randomly in two groups and assigned with the treatment (drug or placebo) keeping investigator and subject unaware of the knowledge, which group received drug and which placebo, of the treatment. Placebo is a formulation without active pharmaceutical ingredient but with similar appearance as that of the drug.
7. **Prophylactic:** Treatment offered to reduce the extent injury or alleviate disease.
8. **Ergogenic aids for athletes**: They are any training technique or nutritional supplement or psychological technique that helps to enhance athlete's performance, improve capacity to exercise and adapt to various training schedules.
9. **Glasgow Outcome Scores (GCS):** It is considered as a scale of scores predicting and assessing the impact of trauma on brain functions. The GCS examines an individual's post-injury verbal and motor responses and eye movement. The three scores give basic measure of an individual's functional status and classify scores of based on the severity of the injury. The score of 13–15 is considered as a mild TBI; scores of 9–12 refers to moderate injury while the scores of 8 or below indicates a severe damage to brain causing major disturbances in normal brain functioning after TBI.
10. **Translation of pre-clinical outcomes to clinic:** Transfer of the technical and therapeutic advances achieved by using experimental animal models (pre-clinical efficacy) to humans for the treatment of diseases.

Key facts of dietary supplementation for traumatic brain injury: Omega-3 fatty acids

- Omega-3 fatty acids are polyunsaturated fatty acids that regulate signal transduction, gene expression and protect neurons.
- It normalizes the levels of brain-derived neurotrophic factor (BDNF), synapsin I, and cAMP-responsive element-binding protein of the rats exposed to brain injury.

- Omega-3 supplementation in rats augment brain repair and improve long-term functional recovery after cerebral ischemia.
- DHA and eicosapentaenoic acid (EPA) are the most promising omega-3 fatty acids for neuro-restorative capacities in TBI.
- However, there are a few clinical evidences for its functional improvements in patients with TBI.

Key facts of dietary supplementation for traumatic brain injury: Sports induced TBI

- Sports and recreational injury in the United States during 2011–14 estimated that 8.6 million episodes of traumatic brain damage every year.
- Athletes who encounter TBI are those who participate in contact-collision sports like boxing, wrestling, American football, rugby, ice hockey, and basketball.
- Severe TBI precipitates into progressive musculoskeletal disorders experienced by athlete; neurodegeneration associated with it may have interconnection with Parkinson's disease, Alzheimer's disease (AD), and dementia.
- The neurochemical changes, due to mild TBI, get restored within 8–10 days.
- A subsequent trauma which initiates cascading alteration in energy homeostasis, ionic exchange, and protein accumulation may be responsible for long-term consequences of severe TBI.

Summary points

- Newer research have shown that consumption of certain diets or their associated physiologically active constituents may be linked to disease risk reduction.
- This chapter focuses on the use of dietary supplements like eicosapentaenoic acid, docosahexaenoic acid, creatine, vitamins, zinc, magnesium and their clinical interventions to alleviate brain damage due to TBI and accelerate recovery from any behavioral or muscular distress.
- Docosahexaenoic acid has been the most promising omega-3 fatty acid for treatment of TBI.
- There appear to be no adverse effects to any of these dietary supplements, when used appropriately in the treatment of TBI.
- The major challenges of the future are performing well-designed clinical studies in patients to confirm anticipated effects, to define the optimal doses, to compare single verses multiple dietary components, and to assess possible side effects.

References

Alves, C. R., Merege Filho, C. A., Benatti, F. B., Brucki, S., Pererira, R. M., Sa Pinto, A. L., et al. (2013). Creatine supplementation associated or not with strength training upon emotional and cognitive measures in older women: A randomized double-blind study. *PLoS One*, *8*(10), e76301.

Amen, D. G., Wu, J. C., Taylor, D., & Willeumier, K. (2011). Reversing brain damage in former NFL players: Implications for traumatic brain injury and substance abuse rehabilitation. *Journal of Psychoactive Drugs*, *43*(1), 1–5.

Aminmansour, B., Nikbakht, H., Ghorbani, A., Rezvani, M., Rahmani, P., Torkashvand, M., et al. (2012). Comparison of the administration of progesterone versus progesterone and vitamin D in improvement of outcomes in patients with traumatic brain injury: A randomized clinical trial with placebo group. *Advanced Biomedical Research*, *1*, 58.

Andres, R. H., Ducray, A. D., Schlattner, U., Wallimann, T., & Widmer, H. R. (2008). Functions and effects of creatine in the central nervous system. *Brain Research Bulletin*, *76*(4), 329–343.

Arlt, S., Müller-Thomsen, T., Beisiegel, U., & Kontush, A. (2012). Effect of one-year vitamin C- and E-supplementation on cerebrospinal fluid oxidation parameters and clinical course in Alzheimer's disease. *Neurochemical Research*, *37*(12), 2706–2714.

Ashbaugh, A., & McGrew, C. (2016). The role of nutritional supplements in sports concussion treatment. *Current Sports Medicine Reports*, *15*(1), 16–19.

Bacci-Ballerini, F., López Anguera, A., Alcaraz, P., & Hernández Reyes, N. (1983). Treatment of postconcussion syndrome with S-adenosylmethionine. *Medicina Clínica (Barcelona)*, *80*(4), 161–164.

Barlow, K. M., Brooks, B. L., MacMaster, F. P., Kirton, A., Seeger, T., Esser, M., et al. (2014). A double-blind, placebo-controlled intervention trial of 3 and 10 mg sublingual melatonin for post-concussion syndrome in youths (PLAYGAME): Study protocol for a randomized controlled trial. *Trials*, *15*, 271.

Başkaya, M. K., Doğan, A., Rao, A. M., & Dempsey, R. J. (2000). Neuroprotective effects of citicoline on brain edema and blood-brain barrier breakdown after traumatic brain injury. *Journal of Neurosurgery*, *92*(3), 448–452.

Benton, D., & Donohoe, R. (2011). The influence of creatine supplementation on the cognitive functioning of vegetarians and omnivores. *British Journal of Nutrition*, *105*(7), 1100–1105.

Blaylock, R. L., & Maroon, J. (2011). Immunoexcitotoxicity as a central mechanism in chronic traumatic encephalopathy-A unifying hypothesis. *Surgical Neurology International*, *2*, 107.

Blondeau, N., Lipsky, R. H., Bourourou, M., Duncan, M. W., Gorelick, P. B., & Marini, A. M. (2015). Alpha-linolenic acid: An omega-3 fatty acid with neuroprotective properties-ready for use in the stroke clinic? *BioMed Research International*, *2015*, 519830.

Briones, T. L., Woods, J., & Rogozinska, M. (2013). Decreased neuroinflammation and increased brain energy homeostasis following environmental enrichment after mild traumatic brain injury is associated with improvement in cognitive function. *Acta Neuropathologica Communications*, *1*(1), 57.

Buell, J. S., Dawson-Hughes, B., Scott, T. M., Weiner, D. E., Dallal, G. E., Qui, W. Q., et al. (2010). 25-Hydroxyvitamin D, dementia, and cerebrovascular pathology in elders receiving home services. *Neurology*, *74*, 18–26.

Chen, C. C., Wei, S. T., Tsaia, S. C., Chen, X. X., & Cho, D. Y. (2013). Cerebrolysin enhances cognitive recovery of mild traumatic brain injury patients: Double-blind, placebo-controlled, randomized study. *British Journal of Neurosurgery*, *27*(6), 803–807.

Chiang, Y. H., Chao, D. P., Chu, S. F., Lin, H. W., Huang, S. Y., Yeh, Y. S., et al. (2012). Early enteral nutrition and clinical outcomes of severe traumatic brain injury patients in acute stage: A multi-center cohort study. *Journal of Neurotrauma*, *29*(1), 75–80.

Cook, A. M., Peppard, A., & Magnuson, B. (2008). Nutrition considerations in traumatic brain injury. *Nutrition in Clinical Practice*, *23*(6), 608–620.

Davalos, D., Grutzendler, J., Yang, G., Kim, J. V., Zuo, Y., Jung, S., et al. (2005). ATP mediates rapid microglial response to local brain injury in vivo. *Nature Neuroscience*, *8*(6), 752–758.

Davidson, M. H., Stein, E. A., Bays, H. E., Maki, K. C., Doyle, R. T., Shalwitz, R. A., et al. (2007). Efficacy and tolerability of adding prescription omega-3 fatty acids 4 g/d to simvastatin 40 mg/d in hypertriglyceridemic patients: An 8-week, randomized, double-blind, placebo-controlled study. *Clinical Therapeutics*, *29*, 1354–1367.

Dean, P. J. A., Arikan, G., Opitz, B., & Sterr, A. (2017). Potential for use of creatine supplementation following mild traumatic brain injury. *Concussion*, *2*(2), CNC34.

Erdman, J., Maria, O., & Pillsbury, L. (2011). *Nutrition and traumatic brain injury: Improving acute and subacute health outcomes in military personnel*. Washington, DC: National Academies Press.

Freire Royes, L. F., & Cassol, G. (2016). The effects of creatine supplementation and physical exercise on traumatic brain injury. *Mini Reviews in Medicinal Chemistry*, *16*(1), 29–39.

Gardner, R. C., & Yaffe, K. (2015). Epidemiology of mild traumatic brain injury and neurodegenerative disease. *Molecular and Cellular Neurosciences*, *66*, 75–80.

Giza, C. C., & Hovda, D. A. (2001). The neurometabolic cascade of Concussion. *Journal of Athletic Training*, *36*(3), 228–235.

Goldman, S. M., Tanner, C. M., Oakes, D., Bhudhikanok, G. S., Gupta, A., & Langston, J. W. (2006). Head injury and Parkinson's disease risk in twins. *Annals of Neurology*, *60*(1), 65–72.

Gualano, B., Roschel, H., Lancha-Jr, A. H., Brightbill, C. E., & Rawson, E. S. (2012). In sickness and in health: The widespread application of creatine supplementation. *Amino Acids*, *43*(2), 519–529.

Hasadsri, L., Wang, B. H., Lee, J. V., Erdman, J. W., Llano, D. A., Barbey, A. K., et al. (2013). Omega-3 fatty acids as a putative treatment for traumatic brain injury. *Journal of Neurotrauma*, *30*(11), 897–906.

Havenetidis, K. (2016). The use of creatine supplements in the military. *Journal of the Royal Army Medical Corps*, *162*(4), 242–248.

Hering, H., Lin, C. C., & Sheng, M. (2003). Lipid rafts in the maintenance of synapses, dendritic spines, and surface AMPA receptor stability. *Journal of Neuroscience*, *23*, 3262–3271.

Innis, S. M. (2008). Dietary omega 3 fatty acids and the developing brain. *Brain Research*, *1237*, 35–43.

Kelley, D. S., Taylor, P. C., Nelson, G. J., & Mackey, B. E. (1998). Dietary docosahexaenoic acid and immunocompetence in young healthy men. *Lipids*, *33*, 559–566.

Kelley, D. S., Taylor, P. C., Nelson, G. J., Schmidt, P. C., Ferretti, A., Erickson, K. L., et al. (1999). Docosahexaenoic acid ingestion inhibits natural killer cell activity and production of inflammatory mediators in young healthy men. *Lipids*, *34*, 317–324.

Kew, S., Mesa, M. D., Tricon, S., Buckley, R., Minihane, A. M., & Yaqoob, P. (2004). Effects of oils rich in eicosapentaenoic and docosahexaenoic acids on immune cell composition and function in healthy humans. *American Journal of Clinical Nutrition*, *79*, 674–681.

Knapik, J. J., Steelman, R. A., Hoedebecke, S. S., Austin, K. G., Farina, E. K., & Lieberman, H. R. (2016). Prevalence of dietary supplement use by athletes: Systematic review and meta-analysis. *Sports Medicine*, *46*(1), 103–123.

Kotani, S., Sakaguchi, E., Warashina, S., Matsukawa, N., Ishikura, Y., Kiso, Y., et al. (2006). Dietary supplementation of arachidonic and docosahexaenoic acids improves cognitive dysfunction. *Neuroscience Research*, *56*(2), 159–164.

Kreider, R. B., Kalman, D. S., Antonio, J., Ziegenfuss, T. N., Wildman, R., Collins, R., et al. (2017). International Society of Sports Nutrition position stand: Safety and efficacy of creatine supplementation in exercise, sport, and medicine. *Journal of the International Society of Sports Nutrition*, *14*, 18.

Lai, J., Moxey, A., Nowak, G., Vashum, K., Bailey, K., & McEvoy, M. (2012). The efficacy of zinc supplementation in depression: Systematic review of randomised controlled trials. *Journal of Affective Disorders*, *136*, e31–e39.

Leon-Carrion, J., Dominguez-Roldan, J. M., Murillo-Cabezas, F., del Rosario Dominguez-Morales, M., & Munoz-Sanchez, M. A. (2000). The role of citicholine in neuropsychological training after traumatic brain injury. *Neuropsychological Rehabilitation*, *4*(1), 33–40.

Lu, L., Tang, L., Wei, W., Hong, Y., Chen, H., Ying, W., et al. (2014). Nicotinamide mononucleotide improves energy activity and survival rate in an in vitro model of Parkinson's disease. *Experimental and Therapeutic Medicine*, *8*(3), 943–950.

Maroon, J. C., Lepere, D. B., Blaylock, R. L., & Bost, J. W. (2012). Postconcussion syndrome: A review of pathophysiology and potential nonpharmacological approaches to treatment. *Physician and Sportsmedicine*, *40*(4), 73–87.

Mori, T. A., & Beilin, L. J. (2004). Omega-3 fatty acids and inflammation. *Current Atherosclerosis Reports*, *6*(6), 461–467.

Mori, T. A., Woodman, R. J., Burke, V., Puddey, I. B., Croft, K. D., & Beilin, L. J. (2003). Effect of eicosapentaenoic acid and docosahexaenoic acid on oxidative stress and inflammatory markers in treated-hypertensive type 2 diabetic subjects. *Free Radical Biology & Medicine*, *35*, 772–781.

Oliver, J. M., Anzalone, A. J., & Turner, S. M. (2018). Protection before impact: The potential neuroprotective role of nutritional supplementation in sports-related head trauma. *Sports Medicine*, *48*(Suppl. 1), 39–52.

Oliver, J. M., Jones, M. T., Kirk, K. M., Gable, D. A., Repshas, J. T., Johnson, T. A., et al. (2016). Effect of docosahexaenoic acid on a biomarker of head trauma in American football. *Medicine and Science in Sports and Exercise*, *48*(6), 974–982.

Pandi-Perumal, S. R., BaHammam, A. S., Brown, G. M., Spence, D. W., Bharti, V. K., Kaur, C., et al. (2013). Melatonin antioxidative defense: Therapeutical implications for aging and neurodegenerative processes. *Neurotoxicity Research*, *23*(3), 267–300.

Petraglia, A. L., Winkler, E. A., & Bailes, J. E. (2011). Stuck at the bench: Potential natural neuroprotective compounds for concussion. *Surgical Neurology International*, *2*, 146.

Rae, C. D. (2014). A guide to the metabolic pathways and function of metabolites observed in human brain 1H magnetic resonance spectra. *Neurochemical Research*, *39*(1), 1–36.

Rae, C. D., & Broer, S. (2015). Creatine as a booster for human brain function. How might it work? *Neurochemistry International*, *89*, 249–259.

Razmkon, A., Sadidi, A., Sherafat-Kazemzadeh, E., Mehrafshan, A., Jamali, M., Malekpour, B., et al. (2011). Administration of vitamin C and vitamin E in severe head injury: A randomized double-blind controlled trial. *Clinical Neurosurgery*, *58*, 133–137.

Riesberg, L. A., Weed, S. A., Mcdonald, T. L., Eckerson, J. M., & Drescher, K. M. (2016). Beyond muscles: The untapped potential of creatine. *International Immunopharmacology*, *37*, 31–42.

Rosales-Corral, S. A., Lopez-Armas, G., Cruz-Ramos, J., Melnikov, V. G., Tan, D. X., Manchester, L. C., et al. (2012). Alterations in lipid levels of mitochondrial membranes induced by amyloid-β: A protective role of melatonin. *International Journal of Alzheimer's Disease*, *2012*, 459806.

Sakellaris, G., Kotsiou, M., Tamiolaki, M., Kalostos, G., Tsapaki, E., Spanaki, M., et al. (2006). Prevention of complications related to traumatic brain injury in children and adolescents with creatine administration: An open label randomized pilot study. *Journal of Trauma*, *61*(2), 322–329.

Sakellaris, G., Nasis, G., Kotsiou, M., Tamiolaki, M., Charissis, G., & Evangeliou, A. (2008). Prevention of traumatic headache, dizziness and fatigue with creatine administration. A pilot study. *Acta Paediatrica*, *97*(1), 31–34.

Sakellaris, G. S., Partalis, N. I., Nasis, G. D., Kotslou, M. E., Tamolaki, M. D., Bouloukak, E. H., et al. (2012). Outcome of traumatic dysarthria and lingual problems of understanding with creatine administration. An open label randomized pilot study. *Journal of Trauma & Treatment*, *1*(3), 1–4.

Scrimgeour, A. G., & Condlin, M. L. (2014). Nutritional treatment for traumatic brain injury. *Journal of Neurotrauma*, *31*(11), 989–999.

Seifman, M. A., Gomes, K., Nguyen, P. N., Bailey, M., Rosenfeld, J. V., Cooper, D. J., et al. (2014). Measurement of serum melatonin in intensive care unit patients: Changes in traumatic brain injury, trauma, and medical conditions. *Frontiers in Neurology*, *5*, 237.

Sen, A. P., & Gulati, A. (2010). Use of magnesium in traumatic brain injury. *Neurotherapeutics*, *7*(1), 91–99.

Singh, M. (2005). Essential fatty acids, DHA and human brain. *Indian Journal of Pediatrics*, *72*, 239–242.

Sydenham, E., Dangour, A. D., & Lim, W. S. (2012). Omega 3 fatty acid for the prevention of cognitive decline and dementia. *Cochrane Database of Systematic Reviews*, *6*, CD005379.

Theadom, A., Mahon, S., Barker-Collo, S., McPherson, K., Rush, E., Vandal, A. C., et al. (2013). Enzogenol for cognitive functioning in traumatic brain injury: A pilot placebo-controlled RCT. *European Journal of Neurology*, *20*(8), 1135–1144.

Trojian, T. H., Wang, D. H., & Leddy, J. J. (2017). Nutritional supplements for the treatment and prevention of sports-related concussion-evidence still lacking. *Current Sports Medicine Reports*, *16*(4), 247–255.

Tsui-Pierchala, B. A., Encinas, M., Milbrandt, J., & Johnson, E. M. (2002). Lipid rafts in neuronal signaling and function. *Trends in Neurosciences*, *25*, 412–417.

Turner, C. E., Byblow, W. D., & Gant, N. (2015). Creatine supplementation enhances corticomotor excitability and cognitive performance during oxygen deprivation. *Journal of Neuroscience*, *35*(4), 1773–1780.

USFDA. (1997). *Substances affirmed as generally recognized as safe: Menhaden oil: Final Rule: Federal Registry* (pp. 30751–30757).

Valentine, R. C., & Valentine, D. L. (2004). Omega-3 fatty acids in cellular membranes: A unified concept. *Progress in Lipid Research*, *43*, 383–402.

Wahls, T., Rubenstein, L., Hall, M., & Snetselaar, L. (2014). Assessment of dietary adequacy for important brain micronutrients in patients presenting to a traumatic brain injury clinic for evaluation. *Nutritional Neuroscience*, *17*(6), 252–259.

Williams, A., & Ramsden, D. (2005). Nicotinamide: A double edged sword. *Parkinsonism & Related Disorders*, *11*(7), 413–420.

Witham, M. D., Dove, F. J., Sugden, J. A., Doney, A. S., & Struthers, A. D. (2012). The effect of vitamin D replacement on markers of vascular health in stroke patients – A randomised controlled trial. *Nutrition, Metabolism, and Cardiovascular Diseases*, *22*, 864–870.

Wohl, D. A., Tien, H. C., Busby, M., Cunningham, C., Macintosh, B., Mapravnil, S., et al. (2005). Randomized study of the safety and efficacy of fish oil (omega-3 fatty acid) supplementation with dietary and exercise counseling for the treatment of antiretroviral therapy-associated hypertriglyceridemia. *Clinical Infectious Diseases*, *41*, 1498–1504.

Wu, W. H., Lu, S. C., Wang, T. F., Jou, H. J., & Wang, T. A. (2006). Effects of docosahexaenoic acid supplementation on blood lipids, estrogen metabolism, and in vivo oxidative stress in postmenopausal vegetarian women. *European Journal of Clinical Nutrition*, *60*, 386–392.

Young, B., Ott, L., Kasarskis, E., Rapp, R., Moles, K., Dempsey, R. J., et al. (1996). Zinc supplementation is associated with improved neurologic recovery rate and visceral protein levels of patients with severe closed head injury. *Journal of Neurotrauma*, *13*, 25–34.

Zafonte, R., Friedewald, W. T., Lee, S. M., Levin, B., Diaz-Arrastia, R., Ansel, B., et al. (2009). The citicoline brain injury treatment (COBRIT) trial: Design and methods. *Journal of Neurotrauma*, *26*(12), 2207–2216.

Section E

Rehabilitation in traumatic brain injury

Chapter 39

Virtual reality and cognitive rehabilitation after traumatic brain injury

Maria Grazia Maggio[a] and Rocco Salvatore Calabrò[b]
[a]*Cognitive Rehabilitation and Relational Psychotherapy, Messina, Sicily, Italy,* [b]*IRCCS Centro Neurolesi "Bonino-Pulejo", Messina, Italy*

List of abbreviations

CAREN	Computer-Assisted Rehabilitation Environment
CR	cognitive rehabilitation
Pc-based software	computer-assisted rehabilitation
TBI	traumatic brain injury
VR	virtual reality

Introduction

Traumatic brain injury (TBI) is a clinical condition characterized by brain damage due to an external, rapid, and violent force. The injury can lead to alterations in the state of consciousness, with possible deficits in cognitive, neuro-sensory, and emotional-behavioral functions (Podell, Gifford, Bougakov, & Goldberg, 2010; Saatian, Ahmadpoor, Mohammadi, & Mazloumi, 2018). In particular, cognitive dysfunction is the leading cause of post-TBI disability, which interferes with work, relationships, leisure, and activities of daily living, causing high personal and economic consequences that are difficult to quantify. Various studies report that approximately 65% of patients with moderate to severe TBI report long-term cognitive problems (Schretlen & Shapiro, 2003; Whiteneck, Gerhart, & Cusick, 2004). Indeed, the percentage of people with TBI who suffer from a form of cognitive impairment, particularly concerning memory, varies from 20% to 79%, depending on the applied criteria, including severity of the TBI, time of measurement, and tools used (De Luca, Calabrò, & Bramanti, 2018). A review that considered 45 studies on TBI found that about 50% of patients with TBI have cognitive dysfunction (McInnes, Friesen, Mackenzie, Westwood, & Boe, 2017). Cognitive impairment is due both to a direct effect of the brain injury and indirect effect of secondary complications (such as hypoxia and intracranial hypertension). TBI can be related to an unfavorable outcome, even when the dysfunction is apparently mild (Niogi, Mukherjee, Ghajar, et al., 2008). According to various studies, TBI can trigger a gradual worsening in cognitive decline over time, which can cause a long-term cognitive impairment (Prince & Bruhns, 2017; Senathi-Raja, Ponsford, & Schönberger, 2010). As the brain areas most involved are the frontal and temporal lobes, especially in the basal areas and in the subcortical white matter, attention, memory, behavior, affectivity, planning, and organization are the most affected domains, with a significant impact on the quality of life of the patient and his family (Latella, Maggio, De Luca, et al., 2018; Maggio, De Luca, Torrisi, et al., 2018). Executive dysfunction is the most common and disabling deficit following TBI. This can include deficits in reasoning, planning, concept formation, mental flexibility, awareness, and intentional behavior (Lengenfelder, Arjunan, Chiaravalloti, Smith, & DeLuca, 2015; Stuss, 2011).

TBI rarely causes a focal damage: patients are usually subject to widespread injuries, for which an accurate diagnosis and "integrated and holistic" treatment are needed and especially in young people (TBI occurs mainly among 15–30 years) (De Luca, Calabrò, et al., 2018).

Cognitive rehabilitation (CR) programs are essential for the clinical recovery of TBI patients, as they improve functional outcomes and quality of life. These strategies are based on a reparative and compensatory approach, which adapt to the patient's resources and disability, enhancing the process of brain repair and plasticity after injuries (De Luca, Calabrò, et al., 2018).

Diagnosis and Treatment of Traumatic Brain Injury. https://doi.org/10.1016/B978-0-12-823347-4.00014-2

Several researches have highlighted the possible effectiveness of innovative techniques, such as virtual reality (VR), during rehabilitation after TBI (Aida, Chau, & Dunn, 2018; De Luca, Maggio, Maresca, et al., 2019; De Luca, Portaro, Le Cause, et al., 2020; Maggio, De Luca, Molonia, et al., 2019; Maggio, Torrisi, Buda, et al., 2020; Zanier, Zoerle, Di Lernia, & Riva, 2018).

VR is an interactive and virtual environment, created using computerized technologies. VR allows global well-being, thanks to a multi-stimulation that is implemented through auditory and visual feedback, increasing the patient's awareness of their performance. VR has been used in the rehabilitation of various pathologies, with good results in the recovery of neurological patients (Maggio, De Cola, Latella, et al., 2018; Maggio, De Luca, & Manuli, 2020; Maggio, Latella, Maresca, et al., 2019; Maggio, Maresca, De Luca, et al., 2019; Maggio, Maresca, Scarcella, et al., 2019; Maggio, Naro, La Rosa, et al., 2020; Manuli, Maggio, Latella, et al., 2020).

This work aims to evaluate the role of VR in potentiating CR of patients with TBI.

CR strategies

CR is a field of neuroscience that applies to patients with cognitive-behavioral disorders resulting from brain injury or neurodegenerative disease, as per a specific neuropsychological assessment and diagnosis. To be treated, the disorder must cause cognitive impairment and have a significant impact on the patient's life. The goal is to promote functional adaptation, through environmental stimulation adapted to the patient's needs. CR is based on the principle of neuroplasticity, i.e. the ability of our brain to change thanks to interactions with the external environment (Wilson, 2013). In planning the intervention, it is important to understand the individual differences that can influence the process, such as the extent of the injury, age, level of awareness, level of education, family support, and motivation. Cognitive (i.e., memory, attention, agnosia, apraxia), emotional/motivational (apathy, anxiety) and executive functions (planning, self-management, control, disinhibition, rigidity) may benefit from CR.

The rehabilitation protocol consists of several phases. Initially, the definition and analysis of the neuropsychological disorder are carried out, through medical/psychological history, interviews with patients and family members, assessment of the disorder, and clinical observation. Subsequently, the rehabilitation program is implemented, promoting autonomy, and specific-ecological objectives. CR interventions are applicable in all phases of post-injury recovery and in different contexts. In addition, they can be administered in different ways, both individually and collectively, by healthcare professionals skilled in different disciplines (Podell et al., 2010). In current clinical practice, there are two basic types of CR: restorative and compensatory rehabilitation (De Luca, Calabrò, et al., 2018; Morganti, 2004). The first uses systematic cognitive processes to teach people performing functional behaviors in activities of daily living. Therefore, it allows the development of the lost function through specialized cognitive exercises. On the other hand, compensatory methods help the patient to improve the use of aids and tools to overcome the impairment. These two methodologies can be used in combination or alone (Salazar, Warden, Schwab, et al., 2000). CR techniques can be classified into two main categories: conventional (paper/pencil exercises) and innovative methods (i.e., pc-based methods, virtual reality, robotics, telemedicine), both of all based on the use of cognitive strategies to retrain or compensate for the patient's cognitive-behavioral deficits. The innovative methods use multimedia and computer resources with hardware and software systems to reactivate the compromised neuropsychological functions (Manuli, Maggio, Tripoli, et al., 2020).

The use of computers in the field of neuroscience has been widespread since a century ago (Riva, 1997). Currently, there is a rapid spread of ICT (information and communication technology) applications designed to help the neurological population, their families, and their caregivers (Aulisio, Han, & Glueck, 2020; Riva, 2005), and several exploratory research on CR for the training of memory, attention, language, executive functions, and other cognitive domains have been carried out with positive results. Moreover, various authors have underlined the potential usefulness of the non-invasive, non-pharmacological CR intervention in all phases of TBI, especially performed using VR (Aida et al., 2018; Aulisio et al., 2020; De Luca et al., 2019; De Luca, Portaro, et al., 2020; Maggio, De Luca, et al., 2019; Maggio, Torrisi, et al., 2020; Manivannan, Al-Amri, Postans, et al., 2019; Zanier et al., 2018).

Virtual reality

VR is a relatively new field of ICT recently applied to neuropsychology. VR is the multi-sensory and interactive simulation of ecological scenarios, which reproduce real objects and events with which it is possible to interact (Maggio, De Cola, et al., 2018). These devices use specific software through input-output peripherals that allows complex and engaging experiences (Chuah, 2018; Howard, 2017). This tool is capable of promoting not only the improvement of the body functions of people with motor and cognitive disabilities, but also their degree of well-being and participation (Calabrò et al., 2017;

De Luca, Russo, et al., 2018). In fact, the experience with VR allows the patient to be the real center of the training also thanks to two key concepts, namely immersion and presence (Bohil, Alicea, & Biocca, 2011; Riva, 2005). Immersion is the objective perception of a sense of sensory absorption in the three-dimensional environment generated by the computer. The presence, on the other hand, consists of a subjective psychological state for which the user is consciously involved in the virtual context. Based on the degree and type of involvement and presence that VR arouses in the user, it is possible to distinguish three types: immersive VR, non-immersive VR, and semi-immersive VR (Man, 2010).

Immersive VR, such as "the Computer-Assisted Rehabilitation Environment (CAREN—Motek Medical B.V.)" (see Fig. 1), sensorially absorbs the user internally, isolating him/her from the external environment, giving the full sensation and illusion of being inside the virtual world. It generally uses some sound and visual diffusion devices, such as a helmet, capable of isolating from surrounding noises and sounds, allowing the user to concentrate fully on the stimuli presented in the 3D screen.

Semi-immersive VR allows a three-dimensional vision to the patient, who has the impression of seeing the world through a window or a wall. An example is the BTs-Nirvana (BTS Bioengineering, Milan) (see Fig. 2).

Finally, non-immersive VR produces in the user the role of external observer of a scenario proposed on a monitor, such as PC-based software (see Fig. 3).

These perceptual and experiential aspects activated by VR are relevant above all in the rehabilitation field.

VR promotes motor awareness and emotional involvement, which can foster functional execution and control of complex behavioral sequences (Aida et al., 2018; Aulisio et al., 2020; De Luca et al., 2019; De Luca, Portaro, et al., 2020; Maggio, De Luca, et al., 2019; Maggio, Torrisi, et al., 2020; Manivannan et al., 2019; Zanier et al., 2018). This activates the patient's motor and cognitive functions, promoting functional recovery (Maggio, Torrisi, et al., 2020). In fact, VR systems by means of increased sensory feedback can induce changes in the processes of neural plasticity (Calabrò et al., 2017). These effects may be due both to the reactivation/amplification of brain neurotransmission and to the involvement of mirror neurons, thanks to the visual-motor information coming from the observation of the stimuli presented with the VR. This allows for the integration of perception, cognition, and action, with recall of memorized motor plans and reinforcement learning, allowing for greater efficiency in the performance of the subject (Calabrò et al., 2017; Maggio, Torrisi, et al., 2020). Furthermore, the increased feedback obtained through exercises carried out in a virtual environment, develop the "awareness of the results" of the movements performed as well as the "awareness of the quality" of the movements themselves, with positive repercussions at cognitive and motor level (Aida et al., 2018; Aulisio et al., 2020; De Luca et al., 2019;

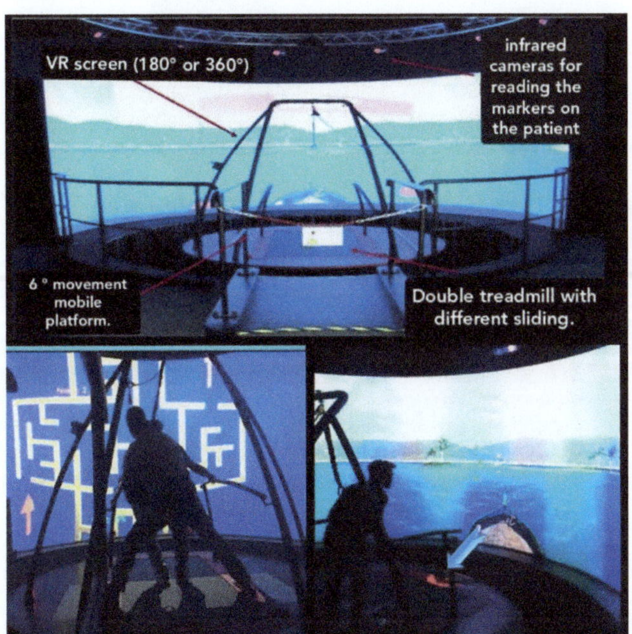

FIG. 1 The Computer-Assisted Rehabilitation Environment (CAREN). This tool is a simulator equipped with a VR screen (up to 360 degrees) and it is composed of a series of software and peripherals, including sensorized treadmill (with foot pressure sensors), high-speed infrared cameras that map the position of the reflective markers positioned on the patient and a mobile platform. Thus, it is an immersive and fully responsive VR simulator useful for enhancing motor and cognitive performance in neurological and orthopedic patients.

FIG. 2 BTs Nirvana. The BTS-Nirvana system consists of computer software, two markerless infrared sensors, a video camera, and a projector connected to a large screen. This device allows the patient to perform motor and cognitive exercises, interacting with virtual scenarios and audiovisual stimuli through movement, allowing sensory involvement with positive effects for the rehabilitation outcomes.

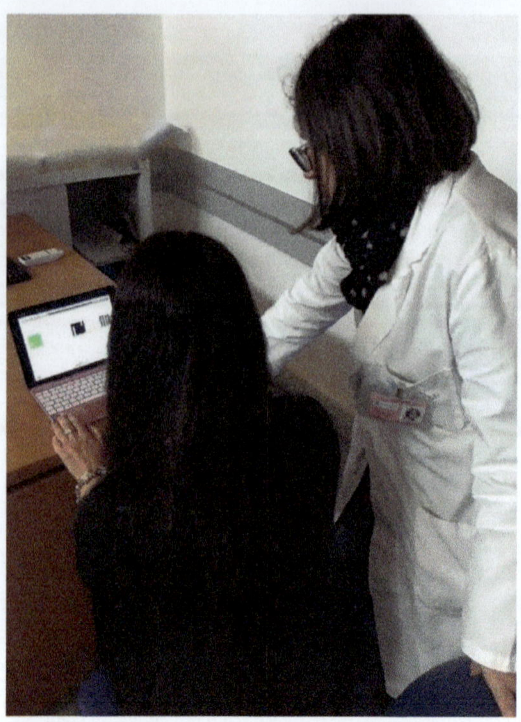

FIG. 3 Computer-assisted rehabilitation (Pc-Based Software). Computer-assisted rehabilitation provides audio-visual feedback that motivates people to engage in therapy and increases awareness of the results achieved. It is a non-immersive VR that uses the computer, with a touch screen. It consists of a series of specific exercises aimed at improving the cognitive domains, in a flexible and adaptable way to the patient's abilities.

De Luca, Portaro, et al., 2020; Maggio, De Luca, et al., 2019; Maggio, Torrisi, et al., 2020; Manivannan et al., 2019; Zanier et al., 2018). VR has been used in the rehabilitation of various pathologies, including TBI, to activate and train memory functions and other cognitive abilities, with an improvement also at motor level, including gait and in upper limb function.

VR advantages and application in CR

Based on the characteristics and potential of VR, the use of VR technology in CR could offer valuable help to qualitatively and quantitatively improve cognitive domains, behavioral skills and QoL of neurological patients (Aida et al., 2018; Aulisio et al., 2020; De Luca et al., 2019; De Luca, Portaro, et al., 2020; Maggio, De Luca, et al., 2019; Maggio, Torrisi, et al., 2020; Manivannan et al., 2019; Zanier et al., 2018). Implementing VR-based CR tools may lead to significant benefits, creating new perspectives in rehabilitation. First of all, VR allows the active involvement of clinicians, users, and caregivers. Indeed, they all are part of the VR system design process, which has a high degree of customization to make it suitable for users (i.e., patients) and their needs. In addition, the VR system is flexible with advanced features as well as adaptable and easy-to-use multi-sensory interfaces. The system also allows the simulation of "real" environments in which it is possible to interact in a protected way, favoring the activation of the subject at several levels: cognitive, motor, emotional, and social. This interaction is related to an increased feedback, which allows the subjects to be more aware of their performance, with greater motor, cognitive and emotional control (De Luca et al., 2019; De Luca, Portaro, et al., 2020; Maggio, De Luca, et al., 2019; Maggio, Torrisi, et al., 2020; Manivannan et al., 2019; Zanier et al., 2018). Finally, the system allows data collection during the CR training activities, evaluating the effectiveness of the program in real time.

VR technologies applied to the CR field have three stages of implementation (García-Betances, Jiménez-Mixco, Arredondo, & Cabrera-Umpiérrez, 2015). The first phase involves the collection of data relating to users, such as patients, clinical specialists and caregivers. This phase also aims at the analysis of the ongoing literature, in order to identify and define the type of training most suitable for the user, both for their cognitive profiles and their needs, promoting the acceptance of the VR systems. In the second phase, the development of the VR system takes place. This phase is based on the selection of the CR methods and activities to be used, as well as on the planning of the type of rehabilitation, the system, and the virtual scenarios. In addition, the evaluation of both patient performance and system efficiency during CR training activities is envisaged.

In the last phase, the system is evaluated and its effectiveness analyzed. The data collected by the system on the patient outcomes is used and analyzed, together with the assessments of medical and technical specialists. Finally, the therapists complete a critical and multi-disciplinary analysis of the effectiveness of the VR-based CR training system and its procedures treatments.

VR in TBI CR: Current knowledge

VR technology is advantageous and useful for the treatment of patients with TBI (Aida et al., 2018; Aulisio et al., 2020; De Luca et al., 2019; De Luca, Portaro, et al., 2020; Maggio, De Luca, et al., 2019; Maggio, Torrisi, et al., 2020; Manivannan et al., 2019; Zanier et al., 2018). Indeed, TBI causes very heterogeneous symptoms and clinical manifestations (especially as a result of frontal lobe dysfunction), with a significant negative impact on daily life.

Current VR therapeutic applications for TBI focus primarily on physical rehabilitation to stimulate balance skills, muscle strength and real-life activity in low-risk environments (Calabrò et al., 2017; Pietrzak, Pullman, & McGuire, 2014; Ustinova, Perkins, Leonard, & Hausbeck, 2014). Few studies have currently been carried out on CR in TBI, focusing on executive functions and attention skills.

VR programs in CR of executive functions and attention abilities have proven effective. Grealy, Johnson, and Rushton (1999) found that VR rehabilitation increased the performance of adult patients in visual and verbal learning tasks, as compared to patients treated with traditional rehabilitation programs. These results were confirmed by Jacoby et al. (2013), who observed better results in executive function tests in adults with TBI treated with VR than in patients not performing VR. Caglio, Latini-Corazzini, D'Agata, et al. (2012) found that adult patients using VR for navigation tasks increased their capacity for working memory, planning and cognitive/behavioral control. Dahdah et al. (2017) carried out a study on 15 TBI patients using a VR version of the Stroop. The authors showed that immersive VR can improve executive functions and speed of information processing in the sub-acute phase of TBI. Ettenhofer, Guise, Brandler, et al. (2019) created a virtual environment for safe driving, confirming in their pilot study on 11 patients that the virtual environment can foster cognitive skills, in particular working memory and visual search/selective attention, as well as autonomy in daily life including driving skills. The same authors conducted a similar study examining fMRI data after a 4-week VR driving intervention on subjects with TBI. The authors found improvements in working memory and processing speed in those who received such intervention; moreover, CR was associated with greater efficiency of brain networks (Gimbel, Ettenhofer, Cordero, Roy, & Chan, 2020). Moreover, Maggio, Torrisi, et al. (2020) in a study of 56 patients with TBI observed that the use of the Lokomat robotic device together with VR can be useful for improving global cognitive, executive and attention functions, as compared to subjects who trained with VR-Lokomat. De Luca et al. (2019), in a randomized clinical trial with 100 TBI patients, observed that VR rehabilitation can increase cognitive flexibility, shifting ability and selective attention compared to a traditional rehabilitation. Another study conducted by the same authors (De Luca et al., 2020) found this tool may be useful and suitable also for rehabilitation at a young age. The authors used immersive VR, observing an improvement in specific cognitive and motor domains, such as attention span, visual-executive processes, emotional awareness and balance. Regarding specific programs on attentional components, Larson, Ramaiya, Zollman, et al. (2011) used virtual training attention exercises on 15 TBI patients, noting that VR was engaging and well tolerated by the patients, with significant improvements in cognitive function. Dvorkin, Ramaiya, Larson, et al. (2013) used interactive visuomotor environments to support attention on visuomotor tasks on 21 patients with TBI in the early stages of recovery. The tool was well tolerated and significantly enhanced patient attention. However, VR also appears to be effective in enhancing other cognitive and behavioral skills. In one study, Knight, Titov, and Crawford (2006) created a virtual road, which included a low-distraction zone and a high-distraction zone, with both visual and auditory stimuli. Participants had to remember target stimuli present in this path. The VR tool has been shown to be useful in stimulating memory skills and resistance to interfering stimuli. Other authors have demonstrated the effectiveness of VR beyond cognitive function by emphasizing the possibility of enhancing mood, coping strategies, and social cognition (De Luca, Portaro, et al., 2020; Maggio, Latella, et al., 2019; Maggio, Maresca, et al., 2019; Maggio, Torrisi, et al., 2020; Manivannan et al., 2019; Westerhof-Evers, Visser-Keizer, Fasotti, & Spikman, 2019). Furthermore, it has been shown that the tool is generally well tolerated by subjects with TBI who have good usability of VR (De Luca, Maggio, et al., 2020).

Conclusions

The current studies demonstrated that VR has the potential to provide an effective rehabilitation tool for the treatment of cognitive impairments in TBI patients. However, these promising results should be further confirmed by studies with larger samples and longer follow-up as methodological quality of evidence is still poor. Finally, VR may be coupled to telerehabilitation that may be useful to bridge the gap between the patients living in rural areas and the main hospitals and to avoid, especially during pandemics, like COVID-19 is, contagion (Calabrò, Bramanti, Garzon, et al., 2018; Maggio, De Luca, Manuli, & Calabrò, 2020).

Applications to other areas of neuroscience

In the current context, the demand to improve the quality and effectiveness of the healthcare and social support services has increased, in order to respond to the multiple needs of the population with neurological diseases and their families. For this reason, the implementation of innovative technologies could favor the management and rehabilitation of various disabilities in this population. To this end, VR can represent a promising methodological approach for cognitive, behavioral, and motor neurorehabilitation. In particular, VR can be used to enhance the effects of conventional therapies, promoting longer rehabilitation sessions, and a reduction in overall hospitalization times (Calabrò et al., 2017; Maggio, De Luca, Torrisi, et al., 2018). The literature has shown that interactive VR rehabilitation is an effective treatment for increasing neuroplasticity, promoting the recovery of different neurological patients. VR has been used in the rehabilitation pathways of various pathologies at different ages, and it is currently being studied by several authors, focusing on cognitive processes, including attention and executive functions (Maggio, De Cola, et al., 2018; Maggio, Torrisi, et al., 2020; Optale, Urgesi, Busato, et al., 2010), memory (Kourtesis, Collina, Doumas, & MacPherson, 2020), language (Maresca, Maggio, Latella, et al., 2019), spatial orientation (Kober, Wood, Hofer, et al., 2013); motor rehabilitation (Calabrò et al., 2017; Calabrò, Naro, Cimino, et al., 2020), and for instrumental activities of daily life, including the use of wheelchairs (Corregidor-Sánchez et al., 2020). In recent years, some VR systems and applications have been developed for patients with MCI and dementia, stroke, musculoskeletal recovery, multiple sclerosis, and others disorders (Calabrò et al., 2017; Calabrò et al., 2020; De Luca et al., 2019; Maggio, Latella, et al., 2019; Maggio, Maresca, De Luca, et al., 2019; Maggio, Maresca, Scarcella, et al., 2019). Immersive VR environments have also been used for neuropsychological assessment and CR therapies such as phobias, stress, anxiety, and memory problems (Kim, Eom, Kwon, Kyeong, & Kim, 2020; Marloth, Chandler, & Vogeley, 2020). In fact, as previously discussed, VR allows concrete, ecological, and realistic experiences, which allow greater control of the individual over different sensor-motor, cognitive and social domains, which are difficult to reproduce in the clinical setting. Furthermore, the experience implemented in VR environments is highly positive and motivating for the patient, with positive repercussions on his compliance. This allows for the reproducibility of experiences and the generalizability of skills learned in real life.

Mini-dictionary of terms

Attention: It is a cognitive process that allows the individual to select information from the environment based on their relevance for purposes and for their perceptual salience. In addition, it allows the regulation of mental processes, establishing the necessary cognitive resources, performing tasks, and maintaining concentration.

Cognitive rehabilitation: CR is a field of neuroscience applied to patients with cognitive-behavioral disorders, resulting from brain injury or neurodegenerative diseases. The goal is to promote functional adaptation, through environmental stimulation adapted to the patient's needs.

Executive function: These are cognitive processes that consist of a complex system of **functional** modules of the mind. They regulate the planning, control, and coordination processes of the cognitive system and govern the activation and modulation of cognitive schemes and processes.

ICT (information and communication technology): Methods and techniques used for the secure **transmission**, reception, and processing of data and information.

Immersive VR: Virtual reality system that sensorial absorbs the user internally, isolating him from the **external** environment, giving the full sensation and illusion of being inside the virtual world. It typically uses some sound and vision diffusion devices, such as a helmet, which can isolate from surrounding noises and sounds, allowing the user to fully focus on the stimuli presented in the 3D screen.

Neural plasticity: The ability of the brain to modify its own structure and functionality according to the activity of its neurons, as well as related to stimuli received from the external environment, in reaction to traumatic injuries or pathological changes, and to the development process of individuals.

Non-immersive VR: Virtual reality system that produces in the user the role of external observer of a **scenario** proposed on a monitor.

Semi-immersive VR: Virtual reality system that allows a three-dimensional vision to the patient, who has **the** impression of seeing the world through a window or a wall.

Traumatic brain injury: A clinical condition characterized by brain damage due to an external, rapid, and violent force. The injury can lead to alterations in the state of consciousness, with possible deficits in cognitive, neuro-sensory, and emotional-behavioral functions.

Virtual reality: VR is a relatively new field of ICT, recently applied to neuropsychology. VR is the multi-sensory and interactive simulation of ecological scenarios, which reproduce real objects and events with which it is possible to interact. These devices use specific software through input-output peripherals that enable complex and engaging experiences.

Key facts of virtual reality

- VR is a relatively new field of ICT recently applied to neuropsychology.
- VR is the multi-sensory and interactive simulation of ecological scenarios, with which it is possible to interact.
- This tool is capable of promoting not only motor and cognitive disabilities, but also well-being and social participation.
- VR allows the patient to experience a sense of immersion and presence.
- The increased feedback of the virtual environment may develop the "awareness of the results" of the movements as well as the "awareness of the performance" with positive repercussions at cognitive and motor level.
- VR has been used in the rehabilitation of various pathologies, including TBI, to activate and train cognitive abilities.

Key fact for cognitive rehabilitation

- CR is a field of neuroscience that applies to patients with cognitive-behavioral disorders resulting from brain injury or neurodegenerative disease.
- The goal of CR is to promote functional adaptation of the patient, through environmental stimulation.
- CR is based on the principle of neuroplasticity.
- In planning the intervention, it is important to understand the individual differences that can influence the process.
- CR techniques can be classified into two main categories: conventional and innovative methods.
- Various authors have underlined the potential usefulness of the non-pharmacological CR intervention in all phases of TBI, especially performed using VR.

Summary points

- Cognitive dysfunction is the leading cause of post-TBI disability.
- Cognitive impairment interferes with work, relationships, leisure, and activities of daily living, causing severe personal and economic consequences in patients and their families.
- Cognitive rehabilitation programs are essential for the clinical recovery of TBI patients, as they improve functional outcomes and quality of life.
- Cognitive rehabilitation strategies are based on a restorative and compensatory approach, which adapts to the patient's resources and disability.
- Cognitive rehabilitation improves the brain's repair process and plasticity after injuries.
- Several studies have highlighted the possible beneficial effects of innovative techniques, such as virtual reality, during rehabilitation after a head injury.
- This work demonstrated that VR has the potential to provide an effective rehabilitation tool for the treatment of cognitive impairments in TBI patients.

References

Aida, J., Chau, B., & Dunn, J. (2018). Immersive virtual reality in traumatic brain injury rehabilitation: A literature review. *NeuroRehabilitation*, *42*(4), 441–448.

Aulisio, M. C., Han, D. Y., & Glueck, A. C. (2020). Virtual reality gaming as a neurorehabilitation tool for brain injuries in adults: A systematic review. *Brain Injury*, *34*(10), 1322–1330.

Bohil, C. J., Alicea, B., & Biocca, F. A. (2011). Virtual reality in neuroscience research and therapy. *Nature Reviews. Neuroscience*, *12*(12), 752–762.

Caglio, M., Latini-Corazzini, L., D'Agata, F., et al. (2012). Virtual navigation for memory rehabilitation in a traumatic brain injured patient. *Neurocase*, *18*(2), 123–131.

Calabrò, R. S., Bramanti, A., Garzon, M., et al. (2018). Telerehabilitation in individuals with severe acquired brain injury: Rationale, study design, and methodology. *Medicine*, *97*(50), e13292.

Calabrò, R. S., Naro, A., Cimino, V., et al. (2020). Improving motor performance in Parkinson's disease: A preliminary study on the promising use of the computer assisted virtual reality environment (CAREN). *Neurological Sciences*, *41*(4), 933–941.

Calabrò, R. S., Naro, A., Russo, M., Leo, A., De Luca, R., Balletta, T., et al. (2017). The role of virtual reality in improving motor performance as revealed by EEG: A randomized clinical trial. *Journal of Neuroengineering and Rehabilitation*, *14*(1), 53.

Chuah, S. H. W. (2018). *Why and Who Will Adopt Extended Reality Technology? Literature Review, Synthesis, and Future Research Agenda.* (December 13, 2018). Available at SSRN: https://ssrn.com/abstract=3300469. or https://doi.org/10.2139/ssrn.3300469.

Corregidor-Sánchez, A. I., Segura-Fragoso, A., Criado-Álvarez, J. J., Rodríguez-Hernández, M., Mohedano-Moriano, A., & Polonio-López, B. (2020). Effectiveness of virtual reality systems to improve the activities of daily life in older people. *International Journal of Environmental Research and Public Health*, *17*(17), E6283.

Dahdah, M. N., Bennett, M., Prajapati, P., Parsons, T. D., Sullivan, E., & Driver, S. (2017). Application of virtual environments in a multi-disciplinary day neurorehabilitation program to improve executive functioning using the Stroop task. *NeuroRehabilitation*, *41*(4), 721–734.

De Luca, R., Calabrò, R. S., & Bramanti, P. (2018). Cognitive rehabilitation after severe acquired brain injury: Current evidence and future directions. *Neuropsychological Rehabilitation*, *28*(6), 879–898.

De Luca, R., Maggio, M. G., Maresca, G., et al. (2019). Improving cognitive function after traumatic brain injury: A clinical trial on the potential use of the semi-immersive virtual reality. *Behavioural Neurology*, *2019*, 9268179.

De Luca, R., Maggio, M. G., Naro, A., Portaro, S., Cannavò, A., & Calabrò, R. S. (2020). Can patients with severe traumatic brain injury be trained with cognitive telerehabilitation? An inpatient feasibility and usability study. *Journal of Clinical Neuroscience*, *79*, 246–250.

De Luca, R., Portaro, S., Le Cause, M., et al. (2020). Cognitive rehabilitation using immersive virtual reality at young age: A case report on traumatic brain injury. *Applied Neuropsychology: Child*, *9*(3), 282–287.

De Luca, R., Russo, M., Naro, A., et al. (2018). Effects of virtual reality-based training with BTs-Nirvana on functional recovery in stroke patients: Preliminary considerations. *International Journal of Neuroscience*, *128*(9), 791–796.

Dvorkin, A. Y., Ramaiya, M., Larson, E. B., et al. (2013). A "virtually minimal" visuo-haptic training of attention in severe traumatic brain injury. *Journal of Neuroengineering and Rehabilitation*, *10*, 92.

Ettenhofer, M. L., Guise, B., Brandler, B., et al. (2019). Neurocognitive Driving Rehabilitation in Virtual Environments (NeuroDRIVE): A pilot clinical trial for chronic traumatic brain injury. *NeuroRehabilitation*, *44*(4), 531–544.

García-Betances, R. I., Jiménez-Mixco, V., Arredondo, M. T., & Cabrera-Umpiérrez, M. F. (2015). Using virtual reality for cognitive training of the elderly. *American Journal of Alzheimer's Disease and Other Dementias*, *30*(1), 49–54.

Gimbel, S. I., Ettenhofer, M. L., Cordero, E., Roy, M., & Chan, L. (2020). Brain bases of recovery following cognitive rehabilitation for traumatic brain injury: A preliminary study. *Brain Imaging and Behavior*. https://doi.org/10.1007/s11682-020-00269-8.

Grealy, M. A., Johnson, D. A., & Rushton, S. K. (1999). Improving cognitive function after brain injury: The use of exercise and virtual reality. *Archives of Physical Medicine and Rehabilitation*, *80*(6), 661–667.

Howard, M. C. (2017). A meta-analysis and systematic literature review of virtual reality rehabilitation programs. *Computers in Human Behavior*, *70*, 317–327.

Jacoby, M., Averbuch, S., Sacher, Y., Katz, N., Weiss, P. L., & Kizony, R. (2013). Effectiveness of executive functions training within a virtual supermarket for adults with traumatic brain injury: A pilot study. *IEEE Transactions on Neural Systems and Rehabilitation Engineering*, *21*(2), 182–190.

Kim, M. K., Eom, H., Kwon, J. H., Kyeong, S., & Kim, J. J. (2020). Neural effects of a short-term virtual reality self-training program to reduce social anxiety. *Psychological Medicine*. https://doi.org/10.1017/S0033291720003098.

Knight, R. G., Titov, N., & Crawford, M. (2006). The effects of distraction on prospective remembering following traumatic brain injury assessed in a simulated naturalistic environment. *Journal of the International Neuropsychological Society*, *12*(1), 8–16.

Kober, S. E., Wood, G., Hofer, D., et al. (2013). Virtual reality in neurologic rehabilitation of spatial disorientation. *Journal of Neuroengineering and Rehabilitation*, *10*, 17.

Kourtesis, P., Collina, S., Doumas, L., & MacPherson, S. E. (2020). Validation of the Virtual Reality Everyday Assessment Lab (VR-EAL): An immersive virtual reality neuropsychological battery with enhanced ecological validity. *Journal of the International Neuropsychological Society*, *27*, 1–16.

Larson, E. B., Ramaiya, M., Zollman, F. S., et al. (2011). Tolerance of a virtual reality intervention for attention remediation in persons with severe TBI. *Brain Injury*, *25*(3), 274–281.

Latella, D., Maggio, M. G., De Luca, R., et al. (2018). Changes in sexual functioning following traumatic brain injury: An overview on a neglected issue. *Journal of Clinical Neuroscience*, *58*, 1–6.

Lengenfelder, J., Arjunan, A., Chiaravalloti, N., Smith, A., & DeLuca, J. (2015). Assessing frontal behavioral syndromes and cognitive functions in traumatic brain injury. *Applied Neuropsychology. Adult*, *22*(1), 7–15.

Maggio, M. G., De Cola, M. C., Latella, D., et al. (2018). What about the role of virtual reality in Parkinson disease's cognitive rehabilitation? Preliminary findings from a randomized clinical trial. *Journal of Geriatric Psychiatry and Neurology*, *31*(6), 312–318.

Maggio, M. G., De Luca, R., Manuli, A., et al. (2020). Do patients with multiple sclerosis benefit from semi-immersive virtual reality? A randomized clinical trial on cognitive and motor outcomes. *Applied Neuropsychology. Adult*, 1–7. https://doi.org/10.1080/23279095.2019.1708364.

Maggio, M. G., De Luca, R., Manuli, A., & Calabrò, R. S. (2020). The five 'W' of cognitive telerehabilitation in the Covid-19 era. *Expert Review of Medical Devices*, *17*(6), 473–475.

Maggio, M. G., De Luca, R., Molonia, F., et al. (2019). Cognitive rehabilitation in patients with traumatic brain injury: A narrative review on the emerging use of virtual reality. *Journal of Clinical Neuroscience*, *61*, 1–4.

Maggio, M. G., De Luca, R., Torrisi, M., et al. (2018). Is there a correlation between family functioning and functional recovery in patients with acquired brain injury? An exploratory study. *Applied Nursing Research*, *41*, 11–14.

Maggio, M. G., Latella, D., Maresca, G., et al. (2019). Virtual reality and cognitive rehabilitation in people with stroke: An overview. *Journal of Neuroscience Nursing*, *51*(2), 101–105.

Maggio, M. G., Maresca, G., De Luca, R., et al. (2019). The growing use of virtual reality in cognitive rehabilitation: Fact, fake or vision? A scoping review. *Journal of the National Medical Association*, *111*(4), 457–463.

Maggio, M. G., Maresca, G., Scarcella, I., et al. (2019). Virtual reality-based cognitive rehabilitation in progressive supranuclear palsy. *Psychogeriatrics*, *19*(6), 616–618.

Maggio, M. G., Naro, A., La Rosa, G., et al. (2020). Virtual reality based cognitive rehabilitation in minimally conscious state: A case report with EEG findings and systematic literature review. *Brain Sciences*, *10*(7), 414.

Maggio, M. G., Torrisi, M., Buda, A., et al. (2020). Effects of robotic neurorehabilitation through lokomat plus virtual reality on cognitive function in patients with traumatic brain injury: A retrospective case-control study. *International Journal of Neuroscience*, *130*(2), 117–123.

Man, D. W. K. (2010). Common issues of virtual reality in neuro-rehabilitation. In J.-J. Kim (Ed.), *Virtual Reality* InTech. http://www.intechopen.com/books/virtualreality/common-issues-of-virtual-reality-in-neuro-rehabilitation.

Manivannan, S., Al-Amri, M., Postans, M., et al. (2019). The effectiveness of virtual reality interventions for improvement of neurocognitive performance after traumatic brain injury: A systematic review. *Journal of Head Trauma Rehabilitation*, *34*(2), E52–E65.

Manuli, A., Maggio, M. G., Latella, D., et al. (2020). Can robotic gait rehabilitation plus Virtual Reality affect cognitive and behavioural outcomes in patients with chronic stroke? A randomized controlled trial involving three different protocols. *Journal of Stroke and Cerebrovascular Diseases*, *29*(8), 104994.

Manuli, A., Maggio, M. G., Tripoli, D., et al. (2020). Patients' perspective and usability of innovation technology in a new rehabilitation pathway: An exploratory study in patients with multiple sclerosis. *Multiple Sclerosis and Related Disorders*, *44*, 102312.

Maresca, G., Maggio, M. G., Latella, D., et al. (2019). Toward improving poststroke aphasia: A pilot study on the growing use of telerehabilitation for the continuity of care. *Journal of Stroke and Cerebrovascular Diseases*, *28*(10), 104303.

Marloth, M., Chandler, J., & Vogeley, K. (2020). Psychiatric interventions in virtual reality: Why we need an ethical framework. *Cambridge Quarterly of Healthcare Ethics*, *29*(4), 574–584.

McInnes, K., Friesen, C. L., Mackenzie, D. E., Westwood, D. A., & Boe, S. G. (2017). Mild traumatic brain injury (mTBI) and chronic cognitive impairment: A scoping review. *PLoS One*, *12*, e0174847.

Morganti, F. (2004). Virtual interaction in cognitive neuropsychology. *Studies in Health Technology and Informatics*, *99*, 55–70.

Niogi, S. N., Mukherjee, P., Ghajar, J., et al. (2008). Extent of microstructural white matter injury in postconcussive syndrome correlates with impaired cognitive reaction time: A 3T diffusion tensor imaging study of mild traumatic brain injury. *AJNR. American Journal of Neuroradiology*, *29*(5), 967–973.

Optale, G., Urgesi, C., Busato, V., et al. (2010). Controlling memory impairment in elderly adults using virtual reality memory training: A randomized controlled pilot study. *Neurorehabilitation and Neural Repair*, *24*(4), 348–357.

Pietrzak, E., Pullman, S., & McGuire, A. (2014). Using virtual reality and videogames for traumatic brain injury rehabilitation: A structured literature review. *Games for Health Journal*, *3*(4), 202–214.

Podell, K., Gifford, K., Bougakov, D., & Goldberg, E. (2010). Neuropsychological assessment in traumatic brain injury. *Psychiatric Clinics of North America*, *33*(4), 855–876.

Prince, C., & Bruhns, M. E. (2017). Evaluation and treatment of mild traumatic brain injury: The role of neuropsychology. *Brain Sciences*, *7*(8), 105.

Riva, G. (1997). Virtual reality as an assessment tool in psychology. *Studies in Health Technology and Informatics*, *44*, 71–79.

Riva, G. (2005). Virtual reality in psychotherapy: Review. *Cyberpsychology & Behavior*, *8*(3), 220–240.

Saatian, M., Ahmadpoor, J., Mohammadi, Y., & Mazloumi, E. (2018). Epidemiology and pattern of traumatic brain injury in a developing country regional trauma center. *Bulletin of Emergency and Trauma*, *6*(1), 45–53.

Salazar, A., Warden, D., Schwab, K., et al. (2000). Cognitive rehabilitation for traumatic brain injury. A randomized trial. *JAMA*, *283*, 3075–3081. 3123–3124.

Schretlen, D. J., & Shapiro, A. M. (2003). A quantitative review of the effects of traumatic brain injury on cognitive functioning. *International Review of Psychiatry*, *15*, 341–349.

Senathi-Raja, D., Ponsford, J., & Schönberger, M. (2010). Impact of age on long-term cognitive function after traumatic brain injury. *Neuropsychology*, *24*, 336–344.

Stuss, D. T. (2011). Traumatic brain injury: Relation to executive dysfunction and the frontal lobes. *Current Opinion in Neurology*, *24*(6), 584–589.

Ustinova, K. I., Perkins, J., Leonard, W. A., & Hausbeck, C. J. (2014). Virtual reality game-based therapy for treatment of postural and co-ordination abnormalities secondary to TBI: A pilot study. *Brain Injury*, *28*(4), 486–495.

Westerhof-Evers, H. J., Visser-Keizer, A. C., Fasotti, L., & Spikman, J. M. (2019). Social cognition and emotion regulation: A multifaceted treatment (T-ScEmo) for patients with traumatic brain injury. *Clinical Rehabilitation*, *33*(5), 820–833.

Whiteneck, G. G., Gerhart, K. A., & Cusick, C. P. (2004). Identifying environment factors that influence the outcomes of people with traumatic brain injury. *Journal of Head Trauma Rehabilitation*, *19*, 191–204.

Wilson, B. A. (2013). Neuropsychology rehabilitation: Theory and practice. *South African Journal of Psychiatry*, *43*, 267–277.

Zanier, E. R., Zoerle, T., Di Lernia, D., & Riva, G. (2018). Virtual reality for traumatic brain injury. *Frontiers in Neurology*, *9*, 345.

Chapter 40

How the elderly and young compare in response to traumatic brain injury rehabilitation

Doo Young Kim[a] and Sung-Bom Pyun[b,c,d]

[a]*Department of Physical Medicine and Rehabilitation, Catholic Kwandong University International St Mary's Hospital, Incheon, Republic of Korea,* [b]*Department of Physical Medicine and Rehabilitation, College of Medicine, Korea University, Seoul, Republic of Korea,* [c]*Department of Biomedical Sciences, College of Medicine, Korea University, Seoul, Republic of Korea,* [d]*Brain Engineering Convergence Research Center, Korea University, Seoul, Republic of Korea*

List of abbreviations

ADL	activities of daily living
BBS	Berg Balance Scale
CNT	computerized neurocognitive tests
DTI	diffusion tensor imaging
FAB	frontal assessment battery
FIM	functional independence measure
fMRI	functional magnetic resonance imaging
GCS	Glasgow Coma Scale
GOAT	Galveston Orientation and Amnesia Test
GOS	Glasgow Outcome Scale
GOS-E	Glasgow Outcome Scale-Extended
ICF	International Classification of Functioning, Disability, and Health
LOTCA	Lowenstein Occupational Therapist Cognitive Assessment
MBI	Modified Barthel Index
MMSE	mini-mental state examination
MoCA	Montreal Cognitive Assessment
TBI	traumatic brain injury
TCT	trunk control test
WAB	Western Aphasia Battery

Introduction

The incidence of traumatic brain injury (TBI) is the highest in young and older (≥ 65 years old) aged individuals, but the overall incidence is a multi-modal distribution across the age spectrum (Dewan et al., 2018; Li, Zhao, Yu, & Zhang, 2016). As the life expectancy of humans has increased over the years, the number of elderly patients with TBI is increasing concurrently (Li et al., 2017). Since these patients differ from their younger counterparts, special consideration is required when comparing their response to rehabilitation (Kim, Kim, & Pyun, 2019).

Characteristics of young and elderly patients with TBI

TBI has a high incidence rate in both young and old people (≥ 65 years old). Motor vehicle crashes are the main cause of TBI in young people; in contrast, fall is the main cause in the elderly population. TBI is a wide spectrum disorder causing various functional impairments, such as physical, cognitive, communicative, and behavioral problems caused by extensive

Diagnosis and Treatment of Traumatic Brain Injury. https://doi.org/10.1016/B978-0-12-823347-4.00002-6

brain insult, requiring multi-disciplinary comprehensive rehabilitation. As the brain and physical functions deteriorate during the aging process, the responses to rehabilitation and final functional outcome may vary between elderly and young patients with TBI (Cifu et al., 1996; Kim et al., 2019; Marquez de la Plata et al., 2008). Considering pathophysiological responses, the risk of subdural hemorrhage after head trauma increases in the elderly patients because their bridge vein is more affected by shearing forces due to natural atrophy of the brain, and the subdural space is relatively larger than that of younger counterparts (Ellis, 1990; Howard, Gross, Dacey, & Winn, 1989). Furthermore, cerebral atrophy, nerve shrinkage, reduced synapses, and reduced plasticity may occur in the brain as aging progresses (Fjell & Walhovd, 2010). Therefore, underlying decrease in cognitive function and cognitive reserve even before the TBI leads to severe cognitive impairments and poor functional outcomes. In addition, an elderly person is often frail with decreased resilience to external stress (Fried, Ferrucci, Darer, Williamson, & Anderson, 2004), and the recovery of overall physical function may be limited.

Several studies have reported differences in functional skills, cognitive abilities, length of hospital stays, medical expenses, and discharge destinations between elderly and young patients with TBI (Cifu et al., 1996; Krishnamoorthy, Distelhorst, Vavilala, & Thompson, 2015; LeBlanc, de Guise, Gosselin, & Feyz, 2006; Marquez de la Plata et al., 2008; Susman et al., 2002). These differences are caused by not only physical deterioration or cognitive decline due to aging, but also by accompanying medical illness. The underlying medical problems, such as cardiovascular disorders, hypertension, and diabetes, are often associated with decreased function and increased mortality after TBI (Herou, Romner, & Tomasevic, 2015).

Assessment of young and elderly patients with TBI

International classification of functioning, disability, and health model for TBI

The International Classification of Functioning, Disability, and Health (ICF) is a classification of health and health-related domains. As the functioning and disability of an individual occurs in a context, ICF includes a list of additional environmental factors. It is a framework of the World Health Organization for measuring health and disability at both individual and population levels. The ICF categorizes health conditions as body structure and function, activity, and participation (Fig. 1). TBI causes a variety of brain structural damages and functional impairments of multiple domains that finally limit the activities of daily livings (ADL) and restricts social participation. For the successful management of TBI, the ICF recommends an individualized rehabilitation approach. Injury to the brain structure causes impairment of body functions, such as physical, cognitive, and communicative functions as well as emotional and behavioral problems. These impairments of body functions prevent independent ADL; common cognitive and behavioral problems after TBI, a common complication, are major causes of patients' inability to participate in society, increasing the burden on caregivers. Therefore, comprehensive evaluation of the severity of functional impairments, limitation of activities, and social participation in patients with TBI provide important clues in identifying the patient's problems and needs and in setting goals for rehabilitation. It is also useful for discharge planning, long-term planning at the community level, and family counseling through evaluation over time after brain injury.

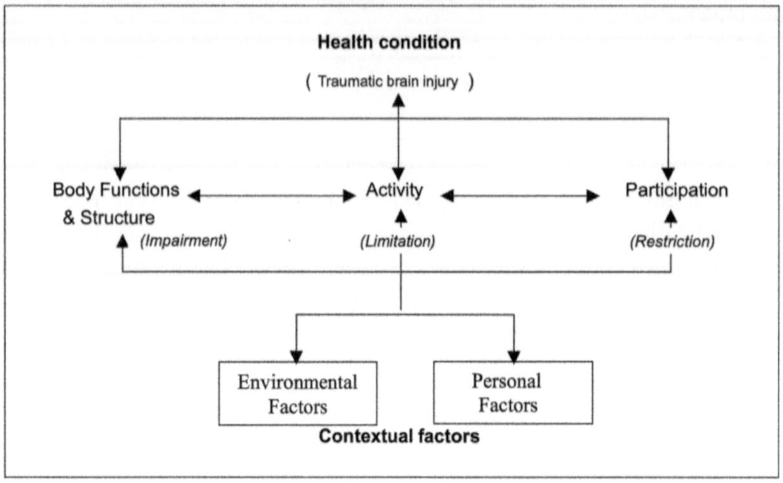

FIG. 1 International Functional, Disability, and Health Classification (ICF) model for traumatic brain injury. For successful treatment of traumatic brain injury, the ICF recommends a rehabilitation approach.

Initial assessment of young and elderly patients with TBI

Initial assessment after TBI includes neurological examination of neuromuscular, sensory, pupillary, cognition, communication, and behavioral impairments. The Glasgow Coma Scale (GCS) is the most commonly used clinical assessment tool for initial severity of brain injury in both young and elderly patients with TBI. The initial GCS score indicates the severity of TBI with 13–15 being mild, 9–12 being moderate, and 3–8 being severe. In patients with TBI, a low GCS score after brain injury is the most reliable predictor of poor prognosis, regardless of age (Table 1) (Kim & Pyun, 2019; Whyte et al., 2009). In severe brain injury with consciousness disorder, the JFK Coma Recovery Scale-Revised (CRS-R) (Giacino, Kalmar, & Whyte, 2004) originally developed in 1991 is a useful tool for differential diagnosis between vegetative state and minimal conscious state. In addition to GCS, post-traumatic amnesia is an important prognostic indicator for the assessment of severity of TBI. It can be assessed using the Galveston Orientation and Amnesia Test (GOAT) that has proven reliability and validity (Bode, Heinemann, & Semik, 2000).

Neuroimaging findings, including brain computed tomography (CT) and magnetic resonance imaging (MRI), are diagnostic tools and strong predictors of outcome for mild to severe TBI. Recent studies have investigated the severity of TBI using advanced neuroimaging techniques, such as diffusion tensor imaging (DTI) and resting-state functional magnetic resonance imaging (fMRI), for white matter integrity and diffuse dysconnectivity reflecting traumatic axonal injury (Benson et al., 2007; Cappa, Conger, & Conger, 2011; Spitz, Maller, O'Sullivan, & Ponsford, 2013; Tang et al., 2012). In addition, some studies evaluated the severity of TBI using biomarkers, such as metabolomics profiling (Table 2) (Bakay, Sweeney, & Wood, 1986; Posti, Dickens, Orešič, Hyötyläinen, & Tenovuo, 2017; Zetterberg, Smith, & Blennow, 2013).

These initial-injury evaluations have a significant impact on the long-term as well as the short-term prognosis for survival. In addition, the initial severity impacts the prognosis of the patient's physical impairment and disability. The severity and site of injury greatly affect the prognosis after TBI. Therefore, to compare the effects of rehabilitation treatment in patients with TBI, the severity of the initial brain injury should be considered.

TABLE 1 Glasgow Coma Scale.

Feature	Response	Score
Best eye response	Open spontaneously	4
	Open to verbal command	3
	Open to pain	2
	No eye opening	1
Best verbal response	Oriented	5
	Confused	4
	Inappropriate words	3
	Incomprehensible sounds	2
	No verbal response	1
Best motor response	Obeys commands	6
	Pain localization	5
	Withdrawal from pain	4
	Flexion to pain	3
	Extension to pain	2
	No motor response	1

The initial Glasgow Coma Scale indicates the severity of traumatic brain injury with a score of 13–15 being mild, 9–12 being moderate, and 3–8 being severe.

TABLE 2 Initial assessment of traumatic brain injury.

	Classification	Tool
Initial severity	Initial evaluation	Pupillary reflex Neurological examination Duration of unconsciousness Multiple trauma
	Clinical assessment	Glasgow Coma Scale (GCS) Galveston Orientation and Amnesia Test (GOAT) The JFK Coma Recovery Scale-Revised (CRS-R)
	Imaging	Conventional Brain CT or MRI Diffusion tensor imaging (DTI) Functional magnetic resonance imaging (fMRI)
	Others	Biomarkers

Clinical assessment tools, imaging techniques, and biomarkers are used to evaluate the initial severity of traumatic brain injury.

Assessment of functional impairments in young and elderly patients with TBI

After acute management of TBI, comprehensive evaluation of the patients is needed for planning and monitoring their rehabilitative management. To assess the approximate outcome of TBI, the overall score of the Glasgow Outcome Scale (GOS) or Glasgow Outcome Scale-Extended (GOS-E) is used. The Rancho Los Amigos Scale is used to express the gross level of cognitive and behavioral functions (Table 3). However, it is difficult to compare the response to rehabilitative intervention because these assessment tools have poor accuracy for neurological conditions and functional performance (Maas et al., 2010; Weir et al., 2012). Thus, an objective and accurate assessment that can measure the physical, cognitive, and behavioral functions is needed to determine the response to rehabilitation in patients with TBI. Considering that the functions deteriorate naturally in the elderly over time, such an assessment tool to measure physical, cognitive, and communication functions are imperative to determine the rehabilitation response (Table 4).

Assessment of physical function in young and elderly patients with TBI

Among the physical functions, locomotion is one of the most important for ADL. For this, it is essential that the body balance should be well maintained through good coordination between muscle strength and sensory function. The Berg Balance Scale (BBS) and Trunk Control Test (TCT) can be used as quantitative evaluation tools for balance function. The BBS consists of 16 items and the maximal score is 56 points; the higher the score, the better the balancing ability. The cut-off values that predict the risk of falls according to the BBS is as follows: 41–56 points for low fall risk, 21–40 points for medium fall risk, and 0–20 points for high fall risk (Berg, Wood-Dauphinee, Williams, & Maki, 1992). The TCT contains four items with a total score of 100 points: rolling to weak side, rolling to strong side, balancing in sitting position, and sitting up from the lying down position. Among the functions of locomotion, if walking is possible, gait function can be assessed using the 10-m Walk Test and Timed Up and Go Test (TUG). The 10-m Walk Test measures the time a patient takes to walk 10 m. The TUG measures the time a sitting patient takes to stand up, turn around an obstacle 3 m in front, and return to the chair. If a patient has impaired functions of balance, muscle strength of the lower extremities, and gait pattern, they will take longer time to complete the tasks; thus, shortening of this time after rehabilitation treatment

TABLE 3 Gross outcome assessment of traumatic brain injury.

Classification	Tool
Overall outcome	Glasgow Outcome Scale (GOS) Glasgow Outcome Scale-Extended (GOS-E)
Arousal and cognitive function	Rancho Los Amigos Scale

Clinical assessment tools to evaluate gross outcome of traumatic brain injury include the Glasgow Outcome Scale, Glasgow Outcome Scale-Extended, and Rancho Los Amigos Scale.

TABLE 4 Comprehensive assessment of traumatic brain injury.

	Classification	Evaluation tests
Comprehensive Assessment	Physical function	Berg Balance Scale (BBS) Trunk Control Test (TCT) 10-m Walk Test Timed Up and Go Test (TUG)
	Cognitive function	Mini-Mental State Examination (MMSE) Montreal Cognitive Assessment (MoCA) Frontal Assessment Battery (FAB) Computerized neurocognitive tests (CNT) Lowenstein Occupational Therapist Cognitive Assessment (LOTCA)
	Communication function	Western Aphasia Battery (WAB)
	Activities of daily livings	Modified Barthel index (MBI) Functional Independence Measure (FIM)

To evaluate the precise outcome of traumatic brain injury, physical function, cognitive function, communication function, and activities of daily living are assessed individually.

indicates its effectiveness. As the elderly often have problems with balancing due to muscle weakness, decreased sensation, and/or orthostatic hypotension, the patient's age should be considered when evaluating physical functions. There are norm data for each age for physical function in these tools. The effect of rehabilitative management can be compared with the distribution relative to the norm values for each age. In particular, the elderly patient with TBI has a lower balance score than the younger counterpart of the same phase after onset of TBI, but the amount of recovery is similar when rehabilitation is performed for the same period (Kim et al., 2019; Perry, Billek-Sawhney, Woollard, Little, & Szalinski, 2019). Polypharmacy, frailty, orthostatic hypotension, and sensory disturbance (proprioception, visual acuity, vestibular function) could adversely affect balance, which is common in the elderly population (Clegg, Young, Iliffe, Rikkert, & Rockwood, 2013; Merzo, Lenell, Nyholm, Enblad, & Lewén, 2016). Therefore, to compare the response to rehabilitation in the physical functions of elderly and young patients with TBI, it is necessary to consider these conditions. In addition, it is appropriate to assess and to compare the amount of functional gain while examining pre-morbidity and adjusting for aging-related decline.

Assessment of cognitive and language functions in young and elderly patients with TBI

Arousal disorders or cognitive impairments, such as consciousness, attention, memory, frontal executive function, and communication disorders, are common sequelae after TBI. The Mini-Mental State Examination (MMSE) and Montreal Cognitive Assessment (MoCA) are used as screening tests to evaluate general cognitive function (Cockrell & Folstein, 2002; McLennan, Mathias, Brennan, & Stewart, 2011). The Frontal Assessment Battery (FAB) is used to assess frontal lobe dysfunction. The MMSE is simple cognitive function test that measures orientation, registration, attention and calculation, recall, language, and copying ability on a 30-point scale. The MoCA is a similar cognitive function test that evaluates visuospatial and executive functions, naming, memory, attention, language, abstraction, delayed recall, and orientation on a 30-point scale. Frontal lobe damage commonly occurs after TBI; the FAB measures frontal executive function, conceptualization, mental flexibility, programming, sensitivity to interference, inhibitory control, and environmental autonomy on an 18-point scale. Each cognitive domain can be precisely assessed using domain-specific cognitive function tests or comprehensive cognitive evaluations, such as computerized neurocognitive function tests and the Lowenstein Occupational Therapist Cognitive Assessment (LOTCA). These can evaluate the degree of preserved or impaired cognitive function and the response to cognitive training or rehabilitation (Arrieux, Cole, & Ahrens, 2017). The domain-specific cognitive function tests include visual attention and working memory test (forward and backward digit span test, forward and backward visual span test), vigilance test (auditory and visual continuous performance test), memory test (verbal and visual learning test), selective attention and inhibitory control test (Stoop test), mental flexibility and executive function test (Trail making A and B), conceptualization and thinking set conversion test (card sorting test), and non-verbal intelligence test (Raven's colored progressive matrices). The LOTCA assesses neurological cognitive impairment through evaluation of orientation, visual and spatial perception, praxis, visuomotor organization, and thinking operations. Cognitive function degrades more because of age than physical function. The normal values for age of each tool have been investigated. When comparing the response to rehabilitation with cognitive function in elderly and young patients with TBI, the reference values according to age groups should be considered. In case of senile dementia, cognitive decline may occur in the same domain of TBI. Thus, it is necessary to measure this function accurately to investigate its correlation to the response to in elderly patients with TBI. Due to aging-related cognitive decline, the final cognitive outcome may be low even after successful recovery and rehabilitation after TBI. Therefore, it is necessary to compare the effect of loss of cognitive function on cognitive recovery after treatment.

Communication disorders, such as cognitive communication disorders or aphasia, may coincide after TBI. In these cases, a combination of language and cognitive assessments are used evaluating communication disorders. In language assessment, the Western Aphasia Battery (WAB) is commonly used and is designed to assess key clinical aspects of linguistic components and non-verbal cognitive functions. The WAB provides information on the presence of aphasia, severity, and its type. Aphasia Quotient is a summary score indicating the overall severity of language impairment (fluency, comprehension, repetition, and naming), and Language Quotient is an extended summary score that evaluates reading and writing. If brain injury involves classical language-related areas, such as Wernicke's area or Broca's area, typical aphasia symptoms can occur. However, in cases of TBI, cognitive communication disorders due to attention, memory, and executive dysfunction are more frequent. These communication disorders can be affected by a decline in cognitive and linguistic function with aging.

Assessment of ADL in young and elderly patients with TBI

The ADL is a comprehensive concept of the patient's functional abilities that indicates not only physical dependence but also combined cognitive and communication functions. A decline in physical performance or cognitive function due to aging negatively affects ADL and increases dependency. The Modified Barthel Index (MBI) and Functional Independence Measure (FIM) are primarily used to assess ADL in patients with TBI. MBI is a measure of the degree of independence of a person with chronic conditions in their daily life. Through direct observation and interviews in the patient's daily life, their level of dependency is evaluated in five categories: complete independence, minimal dependence, moderate dependence, maximal dependence, and total dependence. The score of each item is distributed according to the difference in function, and the scores range from 0 (completely dependent) to 100 (completely independent) points. The assessment area consists of personal hygiene, bathing, feeding, toileting, stair-climbing, eating, dressing, bladder and bowel control, ambulation or wheelchair locomotion, and chair/bed transfer. The FIM is an 18-item clinician's reporting scale that evaluates self-care, continence, mobility, transfer, communication, and cognition. Each of the 18 items is rated on a scale of 1 to 7 based on the level of independence: 1 point indicates a state in which the patient is completely dependent on others for performing the item, and 7 points indicate that the patient can independently perform the item. The FIM is more advantageous in patients with TBI because it can evaluate communication and cognitive functions that are not available in MBI.

Comparison of responses to rehabilitation in the elderly and young patients with TBI

Studies comparing the outcomes between elderly and young patients with TBI reported differences in physical functions, especially for balance and gait, cognitive function, and ADL (Cifu et al., 1996; Kim et al., 2019; Krishnamoorthy et al., 2015; LeBlanc et al., 2006; Marquez de la Plata et al., 2008; Susman et al., 2002). Elderly patients with TBI had initial lower scores for balance than the younger counterparts, but the extent of recovery was similar when rehabilitation was performed under similar conditions (Kim et al., 2019; Perry et al., 2019). Additionally, elderly patients with TBI had lower scores for both pre- and post-rehabilitation ADL than the younger patients with TBI; the degree of improvement in ADL during rehabilitation was similar in both groups suggesting comparable rehabilitation potential of the elderly (Kim et al., 2019; LeBlanc et al., 2006). Thus, when evaluating only the endpoint scores for response to rehabilitation, we may underestimate the possibility of recovery potential in the elderly patients with TBI. To compare the response to rehabilitation between these two groups, we should consider associated common problems in the elderly population, such as sensory disturbance, medication history, and frailty. Pre-morbid polypharmacy, orthostatic hypotension, and sensory disturbances (proprioception, visual acuity, and vestibular function), which are common in the elderly population, could adversely affect their balance (Clegg et al., 2013; Kojima et al., 2012; Merzo et al., 2016). From the viewpoint of pathophysiology of TBI in the elderly, physical, cognitive, and communication functions and the basic dependency for ADL differ from that of younger patients with TBI; hence, a comprehensive and accurate evaluation is necessary to compare their responses to rehabilitation. Since the pre-morbid status of the elderly is relatively low compared with that of the younger counterparts, the response to rehabilitation should not be compared to only the final outcome. Although the final functional outcome scores are usually lower in elderly patients than in younger patients with TBI, the elderly patients also have good potential

TABLE 5 Summary of differences between the elderly and young patients affected by brain injury.

Factors	Young	Elderly
Incidence of traumatic brain injury (most common cause)	Higher (motor vehicle crash)	Higher (fall)
Risk of subdural hemorrhage after head trauma	Lower	Higher
Impact on bridge vein by shearing forces	Lower	Higher
Co-morbidity affecting function	Low	High
Functional scores before rehabilitation	Higher	Lower
Functional scores after rehabilitation	Higher	Lower
Potential to recover after rehabilitation	Good	Good

There are several types of differences in traumatic brain injury between the elderly and young patients, but the potential for rehabilitation is similar.

to recover after rehabilitation. Therefore, it is appropriate to evaluate and to compare the extent of functional gain while investigating pre-morbid conditions and adjusting functional decline with aging (Table 5).

Application to other areas of neuroscience

In this chapter, we compared the responses to rehabilitation treatment between elderly and young patients with TBI. From the perspective of pathophysiology and pre-morbid physical, cognitive, and ADL performances, TBI can cause a difference in the responses to rehabilitation between the elderly and young patients. Although the final functional outcome scores are usually lower in elderly patients than in young patients with TBI, elderly patients also have good potential to recover after rehabilitation. Therefore, it is necessary to evaluate and to compare the extent of functional gain, while investigating pre-morbid conditions and adjusting functional decline with aging.

In addition to TBI, stroke and spinal cord injuries result in a decline in physical, cognitive, and ADL performances that are the main targets of rehabilitation treatment. Aging also causes this decline in multiple functions. Although dysfunction due to central nervous system injury and decline due to aging may appear similar, the two mechanisms differ. Therefore, the changes that occur during the aging process must be considered in comparative studies of the rehabilitation responses between the elderly and younger groups.

Mini-dictionary of terms

- Traumatic brain injury (TBI): TBI refers to damage to the brain caused by trauma, and includes axonal injury as well as traumatic intracranial hemorrhage.
- Frailty: A condition in which the resilience to external stress is decreased.
- Diffusion tensor imaging (DTI): DTI is a technique that shows the orientation of water molecules and helps to check whether axonal injury was coexisted.
- Functional magnetic resonance imaging (fMRI): fMRI identifies active brain regions and displays brain-to-brain connectivity in resting status.
- Glasgow Coma Scale (GCS): The GCS score is the most powerful prognostic predictor of traumatic brain injury, but it is too simple to reflect all neurological abnormalities.
- Frontal lobe function: The functions of the frontal lobe include movement, emotion, behavior, language, and executive functions.
- Communication disorder: In communication disorders, aphasia and cognitive communication disorders must be distinguished. The most important difference is that in aphasia, there is a disorder in the language component.
- Computerized neurocognitive tests (CNT): The advantage of CNTs is that comprehensive cognitive function tests, including concentration and executive functions, can be performed simply through a computer.
- Lowenstein Occupational Therapist Cognitive Assessment (LOTCA): LOTCA evaluates the performance of various cognitive functions in depth.
- Activities of daily living (ADL): ADL refers to the daily activities, including self-care, grooming and dressing, and locomotion such as walking and transfer.

Key facts on DTI

- DTI can confirm the injury of white matter by imaging the fraction of water molecules.
- Using the DTI test, several quantified parameters indicating the integrity of white matter are calculated.
- DTI can be reconstructed with visualized images for the integrity of white matter.
- Among the parameters calculated by DTI, fraction anisotropy is closely related to functional recovery.
- According to recent studies, DTI is emerging as a prognostic factor for neurological recovery of TBI.

Key facts on fMRI

- fMRI measures the activity when the brain performs functions or is at rest.
- fMRI can also measure the connectivity between brain regions when the brain is in a resting state or active.
- The degree of brain activation or connectivity can be estimated to impair brain function.
- It helps to determine the extent of loss of function that occurs in brain lesions observed on MRI.
- fMRI is used to predict the severity and prognosis of brain injury.

Summary points

- Traumatic brain injury (TBI) has a high incidence rate in both young and old people. Motor vehicle crashes are the main cause in young people, but fall is the main cause in elderly people.
- The severity of TBI is more impacted by the severity of trauma than by age.
- As the physical, cognitive, and activities of daily living performances of TBI patients are also affected by aging, the scores for both before and after rehabilitation treatment are low in the elderly.
- Among the communication functions, linguistic components are less susceptible to aging, so accurate assessments are required.
- From the perspective of pathophysiology and low pre-morbid physical, cognitive, and activities of daily living (ADL) performances, TBI can cause a difference in responses to rehabilitation in the elderly and younger patients.
- Although the final functional outcome scores are usually lower in elderly patients than in young patients with TBI, elderly patients also have good potential to recover after rehabilitation.
- Therefore, it is necessary to evaluate and to compare the extent of functional gain, while investigating pre-morbid conditions and adjusting functional decline with aging.

References

Arrieux, J. P., Cole, W. R., & Ahrens, A. P. (2017). A review of the validity of computerized neurocognitive assessment tools in mild traumatic brain injury assessment. *Concussion*, *2*(1), CNC31.

Bakay, R., Sweeney, K. M., & Wood, J. H. (1986). Pathophysiology of cerebrospinal fluid in head injury: Part 2. Biochemical markers for central nervous system trauma. *Neurosurgery*, *18*(3), 376–382.

Benson, R. R., Meda, S. A., Vasudevan, S., Kou, Z., Govindarajan, K. A., Hanks, R. A., et al. (2007). Global white matter analysis of diffusion tensor images is predictive of injury severity in traumatic brain injury. *Journal of Neurotrauma*, *24*(3), 446–459.

Berg, K. O., Wood-Dauphinee, S. L., Williams, J. I., & Maki, B. (1992). Measuring balance in the elderly: Validation of an instrument. *Canadian Journal of Public Health*, *83*(Suppl 2), S7–S11.

Bode, R. K., Heinemann, A. W., & Semik, P. (2000). Measurement properties of the Galveston Orientation and Amnesia Test (GOAT) and improvement patterns during inpatient rehabilitation. *The Journal of Head Trauma Rehabilitation*, *15*(1), 637–655.

Cappa, K. A., Conger, J. C., & Conger, A. J. (2011). Injury severity and outcome: A meta-analysis of prospective studies on TBI outcome. *Health Psychology*, *30*(5), 542.

Cifu, D. X., Kreutzer, J. S., Marwitz, J. H., Rosenthal, M., Englander, J., & High, W. (1996). Functional outcomes of older adults with traumatic brain injury: A prospective, multicenter analysis. *Archives of Physical Medicine and Rehabilitation*, *77*(9), 883–888.

Clegg, A., Young, J., Iliffe, S., Rikkert, M. O., & Rockwood, K. (2013). Frailty in elderly people. *Lancet*, *381*(9868), 752–762.

Cockrell, J. R., & Folstein, M. F. (2002). Mini-mental state examination. In *Principles and practice of geriatric psychiatry* (pp. 140–141). John Wiley & Sons Ltd.

Dewan, M. C., Rattani, A., Gupta, S., Baticulon, R. E., Hung, Y. C., Punchak, M., et al. (2018). Estimating the global incidence of traumatic brain injury. *Journal of Neurosurgery*, *130*(4), 1080–1097.

Ellis, G. (1990). Subdural hematoma in the elderly. *Emergency Medicine Clinics of North America*, *8*(2), 281–294.

Fjell, A. M., & Walhovd, K. B. (2010). Structural brain changes in aging: Courses, causes, and cognitive consequences. *Reviews in the Neurosciences*, *21*(3), 187–222.

Fried, L. P., Ferrucci, L., Darer, J., Williamson, J. D., & Anderson, G. (2004). Untangling the concepts of disability, frailty, and comorbidity: Implications for improved targeting and care. *The Journals of Gerontology Series A: Biological Sciences and Medical Sciences*, *59*(3), M255–M263.

Giacino, J. T., Kalmar, K., & Whyte, J. (2004). The JFK Coma Recovery Scale-Revised: Measurement characteristics and diagnostic utility. *Archives of Physical Medicine and Rehabilitation*, *85*(12), 2020–2029.

Herou, E., Romner, B., & Tomasevic, G. (2015). Acute traumatic brain injury: Mortality in the elderly. *World Neurosurgery*, *83*(6), 996–1001. https://doi.org/10.1016/j.wneu.2015.02.023.

Howard, M. A., Gross, A. S., Dacey, R. G., & Winn, H. R. (1989). Acute subdural hematomas: An age-dependent clinical entity. *Journal of Neurosurgery*, *71*(6), 858–863.

Kim, D. Y., Kim, M. H., & Pyun, S. B. (2019). Comparison of functional outcomes between elderly and young patients with traumatic brain injury in a subacute rehabilitation unit. *Topics in Geriatric Rehabilitation*, *35*(2), 141–148.

Kim, D. Y., & Pyun, S. B. (2019). Prediction of functional outcome and discharge destination in patients with traumatic brain injury after post-acute rehabilitation. *International Journal of Rehabilitation Research*, *42*(3), 256–262.

Kojima, T., Akishita, M., Nakamura, T., Nomura, K., Ogawa, S., Iijima, K., et al. (2012). Polypharmacy as a risk for fall occurrence in geriatric outpatients. *Geriatrics & Gerontology International*, *12*(3), 425–430.

Krishnamoorthy, V., Distelhorst, J. T., Vavilala, M. S., & Thompson, H. (2015). Traumatic brain injury in the elderly: Burden, risk factors, and prevention. *Journal of Trauma Nursing*, *22*(4), 204–208. https://doi.org/10.1097/jtn.0000000000000135.

LeBlanc, J., de Guise, E., Gosselin, N., & Feyz, M. (2006). Comparison of functional outcome following acute care in young, middle-aged and elderly patients with traumatic brain injury. *Brain Injury*, *20*(8), 779–790. https://doi.org/10.1080/02699050600831835.

Li, M., Zhao, Z., Yu, G., & Zhang, J. (2016). Epidemiology of traumatic brain injury over the world: A systematic review. *General Medicine: Open Access*, *4*(5), e275.

Li, Y., Zhou, J., Chen, F., Zhang, J., Qiu, J., & Gu, J. (2017). Epidemiology of traumatic brain injury older inpatients in Chinese military hospitals, 2001-2007. *Journal of Clinical Neuroscience*, *44*, 107–113. https://doi.org/10.1016/j.jocn.2017.05.026.

Maas, A. I., Steyerberg, E. W., Marmarou, A., McHugh, G. S., Lingsma, H. F., Butcher, I., et al. (2010). IMPACT recommendations for improving the design and analysis of clinical trials in moderate to severe traumatic brain injury. *Neurotherapeutics*, *7*(1), 127–134.

Marquez de la Plata, C. D., Hart, T., Hammond, F. M., Frol, A. B., Hudak, A., Harper, C. R., et al. (2008). Impact of age on long-term recovery from traumatic brain injury. *Archives of Physical Medicine and Rehabilitation*, *89*(5), 896–903. https://doi.org/10.1016/j.apmr.2007.12.030.

McLennan, S. N., Mathias, J., Brennan, L., & Stewart, S. (2011). Validity of the Montreal Cognitive Assessment (MoCA) as a screening test for mild cognitive impairment (MCI) in a cardiovascular population. *Journal of Geriatric Psychiatry and Neurology*, *24*(1), 33–38.

Merzo, A., Lenell, S., Nyholm, L., Enblad, P., & Lewén, A. (2016). Promising clinical outcome of elderly with TBI after modern neurointensive care. *Acta Neurochirurgica*, *158*(1), 125–133.

Perry, S. B., Billek-Sawhney, B., Woollard, J., Little, S., & Szalinski, A. (2019). Balance and mobility improvements during inpatient rehabilitation are similar in young-old, mid-old, and old-old adults with traumatic brain injury. *The Journal of Head Trauma Rehabilitation*, *34*(2), E66–E73.

Posti, J. P., Dickens, A. M., Orešič, M., Hyötyläinen, T., & Tenovuo, O. (2017). Metabolomics profiling as a diagnostic tool in severe traumatic brain injury. *Frontiers in Neurology*, *8*, 398.

Spitz, G., Maller, J. J., O'Sullivan, R., & Ponsford, J. L. (2013). White matter integrity following traumatic brain injury: The association with severity of injury and cognitive functioning. *Brain Topography*, *26*(4), 648–660.

Susman, M., DiRusso, S. M., Sullivan, T., Risucci, D., Nealon, P., Cuff, S., et al. (2002). Traumatic brain injury in the elderly: Increased mortality and worse functional outcome at discharge despite lower injury severity. *Journal of Trauma and Acute Care Surgery*, *53*(2), 219–224.

Tang, C., Eaves, E., Dams-O'Connor, K., Ho, L., Leung, E., Wong, E., et al. (2012). Diffuse disconnectivity in traumatic brain injury: A resting state fMRI and DTI study. *Translational Neuroscience*, *3*(1), 9–14.

Weir, J., Steyerberg, E. W., Butcher, I., Lu, J., Lingsma, H. F., McHugh, G. S., et al. (2012). Does the extended Glasgow Outcome Scale add value to the conventional Glasgow Outcome Scale? *Journal of Neurotrauma*, *29*(1), 53–58.

Whyte, J., Gosseries, O., Chervoneva, I., DiPasquale, M. C., Giacino, J., Kalmar, K., et al. (2009). Predictors of short-term outcome in brain-injured patients with disorders of consciousness. *Progress in Brain Research*, *177*, 63–72. https://doi.org/10.1016/s0079-6123(09)17706-3.

Zetterberg, H., Smith, D. H., & Blennow, K. (2013). Biomarkers of mild traumatic brain injury in cerebrospinal fluid and blood. *Nature Reviews. Neurology*, *9*(4), 201.

Chapter 41

Rehabilitation of social cognition after traumatic brain injury

Garzon Heredia Alicia[a,b], Allain Philippe[a,b], Azouvi Philippe[c,d], and Vallat-Azouvi Claire[e]

[a]*Department of Psychology des Pays de la Loire, SFR Confluences, University of Angers, University of Nantes, Maison de la recherche Germaine Tillion, Angers Cedex 01, France,* [b]*Department of Neurology, CHU Angers, Angers, France,* [c]*AP-HP, Raymond Poincare Hospital, Service de MPR, Garches, France,* [d]*INSERM DevPsy Team, Centre de Recherche en Épidémiologie et Santé des Populations, UMR 1018, Paris-Saclay University, UVSQ, Versailles, France,* [e]*Department of Fonctionnement et Dysfonctionnement Cognitifs : les âges de la vie (DYSCO), University of Paris 8, Saint-Denis, France*

List of abbreviations

SCIT	Social Cognition and Interaction Training
CET	Cognitive Enhancement Therapy
ToM	Theory of Mind
CPT	Cognitive Pragmatic Treatment
fMRI	Functional Magnetic Resonance Imaging
RCT	Randomized Controlled-Trial

Introduction

The interest in rehabilitation of social cognition seems to have increased in recent decades, with a concentration of studies in autistic syndrome and psychiatric disorders, especially schizophrenia and bipolar disorder.

In the scientific literature, there are different types of rehabilitation programs of social cognition which can be distinguished into three categories, depending on the ability or abilities targeted. Thus, there are interventions specifically targeting one or more domains of social cognition, as training in emotion recognition. There are broad-based interventions targeting social cognition globally through its different components. Finally, there are non-specific interventions, which target skills that do not belong to social cognition but have an impact on it, as the social communication interventions.

Before discussing the therapies offered to patients with brain injuries, we will briefly outline the programs developed for other populations.

First historical programs: Rehabilitation of social cognition in other pathologies

Publications on social cognition remediation and training programs started in the 1980s and have increased exponentially since 2013 (Fernandez-Sotos et al., 2019). Studies with a high level of scientific evidence also increased. Most of the articles with high level of scientific evidence focused on targeted rehabilitation, mainly on theory of mind and emotional processing, defined as the abilities to identify, facilitate, regulate, understand and manage emotions. The pathologies most studied in these publications are schizophrenia spectrum disorders, autism spectrum disorders, depressive and anxiety disorders and personality disorders. Some studies included patients with eating disorders, attention deficit and hyperactivity disorders, or intellectual disabilities. Finally, a total of 214 articles have been published between 2006 and 2016 on rehabilitation of social cognition in these populations (Fernandez-Sotos et al., 2019).

According to Fernandez-Sotos et al. (2019), the interest for social cognition rehabilitation in schizophrenia obeys practical reasons: the presence of deficits in quality of life and everyday functioning in most patients, the perennial character of their deficits, the impact of disorders on their family, economy and personal life, and the importance of the improvement in social cognition in quality of life and everyday functioning. Similar reasons could be applied for patients with autistic syndrome.

Diagnosis and Treatment of Traumatic Brain Injury. https://doi.org/10.1016/B978-0-12-823347-4.00021-X

To illustrate these interventions, one of the most replicated social cognition programs for patients with schizophrenia is the Social Cognition and Interaction Training (SCIT), developed by Penn et al. (2005). This training targets emotion perception, attributional style and theory of mind during 20 or 24 weeks. Appeared some years earlier, the Cognitive Enhancement Therapy (CET) designed by Hogarty and Flesher (1999) has attracted a lot of attention. This program targeted perspective taking and the ability to evaluate the social context during 24 weeks. Both programs had promising results with improvements in trained areas. A few programs have been developed and studied more recently for schizophrenia (Brunet-Gouet, Oker, Martin, Grynszpan, & Jackson, 2016; Gaudelus et al., 2016).

Rehabilitation of social cognition in traumatic patients with brain injury

Two large reviews of the literature have appeared recently concerning rehabilitation of social cognition for patients with traumatic brain injury. Cassel, McDonald, Kelly, and Togher (2019) analyzed 12 publications concerning this subject. In the same year, Vallat-Azouvi, Azouvi, Le-Bornec, and Brunet-Gouet analyzed 16 articles. Compared to other pathologies like schizophrenia and autistic syndrome, with 214 articles published between 2006 and 2016, the traumatic brain injury population has received less interest in the rehabilitation of social cognition.

For eight publications, the level of scientific evidence was considered as "class I" according to the classification of Cicerone et al. (2011, 2019); Class I: randomized controlled trials. Class II: non-randomized cohort or case–control studies. Class III: single case or multiple single case studies and cohort studies without control. In this way, the randomized controlled design was the most used design (Vallat-Azouvi, Azouvi, Le-Bornec, & Brunet-Gouet, 2019). The functions targeted by these programs were generally theory of mind, emotion perception and social communication. Little or no studies targeted emotional regulation, emotion awareness or empathy. Most of the rehabilitation interventions targeted only one domain of social cognition, principally emotion recognition and theory of mind. According to Cassel et al. (2019), the main behavior and cognitive rehabilitation techniques used in these interventions was repetition, positive reinforcement, errorless learning, self-instructional training and vanishing cues.

The results of these studies are encouraging, despite the small number of patients included for some papers, given that most of interventions have obtained improvements in at least one domain of social cognition. Some interventions have shown increased levels of quality or life or functioning. The recommendations of the INCOG expert group published in 2014 by Tate et al. (2014) and Togher et al. (2014) did not take into account the rehabilitation of social cognition, for which there was still little literature at that time, apart from approaches focused on social communication. In their recent systematic review of the literature from 2009 through 2014, Cicerone et al. (2019) included recommendations for communication and social cognition based on a few studies (two class I studies, one class II studies and five class III studies on the area of communication, mostly after traumatic brain injury and one class III including 5 patients with cerebrovascular diseases). Specific interventions for functional communication deficits were recommended as a Practice Standard, including pragmatic conversational skills, following traumatic brain injury. One class I (Neumann, Babbage, Zupan, & Willer, 2015) and one class III (Radice-Neumann, Zupan, Tomita, & Willer, 2009) study suggested that specific intervention to improve the recognition of emotion from facial expressions may be effectively incorporated as a specific component of the Practice Standard for treating functional communication deficits after traumatic brain injury.

In the rest of this chapter, we will first detail the protocols addressing a sub-domain of social cognition, second, the rehabilitation programs of interpersonal and communication skills of theory of mind and third, the multi-component rehabilitation programs of social cognition.

Table 1 summarizes the different protocols presented in this chapter as well as the effects shown in the validation studies.

Rehabilitation programs targeting one domain of social cognition

Treatment of emotion perception

Bornhofen and McDonald (2008a, 2008b): Facial emotion recognition

Bornhofen and McDonald (2008a, 2008b) have developed one of the first programs targeting facial emotion recognition deficits for people with traumatic brain injury, more than 9 months after the trauma. The intervention was tested in 12 participants with a randomized controlled trial design. The program takes place over 8 weeks, at a frequency of 2 sessions of 1.5 h per week. One therapist guides a group of 2 or 3 participants through the different activities.

The intervention involves the recognition of static and dynamic facial expressions in a program of 4 phases structured in a hierarchy of progressive difficulty:

TABLE 1 Characteristic of treatments of social cognition for traumatic brain injury patients.

Authors	Training target	Design	N =	Results	Generalization	Class cicerone
Bornhofen and McDonald (2008a)	Combination of treatment addressing Social Perception	RCT	12	Improvement of judgment of emotional stimuli and in making social inferences	NA	I
Bornhofen and McDonald (2008b)	Social perception	Comparison of errorless learning and self-instruction	18	Modest improvement in emotion perception	NA	
Radice-Neumann et al. (2009)	FAR	RCT	19	Improved FAR and ability to infer emotions from context	Improvement of socio-emotional behavior	I
Neumann et al. (2015)	FAR	RCT	71	Improved FAR	No	I
McDonald, Bornhofen, and Hunt (2009)	FAR	Comparison of focusing attention and mimicry	22	Limited effect of mimic strategy	NA	I
McDonald et al. (2013)	Recognition of emotional prosody	RCT	20	No significant effect	No	I
Guercio, Podolska-Schroeder, and Rehfeldt (2004)	FAR	Single-case	3	Improved FAR	NA	III
Williamson and Isaki (2015)	FAR	Single-case	2	Improved FAR	NA	III
Lundgren and Brownell (2015)	ToM	Single-case	3	Improved ToM	NA	III
Helffenstein and Wechsler (1982)	Social perception	RCT	16	Improvement in different domains	Improvement in behavior	I
Dahlberg et al. (2007)	Social communication skills	RCT	52	Improvement in communication	Improvement in social communication skills	I
Finch, Cornwell, Copley, Doig, and Fleming (2017)	Social communication	Pre-post	8	Improvement of goal attainment scaling	NA	II
Winegardner, Keohane, Prince, and Neumann (2016)	Perspective-taking and negative attribution bias	Single-case	2	Limited changes on objective measures	Decreased aggression according to relatives	III
Gabbatore et al. (2015)	Communicative pragmatic abilities (including ToM)	Pre-post	15	Improvement in all communication modalities	NA	II

Continued

TABLE 1 Characteristic of treatments of social cognition for traumatic brain injury patients—cont'd

Authors	Training target	Design	N=	Results	Generalization	Class cicerone
Westerhof-Evers et al. (2017)	Multi-faceted Treatment for Social cognition and emotion regulation	RCT	61	Improvement FAR and theory of mind	Improvement on proxy-rated empathy, societal participation and goal attainment	I
McDonald et al. (2008)	Emotion perception and social skills	RCT	51	Improvement on partner directed behavior	No effect on other outcome measures	I

N, number of participants; *FAR*, facial affect recognition; *ToM*, theory of mind; *NA*, non available; *RCT*, randomized controlled-trials.
We highlight the number of study participants, the design of study, the results including generalization, and the class obtained according to cicerone.

- *First phase:* the objective of this phase is to interpret typical emotional contexts using the preserved semantic knowledge of the patients. The exercises are based on emotions frequently associated with everyday life situations.
- *Second phase:* this phase focuses on judging static visual clues using exercises in the form of drawings or photos.
- *Third phase:* in the third phase, the participant has to judge dynamic visual cues. At the beginning, the exercises included a single modality, visual or verbal, to gradually integrate the other one. In this phase, participant is invited to play games role, to view video sequences with one or two modalities and to observe the model of the therapist.
- *Fourth phase:* In the last phase, the exercises target inferences about emotional behavior and situational cues. In this phase, participant learns to differentiate sarcasm from lie through the therapist's model, role-play and video footage.

Some of the given exercises were adapted from Spence (1995) and Peterson and Gannoni (1988) for training social skills in children and teenagers. The exercises used mirror imitation, photos, drawings, videos and role plays (Fig. 1). Most activities were structured in the form of games to increase motivation. The capacities developed are inspired by the work of Ekman (2003a, 2003b), Ekman and Friesen (1975), Feldman, Philippot, and Custrini (1991) and Mehrabian (1972). This intervention program used errorless learning techniques, self-instruction, imitation, and regular or immediate feedback. Numerous simulation tasks were dispersed in the program to increase the encoding. Regular reviews in session and at home promoted the storage of information.

FIG. 1 Stimuli from "Reading a smile (And other great expressions): an emotion perception treatment program" (Bornhofen & McDonald, 2009). These verbal and visual emotional stimuli were used in the emotion perception rehabilitation from Bornhofen and McDonald. The cards represent actors expressing an emotion, which was written below for easy identification by the patient. Thanks to Bornhofen, C., & McDonald, S. for allowing us the use of their image.

The study results showed improvement in dynamic and static emotional perception immediately after the intervention and after one month (Bornhofen & McDonald, 2008a, 2008b). The sarcasm-lie differentiation was also improved. However, the authors found no change for sarcasm-sincerity differentiation or for social functioning.

Radice-Neumann et al. (2009) and Neumann et al. (2015): Face intervention program

The "Face Intervention Program" designed by Radice-Neumann et al. (2009) was replicated by Neumann et al. (2015), and the effectiveness has been assessed in two randomized controlled trials with contrasting results. This program targeted the ability to recognize facial emotions in a sample. The training program is an individual and computerized program, completed by the assistance of a therapist.

The total duration of the training program is 9 h, spread over 2 or 3 weeks. Participant has 2 or 3 trainings a week with sessions of 1 or 1.5 h. The intervention seeks to improve 3 key skills in recognizing emotions:

- Pay attention to facial features indicative of emotions at the individual and global level.
- Associate these clues with personal experiences to associate them to emotions of others.
- Develop a conceptual understanding of emotions.

The goal of treatment, achieved through these skills, is to help participant to recognize the feelings of others and improve socialization. The participant should complete and pass 85% of the exercises to move to the next level. Otherwise, he has to repeat the exercise until success. If the participant took a long time to pass a level, the therapist could guide him through the next level. However, in the next session, he has to validate the level failed.

This program has two part-interventions. In the first step, the participant is stimulated to focus attention on emotionally relevant facial features. In the second part, "Part II", the participant is stimulated to focus attention on his own internal emotions, given that self-emotion processing is important for identification of feelings of others. For the part one, the computerized software offers 7 difficulty levels which contain emotions with obvious expression at the first levels and subtle emotions at the later levels. The 40 faces used for training are productions in 3 dimensions, taken from the Pennsylvania Emotion Recognition Test (PERT; Gur et al., 2002):

- *First step* "Identify the emotion": the software drew colored arrows to focus the participant attention to the facial features rich in emotional information. The participant has to observe these traits and identify the emotion expressed among four propositions. A total of 9 faces are presented and identified. Immediate feedbacks are given, depending on its response. When the participant gives a correct answer, the therapist asks him to describe the traits that helped him identify the emotion. If the patient fails, the software describes the characteristic features of emotion trained and the patient must re-identify the emotion (Fig. 2).

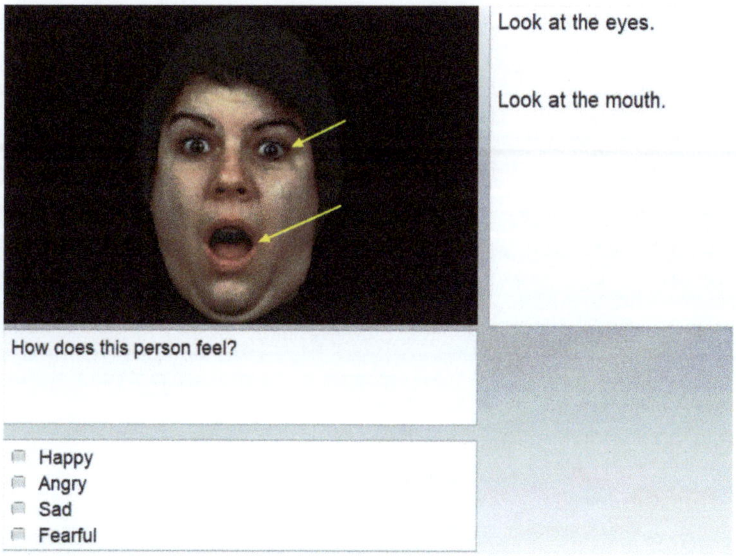

FIG. 2 Exercises from the Face Intervention Program (Radice-Neumann et al., 2009): Emotion identification. This exercise appears in the First step of the rehabilitation program of Radice-Neumann et al. The arrows and indications to the right allow the patient to pay attention to areas of the face rich in emotional information. Thanks to Radice-Neumann, Zupan, Tomita and Willer for allowing us the use of their image.

- *Second step* "Identifying emotion and personal experience"*:* the same 9 faces are presented in random order and the participant should identify them without previous clues (colored arrows) on facial features. The participant is asked to briefly describe a situation in which he felt this emotion.
- *Third step* "Identifying the emotion and imitating the emotion"*:* the software presents the 9 faces in random order, and asks the participant to identify them without clues. The participant is asked to imitate the emotion in front of a mirror.
- *Fourth step* "Resemblance of faces, identify similar characteristics in different faces"*:* this step stimulates the global recognition of emotions through 2 different faces. The participant has to choose an overall description of the emotion between different proposals concerning the characteristic traits, and this for 4 different emotions: joy, sadness, nervousness and fear.
- *Fifth step* "Election of faces, distinguish different emotions of other emotions "*:* the software asks the participant to choose the face representing an emotion given among four propositions. After selecting a face, he is asked to describe the characteristics indicating the given emotion.
- *Sixth step* "Advanced treatment of emotion, deep knowledge emotions"*:* the participant must put himself in the place of the people presenting the emotion on the screen in a total of 8 faces. He is requested to feel as keenly as possible the emotions, in order to be able to answer questions about the situation experienced, the changes physiological feelings, the thoughts that have arisen, the reaction to the situation, the perception of own emotion and the reaction of third parties to its reaction (Fig. 3).

After the administration of the "Part I," the participant could perform an additional exercise of comparison of emotions. This exercise presented two faces at the same time which represented emotions usually confused, such as sadness and nervousness. The participant had to answer different questions about characteristic facial features and on the energy level of the emotion.

In "Part II," the trainer chose 8 emotional events that the participant has described in "Part I." The emotional events correspond to 4 emotions trained in the fourth step (joy, sadness, nervousness and fear). For the most relevant emotional events, the participant is asked to describe the physical and physiological changes of the body which is associated with each emotion. At the end of "Part II", the participant could do a final computer exercise to review the learnings from the "Part I".

The first study (Radice-Neumann et al., 2009) on a sample of 19 patients, 1 year from their head trauma in moderate–severe stage, compared emotion processing from faces to emotion processing from written context. The authors had found a significant improvement in the experimental group on face emotion recognition from faces and ability to infer emotions from context, with a generalization to socio-emotional behavior, including lower levels of proxy-rated aggression. In contrast, these findings were not replicated in a larger study ($n=71$) (Neumann et al., 2015). In this study, the authors showed an improvement of facial affect recognition occurring only in the experimental group, which was maintained after 6 months, after 9 h of computer-based training with a therapist. However, there was no generalization to untrained skills.

The program has the advantages to have a validated manual, software structured by levels of difficulty and an easy application.

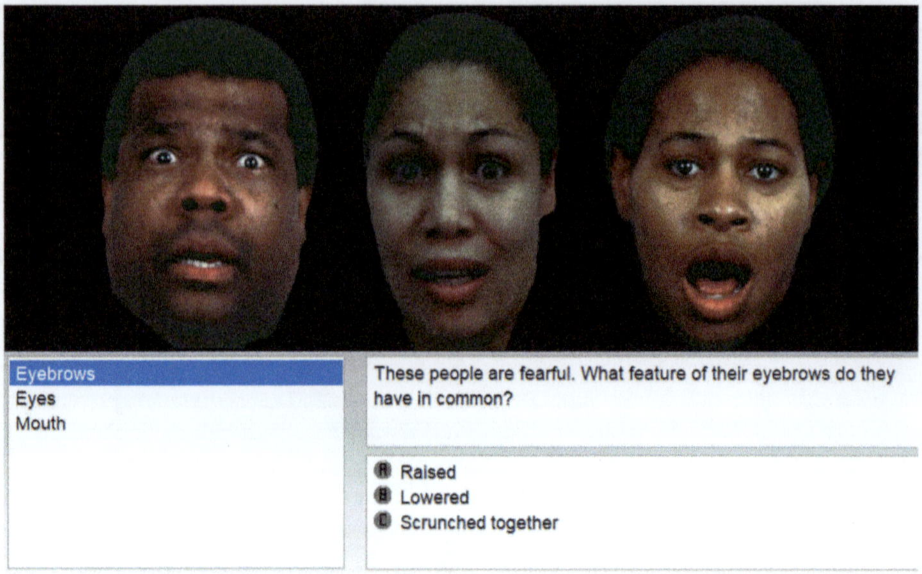

FIG. 3 Exercise from Face Intervention Program (Radice-Neumann et al., 2009): Emotion abstraction. This exercise promotes the abstraction of common features of an emotion from several photos of actors. Thanks to Radice-Neumann, Zupan, Tomita and Willer for allowing us the use of their image.

McDonald et al. (2013): Emotional prosody recognition

This short intervention aims to improve recognition of emotional prosody. It was developed by Bornhofen and McDonald (2008a, 2008b), with studies centered on the rehabilitation of recognition of facial emotions, voices and gestures in severe and chronic head trauma. The program has been validated in people who have suffered brain damage from head trauma, stroke or other medical lesions at an average of 9 years from the lesion. It takes place over 3 sessions of 2 h, with a break during the session.

The program uses exercises of progressive difficulty, guided tasks of perception and distinction of prosodic indices. The activities are structured like a game to increase motivation. The tasks include activities in group, task reviews and home exercises.

The intervention has two objectives to be achieved in the sessions:

- The establishment of a common vocabulary of words and emotional categories, through group debates on the 7 basic emotions and associated behaviors.
- The increase in the ability to distinguish prosody patterns associated with emotions, with simple examples at the beginning which become more complex, with more subtle emotions.

To achieve these goals, the techniques used are vicar learning, peer support, role play and listening to recordings, modeling, errorless learning, positive reinforcement for effort, feedback regular and immediate, the use of verbal context to obtain emotional cues and variety of stimuli used.

The program has been shown to be not effective in a randomized, controlled study with waitlist. But out of 10 patients, 6 showed a significant improvement in recognition of emotional prosody 1 month after the procedure. Some patients reported subjective improvement in understanding emotions in lifelong conversations daily, but no objective measure was used to quantify this change.

Guercio et al. (2004): Facial emotion recognition

This behavioral-inspired intervention leads to the recognition of facial emotions through the "match-to-sample" learning method or "correspondence to the stimulus." It consists of learning by trial and error, by selecting the correspondence to a stimulus presented among 3 options. In this case, Guercio et al. (2004) use labels with the names of the emotions as basic stimuli (Stimulus A), Ekman faces representing emotions (Stimulus B) and photos of strangers showing the same emotions (Stimulus C). Photos used for C stimuli were created with actors who had different age and sex compared to actors on stimuli B.

The program training is computerized and does not require the intervention of a therapist. The learning method uses the "one-to-many" paradigm to train associations between stimuli. Thus, the participants go through 4 modules combining the following stimuli: B-A, C-A, B-C and C-B. When the participants chose an option among the three offered, immediate feedback appears on the screen. To move on to the next module, the participant has to pass 8 of 9 items. Each module lasts between 15 and 30 min, and stimuli are separated from feedback with a 3 s inter-stimuli phase.

The program is individual and has been tried on single case studies with 3 participants (Guercio et al., 2004) who suffered a head trauma from 3 to 6 months before the program admission. This intervention has been shown to be effective, in this study, on recognition facial emotion immediately after the operation, but no follow-up was studied. The trained emotions were happiness, sadness and anger.

Williamson and Isaki (2015): Modified facial affect recognition

Williamson and Isaki (2015) have created a modified facial affect recognition training tested in two adults with moderate to severe chronic traumatic brain injury, in cases study paradigm. The participants were two males, 53 and 44 years old who had suffered from moderate and severe TBI and carried out this protocol respectively 12 and 7 years post-injury. The total duration of the intervention was 6 weeks, with 60 min sessions, one or two sessions per week.

The intervention took place with telepractice thanks to the VSee web-based teleconferencing program, installed on personal computers of the participants. The therapist and the participants saw each other through webcam. The main material used was cards with static emotion pictures, representing facial expression of happiness, sadness, angry, surprise and disgust with different gender, ethnic group and intensity of emotion. Participants were asked to identify those emotions and personally reflecting on those emotions. Also, they were asked to identify sarcasms and emotions in social stories and role-playing.

The main objective of the training was to identify emotions, but the program targets social communication skills too. The intervention had three parts, among which the first one needed two sessions a week during 3 weeks. The frequency of sessions decreased for the second and third part, to one session a week:

- *First part*: The therapist presents 2D images of facial expression emotion cards representing angry, sad, bored, thoughtful, happy, uncomfortable, fearful, and disgusted. For each session, two new emotions were presented to the participant. Guided by the therapist, the participant has to identify emotions using visual and verbal cues.
- *Second part:* participant was asked to describe the physical and psychological changes that occurred in social situations corresponding to various emotions.
- *Third part*: through role-playing, the therapist presents to the participant various social stories including by sarcasm. The participant was asked to identify sarcasm, the facial emotion expressions and the real emotion that the character was feeling. In the final sessions, the participant reviewed all emotions cards. In these sessions, executive function and theory of mind were both addressed.

After this training, participants have shown an improvement of facial affect and above all they have shown improvements in recognizing emotions which were impaired before the training. The authors evaluated performances before and after training, but no follow-up measures were taken.

Treatment of theory of mind (ToM)

Lundgren, Brownell, Cayer-Meade, and Spitzer (2007): Decoding first and second order beliefs and intentional deception

Lundgren et al. (2007) have developed a theory of mind training program for patients suffering from head injury or right hemispherical stroke. This program is inspired by the theory of mind tasks existing for the autistic population (Wellman et al., 2002) and for patients with right hemisphere lesions (Lundgren, Brownell, Cayer-Meade, & Roy, 2006a, 2006b). The intervention trains participants to decode first and second order beliefs, as well as intentional deception.

The intervention uses comics with speech bubbles making the characters thinking explicit to facilitate access to their state of mind. This method promotes the adoption of the perspective of one or several characters in order to evaluate their thoughts, to understand the differences between their thoughts and their possible actions, and to inhibit egocentric thoughts.

The program had 4 phases preceded by a warm-up phase. There are 2 phases dedicated to first-order thoughts and 2 phases dedicated to second-order thoughts. To advance among the tasks of progressive difficulty, the participant has to obtain a 90% or 100% success in the module. In this way, the duration of the training varied depending on participant performance. In all phases, a therapist guided the participant to use the characters bubbles to predict their behavior.

This intervention has been tried on 3 post-right hemispherical stroke patients and 3 patients with head trauma in gradually published case studies. The pilot study (Lundgren et al., 2007) observed, in two participants following a right hemispherical stroke, a good adherence to the program, as well as improved performance in theory of mind in the later stages of the intervention. The second study (Lundgren & Brownell, 2011) had found an improvement in theory of mind applied to comics, and a slight improvement in a battery of affection for a post-stroke participant with right hemispherical damage. The third and last published study (Lundgren & Brownell, 2015) found, in 3 head trauma victims, a gradual improvement in theory of mind and executive functioning.

Rehabilitation programs of interpersonal and communication skills of theory of mind

Helffenstein and Wechsler (1982): Interpersonal process recall remediation (IPR)

Helffenstein and Wechsler (1982) developed the Interpersonal Process Recall (IPR) Remediation, a training program for the interpersonal and communicational skills. The participants had received 20 h of training in sessions of 55–65 min. In the first 10–15 min, participants saw a videotaped unstructured interaction between the subject and one staff member. If one skill was targeted, the videotaped interaction was structured. After that, the participant and 2 staff members reviewed and processed the tape during 45–50 min. Helped by the professionals, the participant developed a more adaptative or appropriate means of interaction. To do that, the staff gave modeling and feedback to the participants, who rehearsed the new skill.

This study had a randomized controlled design with a control group. The participants were 13 males and 3 females from 17 to 35 years, 2 years after their traumatic brain injury. After this intervention, the authors found significantly changes in anxiety trait and increase of self-concept (including the social-self and moral-self). Also, the results have shown improvements in interpersonal and communication skills and increase in the frequency of their adaptative and effective communication skills. These changes were maintained 1 month after the remediation period, but limited to a few skills.

Dahlberg et al. (2007): Social communication skills treatment in group design

Dahlberg et al. (2007) have put into practice the social communication skills treatment from the workbook of Hawley and Newman (2006) in a randomized controlled trial including 25 participants with moderate to severe traumatic brain injury. This study used a group approach, with therapy groups of 8 participants which met for 1.5 h once a week for 12 weeks.

This treatment took into account the personal goals of participants, focus on their self-awareness and self-assessment. The strategies used in this training program were the group feedback, problem solving, practice of new skills and video feedback. The exercises encouraged generalization through homework and family and friend involvement.

The results have shown significant improvement in social communication skills that were maintained after 6 months follow-up. Participants have improved their participation in conversation, role of listener and speaker, social style and quantity and clarity of expression. Some of these improvements remained stable after 9 months. Further, a positive effect was observed on the life satisfaction of participants.

Finch et al. (2017): Remediation of social communication program with group or individual design

The remediation of social communication program, developed by Finch et al. (2017), has been tested in eight participants with traumatic brain injury, among whom 7 had severe trauma and one had a moderate trauma. The time after trauma varied between 4 months to 4 years and 8 months, but only 2 participants (with 4 and 6 months after trauma) were in the period of spontaneous recovery.

During the training, the participants were involved in group sessions with or without individual sessions, twice a week for 1 h. A speech pathologist leaded the sessions using metacognitive strategies and goal-based intervention targeting the social communication skills. The group sessions allowed the group feedback and the opportunity to practice the skills trained during the individual sessions in a safe environment.

Six of the participants presented changes in their self-reported communication skills which approached statistical significance after the remediation. The evaluation of communicative behaviors showed improvements which were maintained 1 month after remediation, especially for aesthetic scales including gesture, humor, speech rate, volume and pauses during communication.

Winegardner et al. (2016): Perspective group training

Based on the principle that anger responses in the traumatic brain injury population could be a consequence of poor perspective-taking, through a distorted analysis of others intention, Winegardner et al. (2016) had developed a perspectives training to treat anger. This "Perspective group" was applied once a week for 1 h for 6 weeks, into two case studies. The participant were two men with brain injury, one after severe traumatic brain injury 14 years before, and the other after subarachnoid hemorrhage while undergoing neurosurgical intervention 11 years before.

The treatment includes exercises with hypothetical situations in video scenarios, role-playing and real-life situations. Guided by one psychologist and two speech and language therapist, the participant used cognitive reappraisal strategies and perceptual positioning to modify their negative attributions.

After completing the intervention, authors reported diminished aggression but minimal changes were observed on perspective-taking. However, the authors emphasized that participants' spouses have reported behavioral changes related to better perspective taking and decreased aggression. Performances in the theory of mind tasks, on the other hand, remained stable after rehabilitation.

Gabbatore et al. (2015): Cognitive pragmatic treatment

The Cognitive Pragmatic Treatment (CPT) is a rehabilitation program from Gabbatore et al. (2015) which addresses the communicative abilities. The authors tested this treatment in eight participants with severe traumatic brain injury, one to 16 years after injury. After the training period, six of these participants were evaluated with functional magnetic resonance imaging (fMRI).

The training takes place twice a week for 3 months, during sessions of 1.5 h. Each session was dedicated improving one communication modality, as linguistic, extralinguistic or social appropriateness. One trained psychotherapist guided the group of five patients through the exercises.

The results after training were significantly better than at baseline in communicative abilities for production and comprehension. These improvements were maintained after 3 months. Thus, the fMRI has shown changes in the brain with a significant increase in functional activity in the bilateral pre-central gyrus, the right middle and superior temporal gyri, in the right cingulate gyrus and in the left inferior parietal lobe. In order to measure these changes, the authors had evaluated the amplitude of low frequency fluctuation (ALFF) index.

Multi-component rehabilitation programs of social cognition

Westerhof-Evers et al. (2017): T-SCEmo program

The T-ScEmo program, recently developed by Westerhof-Evers et al. (2017), targets deficits in facial emotion recognition, empathy, theory of mind and behavioral regulation in head trauma patients. The authors tested this intervention in a multicenter rehabilitation RCT study including 60 patients with traumatic brain injuries with ventro-median lesions at a chronic state.

The intervention began with psychoeducation sessions to improve the awareness of the difficulties associated with deficits in social cognition and to motivate the patient to participate in training. Then, the program was divided into 3 modules concerning the targeted domains of social cognition:

- *First module "Emotion perception":* It involved encoding facial emotions through computer exercises, imitation and simulation. The participant learns the parts of the face that are fundamental for recognizing emotions using the computer software. He is asked to identify the characteristic cues of primary and secondary emotions. The three strategies used to favor emotion perception are facial-feature processing, mimicry and the experience of one own emotions.
- *Second module* "Perspective taking and theory of mind": It included empathy and taking the perspective necessary to optimize the interpretation and understanding of social information. In this module, the patients learned the pattern of interdependence between thoughts, actions and emotions of the cognitive behavioral therapy. The exercises focus on the notion that own ideas and emotions may differ from those of others, and that these two perspectives may coexist. The participant carries out perspective-taking and learning exercises on how to openly inquire about the ideas or emotions of others. To do that, the participant used hypothetical and real-life personal incidents. This module included psychoeducation and self-monitoring strategies.
- *Third module "Social behavior":* It targeted regulation of social behavior through the training of basic social skills as listening, turn taking or giving compliments. The participant learned to control uninhibited behavior and to manifest behaviors socially appropriate. To do this, there were exercises to detect unreasonable social behavior and to identify factors favoring them. In this module, the participant was trained in cognitive strategies to manage behavior and strategies to limit damage in case of inadequate reactions.

The control group received computerized cognitive function training. Compared with the control group, the experimental group improved significantly in the perception of emotions, theory of mind and empathic behavior. The intervention has shown beneficial effects on social behavior in everyday life, with efficacy in societal participation, quality of life and quality of life partner relationship. All these changes were stable during 5 months after training.

McDonald et al. (2008): Social skills treatment-perception of emotions, emotional regulation and social behavior

This social skills treatment trains several aspects of social cognition: perception of emotions, emotional regulation and social behavior. It was created by McDonald et al. (2008) based on the results of previous studies (Bornhofen & McDonald, 2008a, 2008b; Flanagan, McDonald, & Togher, 1995). The intervention was tested in a population suffering from head trauma or anoxia, with severe lesions for more than 12 months.

The duration of the treatment was 12 weeks. There were two sessions per week: one collective session of 3 h, and one individual session of 1 h.

- *The training group:* It was composed by 3–5 participants and 2 therapists, and was entirely dedicated to training social behavior and social perception. During 2 h, the participants worked on different aspects of their social behavior, such as assertiveness, starting a conversation, or making compliments. This part of the session began with a warm-up game, a review of chores for the home and introduction to the topic of the session. Participants were looking for set of potential solutions to social situations and therapists suggested models of appropriate and inappropriate behavior, as well as a role play to develop skills. The remaining hour was dedicated to training social perception. With the therapists, participants decoded facial expressions, voices and gestures and made inferences about the nature of the exchange or the predisposition of the interlocutor. The exercises had progressive difficulty and ranged from static (photos) to dynamic emotions (videos, recordings). The exercises used discrimination, games role playing, mirror imitation, and social skills games.

- *The individual training:* guided by a clinical psychologist, these sessions targeted emotional regulation and helped participants understanding and identifying their self-esteem, signs of anxiety and depression. The session used cognitive-conductive orientation techniques such as relaxation, cognitive strategies, gradual exposure and training in assertiveness.

The entire program is based in positive reinforcement techniques, immediate feedback, repetition and errorless learning. It has been validated in a randomized study with control group and waitlist. The effects obtained were modest, but authors have observed an improvement in social behavior. Nevertheless, social perception and emotional regulation did not improve. The advantage of this program is its combination of individual and collective care that allows the training of the targeted social skills in a social situation. It counts with a manual for two of its three targeted skills.

Conclusion

This review highlights the growing interest in the rehabilitation of social cognition in brain-injured people, despite the significant difference in publications with other populations, such as people suffering from schizophrenia or autistic spectrum disorder. In this way, five interventions targeting emotion perception (Bornhofen & McDonald, 2008a, 2008b; Guercio et al., 2004; McDonald et al., 2009; McDonald et al., 2013; Neumann et al., 2015; Radice-Neumann et al., 2009; Williamson & Isaki, 2015) and one intervention targeting theory of mind (Lundgren & Brownell, 2015) has been developed since 2008 to today. From 1982, five programs targeting interpersonal and communication skills of theory of mind has been tested (Dahlberg et al., 2007; Finch et al., 2017; Gabbatore et al., 2015; Helffenstein & Wechsler, 1982; Winegardner et al., 2016). Very few authors have attempted to create comprehensive social cognition rehabilitation programs targeting the full range of abilities. This is the case for McDonald et al. (2008) and Westerhof-Evers et al. (2017), whose last team obtained improvement in two domains of social cognition with generalization to social participation, proxy-rated empathy and goal attainment. Progress in social cognition rehabilitation seems evident, but there is still a lot of work to ensure effective rehabilitation of people with traumatic brain injury. Among these 16 studies, only 9 were randomized-controlled trial studies with a class I from the Cicerone classification (Table 1). The generalization effect remains rare in these studies, and we found little or no interventions targeting empathy, emotional awareness or emotional regulation. A comprehensive rehabilitation program is currently being developed by Vallat-Azouvi and colleagues. The Cogniso-TC program targets the capacities of social cognition in a specific way according to the needs of the individual.

Despite the still relatively low level of evidence for the population of patients with non-degenerative acquired brain injury, and then the relatively low level of evidence, the few published studies seem to show effectiveness in different areas of SC. Interest is growing in clinical practice and a number of protocols or materials are emerging and are being used in the clinic. The clinician will have to rely on methods validated for this population in order to guarantee maximum effectiveness. Methodologies, both from the point of view of the targeted components (specific components or comprehensive rehabilitation) or the procedures used (individual or group rehabilitation) are still debated in the literature.

Applications to other areas of neuroscience

In this chapter, we present a review of the literature concerning the social cognition rehabilitation in traumatic brain-injured patients, including specific and broad rehabilitation programs, with some programs also addressing communication skills.

Some of these programs look like those developed for patients suffering from schizophrenia and autism spectrum disorders. They use the same stimuli or the same cognitive techniques. For example, various emotion recognition programs which have shown its effectiveness in the schizophrenia population (see for example the Micro-Expression Training Tool by Ekman, 2003a, 2003b; Attention-Shaping by Combs, Tosheva, Wanner, & Basso, 2009) used similar exercises to the Face Intervention Program from Radice-Neumann et al. (2009) to draw attention to the more informative parts of the face. A number of adaptation will be necessary for some components of social cognition for head trauma patients.

Indeed, apart from the specificities related to each population, the learning and rehabilitation techniques, as well as the structure of certain exercises, seem to be effective for several different populations. In this way, perhaps some of these programs could be adapted for application to the stroke population, which currently has no specific program for the rehabilitation of social cognition.

Two studies seem to have asked this question before and have already included post-stroke patients in their validation studies. The first one is the study by Lundgren et al. (2007), which included 3 case studies on post-stroke with right hemispherical lesions in their study about the rehabilitation of theory of mind. The second one is the study by McDonald et al. (2013). This study included 3 post-stroke patients, who have benefited from the rehabilitation of emotion recognition in prosody. These authors did not mention any difference in the results of patients with the two pathologies, which suggests that they obtained the same conclusions for both.

We do not yet have enough evidence to believe that the studies cited in this chapter could be effective for people with stroke, but the idea of this application is a lead to be exploited in future research.

Mini-dictionary of terms

Social cognition: Socio-emotional abilities and experiences which regulate the relationships between individuals, and which can explain individual human behaviors or behaviors in group.
Rehabilitation: Techniques and methods that serve to recover, cognitive processes, functions or activities that has diminished or has been lost due to an accident or illness. The techniques can be based on restoration, or reorganization or compensation hypotheses.
Generalization: Improvements of skills in other cognitive functions that have not been specifically trained in the rehabilitation program.
Transfer in daily life: Functional improvement in everyday life, related with the processes involved in rehabilitation.
Errorless learning techniques: Techniques based on a frequent repetition of the correct answers; this technique prevent patients from giving wrong answers.
Self-instruction technique: Cognitive learning strategy to control one' own behavior through auto-repeated instructions.
Modeling: Technique where the learned person (for example, another people in a group, a therapist or a teacher) shows a behavior that the patient or student will have to copy. This technique is not systematically conscious but induced by the therapist in a group dynamic.
Problem-solving strategy: Method used to identify step by step the key elements of the problem in order to solve it in different steps. This technique is mainly based on verbalization and feedback.

Key facts of rehabilitation in social cognition

Rehabilitation in social cognition:

- Social cognition rehabilitation is in its infancy in the field of head trauma.
- Some programs are under development.
- Results of studies showed, in general, effectiveness in improving certain capacities.
- The studies designs are heterogeneous: randomized-controlled trial, cohort studies, comparative studies and single case studies.
- Some studies included a follow-up to assess the maintenance of improvement.
- Very few studies assess the generalization of acquired knowledge in everyday life.

Cognitive rehabilitation strategies in social cognition:

- Some programs are individual, in order to target specifically the impaired abilities of the patient, or collective, in order to train the social abilities in a social group. Few studies use or compare both approaches.
- The most used strategies are errorless learning, regular feedback, repetition, positive reinforcement, modeling, self-instruction, imitation and problem-solving strategies.
- Regarding studies using role-playing as a tool, some recorded or filmed the participants, others did not.
- Some programs have homework to promote generalization in everyday life.

References

Bornhofen, C., & McDonald, S. (2008a). Treating deficits in emotion perception following traumatic brain injury. *Neuropsychological Rehabilitation*, *18*(1), 22–44.

Bornhofen, C., & McDonald, S. (2008b). Comparing strategies for treating emotion perception deficits in traumatic brain injury. *The Journal of Head Trauma Rehabilitation*, *23*(2), 103–115.

Bornhofen, C., & McDonald, S. (2009). *Reading a smile (and other great expressions): An emotion perception treatment program*. Randwick: Australian Society for the Study of Brain Impairment.

Brunet-Gouet, E., Oker, A., Martin, J. C., Grynszpan, O., & Jackson, P. L. (2016). Advances in virtual agents and affective computing for the understanding and remediation of social cognitive disorders. *Frontiers in Human Neuroscience*, *9*, 697.

Cassel, A., McDonald, S., Kelly, M., & Togher, L. (2019). Learning from the minds of others: A review of social cognition treatments and their relevance to traumatic brain injury. *Neuropsychological Rehabilitation*, *29*(1), 22–55.

Cicerone, K. D., Goldin, Y., Ganci, K., Rosenbaum, A., Wethe, J. V., Langenbahn, D. M., et al. (2019). Evidence-based cognitive rehabilitation: Systematic review of the literature from 2009 through 2014. *Archives of Physical Medicine and Rehabilitation*, *100*(8), 1515–1533.

Cicerone, K. D., Langenbahn, D. M., Braden, C., Malec, J. F., Kalmar, K., Fraas, M., et al. (2011). Evidence-based cognitive rehabilitation: Updated review of the literature from 2003 through 2008. *Archives of Physical Medicine and Rehabilitation*, *92*(4), 519–530.

Combs, D. R., Tosheva, A., Wanner, J., & Basso, M. R. (2009). Remediation of emotion perception deficits in schizophrenia: The use of attentional prompts. *Schizophrenia Research*, *87*, 340–341.

Dahlberg, C. A., Cusick, C. P., Hawley, L. A., Newman, J. K., Morey, C. E., Harrison-Felix, C. L., et al. (2007). Treatment efficacy of social communication skills training after traumatic brain injury: A randomized treatment and deferred treatment-controlled trial. *Archives of Physical Medicine and Rehabilitation*, *88*(12), 1561–1573.

Ekman, P. (2003a). *Emotions revealed: Recognising faces and feelings to improve communication and emotional life*. New York: Times Books.

Ekman, P. (2003b). *Emotions revelated: Recognizing faces and feelings to improve communication and emotional life*. New York: Henry Holt and Company.

Ekman, P., & Friesen, W. V. (1975). *Unmasking the face*. Englewood Cliffs, NJ: Prentice-Hall.

Feldman, R. S., Philippot, P., & Custrini, R. J. (1991). Social competence and nonverbal behaviour. In R. S. Feldman, & B. Rime (Eds.), *Fundamentals of nonverbal behavior* (pp. 329–350). Cambridge: Cambridge University Press.

Fernandez-Sotos, P., Torio, I., Fernandez-Caballero, A., Navarro, E., Gonzalez, P., Dompablo, M., et al. (2019). Social cognition remediation interventions: A systematic mapping review. *PLoS One*, *14*(6), e0218720.

Finch, E., Cornwell, P., Copley, A., Doig, E., & Fleming, J. (2017). Remediation of social communication impairments following traumatic brain injury using metacognitive strategy intervention: A pilot study. *Brain Injury*, *31*(13–14), 1830–1839.

Flanagan, S., McDonald, S., & Togher, L. (1995). Evaluation of the BRISS as a measure of social skills in the traumatically brain injured. *Brain Injury*, *9*, 321–338.

Gabbatore, I., Sacco, K., Angeleri, R., Zettin, M., Bara, B. G., & Bosco, F. M. (2015). Cognitive pragmatic treatment: a rehabilitative program for traumatic brain injury individuals. *The Journal of Head Trauma Rehabilitation*, *30*(5), E14–E28.

Gaudelus, B., Virgile, J., Geliot, S., The GAÏA/RECOS Team, & Franck, N. (2016). Improving facial emotion recognition in schizophrenia: A controlled study comparing specific and attentional focused cognitive remediation. *Frontiers in Psychiatry*, *7*, 105.

Guercio, J. M., Podolska-Schroeder, H., & Rehfeldt, R. A. (2004). Using stimulus equivalence technology to teach emotion recognition to adults with acquired brain injury. *Brain Injury*, *18*(6), 593–601.

Gur, R. C., Sara, R., Hagendoorn, M., Marom, O., Hughett, P., Macy, L., et al. (2002). A method for obtaining 3-dimensional facial expressions and its standardization for use in neurocognitive studies. *Journal of Neuroscience Methods*, *115*(2), 137–143.

Hawley, L., & Newman, J. (2006). *Social skills and traumatic brain injury: A workbook for group treatment*. Denver: Authors.

Helffenstein, D. A., & Wechsler, F. S. (1982). The use of interpersonal process recall (IPR) in the remediation of interpersonal and communication skill deficits in the newly brain-injured. *Clinical Neuropsychology*, *4*(3), 139–142.

Hogarty, G. E., & Flesher, S. (1999). Practice principles of cognitive enhancement therapy for schizophrenia. *Schizophrenia Bulletin*, *25*(4), 693–708.

Lundgren, K., & Brownell, H. (2011). Remediation of theory of mind impairments in brain-injured adults. In J. Guendouzi, F. Loncke, & M. J. Williams (Eds.), *The handbook of psycholinguistic and cognitive processes: Perspectives in communication disorders* (pp. 579–602). Hove: Psychology Press.

Lundgren, K., & Brownell, H. (2015). Selective training of theory of mind in traumatic brain injury: A series of single subject training studies. *The Open Behavioral Science Journal*, *9*, 1–11.

Lundgren, K., Brownell, H., Cayer-Meade, C., & Roy, S. (2006a). Remediation of metaphor comprehension deficit: A pilot study. *Brain and Language*, *99*(1–2), 69–70.

Lundgren, K., Brownell, H., Cayer-Meade, C., & Roy, S. (2006b). Remediation of metaphor comprehension deficit: A pilot study. In *Poster presentation at the annual convention of the American Speech-Language Hearing Association*. FL: Miami.

Lundgren, K., Brownell, H., Cayer-Meade, C., & Spitzer, J. (2007). Training theory of mind following right hemisphere damage: A pilot study. *Brain and Language*, *103*(1), 209–210.

McDonald, S., Bornhofen, C., & Hunt, C. (2009). Addressing deficits in emotion recognition after severe traumatic brain injury: the role of focused attention and mimicry. *Neuropsychol Rehabilitation*, *19(3)*, 321–339.

McDonald, S., Tate, R., Togher, L., Bornhofen, C., Long, E., Gertler, P., et al. (2008). Social skills treatment for people with severe, chronic acquired brain injuries: A multicenter trial. *Archives of Physical Medicine and Rehabilitation*, *89*(9), 1648–1659.

McDonald, S., Togher, L., Tate, R., Randall, R., English, T., & Gowland, A. (2013). A randomised controlled trial evaluating a brief intervention for deficits in recognising emotional prosody following severe ABI. *Neuropsychological Rehabilitation*, *23*(2), 267–286.

Mehrabian, A. (1972). *Nonverbal communication*. Chicago: Aldine-Atherson.

Neumann, D., Babbage, D. R., Zupan, B., & Willer, B. (2015). A randomized controlled trial of emotion recognition training after traumatic brain injury. *The Journal of Head Trauma Rehabilitation*, *30*(3), E12–E23.

Penn, D., Roberts, D. L., Munt, E. D., Silverstein, E., Jones, N., & Sheitman, B. (2005). A pilot study of social cognition and interaction training (SCIT) for schizophrenia. *Schizophrenia Research*, *80*(2–3), 357–359.

Peterson, L., & Gannoni, A. (1988). *Manual for social skills training in young people with parent and teacher programmes.* Beverley, South Africa: Kitchener Press.

Radice-Neumann, D., Zupan, B., Tomita, M., & Willer, B. (2009). Training emotional processing in persons with brain injury. *The Journal of Head Trauma Rehabilitation*, *24*(5), 313–323.

Spence, S. H. (1995). *Social skills training: Enhancing social competence with children and adolescents*. Windsor, UK: Nfer-Nelson.

Tate, R., Kennedy, M., Ponsford, J., Douglas, J., Velikonja, D., Bayley, M., & Stergiou-Kita, M. (2014). INCOG recommendations for management of cognition following traumatic brain injury, Part III: Executive function and self-awareness. *The Journal of Head Trauma Rehabilitation*, *29*(4), 338–352.

Togher, L., Wiseman-Hakes, C., Douglas, J., Stergiou-Kita, M., Ponsford, J., Teasell, R., … Turkstra, L. (2014). INCOG recommendations for management of cognition following traumatic brain injury, Part IV: Cognitive communication. *The Journal of Head Trauma Rehabilitation*, *29*(4), 353–368.

Vallat-Azouvi, C., Azouvi, P., Le-Bornec, G., & Brunet-Gouet, E. (2019). Treatment of social cognition impairments in patients with traumatic brain injury: A critical review. *Brain Injury*, *33*(1), 87–93.

Wellman, H. M., Baron-Cohen, S., Caswell, R., Gomez, J. C., Swettenham, J., Toye, E., et al. (2002). Thought-bubbles help children with autism acquire an alternative to a theory of mind. *Autism*, *6*(4), 343–363.

Westerhof-Evers, H. J., Visser-Keizer, A. C., Fasotti, L., Schönherr, M. C., Vink, M., van der Naalt, J., et al. (2017). Effectiveness of a treatment for impairments in social cognition and emotion regulation (T-ScEmo) after traumatic brain injury: A randomized controlled trial. *The Journal of Head Trauma Rehabilitation*, *32*(5), 296–307.

Williamson, J., & Isaki, E. (2015). Facial affect recognition training through telepractice: Two case studies of individuals with chronic traumatic brain injury. *International Journal of Telerehabilitation*, *7*(1), 13.

Winegardner, J., Keohane, C., Prince, L., & Neumann, D. (2016). Perspective training to treat anger problems after brain injury: Two case studies. *NeuroRehabilitation*, *39*(1), 153–162.

Chapter 42

Psycho-educational intervention on caregivers within the rehabilitation process: From the post-acute to the homecoming phases

Umberto Bivona[a,b], Eva Azicnuda[b], Marta Rapiti[b], and Daniela Silvestro[b]
[a]*Post-Coma Unit, IRCCS Fondazione Santa Lucia, Rome, Italy,* [b]*IRCCS Fondazione Santa Lucia, Rome, Italy*

List of abbreviations

ADL activities of daily life
CINRS Caregiving Impact on Neuro-Rehabilitation Scale
sABI severe acquired brain injury
sTBI severe traumatic brain injury
TBI traumatic brain injury
ToM theory of mind

Introduction

Traumatic Brain Injury (TBI) is still considered as a "silent epidemic" since people are mostly unaware of the magnitude and socio-economic consequences of the brain injury (Roozenbeek, Maas, & Menon, 2013). Indeed, as reported elsewhere in the present Handbook, severe TBI (sTBI) can cause a wide range of neuromotor, neuropsychological and psycho-behavioral sequelae (Arciniegas & Wortzel, 2014; Ciurli, Formisano, Bivona, Cantagallo, & Angelelli, 2011; Rapoport, Mccullagh, Streiner, & Feinstein, 2003), with high unemployment rates (Forslund et al., 2014; Kreutzer et al., 2003; Scaratti et al., 2017) and difficulties in living independently (Kersel, Marsh, Havill, & Sleigh, 2001), fulfilling domestic roles, including housekeeping and parenting (Kieffer-Kristensen, Siersma, & Teasdale, 2013). Moreover, TBI can often cause reduced ability to participate in leisure activities and social isolation (Douglas, 2020; Jourdan et al., 2016). Of note, these difficulties can last up to 10 years post-injury (Douglas, 2017; Ponsford et al., 2014), if not even never-ending (Salas, Casassus, Rowlands, Pimm, & Flanagan, 2018).

As such, sTBI implies an *inter*-professional and long lasting neuro-rehabilitative process, but even more importantly a constant support by at least one informal caregiver (Spoorenberg et al., 2013).

Indeed, since Lezak (1988) stated that brain damage is "a family affair," which involves not only the individual who suffered the brain injury, but also the whole family system (see Fig. 1). In her pioneering work, the author underlined how dramatic could be facing with a person who, mainly due to the post-injury emotional–behavioral changes, in most of the cases seems even another person. Indeed, caregivers may suffer many of the core problems arising from the survivor's dependency, as well as from social reactions to the survivors' disability (Lezak, 1988).

Of note, while caregivers have to face their own burden and concern, they also have an active role in the rehabilitation process (Bivona et al., 2020; Farnese et al., 2020). Indeed, they play a critical role as an extension of the health care system, even if they often need formal training and support to care for persons with sTBI (Ramkumar & Elliot, 2010).

This implies the need of a dual intervention on caregivers, aimed at both sustaining them according to a classical psychotherapeutic approach, as well as educating them on how interacting at best with their loved one, in order to facilitate the neuro-rehabilitation process.

Diagnosis and Treatment of Traumatic Brain Injury. https://doi.org/10.1016/B978-0-12-823347-4.00006-3

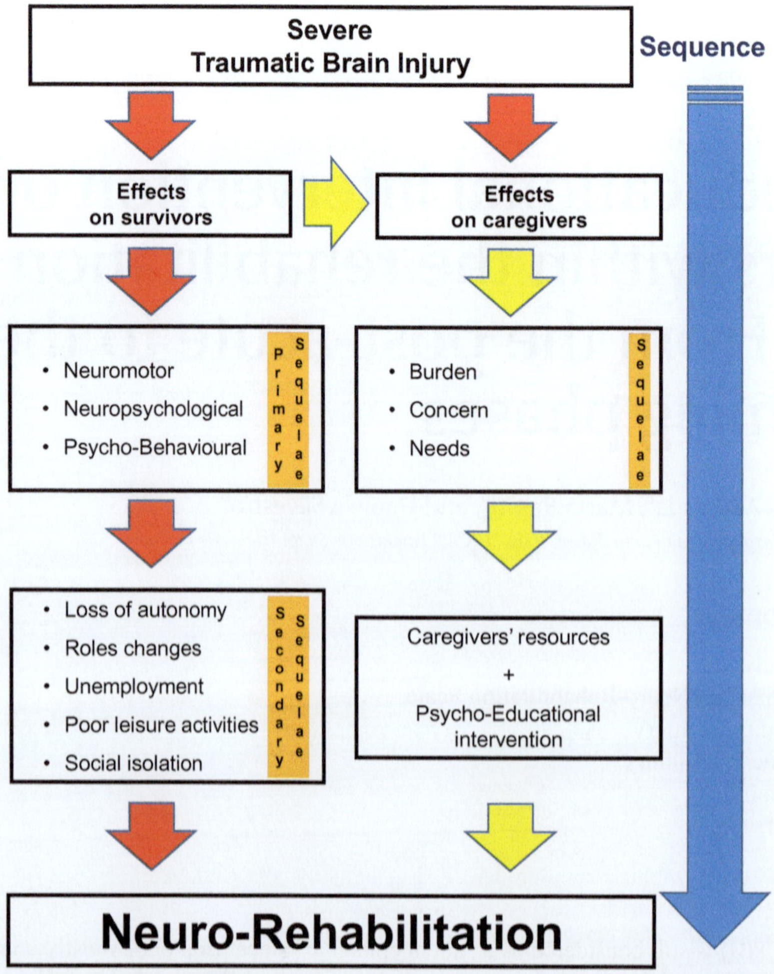

FIG. 1 Flow chart showing the possible direct and indirect effects of a sTBI on both survivors and their caregivers. The figure includes also the possible primary and secondary sequelae of the brain damage on survivors, as well as the main psychological sequelae on caregiver. Therefore, the Neuro-Rehabilitation on survivors after sTBI must address its sequelae after including also caregivers, through a Psycho-Educational support. sTBI = severe traumatic brain injury.

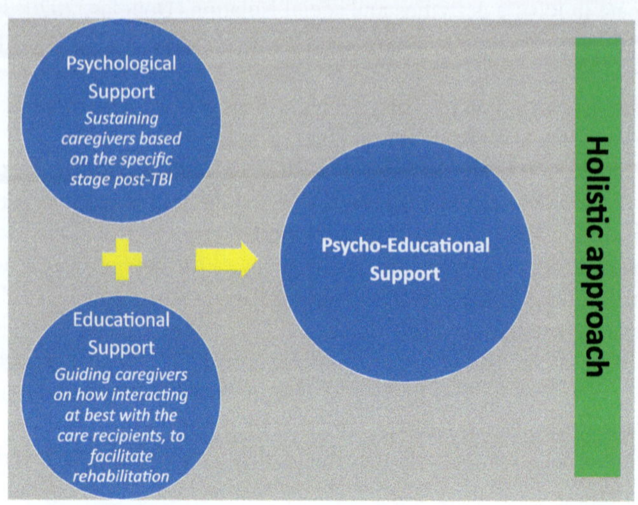

FIG. 2 The Psycho-Educational Intervention on caregivers after sTBI, within a holistic and competence-based approach, stressing health, collaboration, coping, and empowerment. sTBI = severe traumatic brain injury.

The present chapter aims at describing the main characteristics of this "psycho-educational" support, provided by a psychotherapist expert in neuro-rehabilitation to caregivers of survivors of sTBI (see Fig. 2). This dual intervention reflects a holistic and competence-based approach, stressing health, collaboration, coping, and empowerment (Coetzer, 2007; Gómez-de-Regil, Estrella-Castillo, & Vega-Cauich, 2019).

We tried to address all the possible features and issues of such support, according to the different phases after coma recovery, from the post-acute to the homecoming phases.

Caregivers as primary resources within the neuro-rehabilitation process

One of the challenges of providing healthcare services is to enhance its value by integrating caregivers into the care process, both concretely managing their survivor's health conditions and treatment (i.e., *co-executing*) and participating in the whole healthcare process (i.e., *co-planning*) (Farnese et al., 2020). Indeed, clinicians who work in the field of neuro-rehabilitation know very well the importance of an appropriate caregiving approach in helping and accelerating the whole rehabilitation process of persons with sTBI.

However, caregivers can be a double-edged sword. Indeed, it is very common to observe the negative influence of an inappropriate caregiving, that in many cases hamper the neuro-rehabilitation process (Bivona et al., 2020). With this regard, particular attention deserves the different approach often observed in the clinical practice based on the kinship with the care recipient. For example, since due to the brain damage survivors often lose their role in the family system, the balance of functions within it may upset, with predictable consequence even on the couple sexual life (Bivona et al., 2016; Kratz, Sander, Brickell, Lange, & Carlozzi, 2017). Particularly when the survivors is a man, it is common to observe that mothers, more than spouses, even when the care recipient becomes (potentially) independent in many activities of daily life (ADL), may tend to replace him, hampering for instance his/her independence during meals, personal hygiene and care, moving from one place to another, etc. Consequently, many conflicts or alienation between the family of origin and the nuclear family of the survivor can arise, with predictable consequences on caregiving itself (Segev, Levinger, & Hochman, 2018).

This clinical observation, likely related to a tendency, in mothers, to regress in managing their loved one to the typical early relationship mother–child, is in line with studies which demonstrated that spouse caregivers experience more psychosocial distress than parent caregivers (Perlesz, Kinsella, & Crowe, 2000; Verhaeghe, Defloor, & Grypdonck, 2005), as well as lower perceived health status (McPherson, Pentland, & McNaughton, 2000). Indeed, the brain damage often implies interacting with the care recipient as if he/she was a child/adolescent, which for the healthy partners can lead to see caregiving as a new parent/child relationship, with easily imaginable consequences. However, other authors have not found substantial differences between caregivers in relation to kinship (Ponsford, Olver, Ponsford, & Nelms, 2003), while others found that parents even more than partners reported more concerns for the future (Kratz et al., 2017).

On the other hand, adequate caregivers allow reaching more quickly the neuro-rehabilitative aims, for instance when they just observe and supervise survivors in every activity they are already able to do, even if this often implies for caregivers struggling with their natural tendency to help their loved ones, feeling sense of guilty, or engage in conflicting or unsatisfactory relations with the care recipient.

Indeed, many studies have demonstrated the facilitating role of caregivers in the field of severe acquired brain injury (sABI). For instance, it has been demonstrated that survivors with TBI and caregivers who effectively used problem-solving and behavioral coping strategies, showed also low levels of depression (Leach, Frank, Bouman, & Farmer, 1994). Even caregiver empowerment program positively influenced functional capacity and quality of life of the survivors with stroke (Dharma, Damhudi, Yardes, & Haeriyanto, 2018).

With this regard, it is fundamental provide caregivers with honest information and education regarding head injury (Campbell, 1988). In this vein, a recent study (Farnese et al., 2020) showed that sharing knowledge was positively related with caregivers' co-executing and co-planning (i.e., caregivers' participation in the whole healthcare process). Co-executing was also associated with caregivers' satisfaction, according to the staff perspective.

Also Bivona et al., recently investigated the possible relationships between the psychological status of the caregiver, specific features of caregiving as perceived by the cognitive therapist in neuro-rehabilitation, the caregivers' subjective approach to neuro-rehabilitation, and the functional outcome of the survivor. Their study, conducted by means of a new scale, namely the Caregiving Impact on Neuro-Rehabilitation Scale (CINRS), explored the impact of caregiving in neuro-rehabilitation in terms of their *participation* to the neuro-rehabilitation sessions, as well as of their tendency to *follow the suggestions by the therapist*, useful to help survivors in generalizing to the ADL what they learned in neuro-rehabilitation (Bivona et al., 2020). Results showed a close positive relationship between both caregivers' approach and psychological status, and the impact of caregiving in neuro-rehabilitation, as well as the functional outcome of survivors. These findings suggest the existence of a virtuous circle (see Fig. 3), according to which a better caregiving is associated to a better functional outcome of the survivors, which in turn positively influences the caregivers' psychological well-being (Bivona et al., 2020).

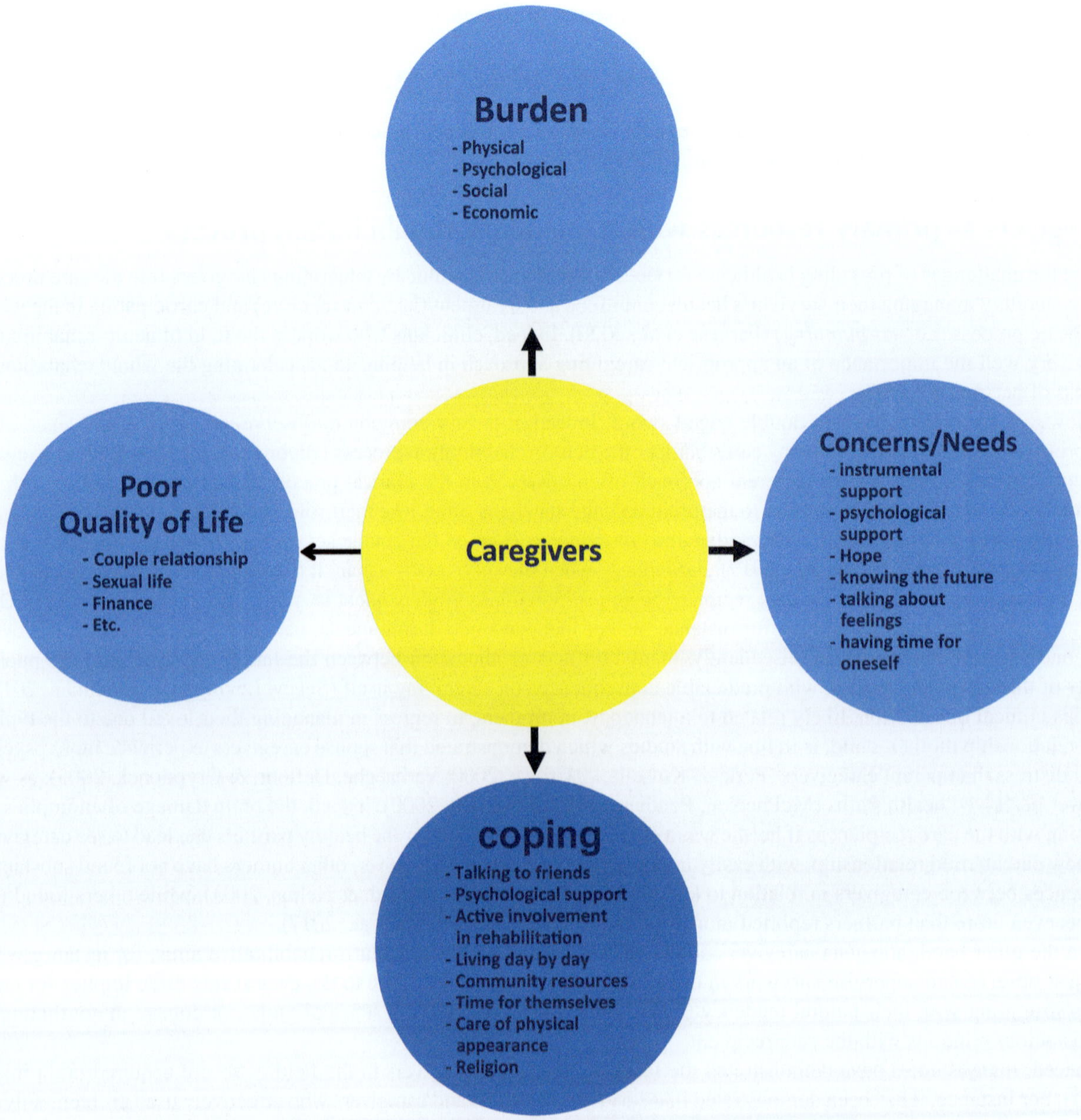

FIG. 3 Virtuous-Circle Model on the impact of Psycho-Educational intervention on both caregivers psychological well-being and type of caregiving style, which in turn can be strictly associated to patient rehabilitation outcome. Circularly, it is also expected that a better patients' rehabilitation outcome could increase caregivers' psychological well-being.

In summary, the above studies underline the usefulness of an educational intervention on caregivers, aimed at facilitating the functional recovery of their loved one (for example, avoiding substitutive behaviors, or helping survivors to generalize what learned in neuro-rehabilitation). Accordingly, in the neuro-rehabilitation educational context it is fundamental that clinicians guide caregivers, preventing dysfunctional behaviors and poor compliance in them. However, further studies on larger samples of caregivers are needed to better evaluate the real impact of an adequate caregiving on the survivor's neuro-rehabilitation outcome, as well as to structure specific guidelines addressed to caregivers.

Caregiver as victims themselves of the brain injury occurred to their loved one

Severe TBI may have an immediate and *direct* effect on family, due to the dramatic experience that is, for instance, receiving an unexpected call by an Emergency Room, or being the direct witness of a sudden stroke occurred to a loved one. This shocking initial emotional burden should be considered in addition to all the *indirect* effects of TBI.

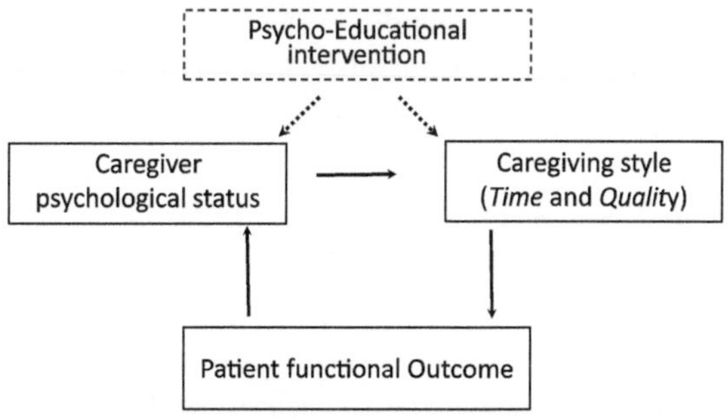

FIG. 4 The figure shows how providing care to an individual with severe TBI can lead to poor quality of life, as well as to high burden and concerns in relation to their needs. Effective coping strategies are fundamental to overcome the sTBI-related distress.

Indeed, as stated above, it is well known that providing care to an individual with moderate to severe ABI is related in the caregivers to increased psychosocial distress and poor functioning (Kreutzer et al., 2009; Lezak, 1988; Ponsford & Schonberger, 2010; Schulz & Sherwood, 2008), stressful changes in lifestyle and habits (D'Ippolito et al., 2018), worsening in quality of life (Formisano et al., 2015; Perlesz et al., 2000) including sexual life (Bivona et al., 2016) and quality of relationship with the survivors due to impairment in social cognition competences (Bivona et al., 2015), as well as work and financial difficulties (Van Houtven, Coe, & Skira, 2013; Wolff, Spillman, Freedman, & Kasper, 2016), and burden (Kreutzer et al., 2009).

Accordingly, caregivers are likely to suffer high burden and concerns in relation to their needs, whereby coping strategies are fundamental to functionally overcome this dramatic phase of their life (see Fig. 4).

Caregivers' burden

Caregivers' burden can be the result of emotional or physical diseases (Coe & Van Houtven, 2009; Schulz & Sherwood, 2008), and social life and financial status negative changes (Van Houtven et al., 2013), as a result of caring for the care recipient (Bivona et al., 2020; Turner et al., 2007), which can accumulate over a long period of time (Kasuya, Polgar-Bailey, Robbyn, & Takeuchi, 2000).

As previously stated, the kinship with the survivor can differently affect caregiver's level of psychological well-being. In a recent qualitative study (Kratz et al., 2017) emerged that parent caregivers were distressed about conflict between the survivor and his/her siblings; conversely, survivors' partners were much more stressed by the changes in the relationships between their partner (i.e., the survivor) and their children. Moreover, they felt guilt due to making decisions alone and, accordingly, to cutting out the survivors by the decision making process. In this latter case, it is particular needed for caregivers partners receiving specific training on problem solving, which has been demonstrated to be a useful factor in decreasing depression and improving perceived health (Rivera, Elliott, Berry, & Grant, 2008).

Of note, partners caregivers seemed to be emotionally wounded by survivors' aggressiveness (Kratz et al., 2017), and in many cases the post-traumatic changes can even lead to separation or divorce (Kreutzer, Sima, Marwitz, & Lukow II, 2016). This is likely due to interacting with a person (the care recipient) who in many cases after the sTBI is so different than before the accident to seem even a stranger. Conversely, parents, despite also suffered aggression and abuse, did not describe them in terms of a breaking down of the relationship and intimacy (Kratz et al., 2017). Once again, this could be due to the natural link between the care recipients' and their parents.

Some specific variables, i.e., poor social networks, loneliness, and caring for a survivor with more severe disability, have been related to caregiver burden (Manskow et al., 2015). Moreover, experiencing loneliness was an independent predictor of increased burden from 1 to 2 years post-injury, while life satisfaction was lower at the 2 year follow-up than at 1 year (Manskow et al., 2017). In line with these results, other authors in the same follow-up times found that greater life dissatisfaction have been related to more emotional and behavioral problems, and lower level of functional outcome in the survivors (Livingston et al., 2010).

A moderate relationship has been found between survivors' cognitive (Powell et al., 2017), behavioral and emotional changes, and relatives' coping with stress as well as their burden and psychological distress (Powell et al., 2017; Wells, Dywan, & Dumas, 2005). Also social cognition (in terms of theory of mind - ToM) deficit in survivors with sABI, rather than their cognitive impairment itself, has been linked to caregivers' poor quality of life (Bivona et al., 2015). Therefore, if for a caregiver can be basically difficult facing the frequent post-sTBI psycho-behavioral changes in their loved one (for instance in terms of irritability, aggressiveness, hypersexuality, etc.), this latter findings suggest how further difficult can be for them a coping process when the survivors are even not able to put themselves in their caregivers' shoes.

As for severity of the brain injury, some authors found a close relationship with psychological status of caregivers (Livingston et al., 2010), even if different data emerged from other studies (Liu, Zhu, Liu, & Guo, 2015; Qadeer et al., 2017; Kreitzer et al., 2020.

Caregivers' concerns and needs across the different stages post-TBI

Caregivers need in general instrumental support, which may be met through established community, government, or hospital-based programs (Kratz et al., 2017). Also, getting psychological support, feeling hope, knowing what the future holds for the survivor, and talking about feelings have been reported as the most important psychological needs, while having time for oneself was the most important social need (Verhaeghe et al., 2005).

Similar to burden, caregivers' concerns and needs might change over time according to the recovery from TBI. Indeed, in the acute and post-acute phases some authors found that caregivers mainly need understanding the injuries, treatments and consequences of ABI, as well as getting guidance (e.g., having more communication with providers) and quality of services (Rotondi, Sinkul, Balzer, Harris, & Moldova, 2007). Other needs, namely knowing the prognosis, having questions answered honestly, and being assured that the survivor was getting the best care, have also been identified in the same phase (Lefebvre & Levert, 2012).

Later on, during the outpatient rehabilitation phase, caregivers seem to need having guidance (eg, with respect to the safety of the survivor at home, or on how to respond to survivor's emotions and behaviors, dealing with role changes, and finding the best providers) (Rotondi et al., 2007).

In the long term, caregivers have reported the need of a reassessment of the survivor's abilities, financial assistance/counseling, and help with family emotional and mental health. In another recent study focused in the first few months after the TBI survivor was discharged to the community, Powell et al. (2017) found that more than one-third of caregivers reported being concerned about managing of their own emotional adjustment, finding strategies for getting things done, as well as managing survivors' emotions and behaviors, and engaging in healthful habits. The authors underline that these need and concerns can be addressed during the neuro-rehabilitation of the care recipient, even if other ways to ensure that caregiver concerns are routinely addressed over the long term as issues arise should also be considered (Powell et al., 2017).

In fact, understanding the needs and concerns of caregivers is crucial to assist at best them in meeting these new demands (Powell et al., 2017), as well as in coping with all consequences of TBI, by searching at new functional and satisfying aims, despite the dramatic changed to whom they are forced due to the brain damage occurred to their loved one.

Caregivers' coping strategies

Talking to friends, participating in psychological support groups, being actively involved in the survivor's rehabilitation, living day by day and focusing on the here and now as well as on positive aspects, identifying and using community resources, devoting time to themselves and taking care of their physical appearance, engaging in prayer, have been reported among the main and effective coping strategies for caregivers of survivors of TBI (Willer, Allen, Liss, & Zicht, 1991). Spirituality, in particular, has been recently found having a direct negative link with their burden, as well as a direct positive association with hope, which, in succession, had a positive link with resilience (Simpson, Anderson, Jones, Genders, & Gopinath, 2020).

Caregivers' burden may be mediated also by an adequate social support, which is considered a useful resource to improve caregivers' problem-solving skills (see the review by Baker, Barker, Sampson, & Martin, 2017 for details). Congruently, also day programs (Gerber & Gargaro, 2015) have been developed to provide social, recreational and skill training activities for persons living with an ABI (designed to reduce participants' social isolation, increase participation in community activities and increase social and leisure skills). Indeed, the authors demonstrated that participation in this program helped reducing both burden on family caregivers, and social isolation for ABI participants (Gerber & Gargaro, 2015).

More in general, family support seems to play a crucial role in maintaining employment after TBI (West & Hughes, 1990), while for persons with severe injury, greater perceived social support in caregivers was predictive of greater social integration and productivity (Sady et al., 2010).

Caregiver supportive services can be another important resource of the community, since they can allow caregivers facing at best the negative consequences of TBI, reducing their burden and mental distress (Dyer, Kansagara, McInnes, Freeman, & Woods, 2012), as well as improving survivor level of functioning (Meis et al., 2013). Therefore, standardized support and training may be important to optimize survivor and caregiver well-being and functioning (Wolff et al., 2016).

Conclusion

Given the above background, that is considering, by one side, the relevant role of caregivers in neuro-rehabilitation, and remember, by the other, how caregivers themselves can be distressed by the burden related to the caring process, we would share our multi-annual clinical experience at the Caregivers Support Service of the Santa Lucia Foundation rehabilitation hospital in Rome (Italy), dedicated to caregivers of survivors with sABI.

Within this Service, when caregiver require our support (that is, when they *actively* accept our invitation to be psychologically assisted), we adopt a model aimed at: a) increasing their self-awareness and knowledge of needs and expectations regarding neuro-rehabilitation of their loved ones; b) establishing, as a consequence, realistic neuro-rehabilitative aims; c) focusing on possible sABI-related emotional distress (e.g., depression, anger, feelings of guilt, anguish and anxiety about the future, etc.) (Silvestro et al., 2016).

We also differentiate the intervention according to three different phases post-sABI (see Fig. 5). Indeed, based on the above mentioned scientific literature, during the early post-acute phase the psycho-educational intervention is mainly focused on educational aspects, i.e., primarily on how caregivers can interact at best with their loved one. In fact, as soon as the specific neuro-rehabilitation project on the survivors begins, caregivers mainly need clinical information on them, as well as any information by therapists useful to improve the health of their loved one.

In fact, they show considering themselves "as a part of the loved one," losing as a consequence the dimension of their Self. For instance, they usually talk exclusively about the survivor, who represents the "thermometer of their psychological health." Their needs seem only oriented towards the current situation of the survivor, as well as to face the fear for the future. At this stage, denial can be one of the main (and effective) defensive mechanisms.

In a further phase, about 6 months after the beginning of rehabilitation, some of the caregivers' needs, which seem now at least partially unrelated to their loved one, can arise, such as caring for themselves, and resume social relationships. They now seem living a growing emotional separation from the survivor. At this stage, the psycho-educational intervention can be focused not only on its educational aspect directly involving the survivors, but even on the caregivers' psychological disease itself, assisting them both to face the "emotional detachment" by their loved one and to elaborate the partial grief with respect to the survivors' part of the Self that the brain injury has taken away.

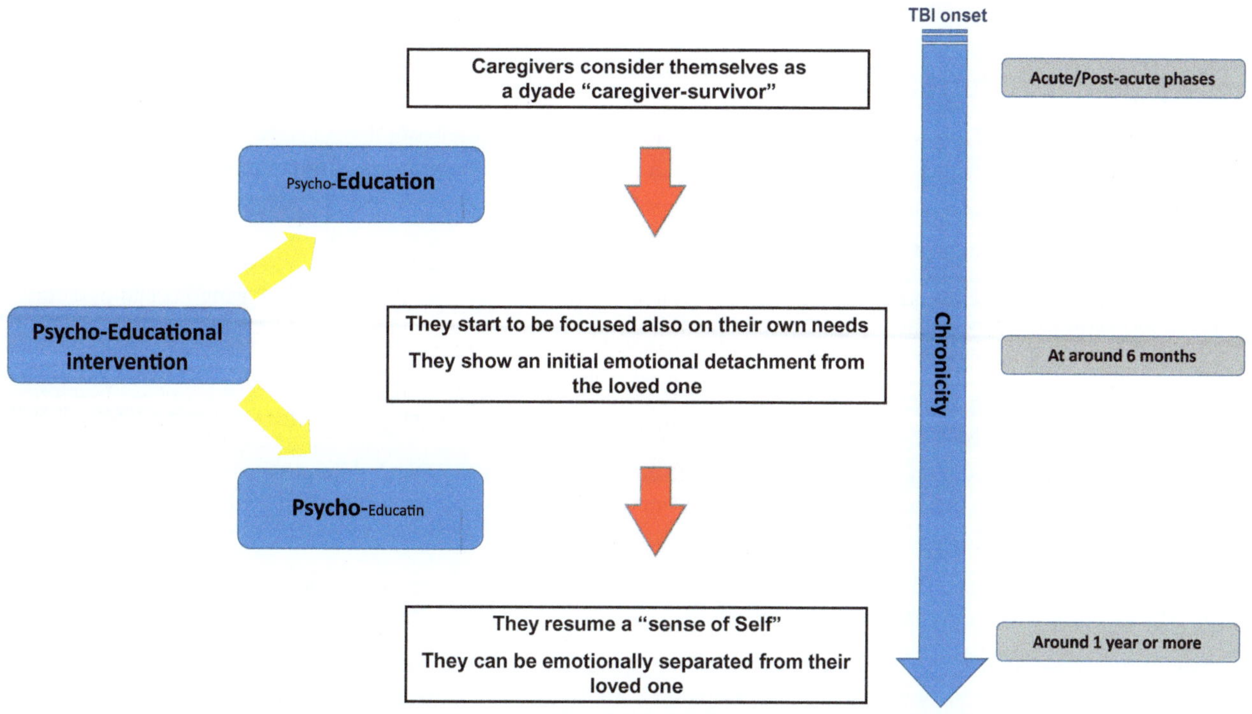

FIG. 5 Evolution of the caregivers' psychological status of the self according to the chronicity from the sTBI onset. Starting from considering themselves as a "part of the loved one," they can feel being emotionally separated from their loved ones only at the more advanced stages post-injury. Accordingly, the intervention by the psychotherapist is mainly focused on educational aspects in the early stages post-TBI, becoming mainly psychological in the advanced ones.

Finally, in the chronic phase at around 1 year after the ABI, caregivers seem, instead, increasingly be focused on themselves, and much more prone to accept and elaborate their emotions and difficulties resulting from the injury. Denial of the consequences of the TBI is now more an obstacle for caregivers then a resource. In this stage the intervention is mainly psychological, since we aim to break down this defensive barrier helping caregivers to focus on the survivors' resources, e.g., the cognitive functions spared by the injury and their own personal coping strategies (Silvestro et al., 2016).

Accordingly, in every stage post-injury, it is fundamental to identify the possible caregivers, which should be fully considered a member of the rehabilitation team, and as such a resource. However, it is just as important to identify also those relatives that should not be able (or not be ready, in a particular phase post-injury) to be part of the team. Indeed, it is also fundamental identifying all cases of possible unadequate caregivers, in order to detect and prevent all possible dysfunctional behaviors that can hinder neuro-rehabilitation, that is a complex and difficult process aiming at giving back at best an individual with a sTBI to his/her life.

Applications to other areas of neuroscience

In the present chapter, we aimed at describing the psycho-educational support to caregivers of survivors of a sTBI, which integrates psychotherapeutic and educational interventions reflecting a holistic and competence-based approach. Indeed, as largely stated in our chapter, the current neuro-rehabilitative approach to the survivors of sTBI requires a holistic perspective, according to which the care recipients are seen within a whole system including their caregivers (i.e., firstly their family members).

Indeed, if by one side caregivers are victims themselves of the accident occurred to their loved one (Reinhard, Given, Petlick, & Bemis, 2008), and as such they require a specific psychological support, they can also be fully considered as informal members of the rehabilitation team, who can effectively facilitate the rehabilitation process.

In fact, it is well known in the scientific literature how indispensable is the caregiver in many neurological diseases, for instance in assisting survivors when the care recipient is an individual who suffers from advanced stages of degenerative diseases, such as dementia or auto-immune diseases (Tramonti et al., 2019). However, it is worth noting that, since sTBI mainly involves younger population with longer life expectance, caregiving in this field is a more complex phenomenon, with different facets according to the chronicity and severity of the brain damage. With this regard, the role of caregivers in facilitating the neuro-rehabilitation process after sTBI is rather more comparable to caregiving in pediatric neurological and neuropsychological disease of traumatic, non-traumatic and genetic syndromes, where caregiving is as relevant as severe are the survivors' cognitive and behavioral, and more in general, functional features.

Mini-dictionary of terms

Acquired brain injury (ABI): A wide spectrum of traumatic and non-traumatic brain damages. Non-traumatic ABI can be caused by focal brain lesions, anoxia, tumors, aneurysm, vascular malformations, and infections of the brain.

Activities of daily life (ADL): The most basic functions of living independently, including feeding ourselves, bathing, dressing, money management, etc.

Caregiver burden: Physical, emotional, financial, psychological fatigue that caregivers feel in caring on and assisting a loved one with neurological and neuropsychological disease.

Caregiving Impact on Neuro-Rehabilitation Scale (CINRS): A new Scale that explores the impact of caregiving in neuro-rehabilitation. The CINRS consists of two sections: (1) CINRS-A, which specifically investigated the amount of time spent by the caregiver in participating to the neuro-rehabilitation sessions, as well as the quality of the caregiver assistance to the survivor (in terms of trust, compliance, and cooperation to the neuro-rehabilitation team and project); and (2) CINRS-B, by which the therapist judged the general improvement of the survivor at the end of the neuro-rehabilitation project.

Coping strategies: A basic and fundamental process useful for adaptation and survival, which can improve or accentuate the effects of mental and/or physical stress on health, facilitating the development of psychopathology and resilience.

Glasgow Coma Scale (GCS): Probably the most used measure to assess the severity of TBI, based on three aspects of responsiveness: eye-opening, motor, and verbal responses. A GCS ≤ 8 in index of severe TBI.

Holistic approach: Providing support after considering the individual as a system, including his/her physical, emotional, social and spiritual well-being.

Informal caregiver (namely, "caregiver" in the present chapter): Any relative, partner, friend or neighbor who, voluntarily and without receiving any payment, provides care and support to a loved one who is not self-sufficient in his/her family or social ***network.***

Severe TBI (sTBI): A TBI that implies a GCS score $\leq$8in the acute phase.

Traumatic brain injury (TBI): A non-degenerative acquired insult to the brain, secondary to any applied external mechanical force. TBI can be caused by a blow to the hand (closed head injury) or by an object that pierces the skull and enters brain tissue (open head injury).

Key facts of psycho-educational intervention

- Severe TBI requires an inter-professional and long lasting neuro-rehabilitative process, but even more importantly a constant support by at least one informal caregiver
- Psycho-educational intervention is provided to caregivers both sustaining them at a psychological level, and guiding them on how managing at best the relationship with the care recipient
- Psycho-educational reflects a paradigm shift to a more holistic and competence-based approach, stressing health, collaboration, coping, and empowerment
- Psycho-educational techniques are used to help removing barriers to comprehending and digesting complex and emotionally loaded information and to develop strategies to use the information in a proactive fashion
- It has been demonstrated that a better caregivers' approach to neuro-rehabilitation is associated to a better functional outcome of survivors with sABI

Summary points

1. Brain Damage involves the whole family system at different levels
2. Appropriate caregiving approach is essential in neuro-rehabilitation
3. Psycho-educational intervention both improves caregivers' well-being and taking care of survivors
4. Better caregivers' approach is associated to better functional outcome in survivors
5. Caregiver support differs according to different phases post-sABI

References

Arciniegas, D. B., & Wortzel, H. S. (2014). Emotional and behavioral dyscontrol after traumatic brain injury. *Psychiatric Clinics of North America*, *37*(1), 31–53.

Baker, A., Barker, S., Sampson, A., & Martin, A. (2017). Caregiver outcomes and interventions: A systematic scoping review of the traumatic brain injury and spinal cord injury literature. *Clinical Rehabilitation*, *31*(1), 45–60.

Bivona, U., Antonucci, G., Contrada, M., Rizza, F., Leoni, F., Zasler, N. D., et al. (2016). A biopsychosocial analysis of sexuality in adult males and their partners after severe traumatic brain injury. *Brain Injury*, *30*(9), 1082–1095.

Bivona, U., Formisano, R., De Laurentiis, S., Accetta, N., Di Cosimo, M. R., Massicci, R., et al. (2015). Theory of mind impairment after severe traumatic brain injury and its relationship with caregivers'quality of life. *Restorative Neurology and Neuroscience*, *33*(3), 335–345.

Bivona, U., Villalobos, D., De Luca, M., Zilli, F., Ferri, G., Lucatello, S., et al. (2020). Psychological status and role of caregivers in the neuro-rehabilitation of patients with severe Acquired Brain Injury (ABI). *Brain Injury*, *34*(13–14), 1714–1722. https://doi.org/10.1080/02699052.2020.1812002. 33190555.

Campbell, C. H. (1988). Needs of relatives and helpfulness of support groups in severe head injury. *Rehabilitation Nursing*, *13*(6), 320–325.

Ciurli, P., Formisano, R., Bivona, U., Cantagallo, A., & Angelelli, P. (2011). Neuropsychiatric disorders in persons with severe traumatic brain injury: Prevalence, phenomenology, and relationship with demographic, clinical, and functional features. *The Journal of Head Trauma Rehabilitation*, *26*(2), 116–126.

Coe, N. B., & Van Houtven, C. H. (2009). Caring for mom and neglecting yourself? The health effects of caring for an elderly parent. *Health Economics*, *18*(9), 991–1010.

Coetzer, R. (2007). Psychotherapy following traumatic brain injury: Integrating theory and practice. *The Journal of Head Trauma Rehabilitation*, *22*(1), 39–47.

D'Ippolito, M., Aloisi, M., Azicnuda, E., Silvestro, D., Giustini, M., Verni, F., et al. (2018). Changes in caregivers lifestyle after severe acquired brain injury: A preliminary investigation. *BioMed Research International*, *2018*, 2824081.

Dharma, K. K., Damhudi, D., Yardes, N., & Haeriyanto, S. (2018). Increase in the functional capacity and quality of life among stroke patients by family caregiver empowerment program based on adaptation model. *International Journal of Nursing Sciences*, *5*(4), 357–364.

Douglas, J. (2017). "So that's the way it is for me – always being left out." Acquired pragmatic language impairment and social functioning following traumatic brain injury. *Brain Impairment*, *18*, 321–331.

Douglas, J. (2020). Loss of friendship following traumatic brain injury: A model grounded in the experience of adults with severe injury. *Neuropsychological Rehabilitation*, *30*(7), 1277–1302.

Dyer, E. A., Kansagara, D., McInnes, D. K., Freeman, M., & Woods, S. (2012). Mobile applications and internet-based approaches for supporting non-professional caregivers: A systematic review. In *VA Evidence-based Synthesis Program Reports* Department of Veterans (US).

Farnese, M. L., Girardi, G., Fida, R., Bivona, U., Bartolo, M., De Tanti, A., et al. (2020). Caregivers' engagement during in-hospital care of sABI's patients: Evaluation of informal co-production from the health providers' perspective. *Health & Social Care in the Community*, *28*(6), 2086–2094.

Formisano, R., De Laurentiis, S., Accetta, N., Di Cosimo, M. R., Massicci, R., Ciurli, P., et al. (2015). Quality of life as perceived by the patients with TBI and their caregivers. *Journal of the Neurological Sciences*, *357*, e350–e356.

Forslund, M. V., Arango-Lasprilla, J. C., Roe, C., Perrin, P. B., Sigurdardottir, S., & Andelic, N. (2014). Multi-level modelling of employment probability trajectories and employment stability at 1, 2 and 5 years after traumatic brain injury. *Brain Injury*, *28*(7), 980–986.

Gerber, G. J., & Gargaro, J. (2015). Participation in a social and recreational day programme increases community integration and reduces family burden of persons with acquired brain injury. *Brain Injury*, *29*(6), 722–729.

Gómez-de-Regil, L., Estrella-Castillo, D. F., & Vega-Cauich, J. (2019). Psychological intervention in traumatic brain injury patients. *Behavioural Neurology*, *2019*(1), 1–8.

Jourdan, C., Bayen, E., Pradat-Diehl, P., Ghout, I., Darnoux, E., Azerad, S., et al. (2016). A comprehensive picture of 4-year outcome of severe brain injuries. Results from the PariS-TBI study. *Annals of Physical and Rehabilitation Medicine*, *59*(2), 100–106.

Kasuya, R. T., Polgar-Bailey, M. P., Robbyn, M. P. H., & Takeuchi, M. S. W. (2000). Caregiver burden and burnout. A guide for primary care physicians. *Postgraduate Medicine*, *108*, 119–123.

Kersel, D. A., Marsh, N. V., Havill, J. H., & Sleigh, J. W. (2001). Psychosocial functioning during the year following severe traumatic brain injury. *Brain Injury*, *15*, 683–696.

Kieffer-Kristensen, R., Siersma, V. D., & Teasdale, T. W. (2013). Family matters: Parental-acquired brain injury and child functioning. *NeuroRehabilitation*, *32*(1), 59–68.

Kratz, A. L., Sander, A. M., Brickell, T. A., Lange, R. T., & Carlozzi, N. E. (2017). Traumatic brain injury caregivers: A qualitative analysis of spouse and parent perspectives on quality of life. *Neuropsychological Rehabilitation*, *27*(1), 16–37.

Kreitzer, N., Bakas, T., Kurowski, B., Lindsell, C., Ferioli, S., Foreman, B., et al. (2020). The experience of caregivers following a moderate to severe traumatic brain injury requiring ICU admission. *The Journal of Head Trauma Rehabilitation*, *35*(3), E299–E309.

Kreutzer, J. S., Marwitz, J. H., Walker, W., Sander, A., Sherer, M., Bogner, J., et al. (2003). Moderating factors in return to work and job stability after traumatic brain injury. *The Journal of Head Trauma Rehabilitation*, *18*, 128–138.

Kreutzer, J. S., Rapport, L. J., Marwitz, J. H., Harrison-Felix, C., Hart, T., Glenn, M., et al. (2009). Caregivers' well-being after traumatic brain injury: A multicenter prospective investigation. *Archives of Physical Medicine and Rehabilitation*, *90*(6), 939–946.

Kreutzer, J. S., Sima, A. P., Marwitz, J. H., & Lukow, H. R., II. (2016). Marital instability after brain injury: An exploratory analysis. *NeuroRehabilitation*, *38*(3), 271–279.

Leach, L. R., Frank, R. G., Bouman, D. E., & Farmer, J. (1994). Family functioning, social support and depression after traumatic brain injury. *Brain Injury*, *8*(7), 599–606.

Lefebvre, H., & Levert, M. J. (2012). The close relatives of people who have had a traumatic brain injury and their special needs. *Brain Injury*, *26*(9), 1084–1097.

Lezak, M. D. (1988). Brain damage is a family affair. *Journal of Clinical and Experimental Neuropsychology*, *10*(1), 111–123.

Liu, W., Zhu, J., Liu, J., & Guo, Q. (2015). Psychological state and needs of family member caregivers for victims of traumatic brain injury: A cross-sectional descriptive study. *International Journal of Nursing Sciences*, *2*(3), 231–236.

Livingston, L. A., Kennedy, R. E., Marwitz, J. H., Arango-Lasprilla, J. C., Rapport, L. J., Bushnik, T., et al. (2010). Predictors of family caregivers' life satisfaction after traumatic brain injury at one and two years postinjury: A longitudinal multi-center investigation. *NeuroRehabilitation*, *27*(1), 73–81.

Manskow, U. S., Friborg, O., Røe, C., Braine, M., Damsgard, E., & Anke, A. (2017). Patterns of change and stability in caregiver burden and life satisfaction from 1 to 2 years after severe traumatic brain injury: A Norwegian longitudinal study. *NeuroRehabilitation*, *40*(2), 211–222.

Manskow, U. S., Sigurdardottir, S., Røe, C., Andelic, N., Skandsen, T., Damsgård, E., et al. (2015). Factors affecting caregiver burden 1 year after severe traumatic brain injury: A prospective Nationwide multicenter study. *The Journal of Head Trauma Rehabilitation*, *30*(6), 411–423.

McPherson, K. M., Pentland, B., & McNaughton, H. K. (2000). Brain injury – The perceived health of carers. *Disability and Rehabilitation*, *22*(15), 683–689.

Meis, L. A., Griffin, J. M., Greer, N., Jensen, A. C., Macdonald, R., Carlyle, M., et al. (2013). Couple and family involvement in adult mental health treatment: A systematic review. *Clinical Psychology Review*, *33*(2), 275–286.

Perlesz, A., Kinsella, G., & Crowe, S. (2000). Psychological distress and family satisfaction following traumatic brain injury: Injured individuals and their primary, secondary, and tertiary carers. *The Journal of Head Trauma Rehabilitation*, *15*(3), 909–929.

Ponsford, J., Downing, M., Olver, J., Ponsford, M., Acher, R., Carty, M., et al. (2014). Longitudinal follow-up of patients with traumatic brain injury: Outcome at two, five, and ten years post-injury. *Journal of Neurotrauma*, *31*, 64–77.

Ponsford, J., Olver, J., Ponsford, M., & Nelms, R. (2003). Long-term adjustment of families following traumatic brain injury where comprehensive rehabilitation has been provided. *Brain Injury*, *17*(6), 453–468.

Ponsford, J., & Schonberger, M. (2010). Family functioning and emotional state two and five years after traumatic brain injury. *Journal of the International Neuropsychological Society*, *16*(2), 306–317.

Powell, J. M., Wise, E. K., Brockway, J. A., Fraser, R., Temkin, N., & Bell, K. R. (2017). Characteristics and concerns of caregivers of adults with traumatic brain injury. *The Journal of Head Trauma Rehabilitation*, *32*(1), E33–E41(9).

Qadeer, A., Khalid, U., Amin, M., Murtaza, S., Khaliq, M. F., & Shoaib, M. (2017). Caregiver's burden of the patients with traumatic brain injury. *Cureus*, *9*(8), e1590.

Ramkumar, N. A., & Elliot, T. R. (2010). Family caregiving of persons following neurotrauma: Issues in research, service and policy. *NeuroRehabilitation*, *27*, 105–112.

Rapoport, M. J., Mccullagh, S., Streiner, D., & Feinstein, A. (2003). The clinical significance of major depression following mild traumatic brain injury. *Psychosomatics*, *44*(1), 31–37.

Reinhard, S. C., Given, B., Petlick, N. H., & Bemis, A. (2008). Supporting family caregivers in providing care. In R. G. Hughes (Ed.), *Patient safety and quality: An evidence-based handbook for nurses* Agency for Healthcare Research and Quality (US).

Rivera, P. A., Elliott, T. R., Berry, J. W., & Grant, J. S. (2008). Problem-solving training for family caregivers of persons with traumatic brain injuries: A randomized controlled trial. *Archives of Physical Medicine and Rehabilitation*, *89*(5), 931–941.

Roozenbeek, B., Maas, A. I., & Menon, D. K. (2013). Changing patterns in the epidemiology of traumatic brain injury. *Nature Reviews Neurology*, *9*(4), 231–236.

Rotondi, A. J., Sinkul, J., Balzer, K., Harris, J., & Moldova, R. (2007). A qualitative needs assessment of persons who have experienced traumatic injury and their primary family caregivers. *The Journal of Head Trauma Rehabilitation*, *22*(1), 14–25.

Sady, M. D., Sander, A. M., Clark, A. N., Sherer, M., Nakase-Richardson, R., & Malec, J. F. (2010). Relationship of preinjury caregiver and family functioning to community integration in adults with traumatic brain injury. *Archives of Physical Medicine and Rehabilitation*, *91*(10), 1542–1550.

Salas, C. E., Casassus, M., Rowlands, L., Pimm, S., & Flanagan, D. (2018). "relating through sameness": A qualitative study of friendship and social isolation in chronic traumatic brain injury. *Neuropsychological Rehabilitation*, *28*(7), 1161–1178.

Scaratti, C., Leonardi, M., Sattin, D., Schiavolin, S., Willems, M., & Raggi, A. (2017). Work-related difficulties in patients with traumatic brain injury: A systematic review on predictors and associated factors. *Disability and Rehabilitation*, *39*(9), 847–855.

Schulz, R., & Sherwood, P. R. (2008). Physical and mental health effects of family caregiving. *American Journal of Nursing*, *108*(9 Suppl), 23–27.

Segev, E., Levinger, M., & Hochman, Y. (2018). "Shared Destiny": The dynamics of relationships in families of patients with brain injury. *Qualitative Health Research*, *28*(1), 112–125.

Silvestro, D., Azicnuda, E., D'Ippolito, M., Giustini, M., Formisano, R., & Bivona, U. (2016). Beyond the classical psychotherapeutic setting to better provide support for caregivers of persons with severe acquired brain injury: Some clinical evidence. *Journal of Psychology and Psychotherapy*, *6*(2), 1–4.

Simpson, G. K., Anderson, M. I., Jones, K. F., Genders, M., & Gopinath, B. (2020). Do spirituality, resilience and hope mediate outcomes among family caregivers after traumatic brain injury or spinal cord injury? A structural equation modelling approach. *NeuroRehabilitation*, *46*(1), 3–15.

Spoorenberg, S. L., Uittenbroek, R. J., Middel, B., Kremer, B. P., Reijneveld, S. A., & Wynia, K. (2013). Embrace, a model for integrated elderly care: Study protocol of a randomized controlled trial on the effectiveness regarding patient outcomes, service use, costs, and quality of care. *BMC Geriatrics*, *13*, 62.

Tramonti, F., Bonfiglio, L., Bongioanni, P., Belviso, C., Fanciullacci, C., Rossi, B., et al. (2019). Caregiver burden and family functioning in different neurological diseases. *Psychology, Health & Medicine*, *24*(1), 27–34.

Turner, B., Fleming, J., Cornwell, P., Worrall, L., Ownsworth, T., Haines, T., et al. (2007). A qualitative study of the transition from hospital to home for individuals with acquired brain injury and their family caregivers. *Brain Injury*, *21*(11), 1119–1130.

Van Houtven, C. H., Coe, N. B., & Skira, M. M. (2013). The effect of informal care on work and wages. *Journal of Health Economics*, *32*(1), 240–252.

Verhaeghe, S., Defloor, T., & Grypdonck, M. (2005). Stress and coping among families of patients with traumatic brain injury: A review of the literature. *Journal of Clinical Nursing*, *14*(8), 1004–1012.

Wells, R., Dywan, J., & Dumas, J. (2005). Life satisfaction and distress in family caregivers as related to specific behavioural changes after traumatic brain injury. *Brain Injury*, *19*, 1105–1115.

West, M., & Hughes, T. (1990). Supported employment phase III: Job retention techniques. In P. Wehman, & J. S. Kreutzer (Eds.), *Vocational rehabilitation for persons with traumatic brain injury*. Rockville, MD: Aspen Publishers, Inc.

Willer, B. S., Allen, K. M., Liss, M., & Zicht, M. S. (1991). Problems and coping strategies of individuals with traumatic brain injury and their spouses. *Archives of Physical Medicine and Rehabilitation*, *72*, 460–464.

Wolff, J. L., Spillman, B. C., Freedman, V. A., & Kasper, J. D. (2016). A national profile of family and unpaid caregivers who assist older adults with health care activities. *JAMA Internal Medicine*, *176*(3), 372–379.

Chapter 43

Recommended resources and sites for the neuroscience of traumatic brain injury

Rajkumar Rajendram[a,b,c], Vinood B. Patel[d], and Victor R. Preedy[e]

[a]College of Medicine, King Saud bin Abdulaziz University for Health Sciences, Riyadh, Saudi Arabia, [b]Department of Medicine, King Abdulaziz Medical City, King Abdullah International Medical Research Center, Ministry of National Guard Health Affairs, Riyadh, Saudi Arabia, [c]Stoke Mandeville Hospital, Buckinghamshire Healthcare NHS Trust, Aylesbury, United Kingdom, [d]School of Life Sciences, University of Westminster, London, United Kingdom, [e]Diabetes and Nutritional Sciences Research Division, Faculty of Life Science and Medicine, King's College London, London, United Kingdom

List of abbreviations

ABI acquired brain injury
BCE before the common (or current) era
CNS central nervous system
TBI traumatic brain injury

Introduction

The Edwin Smith Papyrus, written around 2500 years BCE, contains some of the earliest documented reports of patients with traumatic brain injury (TBI) (Sanchez & Burridge, 2007). When this papyrus was written, combat was probably the most common cause of TBI. In the 21st century, while violence remains a common cause of TBI, others include motor vehicle incidents, sports, and construction (Faul & Coronado, 2015).

As such, the incidence of TBI in men is double that of women, and TBI frequently leads to permanent, devastating disabilities in young, previously independent individuals (Faul & Coronado, 2015). Yet, unfortunately, when treating TBI, many healthcare professionals still have a degree of therapeutic nihilism. This is similar to the therapeutic nihilism associated with spinal cord injury (Rajendram, Patel and Preedy, 2021a).

Yet, because there are many possible modes of injury, TBI is not a homogenous entity. Furthermore, three discrete phases of TBI are now recognized (Prasetyo, 2020). The physical damage to the brain parenchyma at impact is considered to be the primary injury (Prasetyo, 2020). It results in immediate cell death from contusions, lacerations, diffuse axonal injuries, cerebral edema, and intracranial bleeding (Prasetyo, 2020). The complex cascades of biochemical processes initiated by the primary brain injury cause a delayed (i.e., secondary) injury (Prasetyo, 2020). This processes which cause this preventable damage may persist for several days after the primary insult.

The concepts of primary and secondary TBI arose in the 1990s. It was recognized that following the primary insult, secondary insults (e.g., hypoxia or hypotension) could exacerbate TBI (Patel et al., 2005). This concept has defined the monitoring and treatment of TBI for the last 30 years. The concept of tertiary brain injury is more recent (Fleiss & Gressens, 2012; Prasetyo, 2020).

Persisting inflammation and epigenetic phenomena may block the maturation of oligodendrocytes, impair neurogenesis, slow the growth of axons, and alter synapse formation (Prasetyo, 2020). The ongoing chronic tertiary phase of TBI with progressive neurodegeneration and delayed cell death may manifest as worsening cognitive dysfunction, or seizures (Fleiss & Gressens, 2012; Prasetyo, 2020).

For more than a century, it was thought that victims of TBI were unable to regain neurological function because the damaged neurons of the central nervous system (CNS) could not regenerate (Cajal, 1928). Thus, many research groups have attempted to induce neuronal regeneration. These studies have advanced the neuroscience of TBI by demonstrating that the CNS has greater plasticity than originally thought (Barnabe-Heider & Frisen, 2008).

Diagnosis and Treatment of Traumatic Brain Injury. https://doi.org/10.1016/B978-0-12-823347-4.00043-9

The ability to stimulate neuronal regeneration seems almost within our grasp. Yet for the sequelae of TBI to become cured, extensive research is still required. Regardless, even now, scientists struggle to remain up to date. To assist colleagues who are interested in understanding more about the neuroscience of anesthesia and analgesia, we have therefore produced tables containing up-to-date resources in this chapter. The experts who assisted with the compilation of these tables of resources are acknowledged below.

Resources

Tables 1–6 list the most up-to-date information on the regulatory bodies (Table 1), journals (Table 2), books (Table 3), professional societies (Table 4), emerging technologies, platforms (Table 5), and other resources (Table 6) that are relevant to an evidence-based approach to the neuroscience of traumatic brain injury.

Other resources

Sources for the historical aspects of brain injury can located at the Wellcome Collection (https://wellcomecollection.org/collections). Also, The British Library (https://www.bl.uk/) is also a good source of material, some of which is available

TABLE 1 Regulatory bodies and relevant organizations.

Regulatory body/Organization	Web address
Acquired Brain Injury System Navigation of Southeastern Ontario	http://braininjuryhelp.ca/
Adapt Functional Movement Center	https://adaptmovement.org/
Administration for Community Living	https://acl.gov/programs/post-injury-support/traumatic-brain-injury-tbi
American Association of Neurological Surgeons	https://www.aans.org/Patients/Neurosurgical-Conditions-and-Treatments/Traumatic-Brain-Injury
Brain Injury Association of America	https://www.biausa.org/
Brain Injury Association of Durham Region	https://biad.ca/
Brain Injury Association of Fort Erie	https://braininjuryfe.wixsite.com/biafe
Brain Injury Association of London and Region	http://www.braininjurylondon.on.ca/
Brain Injury Association of Niagara	https://www.bianiagara.org/
Brain Injury Association of Peel and Halton	https://biaph.com/
Brain Injury Association of Quinte District	https://biaqd.ca/
Brain Injury Association of Sault Ste. Marie and District	https://www.soobraininjury.com/
Brain Injury Association of the Ottawa Valley	http://biaov.org/
Brain Injury Association of Thunder Bay and Area[a]	http://www.bisno.org/
Brain Injury Guidelines.Org	https://braininjuryguidelines.org/
Brain Injury Services/Step Up! Abi Recovery[b]	http://www.braininjuryservices.com/
Brain Trauma Foundation	https://www.braintrauma.org/
Canadian High-Resolution Traumatic Brain Injury (Cahr-Tbi) Research Collaborative	https://pubmed.ncbi.nlm.nih.gov/32174295/
Center of Excellence for Medical Multimedia (CEMM)	https://tbi.cemmlibrary.org/
Centers for Disease Control and Prevention	https://www.cdc.gov/traumaticbraininjury/index.html
Clinicaltrials.Gov	https://clinicaltrials.gov
Collaborative European Neurotrauma Effectiveness Research in Traumatic Brain Injury (CENTER-TBI)	https://www.center-tbi.eu/

Continued

TABLE 1 Regulatory bodies and relevant organizations—cont'd

Regulatory body/Organization	Web address
Division of Neurosurgery, Department Of Surgery, Faculty of Medicine, Prince of Songkla University, Hat Yai, Songkhla, 90110, Thailand	https://sites.google.com/psu.ac.th/psuneurosurgery
European Brain Injury Consortium	http://www.ebic.nl/
Evidenced-Based Review of Moderate to Severe Acquired Traumatic Brain Injury	https://erabi.ca
Giving Compass	https://givingcompass.org/article/understanding-traumatic-brain-injury/?gclid=CjwKCAjw3pWDBhB3EiwAV1c5rBdW9R7C2hwSThu2SCJluzx-72rXU3tZEajJ9NrlxwYMl1V6ltMurhoCDP8QAvD_BwE
Global Neuro Foundation	https://globalneuro.org/EN/home.html
Hamilton Brain Injury Association	https://hbia.space/
Hamilton Brain Injury Association	https://hbia.space/
Head Injury Rehabilitation Ontario	https://hiro.ca/
Hope After Brain Injury	https://www.hopeafterbraininjury.org/
Indian Head Injury Foundation	https://indianheadinjuryfoundation.org/traumatic-brain-injury/
Institut National D'excellence En Santé Et En Services Sociaux (INESSS)	https://www.inesss.qc.ca/
International Brain Injury Association	https://www.internationalbrain.org
International Initiative for Traumatic Brain Injury Research (Intbir)	http://intbir.nih.gov
Johns Hopkins University	https://www.hopkinsmedicine.org/health/conditions-and-diseases/traumatic-brain-injury
Learning Services Corporation	https://learningservices.com/resources/brain-injury-resources/
March of Dimes	https://www.marchofdimes.ca/
Mayo Clinic	https://www.mayoclinic.org/diseases-conditions/traumatic-brain-injury/symptoms-causes/syc-20378557
Military Health System	https://health.mil/About-MHS/OASDHA/Defense-Health-Agency/Research-and-Development/Traumatic-Brain-Injury-Center-of-Excellence
National Center for Biotechnology Information (NCBI)	https://pubmed.ncbi.nlm.nih.gov/
National Institute of Neurological Disorders and Stroke (NINDS)	https://www.ninds.nih.gov
National Institute on Alcohol Abuse and Alcoholism (NIAAA)	https://www.arcr.niaaa.nih.gov/arcr392/article06.htm
National Institutes of Health (NIH)	https://www.nih.gov/about-nih/what-we-do/nih-turning-discovery-into-health/traumatic-brain-injury-tbi
Neurologic Rehabilitation Institute of Ontario[c]	https://www.nrio.com
New Beginnings ABI & Stroke Recovery Association (Chatham Sarnia)[d]	http://newbeginnings-cksl.com/
New York University	https://med.nyu.edu/departments-institutes/population-health/divisions-sections-centers/medical-ethics/education/high-school-bioethics-project/learning-scenarios/the-nfl-brain-injury
Ontario Brain Injury Association	http://obia.ca/
Ontario Neurotrauma Foundation	onf.org

Continued

TABLE 1 Regulatory bodies and relevant organizations—cont'd

Regulatory body/Organization	Web address
Seizure & Brain Injury Centre	http://seizurebraininjurycentre.com/
The Enhancing Neuroimaging Genetics Through Meta-Analysis (ENIGMA) Brain Injury Group	http://enigma.ini.usc.edu/ongoing/enigma-tbi/
The University of Alabama At Birmingham	https://www.uab.edu/medicine/tbi/
Tony Meduri Traumatic Brain Injury Fund Inc.	https://www.tbifundinc.com/
Transforming Research and Clinical Knowledge in Traumatic Brain Injury (TRACK-TBI)	https://tracktbi.ucsf.edu
Traumatic Brain Injury. Family Caregiver Alliance	https://www.caregiver.org/traumatic-brain-injury
United States Department of Defense	https://dod.defense.gov/News/Special-Reports/0315_tbi/
United States Department of Veterans Affairs	https://www.research.va.gov/topics/tbi.cfm
Upsurgeon	https://www.upsurgeon.com/?v=cd32106bcb6d
World Federation of Neurosurgical Societies (WFNS)	https://wfns.org/
World Health Organization (Who)	https://www.who.int/

[a] *This site may have moved since submission.*
[b] *This is redirected to Head Injury Rehabilitation Ontario.*
[c] *Queries are redirected to another site.*
[d] *The site was not active at the time of going to press.*
This table lists the regulatory bodies and organizations involved with the neuroscience of traumatic brain injury and associated specialties or interests. The links were accurate at the time of going to press but may have moved or altered. In these cases, the use of the "Search" tabs should be explored at the parent web address.

TABLE 2 Relevant journals publishing original research and review articles related to traumatic brain injury.

Journal of Neurotrauma
Brain Injury
World Neurosurgery
Journal of Head Trauma Rehabilitation
Frontiers in Neurology
Journal of Trauma and Acute Care Surgery
PLoS One
Scientific Reports
Neurocritical Care
Archives of Physical Medicine and Rehabilitation
International Journal of Molecular Sciences
Neural Regeneration Research
Journal of Neurosurgery
Experimental Neurology
Neuropsychological Rehabilitation
Frontiers in Neuroscience
Disability and Rehabilitation
Military Medicine
Acta Neurochirurgica

Continued

TABLE 2 Relevant journals publishing original research and review articles related to traumatic brain injury—cont'd

Neurorehabilitation
Brain Research
Critical Care Medicine
Injury
Neurology
Journal of Clinical Neuroscience
Acta Neurochirurgica Supplementum
Brain Sciences
BMJ Open
Journal of Surgical Research
Child S Nervous System

Journals publishing original research and review articles related to the traumatic brain injury. Included in this list are the top 30 journals which have published the greatest number of articles on this topic over the past 5 years. Data derived from Scopus.

TABLE 3 Relevant books.

Book	Authors	Publisher	Year
Biomarkers for Traumatic Brain Injury	Wu AHB, Peacock WF.	Academic Press	2020
Biomarkers for Traumatic Brain Injury (No. 24).	Dambinova SA, Hayes RL, Wang KK.	Royal Society of Chemistry.	2012
Biomarkers of Brain Injury and Neurological Disorders	Wang K, Zhang Z, Kobeissy F.	CRC Press	2014
Brain Injury Medicine	Eapen BC, Cifu DX.	Elsevier	2020
Brain Injury Medicine: Principles and Practice	Zasler N, Katz D, Zafonte R.	Demos Medical Publishing; 1 edition	2006
Brain Injury Medicine Principles and Practice 2nd Ed	Zasler ND, Katz DI, Zafonte RD, Arciniegas DB, Bullock MR, Kreutzer JS.	Demos Medical Publishing	2013
Brain Injury Medicine Principles and Practice 3rd Ed	Zasler N, Katz D, Zafonte R.	Demos Medical Publishing	2021
Brain Neurotrauma: Molecular, Neuropsychological, and Rehabilitation Aspects	Kobeissy FH.	CRC Press	2015
Cognitive Rehabilitation and Neuroimaging. Examining the Evidence from Brain to Behavior	DeLuca J, Chiaravalloti N, Weber E.	Springer Nature Switzerland AG	2020
Cognitive Rehabilitation for Pediatric Neurological Disorders	Locascio G, Slomine B.	Cambridge University Press	2018
Decompressive Craniectomy	Aarabi B	Nova Biomedical	2018
Future of Neurotraumatology	Tunthanathip T.	Book Unit, Faculty of Medicine, Prince of Songkla University	2019
Handbook on the Neuropsychology of Traumatic Brain Injury	Sherer M.	Springer	2020
Hypopituitarism Following Traumatic Brain Injury: Neuroendocrine Dysfunction and Head Trauma	Agha A.	LAP LAMBERT Academic Publishing	2010
Injury Models of the Central Nervous System: Methods and Protocols	Kobeissy FH, Dixon CE, Hayes RL, Mondello S.	Humana Press Inc.	2018

Continued

TABLE 3 Relevant books—cont'd

Book	Authors	Publisher	Year
Management of Adults with Traumatic Brain Injury	Arciniegas DB.	American Psychiatric Pub.	2013
Management of Severe Traumatic Brain Injury Evidence, Tricks, and Pitfalls	Sundstrøm T, Grände PO, Juul N, Kock-Jensen C, Romner B, Wester K.	Springer Berlin Heidelberg	2012
Management of Severe Traumatic Brain Injury: Evidence, Tricks, And Pitfalls. 2nd Ed	Sundstrøm T, Grände PO, Luoto T, Rosenlund C, Undén J, Wester KG.	Springer	2020
Manual of Traumatic Brain Injury: Assessment and Management	Zollman FS.	Demos Medical	2011
Neurosensory Disorders in Mild Traumatic Brain Injury	Balaban C.	Elsevier Science	2018
Neurotrauma: A Comprehensive Textbook on Traumatic Brain Injury and Spinal Cord Injury	Wang KKW.	Oxford University Press	2018
New Therapeutics for Traumatic Brain Injury: Prevention of Secondary Brain Damage And Enhancement of Repair and Regeneration	Heidenreich K.	Elsevier Science Publishing Co Inc.	2017
Pediatric Neurocritical Care	Friess S, Kilbaugh T, Huh J.	Demos Medical	2013
Rehabilitation after Traumatic Brain Injury	Eapen BC, Cifu DX.	Academic Press	2018
Severe Traumatic Brain Injury (in Songklanagarind Surgical Review: Emergency Surgery)	Tunthanathip T.	Book unit, Faculty of Medicine, Prince of Songkla University	2014
Textbook of Traumatic Brain Injury, 3rd Edition	Silver JM, McAllister TW, Arciengas DB.	American Psychiatric Publishing, Inc.	2018
The American Psychiatric Association Publishing Textbook of Neuropsychiatry and Clinical Neurosciences	Yudofsky SC, Hales RE.	American Psychiatric Publishing Inc	2018
The Brain Injury Workbook: Exercises for Cognitive Rehabilitation. 2nd Edition	Powell T.	Taylor & Francis Ltd	2017
The Clinical Problem of Traumatic Head Injury.	Diaz-Arrastia R, Vos PE.	Traumatic Brain Injury, 237.	2014
The Invisible Brain Injury: Cognitive Impairments in Traumatic Brain Injury, Stroke and Other Acquired Brain Pathologies	Atienza AL.	Routledge	2019
The Traumatized Brain: A Family Guide to Understanding Mood, Memory, and Behavior After Brain Injury (A Johns Hopkins Press Health Book)	Rao V.	Johns Hopkins University Press; Illustrated	2015
Topics in Cognitive Rehabilitation in the Tbi Post-Hospital Phase	Anghinah R.	Springer	2018
Translational Research in Traumatic Brain Injury	Laskowitz D, Grant G.	CRC Press.	2016
Trauma. La Gestione Del Paziente	Donelli FM.	Griffin	2018
Traumatic Brain Injury	Vos P, Diaz-Arrastia R.	Wiley	2014
Traumatic Brain Injury	Grafman J, Salazar A.	Elsevier	2015
Traumatic Brain Injury in Sports	Echemendia R, Lovell M, Barth J, Collins M.	CRC Press	2020
Traumatic Brain Injury Pathobiology, Advanced Diagnostics and Acute Management	Long J, Gorbunov N.	IntechOpen	2018
Traumatic Brain Injury Rehabilitation	Lefaivre C.	Routledge	2014

Continued

TABLE 3 Relevant books—cont'd

Book	Authors	Publisher	Year
Traumatic Brain Injury, Part I	Salazar A.	Elsevier Science	2015
Traumatic Brain Injury A Clinician's Guide To Diagnosis, Management, and Rehabilitation	Tsao JW.	Springer New York	2012
Traumatic Brain Injury: A Multidisciplinary Approach 2nd Ed	Whitfield PC, Welbourne J, Thomas E, Summers F, Whyte M, Hutchinson PJ.	Cambridge University Press	2020
Traumatic Brain Injury: Pathobiology, Advanced Diagnostics and Acute Management	Gorbunov N, Joseph L.	BoD–Books on Demand	2018
Traumatic Brain Injury: Rehabilitation, Treatment, and Case Management, 4th Edition	Ashley MJ, Hovda DA.	CRC Press	2017
Understanding Mild Traumatic Brain Injury (Mtbi): An Insightful Guide to Symptoms, Treatments, and Redefining Recovery	Whittemore LL.	Clear Focus Press	2010
Understanding Traumatic Brain Injury: Current Research and Future Directions	Levin HS, Shum DHK, Chan RCK.	Oxford University Press	2014

This table lists books on traumatic brain injury as recommended by authors.

TABLE 4 Professional societies and other organizations relevant to brain injury.

Society/Organization	Web address
Adapt Functional Movement Center	https://adaptmovement.org/conditions/?gclid=CjwKCAjw3pWDBhB3EiwAV1c5rKJg4guJaVYDEWA4ICPOq-pq7DqaL62pyG7CuCEjr_-yT9qSiQqg7BoCWqcQAvD_BwE
American Academy of Neurology	www.aan.com
American Association for The Surgery of Trauma	https://www.aast.org/
American Association of Neurological Surgeons	www.aans.org
American Headache Society	https://americanheadachesociety.org/news/posttraumatic-headaches-persist-five-years-after-traumatic-brain-injury/
American Medical Association	https://www.ama-assn.org/delivering-care/population-care/veterans-health-resources-medical-professionals
American Neurological Association	www.myana.org
American Society of Neurorehabilitation	https://www.asnr.com/i4a/pages/index.cfm?pageid=1
American Society of Neuroscience	https://www.sfn.org/
Association De Neuro-Anesthésie Réanimation De Langue Française	http://www.anarlf.eu
Association of Academic Physiatrists	https://www.physiatry.org/
Brain Injury Association of America	https://www.biausa.org
Brain Injury Association of California	www.biacal.org
Brain Injury New Zealand	https://www.brain-injury.nz/
Brain Society of California	https://brainsocal.org/
Brain Trauma Foundation	https://www.braintrauma.org
Chest Wall Injury Society	https://cwisociety.org/

Continued

TABLE 4 Professional societies and other organizations relevant to brain injury—cont'd

Society/Organization	Web address
Eastern Association for The Surgery of Trauma	https://www.east.org/
European Brain Injury Society	https://www.ebissociety.org/en/
European Federation of Neurorehabilitation Societies	https://efnr.org/
Family Caregiver Alliance	https://www.caregiver.org/traumatic-brain-injury
Federation of European Societies of Neuropsychology	https://www.fesn.eu/
Florence Health	https://huddle.florence-health.com/discover/content/article/pediatric-and-law-experts-define-abusive-head-trauma/
Headway—The Brain Injury Association	https://www.headway.org.uk/
International Association of Functional Neurology and Rehabilitation	www.iafnr.org
International Brain Injury Association	https://www.internationalbrain.org/
International Neurotrauma Society	http://www.internationalneurotraumasociety.org/
Japan Society for Higher Brain Dysfunction	http://www.higherbrain.or.jp
Japanese Association Of Rehabilitation Medicine	https://www.jarm.or.jp/english/
National Neurotrauma Society	https://www.neurotrauma.org/
Neurocritical Care Society	https://www.neurocriticalcare.org/home
Neuro-Developmental Treatment Association	www.ndta.org
Ontario Brain Injury Association	http://obia.ca/
Societe Française D'anesthésie Réanimation	https://sfar.org
Society for Cognitive Rehabilitation	https://www.societyforcognitiverehab.org/
Society for Neuroscience	https://www.sfn.org/
Society of Critical Care Medicine	https://sccm.org/
Taiwan Neuroscience Society	https://neuroscience.org.tw/
Taiwan Neurotrauma Society	https://www.taiwanneurotraumasociety.com/
Victoria Brain Injury Society	https://vbis.ca/
Western Trauma Association	https://www.westerntrauma.org/

This table lists some professional societies and organizations devoted to understanding the neuroscience of traumatic brain injury and its sequalae. Please note, occasionally the location of the websites or web address changes. In these cases the use of the "Search" tabs should be explored at the parent address or site. See also Tables 1 and 6.

TABLE 5 Emerging techniques and platforms relevant to traumatic brain injury.

Organization, Company or Society Name	Web address
Acquired Brain Injury (ABI) Wellness	https://abiwellness.com/
Aging and Neurotrauma Laboratory (Irimia Lab)	www.andrei-irimia.com
American Brain Coalition	https://www.americanbraincoalition.org/default.aspx
Arrowhead Traumatic Brain Injury Conference	www.tbiconference.com
Australia-Europe Neurotrauma Effectiveness Research in Traumatic Brain Injury (Ozenter-TBI)	https://www.monash.edu/medicine/sphpm/anzicrc/research/ozenter-study

Continued

TABLE 5 Emerging techniques and platforms relevant to traumatic brain injury—cont'd

Organization, Company or Society Name	Web address
Banyan Biomarkers	https://www.banyanbio.com
Brain Injury Association of America (BIAA)	https://www.biausa.org/
Brain Injury Professional	https://braininjuryprofessional.com/
Brain Trauma Foundation	https://braintrauma.org/
Brainline military	https://www.brainline.org/military-veterans
Clinicaltrials.Gov	https://clinicaltrials.gov/
Collaborative European Neurotrauma Effectiveness Research in TBI	https://www.center-tbi.eu/
Corticosteroid Randomisation after Significant Head Injury (CRASH)	This site has been designated as restricted at the time of production but other information can be found here https://datacompass.lshtm.ac.uk/id/eprint/35/
Defense Centers of Excellence for Psychological Health and Traumatic Brain Injury	http://dvbic.dcoe.mil/
European Brain Injury Consortium	http://www.ebic.nl
Evalu8Now	https://www.createabilityinc.com/evalu8now/
Evolv	https://evolvrehab.com/virtualrehab/
Federal Interagency Traumatic Brain Injury Research (FITBIR) Informatics System	https://fitbir.nih.gov
Guttmann Neuropersonal Trainer	https://gnpt.es/
Health Technologies for Assessment and Intervention of Traumatic Brain Injury	https://braininjuryprofessional.com/2020/11/technology-evalu8now/
Helsinki Score	http://links.lww.com/NEU/A676
Initiative for Neural Science, Disease & Engineering (Inscide)	https://sites.tufts.edu/INSciDE/tbi/
Intensive Care Monitor +	https://icmplus.neurosurg.cam.ac.uk/
International Mission for Prognosis and Analysis of Clinical Trials (IMPACT) Traumatic Brain Injury	http://www.tbi-impact.org/?p=impact/calc
Mossrehab	https://www.mossrehab.com/technology
Myriad-Rules Based Medicine	https://myriadrbm.com
National Association of State Head Injury Administrators	https://www.nashia.org/
Neuronup (Web Platform for Cognitive Rehabilitation)	https://www.neuronup.com/en
Neurovive Pharmaceuticals—Neurostat Technology	http://abliva.com
Nijmegen	http://www.tbi-prognosis.com/
Nomogram Predicting Intracranial Injury of Pediatric Traumatic Brain Injury	https://thara.shinyapps.io/tsus4/
North American Brain Injury Society[a]	http://www.nabis.org/
Optimal Cerebral Perfusion Pressure	https://cppopt.org/
Pate's Rehabilitation-Brain Injury Treatment Technologies	https://www.paterehab.com/brain-injury-treatment-technologies/
Pediatric Neurocritical Care Research Group	http://www.pncrg.org/
Prognostic Calculator International Mission for Prognosis and Analysis of Clinical Trials (IMPACT) Traumatic Brain Injury (TBI)	http://www.tbi-impact.org/?p=impact/calc
Robolab Therapist	https://robolab.unex.es/index.php/therapist/

Continued

TABLE 5 Emerging techniques and platforms relevant to traumatic brain injury—cont'd

Organization, Company or Society Name	Web address
T3 V3 Driving Simulator and Neurodrive Program (A Novel Rehabilitative Intervention for TBI Involving Clinician-Guided Cognitive and Driving Exercises Conducted Within A Virtual Reality Environment)	http://www.t3rs.net/en/ (driving simulator)
The Center of Excellence for Medical Multimedia (CEMM)	https://tbi.cemmlibrary.org/
The International Brain Injury Association.	https://www.internationalbrain.org/
The Society for Cognitive Rehabilitation	https://www.societyforcognitiverehab.org/
The Society for Neuroscience	http://www.sfn.org/
Therapist Robolab	http://www.therapist.uma.es/
Transcranial Random Noise Stimulation (Trns) in Cognitive Rehabilitation of TBI Patients (RCT Carried out at Neurorehabilitation Hospital "Institut Guttmann", Barcelona, Spain)	https://clinicaltrials.gov/ct2/show/NCT03696381
Transforming Research and Clinical Knowledge in Traumatic Brain Injury (TRACK-TBI)	https://tracktbi.ucsf.edu
Traumatic Brain Injury Center of Excellence, The Military Health System	https://health.mil/About-MHS/OASDHA/Defense-Health-Agency/Research-and-Development/Traumatic-Brain-Injury-Center-of-Excellence
United States Brain Injury Alliance	https://usbia.org/
Wakemed	https://www.wakemed.org/care-and-services/brain-and-spine/rehab-and-physical-therapy/brain-injury-program/advanced-technology

[a] This site could not be accessed when the table was compiled but indirect access was available via The International Brain Injury Association.
This table lists some emerging technologies, centers, and platforms relevant to the neuroscience of traumatic brain injury. Please note, occasionally the location of the websites or web address changes. In these cases, the use of the "Search" tabs should be explored at the parent address or site.

TABLE 6 Other sites and sources of interest relevant to the neuroscience of traumatic brain injury.

Site or source name	Web address
American Association of Neurological Surgeons	https://www.aans.org/Patients/Neurosurgical-Conditions-and-Treatments/Traumatic-Brain-Injury
Aurora Health Care	https://www.aurorahealthcare.org/services/neuroscience/brain-skull-base-care/brain-injury-trauma
Barcelona Brain Health Initiative	https://bbhi.cat/en/
Brain Injury Association of America	https://www.biausa.org/
Brain Injury Canada	https://braininjurycanada.ca/
Brain Injury Professional	https://braininjuryprofessional.com/about-bip/
Brain Trauma Foundation	https://www.braintrauma.org/
Brainline	https://www.brainline.org
Canadian High-Resolution Traumatic Brain Injury (Cahr-Tbi) Research Collaborative	https://pubmed.ncbi.nlm.nih.gov/32174295/
Clinical Practice Guideline for the Rehabilitation of Adults with Moderate to Severe TBI; Institut National D'excellence En Santé Et En Services Sociaux (INESSS)	https://www.inesss.qc.ca/fileadmin/doc/INESSS/Rapports/Traumatologie/Section_2_REC_complete_ENG_final.pdf
Department of Health, Pennsylvania Government	https://www.health.pa.gov/topics/disease/Pages/Traumatic-Brain-Injury.aspx

Continued

TABLE 6 Other sites and sources of interest relevant to the neuroscience of traumatic brain injury—cont'd

Site or source name	Web address
Evidence-Based Review of Moderate-to-Severe Acquired Brain Injury	https://erabi.ca/
Family Caregiver Alliance	https://www.caregiver.org/traumatic-brain-injury
Just Care—Traumatic Brain Injury Association	http://justcarengo.org
Kessler Foundation—Changing The Lives of People With Disabilities	https://kesslerfoundation.org/research/center-traumatic-brain-injury
Machine Learning Application for Predicting Intracranial Injury of Pediatric Traumatic Brain Injury (Python, Streamlit)	https://ctpedtbi.herokuapp.com/
Mcgill University Health Centre	https://muhc.ca/tbi/page/learn-videos
Microdialysis	https://www.mdialysis.com/neuro-monitoring/
National Association of State Head Injury Administrators	https://www.nashia.org/
New York State, Department of Health	https://www.health.ny.gov/prevention/injury_prevention/traumatic_brain_injury/resources.htm
Newswise (Association of Academic Physicians Article)	https://www.newswise.com/articles/alcohol-has-a-dose-dependent-effect-on-the-outcomes-of-traumatic-brain-injury-patients
Northern Brain Injury Association	https://www.nbia.ca/resources/
Onemind-Aurora Initiative	https://onemind.org/aurora/
Oregon Health & Science University	https://www.ohsu.edu/brain-institute/traumatic-brain-injury
Pediatric Critical Care Medicine	Pediatric Critical Care Medicine (lww.com)
Physiopedia	https://www.physio-pedia.com/Discharge_Management_for_Traumatic_Brain_Injury
University of Alabama at Birmingham Traumatic Brain Injury Model System (UAB-TBIMS)	https://www.uab.edu/medicine/tbi/uab-tbi-information-network/bims-infosheets
Ward J, Williams J, Manchester S. (2017). '111 NFL Brains. All But One Had CTE', The New York Times.	https://www.nytimes.com/interactive/2017/07/25/sports/football/nfl-cte.html

This table lists some other sites of interest relevant to the neuroscience of traumatic brain injury. Please note, occasionally the location of the websites or web address changes. In these cases, the use of the "Search" tabs should be explored at the parent address or site.

online. Because traumatic brain injury and its consequences have some features related to other conditions, we also cite other chapters on resources (recommended by authors and practitioners) as follows:

Recommended Resources and Sites for the Neuroscience of Spinal Cord Injury (Rajendram et al., 2021a).

Recommended Research and Resources on the Neuroscience of Parkinson's Disease (Rajendram, Patel, & Preedy, 2020a).

Recommended Resources for Pathology: Oxidative Stress and Dietary Antioxidants (Rajendram, Patel, & Preedy, 2020b).

The Physiology, Neurobiology and Behavior of Pain: Recommended Resources (Rajendram, Patel, & Preedy, 2021b).

Recommended Resources on the Neuroscience of Development (Rajendram, Patel, & Preedy, 2021c).

Recommended Resources on the Neuroscience of Aging (Rajendram, Patel, & Preedy, 2021d).

Mini-dictionary of terms

- **Neuronal plasticity**: The ability of neurons to adapt, reorganize, or grow.
- **Neuronal regeneration**: The development of new axons, synapses neurons, or glia.
- **Neurorehabilitation**: The process which aims to improve the functional outcome after neuronal insults including traumatic brain injury.

- **Parenchyma**: The functional components as opposed to the supportive or structural elements.
- **Therapeutic nihilism**: The belief that the outcome of a patient cannot be improved.

Key facts of traumatic brain injury

- Traumatic brain injury: Any insult to the brain that temporarily or permanently affects its function.
- Traumatic brain injury is not a homogenous entity.
- There are three phases of traumatic brain injury (primary, secondary, and tertiary).
- Failure of neuronal regeneration is thought to prevent functional recovery post traumatic brain injury.
- There is still much to learn about the neuroscience of traumatic brain injury before the sequelae can be reversed.

Summary points

- Patients with traumatic brain injury often survive to discharge home and their quality of life has been improved by technological advances.
- While the sequelae of traumatic brain injury cannot be cured prognostic pessimism is unwarranted.
- Scientists are striving to promote functional recovery after traumatic brain injury by stimulating neuronal regeneration.
- Central neurons have greater plasticity than previously thought and so it seems that the control of neuronal regeneration is feasible.
- Progress has been made in the understanding the neuroscience of traumatic brain injury. Those involved in this field are struggling to remain up to date.

Acknowledgments

We would like to thank the following authors for contributing to the development of this resource. We apologize if some suggested material was not included in this chapter or has been moved to different sections.

Abdelhady S, Kobeissy F, Aftab K, Biroli F, Bunevicius A, Chen L, Chopp M, Coll-Andreu M, Dagher J, Devarajan A, Dhote V, Dobrachinski F, Edwards S, Enam SA, Erikkson E, Figaji A, Gasenzer ER, Green B, Hacioglu A, Hiraoka T, Huh J, Irimia A, Kanat A, Ketharanathan N, Mahmood A, Mohan Maruga Raja MK, Mollayeva T, Palazzo E, Perez PK, Rohlwink U, Roquilly A, Rossini Z, Siraj MA, Tunthanatip T, Wu A, Xiong Y, Zeiler F.

References

Barnabe-Heider, F., & Frisen, J. (2008). Stem cells for spinal cord repair. *Cell Stem Cell*, *3*, 16–24.

Cajal, S. R.y. (1928). *Degeneration and regeneration of the nervous system*. London: Oxford University Press.

Faul, M., & Coronado, V. (2015). Epidemiology of traumatic brain injury. *Handbook of Clinical Neurology*, *127*, 3–13. https://doi.org/10.1016/B978-0-444-52892-6.00001-5.

Fleiss, B., & Gressens, P. (2012). Tertiary mechanisms of brain damage: A new hope for treatment of cerebral palsy? *Lancet Neurology*, *11*, 556–566.

Patel, H. C., Bouamra, O., Woodford, M., King, A. T., Yates, D. W., & Lecky, F. E. (2005). Trends in head injury outcome from 1989 to 2003 and the effect of neurosurgical care: An observational study. *Lancet*, *366*(9496), 1538–1544.

Prasetyo, E. (2020). The primary, secondary, and tertiary brain injury. *Critical Care and Shock*, *23*, 4–13.

Rajendram, R., Patel, V. B., & Preedy, V. R. (2020a). Recommended research and resources on the neuroscience of Parkinson's disease. In V. B. Patel, & V. R. Preedy (Eds.), *Neuroscience of Parkinson's disease*. UK: Elsevier.

Rajendram, R., Patel, V. B., & Preedy, V. R. (2020b). Recommended resources for pathology: Oxidative stress and dietary antioxidants. In V. R. Preedy (Ed.), *Pathology: Oxidative stress and dietary antioxidants*. USA: Elsevier.

Rajendram, R., Patel, V. B., & Preedy, V. R. (2021a). Recommended resources and sites for the neuroscience of spinal cord injury. In R. Rajendram, V. B. Patel, & V. R. Preedy (Eds.), *Neuroscience of spinal cord injury*. USA: Elsevier.

Rajendram, R., Patel, V. B., & Preedy, V. R. (2021b). The physiology, neurobiology and behaviour of pain: Recommended resources. In R. Rajendram, C. R. Martin, V. B. Patel, & V. R. Preedy (Eds.), *The physiology, neurobiology and behaviour of pain*. UK: Elsevier.

Rajendram, R., Patel, V. B., & Preedy, V. R. (2021c). Recommended resources on the neuroscience of development. In C. R. Martin, & V. R. Preedy (Eds.), *Neuroscience of development disease*. UK: Elsevier.

Rajendram, R., Patel, V. B., & Preedy, V. R. (2021d). Recommended resources on the neuroscience of aging. In V. B. Patel, & V. R. Preedy (Eds.), *Neuroscience of aging*. UK: Elsevier.

Sanchez, G. M., & Burridge, A. L. (2007). Decision making in head injury management in the Edwin Smith Papyrus. *Neurosurgical Focus*, *23*(1). https://doi.org/10.3171/foc.2007.23.1.5, E5.

Index

Note: Page numbers followed by *f* indicate figures and *t* indicate tables.

Q

R

W

Y

Z